Advanced Multivariate Statistics with Matrices

Advanced Multivariate Statistics with Matrices

by

Tõnu Kollo
University of Tartu,
Tartu, Estonia

and

Dietrich von Rosen
Swedish University of Agricultural Sciences,
Uppsala, Sweden

Springer

A C.I.P. Catalogue record for this book is available from the Library of Congress.

ISBN-10 1-4020-3418-0 (HB) Springer Dordrecht, Berlin, Heidelberg, New York
ISBN-10 1-4020-3419-9 (e-book) Springer Dordrecht, Berlin, Heidelberg, New York
ISBN-13 978-1-4020-3418-3 (HB) Springer Dordrecht, Berlin, Heidelberg, New York
ISBN-13 978-1-4020-3419-0 (e-book) Springer Dordrecht, Berlin, Heidelberg, New York

---

Published by Springer,
P.O. Box 17, 3300 AA Dordrecht, The Netherlands.

Printed on acid-free paper

Printed in the Netherlands.

To

Imbi, Kaarin, Ülle, Ardo
Tõnu

Tatjana, Philip, Sophie, Michael
Dietrich

TABLE OF CONTENTS

PREFACE

The book presents important tools and techniques for treating problems in modern multivariate statistics in a systematic way. The ambition is to indicate new directions as well as to present the classical part of multivariate statistical analysis in this framework. The book has been written for graduate students and statisticians who are not afraid of matrix formalism. The goal is to provide them with a powerful toolkit for their research and to give necessary background and deeper knowledge for further studies in different areas of multivariate statistics. It can also be useful for researchers in applied mathematics and for people working on data analysis and data mining who can find useful methods and ideas for solving their problems.

It has been designed as a textbook for a two semester graduate course on multivariate statistics. Such a course has been held at the Swedish Agricultural University in 2001/02. On the other hand, it can be used as material for series of shorter courses. In fact, Chapters 1 and 2 have been used for a graduate course "Matrices in Statistics" at University of Tartu for the last few years, and Chapters 2 and 3 formed the material for the graduate course "Multivariate Asymptotic Statistics" in spring 2002. An advanced course "Multivariate Linear Models" may be based on Chapter 4.

A lot of literature is available on multivariate statistical analysis written for different purposes and for people with different interests, background and knowledge. However, the authors feel that there is still space for a treatment like the one presented in this volume. Matrix algebra and theory of linear spaces are continuously developing fields, and it is interesting to observe how statistical applications benefit from algebraic achievements. Our main aim is to present tools and techniques whereas development of specific multivariate methods has been somewhat less important. Often alternative approaches are presented and we do not avoid complicated derivations.

Besides a systematic presentation of basic notions, throughout the book there are several topics which have not been touched or have only been briefly considered in other books on multivariate analysis. The internal logic and development of the material in this book is the following. In Chapter 1 necessary results on matrix algebra and linear spaces are presented. In particular, lattice theory is used. There are three closely related notions of matrix algebra which play a key role in the presentation of multivariate statistics: Kronecker product, vec-operator and the concept of matrix derivative. In Chapter 2 the presentation of distributions is heavily based on matrix algebra, what makes it possible to present complicated expressions of multivariate moments and cumulants in an elegant and compact way. The very basic classes of multivariate and matrix distributions, such as normal, elliptical and Wishart distributions, are studied and several relations and characteristics are presented of which some are new. The choice of the material in Chapter 2 has been made having in mind multivariate asymptotic distributions and multivariate expansions in Chapter 3. This Chapter presents general formal density expansions which are applied in normal and Wishart approximations. Finally, in Chapter 4 the results from multivariate distribution theory and approximations are used in presentation of general linear models with a special emphasis on the Growth Curve model.

The authors are thankful to the Royal Swedish Academy of Sciences and to the Swedish Institute for their financial support. Our sincere gratitude belongs also to the University of Tartu, Uppsala University and the Swedish Agricultural University for their support. Dietrich von Rosen gratefully acknowledges the support from the Swedish Natural Sciences Research Council, while Tõnu Kollo is indebted to the Estonian Science Foundation. Grateful thanks to Professors Heinz Neudecker, Kenneth Nordström and Muni Srivastava. Some results in the book stem from our earlier cooperation. Also discussions with Professors Kai-Tai Fang and Björn Holmquist have been useful for presentation of certain topics. Many thanks to our colleagues for support and stimulating atmosphere. Last but not least we are grateful to all students who helped improve the presentation of the material during the courses held on the material.

Uppsala
November 2004
Tõnu Kollo
Dietrich von Rosen

INTRODUCTION

In 1958 the first edition of *An Introduction to Multivariate Statistical Analysis* by T. W. Anderson appeared and a year before S. N. Roy had published *Some Aspects of Multivariate Analysis.* Some years later, in 1965, *Linear Statistical Inference and Its Applications* by C. R. Rao came out. During the following years several books on multivariate analysis appeared: Dempster (1969), Morrison (1967), Press (1972), Kshirsagar (1972). The topic became very popular in the end of 1970s and the beginning of 1980s. During a short time several monographs were published: Giri (1977), Srivastava & Khatri (1979), Mardia, Kent & Bibby (1979), Muirhead (1982), Takeuchi, Yanai & Mukherjee (1982), Eaton (1983), Farrell (1985) and Siotani, Hayakawa, & Fujikoshi (1985). All these books made considerable contributions to the area though many of them focused on certain topics. In the last 20 years new results in multivariate analysis have been so numerous that it seems impossible to cover all the existing material in one book. One has to make a choice and different authors have made it in different directions. The first class of books presents introductory texts of first courses on undergraduate level (Srivastava & Carter, 1983; Flury, 1997; Srivastava, 2002) or are written for non-statisticians who have some data they want to analyze (Krzanowski, 1990, for example). In some books the presentation is computer oriented (Johnson, 1998; Rencher, 2002), for example). There are many books which present a thorough treatment on specific multivariate methods (Greenacre, 1984; Jolliffe, 1986; McLachlan, 1992; Lauritzen, 1996; Kshirsagar & Smith, 1995, for example) but very few are presenting foundations of the topic in the light of newer matrix algebra. We can refer to Fang & Zhang (1990) and Bilodeau & Brenner (1999), but there still seems to be space for a development. There exists a rapidly growing area of linear algebra related to mathematical statistics which has not been used in full range for a systematic presentation of multivariate statistics. The present book tries to fill this gap to some extent. Matrix theory, which is a cornerstone of multivariate analysis starting from T. W. Anderson, has been enriched during the last few years by several new volumes: Bhatia (1997), Harville (1997), Schott (1997b), Rao & Rao (1998), Zhang (1999). These books form a new basis for presentation of multivariate analysis.

The Kronecker product and vec-operator have been used systematically in Fang & Zhang (1990), but these authors do not tie the presentation to the concept of the matrix derivative which has become a powerful tool in multivariate analysis. Magnus & Neudecker (1999) has become a common reference book on matrix differentiation. In Chapter 2, as well as Chapter 3, we derive most of the results using matrix derivatives. When writing the book, our main aim was to answer the questions "Why?" and "In which way?", so typically the results are presented with proofs or sketches of proofs. However, in many situations the reader can also find an answer to the question "How?".

Before starting with the main text we shall give some general comments and

remarks about the notation and abbreviations.
Throughout the book we use boldface transcription for matrices and vectors. Matrices will be denoted by capital letters and vectors by ordinary small letters of Latin or Greek alphabets. Random variables will be denoted by capital letters from the end of the Latin alphabet. Notion appearing in the text for the first time is printed in *italics*. The end of proofs, definitions and examples is marked by ▮
To shorten proofs we use the following abbreviation

$$\underset{(1.3.2)}{=}$$

which should be red as "the equality is obtained by applying formula (1.3.2)". We have found it both easily understandable and space preserving. In numeration of Definitions, Theorems, Propositions and Lemmas we use a three position system. Theorem 1.2.10 is the tenth theorem of Chapter 1, Section 2. For Corollaries four integers are used: Corollary 1.2.3.1 is the first Corollary of Theorem 1.2.3. In a few cases when we have Corollaries of Lemmas, the capital L has been added to the last number, so Corollary 1.2.3.1L is the first corollary of Lemma 1.2.3. We end the Introduction with the List of Notation, where the page number indicates the first appearance or definition.

LIST OF NOTATION

$\circ$ – elementwise or Hadamard product, p. 3
$\otimes$ – Kronecker or direct product, tensor product, p. 81, 41
$\oplus$ – direct sum, p. 27
$\boxplus$ – orthogonal sum, p. 27
$\mathbf{A}$ – matrix, p. 2
$\mathbf{a}$ – vector, p. 2
c – scalar, p. 3
$\mathbf{A}'$ – transposed matrix, p. 4
$\mathbf{I}_p$ – identity matrix, p. 4
$\mathbf{A}_d$ – diagonalized matrix $\mathbf{A}$, p. 6
$\mathbf{a}_d$ – diagonal matrix, $\mathbf{a}$ as diagonal, p. 6
diag$\mathbf{A}$ – vector of diagonal elements of $\mathbf{A}$, p. 6
$|\mathbf{A}|$ – determinant of $\mathbf{A}$, p. 7
$r(\mathbf{A})$ – rank of $\mathbf{A}$, p. 9
p.d. – positive definite, p. 12
$\mathbf{A}^-$ – generalized inverse, g-inverse, p. 15
$\mathbf{A}^+$ – Moore-Penrose inverse, p. 17
$\mathbf{A}(K)$ – patterned matrix (pattern K), p. 97
$\mathbf{K}_{p,q}$ – commutation matrix, p. 79
vec – vec-operator, p. 89
$\mathbf{A}^{\otimes k}$ – k-th Kroneckerian power, p. 84
$V^k(\mathbf{A})$ – vectorization operator, p. 115
$R^k(\mathbf{A})$ – product vectorization operator, p. 115

$\dfrac{d\mathbf{Y}}{d\mathbf{X}}$ – matrix derivative, p. 127
m.i.v. – mathematically independent and variable, p. 126
$\mathbf{J}(\mathbf{Y} \to \mathbf{X})$ – Jacobian matrix, p. 156
$|\mathbf{J}(\mathbf{Y} \to \mathbf{X})|_+$ – Jacobian, p. 156
$\mathbb{A}^\perp$ – orthocomplement, p. 27
$\mathbb{B}\perp\mathbb{A}$ – perpendicular subspace, p. 27
$\mathbb{B}|\mathbb{A}$ – commutative subspace, p. 31
$\mathcal{R}(\mathbf{A})$ – range space, p. 34
$\mathcal{N}(\mathbf{A})$ – null space, p. 35
$\mathcal{C}(\mathbf{C})$ – column space, p. 48
$\mathbf{X}$ – random matrix, p. 171
$\mathbf{x}$ – random vector, p. 171
X – random variable, p. 171
$f_\mathbf{x}(\mathbf{x})$ – density function, p. 174
$F_\mathbf{x}(\mathbf{x})$ – distribution function, p. 174
$\varphi_\mathbf{x}(\mathbf{t})$ – characteristic function, p. 174
$E[\mathbf{x}]$ – expectation, p. 172
$D[\mathbf{x}]$ – dispersion matrix, p. 173
$c_k[\mathbf{x}]$ – k–th cumulant, p. 181
$m_k[\mathbf{x}]$ – k–th moment, p. 175
$\overline{m}_k[\mathbf{x}]$ – k–th central moment, p. 175
$mc_k[\mathbf{x}]$ – k–th minimal cumulant, p. 185
$mm_k[\mathbf{x}]$ – k–th minimal moment, p. 185
$m\overline{m}_k[\mathbf{x}]$ – k–th minimal central moment, p. 185
$\mathbf{S}$ – sample dispersion matrix, p. 284
$\mathbf{R}$ – sample correlation matrix, p. 289
Ω – theoretical correlation matrix, p. 289
$N_p(\boldsymbol{\mu}, \boldsymbol{\Sigma})$ – multivariate normal distribution, p. 192
$N_{p,n}(\boldsymbol{\mu}, \boldsymbol{\Sigma}, \boldsymbol{\Psi})$ – matrix normal distribution, p. 192
$E_p(\boldsymbol{\mu}, \mathbf{V})$ – elliptical distribution, p. 224
$W_p(\boldsymbol{\Sigma}, n)$ – central Wishart distribution, p. 237
$W_p(\boldsymbol{\Sigma}, n, \boldsymbol{\Delta})$ – noncentral Wishart distribution, p. 237
$M\beta_I(p, m, n)$ – multivariate beta distribution, type I, p. 249
$M\beta_{II}(p, m, n)$ – multivariate beta distribution, type II, p. 250
$\xrightarrow{\mathcal{D}}$ – convergence in distribution, week convergence, p. 277
$\xrightarrow{\mathcal{P}}$ – convergence in probability, p. 278
$O_P(\cdot)$ – p. 278
$o_P(\cdot)$ – p. 278
$(\mathbf{X})()'$ – $(\mathbf{X})(\mathbf{X})'$, p. 355

CHAPTER I

Basic Matrix Theory and Linear Algebra

Matrix theory gives us a language for presenting multivariate statistics in a nice and compact way. Although matrix algebra has been available for a long time, a systematic treatment of multivariate analysis through this approach has been developed during the last three decades mainly. Differences in presentation are visible if one compares the classical book by Wilks (1962) with the book by Fang & Zhang (1990), for example. The relation between matrix algebra and multivariate analysis is mutual. Many approaches in multivariate analysis rest on algebraic methods and in particular on matrices. On the other hand, in many cases multivariate statistics has been a starting point for the development of several topics in matrix algebra, e.g. properties of the commutation and duplication matrices, the star-product, matrix derivatives, etc. In this chapter the style of presentation varies in different sections. In the first section we shall introduce basic notation and notions of matrix calculus. As a rule, we shall present the material without proofs, having in mind that there are many books available where the full presentation can be easily found. From the classical books on matrix theory let us list here Bellmann (1970) on basic level, and Gantmacher (1959) for advanced presentation. From recent books at a higher level Horn & Johnson (1990, 1994) and Bhatia (1997) could be recommended. Several books on matrix algebra have statistical orientation: Graybill (1983) and Searle (1982) at an introductory level, Harville (1997), Schott (1997b), Rao & Rao (1998) and Zhang (1999) at a more advanced level. Sometimes the most relevant work may be found in certain chapters of various books on multivariate analysis, such as Anderson (2003), Rao (1973a), Srivastava & Khatri (1979), Muirhead (1982) or Siotani, Hayakawa & Fujikoshi (1985), for example. In fact, the first section on matrices here is more for notation and a refreshment of the readers' memory than to acquaint them with novel material. Still, in some cases it seems that the results are generally not used and well-known, and therefore we shall add the proofs also. Above all, this concerns generalized inverses.

In the second section we give a short overview of some basic linear algebra accentuating lattice theory. This material is not so widely used by statisticians. Therefore more attention has been paid to the proofs of statements. At the end of this section important topics for the following chapters are considered, including column vector spaces and representations of linear operators in vector spaces. The third section is devoted to partitioned matrices. Here we have omitted proofs only in those few cases, e.g. properties of the direct product and the vec-operator, where we can give references to the full presentation of the material somewhere else. In the last section, we examine matrix derivatives and their properties. Our treatment of the topic differs somewhat from a standard approach and therefore the material is presented with full proofs, except the part which gives an overview of the Fréchet derivative.

1.1 MATRIX ALGEBRA

1.1.1 *Operations and notation*

The *matrix* $\mathbf{A}$ of size $m \times n$ is a rectangular table of elements a_{ij} ($i = 1, \ldots, m$; $j = 1, \ldots, n$):

$$\mathbf{A} = \begin{pmatrix} a_{11} & \ldots & a_{1n} \\ \vdots & \ddots & \vdots \\ a_{m1} & \ldots & a_{mn} \end{pmatrix},$$

where the element a_{ij} is in the $i-$th row and $j-$th column of the table. For indicating the element a_{ij} in the matrix $\mathbf{A}$, the notation $(\mathbf{A})_{ij}$ will also be used. A matrix $\mathbf{A}$ is presented through its elements as $\mathbf{A} = (a_{ij})$. Elements of matrices can be of different nature: real numbers, complex numbers, functions, etc. At the same time we shall assume in the following that the used operations are defined for the elements of $\mathbf{A}$ and we will not point out the necessary restrictions for the elements of the matrices every time. If the elements of $\mathbf{A}$, of size $m \times n$, are real numbers we say that $\mathbf{A}$ is a real matrix and use the notation $\mathbf{A} \in \mathbb{R}^{m \times n}$. Furthermore, if not otherwise stated, the elements of the matrices are supposed to be real. However, many of the definitions and results given below also apply to complex matrices. To shorten the text, the following phrases are used synonymously:

- matrix $\mathbf{A}$ of size $m \times n$;
- $m \times n-$matrix $\mathbf{A}$;
- $\mathbf{A} : m \times n$.

An $m \times n-$matrix $\mathbf{A}$ is called *square*, if $m = n$. When $\mathbf{A}$ is $m \times 1-$matrix, we call it a *vector* :

$$\mathbf{A} = \begin{pmatrix} a_{11} \\ \vdots \\ a_{m1} \end{pmatrix}.$$

If we omit the second index 1 in the last equality we denote the obtained $m-$vector by $\mathbf{a}$:

$$\mathbf{a} = \begin{pmatrix} a_{1} \\ \vdots \\ a_{m} \end{pmatrix}.$$

Two matrices $\mathbf{A} = (a_{ij})$ and $\mathbf{B} = (b_{ij})$ are *equal*,

$$\mathbf{A} = \mathbf{B},$$

if $\mathbf{A}$ and $\mathbf{B}$ are of the same order and all the corresponding elements are equal: $a_{ij} = b_{ij}$. The matrix with all elements equal to 1 will be denoted by $\mathbf{1}$ and, if necessary, the order will be indicated by an index, i.e. $\mathbf{1}_{m \times n}$. A vector of m ones

will be written $\mathbf{1}_m$. Analogously, $\mathbf{0}$ will denote a matrix where all elements are zeros.
The *product of* $\mathbf{A} : m\times n$ *by a scalar* c is an $m\times n-$matrix $c\mathbf{A}$, where the elements of $\mathbf{A}$ are multiplied by c:

$$c\mathbf{A} = (ca_{ij}).$$

Under the scalar c we understand the element of the same nature as the elements of $\mathbf{A}$. So for real numbers a_{ij} the scalar c is a real number, for $a_{ij}-$functions of complex variables the scalar c is also a function from the same class.
The *sum* of two matrices is given by

$$\mathbf{A}+\mathbf{B} = (a_{ij}+b_{ij}), \quad i = 1,\dots,m,\ j = 1,\dots,n.$$

These two fundamental operations, i.e. the sum and multiplication by a scalar, satisfy the following main properties:

$$\begin{aligned}
\mathbf{A}+\mathbf{B} &= \mathbf{B}+\mathbf{A};\\
(\mathbf{A}+\mathbf{B})+\mathbf{C} &= \mathbf{A}+(\mathbf{B}+\mathbf{C});\\
\mathbf{A}+(-1)\mathbf{A} &= \mathbf{0};\\
(c_1+c_2)\mathbf{A} &= c_1\mathbf{A}+c_2\mathbf{A};\\
c(\mathbf{A}+\mathbf{B}) &= c\mathbf{A}+c\mathbf{B};\\
c_1(c_2\mathbf{A}) &= (c_1c_2)\mathbf{A}.
\end{aligned}$$

Multiplication of matrices is possible if the number of columns in the first matrix equals the number of rows in the second matrix. Let $\mathbf{A} : m\times n$ and $\mathbf{B} : n\times r$, then the *product* $\mathbf{C} = \mathbf{AB}$ of the matrices $\mathbf{A} = (a_{ij})$ and $\mathbf{B} = (b_{kl})$ is the $m\times r-$matrix $\mathbf{C} = (c_{ij})$, where

$$c_{ij} = \sum_{k=1}^{n} a_{ik}b_{kj}.$$

Multiplication of matrices is not commutative in general, but the following properties hold, provided that the sizes of the matrices are of proper order:

$$\begin{aligned}
\mathbf{A}(\mathbf{BC}) &= (\mathbf{AB})\mathbf{C};\\
\mathbf{A}(\mathbf{B}+\mathbf{C}) &= \mathbf{AB}+\mathbf{AC};\\
(\mathbf{A}+\mathbf{B})\mathbf{C} &= \mathbf{AC}+\mathbf{BC}.
\end{aligned}$$

Similarly to the summation of matrices, we have the operation of elementwise multiplication, which is defined for matrices of the same order. The *elementwise product* $\mathbf{A}\circ\mathbf{B}$ of $m\times n-$matrices $\mathbf{A} = (a_{ij})$ and $\mathbf{B} = (b_{ij})$ is the $m\times n-$matrix

$$\mathbf{A}\circ\mathbf{B} = (a_{ij}b_{ij}), \quad i = 1,\dots,m,\ j = 1,\dots,n. \tag{1.1.1}$$

This product is also called *Hadamard* or *Schur product.* The main properties of this elementwise product are the following:

$$\begin{aligned}
\mathbf{A}\circ\mathbf{B} &= \mathbf{B}\circ\mathbf{A};\\
\mathbf{A}\circ(\mathbf{B}\circ\mathbf{C}) &= (\mathbf{A}\circ\mathbf{B})\circ\mathbf{C};\\
\mathbf{A}\circ(\mathbf{B}+\mathbf{C}) &= \mathbf{A}\circ\mathbf{B}+\mathbf{A}\circ\mathbf{C}.
\end{aligned}$$

The *transposed* matrix of the $m \times n$–matrix $\mathbf{A} = (a_{ij})$ is defined as an $n \times m$–matrix $\mathbf{A}'$, where the element a_{ij} of $\mathbf{A}$ is in the i–th column and j–th row $(i = 1, \ldots, m;\ j = 1, \ldots, n)$. The transposing operation satisfies the following basic relations:

$$\begin{aligned}(\mathbf{A}')' &= \mathbf{A};\\ (\mathbf{A}+\mathbf{B})' &= \mathbf{A}'+\mathbf{B}';\\ (\mathbf{AB})' &= \mathbf{B}'\mathbf{A}';\\ (\mathbf{A}\circ\mathbf{B})' &= \mathbf{A}'\circ\mathbf{B}'.\end{aligned}$$

Any matrix $\mathbf{A} : m \times n$ can be written as the sum:

$$\mathbf{A} = \sum_{i=1}^{m}\sum_{j=1}^{n} a_{ij}\mathbf{e}_i\mathbf{d}_j', \tag{1.1.2}$$

where $\mathbf{e}_i$ is the m–vector with 1 in the i–th position and 0 in other positions, and $\mathbf{d}_j$ is the n–vector with 1 in the j–th position and 0 elsewhere. The vectors $\mathbf{e}_i$ and $\mathbf{d}_j$ are so-called canonical basis vectors. When proving results for matrix derivatives, for example, the way of writing $\mathbf{A}$ as in (1.1.2) will be of utmost importance. Moreover, when not otherwise stated, $\mathbf{e}_i$ will always denote a canonical basis vector. Sometimes we have to use superscripts on the basis vectors, i.e. $\mathbf{e}_i^1$, $\mathbf{e}_j^2$, $\mathbf{d}_k^1$ and so on.
The role of the unity among matrices is played by the *identity matrix* $\mathbf{I}_m$;

$$\mathbf{I}_m = \begin{pmatrix} 1 & 0 & \ldots & 0\\ 0 & 1 & \ldots & 0\\ \vdots & \vdots & \ddots & \vdots\\ 0 & 0 & \ldots & 1\end{pmatrix} = (\delta_{ij}), \quad (i, j = 1, \ldots, m), \tag{1.1.3}$$

where δ_{ij} is Kronecker's delta, i.e.

$$\delta_{ij} = \begin{cases} 1, & i = j,\\ 0, & i \neq j.\end{cases}$$

If not necessary, the index m in $\mathbf{I}_m$ is dropped. The identity matrix satisfies the trivial equalities

$$\mathbf{I}_m\mathbf{A} = \mathbf{A}\mathbf{I}_n = \mathbf{A},$$

for $\mathbf{A} : m \times n$. Furthermore, the canonical basis vectors $\mathbf{e}_i$, $\mathbf{d}_j$ used in (1.1.2) are identical to the i–th column of $\mathbf{I}_m$ and the j–th column of $\mathbf{I}_n$, respectively. Thus, $\mathbf{I}_m = (\mathbf{e}_1, \ldots, \mathbf{e}_m)$ is another way of representing the identity matrix.
There exist some classes of matrices, which are of special importance in applications. A square matrix $\mathbf{A}$ is *symmetric*, if

$$\mathbf{A} = \mathbf{A}'.$$

A square matrix $\mathbf{A}$ is *skew-symmetric*, if

$$\mathbf{A} = -\mathbf{A}'.$$

A real square matrix $\mathbf{A}$ is *orthogonal*, if

$$\mathbf{A}'\mathbf{A} = \mathbf{I}.$$

Then also $\mathbf{AA}' = \mathbf{I}$. A matrix $\mathbf{A} \in \mathbb{R}^{m\times n}$ is *semiorthogonal* if

$$\mathbf{A}'\mathbf{A} = \mathbf{I}_n \quad \text{or} \quad \mathbf{AA}' = \mathbf{I}_m.$$

The two m-vectors $\mathbf{a}$ and $\mathbf{b}$ are *orthogonal* if

$$\mathbf{a}'\mathbf{b} = 0.$$

Obviously then also $\mathbf{b}'\mathbf{a} = 0$. The orthogonal $m \times m-$matrix $\mathbf{A}$ defines a rotation or reflection in $\mathbb{R}^m$. If we denote

$$\mathbf{y} = \mathbf{Ax},$$

then it follows from the orthogonality of $\mathbf{A}$ that the vectors $\mathbf{x}$ and $\mathbf{y}$ are of the same length, i.e.

$$\mathbf{x}'\mathbf{x} = \mathbf{y}'\mathbf{y}.$$

A square matrix $\mathbf{A}$ is *idempotent* if

$$\mathbf{A}^2 = \mathbf{A}.$$

A square matrix $\mathbf{A}$ is *normal* if

$$\mathbf{AA}' = \mathbf{A}'\mathbf{A}.$$

Note that symmetric, skew-symmetric and orthogonal matrices are all normal matrices.

A square matrix $\mathbf{A} : m \times m$ is a *Toeplitz* matrix, if $a_{ij} = \alpha_{i-j}$ for any $i, j = 1, \ldots, m$. It means that in a Toeplitz matrix $\mathbf{A}$ the elements satisfy $a_{ij} = a_{kl}$ when $i - j = k - l$. Different subclasses of Toeplitz matrices are examined in Basilevsky (1983), for example.

An $m \times m-$matrix $\mathbf{A}$ is an *upper triangular* matrix if all elements of $\mathbf{A}$ below the main diagonal are zeros:

$$\mathbf{A} = \begin{pmatrix} a_{11} & a_{12} & \ldots & a_{1m} \\ 0 & a_{22} & \ldots & a_{2m} \\ \vdots & \vdots & \ddots & \vdots \\ 0 & 0 & \ldots & a_{mm} \end{pmatrix}.$$

If all the elements of $\mathbf{A}$ above its main diagonal are zeros, then $\mathbf{A}$ is a *lower triangular* matrix. Clearly, if $\mathbf{A}$ is an upper triangular matrix , then $\mathbf{A}'$ is a lower triangular matrix.

The *diagonalization* of a square matrix $\mathbf{A}$ is the operation which replaces all the elements outside the main diagonal of $\mathbf{A}$ by zeros. A diagonalized matrix will be denoted by $\mathbf{A}_d$, i.e.

$$\mathbf{A}_d = \begin{pmatrix} a_{11} & 0 & \dots & 0 \\ 0 & a_{22} & \dots & 0 \\ \vdots & \vdots & \ddots & \vdots \\ 0 & 0 & \dots & a_{mm} \end{pmatrix}. \tag{1.1.4}$$

Similarly, $\mathbf{a}_d$ denotes the diagonal matrix obtained from the vector $\mathbf{a} = (a_1, a_2, \dots, a_m)'$:

$$\mathbf{a}_d = \begin{pmatrix} a_1 & 0 & \dots & 0 \\ 0 & a_2 & \dots & 0 \\ \vdots & \vdots & \ddots & \vdots \\ 0 & 0 & \dots & a_m \end{pmatrix}.$$

An extension of this notion will also be used. From square matrices $\mathbf{A}_1$, $\mathbf{A}_2$, $\dots$, $\mathbf{A}_m$ a block diagonal matrix $\mathbf{A}_{[d]}$ is created similarly to $\mathbf{a}_d$:

$$\mathbf{A}_{[d]} = \begin{pmatrix} \mathbf{A}_1 & \mathbf{0} & \dots & \mathbf{0} \\ \mathbf{0} & \mathbf{A}_2 & \dots & \mathbf{0} \\ \vdots & \vdots & \ddots & \vdots \\ \mathbf{0} & \mathbf{0} & \dots & \mathbf{A}_m \end{pmatrix}.$$

Furthermore, we need a notation for the vector consisting of the elements of the main diagonal of a square matrix $\mathbf{A} : m \times m$:

$$\mathrm{diag}\mathbf{A} = (a_{11}, a_{22}, \dots, a_{mm})'.$$

In few cases when we consider complex matrices, we need the following notions. If $\mathbf{A} = (a_{ij}) \in \mathbb{C}^{m\times n}$, then the matrix $\overline{\mathbf{A}} : m \times n$ denotes the *conjugate matrix* of $\mathbf{A}$, where $(\overline{\mathbf{A}})_{ij} = \overline{a}_{ij}$ is the complex conjugate of a_{ij}. The matrix $\mathbf{A}^* : n \times m$ is said to be the *conjugate transpose* of $\mathbf{A}$, if $\mathbf{A}^* = (\overline{\mathbf{A}})'$. A matrix $\mathbf{A} : n \times n$ is *unitary*, if $\mathbf{A}\mathbf{A}^* = \mathbf{I}_n$. In this case $\mathbf{A}^*\mathbf{A} = \mathbf{I}_n$ also holds.

1.1.2 *Determinant, Inverse*

An important characteristic of a square $m \times m-$matrix $\mathbf{A}$ is its *determinant* $|\mathbf{A}|$, which is defined by the equality

$$|\mathbf{A}| = \sum_{(j_1,\dots,j_m)} (-1)^{N(j_1,\dots,j_m)} \prod_{i=1}^{m} a_{ij_i}, \tag{1.1.5}$$

where summation is taken over all different permutations $(j_1,\dots,j_m)$ of the set of integers $\{1,2,\dots,m\}$, and $N(j_1,\dots,j_m)$ is the number of inversions of the permutation $(j_1,\dots,j_m)$. The inversion of a permutation $(j_1,\dots,j_m)$ consists of interchanging two indices so that the larger index comes after the smaller one. For example, if $m=5$ and

$$(j_1,\dots,j_5) = (2,1,5,4,3),$$

then

$$N(2,1,5,4,3) = 1 + N(1,2,5,4,3) = 3 + N(1,2,4,3,5) = 4 + N(1,2,3,4,5) = 4,$$

since $N(1,2,3,4,5) = 0$. Calculating the number of inversions is a complicated problem when m is large. To simplify the calculations, a technique of finding determinants has been developed which is based on minors. The *minor of an element* a_{ij} is the determinant of the $(m-1)\times(m-1)-$submatrix $\mathbf{A}_{(ij)}$ of a square matrix $\mathbf{A} : m\times m$ which is obtained by crossing out the $i-$th row and $j-$th column from $\mathbf{A}$. Through minors the expression of the determinant in (1.1.5) can be presented as

$$|\mathbf{A}| = \sum_{j=1}^{m} a_{ij}(-1)^{i+j}|\mathbf{A}_{(ij)}| \quad \text{for any } i, \tag{1.1.6}$$

or

$$|\mathbf{A}| = \sum_{i=1}^{m} a_{ij}(-1)^{i+j}|\mathbf{A}_{(ij)}| \quad \text{for any } j. \tag{1.1.7}$$

The expression $(-1)^{i+j}|\mathbf{A}_{(ij)}|$ is called *cofactor* of a_{ij}. Any element of $\mathbf{A}$ is a minor of order 1. Fixing r rows, e.g. $(i_1,\dots,i_r)$, and r columns, e.g. $(j_1,\dots,j_r)$, of a matrix $\mathbf{A}$ gives us a square submatrix of order r. The determinant of this matrix is called a *minor of order* r. If $(i_1,\dots,i_r) = (j_1,\dots,j_r)$, we have a *principal submatrix* of a square matrix $\mathbf{A}$ and the determinant of a principal submatrix is called a *principal minor of order* r of $\mathbf{A}$. The sum of the principal minors of order r of $\mathbf{A}$ is denoted $\mathrm{tr}_r\mathbf{A}$ and known as the *$r-$th trace* of $\mathbf{A}$. By convention $\mathrm{tr}_0\mathbf{A} = 1$. The most fundamental properties of the determinant are summarized in

Proposition 1.1.1.

(i) *For* $\mathbf{A} : m\times m$

$$|\mathbf{A}| = |\mathbf{A}'|.$$

(ii) *If* $|\mathbf{A}|$ *and* $|\mathbf{B}|$ *are non-zero,*

$$|\mathbf{AB}| = |\mathbf{A}||\mathbf{B}|.$$

(iii) *For* $\mathbf{A} : m \times n$ *and* $\mathbf{B} : n \times m$

$$|\mathbf{I}_m + \mathbf{AB}| = |\mathbf{I}_n + \mathbf{BA}|.$$

∎

When $|\mathbf{A}| \neq 0$, the $m \times m-$matrix $\mathbf{A}$ is called *non-singular* and then a unique *inverse* of $\mathbf{A}$ exists. Otherwise, when $|\mathbf{A}| = 0$, the matrix $\mathbf{A} : m \times m$ is *singular*. The inverse is denoted $\mathbf{A}^{-1}$ and it is defined by the equation

$$\mathbf{AA}^{-1} = \mathbf{I}_m = \mathbf{A}^{-1}\mathbf{A}.$$

Explicitly we can express a general element of the inverse matrix in the following way

$$(\mathbf{A}^{-1})_{ij} = \frac{(-1)^{i+j}|\mathbf{A}_{(ji)}|}{|\mathbf{A}|}. \tag{1.1.8}$$

For the inverse the main properties are given in

Proposition 1.1.2. *Suppose that all the inverses given below exist. Then*

(i) $$(\mathbf{AB})^{-1} = \mathbf{B}^{-1}\mathbf{A}^{-1};$$

(ii) $$(\mathbf{A}^{-1})' = (\mathbf{A}')^{-1};$$

(iii) $$|\mathbf{A}|^{-1} = |\mathbf{A}^{-1}|.$$

∎

1.1.3 *Rank, Trace*

The vectors $\mathbf{x}_1, \ldots, \mathbf{x}_r$ are said to be *linearly independent*, if

$$\sum_{i=1}^{r} c_i \mathbf{x}_i = \mathbf{0}$$

implies that $c_i = 0$, $i = 1, 2, \ldots, r$. When $\mathbf{z}$ is an m-vector and the $m \times m-$matrix $\mathbf{A}$ is non-singular ($|\mathbf{A}| \neq 0$), then the only solution to the equation

$$\mathbf{Az} = \mathbf{0}, \tag{1.1.9}$$

is the trivial solution $\mathbf{z} = \mathbf{0}$. This means that the rows (columns) of $\mathbf{A}$ are linearly independent. If $|\mathbf{A}| = 0$, then exists at least one non-trivial solution $\mathbf{z} \neq \mathbf{0}$ to equation (1.1.9) and the rows (columns) are dependent. The rank of a matrix is closely related to the linear dependence or independence of its row and column vectors.

Definition 1.1.1. *A matrix* $\mathbf{A} : m \times n$ *is of rank* r, *if the maximum number of linear independent columns of* $\mathbf{A}$ *equals* r. ∎

The rank of a matrix $\mathbf{A}$ is denoted by $r(\mathbf{A})$ and it can be characterized in many different ways. Let us present the most important properties in the next proposition. In some statements given below we will use the notation $\mathbf{A}^o$ for a matrix such that $\mathbf{A}^{o'}\mathbf{A} = \mathbf{0}$ and $r(\mathbf{A}^o) = m - r(\mathbf{A})$ if $\mathbf{A} : m \times n$. Also the notation $(\mathbf{A} : \mathbf{B})$ will be used for a partitioned matrix consisting of two blocks $\mathbf{A}$ and $\mathbf{B}$. Partitioned matrices are discussed in detail in Section 1.3. However, if $\mathbf{a}_i$, $i = 1, 2, \ldots, p$, are the p columns of a matrix $\mathbf{A}$ we sometimes write $\mathbf{A} = (\mathbf{a}_1, \mathbf{a}_2, \ldots, \mathbf{a}_p)$ instead of $\mathbf{A} = (\mathbf{a}_1 : \mathbf{a}_2 : \ldots : \mathbf{a}_p)$.

Proposition 1.1.3.

(i) *The rank of a matrix equals the maximum number of linearly independent rows of* $\mathbf{A}$, *i.e.*
$$r(\mathbf{A}) = r(\mathbf{A}').$$

(ii) *For* $\mathbf{A} : m \times n$
$$r(\mathbf{A}) \leq \min(m, n).$$

(iii) *The rank of a* $\mathbf{A}$ *equals the order of its largest nonzero minor.*

(iv) *For arbitrary matrices* $\mathbf{A}$ *and* $\mathbf{B}$ *of proper sizes*
$$r(\mathbf{AB}) \leq \min(r(\mathbf{A}), r(\mathbf{B})).$$

(v) *For arbitrary matrices* $\mathbf{A}$ *and* $\mathbf{B}$ *of proper sizes*
$$r(\mathbf{A} + \mathbf{B}) \leq r(\mathbf{A}) + r(\mathbf{B}).$$

(vi) *Let* $\mathbf{A}$, $\mathbf{B}$ *and* $\mathbf{C}$ *be of proper sizes and let* $\mathbf{A}$ *and* $\mathbf{C}$ *be non-singular. Then*
$$r(\mathbf{ABC}) = r(\mathbf{B}).$$

(vii) *Let* $\mathbf{A}$: $m \times n$ *and* $\mathbf{B}$ *satisfy* $\mathbf{AB} = \mathbf{0}$. *Then*
$$r(\mathbf{B}) \leq n - r(\mathbf{A}).$$

(viii) *Let* $\mathbf{A}$ *and* $\mathbf{B}$ *be of proper sizes. Then*
$$r(\mathbf{A} : \mathbf{B}) = r(\mathbf{A}'\mathbf{B}^o) + \mathbf{r}(\mathbf{B}).$$

(ix) *Let* $\mathbf{A}$ *and* $\mathbf{B}$ *be of proper sizes. Then*
$$r(\mathbf{A} : \mathbf{B}) = r(\mathbf{A}) + r(\mathbf{B}) - r((\mathbf{A}^o : \mathbf{B}^o)^o).$$

(x) *Let* $\mathbf{A}$: $n \times m$ *and* $\mathbf{B}$: $m \times n$. *Then*

$$r(\mathbf{A} - \mathbf{ABA}) = r(\mathbf{A}) + r(\mathbf{I}_m - \mathbf{BA}) - m = r(\mathbf{A}) + r(\mathbf{I}_n - \mathbf{AB}) - n.$$

∎

Definition 1.1.1 is very seldom used when finding the rank of a matrix. Instead the matrix is transformed by elementary operations to a canonical form, so that we can obtain the rank immediately. The following actions are known as *elementary operations*:

1) interchanging two rows (or columns) of $\mathbf{A}$;
2) multiplying all elements of a row (or column) of $\mathbf{A}$ by some nonzero scalar;
3) adding to any row (or column) of $\mathbf{A}$ any other row (or column) of $\mathbf{A}$ multiplied by a nonzero scalar.

The elementary operations can be achieved by pre- and postmultiplying $\mathbf{A}$ by appropriate non-singular matrices. Moreover, at the same time it can be shown that after repeated usage of elementary operations any $m \times n$−matrix $\mathbf{A}$ can be transformed to a matrix with $\mathbf{I}_r$ in the upper left corner and all the other elements equal to zero. Combining this knowledge with Proposition 1.1.3 (vi) we can state the following.

Theorem 1.1.1. *For every matrix* $\mathbf{A}$ *there exist non-singular matrices* $\mathbf{P}$ *and* $\mathbf{Q}$ *such that*

$$\mathbf{PAQ} = \begin{pmatrix} \mathbf{I}_r & \mathbf{0} \\ \mathbf{0} & \mathbf{0} \end{pmatrix},$$

where $r = r(\mathbf{A})$ *and* $\mathbf{P}, \mathbf{Q}$ *can be presented as products of matrices of elementary operations.* ∎

Later in Proposition 1.1.6 and §1.2.10 we will give somewhat more general results of the same character.

The sum of the diagonal elements of a square matrix is called *trace* and denoted $\mathrm{tr}\mathbf{A}$, i.e. $\mathrm{tr}\mathbf{A} = \sum_i a_{ii}$. Note that $\mathrm{tr}\mathbf{A} = \mathrm{tr}_1\mathbf{A}$. Properties of the trace function will be used repeatedly in the subsequent.

Proposition 1.1.4.

(i) *For* $\mathbf{A}$ *and* $\mathbf{B}$ *of proper sizes*

$$\mathrm{tr}(\mathbf{AB}) = \mathrm{tr}(\mathbf{BA}).$$

(ii) *For* $\mathbf{A}$ *and* $\mathbf{B}$ *of proper sizes and* $\mathbf{B}$ *non-singular*

$$\mathrm{tr}\mathbf{A} = \mathrm{tr}(\mathbf{B}^{-1}\mathbf{AB}).$$

(iii) *For* $\mathbf{A}$ *and* $\mathbf{C}$ *of proper sizes and* $\mathbf{C}$ *orthogonal*

$$\mathrm{tr}\mathbf{A} = \mathrm{tr}(\mathbf{C}'\mathbf{AC}).$$

(iv) *For* $\mathbf{A}$ *and* $\mathbf{B}$ *of proper sizes and constants* $a, b \in \mathbb{R}$

$$\mathrm{tr}(a\mathbf{A} + b\mathbf{B}) = a\mathrm{tr}\mathbf{A} + b\mathrm{tr}\mathbf{B}.$$

(v) *If* $\mathbf{A}$ *satisfies* $\mathbf{A}^2 = n\mathbf{A}$, $n \in \mathbb{N}$, *then*

$$\mathrm{tr}\mathbf{A} = nr(\mathbf{A}).$$

If $\mathbf{A}$ *is idempotent, then*

$$\mathrm{tr}\mathbf{A} = r(\mathbf{A}).$$

(vi) *For any* $\mathbf{A}$

$$\mathrm{tr}(\mathbf{A}'\mathbf{A}) = 0$$

if and only if $\mathbf{A} = \mathbf{0}$.

(vii) *For any* $\mathbf{A}$

$$\mathrm{tr}(\mathbf{A}') = \mathrm{tr}\mathbf{A}.$$

(viii) *For any* $\mathbf{A}$

$$\mathrm{tr}(\mathbf{A}\mathbf{A}') = \mathrm{tr}(\mathbf{A}'\mathbf{A}) = \sum_{i=1}^{m}\sum_{j=1}^{n} a_{ij}^2.$$

(ix) *For symmetric* $\mathbf{A}$ *and* $\mathbf{B}$

$$\mathrm{tr}\{(\mathbf{A}\mathbf{B})^2\} \leq \mathrm{tr}(\mathbf{A}^2\mathbf{B}^2).$$

(x) *For* $\mathbf{A}$ *and* $\mathbf{x}$ *of proper sizes*

$$\mathbf{x}'\mathbf{A}\mathbf{x} = \mathrm{tr}(\mathbf{A}\mathbf{x}\mathbf{x}').$$

(xi) *For any symmetric* $\mathbf{A}$ *such that* $\mathbf{A} \neq \mathbf{0}$

$$r(\mathbf{A}) \geq \frac{(\mathrm{tr}\mathbf{A})^2}{\mathrm{tr}(\mathbf{A}^2)}.$$

∎

There are several integral representations available for the determinant and here we present one, which ties together the notion of the trace function, inverse matrix and determinant. Sometimes the relation in the next theorem is called *Aitken's integral* (Searle, 1982).

Theorem 1.1.2. *Let* $\mathbf{A} \in \mathbb{R}^{m \times m}$ *be non-singular, then*

$$|\mathbf{A}|^{-1} = \frac{1}{(2\pi)^{m/2}} \int_{\mathbb{R}^m} e^{-1/2\mathrm{tr}(\mathbf{A}\mathbf{A}'\mathbf{y}\mathbf{y}')} d\mathbf{y}, \qquad (1.1.10)$$

where $\int_{\mathbb{R}^m}$ *denotes the multiple integral* $\underbrace{\int_{\mathbb{R}} \cdots \int_{\mathbb{R}}}_{m \text{ times}}$ *and* $d\mathbf{y}$ *is the Lebesque measure:* $d\mathbf{y} = \prod_{i=1}^{m} dy_i$. ∎

REMARK: In (1.1.10) the multivariate normal density function is used which will be discussed in Section 2.2.

1.1.4 *Positive definite matrices*

Definition 1.1.2. *A symmetric $m \times m-$matrix $\mathbf{A}$ is positive (negative) definite if $\mathbf{x}'\mathbf{A}\mathbf{x} > 0$ (< 0) for any vector $\mathbf{x} \neq \mathbf{0}$.* ∎

When $\mathbf{A}$ is positive (negative) definite, we denote this by $\mathbf{A} > 0$ (< 0), and sometimes the abbreviation p.d. (n.d.) is used. A symmetric matrix $\mathbf{A}$ is *positive semidefinite* (p.s.d.), if $\mathbf{x}'\mathbf{A}\mathbf{x} \geq 0$ for any $\mathbf{x}$ and $\mathbf{x}'\mathbf{A}\mathbf{x} = 0$ for at least one $\mathbf{x} \neq \mathbf{0}$. A matrix $\mathbf{A}$ is called *negative semidefinite* (n.s.d.), if $\mathbf{x}'\mathbf{A}\mathbf{x} \leq 0$ for any $\mathbf{x}$ and there exists at least one $\mathbf{x} \neq \mathbf{0}$ for which $\mathbf{x}'\mathbf{A}\mathbf{x} = 0$. To denote p.s.d. (n.s.d.) we use $\mathbf{A} \geq 0$ $(\mathbf{A} \leq 0)$. The notion *non-negative definite* (n.n.d.) is used for the matrix $\mathbf{A}$ if $\mathbf{A} \geq 0$ or $\mathbf{A} > 0$, and analogously the matrix $\mathbf{A}$ is *non-positive definite* (n.p.d.) if $\mathbf{A} \leq 0$ or $\mathbf{A} < 0$.
The basic properties of the different types of matrices mentioned above are very similar. As an example we shall present the basic properties of positive definite matrices.

Proposition 1.1.5.

(i) *A matrix $\mathbf{A} \in \mathbb{R}^{m\times m}$ is positive definite if and only if $|\mathbf{A}_i| > 0$ for $i = 1, \dots, m$, where $\mathbf{A}_i$ is $i \times i-$matrix consisting of the elements of the first i rows and columns of $\mathbf{A}$.*

(ii) *If $\mathbf{A} > 0$, then $\mathbf{A}^{-1} > 0$.*

(iii) *A symmetric matrix is positive definite if and only if all its eigenvalues are > 0.*

(iv) *For any $\mathbf{A}$, the matrix $\mathbf{A}\mathbf{A}'$ is n.n.d.*

(v) *If $\mathbf{A}$ is n.n.d., then $\mathbf{A}$ is non-singular if and only if $\mathbf{A} > 0$.*

(vi) *If $\mathbf{A} : m \times m$ is positive definite and $\mathbf{B} : n \times m$ is of rank r, $n \leq m$, then $\mathbf{B}\mathbf{A}\mathbf{B}' > 0$ if and only if $r = n$. $\mathbf{B}\mathbf{A}\mathbf{B}' \geq 0$ if $r < n$.*

(vii) *If $\mathbf{A} > 0$, $\quad \mathbf{B} > 0$, $\quad \mathbf{A} - \mathbf{B} > 0$, then $\mathbf{B}^{-1} - \mathbf{A}^{-1} > 0$ and $|\mathbf{A}| > |\mathbf{B}|$.*

(viii) *If $\mathbf{A} > 0$ and $\mathbf{B} > 0$, then $|\mathbf{A} + \mathbf{B}| \geq |\mathbf{A}| + |\mathbf{B}|$.* ∎

There exist many related results. Rao (1973a) is a suitable reference.

1.1.5 *Factorizations*

A basic property which also may serve as a definition of a positive definite matrix is given in

Theorem 1.1.3.

(i) *The matrix $\mathbf{A}$ is positive definite if and only if $\mathbf{A} = \mathbf{X}\mathbf{X}'$ for some non-singular $\mathbf{X}$.*

(ii) *The matrix $\mathbf{A}$ is non-negative definite if and only if $\mathbf{A} = \mathbf{X}\mathbf{X}'$ for some $\mathbf{X}$.* ∎

The next theorem is a key result for obtaining several complicated distribution results.

Theorem 1.1.4. *(Cholesky decomposition) Let $\mathbf{W} : n \times n$ be positive definite. Then there exists a unique lower triangular matrix $\mathbf{T}$ with positive diagonal elements such that $\mathbf{W} = \mathbf{T}\mathbf{T}'$.*

PROOF: We are going to prove this theorem with the help of an induction argument. If $n = 1$, the theorem is obviously true. Since $\mathbf{W}$ is p.d. we can always find an $\mathbf{X}$ such that $\mathbf{W} = \mathbf{XX}'$. For $n = 2$ it follows that

$$\mathbf{X} = \begin{pmatrix} x_{11} & x_{12} \\ x_{21} & x_{22} \end{pmatrix}$$

and then

$$\mathbf{XX}' = \begin{pmatrix} x_{11}^2 + x_{12}^2 & x_{11}x_{21} + x_{12}x_{22} \\ x_{21}x_{11} + x_{22}x_{12} & x_{21}^2 + x_{22}^2 \end{pmatrix}.$$

Since $\mathbf{W} > 0$, $\mathbf{X}$ is of full rank; $r(\mathbf{X}) = 2$. Moreover, for a lower triangular matrix $\mathbf{T}$

$$\mathbf{TT}' = \begin{pmatrix} t_{11}^2 & t_{11}t_{21} \\ t_{21}t_{11} & t_{21}^2 + t_{22}^2 \end{pmatrix}.$$

Hence, if

$$\begin{aligned} t_{11} &= (x_{11}^2 + x_{12}^2)^{1/2}, \\ t_{21} &= (x_{11}^2 + x_{12}^2)^{-1/2}(x_{21}x_{11} + x_{22}x_{12}), \\ t_{22} &= (x_{21}^2 + x_{22}^2 - (x_{11}^2 + x_{12}^2)^{-1}(x_{21}x_{11} + x_{22}x_{12})^2)^{1/2}, \end{aligned}$$

$\mathbf{W} = \mathbf{TT}'$, where $\mathbf{T}$ has positive diagonal elements. Since $r(\mathbf{X}) = r(\mathbf{T}) = 2$, $\mathbf{W}$ is p.d.. There exist no alternative expressions for $\mathbf{T}$ such that $\mathbf{XX}' = \mathbf{TT}'$ with positive diagonal elements of $\mathbf{T}$. Now let us suppose that we can find the unique lower triangular matrix $\mathbf{T}$ for any p.d. matrix of size $(n-1) \times (n-1)$. For $\mathbf{W} \in \mathbb{R}^{n \times n}$,

$$\mathbf{W} = \mathbf{XX}' = \begin{pmatrix} x_{11}^2 + \mathbf{x}_{12}\mathbf{x}_{12}' & x_{11}\mathbf{x}_{21}' + \mathbf{x}_{12}\mathbf{X}_{22}' \\ \mathbf{x}_{21}x_{11} + \mathbf{X}_{22}\mathbf{x}_{12}' & \mathbf{x}_{21}\mathbf{x}_{21}' + \mathbf{X}_{22}\mathbf{X}_{22}' \end{pmatrix},$$

where

$$\mathbf{X} = \begin{pmatrix} x_{11} & \mathbf{x}_{12} \\ \mathbf{x}_{21} & \mathbf{X}_{22} \end{pmatrix}$$

is of full rank. For the lower triangular matrix

$$\mathbf{T} = \begin{pmatrix} t_{11} & \mathbf{0} \\ \mathbf{t}_{21} & \mathbf{T}_{22} \end{pmatrix}$$

we have

$$\mathbf{TT}' = \begin{pmatrix} t_{11}^2 & t_{11}\mathbf{t}_{21}' \\ \mathbf{t}_{21}t_{11} & \mathbf{t}_{21}\mathbf{t}_{21}' + \mathbf{T}_{22}\mathbf{T}_{22}' \end{pmatrix}.$$

Hence,

$$\begin{aligned} t_{11} &= (x_{11}^2 + \mathbf{x}_{12}\mathbf{x}_{12}')^{1/2}; \\ \mathbf{t}_{21} &= (x_{11}^2 + \mathbf{x}_{12}\mathbf{x}_{12}')^{-1/2}(\mathbf{x}_{21}x_{11} + \mathbf{X}_{22}\mathbf{x}_{12}'). \end{aligned}$$

Below it will be shown that the matrix

$$\mathbf{x}_{21}\mathbf{x}_{21}' + \mathbf{X}_{22}\mathbf{X}_{22}' - (\mathbf{x}_{21}x_{11} + \mathbf{X}_{22}\mathbf{x}_{12}')(x_{11}^2 + \mathbf{x}_{12}\mathbf{x}_{12}')^{-1}(x_{11}\mathbf{x}_{12} + \mathbf{x}_{12}\mathbf{X}_{22}') \quad (1.1.11)$$

is positive definite. Then, by assumption, we can always choose a unique lower triangular matrix $\mathbf{T}_{22}$ such that (1.1.11) equals $\mathbf{T}_{22}\mathbf{T}'_{22}$. Therefore, with the above given choices of elements

$$\mathbf{T} = \begin{pmatrix} t_{11} & \mathbf{0} \\ \mathbf{t}_{21} & \mathbf{T}_{22} \end{pmatrix}$$

is lower triangular and $\mathbf{W} = \mathbf{TT}'$. Now we have only to show that (1.1.11) is p.d. This follows, since (1.1.11) is equal to the product

$$(\mathbf{x}_{21} : \mathbf{X}_{22})\mathbf{P}(\mathbf{x}_{21} : \mathbf{X}_{22})', \tag{1.1.12}$$

where

$$\mathbf{P} = \mathbf{I} - (x_{11} : \mathbf{x}_{12})'((x_{11} : \mathbf{x}_{12})(x_{11} : \mathbf{x}_{12})')^{-1}(x_{11} : \mathbf{x}_{12})$$

is idempotent and symmetric. Furthermore, we have to show that (1.1.12) is of full rank, i.e. $n-1$. Later, in Proposition 1.2.1, it will be noted that for symmetric idempotent matrices $\mathbf{P}^o = \mathbf{I} - \mathbf{P}$. Thus, using Proposition 1.1.3 (viii) we obtain

$$\begin{aligned} r((\mathbf{x}_{21} : \mathbf{X}_{22})\mathbf{P}(\mathbf{x}_{21} : \mathbf{X}_{22})') &= r(\mathbf{P}(\mathbf{x}_{21} : \mathbf{X}_{22})') \\ &= r\begin{pmatrix} x_{11} & \mathbf{x}'_{21} \\ \mathbf{x}'_{12} & \mathbf{X}'_{22} \end{pmatrix} - r\begin{pmatrix} x_{11} \\ \mathbf{x}'_{12} \end{pmatrix} = n-1. \end{aligned}$$

Hence, the expression in (1.1.12) is p.d. and the theorem is established. ∎

Two types of rank factorizations, which are very useful and of principal interest, are presented in

Proposition 1.1.6.

(i) *Let* $\mathbf{A} \in \mathbb{R}^{m\times n}$ *of rank* r. *Then there exist two non-singular matrices* $\mathbf{G} \in \mathbb{R}^{m\times m}$ *and* $\mathbf{H} \in \mathbb{R}^{n\times n}$ *such that*

$$\mathbf{A} = \mathbf{G}^{-1}\begin{pmatrix} \mathbf{I}_r & \mathbf{0} \\ \mathbf{0} & \mathbf{0} \end{pmatrix}\mathbf{H}^{-1}$$

and

$$\mathbf{A} = \mathbf{KL},$$

where $\mathbf{K} \in \mathbb{R}^{m\times r}$, $\mathbf{L} \in \mathbb{R}^{r\times n}$, $\mathbf{K}$ *consists of the first* r *columns of* $\mathbf{G}^{-1}$ *and* $\mathbf{L}$ *of the first* r *rows of* $\mathbf{H}^{-1}$.

(ii) *Let* $\mathbf{A} \in \mathbb{R}^{m\times n}$ *of rank* r. *Then there exist a triangular matrix* $\mathbf{T} \in \mathbb{R}^{m\times m}$ *and an orthogonal matrix* $\mathbf{H} \in \mathbb{R}^{n\times n}$ *such that*

$$\mathbf{A} = \mathbf{T}\begin{pmatrix} \mathbf{I}_r & \mathbf{0} \\ \mathbf{0} & \mathbf{0} \end{pmatrix}\mathbf{H}$$

and

$$\mathbf{A} = \mathbf{KL},$$

where $\mathbf{K} \in \mathbb{R}^{m\times r}$, $\mathbf{L} \in \mathbb{R}^{r\times n}$, $\mathbf{K}$ *consists of the first* r *columns of* $\mathbf{T}$ *and* $\mathbf{L}$ *of the first* r *rows of* $\mathbf{H}$. ∎

For normal matrices, in particular, other useful factorizations are available which are based on eigenvalues and eigenvectors. These, as well as the Jordan factorization, are presented in §1.2.10. It concerns especially Theorem 1.2.39 and Theorem 1.2.41 – Theorem 1.2.44.

1.1.6 *Generalized inverse*

The notion of the generalized inverse of a matrix (shortly g-inverse) is not usually included into a basic course on linear algebra, and therefore we will pay more attention to it here. Classical reference books on generalized inverses are Rao & Mitra (1971), Pringle & Rayner (1971), Campbell & Meyer (1991) and Ben-Israel & Greville (2003). When $\mathbf{A}$ is a non-singular $n \times n-$matrix, the linear equation in $\mathbf{x}$,

$$\mathbf{Ax} = \mathbf{b},$$

can be solved easily by inverting the matrix $\mathbf{A}$, i.e.

$$\mathbf{x} = \mathbf{A}^{-1}\mathbf{b}.$$

Behind the above equation there is a system of linear equations in n variables. Every equation can be considered as a condition on the n variables. However, in many statistical problems it is not so common that the number of variables equals the number of conditions which have to be satisfied. Still, we are interested in solving an equation which now can be stated as

$$\mathbf{Ax} = \mathbf{b}, \tag{1.1.13}$$

where $\mathbf{A} \in \mathbb{R}^{m\times n}$ or $\mathbf{A} \in \mathbb{R}^{m\times m}$ is singular, i.e. $|\mathbf{A}| = 0$, which means that some rows (columns) are linearly dependent on others. Generalized inverses have been introduced in order to solve the system of linear equations in the general case, i.e. to solve (1.1.13).

We say that the *equation* $\mathbf{Ax} = \mathbf{b}$ *is consistent*, if there exists at least one $\mathbf{x}_0$ such that $\mathbf{Ax}_0 = \mathbf{b}$.

Definition 1.1.3. *Let* $\mathbf{A}$ *be an* $m \times n-$*matrix. An* $n \times m-$*matrix* $\mathbf{A}^-$ *is called generalized inverse matrix of the matrix* $\mathbf{A}$, *if*

$$\mathbf{x} = \mathbf{A}^{-}\mathbf{b}$$

is a solution to all consistent equations $\mathbf{Ax} = \mathbf{b}$ *in* $\mathbf{x}$. ∎

There is a fairly nice geometrical interpretation of g-inverses (e.g. see Kruskal, 1973). Let us map $\mathbf{x} \in \mathbb{R}^n$ with the help of a matrix $\mathbf{A}$ to $\mathbb{R}^m$, $m \le n$, i.e. $\mathbf{Ax} = \mathbf{y}$. The problem is, how to find an inverse map from $\mathbf{y}$ to $\mathbf{x}$. Obviously there must exist such a linear map. We will call a certain inverse map for a g-inverse $\mathbf{A}^-$. If $m < n$, the map cannot be unique. However, it is necessary that $\mathbf{AA}^-\mathbf{y} = \mathbf{Ax} = \mathbf{y}$, i.e. $\mathbf{A}^-\mathbf{y}$ will always be mapped on $\mathbf{y}$. The following theorem gives a necessary and sufficient condition for a matrix to be a generalized inverse.

Theorem 1.1.5. *An $n\times m-$matrix $\mathbf{A}^-$ is a generalized inverse matrix of $\mathbf{A} : m\times n$ if and only if*

$$\mathbf{A}\mathbf{A}^-\mathbf{A} = \mathbf{A}. \tag{1.1.14}$$

PROOF: Let us first prove that if $\mathbf{A}^-$ is a generalized inverse, then (1.1.14) holds. For every $\mathbf{z}$ there exists a $\mathbf{b}$ such that $\mathbf{Az} = \mathbf{b}$. Thus, a consistent equation has been constructed and according to the definition of a g-inverse, for every $\mathbf{z}$,

$$\mathbf{z} = \mathbf{A}^-\mathbf{b} = \mathbf{A}^-\mathbf{Az}.$$

Hence, $\mathbf{Az} = \mathbf{AA}^-\mathbf{Az}$ and since $\mathbf{z}$ is arbitrary, (1.1.14) holds.
Let us now prove sufficiency. Suppose that (1.1.14) is valid and that the equation (1.1.13) has a solution for a specific $\mathbf{b}$. Therefore, a vector $\mathbf{w}$ exists which satisfies

$$\mathbf{Aw} = \mathbf{b}.$$

Because $\mathbf{A} = \mathbf{AA}^-\mathbf{A}$, it follows that

$$\mathbf{b} = \mathbf{Aw} = \mathbf{AA}^-\mathbf{Aw} = \mathbf{AA}^-\mathbf{b},$$

which means that $\mathbf{A}^-\mathbf{b}$ is a solution to (1.1.13). ∎

Corollary 1.1.5.1. *All g-inverses $\mathbf{A}^-$ of $\mathbf{A}$ are generated by*

$$\mathbf{A}^- = \mathbf{A}_0^- + \mathbf{Z} - \mathbf{A}_0^-\mathbf{AZAA}_0^-,$$

where $\mathbf{Z}$ is an arbitrary matrix and $\mathbf{A}_0^-$ is a specific g-inverse.

PROOF: By Theorem 1.1.5, $\mathbf{A}^-$ is a g-inverse since $\mathbf{AA}^-\mathbf{A} = \mathbf{A}$. Moreover, if $\mathbf{A}_0^-$ is a specific g-inverse, choose $\mathbf{Z} = \mathbf{A}^- - \mathbf{A}_0^-$. Then

$$\mathbf{A}_0^- + \mathbf{Z} - \mathbf{A}_0^-\mathbf{A}_0\mathbf{Z}\mathbf{A}_0\mathbf{A}_0^- = \mathbf{A}^-.$$

∎

The necessary and sufficient condition (1.1.14) in Theorem 1.1.5 can also be used as a definition of a generalized inverse of a matrix. This way of defining a g-inverse has been used in matrix theory quite often. Unfortunately a generalized inverse matrix is not uniquely defined. Another disadvantage is that the operation of a generalized inverse is not transitive: when $\mathbf{A}^-$ is a generalized inverse of $\mathbf{A}$, $\mathbf{A}$ may not be a generalized inverse of $\mathbf{A}^-$. This disadvantage can be overcome by defining a reflexive generalized inverse.

Definition 1.1.4. *A generalized inverse matrix* $\mathbf{A}^-$ *is a reflexive generalized inverse matrix, if*

$$\mathbf{A}^- = \mathbf{A}^-\mathbf{A}\mathbf{A}^-. \tag{1.1.15}$$

∎

Theorem 1.1.6. *A g-inverse* $\mathbf{A}^- : n \times m$ *is a reflexive g-inverse if and only if*

$$r(\mathbf{A}^-) = r(\mathbf{A}).$$

PROOF: By definition of a reflexive g-inverse, $r(\mathbf{A}^-) = r(\mathbf{A}^-\mathbf{A})$ as well as $r(\mathbf{A}) = r(\mathbf{A}\mathbf{A}^-)$. From Proposition 1.1.4 (v) follows that

$$r(\mathbf{A}\mathbf{A}^-) = \operatorname{tr}(\mathbf{A}\mathbf{A}^-) = \operatorname{tr}(\mathbf{A}^-\mathbf{A}) = r(\mathbf{A}^-\mathbf{A})$$

and thus reflexivity implies $r(\mathbf{A}) = r(\mathbf{A}^-)$.
For proving sufficiency, let us take $\mathbf{A} \in \mathbb{R}^{m\times n}$ and utilize Proposition 1.1.3 (x). Then

$$\begin{aligned} r(\mathbf{A}^- - \mathbf{A}^-\mathbf{A}\mathbf{A}^-) &= r(\mathbf{A}^-) + r(\mathbf{I}_m - \mathbf{A}\mathbf{A}^-) - m = r(\mathbf{A}^-) - r(\mathbf{A}\mathbf{A}^-) \\ &= r(\mathbf{A}^-) - r(\mathbf{A}) = 0 \end{aligned}$$

which establishes the theorem. ∎

Note that for a general g-inverse $r(\mathbf{A}) \leq r(\mathbf{A}^-)$, whereas reflexivity implies equality of the ranks. To obtain a uniquely defined generalized inverse matrix we have to add two more conditions to (1.1.14) and (1.1.15).

Definition 1.1.5. *An* $n \times m-$*matrix* $\mathbf{A}^+$ *is called the Moore-Penrose generalized inverse matrix, if the following equalities are satisfied:*

$$\mathbf{A}\mathbf{A}^+\mathbf{A} = \mathbf{A}, \tag{1.1.16}$$
$$\mathbf{A}^+\mathbf{A}\mathbf{A}^+ = \mathbf{A}^+, \tag{1.1.17}$$
$$(\mathbf{A}\mathbf{A}^+)' = \mathbf{A}\mathbf{A}^+, \tag{1.1.18}$$
$$(\mathbf{A}^+\mathbf{A})' = \mathbf{A}^+\mathbf{A}. \tag{1.1.19}$$

∎

Uniqueness of the Moore-Penrose inverse is proved in the next theorem.

Theorem 1.1.7. *For* $\mathbf{A} : m \times n$ *the Moore-Penrose inverse matrix* $\mathbf{A}^+ : n \times m$ *is uniquely defined.*

PROOF: Let us assume on the contrary, that there exist $\mathbf{B}$ and $\mathbf{C}$, both satisfying (1.1.16) – (1.1.19), with $\mathbf{B} \neq \mathbf{C}$. We will show that this assumption leads to a contradiction. Namely, by using (1.1.16) – (1.1.19) to both matrices $\mathbf{B}$ and $\mathbf{C}$ repeatedly, the following sequence of equalities emerges:

$$\begin{aligned} \mathbf{B} &= \mathbf{BAB} = \mathbf{B}(\mathbf{AB})' = \mathbf{BB}'\mathbf{A}' = \mathbf{BB}'(\mathbf{ACA})' = \mathbf{BB}'\mathbf{A}'(\mathbf{AC})' = \mathbf{BABAC} \\ &= \mathbf{BAC} = (\mathbf{BA})'\mathbf{CAC} = \mathbf{A}'\mathbf{B}'(\mathbf{CA})'\mathbf{C} = \mathbf{A}'\mathbf{C}'\mathbf{C} = (\mathbf{CA})'\mathbf{C} = \mathbf{C}. \end{aligned}$$

∎

The definition of a generalized inverse matrix and, in particular, the definition of the Moore-Penrose inverse are not constructive. They do not tell us how to find these matrices. Here we present one way to obtain them. Let $\mathbf{A}$ be an $m \times n-$matrix of rank r. By interchanging rows and columns of $\mathbf{A}$ we can reach the situation when the $r \times r$ submatrix in the upper left corner of $\mathbf{A}$ is non-singular. Therefore, without loss of generality, it can be assumed that

$$\mathbf{A} = \begin{pmatrix} \mathbf{B} & \mathbf{C} \\ \mathbf{D} & \mathbf{E} \end{pmatrix},$$

where $\mathbf{B} : r \times r$ is non-singular and $(\mathbf{D} : \mathbf{E})'$ is linearly dependent on $(\mathbf{B} : \mathbf{C})'$, i.e $(\mathbf{D} : \mathbf{E})' = (\mathbf{B} : \mathbf{C})'\mathbf{Q}$ for some matrix $\mathbf{Q}$. Thus, $\mathbf{D}' = \mathbf{B}'\mathbf{Q}$ and $\mathbf{E}' = \mathbf{C}'\mathbf{Q}$. Then it is possible to show, by utilizing elementary rules for multiplying partitioned matrices, that

$$\mathbf{A}^{-} = \begin{pmatrix} \mathbf{B}^{-1} & \mathbf{0} \\ \mathbf{0} & \mathbf{0} \end{pmatrix}.$$

In the literature many properties of generalized inverse matrices can be found. Here is a short list of properties for the Moore-Penrose inverse.

Proposition 1.1.7.

(i) *Let $\mathbf{A}$ be non-singular. Then*

$$\mathbf{A}^{+} = \mathbf{A}^{-1}.$$

(ii)

$$(\mathbf{A}^{+})^{+} = \mathbf{A}.$$

(iii)

$$(\mathbf{A}')^{+} = (\mathbf{A}^{+})'.$$

(iv) *Let $\mathbf{A}$ be symmetric and idempotent. Then*

$$\mathbf{A}^{+} = \mathbf{A}.$$

(v) *The matrices $\mathbf{A}\mathbf{A}^{+}$ and $\mathbf{A}^{+}\mathbf{A}$ are idempotent.*

(vi) *The matrices $\mathbf{A}, \mathbf{A}^{+}, \mathbf{A}\mathbf{A}^{+}$ and $\mathbf{A}^{+}\mathbf{A}$ have the same rank.*

(vii)

$$\mathbf{A}'\mathbf{A}\mathbf{A}^{+} = \mathbf{A}' = \mathbf{A}^{+}\mathbf{A}\mathbf{A}'.$$

(viii)

$$\mathbf{A}'(\mathbf{A}^{+})'\mathbf{A}^{+} = \mathbf{A}^{+} = \mathbf{A}^{+}(\mathbf{A}^{+})'\mathbf{A}'.$$

(ix)

$$(\mathbf{A}'\mathbf{A})^{+} = \mathbf{A}^{+}(\mathbf{A}^{+})'.$$

(x)

$$\mathbf{A}\mathbf{A}^{+} = \mathbf{A}(\mathbf{A}'\mathbf{A})^{-}\mathbf{A}'.$$

(xi)

$$\mathbf{A}^{+} = \mathbf{A}'(\mathbf{A}\mathbf{A}')^{-1}, \text{ if } \mathbf{A} \text{ has full row rank.}$$

(xii)

$$\mathbf{A}^{+} = (\mathbf{A}'\mathbf{A})^{-1}\mathbf{A}', \text{ if } \mathbf{A} \text{ has full column rank.}$$

(xiii)

$$\mathbf{A} = \mathbf{0} \text{ if and only if } \mathbf{A}^{+} = \mathbf{0}.$$

(xiv)
$$\mathbf{AB} = \mathbf{0} \textit{ if and only if } \mathbf{B}^+\mathbf{A}^+ = \mathbf{0}.$$

∎

1.1.7 *Problems*

1. Under which conditions is multiplication of matrices commutative?
2. The sum and the elementwise product of matrices have many properties which can be obtained by changing the operation "+" to "$\circ$" in the formulas. Find two examples when this is not true.
3. Show that $\mathbf{A}^{-1}\mathbf{A} = \mathbf{I}$ when $\mathbf{A}\mathbf{A}^{-1} = \mathbf{I}$.
4. Prove formula (1.1.8).
5. Prove statements (v) and (xi) of Proposition 1.1.4.
6. Prove statements (iv), (vii) and (viii) of Proposition 1.1.5.
7. Let $\mathbf{A}$ be a square matrix of order n. Show that
$$|\mathbf{A} + \lambda\mathbf{I}_n| = \sum_{i=0}^{n} \lambda^i \mathrm{tr}_{n-i}\mathbf{A}.$$
8. Let
$$\mathbf{A} = \begin{pmatrix} 0 & 1 & 2 \\ 1 & 2 & 1 \end{pmatrix}.$$
Find $\mathbf{A}^-$.
9. When is the following true: $\mathbf{A}^+ = (\mathbf{A}'\mathbf{A})^-\mathbf{A}'$?
10. Give an example of a g-inverse which is not a reflexive generalized inverse.
11. Find the Moore-Penrose inverse of $\mathbf{A}$ in Problem 8.
12. For $\mathbf{A}$ in Problem 8 find a reflexive inverse which is not the Moore-Penrose inverse.
13. Is $\mathbf{A}^+$ symmetric if $\mathbf{A}$ is symmetric?
14. Let $\mathbf{\Gamma}$ be an orthogonal matrix. Show that $\mathbf{\Gamma} \circ \mathbf{\Gamma}$ is *doubly stochastic*, i.e. the sum of the elements of each row and column equals 1.
15. (Hadamard inequality) For non-singular $\mathbf{B} = (b_{ij}) : n \times n$ show that
$$|\mathbf{B}|^2 \leq \prod_{i=1}^{n}\sum_{j=1}^{n} b_{ij}^2.$$

1.2 ALGEBRA OF SUBSPACES

1.2.1 *Introduction*

In statistics and, particularly, multivariate statistical analysis there is a wide range of applications of *vector* (*linear*) *spaces.* When working with linear models or linearizations, it is often natural to utilize vector space theory. Depending on the level of abstraction a statistician needs certain tools. They may vary from column vector spaces associated with matrices to abstract coordinate free vector space relations. A fairly universal set of relations between subspaces in a finite-dimensional vector space is provided by *lattice theory.* In order to introduce the reader to this topic we recall the definition of a vector (linear) space $\mathbb{V}$:
Let $\mathbf{x}$, $\mathbf{y}$, $\mathbf{z}$, $\ldots$ belong to $\mathbb{V}$, where the operations " + " (sum of vectors) and " $\cdot$ " (multiplication by scalar) are defined so that $\mathbf{x}+\mathbf{y} \in \mathbb{V}$, $\alpha\mathbf{x} = \alpha \cdot \mathbf{x} \in \mathbb{V}$, where α belongs to some field K and

$$\mathbf{x}+\mathbf{y} = \mathbf{y}+\mathbf{x},$$
$$(\mathbf{x}+\mathbf{y})+\mathbf{z} = \mathbf{x}+(\mathbf{y}+\mathbf{z}),$$
there exists a unique null vector $\mathbf{0}$ in the space so that, for all $\mathbf{x} \in \mathbb{V}$, $\mathbf{x}+\mathbf{0} = \mathbf{x}$,
for every $\mathbf{x} \in \mathbb{V}$ there exists a unique $-\mathbf{x} \in \mathbb{V}$ so that $\mathbf{x}+(-\mathbf{x}) = \mathbf{0}$,
$$1 \cdot \mathbf{x} = \mathbf{x},$$
$$\alpha(\beta\mathbf{x}) = (\alpha\beta)\mathbf{x}, \quad \alpha, \beta \in K,$$
$$(\alpha+\beta)\mathbf{x} = \alpha\mathbf{x}+\beta\mathbf{x},$$
$$\alpha(\mathbf{x}+\mathbf{y}) = \alpha\mathbf{x}+\alpha\mathbf{y}.$$

If these conditions are satisfied we say that we have a vector space $\mathbb{V}$ over the field K.
However, the starting point of this paragraph is the observation that the totality of subspaces of $\mathbb{V}$ forms a modular lattice with respect to set inclusion (see the next paragraph for a definition) which is sometimes called the *Dedekind lattice.* The modular lattice structure implies some very fundamental algebraic rules. It is interesting to see that these rules generate new results which are useful for applications as well as for getting better insight of the problems under consideration. In particular, we are going to deal with decompositions of subspaces.
Usually, in multivariate analysis inner product spaces are used, and then it is possible to apply a specific part of the lattice theory, namely the theory of orthomodular lattices. From the definition of an orthomodular lattice several fundamental vector space decompositions follow, and under the inner product assumptions one really appreciates the elegance and power of the theory. Furthermore, in order to treat multivariate problems, we extend some of the results to tensor products of linear spaces. In this section it is supposed that the reader is familiar with a basic course of linear algebra. The proofs given here will be somewhat more condensed than in other parts of this book. Finally, we mention that at the end of this section we will apply some of the vector space results when considering eigenvalues and eigenvectors. In particular, invariant spaces and normal operators (matrices) will be considered.

1.2.2 *Lattices and algebra of subspaces*

This paragraph contains certain fundamental laws as well as principles of the algebra of subspaces when treating subspaces as elements of the lattice of subspaces. We also fix some basic terminology as well as useful definitions. The reader, who is interested in getting deeper insight into lattice theory, is referred to the classical book by Birkhoff (1967) or one at a more advanced level by Grätzer (1998).
There are many examples of sets A where some ordering $\leq$ is defined and where the following three properties hold for all $a, b, c \in A$:

(i) $a \leq a$;
(ii) if $a \leq b$ and $b \leq a$ then $a = b$;
(iii) if $a \leq b$ and $b \leq c$ then $a \leq c$.

The relations (i) – (iii) are called *partial order relations.* A set A equipped with relations (i) – (iii) is called a *partially ordered set* and is denoted by $< A, \leq >$. The notation $a \geq b$ (meaning $b \leq a$) can also be regarded as a definition of a partial order relation. The relation $\geq$ satisfies also (i) – (iii), if $\leq$ does, and so $< A, \geq >$ is a partially ordered set. Then $< A, \geq >$ is called the *dual* of $< A, \leq >$. Let φ be a statement about $< A, \leq >$. If in φ we change all occurrences of $\leq$ by $\geq$ we get the *dual of* φ. The importance of duals follows from the Duality Principle which we formulate following Grätzer (1998). The Principle is often used in lattice theory to shorten proofs.

Duality Principle. *If a statement is true in all partially ordered sets, then its dual is also true in all partially ordered sets.*

Let $< A, \leq >$ form a partially ordered set and $H \subseteq A$, $a \in A$. Then a is an *upper bound of* H, if and only if $h \leq a$, for all $h \in H$. An upper bound is *the least upper bound* (l.u.b.) of H, or *supremum of* H, if and only if, for any upper bound b of H, we have $a \leq b$. We shall then write $a = \sup H$. The concepts of *lower bound* and *greatest lower bound* (g.l.b.), or *infimum*, are similarly defined. The latter is denoted by $\inf H$. Let $\emptyset$ be the empty set. Observe that $\inf \emptyset$ exists, if and only if A has a largest element. At the same time $\sup \emptyset$ exists, if and only if A has a smallest element.

Definition 1.2.1. *A partially ordered set* $< A, \leq >$ *is a lattice if* $\inf\{a, b\}$ *and* $\sup\{a, b\}$ *exist, for all* $a, b \in A$. ∎

In the rest of this paragraph we are going to work with subsets of vector spaces. When considering vector spaces some authors use a coordinate free approach, whereas others prefer to work with methods which depend on the choice of basis. Since the theory of lattices will be employed, the main results of this section are given in a spirit of a coordinate free approach. More precisely, we consider the elements of the space together with the axioms which build up the space. It is noted that later when working explicitly with matrix derivatives and approximations, we switch over to an approach based on coordinates.
We consider a finite-dimensional vector space $\mathbb{V}$, and denote its subspaces by $\mathbb{A}$, $\mathbb{B}$, $\mathbb{C}$, ... (possibly indexed). Moreover, $\mathbf{\Lambda}$ stands for the totality of subspaces of

$\mathbb{V}$. The following presentation was initiated by Nordström & von Rosen (1987). It is obvious that $\mathbf{\Lambda}$ is partially ordered with respect to set-inclusion $\subseteq$, which means that the following lemma is true.

Lemma 1.2.1. *Let $\mathbb{A}$, $\mathbb{B}$ and $\mathbb{C}$ be arbitrary elements in $\mathbf{\Lambda}$. Then*

(i) $\mathbb{A} \subseteq \mathbb{A}$; reflexivity

(ii) *if* $\mathbb{A} \subseteq \mathbb{B}$ *and* $\mathbb{B} \subseteq \mathbb{A}$ *then* $\mathbb{A} = \mathbb{B}$; antisymmetry

(iii) *if* $\mathbb{A} \subseteq \mathbb{B}$ *and* $\mathbb{B} \subseteq \mathbb{C}$ *then* $\mathbb{A} \subseteq \mathbb{C}$. transitivity

∎

For given subspaces $\mathbb{A}$ and $\mathbb{B}$, an upper bound is an element in $\mathbf{\Lambda}$ which includes every element in $\{\mathbb{A}, \mathbb{B}\}$. As noted before, the least upper bound (l.u.b.) is an upper bound which is included in every upper bound. From Lemma 1.2.1 it follows that the l.u.b. is unique. Analogously, we have a lower bound and a unique greatest lower bound (g.l.b.). For arbitrary subspaces $\mathbb{A}$ and $\mathbb{B}$, the intersection $\mathbb{A} \cap \mathbb{B} = \{\mathbf{x} : \mathbf{x} \in \mathbb{A}, \mathbf{x} \in \mathbb{B}\}$ and the sum $\mathbb{A} + \mathbb{B} = \{\mathbf{y} + \mathbf{z} : \mathbf{y} \in \mathbb{A}, \mathbf{z} \in \mathbb{B}\}$ act as g.l.b. and l.u.b. of $\{\mathbb{A}, \mathbb{B}\}$, respectively (Jacobson, 1953, p. 26). For instance, $\mathbb{A} \cap \mathbb{B}$ is included in $\mathbb{A}$ and $\mathbb{B}$, and any space included in $\mathbb{A}$ and $\mathbb{B}$ is also included in $\mathbb{A} \cap \mathbb{B}$. Since $\mathbf{\Lambda}$ is partially ordered and a l.u.b. as well as a g.l.b. exist, $\mathbf{\Lambda}$, or strictly speaking, the ordered pair $(\cap, \subseteq)$, forms a *lattice of subspaces* (e.g. see Birkhoff, 1967, p. 6). This implies that it is possible to use lattice theory when studying intersections and sums of subspaces. Moreover, if $\{\mathbb{A}_i\}$ is any sequence of subsets of $\mathbf{\Lambda}$, the subspaces $\cap_i \mathbb{A}_i$ and $\sum_i \mathbb{A}_i$ act as g.l.b. and l.u.b. for $\{\mathbb{A}_i\}$, respectively (e.g. see Birkhoff, 1967, p. 6).

From Lemma 1.2.1 it follows that if the relation of set-inclusion is reversed, $\mathbf{\Lambda}$ is again partially ordered. Thus, interchanging the compositions $\cap$ and $+$ we get the dual lattice of $\mathbf{\Lambda}$. Hence, to any statement concerning elements of $\mathbf{\Lambda}$, a dual statement is obtained replacing compositions and relations with their duals. Consequently, only one statement in the dual pair of statements needs to be proved. In the following theorem we have brought together some of the most basic algebraic laws for sums and intersections of subspaces.

Theorem 1.2.1. *Let $\mathbb{A}$, $\mathbb{B}$ and $\mathbb{C}$ be arbitrary elements of the subspace lattice $\mathbf{\Lambda}$. Then the following laws hold:*

(i) $\mathbb{A} \cap \mathbb{A} = \mathbb{A}, \quad \mathbb{A} + \mathbb{A} = \mathbb{A};$ idempotent laws

(ii) $\mathbb{A} \cap \mathbb{B} = \mathbb{B} \cap \mathbb{A}, \quad \mathbb{A} + \mathbb{B} = \mathbb{B} + \mathbb{A};$ commutative laws

(iii) $\mathbb{A} \cap (\mathbb{B} \cap \mathbb{C}) = (\mathbb{A} \cap \mathbb{B}) \cap \mathbb{C},$ associative laws

$\mathbb{A} + (\mathbb{B} + \mathbb{C}) = (\mathbb{A} + \mathbb{B}) + \mathbb{C};$

(iv) $\mathbb{A} \cap (\mathbb{A} + \mathbb{B}) = \mathbb{A} + (\mathbb{A} \cap \mathbb{B}) = \mathbb{A};$ absorptive laws

(v) $\mathbb{A} \subseteq \mathbb{B} \Leftrightarrow \mathbb{A} \cap \mathbb{B} = \mathbb{A} \Leftrightarrow \mathbb{A} + \mathbb{B} = \mathbb{B};$ consistency laws

(vi) $\mathbb{A} \subseteq \mathbb{B} \Rightarrow (\mathbb{A} \cap \mathbb{C}) \subseteq \mathbb{B} \cap \mathbb{C};$ isotonicity of compositions

(vii) $\mathbb{A} \cap (\mathbb{B} + \mathbb{C}) \supseteq (\mathbb{A} \cap \mathbb{B}) + (\mathbb{A} \cap \mathbb{C}),$ distributive inequalities

$\mathbb{A} + (\mathbb{B} \cap \mathbb{C}) \subseteq (\mathbb{A} + \mathbb{B}) \cap (\mathbb{A} + \mathbb{C});$

(viii) $\mathbb{C} \subseteq \mathbb{A} \Rightarrow \mathbb{A} \cap (\mathbb{B} + \mathbb{C}) = (\mathbb{A} \cap \mathbb{B}) + \mathbb{C},$ modular laws

$\mathbb{A} \subseteq \mathbb{C} \Rightarrow \mathbb{A} + (\mathbb{B} \cap \mathbb{C}) = (\mathbb{A} + \mathbb{B}) \cap \mathbb{C}.$

PROOF: The statements in (i) – (vii) hold in any lattice and the proofs of (i) – (vi) are fairly straightforward. We are only going to prove the first statements of (vii) and (viii) since the second parts can be obtained by dualization. Now, concerning (vii) we have

$$(\mathbb{A} \cap \mathbb{B}) + (\mathbb{A} \cap \mathbb{C}) \subseteq \mathbb{A} \cap (\mathbb{B} + \mathbb{C}) + \mathbb{A} \cap (\mathbb{B} + \mathbb{C})$$

and the proof follows from (i).
For the proof of (viii) we note that, by assumption and (vii),

$$(\mathbb{A} \cap \mathbb{B}) + \mathbb{C} \subseteq \mathbb{A} \cap (\mathbb{B} + \mathbb{C}).$$

For the opposite relation let $\mathbf{x} \in \mathbb{A} \cap (\mathbb{B} + \mathbb{C})$. Thus, $\mathbf{x} = \mathbf{y} + \mathbf{z}$ for some $\mathbf{y} \in \mathbb{B}$ and $\mathbf{z} \in \mathbb{C}$. According to the assumption $\mathbb{C} \subseteq \mathbb{A}$. Using the definition of a vector space $\mathbf{y} = \mathbf{x} - \mathbf{z} \in \mathbb{A}$ implies $\mathbf{y} \in \mathbb{A} \cap \mathbb{B}$. Hence $\mathbf{x} \in (\mathbb{A} \cap \mathbb{B}) + \mathbb{C}$. ∎

The set $\mathbf{\Lambda}$ forms a *modular lattice of subspaces* since the properties (viii) of the theorem above are satisfied. In particular, we note that strict inequalities may hold in (vii). Thus, $\mathbf{\Lambda}$ is not a *distributive lattice* of subspaces, which is somewhat unfortunate as distributivity is by far a more powerful property than modularity. In the subsequent we have collected some useful identities in a series of corollaries.

Corollary 1.2.1.1. *Let* $\mathbb{A}$, $\mathbb{B}$ *and* $\mathbb{C}$ *be arbitrary elements of the subspace lattice* $\mathbf{\Lambda}$. *Then*

(i) $(\mathbb{A} \cap (\mathbb{B} + \mathbb{C})) + \mathbb{B} = ((\mathbb{A} + \mathbb{B}) \cap \mathbb{C}) + \mathbb{B},$

$(\mathbb{A} + (\mathbb{B} \cap \mathbb{C})) \cap \mathbb{B} = ((\mathbb{A} \cap \mathbb{B}) + \mathbb{C}) \cap \mathbb{B};$

(ii) $(\mathbb{A} \cap (\mathbb{B} + \mathbb{C})) + (\mathbb{B} \cap \mathbb{C}) = (\mathbb{A} + (\mathbb{B} \cap \mathbb{C})) \cap (\mathbb{B} + \mathbb{C}),$

$((\mathbb{A} \cap \mathbb{B}) + \mathbb{C}) \cap (\mathbb{A} + \mathbb{B}) = ((\mathbb{A} + \mathbb{B}) \cap \mathbb{C}) + (\mathbb{A} \cap \mathbb{B});$

(iii) $\mathbb{A} \cap (\mathbb{B} + (\mathbb{A} \cap \mathbb{C})) = (\mathbb{A} \cap \mathbb{B}) + (\mathbb{A} \cap \mathbb{C}),$ modular identities

$\mathbb{A} + (\mathbb{B} \cap (\mathbb{A} + \mathbb{C})) = (\mathbb{A} + \mathbb{B}) \cap (\mathbb{A} + \mathbb{C});$

(iv) $\mathbb{A} \cap (\mathbb{B} + \mathbb{C}) = \mathbb{A} \cap ((\mathbb{B} \cap (\mathbb{A} + \mathbb{C})) + \mathbb{C}),$ shearing identities

$\mathbb{A} + (\mathbb{B} \cap \mathbb{C}) = \mathbb{A} + ((\mathbb{B} + (\mathbb{A} \cap \mathbb{C})) \cap \mathbb{C}).$

PROOF: To prove the first part of (i), apply Theorem 1.2.1 (viii) to the left and right hand side, which shows that both sides equal to $(\mathbb{A}+\mathbb{B})\cap(\mathbb{B}+\mathbb{C})$. The second part of (i) is a dual relation. Since $\mathbb{B}\cap\mathbb{C}\subseteq\mathbb{B}+\mathbb{C}$, the first part of (ii) follows by applying Theorem 1.2.1 (viii) once more, and the latter part of (ii) then follows by virtue of symmetry. Moreover, (iii) implies (iv), and (iii) is obtained from Theorem 1.2.1 (viii) by noting that $\mathbb{A}\cap\mathbb{C}\subseteq\mathbb{A}$. ∎

Note that the modular identities as well as the shearing identities imply the modular laws in Theorem 1.2.1. Thus these relations give equivalent conditions for $\mathbf{\Lambda}$ to be a modular lattice. For example, if $\mathbb{C}\subseteq\mathbb{A}$, Corollary 1.2.1.1 (iii) reduces to Theorem 1.2.1 (viii).

Corollary 1.2.1.2. *Let* $\mathbb{A}$, $\mathbb{B}$ *and* $\mathbb{C}$ *be arbitrary elements of the subspace lattice* $\mathbf{\Lambda}$. *Then*

(i)
$$(\mathbb{A}\cap\mathbb{B})+(\mathbb{A}\cap\mathbb{C})+(\mathbb{B}\cap\mathbb{C})=(\mathbb{A}+(\mathbb{B}\cap\mathbb{C}))\cap(\mathbb{B}+(\mathbb{A}\cap\mathbb{C}))$$
$$=\{\mathbb{A}\cap(\mathbb{B}+\mathbb{C})+(\mathbb{B}\cap\mathbb{C})\}\cap\{(\mathbb{B}\cap(\mathbb{A}+\mathbb{C}))+(\mathbb{A}\cap\mathbb{C})\},$$
$$(\mathbb{A}+\mathbb{B})\cap(\mathbb{A}+\mathbb{C})\cap(\mathbb{B}+\mathbb{C})=(\mathbb{A}\cap(\mathbb{B}+\mathbb{C}))+(\mathbb{B}\cap(\mathbb{A}+\mathbb{C}))$$
$$=\{(\mathbb{A}+(\mathbb{B}\cap\mathbb{C})\cap(\mathbb{B}+\mathbb{C})\}+\{(\mathbb{B}+(\mathbb{A}\cap\mathbb{C}))\cap(\mathbb{A}+\mathbb{C})\};$$

(ii)
$$\{(\mathbb{A}\cap\mathbb{B})+(\mathbb{A}\cap\mathbb{C})\}\cap\{(\mathbb{A}\cap\mathbb{B})+(\mathbb{B}\cap\mathbb{C})\}=\mathbb{A}\cap\mathbb{B},$$
$$\{(\mathbb{A}+\mathbb{B})\cap(\mathbb{A}+\mathbb{C})\}+\{(\mathbb{A}+\mathbb{B})\cap(\mathbb{B}+\mathbb{C})\}=\mathbb{A}+\mathbb{B}.$$

PROOF: By virtue of Theorem 1.2.1 (viii)

$$\begin{aligned}(\mathbb{A}\cap\mathbb{B})+(\mathbb{A}\cap\mathbb{C})+(\mathbb{B}\cap\mathbb{C})&=\mathbb{A}\cap(\mathbb{B}+(\mathbb{A}\cap\mathbb{C}))+(\mathbb{B}\cap\mathbb{C})\\&=(\mathbb{A}+(\mathbb{B}\cap\mathbb{C}))\cap(\mathbb{B}+(\mathbb{A}\cap\mathbb{C})).\end{aligned}$$

Applying Corollary 1.2.1.1 (ii) we also have

$$\begin{aligned}&(\mathbb{A}\cap(\mathbb{B}+\mathbb{C})+(\mathbb{B}\cap\mathbb{C}))\cap(\mathbb{B}\cap(\mathbb{A}+\mathbb{C})+(\mathbb{A}\cap\mathbb{C}))\\&\quad=(\mathbb{A}+(\mathbb{B}\cap\mathbb{C}))\cap(\mathbb{B}+\mathbb{C})\cap(\mathbb{B}+(\mathbb{A}\cap\mathbb{C}))\cap(\mathbb{A}+\mathbb{C})\\&\quad=(\mathbb{A}+(\mathbb{B}\cap\mathbb{C}))\cap(\mathbb{B}+(\mathbb{A}\cap\mathbb{C}))\end{aligned}$$

establishing the first part of (i). The latter part of (i) is a dual statement. Applying Corollary 1.2.1.1 (iii) yields

$$\begin{aligned}&((\mathbb{A}\cap\mathbb{B})+(\mathbb{A}\cap\mathbb{C}))\cap((\mathbb{A}\cap\mathbb{B})+(\mathbb{B}\cap\mathbb{C}))\\&\quad=\mathbb{A}\cap(\mathbb{B}+(\mathbb{A}\cap\mathbb{C}))\cap\mathbb{B}\cap(\mathbb{A}+(\mathbb{B}\cap\mathbb{C}))=\mathbb{A}\cap\mathbb{B}\end{aligned}$$

which establishes the first part of (ii). The second part is again dual. ∎

A different kind of consequence of the modularity of $\mathbf{\Lambda}$ is the following.

Corollary 1.2.1.3. *Equality in Theorem 1.2.1* (vii) *holds if and only if*

$$(\mathbb{A} \cap \mathbb{B}) + (\mathbb{B} \cap \mathbb{C}) + (\mathbb{C} \cap \mathbb{A}) = (\mathbb{A} + \mathbb{B}) \cap (\mathbb{B} + \mathbb{C}) \cap (\mathbb{C} + \mathbb{A}). \qquad \text{median law}$$

PROOF: Distributivity implies that both sides in the median law equal to

$$\mathbb{A} \cap (\mathbb{B} + \mathbb{C}) + (\mathbb{B} \cap \mathbb{C}).$$

Conversely, if the median law holds,

$$\begin{aligned}(\mathbb{A} \cap \mathbb{B}) + (\mathbb{A} \cap \mathbb{C}) &= \mathbb{A} \cap (\mathbb{B} + \mathbb{A} \cap \mathbb{C}) = \mathbb{A} \cap (\mathbb{B} + \mathbb{A} \cap \mathbb{B} + \mathbb{B} \cap \mathbb{C} + \mathbb{A} \cap \mathbb{C}) \\ &= \mathbb{A} \cap (\mathbb{B} + (\mathbb{A} + \mathbb{B}) \cap (\mathbb{B} + \mathbb{C}) \cap (\mathbb{C} + \mathbb{A})) \\ &= \mathbb{A} \cap (\mathbb{B} + (\mathbb{B} + \mathbb{C}) \cap (\mathbb{C} + \mathbb{A})) = \mathbb{A} \cap (\mathbb{B} + \mathbb{C}).\end{aligned}$$

∎

Corollary 1.2.1.4. *Let $\mathbb{B}$ and $\mathbb{C}$ be comparable elements of $\mathbf{\Lambda}$, i.e. subspaces such that $\mathbb{B} \subseteq \mathbb{C}$ or $\mathbb{C} \subseteq \mathbb{B}$ holds, and let $\mathbb{A}$ be an arbitrary subspace. Then*

$$\mathbb{A}+\mathbb{B} = \mathbb{A} + \mathbb{C}, \;\; \mathbb{A} \cap \mathbb{B} = \mathbb{A} \cap \mathbb{C} \quad \Rightarrow \quad \mathbb{B} = \mathbb{C}. \qquad \text{cancellation law}$$

PROOF: Suppose, for example, that $\mathbb{B} \subseteq \mathbb{C}$ holds. Applying (iv) and (viii) of Theorem 1.2.1 we have

$$\mathbb{B} = (\mathbb{A} \cap \mathbb{B}) + \mathbb{B} = (\mathbb{A} \cap \mathbb{C}) + \mathbb{B} = (\mathbb{A} + \mathbb{B}) \cap \mathbb{C} = (\mathbb{A} + \mathbb{C}) \cap \mathbb{C} = \mathbb{C}.$$

∎

Theorem 1.2.1 as well as its corollaries were mainly dealing with so-called subspace polynomials, expressions involving $\cap$ and $+$, and formed from three elements of $\mathbf{\Lambda}$. Considering subspace polynomials formed from a finite set of subspaces, the situation is, in general, much more complicated.

1.2.3 *Disjointness, orthogonality and commutativity*

This paragraph is devoted to the concepts of disjointness (independency), orthogonality and commutativity of subspaces. It means that we have picked out those subspaces of the lattice which satisfy certain relations and for disjointness and orthogonality we have additionally assumed the existence of an inner product. Since the basic properties of disjoint subspaces are quite well-known and since orthogonality and commutativity have much broader impact on statistics we are concentrated on these two concepts.

Among the different definitions in the literature, the following given by Jacobson (1953, p. 28) appears rather intuitive.

Definition 1.2.2. *Let $\{\mathbb{A}_i\}$, $i = 1, \ldots, n$, be a finite set of subspaces of $\mathbf{\Lambda}$. The subspaces $\{\mathbb{A}_i\}$ are said to be disjoint if and only if $\mathbb{A}_i \cap (\sum_{j \neq i} \mathbb{A}_j) = \{\mathbf{0}\}$, for all values of i.* ∎

Although Definition 1.2.2 seems to be natural there are many situations where equivalent formulations suit better. We give two interesting examples of the reformulations in the next lemma. Other equivalent conditions can be found in Jacobson (1953, pp. 28–30), for example.

Lemma 1.2.2. *The subspaces $\{\mathbb{A}_i\}$, $i = 1, \ldots, n$, are disjoint if and only if any one of the following equivalent conditions hold:*

(i) $(\sum_i \mathbb{A}_i) \cap (\sum_j \mathbb{A}_j) = \{\mathbf{0}\}$, *$i \in I$, $j \in J$, for all disjoint subsets I and J of the finite index set;*

(ii) $\mathbb{A}_i \cap (\sum_j \mathbb{A}_j) = \{\mathbf{0}\}$, for all $i > j$.

PROOF: Obviously (i) is sufficient. To prove necessity we use an induction argument. Observe that if I consists of a single element, (i) follows by virtue of Theorem 1.2.1 (vi). Now assume that $\mathbb{B}_{i^\star} = \sum_{i \neq i^\star} \mathbb{A}_i$, $i \in I$ satisfies (i), where $i^\star$ is a fixed element in I. Then, for $i, i^\star \in I$ and $j \in J$

$$\begin{aligned}(\sum_i \mathbb{A}_i) \cap (\sum_j \mathbb{A}_j) &= (\mathbb{A}_{i^\star} + \mathbb{B}_{i^\star}) \cap (\sum_j \mathbb{A}_j) \\ &= (\mathbb{A}_{i^\star} + (\mathbb{B}_{i^\star} \cap (\mathbb{A}_{i^\star} + \sum_j \mathbb{A}_j))) \cap (\sum_j \mathbb{A}_j) = \mathbb{A}_{i^\star} \cap (\sum_j \mathbb{A}_j) = \{\mathbf{0}\},\end{aligned}$$

where the shearing identity (Corollary 1.2.1.1 (iv)), the assumption on $\mathbb{B}_{i^\star}$ and the disjointness of $\{\mathbb{A}_i\}$ have been used. Hence, necessity of (i) is established. Condition (ii) is obviously necessary. To prove sufficiency, assume that (ii) holds. Applying the equality

$$\mathbb{A} \cap (\mathbb{B} + \mathbb{C}) = \mathbb{A} \cap (\mathbb{B} + (\mathbb{C} \cap \mathbb{D})),$$

for any $\mathbb{D}$ such that $\mathbb{A} + \mathbb{B} \subseteq \mathbb{D} \subseteq \mathbb{V}$, and remembering that $\mathbb{V}$ represents the whole space, we obtain

$$\begin{aligned}\mathbb{A}_i \cap (\sum_{j \neq i} \mathbb{A}_j) &= \mathbb{A}_i \cap (\mathbb{A}_n + \sum_{j \neq i,n} \mathbb{A}_j) \\ &= \mathbb{A}_i \cap ((\mathbb{A}_n \cap (\sum_{j \neq n} \mathbb{A}_j)) + \sum_{j \neq i,n} \mathbb{A}_j) = \mathbb{A}_i \cap (\sum_{j \neq i,n} \mathbb{A}_j).\end{aligned}$$

Similarly,

$$\begin{aligned}\mathbb{A}_i \cap (\sum_{j \neq i} \mathbb{A}_j) &= \mathbb{A}_i \cap (\mathbb{A}_{i+1} + \sum_{j<i} \mathbb{A}_j) \\ &= \mathbb{A}_i \cap ((\mathbb{A}_{i+1} \cap (\sum_{j \neq i} \mathbb{A}_j)) + \sum_{j \leq i} \mathbb{A}_j) = \mathbb{A}_i \cap (\sum_{j<i} \mathbb{A}_j) = \{\mathbf{0}\}.\end{aligned}$$

Since the argument holds for arbitrary $i > 1$, sufficiency of (ii) is established. ∎

It is interesting to note that the "sequential disjointness" of $\{\mathbb{A}_i\}$, given in Lemma 1.2.2 (ii) above, is sufficient for disjointness of $\{\mathbb{A}_i\}$. This possibility of defining the disjointness of $\{\mathbb{A}_i\}$ in terms of fewer relations expressed by Lemma 1.2.2 (ii) heavily depends on the modularity of $\mathbf{\Lambda}$, as can be seen from the proof.

Definition 1.2.3. *If* $\{\mathbb{A}_i\}$ *are disjoint and* $\mathbb{A} = \sum_i \mathbb{A}_i$*, we say that* $\mathbb{A}$ *is the direct sum (internal) of the subspaces* $\{\mathbb{A}_i\}$*, and write* $\mathbb{A} = \oplus_i \mathbb{A}_i$. ∎

An important result for disjoint subspaces is the implication of the cancellation law (Corollary 1.2.1.4), which is useful when comparing various vector space decompositions. The required results are often immediately obtained by the next theorem. Note that if the subspaces are not comparable the theorem may fail.

Theorem 1.2.2. *Let* $\mathbb{B}$ *and* $\mathbb{C}$ *be comparable subspaces of* $\mathbf{\Lambda}$*, and* $\mathbb{A}$ *an arbitrary subspace. Then*

$$\mathbb{A} \oplus \mathbb{B} = \mathbb{A} \oplus \mathbb{C} \Rightarrow \mathbb{B} = \mathbb{C}.$$ ∎

Now we start to discuss orthogonality and suppose that the finite-dimensional vector space $\mathbb{V}$ over an arbitrary field K of characteristic 0 is equipped with a non-degenerate inner product, i.e. a symmetric positive bilinear functional on the field K. Here we treat arbitrary finite-dimensional spaces and suppose only that there exists an inner product. Later we consider real or complex spaces and in the proofs the defining properties of the inner products are utilized. Let $P(\mathbb{V})$ denote the *power set* (collection of all subsets) of $\mathbb{V}$. Then, to every set A in $P(\mathbb{V})$ corresponds a unique *perpendicular subspace* (relative to the inner product) $\perp (A)$. The map $\perp: P(\mathbb{V}) \to \mathbf{\Lambda}$ is obviously onto, but not one-to-one. However, restricting the domain of $\perp$ to $\mathbf{\Lambda}$, we obtain the bijective orthocomplementation map $\perp|_{\mathbf{\Lambda}} : \mathbf{\Lambda} \to \mathbf{\Lambda}$, $\perp|_{\mathbf{\Lambda}}(\mathbb{A}) = \mathbb{A}^{\perp}$.

Definition 1.2.4. *The subspace* $\mathbb{A}^{\perp}$ *is called the orthocomplement of* $\mathbb{A}$. ∎

It is interesting to compare the definition with Lemma 1.2.2 (ii) and it follows that orthogonality is a much stronger property than disjointness. For $\{\mathbb{A}_i\}$ we give the following definition.

Definition 1.2.5. *Let* $\{\mathbb{A}_i\}$ *be a finite set of subspaces of* $\mathbb{V}$.

(i) *The subspaces* $\{\mathbb{A}_i\}$ *are said to be orthogonal, if and only if* $\mathbb{A}_i \subseteq \mathbb{A}_j^{\perp}$ *holds, for all* $i \neq j$*, and this will be denoted* $\mathbb{A}_i \perp \mathbb{A}_j$.

(ii) *If* $\mathbb{A} = \sum_i \mathbb{A}_i$ *and the subspaces* $\{\mathbb{A}_i\}$ *are orthogonal, we say that* $\mathbb{A}$ *is the orthogonal sum of the subspaces* $\{\mathbb{A}_i\}$ *and write* $\mathbb{A} = \boxplus_i \mathbb{A}_i$. ∎

For the orthocomplements we have the following well-known facts.

Theorem 1.2.3. *Let* $\mathbb{A}$ *and* $\mathbb{B}$ *be arbitrary elements of* $\mathbf{\Lambda}$*. Then*

(i) $\mathbb{A} \cap \mathbb{A}^{\perp} = \{\mathbf{0}\}, \quad \mathbb{A} \boxplus \mathbb{A}^{\perp} = \mathbb{V}, \quad (\mathbb{A}^{\perp})^{\perp} = \mathbb{A}$; projection theorem

(ii) $(\mathbb{A} \cap \mathbb{B})^{\perp} = \mathbb{A}^{\perp} + \mathbb{B}^{\perp}, \quad (\mathbb{A} + \mathbb{B})^{\perp} = \mathbb{A}^{\perp} \cap \mathbb{B}^{\perp}$; de Morgan's laws

(iii) $(\mathbb{A} \subseteq \mathbb{B}) \Rightarrow \mathbb{B}^{\perp} \subseteq \mathbb{A}^{\perp}$. antitonicity of orthocomplementation

∎

By virtue of Theorem 1.2.3 (i) the orthocomplement of $\mathbb{A}$ is the perpendicular direct complement of $\mathbb{A}$. Hence, $\mathbf{\Lambda}$ is self-dual (with orthocomplementation as

dual automorphism) which is an evident but important observation since dual statements for orthocomplements in the lemmas and theorems given below may be formulated.
Theorem 1.2.3 (i) and (ii) imply that inner product spaces satisfy the conditions for a lattice to be an ortholattice of subspaces. However, $\mathbf{\Lambda}$ is modular and thus belongs to the class of modular ortholattices. Moreover, it can be shown that $\mathbf{\Lambda}$ also is an *orthomodular lattice* of subspaces, i.e. the elements of $\mathbf{\Lambda}$ satisfy the next lemma.

Lemma 1.2.3. *Let $\mathbb{A}$ and $\mathbb{B}$ be arbitrary elements of $\mathbf{\Lambda}$. The following two conditions always hold:*

(i) $$\mathbb{A} \subseteq \mathbb{B} \Rightarrow \mathbb{B} = \mathbb{A} \boxplus (\mathbb{A}^\perp \cap \mathbb{B}); \qquad \text{orthomodular law}$$

(ii) $$\mathbb{A} = (\mathbb{A} \cap \mathbb{B}) \boxplus (\mathbb{A} \cap \mathbb{B}^\perp) \quad \text{symmetricity of commutativity}$$

if and only if

$$\mathbb{B} = (\mathbb{A} \cap \mathbb{B}) \boxplus (\mathbb{A}^\perp \cap \mathbb{B}).$$ ∎

Note that in general the two conditions of the lemma are equivalent for any ortholattice. The concept of commutativity is defined later in this paragraph. Of course, Lemma 1.2.3 can be obtained in an elementary way without any reference to lattices. The point is to observe that $\mathbf{\Lambda}$ constitutes an orthomodular lattice, since it places concepts and results of the theory of orthomodular lattices at our disposal. For a collection of results and references, the books by Kalmbach (1983) and Beran (1985) are recommended. Two useful results, similar to Theorem 1.2.2, are presented in the next theorem.

Theorem 1.2.4. *Let $\mathbb{A}$, $\mathbb{B}$ and $\mathbb{C}$ be arbitrary subspaces of $\mathbf{\Lambda}$. Then*

(i) $$\mathbb{A} \boxplus \mathbb{B} = \mathbb{A} \boxplus \mathbb{C} \Leftrightarrow \mathbb{B} = \mathbb{C};$$

(ii) $$\mathbb{A} \boxplus \mathbb{B} \subseteq \mathbb{A} \boxplus \mathbb{C} \Leftrightarrow \mathbb{B} \subseteq \mathbb{C}.$$

PROOF: Since $\mathbb{A} \subseteq \mathbb{B}^\perp$ and $\mathbb{A} \subseteq \mathbb{C}^\perp$ we get from Lemma 1.2.3 (i)

$$\mathbb{B}^\perp = \mathbb{A} \boxplus (\mathbb{A}^\perp \cap \mathbb{B}^\perp) = \mathbb{A}) \boxplus (\mathbb{A}^\perp \cap \mathbb{C}^\perp) = \mathbb{C}^\perp,$$

where in the second equality de Morgan's law, i.e Theorem 1.2.3 (ii), has been applied to $\mathbb{A} \boxplus \mathbb{B} = \mathbb{A} \boxplus \mathbb{C}$. Thus (i) is proved, and (ii) follows analogously. ∎

Note that we do not have to assume $\mathbb{B}$ and $\mathbb{C}$ to be comparable. It is orthogonality between $\mathbb{A}$ and $\mathbb{C}$, and $\mathbb{B}$ and $\mathbb{C}$, that makes them comparable. Another important property of orthogonal subspaces not shared by disjoint subspaces may also be worth observing.

Theorem 1.2.5. *Let $\mathbb{B}$ and $\{\mathbb{A}_i\}$ be arbitrary subspaces of $\mathbf{\Lambda}$ such that $\mathbb{B} \perp \mathbb{A}_i$ for all i. Then $\mathbb{B} \perp \sum_i \mathbb{A}_i$.* ∎

This fact about orthogonal subspaces does not hold for disjoint subspaces, i.e. $\mathbb{B} \cap \mathbb{A}_i = \{\mathbf{0}\}$, for all i, does not imply $\mathbb{B} \cap (\sum_i \mathbb{A}_i) = \{\mathbf{0}\}$. Basically, this stems from the non-distributive character of $\mathbf{\Lambda}$ exhibited in Theorem 1.2.1.
One of the main results in this section will be given in the following theorem. The decompositions of vector spaces belong to those constructions which are commonly applied in statistics. We are going to use these results in the following paragraphs. There are many other decompositions available but the reader who grasps the course of derivation can easily find alternative results.

Theorem 1.2.6. *Let $\mathbb{A}$, $\mathbb{B}$ and $\mathbb{C}$ be arbitrary elements of $\mathbf{\Lambda}$. Then*

(i) $\mathbb{A} = (\mathbb{A} \cap \mathbb{B}) \boxplus \mathbb{A} \cap (\mathbb{A}^\perp + \mathbb{B}^\perp)$;

(ii) $\mathbb{A} + \mathbb{B} = \mathbb{A} \boxplus (\mathbb{A} + \mathbb{B}) \cap \mathbb{A}^\perp$;

(iii) $\mathbb{A}^\perp = (\mathbb{A} + \mathbb{B})^\perp \boxplus (\mathbb{A} + \mathbb{B}) \cap \mathbb{A}^\perp$;

(iv) $\mathbb{V} = ((\mathbb{A} + \mathbb{B}) \cap \mathbb{A}^\perp \oplus (\mathbb{A} + \mathbb{B}) \cap \mathbb{B}^\perp) \boxplus (\mathbb{A} + \mathbb{B})^\perp \boxplus (\mathbb{A} \cap \mathbb{B})$;

(v) $\mathbb{V} = \mathbb{A} \cap (\mathbb{B} + \mathbb{C})^\perp \boxplus \mathbb{A} \cap \mathbb{B}^\perp \cap (\mathbb{A}^\perp + \mathbb{B} + \mathbb{C}) \boxplus \mathbb{A} \cap (\mathbb{A}^\perp + \mathbb{B}) \boxplus \mathbb{A}^\perp$.

PROOF: All statements, more or less trivially, follow from Lemma 1.2.3 (i). For example, the lemma immediately establishes (iii), which in turn together with Theorem 1.2.3 (i) verifies (iv). ∎

When obtaining the statements of Theorem 1.2.6, it is easy to understand that results for orthomodular lattices are valuable tools. An example of a more traditional approach goes as follows: firstly show that the subspaces are disjoint and thereafter check the dimension which leads to a more lengthy and technical treatment.
The next theorem indicates that orthogonality puts a strong structure on the subspaces. Relation (i) in it explains why it is easy to apply geometrical arguments when orthogonal subspaces are considered, and (ii) shows a certain distributive property.

Theorem 1.2.7. *Let $\mathbb{A}$, $\mathbb{B}$ and $\mathbb{C}$ be arbitrary subspaces of $\mathbf{\Lambda}$.*

(i) *If $\mathbb{B} \perp \mathbb{C}$ and $\mathbb{A} \perp \mathbb{C}$, then $\mathbb{A} \cap (\mathbb{B} \boxplus \mathbb{C}) = \mathbb{A} \cap \mathbb{B}$.*

(ii) $\mathbb{A} \cap (\mathbb{A}^\perp + \mathbb{B} + \mathbb{C}) = \mathbb{A} \cap (\mathbb{A}^\perp + \mathbb{B}) + \mathbb{A} \cap (\mathbb{A}^\perp + \mathbb{C})$.

PROOF: By virtue of Theorem 1.2.1 (v) and the modular equality (Corollary 1.2.1.1 (iii)),

$$\mathbb{A} \cap (\mathbb{B} + \mathbb{C}) = \mathbb{A} \cap \mathbb{C}^\perp \cap ((\mathbb{B} \cap \mathbb{C}^\perp) \boxplus \mathbb{C}) = \mathbb{A} \cap \mathbb{B} \cap \mathbb{C}^\perp = \mathbb{A} \cap \mathbb{B}$$

establishes (i). The relation in (ii) is verified by using the median law (Corollary 1.2.1.3), Theorem 1.2.6 (i) and some calculations. ∎

Corollary 1.2.7.1. *Let $\mathbb{A}$, $\mathbb{B}$ and $\mathbb{C}$ be arbitrary subspaces of $\mathbf{\Lambda}$. If $\mathbb{B} \perp \mathbb{C}$, $\mathbb{A} \perp \mathbb{C}$ and $\mathbb{A} \subseteq (\mathbb{B} \boxplus \mathbb{C})$, then $\mathbb{A} \subseteq \mathbb{B}$.* ▮

Let $\mathbb{V}_1$ and $\mathbb{V}_2$ be two disjoint subspaces. Then every vector $\mathbf{z} \in \mathbb{V}_1 \oplus \mathbb{V}_2$ can be written in a unique way as a sum $\mathbf{z} = \mathbf{x}_1 + \mathbf{x}_2$, where $\mathbf{x}_1 \in \mathbb{V}_1$ and $\mathbf{x}_2 \in \mathbb{V}_2$. To see this, suppose that

$$\begin{aligned} \mathbf{z} &= \mathbf{x}_1 + \mathbf{x}_2, \qquad \mathbf{x}_1 \in \mathbb{V}_1, \quad \mathbf{x}_2 \in \mathbb{V}_2; \\ \mathbf{z} &= \mathbf{x}_3 + \mathbf{x}_4, \qquad \mathbf{x}_3 \in \mathbb{V}_1, \quad \mathbf{x}_4 \in \mathbb{V}_2. \end{aligned}$$

Then

$$\mathbf{0} = \mathbf{x}_1 - \mathbf{x}_3 + \mathbf{x}_2 - \mathbf{x}_4$$

which means that $\mathbf{x}_1 - \mathbf{x}_3 = \mathbf{x}_4 - \mathbf{x}_2$. However, since $\mathbf{x}_1 - \mathbf{x}_3 \in \mathbb{V}_1$, $\mathbf{x}_4 - \mathbf{x}_2 \in \mathbb{V}_2$, $\mathbb{V}_1$ and $\mathbb{V}_2$ are disjoint, this is only possible if $\mathbf{x}_1 = \mathbf{x}_3$ and $\mathbf{x}_2 = \mathbf{x}_4$.

Definition 1.2.6. *Let $\mathbb{V}_1$ and $\mathbb{V}_2$ be disjoint and $\mathbf{z} = \mathbf{x}_1 + \mathbf{x}_2$, where $\mathbf{x}_1 \in \mathbb{V}_1$ and $\mathbf{x}_2 \in \mathbb{V}_2$. The mapping $P\mathbf{z} = \mathbf{x}_1$ is called a projection of $\mathbf{z}$ on $\mathbb{V}_1$ along $\mathbb{V}_2$, and P is a projector. If $\mathbb{V}_1$ and $\mathbb{V}_2$ are orthogonal we say that we have an orthogonal projector.* ▮

In the next proposition the notions *range space* and *null space* appear. These will be defined in §1.2.4.

Proposition 1.2.1. *Let P be a projector on $\mathbb{V}_1$ along $\mathbb{V}_2$. Then*

(i) *P is a linear transformation;*

(ii) *$PP = P$, i.e. P is idempotent;*

(iii) *$I - P$ is a projector on $\mathbb{V}_2$ along $\mathbb{V}_1$ where I is the identity mapping defined by $I\mathbf{z} = \mathbf{z}$;*

(iv) *the range space $\mathcal{R}(P)$ is identical to $\mathbb{V}_1$, the null space $\mathcal{N}(P)$ equals $\mathcal{R}(I - P)$;*

(v) *if P is idempotent, then P is a projector;*

(vi) *P is unique.*

PROOF: (i): $P(a\mathbf{z}_1 + b\mathbf{z}_2) = aP(\mathbf{z}_1) + bP(\mathbf{z}_2)$.
(ii): For any $\mathbf{z} = \mathbf{x}_1 + \mathbf{x}_2$ such that $\mathbf{x}_1 \in \mathbb{V}_1$ and $\mathbf{x}_2 \in \mathbb{V}_2$ we have

$$P^2\mathbf{z} = P(P\mathbf{z}) = P\mathbf{x}_1 = \mathbf{x}_1.$$

It is worth observing that statement (ii) can be used as a definition of projector. The proof of statements (iii) – (vi) is left to a reader as an exercise. ▮

The rest of this paragraph is devoted to the concept of commutativity. Although it is often not explicitly treated in statistics, it is one of the most important concepts in linear models theory. It will be shown that under commutativity a linear space can be decomposed into orthogonal subspaces. Each of these subspaces has a corresponding linear model, as in analysis of variance, for example.

Definition 1.2.7. *The subspaces $\{\mathbb{A}_i\}$ are said to be commutative, which will be denoted $\mathbb{A}_i|\mathbb{A}_j$, if for $\forall i, j$,*

$$\mathbb{A}_i = (\mathbb{A}_i \cap \mathbb{A}_j) \boxplus (\mathbb{A}_i \cap \mathbb{A}_j^\perp).$$

∎

Note that $\mathbb{A}_i = \mathbb{A}_i \cap \mathbb{V} = \mathbb{A}_i \cap (\mathbb{A}_j^\perp \boxplus \mathbb{A}_j)$, where $\mathbb{V}$ stands for the whole space. According to Definition 1.2.7 a distributive property holds under commutativity, which leads us to easily interpretable decompositions.

Now we present some alternative characterizations of commutativity.

Theorem 1.2.8. *The subspaces $\{\mathbb{A}_i\}$ are commutative if and only if any of the following equivalent conditions hold:*

(i) $$\mathbb{A}_i \cap (\mathbb{A}_i \cap \mathbb{A}_j)^\perp = \mathbb{A}_i \cap \mathbb{A}_j^\perp, \qquad \forall i, j;$$

(ii) $$\mathbb{A}_i \cap (\mathbb{A}_i \cap \mathbb{A}_j)^\perp \perp \mathbb{A}_j \cap (\mathbb{A}_i \cap \mathbb{A}_j)^\perp, \qquad \forall i, j;$$

(iii) $$\mathbb{A}_i \cap (\mathbb{A}_i \cap \mathbb{A}_j)^\perp \subseteq \mathbb{A}_j^\perp, \qquad \forall i, j.$$

PROOF: First it is proved that $\{\mathbb{A}_i\}$ are commutative if and only if (i) holds, thereafter the equivalence between (i) and (ii) is proved, and finally (iii) is shown to be equivalent to (i).

From Lemma 1.2.3 (i) it follows that we always have

$$\mathbb{A}_i = (\mathbb{A}_i \cap \mathbb{A}_j) \boxplus (\mathbb{A}_i \cap \mathbb{A}_j)^\perp \cap \mathbb{A}_i.$$

Thus, by definition of commutativity and Theorem 1.2.4, commutativity implies (i). For the opposite relation note that (i) via Lemma 1.2.3 (i) implies commutativity. Turning to the equivalence between (i) and (ii), let us for notational convenience put

$$\mathbb{A} = \mathbb{A}_i \cap (\mathbb{A}_i \cap \mathbb{A}_j)^\perp, \quad \mathbb{B} = \mathbb{A}_j^\perp \boxplus (\mathbb{A}_i \cap \mathbb{A}_j), \quad \mathbb{C} = \mathbb{A}_i \cap \mathbb{A}_j^\perp.$$

We are going to show that $\mathbb{A} = \mathbb{C}$. If (ii) is true, $\mathbb{A} \subseteq \mathbb{B}$. Therefore, Lemma 1.2.3 (i) implies $\mathbb{B} = \mathbb{A} \boxplus \mathbb{A}^\perp \cap \mathbb{B}$, and we always have that $\mathbb{B} = \mathbb{C} \boxplus \mathbb{C}^\perp \cap \mathbb{B}$. However, $\mathbb{C}^\perp \cap \mathbb{B} = (\mathbb{A}^\perp + \mathbb{B}^\perp) \cap \mathbb{B} = \mathbb{A}^\perp \cap \mathbb{B}$, giving us $\mathbb{A} = \mathbb{C}$. Thus (ii) implies (i). The converse is trivial. Turning to the equivalence between (i) and (iii), it follows from Theorem 1.2.1 (v) that if (iii) holds,

$$\mathbb{A}_i \cap (\mathbb{A}_i \cap \mathbb{A}_j)^\perp = \mathbb{A}_i \cap (\mathbb{A}_i \cap \mathbb{A}_j)^\perp \cap \mathbb{A}_j^\perp = \mathbb{A}_i \cap \mathbb{A}_j^\perp.$$

The converse is obvious. ∎

The statement in Theorem 1.2.8 (ii) expresses orthogonality of $\mathbb{A}_i$ and $\mathbb{A}_j$ modulo $\mathbb{A}_i \cap \mathbb{A}_j$. If $\mathbb{A}_i$ and $\mathbb{A}_j$ are subspaces corresponding to factors i and j in an analysis of variance model, then Theorem 1.2.8 (ii) gives us the usual condition for orthogonality of i and j (e.g. see Tjur, 1984). We may also note that a pair of subspaces $\mathbb{A}_i$ and $\mathbb{A}_j$ satisfying Theorem 1.2.8 (ii) is referred to in the literature as "orthogonally incident" (Afriat, 1957) or "geometrically orthogonal" (Tjur, 1984). Furthermore, there is a close connection between orthogonal projectors and commutativity. It is interesting to note that the orthomodular lattice structure of $\mathbf{\Lambda}$ is carried over in a natural way to the set of idempotent and self-adjoint (see Definition 1.2.6) linear operators defined on $\mathbb{V}$. An idempotent linear operator is a projector according to Proposition 1.2.1, and a self-adjoint projector is called an orthogonal projector since $(I - P)$ is orthogonal to P and projects on the orthogonal complement to the space which P projects on.

Theorem 1.2.9. *Let P_i and P_{ij} denote the orthogonal projectors on $\mathbb{A}_i$ and $\mathbb{A}_i \cap \mathbb{A}_j$, respectively. The subspaces $\{\mathbb{A}_i\}$ are commutative if and only if any of the following two equivalent conditions hold:*

(i)
$$P_iP_j = P_jP_i, \qquad \forall i,j;$$

(ii)
$$P_iP_j = P_{ij}, \qquad \forall i,j.$$

PROOF: Suppose that $\mathbb{A}_i$ and $\mathbb{A}_j$ commute. Since

$$\mathbb{A}_i = (\mathbb{A}_i \cap \mathbb{A}_j) \boxplus (\mathbb{A}_i \cap \mathbb{A}_j^{\perp})$$

we have

$$P_i = P_{ij} + Q,$$

where Q is an orthogonal projector on $\mathbb{A}_i \cap \mathbb{A}_j^{\perp}$. Thus, $P_jQ = 0$, $P_jP_{ij} = P_{ij}$ and we obtain that $P_jP_i = P_{ij}$. Similarly, we may show via Lemma 1.2.3 (ii) that $P_iP_j = P_{ij}$. Hence, commutativity implies both (i) and (ii) of the theorem. Moreover, since P_i, P_j and P_{ij} are orthogonal (self-adjoint),

$$P_iP_j = P_{ij} = P_jP_i.$$

Hence, (ii) leads to (i). Finally we show that (ii) implies commutativity. From Theorem 1.2.6 (i) it follows that

$$\mathbb{A}_i = \mathbb{A}_i \cap \mathbb{A}_j \boxplus \mathbb{A}_i \cap (\mathbb{A}_i \cap \mathbb{A}_j)^{\perp}$$

and for projectors we have

$$P_i = P_{ij} + Q,$$

where Q is an orthogonal projector on $\mathbb{A}_i \cap (\mathbb{A}_i \cap \mathbb{A}_j)^{\perp}$. Thus,

$$P_jP_i = P_{ij} + P_jQ$$

and if (ii) holds, $P_jQ = 0$ and therefore Theorem 1.2.8 (iii) implies commutativity. ∎

There are many other possible characterizations of commutativity beside those presented in Theorem 1.2.8 and Theorem 1.2.9. A summary of the topic with statistical applications has been given by Baksalary (1987). In particular, Baksalary presented 46 alternative characterizations including those mentioned in this paragraph. Another useful theorem can easily be established by employing Theorem 1.2.9 (i).

Theorem 1.2.10. *Let* $\mathbb{A}$ *and* $\mathbb{B}$ *be subspaces of* $\mathbb{V}$. *Then*

(i)
$$\mathbb{A} \subseteq \mathbb{B} \Rightarrow \mathbb{A}|\mathbb{B};$$

(ii)
$$\mathbb{A} \perp \mathbb{B} \Rightarrow \mathbb{A}|\mathbb{B};$$

(iii)
$$\mathbb{A}|\mathbb{B} \Rightarrow \mathbb{A}|\mathbb{B}^{\perp};$$

(iv)
$$\mathbb{A}|\mathbb{B} \Rightarrow \mathbb{B}|\mathbb{A}.$$

∎

Note that inclusion as well as orthogonality implies commutativity, and that (iv) is identical to Lemma 1.2.3 (ii). Moreover, to clarify the implication of commutativity we make use of the following version of a general result.

Theorem 1.2.11. *Let* $\{\mathbb{A}_i\}$ *be a finite set of subspaces of* $\mathbb{V}$, *and* $\mathbb{B}$ *a subspace of* $\mathbb{V}$ *that commutes with each of the subspaces* $\mathbb{A}_i$. *Then*

(i) $\mathbb{B}|\sum_i \mathbb{A}_i$ *and* $\mathbb{B} \cap (\sum_i \mathbb{A}_i) = \sum_i (\mathbb{B} \cap \mathbb{A}_i)$;

(ii) $\mathbb{B}| \cap_i \mathbb{A}_i$ *and* $\mathbb{B} + (\cap_i \mathbb{A}_i) = \cap_i (\mathbb{B} + \mathbb{A}_i)$.

PROOF: Setting $\mathbb{A} = \sum_i \mathbb{A}_i$ the first part of (i) is proved if we are able to show that $\mathbb{A} = (\mathbb{A} \cap \mathbb{B}) \boxplus (\mathbb{A} \cap \mathbb{B}^{\perp})$. To this end it is obviously sufficient to prove that

$$\mathbb{A} \subseteq (\mathbb{A} \cap \mathbb{B}) \boxplus (\mathbb{A} \cap \mathbb{B}^{\perp}).$$

By virtue of the assumed commutativity we have

$$\mathbb{A}_i = (\mathbb{A}_i \cap \mathbb{B}) \boxplus (\mathbb{A}_i \cap \mathbb{B}^{\perp})$$

for all i implying

$$\mathbb{A} = (\sum_i (\mathbb{A}_i \cap \mathbb{B})) \boxplus (\sum_i (\mathbb{A}_i \cap \mathbb{B}^{\perp})). \tag{1.2.1}$$

Applying Theorem 1.2.1 (vii) yields

$$\sum_i (\mathbb{A}_i \cap \mathbb{B}) \subseteq \mathbb{A} \cap \mathbb{B} \text{ and } \sum_i (\mathbb{A}_i \cap \mathbb{B}^{\perp}) \subseteq \mathbb{A} \cap \mathbb{B}^{\perp}. \tag{1.2.2}$$

Combining (1.2.1) and (1.2.2) gives us the first part of (i). To prove the second part of (i) note that (1.2.2) gives

$$\mathbb{A} \cap \mathbb{B} \subseteq (\sum_i (\mathbb{A}_i \cap \mathbb{B}) \boxplus (\mathbb{A} \cap \mathbb{B}^{\perp}))$$

and Theorem 1.2.5 implies that $\mathbb{A} \cap \mathbb{B}$ is orthogonal to $\sum_i (\mathbb{A}_i \cap \mathbb{B}^{\perp})$. Thus, from Corollary 1.2.7.1 it follows that $\mathbb{A} \cap \mathbb{B} \subseteq \sum_i (\mathbb{A}_i \cap \mathbb{B})$, and then utilizing (1.2.2) the second part of (i) can be verified. Writing $\cap_i \mathbb{A}_i$ as $(\sum_i \mathbb{A}_i^{\perp})^{\perp}$ and using Theorem 1.2.10 (iii) together with the first part of (i) proves that $\mathbb{B}| \cap_i \mathbb{A}_i$. The second part of (ii) follows similarly. ∎

Corollary 1.2.11.1. *Let $\{\mathbb{A}_i\}$ and $\{\mathbb{B}_j\}$ be finite sets of subspaces of $\mathbb{V}$ such that $\mathbb{A}_i|\mathbb{B}_j$, for all i, j. Then*

(i) $$(\sum_i \mathbb{A}_i) \cap (\sum_j \mathbb{B}_j) = \sum_{ij}(\mathbb{A}_i \cap \mathbb{B}_j);$$

(ii) $$(\bigcap_i \mathbb{A}_i) + (\bigcap_j \mathbb{B}_j) = \bigcap_{ij}(\mathbb{A}_i + \mathbb{B}_j).$$

∎

Corollary 1.2.11.2. *Let $\{\mathbb{A}_i\}$, $i = 1, \ldots, n$, be a finite set of subspaces of $\mathbb{V}$, and $\mathbb{B}$ a subspace of $\mathbb{V}$. The following conditions are equivalent*

(i) $\mathbb{B}|\mathbb{A}_i, \quad \forall i;$

(ii) $\mathbb{B} = \boxplus_{i \leq n} (\mathbb{B} \cap \mathbb{A}_i) \boxplus \bigcap_{i \leq n} \mathbb{A}_i^{\perp} \cap \mathbb{B};$

(iii) $\mathbb{V} = \boxplus_{i \leq n} (\mathbb{B} \cap \mathbb{A}_i) \boxplus (\bigcap_{i \leq n} \mathbb{A}_i^{\perp} \cap \mathbb{B}) \boxplus \boxplus_{i \leq n} (\mathbb{B}^{\perp} \cap \mathbb{A}_i) \boxplus (\bigcap_{i \leq n} A_i^{\perp} \cap \mathbb{B}^{\perp}).$

PROOF: We just prove that (ii) implies (i).

$$\begin{aligned}
\mathbb{B} \cap \mathbb{A}_j^{\perp} &= (\boxplus_i (\mathbb{B} \cap \mathbb{A}_i) \boxplus \bigcap_i \mathbb{A}_i^{\perp} \cap \mathbb{B}) \cap \mathbb{A}_j^{\perp} \\
&= \boxplus_{i \neq j} (\mathbb{B} \cap \mathbb{A}_i) \boxplus (\mathbb{B} \cap \mathbb{A}_j \boxplus \bigcap_i \mathbb{A}_i^{\perp} \cap \mathbb{B}) \cap \mathbb{A}_j^{\perp} \\
&= \boxplus_{i \neq j} (\mathbb{B} \cap \mathbb{A}_i) \boxplus \bigcap_i \mathbb{A}_i^{\perp} \cap \mathbb{B}.
\end{aligned}$$

By adding $\mathbb{B} \cap \mathbb{A}_j$ to this expression, by assumption we obtain $\mathbb{B}$ again. Hence,

$$\mathbb{B} = \mathbb{B} \cap \mathbb{A}_j \boxplus \mathbb{B} \cap \mathbb{A}_j^{\perp}.$$

∎

These two corollaries clearly spell out the implications of commutativity of subspaces. From Corollary 1.2.11.1 we see explicitly that distributivity holds under commutativity whereas Corollary 1.2.11.2 exhibits simple orthogonal decompositions of a vector space that are valid under commutativity. In fact, Corollary 1.2.11.2 shows that when decomposing $\mathbb{V}$ into orthogonal subspaces, commutativity is a necessary and sufficient condition.

1.2.4 *Range spaces*

In this paragraph we will discuss linear transformations from a vector space $\mathbb{V}$ to another space $\mathbb{W}$. The *range space* of a linear transformation $A : \mathbb{V} \to \mathbb{W}$ will be denoted $\mathcal{R}(A)$ and is defined by

$$\mathcal{R}(A) = \{\mathbf{x} : \mathbf{x} = A\mathbf{y}, \mathbf{y} \in \mathbb{V}\},$$

whereas the *null space* $\mathcal{N}(A)$ of A is defined by

$$\mathcal{N}(A) = \{\mathbf{y} : A\mathbf{y} = \{\mathbf{0}\}, \mathbf{y} \in \mathbb{V}\}.$$

It is supposed that the spaces are defined over the real or complex fields and are equipped with an inner product. Although several of the lemmas and theorems given below hold for transformations on finite-dimensional vector spaces defined over arbitrary fields, some of the theorems rest on the fact that if $A : \mathbb{V} \to \mathbb{W}$, the adjoint transformation A' is a map from $\mathbb{W}$ to $\mathbb{V}$. All transformations in this paragraph can be identified via corresponding matrices, relative to a given basis. Thus the results of this paragraph hold for spaces generated by columns of matrices, i.e. we consider column vector spaces without any particular reference to a fixed basis, and therefore so-called column vector spaces can be identified with their corresponding range space.

Definition 1.2.8. *An inner product* $(\bullet, \bullet)$ *in a complex or real space is a scalar valued function of the ordered pairs of vectors* $\mathbf{x}$ *and* $\mathbf{y}$*, such that*

$$\begin{aligned} (\mathbf{x}, \mathbf{y}) &= \overline{(\mathbf{y}, \mathbf{x})}; && \text{hermitian (symmetric)} \\ (a\mathbf{x}_1 + b\mathbf{x}_2, \mathbf{y}) &= a(\mathbf{x}_1, \mathbf{y}) + b(\mathbf{x}_2, \mathbf{y}); && \text{bilinear} \\ (\mathbf{x}, \mathbf{x}) &\geq 0, \quad (\mathbf{x}, \mathbf{x}) = 0 \quad \text{if and only if } \mathbf{x} = \mathbf{0}, && \text{positive} \end{aligned}$$

where $\overline{}$ *denotes complex conjugate.* ∎

Definition 1.2.9. *For a space (complex or real) with an inner product the adjoint transformation* $A' : \mathbb{W} \to \mathbb{V}$ *of the linear transformation* $A : \mathbb{V} \to \mathbb{W}$ *is defined by* $(A\mathbf{x}, \mathbf{y})_{\mathbb{W}} = (\mathbf{x}, A'\mathbf{y})_{\mathbb{V}}$*, where indices show to which spaces the inner products belong.* ∎

Note that the adjoint transformation is unique. Furthermore, since two different inner products, $(\mathbf{x}, \mathbf{y})_1$ and $(\mathbf{x}, \mathbf{y})_2$, on the same space induce different orthogonal bases and since a basis can always be mapped to another basis by the aid of a unique non-singular linear transformation A we have $(\mathbf{x}, \mathbf{y})_1 = (A\mathbf{x}, A\mathbf{y})_2$ for some transformation A. If $B = A'A$, we obtain $(\mathbf{x}, \mathbf{y})_1 = (B\mathbf{x}, \mathbf{y})_1$ and B is positive definite, i.e. B is self-adjoint and $(B\mathbf{x}, \mathbf{x}) > 0$, if $\mathbf{x} \neq \mathbf{0}$. A transformation $A : \mathbb{V} \to \mathbb{V}$ is self-adjoint if $A = A'$. Thus, if we fix some inner product we can always express every other inner product in relation to the fixed one, when going over to coordinates. For example, if $(\mathbf{x}, \mathbf{y}) = \mathbf{x}'\mathbf{y}$, which is referred to as the *standard inner product*, then any other inner product for some positive definite transformation B is defined by $\mathbf{x}'B\mathbf{y}$. The results of this paragraph will be given without a particular reference to any special inner product but from the discussion above it follows that it is possible to obtain explicit expressions for any inner product. Finally we note that any positive definite transformation B can be written $B = A'A$, for some A (see also Theorem 1.1.3).
In the subsequent, if not necessary, we do not indicate those spaces on which and from which the linear transformations act. It may be interesting to observe that if the space is equipped with a standard inner product and if we are talking in terms of complex matrices and column vector spaces, the adjoint operation can

be replaced by the conjugate transposing which in the real case is identical to the transposing operator. Our first lemma of this paragraph presents a somewhat trivial but useful result, and the proof of the lemma follows from the definition of a range space.

Lemma 1.2.4. *Let A, B and C be any linear transformations such that AB and AC are defined. Then*

(i) $$\mathcal{R}(AB) = \mathcal{R}(AC) \quad \text{if} \quad \mathcal{R}(B) = \mathcal{R}(C);$$

(ii) $$\mathcal{R}(AB) \subseteq \mathcal{R}(AC) \quad \text{if} \quad \mathcal{R}(B) \subseteq \mathcal{R}(C).$$

∎

The next two lemmas comprise standard results which are very useful. In the proofs we accustom the reader with the technique of using inner products and adjoint transformations.

Lemma 1.2.5. *Let A be an arbitrary linear transformation. Then*

$$\mathcal{N}(A') = \mathcal{R}(A)^{\perp}$$

PROOF: Suppose $\mathbf{y} \in \mathcal{R}(A)^{\perp}$. By definition of $\mathcal{R}(A)$ for any $\mathbf{z}$ we have a vector $\mathbf{x} = A\mathbf{z} \in \mathcal{R}(A)$. Hence,

$$0 = (\mathbf{x}, \mathbf{y}) = (A\mathbf{z}, \mathbf{y}) = (\mathbf{z}, A'\mathbf{y}) \Rightarrow A'\mathbf{y} = \mathbf{0}.$$

Implication holds also in opposite direction and thus the lemma is proved. ∎

Lemma 1.2.6. *Let A be an arbitrary linear transformation. Then*

$$\mathcal{R}(AA') = \mathcal{R}(A).$$

PROOF: From Lemma 1.2.4 (ii) it follows that it is sufficient to show that $\mathcal{R}(A) \subseteq \mathcal{R}(AA')$. For any $\mathbf{y} \in \mathcal{R}(AA')^{\perp} = \mathcal{N}(AA')$ we obtain

$$0 = (AA'\mathbf{y}, \mathbf{y}) = (A'\mathbf{y}, A'\mathbf{y}).$$

Thus, $A'\mathbf{y} = \mathbf{0}$ leads us to $\mathbf{y} \in \mathcal{N}(A') = \mathcal{R}(A)^{\perp}$. Hence $\mathcal{R}(AA')^{\perp} \subseteq \mathcal{R}(A)^{\perp}$ which is equivalent to $\mathcal{R}(A) \subseteq \mathcal{R}(AA')$. ∎

In the subsequent, for an arbitrary transformation A, the transformation A^o denotes any transformation such that $\mathcal{R}(A)^{\perp} = \mathcal{R}(A^o)$ (compare with $\mathbf{A}^o$ in Proposition 1.1.3). Note that A^o depends on the inner product. Moreover, in the next theorem the expression $A'B^o$ appears and for an intuitive geometrical understanding it may be convenient to interpret $A'B^o$ as a transformation A' from (restricted to) the null space of B'.

Theorem 1.2.12. *For any transformations A, B assume that $A'B^o$ is well defined. Then the following statements are equivalent:*

(i)
$$\mathcal{R}(A') \subseteq \mathcal{R}(A'B^o);$$

(ii)
$$\mathcal{R}(A') = \mathcal{R}(A'B^o);$$

(iii)
$$\mathcal{R}(A) \cap \mathcal{R}(B) = \{\mathbf{0}\}.$$

PROOF: First note that if $\mathcal{R}(A) \subseteq \mathcal{R}(B)$ or $\mathcal{R}(B) \subseteq \mathcal{R}(A)$ hold, the theorem is trivially true. The equivalence between (i) and (ii) is obvious. Now suppose that $\mathcal{R}(A) \cap \mathcal{R}(B) = \{\mathbf{0}\}$ holds. Then, for $\mathbf{y} \in \mathcal{R}(A'B^o)^\perp$ and arbitrary $\mathbf{x}$

$$0 = (\mathbf{x}, B^{o'}A\mathbf{y}) = (B^o\mathbf{x}, A\mathbf{y}).$$

Hence, within $\mathcal{R}(A'B^o)^\perp$ we have that $\mathcal{R}(A)$ and $\mathcal{R}(B^o)$ are orthogonal. Thus, $\mathcal{R}(A) \subseteq \mathcal{R}(B)$ within $\mathcal{R}(A'B^o)^\perp$ contradicts the assumption, unless $A\mathbf{y} = \mathbf{0}$ for all $\mathbf{y} \in \mathcal{R}(A'B^o)^\perp$. Hence $\mathcal{R}(A') \subseteq \mathcal{R}(A'B^o)$.
For the converse, suppose that there exists a vector $\mathbf{x} \in \mathcal{R}(A) \cap \mathcal{R}(B)$ implying the existence of vectors $\mathbf{z}_1$ and $\mathbf{z}_2$ such that $\mathbf{x} = A\mathbf{z}_1$ and $\mathbf{x} = B\mathbf{z}_2$. For all $\mathbf{y}$, of course, $(\mathbf{x}, B^o\mathbf{y}) = 0$. Hence,

$$0 = (A\mathbf{z}_1, B^o\mathbf{y}) = (\mathbf{z}_1, A'B^o\mathbf{y})$$

implies $\mathbf{z}_1 \in \mathcal{R}(A'B^o)^\perp = \mathcal{R}(A')^\perp$. Therefore we obtain $A\mathbf{z}_1 = \mathbf{0}$, and thus $\mathbf{x} = \mathbf{0}$.

▮

An interesting consequence of the equivalence between (ii) and (iii) is given in the next corollary. Let us remind the reader that AA' is always a non-negative definite transformation and that any non-negative definite transformation can be written AA' for some A.

Corollary 1.2.12.1. *In the notation of Theorem 1.2.12*

$$\mathcal{R}(AA'B^o) \cap \mathcal{R}(B) = \{\mathbf{0}\}.$$

▮

Fairly often it is meaningful to utilize results concerning the dimensionality of the range space when establishing theorems. The dimensionality is given by the number of linearly independent elements (basis vectors) which via linear operations generate all other elements of that particular space. In the next lemma we give some fundamental results which will be used later. The dimensionality of a space $\mathbb{V}$ is denoted by dim$\mathbb{V}$.

Lemma 1.2.7. *Let A and B be arbitrary linear transformations such that the relations are well defined. Then*

(i) $\dim(\mathcal{R}(A)) = \dim(\mathcal{R}(A'))$;

(ii) $\dim(\mathcal{R}(A)+\mathcal{R}(B)) = \dim(\mathcal{R}(A)) + \dim(\mathcal{R}(B)) - \dim(\mathcal{R}(A)\cap\mathcal{R}(B))$;

(iii) $\dim(\mathcal{R}(A)+\mathcal{R}(B)) = \dim(\mathcal{R}(A'B^o)) + \dim(\mathcal{R}(B))$;

(iv) $\mathcal{R}(A) \subseteq \mathcal{R}(B)$ *if and only if* $\dim(\mathcal{R}(A)+\mathcal{R}(B)) = \dim(\mathcal{R}(B))$.

∎

Theorem 1.2.13. *Let A, B and C be arbitrary linear transformations such that the products of the transformations are well defined. Then*

(i)
$$\mathcal{R}(AA'B) = \mathcal{R}(AA'C) \Leftrightarrow \mathcal{R}(A'B) = \mathcal{R}(A'C);$$

(ii)
$$\mathcal{R}(AA'B) \subseteq \mathcal{R}(AA'C) \Leftrightarrow \mathcal{R}(A'B) \subseteq \mathcal{R}(A'C).$$

PROOF: We will just prove (ii) since (i) can be verified by copying the given proof. Suppose that $\mathcal{R}(AA'B) \subseteq \mathcal{R}(AA'C)$ holds, and let H be a transformation such that $\mathcal{R}(H) = \mathcal{R}(B)+\mathcal{R}(C)$. Then, applying Lemma 1.2.7 (i) and (iv) and Lemma 1.2.6 we obtain that

$$\begin{aligned}\dim(\mathcal{R}(A'C)) &= \dim(\mathcal{R}(C'A)) = \dim(\mathcal{R}(C'AA')) = \dim(\mathcal{R}(AA'C))\\ &= \dim(\mathcal{R}(AA'C)+\mathcal{R}(AA'B)) = \dim(\mathcal{R}(H'AA'))\\ &= \dim(\mathcal{R}(H'A)) = \dim(\mathcal{R}(A'H)) = \dim(\mathcal{R}(A'C)+\mathcal{R}(A'B)).\end{aligned}$$

The converse follows by starting with $\dim(\mathcal{R}(AA'C))$ and then copying the above given procedure. ∎

Theorem 1.2.14. *Let A, B and C be linear transformations such that the spaces of the theorem are well defined. If*

$$\mathcal{R}(A') \oplus \mathcal{R}(B) \subseteq \mathcal{R}(C) \quad \textit{and} \quad \mathcal{R}(A) \oplus \mathcal{R}(B) = \mathcal{R}(C),$$

then

$$\mathcal{R}(A') \oplus \mathcal{R}(B) = \mathcal{R}(C).$$

PROOF: Lemma 1.2.7 (i) implies that

$$\dim(\mathcal{R}(A') \oplus \mathcal{R}(B)) = \dim(\mathcal{R}(A) \oplus \mathcal{R}(B)) = \dim(\mathcal{R}(C)).$$

∎

The next theorem is also fairly useful.

Theorem 1.2.15. *Let A and B be arbitrary linear transformations such that the products of the transformations are well defined. Then*

$$\mathcal{R}(A(A'B^o)^o) = \mathcal{R}(A) \cap \mathcal{R}(B).$$

PROOF: Since $\mathcal{R}(A)^\perp$ and $\mathcal{R}(B)^\perp$ are orthogonal to $\mathcal{R}(A(A'B^o)^o)$, it follows from Theorem 1.2.5 that the sum is also orthogonal to $\mathcal{R}(A(A'B^o)^o)$. Thus, $\mathcal{R}(A(A'B^o)^o) \subseteq \mathcal{R}(A) \cap \mathcal{R}(B)$. Therefore, applying Lemma 1.2.7 (iii) confirms that the spaces must be identical. ∎

Many decompositions of vector spaces are given in forms which use range spaces and projectors. Therefore the next theorem due to Shinozaki & Sibya (1974) is important, since it clearly spells out the relation between the range space of the projection and the intersections of certain subspaces.

Theorem 1.2.16. *Let P be an arbitrary projector and A a linear transformation such that PA is defined. Then*

$$\mathcal{R}(PA) = \mathcal{R}(P) \cap (\mathcal{N}(P) + \mathcal{R}(A)).$$

PROOF: Suppose that $\mathcal{N}(P) + \mathcal{R}(A) = \mathcal{N}(P) + \mathcal{R}(PA)$ holds. In that case the result follows immediately from Theorem 1.2.1 (viii), since $\mathcal{R}(PA) \subseteq \mathcal{R}(P)$. The assumption is true since

$$\mathcal{R}(A) = \mathcal{R}((I-P)A + PA) \subseteq \mathcal{R}((I-P)A) + \mathcal{R}(PA) \subseteq \mathcal{N}(P) + \mathcal{R}(PA)$$

and $\mathcal{R}(PA) = \mathcal{R}(A - (I-P)A) \subseteq \mathcal{R}(A) + \mathcal{N}(P)$. ∎

As already noted above in Theorem 1.2.9, if $P = P'$, then P is said to be an *orthogonal projector*. In this case $\mathcal{N}(P) = \mathcal{R}(P)^{\perp}$ leads to the equality $\mathcal{R}(PA) = \mathcal{R}(P) \cap (\mathcal{R}(P)^{\perp} + \mathcal{R}(A))$ which is a more commonly applied relation.

Corollary 1.2.16.1. *Let P be an arbitrary projector and A a linear transformation such that PA is defined with $\mathcal{N}(P) \subseteq \mathcal{R}(A)$. Then*

$$\mathcal{R}(PA) = \mathcal{R}(P) \cap \mathcal{R}(A).$$

∎

The rest of this paragraph is devoted to two important decompositions of vector spaces. Our next theorem was brought forward by Stein (1972, p. 114) and Rao (1974).

Theorem 1.2.17. *Let A and B be arbitrary transformations such that $A'B^o$ is defined. Then*

$$\mathcal{R}(A) + \mathcal{R}(B) = \mathcal{R}(AA'B^o) \oplus \mathcal{R}(B).$$

PROOF: Corollary 1.2.12.1 states that $\mathcal{R}(AA'B^o)$ and $\mathcal{R}(B)$ are disjoint. Thus it is obvious that

$$\mathcal{R}(AA'B^o) \oplus \mathcal{R}(B) \subseteq \mathcal{R}(A) + \mathcal{R}(B). \qquad (1.2.3)$$

For any $\mathbf{y} \in (\mathcal{R}(AA'B^o) \oplus \mathcal{R}(B))^{\perp}$ we have $\mathbf{y} = B^o\mathbf{z}_1$ for some $\mathbf{z}_1$, since $\mathbf{y} \in \mathcal{R}(B)^{\perp}$. Hence, for all $\mathbf{z}_2$,

$$0 = (AA'B^o\mathbf{z}_2, \mathbf{y}) = (AA'B^o\mathbf{z}_2, B^o\mathbf{z}_1) = (B^{o'}AA'B^o\mathbf{z}_2, \mathbf{z}_1)$$

implies that, for all $\mathbf{z}_3$,

$$0 = (B^{o'}A\mathbf{z}_3, \mathbf{z}_1) = (A\mathbf{z}_3, B^o\mathbf{z}_1) = (A\mathbf{z}_3, \mathbf{y}).$$

Therefore, $\mathcal{R}(A) \subseteq \mathcal{R}(AA'B^o) \oplus \mathcal{R}(B)$ and the opposite inclusion to (1.2.3) is established. ∎

Corollary 1.2.17.1. *If $\mathcal{R}(B) \subseteq \mathcal{R}(A)$, then*

$$\mathcal{R}(A) = \mathcal{R}(AA'B^o) \oplus \mathcal{R}(B).$$

∎

With the help of Theorem 1.2.16 and Theorem 1.2.6 (ii) we can establish equality in (1.2.3) in an alternative way. Choose B^o to be an orthogonal projector on $\mathcal{R}(B)^\perp$. Then it follows that

$$\mathcal{R}((AA'B^o)') \boxplus \mathcal{R}(B) = \mathcal{R}(A) + \mathcal{R}(B).$$

Therefore, Theorem 1.2.14 implies that equality holds in (1.2.3). This shows an interesting application of how to prove a statement by the aid of an adjoint transformation.

Baksalary & Kala (1978), as well as several other authors, use a decomposition which is presented in the next theorem.

Theorem 1.2.18. *Let A, B and C be arbitrary transformations such that the spaces are well defined, and let P be an orthogonal projector on $\mathcal{R}(C)$. Then*

$$\mathbb{V} = \mathbb{B}_1 \boxplus \mathbb{B}_2 \boxplus \mathcal{R}(PA) \boxplus \mathcal{R}(P)^\perp,$$

where

$$\begin{aligned}\mathbb{B}_1 &= \mathcal{R}(P) \cap (\mathcal{R}(PA) + \mathcal{R}(PB))^\perp,\\ \mathbb{B}_2 &= (\mathcal{R}(PA) + \mathcal{R}(PB)) \cap \mathcal{R}(PA)^\perp.\end{aligned}$$

PROOF: Using Theorem 1.2.16 and Corollary 1.2.1.1 (iii) we get more information about the spaces:

$$\begin{aligned}\mathbb{B}_1 &= \mathcal{R}(C) \cap (\mathcal{R}(C) \cap (\mathcal{R}(C)^\perp + \mathcal{R}(A) + \mathcal{R}(B)))^\perp\\ &= \mathcal{R}(C) \cap (\mathcal{R}(C) \cap (\mathcal{R}(C)^\perp + (\mathcal{R}(C) \cap (\mathcal{R}(A) + \mathcal{R}(B))^\perp))\\ &= \mathcal{R}(C) \cap (\mathcal{R}(A) + \mathcal{R}(B))^\perp,\end{aligned}$$

$$\begin{aligned}\mathbb{B}_2 &= \mathcal{R}(C) \cap (\mathcal{R}(C)^\perp + \mathcal{R}(A) + \mathcal{R}(B)) \cap (\mathcal{R}(C) \cap (\mathcal{R}(C)^\perp + \mathcal{R}(A)))^\perp\\ &= \mathcal{R}(C) \cap (\mathcal{R}(C)^\perp + \mathcal{R}(A) + \mathcal{R}(B)) \cap (\mathcal{R}(C)^\perp + (\mathcal{R}(C) \cap \mathcal{R}(A)^\perp))\\ &= \mathcal{R}(C) \cap \mathcal{R}(A)^\perp \cap (\mathcal{R}(C)^\perp + \mathcal{R}(A) + \mathcal{R}(B)).\end{aligned}$$

By virtue of these relations the statement of the theorem can be written as

$$\begin{aligned}\mathbb{V} = {} & \mathcal{R}(C) \cap (\mathcal{R}(A) + \mathcal{R}(B))^\perp \boxplus \mathcal{R}(C) \cap \mathcal{R}(A)^\perp \cap (\mathcal{R}(C)^\perp + \mathcal{R}(A)\\ & + \mathcal{R}(B)) \boxplus \mathcal{R}(C) \cap (R(C)^\perp + \mathcal{R}(A)) \boxplus \mathcal{R}(C)^\perp\end{aligned}$$

which is identical to Theorem 1.2.6 (v). ∎

1.2.5 *Tensor spaces*

In the present paragraph we will give a brief introduction to tensor spaces. The purpose is to present some basic results and indicate the relationship with the Kronecker product. For a more extensive presentation we refer to Greub (1978). We are going to use the notion of a *bilinear* map. Briefly speaking, $\rho(\mathbf{x}, \mathbf{y})$ is a bilinear map if it is linear in each argument. The underlying field is supposed to have characteristic 0.

Definition 1.2.10. *Let $\rho : \mathbb{V}\times\mathbb{W} \rightarrow \mathbb{V}\otimes\mathbb{W}$ be a bilinear map from the Cartesian product of the vector spaces $\mathbb{V}$ and $\mathbb{W}$ to a vector space $\mathbb{V}\otimes\mathbb{W}$, satisfying the following two conditions:*

(i) *if* $\dim\mathbb{V} = n$ *and* $\dim\mathbb{W} = m$, *then* $\dim(\mathbb{V}\otimes\mathbb{W}) = mn$;

(ii) *the set of all vectors* $\rho(\mathbf{x}, \mathbf{y})$, *where* $\mathbf{x} \in \mathbb{V}$, $\mathbf{y} \in \mathbb{W}$, *generates* $\mathbb{V}\otimes\mathbb{W}$.

Then the space $\mathbb{V}\otimes\mathbb{W}$ *is called the tensor product of* $\mathbb{V}$ *and* $\mathbb{W}$. ∎

The space $\mathbb{V}\otimes\mathbb{W}$ is uniquely determined up to an isomorphism. It can be shown (Greub, 1978, p. 9) that there always exists a bilinear map satisfying the conditions given in Definition 1.2.10. Furthermore, let $\mathbb{A} \subseteq \mathbb{V}$, $\mathbb{B} \subseteq \mathbb{W}$ and let ρ_1 be the restriction of ρ to $\mathbb{A} \times \mathbb{B}$. Then ρ_1 generates a vector space (tensor product) $\mathbb{A}\otimes\mathbb{B} \subseteq \mathbb{V}\otimes\mathbb{W}$ and for any $\mathbf{x} \in \mathbb{A}$ and $\mathbf{y} \in \mathbb{B}$, $\rho_1(\mathbf{x}, \mathbf{y}) = \rho(\mathbf{x}, \mathbf{y})$ (Greub 1978, p. 13).

In the next theorem we give the most fundamental relations for subspaces $\mathbb{A}\otimes\mathbb{B}$ of $\mathbb{V}\otimes\mathbb{W}$. Let $\mathbf{\Lambda}_1$ and $\mathbf{\Lambda}_2$ be the lattices of subspaces of $\mathbb{V}$ and $\mathbb{W}$, respectively.

Theorem 1.2.19. *Let* $\mathbb{A}$, $\mathbb{A}_i \in \mathbf{\Lambda}_1$, *i=1,2, and let* $\mathbb{B}$, $\mathbb{B}_i \in \mathbf{\Lambda}_2$, *i=1,2,. Then*

(i) $\mathbb{A}\otimes\mathbb{B} = \{\mathbf{0}\}$ *if and only if* $\mathbb{A} = \{\mathbf{0}\}$ *or* $\mathbb{B} = \{\mathbf{0}\}$;

(ii) $\mathbb{A}_1\otimes\mathbb{B}_1 \subseteq \mathbb{A}_2\otimes\mathbb{B}_2$ *if and only if* $\mathbb{A}_1 \subseteq \mathbb{A}_2$ *and* $\mathbb{B}_1 \subseteq \mathbb{B}_2$, *under the condition that* $\mathbb{A}_2 \neq \{\mathbf{0}\}, \mathbb{B}_2 \neq \{\mathbf{0}\}$;

(iii) $(\mathbb{A}_1 + \mathbb{A}_2)\otimes(\mathbb{B}_1 + \mathbb{B}_2) = (\mathbb{A}_1\otimes\mathbb{B}_1) + (\mathbb{A}_1\otimes\mathbb{B}_2) + (\mathbb{A}_2\otimes\mathbb{B}_1) + (\mathbb{A}_2\otimes\mathbb{B}_2)$;

(iv) $(\mathbb{A}_1\otimes\mathbb{B}_1)\cap(\mathbb{A}_2\otimes\mathbb{B}_2) = (\mathbb{A}_1\cap\mathbb{A}_2)\otimes(\mathbb{B}_1\cap\mathbb{B}_2)$;

(v) $(\mathbb{A}_1\otimes\mathbb{B}_1)\cap(\mathbb{A}_2\otimes\mathbb{B}_2) = \{\mathbf{0}\}$ *if and only if* $(\mathbb{A}_1\cap\mathbb{A}_2) = \{\mathbf{0}\}$ *or* $(\mathbb{B}_1\cap\mathbb{B}_2) = \{\mathbf{0}\}$.

PROOF: Since by Definition 1.2.10 (i) $\dim(\mathbb{A}\otimes\mathbb{B}) = \dim(\mathbb{A}) \times \dim(\mathbb{B})$, (i) is established. The statement in (ii) follows, because if ρ is generating $\mathbb{A}_2\otimes\mathbb{B}_2$, a restriction ρ_1 to ρ generates $\mathbb{A}_1\otimes\mathbb{B}_1$. The relation in (iii) immediately follows from bilinearity and (v) is established with the help of (i) and (iv). Thus the proof of the theorem is completed if (iv) is verified.

Note that (iv) holds if any of the subspaces are equal to zero, and in the subsequent this case will be excluded. First the elements in $\mathbb{V}\times\mathbb{W}$ are ordered in such a manner that $\mathbb{A}_1 \times \mathbb{B}_1 \subseteq \mathbb{A}_2 \times \mathbb{B}_2$ if and only if $\mathbb{A}_1 \subseteq \mathbb{A}_2$ and $\mathbb{B}_1 \subseteq \mathbb{B}_2$. The reason for doing this is that from (ii) it follows that the proposed ordering implies that any bilinear map $\rho : \mathbb{V}\times\mathbb{W} \rightarrow \mathbb{V}\otimes\mathbb{W}$ is isoton. According to Birkhoff (1967, p. 8) a consequence of this ordering of subspaces in $\mathbb{V}\times\mathbb{W}$ is that the totality of subspaces in $\mathbb{V}\times\mathbb{W}$ forms a direct product lattice, and that the g.l.b. of $\{\mathbb{A}_1 \times \mathbb{B}_1, \mathbb{A}_2 \times \mathbb{B}_2\}$ is given by $\mathbb{A}_1 \cap \mathbb{A}_2 \times \mathbb{B}_1 \cap \mathbb{B}_2$. This fact is of utmost importance.

Now, since trivially

$$(\mathbb{A}_1\cap\mathbb{A}_2)\otimes(\mathbb{B}_1\cap\mathbb{B}_2) \subseteq (\mathbb{A}_1\otimes\mathbb{B}_1)\cap(\mathbb{A}_2\otimes\mathbb{B}_2)$$

we obtain that $(\mathbb{A}_1\cap\mathbb{A}_2)\otimes(\mathbb{B}_1\cap\mathbb{B}_2)$ is a lower bound. Moreover, there exists a bilinear map

$$\rho : (\mathbb{A}_1\cap\mathbb{A}_2)\times(\mathbb{B}_1\cap\mathbb{B}_2) \rightarrow (\mathbb{A}_1\cap\mathbb{A}_2)\otimes(\mathbb{B}_1\cap\mathbb{B}_2).$$

Therefore, since $(\mathbb{A}_1 \cap \mathbb{A}_2) \times (\mathbb{B}_1 \cap \mathbb{B}_2)$ and $(\mathbb{A}_1 \otimes \mathbb{A}_2) \cap (\mathbb{B}_1 \otimes \mathbb{B}_2)$ are g.l.b., it follows from the isotonicity of bilinear maps that (iv) is verified if we are able to show that the following inclusion

$$(\mathbb{A}_1 \cap \mathbb{A}_2) \otimes (\mathbb{B}_1 \cap \mathbb{B}_2) \subseteq \sum_i \mathbb{C}_i \otimes \mathbb{D}_i \subsetneq (\mathbb{A}_1 \otimes \mathbb{B}_1) \cap (\mathbb{A}_2 \otimes \mathbb{B}_2). \tag{1.2.4}$$

is impossible for any choices of $\mathbb{C}_i$ and $\mathbb{D}_i$ in (1.2.4). We must consider this, since subspaces of $\mathbb{A} \otimes \mathbb{B}$ may not equal $\mathbb{C} \otimes \mathbb{D}$ for some $\mathbb{C}$ and $\mathbb{D}$, i.e. $\sum_i \mathbb{C}_i \otimes \mathbb{D}_i$ may not equal $\mathbb{C} \otimes \mathbb{D}$. However, if (1.2.4) holds, $\mathbb{C}_i \otimes \mathbb{D}_i$ is included in $\mathbb{A}_1 \otimes \mathbb{B}_1$ and $\mathbb{A}_2 \otimes \mathbb{B}_2$, implying that $\mathbb{C}_i \subseteq \mathbb{A}_1$, $\mathbb{D}_i \subseteq \mathbb{B}_1$, $\mathbb{C}_i \subseteq \mathbb{A}_2$ and $\mathbb{D}_i \subseteq \mathbb{B}_2$, which in turn leads to $\mathbb{C}_i \subseteq \mathbb{A}_1 \cap \mathbb{A}_2$ and $\mathbb{D}_i \subseteq \mathbb{B}_1 \cap \mathbb{B}_2$. Thus,

$$\sum_i \mathbb{C}_i \otimes \mathbb{D}_i \subseteq (\mathbb{A}_1 \cap \mathbb{A}_2) \otimes (\mathbb{B}_1 \cap \mathbb{B}_2)$$

and therefore equality must hold in (1.2.4), which establishes (iv). ∎

The above theorem can easily be extended to multilinear forms, i.e. vector spaces, generated by multilinear mappings (see e.g. Greub, 1978, p. 26, or Marcus, 1973). Furthermore, there are several other ways of proving the theorem (e.g. see Greub, 1978; Chapter I). These proofs usually do not use lattice theory. In particular, product lattices are not considered.

In the subsequent suppose that $\mathbb{V}$ and $\mathbb{W}$ are inner product spaces. Next an inner product on the tensor space $\mathbb{V} \otimes \mathbb{W}$ is defined. The definition is the one which is usually applied and it is needed because the orthogonal complement to a tensor product is going to be considered.

Definition 1.2.11. *Let $\rho : \mathbb{V} \times \mathbb{W} \to \mathbb{V} \otimes \mathbb{W}$ and $(\bullet, \bullet)_{\mathbb{V}}$, $(\bullet, \bullet)_{\mathbb{W}}$ be inner products on $\mathbb{V}$ and $\mathbb{W}$, respectively. The inner product on $\mathbb{V} \times \mathbb{W}$ is defined by*

$$(\rho(\mathbf{x}_1, \mathbf{y}_1), \rho(\mathbf{x}_2, \mathbf{y}_2)_{\mathbb{V}\otimes\mathbb{W}} = (\mathbf{x}_1, \mathbf{x}_2)_{\mathbb{V}}(\mathbf{y}_1, \mathbf{y}_2)_{\mathbb{W}}.$$

∎

The next theorem gives us two well-known relations for tensor products as well as extends some of the statements in Theorem 1.2.6. Other statements of Theorem 1.2.6 can be extended analogously.

Theorem 1.2.20. *Let $\mathbb{A}$, $\mathbb{A}_i \in \mathbf{\Lambda}_1$ and let $\mathbb{B}$, $\mathbb{B}_i \in \mathbf{\Lambda}_2$.*

(i) *Suppose that $\mathbb{A}_1 \otimes \mathbb{B}_1 \neq \{\mathbf{0}\}$ and $\mathbb{A}_2 \otimes \mathbb{B}_2 \neq \{\mathbf{0}\}$. Then*

$$\mathbb{A}_1 \otimes \mathbb{B}_1 \perp \mathbb{A}_2 \otimes \mathbb{B}_2 \quad \text{if and only if } \mathbb{A}_1 \perp \mathbb{A}_2 \text{ or } \mathbb{B}_1 \perp \mathbb{B}_2;$$

(ii)
$$\begin{aligned}(\mathbb{A} \otimes \mathbb{B})^\perp &= (\mathbb{A}^\perp \otimes \mathbb{B}) \boxplus (\mathbb{A} \otimes \mathbb{B}^\perp) \boxplus (\mathbb{A}^\perp \otimes \mathbb{B}^\perp)\\ &= (\mathbb{A}^\perp \otimes \mathbb{W}) \boxplus (\mathbb{A} \otimes \mathbb{B}^\perp)\\ &= (\mathbb{A}^\perp \otimes \mathbb{B}) \boxplus (\mathbb{V} \otimes \mathbb{B}^\perp);\end{aligned}$$

(iii) $(\mathbb{A}_1 \otimes \mathbb{B}_1) = (\mathbb{A}_1 \cap \mathbb{A}_2 \otimes \mathbb{B}_1 \cap \mathbb{B}_2) \boxplus (\mathbb{A}_1 \cap (\mathbb{A}_1^\perp + \mathbb{A}_2^\perp) \otimes \mathbb{B}_1 \cap \mathbb{B}_2)$

$$\boxplus(\mathbb{A}_1 \cap \mathbb{A}_2 \otimes \mathbb{B}_1 \cap (\mathbb{B}_1^\perp + \mathbb{B}_2^\perp))$$

$$\boxplus(\mathbb{A}_1 \cap (\mathbb{A}_1^\perp + \mathbb{A}_2^\perp) \otimes \mathbb{B}_1 \cap (\mathbb{B}_1^\perp + \mathbb{B}_2^\perp));$$

(iv) $$((\mathbb{A}_1 \otimes \mathbb{B}_1) + (\mathbb{A}_2 \otimes \mathbb{B}_2))^\perp$$

$$=\mathbb{A}_1^\perp \cap (\mathbb{A}_1 + \mathbb{A}_2) \otimes \mathbb{B}_2^\perp \cap (\mathbb{B}_1 + \mathbb{B}_2) + (\mathbb{A}_1 + \mathbb{A}_2)^\perp \otimes \mathbb{B}_2^\perp \cap (\mathbb{B}_1 + \mathbb{B}_2)$$

$$+ (\mathbb{A}_1 + \mathbb{A}_2)^\perp \otimes \mathbb{B}_1 \cap \mathbb{B}_2 + \mathbb{A}_2^\perp \cap (\mathbb{A}_1 + \mathbb{A}_2) \otimes \mathbb{B}_1^\perp \cap (\mathbb{B}_1 + \mathbb{B}_2)$$

$$+ (\mathbb{A}_1 + \mathbb{A}_2)^\perp \otimes \mathbb{B}_1^\perp \cap (\mathbb{B}_1 + \mathbb{B}_2) + \mathbb{A}_1^\perp \cap (\mathbb{A}_1 + \mathbb{A}_2) \otimes (\mathbb{B}_1 + \mathbb{B}_2)^\perp$$

$$+ \mathbb{A}_1 \cap \mathbb{A}_2 \otimes (\mathbb{B}_1 + \mathbb{B}_2)^\perp + \mathbb{A}_1^\perp \cap (\mathbb{A}_1 + \mathbb{A}_2) \otimes (\mathbb{B}_1 + \mathbb{B}_2)^\perp$$

$$+ (\mathbb{A}_1 + \mathbb{A}_2)^\perp \otimes (\mathbb{B}_1 + \mathbb{B}_2)^\perp$$

$$= \mathbb{A}_2^\perp \cap (\mathbb{A}_1 + \mathbb{A}_2) \otimes \mathbb{B}_1^\perp + \mathbb{A}_1 \cap \mathbb{A}_2 \otimes (\mathbb{B}_1 + \mathbb{B}_2)^\perp$$

$$+ \mathbb{A}_1^\perp \cap (\mathbb{A}_1 + \mathbb{A}_2) \otimes \mathbb{B}_2^\perp + (\mathbb{A}_1 + \mathbb{A}_2)^\perp \otimes \mathbb{W}$$

$$= \mathbb{A}_1^\perp \otimes \mathbb{B}_2^\perp \cap (\mathbb{B}_1 + \mathbb{B}_2) + (\mathbb{A}_1 + \mathbb{A}_2)^\perp \otimes \mathbb{B}_1 \cap \mathbb{B}_2$$

$$+ \mathbb{A}_2^\perp \otimes \mathbb{B}_1^\perp \cap (\mathbb{B}_1 + \mathbb{B}_2) + \mathbb{V} \otimes (\mathbb{B}_1 + \mathbb{B}_2)^\perp.$$

PROOF: The statement in (i) follows from the definition of the inner product given in Definition 1.2.11, and statement (ii) from utilizing (i) and the relation

$$\begin{aligned}\mathbb{V} \otimes \mathbb{W} &= (\mathbb{A} \boxplus \mathbb{A}^\perp) \otimes (\mathbb{B} \boxplus \mathbb{B}^\perp)\\ &= (\mathbb{A} \otimes \mathbb{B}) \boxplus (\mathbb{A} \otimes \mathbb{B}^\perp) \boxplus (\mathbb{A}^\perp \otimes \mathbb{B}) \boxplus (\mathbb{A}^\perp \otimes \mathbb{B}^\perp),\end{aligned}$$

which is verified by Theorem 1.2.19 (iii). To prove (iii) and (iv) we use Theorem 1.2.19 together with Theorem 1.2.6. ∎

The next theorem is important when considering so-called growth curve models in Chapter 4.

Theorem 1.2.21. *Let $\mathbb{A}_1$, $i = 1, 2, \ldots, s$, be arbitrary elements of $\mathbf{\Lambda}_1$ such that $\mathbb{A}_s \subseteq \mathbb{A}_{s-1} \subseteq \cdots \subseteq \mathbb{A}_1$, and let $\mathbb{B}_i$, $i = 1, 2, \ldots, s$, be arbitrary elements of $\mathbf{\Lambda}_2$. Denote $\mathbb{C}_j = \sum_{1 \le i \le j} \mathbb{B}_i$. Then*

(i) $$(\sum_i \mathbb{A}_i \otimes \mathbb{B}_i)^\perp = (\mathbb{A}_s \otimes \mathbb{C}_s^\perp) \boxplus_{1 \le j \le s-1} (\mathbb{A}_j \cap \mathbb{A}_{j+1}^\perp \otimes \mathbb{C}_j^\perp) \boxplus (\mathbb{A}_1^\perp \otimes \mathbb{W});$$

(ii) $$(\sum_i \mathbb{A}_i \otimes \mathbb{B}_i)^\perp = (\mathbb{V} \otimes \mathbb{C}_s^\perp) \boxplus_{2 \le j \le s} (\mathbb{A}_j \perp \otimes \mathbb{C}_j \cap \mathbb{C}_{j-1}^\perp) \boxplus (\mathbb{A}_1^\perp \mathbb{B}_1).$$

PROOF: Since inclusion of subspaces implies commutativity of subspaces, we obtain

$$\mathbb{A}_i = \mathbb{A}_s \boxplus_{i \le j \le s-1} \mathbb{A}_j \cap \mathbb{A}_{j+1}^\perp, \quad i = 1, 2, \ldots, s-1,$$

and thus

$$\begin{aligned}\sum_i \mathbb{A}_i \otimes \mathbb{B}_i =& \mathbb{A}_s \otimes \mathbb{C}_s \boxplus \sum_{i \leq s-1} \boxplus_{i \leq j \leq s-1} \mathbb{A}_j \cap \mathbb{A}_{j+1}^{\perp} \otimes \mathbb{B}_i \\ =& \mathbb{A}_s \otimes \mathbb{C}_s \sum_{1 \leq j \leq s-1} \mathbb{A}_j \cap \mathbb{A}_{j+1}^{\perp} \otimes \mathbb{C}_j, \end{aligned} \tag{1.2.5}$$

which is obviously orthogonal to the left hand side in (i). Moreover, summing the left hand side in (i) with (1.2.5) gives the whole space which establishes (i). The statement in (ii) is verified in a similar fashion by noting that $\mathbb{C}_{j-1}^{\perp} = \mathbb{C}_j^{\perp} \boxplus \mathbb{C}_j \cap \mathbb{C}_{j-1}^{\perp}$. ∎

Theorem 1.2.22. *Let $\mathbb{A}_i \in \mathbf{\Lambda}_1$ and $\mathbb{B}_i \in \mathbf{\Lambda}_2$ such that $(\mathbb{A}_1 \otimes \mathbb{B}_1) \cap (\mathbb{A}_2 \otimes \mathbb{B}_2) \neq \{\mathbf{0}\}$. Then*

$$\mathbb{A}_1 \otimes \mathbb{B}_1 | \mathbb{A}_2 \otimes \mathbb{B}_2 \quad \textit{if and only if} \quad \mathbb{A}_1 | \mathbb{A}_2 \quad \textit{and} \quad \mathbb{B}_1 | \mathbb{B}_2.$$

PROOF: In order to characterize commutativity Theorem 1.2.8 (i) is used. Suppose that $\mathbb{A}_1 | \mathbb{A}_2$ and $\mathbb{B}_1 | \mathbb{B}_2$ hold. Commutativity follows if

$$(\mathbb{A}_1 \otimes \mathbb{B}_1)^{\perp} + (\mathbb{A}_1 \otimes \mathbb{B}_1) \cap (\mathbb{A}_2 \otimes \mathbb{B}_2) = (\mathbb{A}_1 \otimes \mathbb{B}_1)^{\perp} + (\mathbb{A}_2 \otimes \mathbb{B}_2). \tag{1.2.6}$$

Applying Theorem 1.2.6 (i) and Theorem 1.2.19 (iii) yields

$$\begin{aligned}\mathbb{A}_2 \otimes \mathbb{B}_2 =& (\mathbb{A}_1 \cap \mathbb{A}_2 \otimes \mathbb{B}_1 \cap \mathbb{B}_2) \boxplus (\mathbb{A}_2 \cap (\mathbb{A}_1 \cap \mathbb{A}_2)^{\perp} \otimes \mathbb{B}_2) \\ & \boxplus (\mathbb{A}_1 \cap \mathbb{A}_2 \otimes \mathbb{B}_2 \cap (\mathbb{B}_1 \cap \mathbb{B}_2)^{\perp}). \end{aligned} \tag{1.2.7}$$

If $\mathbb{A}_1 | \mathbb{A}_2$ and $\mathbb{B}_1 | \mathbb{B}_2$, Theorem 1.2.8 (i) and Theorem 1.2.20 (ii) imply

$$\begin{aligned}(\mathbb{A}_2 \cap & (\mathbb{A}_1 \cap \mathbb{A}_2)^{\perp} \otimes \mathbb{B}_2) \boxplus (\mathbb{A}_1 \cap \mathbb{A}_2 \otimes (\mathbb{B}_2 \cap (\mathbb{B}_1 \cap \mathbb{B}_2)^{\perp}) \\ \subseteq & (\mathbb{A}_1 \otimes \mathbb{B}_1)^{\perp} = \mathbb{A}_1^{\perp} \otimes \mathbb{W} \boxplus \mathbb{A}_1 \otimes \mathbb{B}_1^{\perp}, \end{aligned}$$

where $\mathbb{W}$ represents the whole space. Hence (1.2.6) holds. For the converse, let (1.2.6) be true. Using the decomposition given in (1.2.7) we get from Theorem 1.2.1 (v) and Theorem 1.2.20 (ii) that ($\mathbb{V}$ and $\mathbb{W}$ represent the whole spaces)

$$\begin{aligned}\mathbb{A}_1 \cap \mathbb{A}_2 \otimes (\mathbb{B}_2 \cap (\mathbb{B}_1 \cap \mathbb{B}_2)^{\perp}) \subseteq & (\mathbb{A}_1 \otimes \mathbb{B}_1)^{\perp} = \mathbb{A}_1^{\perp} \otimes \mathbb{W} \boxplus \mathbb{A}_1 \otimes \mathbb{B}_1^{\perp}; \\ \mathbb{A}_2 \cap (\mathbb{A}_1 \cap \mathbb{A}_2)^{\perp} \otimes \mathbb{B}_1 \cap \mathbb{B}_2 \subseteq & (\mathbb{A}_1 \otimes \mathbb{B}_1)^{\perp} = \mathbb{A}_1^{\perp} \otimes \mathbb{B}_1 \boxplus \mathbb{V} \otimes \mathbb{B}_1^{\perp}. \end{aligned}$$

Thus, from Corollary 1.2.7.1, Theorem 1.2.19 (ii) and Theorem 1.2.8 (i) it follows that the converse also is verified. ∎

Note that by virtue of Theorem 1.2.19 (v) and Theorem 1.2.20 (i) it follows that $\mathbb{A}_1 \otimes \mathbb{B}_1$ and $\mathbb{A}_2 \otimes \mathbb{B}_2$ are disjoint (orthogonal) if and only if $\mathbb{A}_1$ and $\mathbb{A}_2$ are disjoint (orthogonal) or $\mathbb{B}_1$ and $\mathbb{B}_2$ are disjoint (orthogonal), whereas Theorem 1.2.22 states that $\mathbb{A}_1 \otimes \mathbb{B}_1$ and $\mathbb{A}_2 \otimes \mathbb{B}_2$ commute if and only if $\mathbb{A}_1$ and $\mathbb{B}_1$ commute, and $\mathbb{A}_2$ and $\mathbb{B}_2$ commute.

Finally we present some results for tensor products of linear transformations.

Definition 1.2.12. *Let $A : \mathbb{V}_1 \to \mathbb{W}_1$, $B : \mathbb{V}_2 \to \mathbb{W}_2$ be linear maps, and let $\rho_1 : \mathbb{V}_1 \times \mathbb{V}_2 \to \mathbb{V}_1 \otimes \mathbb{V}_2$ and $\rho_2 : \mathbb{W}_1 \times \mathbb{W}_2 \to \mathbb{W}_1 \otimes \mathbb{W}_2$ be bilinear maps. The tensor product $A \otimes B : \mathbb{V}_1 \otimes \mathbb{V}_2 \to \mathbb{W}_1 \otimes \mathbb{W}_2$ is a linear map determined by*

$$A \otimes B\rho_1(\mathbf{x},\mathbf{y}) = \rho_2(A\mathbf{x}, B\mathbf{y}), \quad \forall \mathbf{x} \in \mathbb{V}_1, \forall \mathbf{y} \in \mathbb{V}_2.$$

∎

Note that by definition

$$\mathcal{R}(A \otimes B) = \mathcal{R}(A) \otimes \mathcal{R}(B),$$

which means that the range space of the tensor product of linear mappings equals the tensor product of the two range spaces. From this observation we can see that Theorem 1.2.19 and Theorem 1.2.20 can be utilized in this context. For example, from Theorem 1.2.19 (ii) it follows that

$$\mathcal{R}(A_1 \otimes B_1) \subseteq \mathcal{R}(A_2 \otimes B_2)$$

if and only if $\mathcal{R}(A_1) \subseteq \mathcal{R}(A_2)$ and $\mathcal{R}(B_1) \subseteq \mathcal{R}(B_2)$. Moreover, Theorem 1.2.20 (ii) yields

$$\mathcal{R}(A \otimes B)^{\perp} = (\mathcal{R}(A)^{\perp} \otimes \mathcal{R}(B)) \boxplus (\mathcal{R}(A) \otimes \mathcal{R}(B)^{\perp}) \boxplus (\mathcal{R}(A)^{\perp} \otimes \mathcal{R}(B)^{\perp}).$$

Other statements in Theorem 1.2.20 could also easily be converted to hold for range spaces, but we leave it to the reader to find out the details.

1.2.6 *Matrix representation of linear operators in vector spaces*

Let $\mathbb{V}$ and $\mathbb{W}$ be finite-dimensional vector spaces with basis vectors $\mathbf{e}_i$, $i \in I$ and $\mathbf{d}_j$, $j \in J$, where I consists of n and J of m elements. Every vector $\mathbf{x} \in \mathbb{V}$ and $\mathbf{y} \in \mathbb{W}$ can be presented through the basis vectors:

$$\mathbf{x} = \sum_{i \in I} x_i \mathbf{e}_i, \qquad \mathbf{y} = \sum_{j \in J} y_j \mathbf{d}_j.$$

The coefficients x_i and y_j are called the *coordinates* of $\mathbf{x}$ and $\mathbf{y}$, respectively. Let $A : \mathbb{V} \to \mathbb{W}$ be a linear operator (transformation) where the coordinates of $A\mathbf{e}_i$ will be denoted a_{ji}, $j \in J$, $i \in I$. Then

$$\sum_{j \in J} y_j \mathbf{d}_j = \mathbf{y} = A\mathbf{x} = \sum_{i \in I} x_i A\mathbf{e}_i = \sum_{i \in I} x_i \sum_{j \in J} a_{ji} \mathbf{d}_j = \sum_{j \in J} (\sum_{i \in I} a_{ji} x_i) \mathbf{d}_j,$$

i.e. the coordinates y_j of $A\mathbf{x}$ are determined by the coordinates of the images $A\mathbf{e}_i$, $i \in I$, and $\mathbf{x}$. This implies that to every linear operator, say A, there exists a matrix $\mathbf{A}$ formed by the coordinates of the images of the basis vectors. However, the matrix can be constructed in different ways, since the construction depends on the chosen basis as well as on the order of the basis vectors.

If we have Euclidean spaces $\mathbb{R}^n$ and $\mathbb{R}^m$, i.e. ordered n and m tuples, then the most natural way is to use the original bases, i.e. $\mathbf{e}_i = (\delta_{ik})_k$, $i, k \in I$; $\mathbf{d}_j = (\delta_{jl})_l$, $j, l \in J$, where δ_{rs} is the Kronecker delta, i.e. it equals 1 if $r = s$ and 0 otherwise. In order to construct a matrix corresponding to an operator A, we have to choose a way of storing the coordinates of $A\mathbf{e}_i$. The two simplest possibilities would be to take the $i-$th row or the $i-$th column of a matrix and usually the $i-$th column is chosen. In this case the matrix which represents the operator $A : \mathbb{R}^n \to \mathbb{R}^m$ is a matrix $\mathbf{A} = (a_{ji}) \in \mathbb{R}^{m\times n}$, where $a_{ji} = (A\mathbf{e}_i)_j$, $j = 1, \dots, m$; $i = 1, \dots, n$. For calculating the coordinates of $A\mathbf{x}$ we can utilize the usual product of a matrix and a vector. On the other hand, if the matrix representation of A is built up in such a way that the coordinates of $A\mathbf{e}_i$ form the $i-$th row of the matrix, we get an $n \times m-$matrix, and we find the coordinates of $A\mathbf{x}$ as a product of a row vector and a matrix.

We are interested in the case where the vector spaces $\mathbb{V}$ and $\mathbb{W}$ are spaces of $p\times q-$ and $r \times s-$arrays (matrices) in $\mathbb{R}^{p\times q}$ and $\mathbb{R}^{r\times s}$, respectively. The basis vectors are denoted $\mathbf{e}_i$ and $\mathbf{d}_j$ so that

$$i \in I = \{(1,1), (1,2), \dots, (p,q)\},$$

$$j \in J = \{(1,1), (1,2), \dots (r,s)\}.$$

In this case it is convenient to use two indices for the basis, i.e. instead of $\mathbf{e}_i$ and $\mathbf{d}_j$, the notation $\mathbf{e}_{vw}$ and $\mathbf{d}_{tu}$ will be used:

$$\mathbf{e}_i \longrightarrow \mathbf{e}_{vw},$$

$$\mathbf{d}_j \longrightarrow \mathbf{d}_{tu},$$

where

$$\mathbf{e}_{vw} = (\delta_{vk}, \delta_{wl}), \qquad v, k = 1, \dots, p; \quad w, l = 1, \dots, q;$$

$$\mathbf{d}_{tu} = (\delta_{tm}, \delta_{un}), \qquad t, m = 1, \dots, r; \quad u, n = 1, \dots, s.$$

Let A be a linear map: $A : \mathbb{R}^{p\times q} \longrightarrow \mathbb{R}^{r\times s}$. The coordinate y_{tu} of the image $A\mathbf{X}$ is found by the equality

$$y_{tu} = \sum_{k=1}^{p} \sum_{l=1}^{q} a_{tukl} x_{kl}. \tag{1.2.8}$$

In order to find the matrix corresponding to the linear transformation A we have to know again the coordinates of the images of the basis vectors. The way of ordering the coordinates into a matrix is a matter of taste. In the next theorem we present the two most frequently used orderings. The proof of the theorem can be found in Parring (1992).

Theorem 1.2.23. *Let $A : \mathbb{U} \to \mathbb{V}$, $\mathbb{U} \in \mathbb{R}^{p\times q}$, $\mathbb{V} \in \mathbb{R}^{r\times s}$ be a linear transformation.*

(i) *If the $tu-$th coordinates of the basis vectors $\mathbf{e}_{vw}$ form a $p \times q-$block $\mathbf{A}_{tu}$:*

$$\mathbf{A}_{tu} = (a_{tuvw}), \qquad v = 1, \dots, p; \; w = 1, \dots, q, \tag{1.2.9}$$

the representation of the transformation A is a partitioned matrix which equals

$$\mathbf{A} = \begin{pmatrix} \mathbf{A}_{11} & \dots & \mathbf{A}_{1s} \\ \vdots & \ddots & \vdots \\ \mathbf{A}_{r1} & \dots & \mathbf{A}_{rs} \end{pmatrix}.$$

Then the coordinates of the image $\mathbf{Y} = A\mathbf{X}$ are given by

$$y_{tu} = \operatorname{tr}(\mathbf{A}'_{tu}\mathbf{X}). \tag{1.2.10}$$

(ii) *If the representation of the transformation A is a partitioned matrix which equals*

$$\mathbf{A} = \begin{pmatrix} \mathbf{A}_{11} & \dots & \mathbf{A}_{1q} \\ \vdots & \ddots & \vdots \\ \mathbf{A}_{p1} & \dots & \mathbf{A}_{pq} \end{pmatrix},$$

where $\mathbf{A}_{kl} : r \times s$ consists of all coordinates of the basis vector $\mathbf{e}_{kl}$

$$\mathbf{A}_{kl} = (a_{tukl}), \qquad t = 1, \dots, r;\ u = 1, \dots, s, \tag{1.2.11}$$

the coordinates of $\mathbf{Y} = A\mathbf{X}$ are given by

$$y_{tu} = \left(\sum_{k=1}^{p} \sum_{l=1}^{q} x_{kl} \mathbf{A}_{kl} \right)_{tu}. \tag{1.2.12}$$

∎

REMARK: In (1.2.12) the coordinate y_{tu} can be considered as an element of the *star-product* $\mathbf{X} \star \mathbf{A}$ of the matrices $\mathbf{X}$ and $\mathbf{A}$ which was introduced by MacRae (1974).

In the next theorem we briefly consider bilinear maps and tensor products.

Theorem 1.2.24. *Let $A : \mathbb{V}_1 \to \mathbb{W}_1$, $\mathbb{V}_1 \in \mathbb{R}^n$, $\mathbb{W}_1 \in \mathbb{R}^r$, $B : \mathbb{V}_2 \to \mathbb{W}_2$, $\mathbb{V}_2 \in \mathbb{R}^q$ and $\mathbb{W}_2 \in \mathbb{R}^s$. Then in Definition 1.2.10 $\mathbf{A} \otimes \mathbf{B}$ and $\operatorname{vec}\mathbf{A}\operatorname{vec}'\mathbf{B}$ are both representations of the linear map $A \otimes B$, where $\otimes$ in $\mathbf{A} \otimes \mathbf{B}$ denotes the Kronecker product which is defined in §1.3.3, and vec is the vec-operator defined in §1.3.4.*

PROOF: Let $\rho_1 : \mathbb{V}_1 \times \mathbb{V}_2 \to \mathbb{V}_1 \otimes \mathbb{V}_2$ and $\rho_2 : \mathbb{W}_1 \times \mathbb{W}_2 \to \mathbb{W}_1 \otimes \mathbb{W}_2$ be bilinear maps. Then, by Definition 1.2.10 and bilinearity,

$$A \otimes B \rho_1(\mathbf{x}^1, \mathbf{x}^2) = \rho_2(A\mathbf{x}^1, B\mathbf{x}^2) = \sum_{j_1 \in J_1} \sum_{j_2 \in J_2} \sum_{i_1 \in I_1} \sum_{i_2 \in I_2} a_{j_1 i_1} b_{j_2 i_2} x^1_{i_1} x^2_{i_2} \rho_2(d^1_{j_1}, d^2_{j_2}),$$

where

$$A\mathbf{x}^1 = \sum_{i_1 \in I_1} a_{j_1 i_1} x^1_{i_1} d^1_{j_1}$$

and

$$B\mathbf{x}^2 = \sum_{i_2 \in I_2} b_{j_2 i_2} x^2_{i_2} d^2_{j_2}.$$

If we order the elements $\{a_{i_1j_1}b_{i_2j_2}\}$ as

$$\begin{pmatrix} a_{11}b_{11} & \cdots & a_{11}b_{1s} & \cdots & a_{1r}b_{11} & \cdots & a_{1r}b_{1s} \\ \vdots & \ddots & \vdots & \ddots & \vdots & \ddots & \vdots \\ a_{11}b_{q1} & \cdots & a_{11}b_{qs} & \cdots & a_{1r}b_{q1} & \cdots & a_{1r}b_{qs} \\ \vdots & \ddots & \vdots & \ddots & \vdots & \ddots & \vdots \\ a_{p1}b_{11} & \cdots & a_{p1}b_{1s} & \cdots & a_{pr}b_{11} & \cdots & a_{pr}b_{1s} \\ \vdots & \ddots & \vdots & \ddots & \vdots & \ddots & \vdots \\ a_{p1}b_{q1} & \cdots & a_{p1}b_{qs} & \cdots & a_{pr}b_{q1} & \cdots & a_{pr}b_{qs} \end{pmatrix}$$

the representation can, with the help of the Kronecker product in §1.3.3, be written as $\mathbf{A} \otimes \mathbf{B}$. Moreover, if the elements $\{a_{i_1j_1}b_{i_2j_2}\}$ are ordered as

$$\begin{pmatrix} a_{11}b_{11} & \cdots & a_{11}b_{q1} & a_{11}b_{12} & \cdots & a_{11}b_{q2} & \cdots & a_{11}b_{qs} \\ \vdots & \ddots & \vdots & \vdots & \ddots & \vdots & \ddots & \vdots \\ a_{p1}b_{11} & \cdots & a_{p1}b_{q1} & a_{p1}b_{12} & \cdots & a_{p1}b_{q2} & \cdots & a_{p1}b_{qs} \\ a_{12}b_{11} & \cdots & a_{12}b_{q1} & a_{12}b_{12} & \cdots & a_{12}b_{q2} & \cdots & a_{12}b_{qs} \\ \vdots & \ddots & \vdots & \vdots & \ddots & \vdots & \ddots & \vdots \\ a_{p2}b_{11} & \cdots & a_{p2}b_{q1} & a_{p2}b_{12} & \cdots & a_{p2}b_{q2} & \cdots & a_{p2}b_{qs} \\ \vdots & \ddots & \vdots & \vdots & \ddots & \vdots & \ddots & \vdots \\ a_{pr}b_{11} & \cdots & a_{pr}b_{q1} & a_{pr}b_{12} & \cdots & a_{pr}b_{q2} & \cdots & a_{pr}b_{qs} \end{pmatrix}$$

we have, according to §1.3.4, the matrix which equals $\text{vec}\mathbf{A}\text{vec}'\mathbf{B}$. ∎

REMARK: If in (1.2.9) $a_{tuvw} = a_{tu}b_{vw}$, it follows that

$$\mathbf{A} = \begin{pmatrix} \mathbf{A}_{11} & \dots & \mathbf{A}_{1s} \\ \vdots & \ddots & \vdots \\ \mathbf{A}_{r1} & \dots & \mathbf{A}_{rs} \end{pmatrix} = \mathbf{A} \otimes \mathbf{B}.$$

When comparing $A : \mathbb{V} \to \mathbb{W}$ in Theorem 1.2.23 with the tensor map $A \otimes B : \mathbb{V}_1 \otimes \mathbb{V}_2 \to \mathbb{W}_1 \otimes \mathbb{W}_2$, where $\dim\mathbb{V} = \dim(\mathbb{V}_1 \otimes \mathbb{V}_2)$ and $\dim\mathbb{W} = \dim(\mathbb{W}_1 \otimes \mathbb{W}_2)$, some insight in the consequences of using bilinear maps instead of linear maps is obtained.

1.2.7 *Column vector spaces*

The vector space generated by the columns of an arbitrary matrix $\mathbf{A} : p \times q$ is denoted $\mathcal{C}(\mathbf{A})$:

$$\mathcal{C}(\mathbf{A}) = \{\mathbf{a} : \mathbf{a} = \mathbf{A}\mathbf{z},\ \mathbf{z} \in \mathbb{R}^q\}.$$

Furthermore, the orthogonal complement to $\mathcal{C}(\mathbf{A})$ is denoted $\mathcal{C}(\mathbf{A})^\perp$, and a matrix which columns generate the orthogonal complement to $\mathcal{C}(\mathbf{A})$ is denoted $\mathbf{A}^o$, i.e. $\mathcal{C}(\mathbf{A}^o) = \mathcal{C}(\mathbf{A})^\perp$. The matrix $\mathbf{A}^o$ shares the property with $\mathbf{A}^-$ of not being unique. For example, we can choose $\mathbf{A}^o = \mathbf{I} - (\mathbf{A}')^-\mathbf{A}'$ or $\mathbf{A}^o = \mathbf{I} - \mathbf{A}(\mathbf{A}'\mathbf{A})^-\mathbf{A}'$ as well as in some other way.

From the definition of a column vector space and the definition of a range space given in §1.2.4, i.e. $\mathcal{R}(A) = \{\mathbf{x} : \mathbf{x} = A\mathbf{y}, \mathbf{y} \in \mathbb{V}\}$, it follows that any column vector space can be identified by a corresponding range space. Hence, all results in §1.2.4 that hold for range spaces will also hold for column vector spaces. Some of the results in this subsection will be restatements of results given in §1.2.4, and now and then we will refer to that paragraph.
The first proposition presents basic relations for column vector spaces.

Proposition 1.2.2. *For column vector spaces the following relations hold.*

(i) $C(\mathbf{A}) \subseteq C(\mathbf{B})$ *if and only if* $\mathbf{A} = \mathbf{BQ}$, *for some matrix* $\mathbf{Q}$.

(ii) *If* $C(\mathbf{A}+\mathbf{BE}) \subseteq C(\mathbf{B})$, *for some matrix* $\mathbf{E}$ *of proper size, then* $C(\mathbf{A}) \subseteq C(\mathbf{B})$.
If $C(\mathbf{A}) \subseteq C(\mathbf{B})$, *then* $C(\mathbf{A}+\mathbf{BE}) \subseteq C(\mathbf{B})$, *for any matrix* $\mathbf{E}$ *of proper size.*

(iii) $C(\mathbf{A}'\mathbf{B}_1) \subseteq C(\mathbf{A}'\mathbf{B}_2)$ *if* $C(\mathbf{B}_1) \subseteq C(\mathbf{B}_2)$;
$C(\mathbf{A}'\mathbf{B}_1) = C(\mathbf{A}'\mathbf{B}_2)$ *if* $C(\mathbf{B}_1) = C(\mathbf{B}_2)$.

(iv) $C(\mathbf{A}'\mathbf{B}) = C(\mathbf{A}')$ *if* $C(\mathbf{A}) \subseteq C(\mathbf{B})$.

(v) $C(\mathbf{A}) \cap C(\mathbf{B}) = C((\mathbf{A}^o : \mathbf{B}^o)^o)$.

(vi) *For any* $\mathbf{A}^-$
$$\mathbf{CA}^-\mathbf{A} = \mathbf{C}$$
if and only if $C(\mathbf{C}') \subseteq C(\mathbf{A}')$.

(vii) $C(\mathbf{A}') = C(\mathbf{A}'\mathbf{B})$ *if and only if* $r(\mathbf{A}'\mathbf{B}) = r(\mathbf{A}')$.

(viii) *Let* $\mathbf{A} \in \mathbb{R}^{p\times q}$, $\mathbf{S} > 0$ *and* $r(\mathbf{H}) = p$. *Then* $C(\mathbf{A}') = C(\mathbf{A}'\mathbf{H}) = C(\mathbf{A}'\mathbf{SA})$.

(ix) *Let* $\mathbf{A} \in \mathbb{R}^{p\times q}$ *and* $\mathbf{S} > 0$. *Then*
$\mathbf{C}(\mathbf{A}'\mathbf{SA})^-\mathbf{A}'\mathbf{SA} = \mathbf{C}$ *if and only if* $C(\mathbf{C}') \subseteq C(\mathbf{A}')$;
$\mathbf{A}(\mathbf{A}'\mathbf{SA})^-\mathbf{A}'\mathbf{SB} = \mathbf{B}$ *if and only if* $C(\mathbf{B}) \subseteq C(\mathbf{A})$;
$\mathbf{CAB}(\mathbf{CAB})^-\mathbf{C} = \mathbf{C}$ *if* $r(\mathbf{CAB}) = r(\mathbf{C})$.

(x) $\mathbf{CA}^-\mathbf{B}$ *is invariant under choice of g-inverse if and only if* $C(\mathbf{C}') \subseteq C(\mathbf{A}')$ *and* $C(\mathbf{B}) \subseteq C(\mathbf{A})$.

(xi) *Let* $\mathbf{S} > 0$, *then* $\mathbf{C}_1(\mathbf{A}'\mathbf{SA})^-\mathbf{C}_2$ *is invariant under any choice of* $(\mathbf{A}'\mathbf{SA})^-$ *if and only if* $C(\mathbf{C}_1') \subseteq C(\mathbf{A}')$ *and* $C(\mathbf{C}_2) \subseteq C(\mathbf{A}')$.

(xii) *If* $C(\mathbf{C}') \subseteq C(\mathbf{A}')$ *and* $\mathbf{S} > 0$, *then*
$$C(\mathbf{C}) = C(\mathbf{C}(\mathbf{A}'\mathbf{SA})^-) = C(\mathbf{C}(\mathbf{A}'\mathbf{SA})^-\mathbf{A}'\mathbf{S}).$$

(xiii) $C(\mathbf{AB}) \subseteq C(\mathbf{AB}^o)$ *if and only if* $C(\mathbf{A}) = C(\mathbf{AB}^o)$. ∎

In the next theorem we present one of the most fundamental relations for the treatment of linear models with linear restrictions on the parameter space. We will apply this result in Chapter 4.

Theorem 1.2.25. *Let* $\mathbf{S}$ *be p.d.,* $\mathbf{A}$ *and* $\mathbf{B}$ *be arbitrary matrices of proper sizes, and* $\mathbf{C}$ *any matrix such that* $\mathcal{C}(\mathbf{C}) = \mathcal{C}(\mathbf{A}') \cap \mathcal{C}(\mathbf{B})$. *Then*

$$\begin{aligned}\mathbf{A}(\mathbf{A}'\mathbf{S}\mathbf{A})^-\mathbf{A}' - \mathbf{A}\mathbf{B}^o(\mathbf{B}^{o'}\mathbf{A}'\mathbf{S}\mathbf{A}\mathbf{B}^o)^-\mathbf{B}^{o'}\mathbf{A}'&\\ =\mathbf{A}(\mathbf{A}'\mathbf{S}\mathbf{A})^-\mathbf{C}(\mathbf{C}'(\mathbf{A}'\mathbf{S}\mathbf{A})^-\mathbf{C})^-\mathbf{C}'(\mathbf{A}'\mathbf{S}\mathbf{A})^-\mathbf{A}'.&\end{aligned}$$

PROOF: We will prove the theorem when $\mathbf{S} = \mathbf{I}$. The general case follows immediately by changing $\mathbf{A}$ to $\mathbf{S}^{1/2}\mathbf{A}$. Now, if $\mathbf{S} = \mathbf{I}$ we see that the statement involves three orthogonal projectors (symmetric idempotent matrices):

$$\mathbf{A}(\mathbf{A}'\mathbf{A})^-\mathbf{A}',$$

$$\mathbf{A}\mathbf{B}^o(\mathbf{B}^{o'}\mathbf{A}'\mathbf{A}\mathbf{B}^o)^-\mathbf{B}^{o'}\mathbf{A}'$$

and

$$\mathbf{A}(\mathbf{A}'\mathbf{A})^-\mathbf{C}(\mathbf{C}'(\mathbf{A}'\mathbf{A})^-\mathbf{C})^-\mathbf{C}'(\mathbf{A}'\mathbf{A})^-\mathbf{A}'.$$

Let $\mathbf{P} = \mathbf{A}(\mathbf{A}'\mathbf{A})^-\mathbf{A}'$ and then

$$\mathcal{C}(\mathbf{A}) \cap \mathcal{C}(\mathbf{A}\mathbf{B}^o)^\perp = \mathcal{C}(\mathbf{P}) \cap \mathcal{C}(\mathbf{A}\mathbf{B}^o)^\perp.$$

By virtue of Theorem 1.2.15,

$$\mathcal{C}(\mathbf{P}(\mathbf{P}\mathbf{A}\mathbf{B}^o)^o) = \mathcal{C}(\mathbf{P}(\mathbf{A}\mathbf{B}^o)^o) = \mathcal{C}(\mathbf{A}(\mathbf{A}'\mathbf{A})^-\mathbf{C}).$$

Since $\mathcal{C}(\mathbf{A}\mathbf{B}^o) \subseteq \mathcal{C}(\mathbf{A})$, it follows from Theorem 1.2.6 (ii) that,

$$\mathcal{C}(\mathbf{A}) = \mathcal{C}(\mathbf{A}\mathbf{B}^o) \boxplus \mathcal{C}(\mathbf{A}) \cap \mathcal{C}(\mathbf{A}\mathbf{B}^o)^\perp = \mathcal{C}(\mathbf{A}\mathbf{B}^o) \boxplus \mathcal{C}(\mathbf{A}(\mathbf{A}'\mathbf{A})^-\mathbf{C})$$

and the given projections are projections on the subspaces $\mathcal{C}(\mathbf{A})$, $\mathcal{C}(\mathbf{A}\mathbf{B}^o)$ and $\mathcal{C}(\mathbf{A}(\mathbf{A}'\mathbf{A})^-\mathbf{C})$. ∎

Corollary 1.2.25.1. *For* $\mathbf{S} > 0$ *and an arbitrary matrix* $\mathbf{B}$ *of proper size*

$$\mathbf{S}^{-1} - \mathbf{B}^o(\mathbf{B}^{o'}\mathbf{S}\mathbf{B}^o)^-\mathbf{B}^{o'} = \mathbf{S}^{-1}\mathbf{B}(\mathbf{B}'\mathbf{S}^{-1}\mathbf{B})^-\mathbf{B}'\mathbf{S}^{-1}.$$

∎

The matrix $\mathbf{A}\mathbf{A}'(\mathbf{A}\mathbf{A}' + \mathbf{B}\mathbf{B}')^-\mathbf{B}\mathbf{B}'$ is called *parallel sum* of $\mathbf{A}\mathbf{A}'$ and $\mathbf{B}\mathbf{B}'$. Two basic properties of the parallel sum are given in

Lemma 1.2.8. *Let all the matrix operations be well defined. Then*

(i) $\mathbf{A}\mathbf{A}'(\mathbf{A}\mathbf{A}' + \mathbf{B}\mathbf{B}')^-\mathbf{B}\mathbf{B}' = \mathbf{B}\mathbf{B}'(\mathbf{A}\mathbf{A}' + \mathbf{B}\mathbf{B}')^-\mathbf{A}\mathbf{A}'$;

(ii) $(\mathbf{A}\mathbf{A}'(\mathbf{A}\mathbf{A}' + \mathbf{B}\mathbf{B}')^-\mathbf{B}\mathbf{B}')^- = (\mathbf{A}\mathbf{A}')^- + (\mathbf{B}\mathbf{B}')^-$.

PROOF: Since $\mathcal{C}(\mathbf{B}) = \mathcal{C}(\mathbf{B}\mathbf{B}') \subseteq \mathcal{C}(\mathbf{A} : \mathbf{B}) = \mathcal{C}(\mathbf{A}\mathbf{A}' + \mathbf{B}\mathbf{B}')$ we have

$$\mathbf{B}\mathbf{B}' = (\mathbf{A}\mathbf{A}' + \mathbf{B}\mathbf{B}')(\mathbf{A}\mathbf{A}' + \mathbf{B}\mathbf{B}')^-\mathbf{B}\mathbf{B}' = \mathbf{B}\mathbf{B}'(\mathbf{A}\mathbf{A}' + \mathbf{B}\mathbf{B}')^-(\mathbf{A}\mathbf{A}' + \mathbf{B}\mathbf{B}'),$$

where in the second equality it has been used that $\mathcal{C}(\mathbf{B}) \subseteq \mathcal{C}(\mathbf{AA}'+\mathbf{BB}')$. This implies (i).
For (ii) we utilize (i) and observe that

$$\begin{aligned}
&\mathbf{AA}'(\mathbf{AA}'+\mathbf{BB}')^-\mathbf{BB}'\{(\mathbf{AA}')^- + (\mathbf{BB}')^-\}\mathbf{AA}'(\mathbf{AA}'+\mathbf{BB}')^-\mathbf{BB}'\\
&\quad=\mathbf{BB}'(\mathbf{AA}'+\mathbf{BB}')^-\mathbf{AA}'(\mathbf{AA}'+\mathbf{BB}')^-\mathbf{BB}'\\
&\qquad+\mathbf{AA}'(\mathbf{AA}'+\mathbf{BB}')^-\mathbf{BB}'(\mathbf{AA}'+\mathbf{BB}')^-\mathbf{AA}'\\
&\quad=\mathbf{BB}'(\mathbf{AA}'+\mathbf{BB}')^-\mathbf{AA}'(\mathbf{AA}'+\mathbf{BB}')^-(\mathbf{AA}'+\mathbf{BB}')\\
&\quad=\mathbf{BB}'(\mathbf{AA}'+\mathbf{BB}')^-\mathbf{AA}'.
\end{aligned}$$

∎

In the next theorem some useful results for the intersection of column subspaces are collected.

Theorem 1.2.26. *Let* $\mathbf{A}$ *and* $\mathbf{B}$ *be matrices of proper sizes. Then*

(i) $\mathcal{C}(\mathbf{A}) \cap \mathcal{C}(\mathbf{B}) = \mathcal{C}(\mathbf{A}^o : \mathbf{B}^o)^\perp$;

(ii) $\mathcal{C}(\mathbf{A}) \cap \mathcal{C}(\mathbf{B}) = \mathcal{C}(\mathbf{A}(\mathbf{A}'\mathbf{B}^o)^o)$;

(iii) $\mathcal{C}(\mathbf{A}) \cap \mathcal{C}(\mathbf{B}) = \mathcal{C}(\mathbf{AA}'(\mathbf{AA}'+\mathbf{BB}')^-\mathbf{BB}') = \mathcal{C}(\mathbf{BB}'(\mathbf{AA}'+\mathbf{BB}')^-\mathbf{AA}')$.

PROOF: The results (i) and (ii) have already been given by Theorem 1.2.3 (ii) and Theorem 1.2.15. For proving (iii) we use Lemma 1.2.8. Clearly, by (i) of the lemma

$$\mathcal{C}(\mathbf{BB}'(\mathbf{AA}'+\mathbf{BB}')^-\mathbf{AA}') \subseteq \mathcal{C}(\mathbf{A}) \cap \mathcal{C}(\mathbf{B}).$$

Since

$$\begin{aligned}
&\mathbf{AA}'(\mathbf{AA}'+\mathbf{BB}')^-\mathbf{BB}'\{(\mathbf{AA}')^- + (\mathbf{BB}')^-\}\mathbf{A}(\mathbf{A}'\mathbf{B}^o)^o\\
&= \mathbf{AA}'(\mathbf{AA}'+\mathbf{BB}')^-\mathbf{B}(\mathbf{B}'\mathbf{A}^o)^o + \mathbf{BB}'(\mathbf{AA}'+\mathbf{BB}')^-\mathbf{A}(\mathbf{A}'\mathbf{B}^o)^o\\
&= (\mathbf{AA}'+\mathbf{BB}')(\mathbf{AA}'+\mathbf{BB}')^-\mathbf{A}(\mathbf{A}'\mathbf{B}^o)^o = \mathbf{A}(\mathbf{A}'\mathbf{B}^o)^o,
\end{aligned}$$

it follows that

$$\mathcal{C}(\mathbf{A}) \cap \mathcal{C}(\mathbf{B}) \subseteq \mathcal{C}(\mathbf{AA}'(\mathbf{AA}'+\mathbf{BB}')^-\mathbf{BB}'),$$

and (iii) is established. ∎

For further results on parallel sums of matrices as well as on *parallel differences*, we refer to Rao & Mitra (1971).

1.2.8 *Eigenvalues and eigenvectors*

Let $\mathbf{A}$ be an $m \times m-$matrix. We are interested in vectors $\mathbf{x} \neq \mathbf{0} \in \mathbb{R}^m$, whose direction does not change when multiplied with $\mathbf{A}$, i.e.

$$\mathbf{Ax} = \lambda\mathbf{x}. \tag{1.2.13}$$

This means that the matrix $(\mathbf{A} - \lambda\mathbf{I})$ must be singular and the equation in $\mathbf{x}$ given in (1.2.13) has a nontrivial solution, if

$$|\mathbf{A} - \lambda\mathbf{I}| = 0. \tag{1.2.14}$$

The equation in (1.2.14) is called *characteristic equation* of the matrix $\mathbf{A}$. The left hand side of (1.2.14) is a polynomial in λ of $m-$th degree (see also Problem 7 in §1.1.7) with m roots, of which some are possibly complex and some may be identical to each other.

Definition 1.2.13. *The values λ_i, which satisfy (1.2.14), are called eigenvalues or latent roots of the matrix* $\mathbf{A}$. *The vector* $\mathbf{x}_i$, *which corresponds to the eigenvalue* λ_i *in (1.2.13), is called eigenvector or latent vector of* $\mathbf{A}$ *which corresponds to* λ_i.

Eigenvectors are not uniquely defined. If $\mathbf{x}_i$ is an eigenvector, then $c\mathbf{x}_i$, $c \in \mathbb{R}$, is also an eigenvector. We call an eigenvector *standardized* if it is of unit-length. Eigenvalues and eigenvectors have many useful properties. In the following we list those which are the most important and which we are going to use later. Observe that many of the properties below follow immediately from elementary properties of the determinant. Some of the statements will be proven later.

Proposition 1.2.3.

(i) *If* $\mathbf{B} = \mathbf{C}\mathbf{A}\mathbf{C}^{-1}$, *where* $\mathbf{A}, \mathbf{B}, \mathbf{C}$ *are* $m \times m-$*matrices, then* $\mathbf{A}$ *and* $\mathbf{B}$ *have the same eigenvalues.*

(ii) *If* $\mathbf{A}$ *is a real symmetric matrix, then all its eigenvalues are real.*

(iii) *The matrices* $\mathbf{A}$ *and* $\mathbf{A}'$ *have the same eigenvalues.*

(iv) *The eigenvectors of* $\mathbf{A}$ *and* $\mathbf{A} + c\mathbf{I}$ *are the same for all constants* c.

(v) *Let* $\mathbf{A}$ *and* $\mathbf{B}$ *be* $m \times m-$*matrices with* $\mathbf{A}$ *being non-singular. Then the matrices* $\mathbf{AB}$ *and* $\mathbf{BA}$ *have the same eigenvalues.*

(vi) *If* $\lambda_1, \ldots, \lambda_m$ *are the eigenvalues of a non-singular matrix* $\mathbf{A}$, *then* $\lambda_1^{-1}, \ldots, \lambda_m^{-1}$ *are the eigenvalues of* $\mathbf{A}^{-1}$.

(vii) *If* $\mathbf{A}$ *is an orthogonal matrix, then the modulus of each eigenvalue of* $\mathbf{A}$ *equals one.*

(viii) *All eigenvalues of a symmetric idempotent matrix* $\mathbf{A}$ *equal one or zero.*

(ix) *If* $\mathbf{A}$ *is a triangular matrix (upper or lower), then its eigenvalues are identical to the diagonal elements.*

(x) The trace of $\mathbf{A} : m \times m$ equals the sum of the eigenvalues λ_i of $\mathbf{A}$, i.e.

$$\mathrm{tr}\mathbf{A} = \sum_{i=1}^{m} \lambda_i. \tag{1.2.15}$$

(xi) The determinant of $\mathbf{A} : m \times m$ equals the product of the eigenvalues λ_i of $\mathbf{A}$, i.e.

$$|\mathbf{A}| = \prod_{i=1}^{m} \lambda_i.$$

(xii) If $\mathbf{A}$ is an $m \times m-$matrix, then

$$\mathrm{tr}(\mathbf{A}^k) = \sum_{i=1}^{m} \lambda_i^k,$$

where $\lambda_1, \ldots, \lambda_m$ are the eigenvalues of $\mathbf{A}$.

(xiii) If $\mathbf{A}$ is a symmetric $m \times m-$matrix with the eigenvalues λ_i, then

$$\sum_{i=1}^{m} \lambda_i^2 = \sum_{i=1}^{m}\sum_{j=1}^{m} a_{ij}^2.$$

(xiv) If all eigenvalues of $m \times m-$matrix $\mathbf{A}$ are real and k of them are non-zero, then

$$(\mathrm{tr}\mathbf{A})^2 \leq k\mathrm{tr}(\mathbf{A}^2).$$

(xv) Let $\mathbf{A} : m \times m$ have rank r and let the number of non-zero eigenvalues of $\mathbf{A}$ be k. Then $r \geq k$.

(xvi) Let $\lambda_1 \geq \lambda_2 \geq \cdots \geq \lambda_m$ be the eigenvalues of $\mathbf{A} > 0$, and let $\mathbf{B}$: $m \times k$ be of rank k such that $\mathrm{diag}(\mathbf{B}'\mathbf{B}) = (g_1, \ldots, g_k)'$. Then

$$\max_{\mathbf{B}} |\mathbf{B}'\mathbf{A}\mathbf{B}| = \prod_{i=1}^{k} \lambda_i g_i.$$

(xvii) Let $\lambda_1 \geq \lambda_2 \geq \cdots \geq \lambda_m$ be the eigenvalues of $\mathbf{A} > 0$, and let $\mathbf{B}$: $m \times k$ be of rank k such that $\mathbf{B}'\mathbf{B}$ is diagonal with diagonal elements $g_1, \ldots, g_k$. Then

$$\min_{\mathbf{B}} |\mathbf{B}'\mathbf{A}\mathbf{B}| = \prod_{i=1}^{k} \lambda_{m+1-i} g_i.$$

(xviii) Let $\lambda_1 \geq \lambda_2 \geq \cdots \geq \lambda_m$ be the eigenvalues of a symmetric matrix $\mathbf{A}$: $m \times m$, and let $\mu_1 \geq \mu_2 \geq \cdots \geq \mu_m$ be the eigenvalues of $\mathbf{B} > 0 : m \times m$. Then

$$\sum_{i=1}^{m} \frac{\lambda_i}{\mu_i} \leq \mathrm{tr}(\mathbf{A}\mathbf{B}^{-1}) \leq \sum_{i=1}^{m} \frac{\lambda_i}{\mu_{m-i+1}}.$$

(xix) Let $\lambda_1 \geq \lambda_2 \geq \cdots \geq \lambda_m$ be the eigenvalues of a symmetric matrix $\mathbf{A}$: $m \times m$. Then, for any symmetric idempotent matrix $\mathbf{B} : m \times m$ of rank r,

$$\sum_{i=1}^{r} \lambda_{m-i+1} \leq \mathrm{tr}(\mathbf{A}\mathbf{B}) \leq \sum_{i=1}^{r} \lambda_i.$$

PROOF: The properties (xvi) and (xvii) are based on the Poincare separation theorem (see Rao, 1973a, pp. 65–66). The proof of (xviii) and (xix) are given in Srivastava & Khatri (1979, Theorem 1.10.2). ∎

In the next theorem we will show that eigenvectors corresponding to different eigenvalues are linearly independent. Hence eigenvectors can be used as a basis.

Theorem 1.2.27. *Let $\lambda_1, \lambda_2, \dots, \lambda_m$, $\lambda_i \neq \lambda_j$, $i \neq j$, be eigenvalues of a matrix $\mathbf{A}$ and $\mathbf{x}_i$, $i = 1, 2, \dots, m$, be the corresponding eigenvectors. Then the vectors $\{\mathbf{x}_i\}$ are linearly independent.*

PROOF: Suppose that

$$\sum_{i=1}^{m} c_i \mathbf{x}_i = \mathbf{0}.$$

Multiplying this equality with the product

$$\prod_{\substack{k=1 \\ k \neq j}}^{m} (\mathbf{A} - \lambda_k \mathbf{I})$$

we get $c_j = 0$, since

$$c_j(\lambda_j - \lambda_m)(\lambda_j - \lambda_{m-1}) \times \cdots \times (\lambda_j - \lambda_{j+1})(\lambda_j - \lambda_{j-1}) \times \cdots \times (\lambda_j - \lambda_1)\mathbf{x}_j = \mathbf{0}.$$

Thus,

$$\sum_{\substack{i=1 \\ i \neq j}}^{m} c_i \mathbf{x}_i = \mathbf{0}.$$

By repeating the same procedure for $j = 1, \cdots, m$, we obtain that the coefficients satisfy $c_m = c_{m-1} = \dots = c_1 = 0$. ∎

If $\mathbf{A}$ is symmetric we can strengthen the result.

Theorem 1.2.28. *Let $\lambda_1, \lambda_2, \dots, \lambda_m$, $\lambda_i \neq \lambda_j$, $i \neq j$, be eigenvalues of a symmetric matrix $\mathbf{A}$ and $\mathbf{x}_i$, $i = 1, 2, \dots, m$, be the corresponding eigenvectors. Then, the vectors $\{\mathbf{x}_i\}$ are orthogonal.*

PROOF: Let λ and μ be two different eigenvalues and consider

$$\begin{aligned} \mathbf{Ax} &= \lambda \mathbf{x} \\ \mathbf{Ay} &= \mu \mathbf{y}. \end{aligned}$$

Then $\mathbf{y}'\mathbf{Ax} = \lambda \mathbf{y}'\mathbf{x}$ and $\mathbf{x}'\mathbf{Ay} = \mu \mathbf{x}'\mathbf{y}$ which imply

$$0 = \mathbf{y}'\mathbf{Ax} - \mathbf{x}'\mathbf{Ay} = (\lambda - \mu)\mathbf{y}'\mathbf{x}$$

and thus those eigenvectors which correspond to different eigenvalues are orthogonal. ∎

Next we are going to present a result which is connected to the well-known Cayley-Hamilton theorem. One elegant way of proving the theorem is given by Rao (1973a, p. 44). Here we will show an alternative way of proving a related result which gives information about eigenvalues and eigenvectors of square matrices in an elementary fashion.

Theorem 1.2.29. *Let the matrix* $\mathbf{A}$ *be of size* $m \times m$ *and* $r(\mathbf{A}) = r$. *Then*

$$\mathbf{A}^{r+1} = \sum_{i=1}^{r} c_{i-1}\mathbf{A}^i,$$

for some known constants c_i.

PROOF: Let $\mathbf{D} : r \times r$ be a diagonal matrix with diagonal elements d_i such that $d_i \neq d_j$ if $i \neq j$ and $d_i \neq 0$, $i, j = 1, \ldots, r$. If we can find a solution to

$$\begin{pmatrix} d_1^r \\ d_2^r \\ \vdots \\ d_r^r \end{pmatrix} = \begin{pmatrix} 1 & d_1 & d_1^2 & \ldots & d_1^{r-1} \\ 1 & d_2 & d_2^2 & \ldots & d_2^{r-1} \\ \vdots & \vdots & \vdots & \ddots & \vdots \\ 1 & d_r & d_r^2 & \ldots & d_r^{r-1} \end{pmatrix} \begin{pmatrix} c_0 \\ c_1 \\ \vdots \\ c_{r-1} \end{pmatrix} \tag{1.2.16}$$

we may write

$$\mathbf{D}^r = \sum_{i=0}^{r-1} c_i \mathbf{D}^i.$$

However, the Vandermonde determinant

$$\begin{vmatrix} 1 & d_1 & d_1^2 & \ldots & d_1^{r-1} \\ 1 & d_2 & d_2^2 & \ldots & d_2^{r-1} \\ \vdots & \vdots & \vdots & \ddots & \vdots \\ 1 & d_r & d_r^2 & \ldots & d_r^{r-1} \end{vmatrix} = \prod_{i<j}(d_j - d_i), \tag{1.2.17}$$

differs from 0, since by assumption $d_i \neq d_j$. Thus, (1.2.16) has a unique solution. By definition of eigenvalues and eigenvectors,

$$\mathbf{AZ} = \mathbf{Z\Lambda}, \tag{1.2.18}$$

where $\mathbf{Z} : m \times r$ consists of r linearly independent eigenvectors and the diagonal matrix $\mathbf{\Lambda}$: $r \times r$ of different non-zero eigenvalues. From (1.2.18) and Definition 1.1.3 it follows that

$$\mathbf{A} = \mathbf{Z\Lambda Z}^-$$

and since $\mathbf{\Lambda Z}^-\mathbf{Z} = \mathbf{\Lambda}$, which holds because $\mathcal{C}(\mathbf{\Lambda}) = \mathcal{C}(\mathbf{Z}')$,

$$\mathbf{A}^r = \mathbf{Z\Lambda}^r\mathbf{Z}^-.$$

Thus, since $\mathbf{\Lambda}$ is a diagonal matrix, $\mathbf{\Lambda}^r = \sum_{i=0}^{r-1} c_i \mathbf{\Lambda}^i$. Then,

$$\mathbf{A}^r = \mathbf{Z\Lambda}^r\mathbf{Z}^- = c_0\mathbf{ZZ}^- + \sum_{i=1}^{r-1} c_i \mathbf{Z\Lambda}^i\mathbf{Z}^- = c_0\mathbf{ZZ}^- + \sum_{i=1}^{r-1} c_i\mathbf{A}^i.$$

Postmultiplying the obtained equality by $\mathbf{A} = \mathbf{Z\Lambda Z}^-$ yields

$$\mathbf{A}^{r+1} = c_0\mathbf{A} + \sum_{i=1}^{r-1} c_i\mathbf{A}^{i+1},$$

which establishes the theorem. ∎

REMARK: In the above proof we have supposed that all eigenvalues are distinct. However, since we can always use $\mathbf{\Lambda}^r = \sum_{i=0}^{r-1} c_i\mathbf{\Lambda}^i$, this is not a significant restriction.

Corollary 1.2.29.1. *If* $\mathbf{A} : m \times m$ *is non-singular, then*

$$\mathbf{A}^m = \sum_{i=0}^{m-1} c_i \mathbf{A}^i$$

and

$$c_0 \mathbf{A}^{-1} = \mathbf{A}^{m-1} + \sum_{i=1}^{m-1} c_i \mathbf{A}^{i-1}.$$

∎

Let λ be an eigenvalue, of multiplicity r, of a symmetric matrix $\mathbf{A}$. Then there exist r orthogonal eigenvectors corresponding to λ. This fact will be explained in more detail in §1.2.9 and §1.2.10. The linear space of all linear combinations of the eigenvectors is called the *eigenspace*, which corresponds to λ, and is denoted by $\mathbb{V}(\lambda)$. The dimension of $\mathbb{V}(\lambda)$ equals the multiplicity of λ. If λ and μ are two different eigenvalues of a symmetric $\mathbf{A}$, then $\mathbb{V}(\lambda)$ and $\mathbb{V}(\mu)$ are orthogonal subspaces in $\mathbb{R}^m$ (see Theorem 1.2.28). If $\lambda_1, \ldots, \lambda_k$ are all different eigenvalues of $\mathbf{A}$ with multiplicities $m_1, \ldots, m_k$, then the space $\mathbb{R}^m$ can be presented as an orthogonal sum of subspaces $\mathbb{V}(\lambda_i)$:

$$\mathbb{R}^m = \boxplus_{i=1}^{k} \mathbb{V}(\lambda_i)$$

and therefore any vector $\mathbf{x} \in \mathbb{R}^m$ can be presented, in a unique way, as a sum

$$\mathbf{x} = \sum_{i=1}^{k} \mathbf{x}_i, \qquad (1.2.19)$$

where $\mathbf{x}_i \in \mathbb{V}(\lambda_i)$.

The *eigenprojector* $\mathbf{P}_{\lambda_i}$ of the matrix $\mathbf{A}$ which corresponds to λ_i is an $m \times m-$matrix, which transforms the space $\mathbb{R}^m$ onto the space $\mathbb{V}(\lambda_i)$. An arbitrary vector $\mathbf{x} \in \mathbb{R}^m$ may be transformed by the projector $\mathbf{P}_{\lambda_i}$ to the vector $\mathbf{x}_i$ via

$$\mathbf{P}_{\lambda_i}\mathbf{x} = \mathbf{x}_i,$$

where $\mathbf{x}_i$ is the $i-$th term in the sum (1.2.19). If V is a subset of the set of all different eigenvalues of $\mathbf{A}$, e.g.

$$V = \{\lambda_1, \ldots, \lambda_n\},$$

then the eigenprojector $\mathbf{P}_V$, which corresponds to the eigenvalues $\lambda_i \in V$, is of the form

$$\mathbf{P}_V = \sum_{\lambda_i \in V} \mathbf{P}_{\lambda_i}.$$

The eigenprojector $\mathbf{P}_{\lambda_i}$ of a symmetric matrix $\mathbf{A}$ can be presented through the standardized eigenvectors $\mathbf{y}_i$, i.e.

$$\mathbf{P}_{\lambda_i} = \sum_{j=1}^{m_i} \mathbf{y}_j \mathbf{y}_j'.$$

Basic properties of eigenprojectors of symmetric matrices are given by the following equalities:

$$\begin{aligned}\mathbf{P}_{\lambda_i}\mathbf{P}_{\lambda_i} &= \mathbf{P}_{\lambda_i};\\ \mathbf{P}_{\lambda_i}\mathbf{P}_{\lambda_j} &= \mathbf{0}, \quad i \neq j;\\ \sum_{i=1}^{k}\mathbf{P}_{\lambda_i} &= \mathbf{I}.\end{aligned}$$

These relations rely on the fact that $\mathbf{y}_j'\mathbf{y}_j = 1$ and $\mathbf{y}_i'\mathbf{y}_j = 0$, $i \neq j$. The eigenprojectors $\mathbf{P}_{\lambda_i}$ enable us to present a symmetric matrix $\mathbf{A}$ through its *spectral decomposition*:

$$\mathbf{A} = \sum_{i=1}^{k}\lambda_i\mathbf{P}_{\lambda_i}.$$

In the next theorem we shall construct the Moore-Penrose inverse for a symmetric matrix with the help of eigenprojectors.

Theorem 1.2.30. *Let* $\mathbf{A}$ *be a symmetric* $n \times n-$*matrix. Then*

$$\begin{aligned}\mathbf{A} &= \mathbf{P}\boldsymbol{\Lambda}\mathbf{P}';\\ \mathbf{A}^+ &= \mathbf{P}\boldsymbol{\Lambda}^+\mathbf{P}',\end{aligned}$$

where $\boldsymbol{\Lambda}^+$ *is the diagonal matrix with elements*

$$(\boldsymbol{\Lambda}^+)_{ii} = \begin{cases}\lambda_i^{-1}, & \text{if} \quad \lambda_i \neq 0;\\ 0, & \text{if} \quad \lambda_i = 0,\end{cases}$$

$\mathbf{P}$ *is the orthogonal matrix which consists of standardized orthogonal eigenvectors of* $\mathbf{A}$*, and* λ_i *is an eigenvalue of* $\mathbf{A}$ *with the corresponding eigenvectors* $\mathbf{P}_{\lambda_i}$*.*

PROOF: The first statement is a reformulation of the spectral decomposition. To prove the theorem we have to show that the relations (1.1.16) - (1.1.19) are fulfilled. For (1.1.16) we get

$$\mathbf{A}\mathbf{A}^+\mathbf{A} = \mathbf{P}\boldsymbol{\Lambda}\mathbf{P}'\mathbf{P}\boldsymbol{\Lambda}^+\mathbf{P}'\mathbf{P}\boldsymbol{\Lambda}\mathbf{P}' = \mathbf{P}\boldsymbol{\Lambda}\mathbf{P}' = \mathbf{A}.$$

The remaining three equalities follow in a similar way:

$$\begin{aligned}\mathbf{A}^+\mathbf{A}\mathbf{A}^+ =&\mathbf{P}\boldsymbol{\Lambda}^+\mathbf{P}'\mathbf{P}\boldsymbol{\Lambda}\mathbf{P}'\mathbf{P}\boldsymbol{\Lambda}^+\mathbf{P}' = \mathbf{A}^+;\\ (\mathbf{A}\mathbf{A}^+)' =&(\mathbf{P}\boldsymbol{\Lambda}\mathbf{P}'\mathbf{P}\boldsymbol{\Lambda}^+\mathbf{P}')' = \mathbf{P}\boldsymbol{\Lambda}^+\mathbf{P}'\mathbf{P}\boldsymbol{\Lambda}\mathbf{P}' = \mathbf{P}\boldsymbol{\Lambda}^+\boldsymbol{\Lambda}\mathbf{P}' = \mathbf{P}\boldsymbol{\Lambda}\boldsymbol{\Lambda}^+\mathbf{P}' = \mathbf{A}\mathbf{A}^+;\\ (\mathbf{A}^+\mathbf{A})' =&(\mathbf{P}\boldsymbol{\Lambda}^+\mathbf{P}'\mathbf{P}\boldsymbol{\Lambda}\mathbf{P}')' = \mathbf{P}\boldsymbol{\Lambda}^+\mathbf{P}'\mathbf{P}\boldsymbol{\Lambda}\mathbf{P}' = \mathbf{A}^+\mathbf{A}.\end{aligned}$$

∎

Therefore, for a symmetric matrix there is a simple way of constructing the Moore-Penrose inverse of $\mathbf{A}$, i.e. the solution is obtained after solving an eigenvalue problem.

1.2.9. *Eigenstructure of normal matrices*

There exist two closely connected notions: eigenvectors and invariant linear subspaces. We have chosen to present the results in this paragraph using matrices and column vector spaces but the results could also have been stated via linear transformations.

Definition 1.2.14. *A subspace* $\mathcal{C}(\mathbf{M})$ *is* **A**-*invariant, if*

$$\mathcal{C}(\mathbf{AM}) \subseteq \mathcal{C}(\mathbf{M}).$$

∎

It follows from Theorem 1.2.12 that equality in Definition 1.2.14 holds if and only if $\mathcal{C}(\mathbf{A}') \cap \mathcal{C}(\mathbf{M})^{\perp} = \{\mathbf{0}\}$. Moreover, note that Definition 1.2.14 implicitly implies that $\mathbf{A}$ is a square matrix. For results concerning invariant subspaces we refer to the book by Gohberg, Lancaster & Rodman (1986).

Theorem 1.2.31. *The space* $\mathcal{C}(\mathbf{M})$ *is* **A**-*invariant if and only if* $\mathcal{C}(\mathbf{M})^{\perp}$ *is* $\mathbf{A}'$-*invariant.*

PROOF: Suppose that $\mathcal{C}(\mathbf{AM}) \subseteq \mathcal{C}(\mathbf{M})$. Then, $\mathbf{M}^{o'}\mathbf{AM} = \mathbf{0} = \mathbf{M}'\mathbf{A}'\mathbf{M}^{o}$ which implies that $\mathcal{C}(\mathbf{A}'\mathbf{M}^{o}) \subseteq \mathcal{C}(\mathbf{M}^{o})$. The converse follows immediately. ∎

The next theorem will connect invariant subspaces and eigenvectors.

Theorem 1.2.32. *Let* $\mathcal{C}(\mathbf{A}') \cap \mathcal{C}(\mathbf{M})^{\perp} = \{\mathbf{0}\}$ *and* $\dim \mathcal{C}(\mathbf{M}) = s$. *Then* $\mathcal{C}(\mathbf{M})$ *is* **A**-*invariant if and only if there exist* s *linearly independent eigenvectors* $\mathbf{x}_1, \mathbf{x}_2, \ldots, \mathbf{x}_s$ *of* **A** *such that*

$$\mathcal{C}(\mathbf{M}) = \mathcal{C}(\mathbf{x}_1) \oplus \mathcal{C}(\mathbf{x}_2) \oplus \cdots \oplus \mathcal{C}(\mathbf{x}_s).$$

PROOF: If $\mathcal{C}(\mathbf{M})$ is generated by eigenvectors of $\mathbf{A}$ then it is clear that $\mathcal{C}(\mathbf{M})$ is $\mathbf{A}$-invariant. Suppose that $\mathcal{C}(\mathbf{M})$ is $\mathbf{A}$-invariant and let $\mathbf{P}$ be a projector on $\mathcal{C}(\mathbf{M})$, then $\mathcal{C}(\mathbf{P}) = \mathcal{C}(\mathbf{M})$. By assumptions and Theorem 1.2.12

$$\mathcal{C}(\mathbf{PAP}) = \mathcal{C}(\mathbf{PAM}) = \mathcal{C}(\mathbf{AM}) = \mathcal{C}(\mathbf{M}) \tag{1.2.20}$$

which implies that the eigenvectors of $\mathbf{PAP}$ span $\mathcal{C}(\mathbf{M})$, as well as that $\mathbf{PAP} = \mathbf{AP}$. However, for any eigenvector $\mathbf{x}$ of $\mathbf{PAP}$ with the corresponding eigenvalue λ,

$$(\mathbf{A} - \lambda\mathbf{I})\mathbf{x} = (\mathbf{A} - \lambda\mathbf{I})\mathbf{Px} = (\mathbf{AP} - \lambda\mathbf{I})\mathbf{Px} = (\mathbf{PAP} - \lambda\mathbf{I})\mathbf{x} = \mathbf{0}$$

and hence all eigenvectors of $\mathbf{PAP}$ are also eigenvectors of $\mathbf{A}$. This means that there exist s eigenvectors of $\mathbf{A}$ which generate $\mathcal{C}(\mathbf{M})$. ∎

REMARK: Observe that $\mathbf{x}$ may be complex and therefore we have to assume an underlying complex field, i.e. $\mathbf{M} = \sum_i c_i\mathbf{x}_i$, where c_i as well as $\mathbf{x}_i$ may be complex. Furthermore, it follows from the proof that if $\mathcal{C}(\mathbf{M})$ is $\mathbf{A}$-invariant, any eigenvector of $\mathbf{A}$ is included in $\mathcal{C}(\mathbf{M})$.

If $\mathcal{C}(\mathbf{A}') \cap \mathcal{C}(\mathbf{M})^{\perp} = \{\mathbf{0}\}$ does not hold, it follows from the modular equality in Corollary 1.2.1.1 (iii) that

$$\mathcal{C}(\mathbf{M}) = \mathcal{C}(\mathbf{AM}) \boxplus \mathcal{C}(\mathbf{M}) \cap \mathcal{C}(\mathbf{AM})^{\perp}. \tag{1.2.21}$$

Let $\mathbf{P}$ be a projector on $\mathcal{C}(\mathbf{M})$, as in the proof of Theorem 1.2.32. Since $\mathcal{C}(\mathbf{PAP}) = \mathcal{C}(\mathbf{AM})$, we note that

$$\mathcal{C}(\mathbf{AM}) = \mathcal{C}(\mathbf{x}_1) \oplus \mathcal{C}(\mathbf{x}_2) \oplus \cdots \oplus \mathcal{C}(\mathbf{x}_r),$$

where $\mathbf{x}_i$, $i = 1, 2, \ldots, r$, are eigenvectors of $\mathbf{A}$. Since in most statistical applications $\mathbf{A}$ will be symmetric, we are going to explore (1.2.21) under this assumption. It follows from Theorem 1.2.15 that

$$\mathcal{C}(\mathbf{M}) \cap \mathcal{C}(\mathbf{AM})^{\perp} = \mathcal{C}(\mathbf{M}(\mathbf{M}'\mathbf{AM})^{o}) \tag{1.2.22}$$

and then once again, applying Theorem 1.2.15,

$$\mathcal{C}(\mathbf{AM}(\mathbf{M}'\mathbf{AM})^{o}) = \mathcal{C}(\mathbf{AM}) \cap \mathcal{C}(\mathbf{M})^{\perp} \subseteq \mathcal{C}(\mathbf{M}) \cap \mathcal{C}(\mathbf{M})^{\perp} = \{\mathbf{0}\},$$

which implies that $\mathcal{C}(\mathbf{M}(\mathbf{M}'\mathbf{AM})^{o}) \subseteq \mathcal{C}(\mathbf{A})^{\perp}$. Choose $\mathbf{A}^{o}$ to be an orthogonal projector on $\mathcal{C}(\mathbf{A})^{\perp}$, i.e. $\mathbf{A}^{o} = \mathbf{I} - \mathbf{A}(\mathbf{AA})^{-}\mathbf{A}$. Then, by Theorem 1.2.16 and Theorem 1.2.1 (viii),

$$\begin{aligned} &\mathcal{C}(\mathbf{A}^{o}\mathbf{M}(\mathbf{M}'\mathbf{AM})^{o}) \\ &\quad = \mathcal{C}(\mathbf{A})^{\perp} \cap (\mathcal{C}(\mathbf{A}) + \mathcal{C}(\mathbf{M}(\mathbf{M}'\mathbf{AM})^{o})) = \mathcal{C}(\mathbf{A})^{\perp} \cap \mathcal{C}(\mathbf{M}(\mathbf{M}'\mathbf{AM})^{o}) \\ &\quad \underset{(1.2.22)}{=} \mathcal{C}(\mathbf{M}) \cap \mathcal{C}(\mathbf{AM})^{\perp} \cap \mathcal{C}(\mathbf{A})^{\perp} \subseteq \mathcal{C}(\mathbf{M}) \cap \mathcal{C}(\mathbf{AM})^{\perp}. \end{aligned}$$

Therefore $\mathcal{C}(\mathbf{M}) \cap \mathcal{C}(\mathbf{AM})^{\perp}$ is $\mathbf{A}^{o}$-invariant and is included in $\mathcal{C}(\mathbf{A}^{o})$. Hence we have established the following theorem.

Theorem 1.2.33. *Suppose that* $\mathbf{A}$ *is a symmetric matrix. Let* $\mathcal{C}(\mathbf{M})$ *be* $\mathbf{A}$*-invariant,* $\dim \mathcal{C}(\mathbf{M}) = s$, $\mathbf{x}_1, \ldots, \mathbf{x}_t$ *linearly independent eigenvectors of* $\mathbf{A}$, *and* $\mathbf{y}_1, \ldots, \mathbf{y}_u$ *eigenvectors of* $\mathbf{I} - \mathbf{A}(\mathbf{AA})^{-}\mathbf{A}$. *Then*

$$\mathcal{C}(\mathbf{M}) = \mathcal{C}(\mathbf{x}_1) \boxplus \mathcal{C}(\mathbf{x}_2) \boxplus \cdots \boxplus \mathcal{C}(\mathbf{x}_v) \boxplus \mathcal{C}(\mathbf{y}_1) \boxplus \mathcal{C}(\mathbf{y}_2) \boxplus \cdots \boxplus \mathcal{C}(\mathbf{y}_w),$$

where $v \leq t$, $w \leq u$ *and* $v + w = s$. ∎

Suppose that $\mathbf{A}$ is a square matrix. Next we will consider the smallest $\mathbf{A}$-invariant subspace with one generator which is sometimes called *Krylov subspace* or *cyclic invariant subspace.* The space is given by

$$\mathcal{C}(\mathbf{x}, \mathbf{Ax}, \mathbf{A}^2\mathbf{x}, \ldots, \mathbf{A}^{a-1}\mathbf{x}), \quad a \leq p, \ \mathbf{x} \in \mathbb{R}^p.$$

In particular, a proof is given that this space is $\mathbf{A}$-invariant. We are going to study this space by generating a specific basis. In statistics we are interested in the basis which is generated by a partial least squares algorithm (PLS). For details about the algorithm and PLS we refer to Helland (1988, 1990). Let

$$\mathbf{G}_a = (\mathbf{g}_1, \mathbf{g}_2, \ldots, \mathbf{g}_a),$$

where $\mathbf{g}_i$, $i = 1, 2, \ldots, p$, are recursively defined vectors:

$$\begin{aligned} \mathbf{g}_1 &= \mathbf{x}; \\ \mathbf{g}_{a+1} &= (\mathbf{I} - \mathbf{A}_{a-1}\mathbf{g}_a(\mathbf{g}_a'\mathbf{A}_{a-1}\mathbf{g}_a)^{-}\mathbf{g}_a')\mathbf{g}_a; && (1.2.23) \\ \mathbf{A}_a &= \mathbf{A}_{a-1} - \mathbf{A}_{a-1}\mathbf{g}_a(\mathbf{g}_a'\mathbf{A}_{a-1}\mathbf{g}_a)^{-}\mathbf{g}_a'\mathbf{A}_{a-1}, \quad \mathbf{A}_0 = \mathbf{A}. && (1.2.24) \end{aligned}$$

In (1.2.23) and (1.2.24) as well in the applications it will be supposed that $\mathbf{g}'_a\mathbf{A}_{a-1}\mathbf{g}_a \neq 0$ when $\mathbf{A}_{a-1}\mathbf{g}_a \neq \mathbf{0}$. In statistics $\mathbf{A}$ is usually positive definite and therefore this assumption holds. First of all it will be shown that the vectors $\mathbf{g}_i$ form a basis. Note that the matrix

$$\mathbf{A}_{a-1}\mathbf{g}_a(\mathbf{g}'_a\mathbf{A}_{a-1}\mathbf{g}_a)^{-}\mathbf{g}'_a$$

is idempotent, and by Theorem 1.2.16 it follows that

$$\begin{aligned} C(\mathbf{g}_{a+1}) &= C(\mathbf{g}_a)^{\perp} \cap \{C(\mathbf{A}_{a-1}\mathbf{g}_a) + C(\mathbf{g}_a)\} \\ &= C(\mathbf{g}_a)^{\perp} \cap \{C(\mathbf{g}_{a-1})^{\perp} \cap (C(\mathbf{A}_{a-2}\mathbf{g}_{a-1}) + C(\mathbf{A}_{a-2}\mathbf{g}_a)) \\ &\qquad + C(\mathbf{g}_{a-1})^{\perp} \cap (C(\mathbf{A}_{a-2}\mathbf{g}_{a-1}) + C(\mathbf{g}_{a-1}))\}. \end{aligned}$$

Moreover, using the distributive inequalities in Theorem 1.2.1 (vii) we obtain

$$\begin{aligned} C(\mathbf{g}_{a+1}) &\subseteq C(\mathbf{g}_a)^{\perp} \cap C(\mathbf{g}_{a-1})^{\perp} \cap \{C(\mathbf{A}_{a-2}(\mathbf{g}_a : \mathbf{g}_{a-1})) + C(\mathbf{g}_{a-1})\} \\ &\subseteq \ldots \subseteq C(\mathbf{G}_a)^{\perp} \cap \{C(\mathbf{A}\mathbf{G}_a) + C(\mathbf{g}_1)\}. \end{aligned}$$

Thus, $C(\mathbf{g}_{a+1}) \subseteq C(\mathbf{G}_a)^{\perp}$, which means that $\mathbf{G}_a$ defines an orthogonal basis and $\dim\mathbf{G}_a = a$. Now we are going to show that

$$C(\mathbf{G}_a) = C(\mathbf{g}_1, \mathbf{A}\mathbf{g}_1, \ldots, \mathbf{A}^{a-1}\mathbf{g}_1). \tag{1.2.25}$$

For $a = 1$ equality (1.2.25) is obviously true as well as for $a = 2$, since $C(\mathbf{g}_1) \subseteq C(\mathbf{A}\mathbf{g}_1) + C(\mathbf{g}_1)$ implies (modular laws) that

$$C(\mathbf{g}_2, \mathbf{g}_1) = C(\mathbf{g}_1)^{\perp} \cap \{C(\mathbf{A}\mathbf{g}_1) + C(\mathbf{g}_1)\} + C(\mathbf{g}_1) = C(\mathbf{A}\mathbf{g}_1, \mathbf{g}_1).$$

In order to verify (1.2.25) we will use induction. Suppose that

$$C(\mathbf{G}_{a-1}) = C(\mathbf{g}_1, \mathbf{A}\mathbf{g}_1, \ldots, \mathbf{A}^{a-2}\mathbf{g}_1)$$

holds, and then

$$\begin{aligned} C(\mathbf{G}_a) &= C(\mathbf{g}_a) + C(\mathbf{G}_{a-1}) \\ &= C(\mathbf{g}_{a-1})^{\perp} \cap \{C(\mathbf{A}_{a-2}\mathbf{g}_{a-1}) + C(\mathbf{g}_{a-1})\} + C(\mathbf{G}_{a-1})\} \\ &\subseteq \ldots \subseteq C(\mathbf{G}_{a-1})^{\perp} \cap \{C(\mathbf{A}\mathbf{G}_{a-1}) + C(\mathbf{g}_1)\} + C(\mathbf{G}_{a-1}) \\ &= C(\mathbf{G}_{a-1})^{\perp} \cap \{C(\mathbf{G}_{a-1}) + C(\mathbf{A}^{a-1}\mathbf{g}_1)\} + C(\mathbf{G}_{a-1}) \\ &= C(\mathbf{g}_1, \mathbf{A}\mathbf{g}_1, \ldots, \mathbf{A}^{a-1}\mathbf{g}_1). \end{aligned}$$

However, since $\dim C(\mathbf{G}_a) = a$ and $\dim C(\mathbf{g}_1, \mathbf{A}\mathbf{g}_1, \ldots, \mathbf{A}^{a-1}\mathbf{g}_1) \leq a$, we may conclude that

$$C(\mathbf{G}_a) = C(\mathbf{g}_1, \mathbf{A}\mathbf{g}_1, \ldots, \mathbf{A}^{a-1}\mathbf{g}_1)$$

and that the vectors $\mathbf{g}_1, \mathbf{A}\mathbf{g}_1, \ldots, \mathbf{A}^{a-1}\mathbf{g}_1$ are linearly independent.

Let us consider the case when $C(\mathbf{g}_1, \mathbf{A}\mathbf{g}_1, \ldots, \mathbf{A}^{a-1}\mathbf{g}_1)$ is $\mathbf{A}$-invariant. Suppose that $\mathbf{A}_{a-1}\mathbf{g}_a = \mathbf{0}$, which, by (1.2.23), implies $\mathbf{g}_{a+1} = \mathbf{g}_a$ and $\mathbf{A}_a = \mathbf{A}_{a-1}$. Observe that

$$\begin{aligned} C(\mathbf{A}'_{a-1}) &= C(\mathbf{g}_{a-1})^{\perp} \cap \{C(\mathbf{A}'_{a-2}\mathbf{g}_a) + C(\mathbf{A}'_{a-2})\} \\ &= C(\mathbf{g}_a)^{\perp} \cap C(\mathbf{A}'_{a-2}) = \cdots = C(\mathbf{G}_{a-1})^{\perp} \cap C(\mathbf{A}'). \end{aligned}$$

Since $\mathbf{A}_{a-1}\mathbf{g}_a = \mathbf{0}$ yields

$$C(\mathbf{g}_a) \subseteq C(\mathbf{G}_{a-1}) + C(\mathbf{A}')^{\perp}$$

it follows that

$$C(\mathbf{A}\mathbf{g}_a) \subseteq C(\mathbf{A}\mathbf{G}_{a-1}) = (\mathbf{A}\mathbf{g}_1, \ldots, \mathbf{A}^{a-1}\mathbf{g}_1).$$

Thus, $C(\mathbf{g}_1, \mathbf{A}\mathbf{g}_1, \ldots, \mathbf{A}^{a-1}\mathbf{g}_1)$ is $\mathbf{A}$-invariant, i.e. $\mathbf{G}_a$ is $\mathbf{A}$-invariant. Furthermore, if $\mathbf{g}_{a+1} = \mathbf{0}$, then

$$C(\mathbf{A}\mathbf{G}_a) \subseteq C(\mathbf{G}_{a+1}) = C(\mathbf{G}_a),$$

since $C(\mathbf{G}_a) = C(\mathbf{g}_1, \mathbf{A}\mathbf{g}_1, \ldots, \mathbf{A}^{a-1}\mathbf{g}_1)$, which means that $\mathbf{G}_a$ is $\mathbf{A}$-invariant. On the other hand, if $C(\mathbf{A}\mathbf{G}_a) \subseteq C(\mathbf{G}_a)$ and $\mathbf{A}_{a-1}\mathbf{g}_a \neq \mathbf{0}$, then $C(\mathbf{g}_{a+1}) \subseteq C(\mathbf{G}_a)^{\perp} \cap C(\mathbf{G}_a) = \{\mathbf{0}\}$. Hence we have proved the following theorem.

Theorem 1.2.34. *Let $\mathbf{g}_i$ be a $p-$vector and $\mathbf{A}_{i-1} : p \times p$, $i = 1, 2, \ldots, a+1$, $a \leq p$, be given by (1.2.23) and (1.2.24), respectively. Suppose that $\mathbf{g}'_a\mathbf{A}_{a-1}\mathbf{g}_a \neq 0$ when $\mathbf{A}_{a-1}\mathbf{g}_a \neq \mathbf{0}$.*

(i) *If $\mathbf{A}_{a-1}\mathbf{g}_a = \mathbf{0}$ for some $a \leq p$, then $C(\mathbf{g}_1, \mathbf{A}\mathbf{g}_1, \ldots, \mathbf{A}^{a-1}\mathbf{g}_1)$ is $\mathbf{A}$-invariant.*

(ii) *If $\mathbf{A}_{a-1}\mathbf{g}_a \neq \mathbf{0}$, the vectors $\mathbf{g}_j = \mathbf{0}$, for $j = a+1, a+2, \ldots$, if and only if $C(\mathbf{g}_1, \mathbf{A}\mathbf{g}_1, \ldots, \mathbf{A}^{a-1}\mathbf{g}_1)$ is $\mathbf{A}$-invariant.* ∎

Corollary 1.2.34.1. *If $\mathbf{g}_1 \in \mathbb{R}^p$, then $\mathbf{g}_{p+1} = \mathbf{0}$ and $C(\mathbf{g}_1, \mathbf{A}\mathbf{g}_1, \ldots, \mathbf{A}^{p-1}\mathbf{g}_1)$ is $\mathbf{A}$-invariant.* ∎

To prove Corollary 1.2.34.1 we may alternatively use Theorem 1.2.29 where it was shown that there exist constants c_i, $i = 1, 2, \ldots n \leq p$, such that

$$\sum_{i=1}^{n} c_i \mathbf{A}^i = \mathbf{0}.$$

Thus, $\mathbf{A}^n\mathbf{g}_1$ is a linear function of $\mathbf{g}_1, \mathbf{A}\mathbf{g}_1, \ldots, \mathbf{A}^{n-1}\mathbf{g}_1$.
An important special case when considering eigenspaces is given in the next statement.

Theorem 1.2.35. *If* $\mathbf{g}_1 \in \mathbb{R}^p$, $\mathcal{C}(\mathbf{g}_1) \subseteq \mathcal{C}(\mathbf{A})$, $\mathbf{A} : p \times p$ *and* $\mathcal{C}(\mathbf{A})$ *is generated by* $a \leq p$ *eigenvectors of* $\mathbf{A}$, *then*

$$\mathcal{C}(\mathbf{g}_1, \mathbf{A}\mathbf{g}_1, \ldots, \mathbf{A}^{a-1}\mathbf{g}_1) = \mathcal{C}(\mathbf{A}).$$

PROOF: Obviously $\mathcal{C}(\mathbf{g}_1, \mathbf{A}\mathbf{g}_1, \ldots, \mathbf{A}^{a-1}\mathbf{g}_1) \subseteq \mathcal{C}(\mathbf{A})$. According to the Remark after Theorem 1.2.32 and Corollary 1.2.34.1 any eigenvector of $\mathbf{A}$ belongs to $\mathcal{C}(\mathbf{g}_1, \mathbf{A}\mathbf{g}_1, \ldots, \mathbf{A}^{a-1}\mathbf{g}_1)$. Thus, $\mathcal{C}(\mathbf{A}) \subseteq \mathcal{C}(\mathbf{g}_1, \mathbf{A}\mathbf{g}_1, \ldots, \mathbf{A}^{a-1}\mathbf{g}_1)$. ∎

Up to now we have not examined complex matrices. In Theorem 1.2.32 – Theorem 1.2.35 we have presented the results for real matrices. However, all the results are valid for complex matrices if we interpret transposed matrices as conjugate transposed matrices. In the following we shall consider complex eigenvalues and eigenvectors. Necessary notation was given at the end of §1.1.1. Our intention is to utilize the field of complex numbers at a minimum level. Therefore, several results given below can be extended but we leave this to an interested reader.

The class of normal matrices will be studied in some detail. Remember that according to the definition a normal matrix $\mathbf{A}$ satisfies the equality

$$\mathbf{A}\mathbf{A}' = \mathbf{A}'\mathbf{A}. \tag{1.2.26}$$

Let $\mathbf{x}$ be an eigenvector of $\mathbf{A} \in \mathbb{R}^{p\times p}$ and λ the corresponding eigenvalue. Then, by (1.2.13)

$$\mathbf{A}\mathbf{x} = \lambda\mathbf{x}$$

holds. Here $\mathbf{x}$ and λ may very well be complex. As noted before, the possible values of λ are determined via the characteristic equation (1.2.14). If (1.2.13) holds, then

$$\mathbf{A}\mathbf{A}'\mathbf{x} = \mathbf{A}'\mathbf{A}\mathbf{x} = \lambda\mathbf{A}'\mathbf{x}$$

and thus $\mathbf{A}'\mathbf{x}$ is also an eigenvector. Furthermore, in the same way we may show that $(\mathbf{A}')^2\mathbf{x}, (\mathbf{A}')^3\mathbf{x}$, etc. are eigenvectors, and finally we can get a finite sequence of eigenvectors,

$$\mathbf{x}, \mathbf{A}'\mathbf{x}, (\mathbf{A}')^2\mathbf{x}, \ldots, (\mathbf{A}')^{p-1}\mathbf{x}, \tag{1.2.27}$$

which corresponds to λ. From Corollary 1.2.34.1 it follows that the vectors in (1.2.27) span an $\mathbf{A}'$-invariant subspace. Let

$$\mathbf{z} \in \mathcal{C}(\mathbf{x}, \mathbf{A}'\mathbf{x}, (\mathbf{A}')^2\mathbf{x}, \ldots, (\mathbf{A}')^{p-1}\mathbf{x}),$$

where $\mathbf{x}$ is an eigenvector of $\mathbf{A}$. Then, $\mathbf{z} = \sum_{i=0}^{p-1} c_i(\mathbf{A}')^i\mathbf{x}$ for some c_i and

$$\mathbf{A}\mathbf{z} = \sum_{i=0}^{p-1} c_i\mathbf{A}(\mathbf{A}')^i\mathbf{x} = \sum_{i=0}^{p-1} c_i(\mathbf{A}')^i\mathbf{A}\mathbf{x} = \lambda\mathbf{z},$$

which means that any vector in the space $\mathcal{C}(\mathbf{x}, \mathbf{A}'\mathbf{x}, (\mathbf{A}')^2\mathbf{x}, \ldots, (\mathbf{A}')^{p-1}\mathbf{x})$ is an eigenvector of $\mathbf{A}$. In particular, since $\mathcal{C}(\mathbf{x}, \mathbf{A}'\mathbf{x}, (\mathbf{A}')^2\mathbf{x}, \ldots, (\mathbf{A}')^{p-1}\mathbf{x})$ is $\mathbf{A}'$-invariant, there must, according to Theorem 1.2.32, be at least one eigenvector of $\mathbf{A}'$ which belongs to

$$\mathcal{C}(\mathbf{x}, \mathbf{A}'\mathbf{x}, (\mathbf{A}')^2\mathbf{x}, \ldots, (\mathbf{A}')^{p-1}\mathbf{x}),$$

which then also is an eigenvector of $\mathbf{A}$. Denote the joint eigenvector by $\mathbf{y}_1$ and study $\mathcal{C}(\mathbf{y}_1)^\perp$. By Theorem 1.2.31 it follows that $\mathcal{C}(\mathbf{y}_1)^\perp$ is $\mathbf{A}$-invariant as well as $\mathbf{A}'$-invariant. Thus we may choose an eigenvector of $\mathbf{A}$, say $\mathbf{x}_1$, which belongs to $\mathcal{C}(\mathbf{y}_1)^\perp$. Therefore, by repeating the above procedure, we can state that the space $\mathcal{C}(\mathbf{x}_1, \mathbf{A}'\mathbf{x}_1, (\mathbf{A}')^2\mathbf{x}_1, \ldots, (\mathbf{A}')^{p-1}\mathbf{x}_1)$ is $\mathbf{A}'$-invariant as well as belongs to $\mathcal{C}(\mathbf{y}_1)^\perp$, and we may find a common eigenvector, say $\mathbf{y}_2$, of $\mathbf{A}$ and $\mathbf{A}'$ which is orthogonal to the common eigenvector $\mathbf{y}_1$. In the next step we start with the space

$$\mathcal{C}(\mathbf{y}_1, \mathbf{y}_2)^\perp$$

which is $\mathbf{A}$-invariant as well as $\mathbf{A}'$-invariant, and by further proceeding in this way the next theorem can be established.

Theorem 1.2.36. *If $\mathbf{A}$ is normal, then there exists a set of orthogonal eigenvectors of $\mathbf{A}$ and $\mathbf{A}'$ which spans the whole space $\mathcal{C}(\mathbf{A}) = \mathcal{C}(\mathbf{A}')$.* ∎

Suppose that we have a system of orthogonal eigenvectors of a matrix $\mathbf{A} : p \times p$, where the eigenvectors $\mathbf{x}_i$ span the whole space and are collected into an eigenvector matrix $\mathbf{X} = (\mathbf{x}_1, \ldots, \mathbf{x}_p)$. Let the eigenvectors $\mathbf{x}_i$, corresponding to the eigenvalues λ_i, be ordered in such a way that

$$\mathbf{AX} = \mathbf{X\Lambda}, \quad \mathbf{X} = (\mathbf{X}_1 : \mathbf{X}_2), \quad \mathbf{X}_1^\star\mathbf{X}_1 = \mathbf{D}, \quad \mathbf{X}_1^\star\mathbf{X}_2 = \mathbf{0}, \quad \mathbf{\Lambda} = \begin{pmatrix} \mathbf{\Lambda}_1 & \mathbf{0} \\ \mathbf{0} & \mathbf{0} \end{pmatrix},$$

where the partition of $\mathbf{\Lambda}$ corresponds to the partition of $\mathbf{X}$, $\mathbf{X}$ is non-singular (in fact, unitary), $\mathbf{D}$ and $\mathbf{\Lambda}_1$ are both non-singular diagonal matrices with the difference that $\mathbf{D}$ is real whereas $\mathbf{\Lambda}_1$ may be complex. Remember that $\mathbf{A}^\star$ denotes the conjugate transpose. Since

$$\mathbf{X}^\star\mathbf{A}'\mathbf{X} = \mathbf{\Lambda}^\star\mathbf{X}^\star\mathbf{X} = \mathbf{\Lambda}_1^\star\mathbf{D} = \mathbf{D\Lambda}_1^\star = \mathbf{X}^\star\mathbf{X\Lambda}^\star$$

and because $\mathbf{X}$ is non-singular it follows that

$$\mathbf{A}'\mathbf{X} = \mathbf{X\Lambda}^\star.$$

Thus

$$\mathbf{AA}'\mathbf{X} = \mathbf{AX\Lambda}^\star = \mathbf{X\Lambda\Lambda}^\star$$

and

$$\mathbf{A}'\mathbf{AX} = \mathbf{A}'\mathbf{X\Lambda} = \mathbf{X\Lambda}^\star\mathbf{\Lambda} = \mathbf{X\Lambda\Lambda}^\star = \mathbf{AX\Lambda}^\star = \mathbf{AA}'\mathbf{X}$$

which implies $\mathbf{AA}' = \mathbf{A}'\mathbf{A}$ and the next theorem has been established.

Theorem 1.2.37. *If there exists a system of orthogonal eigenvectors of $\mathbf{A}$ which spans $\mathcal{C}(\mathbf{A})$, then $\mathbf{A}$ is normal.* ∎

An implication of Theorem 1.2.36 and Theorem 1.2.37 is that a matrix $\mathbf{A}$ is normal if and only if there exists a system of orthogonal eigenvectors of $\mathbf{A}$ which spans $\mathcal{C}(\mathbf{A})$. Furthermore, by using the ideas of the proof of Theorem 1.2.37 the next result follows.

Theorem 1.2.38. *If* $\mathbf{A}$ *and* $\mathbf{A}'$ *have a common eigenvector* $\mathbf{x}$ *and* λ *is the corresponding eigenvalue of* $\mathbf{A}$, *then* $\overline{\lambda}$ *is the eigenvalue of* $\mathbf{A}'$.

PROOF: By assumption

$$\mathbf{A}\mathbf{x} = \lambda\mathbf{x},$$
$$\mathbf{A}'\mathbf{x} = \mu\mathbf{x}$$

and then $\mathbf{x}^\star\mathbf{A}\mathbf{x} = \lambda\mathbf{x}^\star\mathbf{x}$ implies

$$\mu\mathbf{x}^\star\mathbf{x} = \mathbf{x}^\star\mathbf{A}'\mathbf{x} = (\mathbf{x}^\star\mathbf{A}\mathbf{x})^\star = \overline{\lambda}\mathbf{x}^\star\mathbf{x}.$$

∎

Corollary 1.2.38.1. *The eigenvalues and eigenvectors of a symmetric matrix are real.* ∎

Corollary 1.2.38.2. *If the eigenvalues of a normal matrix* $\mathbf{A}$ *are all real then* $\mathbf{A}$ *is symmetric.* ∎

1.2.10 *Eigenvalue-based factorizations*

Let $\mathbf{A}$ be a real $n \times n-$matrix where the non-zero eigenvalues are denoted λ_j and the corresponding unit length eigenvectors $\mathbf{x}_j,\ j = 1,\ldots,n_1$. Since $\mathbf{A}$ is real, it follows that $\mathbf{x}_j$ is real if and only if λ_j is real. Furthermore, it is possible to show that if $\mathbf{x}_j$ is a complex eigenvector, $\overline{\mathbf{x}}_j$ is also an eigenvector. Let $\mathbf{x}_{2j-1} = \mathbf{u}_j + i\mathbf{v}_j$, $\mathbf{x}_{2j} = \mathbf{u}_j - i\mathbf{v}_j,\ j = 1,2,\ldots,q$, be the complex eigenvectors corresponding to the complex eigenvalues $\lambda_{2j-1} = \mu_j + i\delta_j,\ \lambda_{2j} = \mu_j - i\delta_j,\ j = 1,2,\ldots,q$, and $\mathbf{x}_j,\ j = 2q+1,\ldots,n_1$, be real eigenvectors corresponding to the real non-zero eigenvalues λ_j. Moreover, let $\lambda_{n_1+1},\ldots,\lambda_n$ be all zero eigenvalues. Let

$$\mathbf{E} = \left\{\begin{pmatrix} \mu_1 & \delta_1 \\ -\delta_1 & \mu_1 \end{pmatrix}, \begin{pmatrix} \mu_2 & \delta_2 \\ -\delta_2 & \mu_2 \end{pmatrix}, \ldots, \begin{pmatrix} \mu_q & \delta_q \\ -\delta_q & \mu_q \end{pmatrix}, \lambda_{2q+1},\ldots,\lambda_{n_1}, \mathbf{0}_{n-n_1}\right\}_{[d]} \tag{1.2.28}$$

and

$$\mathbf{Q}_1 = (\mathbf{u}_1, \mathbf{v}_1, \mathbf{u}_2, \mathbf{v}_2, \ldots, \mathbf{x}_{2q+1}, \ldots, \mathbf{x}_{n_1}), \tag{1.2.29}$$

where it is supposed that the eigenvectors are of unit length. Next a theorem is proven which gives us an eigenvalue-based factorization of a normal matrix.

Theorem 1.2.39. *Any normal matrix* $\mathbf{A} : n \times n$ *can be presented as a product*

$$\mathbf{A} = \mathbf{Q}\mathbf{E}\mathbf{Q}',$$

where $\mathbf{E}$ *is given by (1.2.28) and* $\mathbf{Q}$ *is an orthogonal matrix.*

PROOF: Suppose that $\mathbf{A} : n \times n$ is normal and $r(\mathbf{A}) = n_1$. Then one can always find a system of eigenvectors $\mathbf{x}_i$ of $\mathbf{A}$ corresponding to the non-zero eigenvalues $\lambda_i,\ i = 1,\ldots,n_1$, in such a way that

$$\mathbf{A}\mathbf{X} = \mathbf{X}\mathbf{\Lambda}, \quad \mathbf{X} = (\mathbf{X}_1 : \mathbf{X}_2), \quad \mathbf{X}_1^\star\mathbf{X}_1 = \mathbf{I}_{n_1}, \quad \mathbf{X}_1^\star\mathbf{X}_2 = \mathbf{0},$$

where $\mathbf{\Lambda} = (\mathbf{\Lambda}_1, \mathbf{0}_{n-n_1})_{[d]}$, with $\mathbf{\Lambda}_1 = (\lambda_1, \ldots, \lambda_{n_1})_d$ and $\mathbf{X} = (\mathbf{x}_1, \ldots, \mathbf{x}_{n_1} : \mathbf{X}_2)$. Here $\mathbf{X}_1$ is of size $n \times n_1$, $\mathbf{X}_2$ is of size $n \times (n - n_1)$ and $\mathbf{X}$ is non-singular. Put $\mathbf{Q} = (\mathbf{Q}_1 : \mathbf{Q}_2)$, where $\mathbf{Q}_2'\mathbf{Q}_1 = \mathbf{0}$, $\mathbf{Q}_2'\mathbf{Q}_2 = \mathbf{I}$, $\mathbf{Q}_1$ is defined by (1.2.29), and $\mathbf{Q}$ is non-singular. Let the eigenvectors be standardized, i.e.

$$\mathbf{I} = \mathbf{X}_1^{\star}\mathbf{X}_1 = \mathbf{Q}_1'\mathbf{Q}_1.$$

Therefore, the equation $\mathbf{AX} = \mathbf{X\Lambda}$ is identical to

$$\mathbf{AQ} = \mathbf{QE},$$

which means that $\mathbf{A} = \mathbf{QEQ}^{-1}$. However, since $\mathbf{Q}$ is of full rank, $\mathbf{Q}' = \mathbf{Q}^{-1}$, and thus $\mathbf{QQ}' = \mathbf{Q}'\mathbf{Q} = \mathbf{I}$, i.e. $\mathbf{Q}$ is orthogonal. ∎

From the theorem two important corollaries follow, of which the first one already has been presented in Theorem 1.2.30.

Corollary 1.2.39.1. *Let $\mathbf{A} : n \times n$ be a real symmetric matrix. Then there exists an orthogonal matrix $\mathbf{Q}$ such that*

$$\mathbf{Q}'\mathbf{AQ} = \mathbf{\Lambda}$$

and

$$\mathbf{A} = \mathbf{Q\Lambda Q}',$$

where

$$\mathbf{\Lambda} = (\lambda_1, \ldots, \lambda_{n_1}, 0, \ldots, 0)_d.$$

PROOF: If $\mathbf{A}$ is symmetric, $\mathbf{E}$ in (1.2.28) must also be symmetric, and therefore all the eigenvalues λ_i have to be real. ∎

Corollary 1.2.39.2. *Let $\mathbf{A} : n \times n$ be a real skew-symmetric matrix. Then there exists an orthogonal matrix $\mathbf{Q}$ such that*

$$\mathbf{Q}'\mathbf{AQ} = \mathbf{\Lambda}$$

and

$$\mathbf{Q\Lambda Q}' = \mathbf{A},$$

where

$$\mathbf{\Lambda} = \left\{ \begin{pmatrix} 0 & \delta_1 \\ -\delta_1 & 0 \end{pmatrix}, \begin{pmatrix} 0 & \delta_2 \\ -\delta_2 & 0 \end{pmatrix}, \ldots, \begin{pmatrix} 0 & \delta_q \\ -\delta_q & 0 \end{pmatrix}, \mathbf{0}_{n-2q} \right\}_{[d]}.$$

PROOF: If $\mathbf{A}$ is skew-symmetric, $\mathbf{E}$ in (1.2.28) must also be skew-symmetric. Hence, for the complex eigenvalues $\mu_j = 0$, $j = 1, 2, \ldots, 2q$, and all real eigenvalues have to be equal to zero. ∎

Corollary 1.2.39.3. *If* $\mathbf{A}$ *is normal and* $\mathbf{x}$ *is an eigenvector of* $\mathbf{A}$, *then* $\mathbf{x}$ *is also an eigenvector of* $\mathbf{A}'$.

PROOF: From the theorem it follows that $\mathbf{A} = \mathbf{QEQ}'$. Then $\mathbf{Ax} = \lambda\mathbf{x}$ implies that

$$\mathbf{EQ}'\mathbf{x} = \lambda\mathbf{Q}'\mathbf{x}$$

and thus $\mathbf{Q}'\mathbf{x}$ is an eigenvector of $\mathbf{E}$. Furthermore, by the structure in $\mathbf{E}$ any eigenvector of $\mathbf{E}$ is also an eigenvector of $\mathbf{E}'$. Therefore, according to Theorem 1.2.36, $\mathbf{E}'\mathbf{Q}'\mathbf{x} = \overline{\lambda}\mathbf{Q}'\mathbf{x}$, which gives us

$$\mathbf{A}'\mathbf{x} = \mathbf{QE}'\mathbf{Q}'\mathbf{x} = \overline{\lambda}\mathbf{x}.$$

∎

In the following we will perform a brief study of normal matrices which commute, i.e. $\mathbf{AB} = \mathbf{BA}$.

Lemma 1.2.9. *If for normal matrices* $\mathbf{A}$ *and* $\mathbf{B}$ *the equality* $\mathbf{AB} = \mathbf{BA}$ *holds, then*

$$\mathbf{A}'\mathbf{B} = \mathbf{BA}'$$

and

$$\mathbf{AB}' = \mathbf{B}'\mathbf{A}.$$

PROOF: Let

$$\mathbf{V} = \mathbf{A}'\mathbf{B} - \mathbf{BA}'$$

and consider $\mathbf{VV}'$:

$$\mathbf{VV}' = \mathbf{A}'\mathbf{BB}'\mathbf{A} - \mathbf{BA}'\mathbf{B}'\mathbf{A} - \mathbf{A}'\mathbf{BAB}' + \mathbf{BA}'\mathbf{AB}'$$

which by assumption is identical to

$$\begin{aligned}\mathbf{VV}' &= \mathbf{A}'\mathbf{BB}'\mathbf{A} - \mathbf{BB}'\mathbf{A}'\mathbf{A} - \mathbf{A}'\mathbf{ABB}' + \mathbf{BA}'\mathbf{AB}' \\ &= \mathbf{A}'\mathbf{BB}'\mathbf{A} - \mathbf{BB}'\mathbf{AA}' - \mathbf{A}'\mathbf{AB}'\mathbf{B} + \mathbf{BA}'\mathbf{AB}'.\end{aligned}$$

Thus

$$\mathrm{tr}(\mathbf{VV}') = 0,$$

which by Proposition 1.1.4 (vi) implies that $\mathbf{V} = \mathbf{0}$. ∎

Lemma 1.2.10. *If for normal matrices* $\mathbf{A}$ *and* $\mathbf{B}$ *the equality* $\mathbf{AB} = \mathbf{BA}$ *holds, then* $C(\mathbf{AB}) = C(\mathbf{A}) \cap C(\mathbf{B})$.

PROOF: It is obvious that $C(\mathbf{AB}) \subseteq C(\mathbf{A}) \cap C(\mathbf{B})$. Then, by Theorem 1.2.6 (i),

$$C(\mathbf{A}) \cap C(\mathbf{B}) = C(\mathbf{AB}) \boxplus C(\mathbf{A}) \cap C(\mathbf{B}) \cap C(\mathbf{AB})^{\perp}. \tag{1.2.30}$$

We are going to show that if $\mathbf{AB} = \mathbf{BA}$, then

$$C(\mathbf{A}) \cap C(\mathbf{B}) \cap C(\mathbf{AB})^{\perp} = \{\mathbf{0}\}.$$

Theorem 1.2.15 implies that

$$\mathcal{C}(\mathbf{A}) \cap \mathcal{C}(\mathbf{B}) \cap \mathcal{C}(\mathbf{AB})^{\perp} = \mathcal{C}(\mathbf{A}(\mathbf{A}'\mathbf{B}^o)^o((\mathbf{A}'\mathbf{B}^o)^{o'}\mathbf{A}'\mathbf{AB})^o). \qquad (1.2.31)$$

Moreover, using Theorem 1.2.15 again with the assumption of the lemma, it follows that

$$\begin{aligned}\mathcal{C}((\mathbf{A}'\mathbf{B}^o)^{o'}\mathbf{A}'\mathbf{AB}) &= \mathcal{C}((\mathbf{A}'\mathbf{B}^o)^{o'}\mathbf{BA}'\mathbf{A}) = \mathcal{C}((\mathbf{A}'\mathbf{B}^o)^{o'}\mathbf{BA}') \\ &= \mathcal{C}((\mathbf{A}'\mathbf{B}^o)^{o'}\mathbf{A}'\mathbf{B}).\end{aligned}$$

Thus the right hand side of (1.2.31) equals

$$\mathcal{C}(\mathbf{A}) \cap \mathcal{C}(\mathbf{B}) \cap \mathcal{C}(\mathbf{B})^{\perp} = \{\mathbf{0}\},$$

which via (1.2.30) establishes the lemma. ∎

Lemma 1.2.11. *If for normal matrices* $\mathbf{A}$ *and* $\mathbf{B}$ *the equality* $\mathbf{AB} = \mathbf{BA}$ *holds, then* $\mathcal{C}(\mathbf{A}) | \mathcal{C}(\mathbf{B})$, *i.e. the spaces* $\mathcal{C}(\mathbf{A})$ *and* $\mathcal{C}(\mathbf{B})$ *commute.*

PROOF: From Theorem 1.2.8 (iii) it follows that we have to show that

$$\mathcal{C}(\mathbf{A}) \cap \mathcal{C}(\mathbf{A} \cap \mathbf{B})^{\perp} \subseteq \mathcal{C}(\mathbf{B})^{\perp},$$

which by Lemma 1.2.10 is equivalent to

$$\mathbf{B}'\mathbf{A}(\mathbf{A}'\mathbf{AB})^o = \mathbf{0}.$$

By Lemma 1.2.9 this is true since

$$\mathcal{C}(\mathbf{A}'\mathbf{AB}) = \mathcal{C}(\mathbf{BA}'\mathbf{A}) = \mathcal{C}(\mathbf{BA}') = \mathcal{C}(\mathbf{A}'\mathbf{B}).$$

∎

Theorem 1.2.40. *Let* $\mathbf{A} \in \mathbb{R}^{p\times p}$ *and* $\mathbf{B} \in \mathbb{R}^{p\times p}$ *be normal matrices such that* $\mathbf{AB} = \mathbf{BA}$. *Then there exists a set of orthogonal eigenvectors of* $\mathbf{AB}$ *which also are eigenvectors of* $\mathbf{A}$ *and* $\mathbf{B}$ *which generate* $\mathcal{C}(\mathbf{AB})$.

PROOF: Let $\mathbf{y}_1$ be an eigenvector of $\mathbf{AB}$, i.e. $\mathbf{ABy}_1 = \lambda\mathbf{y}_1$. Then $\mathbf{y}_1, \mathbf{A}'\mathbf{y}_1, \ldots, (\mathbf{A}')^{a-1}\mathbf{y}_1$, for some $a \leq p$, are eigenvectors of $\mathbf{AB}$. The set is $\mathbf{A}$-invariant and thus there is an eigenvector of $\mathbf{A}'$ and, because of normality, also of $\mathbf{A}$ which belongs to

$$\mathcal{C}(\mathbf{y}_1, \mathbf{A}'\mathbf{y}_1, \ldots, (\mathbf{A}')^{a-1}\mathbf{y}_1).$$

Denote this vector by $\mathbf{x}_1$. Thus,

$$\mathbf{x}_1 = \sum_{i=1}^{a} c_i(\mathbf{A}')^{i-1}\mathbf{y}_1$$

for some constants c_i. This implies that $\mathbf{ABx}_1 = \lambda\mathbf{x}_1$, which means that $\mathbf{x}_1$ is an eigenvector of $\mathbf{AB}$ as well as to $\mathbf{A}$. Furthermore, $\mathbf{x}_1, \mathbf{B}'\mathbf{x}_1, \ldots, (\mathbf{B}')^b\mathbf{x}_1$, for some $b \leq p$, are all eigenvectors of $\mathbf{AB}$. The space generated by this sequence is $\mathbf{B}'$-invariant and, because of normality, there is an eigenvector, say $\mathbf{z}_1$, to $\mathbf{B}$ which belongs to

$$\mathcal{C}(\mathbf{x}_1, \mathbf{B}'\mathbf{x}_1, \ldots, (\mathbf{B}')^{b-1}\mathbf{x}_1).$$

Hence,

$$\mathbf{z}_1 = \sum_{j=1}^{b} d_j (\mathbf{B}')^{j-1}\mathbf{x}_1 = \sum_{j=1}^{b}\sum_{i=1}^{a} d_j c_i (\mathbf{B}')^{j-1}(\mathbf{A}')^{i-1}\mathbf{y}_1$$

for some constants d_j. It follows immediately that $\mathbf{z}_1$ is an eigenvector of $\mathbf{AB}$, $\mathbf{A}$ and $\mathbf{B}$. Now, as in the proof of Theorem 1.2.36, study $\mathcal{C}(\mathbf{z}_1)^\perp$ which among others is $(\mathbf{AB})'$-invariant and $\mathbf{AB}$-invariant. Thus there exists an eigenvector of $\mathbf{AB}$, say $\mathbf{y}_2$, which belongs to $\mathcal{C}(\mathbf{z}_1)^\perp$. Thus,

$$\mathcal{C}(\mathbf{y}_2, \mathbf{A}'\mathbf{y}_2, \ldots, (\mathbf{A}')^{a-1}\mathbf{y}_2)$$

is $\mathbf{A}'$-invariant and orthogonal to $\mathcal{C}(\mathbf{z}_1)$. Now we can continue as before and end up with

$$\mathbf{z}_2 = \sum_{j=1}^{b}\sum_{i=1}^{a} d_j c_i (\mathbf{B}')^{j-1}(\mathbf{A}')^{i-1}\mathbf{y}_2,$$

which is an eigenvector of $\mathbf{AB}$, $\mathbf{A}$ and $\mathbf{B}$. Clearly, $\mathbf{z}_1'\mathbf{z}_2 = \mathbf{0}$ since $\mathbf{y}_1$ is an eigenvector of $\mathbf{B}$ and $\mathbf{A}$. We may continue by considering $\mathcal{C}(\mathbf{z}_1, \mathbf{z}_2)^\perp$. ∎

Now, if $\mathbf{A}$ and $\mathbf{B}$ commute, we have by Definition 1.2.5 that

$$\mathcal{C}(\mathbf{A}) = \mathcal{C}(\mathbf{A}) \cap \mathcal{C}(\mathbf{B}) \boxplus \mathcal{C}(\mathbf{A}) \cap \mathcal{C}(\mathbf{B})^\perp,$$
$$\mathcal{C}(\mathbf{B}) = \mathcal{C}(\mathbf{A}) \cap \mathcal{C}(\mathbf{B}) \boxplus \mathcal{C}(\mathbf{A})^\perp \cap \mathcal{C}(\mathbf{B}).$$

Furthermore, applying Lemma 1.2.9 gives that $\mathbf{AB}$ is normal and thus there exists a system of orthogonal eigenvectors $\mathbf{x}_1, \mathbf{x}_2, \ldots, \mathbf{x}_{r(\mathbf{AB})}$ which spans $\mathcal{C}(\mathbf{A}) \cap \mathcal{C}(\mathbf{B})$. By Theorem 1.2.40 these vectors are also eigenvectors of $\mathcal{C}(\mathbf{A})$ and $\mathcal{C}(\mathbf{B})$. Since $\mathbf{A}$ is normal, orthogonal eigenvectors $\mathbf{u}_{r(\mathbf{AB})+1}, \mathbf{u}_{r(\mathbf{AB})+2}, \ldots, \mathbf{u}_{r(\mathbf{A})}$ which are orthogonal to $\mathbf{x}_1, \mathbf{x}_2, \ldots, \mathbf{x}_{r(\mathbf{AB})}$ can additionally be found, and both sets of these eigenvectors span $\mathcal{C}(\mathbf{A})$. Let us denote one of these eigenvectors by $\mathbf{y}$. Thus, since $\mathbf{y}$ is also an eigenvector of $\mathbf{A}'$,

$$\mathbf{A}'\mathbf{y} = \lambda\mathbf{y}, \tag{1.2.32}$$
$$\mathbf{A}'\mathbf{B}'\mathbf{y} = \mathbf{0} \tag{1.2.33}$$

for some λ. However, (1.2.32) and (1.2.33) imply that

$$\mathbf{0} = \mathbf{A}'\mathbf{B}'\mathbf{y} = \mathbf{B}'\mathbf{A}'\mathbf{y} = \lambda\mathbf{B}'\mathbf{y}.$$

Hence $\mathbf{y}$ is orthogonal to $\mathbf{B}$ and we have found that $\mathbf{u}_{r(\mathbf{AB})+1}, \mathbf{u}_{r(\mathbf{AB})+2}, \ldots, \mathbf{u}_{r(\mathbf{A})}$ generate

$$\mathcal{C}(\mathbf{A}) \cap \mathcal{C}(\mathbf{B})^\perp.$$

Indeed, this result is also established by applying Theorem 1.2.4 (i). Furthermore, in the same way we can find orthogonal eigenvectors $\mathbf{v}_{r(\mathbf{AB})+1}, \mathbf{v}_{r(\mathbf{AB})+2}, \ldots, \mathbf{v}_{r(\mathbf{B})}$ of $\mathbf{B}$, which generate

$$C(\mathbf{B}) \cap C(\mathbf{A})^{\perp}.$$

Thus, if we put the eigenvectors together and suppose that they are standardized, we get a matrix

$$\mathbf{X}_1 = (\mathbf{x}_1, \mathbf{x}_2, \ldots, \mathbf{x}_{r(\mathbf{AB})}, \mathbf{u}_{r(\mathbf{AB})+1}, \mathbf{u}_{r(\mathbf{AB})+2}, \\ \ldots, \mathbf{u}_{r(\mathbf{A})}, \mathbf{v}_{r(\mathbf{AB})+1}, \mathbf{v}_{r(\mathbf{AB})+2}, \ldots, \mathbf{v}_{r(\mathbf{B})}),$$

where $\mathbf{X}_1^{\star}\mathbf{X}_1 = \mathbf{I}$. Consider the matrix $\mathbf{X} = (\mathbf{X}_1, \mathbf{X}_2)$ where $\mathbf{X}_2$ is such that $\mathbf{X}_2^{\star}\mathbf{X}_1 = \mathbf{0}$, $\mathbf{X}_2^{\star}\mathbf{X}_2 = \mathbf{I}$. We obtain that if $\mathbf{X}$ is of full rank, it is also orthogonal and satisfies

$$\mathbf{AX} = \mathbf{X}\boldsymbol{\Delta}_1, \qquad \boldsymbol{\Delta}_1 = \begin{pmatrix} \boldsymbol{\Lambda}_1 & \mathbf{0} \\ \mathbf{0} & \mathbf{0} \end{pmatrix},$$

$$\mathbf{BX} = \mathbf{X}\boldsymbol{\Delta}_2, \qquad \boldsymbol{\Delta}_2 = \begin{pmatrix} \boldsymbol{\Lambda}_2 & \mathbf{0} \\ \mathbf{0} & \mathbf{0} \end{pmatrix},$$

where $\boldsymbol{\Lambda}_1$ and $\boldsymbol{\Lambda}_2$ are diagonal matrices consisting of the eigenvalues of $\mathbf{A}$ and $\mathbf{B}$, respectively. From the proof of Theorem 1.2.39 it follows that we have an orthogonal matrix $\mathbf{Q}$ and

$$\mathbf{E}_1 = \left\{ \begin{pmatrix} \mu_1 & \delta_1 \\ -\delta_1 & \mu_1 \end{pmatrix}, \ldots, \begin{pmatrix} \mu_q & \delta_q \\ -\delta_q & \mu_q \end{pmatrix}, \lambda_{2q+1}, \ldots, \lambda_{n_1} \right\}_{[d]},$$

$$\mathbf{E}_2 = \left\{ \begin{pmatrix} \mu_1' & \delta_1' \\ -\delta_1' & \mu_1' \end{pmatrix}, \ldots, \begin{pmatrix} \mu_q' & \delta_q' \\ -\delta_q' & \mu_q' \end{pmatrix}, \lambda_{2q+1}', \ldots, \lambda_{n_1}' \right\}_{[d]}$$

such that

$$\mathbf{A} = \mathbf{Q}\mathbf{E}_1\mathbf{Q}'$$
$$\mathbf{B} = \mathbf{Q}\mathbf{E}_2\mathbf{Q}'.$$

The procedure above can immediately be extended to an arbitrary sequence of normal matrices $\mathbf{A}_i$, $i = 1, 2, \ldots$, which commute pairwise, and we get an important factorization theorem.

Theorem 1.2.41. *Let $\{\mathbf{A}_i\}$ be a sequence of normal matrices which commute pairwise, i.e. $\mathbf{A}_i\mathbf{A}_j = \mathbf{A}_j\mathbf{A}_i$, $i \neq j$, $i, j = 1, 2, \ldots$. Then there exists an orthogonal matrix $\mathbf{Q}$ such that*

$$\mathbf{A}_i = \mathbf{Q}\left\{ \begin{pmatrix} \mu_1^{(i)} & \delta_1^{(i)} \\ -\delta_1^{(i)} & \mu_1^{(i)} \end{pmatrix}, \ldots, \begin{pmatrix} \mu_q^{(i)} & \delta_q^{(i)} \\ -\delta_q^{(i)} & \mu_q^{(i)} \end{pmatrix}, \lambda_{2q+1}^{(i)}, \ldots, \lambda_{n_1}^{(i)}, \mathbf{0}_{n-n_1} \right\}_{[d]} \mathbf{Q}',$$

where $\lambda_k^{(i)}$ and $\mu_k^{(i)} \pm i\delta_k^{(i)}$ stand for non-zero eigenvalues of $\mathbf{A}_i$. ∎

REMARK: For some i some of the blocks

$$\begin{pmatrix} \mu_j^{(i)} & \delta_j^{(i)} \\ -\delta_j^{(i)} & \mu_j^{(i)} \end{pmatrix}$$

or $\lambda_j^{(i)}$ may be zero.

Theorem 1.2.42. *Let* $\mathbf{A} > 0$ *and* $\mathbf{B}$ *symmetric. Then, there exist a non-singular matrix* $\mathbf{T}$ *and diagonal matrix* $\mathbf{\Lambda} = (\lambda_1, \ldots, \lambda_m)_d$ *such that*

$$\mathbf{A} = \mathbf{TT}', \qquad \mathbf{B} = \mathbf{T\Lambda T}',$$

where λ_i, $i = 1, 2, \ldots m$, *are eigenvalues of* $\mathbf{A}^{-1}\mathbf{B}$.

PROOF: Let $\mathbf{A} = \mathbf{XX}'$ and consider the matrix $\mathbf{X}^{-1}\mathbf{B}(\mathbf{X}')^{-1}$. Since, $\mathbf{X}^{-1}\mathbf{B}(\mathbf{X}')^{-1}$ is symmetric, it follows by Corollary 1.2.39.1 that there exists an orthogonal matrix $\mathbf{Q}$ such that

$$\mathbf{X}^{-1}\mathbf{B}(\mathbf{X}')^{-1} = \mathbf{Q\Lambda Q}'.$$

Thus $\mathbf{A} = \mathbf{XQQ}'\mathbf{X}'$ and $\mathbf{B} = \mathbf{XQ\Lambda Q}'\mathbf{X}'$. If we put $\mathbf{T} = \mathbf{XQ}$, the theorem is established. ∎

Corollary 1.2.42.1. *Let* $\mathbf{A} > 0$ *and* $\mathbf{B}$ *p.s.d. Then there exist a non-singular matrix* $\mathbf{T}$ *and diagonal matrix* $\mathbf{\Lambda} = (\lambda_1, \ldots, \lambda_m)_d$ *such that*

$$\mathbf{A} = \mathbf{TT}', \qquad \mathbf{B} = \mathbf{T\Lambda T}',$$

where $\lambda_i \geq 0$, $i = 1, 2, \ldots m$, *are eigenvalues of* $\mathbf{A}^{-1}\mathbf{B}$. ∎

It follows from the proof of Theorem 1.2.42 that we can change the theorem and establish

Theorem 1.2.43. *Let* $\mathbf{A} > 0$ *and* $\mathbf{B}$ *be normal. Then there exists a matrix* $\mathbf{T} > 0$ *and an orthogonal matrix* $\mathbf{Q}$ *such that* $\mathbf{A} = \mathbf{QTQ}'$ *and* $\mathbf{B} = \mathbf{QEQ}'$, *where*

$$\mathbf{E} = \left\{ \begin{pmatrix} \mu_1 & \delta_1 \\ -\delta_1 & \mu_1 \end{pmatrix}, \begin{pmatrix} \mu_2 & \delta_2 \\ -\delta_2 & \mu_2 \end{pmatrix}, \ldots, \begin{pmatrix} \mu_q & \delta_q \\ -\delta_q & \mu_q \end{pmatrix}, \mathbf{0} \right\}_{[d]}.$$

∎

Corollary 1.2.43.1. *Let* $\mathbf{A} > 0$ *and* $\mathbf{B}$ *be skew-symmetric. Then there is a matrix* $\mathbf{T} > 0$ *and an orthogonal matrix* $\mathbf{Q}$ *such that* $\mathbf{A} = \mathbf{QTQ}'$ *and* $\mathbf{B} = \mathbf{Q\Lambda Q}'$, *where*

$$\mathbf{\Lambda} = \left\{ \begin{pmatrix} 0 & \delta_1 \\ -\delta_1 & 0 \end{pmatrix}, \begin{pmatrix} 0 & \delta_2 \\ -\delta_2 & 0 \end{pmatrix}, \ldots, \begin{pmatrix} 0 & \delta_q \\ -\delta_q & 0 \end{pmatrix}, \mathbf{0} \right\}_{[d]}.$$

∎

Another type of factorizations are given by the Schur and Jordan factorizations which are also called Schur and Jordan decompositions.

Theorem 1.2.44. *(Schur factorization) For* $\mathbf{A} \in \mathbb{R}^{m \times m}$ *there exists a unitary matrix* $\mathbf{U}$ *such that*

$$\mathbf{U}^*\mathbf{AU} = \mathbf{H},$$

where $\mathbf{H}$ *is an upper triangular matrix with eigenvalues of* $\mathbf{A}$ *as its main diagonal elements.*

PROOF: Details of the proof of the theorem can be found in Bellmann (1970, pp 202-203). ∎

In the special case, when $\mathbf{A}$ is symmetric, the statement of Theorem 1.2.44 coincides with Corollary 1.2.39.1. The last theorem of this section is connected to so-called *Jordan normal forms* (see Gantmacher, 1959).

Theorem 1.2.45. *(Jordan factorization) Let* $\mathbf{A} \in \mathbb{R}^{m\times m}$ *and* $\mathbf{J}_k(\lambda) : k \times k$ *denotes the upper triangular matrix of the form*

$$\mathbf{J}_{k_i}(\lambda) = \begin{pmatrix} \lambda_i & 1 & 0 & \dots & 0 \\ 0 & \lambda_i & 1 & \dots & 0 \\ \vdots & \vdots & \vdots & & \vdots \\ 0 & 0 & 0 & \dots & 1 \\ 0 & 0 & 0 & \dots & \lambda_i \end{pmatrix}.$$

Then there exists a non-singular matrix $\mathbf{H} : m \times m$ *such that*

$$\mathbf{A} = \mathbf{HJH}^{-1},$$

where

$$\mathbf{J} = (\mathbf{J}_{k_1}, \mathbf{J}_{k_2}, \dots, \mathbf{J}_{k_r})_{[d]}$$

with $k_1 + k_2 + \cdots + k_r = m$. *The values* λ_i *are eigenvalues of* $\mathbf{A}$ *which do not have to be distinct.* ∎

1.2.11 *Problems*

1. Show that $C(\mathbf{A}') = C(\mathbf{A}^+)$.
2. Verify (iii) – (vi) in Proposition 1.2.1.
3. Show that an orthogonal projector is self-adjoint.
4. Prove (v), (xi) and (xiii) in Proposition 1.2.2.
5. Verify relation (1.2.17). Determine the following two determinants

$$\begin{vmatrix} 1+d_1+d_1^2+d_1^3 & d_1+d_1^2+d_1^3 & d_1^2+d_1^3 & d_1^3 \\ 1+d_2+d_2^2+d_2^3 & d_2+d_2^2+d_2^3 & d_2^2+d_2^3 & d_2^3 \\ 1+d_3+d_3^2+d_3^3 & d_3+d_3^2+d_3^3 & d_3^2+d_3^3 & d_3^3 \\ 1+d_4+d_4^2+d_4^3 & d_4+d_4^2+d_4^3 & d_4^2+d_4^3 & d_4^3 \end{vmatrix} \text{ and } \begin{vmatrix} 2 & d_1 & d_1^2 & d_1^3 \\ 1 & d_2 & d_2^2 & d_2^3 \\ 1 & d_3 & d_3^2 & d_3^3 \\ 1 & d_4 & d_4^2 & d_4^3 \end{vmatrix}.$$

6. Show that Gram-Schmidt orthogonalization means that we are using the following decomposition:

$$\begin{aligned} &C(\mathbf{x}_1 : \mathbf{x}_2 : \dots : \mathbf{x}_m) \\ &= C(\mathbf{x}_1) + C(\mathbf{x}_1)^\perp \cap C(\mathbf{x}_1 : \mathbf{x}_2) + C(\mathbf{x}_1 : \mathbf{x}_2)^\perp \cap C(\mathbf{x}_1 : \mathbf{x}_2 : \mathbf{x}_3) + \cdots \\ &\quad \cdots + C(\mathbf{x}_1 : \mathbf{x}_2 : \dots : \mathbf{x}_{m-1})^\perp \cap C(\mathbf{x}_1 : \mathbf{x}_2 : \dots : \mathbf{x}_m), \end{aligned}$$

   where $\mathbf{x}_i \in \mathbb{R}^p$
7. Prove Theorem 1.2.21 (i) when it is assumed that instead of $\mathbb{A}_s \subseteq \mathbb{A}_{s-1} \subseteq \ldots \subseteq \mathbb{A}_1$ the subspaces commute.
8. Take 3×3 matrices $\mathbf{A}$ and $\mathbf{B}$ which satisfy the assumptions of Theorem 1.2.42–Theorem 1.2.45 and construct all four factorizations.
9. What happens in Theorem 1.2.35 if $\mathbf{g}_1$ is supposed to be complex? Study it with the help of eigenvectors and eigenvalues of the normal matrix

$$\begin{pmatrix} 1 & 1 \\ -1 & 1 \end{pmatrix}.$$

10. Prove Theorem 1.2.44.

1.3 PARTITIONED MATRICES

1.3.1 *Basic notation and relations*

Statistical data is often obtained as measurements from some repeated procedure. For example, we have often independent, identically distributed observation vectors on several experimental units or we may have different types of information available about our data, like within individuals and between individuals information in repeated measurements. In both cases we will present the information (data and design) in form of matrices which have certain pattern structure or consist of certain blocks because of the underlying procedure. Moreover, certain useful statistical quantities lead us to matrices of specific structures. An obvious example is the dispersion matrix which is defined as a symmetric characteristic of dependency. In Section 1.1 we listed several structures of matrices like symmetric, skew-symmetric, diagonal, triangular displaying structure in an explicit way, or normal, positive definite, idempotent, orthogonal matrices which have certain implicit structure in them. In this section we give a general overview of techniques for handling structures in matrices.

Definition 1.3.1. *A matrix* $\mathbf{A} : p \times q$ *is called partitioned matrix (or block-matrix) if it consists of* uv *submatrices* $\mathbf{A}_{ij} : p_i \times q_j$ *(blocks) so that*

$$\mathbf{A} = \begin{pmatrix} \mathbf{A}_{11} & \mathbf{A}_{12} & \cdots & \mathbf{A}_{1v} \\ \vdots & \vdots & \ddots & \vdots \\ \mathbf{A}_{u1} & \mathbf{A}_{u2} & \cdots & \mathbf{A}_{uv} \end{pmatrix}, \qquad \sum_{i=1}^{u} p_i = p; \sum_{j=1}^{v} q_j = q.$$

∎

A partitioned matrix will often be denoted

$$\mathbf{A} = [\mathbf{A}_{ij}] \qquad i = 1, \ldots, u; \; j = 1, \ldots, v.$$

If it is necessary, dots are used to separate blocks in a partitioned matrix.
Following Anderson (2003) we shall use double indices for indicating rows and columns of a partitioned matrix. A row of the partitioned matrix $\mathbf{A}$ is denoted by an index (k,l), if this is the $(\sum_{i=1}^{k-1} p_i + l)-$th row of $\mathbf{A}$, i.e. this is the $l-$th row of the $k-$th block-row $(\mathbf{A}_{k1} : \ldots : \mathbf{A}_{kv})$.
A column of $\mathbf{A}$ is denoted by an index (g,h), if this column is the $(\sum_{i=1}^{g-1} q_i + h)-$th column of the matrix $\mathbf{A}$, i.e. it is the $h-$th column in the $g-$th column of the blocks

$$\begin{pmatrix} \mathbf{A}_{1g} \\ \vdots \\ \mathbf{A}_{ug} \end{pmatrix}.$$

The element of the partitioned matrix $\mathbf{A}$ in the $(k,l)-$th row and $(g,h)-$th column is denoted by $a_{(k,l)(g,h)}$ or $(\mathbf{A})_{(k,l)(g,h)}$.
Using standard numeration of rows and columns of a matrix we get the relation

$$a_{(k,l)\,(g,h)} = a_{\sum_{i=1}^{k-1} p_i + l, \sum_{j=1}^{g-1} q_j + h}. \tag{1.3.1}$$

In a special case, when $u = 1$, the $p\times q-$matrix $\mathbf{A}$ is a partitioned matrix of column blocks which is divided into submatrices $\mathbf{A}_j : p \times q_j$ so that $q_j > 0, \sum_{j=1}^{v} q_j = q$:

$$\mathbf{A} = (\mathbf{A}_1 : \mathbf{A}_2 : \ldots : \mathbf{A}_v).$$

We denote this matrix

$$\mathbf{A} = [\mathbf{A}_{1j}], \quad j = 1, \ldots, v.$$

For indicating the $h-$th column of a submatrix $\mathbf{A}_g$ we use the index (g, h), and the element of $\mathbf{A}$ in the $k-$th row and the $(g, h)-$th column is denoted by $a_{k(g,h)}$. If $\mathbf{A} : p \times q$ is divided into submatrices $\mathbf{A}_i$ so that $\mathbf{A}_i$ is a $p_i \times q$ submatrix, where $p_i > 0, \sum_{i=1}^{u} p_i = p$, we have a partitioned matrix of row blocks

$$\mathbf{A} = \begin{pmatrix} \mathbf{A}_1 \\ \vdots \\ \mathbf{A}_u \end{pmatrix},$$

which is denoted by

$$\mathbf{A} = [\mathbf{A}_{i1}] \quad i = 1, \ldots, u.$$

The index of the $l-$th row of the submatrix $\mathbf{A}_k$ is denoted by (k, l), and the element in the $(k, l)-$th row and the $g-$th column is denoted by $a_{(k,l)g}$.
A partitioned matrix $\mathbf{A} = [\mathbf{A}_{ij}]$, $i = 1, \ldots, u$, $j = 1, \ldots, v$ is called *block-diagonal* if $\mathbf{A}_{ij} = \mathbf{0}$, $i \neq j$. The partitioned matrix, obtained from $\mathbf{A} = [\mathbf{A}_{ij}]$, $i = 1, \ldots, u$, $j = 1, \ldots, v$, by block-diagonalization, i.e. by putting $\mathbf{A}_{ij} = \mathbf{0}$, $i \neq j$, has been denoted by $\mathbf{A}_{[d]}$ in §1.1.1. The same notation $\mathbf{A}_{[d]}$ is used when a block-diagonal matrix is constructed from u blocks of matrices.
So, whenever a block-structure appears in matrices, it introduces double indices into notation. Now we shall list some basic properties of partitioned matrices.

Proposition 1.3.1. *For partitioned matrices* $\mathbf{A}$ *and* $\mathbf{B}$ *the following basic properties hold:*
(i)

$$c\mathbf{A} = [c\mathbf{A}_{ij}], \qquad i = 1, \ldots, u;\ j = 1, \ldots, v,$$

where $\mathbf{A} = [\mathbf{A}_{ij}]$ *and* c *is a constant.*
(ii) *If* $\mathbf{A}$ *and* $\mathbf{B}$ *are partitioned matrices with blocks of the same dimensions, then*

$$\mathbf{A} + \mathbf{B} = [\mathbf{A}_{ij} + \mathbf{B}_{ij}], \qquad i = 1, \ldots, u;\ j = 1, \ldots, v,$$

i.e.

$$[(\mathbf{A} + \mathbf{B})_{ij}] = [\mathbf{A}_{ij}] + [\mathbf{B}_{ij}].$$

(iii) *If* $\mathbf{A} = [\mathbf{A}_{ij}]$ *is an* $p \times q-$*partitioned matrix with blocks*

$$\mathbf{A}_{ij} : p_i \times q_j, \qquad (\sum_{i=1}^{u} p_i = p; \sum_{j=1}^{v} q_j = q)$$

and $\mathbf{B} = [\mathbf{B}_{jl}] : q \times r-$*partitioned matrix, where*

$$\mathbf{B}_{jl} : q_j \times r_l, \qquad (\sum_{l=1}^{w} r_l = r),$$

then the $p \times r-$*partitioned matrix* $\mathbf{AB} = [\mathbf{C}_{il}]$ *consists of* $p_i \times r_l-$*submatrices*

$$\mathbf{C}_{il} = \sum_{j=1}^{v} \mathbf{A}_{ij}\mathbf{B}_{jl}, \quad i = 1, \dots, u;\ l = 1, \dots, w.$$

∎

In many applications a block-structure is given by

$$\mathbf{A} = \begin{pmatrix} \mathbf{A}_{11} & \mathbf{A}_{12} \\ \mathbf{A}_{21} & \mathbf{A}_{22} \end{pmatrix}. \tag{1.3.2}$$

In this case there exist useful formulas for the inverse as well as the determinant of the matrix.

Proposition 1.3.2. *Let a partitioned non-singular matrix* $\mathbf{A}$ *be given by (1.3.2). If* $\mathbf{A}_{22}$ *is non-singular, then*

$$|\mathbf{A}| = |\mathbf{A}_{22}||\mathbf{A}_{11} - \mathbf{A}_{12}\mathbf{A}_{22}^{-1}\mathbf{A}_{21}|;$$

if $\mathbf{A}_{11}$ *is non-singular, then*

$$|\mathbf{A}| = |\mathbf{A}_{11}||\mathbf{A}_{22} - \mathbf{A}_{21}\mathbf{A}_{11}^{-1}\mathbf{A}_{12}|.$$

PROOF: It follows by definition of a determinant (see §1.1.2) that

$$\begin{vmatrix} \mathbf{T}_{11} & \mathbf{T}_{12} \\ \mathbf{0} & \mathbf{T}_{22} \end{vmatrix} = |\mathbf{T}_{11}||\mathbf{T}_{22}|$$

and then the results follow by noting that

$$|\mathbf{A}_{22}||\mathbf{A}_{11} - \mathbf{A}_{12}\mathbf{A}_{22}^{-1}\mathbf{A}_{21}| = \begin{vmatrix} \mathbf{I} & -\mathbf{A}_{12}\mathbf{A}_{22}^{-1} \\ \mathbf{0} & \mathbf{I} \end{vmatrix}\begin{vmatrix} \mathbf{A}_{11} & \mathbf{A}_{12} \\ \mathbf{A}_{21} & \mathbf{A}_{22} \end{vmatrix}\begin{vmatrix} \mathbf{I} & \mathbf{0} \\ -\mathbf{A}_{22}^{-1}\mathbf{A}_{21} & \mathbf{I} \end{vmatrix}.$$

∎

Proposition 1.3.3. *Let a non-singular* $\mathbf{A}$ *be partitioned according to (1.3.2). Then, when all below given inverses exist,*

$$\mathbf{A}^{-1} = \begin{pmatrix} \mathbf{C}_{22}^{-1} & -\mathbf{A}_{11}^{-1}\mathbf{A}_{12}\mathbf{C}_{11}^{-1} \\ -\mathbf{A}_{22}^{-1}\mathbf{A}_{21}\mathbf{C}_{22}^{-1} & \mathbf{C}_{11}^{-1} \end{pmatrix},$$

where

$$\begin{aligned} \mathbf{C}_{11} &= \mathbf{A}_{22} - \mathbf{A}_{21}\mathbf{A}_{11}^{-1}\mathbf{A}_{12}, \\ \mathbf{C}_{22} &= \mathbf{A}_{11} - \mathbf{A}_{12}\mathbf{A}_{22}^{-1}\mathbf{A}_{21}. \end{aligned}$$

∎

The matrices $\mathbf{C}_{11}$ and $\mathbf{C}_{22}$ in Proposition 1.3.3 are called *Schur complements* of $\mathbf{A}_{11}$ and $\mathbf{A}_{22}$, respectively. There exist a wide range of applications for Schur complements in statistics (see Ouellette, 1981).

Proposition 1.3.4. *Let*

$$\mathbf{A} = \begin{pmatrix} \mathbf{A}_{11} & \mathbf{A}_{12} \\ \mathbf{A}_{21} & \mathbf{A}_{22} \end{pmatrix} \quad \text{and} \quad \mathbf{A}^{-1} = \begin{pmatrix} \mathbf{A}^{11} & \mathbf{A}^{12} \\ \mathbf{A}^{21} & \mathbf{A}^{22} \end{pmatrix}$$

so that the dimensions of $\mathbf{A}_{11}$, $\mathbf{A}_{12}$, $\mathbf{A}_{21}$, $\mathbf{A}_{22}$ *correspond to the dimensions of* $\mathbf{A}^{11}$, $\mathbf{A}^{12}$, $\mathbf{A}^{21}$, $\mathbf{A}^{22}$, *respectively. Then*

(i) $(\mathbf{A}^{11})^{-1}\mathbf{A}^{12} = -\mathbf{A}_{12}\mathbf{A}_{22}^{-1}$;

(ii) $C(\mathbf{A}_{12}) \subseteq C(\mathbf{A}_{11} - \mathbf{A}_{12}\mathbf{A}_{22}^{-1}\mathbf{A}_{21})$;

(iii) *if* $\mathbf{A} > 0$, *then* $C(\mathbf{A}_{12}) \subseteq C(\mathbf{A}_{11})$. ∎

Proposition 1.3.5. *If the Schur complements* $\mathbf{C}_{ii}$ $(i = 1, 2)$ *in Proposition 1.3.3 are non-singular, then*

(i)
$$\mathbf{C}_{11}^{-1} = \mathbf{A}_{22}^{-1} + \mathbf{A}_{22}^{-1}\mathbf{A}_{21}(\mathbf{A}_{11} - \mathbf{A}_{12}\mathbf{A}_{22}^{-1}\mathbf{A}_{21})^{-1}\mathbf{A}_{12}\mathbf{A}_{22}^{-1},$$
$$\mathbf{C}_{22}^{-1} = \mathbf{A}_{11}^{-1} + \mathbf{A}_{11}^{-1}\mathbf{A}_{12}(\mathbf{A}_{22} - \mathbf{A}_{21}\mathbf{A}_{11}^{-1}\mathbf{A}_{12})^{-1}\mathbf{A}_{21}\mathbf{A}_{11}^{-1};$$

(ii)
$$\begin{aligned}(\mathbf{B}_1' : \mathbf{B}_2') &\begin{pmatrix} \mathbf{A}_{11} & \mathbf{A}_{12} \\ \mathbf{A}_{21} & \mathbf{A}_{22} \end{pmatrix}^{-1} \begin{pmatrix} \mathbf{B}_1 \\ \mathbf{B}_2 \end{pmatrix} \\ &= (\mathbf{B}_1 - \mathbf{A}_{12}\mathbf{A}_{22}^{-1}\mathbf{B}_2)'(\mathbf{A}_{11} - \mathbf{A}_{12}\mathbf{A}_{22}^{-1}\mathbf{A}_{21})^{-1}(\mathbf{B}_1 - \mathbf{A}_{12}\mathbf{A}_{22}^{-1}\mathbf{B}_2) \\ &\quad + \mathbf{B}_2'\mathbf{A}_{22}^{-1}\mathbf{B}_2,\end{aligned}$$

where $\mathbf{B}_1$ *and* $\mathbf{B}_2$ *are matrices of proper sizes.* ∎

Related to Proposition 1.3.5 is the so-called *Inverse binomial theorem.*

Proposition 1.3.6 *(Inverse binomial theorem).* *Suppose that all included inverses exist. Then*

$$(\mathbf{A} + \mathbf{BCD})^{-1} = \mathbf{A}^{-1} - \mathbf{A}^{-1}\mathbf{B}(\mathbf{DA}^{-1}\mathbf{B} + \mathbf{C}^{-1})^{-1}\mathbf{DA}^{-1}.$$

PROOF: Premultiplication by $\mathbf{A} + \mathbf{BCD}$ gives

$$\begin{aligned}(\mathbf{A} + \mathbf{BCD})&(\mathbf{A}^{-1} - \mathbf{A}^{-1}\mathbf{B}(\mathbf{DA}^{-1}\mathbf{B} + \mathbf{C}^{-1})^{-1}\mathbf{DA}^{-1}) \\ &= \mathbf{I} - \mathbf{B}(\mathbf{DA}^{-1}\mathbf{B} + \mathbf{C}^{-1})^{-1}\mathbf{DA}^{-1} + \mathbf{BCDA}^{-1} \\ &\quad - \mathbf{BCDA}^{-1}\mathbf{B}(\mathbf{DA}^{-1}\mathbf{B} + \mathbf{C}^{-1})^{-1}\mathbf{DA}^{-1} \\ &= \mathbf{I} - \mathbf{B}(\mathbf{CDA}^{-1}\mathbf{B} + \mathbf{I})(\mathbf{DA}^{-1}\mathbf{B} + \mathbf{C}^{-1})^{-1}\mathbf{DA}^{-1} + \mathbf{BCDA}^{-1} \\ &= \mathbf{I} - \mathbf{BCDA}^{-1} + \mathbf{BCDA}^{-1} = \mathbf{I}.\end{aligned}$$

∎

Propositions 1.3.3 and 1.3.6 can be extended to situations when the inverses do not exist. In such case some additional subspace conditions have to be imposed.

Proposition 1.3.7.
(i) *Let* $C(\mathbf{B}) \subseteq C(\mathbf{A})$, $C(\mathbf{D}') \subseteq C(\mathbf{A}')$ *and* $\mathbf{C}$ *be non-singular. Then*

$$(\mathbf{A}+\mathbf{BCD})^- = \mathbf{A}^- - \mathbf{A}^-\mathbf{B}(\mathbf{DA}^-\mathbf{B}+\mathbf{C}^{-1})^-\mathbf{DA}^-.$$

(ii) *Let* $C(\mathbf{B}) \subseteq C(\mathbf{A})$, $C(\mathbf{C}') \subseteq C(\mathbf{A}')$. *Then*

$$\begin{pmatrix} \mathbf{A} & \mathbf{B} \\ \mathbf{C} & \mathbf{D} \end{pmatrix}^- = \begin{pmatrix} \mathbf{A}^- & \mathbf{0} \\ \mathbf{0} & \mathbf{0} \end{pmatrix} + \begin{pmatrix} -\mathbf{A}^-\mathbf{B} \\ \mathbf{I} \end{pmatrix} (\mathbf{D}-\mathbf{CA}^-\mathbf{B})^-(-\mathbf{CA}^- : \mathbf{I}).$$

∎

The next two results concern with the g-inverse of $(\mathbf{A} : \mathbf{B})$.

Proposition 1.3.8. *One choice of g-inverse of the partitioned matrix* $(\mathbf{A} : \mathbf{B})$ *is given by*

$$(\mathbf{A} : \mathbf{B})^- = \begin{pmatrix} \mathbf{A}^+ - \mathbf{A}^+\mathbf{BR}^+ \\ \mathbf{R}^+ \end{pmatrix},$$

where

$$\mathbf{R} = (\mathbf{I} - \mathbf{AA}^+)\mathbf{B}.$$

PROOF: First it is noted that

$$\begin{aligned}(\mathbf{A} : \mathbf{B})(\mathbf{A} : \mathbf{B})^- &= \mathbf{AA}^+ - \mathbf{AA}^+\mathbf{BR}^+ + \mathbf{BR}^+ = \mathbf{AA}^+ + (\mathbf{I} - \mathbf{AA}^+)\mathbf{BR}^+ \\ &= \mathbf{AA}^+ + \mathbf{RR}^+\end{aligned}$$

and by (1.1.18)

$$\mathbf{RR}^+\mathbf{A} = \mathbf{0}, \quad \mathbf{RR}^+\mathbf{B} = \mathbf{RR}^+\mathbf{R} = \mathbf{R}, \quad \mathbf{AA}^+\mathbf{B} + \mathbf{RR}^+\mathbf{B} = \mathbf{B}.$$

Thus,

$$(\mathbf{A} : \mathbf{B})(\mathbf{A} : \mathbf{B})^-(\mathbf{A} : \mathbf{B}) = (\mathbf{A} : \mathbf{B}).$$

∎

Observe that the g-inverse in Proposition 1.3.8 is reflexive since $\mathbf{R}^+\mathbf{A} = \mathbf{0}$ implies

$$\begin{aligned}(\mathbf{A} : \mathbf{B})^-(\mathbf{A} : \mathbf{B})(\mathbf{A} : \mathbf{B})^- &= \begin{pmatrix} \mathbf{A}^+ - \mathbf{A}^+\mathbf{BR}^+ \\ \mathbf{R}^+ \end{pmatrix}(\mathbf{AA}^+ + \mathbf{RR}^+) \\ &= \begin{pmatrix} \mathbf{A}^+ - \mathbf{A}^+\mathbf{BR}^+ \\ \mathbf{R}^+ \end{pmatrix} = (\mathbf{A} : \mathbf{B})^-.\end{aligned}$$

However, it is easy to show that the g-inverse is not a Moore-Penrose inverse. Therefore, in order to obtain a Moore-Penrose inverse, the proposition has to be modified somewhat (Cline, 1964).

Proposition 1.3.9. *The Moore-Penrose inverse of the partitioned matrix* $(\mathbf{A}:\mathbf{B})$ *equals*

$$(\mathbf{A}:\mathbf{B})^+ = \begin{pmatrix} \mathbf{A}^+ - \mathbf{A}^+\mathbf{B}\mathbf{H} \\ \mathbf{H} \end{pmatrix},$$

where

$$\begin{aligned}\mathbf{H} =& \mathbf{R}^+ + (\mathbf{I}-\mathbf{R}^+\mathbf{R})\mathbf{Z}\mathbf{B}'\mathbf{A}^{+'}\mathbf{A}^+(\mathbf{I}-\mathbf{B}\mathbf{R}^+),\\ \mathbf{R} =& (\mathbf{I}-\mathbf{A}\mathbf{A}^+)\mathbf{B},\\ \mathbf{Z} =& \{\mathbf{I}+(\mathbf{I}-\mathbf{R}^+\mathbf{R})\mathbf{B}'\mathbf{A}^{+'}\mathbf{A}^+\mathbf{B}(\mathbf{I}-\mathbf{R}^+\mathbf{R})\}^{-1}.\end{aligned}$$

∎

Some simple block structures have nice mathematical properties. For example, consider

$$\begin{pmatrix} \mathbf{A} & \mathbf{B} \\ -\mathbf{B} & \mathbf{A} \end{pmatrix} \tag{1.3.3}$$

and multiply two matrices of the same structure:

$$\begin{pmatrix} \mathbf{A}_1 & \mathbf{B}_1 \\ -\mathbf{B}_1 & \mathbf{A}_1 \end{pmatrix}\begin{pmatrix} \mathbf{A}_2 & \mathbf{B}_2 \\ -\mathbf{B}_2 & \mathbf{A}_2 \end{pmatrix} = \begin{pmatrix} \mathbf{A}_1\mathbf{A}_2-\mathbf{B}_1\mathbf{B}_2 & \mathbf{A}_1\mathbf{B}_2+\mathbf{B}_1\mathbf{A}_2 \\ -\mathbf{B}_1\mathbf{A}_2-\mathbf{A}_1\mathbf{B}_2 & \mathbf{A}_1\mathbf{A}_2-\mathbf{B}_1\mathbf{B}_2. \end{pmatrix}. \tag{1.3.4}$$

The interesting fact is that the matrix on the right hand side of (1.3.4) is of the same type as the matrix in (1.3.3). Hence, a class of matrices can be defined which is closed under multiplication, and the matrices (1.3.3) may form a group under some assumptions on $\mathbf{A}$ and $\mathbf{B}$. Furthermore, consider the complex matrices $\mathbf{A}_1+i\mathbf{B}_1$ and $\mathbf{A}_2+i\mathbf{B}_2$, where i is the imaginary unit, and multiply them. Then

$$(\mathbf{A}_1+i\mathbf{B}_1)(\mathbf{A}_2+i\mathbf{B}_2) = \mathbf{A}_1\mathbf{A}_2-\mathbf{B}_1\mathbf{B}_2+i(\mathbf{A}_1\mathbf{B}_2+\mathbf{B}_1\mathbf{A}_2) \tag{1.3.5}$$

and we see that multiplication in (1.3.5) is equivalent to multiplication in (1.3.4). Indeed, we have obtained a generalized version of the fact that the space of complex $p-$vectors is isomorphic to a $2p-$dimensional real-valued space. Now the above given ideas will be extended somewhat, and the sum

$$\mathbf{A}+i\mathbf{B}+j\mathbf{C}+k\mathbf{D} \tag{1.3.6}$$

will be considered, where $i^2=j^2=k^2=-1$, $ij=-ji=k$, $jk=-kj=i$ and $ki=-ik=j$. Hence, we have introduced an algebraic structure which is usually called quaternions structure and its elements *quaternions*. For a general and easily accessible discussion of hypercomplex number, of which quaternions and the complex numbers are special cases, we refer to Kantor & Solodovnikov (1989). The reason why quaternions are discussed here is that multiplication of elements of the form (1.3.6) is equivalent to multiplication of matrices with a structure given in the next block matrix:

$$\begin{pmatrix} \mathbf{A} & \mathbf{B} & \mathbf{C} & \mathbf{D} \\ -\mathbf{B} & \mathbf{A} & -\mathbf{D} & \mathbf{C} \\ -\mathbf{C} & \mathbf{D} & \mathbf{A} & -\mathbf{B} \\ -\mathbf{D} & -\mathbf{C} & \mathbf{B} & \mathbf{A} \end{pmatrix}.$$

Now we briefly consider some more general structures than complex numbers and quaternions. The reason is that there exist many statistical applications of these structures. In particular, this is the case when we consider variance components models or covariance structures in multivariate normal distributions (e.g. see Andersson, 1975).

Suppose that we have a linear space $\mathbb{L}$ of matrices, i.e. for any $\mathbf{A}, \mathbf{B} \in \mathbb{L}$ we have $\alpha\mathbf{A} + \beta\mathbf{B} \in \mathbb{L}$. Furthermore, suppose that the identity $\mathbf{I} \in \mathbb{L}$. First we are going to prove an interesting theorem which characterizes squares of matrices in linear spaces via a *Jordan product* $ab + ba$ (usually the product is defined as $\frac{1}{2}(ab + ba)$ where a and b belong to a vector space).

Theorem 1.3.1. *For all* $\mathbf{A}, \mathbf{B} \in \mathbb{L}$

$$\mathbf{AB} + \mathbf{BA} \in \mathbb{L},$$

if and only if $\mathbf{C}^2 \in \mathbb{L}$, *for any* $\mathbf{C} \in \mathbb{L}$.

PROOF: If $\mathbf{A}^2, \mathbf{B}^2 \in \mathbb{L}$ we have, since $\mathbf{A} + \mathbf{B} \in \mathbb{L}$,

$$\mathbf{AB} + \mathbf{BA} = (\mathbf{A} + \mathbf{B})^2 - \mathbf{A}^2 - \mathbf{B}^2 \in \mathbb{L}.$$

For the converse, since $\mathbf{I} + \mathbf{A} \in \mathbb{L}$, it is noted that

$$2\mathbf{A} + 2\mathbf{A}^2 = \mathbf{A}(\mathbf{I} + \mathbf{A}) + (\mathbf{I} + \mathbf{A})\mathbf{A} \in \mathbb{L}$$

and by linearity it follows that also $\mathbf{A}^2 \in \mathbb{L}$. ∎

Let G_1 be the set of all invertable matrices in $\mathbb{L}$, and G_2 the set of all inverses of the invertable matrices, i.e.

$$G_1 = \{\mathbf{\Sigma} : |\mathbf{\Sigma}| \neq 0, \mathbf{\Sigma} \in \mathbb{L}\}, \tag{1.3.7}$$

$$G_2 = \{\mathbf{\Sigma}^{-1} : \mathbf{\Sigma} \in G_1\}. \tag{1.3.8}$$

Theorem 1.3.2. *Let the sets* G_1 *and* G_2 *be given by (1.3.7) and (1.3.8), respectively. Then* $G_1 = G_2$, *if and only if*

$$\mathbf{AB} + \mathbf{BA} \in \mathbb{L}, \quad \forall \mathbf{A}, \mathbf{B} \in \mathbb{L}, \tag{1.3.9}$$

or

$$\mathbf{A}^2 \in \mathbb{L}, \quad \forall \mathbf{A} \in \mathbb{L}. \tag{1.3.10}$$

PROOF: The equivalence between (1.3.9) and (1.3.10) was shown in Theorem 1.3.1. Consider the sets G_1 and G_2, given by (1.3.7) and (1.3.8), respectively. Suppose that $G_2 \subseteq \mathbb{L}$, i.e. every matrix in G_2 belongs to $\mathbb{L}$. Then $G_1 = G_2$. Furthermore, if $G_2 \subseteq \mathbb{L}$, for any $\mathbf{A} \in G_1$,

$$\mathbf{A}^2 = \mathbf{A} - ((\mathbf{A} + \mathbf{I})^{-1} - \mathbf{A}^{-1})^{-1} \in G_1 = G_2.$$

Now suppose that $\mathbf{A}^2 \in \mathbb{L}$ when $\mathbf{A} \in G_1$. Then, by Theorem 1.3.1,

$$\mathbf{A}^2(\mathbf{A}+\mathbf{I}) + (\mathbf{A}+\mathbf{I})\mathbf{A}^2 \in \mathbb{L},$$

since $\mathbf{I}+\mathbf{A} \in \mathbb{L}$. Thus $\mathbf{A}^3 \in \mathbb{L}$. By induction it follows that $\mathbf{A}^i \in \mathbb{L}$, $i = 0, 1, \ldots$. From Corollary 1.2.29 (ii) we obtain that if $\mathbf{A} : n \times n$, then there exist constants c_i such that

$$c_0\mathbf{A}^{-1} = \mathbf{A}^{n-1} - \sum_{i=1}^{n-1} c_i\mathbf{A}^{i-1} \in \mathbb{L},$$

i.e. $G_2 \subseteq \mathbb{L}$ and the theorem is established. ∎

The results in Theorem 1.3.1 and Theorem 1.3.2 have interesting statistical implications. Under so-called Jordan algebra factorizations, Tolver Jensen (1988) showed how to use similar results in a statistical context.

1.3.2 *The commutation matrix*

The notion of interest in this section was probably first introduced by Murnaghan (1938) as *permutation matrix*. Commonly the notion appears under this name in the literature. On the other hand, in recent publications on mathematical statistics more often the word "commutation" matrix has been used and we are also going to follow this tradition. However, as noted by some authors, it would be more appropriate to call the matrix a transposition matrix.

Definition 1.3.2. *The partitioned matrix $\mathbf{K}_{p,q} : pq \times pq$ consisting of $q \times p-$blocks is called commutation matrix, if*

$$(\mathbf{K}_{p,q})_{(i,j)(g,h)} = \begin{cases} 1; & g = j, h = i, \quad i, h = 1, \ldots, p; \quad j, g = 1, \ldots, q, \\ 0; & \text{otherwise.} \end{cases} \tag{1.3.11}$$

∎

From the definition it follows that one element in each column and row of $\mathbf{K}_{p,q}$ equals one and the other elements are zeros. As an example we shall write out the matrix $\mathbf{K}_{2,3}$.

Example 1.3.1.

$$\mathbf{K}_{2,3} = \begin{pmatrix} 1 & 0 & \vdots & 0 & 0 & \vdots & 0 & 0 \\ 0 & 0 & \vdots & 1 & 0 & \vdots & 0 & 0 \\ 0 & 0 & \vdots & 0 & 0 & \vdots & 1 & 0 \\ \cdots & \cdots & \cdots & \cdots & \cdots & \cdots & \cdots & \\ 0 & 1 & \vdots & 0 & 0 & \vdots & 0 & 0 \\ 0 & 0 & \vdots & 0 & 1 & \vdots & 0 & 0 \\ 0 & 0 & \vdots & 0 & 0 & \vdots & 0 & 1 \end{pmatrix}$$

∎

The commutation matrix can also be described in the following way: in the $(i, j)-$th block of $\mathbf{K}_{p,q}$ the $(j, i)-$th element equals one, while all other elements in that block are zeros. The commutation matrix is studied in the paper by Magnus & Neudecker (1979), and also in their book (Magnus & Neudecker, 1999). Here we shall give the main properties of the commutation matrix.

Proposition 1.3.10. *Let* $\mathbf{K}_{p,q}$ *be the commutation matrix. Then*

(i) $\mathbf{K}_{p,q} = \mathbf{K}'_{q,p}$;

(ii) $\mathbf{K}_{p,q}\mathbf{K}_{q,p} = \mathbf{I}_{pq}$;

(iii) $\mathbf{K}_{p,1} = \mathbf{K}_{1,p} = \mathbf{I}_p$;

(iv) $|\mathbf{K}_{p,q}| = \pm 1$. ∎

According to (i) and (ii) of the proposition, it follows that $\mathbf{K}_{p,q}$ is an orthogonal matrix. In the proof of the next proposition and other places it is convenient to use the indicator function $1_{\{a=b\}}$, i.e.

$$1_{\{a=b\}} = \begin{cases} 1 & a = b, \\ 0 & \text{otherwise.} \end{cases}$$

Proposition 1.3.11.

(i) *Let a partitioned matrix* $\mathbf{A} : m \times n$ *consist of* $r \times s-$*blocks:*

$$\mathbf{A} = [\mathbf{A}_{ij}] \quad i = 1, \ldots, p; \; j = 1, \ldots, q.$$

Then the partitioned matrix $\mathbf{AK}_{s,q}$ *consists of* $r \times q-$*blocks and the* $(g,h)-th$ *column of the product is the* $(h,g)-th$ *column of* $\mathbf{A}$.

(ii) *Let a* $m \times n-$*partitioned matrix* $\mathbf{A}$ *consist of* $r \times s-$*blocks*

$$\mathbf{A} = [\mathbf{A}_{ij}] \qquad i = 1, \ldots, p; \; j = 1, \ldots, q.$$

Then the partitioned matrix $\mathbf{K}_{p,r}\mathbf{A}$ *consists of* $p \times s-$*blocks and the* $(i,j)-th$ *row of the product matrix is the* $(j,i)-th$ *row of* $\mathbf{A}$.

PROOF: To prove (i) we have to show that

$$(\mathbf{AK}_{s,q})_{(i,j)(g,h)} = (\mathbf{A})_{(i,j)(h,g)}$$

for any $i = 1, \ldots, p; \; j = 1, \ldots, r; \; g = 1, \ldots, q, \; h = 1, \ldots, s$. By Proposition 1.3.1 (iii) we have

$$\begin{aligned} (\mathbf{AK}_{s,q})_{(i,j)(g,h)} &= \sum_{k=1}^{q}\sum_{l=1}^{s}(\mathbf{A})_{(i,j)(k,l)}(\mathbf{K})_{(k,l)(g,h)} \\ &\underset{(1.3.11)}{=} \sum_{k=1}^{q}\sum_{l=1}^{s}(\mathbf{A})_{(i,j)(k,l)}1_{\{g=l\}}1_{\{h=k\}} = (\mathbf{A})_{(i,j)(h,g)}. \end{aligned}$$

Thus, statement (i) is proved. The proof of (ii) is similar and is left as an exercise to the reader. ∎

Some important properties of the commutation matrix, in connection with the direct product (Kronecker product) and the vec-operator, will appear in the following paragraphs

1.3.3 *Direct product*

The direct product is one of the key tools in matrix theory which is applied to multivariate statistical analysis. The notion is used under different names. The classical books on matrix theory use the name "direct product" more often (Searle, 1982; Graybill, 1983, for example) while in the statistical literature and in recent issues the term "Kronecker product" is more common (Schott, 1997b; Magnus & Neudecker, 1999; and others). We shall use them synonymously throughout the text.

Definition 1.3.3. *Let* $\mathbf{A} = (a_{ij})$ *be a* $p\times q-$*matrix and* $\mathbf{B} = (b_{ij})$ *an* $r\times s-$*matrix. Then the* $pr \times qs-$*matrix* $\mathbf{A}\otimes\mathbf{B}$ *is called a direct product (Kronecker product) of the matrices* $\mathbf{A}$ *and* $\mathbf{B}$, *if*

$$\mathbf{A}\otimes\mathbf{B} = [a_{ij}\mathbf{B}], \quad i = 1,\dots,p;\ j = 1,\dots,q, \tag{1.3.12}$$

where

$$a_{ij}\mathbf{B} = \begin{pmatrix} a_{ij}b_{11} & \dots & a_{ij}b_{1s} \\ \vdots & \ddots & \vdots \\ a_{ij}b_{r1} & \dots & a_{ij}b_{rs} \end{pmatrix}.$$

∎

This definition somewhat overrules a principle of symmetry. We could also define $\mathbf{A}\otimes\mathbf{B}$ as consisting of blocks $[\mathbf{A}b_{kl}]$, $k = 1,\dots,r;\ l = 1,\dots,s$ (Graybill, 1983, for example). Sometimes one distinguishes between the two versions, i.e. by defining the right and left Kronecker product. It is easy to see that in both cases we have notions with similar properties which are equally useful in applications. By tradition it has happened that definition (1.3.12) is used more often. Moreover, the so-called *half Kronecker products* (Holmquist, 1985b) are also related to this notion.
Here a list of basic properties of the direct product will be given where we shall not indicate the sizes of the matrices if these coincide with those given in Definition 1.3.3.

Proposition 1.3.12.

(i)
$$(\mathbf{A}\otimes\mathbf{B})_{r(k-1)+l,s(g-1)+h} = (\mathbf{A}\otimes\mathbf{B})_{(k,l)(g,h)}.$$

(ii)
$$(\mathbf{A}\otimes\mathbf{B})_{(k,l)(g,h)} = a_{kg}b_{lh}. \tag{1.3.13}$$

(iii)
$$(\mathbf{A}\otimes\mathbf{B})' = \mathbf{A}'\otimes\mathbf{B}'.$$

(iv) *Let* $\mathbf{A},\mathbf{B} : p\times q$ *and* $\mathbf{C},\mathbf{D} : r\times s$. *Then*

$$(\mathbf{A}+\mathbf{B})\otimes(\mathbf{C}+\mathbf{D}) = \mathbf{A}\otimes\mathbf{C}+\mathbf{A}\otimes\mathbf{D}+\mathbf{B}\otimes\mathbf{C}+\mathbf{B}\otimes\mathbf{D}.$$

(v) *Let* $\mathbf{A} : p\times q$, $\mathbf{B} : r\times s$ *and* $\mathbf{C} : t\times u$. *Then*

$$\mathbf{A}\otimes(\mathbf{B}\otimes\mathbf{C}) = (\mathbf{A}\otimes\mathbf{B})\otimes\mathbf{C}.$$

(vi) *Let* $\mathbf{A} : p\times q$ *and* $\mathbf{B} : q\times w$ *and* $\mathbf{C} : r\times s$, $\mathbf{D} : s\times t$. *Then*

$$(\mathbf{A}\otimes\mathbf{C})(\mathbf{B}\otimes\mathbf{D}) = (\mathbf{AB})\otimes(\mathbf{CD}). \tag{1.3.14}$$

(vii) *Let* $\mathbf{A} : p\times p$ *and* $\mathbf{B} : q\times q$ *be non-singular matrices. Then*

$$(\mathbf{A}\otimes\mathbf{B})^{-1} = \mathbf{A}^{-1}\otimes\mathbf{B}^{-1}.$$

In the singular case one choice of a g-inverse is given by

$$(\mathbf{A}\otimes\mathbf{B})^{-}=\mathbf{A}^{-}\otimes\mathbf{B}^{-}.$$

(viii) *Let* $\mathbf{A}: p\times q$ *and* $\mathbf{B}: r\times s$. *Then*

$$\mathbf{A}\otimes\mathbf{B}=\mathbf{K}_{p,r}(\mathbf{B}\otimes\mathbf{A})\mathbf{K}_{s,q}. \tag{1.3.15}$$

(ix) *If* $\mathbf{A}$ *is* $p\times p-$*matrix and* $\mathbf{B}$ *is* $q\times q-$*matrix, then*

$$|\mathbf{A}\otimes\mathbf{B}|=|\mathbf{A}|^{q}|\mathbf{B}|^{p}.$$

(x) $r(\mathbf{A}\otimes\mathbf{B})=r(\mathbf{A})r(\mathbf{B})$.
(xi) $\mathbf{A}\otimes\mathbf{B}=\mathbf{0}$ *if and only if* $\mathbf{A}=\mathbf{0}$ *or* $\mathbf{B}=\mathbf{0}$.
(xii) *If* $\mathbf{a}$ *is a* $p-$*vector and* $\mathbf{b}$ *is a* $q-$*vector, then the outer product*

$$\mathbf{a}\mathbf{b}'=\mathbf{a}\otimes\mathbf{b}'=\mathbf{b}'\otimes\mathbf{a}. \tag{1.3.16}$$

(xiii) *Any eigenvalue of* $\mathbf{A}\otimes\mathbf{B}$ *equals the product of an eigenvalue of* $\mathbf{A}$ *with an eigenvalue of* $\mathbf{B}$. ∎

The direct product gives us another possibility to present the commutation matrix of the previous paragraph. The alternative form via basis vectors often gives us easy proofs for relations where the commutation matrix is involved. It is also convenient to use a similar representation of the direct product of matrices.

Proposition 1.3.13.
(i) *Let* $\mathbf{e}_i$ *be the* $i-$*th column vector of* $\mathbf{I}_p$ *and* $\mathbf{d}_j$ *the* $j-$*th column vector of* $\mathbf{I}_q$. *Then*

$$\mathbf{K}_{p,q}=\sum_{i=1}^{p}\sum_{j=1}^{q}(\mathbf{e}_i\mathbf{d}_j')\otimes(\mathbf{d}_j\mathbf{e}_i'). \tag{1.3.17}$$

(ii) *Let* $\mathbf{e}_{i_1}$, $\mathbf{e}_{i_2}$ *be the* i_1-*th and* i_2-*th column vectors of* $\mathbf{I}_p$ *and* $\mathbf{I}_r$, *respectively, and* $\mathbf{d}_{j_1}$, $\mathbf{d}_{j_2}$ *the* j_1-*th and* j_2-*th column vectors of* $\mathbf{I}_q$ *and* $\mathbf{I}_s$, *respectively. Then for* $\mathbf{A}: p\times q$ *and* $\mathbf{B}: r\times s$

$$\mathbf{A}\otimes\mathbf{B}=\sum_{\substack{1\le i_1\le p,\\ 1\le j_1\le q}}\sum_{\substack{1\le i_2\le r,\\ 1\le j_2\le s}} a_{i_1j_1}b_{i_2j_2}(\mathbf{e}_{i_1}\mathbf{d}_{j_1}')\otimes(\mathbf{e}_{i_2}\mathbf{d}_{j_2}').$$

PROOF: (i): It will be shown that the corresponding elements of the matrix given on the right hand side of (1.3.17) and the commutation matrix $\mathbf{K}_{p,q}$ are identical:

$$\left(\sum_{i=1}^{p}\sum_{j=1}^{q}(\mathbf{e}_i\mathbf{d}_j')\otimes(\mathbf{d}_j\mathbf{e}_i')\right)_{(k,l)(g,h)}\underset{(1.3.13)}{=}\sum_{i=1}^{p}\sum_{j=1}^{q}(\mathbf{e}_i\mathbf{d}_j')_{kg}(\mathbf{d}_j\mathbf{e}_i')_{lh}$$

$$=\sum_{i=1}^{p}\sum_{j=1}^{q}(\mathbf{e}_k\mathbf{d}_g')_{kg}(\mathbf{d}_l\mathbf{e}_h')_{lh}\underset{(1.3.11)}{=}(\mathbf{K}_{p,q})_{(k,l)(g,h)}.$$

(ii): For the matrices **A** and **B**, via (1.1.2), the following representations through basis vectors are obtained:

$$\mathbf{A} = \sum_{i_1=1}^{p}\sum_{j_1=1}^{q} a_{i_1j_1}(\mathbf{e}_{i_1}\mathbf{d}'_{j_1}), \qquad \mathbf{B} = \sum_{i_2=1}^{r}\sum_{j_2=1}^{s} b_{i_2j_2}(\mathbf{e}_{i_2}\mathbf{d}'_{j_2}).$$

Then

$$\mathbf{A}\otimes\mathbf{B} = \sum_{\substack{1\leq i_1\leq p,\\ 1\leq j_1\leq q}}\ \sum_{\substack{1\leq i_2\leq r,\\ 1\leq j_2\leq s}} a_{i_1j_1}b_{i_2j_2}(\mathbf{e}_{i_1}\mathbf{d}'_{j_1})\otimes(\mathbf{e}_{i_2}\mathbf{d}'_{j_2}).$$

∎

To acquaint the reader with different techniques, the statement (1.3.15) is now verified in two ways: using basis vectors and using double indexation of elements. Proposition 1.3.13 gives

$$\begin{aligned}
&\mathbf{K}_{p,r}(\mathbf{B}\otimes\mathbf{A})\mathbf{K}_{s,q}\\
&\quad= \sum_{\substack{1\leq i_1\leq p,\\ 1\leq j_1\leq q}}\ \sum_{\substack{1\leq i_2\leq r,\\ 1\leq j_2\leq s}} (\mathbf{e}_{i_1}\mathbf{e}'_{i_2}\otimes\mathbf{e}_{i_2}\mathbf{e}'_{i_1})(\mathbf{B}\otimes\mathbf{A})(\mathbf{d}_{j_2}\mathbf{d}'_{j_1}\otimes\mathbf{d}_{j_1}\mathbf{d}'_{j_2})\\
&\quad= \sum_{\substack{1\leq i_1\leq p,\\ 1\leq j_1\leq q}}\ \sum_{\substack{1\leq i_2\leq r,\\ 1\leq j_2\leq s}} b_{i_2j_2}a_{i_1j_1}(\mathbf{e}_{i_1}\mathbf{e}'_{i_2}\otimes\mathbf{e}_{i_2}\mathbf{e}'_{i_1})(\mathbf{e}_{i_2}\mathbf{d}'_{j_2}\otimes\mathbf{e}_{i_1}\mathbf{d}'_{j_1})(\mathbf{d}_{j_2}\mathbf{d}'_{j_1}\otimes\mathbf{d}_{j_1}\mathbf{d}'_{j_2})\\
&\quad= \sum_{\substack{1\leq i_1\leq p,\\ 1\leq j_1\leq q}}\ \sum_{\substack{1\leq i_2\leq r,\\ 1\leq j_2\leq s}} b_{i_2j_2}a_{i_1j_1}\mathbf{e}_{i_1}\mathbf{d}'_{j_1}\otimes\mathbf{e}_{j_1}\mathbf{d}'_{j_2} = \mathbf{A}\otimes\mathbf{B}.
\end{aligned}$$

The same result is obtained by using Proposition 1.3.12 in the following way:

$$\begin{aligned}
(\mathbf{K}_{p,r}(\mathbf{B}\otimes\mathbf{A})\mathbf{K}_{s,q})_{(ij)(gh)} &= \sum_{k,l}(\mathbf{K}_{p,r})_{(ij)(kl)}\,((\mathbf{B}\otimes\mathbf{A})\mathbf{K}_{s,q})_{(kl)(gh)}\\
= \sum_{m,n}(\mathbf{B}\otimes\mathbf{A})_{(ji)(mn)}(\mathbf{K}_{s,q})_{(mn)(gh)} &= (\mathbf{B}\otimes\mathbf{A})_{(ji)(hg)} \underset{(1.3.13)}{=} b_{jh}a_{ig}\\
\underset{(1.3.13)}{=} (\mathbf{A}\otimes\mathbf{B})_{(ij)(gh)}.
\end{aligned}$$

In §1.2.5 it was noted that the range space of the tensor product of two mappings equals the tensor product of the range spaces of respective mapping. Let us now consider the column vector space of the Kronecker product of the matrices **A** and **B**, i.e. $\mathcal{C}(\mathbf{A}\otimes\mathbf{B})$. From Definition 1.2.8 of a tensor product (see also Takemura, 1983) it follows that a tensor product of $\mathcal{C}(\mathbf{A})$ and $\mathcal{C}(\mathbf{B})$ is given by

$$\mathcal{C}(\mathbf{A})\otimes\mathcal{C}(\mathbf{B}) = \mathcal{C}(\mathbf{a}_1\otimes\mathbf{b}_1 : \ldots : \mathbf{a}_1\otimes\mathbf{b}_s : \mathbf{a}_2\otimes\mathbf{b}_1 : \ldots : \mathbf{a}_1\otimes\mathbf{b}_s : \ldots : \mathbf{a}_r\otimes\mathbf{b}_s),$$

where $\mathbf{A} = (\mathbf{a}_1 : \ldots : \mathbf{a}_r)$ and $\mathbf{B} = (\mathbf{b}_1 : \ldots : \mathbf{b}_s)$. This means that the tensor product of $\mathcal{C}(\mathbf{A})$ and $\mathcal{C}(\mathbf{B})$ includes all combinations of the direct products of the columns of **A** with the columns of **B**. However,

$$\mathcal{C}((\mathbf{A}_1 : \mathbf{A}_2)\otimes(\mathbf{B}_1 : \mathbf{B}_2)) = \mathcal{C}((\mathbf{A}_1\otimes(\mathbf{B}_1 : \mathbf{B}_2)) : (\mathbf{A}_2\otimes(\mathbf{B}_1 : \mathbf{B}_2)))$$

and since by definition of the Kronecker product the columns of $\mathbf{A}_i \otimes (\mathbf{B}_1 : \mathbf{B}_2)$, $i = 1, 2$, are identical to those of $(\mathbf{A}_i \otimes \mathbf{B}_1) : (\mathbf{A}_i \otimes \mathbf{B}_2)$, although differently arranged, we have established that

$$C((\mathbf{A}_1 : \mathbf{A}_2) \otimes (\mathbf{B}_1 : \mathbf{B}_2)) = C((\mathbf{A}_1 \otimes \mathbf{B}_1) : (\mathbf{A}_2 \otimes \mathbf{B}_1) : (\mathbf{A}_1 \otimes \mathbf{B}_2) : (\mathbf{A}_2 \otimes \mathbf{B}_2)).$$

Thus, according to the definition of a tensor product of linear spaces, the column vector space of a Kronecker product of matrices is a tensor product, and we may write

$$C(\mathbf{A}) \otimes C(\mathbf{B}) = C(\mathbf{A} \otimes \mathbf{B}).$$

Moreover, when later in §1.3.5 considering column spaces of tensor products, the results of §1.2.5 are at our disposal which is of utmost importance.
Now, the direct product will be considered in a special case, namely when a vector is multiplied by itself several times.

Definition 1.3.4. *We call a p^k−vector $\mathbf{a}^{\otimes k}$, the k−th power of the p−vector $\mathbf{a}$, if $\mathbf{a}^{\otimes 0} = 1$ and*

$$\mathbf{a}^{\otimes k} = \underbrace{\mathbf{a} \otimes \cdots \otimes \mathbf{a}}_{k\ times}.$$

In general, for any matrix $\mathbf{A}$ the Kroneckerian power is given by

$$\mathbf{A}^{\otimes k} = \underbrace{\mathbf{A} \otimes \cdots \otimes \mathbf{A}}_{k\ times}.$$

∎

Furthermore, Proposition 1.3.12 (v) implies that $\mathbf{A}^{\otimes k}\mathbf{B}^{\otimes k} = (\mathbf{AB})^{\otimes k}$. In particular, it is noted that

$$\mathbf{a}^{\otimes k} \otimes \mathbf{a}^{\otimes j} = \mathbf{a}^{\otimes (k+j)}, \quad k, j \in \mathbb{N}. \tag{1.3.18}$$

The following statement makes it possible to identify where in a long vector of the Kroneckerian power a certain element of the product is situated.

Lemma 1.3.1. *Let $\mathbf{a} = (a_1 \ldots, a_p)'$ be a p−vector. Then for any $i_1, \ldots, i_k \in \{1, \ldots, p\}$ the following equality holds:*

$$a_{i_1} a_{i_2} \ldots a_{i_k} = (\mathbf{a}^{\otimes k})_j, \tag{1.3.19}$$

where

$$j = (i_1 - 1)p^{k-1} + (i_2 - 1)p^{k-2} + \ldots + (i_{k-1} - 1)p + i_k. \tag{1.3.20}$$

PROOF: We are going to use induction. For $k = 1$ the equality holds trivially, and for $k = 2$ it follows immediately from Proposition 1.3.12 (i). Let us assume, that the statements (1.3.19) and (1.3.20) are valid for $k = n - 1$:

$$a_{i_1} \ldots a_{i_{n-1}} = (\mathbf{a}^{\otimes (n-1)})_i, \tag{1.3.21}$$

where

$$i = (i_1 - 1)p^{n-2} + (i_2 - 1)p^{n-3} + \ldots + (i_{k-2} - 1)p + i_{n-1} \tag{1.3.22}$$

with $i \in \{1, \ldots, p^{n-1}\}$; $i_1, \ldots, i_{n-1} \in \{1, \ldots, p\}$. It will be shown that (1.3.19) and (1.3.20) also take place for $k = n$. From (1.3.18)

$$\mathbf{a}^{\otimes n} = \mathbf{a}^{\otimes(n-1)} \otimes \mathbf{a}.$$

For these two vectors Proposition 1.3.12 (xii) yields

$$(\mathbf{a}^{\otimes n})_j = (\mathbf{a}^{\otimes(n-1)})_i a_{i_n}, \tag{1.3.23}$$
$$j = (i-1)p + i_n, \tag{1.3.24}$$

with $j \in \{1, \ldots, p^n\}$; $i \in \{1, \ldots, p^{n-1}\}$; $i_n \in \{1, \ldots, p\}$. Replacing the vector $\mathbf{a}^{\otimes(n-1)}$ in (1.3.23) by the expression (1.3.21), and in (1.3.24) the index i by formula (1.3.22) we get the desired result. ▮

From Lemma 1.3.1 it follows that some coordinates of $\mathbf{a}^{\otimes k}$ are always equal. If all the indices i_j are different ($j = 1, \ldots, k$), then at least $k!$ coordinates of $\mathbf{a}^{\otimes k}$ are equal, for example. This means that there exist permutations of the coordinates of $\mathbf{a}^{\otimes k}$ which do not change the vector. These permutations can be presented through the commutation matrix $\mathbf{K}_{p,p}$.

Example 1.3.2. By changing the order of multipliers $\mathbf{a}$ in the product

$$\mathbf{a}^{\otimes 2} = \mathbf{a} \otimes \mathbf{a},$$

it follows via Proposition 1.3.12 (viii) that for $\mathbf{a} : p \times 1$

$$\mathbf{a}^{\otimes 2} = \mathbf{K}_{p,p}\mathbf{a}^{\otimes 2}.$$

For the powers of higher order many more equivalent permutations do exist. In the case of the third power we have the following relations

$$\begin{aligned}
\mathbf{a}^{\otimes 3} &= \mathbf{K}_{p^2,p}\mathbf{a}^{\otimes 3};\\
\mathbf{a}^{\otimes 3} &= \mathbf{K}_{p,p^2}\mathbf{a}^{\otimes 3};\\
\mathbf{a}^{\otimes 3} &= (\mathbf{I}_p \otimes \mathbf{K}_{p,p})\mathbf{a}^{\otimes 3};\\
\mathbf{a}^{\otimes 3} &= (\mathbf{K}_{p,p} \otimes \mathbf{I}_p)\mathbf{K}_{p^2,p}\mathbf{a}^{\otimes 3};\\
\mathbf{a}^{\otimes 3} &= (\mathbf{K}_{p,p} \otimes \mathbf{I}_p)\mathbf{a}^{\otimes 3}.
\end{aligned}$$

Observe that the relations above are permutations of basis vectors (permutations acting on tensor products). This follows from the equality

$$\mathbf{a}^{\otimes 3} = \sum_{i_1 i_2 i_3} a_{i_1} a_{i_2} a_{i_3} (\mathbf{e}_{i_1} \otimes \mathbf{e}_{i_2} \otimes \mathbf{e}_{i_3}).$$

Thus

$$
\begin{array}{rll}
\mathbf{K}_{p^2,p}\mathbf{a}^{\otimes 3} & \text{corresponds to the permutation} & \mathbf{e}_{i_2}\otimes\mathbf{e}_{i_3}\otimes\mathbf{e}_{i_1},\\
\mathbf{K}_{p,p^2}\mathbf{a}^{\otimes 3} & \text{corresponds to the permutation} & \mathbf{e}_{i_3}\otimes\mathbf{e}_{i_1}\otimes\mathbf{e}_{i_2},\\
(\mathbf{I}_p\otimes\mathbf{K}_{p,p})\mathbf{a}^{\otimes 3} & \text{corresponds to the permutation} & \mathbf{e}_{i_1}\otimes\mathbf{e}_{i_3}\otimes\mathbf{e}_{i_2},\\
(\mathbf{K}_{p,p}\otimes\mathbf{I}_p)\mathbf{K}_{p^2,p}\mathbf{a}^{\otimes 3} & \text{corresponds to the permutation} & \mathbf{e}_{i_2}\otimes\mathbf{e}_{i_1}\otimes\mathbf{e}_{i_3},\\
(\mathbf{K}_{p,p}\otimes\mathbf{I}_p)\mathbf{a}^{\otimes 3} & \text{corresponds to the permutation} & \mathbf{e}_{i_3}\otimes\mathbf{e}_{i_2}\otimes\mathbf{e}_{i_1},
\end{array}
$$

and together with the identity permutation $\mathbf{I}_{p^3}$ all 3! possible permutations are given. Hence, as noted above, if $i_1 \neq i_2 \neq i_3$, the product $a_{i_1}a_{i_2}a_{i_3}$ appears in 3! different places in $\mathbf{a}^{\otimes 3}$. All representations of $\mathbf{a}^{\otimes 3}$ follow from the basic properties of the direct product. The last one is obtained, for example, from the following chain of equalities:

$$
\mathbf{a}^{\otimes 3} \underset{(1.3.15)}{=} \mathbf{K}_{p,p}(\mathbf{a}\otimes\mathbf{a})\otimes\mathbf{a} = \mathbf{K}_{p,p}(\mathbf{a}\otimes\mathbf{a})\otimes(\mathbf{I}_p\mathbf{a}) \underset{(1.3.14)}{=} (\mathbf{K}_{p,p}\otimes\mathbf{I}_p)\mathbf{a}^{\otimes 3},
$$

or by noting that

$$
(\mathbf{K}_{p,p}\otimes\mathbf{I}_p)(\mathbf{e}_{i_1}\otimes\mathbf{e}_{i_2}\otimes\mathbf{e}_{i_3}) = \mathbf{e}_{i_2}\otimes\mathbf{e}_{i_1}\otimes\mathbf{e}_{i_3}.
$$

∎

Below an expression for $(\mathbf{A}+\mathbf{B})^{\otimes k}$ will be found. For small k this is a trivial problem but for arbitrary k complicated expressions are involved. In the next theorem $\prod_{j=0}^{k}\mathbf{A}_j$ stands for the matrix product $\mathbf{A}_0\mathbf{A}_1\cdots\mathbf{A}_k$.

Theorem 1.3.3. *Let $\mathbf{A} : p\times n$ and $\mathbf{B} : p\times n$. Denote $\mathbf{i}_j = (i_1,\ldots,i_j)$,*

$$
\mathbf{L}_s(j,k,\mathbf{i}_j) = \prod_{r=1}^{j}(\mathbf{I}_{s^{r-1}}\otimes\mathbf{K}_{s^{i_r-r},s}\otimes\mathbf{I}_{s^{k-i_r}}), \quad \mathbf{L}_s(0,k,\mathbf{i}_0) = \mathbf{I}_{s^k}
$$

and

$$
J_{j,k} = \{\mathbf{i}_j;\, j\leq i_j\leq k,\, j-1\leq i_{j-1}\leq i_j-1,\ldots,1\leq i_1\leq i_2-1\}.
$$

Let $\sum_{J_{0,k}}\mathbf{Q} = \mathbf{Q}$ for any matrix $\mathbf{Q}$. Then

$$
(\mathbf{A}+\mathbf{B})^{\otimes k} = \sum_{j=0}^{k}\sum_{\mathbf{i}_j\in J_{j,k}}\mathbf{L}_p(j,k,\mathbf{i}_j)(\mathbf{A}^{\otimes j}\otimes\mathbf{B}^{\otimes k-j})\mathbf{L}_n(j,k,\mathbf{i}_j)'.
$$

PROOF: An induction argument will be used. The theorem holds for $k = 1,2,3$. Suppose that the statement is true for $(\mathbf{A}+\mathbf{B})^{\otimes k-1}$. We are going to apply that

$$
\begin{aligned}
&\mathbf{L}_p(j,k-1,\mathbf{i}_j)(\mathbf{A}^{\otimes j}\otimes\mathbf{B}^{\otimes k-1-j})\mathbf{L}_n(j,k-1,\mathbf{i}_j)'\otimes\mathbf{B}\\
&\quad= (\mathbf{L}_p(j,k-1,\mathbf{i}_j)\otimes\mathbf{I}_p)(\mathbf{A}^{\otimes j}\otimes\mathbf{B}^{\otimes k-1-j}\otimes\mathbf{B})(\mathbf{L}_n(j,k-1,\mathbf{i}_j)'\otimes\mathbf{I}_n)
\end{aligned}
$$

and

$$
\begin{aligned}
&\mathbf{L}_p(j,k-1,\mathbf{i}_j)(\mathbf{A}^{\otimes j}\otimes\mathbf{B}^{\otimes k-1-j})\mathbf{L}_n(j,k-1,\mathbf{i}_j)'\otimes\mathbf{A}\\
&\quad=(\mathbf{L}_p(j,k-1,\mathbf{i}_j)\otimes\mathbf{I}_p)(\mathbf{A}^{\otimes j}\otimes\mathbf{B}^{\otimes k-1-j}\otimes\mathbf{A})(\mathbf{L}_n(j,k-1,\mathbf{i}_j)'\otimes\mathbf{I}_n)\\
&\quad=(\mathbf{L}_p(j,k-1,\mathbf{i}_j)\otimes\mathbf{I}_p)(\mathbf{I}_{p^j}\otimes\mathbf{K}_{p^{k-1-j},p})(\mathbf{A}^{\otimes j}\otimes\mathbf{A}\otimes\mathbf{B}^{\otimes k-1-j})\\
&\qquad\times(\mathbf{I}_{n^j}\otimes\mathbf{K}_{n,n^{k-1-j}})(\mathbf{L}_n(j,k-1,\mathbf{i}_j)'\otimes\mathbf{I}_n).
\end{aligned}
$$

By assumption

$$
\begin{aligned}
&(\mathbf{A}+\mathbf{B})^{\otimes k}=\{(\mathbf{A}+\mathbf{B})^{\otimes k-1}\}\otimes(\mathbf{A}+\mathbf{B})\\
&=\{\sum_{j=1}^{k-2}\sum_{J_{j,k-1}}\mathbf{L}_p(j,k-1,\mathbf{i}_j)(\mathbf{A}^{\otimes j}\otimes\mathbf{B}^{\otimes k-1-j})\mathbf{L}_n(j,k-1,\mathbf{i}_j)'\\
&\quad+\mathbf{A}^{\otimes k-1}+\mathbf{B}^{\otimes k-j}\}\otimes(\mathbf{A}+\mathbf{B})\\
&=\mathbf{S}_1+\mathbf{S}_2+\mathbf{A}^{\otimes k-1}\otimes\mathbf{B}+\mathbf{K}_{p^{k-1},p}(\mathbf{A}\otimes\mathbf{B}^{\otimes k-1})\mathbf{K}_{n,n^{k-1}}+\mathbf{A}^{\otimes k}+\mathbf{B}^{\otimes k}, \quad (1.3.25)
\end{aligned}
$$

where

$$
\begin{aligned}
\mathbf{S}_1=&\sum_{j=1}^{k-2}\sum_{J_{j,k-1}}(\mathbf{L}_p(j,k-1,\mathbf{i}_j)\otimes\mathbf{I}_p)(\mathbf{I}_{p^j}\otimes\mathbf{K}_{p^{k-1-j},p})(\mathbf{A}^{\otimes j+1}\otimes\mathbf{B}^{\otimes k-1-j})\\
&\times(\mathbf{I}_{n^j}\otimes\mathbf{K}_{n,n^{k-1-j}})(\mathbf{L}_n(j,k-1,\mathbf{i}_j)'\otimes\mathbf{I}_n)
\end{aligned}
$$

and

$$
\mathbf{S}_2=\sum_{j=1}^{k-2}\sum_{J_{j,k-1}}(\mathbf{L}_p(j,k-1,\mathbf{i}_j)\otimes\mathbf{I}_p)(\mathbf{A}^{\otimes j}\otimes\mathbf{B}^{\otimes k-j})(\mathbf{L}_n(j,k-1,\mathbf{i}_j)'\otimes\mathbf{I}_n).
$$

If the indices in $\mathbf{S}_1$ are altered, i.e. $j+1\to j$, then

$$
\begin{aligned}
\mathbf{S}_1=&\sum_{j=2}^{k-1}\sum_{J_{j-1,k-1}}(\mathbf{L}_p(j-1,k-1,\mathbf{i}_{j-1})\otimes\mathbf{I}_p)(\mathbf{I}_{p^{j-1}}\otimes\mathbf{K}_{p^{k-j},p})(\mathbf{A}^{\otimes j}\otimes\mathbf{B}^{\otimes k-j})\\
&\times(\mathbf{I}_{n^{j-1}}\otimes\mathbf{K}_{n,n^{k-j}})(\mathbf{L}_n(j-1,k-1,\mathbf{i}_{j-1})'\otimes\mathbf{I}_n).
\end{aligned}
$$

Let $i_j=k$, which implies that we may replace $(\mathbf{I}_{s^{j-1}}\otimes\mathbf{K}_{s^{k-j},s})$ by $(\mathbf{I}_{s^{j-1}}\otimes\mathbf{K}_{s^{i_j-j},s})$ and obtain

$$
(\mathbf{L}_s(j-1,k-1,\mathbf{i}_{j-1})\otimes\mathbf{I}_s)(\mathbf{I}_{s^{j-1}}\otimes\mathbf{K}_{s^{k-j},s})=\mathbf{L}_s(j,k,\mathbf{i}_j).
$$

Hence,

$$
\mathbf{S}_1=\sum_{j=2}^{k-1}\sum_{i_j=k}^{k}\sum_{J_{j-1,k-1}}\mathbf{L}_p(j,k,\mathbf{i}_j)(\mathbf{A}^{\otimes j}\otimes\mathbf{B}^{\otimes k-j})\mathbf{L}_n(j,k,\mathbf{i}_j)'.
$$

Furthermore,

$$\mathbf{L}_s(j,k-1,\mathbf{i}_j)\otimes\mathbf{I}_s=\mathbf{L}_s(j,k,\mathbf{i}_j)$$

and

$$\mathbf{S}_2=\sum_{j=1}^{k-2}\sum_{J_{j,k-1}}\mathbf{L}_p(j,k,\mathbf{i}_j)(\mathbf{A}^{\otimes j}\otimes\mathbf{B}^{\otimes k-j})\mathbf{L}_n(j,k,\mathbf{i}_j)'.$$

From the expressions of the index sets $J_{j-1,k-1}$ and $J_{j,k}$ it follows that

$$\mathbf{S}_1+\mathbf{S}_2=\sum_{j=2}^{k-2}\sum_{J_{j,k}}\mathbf{L}_p(j,k,\mathbf{i}_j)(\mathbf{A}^{\otimes j}\otimes\mathbf{B}^{\otimes k-j})\mathbf{L}_n(j,k,\mathbf{i}_j)'+\mathbf{S}_3+\mathbf{S}_4, \quad (1.3.26)$$

where

$$\mathbf{S}_3=\sum_{j=k-1}^{k-1}\sum_{i_j=k}^{k}\sum_{J_{j-1,k-1}}\mathbf{L}_p(j,k,\mathbf{i}_j)(\mathbf{A}^{\otimes j}\otimes\mathbf{B}^{\otimes k-j})\mathbf{L}_n(j,k,\mathbf{i}_j)'$$

and

$$\mathbf{S}_4=\sum_{j=1}^{1}\sum_{J_{j,k-1}}\mathbf{L}_p(j,k,\mathbf{i}_j)(\mathbf{A}^{\otimes j}\otimes\mathbf{B}^{\otimes k-j})\mathbf{L}_n(j,k,\mathbf{i}_j)'.$$

However,

$$\mathbf{S}_3+\mathbf{A}^{\otimes k-1}\otimes\mathbf{B}=\sum_{j=k-1}^{k-1}\sum_{J_{j,k}}\mathbf{L}_p(j,k,\mathbf{i}_j)(\mathbf{A}^{\otimes k-1}\otimes\mathbf{B})\mathbf{L}_n(j,k,\mathbf{i}_j)' \quad (1.3.27)$$

and

$$\mathbf{S}_4+\mathbf{K}_{p^{k-1},p}(\mathbf{A}^{\otimes k-1}\otimes\mathbf{B})\mathbf{K}_{n,n^{k-1}}=\sum_{j=1}^{1}\sum_{J_{j,k}}\mathbf{L}_p(j,k,\mathbf{i}_j)(\mathbf{A}\otimes\mathbf{B}^{k-j})\mathbf{L}_n(j,k,\mathbf{i}_j)'. \quad (1.3.28)$$

Hence, by summing the expressions in (1.3.26), (1.3.27) and (1.3.28), it follows from (1.3.25) that the theorem is established. ∎

The matrix $\mathbf{L}_s(j,k,\mathbf{i}_j)$ in the theorem is a permutation operator acting on $(\mathbf{A}^{\otimes j})\otimes(\mathbf{B}^{\otimes k-j})$. For each j the number of permutations acting on

$$(\mathbf{A}^{\otimes j})\otimes(\mathbf{B}^{\otimes k-j})$$

equals $\binom{k}{j}$. Moreover, $\mathbf{A}^{\otimes k}$ and $\mathbf{B}^{\otimes k}$ are the special cases, when $j=k$ and $j=0$, respectively. Instead of the expression presented in the lemma, we may write

$$(\mathbf{A}+\mathbf{B})^{\otimes k}=\sum_{j=0}^{k}\sum_{\sigma\in S_{k,j}^{p,n}}\sigma(\mathbf{A}^{\otimes j}\otimes\mathbf{B}^{\otimes k-j}),$$

where $S_{k,j}^{p,n}$ is a certain set of permutations. An analogue of Theorem 1.3.3 for the more general relation $(\sum_i\mathbf{A}_i)^{\otimes k}$ has been derived by Holmquist (1985a).

1.3.4 vec-*operator*

Besides the direct product, the vec-operator is the second basic tool from "newer" matrix algebra in multivariate statistical analysis.

Definition 1.3.5. *Let* $\mathbf{A} = (\mathbf{a}_1, \ldots, \mathbf{a}_q)$ *be* $p \times q-$*matrix, where* $\mathbf{a}_i$, $i = 1, \ldots, q$, *is the* $i-$*th column vector. The vectorization operator* $\mathrm{vec}(\cdot)$ *is an operator from* $\mathbb{R}^{p\times q}$ *to* $\mathbb{R}^{pq}$, *with*

$$\mathrm{vec}\mathbf{A} = \begin{pmatrix} \mathbf{a}_1 \\ \vdots \\ \mathbf{a}_q \end{pmatrix}.$$

∎

As for the commutation matrix (Definition 1.3.2 and Proposition 1.3.13), there exist possibilities for alternative equivalent definitions of the vec-operator, for instance:

$$\mathrm{vec} : \mathbb{R}^{p\times q} \to \mathbb{R}^{pq}, \quad \mathrm{vec}(\mathbf{ab}') = \mathbf{b} \otimes \mathbf{a} \quad \forall \mathbf{a} \in \mathbb{R}^p, \forall \mathbf{b} \in \mathbb{R}^q. \tag{1.3.28}$$

According to (1.3.28) we can write, through unit basis vectors $\mathbf{d}_i$ and $\mathbf{e}_j$,

$$\mathrm{vec}\mathbf{A} = \sum_{ij} a_{ij}\mathbf{e}_j \otimes \mathbf{d}_i, \quad \mathbf{e}_j \in \mathbb{R}^q, \ \mathbf{d}_i \in \mathbb{R}^p, \tag{1.3.29}$$

since $\mathbf{A} = \sum_{ij} a_{ij}\mathbf{d}_i\mathbf{e}_j'$. Many results can be proved by combining (1.3.29) and Proposition 1.3.12. Moreover, the vec-operator is linear, and there exists a unique inverse operator vec^{-1} such that for any vectors $\mathbf{e}, \mathbf{d}$

$$\mathrm{vec}^{-1}(\mathbf{e} \otimes \mathbf{d}) = \mathbf{de}'.$$

There exist direct generalizations which act on general Kroneckerian powers of vectors. We refer to Holmquist (1985b), where both generalized commutation matrices and generalized vec-operators are handled.
The idea of representing a matrix as a long vector consisting of its columns appears the first time in Sylvester (1884). The notation "vec" was introduced by Koopmans, Rubin & Leipnik (1950). Its regular use in statistical publications started in the late 1960s. In the following we give basic properties of the vec-operator. Several proofs of the statements can be found in the book by Magnus & Neudecker (1999), and the rest are straightforward consequences of the definitions of the vec-operator and the commutation matrix.

Proposition 1.3.14.
(i) *Let* $\mathbf{A} : p \times q$. *Then*

$$\mathbf{K}_{p,q}\mathrm{vec}\mathbf{A} = \mathrm{vec}\mathbf{A}'. \tag{1.3.30}$$

(ii) *Let* $\mathbf{A} : p \times q$, $\mathbf{B} : q \times r$ *and* $\mathbf{C} : r \times s$. *Then*

$$\mathrm{vec}(\mathbf{ABC}) = (\mathbf{C}' \otimes \mathbf{A})\mathrm{vec}\mathbf{B}. \tag{1.3.31}$$

(iii) *Let* $\mathbf{A} : p \times q$, $\mathbf{B} : q \times r$, $\mathbf{C} : r \times s$ *and* $\mathbf{D} : s \times p$. *Then*

$$\begin{aligned} \mathrm{tr}(\mathbf{AB}) &= \mathrm{vec}'\mathbf{A}'\mathrm{vec}\mathbf{B}, \quad r = p; \qquad (1.3.32)\\ \mathrm{tr}(\mathbf{ABCD}) &= \mathrm{vec}'\mathbf{A}(\mathbf{B} \otimes \mathbf{D}')\mathrm{vec}\mathbf{C}' = \mathrm{vec}'\mathbf{A}\mathrm{vec}(\mathbf{D}'\mathbf{C}'\mathbf{B}');\\ \mathrm{tr}(\mathbf{ABCD}) &= (\mathrm{vec}'(\mathbf{C}') \otimes \mathrm{vec}'\mathbf{A})(\mathbf{I}_r \otimes \mathbf{K}_{s,q} \otimes \mathbf{I}_p)(\mathrm{vec}\mathbf{B} \otimes \mathrm{vec}\mathbf{D}');\\ \mathrm{tr}(\mathbf{ABCD}) &= (\mathrm{vec}'\mathbf{B} \otimes \mathrm{vec}'\mathbf{D})\mathbf{K}_{r,pqs}(\mathrm{vec}\mathbf{A} \otimes \mathrm{vec}\mathbf{C}). \end{aligned}$$

(iv) *Let* $\mathbf{A} : p \times q$ *and* $\mathbf{B} : r \times s$. *Then*

$$\begin{aligned}
\mathrm{vec}(\mathbf{A} \otimes \mathbf{B}) &= (\mathbf{I}_q \otimes \mathbf{K}_{s,p} \otimes \mathbf{I}_r)(\mathrm{vec}\mathbf{A} \otimes \mathrm{vec}\mathbf{B}); \quad (1.3.33)\\
(\mathbf{I}_q \otimes \mathbf{K}_{p,s} \otimes \mathbf{I}_r)\mathrm{vec}(\mathbf{A} \otimes \mathbf{B}) &= (\mathrm{vec}\mathbf{A} \otimes \mathrm{vec}\mathbf{B});\\
\mathbf{K}_{r,qps}(\mathrm{vec}\mathbf{A} \otimes \mathrm{vec}\mathbf{B}) &= \mathrm{vec}(\mathbf{K}_{p,s}(\mathbf{B}' \otimes \mathbf{A}));\\
(\mathbf{I}_r \otimes \mathbf{K}_{p,q} \otimes \mathbf{I}_s)\mathbf{K}_{r,qps}(\mathrm{vec}\mathbf{A} \otimes \mathrm{vec}\mathbf{B}) &= \mathrm{vec}(\mathbf{K}_{q,s}(\mathbf{B}' \otimes \mathbf{A}')).
\end{aligned}$$

(v) *Let* $\mathbf{A} : p \times q$, $\mathbf{B} : r \times s$ *and set*

$$\begin{aligned}
\mathbf{G}_1 =& (\mathbf{I}_q \otimes \mathbf{K}_{s,p} \otimes \mathbf{I}_r)\mathbf{K}_{pq,rs},\\
\mathbf{G}_2 =& (\mathbf{I}_p \otimes \mathbf{K}_{q,r} \otimes \mathbf{I}_s)\mathbf{K}_{pr,qs},\\
\mathbf{G}_3 =& (\mathbf{I}_q \otimes \mathbf{K}_{s,r} \otimes \mathbf{I}_p)\mathbf{K}_{qr,sp}.
\end{aligned}$$

Then

$$\begin{aligned}
&\mathbf{G}_i^{-1} = \mathbf{G}_i' \quad i = 1, 2, 3;\\
&\mathbf{G}_1(\mathrm{vec}\mathbf{B} \otimes \mathrm{vec}\mathbf{A}) = \mathrm{vec}(\mathbf{A} \otimes \mathbf{B});\\
&\mathbf{G}_2\mathrm{vec}(\mathbf{A} \otimes \mathbf{B}) = \mathrm{vec}\mathbf{A}' \otimes \mathrm{vec}\mathbf{B}';\\
&\mathbf{G}_3\mathrm{vec}(\mathbf{K}_{q,r}(\mathbf{B} \otimes \mathbf{A}')) = \mathrm{vec}(\mathbf{K}_{r,p}(\mathbf{A} \otimes \mathbf{B})).
\end{aligned}$$

(vi) *Let* $\mathbf{A}_i : p \times q$, $\mathbf{B}_i : r \times s$, $\mathbf{C}_i : q \times s$, $\mathbf{D}_i : p \times r$, $\mathbf{E}_i : p \times s$ *and* $\mathbf{F}_i : q \times r$. *Then the equation*

$$\sum_i (\mathbf{A}_i \otimes \mathbf{B}_i) = \sum_i \mathrm{vec}\mathbf{D}_i'\mathrm{vec}'(\mathbf{C}_i') + \mathbf{K}_{p,r}(\mathbf{F}_i' \otimes \mathbf{E}_i)$$

is equivalent to

$$\sum_i \mathrm{vec}\mathbf{A}_i\mathrm{vec}'\mathbf{B}_i = \sum_i \mathbf{C}_i \otimes \mathbf{D}_i + \mathbf{K}_{q,p}(\mathbf{E}_i' \otimes \mathbf{F}_i).$$

∎

The relation in (i) is often used as a definition of the commutation matrix. Furthermore, note that the first equality of (v) means that $\mathbf{G}_i$, $i = 1, 2, 3$, are orthogonal matrices, and that from (vi) it follows that the equation

$$\mathbf{A} \otimes \mathbf{B} = \mathrm{vec}\mathbf{D}'\mathrm{vec}'(\mathbf{C}') + \mathbf{K}_{p,r}(\mathbf{F}' \otimes \mathbf{E})$$

is equivalent to

$$\mathrm{vec}\mathbf{A}\mathrm{vec}'\mathbf{B} = \mathbf{C} \otimes \mathbf{D} + \mathbf{K}_{q,p}(\mathbf{E}' \otimes \mathbf{F}).$$

The next property enables us to present the direct product of matrices through the vec-representation of those matrices.

Proposition 1.3.15. *Let* $\mathbf{A} : r \times s$, $\mathbf{B} : s \times t$, $\mathbf{C} : m \times n$ *and* $\mathbf{D} : n \times p$. *Then*

$$\mathbf{AB} \otimes \mathbf{CD} = (\mathbf{I}_{rm} \otimes \mathrm{vec}'\mathbf{D}')(\mathbf{I}_r \otimes \mathrm{vec}\mathbf{C}'\mathrm{vec}'\mathbf{B}' \otimes \mathbf{I}_p)(\mathrm{vec}\mathbf{A}' \otimes \mathbf{I}_{pt}).$$

∎

1.3.5 *Linear equations*

Matrix equations play an important role in many fields. We will give a fairly simple approach to solve linear equations which is based on some results about vector space decompositions of tensor spaces given in §1.2.5. In (1.1.13) we considered $\mathbf{Ax} = \mathbf{b}$ as an equation in $\mathbf{x}$. First it is noted that $\mathbf{Ax} = \mathbf{b}$ has a solution, i.e. it is consistent, if and only if $\mathbf{b} \in \mathcal{C}(\mathbf{A})$. Furthermore, any solution of a linear equation system $\mathbf{Ax} = \mathbf{b}$ consists of two parts, namely, a particular solution $\mathbf{x}_0$ and the general solution of the homogenous equation $\mathbf{Ax} = \mathbf{0}$. By Definition 1.1.3 of a g-inverse we observe that if the equation is consistent, $\mathbf{x} = \mathbf{A}^-\mathbf{b}$ is a particular solution. Moreover, $\mathbf{x} = (\mathbf{I} - \mathbf{A}^-\mathbf{A})\mathbf{z}$, where $\mathbf{z}$ is arbitrary, is a general solution of $\mathbf{Ax} = \mathbf{0}$. To see this, note that by Theorem 1.1.5, $(\mathbf{I} - \mathbf{A}^-\mathbf{A})\mathbf{z}$ is a solution of $\mathbf{Ax} = \mathbf{0}$. For the opposite, suppose that $\underline{\mathbf{x}}$ is a solution of $\mathbf{Ax} = \mathbf{0}$. This implies that $\underline{\mathbf{x}} = (\mathbf{I} - \mathbf{A}^-\mathbf{A})\underline{\mathbf{x}}$ and thus we have found a $\mathbf{z}$ such that $\underline{\mathbf{x}} = (\mathbf{I} - \mathbf{A}^-\mathbf{A})\mathbf{z}$. Hence, it can be stated that the general solution of

$$\mathbf{Ax} = \mathbf{b}$$

equals

$$\mathbf{x} = \mathbf{A}^-\mathbf{b} + (\mathbf{I} - \mathbf{A}^-\mathbf{A})\mathbf{z},$$

which is a well-known and a commonly applied solution. However, $\mathbf{z}$ is of the same size as $\mathbf{x}$, which is unnatural since all the solutions to $\mathbf{Ax} = \mathbf{0}$ can be generated by a smaller number of arbitrary elements than the size of $\mathbf{x}$. Keeping this in mind, it is observed that $\mathbf{Ax} = \mathbf{0}$ means that $\mathbf{x} \in \mathcal{C}(\mathbf{A}')^{\perp}$. Hence, all solutions to $\mathbf{Ax} = \mathbf{0}$ are given by the orthogonal complement to $\mathcal{C}(\mathbf{A}')$, which leads us to the relation

$$\mathbf{x} = (\mathbf{A}')^o\mathbf{z},$$

where $\mathbf{z}$ is arbitrary. By a proper choice of $(\mathbf{A}')^o$ it is possible to generate all solutions with a minimum number of free elements in $\mathbf{z}$.

Now we are going to consider matrix extensions of the above equation:

$$\mathbf{AXB} = \mathbf{C},$$

and

$$\mathbf{A}_i\mathbf{XB}_i = \mathbf{C}_i, \quad i = 1, 2.$$

Furthermore, the solutions of these equations will be utilized when the equation

$$\mathbf{A}_1\mathbf{X}_1\mathbf{B}_1 + \mathbf{A}_2\mathbf{X}_2\mathbf{B}_2 = \mathbf{0}$$

is solved. It is interesting to compare our approach with the one given by Rao & Mitra (1971, Section 2.3) where some related results are also presented.

Theorem 1.3.4. *A representation of the general solution of the consistent equation in* $\mathbf{X}$*:*

$$\mathbf{AXB} = \mathbf{C}$$

is given by any of the following three formulas:

$$\mathbf{X} = \mathbf{X}_0 + (\mathbf{A}')^o\mathbf{Z}_1\mathbf{B}' + \mathbf{A}'\mathbf{Z}_2\mathbf{B}^{o'} + (\mathbf{A}')^o\mathbf{Z}_3\mathbf{B}^{o'};$$

$$\mathbf{X} = \mathbf{X}_0 + (\mathbf{A}')^o\mathbf{Z}_1 + \mathbf{A}'\mathbf{Z}_2\mathbf{B}^{o'};$$

$$\mathbf{X} = \mathbf{X}_0 + \mathbf{Z}_1\mathbf{B}^{o'} + (\mathbf{A}')^o\mathbf{Z}_2\mathbf{B}',$$

where $\mathbf{X}_0$ *is a particular solution and* $\mathbf{Z}_i$, $i = 1, 2, 3$, *stand for arbitrary matrices of proper sizes.*

PROOF: Since $\mathbf{AXB} = \mathbf{0}$ is equivalent to $(\mathbf{B}' \otimes \mathbf{A})\mathrm{vec}\mathbf{X} = \mathbf{0}$, we are going to consider $\mathcal{C}(\mathbf{B} \otimes \mathbf{A}')^\perp$. A direct application of Theorem 1.2.20 (ii) yields

$$\begin{aligned}\mathcal{C}(\mathbf{B} \otimes \mathbf{A}')^\perp &= \mathcal{C}(\mathbf{B}^o \otimes \mathbf{A}') \boxplus \mathcal{C}(\mathbf{B} \otimes (\mathbf{A}')^o) \boxplus \mathcal{C}(\mathbf{B}^o \otimes (\mathbf{A}')^o)\\ &= \mathcal{C}(\mathbf{B}^o \otimes \mathbf{I}) \boxplus \mathcal{C}(\mathbf{B} \otimes (\mathbf{A}')^o)\\ &= \mathcal{C}(\mathbf{B}^o \otimes \mathbf{A}') \boxplus \mathcal{C}(\mathbf{I} \otimes (\mathbf{A}')^o).\end{aligned}$$

Hence

$$\mathrm{vec}\mathbf{X} = (\mathbf{B}^o \otimes \mathbf{A}')\mathrm{vec}\mathbf{Z}_1 + (\mathbf{B} \otimes (\mathbf{A}')^o)\mathrm{vec}\mathbf{Z}_2 + (\mathbf{B}^o \otimes (\mathbf{A}')^o)\mathrm{vec}\mathbf{Z}_3$$

or

$$\mathrm{vec}\mathbf{X} = (\mathbf{B}^o \otimes \mathbf{A}')\mathrm{vec}\mathbf{Z}_1 + (\mathbf{I} \otimes (\mathbf{A}')^o)\mathrm{vec}\mathbf{Z}_2$$

or

$$\mathrm{vec}\mathbf{X} = (\mathbf{B}^o \otimes \mathbf{I})\mathrm{vec}\mathbf{Z}_1 + (\mathbf{B} \otimes (\mathbf{A}')^o)\mathrm{vec}\mathbf{Z}_2,$$

which are equivalent to the statements of the theorem. ∎

Theorem 1.3.5. *The equation* $\mathbf{AXB} = \mathbf{C}$ *is consistent if and only if* $\mathcal{C}(\mathbf{C}) \subseteq \mathcal{C}(\mathbf{A})$ *and* $\mathcal{C}(\mathbf{C}') \subseteq \mathcal{C}(\mathbf{B}')$*. A particular solution of the equation is given by*

$$\mathbf{X}_0 = \mathbf{A}^-\mathbf{C}\mathbf{B}^-.$$

PROOF: By Proposition 1.2.2 (i), $\mathcal{C}(\mathbf{C}) \subseteq \mathcal{C}(\mathbf{A})$ and $\mathcal{C}(\mathbf{C}') \subseteq \mathcal{C}(\mathbf{B}')$ hold if $\mathbf{AXB} = \mathbf{C}$ and so the conditions are necessary. To prove sufficiency assume that $\mathcal{C}(\mathbf{C}) \subseteq \mathcal{C}(\mathbf{A})$ and $\mathcal{C}(\mathbf{C}') \subseteq \mathcal{C}(\mathbf{B}')$ are true. Then $\mathbf{AXB} = \mathbf{C}$ is consistent since a particular solution is given by $\mathbf{X}_0$ of the theorem. ∎

The next equation has been considered in many papers, among others by Mitra (1973, 1990), Shinozaki & Sibuya (1974) and Baksalary & Kala (1980).

Theorem 1.3.6. *A representation of the general solution of the consistent equations in* $\mathbf{X}$*:*

$$\begin{cases} \mathbf{A}_1\mathbf{X}\mathbf{B}_1 = \mathbf{C}_1 \\ \mathbf{A}_2\mathbf{X}\mathbf{B}_2 = \mathbf{C}_2 \end{cases}$$

is given by

$$\begin{aligned} \mathbf{X} =& \mathbf{X}_0 + \mathbf{T}_3\mathbf{Z}_1\mathbf{S}_1' + \mathbf{T}_4\mathbf{Z}_2\mathbf{S}_1' + \mathbf{T}_4\mathbf{Z}_3\mathbf{S}_2' + \mathbf{T}_1\mathbf{Z}_4\mathbf{S}_3' + \mathbf{T}_4\mathbf{Z}_5\mathbf{S}_3' \\ &+ \mathbf{T}_1\mathbf{Z}_6\mathbf{S}_4' + \mathbf{T}_2\mathbf{Z}_7\mathbf{S}_4' + \mathbf{T}_3\mathbf{Z}_8\mathbf{S}_4' + \mathbf{T}_4\mathbf{Z}_9\mathbf{S}_4' \end{aligned}$$

or

$$\mathbf{X} = \mathbf{X}_0 + (\mathbf{A}_2')^o\mathbf{Z}_1\mathbf{S}_1' + (\mathbf{A}_1' : \mathbf{A}_2')^0\mathbf{Z}_2\mathbf{S}_2' + (\mathbf{A}_1')^o\mathbf{Z}_3\mathbf{S}_3' + \mathbf{Z}_4\mathbf{S}_4'$$

or

$$\mathbf{X} = \mathbf{X}_0 + \mathbf{T}_1\mathbf{Z}_1\mathbf{B}_2^{o'} + \mathbf{T}_2\mathbf{Z}_2(\mathbf{B}_1 : \mathbf{B}_2)^{o'} + \mathbf{T}_3\mathbf{Z}_3\mathbf{B}_1^{o'} + \mathbf{T}_4\mathbf{Z}_4,$$

where $\mathbf{X}_0$ *is a particular solution,* $\mathbf{Z}_i$, $i = 1, \ldots, 9$, *are arbitrary matrices of proper sizes and* $\mathbf{S}_i$, $\mathbf{T}_i$, $i = 1, \ldots, 4$, *are any matrices satisfying the conditions*

$$\begin{aligned} C(\mathbf{S}_1) &= C(\mathbf{B}_1 : \mathbf{B}_2) \cap C(\mathbf{B}_1)^\perp, & C(\mathbf{T}_1) &= C(\mathbf{A}_1' : \mathbf{A}_2') \cap C(\mathbf{A}_1')^\perp, \\ C(\mathbf{S}_2) &= C(\mathbf{B}_1) \cap C(\mathbf{B}_2), & C(\mathbf{T}_2) &= C(\mathbf{A}_1') \cap C(\mathbf{A}_2'), \\ C(\mathbf{S}_3) &= C(\mathbf{B}_1 : \mathbf{B}_2) \cap C(\mathbf{B}_2)^\perp, & C(\mathbf{T}_3) &= C(\mathbf{A}_1' : \mathbf{A}_2') \cap C(\mathbf{A}_2')^\perp, \\ C(\mathbf{S}_4) &= C(\mathbf{B}_1 : \mathbf{B}_2)^\perp, & C(\mathbf{T}_4) &= C(\mathbf{A}_1' : \mathbf{A}_2')^\perp. \end{aligned}$$

Proof: The proof follows immediately from Theorem 1.2.20 (iv) and similar considerations to those given in the proof of Theorem 1.3.4. ∎

Corollary 1.3.6.1. *If* $C(\mathbf{B}_1) \subseteq C(\mathbf{B}_2)$ *and* $C(\mathbf{A}_2') \subseteq C(\mathbf{A}_1')$ *hold, then*

$$\mathbf{X} = \mathbf{X}_0 + (\mathbf{A}_2' : (\mathbf{A}_1')^o)^o\mathbf{Z}_1\mathbf{B}_1^{o'} + \mathbf{A}_2'\mathbf{Z}_2\mathbf{B}_2^{o'} + (\mathbf{A}_1')^o\mathbf{Z}_3.$$

∎

Corollary 1.3.6.2. *If* $\mathbf{A}_1 = \mathbf{I}$ *and* $\mathbf{B}_2 = \mathbf{I}$ *hold, then*

$$\mathbf{X} = \mathbf{X}_0 + (\mathbf{A}_2')^o\mathbf{Z}_1\mathbf{B}_1^{0'}.$$

∎

The next theorem is due to Mitra (1973) (see also Shinozaki & Sibuya, 1974).

Theorem 1.3.7. *The equation system*

$$\begin{cases} \mathbf{A}_1\mathbf{X}\mathbf{B}_1 = \mathbf{C}_1 \\ \mathbf{A}_2\mathbf{X}\mathbf{B}_2 = \mathbf{C}_2 \end{cases} \qquad (1.3.34)$$

is consistent if and only if $\mathbf{A}_1\mathbf{X}\mathbf{B}_1 = \mathbf{C}_1$ *and* $\mathbf{A}_2\mathbf{X}\mathbf{B}_2 = \mathbf{C}_2$ *are consistent and*

$$\begin{aligned} &\mathbf{A}_1'\mathbf{A}_1(\mathbf{A}_1'\mathbf{A}_1 + \mathbf{A}_2'\mathbf{A}_2)^-\mathbf{A}_2'\mathbf{C}_2\mathbf{B}_2'(\mathbf{B}_1\mathbf{B}_1' + \mathbf{B}_2\mathbf{B}_2')^-\mathbf{B}_1\mathbf{B}_1' \\ &\quad = \mathbf{A}_2'\mathbf{A}_2(\mathbf{A}_2'\mathbf{A}_2 + \mathbf{A}_1'\mathbf{A}_1)^-\mathbf{A}_1'\mathbf{C}_1\mathbf{B}_1'(\mathbf{B}_2\mathbf{B}_2' + \mathbf{B}_1\mathbf{B}_1')^-\mathbf{B}_2\mathbf{B}_2'. \end{aligned} \qquad (1.3.35)$$

A particular solution is given by

$$\begin{aligned}\mathbf{X}_0 = & (\mathbf{A}_2'\mathbf{A}_2 + \mathbf{A}_1'\mathbf{A}_1)^-\mathbf{A}_1'\mathbf{C}_1\mathbf{B}_1'(\mathbf{B}_2\mathbf{B}_2' + \mathbf{B}_1\mathbf{B}_1')^- \\ & + (\mathbf{A}_2'\mathbf{A}_2 + \mathbf{A}_1'\mathbf{A}_1)^-\mathbf{A}_2'\mathbf{C}_2\mathbf{B}_2'(\mathbf{B}_2\mathbf{B}_2' + \mathbf{B}_1\mathbf{B}_1')^- \\ & + (\mathbf{A}_1'\mathbf{A}_1)^-\mathbf{A}_2'\mathbf{A}_2(\mathbf{A}_2'\mathbf{A}_2 + \mathbf{A}_1'\mathbf{A}_1)^-\mathbf{A}_1'\mathbf{C}_1\mathbf{B}_1'(\mathbf{B}_2\mathbf{B}_2' + \mathbf{B}_1\mathbf{B}_1')^- \\ & + (\mathbf{A}_2'\mathbf{A}_2)^-\mathbf{A}_1'\mathbf{A}_1(\mathbf{A}_2'\mathbf{A}_2 + \mathbf{A}_1'\mathbf{A}_1)^-\mathbf{A}_2'\mathbf{C}_2\mathbf{B}_2'(\mathbf{B}_2\mathbf{B}_2' + \mathbf{B}_1\mathbf{B}_1')^-. \quad (1.3.36)\end{aligned}$$

PROOF: First of all it is noted that the equations given by (1.3.34) are equivalent to

$$\mathbf{A}_1'\mathbf{A}_1\mathbf{X}\mathbf{B}_1\mathbf{B}_1' = \mathbf{A}_1'\mathbf{C}_1\mathbf{B}_1', \quad (1.3.37)$$

$$\mathbf{A}_2'\mathbf{A}_2\mathbf{X}\mathbf{B}_2\mathbf{B}_2' = \mathbf{A}_2'\mathbf{C}_2\mathbf{B}_2'. \quad (1.3.38)$$

After pre- and post-multiplication we obtain the following equations

$$\begin{aligned}&\mathbf{A}_2'\mathbf{A}_2(\mathbf{A}_2'\mathbf{A}_2 + \mathbf{A}_1'\mathbf{A}_1)^-\mathbf{A}_1'\mathbf{A}_1\mathbf{X}\mathbf{B}_1\mathbf{B}_1'(\mathbf{B}_2\mathbf{B}_2' + \mathbf{B}_1\mathbf{B}_1')^-\mathbf{B}_2\mathbf{B}_2' \\ &= \mathbf{A}_2'\mathbf{A}_2(\mathbf{A}_2'\mathbf{A}_2 + \mathbf{A}_1'\mathbf{A}_1)^-\mathbf{A}_1'\mathbf{C}_1\mathbf{B}_1'(\mathbf{B}_2\mathbf{B}_2' + \mathbf{B}_1\mathbf{B}_1')^-\mathbf{B}_2\mathbf{B}_2'\end{aligned}$$

and

$$\begin{aligned}&\mathbf{A}_1'\mathbf{A}_1(\mathbf{A}_2'\mathbf{A}_2 + \mathbf{A}_1'\mathbf{A}_1)^-\mathbf{A}_2'\mathbf{A}_2\mathbf{X}\mathbf{B}_2\mathbf{B}_2'(\mathbf{B}_2\mathbf{B}_2' + \mathbf{B}_1\mathbf{B}_1')^-\mathbf{B}_1\mathbf{B}_1' \\ &= \mathbf{A}_1'\mathbf{A}_1(\mathbf{A}_2'\mathbf{A}_2 + \mathbf{A}_1'\mathbf{A}_1)^-\mathbf{A}_2'\mathbf{C}_2\mathbf{B}_2'(\mathbf{B}_2\mathbf{B}_2' + \mathbf{B}_1\mathbf{B}_1')^-\mathbf{B}_1\mathbf{B}_1'.\end{aligned}$$

Since, by Lemma 1.2.8

$$\mathbf{A}_1'\mathbf{A}_1(\mathbf{A}_2'\mathbf{A}_2 + \mathbf{A}_1'\mathbf{A}_1)^-\mathbf{A}_2'\mathbf{A}_2 = \mathbf{A}_2'\mathbf{A}_2(\mathbf{A}_2'\mathbf{A}_2 + \mathbf{A}_1'\mathbf{A}_1)^-\mathbf{A}_1'\mathbf{A}_1$$

and

$$\mathbf{B}_2\mathbf{B}_2'(\mathbf{B}_2\mathbf{B}_2' + \mathbf{B}_1\mathbf{B}_1')^-\mathbf{B}_1\mathbf{B}_1' = \mathbf{B}_1\mathbf{B}_1'(\mathbf{B}_2\mathbf{B}_2' + \mathbf{B}_1\mathbf{B}_1')^-\mathbf{B}_2\mathbf{B}_2'$$

the condition in (1.3.35) must hold. The choice of g-inverse in the parallel sum is immaterial.

In the next it is observed that from Theorem 1.3.5 it follows that if (1.3.35) holds then

$$\begin{aligned}&(\mathbf{A}_2'\mathbf{A}_2)^-\mathbf{A}_1'\mathbf{A}_1(\mathbf{A}_2'\mathbf{A}_2 + \mathbf{A}_1'\mathbf{A}_1)^-\mathbf{A}_2'\mathbf{C}_2\mathbf{B}_2'(\mathbf{B}_2\mathbf{B}_2' + \mathbf{B}_1\mathbf{B}_1')^-\mathbf{B}_1\mathbf{B}_1' \\ &= (\mathbf{A}_2'\mathbf{A}_2 + \mathbf{A}_1'\mathbf{A}_1)^-\mathbf{A}_1'\mathbf{C}_1\mathbf{B}_1'(\mathbf{B}_2\mathbf{B}_2' + \mathbf{B}_1\mathbf{B}_1')^-\mathbf{B}_2\mathbf{B}_2'\end{aligned}$$

and

$$\begin{aligned}&(\mathbf{A}_1'\mathbf{A}_1)^-\mathbf{A}_2'\mathbf{A}_2(\mathbf{A}_2'\mathbf{A}_2 + \mathbf{A}_1'\mathbf{A}_1)^-\mathbf{A}_1'\mathbf{C}_1\mathbf{B}_1'(\mathbf{B}_2\mathbf{B}_2' + \mathbf{B}_1\mathbf{B}_1')^-\mathbf{B}_2\mathbf{B}_2' \\ &= (\mathbf{A}_2'\mathbf{A}_2 + \mathbf{A}_1'\mathbf{A}_1)^-\mathbf{A}_2'\mathbf{C}_2\mathbf{B}_2'(\mathbf{B}_2\mathbf{B}_2' + \mathbf{B}_1\mathbf{B}_1')^-\mathbf{B}_1\mathbf{B}_1'.\end{aligned}$$

Under these conditions it follows immediately that $\mathbf{X}_0$ given by (1.3.36) is a solution. Let us show that $\mathbf{A}_1'\mathbf{A}_1\mathbf{X}_0\mathbf{B}_1\mathbf{B}_1' = \mathbf{A}_1'\mathbf{C}_1\mathbf{B}_1'$:

$$\begin{aligned}&\mathbf{A}_1'\mathbf{A}_1\mathbf{X}_0\mathbf{B}_1\mathbf{B}_1' \\ &\quad = \mathbf{A}_1'\mathbf{A}_1(\mathbf{A}_2'\mathbf{A}_2 + \mathbf{A}_1'\mathbf{A}_1)^-\mathbf{A}_1'\mathbf{C}_1\mathbf{B}_1'(\mathbf{B}_2\mathbf{B}_2' + \mathbf{B}_1\mathbf{B}_1')^-\mathbf{B}_1\mathbf{B}_1' \\ &\quad\quad + \mathbf{A}_2'\mathbf{A}_2(\mathbf{A}_2'\mathbf{A}_2 + \mathbf{A}_1'\mathbf{A}_1)^-\mathbf{A}_1'\mathbf{C}_1\mathbf{B}_1'(\mathbf{B}_2\mathbf{B}_2' + \mathbf{B}_1\mathbf{B}_1')^-\mathbf{B}_2\mathbf{B}_2' \\ &\quad\quad + \mathbf{A}_1'\mathbf{A}_1(\mathbf{A}_1'\mathbf{A}_1)^-\mathbf{A}_2'\mathbf{A}_2(\mathbf{A}_2'\mathbf{A}_2 + \mathbf{A}_1'\mathbf{A}_1)^-\mathbf{A}_1'\mathbf{C}_1\mathbf{B}_1'(\mathbf{B}_2\mathbf{B}_2' + \mathbf{B}_1\mathbf{B}_1')^-\mathbf{B}_1\mathbf{B}_1' \\ &\quad\quad + \mathbf{A}_1'\mathbf{A}_1(\mathbf{A}_2'\mathbf{A}_2 + \mathbf{A}_1'\mathbf{A}_1)^-\mathbf{A}_2'\mathbf{C}_2\mathbf{B}_2'(\mathbf{B}_2\mathbf{B}_2' + \mathbf{B}_1\mathbf{B}_1')^-\mathbf{B}_2\mathbf{B}_2' \\ &\quad = \mathbf{A}_1'\mathbf{C}_1\mathbf{B}_1'(\mathbf{B}_2\mathbf{B}_2' + \mathbf{B}_1\mathbf{B}_1')^-\mathbf{B}_1\mathbf{B}_1' + \mathbf{A}_1'\mathbf{C}_1\mathbf{B}_1'(\mathbf{B}_2\mathbf{B}_2' + \mathbf{B}_1\mathbf{B}_1')^-\mathbf{B}_2\mathbf{B}_2' \\ &\quad = \mathbf{A}_1'\mathbf{C}_1\mathbf{B}_1'.\end{aligned}$$

∎

Theorem 1.3.8. *A representation of the general solution of*

$$\mathbf{A}_1\mathbf{X}_1\mathbf{B}_1 + \mathbf{A}_2\mathbf{X}_2\mathbf{B}_2 = \mathbf{0}$$

is given by

$$\begin{aligned}
\mathbf{X}_1 &= -\mathbf{A}_1^{-}\mathbf{A}_2(\mathbf{A}_2'\mathbf{A}_1^o : (\mathbf{A}_2')^o)^o\mathbf{Z}_3(\mathbf{B}_2(\mathbf{B}_1')^o)^{o'}\mathbf{B}_2\mathbf{B}_1^{-} + (\mathbf{A}_1')^o\mathbf{Z}_1 + \mathbf{A}_1'\mathbf{Z}_2\mathbf{B}_1^{o'},\\
\mathbf{X}_2 &= (\mathbf{A}_2'\mathbf{A}_1^o : (\mathbf{A}_2')^o)^o\mathbf{Z}_3(\mathbf{B}_2(\mathbf{B}_1')^o)^{o'} + \mathbf{A}_2'\mathbf{A}_1^o\mathbf{Z}_4\mathbf{B}_2^{o'} + (\mathbf{A}_2')^o\mathbf{Z}_5
\end{aligned}$$

or

$$\begin{aligned}
\mathbf{X}_1 &= -\mathbf{A}_1^{-}\mathbf{A}_2(\mathbf{A}_2'\mathbf{A}_1^o)^o(\mathbf{A}_2'\mathbf{A}_1^o)^{o'}\mathbf{A}_2'\mathbf{Z}_6(\mathbf{B}_2(\mathbf{B}_1')^o)^{o'}\mathbf{B}_2\mathbf{B}_1^{-} + (\mathbf{A}_1')^o\mathbf{Z}_1 + \mathbf{A}_1'\mathbf{Z}_2\mathbf{B}_1^{o'},\\
\mathbf{X}_2 &= (\mathbf{A}_2'\mathbf{A}_1^o)^{o'}((\mathbf{A}_2'\mathbf{A}_1^o)^{o'}\mathbf{A}_2')^o\mathbf{Z}_5 + (\mathbf{A}_2'\mathbf{A}_1^o)^o(\mathbf{A}_2'\mathbf{A}_1^o)^{o'}\mathbf{A}_2'\mathbf{Z}_6(\mathbf{B}_2(\mathbf{B}_1')^o)^{o'}\\
&\quad + \mathbf{A}_2'\mathbf{A}_1^o\mathbf{Z}_4\mathbf{B}_2^{o'},
\end{aligned}$$

where $\mathbf{Z}_i$*,* $i = 1, \ldots, 6$*, are arbitrary matrices.*

PROOF: From Theorem 1.3.5 it follows that there exists a solution if and only if

$$C(\mathbf{A}_2\mathbf{X}_2\mathbf{B}_2) \subseteq C(\mathbf{A}_1), \qquad C(\mathbf{B}_2'\mathbf{X}_2'\mathbf{A}_2') \subseteq C(\mathbf{B}_1').$$

These relations are equivalent to

$$\mathbf{A}_1^{o'}\mathbf{A}_2\mathbf{X}_2\mathbf{B}_2 = \mathbf{0}, \tag{1.3.39}$$

$$\mathbf{A}_2\mathbf{X}_2\mathbf{B}_2(\mathbf{B}_1')^o = \mathbf{0} \tag{1.3.40}$$

and from Theorem 1.3.4 and Theorem 1.3.5 it follows that

$$\mathbf{X}_1 = -\mathbf{A}_1^{-}\mathbf{A}_2\mathbf{X}_2\mathbf{B}_2\mathbf{B}_1^{-} + (\mathbf{A}_1')^o\mathbf{Z}_1 + \mathbf{A}_1'\mathbf{Z}_2\mathbf{B}_1^{o'}. \tag{1.3.41}$$

Equations (1.3.39) and (1.3.40) do not depend on $\mathbf{X}_1$, and since the assumptions of Corollary 1.3.6.1 are fulfilled, a general solution for $\mathbf{X}_2$ is obtained, which is then inserted into (1.3.41).
The alternative representation is established by applying Theorem 1.3.4 twice. When solving (1.3.39),

$$\mathbf{X}_2 = (\mathbf{A}_2'\mathbf{A}_1^o)^o\mathbf{Z}_3 + \mathbf{A}_2'\mathbf{A}_1^o\mathbf{Z}_4\mathbf{B}_2^{o'} \tag{1.3.42}$$

is obtained, and inserting (1.3.39) into (1.3.40) yields

$$\mathbf{A}_2(\mathbf{A}_2'\mathbf{A}_1^o)^o\mathbf{Z}_3\mathbf{B}_2(\mathbf{B}_1')^o = \mathbf{0}.$$

Hence,

$$\mathbf{Z}_3 = ((\mathbf{A}_2'\mathbf{A}_1^o)^{o'}\mathbf{A}_2')^o\mathbf{Z}_5 + (\mathbf{A}_2'\mathbf{A}_1^o)^{o'}\mathbf{A}_2'\mathbf{Z}_6(\mathbf{B}_2(\mathbf{B}_1')^o)^{o'},$$

which in turn is inserted into (1.3.42). The solution for $\mathbf{X}_1$ follows once again from (1.3.41). ∎

Next we present the solution of a general system of matrix equations, when a nested subspace condition holds.

Theorem 1.3.9. *Let* $\mathbf{H}_i = (\mathbf{A}_1' : \mathbf{A}_2' : \ldots : \mathbf{A}_i')$. *A representation of the solution of*

$$\mathbf{A}_i\mathbf{X}\mathbf{B}_i = \mathbf{0}, \qquad i = 1, 2, \ldots, s,$$

when the nested subspace condition $\mathcal{C}(\mathbf{B}_s) \subseteq \mathcal{C}(\mathbf{B}_{s-1}) \subseteq \cdots \subseteq \mathcal{C}(\mathbf{B}_1)$ *holds, is given by*

$$\mathbf{X} = \mathbf{H}_s^o\mathbf{Z}_1\mathbf{B}_s' + \sum_{i=1}^{s-1}\mathbf{H}_i^o\mathbf{Z}_{i+1}(\mathbf{B}_i^o : \mathbf{B}_{i+1})^{o'} + \mathbf{Z}_{s+1}\mathbf{B}_1^{o'}$$

or

$$\mathbf{X} = \mathbf{H}_s^o\mathbf{Z}_1 + \sum_{i=2}^{s}(\mathbf{H}_{i-1} : \mathbf{H}_i^o)^o\mathbf{Z}_i\mathbf{B}_i^{o'} + \mathbf{Z}_{s+1}\mathbf{B}_1^{o'},$$

where $\mathbf{Z}_i$ *are arbitrary matrices.*

PROOF: By vectorizing the linear equations the proof follows immediately from Theorem 1.2.21. ∎

Finally the general solution of a linear equation is given in a form which sometimes is convenient to use.

Theorem 1.3.10. *A representation of the general solution of the consistent equation in* $\mathbf{x}$*:*

$$\mathbf{A}\mathbf{x} = \mathbf{b}$$

is given by

$$\mathbf{x} = \mathbf{A}^-\mathbf{b},$$

where $\mathbf{A}^-$ *is an arbitrary g-inverse.*

PROOF: From Theorem 1.3.5 and Theorem 1.3.6 it follows that a general solution is given by

$$\mathbf{x} = \mathbf{A}_0^-\mathbf{b} + (\mathbf{I} - \mathbf{A}_0^-\mathbf{A}_0)\mathbf{q}, \tag{1.3.43}$$

where $\mathbf{A}_0^-$ is a particular g-inverse and $\mathbf{q}$ is an arbitrary vector. Furthermore, since $\mathbf{A}\mathbf{A}^-\mathbf{A} = \mathbf{A}$, all g-inverses to $\mathbf{A}$ can be represented via

$$\mathbf{A}^- = \mathbf{A}_0^- + \mathbf{Z} - \mathbf{A}_0^-\mathbf{A}\mathbf{Z}\mathbf{A}\mathbf{A}_0^-, \tag{1.3.44}$$

where $\mathbf{Z}$ is an arbitrary matrix. Now choose $\mathbf{Z} = \mathbf{q}(\mathbf{b}'\mathbf{b})^{-1}\mathbf{b}'$. Then

$$\mathbf{A}^-\mathbf{b} = \mathbf{A}_0^-\mathbf{b} + \mathbf{Z}\mathbf{b} - \mathbf{A}_0^-\mathbf{A}\mathbf{Z}\mathbf{A}\mathbf{A}_0^-\mathbf{b} = \mathbf{A}_0^-\mathbf{b} + (\mathbf{I} - \mathbf{A}_0^-\mathbf{A})\mathbf{Z}\mathbf{b} = \mathbf{A}_0^-\mathbf{b} + (\mathbf{I} - \mathbf{A}_0^-\mathbf{A})\mathbf{q}.$$

Thus, by a suitable choice of $\mathbf{Z}$ in (1.3.44), all solutions in (1.3.43) are of the form $\mathbf{A}^-\mathbf{b}$. ∎

Observe the difference between this result and previous theorems where it was utilized that there always exists a particular choice of g-inverse. In Theorem 1.3.10 it is crucial that $\mathbf{A}^-$ represents all g-inverses.

1.3.6 *Patterned matrices*

The notion *patterned matrix* has been used differently. Graybill (1983) explains it as: "... by recognizing a particular structure, or pattern, in certain matrices... we call such matrices *patterned matrices*." Nel (1980) uses the notion in a more restricted meaning: "The matrix is said to be *patterned*, if a certain number of its elements are constants or repeated by the absolute value of the elements."

Furthermore, there are problems when we need only some part of a matrix and it is of no importance what kind of relation this part has with the rest of the matrix. For example, when examining the asymptotic distribution of the multiple correlation coefficients, so-called *amputated* matrices were introduced by Parring (1980) in order to cut away certain rows and columns of the original matrix. In this case we only need to identify certain parts of the matrix.

We are going to consider patterned matrices as subsets of matrix elements without tieing the notion of a patterned matrix to any specific relation among the elements of the matrix. We talk about a patterned matrix $\mathbf{A}(K)$ if any element or certain part of the original matrix, defined by an index-set K, has been selected from $\mathbf{A}$, i.e. a certain pattern has been "cut out" from the original matrix. If the selected part consists of constants and repeated by the absolute value of the elements we get Nel's (1980) version of a patterned matrix. In fact, the major part of applications of patterned matrices concern symmetric, skew-symmetric, diagonal, triangular etc. matrices which are all patterned matrices in the sense of Nel (1980).

Definition 1.3.6. *Let* $\mathbf{A}$ *be a* $p \times q-$*matrix and* K *a set of pairs of indices:*

$$K = \{(i,j) : i \in I_K, j \in J_K; I_K \subset \{1,\ldots,p\}; J_K \subset \{1,\ldots,q\}\}. \tag{1.3.45}$$

We call $\mathbf{A}(K)$ *a patterned matrix and the set* K *a pattern of the* $p \times q-$*matrix* $\mathbf{A}$, *if* $\mathbf{A}(K)$ *consists of elements* a_{ij} *of* $\mathbf{A}$ *where* $(i,j) \in K$. ∎

Note that $\mathbf{A}(K)$ is not a matrix in a strict sense since it is not a rectangle of elements. One should just regard $\mathbf{A}(K)$ as a convenient notion for a specific collection of elements. When the elements of $\mathbf{A}(K)$ are collected into one column by columns of $\mathbf{A}$ in a natural order, we get a vector of dimensionality r, where r is the number of pairs in K. Let us denote this vector by $\mathrm{vec}\mathbf{A}(K)$. Clearly, there exists always a matrix which transforms $\mathrm{vec}\mathbf{A}$ into $\mathrm{vec}\mathbf{A}(K)$. Let us denote a $r \times pq-$matrix by $\mathbf{T}(K)$ if it satisfies the equality

$$\mathrm{vec}\mathbf{A}(K) = \mathbf{T}(K)\mathrm{vec}\mathbf{A}, \tag{1.3.46}$$

for $\mathbf{A} : p \times q$ and pattern K defined by (1.3.45). Nel (1980) called $\mathbf{T}(K)$ the *transformation* matrix. If some elements of $\mathbf{A}$ are equal by modulus then the transformation matrix $\mathbf{T}(K)$ is not uniquely defined by (1.3.46). Consider a simple example.

Example 1.3.3. Let $\mathbf{S}$ be a 3×3 symmetric matrix:

$$\mathbf{S} = \begin{pmatrix} s_{11} & s_{12} & 0 \\ s_{12} & s_{22} & s_{23} \\ 0 & s_{23} & s_{33} \end{pmatrix}.$$

If we define a pattern K for the lower triangular part:

$$K = \{(i,j) : i,j = 1,2,3; i \geq j\},$$

then the matrix

$$\mathbf{T}_1 = \begin{pmatrix} 1 & 0 & 0 & \vdots & 0 & 0 & 0 & \vdots & 0 & 0 & 0 \\ 0 & 1 & 0 & \vdots & 0 & 0 & 0 & \vdots & 0 & 0 & 0 \\ 0 & 0 & 0 & \vdots & 0 & 0 & 0 & \vdots & 0 & 0 & 0 \\ 0 & 0 & 0 & \vdots & 0 & 1 & 0 & \vdots & 0 & 0 & 0 \\ 0 & 0 & 0 & \vdots & 0 & 0 & 1 & \vdots & 0 & 0 & 0 \\ 0 & 0 & 0 & \vdots & 0 & 0 & 0 & \vdots & 0 & 0 & 1 \end{pmatrix}$$

is a transformation matrix because it satisfies (1.3.46). However, the matrix

$$\mathbf{T}_2 = \begin{pmatrix} 1 & 0 & 0 & \vdots & 0 & 0 & 0 & \vdots & 0 & 0 & 0 \\ 0 & \frac{1}{2} & 0 & \vdots & \frac{1}{2} & 0 & 0 & \vdots & 0 & 0 & 0 \\ 0 & 0 & \frac{1}{2} & \vdots & 0 & 0 & 0 & \vdots & \frac{1}{2} & 0 & 0 \\ 0 & 0 & 0 & \vdots & 0 & 1 & 0 & \vdots & 0 & 0 & 0 \\ 0 & 0 & 0 & \vdots & 0 & 0 & \frac{1}{2} & \vdots & 0 & \frac{1}{2} & 0 \\ 0 & 0 & 0 & \vdots & 0 & 0 & 0 & \vdots & 0 & 0 & 1 \end{pmatrix} \qquad (1.3.47)$$

also satisfies (1.3.46). Moreover, in (1.3.47) we can replace the third row by zeros and still the matrix will satisfy (1.3.46). ∎

If one looks for the transformation that just picks out "proper" elements from $\mathbf{A}$, the simplest way is to use a matrix which consists of ones and zeros like $\mathbf{T}_1$. If the information about the matrices $\mathbf{A}$ and $\mathbf{A}(K)$ is given by an index-set K, we can find a transformation matrix

$$\mathbf{T}(K) : \text{vec}\mathbf{A} \rightarrow \text{vec}\mathbf{A}(K)$$

solely via a 0-1 construction. Formally, the transformation matrix $\mathbf{T}(K)$ is an $r \times pq-$partitioned matrix of columns consisting of $r \times p-$blocks, and if $(\text{vec}\mathbf{A}(K))_f = a_{ij}$, where $f \in \{1,\ldots,r\}$, $i \in \{1,\ldots,p\}$, $j \in \{1,\ldots,q\}$, the general element of the matrix is given by

$$(\mathbf{T}(K))_{f(g,h)} = \begin{cases} 1 & i = h, j = g, \quad g = 1,\ldots,p;\ h = 1,\ldots,q; \\ 0 & \text{elsewhere}, \end{cases} \qquad (1.3.48)$$

where the numeration of the elements follows (1.3.1). As an example, let us write out the transformation matrix for the upper triangle of an $n \times n-$matrix $\mathbf{A}$ which will be denoted by $\mathbf{G}_n$. From (1.3.48) we get the following block-diagonal partitioned $\frac{1}{2}n(n+1) \times n^2-$matrix $\mathbf{G}_n$:

$$\mathbf{G}_n = ([\mathbf{G}_{11}], \ldots, [\mathbf{G}_{nn}])_{[d]}, \tag{1.3.49}$$

$$\mathbf{G}_{ii} = (\mathbf{e}_1, \ldots, \mathbf{e}_i)' \quad i = 1, \ldots, n, \tag{1.3.50}$$

where, as previously, $\mathbf{e}_i$ is the $i-$th unit vector, i.e. $\mathbf{e}_i$ is the ith column of $\mathbf{I}_n$.
Besides the general notion of a patterned matrix we need to apply it also in the case when there are some relationships among the elements in the matrix. We would like to use the patterned matrix to describe the relationships. Let us start assuming that we have some additional information about the relationships between the elements of our matrix. Probably the most detailed analysis of the subject has been given by Magnus (1983) who introduced linear and affine structures for this purpose. In his terminology (Magnus, 1983) a matrix $\mathbf{A}$ is L-structured, if there exist some, say $(pq - r)$, linear relationships between the elements of $\mathbf{A}$. We are interested in the possibility to "restore" vec$\mathbf{A}$ from vec$\mathbf{A}(K)$. Therefore we need to know what kind of structure we have so that we can utilize this additional information. Before going into details we present some general statements which are valid for all matrices with some linear relationship among the elements.
Suppose that we have a class of $p \times q-$matrices, say $\mathbb{M}_{p,q}$, where some arbitrary linear structure between the elements exists. Typical classes of matrices are symmetric, skew-symmetric, diagonal, triangular, symmetric off-diagonal. etc. For all $\alpha, \beta \in \mathbb{R}$ and arbitrary $\mathbf{A}$, $\mathbf{D} \in \mathbb{M}_{p,q}$

$$\alpha\mathbf{A} + \beta\mathbf{D} \in \mathbb{M}_{p,q},$$

and it follows that $\mathbb{M}_{p,q}$ constitutes a linear space. Consider the vectorized form of the matrices $\mathbf{A} \in \mathbb{M}_{p,q}$, i.e. vec$\mathbf{A}$. Because of the assumed linear structure, the vectors vec$\mathbf{A}$ belong to a certain subspace in $\mathbb{R}^{pq}$, say r-dimensional space and $r < pq$. The case $r = pq$ means that there is no linear structure in the matrix of interest.
Since the space $\mathbb{M}_{p,q}$ is a linear space, we can to the space where vec$\mathbf{A}$ belongs, always choose a matrix $\mathbf{B} : pq \times r$ which consists of r independent basis vectors such that $\mathbf{B}'\mathbf{B} = \mathbf{I}_r$. Furthermore, $r < pq$ rows of $\mathbf{B}$ are linearly independent. Let these r linearly independent rows of $\mathbf{B}$ form a matrix $\underline{\mathbf{B}} : r \times r$ of rank r and thus $\underline{\mathbf{B}}$ is invertable. We intend to find matrices (transformations) $\mathbf{T}$ and $\mathbf{C}$ such that

$$\mathbf{TB} = \underline{\mathbf{B}}, \qquad \mathbf{C}\underline{\mathbf{B}} = \mathbf{B}. \tag{1.3.51}$$

This means that $\mathbf{T}$ selects a unique set of row vectors from $\mathbf{B}$, whereas $\mathbf{C}$ creates the original matrix from $\underline{\mathbf{B}}$. Since $\mathbf{B}'\mathbf{B} = \mathbf{I}_r$, it follows immediately that one solution is given by

$$\mathbf{T} = \underline{\mathbf{B}}\mathbf{B}', \qquad \mathbf{C} = \mathbf{B}\underline{\mathbf{B}}^{-1}.$$

Furthermore, from (1.3.51) it follows that $\mathbf{TC} = \mathbf{I}$, which implies

$$\mathbf{TCT} = \mathbf{T}, \quad \mathbf{CTC} = \mathbf{C},$$

and since $\mathbf{CT} = \mathbf{BB}'$ is symmetric, $\mathbf{C}$ is the Moore-Penrose inverse of $\mathbf{T}$, i.e. $\mathbf{C} = \mathbf{T}^+$.
For any matrix $\mathbf{A} \in \mathbb{M}_{p,q}$ we can always express vec$\mathbf{A} = \mathbf{Bq}$ for some vector $\mathbf{q}$. In the following we shall find the matrix $\mathbf{B}$ such that $\mathbf{T}$ satisfies (1.3.46). Thus $\mathbf{T}$ is a transformation matrix while $\mathbf{T}^+$ realizes the inverse transformation. Furthermore, $\mathbf{T}^+\mathbf{T} = \mathbf{BB}'$ will always act as an orthogonal projector which later in special cases will be studied in more detail. If we would not have required the basis vectors to satisfy $\mathbf{B}'\mathbf{B} = \mathbf{I}_r$, we could also have obtained $\mathbf{T}$ and $\mathbf{C}$ matrices, but then $\mathbf{BB}'$ would not have been an orthogonal projector. In order to consider the inverse transformation which enables us to "restore" vec$\mathbf{A}$ from vec$\mathbf{A}(K)$ we need additional information. The information is given in the form of a pattern identifier induced by the linear structure. Therefore, from now on we will just be focusing on certain patterned matrices that we call linearly structured matrices.

Definition 1.3.7. *A matrix* $\mathbf{A}$ *is linearly structured if the only linear structure between the elements is given by* $|a_{ij}| = |a_{kl}| \neq 0$ *and there exists at least one* $(i,j) \neq (k,l)$ *so that* $|a_{ij}| = |a_{kl}| \neq 0$.
For a linearly structured matrix $\mathbf{A} : p \times q$, *with* r *different by absolute value non-zero elements, a pattern identifier* $k(i,j)$ *is a function*

$$k(\cdot,\cdot) : I \times J \longrightarrow H,$$

$$I = \{1,\ldots,p\}, \quad J = \{1,\ldots,q\}, \quad H = \{1,\ldots,r\}$$

such that for $a_{ij} \neq 0,\ a_{gh} \neq 0$,

$$k(i,j) = k(g,h) \Longleftrightarrow |a_{ij}| = |a_{gh}|.$$

∎

In Definition 1.3.7 we have not mentioned $k(i,j)$ when $a_{ij} = 0$. However, from the sequel it follows that it is completely immaterial which value $k(i,j)$ takes if $a_{ij} = 0$. For simplicity we may put $k(i,j) = 0$, if $a_{ij} = 0$. It follows also from the definition that all possible patterned matrices $\mathbf{A}(K)$ consisting of different non-zero elements of $\mathbf{A}$ have the same pattern identifier $k(i,j)$. In the following we will again use the indicator function $1_{\{a=b\}}$, i.e.

$$1_{\{a=b\}} = \begin{cases} 1, & a = b, \quad a \neq 0, \\ 0, & \text{otherwise}. \end{cases}$$

The next lemma gives a realization of the basis matrix $\mathbf{B}$.

Lemma 1.3.2. *Let* $\mathbf{A} : p \times q$ *be a linearly structured matrix with* r *different by absolute value non-zero elements and pattern identifier* $k(i,j)$. *The basis matrix* $\mathbf{B}$ *generates* vec$\mathbf{A}$, *i.e.* vec$\mathbf{A} = \mathbf{Bq}$ *for some* $\mathbf{q}$, *if*

$$\mathbf{B} = \sum_{i=1}^{p}\sum_{j=1}^{q}(\mathbf{d}_j \otimes \mathbf{e}_i)\mathbf{f}'_{k(i,j)}\frac{\operatorname{sgn}(a_{ij})}{\sqrt{m(i,j)}}, \tag{1.3.52}$$

where

$$m(i,j)=\sum_{r=1}^{p}\sum_{s=1}^{q}1_{\{|a_{ij}|=|a_{rs}|\}} \tag{1.3.53}$$

and the unit basis vectors $\mathbf{d}_j : q\times 1$, $\mathbf{e}_i : p\times 1$ *and* $\mathbf{f}_{k(i,j)} : r\times 1$.

PROOF: The statement follows, since

$$\begin{aligned}\mathrm{vec}\mathbf{A}=&\sum_{i,j}a_{ij}(\mathbf{d}_j\otimes\mathbf{e}_i)\\ =&\sum_{i,j}\sum_{g,h}(\mathbf{d}_j\otimes\mathbf{e}_i)\mathbf{f}'_{k(i,j)}\mathbf{f}_{k(g,h)}\frac{a_{ij}}{|a_{ij}|}|a_{gh}|\frac{1}{\sqrt{m(i,j)}}\frac{1}{\sqrt{m(g,h)}}\\ =&\mathbf{B}\sum_{g,h}\mathbf{f}_{k(g,h)}|a_{gh}|\frac{1}{\sqrt{m(g,h)}}.\end{aligned}$$

∎

Lemma 1.3.3. *Let* $\mathbf{B}$ *be given by (1.3.52). Then, in the notation of Lemma 1.3.2,*

(i) *the column vectors of* $\mathbf{B}$ *are orthonormal, i.e.* $\mathbf{B}'\mathbf{B}=\mathbf{I}_r$*;*

(ii)
$$\mathbf{B}\mathbf{B}'=\sum_{i,j}\sum_{k,l}(\mathbf{d}_j\otimes\mathbf{e}_i)(\mathbf{d}_l\otimes\mathbf{e}_k)'1_{\{|a_{ij}|=|a_{kl}|\}}\frac{1}{m(i,j)}. \tag{1.3.54}$$

PROOF: Since $\mathbf{f}_{k(i,j)}=\mathbf{f}_{k(g,h)}$ if $1_{\{|a_{ij}|=|a_{gh}|\}}=1$, straightforward calculations show that

$$\mathbf{B}'\mathbf{B}=\sum_{i,j}\mathbf{f}_{k(i,j)}\mathbf{f}'_{k(i,j)}\frac{1}{m(i,j)}=\mathbf{I}_r$$

and thus (i) is verified. Statement (ii) follows in a similar manner. ∎

Note that for linearly structured matrices the product $\mathbf{B}\mathbf{B}'$ is independent of the pattern identifier $k(i,j)$.

Lemma 1.3.4. *For a linearly structured matrix* $\mathbf{A} : p\times q$*, with the pattern identifier* $k(i,j)$*, the matrix* $\underline{\mathbf{B}}$ *in (1.3.51) is given by*

$$\underline{\mathbf{B}}=\sum_{i=1}^{p}\sum_{j=1}^{q}\sum_{s=1}^{r}\mathbf{f}_s\mathbf{f}'_s1_{\{k(i,j)=s\}}m(i,j)^{-3/2}, \tag{1.3.55}$$

where $\mathbf{f}_{k(i,j)}$ *and* $m(i,j)$ *are defined in Lemma 1.3.2.* ∎

REMARK: Note that $\underline{\mathbf{B}}$ is diagonal and thus

$$\underline{\mathbf{B}}^{-1}=\sum_{i=1}^{p}\sum_{j=1}^{q}\sum_{s=1}^{r}\mathbf{f}_s\mathbf{f}'_s1_{\{k(i,j)=s\}}m(i,j)^{-1/2}. \tag{1.3.56}$$

From Lemma 1.3.2 and Lemma 1.3.4 it follows that $\mathbf{T}$ and $\mathbf{C}=\mathbf{T}^+$ can easily be established for any linearly structured matrix.

Theorem 1.3.11. *Let* $\mathbf{T} = \underline{\mathbf{B}}\mathbf{B}'$ *and* $\mathbf{T}^+ = \mathbf{B}\underline{\mathbf{B}}^{-1}$ *where* $\mathbf{B}$ *and* $\underline{\mathbf{B}}$ *are given by (1.3.52) and (1.3.55), respectively. Then*

$$\mathbf{T} = \sum_{s=1}^{r}\sum_{i,j}\mathbf{f}_s(\mathbf{d}_j\otimes\mathbf{e}_i)'1_{\{k(i,j)=s\}}m(i,j)^{-1}\mathrm{sgn}(a_{ij}), \tag{1.3.57}$$

$$\mathbf{T}^+ = \sum_{s=1}^{r}\sum_{i,j}(\mathbf{d}_j\otimes\mathbf{e}_i)\mathbf{f}_s'1_{\{k(i,j)=s\}}\mathrm{sgn}(a_{ij}), \tag{1.3.58}$$

where $m(i,j)$ *is defined by (1.3.53) and the basis vectors* $\mathbf{e}_i,\ \mathbf{d}_j,\ \mathbf{f}_{k(i,j)}$ *are defined in Lemma 1.3.2.*

PROOF: From (1.3.52) and (1.3.55) we get

$$\begin{aligned}
\mathbf{T} &= \underline{\mathbf{B}}\mathbf{B}'\\
&= \sum_{i_1,j_1}\sum_{s=1}^{r}\mathbf{f}_s\mathbf{f}_s'1_{\{k(i_1,j_1)=s\}}m(i_1,j_1)^{-3/2}\sum_{i,j}\mathbf{f}_{k(i,j)}(\mathbf{d}_j\otimes\mathbf{e}_i)'\frac{\mathrm{sgn}(a_{ij})}{\sqrt{m(i,j)}}\\
&= \sum_{i_1,j_1}\sum_{i,j}\sum_{s=1}^{r}\mathbf{f}_s(\mathbf{d}_j\otimes\mathbf{e}_i)'1_{\{k(i_1,j_1)=s\}}1_{\{k(i,j)=s\}}\\
&\qquad\times m(i_1,j_1)^{-3/2}m(i,j)^{-1/2}\mathrm{sgn}(a_{ij})\\
&= \sum_{i,j}\sum_{s=1}^{r}\mathbf{f}_s(\mathbf{d}_j\otimes\mathbf{e}_i)'1_{\{k(i,j)=s\}}m(i,j)^{-1}\mathrm{sgn}(a_{ij}),
\end{aligned}$$

since

$$\sum_{i_1,j_1}1_{\{k(i_1,j_1)=s\}} = m(i_1,j_1)$$

and $k(i_1,j_1)=k(i,j)$ implies that $m(i_1,j_1)=m(i,j)$. Thus, the first statement is proved. The second statement follows in the same way from (1.3.52) and (1.3.56):

$$\begin{aligned}
\mathbf{T}^+ &= \mathbf{B}\underline{\mathbf{B}}^{-1}\\
&= \sum_{i,j}(\mathbf{d}_j\otimes\mathbf{e}_i)\mathbf{f}_{k(i,j)}'\frac{\mathrm{sgn}(a_{ij})}{\sqrt{m(i,j)}}\sum_{i_1,j_1}\sum_{s=1}^{r}\mathbf{f}_s\mathbf{f}_s'1_{\{k(i_1,j_1)=s\}}m(i_1,j_1)^{-1/2}\\
&= \sum_{i_1,j_1}\sum_{i,j}\sum_{s=1}^{r}(\mathbf{d}_j\otimes\mathbf{e}_i)\mathbf{f}_s'1_{\{k(i_1,j_1)=s\}}1_{\{k(i,j)=s\}}\\
&\qquad\times m(i_1,j_1)^{-1/2}m(i,j)^{-1/2}\mathrm{sgn}(a_{ij})\\
&= \sum_{i,j}\sum_{s=1}^{r}(\mathbf{d}_j\otimes\mathbf{e}_i)\mathbf{f}_s'1_{\{k(i,j)=s\}}\mathrm{sgn}(a_{ij}).
\end{aligned}$$

∎

The following situation has emerged. For a linearly structured matrix $\mathbf{A}$ the structure is described by the pattern identifier $k(i,j)$ with several possible patterns K.

The information provided by the $k(i,j)$ can be used to eliminate repeated elements from the matrix $\mathbf{A}$ in order to get a patterned matrix $\mathbf{A}(K)$. We know that $\mathrm{vec}\mathbf{A} = \mathbf{Bq}$ for some vector $\mathbf{q}$, where $\mathbf{B}$ is defined in Lemma 1.3.2. Furthermore, for the same $\mathbf{q}$ we may state that $\mathrm{vec}\mathbf{A}(K) = \underline{\mathbf{B}}\mathbf{q}$. The next results are important consequences of the theorem.

Corollary 1.3.11.1. *Let* $\mathbf{A}$ *be a linearly structured matrix and* $\mathbf{B}$ *a basis which generates* $\mathrm{vec}\mathbf{A}$. *Let* $\underline{\mathbf{B}}$ *given in Lemma 1.3.4 generate* $\mathrm{vec}\mathbf{A}(K)$. *Then the transformation matrix* $\mathbf{T}$, *given by (1.3.57), satisfies the relation*

$$\mathrm{vec}\mathbf{A}(K) = \mathbf{T}\mathrm{vec}\mathbf{A} \tag{1.3.59,}$$

and its Moore-Penrose inverse $\mathbf{T}^+$, *given by (1.3.58), defines the inverse transformation*

$$\mathrm{vec}\mathbf{A} = \mathbf{T}^+\mathrm{vec}\mathbf{A}(K) \tag{1.3.60}$$

for any pattern K *which corresponds to the pattern identifier* $k(i,j)$ *used in* $\mathbf{T}$.

PROOF: Straightforward calculations yield that for some $\mathbf{q}$

$$\mathbf{T}\mathrm{vec}\mathbf{A} = \mathbf{TBq} = \underline{\mathbf{B}}\mathbf{q} = \mathrm{vec}\mathbf{A}(K).$$

Furthermore,

$$\mathbf{T}^+\mathrm{vec}\mathbf{A}(K) = \mathbf{T}^+\underline{\mathbf{B}}\mathbf{q} = \mathbf{Bq} = \mathrm{vec}\mathbf{A}.$$

∎

As noted before, the transformation matrix $\mathbf{T}$ which satisfies (1.3.59) and (1.3.60) was called *transition* matrix by Nel (1980). The next two corollaries of Theorem 1.3.11 give us expressions for any element of $\mathbf{T}$ and $\mathbf{T}^+$, respectively. If we need to point out that the transition matrix $\mathbf{T}$ is applied to a certain matrix $\mathbf{A}$ from the considered class of matrices $\mathbb{M}$, we shall write $\mathbf{A}$ as an argument of $\mathbf{T}$, i.e. $\mathbf{T}(\mathbf{A})$, and sometimes we also indicate which class $\mathbf{A}$ belongs to.

Corollary 1.3.11.2. *Suppose* $\mathbf{A}$ *is a* $p \times q$ *linearly structured matrix with the pattern identifier* $k(i,j)$, *and* $\mathbf{A}(K)$ *denotes a patterned matrix where* K *is one possible pattern which corresponds to* $k(i,j)$, *and* $\mathbf{T}(\mathbf{A})$ *is defined by (1.3.57). Then the elements of* $\mathbf{T}(\mathbf{A})$ *are given by*

$$(\mathbf{T}(\mathbf{A}))_{s(j,i)} = \begin{cases} \dfrac{1}{m_s}, & \textit{if} \quad a_{ij} = (\mathrm{vec}\mathbf{A}(K))_s, \\ -\dfrac{1}{m_s}, & \textit{if} \quad a_{ij} = -(\mathrm{vec}\mathbf{A}(K))_s, \\ 0, & \textit{otherwise}, \end{cases} \tag{1.3.61}$$

where $s = 1,\dots,r$, $i = 1,\dots,p$, $j = 1,\dots,q$, $\mathrm{vec}\mathbf{A}(K) : r \times 1$ *is as in Corollary 1.3.11.1, the numeration of the elements follows (1.3.1) and*

$$m_s = \sum_{i=1}^{p}\sum_{j=1}^{q} 1_{\{|(\mathrm{vec}\mathbf{A}(K))_s| = |a_{ij}|\}}$$

∎

Corollary 1.3.11.3. *Suppose* $\mathbf{A} : p \times q$ *is a linearly structured matrix with the pattern identifier* $k(i,j)$, *and* $\mathbf{A}(K)$ *denotes a patterned matrix where* K *is one possible pattern which corresponds to* $k(i,j)$, *and* $\mathbf{T}$ *is defined by (1.3.57). Then the elements of* $\mathbf{T}^+$ *are given by*

$$(\mathbf{T}^+)_{(j,i)s} = \begin{cases} 1 & \textit{if} \quad a_{ij} = (\mathrm{vec}\ \mathbf{A}(K))_s, \\ -1 & \textit{if} \quad a_{ij} = -(\mathrm{vec}\ \mathbf{A}(K))_s, \\ 0 & \textit{otherwise}, \end{cases} \tag{1.3.62}$$

where $s = 1, \ldots, r$, $i = 1, \ldots, p$, $j = 1, \ldots, q$, $\mathrm{vec}\mathbf{A}(K) : r \times 1$ *is as in Corollary 1.3.11.1 and the numeration of the elements follows (1.3.1).* ∎

Example 1.3.4 Let us see what the transition and inverse transition matrices look like in Example 1.3.3. Indeed, we can easily check that $\mathbf{T}_2$ in (1.3.47) is the transition matrix. Direct calculation yields the inverse transition matrix:

$$\mathbf{T}_2^+ = \begin{pmatrix} 1 & 0 & 0 & 0 & 0 & 0 \\ 0 & 1 & 0 & 0 & 0 & 0 \\ 0 & 0 & 1 & 0 & 0 & 0 \\ - & - & - & - & - & - \\ 0 & 1 & 0 & 0 & 0 & 0 \\ 0 & 0 & 0 & 1 & 0 & 0 \\ 0 & 0 & 0 & 0 & 1 & 0 \\ - & - & - & - & - & - \\ 0 & 0 & 1 & 0 & 0 & 0 \\ 0 & 0 & 0 & 0 & 1 & 0 \\ 0 & 0 & 0 & 0 & 0 & 1 \end{pmatrix}.$$

∎

For applications the most important special case is the symmetric matrix and its transition matrix. There are many possibilities of picking out $\frac{1}{2}n(n+1)$ different elements from a symmetric $n \times n-$matrix ($2^{\frac{n(n-1)}{2}}$, in fact). Two most common of them being the lower and the upper triangles. The point of using the transition matrix is that for any choice of elements, the transition matrix will always be the same and is described by the so-called duplication matrix which has been carefully examined by Magnus & Neudecker (1999), as well as by others. In our treatment we shall use somewhat different notation. Let $\mathbf{A}$ be a symmetric $n \times n-$matrix and let the patterned matrix consisting of the elements of its lower triangle be represented by $\mathbf{A}_\triangle$. Denote the corresponding transition matrix by $\mathbf{D}_n : \frac{1}{2}n(n+1) \times n^2$:

$$\mathbf{D}_n \mathrm{vec}\mathbf{A} = \mathrm{vec}\mathbf{A}_\triangle. \tag{1.3.63}$$

Then

$$\mathbf{D}_n^+ \mathrm{vec}\mathbf{A}_\triangle = \mathrm{vec}\mathbf{A}. \tag{1.3.64}$$

The matrix $\mathbf{D}_n^+$ is called *duplication matrix* and its basic properties have been collected in the next proposition (for proofs, see Magnus & Neudecker, 1999).

Proposition 1.3.16. *Let the duplication matrix* $\mathbf{D}_n^+$ *be defined by (1.3.64) and the corresponding transition matrix* $\mathbf{D}_n$ *by (1.3.63). Then*

(i)
$$\mathbf{K}_{n,n}\mathbf{D}_n^+ = \mathbf{D}_n^+;$$

(ii)
$$\mathbf{D}_n^+\mathbf{D}_n = \frac{1}{2}(\mathbf{I}_{n^2} + \mathbf{K}_{n,n});$$

(iii) *for any* $n-$*vector* $\mathbf{b}$ *and* $\mathbf{A} : n \times n$

$$\mathbf{D}_n^+\mathbf{D}_n(\mathbf{b} \otimes \mathbf{A}) = \frac{1}{2}(\mathbf{b} \otimes \mathbf{A} + \mathbf{A} \otimes \mathbf{b});$$

(iv) *for* $\mathbf{A} : n \times n$

$$\mathbf{D}_n^+\mathbf{D}_n(\mathbf{A} \otimes \mathbf{A})\mathbf{D}_n^+ = (\mathbf{A} \otimes \mathbf{A})\mathbf{D}_n^+;$$

(v) *for* $\mathbf{A} : n \times n$

$$\mathbf{D}_n^+\mathbf{D}_n(\mathbf{A} \otimes \mathbf{A})\mathbf{D}_n' = (\mathbf{A} \otimes \mathbf{A})\mathbf{D}_n';$$

(vi) *for non-singular* $\mathbf{A} : n \times n$

$$(\mathbf{D}_n(\mathbf{A} \otimes \mathbf{A})\mathbf{D}_n^+)^{-1} = \mathbf{D}_n(\mathbf{A}^{-1} \otimes \mathbf{A}^{-1})\mathbf{D}_n^+;$$

(vii) *for non-singular* $\mathbf{A} : n \times n$

$$(\mathbf{D}_n^{+\prime}(\mathbf{A} \otimes \mathbf{A})\mathbf{D}_n^+)^{-1} = \mathbf{D}_n(\mathbf{A}^{-1} \otimes \mathbf{A}^{-1})\mathbf{D}_n'.$$

∎

As we saw in Theorem 1.3.11, it was possible to characterize transition matrices mathematically. Unfortunately, in general, not many interesting and easily interpretable properties can be found for these matrices. One reason for this is that $\mathbf{T}$ is a function of the pattern identifier, which means that results depend on $\mathbf{f}_{k(i,j)}$. However, Nel (1980) brought forward the notion of *pattern matrix* which shares some basic properties with $\mathbf{T}^+$. For example, both generate the same subspace. We shall call it *pattern projection matrix.*

Definition 1.3.8. *Let* $\mathbf{T}$ *be the transition matrix for a linearly structured matrix* $\mathbf{A} : p \times q$ *with the pattern identifier* $k(i,j)$. *Then the matrix* $\mathbf{M} : pq \times pq$, *defined by*

$$\mathbf{M} = \mathbf{T}^+\mathbf{T} \tag{1.3.65}$$

is called the pattern projection matrix of $\mathbf{A}$. ∎

Observe that $\mathbf{M}$ is an orthogonal projector and from a geometrical point of view this is an interesting definition. Some basic properties of the pattern matrix $\mathbf{M}$ are collected in the following proposition. The statements all follow in elementary way from Definition 1.3.8 and from the fact that $\mathbf{M}$ is a projector.

Proposition 1.3.17. *Let* $\mathbf{M}$ *be given by (1.3.65). Then*

(i) $\mathbf{M}$ *is symmetric;*

(ii) $\mathbf{M}$ *is idempotent;*

(iii) *if* $\mathbf{A}$ *is* $p \times q-$*matrix with the transition matrix* $\mathbf{T}$, *then*

$$\mathbf{M}\mathrm{vec}\mathbf{A} = \mathrm{vec}\mathbf{A}; \tag{1.3.66}$$

(iv) $\mathbf{T}$ *is invariant under right multiplication and* $\mathbf{T}^+$ *is invariant under left multiplication by* $\mathbf{M}$, *respectively.* ∎

Now some widely used classes of linearly structured $n \times n$ square matrices will be studied. We shall use the following notation:

- s (s_n) for symmetric matrices (s_n if we want to stress the dimensions of the matrix);
- ss (ss_n) – skew-symmetric matrices;
- d (d_n) – diagonal matrices;
- c (c_n) – correlation type matrices or symmetric off-diagonal matrices;
- u (u_n) – upper triangular matrices;
- l (l_n) – lower triangular matrices;
- t (t_n) – Toeplitz matrices.

For instance, $\mathbf{M}(s)$ and $\mathbf{T}(s)$ will be the notation for $\mathbf{M}$ and $\mathbf{T}$ for a symmetric $\mathbf{A}$. In the following propositions the pattern identifier $k(i,j)$ and matrices $\mathbf{B}$, $\mathbf{T}$, $\mathbf{T}^+$ and $\mathbf{M}$ are presented for the listed classes of matrices. Observe that it is supposed that there exist no other relations between the elements than the basic ones which define the class. For example, all diagonal elements in the symmetric class and diagonal class differ. In all these propositions $\mathbf{e}_i$ and $\mathbf{f}_{k(i,j)}$ stand for the $n-$ and $r-$dimensional basis vectors, respectively, where r depends on the structure. Let us start with the class of symmetric matrices.

Proposition 1.3.18. *For the class of symmetric* $n \times n-$*matrices* $\mathbf{A}$ *the following equalities hold:*

$$\begin{aligned} k(i,j) =& n(\min(i,j)-1) - \tfrac{1}{2}\min(i,j)(\min(i,j)-1) + \max(i,j), \\ \mathbf{B}(s_n) =& \frac{1}{\sqrt{2}} \sum_{\substack{i,j=1 \\ i \neq j}}^{n} (\mathbf{e}_j \otimes \mathbf{e}_i)\mathbf{f}'_{k(i,j)}\mathrm{sgn}(a_{ij}) + \sum_{i=1}^{n} (\mathbf{e}_i \otimes \mathbf{e}_i)\mathbf{f}'_{k(i,i)}\mathrm{sgn}(a_{ii}), \end{aligned}$$

$$\mathbf{T}(s_n) = \frac{1}{2}\sum_{\substack{i,j=1\\i<j}}^{n}\sum_{s=1}^{r}\mathbf{f}_s(\mathbf{e}_j\otimes\mathbf{e}_i+\mathbf{e}_i\otimes\mathbf{e}_j)'1_{\{k(i,j)=s\}}\mathrm{sgn}(a_{ij})$$

$$+\sum_{i=1}^{n}\sum_{s=1}^{r}\mathbf{f}_s(\mathbf{e}_i\otimes\mathbf{e}_i)'1_{\{k(i,i)=s\}}\mathrm{sgn}(a_{ii}), \qquad r=\frac{1}{2}n(n+1),$$

$$\mathbf{T}^+(s) = \sum_{\substack{i,j=1\\i<j}}^{n}\sum_{s=1}^{r}(\mathbf{e}_j\otimes\mathbf{e}_i+\mathbf{e}_i\otimes\mathbf{e}_j)\mathbf{f}_s'1_{\{k(i,j)=s\}}\mathrm{sgn}(a_{ij})$$

$$+\sum_{i=1}^{n}\sum_{s=1}^{r}(\mathbf{e}_i\otimes\mathbf{e}_i)\mathbf{f}_s'1_{\{k(i,i)=s\}}\mathrm{sgn}(a_{ii}),$$

$$\mathbf{M}(s) = \frac{1}{2}(\mathbf{I}+\mathbf{K}_{n,n}).$$

PROOF: If we count the different elements row by row we have n elements in the first row, $n-1$ different elements in the second row, $n-2$ elements in the third row, etc. Thus, for a_{ij}, $i\leq j$, we are in the i–th row where the j–th element should be considered. It follows that a_{ij} is the element with number

$$n+n-1+n-2+\cdots+n-i+1+j-i$$

which equals

$$n(i-1)-\frac{1}{2}i(i-1)+j=k(i,j).$$

If we have an element a_{ij}, $i>j$, it follows by symmetry that the expression for $k(i,j)$ is true and we have verified that $k(i,j)$ holds. Moreover, $\mathbf{B}(s)$ follows immediately from (1.3.52). For $\mathbf{T}(s)$ and $\mathbf{T}^+(s)$ we apply Theorem 1.3.11. ∎

Proposition 1.3.19. *For the class of skew-symmetric matrices* $\mathbf{A}: n\times n$ *we have*

$$k(i,j) = n(\min(i,j)-1)-\tfrac{1}{2}\min(i,j)(\min(i,j)-1)+\max(i,j)-\min(i,j),\ i\neq j,$$
$$k(i,i) = 0,$$

$$\mathbf{B}(ss_n) = \frac{1}{\sqrt{2}}\sum_{\substack{i,j=1\\i\neq j}}^{n}(\mathbf{e}_j\otimes\mathbf{e}_i)\mathbf{f}_{k(i,j)}'\mathrm{sgn}(a_{ij}),$$

$$\mathbf{T}(ss_n) = \frac{1}{2}\sum_{\substack{i,j=1\\i<j}}^{n}\sum_{s=1}^{r}\mathbf{f}_s(\mathbf{e}_j\otimes\mathbf{e}_i-\mathbf{e}_i\otimes\mathbf{e}_j)'1_{\{k(i,j)=s\}}\mathrm{sgn}(a_{ij}), \qquad r=\frac{1}{2}n(n-1),$$

$$\mathbf{T}^+(ss) = \sum_{\substack{i,j=1\\i<j}}^{n}\sum_{s=1}^{r}(\mathbf{e}_j\otimes\mathbf{e}_i-\mathbf{e}_i\otimes\mathbf{e}_j)\mathbf{f}_s'1_{\{k(i,j)=s\}}\mathrm{sgn}(a_{ij}),$$

$$\mathbf{M}(ss) = \frac{1}{2}(\mathbf{I}-\mathbf{K}_{n,n}).$$

PROOF: Calculations similar to those in Proposition 1.3.18 show that the relation for $k(i,j)$ holds. The relations for $\mathbf{B}(ss)$, $\mathbf{T}(ss)$ and $\mathbf{T}^+(ss)$ are also obtained in the same way as in the previous proposition. For deriving the last statement we use the fact that $\mathrm{sgn}(a_{ij})\mathrm{sgn}(a_{ij}) = 1$:

$$\mathbf{M}(ss) = \mathbf{B}(ss)\mathbf{B}'(ss) = \frac{1}{\sqrt{2}} \sum_{i,j=1}^{n} (\mathbf{e}_j \otimes \mathbf{e}_i)(\mathbf{e}_j \otimes \mathbf{e}_i - \mathbf{e}_i \otimes \mathbf{e}_j) = \frac{1}{2}(\mathbf{I} - \mathbf{K}_{n,n}).$$

∎

Proposition 1.3.20. *For diagonal matrices $\mathbf{A} : n \times n$ with different diagonal elements we obtain*

$$\begin{aligned}
k(i,i) =& i, \qquad k(i,j) = 0, \quad i \neq j,\\
\mathbf{B}(d_n) =& \sum_{i=1}^{n} (\mathbf{e}_i \otimes \mathbf{e}_i)\mathbf{e}_i',\\
\mathbf{T}(d_n) =& \sum_{i=1}^{n} \mathbf{e}_i(\mathbf{e}_i \otimes \mathbf{e}_i)',\\
\mathbf{T}^+(d_n) =& \sum_{i=1}^{n} (\mathbf{e}_i \otimes \mathbf{e}_i)\mathbf{e}_i',\\
\mathbf{M}(d_n) =& \sum_{i=1}^{n} (\mathbf{e}_i \otimes \mathbf{e}_i)(\mathbf{e}_i \otimes \mathbf{e}_i)' = (\mathbf{K}_{n,n})_d.
\end{aligned}$$

PROOF: In Lemma 1.3.2 and Theorem 1.3.11 we can choose $f_{k(i,j)} = \mathbf{e}_i$ which imply the first four statements. The last statement follows from straightforward calculations, which yield

$$\mathbf{M}(d) = \mathbf{B}(d)\mathbf{B}'(d) = \sum_{i=1}^{n} (\mathbf{e}_i \otimes \mathbf{e}_i)(\mathbf{e}_i \otimes \mathbf{e}_i)' = (\mathbf{K}_{n,n})_d.$$

∎

Observe that we do not have to include $\mathrm{sgn}(a_{ij})$ in the given relation in Proposition 1.3.20.

Proposition 1.3.21. *For any $\mathbf{A} : n \times n$ belonging to the class of symmetric off-diagonal matrices it follows that*

$$\begin{aligned}
k(i,j) =& n(\min(i,j) - 1) - \tfrac{1}{2}\min(i,j)(\min(i,j) - 1) + \max(i,j) - \min(i,j), \ i \neq j,\\
k(i,i) =& 0,
\end{aligned}$$

$$\mathbf{B}(c_n) = \frac{1}{\sqrt{2}} \sum_{\substack{i,j=1\\ i\neq j}}^{n} (\mathbf{e}_j \otimes \mathbf{e}_i)\mathbf{f}_{k(i,j)}'\mathrm{sgn}(a_{ij}),$$

$$\mathbf{T}(c_n) = \frac{1}{2}\sum_{\substack{i,j=1\\ i<j}}^{n}\sum_{s=1}^{r}\mathbf{f}_s(\mathbf{e}_j\otimes\mathbf{e}_i' + \mathbf{e}_i\otimes\mathbf{e}_j)'1_{\{k(i,j)=s\}}\mathrm{sgn}(a_{ij}), \quad r = \frac{1}{2}n(n-1),$$

$$\mathbf{T}^+(c_n) = \sum_{\substack{i,j=1\\ i<j}}^{n}\sum_{i=1}^{n}(\mathbf{e}_j\otimes\mathbf{e}_i + \mathbf{e}_i\otimes\mathbf{e}_j)\mathbf{f}_s'1_{\{k(i,j)=s\}}\mathrm{sgn}(a_{ij}),$$

$$\mathbf{M}(c_n) = \frac{1}{2}(\mathbf{I}+\mathbf{K}_{n,n}) - (\mathbf{K}_{n,n})_d.$$

PROOF: The same elements as in the skew-symmetric case are involved. Therefore, the expression for $k(i,j)$ follows from Proposition 1.3.19. We just prove the last statement since the others follow from Lemma 1.3.2 and Theorem 1.3.11. By definition of $\mathbf{B}(c)$

$$\begin{aligned}\mathbf{M}(c) =&\mathbf{B}(c)\mathbf{B}'(c) = \sum_{\substack{i,j=1\\ i\neq j}}^{n}\sum_{\substack{k,l=1\\ k\neq l}}^{n}\frac{1}{2}(\mathbf{e}_j\otimes\mathbf{e}_i)f_{k(i,j)}'f_{k(k,l)}(\mathbf{e}_l\otimes\mathbf{e}_k)'\mathrm{sgn}(a_{ij})\mathrm{sgn}(a_{kl})\\ =&\frac{1}{2}\sum_{\substack{i,j=1\\ i\neq j}}^{n}(\mathbf{e}_j\otimes\mathbf{e}_i)(\mathbf{e}_j\otimes\mathbf{e}_i + \mathbf{e}_i\otimes\mathbf{e}_j)' = \frac{1}{2}(\mathbf{I}+\mathbf{K}_{n,n}) - (\mathbf{K}_{n,n})_d.\end{aligned}$$

∎

Proposition 1.3.22. *For any $\mathbf{A} : n\times n$ from the class of upper triangular matrices*

$$\begin{aligned}k(i,j) =& n(i-1) - \tfrac{1}{2}i(i-1) + j, \quad i\leq j,\\ k(i,j) =& 0, \quad i>j,\end{aligned}$$

$$\mathbf{B}(u_n) = \sum_{\substack{i,j=1\\ i\leq j}}^{n}(\mathbf{e}_j\otimes\mathbf{e}_i)\mathbf{f}_{k(i,j)}'\mathrm{sgn}(a_{ij}),$$

$$\mathbf{T}(u_n) = \sum_{\substack{i,j=1\\ i\leq j}}^{n}\sum_{s=1}^{r}\mathbf{f}_s(\mathbf{e}_j\otimes\mathbf{e}_i)'1_{\{k(i,j)=s\}}\mathrm{sgn}(a_{ij}), \quad r = \frac{1}{2}n(n+1),$$

$$\mathbf{T}^+(u_n) = \sum_{\substack{i,j=1\\ i\leq j}}^{n}\sum_{s=1}^{r}(\mathbf{e}_j\otimes\mathbf{e}_i)\mathbf{f}_s'1_{\{k(i,j)=s\}}\mathrm{sgn}(a_{ij}),$$

$$\mathbf{M}(u_n) = \sum_{\substack{i,j=1\\ i\leq j}}^{n}(\mathbf{e}_j\otimes\mathbf{e}_i)(\mathbf{e}_j\otimes\mathbf{e}_i)'.$$

PROOF: Here the same elements as in the symmetric case are regarded. The expression for $k(i,j)$ is thus a consequence of Proposition 1.3.18. Other statements could be obtained by copying the proof of Proposition 1.3.21, for example. ∎

Observe the similarity between $\mathbf{T}(u_n)$ in Proposition 1.3.22 and $\mathbf{G}_n$ given by (1.3.49) and (1.3.50). By symmetry it follows that we may state a proposition for lower triangular matrices. Moreover, note the similarity between $\mathbf{T}(l_n)$ and $\mathbf{D}_n$ given by (1.3.63) and (1.3.64).

Proposition 1.3.23. *For any* $\mathbf{A} : n\times n$ *from the class of lower triangular matrices*

$$
\begin{aligned}
k(i,j) =& n(j-1) - \tfrac{1}{2}j(j-1) + i, \quad j \leq i,\\
k(i,j) =& 0, \quad i < j,\\
\mathbf{B}(l_n) =& \sum_{\substack{i,j=1\\ i\geq j}}^{n} (\mathbf{e}_i \otimes \mathbf{e}_j)\mathbf{f}'_{k(i,j)}\mathrm{sgn}(a_{ij}),\\
\mathbf{T}(l_n) =& \sum_{\substack{i,j=1\\ i\geq j}}^{n}\sum_{s=1}^{r} \mathbf{f}_s(\mathbf{e}_i \otimes \mathbf{e}_j)' 1_{\{k(i,j)=s\}}\mathrm{sgn}(a_{ij}), \quad r = \frac{1}{2}n(n+1),\\
\mathbf{T}^+(l_n) =& \sum_{\substack{i,j=1\\ i\geq j}}^{n}\sum_{s=1}^{r} (\mathbf{e}_i \otimes \mathbf{e}_j)\mathbf{f}'_s 1_{\{k(i,j)=s\}}\mathrm{sgn}(a_{ij}),\\
\mathbf{M}(l_n) =& \sum_{\substack{i,j=1\\ i\geq j}}^{n} (\mathbf{e}_i \otimes \mathbf{e}_j)(\mathbf{e}_i \otimes \mathbf{e}_j)'.
\end{aligned}
$$

∎

Proposition 1.3.24. *For any Toeplitz matrix* $\mathbf{A} : n \times n$,

$$
\begin{aligned}
k(i,j) =& n + j - i,\\
\mathbf{B}(t_n) =& \sum_{i,j=1}^{n} (\mathbf{e}_j \otimes \mathbf{e}_i)\mathbf{f}'_{k(i,j)}\frac{1}{\sqrt{n-|j-i|}}\mathrm{sgn}(a_{ij}),\\
\mathbf{T}(t_n) =& \sum_{i,j=1}^{n}\sum_{s=1}^{r} \mathbf{f}_s(\mathbf{e}_j \otimes \mathbf{e}_i)' 1_{\{k(i,j)\}}\frac{1}{n-|j-i|}\mathrm{sgn}(a_{ij}), \quad r = 2n-1,\\
\mathbf{T}^+(t_n) =& \sum_{i,j=1}^{n}\sum_{s=1}^{r} (\mathbf{e}_j \otimes \mathbf{e}_i)\mathbf{f}'_s 1_{\{k(i,j)\}}\mathrm{sgn}(a_{ij}),\\
\mathbf{M}(t_n) =& \sum_{\substack{i,j,k,l=1\\ j-i=l-k}}^{n} (\mathbf{e}_j \otimes \mathbf{e}_i)(\mathbf{e}_l \otimes \mathbf{e}_k)'\frac{1}{n-|j-i|}.
\end{aligned}
$$

PROOF: Since by definition of a Toeplitz matrix (see §1.1.1) $a_{ij} = a_{i-j}$ we have $2n-1$ different elements in $\mathbf{A}$. Because $j-i$ ranges through

$$-(n-1), -(n-2), \ldots, (n-1),$$

the equality for $k(i,j)$ is true. If $j - i = t$ we have $j = t + i$. Since $1 \leq j \leq n$, we have for $t \geq 0$ the restriction $1 \leq i \leq n-t$. If $t \leq 0$ then $1 \leq i \leq n+t$. Therefore, for fixed t, we have $n - |t|$ elements. Hence, $\mathbf{B}(t)$ follows from Lemma 1.3.2. The other statements are obtained by using Theorem 1.3.11 and (1.3.65). ∎

We will now study the pattern projection matrices $\mathbf{M}(\cdot)$ in more detail and examine their action on tensor spaces. For symmetric matrices it was shown that $\mathbf{M}(s) = \frac{1}{2}(\mathbf{I}_{n^2} + \mathbf{K}_{n,n})$. From a tensor space point of view any vectorized symmetric matrix $\mathbf{A} : n \times n$ can be written as

$$\mathrm{vec}\mathbf{A} = \sum_{i,j=1}^{n} a_{ij}(\mathbf{e}_j \otimes \mathbf{e}_i) = \sum_{i,j=1}^{n} a_{ij}\frac{1}{2}((\mathbf{e}_j \otimes \mathbf{e}_i) + (\mathbf{e}_i \otimes \mathbf{e}_j)). \tag{1.3.67}$$

Moreover,

$$\frac{1}{2}(\mathbf{I}_{n^2} + \mathbf{K}_{n,n})\frac{1}{2}((\mathbf{e}_j \otimes \mathbf{e}_i) + (\mathbf{e}_i \otimes \mathbf{e}_j)) = \frac{1}{2}((\mathbf{e}_j \otimes \mathbf{e}_i) + (\mathbf{e}_i \otimes \mathbf{e}_j)), \tag{1.3.68}$$

$$\frac{1}{2}(\mathbf{I}_{n^2} + \mathbf{K}_{n,n})(\mathbf{e}_j \otimes \mathbf{e}_i) = \frac{1}{2}[(\mathbf{e}_j \otimes \mathbf{e}_i) + (\mathbf{e}_i \otimes \mathbf{e}_j)] \tag{1.3.69}$$

and for arbitrary $\mathbf{H} : n \times n$

$$\frac{1}{2}(\mathbf{I}_{n^2} + \mathbf{K}_{n,n})\mathrm{vec}\mathbf{H} = \frac{1}{2}\mathrm{vec}(\mathbf{H} + \mathbf{H}'), \tag{1.3.70}$$

which means that $\mathbf{H}$ has been symmetrized. Indeed, (1.3.67) – (1.3.70) all show, in different ways, how the projector $\frac{1}{2}(\mathbf{I}_{n^2} + \mathbf{K}_{n,n})$ acts. Direct multiplication of terms shows that

$$\frac{1}{2}(\mathbf{I}_{n^2} + \mathbf{K}_{n,n})\frac{1}{2}(\mathbf{I}_{n^2} - \mathbf{K}_{n,n}) = \mathbf{0}$$

and since $\frac{1}{2}(\mathbf{I}_{n^2} - \mathbf{K}_{n,n})$ is the pattern matrix and a projector on the space of skew-symmetric matrices, it follows that vectorized symmetric and skew-symmetric matrices are orthogonal. Furthermore, vectorized skew-symmetric matrices span the orthogonal complement to the space generated by the vectorized symmetric matrices since

$$\mathbf{I}_{n^2} = \frac{1}{2}(\mathbf{I}_{n^2} + \mathbf{K}_{n,n}) + \frac{1}{2}(\mathbf{I}_{n^2} - \mathbf{K}_{n,n}).$$

Similarly to (1.3.67) for symmetric matrices, we have representations of vectorized skew-symmetric matrices

$$\mathrm{vec}\mathbf{A} = \sum_{i,j=1}^{n} a_{ij}(\mathbf{e}_j \otimes \mathbf{e}_i) = \sum_{i,j=1}^{n} a_{ij}\frac{1}{2}((\mathbf{e}_j \otimes \mathbf{e}_i) - (\mathbf{e}_i \otimes \mathbf{e}_j)).$$

Moreover,

$$\frac{1}{2}(\mathbf{I}_{n^2} - \mathbf{K}_{n,n})\frac{1}{2}((\mathbf{e}_j \otimes \mathbf{e}_i) - (\mathbf{e}_i \otimes \mathbf{e}_j)) = \frac{1}{2}((\mathbf{e}_j \otimes \mathbf{e}_i) - (\mathbf{e}_i \otimes \mathbf{e}_j)),$$

$$\frac{1}{2}(\mathbf{I}_{n^2} - \mathbf{K}_{n,n})(\mathbf{e}_j \otimes \mathbf{e}_i) = \frac{1}{2}((\mathbf{e}_j \otimes \mathbf{e}_i) - (\mathbf{e}_i \otimes \mathbf{e}_j))$$

and for arbitrary $\mathbf{H} : n \times n$

$$\frac{1}{2}(\mathbf{I}_{n^2} - \mathbf{K}_{n,n})\text{vec}\mathbf{H} = \frac{1}{2}\text{vec}(\mathbf{H} - \mathbf{H}'),$$

where $\mathbf{H} - \mathbf{H}'$ is obviously skew-symmetric. Furthermore, the whole space can be generated by $(\mathbf{e}_j \otimes \mathbf{e}_i)$ which can also be represented another way:

$$(\mathbf{e}_j \otimes \mathbf{e}_i) = \frac{1}{2}((\mathbf{e}_j \otimes \mathbf{e}_i) + (\mathbf{e}_i \otimes \mathbf{e}_j)) + \frac{1}{2}((\mathbf{e}_j \otimes \mathbf{e}_i) - (\mathbf{e}_i \otimes \mathbf{e}_j)).$$

Next we are going to decompose the space generated by vectorized symmetric matrices into two orthogonal spaces, namely the spaces generated by vectorized diagonal matrices and vectorized symmetric off-diagonal matrices. Everything follows from the relations:

$$\frac{1}{2}(\mathbf{I}_{n^2}+\mathbf{K}_{n,n}) = \frac{1}{2}(\mathbf{I}_{n^2} + \mathbf{K}_{n,n}) - (\mathbf{K}_{n,n})_d + (\mathbf{K}_{n,n})_d;$$
$$(\frac{1}{2}(\mathbf{I}_{n^2} + \mathbf{K}_{n,n}) - (\mathbf{K}_{n,n})_d)(\mathbf{K}_{n,n})_d = \mathbf{0}.$$

Thus, according to Proposition 1.3.20 and Proposition 1.3.21, any matrix can be written as a sum of a symmetric off-diagonal matrix, a diagonal matrix and a skew-symmetric matrix.
One advantage of working with projectors is that it is easy to work with any structure generated by linearly structured matrices as well as to combine these structures.
Moreover, we have seen that the class of skew-symmetric matrices is orthogonal to the class of symmetric matrices. It is fairly easy to generalize this result to arbitrary classes of linearly structured matrices. Consider the basis matrix $\mathbf{B}$ given by (1.3.52). Let

$$\mathbf{q} = \sum_{i=1}^{p}\sum_{j=1}^{q} q_{ij}(\mathbf{d}_j \otimes \mathbf{e}_i),$$

and suppose that $\mathbf{q}'\mathbf{B} = \mathbf{0}$, i.e.

$$\mathbf{0} = \mathbf{q}'\mathbf{B} = \sum_{i=1}^{p}\sum_{j=1}^{q} q_{ij}\mathbf{f}'_{k(i,j)}\frac{\text{sgn}(a_{ij})}{\sqrt{m(i,j)}}, \qquad (1.3.71)$$

where $m(i,j)$ is defined by (1.3.53). Since $k(i,j) \in \{1, 2, \ldots, r\}$, it follows that (1.3.71) is equivalent to

$$\sum_{i=1}^{p}\sum_{\substack{j=1\\k(i,j)=s}}^{q} q_{ij}\frac{\text{sgn}(a_{ij})}{\sqrt{m(i,j)}} = 0, \quad s = 1, 2, \ldots, r. \qquad (1.3.72)$$

This is a linear equation in q_{ij} which can always be solved. If $m(i,j) = m$, we will obtain $m - 1$ different solutions. In particular, if $m = 1$ for (i,j) such that

$k(i,j) = s$, the corresponding $q_{ij} = 0$. If $\mathrm{sgn}(a_{ij}) = 1$ and $m = 2$ for, say (i_1, j_1) and (i_2, j_2), we have

$$q_{i_1,j_1} + q_{i_2,j_2} = 0$$

and therefore $q_{i_1,j_1} = -q_{i_2,j_2}$. Thus, from (1.3.72) we can determine all vectors which are orthogonal to $\mathbf{B}$ and find the orthogonal complement to $\mathbf{B}$.

Let us consider Toeplitz matrices in more detail. Another way to express the basis matrix $\mathbf{B}(t)$ in Proposition 1.3.24 is

$$\mathbf{B}(t) = \sum_{k=1}^{2n-1} \sum_{\substack{i=1 \\ i-|n-k|\geq 1}}^{n} \frac{1}{\sqrt{n-|n-k|}} (\mathbf{e}_{k+i-n} \otimes \mathbf{e}_i)\mathbf{f}_k' \mathrm{sgn}(a_{ik+i-n}).$$

An advantage of this representation is that we can see how the linear space, which corresponds to a Toeplitz matrix, is built up. The representation above decomposes the space into $(2n-1)$ orthogonal spaces of dimension 1. Moreover, since $\mathbf{M}(t) = \mathbf{B}(t)\mathbf{B}(t)'$,

$$\mathbf{M}(t) = \sum_{k=1}^{2n-1} \sum_{\substack{i=1 \\ i-|n-k|\geq 1}}^{n} \sum_{\substack{j=1 \\ j-|n-k|\geq 1}}^{n} \frac{1}{n-|n-k|} (\mathbf{e}_{k+i-n} \otimes \mathbf{e}_i)(\mathbf{e}_{k+j-n} \otimes \mathbf{e}_j)'.$$

We are going to determine the class of matrices which is orthogonal to the class of Toeplitz matrices.

Theorem 1.3.12. *A basis matrix $\mathbf{B}(t)^o$ of the class which is orthogonal to all vectorized $n \times n$ Toeplitz matrices is given by*

$$\mathbf{B}^o(t) = \sum_{k=2}^{2n-1} \sum_{r=1}^{n-2-|n-k|} \sum_{\substack{i=1 \\ i-|n-k|\geq 1}}^{n} (\mathbf{e}_{k+i-n} \otimes \mathbf{e}_i)(\mathbf{g}_k^r)' m_{k,r,i},$$

where $\mathbf{e}_i : n \times 1$, $\mathbf{g}_k^r : (n-1)^2 \times 1$, the $j-th$ element of $\mathbf{g}_k^r$ equals

$$(\mathbf{g}_k^r)_j = \begin{cases} 1, & \text{if } j = \frac{1}{2}(k-1)(k-2) + r, \quad 2 \leq k \leq n,\ 1 \leq r \leq k-1, \\ 1, & \text{if } j = (n-1)^2 - \frac{1}{2}(2n-k-1)(2n-k-2) - r + 1, \\ & \qquad n < k \leq 2n-2,\ 1 \leq r \leq 2n-k-1, \\ 0, & \text{otherwise} \end{cases}$$

and

$$m_{k,r,i} = \begin{cases} -\dfrac{n-|n-k|-r}{\sqrt{(n-|n-k|-r)^2 + n - |n-k| - r}}, & i = 1 + |n-k|, \\ \dfrac{1}{\sqrt{(n-|n-k|-r)^2 + n - |n-k| - r}}, & \\ & i = 2 + |n-k|, 3 + |n-k|, \ldots, n-r+1, \\ 0, & \text{otherwise.} \end{cases}$$

PROOF: First we note that $\mathbf{B}(t)$ is of size $n^2 \times (2n-1)$ and $\mathbf{B}^o(t)$ is built up so that for each k we have $n-1-|n-k|$ columns in $\mathbf{B}^o(t)$ which are specified with the help of the index r. However, by definition of $m_{k,r,i}$, these columns are orthogonal to each other and thus $r(\mathbf{B}^o(t)) = n^2-(2n-1) = (n-1)^2$. Therefore it remains to show that $\mathbf{B}^{o'}(t)\mathbf{B}(t) = \mathbf{0}$ holds. It is enough to do this for a fixed k. Now,

$$\sum_{r=1}^{n-1-|n-k|} \sum_{\substack{i,i_1=1 \\ i,i_1\geq 1+|n-k|}}^{n} \mathbf{g}_k^r(\mathbf{e}_{k+i-n}\otimes \mathbf{e}_i)'(\mathbf{e}_{k+i_1-n}\otimes \mathbf{e}_{i_1})\mathbf{f}_k' m_{k,r,i}\mathrm{sgn}(a_{i_1 k+i_1-n})$$

$$= \sum_{r=1}^{n-1-|n-k|} \sum_{\substack{i=1 \\ i-|n-k|\geq 1}}^{n} \mathbf{g}_k^r\mathbf{f}_k' m_{k,r,i}\mathrm{sgn}(a_{i_1 k+i_1-n}).$$

Since by definition of a Toeplitz matrix $\mathrm{sgn}(a_{ik+i-n}) = \mathrm{sgn}(a_{n-k})$ is independent of i and

$$\sum_{\substack{i=1 \\ i-|n-k|\geq 1}}^{n} m_{k,r,i} = 0,$$

we have that $\mathbf{B}^{o'}(t)\mathbf{B}(t) = \mathbf{0}$. ∎

1.3.7 *Vectorization operators*

In this paragraph we are going to introduce notions which have been designed particularly for symmetric matrices and other matrices which comprise linear structures. These notions enable us to select all nonrepeated elements from the set of all possible moments of a certain order, or all partial derivatives of a given order, for example. The presentation of the material is based on the paper by Kollo & von Rosen (1995b). Let (i,j) stand for the number of combinations given by

$$(i,j) = \binom{i-1+j-1}{i-1}, \quad i,j = 1,2,\ldots, \tag{1.3.73}$$

and

$$(0,j) = (i,0) = 0.$$

Then (i,j) has the following properties which will be used later:

(i) $(i,j) = (j,i)$;

(ii) $(i,j) = (i,j-1) + (i-1,j)$;

(iii) $(i+1,j) = \sum_{k=1}^{j}(i,k)$. (1.3.74)

In particular,

$$(1,j)=1, \quad (2,j)=j, \quad (3,j)=j(j+1)/2, \quad (4,j)=\sum_{k=1}^{j}k(k+1)/2.$$

Moreover, using (ii) it is easy to construct the following table for small values of i and j, which is just a reformulation of Pascal's triangle. One advantage of this table is that (iii) follows immediately.

Table 1.3.1. The combinations (i, j) for $i, j \leq 7$.

	i						
j	1	2	3	4	5	6	7
1	1	1	1	1	1	1	1
2	1	2	3	4	5	6	7
3	1	3	6	10	15	21	28
4	1	4	10	20	35	56	84
5	1	5	15	35	70	126	210
6	1	6	21	56	126	252	462
7	1	7	28	84	210	462	924

In the following a new operator is defined which is called *vectorization operator* and which is fundamental to all results of this section.

Definition 1.3.9. *For any matrix* $\mathbf{A} : (j, n) \times n$, $j = 1, 2, \ldots$ *the vectorization operator is given by*

$$V^j(\mathbf{A}) = (a_{11}, a_{12}, \ldots, a_{(j,2)2}, a_{13}, \ldots, a_{(j,3)3}, \ldots, a_{1n}, \ldots, a_{(j,n)n})', \qquad (1.3.75)$$

where (i,j) is defined by (1.3.73). ∎

The vector $V^j(\mathbf{A})$ consists of $(j + 1, n)$ elements. In particular, for $j = 1, 2, 3$, we have

$$\begin{aligned} V^1(\mathbf{A}) =&(a_{11}, a_{12}, \ldots, a_{1n})', \\ V^2(\mathbf{A}) =&(a_{11}, a_{12}, a_{22}, a_{13}, a_{23}, a_{33}, \ldots, a_{1n}, a_{2n}, \ldots, a_{nn})', \\ V^3(\mathbf{A}) =&(a_{11}, a_{12}, a_{22}, a_{32}, a_{13}, a_{23}, a_{33}, a_{43}, a_{53}, a_{63}, \ldots, a_{\frac{n(n+1)}{2}n})'. \end{aligned}$$

In Definition 1.3.9 the index j is connected to the size of the matrix $\mathbf{A}$. In principle we could have omitted this restriction, and $V^j(\mathbf{A})$ could have been defined as a selection operator, e.g. $V^1(\mathbf{A})$ could have been an operator which picks out the first row of any matrix $\mathbf{A}$ and thereafter transposes it.
Now an operator R^j will be defined which makes it possible to collect, in a suitable way, all monomials obtained from a certain patterned matrix. With *monomials of a matrix* $\mathbf{A} = (a_{ij})$ we mean products of the elements a_{ij}, for example, $a_{11}a_{22}a_{33}$ and $a_{21}a_{32}a_{41}$ are monomials of order 3.

Definition 1.3.10. *For a patterned matrix* $\mathbf{A} : p \times q$, *the product vectorization operator* $R^j(\mathbf{A})$ *is given by*

$$R^j(\mathbf{A}) = V^j(R^{j-1}(\mathbf{A})\text{vec}'\mathbf{A}(K)), \quad j = 1, 2, \ldots, \qquad (1.3.76)$$

where $R^0(\mathbf{A}) = 1$ and $\mathbf{A}(K)$ is given in §1.3.6. ∎

In particular, for a symmetric $\mathbf{A}$, the operator $R^j(\mathbf{A})$ equals

$$R^j(\mathbf{A}) = V^j(R^{j-1}(\mathbf{A})V^2(\mathbf{A})'), \quad j = 1, 2, \ldots \tag{1.3.77}$$

From Definition 1.3.10 it follows that $R^j(\mathrm{vec}\mathbf{A}(K)) = R^j(\mathbf{A}) = R^j(\mathbf{A}(K))$. Moreover, let us see how $R^j(\bullet)$ transforms a $p-$vector $\mathbf{x} = (x_1, \ldots, x_p)'$. From Definition 1.3.10 it follows that

$$R^1(\mathbf{x}) = \mathbf{x}, \quad R^2(\mathbf{x}) = V^2(\mathbf{x}\mathbf{x}').$$

Thus, $R^2(\mathbf{x})$ is a vectorized upper triangle of the matrix $\mathbf{x}\mathbf{x}'$.

$$R^3(\mathbf{x}) = V^3(R^2(\mathbf{x})\mathbf{x}') = V^3(V^2(\mathbf{x}\mathbf{x}')\mathbf{x}'),$$

which forms a $(4,p)-$vector of all different monomials $x_i x_j x_k$, $i, j, k = 1, 2, \ldots$. In order to establish the fact that $R^j(\mathbf{A})$ represents all different monomials of order j, with respect to the elements a_{kl}, a set G_j of monomials of order j is introduced:

$$G_j = \left\{ \prod_{k,l} a_{kl}{}^{i_j(k,l)} : i_j(k,l) \in \{0, 1, \ldots, j\}, \sum_{k,l} i_j(k,l) = j \right\}.$$

To illustrate the set G_j let us look at the following example.

Example 1.3.5 Let

$$\mathbf{X} = \begin{pmatrix} x & 2x \\ e^x & -2 \end{pmatrix}.$$

Then $i_2(k,l) \in \{0, 1, 2\}$, $\quad \sum_{k,l} i_2(k,l) = 2$ and

$$G_2 = \left\{ x^2,\ 4x^2,\ e^{2x},\ 4,\ 2x^2,\ xe^x,\ -2x,\ 2xe^x,\ -4x,\ -2e^x \right\}.$$

∎

Theorem 1.3.13. *$R^j(\mathbf{A})$ consists of all elements in G_j and each element appears only once.*

PROOF: We shall present the framework of the proof and shall not go into details. The theorem is obviously true for $j = 1$ and $j = 2$. Suppose that the theorem holds for $j - 1$, i.e. $R^{j-1}(\mathbf{A})$ consist of elements from G_{j-1} in a unique way. Denote the first (j,k) elements in $R^{j-1}(\mathbf{A})$ by R_k^{j-1}. In the rest of the proof suppose that $\mathbf{A}$ is symmetric. For a symmetric $\mathbf{A} : n \times n$

$$R^{j-1}_{(3,n-1)+n} = R^{j-1}(\mathbf{A}),$$

and by assumption $R_k^{j-1}(\mathbf{A})$, $k = 1, 2, \ldots, (3, n-1)+n$, consist of unique elements. Now,

$$R^j(\mathbf{A}) = \mathrm{vec}((R_1^{j-1})'a_{11}, (R_2^{j-1})'a_{12}, \ldots, (R^{j-1}_{(3,n-1)+n-1})'a_{n-1n}, (R^{j-1}_{(3,n-1)+n})'a_{nn})$$

and since $R^{j-1}_{(3,n-1)+n-1}$ does not include a_{nn}, $R^{j-1}_{(3,n-1)+n-2}$ does not include a_{nn}, a_{n-1n}, $R^{j-1}_{(3,n-1)+n-3}$ does not include $a_{nn}, a_{n-1n}, a_{n-2n}$, etc., all elements appear only once in $R^j(\mathbf{A})$. ∎

The operator $R^j(\mathbf{A})$ will be illustrated by a simple example.

Example 1.3.6 Let

$$\mathbf{A} = \begin{pmatrix} a_{11} & a_{12} \\ a_{12} & a_{22} \end{pmatrix}$$

be a symmetric matrix. Then $\mathrm{vec}'(\mathbf{A}(K)) = (a_{11}, a_{12}, a_{22})$ and

$$\begin{aligned}
R^1(\mathbf{A}) =&\mathrm{vec}(\mathbf{A}(K)),\\
R^2(\mathbf{A}) =&V^2(R^1(\mathbf{A})\mathrm{vec}'(\mathbf{A}(K)))\\
=&V^2\begin{pmatrix} a_{11}^2 & a_{11}a_{12} & a_{11}a_{22} \\ a_{12}a_{11} & a_{12}^2 & a_{12}a_{22} \\ a_{22}a_{11} & a_{22}a_{12} & a_{22}^2 \end{pmatrix} = (a_{11}^2, a_{11}a_{12}, a_{12}^2, a_{11}a_{22}, a_{12}a_{22}, a_{22}^2)',\\
R^3(\mathbf{A}) =&V^3(R^2(\mathbf{A})\mathrm{vec}'(\mathbf{A}(K)))\\
=&V^3\begin{pmatrix} a_{11}^3 & a_{11}^2a_{12} & a_{11}^2a_{22} \\ a_{11}^2a_{12} & a_{11}a_{12}^2 & a_{11}a_{12}a_{22} \\ a_{11}a_{12}^2 & a_{12}^3 & a_{12}^2a_{22} \\ a_{11}^2a_{22} & a_{11}a_{12}a_{22} & a_{11}a_{22}^2 \\ a_{11}a_{12}a_{22} & a_{12}^2a_{22} & a_{12}a_{22}^2 \\ a_{11}a_{22}^2 & a_{12}a_{22}^2 & a_{22}^3 \end{pmatrix}\\
=&(a_{11}^3, a_{11}^2a_{12}, a_{11}a_{12}^2, a_{12}^3, a_{11}^2a_{22}, a_{11}a_{12}a_{22}, a_{12}^2a_{22}, a_{11}a_{22}^2, a_{12}a_{22}^2, a_{22}^3)'.
\end{aligned} \tag{1.3.78}$$

Moreover,

$$\begin{aligned}
&R_1^2 = a_{11}^2,\ \ R_2^2 = (a_{11}^2, a_{11}a_{12}, a_{12}^2)',\ \ R_3^2 = R^2(\mathbf{A}),\ \ R_1^3 = a_{11}^3,\\
&R_2^3 = (a_{11}^3, a_{11}^2a_{12}, a_{11}a_{12}^2, a_{12}^3)' \quad \text{and} \quad R_3^3 = R^3(\mathbf{A}).
\end{aligned}$$

We also see that

$$R^3(\mathbf{A}) = \mathrm{vec}((R_1^2)'a_{11}, (R_2^2)'a_{12}, (R_3^2)'a_{33}).$$

∎

Let, as before, $\mathbf{e}_m$ denote the $m-$th unit vector, i.e. the $m-$th coordinate of $\mathbf{e}_m$ equals 1 and other coordinates equal zero. The next theorem follows from Definition 1.3.9.

Theorem 1.3.14. *Let $\mathbf{A} : (s,n) \times n$ and $\mathbf{e}_m : (s+1,n) - vector$. Then*

$$V^s(\mathbf{A}) = \sum_i \sum_j a_{ij}\mathbf{e}_{(s+1,j-1)+i}, \quad 1 \le i \le (s,j),\, 1 \le j \le n.$$

∎

By using Definition 1.3.10 and Theorem 1.3.14 repeatedly, the next theorem can be established.

Theorem 1.3.15. *For a patterned matrix* $\mathbf{A}$*, let* $\mathrm{vec}\mathbf{A}(K) = \mathbf{c} = (c_i)$, $i = 1, 2, \ldots, n$. *Then*

$$R^j(\mathbf{A}) = \sum_{I_j} (\prod_{r=1}^{j} c_{i_r}) \mathbf{e}_{u_j},$$

where

$$I_j = \{(i_1, i_2, \ldots, i_j) : i_g = 1, 2, \ldots, n; 1 \le u_{k-1} \le (k, i_k), \quad k = 1, 2, \ldots, j\}$$

and

$$u_s = 1 + \sum_{r=2}^{s+1} (r, i_{r-1} - 1), \quad u_0 = 1.$$

PROOF: The theorem obviously holds for $j = 1$. Furthermore,

$$\mathbf{cc}' = \sum_{i_1, i_2} c_{i_1} c_{i_2} \mathbf{e}_{i_1} \mathbf{e}'_{i_2}$$

and since by Theorem 1.3.14

$$V^2(\mathbf{e}_{u_1} \mathbf{e}'_{i_2}) = \mathbf{e}_{u_2},$$

the theorem is true for $j = 2$. Now suppose that the theorem is true for $j - 1$. For j we have

$$R^j(\mathbf{A}) = V^j(R^{j-1}(\mathbf{A})\mathbf{c}')$$

and since

$$V^j(\mathbf{e}_{u_{j-1}} \mathbf{e}'_j) = \mathbf{e}_{u_j}$$

the theorem is established. ∎

If we want to express $R^j(\mathbf{A})$ through the elements of $\mathbf{A}$, instead of coordinates of $\mathrm{vec}\mathbf{A}(K)$, we have to know how $\mathrm{vec}\mathbf{A}(K)$ has been constructed. The index set I_j and c_i are then transformed to the index set K and a_{kl}, respectively. If $\mathbf{A}$ has some known structure we can give more details, i.e. we can present $R^j(\mathbf{A})$ through the elements of $\mathbf{A}$ instead of using $\mathrm{vec}\mathbf{A}(K)$. In particular, if $\mathbf{A}$ is symmetric, we obtain

Theorem 1.3.16. *For symmetric* $\mathbf{A} : n \times n$

$$R^j(\mathbf{A}) = \sum_{I_j} (\prod_{r=1}^{j} a_{i_{2r-1} i_{2r}}) \mathbf{e}_{u_j}$$

where

$$\begin{aligned} I_j =& \{(i_1, i_2, \ldots, i_j) : i_g = 1, 2, \ldots, n; i_{2k-1} \le i_{2k}; \\ & 1 \le u_k \le (k+1, (3, i_{2k+2} - 1) + i_{2k+1}), \quad k = 0, \ldots, j-1\} \end{aligned}$$

and

$$u_j = 1 + \sum_{k=2}^{j+1} (k, (3, i_{2k-2} - 1) + i_{2k-3} - 1), \quad u_0 = 1.$$

∎

For a better understanding of Theorem 1.3.15 and Theorem 1.3.16 we suggest the reader to study Example 1.3.6.

1.3.8 *Problems*

1. Let $\mathbf{V}$ be positive definite and set $\mathbf{S} = \mathbf{V} + \mathbf{BKB}'$, where the matrix $\mathbf{K}$ is such that $\mathcal{C}(\mathbf{S}) = \mathcal{C}(\mathbf{V} : \mathbf{B})$. Show that

$$\mathbf{S} = \mathbf{V}\mathbf{B}^o(\mathbf{B}^{o'}\mathbf{V}\mathbf{B}^o)^-\mathbf{B}^{o'}\mathbf{V} + \mathbf{B}(\mathbf{B}'\mathbf{S}\mathbf{B})^-\mathbf{B}'.$$

2. Extend Theorem 1.3.8 and give necessary and sufficient conditions for

$$\mathbf{A}_1\mathbf{X}_1\mathbf{B}_1 + \mathbf{A}_2\mathbf{X}_2\mathbf{B}_2 = \mathbf{C}$$

   to have a solution. Find the solution (Baksalary & Kala, 1980).
3. Verify Corollary 1.3.6.1.
4. If $(\mathbf{A} : \mathbf{B})$ has full column rank, show that the Moore-Penrose inverse of $(\mathbf{A} : \mathbf{B})$ equals

$$\begin{aligned}(\mathbf{A} : \mathbf{B})^+ &= \begin{pmatrix} (\mathbf{A}'\mathbf{Q}_\mathbf{B}\mathbf{A})^{-1}\mathbf{A}'\mathbf{Q}_\mathbf{B} \\ (\mathbf{B}'\mathbf{B})^{-1}\mathbf{B}'(\mathbf{I} - \mathbf{A}(\mathbf{A}'\mathbf{Q}_\mathbf{B}\mathbf{A})^{-1}\mathbf{A}'\mathbf{Q}_\mathbf{B}) \end{pmatrix} \\ &= \begin{pmatrix} (\mathbf{A}'\mathbf{A})^{-1}\mathbf{A}'(\mathbf{I} - \mathbf{B}(\mathbf{B}'\mathbf{Q}_\mathbf{A}\mathbf{B})^{-1}\mathbf{B}'\mathbf{Q}_\mathbf{A}) \\ (\mathbf{B}'\mathbf{Q}_\mathbf{A}\mathbf{B})^{-1}\mathbf{B}'\mathbf{Q}_\mathbf{A} \end{pmatrix},\end{aligned}$$

   where

$$\begin{aligned}\mathbf{Q}_\mathbf{A} &= \mathbf{I} - \mathbf{A}(\mathbf{A}'\mathbf{A})^{-1}\mathbf{A}', \\ \mathbf{Q}_\mathbf{B} &= \mathbf{I} - \mathbf{B}(\mathbf{B}'\mathbf{B})^{-1}\mathbf{B}'.\end{aligned}$$

5. Prove Proposition 1.3.12 (x) – (xiii).
6. Present the sample dispersion matrix

$$\mathbf{S} = \frac{1}{n-1}\sum_{i=1}^{n}(\mathbf{x}_i - \overline{\mathbf{x}})(\mathbf{x}_i - \overline{\mathbf{x}})'$$

   without using the summation operator.
7. Prove that

$$\mathbf{a}^{\otimes 3} = (\mathbf{K}_{p,p} \otimes \mathbf{I}_p)\mathbf{K}_{p^2,p}\mathbf{a}^{\otimes 3}.$$

8. Prove Proposition 1.3.14 (ii), (iii) and (vi).

9. For any $\mathbf{A}$: $p \times n$ from the class of semiorthogonal matrices, i.e. $\mathbf{AA}' = \mathbf{I}_p$, which will be denoted so_{pn}, there exist $r = np - \frac{1}{2}p(p+1)$ "free" elements. Show that

$$\mathbf{T}^{+}(so_{pn}) = \sum_{\substack{i<j=1 \\ 1\leq i\leq p}}^{n} \sum_{s=1}^{r} (\mathbf{e}_j \otimes \mathbf{d}_i)\mathbf{f}'_s 1_{\{k(i,j)=s\}} \mathrm{sgn}(a_{ij}), \qquad r = np - \frac{1}{2}p(p+1),$$

where $k(i,j) = n(i-1) - \frac{1}{2}i(i+1) + j$, $j > i$ and $k(i,j) = 0$, $i \geq j$. Find also $\mathbf{B}(so_{pn})$, $\mathbf{T}(so_{pn})$ and $\mathbf{M}(so_{pn})$.
10. Prove Proposition 1.3.11 (ii).

1.4 MATRIX DERIVATIVES

1.4.1 *Introduction*

The matrix derivative is one of the key notions in matrix theory for multivariate analysis. Solving extreme value problems, finding maximum likelihood estimators, deriving parameters of multivariate limit distributions and expanding statistics into asymptotic series may all serve as examples where matrix derivatives play an important role. Generally speaking, in all these problems one has to find a derivative of one matrix, say $\mathbf{Y}$, by another matrix $\mathbf{X}$, which is the main issue of this section. Matrix derivative is a comparatively new notion among the tools for multivariate analysis. At the same time the notion has been available for more than 70 years, at least. Probably the first papers on the topic were written by Turnbull (1927, 1930, 1931) who examined a differential operator $\frac{\partial}{\partial \mathbf{X}}$ of the form of a $p \times q-$matrix. Turnbull applied it to Taylor's theorem and for differentiating characteristic functions. Today one can refer to two main types of matrix derivatives which are mathematically equivalent representations of the Fréchet derivative as we will see in the next paragraph. The real starting point for a modern presentation of the topic is the paper by Dwyer & MacPhail (1948). Their ideas were further developed in one direction by Bargmann (1964) and the theory obtained its present form in the paper of MacRae (1974). This work has later on been continued in many other papers. The basic idea behind these papers is to order all partial derivatives $\frac{\partial y_{kl}}{\partial x_{ij}}$ by using the Kronecker product while preserving the original matrix structure of $\mathbf{Y}$ and $\mathbf{X}$. Therefore, these methods are called *Kronecker arrangement methods*. Another group of matrix derivatives is based on vectorized matrices and therefore it is called *vector arrangement methods*. Origin of this direction goes back to 1969 when two papers appeared: Neudecker (1969) and Tracy & Dwyer (1969). Since then many contributions have been made by several authors, among others McDonald & Swaminathan (1973) and Bentler & Lee (1975). Because of a simple chain rule for differentiating composite functions, this direction has become somewhat more popular than the Kroneckerian arrangement and it is used in most of the books on this topic. The first monograph was written by Rogers (1980). From later books we refer to Graham (1981), Magnus & Neudecker (1999) and Kollo (1991). Nowadays the monograph by Magnus & Neudecker has become the main reference in this area.

In this section most of the results will be given with proofs, since this part of the matrix theory is the newest and has not been systematically used in multivariate analysis. Additionally, it seems that the material is not so widely accepted among statisticians.

Mathematically the notion of a matrix derivative is a realization of the Fréchet derivative known from functional analysis. The problems of existence and representation of the matrix derivative follow from general properties of the Fréchet derivative. That is why we are first going to give a short overview of Fréchet derivatives. We refer to §1.2.6 for basic facts about matrix representations of linear operators. However, in §1.4.1 we will slightly change the notation and adopt the terminology from functional analysis. In §1.2.4 we considered linear transformations from one vector space to another, i.e. $A : \mathbb{V} \to \mathbb{W}$. In this section we will consider a linear map $f : \mathbb{V} \to \mathbb{W}$ instead, which is precisely the same if $\mathbb{V}$ and $\mathbb{W}$ are vector spaces.

1.4.2 *Fréchet derivative and its matrix representation*

Relations between the Fréchet derivative and its matrix representations have not been widely considered. We can refer here to Wrobleski (1963) and Parring (1992), who have discussed relations between matrix derivatives, Fréchet derivatives and Gâteaux derivatives when considering existence problems. As the existence of the matrix derivative follows from the existence of the Fréchet derivative, we are going to present the basic framework of the notion in the following. Let f be a mapping from a normed linear space $\mathbb{V}$ to a normed linear space $\mathbb{W}$, i.e.

$$f : \mathbb{V} \longrightarrow \mathbb{W}.$$

The mapping f is *Fréchet differentiable* at $\mathbf{x}$ if the following representation of f takes place:

$$f(\mathbf{x} + \mathbf{h}) = f(\mathbf{x}) + D_{\mathbf{x}}\mathbf{h} + \varepsilon(\mathbf{x}, \mathbf{h}), \tag{1.4.1}$$

where $D_{\mathbf{x}}$ is a linear continuous operator, and uniformly for each $\mathbf{h}$

$$\frac{\| \varepsilon(\mathbf{x}, \mathbf{h}) \|}{\| \mathbf{h} \|} \longrightarrow 0,$$

if $\| \mathbf{h} \| \longrightarrow 0$, where $\| \bullet \|$ denotes the norm in $\mathbb{V}$. The linear continuous operator

$$D_{\mathbf{x}} : \mathbb{V} \longrightarrow \mathbb{W}$$

is called the *Fréchet derivative* (or strong derivative) of the mapping f at $\mathbf{x}$. The operator $D_{\mathbf{x}}$ is usually denoted by $f'(\mathbf{x})$ or $Df(\mathbf{x})$. The term $D_{\mathbf{x}}\mathbf{h}$ in (1.4.1) is called the *Fréchet differential* of f at $\mathbf{x}$. The mapping f is differentiable in the set A if f is differentiable for every $\mathbf{x} \in A$. The above defined derivative comprises the most important properties of the classical derivative of real valued functions (see Kolmogorov & Fomin, 1970, pp. 470–471; Spivak, 1965, pp. 19–22, for example). Here are some of them stated.

Proposition 1.4.1.

(i) *If $f(\mathbf{x}) = const.$, then $f'(\mathbf{x}) = 0$.*

(ii) *The Fréchet derivative of a linear continuous mapping is the mapping itself.*

(iii) *If f and g are differentiable at $\mathbf{x}$, then $(f + g)$ and (cf), where c is a constant, are differentiable at $\mathbf{x}$ and*

$$(f + g)'(\mathbf{x}) = f'(\mathbf{x}) + g'(\mathbf{x}),$$

$$(cf)'(\mathbf{x}) = cf'(\mathbf{x}).$$

(iv) *Let $\mathbb{V}, \mathbb{W}$ and $\mathbb{U}$ be normed spaces and mappings*

$$f : \mathbb{V} \longrightarrow \mathbb{W}; \qquad g : \mathbb{W} \longrightarrow \mathbb{U}$$

such that f is differentiable at $\mathbf{x}$ and g is differentiable at $y = f(\mathbf{x})$. Then the composition $h = g \odot f$; $h : \mathbb{V} \longrightarrow \mathbb{U}$ is differentiable at $\mathbf{x}$ and

$$h'(\mathbf{x}) = g'(\mathbf{y}) \odot f'(\mathbf{x}).$$

∎

It is important to have a derivative where property (iv) holds. This enables us to differentiate composite functions. Another well-known variant, the Gâteaux derivative (or weak derivative) does not have this property, for instance.
From §1.2.6 it follows that being a linear operator, the Fréchet derivative can be represented in a matrix form in the case of finite-dimensional vector spaces. Denote the basis of $\mathbb{W}$ by $\{\mathbf{d}_j\}$. Then the mapping $f : \mathbb{V} \longrightarrow \mathbb{W}$ can be presented, at any point $\mathbf{x}$ of $\mathbb{V}$, by

$$f(\mathbf{x}) = \sum_{j \in J} f_j(\mathbf{x})\mathbf{d}_j.$$

Here J is connected to the set of vectors $\mathbf{x} \in \mathbb{V}$, for which f is differentiable.
If there exists a derivative $f'(\mathbf{x})$, then the partial derivatives $\frac{\partial f_j(\mathbf{x})}{\partial x_i}$, $i \in I, j \in J$ exist and the matrix representing the Fréchet derivative is given by a matrix of partial derivatives (see Spivak, 1965, for example). The opposite is not true, i.e. in general from the existence of partial derivatives the existence of a Fréchet derivative does not follow. Sufficient and necessary conditions for the existence of $f'(\mathbf{x})$ are not easy to use (Spivak, 1965, pp. 20–21): $f'(\mathbf{x})$ exists if and only if every coordinate function f_k is differentiable at $\mathbf{x}$. Fortunately there exist comparatively simple sufficient conditions for the existence of Fréchet derivatives (Spivak, 1965, p. 31).

Theorem 1.4.1. *Let $f : \mathbb{V} \longrightarrow \mathbb{W}$ be a mapping such that all partial derivatives $\frac{\partial f_j(\mathbf{x})}{\partial x_i}$, $i \in I, j \in J$, exist in an open set which includes the point $\mathbf{x}_0$, and are continuous at $\mathbf{x}_0$. Then f is Fréchet differentiable at $\mathbf{x}_0$.* ∎

As we saw in §1.2.6, the form of the matrix which represents the Fréchet derivative as a linear operator has to be fixed by convention. If we have $\mathbb{R}^{p\times q}$ and $\mathbb{R}^{r\times s}$ as $\mathbb{V}$ and $\mathbb{W}$, respectively, then two natural realizations of the Fréchet derivative are given by the relations (1.2.9) and (1.2.11). In both cases the Fréchet differential is given by the same equality

$$Df(\mathbf{x})\mathbf{h} = \sum_{j \in J} (Df_j(\mathbf{x})\mathbf{h})\mathbf{d}_j.$$

Let us examine, what kind of matrix representations of the Fréchet derivative we get when starting from the different orderings (1.2.9) and (1.2.11), respectively. In (1.2.9) the matrix of the linear operator consists of $p \times q-$blocks, where the $(tu)-$th block includes all $(tu)-$th coordinates of the basis vectors in $\mathbb{R}^{p\times q}$. For the Fréchet derivative it means that in the $(tu)-$th block we have partial derivatives of

y_{tu} with respect to the elements of $\mathbf{X}$. It is convenient to present all the elements jointly in a representation matrix $\frac{d\mathbf{Y}}{d\mathbf{X}}$ in the following way (MacRae, 1974):

$$\frac{d\mathbf{Y}}{d\mathbf{X}} = \mathbf{Y} \otimes \frac{\partial}{\partial \mathbf{X}}, \tag{1.4.2}$$

where the partial differentiation operator $\frac{\partial}{\partial \mathbf{X}}$ is of the form

$$\frac{\partial}{\partial \mathbf{X}} = \begin{pmatrix} \frac{\partial}{\partial x_{11}} & \cdots & \frac{\partial}{\partial x_{1q}} \\ \vdots & \ddots & \vdots \\ \frac{\partial}{\partial x_{p1}} & \cdots & \frac{\partial}{\partial x_{pq}} \end{pmatrix}.$$

The Kronecker product $\mathbf{Y} \otimes \frac{\partial}{\partial \mathbf{X}}$ means that the operator of partial differentiation is applied to the elements of the matrix $\mathbf{Y}$ by the rule of the Kronecker product: the tu–th block of $\frac{d\mathbf{Y}}{d\mathbf{X}}$ is of the form

$$\left[\frac{d\mathbf{Y}}{d\mathbf{X}}\right]_{tu} = \begin{pmatrix} \frac{\partial y_{tu}}{\partial x_{11}} & \cdots & \frac{\partial y_{tu}}{\partial x_{1q}} \\ \vdots & \ddots & \vdots \\ \frac{\partial y_{tu}}{\partial x_{p1}} & \cdots & \frac{\partial y_{tu}}{\partial x_{pq}} \end{pmatrix}, \qquad t = 1, \ldots, r;\ u = 1, \ldots, s.$$

Another way of representing (1.4.2) is given by

$$\frac{d\mathbf{Y}}{d\mathbf{X}} = \sum_{i,j,k,l} \frac{\partial y_{ij}}{\partial x_{kl}} \mathbf{r}_i \mathbf{s}_j' \otimes \mathbf{f}_k \mathbf{g}_l',$$

where $\mathbf{r}_i$, $\mathbf{s}_j$, $\mathbf{f}_k$ and $\mathbf{g}_l$ are unit basis vectors, as in (1.1.2), with dimensionalities r, s, p and q, respectively.

If the elements of our linear operator are ordered into a matrix by (1.2.11), then the (uv)–th $r \times s$–block consists of the coordinates of the (uv)–th basis vector. The matrix which represents the Fréchet derivative consists of $r \times s$–blocks where in the (uv)–th block we have the partial derivatives of the elements of $\mathbf{Y}$ by x_{uv}. The set of all partial derivatives is now easy to present as

$$\frac{d\mathbf{Y}}{d\mathbf{X}} = \frac{\partial}{\partial \mathbf{X}} \otimes \mathbf{Y}, \tag{1.4.3}$$

where the operator $\frac{\partial}{\partial \mathbf{X}}$ is given as before. The product in (1.4.3) means that the (ij)–th block of the matrix derivative equals

$$\left[\frac{d\mathbf{Y}}{d\mathbf{X}}\right]_{ij} = \begin{pmatrix} \frac{\partial y_{11}}{\partial x_{ij}} & \cdots & \frac{\partial y_{1s}}{\partial x_{ij}} \\ \vdots & \vdots & \vdots \\ \frac{\partial y_{r1}}{\partial x_{ij}} & \cdots & \frac{\partial y_{rs}}{\partial x_{ij}} \end{pmatrix}, \qquad i = 1, \ldots, p;\ j = 1, \ldots, q.$$

Here
$$\frac{d\mathbf{Y}}{d\mathbf{X}} = \sum_{i,j,k,l} \frac{\partial y_{ij}}{\partial x_{kl}} \mathbf{f}_k \mathbf{g}_l' \otimes \mathbf{r}_i \mathbf{s}_j',$$
where the basis vectors are defined as before.

Before proceeding we note that the Fréchet derivative is a linear map of an element $\mathbf{x}$ in a linear space. By Proposition 1.4.1 (iii) it follows that if we write $f'(\mathbf{x}) = (f', \mathbf{x})$,
$$(c_1 f_1' + c_2 f_2', \mathbf{x}) = c_1(f_1', \mathbf{x}) + c_2(f_2', \mathbf{x}),$$
by linearity of f' and by linearity in the argument
$$(f', c_1\mathbf{x}_1 + c_2\mathbf{x}_2) = c_1(f', \mathbf{x}_1) + c_2(f', \mathbf{x}_2),$$
where c_1 and c_2 are constants. Hence, using Definition 1.2.10 (ii) and assuming that (i) of the definition holds, which can be shown to be the case, it follows that $f'(\mathbf{x})$ generates a tensor space. This explains, via Theorem 1.2.24, in a more abstract way why, for example, (1.4.2) is a representation of the Fréchet derivative.

The next two popular representations of the Fréchet derivative arise from the following argument. Let us define the norm in $\mathbb{R}^{p\times q}$ as
$$\| \mathbf{X} \| = \sqrt{\mathrm{tr}(\mathbf{X}'\mathbf{X})} = \sqrt{\mathrm{vec}'\mathbf{X}\mathrm{vec}\mathbf{X}},$$
where $\mathbf{X} \in \mathbb{R}^{p\times q}$. Then the spaces $\mathbb{R}^{p\times q}$ and $\mathbb{R}^{pq}$ are isometric, which means that a study of
$$f : \mathbb{R}^{p\times q} \longrightarrow \mathbb{R}^{r\times s}$$
can be replaced by a study of
$$f : \mathbb{R}^{pq} \longrightarrow \mathbb{R}^{rs}.$$

Using the vectorized representation of a matrix, we get with the help of the vec-operator a matrix representation of the Fréchet derivative as in Euclidian spaces: coordinates of the i–th basis vector are kept as i–th row or i–th column of the representation matrix. If we keep the coordinates of the image of the vector $\mathbf{e}_i \in \mathbb{V}$ in the i–th row, the matrix $\frac{d\mathbf{Y}}{d\mathbf{X}}$ is given by the equality
$$\frac{d\mathbf{Y}}{d\mathbf{X}} = \frac{\partial}{\partial \mathrm{vec}\mathbf{X}} \mathrm{vec}'\mathbf{Y} = \mathrm{vec}'\mathbf{Y} \otimes \frac{\partial}{\partial \mathrm{vec}\mathbf{X}}, \tag{1.4.4}$$
where
$$\frac{\partial}{\partial \mathrm{vec}\mathbf{X}} = \left(\frac{\partial}{\partial x_{11}}, \ldots, \frac{\partial}{\partial x_{p1}}, \frac{\partial}{\partial x_{12}}, \ldots, \frac{\partial}{\partial x_{p2}}, \ldots, \frac{\partial}{\partial x_{1q}}, \ldots, \frac{\partial}{\partial x_{pq}} \right)'$$
and the direct product in (1.4.4) defines the element in $\mathrm{vec}\mathbf{Y}$ which is differentiated by the operator $\frac{\partial}{\partial \mathrm{vec}\mathbf{X}}$. Through the basis vectors the derivative in (1.4.4) may be presented in the following way:
$$\frac{d\mathbf{Y}}{d\mathbf{X}} = \sum_{i,j,k,l} \frac{\partial y_{ij}}{\partial x_{kl}} (\mathbf{g}_l \otimes \mathbf{f}_k)(\mathbf{s}_j \otimes \mathbf{r}_i)' = \sum_{i,j,k,l} \frac{\partial y_{ij}}{\partial x_{kl}} (\mathbf{g}_l \mathbf{s}_j') \otimes (\mathbf{f}_k \mathbf{s}_i'), \tag{1.4.5}$$

where $\mathbf{f}_k$, $\mathbf{g}_l$, $\mathbf{r}_i$, $\mathbf{s}_j$ are the basis vectors of sizes p, q, r and s, respectively. If the coordinates of the image of $\mathbf{e}_i \in \mathbb{V}$ form the $i-$th column of the representation matrix of the Fréchet derivative, the matrix is given by the equality (Neudecker, 1969)

$$\frac{d\mathbf{Y}}{d\mathbf{X}} = \left(\frac{\partial}{\partial \mathrm{vec}\mathbf{X}}\right)' \otimes \mathrm{vec}\mathbf{Y} \tag{1.4.6}$$

which is identical to

$$\frac{d\mathbf{Y}}{d\mathbf{X}} = \sum_{i,j,k,l} \frac{\partial y_{ij}}{\partial x_{kl}}(\mathbf{s}_j \otimes \mathbf{r}_i)(\mathbf{g}_l \otimes \mathbf{f}_k)' = \sum_{i,j,k,l} \frac{\partial y_{ij}}{\partial x_{kl}}(\mathbf{s}_j\mathbf{g}_l') \otimes (\mathbf{r}_i\mathbf{f}_k'),$$

where the basis vectors are defined as in (1.4.5). The equalities (1.4.2), (1.4.3), (1.4.4) and (1.4.6) represent different forms of matrix derivatives found in the literature. Mathematically it is evident that they all have equal rights to be used in practical calculations but it is worth observing that they all have their pros and cons. The main reason why we prefer the vec-arrangement, instead of the Kroneckerian arrangement of partial derivatives, is the simplicity of differentiating composite functions. In the Kroneckerian approach, i.e. (1.4.2) and (1.4.3), an additional operation of matrix calculus, namely the *star product* (MacRae, 1974), has to be introduced while in the case of vec-arrangement we get a chain rule for differentiating composite functions which is analogous to the univariate chain rule. Full analogy with the classical chain rule will be obtained when using the derivative in (1.4.6). This variant of the matrix derivative is used in the books by Magnus & Neudecker (1999) and Kollo (1991). Unfortunately, when we are going to differentiate characteristic functions by using (1.4.6), we shall get an undesirable result: the first order moment equals the transposed expectation of a random vector. It appears that the only relation from the four different variants mentioned above, which gives the first two moments of a random vector in the form of a vector and a square matrix and at the same time supplies us with the chain rule, is given by (1.4.4). This is the main reason why we prefer this way of ordering the coordinates of the images of the basis vectors, and in the next paragraph that derivative will be exploited.

1.4.3 *Matrix derivatives, properties*

As we have already mentioned in the beginning of this chapter, generally a matrix derivative is a derivative of one matrix $\mathbf{Y}$ by another matrix $\mathbf{X}$. Many authors follow McDonald & Swaminathan (1973) and assume that the matrix $\mathbf{X}$ by which we differentiate is *mathematically independent and variable.* This will be abbreviated m.i.v. It means the following:

a) the elements of $\mathbf{X}$ are non-constant;
b) no two or more elements are functionally dependent.

The restrictiveness of this assumption is obvious because it excludes important classes of matrices, such as symmetric matrices, diagonal matrices, triangular matrices, etc. for $\mathbf{X}$. At the same time the Fréchet derivative, which serves as mathematical basis of matrix derivatives, can be found if we include two natural assumptions.

a) If $x_{ij} = \text{const.}$, then $\dfrac{\partial y}{\partial x_{ij}} = 0$ for any y.

b) If $x_{ij} = x_{kl}$ for some pairs of indices $(i,j) \neq (k,l)$, then $\dfrac{\partial y}{\partial x_{ij}} = \dfrac{\partial y}{\partial x_{kl}}$ for any y which is differentiable by x_{ij}.

Alternatively we could have developed a theory for matrix derivatives where we would have taken into account that matrices with linear relationships among the elements could have been presented via defining a specific basis.

Definition 1.4.1. *Let the elements of* $\mathbf{Y} \in \mathbb{R}^{r\times s}$ *be functions of* $\mathbf{X} \in \mathbb{R}^{p\times q}$. *The matrix* $\dfrac{d\mathbf{Y}}{d\mathbf{X}} \in \mathbb{R}^{pq\times rs}$ *is called matrix derivative of* $\mathbf{Y}$ *by* $\mathbf{X}$ *in a set* A, *if the partial derivatives* $\dfrac{\partial y_{kl}}{\partial x_{ij}}$ *exist, are continuous in* A, *and*

$$\frac{d\mathbf{Y}}{d\mathbf{X}} = \frac{\partial}{\partial \text{vec}\mathbf{X}}\text{vec}'\mathbf{Y}, \tag{1.4.7}$$

where

$$\frac{\partial}{\partial \text{vec}\mathbf{X}} = \left(\frac{\partial}{\partial x_{11}}, \ldots, \frac{\partial}{\partial x_{p1}}, \frac{\partial}{\partial x_{12}}, \ldots, \frac{\partial}{\partial x_{p2}}, \ldots, \frac{\partial}{\partial x_{1q}}, \ldots, \frac{\partial}{\partial x_{pq}}\right)'. \tag{1.4.8}$$

■

It means that (1.4.4) is used for defining the matrix derivative. At the same time we observe that the matrix derivative (1.4.7) remains the same if we change $\mathbf{X}$ and $\mathbf{Y}$ to vec$\mathbf{X}$ and vec$\mathbf{Y}$, respectively. So we have the identity

$$\frac{d\mathbf{Y}}{d\mathbf{X}} \equiv \frac{d\text{vec}'\mathbf{Y}}{d\text{vec}\mathbf{X}}.$$

The derivative used in the book by Magnus & Neudecker (1999) is the transposed version of (1.4.7), i.e. (1.4.6).

We are going to use several properties of the matrix derivative repeatedly. Since there is no good reference volume where all the properties can be found, we have decided to present them with proofs. Moreover, in §1.4.9 we have collected most of them into Table 1.4.2. If not otherwise stated, the matrices in the propositions will have the same sizes as in Definition 1.4.1. To avoid zero columns in the derivatives we assume that $x_{ij} \neq \text{const.}$, $i = 1, \ldots, p$, $j = 1, \ldots, q$.

Proposition 1.4.2. *Let* $\mathbf{X} \in \mathbf{R}^{p\times q}$ *and the elements of* $\mathbf{X}$ *m.i.v.*

(i) *Then*

$$\frac{d\mathbf{X}}{d\mathbf{X}} = \mathbf{I}_{pq}. \tag{1.4.9}$$

(ii) *Let* c *be a constant. Then*

$$\frac{d(c\mathbf{X})}{d\mathbf{X}} = c\mathbf{I}_{pq}. \tag{1.4.10}$$

(iii) *Let* $\mathbf{A}$ *be a matrix of proper size with constant elements. Then*

$$\frac{d\mathbf{A}'\mathbf{X}}{d\mathbf{X}} = \mathbf{I}_q \otimes \mathbf{A}. \tag{1.4.11}$$

(iv) *Let* $\mathbf{A}$ *be a matrix of proper size with constant elements. Then*

$$\frac{d(\mathbf{A}'\text{vec}\mathbf{X})}{d\mathbf{X}} = \mathbf{A}. \tag{1.4.12}$$

(v) *Let* $\mathbf{Z}$ *and* $\mathbf{Y}$ *be of the same size. Then*

$$\frac{d(\mathbf{Y}+\mathbf{Z})}{d\mathbf{X}} = \frac{d\mathbf{Y}}{d\mathbf{X}} + \frac{d\mathbf{Z}}{d\mathbf{X}}. \tag{1.4.13}$$

PROOF: The statements (ii) and (v) follow straightforwardly from the definition in (1.4.7). To show (i), (1.4.5) is used:

$$\begin{aligned}\frac{d\mathbf{X}}{d\mathbf{X}} &= \sum_{ijkl} \frac{\partial x_{ij}}{\partial x_{kl}} (\mathbf{e}_l \otimes \mathbf{d}_k)(\mathbf{e}_j \otimes \mathbf{d}_i)' = \sum_{ij} (\mathbf{e}_j \otimes \mathbf{d}_i)(\mathbf{e}_j \otimes \mathbf{d}_i)' \\ &= \sum_{ij} (\mathbf{e}_j\mathbf{e}_j') \otimes (\mathbf{d}_i\mathbf{d}_i') = \mathbf{I}_p \otimes \mathbf{I}_q = \mathbf{I}_{pq},\end{aligned}$$

which completes the proof of (i).
Statement (iii) follows from the next chain of equalities:

$$\frac{d\mathbf{A}'\mathbf{X}}{d\mathbf{X}} = \frac{d\mathbf{X}}{d\mathbf{X}}(\mathbf{I} \otimes \mathbf{A}) = \mathbf{I} \otimes \mathbf{A}.$$

It remains to prove (iv):

$$\frac{d\mathbf{A}'\text{vec}\mathbf{X}}{d\mathbf{X}} = \frac{d\text{vec}'\mathbf{X}\mathbf{A}}{d\mathbf{X}} = \frac{d\mathbf{X}}{d\mathbf{X}}\mathbf{A} = \mathbf{A}.$$

∎

Proposition 1.4.3. *(chain rule) Let* $\mathbf{Z} : t \times u$ *be a function of* $\mathbf{Y}$ *and* $\mathbf{Y}$ *be a function of* $\mathbf{X}$. *Then*

$$\frac{d\mathbf{Z}}{d\mathbf{X}} = \frac{d\mathbf{Y}}{d\mathbf{X}}\frac{d\mathbf{Z}}{d\mathbf{Y}}. \tag{1.4.14}$$

PROOF: It is well known from mathematical analysis that

$$\frac{\partial z_{ij}}{\partial x_{kl}} = \sum_{mn} \frac{\partial z_{ij}}{\partial y_{mn}} \frac{\partial y_{mn}}{\partial x_{kl}}.$$

Thus, using (1.4.5) with $\mathbf{d}_i^1$, $\mathbf{e}_j^1$, $\mathbf{d}_k^2$, $\mathbf{e}_l^2$, and $\mathbf{d}_m^3$, $\mathbf{e}_n^3$ being basis vectors of size p, q, t, u and r, s, respectively, we have

$$\begin{aligned}
\frac{d\mathbf{Z}}{d\mathbf{X}} &= \sum_{ijkl} \frac{\partial z_{ij}}{\partial x_{kl}} (\mathbf{e}_l^1 \otimes \mathbf{d}_k^1)(\mathbf{e}_j^2 \otimes \mathbf{d}_i^2)' \\
&= \sum_{ijklmn} \frac{\partial z_{ij}}{\partial y_{mn}} \frac{\partial y_{mn}}{\partial x_{kl}} (\mathbf{e}_l^1 \otimes \mathbf{d}_k^1)(\mathbf{e}_n^3 \otimes \mathbf{d}_m^3)'(\mathbf{e}_n^3 \otimes \mathbf{d}_m^3)(\mathbf{e}_j^2 \otimes \mathbf{d}_i^2)' \\
&= \sum_{ijklmnop} \frac{\partial z_{ij}}{\partial y_{mn}} \frac{\partial y_{op}}{\partial x_{kl}} (\mathbf{e}_l^1 \otimes \mathbf{d}_k^1)(\mathbf{e}_p^3 \otimes \mathbf{d}_o^3)'(\mathbf{e}_n^3 \otimes \mathbf{d}_m^3)(\mathbf{e}_j^2 \otimes \mathbf{d}_i^2)' \\
&= \sum_{klop} \frac{\partial y_{op}}{\partial x_{kl}} (\mathbf{e}_l^1 \otimes \mathbf{d}_k^1)(\mathbf{e}_p^3 \otimes \mathbf{d}_o^3)' \sum_{ijmn} \frac{\partial z_{ij}}{\partial y_{mn}} (\mathbf{e}_n^3 \otimes \mathbf{d}_m^3)(\mathbf{e}_j^2 \otimes \mathbf{d}_i^2)' = \frac{d\mathbf{Y}}{d\mathbf{X}} \frac{d\mathbf{Z}}{d\mathbf{Y}}.
\end{aligned}$$

∎

Proposition 1.4.4. *Let* **A** *and* **B** *be constant matrices of proper sizes. Then*

(i)
$$\frac{d(\mathbf{AXB})}{d\mathbf{X}} = \mathbf{B} \otimes \mathbf{A}'; \tag{1.4.15}$$

(ii)
$$\frac{d(\mathbf{AYB})}{d\mathbf{X}} = \frac{d\mathbf{Y}}{d\mathbf{X}}(\mathbf{B} \otimes \mathbf{A}'). \tag{1.4.16}$$

PROOF: To prove (i) we start from definition (1.4.7) and get the following row of equalities:

$$\frac{d(\mathbf{AXB})}{d\mathbf{X}} = \frac{d}{d\mathrm{vec}\mathbf{X}} \otimes \mathrm{vec}'(\mathbf{AXB}) = \frac{d[\mathrm{vec}'\mathbf{X}(\mathbf{B} \otimes \mathbf{A}')]}{d\mathbf{X}} = \frac{d\mathbf{X}}{d\mathbf{X}}(\mathbf{B}\otimes\mathbf{A}') = (\mathbf{B}\otimes\mathbf{A}').$$

For statement (ii) it follows by using the chain rule and (1.4.15) that

$$\frac{d(\mathbf{AYB})}{d\mathbf{X}} = \frac{d\mathbf{Y}}{d\mathbf{X}} \frac{d\mathbf{AYB}}{d\mathbf{Y}} = \frac{d\mathbf{Y}}{d\mathbf{X}}(\mathbf{B} \otimes \mathbf{A}').$$

∎

Proposition 1.4.5. *Let the elements of* **X** *be m.i.v. Then*

$$\frac{d\mathbf{X}'}{d\mathbf{X}} = \mathbf{K}_{q,p}, \qquad \frac{d\mathbf{X}}{d\mathbf{X}'} = \mathbf{K}_{p,q}. \tag{1.4.17}$$

PROOF: Observe that

$$\frac{d\mathbf{X}'}{d\mathbf{X}} = \frac{d\mathrm{vec}\mathbf{X}'}{d\mathrm{vec}\mathbf{X}} \underset{(1.3.30)}{=} \frac{d(\mathbf{K}_{p,q}\mathrm{vec}\mathbf{X})}{d\mathrm{vec}\mathbf{X}} = \mathbf{K}'_{p,q} = \mathbf{K}_{q,p},$$

where the last equality is obtained by Proposition 1.3.10 (i). The second equality in (1.4.17) is proved similarly. ∎

Proposition 1.4.6. *Let* $\mathbf{W}$ *be a function of* $\mathbf{Y} \in \mathbb{R}^{r\times s}$ *and* $\mathbf{Z} \in \mathbb{R}^{s\times n}$*, which both are functions of* $\mathbf{X}$*. Then*

(i)
$$\frac{d\mathbf{W}}{d\mathbf{X}} = \frac{d\mathbf{Y}}{d\mathbf{X}}\frac{d\mathbf{W}}{d\mathbf{Y}}\bigg|_{\mathbf{Z}=\text{const.}} + \frac{d\mathbf{Z}}{d\mathbf{X}}\frac{d\mathbf{W}}{d\mathbf{Z}}\bigg|_{\mathbf{Y}=\text{const.}}; \qquad (1.4.18)$$

(ii)
$$\frac{d(\mathbf{YZ})}{d\mathbf{X}} = \frac{d\mathbf{Y}}{d\mathbf{X}}(\mathbf{Z}\otimes\mathbf{I}_r) + \frac{d\mathbf{Z}}{d\mathbf{X}}(\mathbf{I}_n\otimes\mathbf{Y}'); \qquad (1.4.19)$$

(iii) *if* $z = z(\mathbf{X})$ *is a real function,*

$$\frac{d(\mathbf{Y}z)}{d\mathbf{X}} = \frac{d\mathbf{Y}}{d\mathbf{X}}z + \frac{dz}{d\mathbf{X}}\text{vec}'\mathbf{Y}.$$

PROOF: (i): By virtue of the formula for partial derivatives of composite functions in analysis we have

$$\frac{\partial w_{ij}}{\partial x_{gh}} = \sum_{mn}\frac{\partial y_{mn}}{\partial x_{gh}}\frac{\partial w_{ij}}{\partial y_{mn}}\bigg|_{\mathbf{Z}=\text{const.}} + \sum_{mn}\frac{\partial z_{mn}}{\partial x_{gh}}\frac{\partial w_{ij}}{\partial z_{mn}}\bigg|_{\mathbf{Y}=\text{const.}}.$$

Thus, since

$$\frac{d\mathbf{W}}{d\mathbf{X}} = \sum_{ijkl}\frac{\partial w_{ij}}{\partial x_{kl}}(\mathbf{e}_l^1\otimes\mathbf{d}_k^1)(\mathbf{e}_j^2\otimes\mathbf{d}_i^2)',$$

we may copy the proof of the chain rule given in Proposition 1.4.3, which immediately establishes the statement.
(ii): From (1.4.18) it follows that

$$\frac{d(\mathbf{YZ})}{d\mathbf{X}} = \frac{d\mathbf{Y}}{d\mathbf{X}}\frac{d(\mathbf{YZ})}{d\mathbf{Y}}\bigg|_{\mathbf{Z}=\text{const.}} + \frac{d\mathbf{Z}}{d\mathbf{X}}\frac{d(\mathbf{YZ})}{d\mathbf{Z}}\bigg|_{\mathbf{Y}=\text{const.}}.$$

Thus, (1.4.16) yields

$$\frac{d(\mathbf{YZ})}{d\mathbf{X}} = \frac{d\mathbf{Y}}{d\mathbf{X}}(\mathbf{Z}\otimes\mathbf{I}_r) + \frac{d\mathbf{Z}}{d\mathbf{X}}(\mathbf{I}_n\otimes\mathbf{Y}').$$

When proving (iii) similar calculations are used. ∎

Proposition 1.4.7.

(i) *Let* $\mathbf{Y} \in \mathbb{R}^{r\times r}$ *and* $\mathbf{Y}^0 = \mathbf{I}_r, n \geq 1$. *Then*

$$\frac{d\mathbf{Y}^n}{d\mathbf{X}} = \frac{d\mathbf{Y}}{d\mathbf{X}}\left(\sum_{\substack{i+j=n-1;\\ i,j\geq 0}}\mathbf{Y}^i\otimes(\mathbf{Y}')^j\right). \qquad (1.4.20)$$

(ii) *Let* $\mathbf{X}$ *be non-singular and the elements of* $\mathbf{X}$ *m.i.v. Then*

$$\frac{d\mathbf{X}^{-1}}{d\mathbf{X}} = -\mathbf{X}^{-1} \otimes (\mathbf{X}')^{-1}. \qquad (1.4.21)$$

(iii) *Let* $\mathbf{Y}$ *be non-singular. Then*

$$\frac{d\mathbf{Y}^{-n}}{d\mathbf{X}} = \frac{d\mathbf{Y}^{-1}}{d\mathbf{X}} \left(\sum_{\substack{i+j=n-1;\\ i,j\geq 0}} \mathbf{Y}^{-i} \otimes (\mathbf{Y}')^{-j} \right)$$

$$= -\frac{d\mathbf{Y}}{d\mathbf{X}} \left(\sum_{\substack{i+j=n-1;\\ i,j\geq 0}} \mathbf{Y}^{-i-1} \otimes (\mathbf{Y}')^{-j-1} \right). \qquad (1.4.22)$$

PROOF: (i): To prove the statement we shall use induction. For $n = 1$

$$\frac{d\mathbf{Y}}{d\mathbf{X}} = \frac{d\mathbf{Y}}{d\mathbf{X}}(\mathbf{I}_r \otimes \mathbf{I}_r)$$

and for $n = 2$

$$\frac{d\mathbf{Y}^2}{d\mathbf{X}} = \frac{d\mathbf{Y}}{d\mathbf{X}}(\mathbf{Y} \otimes \mathbf{I}_r) + \frac{d\mathbf{Y}}{d\mathbf{X}}(\mathbf{I}_r \otimes \mathbf{Y}).$$

Let us assume that the statement is valid for $n = k$. Then for $n = k + 1$

$$\frac{d\mathbf{Y}^{k+1}}{d\mathbf{X}} = \frac{d(\mathbf{Y}\mathbf{Y}^k)}{d\mathbf{X}} = \frac{d\mathbf{Y}^k}{d\mathbf{X}}(\mathbf{I} \otimes \mathbf{Y}') + \frac{d\mathbf{Y}}{d\mathbf{X}}(\mathbf{Y}^k \otimes \mathbf{I})$$

$$= \frac{d\mathbf{Y}}{d\mathbf{X}} \left(\sum_{\substack{i+j=k-1;\\ i,j\geq 0}} \mathbf{Y}^{i} \otimes (\mathbf{Y}')^{j} \right) (\mathbf{I} \otimes \mathbf{Y}') + \frac{d\mathbf{Y}}{d\mathbf{X}}(\mathbf{Y}^k \otimes \mathbf{I})$$

$$= \frac{d\mathbf{Y}}{d\mathbf{X}} \sum_{\substack{i+j=k-1;\\ i,j\geq 0}} (\mathbf{Y}^{i} \otimes (\mathbf{Y}')^{j+1} + \mathbf{Y}^k \otimes \mathbf{I}) = \frac{d\mathbf{Y}}{d\mathbf{X}} \left(\sum_{\substack{m+l=k;\\ m,l\geq 0}} \mathbf{Y}^{m} \otimes (\mathbf{Y}')^{l} \right).$$

(ii): Differentiate the identity

$$\mathbf{X}\mathbf{X}^{-1} = \mathbf{I}_p.$$

After using (1.4.19) we get

$$\frac{d\mathbf{X}}{d\mathbf{X}}(\mathbf{X}^{-1} \otimes \mathbf{I}) + \frac{d\mathbf{X}^{-1}}{d\mathbf{X}}(\mathbf{I} \otimes \mathbf{X}') = \mathbf{0}$$

and therefore

$$\frac{d\mathbf{X}^{-1}}{d\mathbf{X}} = -(\mathbf{X}^{-1} \otimes \mathbf{I})(\mathbf{I} \otimes \mathbf{X}')^{-1} = -\mathbf{X}^{-1} \otimes (\mathbf{X}')^{-1}.$$

(iii): Put $\mathbf{Y}^{-1} = \mathbf{Z}$. Then, by (1.4.20),

$$\frac{d\mathbf{Y}^{-n}}{d\mathbf{X}} = \frac{d\mathbf{Z}^{n}}{d\mathbf{X}} = \frac{d\mathbf{Z}}{d\mathbf{X}}\left(\sum_{\substack{i+j=n-1;\\ i,j\geq 0}} \mathbf{Z}^{i}\otimes(\mathbf{Z}')^{j}\right)$$

$$= \frac{d\mathbf{Y}^{-1}}{d\mathbf{X}}\left(\sum_{\substack{i+j=n-1;\\ i,j\geq 0}} (\mathbf{Y}^{-1})^{i}\otimes(\mathbf{Y}')^{-j}\right).$$

Hence, the first equality of statement (iii) follows from the fact that

$$(\mathbf{Y}^{-1})' = (\mathbf{Y}')^{-1}.$$

The second equality follows from the chain rule, (1.4.14) and (1.4.21), which give

$$\frac{d\mathbf{Y}^{-1}}{d\mathbf{X}} = \frac{d\mathbf{Y}}{d\mathbf{X}}\frac{d\mathbf{Y}^{-1}}{d\mathbf{Y}} = -\frac{d\mathbf{Y}}{d\mathbf{X}}\left(\mathbf{Y}^{-1}\otimes(\mathbf{Y}')^{-1}\right).$$

Thus,

$$\frac{d\mathbf{Y}^{-n}}{d\mathbf{X}} = -\frac{d\mathbf{Y}}{d\mathbf{X}}\left(\mathbf{Y}^{-1}\otimes(\mathbf{Y}')^{-1}\right)\left(\sum_{\substack{i+j=n-1;\\ i,j\geq 0}} \mathbf{Y}^{-i}\otimes(\mathbf{Y}')^{-j}\right)$$

and after applying property (1.3.14) for the Kronecker product we obtain the necessary equality. ∎

Proposition 1.4.8.
(i) *Let* $\mathbf{Z}\in\mathbb{R}^{m\times n}$ *and* $\mathbf{Y}\in\mathbb{R}^{r\times s}$. *Then*

$$\frac{d(\mathbf{Y}\otimes\mathbf{Z})}{d\mathbf{X}} = \left(\frac{d\mathbf{Y}}{d\mathbf{X}}\otimes \mathrm{vec}'\mathbf{Z} + \mathrm{vec}'\mathbf{Y}\otimes\frac{d\mathbf{Z}}{d\mathbf{X}}\right)(\mathbf{I}_s\otimes\mathbf{K}_{r,n}\otimes\mathbf{I}_m). \tag{1.4.23}$$

(ii) *Let* $\mathbf{A}$ *be a matrix with constant elements. Then*

$$\frac{d(\mathbf{Y}\otimes\mathbf{A})}{d\mathbf{X}} = (\frac{d\mathbf{Y}}{d\mathbf{X}}\otimes\mathrm{vec}'\mathbf{A})(\mathbf{I}_s\otimes\mathbf{K}_{r,n}\otimes\mathbf{I}_m),\quad \mathbf{Y}\in\mathbb{R}^{r\times s},\mathbf{A}\in\mathbb{R}^{m\times n}; \tag{1.4.24}$$

$$\frac{d(\mathbf{A}\otimes\mathbf{Y})}{d\mathbf{X}} = (\frac{d\mathbf{Y}}{d\mathbf{X}}\otimes\mathrm{vec}'\mathbf{A})\mathbf{K}_{rs,mn}(\mathbf{I}_n\otimes\mathbf{K}_{m,s}\otimes\mathbf{I}_r),\quad \mathbf{Y}\in\mathbb{R}^{r\times s},\mathbf{A}\in\mathbb{R}^{m\times n}. \tag{1.4.25}$$

PROOF: (i): By virtue of equality (1.3.33) it follows that

$$\frac{d(\mathbf{Y}\otimes\mathbf{Z})}{d\mathbf{X}} = \frac{d}{d\mathrm{vec}\mathbf{X}}(\mathrm{vec}\mathbf{Y}\otimes\mathrm{vec}\mathbf{Z})'(\mathbf{I}_s\otimes\mathbf{K}_{r,n}\otimes\mathbf{I}_m).$$

The properties (1.4.16) and (1.4.18) establish

$$\begin{aligned}\frac{d(\mathbf{Y}\otimes\mathbf{Z})}{d\mathbf{X}} &= \left\{\frac{d}{d\mathrm{vec}\mathbf{X}}(\mathrm{vec}'\mathbf{Y}\otimes\mathrm{vec}'\mathbf{Z})\Big|_{\mathbf{Z}=\mathrm{const.}}\right.\\ &\quad\left.+\frac{d}{d\mathrm{vec}\mathbf{X}}(\mathrm{vec}'\mathbf{Y}\otimes\mathrm{vec}'\mathbf{Z})\Big|_{\mathbf{Y}=\mathrm{const.}}\right\}(\mathbf{I}_s\otimes\mathbf{K}_{r,n}\otimes\mathbf{I}_m)\\ &=\left[\frac{d\mathbf{Y}}{d\mathbf{X}}(\mathbf{I}_{rs}\otimes\mathrm{vec}'\mathbf{Z})+\frac{d\mathbf{Z}}{d\mathbf{X}}(\mathrm{vec}'\mathbf{Y}\otimes\mathbf{I}_{mn})\right](\mathbf{I}_s\otimes\mathbf{K}_{r,n}\otimes\mathbf{I}_m). \qquad (1.4.26)\end{aligned}$$

For the two statements in (ii) we note that $\frac{d\mathbf{A}}{d\mathbf{X}}=\mathbf{0}$. If in (1.4.26) $\mathbf{Z}=\mathbf{A}$ is chosen, then (1.4.24) is directly obtained. To prove (1.4.25) we first remark that (1.4.26) reduces to

$$\frac{d(\mathbf{A}\otimes\mathbf{Y})}{d\mathbf{X}}=(\mathrm{vec}'\mathbf{A}\otimes\frac{d\mathbf{Y}}{d\mathbf{X}})(\mathbf{I}_n\otimes\mathbf{K}_{m,s}\otimes\mathbf{I}_r),$$

and then the commutation property of the Kronecker product (1.3.15) gives the statement. ∎

Proposition 1.4.9. *Let all matrices be of proper sizes and the elements of* $\mathbf{X}$ *m.i.v. Then*

(i)
$$\frac{d\mathrm{tr}\mathbf{Y}}{d\mathbf{X}}=\frac{d\mathbf{Y}}{d\mathbf{X}}\mathrm{vec}\mathbf{I}; \qquad (1.4.27)$$

(ii)
$$\frac{d\mathrm{tr}(\mathbf{A}'\mathbf{X})}{d\mathbf{X}}=\mathrm{vec}\mathbf{A}; \qquad (1.4.28)$$

(iii)
$$\frac{d\mathrm{tr}(\mathbf{AXBX}')}{d\mathbf{X}}=\mathrm{vec}(\mathbf{A}'\mathbf{XB}')+\mathrm{vec}(\mathbf{AXB}). \qquad (1.4.29)$$

PROOF: (i): The first statement follows, since

$$\frac{d\mathrm{tr}\mathbf{Y}}{d\mathbf{X}}=\frac{d\mathbf{Y}}{d\mathbf{X}}\frac{d\mathrm{vec}'\mathbf{I}\mathrm{vec}\mathbf{Y}}{d\mathbf{Y}}\underset{(1.4.12)}{=}\frac{d\mathbf{Y}}{d\mathbf{X}}\mathrm{vec}\mathbf{I}.$$

(ii): The second statement holds because

$$\frac{d\mathrm{tr}(\mathbf{A}'\mathbf{X})}{d\mathbf{X}}=\frac{d\mathbf{A}'\mathbf{X}}{d\mathbf{X}}\mathrm{vec}\mathbf{I}\underset{(1.4.11)}{=}(\mathbf{I}\otimes\mathbf{A})\mathrm{vec}\mathbf{I}=\mathrm{vec}\mathbf{A}.$$

(iii): For the third statement we have to perform more calculations:

$$\begin{aligned}&\frac{d\mathrm{tr}(\mathbf{AXBX}')}{d\mathbf{X}}\\ &=\frac{d(\mathbf{AXBX}')}{d\mathbf{X}}\frac{d\mathrm{tr}(\mathbf{AXBX}')}{d(\mathbf{AXBX}')}\underset{(1.4.19)}{=}\left(\frac{d\mathbf{AX}}{d\mathbf{X}}(\mathbf{BX}'\otimes\mathbf{I})+\frac{d\mathbf{BX}'}{d\mathbf{X}}(\mathbf{I}\otimes\mathbf{X}'\mathbf{A}')\right)\mathrm{vec}\mathbf{I}\\ &\underset{\substack{(1.4.11)\\(1.4.17)}}{=}\{(\mathbf{I}\otimes\mathbf{A}')(\mathbf{BX}'\otimes\mathbf{I})+\mathbf{K}_{q,p}(\mathbf{I}\otimes\mathbf{B}')(\mathbf{I}\otimes\mathbf{X}'\mathbf{A}')\}\mathrm{vec}\mathbf{I}\\ &=\{(\mathbf{BX}'\otimes\mathbf{A}')+(\mathbf{B}'\mathbf{X}'\otimes\mathbf{A}')\}\mathrm{vec}\mathbf{I}=\mathrm{vec}(\mathbf{A}'\mathbf{XB}')+\mathrm{vec}(\mathbf{AXB}).\end{aligned}$$

∎

Proposition 1.4.10. *Let* $\mathbf{X}$ *be non-singular and the elements of* $\mathbf{X}$ *m.i.v. Then*

$$\frac{d|\mathbf{X}|}{d\mathbf{X}} = |\mathbf{X}|\mathrm{vec}(\mathbf{X}^{-1})' \tag{1.4.30}$$

and

$$\frac{d|\mathbf{X}|^r}{d\mathbf{X}} = r|\mathbf{X}|^r\mathrm{vec}(\mathbf{X}^{-1})'. \tag{1.4.31}$$

PROOF: The statement (1.4.30) will be proven in two different ways. At first we use the approach which usually can be found in the literature, whereas the second proof follows from the proof of (1.4.31) and demonstrates the connection between the determinant and trace function.

For (1.4.30) we shall use the representation of the determinant given by (1.1.6), which equals

$$|\mathbf{X}| = \sum_j x_{ij}(-1)^{i+j}|\mathbf{X}_{(ij)}|.$$

Note that a general element of the inverse matrix is given by (1.1.8), i.e.

$$(\mathbf{X}^{-1})_{ij} = \frac{(-1)^{i+j}|\mathbf{X}_{(ji)}|}{|\mathbf{X}|},$$

where $|\mathbf{X}_{(ij)}|$ is the minor of x_{ij}. Then

$$\frac{\partial|\mathbf{X}|}{\partial x_{ij}} = (-1)^{i+j}|\mathbf{X}_{(ij)}| = (\mathbf{X}^{-1})_{ji}|\mathbf{X}|.$$

Since the obtained equality holds for all values of the indices $i, j = 1, \ldots, p$, we get the desired result by definition of the matrix derivative:

$$\frac{d|\mathbf{X}|}{d\mathbf{X}} = \mathrm{vec}(\mathbf{X}^{-1})'|\mathbf{X}|.$$

For (1.4.31) we are going to utilize the integral representation of the determinant given in Theorem 1.1.2. This means that at first the derivative

$$\frac{d|\mathbf{X}|^{-1}}{d\mathbf{X}}$$

is considered and thereafter it is utilized that

$$\frac{d|\mathbf{X}|^r}{d\mathbf{X}} = \frac{d|\mathbf{X}^{-1}|^{-r}}{d\mathbf{X}} = -r|\mathbf{X}^{-1}|^{-r-1}\frac{d|\mathbf{X}^{-1}|}{d\mathbf{X}}. \tag{1.4.32}$$

Then

$$\begin{aligned}\frac{d|\mathbf{X}^{-1}|}{d\mathbf{X}} &= \frac{d|\mathbf{X}'\mathbf{X}|^{-1/2}}{d\mathbf{X}} = \frac{d}{d\mathbf{X}}\frac{1}{(2\pi)^{p/2}}\int_{\mathbb{R}^p} e^{-\frac{1}{2}\mathrm{tr}(\mathbf{X}'\mathbf{X}\mathbf{y}\mathbf{y}')}d\mathbf{y}\\ &= -\frac{1}{2}\frac{1}{(2\pi)^{p/2}}\int_{\mathbb{R}^p}\frac{d\mathrm{tr}(\mathbf{X}'\mathbf{X}\mathbf{y}\mathbf{y}')}{d\mathbf{X}}e^{-\frac{1}{2}\mathrm{tr}(\mathbf{X}'\mathbf{X}\mathbf{y}\mathbf{y}')}d\mathbf{y}.\end{aligned}$$

Differentiation under the integral is allowed as we are integrating the normal density function. Using (1.4.28), we obtain that the right hand side equals

$$
\begin{aligned}
-\frac{1}{2}\frac{1}{(2\pi)^{p/2}}\int_{\mathbb{R}^p} 2\mathrm{vec}(\mathbf{X}\mathbf{y}\mathbf{y}')e^{-\frac{1}{2}\mathrm{tr}(\mathbf{X}'\mathbf{X}\mathbf{y}\mathbf{y}')}d\mathbf{y}\\
= -|\mathbf{X}'\mathbf{X}|^{-1/2}\mathrm{vec}(\mathbf{X}(\mathbf{X}'\mathbf{X})^{-1}) = -|\mathbf{X}|^{-1}\mathrm{vec}((\mathbf{X}')^{-1}),
\end{aligned}
$$

and therefore, from (1.4.32), it follows that

$$
\frac{d|\mathbf{X}|^r}{d\mathbf{X}} = r|\mathbf{X}|^r\mathrm{vec}((\mathbf{X}')^{-1}).
$$

In particular, this relation holds for $r = 1$, i.e. (1.4.30) is valid. ∎

1.4.4 *Derivatives of patterned matrices*

In Definition 1.4.1 we included the possibility to differentiate the matrix $\mathbf{Y}$ by $\mathbf{X}$, when some elements in $\mathbf{X}$ can be repeated or are constants. In many situations we are interested in differentiating by nonrepeated and non-constant elements, for example when finding Jacobians. When $\mathbf{X}$ is a partitioned matrix, we may want to find a derivative by a certain block or section of the matrix. In general, we need a matrix derivative of one patterned matrix by another patterned matrix. The notion of patterned matrix was considered in §1.3.6. In this case we consider the matrix derivative as a collection of partial derivatives and not as a representation of a linear map.

Definition 1.4.2. *Let $\mathbf{X} \in \mathbf{R}^{p\times q}$, $\mathbf{Y} \in \mathbf{R}^{r\times s}$, and let $\mathbf{X}(K_1)$ and $\mathbf{Y}(K_2)$ be two patterned matrices with patterns K_1 and K_2, respectively. The matrix derivative $\dfrac{d\mathbf{Y}(K_2)}{d\mathbf{X}(K_1)}$ is defined by the equality*

$$
\frac{d\mathbf{Y}(K_2)}{d\mathbf{X}(K_1)} = \frac{d}{d\mathrm{vec}\mathbf{X}(K_1)}\mathrm{vec}'\mathbf{Y}(K_2), \tag{1.4.33}
$$

with patterns

$$
K_1 = \{(i,j) : i \in I_{K_1}, j \in J_{K_1}; I_{K_1} \subset \{1,\ldots,p\}, J_{K_1} \subset \{1,\ldots,q\}\},
$$

$$
K_2 = \{(i,j) : i \in I_{K_2}, j \in J_{K_2}; I_{K_2} \subset \{1,\ldots,r\}, J_{K_2} \subset \{1,\ldots,s\}\}.
$$

∎

Note that from Definition 1.4.2 the obvious result

$$
\frac{d\mathbf{X}(K)}{d\mathbf{X}(K)} = \mathbf{I}
$$

follows.

Theorem 1.4.2. *Let* $\mathbf{X} \in \mathbf{R}^{p\times q}$, $\mathbf{Y} \in \mathbf{R}^{r\times s}$, *and* $\mathbf{X}(K_1)$, $\mathbf{Y}(K_2)$ *be patterned matrices with the corresponding transformation matrices* $\mathbf{T}(K_1)$ *and* $\mathbf{T}(K_2)$, *given by (1.3.46). Then*

$$\frac{d\mathbf{Y}(K_2)}{d\mathbf{X}(K_1)} = \mathbf{T}(K_1)\frac{d\mathbf{Y}}{d\mathbf{X}}(\mathbf{T}(K_2))'. \tag{1.4.34}$$

PROOF: Combining Definition 1.4.2 with (1.3.44) gives that

$$\frac{d\mathbf{Y}(K_2)}{d\mathbf{X}(K_1)} = \frac{\partial}{\partial(\mathbf{T}(K_1)\mathrm{vec}\mathbf{X})}(\mathbf{T}(K_2)\mathrm{vec}\mathbf{Y})' = \left(\mathbf{T}(K_1)\frac{\partial}{\partial \mathrm{vec}\mathbf{X}}\right)(\mathbf{T}(K_2)\mathrm{vec}\mathbf{Y})'$$
$$= \mathbf{T}(K_1)\frac{d\mathbf{Y}}{d\mathbf{X}}(\mathbf{T}(K_2))'.$$

∎

The idea above is fairly simple. The approach just means that by the transformation matrices we cut out a proper part from the whole matrix of partial derivatives. However, things become more interesting and useful when we consider patterned matrices. In this case we have full information about repeatedness of elements in matrices $\mathbf{X}$ and $\mathbf{Y}$, i.e. the pattern is completely known. We can use the transition matrices $\mathbf{T}(\mathbf{X})$ and $\mathbf{T}(\mathbf{Y})$, defined by (1.3.61).

Theorem 1.4.3. *Let* $\mathbf{X} \in \mathbb{R}^{p\times q}$, $\mathbf{Y} \in \mathbb{R}^{r\times s}$, *and* $\mathbf{X}(K_1)$, $\mathbf{Y}(K_2)$ *be patterned matrices, and let* $\mathbf{T}(\mathbf{X})$ *and* $\mathbf{T}(\mathbf{Y})$ *be the corresponding transition matrices. Then*

$$\frac{d\mathbf{Y}(K_2)}{d\mathbf{X}(K_1)} = \mathbf{T}(\mathbf{X})\frac{d\mathbf{Y}}{d\mathbf{X}}(\mathbf{T}(\mathbf{Y}))' \tag{1.4.35}$$

and

$$\frac{d\mathbf{Y}}{d\mathbf{X}} = \mathbf{T}^{+}(\mathbf{X})\frac{d\mathbf{Y}(K_2)}{d\mathbf{X}(K_1)}(\mathbf{T}^{+}(\mathbf{Y}))', \tag{1.4.36}$$

where $\mathbf{T}(\cdot)$ *and* $\mathbf{T}^{+}(\cdot)$ *are given by Corollaries 1.3.11.2 and 1.3.11.3, respectively.*

PROOF: The statement in (1.4.35) repeats the previous theorem, whereas (1.4.36) follows directly from Definition 1.4.1 and the basic property of $\mathbf{T}^{+}(\cdot)$. ∎

Theorem 1.4.3 shows the correspondence between Definitions 1.4.1 and 1.4.2. In the case when the patterns consist of all different elements of the matrices $\mathbf{X}$ and $\mathbf{Y}$, the derivatives can be obtained from each other via the transition matrices. From all possible patterns the most important applications are related to diagonal, symmetric, skew-symmetric and correlation type matrices. The last one is understood to be a symmetric matrix with constants on the main diagonal. Theorem 1.4.3 gives us also a possibility to find derivatives in the cases when a specific pattern is given for only one of the matrices $\mathbf{Y}$ and $\mathbf{X}$. In the next corollary a derivative by a symmetric matrix is given, which will be used frequently later.

Corollary 1.4.3.1. *Let* $\mathbf{X}$ *be a symmetric* $p\times p-$*matrix and* $\mathbf{X}^{\triangle}$ *denote its upper triangle. Then*

$$\frac{d\mathbf{X}}{d\mathbf{X}^{\triangle}} = \mathbf{G}_p\mathbf{H}_p,$$

where $\mathbf{G}_p$ *is the transformation matrix of the upper triangle defined by (1.3.49) and (1.3.50), and* $\mathbf{H}_p = \mathbf{I}_{p^2} + \mathbf{K}_{p,p} - (\mathbf{K}_{p,p})_d$. ∎

In the following Proposition 1.4.11 we present the derivatives of matrices with some basic dependence structures. The formulae are closely related to the corresponding pattern matrices as can be seen from the expressions in Proposition 1.4.11 and Propositions 1.3.18 – 1.3.21.

Proposition 1.4.11. *Let* $\mathbf{X} \in \mathbf{R}^{p\times p}$. *Then*

(i) *for the diagonal matrix* $\mathbf{X}_d$

$$\frac{d\mathbf{X}_d}{d\mathbf{X}} = (\mathbf{K}_{p,p})_d; \tag{1.4.37}$$

(ii) *for the symmetric matrix* $\mathbf{X}$

$$\frac{d\mathbf{X}}{d\mathbf{X}} = \mathbf{I}_{p^2} + \mathbf{K}_{p,p} - (\mathbf{K}_{p,p})_d; \tag{1.4.38}$$

(iii) *for the skew-symmetric matrix* $\mathbf{X}$

$$\frac{d\mathbf{X}}{d\mathbf{X}} = \mathbf{I}_{p^2} - \mathbf{K}_{p,p}; \tag{1.4.39}$$

(iv) *for the correlation type matrix* $\mathbf{X}$

$$\frac{d\mathbf{X}}{d\mathbf{X}} = \mathbf{I}_{p^2} + \mathbf{K}_{p,p} - 2(\mathbf{K}_{p,p})_d. \tag{1.4.40}$$

∎

1.4.5 *Higher order derivatives*

Higher order derivatives are needed later in series expansions. A natural way of extending the concept of matrix derivative to higher order derivatives is by a recursive definition.

Definition 1.4.3. *The matrix derivative of order k of* $\mathbf{Y}$ *by* $\mathbf{X}$ *is defined as the matrix derivative of the matrix derivative of order* $(k-1)$ *of* $\mathbf{Y}$ *by* $\mathbf{X}$:

$$\frac{d^k\mathbf{Y}}{d\mathbf{X}^k} = \frac{d}{d\mathbf{X}}\left(\frac{d^{k-1}\mathbf{Y}}{d\mathbf{X}^{k-1}}\right), \tag{1.4.41}$$

where $\dfrac{d\mathbf{Y}}{d\mathbf{X}}$ *is defined by (1.4.7).* ∎

The following property shows how the k−th order derivative can be presented non-recursively through the differential operator in (1.4.7). Moreover, in §1.4.3 we considered Fréchet derivatives and noted their relation to tensor products. Remark that one can generalize this result to higher order derivatives and observe that a multilinear mapping with a corresponding higher order tensor product (Kroneckerian power in matrix language) will be obtained.

Theorem 1.4.4. *The $k-$th derivative* $\frac{d^k\mathbf{Y}}{d\mathbf{X}^k}$ *can be written as*

$$\frac{d^k\mathbf{Y}}{d\mathbf{X}^k} = \frac{\partial}{\partial \text{vec}\mathbf{X}}\text{vec}'\mathbf{Y} \otimes \underbrace{\frac{\partial}{\partial \text{vec}'\mathbf{X}} \otimes \cdots \otimes \frac{\partial}{\partial \text{vec}'\mathbf{X}}}_{k-1 \text{ times}}. \tag{1.4.42}$$

PROOF: The statement is proved with the help of induction. For $k = 1$, the relation (1.4.42) is valid due to the definition of the derivative in (1.4.7). Suppose that (1.4.42) is true for $k = n - 1$:

$$\frac{d^{n-1}\mathbf{Y}}{d\mathbf{X}^{n-1}} = \frac{\partial}{\partial \text{vec}\mathbf{X}}\text{vec}'\mathbf{Y} \otimes \underbrace{\frac{\partial}{\partial \text{vec}'\mathbf{X}} \otimes \cdots \otimes \frac{\partial}{\partial \text{vec}'\mathbf{X}}}_{(n-2)\ times}.$$

Let us show that the statement then also holds for $k = n$:

$$\begin{aligned}\frac{d^k\mathbf{Y}}{d\mathbf{X}^k} &= \frac{\partial}{\partial \text{vec}\mathbf{X}}\text{vec}'\frac{d^{n-1}\mathbf{Y}}{d\mathbf{X}^{n-1}} \\ &= \frac{\partial}{\partial \text{vec}\mathbf{X}}\text{vec}'(\frac{\partial}{\partial \text{vec}\mathbf{X}}\text{vec}'\mathbf{Y} \otimes \underbrace{\frac{\partial}{\partial \text{vec}'\mathbf{X}} \otimes \cdots \otimes \frac{\partial}{\partial \text{vec}'\mathbf{X}}}_{(n-2)\ times}) \\ &= \frac{\partial}{\partial \text{vec}\mathbf{X}}\text{vec}'\mathbf{Y} \otimes \underbrace{\frac{\partial}{\partial \text{vec}'\mathbf{X}} \otimes \cdots \otimes \frac{\partial}{\partial \text{vec}'\mathbf{X}}}_{n-1\ times}.\end{aligned}$$

∎

Another important property concerns the representation of the $k-$th order derivative of another derivative of lower order.

Theorem 1.4.5. *The $k-$th order matrix derivative of the $l-$th order matrix derivative* $\frac{d^l\mathbf{Y}}{d\mathbf{X}^l}$ *is the $(k+l)-$th order derivative of $\mathbf{Y}$ by $\mathbf{X}$:*

$$\frac{d^k}{d\mathbf{X}^k}\left(\frac{d^l\mathbf{Y}}{d\mathbf{X}^l}\right) = \frac{d^{k+l}\mathbf{Y}}{d\mathbf{X}^{k+l}}.$$

PROOF: By virtue of Definition 1.4.3 and Theorem 1.4.4

$$\begin{aligned}\frac{d^k}{d\mathbf{X}^k}\left(\frac{d^l\mathbf{Y}}{d\mathbf{X}^l}\right) &= \frac{d^k}{d\mathbf{X}^k}(\frac{\partial}{\partial \text{vec}\mathbf{X}}\text{vec}'\mathbf{Y} \otimes \underbrace{\frac{\partial}{\partial \text{vec}'\mathbf{X}} \otimes \cdots \otimes \frac{\partial}{\partial \text{vec}'\mathbf{X}}}_{l-1\ times}) \\ &= \frac{\partial}{\partial \text{vec}\mathbf{X}}\text{vec}'\mathbf{Y} \otimes \underbrace{\frac{\partial}{\partial \text{vec}'\mathbf{X}} \otimes \cdots \otimes \frac{\partial}{\partial \text{vec}'\mathbf{X}}}_{l\ times} \otimes \underbrace{\frac{\partial}{\partial \text{vec}'\mathbf{X}} \otimes \frac{\partial}{\partial \text{vec}'\mathbf{X}} \otimes \cdots \otimes \frac{\partial}{\partial \text{vec}'\mathbf{X}}}_{k-1\ times} \\ &= \frac{d^{k+l}\mathbf{Y}}{d\mathbf{X}^{k+l}}.\end{aligned}$$

∎

1.4.6 *Higher order derivatives and patterned matrices*

A natural generalization of higher order derivatives to patterned matrices is given below.

Definition 1.4.4. *Let $\mathbf{X}(K_1)$ and $\mathbf{Y}(K_2)$ be patterned matrices with the patterns K_1 and K_2, respectively. The k–th order matrix derivative of $\mathbf{Y}(K_2)$ by $\mathbf{X}(K_1)$ is defined by the equality*

$$\frac{d^k\mathbf{Y}(K_2)}{d\mathbf{X}(K_1)^k} = \frac{d}{d\mathbf{X}(K_1)}\left(\frac{d^{k-1}\mathbf{Y}(K_2)}{d\mathbf{X}(K_1)^{k-1}}\right), \quad k \geq 2, \tag{1.4.43}$$

where $\frac{d\mathbf{Y}(K_2)}{d\mathbf{X}(K_1)}$ is defined by (1.4.33) and the patterns K_1 and K_2 are presented in Definition 1.4.2. ∎

In the next theorem we shall give the relation between higher order derivatives of patterned matrices and higher order derivatives with all repeated elements in it.

Theorem 1.4.6. *Let $\mathbf{X}(K_1)$ and $\mathbf{Y}(K_2)$ be patterned matrices with transition matrices $\mathbf{T}(\mathbf{X})$ and $\mathbf{T}(\mathbf{Y})$, defined by (1.3.61). Then, for $k \geq 2$,*

$$\frac{d^k\mathbf{Y}(K_2)}{d\mathbf{X}(K_1)^k} = \mathbf{T}(\mathbf{X})\frac{d^k\mathbf{Y}}{d\mathbf{X}^k}\left(\mathbf{T}(\mathbf{X})^{\otimes(k-1)} \otimes \mathbf{T}(\mathbf{Y})\right)', \tag{1.4.44}$$

$$\frac{d^k\mathbf{Y}}{d\mathbf{X}^k} = \mathbf{T}^+(\mathbf{X})\frac{d^k\mathbf{Y}(K_2)}{d\mathbf{X}(K_1)^k}\left(\mathbf{T}^+(\mathbf{X})^{\otimes(k-1)} \otimes \mathbf{T}^+(\mathbf{Y})\right)'. \tag{1.4.45}$$

PROOF: To prove the statements, once again an induction argument will be applied. Because of the same structure of statements, the proofs of (1.4.44) and (1.4.45) are similar and we shall only prove (1.4.45). For $k = 1$, since $\mathbf{T}(\mathbf{X})^{\otimes 0} = 1$ and $\mathbf{T}^+(\mathbf{X})^{\otimes 0} = 1$, it is easy to see that we get our statements in (1.4.44) and (1.4.45) from (1.4.35) and (1.4.36), respectively. For $k = 2$, by Definition 1.4.2 and (1.4.36) it follows that

$$\frac{d^2\mathbf{Y}}{d\mathbf{X}^2} = \frac{d}{d\mathbf{X}}\left(\frac{d\mathbf{Y}}{d\mathbf{X}}\right) = \frac{d}{d\mathbf{X}}\left(\mathbf{T}^+(\mathbf{X})\frac{d\mathbf{Y}(K_2)}{d\mathbf{X}(K_1)}\mathbf{T}^+(\mathbf{Y})'\right).$$

Definition 1.4.1 and assumptions about the transition matrices give

$$\begin{aligned}\frac{d^2\mathbf{Y}}{d\mathbf{X}^2} &= \frac{d}{d\text{vec}\mathbf{X}}\text{vec}'\left(\mathbf{T}^+(\mathbf{X})\frac{d\mathbf{Y}(K_2)}{d\mathbf{X}(K_1)}\mathbf{T}^+(\mathbf{Y})'\right)\\ &= \mathbf{T}^+(\mathbf{X})\frac{d}{d\text{vec}\mathbf{X}(K_1)}\text{vec}'\frac{d\mathbf{Y}(K_2)}{d\mathbf{X}(K_1)}(\mathbf{T}^+(\mathbf{Y}) \otimes \mathbf{T}^+(\mathbf{X}))'.\end{aligned}$$

Using property (1.3.14) of the Kronecker product we get the desired equality:

$$\begin{aligned}\frac{d^2\mathbf{Y}}{d\mathbf{X}^2} &= (\mathbf{T}^+(\mathbf{X}) \otimes \text{vec}\frac{d\mathbf{Y}(K_2)}{d\mathbf{X}(K_1)})(\frac{d}{d\text{vec}'\mathbf{X}(K_1)} \otimes (\mathbf{T}^+(\mathbf{X}) \otimes \mathbf{T}^+(\mathbf{Y}))')\\ &= \mathbf{T}^+(\mathbf{X})\frac{d^2\mathbf{Y}(K_2)}{d\mathbf{X}(K_1)^2}(\mathbf{T}^+(\mathbf{X}) \otimes \mathbf{T}^+(\mathbf{Y}))'.\end{aligned}$$

Assume that statement (1.4.45) holds for $k = n-1$:

$$\frac{d^{n-1}\mathbf{Y}}{d\mathbf{X}^{n-1}} = \mathbf{T}^+(\mathbf{X})\frac{d^{n-1}\mathbf{Y}(K_2)}{d\mathbf{X}(K_1)^{n-1}}\left(\mathbf{T}^+(\mathbf{X})^{\otimes(n-2)}\otimes\mathbf{T}^+(\mathbf{Y})\right)'.$$

Let us show that then (1.4.45) is also valid for $k = n$. Repeating the same argumentation as in the case $k = 2$, we get the following chain of equalities

$$\begin{aligned}
\frac{d^n\mathbf{Y}}{d\mathbf{X}^n} =& \frac{d}{d\mathbf{X}}\left(\frac{d^{n-1}\mathbf{Y}}{d\mathbf{X}^{n-1}}\right)\\
=&\mathbf{T}^+(\mathbf{X})\frac{d}{d\mathrm{vec}\mathbf{X}(K_1)}\mathrm{vec}'\{\mathbf{T}^+(\mathbf{X})\frac{d^{n-1}\mathbf{Y}}{d\mathbf{X}^{n-1}}(\mathbf{T}^+(\mathbf{X})^{\otimes(n-2)}\otimes\mathbf{T}^+(\mathbf{Y}))'\}\\
=&\mathbf{T}^+(\mathbf{X})\frac{d}{d\mathrm{vec}\mathbf{X}(K_1)}\mathrm{vec}'\frac{d^{n-1}\mathbf{Y}}{d\mathbf{X}^{n-1}}(\mathbf{T}^+(\mathbf{X})^{\otimes(n-1)}\otimes\mathbf{T}^+(\mathbf{Y}))'\\
=&\mathbf{T}^+(\mathbf{X})\frac{d^n\mathbf{Y}}{d\mathbf{X}^n}(\mathbf{T}^+(\mathbf{X})^{\otimes(n-1)}\otimes\mathbf{T}^+(\mathbf{Y}))'.
\end{aligned}$$

∎

In the case of symmetric matrices we shall reformulate the statement (1.4.45) as a corollary of the theorem.

Corollary 1.4.6.1. *Let $\mathbf{X}\in\mathbb{R}^{p\times p}$, $\mathbf{Y}\in\mathbb{R}^{r\times r}$ be symmetric matrices and $\mathbf{Y}_\triangle$, $\mathbf{X}_\triangle$ the lower triangles of these matrices. Then*

$$\frac{d^k\mathbf{Y}}{d\mathbf{X}^k} = \mathbf{D}_p^+\frac{d^k\mathbf{Y}_\triangle}{d\mathbf{X}_\triangle^k}\left((\mathbf{D}_p^+)^{\otimes(k-1)}\otimes\mathbf{D}_r^+\right)', \tag{1.4.46}$$

where $\mathbf{D}_p^+$ is defined by (1.3.64). ∎

1.4.7 *Differentiating symmetric matrices using an alternative derivative*

In this paragraph we shall consider another variant of the matrix derivative, namely, the derivative given by (1.4.2). To avoid misunderstandings, we shall denote the derivative (1.4.2) by $\frac{\widetilde{d}\mathbf{Y}}{\widetilde{d}\mathbf{X}}$ instead of $\frac{d\mathbf{Y}}{d\mathbf{X}}$ which is used for the matrix derivative defined by (1.4.7). The paragraph is intended for those who want to go deeper into the applications of matrix derivatives. Our main purpose is to present some ideas about differentiating symmetric matrices. Here we will consider the matrix derivative as a matrix representation of a certain operator. The results can not be directly applied to minimization or maximization problems, but the derivative is useful when obtaining moments for symmetric matrices, which will be illustrated later in §2.4.5.

As noted before, the derivative (1.4.2) can be presented as

$$\frac{\widetilde{d}\mathbf{Y}}{\widetilde{d}\mathbf{X}} = \sum_{ijkl}\frac{\partial y_{ij}}{\partial x_{kl}}((\mathbf{r}_i\mathbf{s}_j')\otimes(\mathbf{f}_k\mathbf{g}_l')), \tag{1.4.47}$$

where $\mathbf{r}_i$, $\mathbf{s}_j$, $\mathbf{f}_k$, $\mathbf{g}_l$ are the basis vectors of sizes r, s, p and q, respectively. Assume that $\mathbf{X}$ consists of mathematically independent and variable (m.i.v.) elements. However, when $\mathbf{X} : n \times n$ is symmetric and the elements of the lower triangle of $\mathbf{X}$ are m.i.v., we define, inspired by Srivastava & Khatri (1979, p. 37),

$$\frac{\widetilde{d}\mathbf{Y}}{\widetilde{d}\mathbf{X}} = \sum_{ijkl} \frac{\partial y_{ij}}{\partial x_{kl}} ((\mathbf{r}_i \mathbf{s}_j') \otimes (\epsilon_{kl} \mathbf{f}_k \mathbf{f}_l')), \quad \epsilon_{kl} = \begin{cases} 1, & k = l, \\ \frac{1}{2}, & k \neq l. \end{cases} \tag{1.4.48}$$

In §1.4.4 the idea was to cut out a certain unique part from a matrix. In this paragraph we consider all elements with a weight which depends on the repeatedness of the element in the matrix, i.e. whether the element is a diagonal element or an off-diagonal element. Note that (1.4.48) is identical to

$$\frac{\widetilde{d}\mathbf{Y}}{\widetilde{d}\mathbf{X}} = \frac{1}{2} \sum_{\substack{ij \\ k<l}} \frac{\partial y_{ij}}{\partial x_{kl}} ((\mathbf{r}_i \mathbf{s}_j') \otimes (\mathbf{f}_k \mathbf{f}_l' + \mathbf{f}_l \mathbf{f}_k')) + \sum_{ijk} \frac{\partial y_{ij}}{\partial x_{kk}} ((\mathbf{r}_i \mathbf{s}_j') \otimes \mathbf{f}_k \mathbf{f}_k').$$

A general basic derivative would be

$$\sum_{ijkl} \frac{\partial y_{ij}}{\partial x_{kl}} \mathbf{e}_{ij} \otimes \mathbf{d}_{kl},$$

where $\{\mathbf{e}_{ij}\}$ is a basis for the space which $\mathbf{Y}$ belongs to, and $\{\mathbf{d}_{kl}\}$ is a basis for the space which $\mathbf{X}$ belongs to. In the symmetric case, i.e. when $\mathbf{X}$ is symmetric, we can use the following set of matrices as basis $\mathbf{d}_{kl}$:

$$\frac{1}{2}(\mathbf{f}_k \mathbf{f}_l' + \mathbf{f}_l \mathbf{f}_k'),\ k < l, \quad \mathbf{f}_k \mathbf{f}_k', \quad k, l = 1, \ldots, p.$$

This means that from here we get the derivative (1.4.48). Furthermore, there is a clear connection with §1.3.6. For example, if the basis matrices $\frac{1}{2}(\mathbf{f}_k \mathbf{f}_l' + \mathbf{f}_l \mathbf{f}_k')$, $k < l$, and $\mathbf{f}_k \mathbf{f}_k'$ are vectorized, i.e. consider

$$\frac{1}{2}(\mathbf{f}_l \otimes \mathbf{f}_k + \mathbf{f}_k \otimes \mathbf{f}_l),\ k < l, \quad \mathbf{f}_k \otimes \mathbf{f}_k,$$

then we obtain the columns of the pattern projection matrix $\mathbf{M}(s)$ in Proposition 1.3.18.

Below we give several properties of the derivative defined by (1.4.47) and (1.4.48), which will be proved rigorously only for the symmetric case.

Proposition 1.4.12. *Let the derivatives below be given by either (1.4.47) or (1.4.48). Then*

(i)
$$\frac{\widetilde{d}(\mathbf{Z}\mathbf{Y})}{\widetilde{d}\mathbf{X}} = \frac{\widetilde{d}\mathbf{Z}}{\widetilde{d}\mathbf{X}}(\mathbf{Y} \otimes \mathbf{I}) + (\mathbf{Z} \otimes \mathbf{I})\frac{\widetilde{d}\mathbf{Y}}{\widetilde{d}\mathbf{X}};$$

(ii) *for a scalar function* $y = y(\mathbf{X})$ *and matrix* $\mathbf{Z}$

$$\frac{\widetilde{d}(\mathbf{Z}y)}{\widetilde{d}\mathbf{X}} = \frac{\widetilde{d}\mathbf{Z}}{\widetilde{d}\mathbf{X}}y + \mathbf{Z} \otimes \frac{\widetilde{d}y}{\widetilde{d}\mathbf{X}};$$

(iii) *for constant matrices* $\mathbf{A}$ *and* $\mathbf{B}$ *and* $\mathbf{X} \in \mathbb{R}^{p \times n}$

$$\frac{\widetilde{d}(\mathbf{AYB})}{\widetilde{d}\mathbf{X}} = (\mathbf{A} \otimes \mathbf{I}_p)\frac{\widetilde{d}\mathbf{Y}}{\widetilde{d}\mathbf{X}}(\mathbf{B} \otimes \mathbf{I}_n);$$

(iv) *for* $\mathbf{Y} \in \mathbb{R}^{q \times r}$, $\mathbf{Z} \in \mathbb{R}^{s \times t}$, *being functions of* $\mathbf{X} \in \mathbb{R}^{p \times n}$,

$$\frac{\widetilde{d}(\mathbf{Y} \otimes \mathbf{Z})}{\widetilde{d}\mathbf{X}} = \mathbf{Y} \otimes \frac{\widetilde{d}\mathbf{Z}}{\widetilde{d}\mathbf{X}} + (\mathbf{K}_{q,s} \otimes \mathbf{I}_p)(\mathbf{Z} \otimes \frac{\widetilde{d}\mathbf{Z}}{\widetilde{d}\mathbf{X}})(\mathbf{K}_{t,r} \otimes \mathbf{I}_n);$$

(v) *if* $\mathbf{Y} \in \mathbb{R}^{q \times q}$ *is a function of* $\mathbf{X} \in \mathbb{R}^{p \times n}$, *then*

$$\frac{\widetilde{d}(\mathrm{tr}\mathbf{Y})}{\widetilde{d}\mathbf{X}} = \sum_{m=1}^{q}(\mathbf{e}'_m \otimes \mathbf{I}_p)\frac{\widetilde{d}\mathbf{Y}}{\widetilde{d}\mathbf{X}}(\mathbf{e}_m \otimes \mathbf{I}_n),$$

where $\mathbf{e}_m$ *is the* $m-th$ *column of* $\mathbf{I}_q$;

(vi) *if* $\mathbf{Y} \in \mathbb{R}^{q \times q}$ *is non-singular and a function of* $\mathbf{X} \in \mathbb{R}^{p \times n}$, *then*

$$\frac{\widetilde{d}\mathbf{Y}^{-1}}{\widetilde{d}\mathbf{X}} = -(\mathbf{Y}^{-1} \otimes \mathbf{I}_p)\frac{\widetilde{d}\mathbf{Y}}{\widetilde{d}\mathbf{X}}(\mathbf{Y}^{-1} \otimes \mathbf{I}_n);$$

(vii) *if* $\mathbf{X} \in \mathbb{R}^{p \times n}$, *then*

$$\frac{\widetilde{d}\mathbf{X}}{\widetilde{d}\mathbf{X}} = \begin{cases} \mathrm{vec}\mathbf{I}_p\mathrm{vec}'\mathbf{I}_n & \text{if } \mathbf{X} \text{ is m.i.v.,} \\ \frac{1}{2}\{\mathrm{vec}\mathbf{I}_p\mathrm{vec}'\mathbf{I}_p + \mathbf{K}_{p,p}\} & \text{if } \mathbf{X}: p \times p \text{ is symmetric;} \end{cases}$$

(viii) *if* $\mathbf{X} \in \mathbb{R}^{p \times n}$, *then*

$$\frac{\widetilde{d}\mathbf{X}'}{\widetilde{d}\mathbf{X}} = \begin{cases} \mathbf{K}_{n,p} & \text{if } \mathbf{X} \text{ is m.i.v.,} \\ \frac{1}{2}\{\mathrm{vec}\mathbf{I}_p\mathrm{vec}'\mathbf{I}_p + \mathbf{K}_{p,p}\} & \text{if } \mathbf{X}: p \times p \text{ is symmetric;} \end{cases}$$

(ix) *for a constant matrix* $\mathbf{A}$

$$\frac{\widetilde{d}\mathrm{tr}(\mathbf{A}'\mathbf{X})}{\widetilde{d}\mathbf{X}} = \begin{cases} \mathbf{A} & \text{if } \mathbf{X} \text{ is m.i.v.,} \\ \frac{1}{2}(\mathbf{A} + \mathbf{A}') & \text{if } \mathbf{X} \text{ is symmetric;} \end{cases}$$

(x) *for constant matrices* $\mathbf{A}$ *and* $\mathbf{B}$

$$\frac{\widetilde{d}\mathrm{tr}(\mathbf{XAX}'\mathbf{B})}{\widetilde{d}\mathbf{X}} = \begin{cases} \mathbf{BXA} + \mathbf{B}'\mathbf{XA}' & \text{if } \mathbf{X} \text{ is m.i.v.,} \\ \frac{1}{2}(\mathbf{BXA} + \mathbf{B}'\mathbf{XA}' + \mathbf{A}'\mathbf{XB}' + \mathbf{AXB}) & \\ & \text{if } \mathbf{X} \text{ is symmetric;} \end{cases}$$

(xi) *if* **AXB** *is non-singular with* **A** *and* **B** *being matrices of constants, then*

$$\frac{\widetilde{d}\mathrm{tr}((\mathbf{AXB})^{-1})}{\widetilde{d}\mathbf{X}} = \begin{cases} -\mathbf{A}'(\mathbf{B}'\mathbf{X}'\mathbf{A}')^{-1}(\mathbf{B}'\mathbf{X}'\mathbf{A}')^{-1}\mathbf{B}' & \textit{if } \mathbf{X} \textit{ is m.i.v.,} \\ -\frac{1}{2}\{\mathbf{A}'(\mathbf{B}'\mathbf{X}\mathbf{A}')^{-1}(\mathbf{B}'\mathbf{X}\mathbf{A}')^{-1}\mathbf{B}' & \\ \quad +\mathbf{B}(\mathbf{AXB})^{-1}(\mathbf{AXB})^{-1}\mathbf{A}\} & \textit{if } \mathbf{X} \textit{ is symmetric;} \end{cases}$$

(xii) *for a non-singular* **X** *and any real* r

$$\frac{\widetilde{d}|\mathbf{X}|^r}{\widetilde{d}\mathbf{X}} = r(\mathbf{X}')^{-1}|\mathbf{X}|^r.$$

PROOF: (i): Suppose that **X** is symmetric. Straightforward calculations yield

$$\begin{aligned}
\frac{\widetilde{d}(\mathbf{ZY})}{\widetilde{d}\mathbf{X}} =& \sum_{ijkl} \frac{\partial \sum_m z_{im}y_{mj}}{\partial x_{kl}}(\mathbf{r}_i\mathbf{s}_j') \otimes (\epsilon_{kl}\mathbf{f}_k\mathbf{f}_l') \\
=& \sum_{ijkl}\sum_m \left(\frac{\partial z_{im}}{\partial x_{kl}}y_{mj} + z_{im}\frac{\partial y_{mj}}{\partial x_{kl}}\right)(\mathbf{r}_i\mathbf{s}_j') \otimes (\epsilon_{kl}\mathbf{f}_k\mathbf{f}_l') \\
=& \sum_{ijkl}\sum_m\sum_n \left(\frac{\partial z_{im}}{\partial x_{kl}}y_{nj} + z_{im}\frac{\partial y_{nj}}{\partial x_{kl}}\right)(\mathbf{r}_i\mathbf{s}_m^{1'}\mathbf{s}_n^1\mathbf{s}_j') \otimes (\epsilon_{kl}\mathbf{f}_k\mathbf{f}_l') \\
=& \frac{\widetilde{d}\mathbf{Z}}{\widetilde{d}\mathbf{X}}(\mathbf{Y}\otimes\mathbf{I}) + (\mathbf{Z}\otimes\mathbf{I})\frac{\widetilde{d}\mathbf{Y}}{\widetilde{d}\mathbf{X}},
\end{aligned}$$

since

$$\begin{aligned}
(\mathbf{r}_i\mathbf{s}_m^{1'}\mathbf{s}_n^1\mathbf{s}_j') \otimes (\epsilon_{kl}\mathbf{f}_k\mathbf{f}_l') &= (\mathbf{r}_i\mathbf{s}_m^{1'} \otimes \epsilon_{kl}\mathbf{f}_k\mathbf{f}_l')(\mathbf{s}_n^1\mathbf{s}_j' \otimes \mathbf{I}_n) \\
&= (\mathbf{r}_i\mathbf{s}_m' \otimes \mathbf{I}_p)(\mathbf{s}_n^1\mathbf{s}_j' \otimes \epsilon_{kl}\mathbf{f}_k\mathbf{f}_l').
\end{aligned}$$

(ii): Similarly, for a symmetric **X**,

$$\begin{aligned}
\frac{\widetilde{d}(\mathbf{Z}y)}{\widetilde{d}\mathbf{X}} =& \sum_{ijkl}\frac{\partial z_{ij}y}{\partial x_{kl}}(\mathbf{r}_i\mathbf{s}_j') \otimes (\epsilon_{kl}\mathbf{f}_k\mathbf{f}_l') \\
=& \sum_{ijkl} y\frac{\partial z_{ij}}{\partial x_{kl}}(\mathbf{r}_i\mathbf{s}_j') \otimes (\epsilon_{kl}\mathbf{f}_k\mathbf{f}_l') + \sum_{ijkl} z_{ij}(\mathbf{r}_i\mathbf{s}_j' \otimes \frac{\partial y}{\partial x_{kl}}\epsilon_{kl}\mathbf{f}_k\mathbf{f}_l') \\
=& \frac{\widetilde{d}\mathbf{Z}}{\widetilde{d}\mathbf{X}}y + \mathbf{Z}\otimes\frac{\widetilde{d}\mathbf{Y}}{\widetilde{d}\mathbf{X}}.
\end{aligned}$$

(iii): From (i) it follows that

$$\frac{\widetilde{d}(\mathbf{AYB})}{\widetilde{d}\mathbf{X}} = (\mathbf{A}\otimes\mathbf{I}_p)\frac{\widetilde{d}\mathbf{YB}}{\widetilde{d}\mathbf{X}} = (\mathbf{A}\otimes\mathbf{I}_p)\frac{\widetilde{d}\mathbf{Y}}{\widetilde{d}\mathbf{X}}(\mathbf{B}\otimes\mathbf{I}_n).$$

(iv): By using (i) and Proposition 1.3.12 we get

$$\begin{aligned}\frac{\widetilde{d}(\mathbf{Y}\otimes\mathbf{Z})}{\widetilde{d}\mathbf{X}} &= \frac{\widetilde{d}((\mathbf{I}_q\otimes\mathbf{Z})(\mathbf{Y}\otimes\mathbf{I}_t))}{\widetilde{d}\mathbf{X}}\\ &= \frac{\widetilde{d}(\mathbf{I}_q\otimes\mathbf{Z})}{\widetilde{d}\mathbf{X}}(\mathbf{Y}\otimes\mathbf{I}_t\otimes\mathbf{I}_n) + (\mathbf{I}_q\otimes\mathbf{Z}\otimes\mathbf{I}_p)\frac{\widetilde{d}(\mathbf{Y}\otimes\mathbf{I}_t)}{\widetilde{d}\mathbf{X}}.\end{aligned}\tag{1.4.49}$$

Now

$$\frac{\widetilde{d}(\mathbf{I}_q\otimes\mathbf{Z})}{\widetilde{d}\mathbf{X}}(\mathbf{Y}\otimes\mathbf{I}_t\otimes\mathbf{I}_n) = (\mathbf{I}_q\otimes\frac{\widetilde{d}\mathbf{Z}}{\widetilde{d}\mathbf{X}})(\mathbf{Y}\otimes\mathbf{I}_t\otimes\mathbf{I}_n) = \mathbf{Y}\otimes\frac{\widetilde{d}\mathbf{Z}}{\widetilde{d}\mathbf{X}}\tag{1.4.50}$$

and

$$\begin{aligned}(\mathbf{I}_q\otimes\mathbf{Z}\otimes\mathbf{I}_p)\frac{\widetilde{d}(\mathbf{Y}\otimes\mathbf{I}_t)}{\widetilde{d}\mathbf{X}} &= (\mathbf{I}_q\otimes\mathbf{Z}\otimes\mathbf{I}_p)\frac{\widetilde{d}(\mathbf{K}_{q,t}(\mathbf{I}_t\otimes\mathbf{Y})\mathbf{K}_{t,r})}{\widetilde{d}\mathbf{X}}\\ &= (\mathbf{I}_q\otimes\mathbf{Z}\otimes\mathbf{I}_p)(\mathbf{K}_{q,t}\otimes\mathbf{I}_p)\frac{\widetilde{d}(\mathbf{I}_t\otimes\mathbf{Y})}{\widetilde{d}\mathbf{X}}(\mathbf{K}_{t,r}\otimes\mathbf{I}_n)\\ &= (\mathbf{K}_{q,s}\otimes\mathbf{I}_p)(\mathbf{Z}\otimes\mathbf{I}_q\otimes\mathbf{I}_p)(\mathbf{I}_t\otimes\frac{\widetilde{d}\mathbf{Y}}{\widetilde{d}\mathbf{X}})(\mathbf{K}_{t,r}\otimes\mathbf{I}_n)\\ &= (\mathbf{K}_{q,s}\otimes\mathbf{I}_p)(\mathbf{Z}\otimes\frac{\widetilde{d}\mathbf{Y}}{\widetilde{d}\mathbf{X}})(\mathbf{K}_{t,r}\otimes\mathbf{I}_n).\end{aligned}\tag{1.4.51}$$

Thus, inserting (1.4.50) and (1.4.51) in the right-hand side of (1.4.49) verifies (iv).
(v): Since $\sum_m \mathbf{s}_m\mathbf{s}_m' = \mathbf{I}_q$, it follows that

$$\mathrm{tr}\mathbf{Y} = \mathrm{tr}\left(\sum_m \mathbf{s}_m\mathbf{s}_m'\mathbf{Y}\right) = \sum_m \mathrm{tr}(\mathbf{s}_m\mathbf{s}_m'\mathbf{Y}) = \sum_m \mathrm{tr}(\mathbf{s}_m'\mathbf{Y}\mathbf{s}_m) = \sum_m \mathbf{s}_m'\mathbf{Y}\mathbf{s}_m$$

and then (iii) implies (v).
(vi): Observe that

$$\mathbf{0} = \frac{\widetilde{d}\mathbf{Y}^{-1}\mathbf{Y}}{\widetilde{d}\mathbf{X}} = \frac{\widetilde{d}\mathbf{Y}^{-1}}{\widetilde{d}\mathbf{X}}(\mathbf{Y}\otimes\mathbf{I}) + (\mathbf{Y}^{-1}\otimes\mathbf{I})\frac{\widetilde{d}\mathbf{Y}}{\widetilde{d}\mathbf{X}}.$$

(vii): Here it is noted that

$$\begin{aligned}&\frac{\widetilde{d}\mathbf{X}}{\widetilde{d}\mathbf{X}}\\ &= \begin{cases}\displaystyle\sum_{ijkl}\frac{\partial x_{ij}}{\partial x_{kl}}((\mathbf{r}_i\mathbf{s}_j')\otimes(\mathbf{r}_k\mathbf{s}_l')) = \sum_{kl}((\mathbf{r}_k\mathbf{s}_l')\otimes(\mathbf{r}_k\mathbf{s}_l')),\\ \displaystyle\sum_{ijkl}\frac{\partial x_{ij}}{\partial x_{kl}}((\mathbf{r}_i\mathbf{r}_j')\otimes(\epsilon_{kl}\mathbf{r}_k\mathbf{r}_l')) = \sum_{kl}\frac{1}{2}((\mathbf{r}_k\mathbf{r}_l')\otimes(\mathbf{r}_k\mathbf{r}_l') + (\mathbf{r}_l\mathbf{r}_k')\otimes(\mathbf{r}_k\mathbf{r}_l'))\end{cases}\\ &= \begin{cases}\mathrm{vec}\mathbf{I}_p\mathrm{vec}'\mathbf{I}_n & \text{if } \mathbf{X} \text{ is m.i.v.,}\\ \frac{1}{2}\{\mathrm{vec}\mathbf{I}_p\mathrm{vec}'\mathbf{I}_p + \mathbf{K}_{p,p}\} & \text{if } \mathbf{X}: p\times p \text{ is symmetric.}\end{cases}\end{aligned}$$

(viii): If $\mathbf{X}$ is m.i.v., then

$$\frac{\widetilde{d}\mathbf{X}'}{\widetilde{d}\mathbf{X}} = \sum_{ijkl} \frac{\partial x_{ij}}{\partial x_{kl}}((\mathbf{s}_j\mathbf{r}_i') \otimes (\mathbf{r}_k\mathbf{s}_l')) = \sum_{kl}((\mathbf{s}_l\mathbf{r}_k') \otimes (\mathbf{r}_k\mathbf{s}_l')) = \mathbf{K}_{n,p}.$$

(ix): By virtue of (iii), (v) and (vii)

$$\frac{\widetilde{d}\mathrm{tr}(\mathbf{A}'\mathbf{X})}{\widetilde{d}\mathbf{X}} = \sum_m (\mathbf{s}_m'\mathbf{A}' \otimes \mathbf{I}_p)\frac{\widetilde{d}\mathbf{X}}{\widetilde{d}\mathbf{X}}(\mathbf{s}_m \otimes \mathbf{I}_n)$$
$$= \begin{cases} \sum_m \mathrm{vec}(\mathbf{A}\mathbf{s}_m)\mathrm{vec}'(\mathbf{s}_m) = \sum_m \mathbf{A}\mathbf{s}_m\mathbf{s}_m' = \mathbf{A} & \text{if } \mathbf{X} \text{ is m.i.v.,} \\ \frac{1}{2}\sum_m\{\mathrm{vec}(\mathbf{A}\mathbf{s}_m)\mathrm{vec}'(\mathbf{s}_m) + \mathbf{K}_{1,p}\mathrm{vec}(\mathbf{A}'\mathbf{s}_m)\mathrm{vec}'(\mathbf{s}_m)\} = \frac{1}{2}(\mathbf{A}+\mathbf{A}') & \\ & \text{if } \mathbf{X} \text{ is symmetric.} \end{cases}$$

(x): Similar calculations show that

$$\frac{\widetilde{d}\mathrm{tr}(\mathbf{X}\mathbf{A}\mathbf{X}'\mathbf{B})}{\widetilde{d}\mathbf{X}}$$
$$= \begin{cases} \sum_m \mathbf{s}_m\mathbf{s}_m'\mathbf{B}'\mathbf{X}\mathbf{A} + \mathbf{B}\mathbf{s}_m\mathbf{s}_m'\mathbf{X}\mathbf{A}' \\ \frac{1}{2}\sum_m \mathbf{s}_m\mathbf{s}_m'\mathbf{B}'\mathbf{X}\mathbf{A}' + \mathbf{A}\mathbf{X}\mathbf{B}\mathbf{s}_m\mathbf{s}_m' + \mathbf{A}'\mathbf{X}\mathbf{s}_m\mathbf{s}_m'\mathbf{B}' + \mathbf{B}\mathbf{s}_m\mathbf{s}_m'\mathbf{X}\mathbf{A} \end{cases}$$
$$= \begin{cases} \mathbf{B}\mathbf{X}\mathbf{A} + \mathbf{B}'\mathbf{X}\mathbf{A}' & \text{if } \mathbf{X} \text{ is m.i.v.,} \\ \frac{1}{2}(\mathbf{B}\mathbf{X}\mathbf{A} + \mathbf{B}'\mathbf{X}\mathbf{A}' + \mathbf{A}'\mathbf{X}\mathbf{B}' + \mathbf{A}\mathbf{X}\mathbf{B}) & \text{if } \mathbf{X} \text{ is symmetric.} \end{cases}$$

(xi): From (v), (vi), (vii) and (iii) it follows that

$$\frac{\widetilde{d}\mathrm{tr}((\mathbf{A}\mathbf{X}\mathbf{B})^{-1})}{\widetilde{d}\mathbf{X}}$$
$$= \sum_m (\mathbf{s}_m' \otimes \mathbf{I})(-((\mathbf{A}\mathbf{X}\mathbf{B})^{-1} \otimes \mathbf{I}))\frac{\widetilde{d}\mathbf{A}\mathbf{X}\mathbf{B}}{\widetilde{d}\mathbf{X}}((\mathbf{A}\mathbf{X}\mathbf{B})^{-1} \otimes \mathbf{I})(\mathbf{e}_m \otimes \mathbf{I})$$
$$= -\sum_m ((\mathbf{s}_m'(\mathbf{A}\mathbf{X}\mathbf{B})^{-1}\mathbf{A}) \otimes \mathbf{I})\frac{\widetilde{d}\mathbf{X}}{\widetilde{d}\mathbf{X}}((\mathbf{B}(\mathbf{A}\mathbf{X}\mathbf{B})^{-1}\mathbf{s}_m) \otimes \mathbf{I})$$
$$= \begin{cases} -\sum_m \mathrm{vec}(\mathbf{A}'(\mathbf{B}'\mathbf{X}'\mathbf{A}')^{-1}\mathbf{s}_m)\mathrm{vec}'(\mathbf{B}(\mathbf{A}\mathbf{X}\mathbf{B})^{-1}\mathbf{s}_m) \\ -\sum_m \frac{1}{2}\{\mathrm{vec}(\mathbf{A}'(\mathbf{B}'\mathbf{X}'\mathbf{A}')^{-1}\mathbf{s}_m)\mathrm{vec}'(\mathbf{B}(\mathbf{A}\mathbf{X}\mathbf{B})^{-1}\mathbf{s}_m) \\ \quad +((\mathbf{s}_m'(\mathbf{A}\mathbf{X}\mathbf{B})^{-1}\mathbf{A}) \otimes \mathbf{I})\mathbf{K}_{p,p}((\mathbf{B}(\mathbf{A}\mathbf{X}\mathbf{B})^{-1}\mathbf{s}_m) \otimes \mathbf{I})\} \end{cases}$$
$$= \begin{cases} -\mathbf{A}'(\mathbf{B}'\mathbf{X}'\mathbf{A}')^{-1}(\mathbf{B}'\mathbf{X}'\mathbf{A}')^{-1}\mathbf{B}' & \text{if } \mathbf{X} \text{ is m.i.v.} \\ -\frac{1}{2}\{\mathbf{A}'(\mathbf{B}'\mathbf{X}\mathbf{A}')^{-1}(\mathbf{B}'\mathbf{X}\mathbf{A}')^{-1}\mathbf{B}' + \mathbf{B}(\mathbf{A}\mathbf{X}\mathbf{B})^{-1}(\mathbf{A}\mathbf{X}\mathbf{B})^{-1}\mathbf{A}\} & \\ & \text{if } \mathbf{X} \text{ is symmetric.} \end{cases}$$

(xii): The statement holds for both the m.i.v. and the symmetric matrices and the proof is almost identical to the one of Proposition 1.4.10. ∎

1.4.8 *Minimal derivatives*

Dealing with higher order derivatives, one practical problem is how to select the different partial derivatives from all possible combinations of partial derivatives. Matrix derivatives of any order which consist of all different partial derivatives, where each partial derivative appears only once, will be referred to as *minimal derivatives*. For the first derivatives one solution to the problem was given in §1.4.4 where derivatives by patterned matrices were considered. If one is interested in higher order derivatives, a solution to the problem can be obtained by applying differentiation of patterned matrices iteratively when using different patterns K which depend on the order of the derivative. Parallel to this possibility another approach can be used. In the papers by Kollo & von Rosen (1995a, 1995b) the derivative is defined in a slightly different way from what is standard. The derivative which we are now going to examine is based on the vectorization operators $V(\cdot)$ and $R(\cdot)$ described in §1.3.7.

Let $\mathbf{X}(K)$ be a patterned matrix which consists of all different elements and $\mathrm{vec}\mathbf{X}(K)$ the $k-$vector consisting of the elements of $\mathbf{X}(K)$. In the case of a symmetric matrix $\mathbf{X}$ the vectorization operator $V(\cdot)$ can be used to perform the selection of different elements of $\mathbf{X}$ in the form of the upper triangle of $\mathbf{X}$.

Definition 1.4.5. *Let $\mathbf{X}(K_1)$ and $\mathbf{Y}(K_2)$ be two patterned matrices, where the elements of $\mathbf{Y}(K_2)$ are functions of the elements of $\mathbf{X}(K_1)$. The $j-$th order minimal derivative is given by the equality*

$$\frac{\widehat{d^j}\mathbf{Y}(K_2)}{\widehat{d}\mathbf{X}(K_1)^j} = V^j\left(\frac{d}{d\mathrm{vec}\mathbf{X}(K_1)}\left(\frac{\widehat{d^{j-1}}\mathbf{Y}(K_2)}{\widehat{d}\mathbf{X}(K_1)^{j-1}}\right)\right) \tag{1.4.52}$$

with

$$\frac{\widehat{d}\mathbf{Y}(K_2)}{\widehat{d}\mathbf{X}(K_1)} = V^1\left(\frac{d\mathbf{Y}(K_2)}{d\mathbf{X}(K_1)}\right), \tag{1.4.53}$$

where the derivative $\dfrac{d\mathbf{Y}(K_2)}{d\mathbf{X}(K_1)}$ is defined by (1.4.33) and the vectorization operator V^j by (1.3.75). ∎

If $\mathbf{X}$ and $\mathbf{Y}$ are symmetric matrices and the elements of their upper triangular parts are different, we may choose $\mathrm{vec}\mathbf{X}(K_1)$ and $\mathrm{vec}\mathbf{Y}(K_2)$ to be identical to $V^2(\mathbf{X})$ and $V^2(\mathbf{Y})$. Whenever in the text we refer to a symmetric matrix, we always assume that this assumption is fulfilled. When taking the derivative of a symmetric matrix by a symmetric matrix, we shall use the notation $\dfrac{\widehat{d}\mathbf{Y}}{\widehat{d}\mathbf{X}}$ and will omit the index sets after the matrices. So for symmetric matrices $\mathbf{Y}$ and $\mathbf{X}$ Definition 1.4.5 takes the following form:

$$\frac{\widehat{d^j}\mathbf{Y}}{\widehat{d}\mathbf{X}^j} = V^j\left(\frac{d}{d(V^2(\mathbf{X}))}\left(\frac{\widehat{d^{j-1}}\mathbf{Y}}{\widehat{d}\mathbf{X}^{j-1}}\right)\right), \tag{1.4.54}$$

with

$$\frac{\widehat{d}\mathbf{Y}}{\widehat{d}\mathbf{X}} = V^1\left(\frac{d(V^2(\mathbf{Y}))}{d(V^2(\mathbf{X}))}\right). \tag{1.4.55}$$

The next theorem shows that for symmetric matrices $\frac{\widehat{d^j}\mathbf{Y}}{\widehat{d\mathbf{X}^j}}$ consists of all different partial derivatives of $j-$th order. However, it is worth noting that the theorem is true for matrices with arbitrary structure on the elements.

Theorem 1.4.7. *Let* $\mathbf{X}$, $\mathbf{Y}$ *be symmetric matrices with* $\mathbf{Y} = \mathbf{Y}(\mathbf{X})$. *Then the derivative, defined by (1.4.54), consists of all different partial derivatives* $\frac{\partial^j y_{gh}}{\partial x_{kl}^j}$, *which appear only once.*

PROOF: To prove the theorem it will be shown that (1.4.54) is determined by the same construction as the product vectorizing operator $R^j(\cdot)$ and the statement follows then from the uniqueness of the elements of $R^j(\cdot)$. First let us define, on the basis of $V^j(\cdot)$, a composition operator $Q^j(\cdot)$.

Definition 1.4.6. *For symmetric* $\mathbf{A} : n \times n$

$$Q^j(\mathbf{A}) = V^j(Q^{j-1}(\mathbf{A}) \odot V^2(A)'), \qquad j = 1, 2, \ldots, \tag{1.4.56}$$

where $Q^0(\mathbf{A}) = 1$ *and* $\odot$ *is a composition operation which is specially defined for different forms of* $Q^j(\mathbf{A})$. ∎

There are two important special cases of $Q^j(\mathbf{A})$, which will be referred to in the proof of Theorem 1.4.7. From (1.3.77), which defines the product vectorization operator $R^j(\cdot)$ for symmetric matrices, it follows that the operator $Q^j(\cdot)$ turns into $R^j(\cdot)$, if the composition operation is the usual matrix product operation. When the composition is the differentiation procedure, the operator $Q^j(\cdot)$ turns into the $j-$th order matrix derivative given by (1.4.54). We have shown in Theorem 1.3.13 that $R^j(\mathbf{A})$ consists of all different monomials of order j. Because of the same structure of $Q^j(\cdot)$ for different composition operators, we get the statement of the theorem immediately. ∎

1.4.9 *Tables of derivatives*

In this paragraph the derivatives given by (1.4.7), (1.4.47) and (1.4.48) are summarized and their properties presented in two tables.

Table 1.4.1. Properties of the matrix derivatives given by (1.4.47) and (1.4.48). If $\mathbf{X} : p \times n$ is m.i.v., then (1.4.47) is used, whereas if $\mathbf{X}$ is symmetric, (1.4.48) is used. If dimensions are not given, the results hold for both derivatives.

Differentiated function	Derivative
$\mathbf{Z}+\mathbf{Y}$	$\frac{\widetilde{d}\mathbf{Y}}{\widetilde{d}\mathbf{X}}+\frac{\widetilde{d}\mathbf{Z}}{\widetilde{d}\mathbf{X}}$
$\mathbf{Z}\mathbf{Y}$	$\frac{\widetilde{d}\mathbf{Z}}{\widetilde{d}\mathbf{X}}(\mathbf{Y}\otimes\mathbf{I})+(\mathbf{Z}\otimes\mathbf{I})\frac{\widetilde{d}\mathbf{Y}}{\widetilde{d}\mathbf{X}}$
$\mathbf{Z}y$	$\frac{\widetilde{d}\mathbf{Z}}{\widetilde{d}\mathbf{X}}y+\mathbf{Z}\otimes\frac{\widetilde{d}y}{\widetilde{d}\mathbf{X}}$
$\mathbf{AYB}$	$(\mathbf{A}\otimes\mathbf{I}_p)\frac{\widetilde{d}\mathbf{Y}}{\widetilde{d}\mathbf{X}}(\mathbf{B}\otimes\mathbf{I}_n)$
$\mathbf{Y}\otimes\mathbf{Z}, \quad \mathbf{Y}\in\mathbb{R}^{q\times r}, \mathbf{Z}\in\mathbb{R}^{s\times t}$	$\mathbf{Y}\otimes\frac{\widetilde{d}\mathbf{Z}}{\widetilde{d}\mathbf{X}}+(\mathbf{K}_{q,s}\otimes\mathbf{I}_p)(\mathbf{Z}\otimes\frac{\widetilde{d}\mathbf{Y}}{\widetilde{d}\mathbf{X}})(\mathbf{K}_{t,r}\otimes\mathbf{I}_n)$
$\mathrm{tr}\mathbf{Y}, \quad \mathbf{Y}\in\mathbb{R}^{q\times q}$	$\sum_{m=1}^{q}(\mathbf{e}'_m\otimes\mathbf{I}_p)\frac{\widetilde{d}\mathbf{Y}}{\widetilde{d}\mathbf{X}}(\mathbf{e}_m\otimes\mathbf{I}_n)$
$\mathbf{Y}^{-1}$	$-(\mathbf{Y}^{-1}\otimes\mathbf{I}_p)\frac{\widetilde{d}\mathbf{Y}}{\widetilde{d}\mathbf{X}}(\mathbf{Y}^{-1}\otimes\mathbf{I}_n)$
$\mathbf{X}, \quad \mathbf{X}$ is m.i.v.	$\mathrm{vec}\mathbf{I}_p\mathrm{vec}'\mathbf{I}_n$
$\mathbf{X}, \quad \mathbf{X}$ is symmetric	$\frac{1}{2}(\mathrm{vec}\mathbf{I}_p\mathrm{vec}'\mathbf{I}_p+\mathbf{K}_{p,p})$
$\mathbf{X}', \quad \mathbf{X}$ is m.i.v.	$\mathbf{K}_{n,p}$
$\mathrm{tr}(\mathbf{A}'\mathbf{X}), \quad \mathbf{X}$ is m.i.v.	$\mathbf{A}$
$\mathrm{tr}(\mathbf{A}'\mathbf{X}), \quad \mathbf{X}$ is symmetric	$\frac{1}{2}(\mathbf{A}+\mathbf{A}')$
$\mathrm{tr}(\mathbf{XAX}'\mathbf{B}), \quad \mathbf{X}$ is m.i.v.	$\mathbf{BXA}+\mathbf{B}'\mathbf{XA}'$
$\mathrm{tr}(\mathbf{XAX}'\mathbf{B}), \quad \mathbf{X}$ is symmetric	$\frac{1}{2}(\mathbf{BXA}+\mathbf{B}'\mathbf{XA}'+\mathbf{A}'\mathbf{XB}'+\mathbf{AXB})$
$\lvert\mathbf{X}\rvert^r, \quad r\in\mathbb{R}$	$r(\mathbf{X}')^{-1}\lvert\mathbf{X}\rvert^r$

Table 1.4.2. Properties of the matrix derivative given by (1.4.7). In the table $\mathbf{X} \in \mathbb{R}^{p\times q}$ and $\mathbf{Y} \in \mathbb{R}^{r\times s}$.

Differentiated function	Derivative $\frac{d\mathbf{Y}}{d\mathbf{X}}$	Formula
$\mathbf{X}$	$\mathbf{I}_{pq}$	(1.4.9)
$c\mathbf{X}$	$c\mathbf{I}_{pq}$	(1.4.10)
$\mathbf{A}'\text{vec}\mathbf{X}$	$\mathbf{A}$	(1.4.12)
$\mathbf{Y}+\mathbf{Z}$	$\frac{d\mathbf{Y}}{d\mathbf{X}}+\frac{d\mathbf{Z}}{d\mathbf{X}}$	(1.4.13)
$\mathbf{Z}=\mathbf{Z}(\mathbf{Y}), \mathbf{Y}=\mathbf{Y}(\mathbf{X})$	$\frac{d\mathbf{Z}}{d\mathbf{X}}=\frac{d\mathbf{Y}}{d\mathbf{X}}\frac{d\mathbf{Z}}{d\mathbf{Y}}$	(1.4.14)
$\mathbf{Y}=\mathbf{AXB}$	$\frac{d\mathbf{Y}}{d\mathbf{X}}=\mathbf{B}\otimes\mathbf{A}'$	(1.4.15)
$\mathbf{Z}=\mathbf{AYB}$	$\frac{d\mathbf{Z}}{d\mathbf{X}}=\frac{d\mathbf{Y}}{d\mathbf{X}}(\mathbf{B}\otimes\mathbf{A}')$	(1.4.16)
$\mathbf{X}'$	$\mathbf{K}_{q,p}$	(1.4.17)
$\mathbf{W}=\mathbf{W}(\mathbf{Y}(\mathbf{X}),\mathbf{Z}(\mathbf{X}))$	$\frac{d\mathbf{W}}{d\mathbf{X}}=\frac{d\mathbf{Y}}{d\mathbf{X}}\frac{d\mathbf{W}}{d\mathbf{Y}}\Big\vert_{\mathbf{Z}=\text{const.}}+\frac{d\mathbf{Z}}{d\mathbf{X}}\frac{d\mathbf{W}}{d\mathbf{Z}}\Big\vert_{\mathbf{Y}=\text{const.}}$	(1.4.18)
$\mathbf{W}=\mathbf{YZ},\ \ \mathbf{Z}\in\mathbb{R}^{s\times t}$	$\frac{d\mathbf{W}}{d\mathbf{X}}=\frac{d\mathbf{Y}}{d\mathbf{X}}(\mathbf{Z}\otimes\mathbf{I}_r)+\frac{d\mathbf{Z}}{d\mathbf{X}}(\mathbf{I}_t\otimes\mathbf{Y}')$	(1.4.19)
$\mathbf{Y}^n, \mathbf{Y}^0=\mathbf{I}_r\ n\geq 1$	$\frac{d\mathbf{Y}^n}{d\mathbf{X}}=\frac{d\mathbf{Y}}{d\mathbf{X}}\Big\{\sum_{\substack{i+j=n-1;\\ i,j\geq 0}}\mathbf{Y}^i\otimes(\mathbf{Y}')^j\Big\}$	(1.4.20)
$\mathbf{X}^{-1},\quad \mathbf{X}\in\mathbb{R}^{p\times p}$	$-\mathbf{X}^{-1}\otimes(\mathbf{X}')^{-1},$	(1.4.21)
$\mathbf{Y}^{-n}, \mathbf{Y}^0=\mathbf{I}_r,\ n\geq 1$	$\frac{d\mathbf{Y}^{-n}}{d\mathbf{X}}=-\frac{d\mathbf{Y}}{d\mathbf{X}}\Big\{\sum_{\substack{i+j=n-1;\\ i,j\geq 0}}\mathbf{Y}^{-i-1}\otimes(\mathbf{Y}')^{-j-1}\Big\}$	(1.4.22)
$\mathbf{Y}\otimes\mathbf{Z},\ \ \mathbf{Z}\in\mathbb{R}^{m\times n}$	$\Big\{\frac{d\mathbf{Y}}{d\mathbf{X}}\otimes\text{vec}'\mathbf{Z}+\text{vec}'\mathbf{Y}\otimes\frac{d\mathbf{Z}}{d\mathbf{X}}\Big\}(\mathbf{I}_s\otimes\mathbf{K}_{r,n}\otimes\mathbf{I}_m)$	(1.4.23)
$\mathbf{Y}\otimes\mathbf{A},\ \ \mathbf{A}\in\mathbb{R}^{m\times n}$	$(\frac{d\mathbf{Y}}{d\mathbf{X}}\otimes\text{vec}'\mathbf{A})(\mathbf{I}_s\otimes\mathbf{K}_{r,n}\otimes\mathbf{I}_m)$	(1.4.24)
$\mathbf{A}\otimes\mathbf{Y},\ \ \mathbf{A}\in\mathbb{R}^{m\times n}$	$(\frac{d\mathbf{Y}}{d\mathbf{X}}\otimes\text{vec}'\mathbf{A})\mathbf{K}_{rs,mn}(\mathbf{I}_n\otimes\mathbf{K}_{m,s}\otimes\mathbf{I}_r)$	(1.4.25)
$\text{tr}(\mathbf{A}'\mathbf{X})$	$\text{vec}\mathbf{A}$	(1.4.28)
$\lvert\mathbf{X}\rvert,\quad \mathbf{X}\in\mathbb{R}^{p\times p}$	$\lvert\mathbf{X}\rvert\text{vec}(\mathbf{X}^{-1})'$	(1.4.30)
$\mathbf{X}_d,\quad \mathbf{X}\in\mathbb{R}^{p\times p}$	$(\mathbf{K}_{p,p})_d$	(1.4.37)
$\mathbf{X}-\textit{symmetric}$	$\frac{d\mathbf{X}}{d\mathbf{X}}=\mathbf{I}_{p^2}+\mathbf{K}_{p,p}-(\mathbf{K}_{p,p})_d,\qquad \mathbf{X}\in\mathbb{R}^{p\times p}$	(1.4.38)
$\mathbf{X}-\textit{skew-symmetric}$	$\frac{d\mathbf{X}}{d\mathbf{X}}=\mathbf{I}_{p^2}-\mathbf{K}_{p,p},\qquad \mathbf{X}\in\mathbb{R}^{p\times p}$	(1.4.39)

1.4.10 *Taylor expansion*

A basic tool for further studies in the following chapters is the Taylor series expansion. We will derive a general formula, which is valid for a matrix function with a matrix argument:

$$\mathbf{f} : \mathbb{R}^{p\times q} \longrightarrow \mathbb{R}^{r\times s}.$$

Because of the isometry between the vector spaces $\mathbb{R}^{p\times q}$ and $\mathbb{R}^{pq}$ (see §1.4.2), the study of a mapping $\mathbf{f}$ is equivalent to the study of a function defined on the Euclidian space $\mathbb{R}^{pq}$ which maps on $\mathbb{R}^{rs}$. In this paragraph we are going to convert classical Taylor series expansions of multivariate functions into compact matrix forms.

Let us remind of the one-dimensional case. When f is an ordinary univariate real function, it can be expanded into a Taylor series in a neighborhood $\mathcal{D}$ of a point x_0 if $f(x)$ is $(n+1)$ times differentiable at x_0:

$$f(x) = \sum_{k=0}^{n} \frac{1}{k!} \frac{d^k f(x)}{dx^k}\bigg|_{x=x_0} (x-x_0)^k + r_n, \qquad (1.4.57)$$

where the error term

$$r_n = \frac{1}{(n+1)!} \frac{d^{n+1} f(x)}{dx^{n+1}}\bigg|_{x=\xi} (x-x_0)^{n+1}, \qquad \text{for some } \xi \in \mathcal{D}.$$

When $\mathbf{f}(\mathbf{x})$ is a function from $\mathbb{R}^p$ to $\mathbb{R}^q$, every coordinate $f_i(\mathbf{x})$ of $\mathbf{f}(\mathbf{x})$ can be expanded into a Taylor series in a neighborhood $\mathcal{D}$ of $\mathbf{x}_0$, if all partial derivatives up to the order $(n+1)$ of $\mathbf{f}(\mathbf{x})$ exist and are continuous in $\mathcal{D}$:

$$\begin{aligned} f_i(\mathbf{x}) &= f_i(\mathbf{x}_0) \\ &+ \sum_{i_1=1}^{p} \frac{\partial f_i(\mathbf{x})}{\partial x_{i_1}}\bigg|_{\mathbf{x}=\mathbf{x}_0} (\mathbf{x}-\mathbf{x}_0)_{i_1} + \frac{1}{2!} \sum_{i_1,i_2=1}^{p} \frac{\partial^2 f_i(\mathbf{x})}{\partial x_{i_1}\partial x_{i_2}}\bigg|_{\mathbf{x}=\mathbf{x}_0} (\mathbf{x}-\mathbf{x}_0)_{i_1}(\mathbf{x}-\mathbf{x}_0)_{i_2} \\ &+ \cdots + \frac{1}{n!} \sum_{i_1,\ldots,i_n=1}^{p} \frac{\partial^n f_i(\mathbf{x})}{\partial x_{i_1}\ldots\partial x_{i_n}}\bigg|_{\mathbf{x}=\mathbf{x}_0} (\mathbf{x}-\mathbf{x}_0)_{i_1} \times \cdots \times (\mathbf{x}-\mathbf{x}_0)_{i_n} + r_n^i, \end{aligned} \qquad (1.4.58)$$

where the error term is

$$r_n^i = \frac{1}{(n+1)!} \sum_{i_1,\ldots,i_{n+1}=1}^{p} \frac{\partial^{n+1} f_i(\mathbf{x})}{\partial x_{i_1}\ldots\partial x_{i_{n+1}}}\bigg|_{\mathbf{x}=\xi} (\mathbf{x}-\mathbf{x}_0)_{i_1} \times \cdots \times (\mathbf{x}-\mathbf{x}_0)_{i_{n+1}}, \qquad \text{for some } \xi \in \mathcal{D}.$$

In the following theorem we shall present the expansions of the coordinate functions $f_i(\mathbf{x})$ in a compact matrix form.

Theorem 1.4.8. *If the function* $\mathbf{f}(\mathbf{x})$ *from* $\mathbb{R}^p$ *to* $\mathbb{R}^q$ *has continuous partial derivatives up to the order* $(n+1)$ *in a neighborhood* $\mathcal{D}$ *of a point* $\mathbf{x}_0$*, then the function* $\mathbf{f}(\mathbf{x})$ *can be expanded into the Taylor series at the point* $\mathbf{x}_0$ *in the following way:*

$$\mathbf{f}(\mathbf{x}) = \mathbf{f}(\mathbf{x}_0) + \sum_{k=1}^{n} \frac{1}{k!} \left(\mathbf{I}_q \otimes (\mathbf{x}-\mathbf{x}_0)^{\otimes(k-1)}\right)' \left(\frac{d^k\mathbf{f}(\mathbf{x})}{d\mathbf{x}^k}\right)'\Bigg|_{\mathbf{x}=\mathbf{x}_0} (\mathbf{x}-\mathbf{x}_0) + \mathbf{r}_n, \tag{1.4.59}$$

where the error term is

$$\mathbf{r}_n = \frac{1}{(n+1)!} \left((\mathbf{I}_q \otimes (\mathbf{x}-\mathbf{x}_0)^{\otimes(k-1)}\right)' \left(\frac{d^{n+1}\mathbf{f}(\mathbf{x})}{d\mathbf{x}^{n+1}}\right)'\Bigg|_{\mathbf{x}=\boldsymbol{\xi}} (\mathbf{x}-\mathbf{x}_0),$$

for some $\boldsymbol{\xi} \in \mathcal{D}$, (1.4.60)

and the derivative is given by (1.4.41).

PROOF: The starting point of the proof will be (1.4.58), which holds for the coordinate functions $f_i(\mathbf{x})$ of

$$\mathbf{f}(\mathbf{x}) = \sum_{i=1}^{q} f_i(\mathbf{x})\mathbf{d}_i = (f_1(\mathbf{x}), f_2(\mathbf{x}), \ldots, f_q(\mathbf{x}))'.$$

From Theorem 1.4.4 it follows that

$$\frac{d^k\mathbf{f}(\mathbf{x})}{d\mathbf{x}^k}\Bigg|_{\mathbf{x}=\mathbf{x}_0} = \sum_{i=1}^{q} \sum_{i_1,\ldots,i_k=1}^{p} \frac{\partial^k f_i(\mathbf{x})}{\partial x_{i_1}\ldots\partial x_{i_k}}\Bigg|_{\mathbf{x}=\mathbf{x}_0} \mathbf{e}_{i_k}(\mathbf{d}_i \otimes \mathbf{e}_{i_1} \otimes \cdots \otimes \mathbf{e}_{i_{k-1}})', \tag{1.4.61}$$

where $\mathbf{e}_{i_\bullet}$ and $\mathbf{d}_i$ are unit basis vectors of size p and q, respectively. Premultiplying (1.4.61) with $(\mathbf{x}-\mathbf{x}_0)'$ and postmultiplying it with $\mathbf{I}_q \otimes (\mathbf{x}-\mathbf{x}_0)^{\otimes k-1}$ yields

$$\begin{aligned}(\mathbf{x}-\mathbf{x}_0)' \frac{d^k\mathbf{f}(\mathbf{x})}{d\mathbf{x}^k}\Bigg|_{\mathbf{x}-\mathbf{x}_0} &(\mathbf{I}_q \otimes (\mathbf{x}-\mathbf{x}_0)^{\otimes k-1}) \\ = \sum_{i=1}^{q} \sum_{i_1,\ldots,i_k=1}^{p} &\frac{\partial^k f_i(\mathbf{x})}{\partial x_{i_1}\ldots\partial x_{i_k}}\Bigg|_{\mathbf{x}=\mathbf{x}_0} \mathbf{d}_i'(\mathbf{x}-\mathbf{x}_0)_{i_1} \times \cdots \times (\mathbf{x}-\mathbf{x}_0)_{i_k}.\end{aligned} \tag{1.4.62}$$

Since $\mathbf{f}(\mathbf{x}) = \sum_{i=1}^{q} f_i(\mathbf{x})\mathbf{d}_i$, it follows from (1.4.58) that by (1.4.62) we have the term including the k−th derivative in the Taylor expansion for all coordinates (in transposed form). To complete the proof we remark that for the error term we can copy the calculations given above and thus the theorem is established. ∎

Comparing expansion (1.4.59) with the univariate Taylor series (1.4.57), we note that, unfortunately, a full analogy between the formulas has not been obtained. However, the Taylor expansion in the multivariate case is only slightly more complicated than in the univariate case. The difference is even smaller in the most important case of applications, namely, when the function $f(\mathbf{x})$ is a mapping on the real line. We shall present it as a corollary of the theorem.

Corollary 1.4.8.1. *Let the function $f(\mathbf{x})$ from $\mathbb{R}^p$ to $\mathbb{R}$ have continuous partial derivatives up to order $(n+1)$ in a neighborhood $\mathcal{D}$ of the point $\mathbf{x}_0$. Then the function $f(\mathbf{x})$ can be expanded into a Taylor series at the point $\mathbf{x}_0$ in the following way:*

$$f(\mathbf{x}) = f(\mathbf{x}_0) + \sum_{k=1}^{n} \frac{1}{k!}(\mathbf{x}-\mathbf{x}_0)' \left(\frac{d^k f(\mathbf{x})}{d\mathbf{x}^k}\right)'\Bigg|_{\mathbf{x}=\mathbf{x}_0} (\mathbf{x}-\mathbf{x}_0)^{\otimes(k-1)} + r_n, \quad (1.4.63)$$

where the error term is

$$r_n = \frac{1}{(n+1)!}(\mathbf{x}-\mathbf{x}_0)' \left(\frac{d^{n+1} f(\mathbf{x})}{d\mathbf{x}^{n+1}}\right)'\Bigg|_{\mathbf{x}=\boldsymbol{\xi}} (\mathbf{x}-\mathbf{x}_0)^{\otimes n}, \qquad \text{for some } \boldsymbol{\xi} \in \mathcal{D}, \quad (1.4.64)$$

and the derivative is given by (1.4.41). ∎

There is also another way of presenting (1.4.63) and (1.4.64), which sometimes is preferable in applications.

Corollary 1.4.8.2. *Let the function $f(\mathbf{x})$ from $\mathbb{R}^p$ to $\mathbb{R}$ have continuous partial derivatives up to order $(n+1)$ in a neighborhood $\mathcal{D}$ of the point $\mathbf{x}_0$. Then the function $f(\mathbf{x})$ can be expanded into a Taylor series at the point $\mathbf{x}_0$ in the following way:*

$$f(\mathbf{x}) = f(\mathbf{x}_0) + \sum_{k=1}^{n} \frac{1}{k!}\left((\mathbf{x}-\mathbf{x}_0)'\right)^{\otimes k} \operatorname{vec}\left(\frac{d^k f(\mathbf{x})}{d\mathbf{x}^k}\right)'\Bigg|_{\mathbf{x}=\mathbf{x}_0} + r_n, \quad (1.4.65)$$

where the error term is

$$r_n = \frac{1}{(n+1)!}\left((\mathbf{x}-\mathbf{x}_0)'\right)^{\otimes(n+1)} \operatorname{vec}\left(\frac{d^{n+1} f(\mathbf{x})}{d\mathbf{x}^{n+1}}\right)'\Bigg|_{\mathbf{x}=\boldsymbol{\xi}}, \qquad \text{for some } \boldsymbol{\xi} \in \mathcal{D}, \quad (1.4.66)$$

and the derivative is given by (1.4.41).

PROOF: The statement follows from Corollary 1.4.8.1 after applying property (1.3.31) of the vec-operator to (1.4.63) and (1.4.64). ∎

1.4.11 *Integration by parts and orthogonal polynomials*

A multivariate analogue of the formula of integration by parts is going to be presented. The expression will not look as nice as in the univariate case but later it will be shown that the multivariate version is effective in various applications. In the sequel some specific notation is needed. Let $\int_\Omega \mathbf{F}(\mathbf{X})d\mathbf{X}$ denote an ordinary multiple integral, where $\mathbf{F}(\mathbf{X}) \in \mathbb{R}^{p\times t}$ and integration is performed elementwise of the matrix function $\mathbf{F}(\mathbf{X})$ with matrix argument $\mathbf{X} \in \mathbb{R}^{q\times m}$. In the integral, $d\mathbf{X}$ denotes the Lebesque measure $\prod_{k=1,\dots,q,\, l=1,\dots,m} dx_{kl}$. Furthermore, $\int_{\Omega\setminus x_{ij}}$ means that we are integrating over all variables in $\mathbf{X}$ except x_{ij} and $d\mathbf{X}_{\setminus x_{ij}} = d\mathbf{X}/dx_{ij}$, i.e. the Lebesque measure for all variables in $\mathbf{X}$ except x_{ij}. Finally, let $\partial\Omega$ denote the boundary of Ω. The set Ω may be complicated. For example, the cone over all positive definite matrices is a space which is often utilized. On the other hand, it can also be fairly simple, for instance, $\Omega = \mathbb{R}^{q\times m}$. Before presenting the theorem, let us recall the formula for univariate integration by parts :

$$\int_\Omega \frac{df(x)}{dx} g(x)dx = f(x)g(x)\Big|_{x\in\partial\Omega} - \int_\Omega f(x)\frac{dg(x)}{dx}dx.$$

Theorem 1.4.9. *Let* $\mathbf{F}(\mathbf{X}) \in \mathbb{R}^{p\times t}$, $\mathbf{G}(\mathbf{X}) \in \mathbb{R}^{t\times n}$, $\mathbf{X} \in \mathbb{R}^{q\times m}$ *and suppose that all integrals given below exist. Then*

$$\int_\Omega \frac{d\mathbf{G}(\mathbf{X})}{d\mathbf{X}}(\mathbf{I}_n \otimes \mathbf{F}'(\mathbf{X}))d\mathbf{X} + \int_\Omega \frac{d\mathbf{F}(\mathbf{X})}{d\mathbf{X}}(\mathbf{G}(\mathbf{X}) \otimes \mathbf{I}_p)d\mathbf{X} = \mathbf{Q},$$

where

$$\mathbf{Q} = \sum_{u,v\in I}(\mathbf{e}_v \otimes \mathbf{d}_u)\left(\int_{\Omega\setminus x_{uv}} \frac{d\mathbf{F}(\mathbf{X})\mathbf{G}(\mathbf{X})}{dx_{uv}} d\mathbf{X}_{\setminus x_{uv}}\Big|_{x_{uv}\in\partial\Omega}\right),$$

$I = \{u, v : u = 1,\dots,q, v = 1,\dots m\}$ *and the derivative is given by (1.4.7).*

PROOF: Note that $\dfrac{d\mathbf{F}(\mathbf{X})}{d\mathbf{X}} \in \mathbb{R}^{qm\times pt}$ and $\dfrac{d\mathbf{G}(\mathbf{X})}{d\mathbf{X}} \in \mathbb{R}^{qm\times tn}$. We are going to integrate the derivative

$$\frac{d(\mathbf{F}(\mathbf{X})\mathbf{G}(\mathbf{X}))_{ij}}{dx_{uv}}.$$

The first fundamental theorem in analysis states that

$$\int_\Omega \frac{d(\mathbf{F}(\mathbf{X})\mathbf{G}(\mathbf{X}))_{ij}}{dx_{uv}} dx_{uv} d\mathbf{X}_{\setminus x_{uv}} = \int_{\Omega\setminus x_{uv}} (\mathbf{F}(\mathbf{X})\mathbf{G}(\mathbf{X}))_{ij}\Big|_{x_{uv}\in\partial\Omega} d\mathbf{X}_{\setminus x_{uv}}$$

which implies

$$\int_\Omega \frac{d\mathbf{F}(\mathbf{X})\mathbf{G}(\mathbf{X})}{dx_{uv}} dx_{uv} d\mathbf{X}_{\setminus x_{uv}} = \int_{\Omega\setminus x_{uv}} \mathbf{F}(\mathbf{X})\mathbf{G}(\mathbf{X})\Big|_{x_{uv}\in\partial\Omega} d\mathbf{X}_{\setminus x_{uv}}.$$

Thus, the theorem is verified since

$$\begin{aligned}\mathbf{Q} &= \int_\Omega \frac{d\mathbf{F}(\mathbf{X})\mathbf{G}(\mathbf{X})}{d\mathbf{X}} d\mathbf{X} \\ &\underset{(1.4.16)}{=} \int_\Omega \frac{d\mathbf{G}(\mathbf{X})}{d\mathbf{X}}(\mathbf{I} \otimes \mathbf{F}'(\mathbf{X}))d\mathbf{X} + \int_\Omega \frac{d\mathbf{F}(\mathbf{X})}{d\mathbf{X}}(\mathbf{G}(\mathbf{X}) \otimes \mathbf{I})d\mathbf{X}.\end{aligned}$$

∎

If $\mathbf{F}(\mathbf{X})\mathbf{G}(\mathbf{X})|_{\mathbf{X}_{uv}\in\partial\Omega} = \mathbf{0}$, $u = 1,\dots,q, v = 1,\dots,m$, then $\mathbf{Q} = \mathbf{0}$. Suppose that this holds, then a useful relation is presented in the next corollary based on this fact.

Corollary 1.4.9.1. *If* $\mathbf{F}(\mathbf{X})\mathbf{G}(\mathbf{X})|_{x_{uv}\in\partial\mathbf{\Omega}} = \mathbf{0}$, $u = 1,\dots,q$, $v = 1,\dots,m$, *then*

$$\int_{\mathbf{\Omega}} \frac{d\mathbf{F}(\mathbf{X})}{d\mathbf{X}}(\mathbf{G}(\mathbf{X})\otimes\mathbf{I}_p)d\mathbf{X} = -\int_{\mathbf{\Omega}} \frac{d\mathbf{G}(\mathbf{X})}{d\mathbf{X}}(\mathbf{I}_n\otimes\mathbf{F}'(\mathbf{X}))d\mathbf{X},$$

and the derivative is given by (1.4.7). ∎

In the following some general ideas of generating orthogonal polynomials will be discussed. Suppose that $f(\mathbf{X})\in\mathbb{R}$, $\mathbf{X}\in\mathbb{R}^{q\times m}$ is a function such that

$$\frac{d^k f(\mathbf{X})}{d\mathbf{X}^k} = \mathbf{P}_k(\mathbf{X})f(\mathbf{X}), \tag{1.4.67}$$

where $\mathbf{P}_k(\mathbf{X})$ is a polynomial in $\mathbf{X}$ of order k, i.e. $\sum_{i=0}^k c_i\mathrm{vec}\mathbf{X}(\mathrm{vec}'\mathbf{X})^{\otimes i-1}$, where c_i are known constants. Furthermore, suppose that on $\partial\mathbf{\Omega}$

$$\mathrm{vec}\mathbf{P}_{k-r-1}(\mathbf{X})\mathrm{vec}'\frac{d^r\mathbf{P}_l(\mathbf{X})}{d\mathbf{X}^r}f(\mathbf{X})\Big|_{x_{uv}\in\partial\mathbf{\Omega}} = \mathbf{0};$$
$$r = 0,\dots,l<k,\ u = 1,\dots,q,\ v = 1,\dots,m. \tag{1.4.68}$$

Theorem 1.4.10. *Suppose that (1.4.68) holds. Then, for* $k > l$,

$$\int_{\mathbf{\Omega}} \mathrm{vec}\mathbf{P}_k(\mathbf{X})\mathrm{vec}'\mathbf{P}_l(\mathbf{X})f(\mathbf{X})d\mathbf{X} = \mathbf{0}.$$

PROOF: First it is noted that for $r = 0$ the equality (1.4.68) implies that in Theorem 1.4.9

$$\mathbf{Q} = \mathrm{vec}\int_{\mathbf{\Omega}} \frac{d}{d\mathbf{X}}\{\mathrm{vec}\mathbf{P}_{k-1}(\mathbf{X})\otimes\mathrm{vec}\mathbf{P}_l(\mathbf{X})f(\mathbf{X})\}d\mathbf{X} = \mathbf{0}.$$

Hence, from Corollary 1.4.9.1 it follows that

$$\begin{aligned}\mathbf{0} = \mathrm{vec}&\int_{\mathbf{\Omega}} \mathbf{P}_k(\mathbf{X})(\mathbf{I}\otimes\mathrm{vec}'\mathbf{P}_l(\mathbf{X}))f(\mathbf{X})d\mathbf{X}\\ &+\mathrm{vec}\int_{\mathbf{\Omega}}(\mathrm{vec}\mathbf{P}_{k-1}(\mathbf{X})\otimes\frac{d\mathbf{P}_l(\mathbf{X})}{d\mathbf{X}})f(\mathbf{X})d\mathbf{X},\end{aligned}$$

which is equivalent to

$$\begin{aligned}\mathbf{0} = \mathbf{H}_1^1&\int_{\mathbf{\Omega}} \mathrm{vec}\mathbf{P}_k(\mathbf{X})\otimes\mathrm{vec}\mathbf{P}_l(\mathbf{X})f(\mathbf{X})d\mathbf{X}\\ &-\mathbf{H}_1^2\int_{\mathbf{\Omega}}\mathrm{vec}\mathbf{P}_{k-1}(\mathbf{X})\otimes\mathrm{vec}\frac{d\mathbf{P}_l(\mathbf{X})}{d\mathbf{X}}f(\mathbf{X})d\mathbf{X}\end{aligned}$$

for some known non-singular matrices $\mathbf{H}_1^1$ and $\mathbf{H}_1^2$. Thus,

$$\mathbf{H}_1^1\int_{\mathbf{\Omega}}\mathrm{vec}\mathbf{P}_k(\mathbf{X})\otimes\mathrm{vec}\mathbf{P}_l(\mathbf{X})f(\mathbf{X})d\mathbf{X} = \mathbf{H}_1^2\int_{\mathbf{\Omega}}\mathrm{vec}\mathbf{P}_{k-1}(\mathbf{X})\otimes\mathrm{vec}\frac{d\mathbf{P}_l(\mathbf{X})}{d\mathbf{X}}f(\mathbf{X})d\mathbf{X}. \tag{1.4.69}$$

We are going to proceed in the same manner and note that

$$\begin{aligned}\mathbf{0} &= \mathrm{vec}\int_{\boldsymbol{\Omega}}\frac{d}{d\mathbf{X}}\{\mathrm{vec}\mathbf{P}_{k-2}(\mathbf{X})\otimes\mathrm{vec}\frac{d\mathbf{P}_l(\mathbf{X})}{d\mathbf{X}}f(\mathbf{X})\}d\mathbf{X}\\ &= \mathbf{H}_2^1\int_{\boldsymbol{\Omega}}\{\mathrm{vec}\mathbf{P}_{k-1}(\mathbf{X})\otimes\mathrm{vec}\frac{d\mathbf{P}_l(\mathbf{X})}{d\mathbf{X}}f(\mathbf{X})\}d\mathbf{X}\\ &\quad -\mathbf{H}_2^2\int_{\boldsymbol{\Omega}}\{\mathrm{vec}\mathbf{P}_{k-2}(\mathbf{X})\otimes\mathrm{vec}\frac{d^2\mathbf{P}_l(\mathbf{X})}{d\mathbf{X}^2}f(\mathbf{X})\}d\mathbf{X}\end{aligned}$$

for some non-singular matrices $\mathbf{H}_2^1$ and $\mathbf{H}_2^2$. Thus,

$$\begin{aligned}&\int_{\boldsymbol{\Omega}}\mathrm{vec}\mathbf{P}_k(\mathbf{X})\otimes\mathrm{vec}\mathbf{P}_l(\mathbf{X})f(\mathbf{X})d\mathbf{X}\\ &=(\mathbf{H}_1^1)^{-1}\mathbf{H}_1^2\int_{\boldsymbol{\Omega}}\{\mathrm{vec}\mathbf{P}_{k-1}(\mathbf{X})\otimes\mathrm{vec}\frac{d\mathbf{P}_l(\mathbf{X})}{d\mathbf{X}}f(\mathbf{X})\}d\mathbf{X}\\ &=(\mathbf{H}_1^1)^{-1}\mathbf{H}_1^2(\mathbf{H}_2^1)^{-1}\mathbf{H}_2^2\int_{\boldsymbol{\Omega}}\{\mathrm{vec}\mathbf{P}_{k-2}(\mathbf{X})\otimes\mathrm{vec}\frac{d^2\mathbf{P}_l(\mathbf{X})}{d\mathbf{X}^2}f(\mathbf{X})\}d\mathbf{X}=\cdots\\ &=(\mathbf{H}_1^1)^{-1}\mathbf{H}_1^2(\mathbf{H}_2^1)^{-1}\mathbf{H}_2^2\cdots(\mathbf{H}_l^1)^{-1}\mathbf{H}_l^2\int_{\boldsymbol{\Omega}}\{\mathrm{vec}\mathbf{P}_{k-l}(\mathbf{X})\otimes\mathrm{vec}\frac{d^l\mathbf{P}_l(\mathbf{X})}{d\mathbf{X}^l}f(\mathbf{X})\}d\mathbf{X}.\end{aligned}$$

However, since $\mathbf{P}_l(\mathbf{X})$ is a polynomial of order l, $\dfrac{d^l\mathbf{P}_l(\mathbf{X})}{d\mathbf{X}^l}$ is a constant matrix and since, by assumption given in (1.4.68),

$$\int_{\boldsymbol{\Omega}}\mathbf{P}_{k-l}(\mathbf{X})f(\mathbf{X})d\mathbf{X}=\int_{\boldsymbol{\Omega}}\frac{d}{d\mathbf{X}}\mathbf{P}_{k-l-1}(\mathbf{X})f(\mathbf{X})d\mathbf{X}=\mathbf{0}.$$

It has been shown that $\mathrm{vec}\mathbf{P}_k(\mathbf{X})$ and $\mathrm{vec}\mathbf{P}_l(\mathbf{X})$ are orthogonal with respect to $f(\mathbf{X})$. ∎

Later in our applications, $f(\mathbf{X})$ will be a density function and then the result of the theorem can be rephrased as

$$E[\mathrm{vec}\mathbf{P}_k(\mathbf{X})\otimes\mathrm{vec}\mathbf{P}_l(\mathbf{X})]=\mathbf{0},$$

which is equivalent to

$$E[\mathrm{vec}\mathbf{P}_k(\mathbf{X})\mathrm{vec}'\mathbf{P}_l(\mathbf{X})]=\mathbf{0}, \tag{1.4.70}$$

where $E[\bullet]$ denotes expectation.

1.4.12 *Jacobians*

Jacobians $|\mathbf{J}(\mathbf{Y}\rightarrow\mathbf{X})|_+$ appear in multiple integrals when the original variables are transformed, i.e. assuming that the multiple integral is well defined and that the functions are integrable

$$\int\mathbf{F}(\mathbf{Y})d\mathbf{Y}=\int\mathbf{F}(\mathbf{G}(\mathbf{X}))|\mathbf{J}(\mathbf{Y}\rightarrow\mathbf{X})|_+d\mathbf{X},$$

where $\mathbf{Y} = \mathbf{G}(\mathbf{X})$, for some one-to-one matrix function $\mathbf{G}(\bullet)$. In a statistical context Jacobians are used when density functions are transformed. Observe the well-known relation from calculus

$$|\mathbf{J}(\mathbf{Y} \to \mathbf{X})|_+ = |\mathbf{J}(\mathbf{X} \to \mathbf{Y})|_+^{-1}.$$

The *Jacobian matrix* of the one-to-one transformation from $\mathbf{Y} = \mathbf{G}(\mathbf{X})$ to $\mathbf{X}$ is given by

$$\mathbf{J}(\mathbf{Y} \to \mathbf{X}) = \frac{d\mathrm{vec}\mathbf{Y}(K_1)}{d\mathrm{vec}\mathbf{X}(K_2)},$$

where K_1 and K_2 describe the patterns in $\mathbf{Y}$ and $\mathbf{X}$, respectively. For notational convenience we will often use $\frac{d_{k_1}\mathbf{Y}}{d_{k_2}\mathbf{X}}$ instead of $\frac{d\mathrm{vec}\mathbf{Y}(K_1)}{d\mathrm{vec}\mathbf{X}(K_2)}$ in this paragraph. The absolute value of the determinant of $\mathbf{J}(\mathbf{Y} \to \mathbf{X})$, i.e. $|\mathbf{J}(\mathbf{Y} \to \mathbf{X})|_+$, is called the *Jacobian* of the transformation. Many useful results and references on Jacobians can be found in the books by Srivastava & Khatri (1979), Muirhead (1982), Magnus (1988) and Mathai (1997). In particular we would like to mention the classical references Deemer & Olkin (1951) and Olkin & Sampson (1972). In this paragraph we will present some of the most frequently used Jacobians. The purpose, besides presenting these Jacobians, is to illustrate how to utilize the notions of patterned matrices and matrix derivatives.

Theorem 1.4.11. *Let $\mathbf{Y} = \mathbf{F}(\mathbf{X})$ and $\mathbf{X} = \mathbf{G}(\mathbf{Z})$. Then*

$$\mathbf{J}(\mathbf{Y} \to \mathbf{Z}) = \mathbf{J}(\mathbf{Y} \to \mathbf{X})\mathbf{J}(\mathbf{X} \to \mathbf{Z}).$$

PROOF: The chain rule, given by Proposition 1.4.3, establishes the theorem. ∎

The next theorem is taken from Srivastava & Khatri (1979, Theorem 1.11.2). However, we shall give a different proof.

Theorem 1.4.12. *Define the conditional transformation as*

$$\begin{aligned} y_1 &= f_1(x_1, x_2, \ldots, x_n), \\ y_i &= f_i(y_1, y_2, \ldots, y_{i-1}, x_1, x_2, \ldots, x_n). \qquad i = 1, 2, \ldots, n \end{aligned}$$

Then

$$|\mathbf{J}(y_1, y_2, \ldots, y_n \to x_1, x_2, \ldots, x_n)|_+ = \prod_{i=1}^{n} |J(y_i \to x_i)|_+.$$

PROOF: We will once again apply the chain rule

$$\begin{aligned} \frac{d\mathrm{vec}(y_1, y_2, \ldots, y_n)}{d\mathrm{vec}(x_1, x_2, \ldots, x_n)} &= \frac{d\mathrm{vec}(y_1, y_2, \ldots, y_{n-1}, x_n)}{d\mathrm{vec}(x_1, x_2, \ldots, x_n)} \frac{d\mathrm{vec}(y_1, y_2, \ldots, y_n)}{d\mathrm{vec}(y_1, y_2, \ldots, y_{n-1}, x_n)} \\ &= \begin{pmatrix} \frac{d\mathrm{vec}(y_1, y_2, \ldots, y_{n-1})}{d\mathrm{vec}(x_1, x_2, \ldots, x_{n-1})} & \mathbf{0} \\ \frac{d\mathrm{vec}(y_1, y_2, \ldots, y_{n-1})}{dx_n} & 1 \end{pmatrix} \begin{pmatrix} \mathbf{I} & \mathbf{0} \\ \frac{d\mathrm{vec}(y_1, y_2, \ldots, y_{n-1})}{dx_n} & \frac{dy_n}{dx_n} \end{pmatrix}. \end{aligned}$$

Thus, because of the triangular structure of the variables,

$$|\mathbf{J}(y_1, y_2, \ldots, y_n \to x_1, x_2, \ldots, x_n)|_+ = \left|\frac{d\mathrm{vec}(y_1, y_2, \ldots, y_{n-1})}{d\mathrm{vec}(x_1, x_2, \ldots, x_{n-1})}\right|_+ \left|\frac{dy_n}{dx_n}\right|_+$$

and then it follows that

$$|\mathbf{J}(y_1, y_2, \ldots, y_n \to x_1, x_2, \ldots, x_n)|_+ = \prod_{i=1}^{n} |J(y_i \to x_i)|_+.$$

∎

In the following certain linear transformations will be considered.

Theorem 1.4.13. *Let* $\mathbf{A} : n \times n$ *be non-singular,* $\mathbf{V}$ *symmetric and* $\mathbf{W}$ *skew-symmetric.*

(i) $|\mathbf{J}(\mathbf{AVA}' \to \mathbf{V})|_+ = |\mathbf{A}|_+^{n+1}$,

(ii) $|\mathbf{J}(\mathbf{AWA}' \to \mathbf{W})|_+ = |\mathbf{A}|_+^{n-1}$.

PROOF: Without loss of generality we suppose that the eigenvalues of $\mathbf{A}'\mathbf{A}$ are non-zero and distinct. Now

$$\begin{aligned}\left|\frac{d_k\mathbf{AVA}'}{d_k\mathbf{V}}\right|_+ &= \left|\frac{d\mathbf{V}}{d_k\mathbf{V}}(\mathbf{A}'\otimes\mathbf{A}')\mathbf{T}'(s)\right|_+ = \left|\frac{d_k\mathbf{V}}{d_k\mathbf{V}}(\mathbf{T}^+(s))'(\mathbf{A}'\otimes\mathbf{A}')\mathbf{T}'(s)\right|_+ \\ &= \left|(\mathbf{T}^+(s))'(\mathbf{A}'\otimes\mathbf{A}')\mathbf{T}'(s)\right|_+, \qquad (1.4.71)\end{aligned}$$

where $\mathbf{T}(s)$ and $\mathbf{T}^+(s)$ are given in Proposition 1.3.18 and the index k in the derivative refers to the symmetric structure. By definition of $\mathbf{T}(s)$ the equality (1.4.71) can be further explored:

$$\begin{aligned}&\left|\frac{d_k\mathbf{AVA}'}{d_k\mathbf{V}}\right|_+ \\ &= |\underline{\mathbf{B}}(s)^{-1}\mathbf{B}'(s)(\mathbf{A}'\otimes\mathbf{A}')\mathbf{B}(s)\underline{\mathbf{B}}'|_+ = |\mathbf{B}'(s)(\mathbf{A}'\otimes\mathbf{A}')\mathbf{B}(s)|_+ \\ &= |\mathbf{B}'(s)(\mathbf{A}'\otimes\mathbf{A}')\mathbf{B}(s)\mathbf{B}'(s)(\mathbf{A}\otimes\mathbf{A})\mathbf{B}(s)|^{1/2} = |\mathbf{B}'(s)(\mathbf{A}'\mathbf{A}\otimes\mathbf{A}'\mathbf{A})\mathbf{B}(s)|^{1/2},\end{aligned}$$

since $\mathbf{B}(s)\mathbf{B}'(s) = \frac{1}{2}(\mathbf{I}+\mathbf{K}_{n,n})$. According to Proposition 1.3.12 (xiii) all different eigenvalues of $\mathbf{A}'\mathbf{A}\otimes\mathbf{A}'\mathbf{A}$ are given by $\lambda_i\lambda_j$, $j \geq i = 1, 2, \ldots, n$, where λ_k is an eigenvalue of $\mathbf{A}'\mathbf{A}$. However, these eigenvalues are also eigenvalues of the product $\mathbf{B}'(s)(\mathbf{A}'\mathbf{A}\otimes\mathbf{A}'\mathbf{A})\mathbf{B}(s)$ since, if $\mathbf{u}$ is an eigenvector which corresponds to $\lambda_i\lambda_j$,

$$(\mathbf{A}'\mathbf{A}\otimes\mathbf{A}'\mathbf{A})\mathbf{u} = \lambda_i\lambda_j\mathbf{u}$$

implies that

$$\begin{aligned}\mathbf{B}'(s)(\mathbf{A}'\mathbf{A}\otimes\mathbf{A}'\mathbf{A})\mathbf{B}(s)\mathbf{B}'(s)\mathbf{u} &= \mathbf{B}'(s)\mathbf{B}(s)\mathbf{B}'(s)(\mathbf{A}'\mathbf{A}\otimes\mathbf{A}'\mathbf{A})\mathbf{u} \\ &= \mathbf{B}'(s)(\mathbf{A}'\mathbf{A}\otimes\mathbf{A}'\mathbf{A})\mathbf{u} = \lambda_i\lambda_j\mathbf{B}'(s)\mathbf{u}.\end{aligned}$$

Moreover, since

$$r(\mathbf{B}'(s)(\mathbf{A}'\mathbf{A}\otimes\mathbf{A}'\mathbf{A})\mathbf{B}(s)) = r(\mathbf{B}(s)) = \tfrac{1}{2}n(n+1),$$

all eigenvalues of $\mathbf{B}'(s)(\mathbf{A}'\mathbf{A}\otimes\mathbf{A}'\mathbf{A})\mathbf{B}(s)$ are given by $\lambda_i\lambda_j$. Thus, by Proposition 1.2.3 (xi)

$$\begin{aligned}|\mathbf{B}'(s)(\mathbf{A}'\mathbf{A}\otimes\mathbf{A}'\mathbf{A})\mathbf{B}(s)|^{1/2} &= \prod_{\substack{i,j=1\\ i\leq j}}^{n}|\lambda_i\lambda_j|^{1/2} = \prod_{i=1}^{n}|\lambda_i|^{\frac{1}{2}(n+1)}\\ &= |\mathbf{A}'\mathbf{A}|^{\frac{1}{2}(n+1)} = |\mathbf{A}|_+^{n+1}.\end{aligned}$$

Hence, (i) of the theorem is established.

For (ii) we utilize, as in the symmetric case, that

$$\begin{aligned}\left|\frac{d_k\mathbf{AWA}'}{d_k\mathbf{W}}\right|_+ &= \left|(\mathbf{T}^+(ss))'(\mathbf{A}'\otimes\mathbf{A}')\mathbf{T}'(ss)\right|_+ = |\mathbf{B}'(ss)(\mathbf{A}'\otimes\mathbf{A}')\mathbf{B}(ss)|_+\\ &= |\mathbf{B}'(ss)(\mathbf{A}'\mathbf{A}\otimes\mathbf{A}'\mathbf{A})\mathbf{B}(ss)|^{1/2},\end{aligned}$$

where k now indicates the skew-symmetric structure. By definition of $\mathbf{B}(s)$ and $\mathbf{B}(ss)$

$$\begin{aligned}&|\mathbf{A}'\mathbf{A}\otimes\mathbf{A}'\mathbf{A}|_+\\ &= |(\mathbf{B}(s)\mathbf{B}'(s)+\mathbf{B}(ss)\mathbf{B}'(ss))(\mathbf{A}'\mathbf{A}\otimes\mathbf{A}'\mathbf{A})(\mathbf{B}(s)\mathbf{B}'(s)+\mathbf{B}(ss)\mathbf{B}'(ss))|_+\\ &= |\mathbf{B}(s)\mathbf{B}'(s)(\mathbf{A}'\mathbf{A}\otimes\mathbf{A}'\mathbf{A})\mathbf{B}(s)\mathbf{B}'(s)+\mathbf{B}(ss)\mathbf{B}'(ss)(\mathbf{A}'\mathbf{A}\otimes\mathbf{A}'\mathbf{A})\mathbf{B}(ss)\mathbf{B}'(ss)|_+,\end{aligned} \tag{1.4.72}$$

since $\mathbf{B}'(s)(\mathbf{A}'\mathbf{A}\otimes\mathbf{A}'\mathbf{A})\mathbf{B}(ss)\mathbf{B}'(ss) = \mathbf{B}'(s)\mathbf{B}(ss)\mathbf{B}'(ss)(\mathbf{A}'\mathbf{A}\otimes\mathbf{A}'\mathbf{A}) = \mathbf{0}$. Furthermore, $(\mathbf{B}(s):\mathbf{B}(ss)) : n\times n$ is an orthogonal matrix and therefore (1.4.72) is equivalent to

$$\begin{aligned}&|\mathbf{A}'\mathbf{A}\otimes\mathbf{A}'\mathbf{A}|_+\\ &= \left|(\mathbf{B}(s):\mathbf{B}(ss))\begin{pmatrix}\mathbf{B}'(s)(\mathbf{A}'\mathbf{A}\otimes\mathbf{A}'\mathbf{A})\mathbf{B}(ss) & \mathbf{0}\\ \mathbf{0} & \mathbf{B}'(ss)(\mathbf{A}'\mathbf{A}\otimes\mathbf{A}'\mathbf{A})\mathbf{B}(ss)\end{pmatrix}\right.\\ &\qquad\left.\times\begin{pmatrix}\mathbf{B}'(s)\\ \mathbf{B}'(ss)\end{pmatrix}\right|_+\\ &= |\mathbf{B}'(s)(\mathbf{A}'\mathbf{A}\otimes\mathbf{A}'\mathbf{A})\mathbf{B}(s)|_+|\mathbf{B}'(ss)(\mathbf{A}'\mathbf{A}\otimes\mathbf{A}'\mathbf{A})\mathbf{B}(ss)|_+.\end{aligned}$$

Thus, it follows from (i) that

$$|\mathbf{B}'(ss)(\mathbf{A}'\mathbf{A}\otimes\mathbf{A}'\mathbf{A})\mathbf{B}(ss)|_+^{1/2} = |\mathbf{A}'\mathbf{A}|^n|\mathbf{A}'\mathbf{A}|^{-\frac{1}{2}(n+1)} = |\mathbf{A}|_+^{n-1}.$$

∎

Theorem 1.4.14. *Let* $\mathbf{Y} = \mathbf{AXB}$, *where* $\mathbf{Y} : p \times n$, $\mathbf{X} : p \times n$, $\mathbf{A}$ *and* $\mathbf{B}$ *are non-singular constant matrices. Then*

$$|\mathbf{J}(\mathbf{Y} \to \mathbf{X})|_+ = |\mathbf{B}|_+^p |\mathbf{A}|_+^n.$$

PROOF: The proof follows immediately since

$$\mathrm{vec}'(\mathbf{AXB}) = \mathrm{vec}'\mathbf{X}(\mathbf{B} \otimes \mathbf{A}')$$

and by Proposition 1.3.12 (ix) $|\mathbf{B} \otimes \mathbf{A}'| = |\mathbf{B}|^p |\mathbf{A}|^n$. ∎

Theorem 1.4.15. *Let* $\mathbf{A}$, $\mathbf{B}$ *and* $\mathbf{V}$ *be diagonal matrices, where* $\mathbf{A}$ *and* $\mathbf{B}$ *are constant matrices, and* $\mathbf{W} = \mathbf{AVB}$. *Then*

$$|\mathbf{J}(\mathbf{W} \to \mathbf{V})|_+ = |\mathbf{A}|_+ |\mathbf{B}|_+.$$

PROOF: The proof is a simplified version of Theorem 1.4.13:

$$|\mathbf{J}(\mathbf{W} \to \mathbf{V})|_+ = \left| \frac{d_k \mathbf{AVB}}{d_k \mathbf{V}} \right|_+ = |\mathbf{B}'(d)(\mathbf{B} \otimes \mathbf{A})\mathbf{B}(d)|_+ = |\mathbf{A}|_+ |\mathbf{B}|_+,$$

where Proposition 1.3.20 has been used and the subindex k indicates the diagonal structure. An alternative proof is based on Theorem 1.4.11:

$$|\mathbf{J}(\mathbf{AVB} \to \mathbf{V})|_+ = |\mathbf{J}(\mathbf{AVB} \to \mathbf{Z}; \mathbf{Z} = \mathbf{AV})|_+ |\mathbf{J}(\mathbf{Z} \to \mathbf{V})|_+.$$

∎

Theorem 1.4.16. *Let* $\mathbf{A}$, $\mathbf{B}$ *and* $\mathbf{V}$ *be upper triangular matrices, and put* $\mathbf{W} = \mathbf{AVB}$. *Then*

$$|\mathbf{J}(\mathbf{W} \to \mathbf{V})|_+ = \prod_{i=1}^{n} |b_{ii}|^i |a_{ii}|^{n-i-1}.$$

PROOF: Similarly to the proof of Theorem 1.4.15

$$|\mathbf{J}(\mathbf{W} \to \mathbf{V})|_+ = |\mathbf{B}'(u)(\mathbf{B} \otimes \mathbf{A}')\mathbf{B}(u)|_+.$$

Now from Proposition 1.3.22 it follows that

$$\mathbf{H} = \mathbf{B}'(u)(\mathbf{B} \otimes \mathbf{A}')\mathbf{B}(u) = \sum_{\substack{i_1 \leq j_1, i_2 \leq j_2 \\ j_1 \leq j_2, i_2 \leq i_1}} b_{j_1 j_2} a_{i_2 i_1} \mathbf{f}_{k(i_1,j_1)} \mathbf{f}'_{k(i_2,j_2)} = \begin{pmatrix} \mathbf{H}_{11} & \mathbf{0} \\ \mathbf{H}_{21} & \mathbf{H}_{22} \end{pmatrix},$$

where

$$\begin{aligned}
\mathbf{H}_{11} &= \sum_{1=i_1=i_2 \leq j_1 \leq j_2} b_{j_1 j_2} a_{11} \mathbf{e}_{j_1} \mathbf{e}'_{j_2}, \\
\mathbf{H}_{21} &= \sum_{1=i_2<i_1 \leq j_1 \leq j_2} b_{j_1 j_2} a_{1 i_1} \widetilde{\mathbf{f}}_{k(i_1,j_1)} \mathbf{e}'_{j_2}, \\
\widetilde{\mathbf{f}}_{k(i_1,j_1)} &= (\mathbf{0} : \mathbf{I}_r)\mathbf{f}_{k(i_1,j_1)}, \quad \mathbf{0} : \tfrac{1}{2}n(n+1) \times 1, \quad r = \tfrac{1}{2}n(n+1) - 1, \\
\mathbf{H}_{22} &= \sum_{2=i_2 \leq i_1 \leq j_1 \leq j_2} b_{j_1 j_2} a_{i_2 i_1} \widetilde{\mathbf{f}}_{k(i_1,j_1)} \widetilde{\mathbf{f}}'_{k(i_2,j_2)}.
\end{aligned}$$

Since $\mathbf{H}_{11}$ is an upper triangular matrix, $|\mathbf{H}_{11}|$ is the product of the diagonal elements of $\mathbf{H}_{11}$. Moreover, $\mathbf{H}_{22}$ is of the same type as $\mathbf{H}$ and we can repeat the above given arguments. Thus,

$$|\mathbf{H}|_+ = (\prod_{j=1}^n |b_{jj}a_{11}|)(\prod_{j=2}^n |b_{jj}a_{22}|) \times \cdots \times |b_{nn}a_{nn}| = \prod_{i=1}^n |b_{ii}|^i |a_{ii}|^{n-i+1}.$$

∎

Many more Jacobians for various linear transformations can be obtained. A rich set of Jacobians of various transformations is given by Magnus (1988). Next we consider Jacobians of some non-linear transformations.

Theorem 1.4.17. *Let* $\mathbf{W} = \mathbf{V}^{-1} : n \times n$.

(i) *If the elements of* $\mathbf{V}$ *are functionally independent, then*

$$|\mathbf{J}(\mathbf{W} \to \mathbf{V})|_+ = |\mathbf{V}|_+^{-2n}.$$

(ii) *If* $\mathbf{V}$ *is symmetric, then*

$$|\mathbf{J}(\mathbf{W} \to \mathbf{V})|_+ = |\mathbf{V}|_+^{-(n+1)}.$$

(iii) *If* $\mathbf{V}$ *is skew-symmetric, then*

$$|\mathbf{J}(\mathbf{W} \to \mathbf{V})|_+ = |\mathbf{V}|_+^{-(n-1)}.$$

PROOF: Using Proposition 1.4.7 (ii)

$$\left|\frac{d\mathbf{V}^{-1}}{d\mathbf{V}}\right|_+ = |-\mathbf{V}^{-1} \otimes (\mathbf{V}^{-1})'|_+ = |\mathbf{V}|^{-2n}.$$

Thus (i) is verified.
For (ii) we first note that

$$\begin{aligned}\mathbf{0} =& \frac{d\mathbf{V}^{-1}\mathbf{V}}{d_k\mathbf{V}} = \frac{d\mathbf{V}^{-1}}{d_k\mathbf{V}}(\mathbf{V} \otimes \mathbf{I}) + \frac{d\mathbf{V}}{d_k\mathbf{V}}(\mathbf{I} \otimes \mathbf{V}^{-1}) \\ =& \frac{d_k\mathbf{V}^{-1}}{d_k\mathbf{V}}(\mathbf{T}^+(s))'(\mathbf{V} \otimes \mathbf{I}) + (\mathbf{T}^+(s))'(\mathbf{I} \otimes \mathbf{V}^{-1}),\end{aligned}$$

where k refers to the symmetric structure. Thus,

$$\frac{d_k\mathbf{V}^{-1}}{d_k\mathbf{V}} = -(\mathbf{T}^+(s))'(\mathbf{V}^{-1} \otimes \mathbf{V}^{-1})\mathbf{T}(s)$$

and the Jacobian is similar to (1.4.71) which implies that (ii), as well as (iii) with some modifications, follow from the proof of Theorem 1.4.13. ∎

In the subsequent theorem we are going to express a symmetric matrix with the help of a triangular matrix. Since the number of non-zero and functionally independent elements is the same in both classes, the Jacobian in the next theorem is well defined.

Theorem 1.4.18. *Let* $\mathbf{W} = \mathbf{TT}'$ *where* $\mathbf{T} : n \times n$ *is a non-singular lower triangular matrix with positive diagonal elements. Then*

$$|\mathbf{J}(\mathbf{W} \to \mathbf{T})|_+ = 2^n \prod_{j=1}^{n} t_{jj}^{n-j+1}.$$

PROOF: It is noted that

$$\begin{aligned}\frac{d_{k_1}\mathbf{TT}'}{d_{k_2}\mathbf{T}} &= \frac{d\mathbf{T}}{d_{k_2}\mathbf{T}}(\mathbf{T}' \otimes \mathbf{I})\mathbf{T}'(s) + \frac{d\mathbf{T}}{d_{k_2}\mathbf{T}}\mathbf{K}_{n,n}(\mathbf{I} \otimes \mathbf{T}')\mathbf{T}'(s)\\ &= (\mathbf{T}^+(l))'(\mathbf{T}' \otimes \mathbf{I})(\mathbf{I} + \mathbf{K}_{n,n})\mathbf{T}'(s),\end{aligned}$$

where k_1 and k_2 refer to the symmetric and lower triangular patterns, respectively, and $\mathbf{T}(s)$ and $\mathbf{T}(l)$ are defined in Proposition 1.3.18 and Proposition 1.3.23. Since $(\mathbf{I} + \mathbf{K}_{n,n})\mathbf{T}'(s) = 2\mathbf{T}'(s)$, we consider the product

$$\begin{aligned}&(\mathbf{T}^+(l))'(\mathbf{T}' \otimes \mathbf{I})\mathbf{T}'(s)\\ &= \sum_{\substack{i,j=1\\ i\geq j}}^{n} \sum_{i_1,j_1=1}^{n} \mathbf{f}_{k_2(i,j)}(\mathbf{e}_j \otimes \mathbf{e}_i)'(\mathbf{T}' \otimes \mathbf{I})(\mathbf{e}_{j_1} \otimes \mathbf{e}_{i_1})\mathbf{f}'_{k_1(i_1,j_1)}\\ &= \sum_{\substack{i,j,j_1=1\\ i\geq j}}^{n} t_{j_1 j}\mathbf{f}_{k_2(i,j)}\mathbf{f}'_{k_1(i,j_1)} = \sum_{\substack{i,j,j_1=1\\ j\leq j_1\leq i}}^{n} t_{j_1 j}\mathbf{f}_{k_2(i,j)}\mathbf{f}'_{k_2(i,j_1)} + \sum_{\substack{i,j,j_1=1\\ j\leq i< j_1}}^{n} t_{j_1 j}\mathbf{f}_{k_2(i,j)}\mathbf{f}'_{k_1(i,j_1)},\end{aligned}$$

where the last equality follows because $\mathbf{T}$ is lower triangular and $k_2(i,j) = k_1(i,j)$, if $i \geq j$. Moreover, $k_2(i,j) < k_1(i,j_1)$, if $j \leq i < j_1$ and therefore the product $(\mathbf{T}^+(l))'(\mathbf{T}' \otimes \mathbf{I})\mathbf{T}'(s)$ is an upper triangular matrix. Thus,

$$\begin{aligned}\left|\frac{d_{k_1}\mathbf{TT}'}{d_{k_2}\mathbf{T}}\right|_+ &= |2(\mathbf{T}^+(l))'(\mathbf{T}' \otimes \mathbf{I})\mathbf{T}'(s)|_+ = 2^n \left|\sum_{i\geq j=1}^{n} t_{jj}\mathbf{f}_{k_2(i,j)}\mathbf{f}'_{k_2(i,j)}\right|_+\\ &= 2^n \prod_{j=1}^{n} |t_{jj}|^{n-j+1}.\end{aligned}$$

■

In the next theorem we will use the singular value decomposition. This means that symmetric, orthogonal and diagonal matrices will be considered. Note that the number of functionally independent elements in an orthogonal matrix is the same as the number of functionally independent elements in a skew-symmetric matrix.

Theorem 1.4.19. *Let* $\mathbf{W} = \mathbf{HDH}'$, *where* $\mathbf{H} : n \times n$ *is an orthogonal matrix and* $\mathbf{D} : n \times n$ *is a diagonal matrix. Then*

$$|\mathbf{J}(\mathbf{W} \to (\mathbf{H}, \mathbf{D}))|_+ = \prod_{i>j=1}^{n} |d_i - d_j| \prod_{j=1}^{n} |\mathbf{H}_{(j)}|_+,$$

where

$$\mathbf{H}_{(j)} = \begin{pmatrix} h_{jj} & \dots & h_{jn} \\ \vdots & \ddots & \vdots \\ h_{nj} & \dots & h_{nn} \end{pmatrix}. \tag{1.4.73}$$

PROOF: By definition,

$$|\mathbf{J}(\mathbf{W} \to (\mathbf{H}, \mathbf{D}))|_+ = \begin{vmatrix} \dfrac{d_{k_1}\mathbf{HDH}'}{d_{k_2}\mathbf{H}} \\ \dfrac{d_{k_1}\mathbf{HDH}'}{d_{k_3}\mathbf{D}} \end{vmatrix}_+, \tag{1.4.74}$$

where k_1, k_2 and k_3 refer to the symmetric, orthogonal and diagonal structures, respectively. In particular, if in $\mathbf{H}$ the elements h_{ij}, $i < j$ are chosen so that they are functionally independent, then

$$\begin{aligned} k_2(i,j) =& n(i-1) - \tfrac{1}{2}i(i-1) + j - i, \quad i < j, \\ k_2(i,j) =& 0 \quad i \geq j. \end{aligned}$$

Now

$$\begin{aligned} \frac{d_{k_1}\mathbf{HDH}'}{d_{k_2}\mathbf{H}} =& \frac{d\mathbf{HDH}'}{d_{k_2}\mathbf{H}}\mathbf{T}'(s) = \{\frac{d\mathbf{H}}{d_{k_2}\mathbf{H}}(\mathbf{DH}' \otimes \mathbf{I}) + \frac{d\mathbf{H}'}{d_{k_2}\mathbf{H}}(\mathbf{I} \otimes \mathbf{DH}')\}\mathbf{T}'(s) \\ =& \frac{d\mathbf{H}}{d_{k_2}\mathbf{H}}(\mathbf{I} \otimes \mathbf{H})(\mathbf{D} \otimes \mathbf{I})(\mathbf{H}' \otimes \mathbf{H}')(\mathbf{I} + \mathbf{K}_{n,n})\mathbf{T}'(s). \end{aligned} \tag{1.4.75}$$

As in the previous theorem we observe that $(\mathbf{I} + \mathbf{K}_{n,n})\mathbf{T}'(s) = 2\mathbf{T}'(s)$. Moreover, because

$$\mathbf{0} = \frac{d\mathbf{HH}'}{d_{k_2}\mathbf{H}} = \frac{d\mathbf{H}}{d_{k_2}\mathbf{H}}(\mathbf{H}' \otimes \mathbf{I})(\mathbf{I} + \mathbf{K}_{n,n}), \tag{1.4.76}$$

it follows that

$$\frac{d\mathbf{H}}{d_{k_2}\mathbf{H}}(\mathbf{I} \otimes \mathbf{H}) = \frac{d\mathbf{H}}{d_{k_2}\mathbf{H}}(\mathbf{I} \otimes \mathbf{H})\tfrac{1}{2}(\mathbf{I} - \mathbf{K}_{n,n}).$$

Thus (1.4.75) is equivalent to

$$\begin{aligned} \frac{d_{k_1}\mathbf{HDH}'}{d_{k_2}\mathbf{H}} =& \frac{d\mathbf{H}}{d_{k_2}\mathbf{H}}(\mathbf{I} \otimes \mathbf{H})\tfrac{1}{2}(\mathbf{I} - \mathbf{K}_{n,n})(\mathbf{D} \otimes \mathbf{I})(\mathbf{H}' \otimes \mathbf{H}')(\mathbf{I} + \mathbf{K}_{n,n})\mathbf{T}'(s) \\ =& \frac{d\mathbf{H}}{d_{k_2}\mathbf{H}}(\mathbf{I} \otimes \mathbf{H})\tfrac{1}{2}(\mathbf{D} \otimes \mathbf{I} - \mathbf{I} \otimes \mathbf{D})(\mathbf{H}' \otimes \mathbf{H}')(\mathbf{I} + \mathbf{K}_{n,n})\mathbf{T}'(s) \\ =& \frac{d\mathbf{H}}{d_{k_2}\mathbf{H}}(\mathbf{I} \otimes \mathbf{H})(\mathbf{D} \otimes \mathbf{I} - \mathbf{I} \otimes \mathbf{D})(\mathbf{H}' \otimes \mathbf{H}')\mathbf{T}'(s). \end{aligned}$$

For the second term of the Jacobian it follows that

$$\frac{d_{k_1}\mathbf{HDH}'}{d_{k_3}\mathbf{D}} = \frac{d\mathbf{D}}{d_{k_3}\mathbf{D}}(\mathbf{H}' \otimes \mathbf{H}')\mathbf{T}'(s) = (\mathbf{T}^+(d))'(\mathbf{H}' \otimes \mathbf{H}')\mathbf{T}'(s),$$

where $\mathbf{T}^+(d)$ can be found in Proposition 1.3.20. Since $\mathbf{T}'(s) = \mathbf{B}(s)\underline{\mathbf{B}}'(s)$, we are going to study the product

$$
\begin{aligned}
&|\underline{\mathbf{B}}(s)| \left| \begin{pmatrix} \frac{d_{k_1}\mathbf{HDH}'}{d_{k_2}\mathbf{H}} \\ \frac{d_{k_1}\mathbf{HDH}'}{d_{k_3}\mathbf{D}} \end{pmatrix} (\underline{\mathbf{B}}'(s))^{-1}(\underline{\mathbf{B}}(s))^{-1} \begin{pmatrix} \frac{d_{k_1}\mathbf{HDH}'}{d_{k_2}\mathbf{H}} \\ \frac{d_{k_1}\mathbf{HDH}'}{d_{k_3}\mathbf{D}} \end{pmatrix}' \right|^{1/2} \\
&\qquad = |\underline{\mathbf{B}}(s)| \begin{vmatrix} \mathbf{K}_{11} & \mathbf{K}_{12} \\ \mathbf{K}_{12}' & \mathbf{K}_{22} \end{vmatrix}^{1/2},
\end{aligned}
\tag{1.4.77}
$$

which equals the determinant given in (1.4.74) and where

$$
\begin{aligned}
\mathbf{K}_{11} =& \frac{d\mathbf{H}}{d_{k_2}\mathbf{H}}(\mathbf{I}\otimes\mathbf{H})(\mathbf{D}\otimes\mathbf{I}-\mathbf{I}\otimes\mathbf{D})(\mathbf{H}'\otimes\mathbf{H}')\mathbf{B}(s)\mathbf{B}'(s)(\mathbf{H}\otimes\mathbf{H}) \\
& \times (\mathbf{D}\otimes\mathbf{I}-\mathbf{I}\otimes\mathbf{D})(\mathbf{I}\otimes\mathbf{H}')\left(\frac{d\mathbf{H}}{d_{k_2}\mathbf{H}}\right)' \\
=& \frac{d\mathbf{H}}{d_{k_2}\mathbf{H}}(\mathbf{I}\otimes\mathbf{H})(\mathbf{D}\otimes\mathbf{I}-\mathbf{I}\otimes\mathbf{D})^2(\mathbf{I}\otimes\mathbf{H}')\left(\frac{d\mathbf{H}}{d_{k_2}\mathbf{H}}\right)', \\
\mathbf{K}_{22} =& (\mathbf{T}^+(d))'(\mathbf{H}'\otimes\mathbf{H}')\mathbf{B}(s)\mathbf{B}'(s)(\mathbf{H}\otimes\mathbf{H})\mathbf{T}^+(d) \\
=& (\mathbf{T}^+(d))'\mathbf{B}(s)\mathbf{B}'(s)\mathbf{T}^+(d) = \mathbf{I}, \\
\mathbf{K}_{12} =& \frac{d\mathbf{H}}{d_{k_2}\mathbf{H}}(\mathbf{I}\otimes\mathbf{H})(\mathbf{D}\otimes\mathbf{I}-\mathbf{I}\otimes\mathbf{D})(\mathbf{H}'\otimes\mathbf{H}')\mathbf{B}(s)\mathbf{B}'(s)(\mathbf{H}\otimes\mathbf{H})\mathbf{T}^+(d) \\
=& \frac{d\mathbf{H}}{d_{k_2}\mathbf{H}}(\mathbf{I}\otimes\mathbf{H})(\mathbf{D}\otimes\mathbf{I}-\mathbf{I}\otimes\mathbf{D})\mathbf{T}^+(d) = \mathbf{0}.
\end{aligned}
$$

Thus, (1.4.77) is identical to

$$
|\underline{\mathbf{B}}(s)| \left| \frac{d\mathbf{H}}{d_{k_2}\mathbf{H}}(\mathbf{I}\otimes\mathbf{H})(\mathbf{D}\otimes\mathbf{I}-\mathbf{I}\otimes\mathbf{D})^2(\mathbf{I}\otimes\mathbf{H}')\left(\frac{d\mathbf{H}}{d_{k_2}\mathbf{H}}\right)' \right|^{1/2}. \tag{1.4.78}
$$

Since

$$
\frac{d\mathbf{H}}{d_{k_2}\mathbf{H}}(\mathbf{I}\otimes\mathbf{H})(\mathbf{I}+\mathbf{K}_{n,n}) = \mathbf{0},
$$

we may choose a matrix $\mathbf{Q}$ such that

$$
\frac{d\mathbf{H}}{d_{k_2}\mathbf{H}} = \mathbf{Q}\tfrac{1}{2}(\mathbf{I}-\mathbf{K}_{n,n})(\mathbf{I}\otimes\mathbf{H}')
$$

holds and $\mathbf{QB}(ss) : \frac{1}{2}n(n-1) \times \frac{1}{2}n(n-1)$ is of full rank. Then

$$
\frac{d\mathbf{H}}{d_{k_2}\mathbf{H}}(\mathbf{I}\otimes\mathbf{H}) = \mathbf{QB}(ss)\mathbf{B}'(ss).
$$

Returning to (1.4.78) it follows that this expression is identical to

$$|\underline{\mathbf{B}}(s)||\mathbf{QB}(ss)\mathbf{B}'(ss)\mathbf{Q}'|^{1/2}|\mathbf{B}'(ss)(\mathbf{D}\otimes\mathbf{I}-\mathbf{I}\otimes\mathbf{D})^2\mathbf{B}(ss)|^{1/2}$$
$$=|\underline{\mathbf{B}}(s)|\left|\frac{d\mathbf{H}}{d_{k_2}\mathbf{H}}\left(\frac{d\mathbf{H}}{d_{k_2}\mathbf{H}}\right)'\right|^{1/2}|\mathbf{B}'(ss)(\mathbf{D}\otimes\mathbf{I}-\mathbf{I}\otimes\mathbf{D})^2\mathbf{B}(ss)|^{1/2}. \quad (1.4.79)$$

Each determinant in (1.4.79) will be examined separately. First it is noticed that

$$(\mathbf{D}\otimes\mathbf{I}-\mathbf{I}\otimes\mathbf{D})^2=\sum_{i\neq j}(d_i-d_j)^2\mathbf{e}_i\mathbf{e}_i'\otimes\mathbf{e}_j\mathbf{e}_j'.$$

This implies that

$$\mathbf{B}'(ss)(\mathbf{D}\otimes\mathbf{I}-\mathbf{I}\otimes\mathbf{D})^2\mathbf{B}(ss)=\tfrac{1}{2}\sum_{i\neq j}(d_i-d_j)^2\mathbf{f}_{k_4(i,j)}\mathbf{f}'_{k_4(i,j)}$$
$$=\sum_{i>j}(d_i-d_j)^2\mathbf{f}_{k_4(i,j)}\mathbf{f}'_{k_4(i,j)}, \quad (1.4.80)$$

where $k_4(i,j)$ is given in Proposition 1.3.19. The matrix in (1.4.80) is diagonal and hence

$$|\mathbf{B}'(ss)(\mathbf{D}\otimes\mathbf{I}-\mathbf{I}\otimes\mathbf{D})^2\mathbf{B}(ss)|^{1/2}=(\prod_{i>j}(d_i-d_j)^2)^{1/2}=\prod_{i>j}|d_i-d_j|. \quad (1.4.81)$$

Moreover, using the definition of $\underline{\mathbf{B}}(s)$, given by (1.3.55),

$$|\underline{\mathbf{B}}(s)|=\left|\sum_{i,j}\mathbf{f}_{k_1(i,j)}\mathbf{f}'_{k_1(i,j)}\frac{1}{\sqrt{2}}\right|=2^{\frac{1}{4}n(n-1)}. \quad (1.4.82)$$

It remains to find

$$\left|\frac{d\mathbf{H}}{d_{k_2}\mathbf{H}}\left(\frac{d\mathbf{H}}{d_{k_2}\mathbf{H}}\right)'\right|^{1/2}.$$

We may write

$$\frac{d\mathbf{H}}{d_{k_2}\mathbf{H}}=\sum_{i,j=1}^{n}\sum_{\substack{k,l=1\\k<l}}\frac{\partial h_{ij}}{\partial h_{kl}}\mathbf{f}_{k_2(k,l)}(\mathbf{e}_j\otimes\mathbf{e}_i)'.$$

Furthermore, let su stand for the strict upper triangular matrix, i.e. all elements on the diagonal and below it equal zero. Similarly to the basis matrices in §1.3.6 we may define

$$\mathbf{B}(su)=\sum_{i<j}(\mathbf{e}_j\otimes\mathbf{e}_i)\mathbf{f}'_{k_2(i,j)}.$$

If using $\mathbf{B}(l)$, which was given in Proposition 1.3.23, it follows that $\mathbf{B}'(l)\mathbf{B}(su)=\mathbf{0}$ and

$$(\mathbf{B}(su):\mathbf{B}(l))(\mathbf{B}(su):\mathbf{B}(l))'=\mathbf{B}(su)\mathbf{B}'(su)+\mathbf{B}(l)\mathbf{B}'(l)=\mathbf{I}.$$

Now

$$\frac{d\mathbf{H}}{d_{k_2}\mathbf{H}}\mathbf{B}(su) = \sum_{\substack{i,j\\k<l,i<j}} \sum_{i_1<j_1} \frac{\partial h_{ij}}{\partial h_{kl}}\mathbf{f}_{k_2(k,l)}(\mathbf{e}_j \otimes \mathbf{e}_i)'(\mathbf{e}_{j_1} \otimes \mathbf{e}_{i_1})\mathbf{f}'_{k_2(i_1,j_1)}$$
$$= \sum_{\substack{i,j\\k<l}} \frac{\partial h_{ij}}{\partial h_{kl}}\mathbf{f}_{k_2(k,l)}\mathbf{f}'_{k_2(i,j)} = \sum_{i<j}\mathbf{f}_{k_2(i,j)}\mathbf{f}'_{k_2(i,j)} = \mathbf{I}.$$

Thus,

$$\left|\frac{d\mathbf{H}}{d_{k_2}\mathbf{H}}\left(\frac{d\mathbf{H}}{d_{k_2}\mathbf{H}}\right)'\right| = \left|\frac{d\mathbf{H}}{d_{k_2}\mathbf{H}}(\mathbf{B}(su)\mathbf{B}'(su) + \mathbf{B}(l)\mathbf{B}'(l))\left(\frac{d\mathbf{H}}{d_{k_2}\mathbf{H}}\right)'\right|$$
$$= \left|\mathbf{I} + \frac{d\mathbf{H}}{d_{k_2}\mathbf{H}}\mathbf{B}(l)\mathbf{B}'(l)\left(\frac{d\mathbf{H}}{d_{k_2}\mathbf{H}}\right)'\right|. \qquad (1.4.83)$$

We are going to manipulate $\frac{d\mathbf{H}}{d_{k_2}\mathbf{H}}\mathbf{B}(l)$ and from (1.4.76) it follows that

$$\mathbf{0} = \frac{d\mathbf{H}}{d_{k_2}\mathbf{H}}(\mathbf{I}\otimes\mathbf{H})\mathbf{B}(s) = \frac{d\mathbf{H}}{d_{k_2}\mathbf{H}}(\mathbf{B}(su)\mathbf{B}'(su) + \mathbf{B}(l)\mathbf{B}'(l))(\mathbf{I}\otimes\mathbf{H})\mathbf{B}(s)$$
$$= \mathbf{B}'(su)(\mathbf{I}\otimes\mathbf{H})\mathbf{B}(s) + \frac{d\mathbf{H}}{d_{k_2}\mathbf{H}}\mathbf{B}(l)\mathbf{B}'(l)(\mathbf{I}\otimes\mathbf{H})\mathbf{B}(s),$$

which implies

$$\frac{d\mathbf{H}}{d_{k_2}\mathbf{H}}\mathbf{B}(l) = -\mathbf{B}'(su)(\mathbf{I}\otimes\mathbf{H})\mathbf{B}(s)\{\mathbf{B}'(l)(\mathbf{I}\otimes\mathbf{H})\mathbf{B}(s)\}^{-1}.$$

Therefore the right hand side of (1.4.83) can be simplified:

$$|\mathbf{I}+\mathbf{B}'(su)(\mathbf{I}\otimes\mathbf{H})\mathbf{B}(s)\{\mathbf{B}'(l)(\mathbf{I}\otimes\mathbf{H})\mathbf{B}(s)\}^{-1}\{\mathbf{B}'(s)(\mathbf{I}\otimes\mathbf{H}')\mathbf{B}(l)\}^{-1}$$
$$\times\,\mathbf{B}'(s)(\mathbf{I}\otimes\mathbf{H}')\mathbf{B}(su)|$$
$$= |\mathbf{I} + \{\mathbf{B}'(l)(\mathbf{I}\otimes\mathbf{H})\mathbf{B}(s)\}^{-1}\{\mathbf{B}'(s)(\mathbf{I}\otimes\mathbf{H}')\mathbf{B}(l)\}^{-1}\mathbf{B}'(s)(\mathbf{I}\otimes\mathbf{H}')\mathbf{B}(su)$$
$$\times\,\mathbf{B}'(su)(\mathbf{I}\otimes\mathbf{H})\mathbf{B}(s)|$$
$$= |\mathbf{B}'(l)(\mathbf{I}\otimes\mathbf{H})\mathbf{B}(s)|^{-1},$$

where it has been utilized that $\mathbf{B}(su)\mathbf{B}'(su) = \mathbf{I} - \mathbf{B}(l)\mathbf{B}'(l)$. By definitions of $\mathbf{B}(l)$ and $\mathbf{B}(s)$, in particular, $k_3(i,j) = n(j-1) - \frac{1}{2}j(j-1) + i$, $i \geq j$ and $k_1(i,j) = n(\min(i,j)-1) - \frac{1}{2}\min(i,j)(\min(i,j)-1) + \max(i,j)$,

$$\mathbf{B}'(l)(\mathbf{I}\otimes\mathbf{H})\mathbf{B}(s)$$
$$= \sum_{i\geq j}\mathbf{f}_{k_3(i,j)}(\mathbf{e}_j\otimes\mathbf{e}_i)'(\mathbf{I}\otimes\mathbf{H})\{\sum_{\substack{i_1,j_1\\i_1\neq j_1}}\tfrac{1}{\sqrt{2}}(\mathbf{e}_{j_1}\otimes\mathbf{e}_{i_1})\mathbf{f}'_{k_1(i_1,j_1)} + \sum_{i_1}(\mathbf{e}_{i_1}\otimes\mathbf{e}_{i_1})\mathbf{f}'_{k_1(i_1,i_1)}\}$$
$$= \sum_{\substack{i\geq j\\i_1\neq j}}\tfrac{1}{\sqrt{2}}h_{ii_1}\mathbf{f}_{k_3(i,j)}\mathbf{f}'_{k_1(i_1,j)} + \sum_{i\geq j}h_{ij}\mathbf{f}_{k_3(i,j)}\mathbf{f}'_{k_1(i_1,j)}. \qquad (1.4.84)$$

Now we have to look closer at $k_3(i,j)$ and $k_1(i,j)$. Fix $j = j_0$ and then, if $i_1 \geq j_0$, $\{k_1(i_1,j_0)\}_{i_1\geq j_0} = \{k_3(i,j_0)\}_{i\geq j_0}$ and both $k_1(i,j_0)$ and $k_2(i,j_0)$ are smaller than $n_{j_0} - \frac{1}{2}j_0(j_0-1)$. This means that $\mathbf{f}_{k_3(i,j_0)}\mathbf{f}'_{k_1(i_1,j_0)}$, $i_1 \geq j_0$, $i_1 \geq j_0$ forms a square block. Furthermore, if $i_1 < j_0$, then $\mathbf{f}_{k_1(i_1,j_0)}$ is identical to $\mathbf{f}_{k_1(i_1,j)}$ for $i_1 = j < j_0 = i$. Therefore, $\mathbf{f}_{k_3(i,j_0)}\mathbf{f}'_{k_1(i_1,j_0)}$ stands below the block given by $\mathbf{f}_{k_3(i,i_1)}\mathbf{f}'_{k_1(j_0,i_1)}$. Thus, each j defines a block and the block is generated by $i \geq j$ and $i_1 \geq j$. If $i_1 < j$ then we are "below" the block structure. Thus the determinant of (1.4.84) equals

$$
\begin{aligned}
|\mathbf{B}'(l)(\mathbf{I}\otimes\mathbf{H})\mathbf{B}(s)| &= |\sum_{\substack{i\geq j\\ i_1>j}} \tfrac{1}{\sqrt{2}}h_{ii_1}\mathbf{f}_{k_3(i,j)}\mathbf{f}'_{k_1(i_1,j)} + \sum_{i\geq j} h_{ij}\mathbf{f}_{k_3(i,j)}\mathbf{f}'_{k_1(j,j)}| \\
&= \prod_{j=1}^{n-1}\begin{vmatrix} h_{jj} & \frac{1}{\sqrt{2}}h_{jj+1} & \cdots & \frac{1}{\sqrt{2}}h_{jn} \\ \vdots & \vdots & \ddots & \vdots \\ h_{nj} & \frac{1}{\sqrt{2}}h_{nj+1} & \cdots & \frac{1}{\sqrt{2}}h_{nn}\end{vmatrix}|h_{nn}| = \prod_{j=1}^{n-1}2^{-\frac{1}{2}(n-j)}|\mathbf{H}_{(j)}||h_{nn}| \\
&= 2^{-\frac{1}{4}n(n-1)}\prod_{j=1}^{n-1}|\mathbf{H}_{(j)}||h_{nn}|,
\end{aligned}
\qquad (1.4.85)
$$

where $\mathbf{H}_{(j)}$ is given by (1.4.73). By combining (1.4.80), (1.4.81) and (1.4.85) we have proved the theorem. ∎

Theorem 1.4.20. *Let* $\mathbf{X} : p\times n$, $p \leq n$ *and* $\mathbf{X} = \mathbf{TL}$, *where* $\mathbf{T}$ *is a lower triangular matrix with positive elements, and* $\mathbf{LL}' = \mathbf{I}_p$, *i.e.* $\mathbf{L}$ *is semiorthogonal. Then*

$$
|\mathbf{J}(\mathbf{X}\to\mathbf{T},\mathbf{L})|_+ = \prod_{i=1}^{p} t_{ii}^{n-i}\prod_{i=1}^{p}|\mathbf{L}_i|_+,
$$

where $\mathbf{L}_i = (l_{jk})$ $j,k = 1,2,\ldots,i$ *and the functionally independent elements in* $\mathbf{L}$ *are* $l_{12}, l_{13}, \ldots, l_{1n}, l_{23}, \ldots, l_{2n}, \ldots, l_{p1}, \ldots, l_{pn}$.

PROOF: By definition of the Jacobian

$$
|\mathbf{J}|_+ = \begin{vmatrix} \dfrac{d\mathbf{X}}{d\mathbf{T}} \\ \dfrac{d\mathbf{X}}{d\mathbf{L}} \end{vmatrix}_+,
$$

where

$$
\begin{aligned}
\frac{d\mathbf{X}}{d\mathbf{T}} &= (\mathbf{T}^+(l))'(\mathbf{L}\otimes\mathbf{I}_p), \\
\frac{d\mathbf{X}}{d\mathbf{L}} &= (\mathbf{T}^+(so))'(\mathbf{I}_n\otimes\mathbf{T}'),
\end{aligned}
$$

with $\mathbf{T}^+(l)$ and $\mathbf{T}^+(so)$ given in Proposition 1.3.23 and Problem 9 in §1.3.8, respectively. Now we are going to study $\mathbf{J}$ in detail. Put

$$\mathbf{H}_1 = \sum_{i=1}^{p}\sum_{j=1}^{i}\sum_{k=i+1}^{n} l_{jk}\underline{\mathbf{f}}_s(\mathbf{e}_k \otimes \mathbf{d}_i)',$$

$$\mathbf{H}_2 = \sum_{i=1}^{p}\sum_{j=1}^{i}\sum_{k=1}^{i} l_{jk}\underline{\mathbf{f}}_s(\mathbf{e}_k \otimes \mathbf{d}_i)',$$

$$\mathbf{H}_3 = \sum_{i=1}^{p}\sum_{j=i+1}^{n} t_{ii}\mathbf{f}_s(\mathbf{e}_j \otimes \mathbf{d}_i)',$$

$$\mathbf{H}_4 = \sum_{i=1}^{p}\sum_{j=i+1}^{n}\sum_{m=i+1}^{p} t_{mi}\mathbf{f}_s(\mathbf{e}_j \otimes \mathbf{d}_m)',$$

where $\mathbf{f}_s$ and $\underline{\mathbf{f}}_s$ follow from the definitions of $\mathbf{T}^+(so)$ and $\mathbf{T}^+(l)$, respectively. Observe that using these definitions we get

$$\mathbf{J} = \begin{pmatrix} \mathbf{H}_1 + \mathbf{H}_2 \\ \mathbf{H}_3 + \mathbf{H}_4 \end{pmatrix}. \tag{1.4.86}$$

The reason for using $\mathbf{H}_1 + \mathbf{H}_2$ and $\mathbf{H}_3 + \mathbf{H}_4$ is that only $\mathbf{H}_2$ and $\mathbf{H}_3$ contribute to the determinant. This will be clear from the following.

$\mathbf{J}$ is a huge matrix and its determinant will be explored by a careful investigation of $\mathbf{e}_k \otimes \mathbf{d}_i$, $\mathbf{e}_j \otimes \mathbf{d}_i$, $\underline{\mathbf{f}}_s$ and $\mathbf{f}_s$ in $\mathbf{H}_r$, $r = 1, 2, 3, 4$. First observe that l_{11} is the only non-zero element in the column given by $\mathbf{e}_1 \otimes \mathbf{d}_1$ in $\mathbf{J}$. Let

$$\mathbf{J}_{11} = \mathbf{J}^{(j=i=k=1;\underline{\mathbf{f}}_s,\mathbf{e}_k\otimes\mathbf{d}_i)}_{(j=i=1;\mathbf{e}_j\otimes\mathbf{d}_i)}$$

denote the matrix $\mathbf{J}$ where the row and column in $\mathbf{H}_1$ and $\mathbf{H}_2$, which correspond to $i = j = k = 1$, have been deleted. Note that the notation means the following: The upper index indicates which rows and columns have been deleted from $\mathbf{H}_1$ and $\mathbf{H}_2$ whereas the subindex shows which rows and columns in $\mathbf{H}_3$ and $\mathbf{H}_4$ have been deleted. Thus, by (1.1.6),

$$|\mathbf{J}|_+ = |l_{11}||\mathbf{J}_{11}|_+.$$

However, in all columns of $\mathbf{J}_{11}$ where t_{11} appears, there are no other elements which differ from zero. Let

$$\mathbf{J}_{12} = \mathbf{J}^{(i=1,i<k\leq n;\mathbf{e}_k\otimes\mathbf{d}_i)}_{11(i=1,i<j\leq n;\mathbf{f}_s,\mathbf{e}_j\otimes\mathbf{d}_i)}$$

denote the matrix $\mathbf{J}_{11}$ where the columns in $\mathbf{H}_1$ and $\mathbf{H}_2$ which correspond to $i = 1$ and the rows and columns which correspond to $i = 1, i < j \leq n$ in $\mathbf{H}_3$ and $\mathbf{H}_4$, given in (1.4.86), have been omitted. Hence,

$$|\mathbf{J}|_+ = |l_{11}|t_{11}^{n-1}|\mathbf{J}_{12}|_+.$$

We may proceed in the same manner and obtain

$$|\mathbf{J}|_+ = |l_{11}|t_{11}^{n-1}\begin{vmatrix} l_{11} & l_{12} \\ l_{21} & l_{22} \end{vmatrix}|\mathbf{J}_{21}|_+ = |l_{11}|t_{11}^{n-1}|\mathbf{L}_2|t_{22}^{n-2}|\mathbf{J}_{22}|_+$$
$$= \ldots = \prod_{i=1}^{p}|\mathbf{L}_i|\prod_{i=1}^{p}t_{ii}^{n-i}$$

for some sequence of matrices $\mathbf{J}_{21}$, $\mathbf{J}_{22}$, $\mathbf{J}_{31}$, $\mathbf{J}_{32}$, $\mathbf{J}_{41}$, $\ldots$. For example,

$$\mathbf{J}_{21} = \mathbf{J}_{12(i=2,j\le 2;\mathbf{e}_j\otimes\mathbf{d}_i)}^{(i=2,j\le 2,k\le 2;\mathbf{f}_s,\mathbf{e}_k\otimes\mathbf{d}_i)}$$

and

$$\mathbf{J}_{22} = \mathbf{J}_{21(i=2,i<j\le n;\mathbf{f}_s,\mathbf{e}_j\otimes\mathbf{d}_i)}^{(i=2,i<k\le n;\mathbf{e}_k\otimes\mathbf{d}_i)},$$

where we have used the same notation as when $\mathbf{J}_{11}$ and $\mathbf{J}_{12}$ were defined. The choice of indices in these expressions follows from the construction of $\mathbf{H}_2$ and $\mathbf{H}_3$. ∎

Sometimes it is useful to work with polar coordinates. Therefore, we will present a way of obtaining the Jacobian, when a transformation from the rectangular coordinates $y_1, y_2, \ldots, y_n$ to the polar coordinates $r, \theta_{n-1}, \theta_{n-2}, \ldots, \theta_1$ is performed.

Theorem 1.4.21. *Let*

$$y_1 = r\prod_{i=1}^{n-1}\sin\theta_i, \quad y_2 = r\cos\theta_{n-1}\prod_{i=1}^{n-2}\sin\theta_i, \quad y_3 = r\cos\theta_{n-2}\prod_{i=1}^{n-3}\sin\theta_i, \ldots$$
$$\ldots, y_{n-1} = r\cos\theta_2\sin\theta_1, \quad y_n = r\cos\theta_1.$$

Then

$$|\mathbf{J}(\mathbf{y}\to r,\boldsymbol{\theta})| = r^{n-1}\prod_{i=1}^{n-2}\sin^{n-i-1}\theta_i, \qquad \boldsymbol{\theta} = (\theta_{n-1},\theta_{n-2},\ldots,\theta_1)'.$$

PROOF: We are going to combine Theorem 1.4.11 and Theorem 1.4.12. First we observe that the polar transformation can be obtained via an intermediate transformation in the following way;

$$y_k = y_{k-1}x_k^{-1}, \quad y_1 = x_1, \qquad k = 2, 3, \ldots, n \tag{1.4.87}$$

and

$$x_2 = \tan\theta_{n-1}$$
$$x_k = \tan\theta_{n-k+1}\cos\theta_{n-k+2}, \qquad k = 3, 4, \ldots, n.$$

Thus, by Theorem 1.4.11

$$\mathbf{J}(\mathbf{y}\to r,\boldsymbol{\theta}) = J(\mathbf{y}\to\mathbf{x})J(\mathbf{x}\to r,\boldsymbol{\theta}).$$

Furthermore, from (1.4.87) it follows that $\mathbf{J}(\mathbf{y} \to \mathbf{x})$ is based on a conditional transformation and then Theorem 1.4.12 tells us that

$$\mathbf{J}(\mathbf{y} \to r, \boldsymbol{\theta}) = \prod_{i=1}^{n} J(y_i \to x_i)\mathbf{J}(\mathbf{x} \to r, \boldsymbol{\theta}).$$

From (1.4.87) it follows immediately that

$$\left| \prod_{i=1}^{n} J(y_i \to x_i) \right|_+ = \frac{x_1^{n-1}}{\prod_{i=2}^{n} x_i^{n-i+2}}.$$

This determinant $|\mathbf{J}(\mathbf{x} \to r, \boldsymbol{\theta})|$, equals, by Proposition 1.3.2,

$$|\mathbf{J}(x_2, x_3, \ldots, x_n \to \boldsymbol{\theta})|_+ \prod_{i=1}^{n-1} \sin\theta_i.$$

However, $\mathbf{J}(x_2, x_3, \ldots, x_n \to \boldsymbol{\theta})$ is an upper triangular matrix and the Jacobian equals

$$\frac{1}{\cos\theta_1} \prod_{i=1}^{n-1} \frac{1}{\cos\theta_i}.$$

Thus,

$$\begin{aligned}
|\mathbf{J}(\mathbf{y} \to r, \boldsymbol{\theta})|_+ &= \frac{x_1^{n-1}}{\prod_{i=2}^{n} x_i^{n-i+2}} \prod_{i=1}^{n-1} \sin\theta_i \frac{1}{\cos\theta_1} \prod_{i=1}^{n-1} \frac{1}{\cos\theta_i} \\
&= \frac{x_1^{n-1}}{\prod_{i=2}^{n} x_i^{n-i+1}} \prod_{i=1}^{n-1} \frac{1}{\cos\theta_i}.
\end{aligned}$$

By inserting the expression for x_i, $i = 1, 2, \ldots, n$ into the Jacobian we get

$$\prod_{i=2}^{n} x_i^{n-i+1} = \prod_{i=1}^{n-1} \sin^i\theta_i \prod_{i=1}^{n-1} \frac{1}{\cos\theta_i},$$

and the theorem is verified. ∎

1.4.13 *Problems*

1. Find the derivative $\dfrac{d|\mathbf{Y}\mathbf{X}^2|}{d\mathbf{X}}$, where $\mathbf{Y} = \mathbf{Y}(\mathbf{X})$.
2. Prove that $\dfrac{d\mathbf{X}}{d\mathbf{X}'} = \mathbf{K}_{p,q}$ if $\mathbf{X} : p \times q$.
3. Find $\dfrac{d(\mathbf{A} \circ \mathbf{X})}{d\mathbf{X}}$.
4. Find $\dfrac{d^2(\mathbf{Y}\mathbf{X}^2)}{d\mathbf{X}^2}$.

5. Differentiate $|\mathbf{I}+\mathbf{T}\boldsymbol{\Sigma}|$ using both derivatives given by (1.4.7) and (1.4.48) where $\mathbf{T}$ and $\boldsymbol{\Sigma}$ are symmetric.
6. Derive the first three terms in a Taylor series expansion of $e^{\ln|\mathbf{X}|}$, at the point $\mathbf{X}=\mathbf{I}_p$, where $\mathbf{X}: p\times p$ is non-singular.
7. Find the Jacobian of the transformation $\mathbf{W}=\mathbf{A}\circ\mathbf{V}$ where $\mathbf{A}$ is constant and the size of all included matrices is $p\times p$.
8. Verify that $\boldsymbol{\Sigma}^{-1}\mathbf{x}$ and $\boldsymbol{\Sigma}^{-1}\mathbf{x}\mathbf{x}'\boldsymbol{\Sigma}^{-1}-\boldsymbol{\Sigma}^{-1}$ are orthogonal with respect to the function
$$f(\mathbf{x})=e^{-\frac{1}{2}\mathbf{x}'\boldsymbol{\Sigma}^{-1}\mathbf{x}}.$$
9. Give a detailed proof of statement (iii) of Theorem 1.4.17.
10. Let $\mathbf{T}$ be a Toeplitz matrix. Examine $\dfrac{d\mathbf{T}}{d\mathbf{T}}$.

CHAPTER II

Multivariate Distributions

At first we are going to introduce the main tools for characterization of multivariate distributions: moments and central moments, cumulants, characteristic function, cumulant function, etc. Then attention is focused on the multivariate normal distribution and the Wishart distribution as these are the two basic distributions in multivariate analysis. Random variables will be denoted by capital letters from the end of the Latin alphabet. For random vectors and matrices we shall keep the boldface transcription with the same difference as in the non-random case; small bold letters are used for vectors and capital bold letters for matrices. So $\mathbf{x} = (X_1, \ldots, X_p)'$ is used for a random $p-$vector, and $\mathbf{X} = (X_{ij})$ $(i = 1, \ldots, p;\ j = 1, \ldots, q)$ will denote a random $p \times q-$matrix $\mathbf{X}$.

Nowadays the literature on multivariate distributions is growing rapidly. For a collection of results on continuous and discrete multivariate distributions we refer to Johnson, Kotz & Balakrishnan (1997) and Kotz, Balakrishnan & Johnson (2000). Over the years many specialized topics have emerged, which are all connected to multivariate distributions and their marginal distributions. For example, the theory of copulas (see Nelsen, 1999), multivariate Laplace distributions (Kotz, Kozubowski & Podgórski, 2001) and multivariate $\mathbf{t}-$distributions (Kotz & Nadarajah, 2004) can be mentioned.

The first section is devoted to basic definitions of multivariate moments and cumulants. The main aim is to connect the moments and cumulants to the definition of matrix derivatives via differentiation of the moment generating or cumulant generating functions.

In the second section the multivariate normal distribution is presented. We focus on the matrix normal distribution which is a straightforward extension of the classical multivariate normal distribution. While obtaining moments of higher order, it is interesting to observe the connections between moments and permutations of basis vectors in tensors (Kronecker products).

The third section is dealing with so-called elliptical distributions, in particular with spherical distributions. The class of elliptical distributions is a natural extension of the class of multivariate normal distributions. In Section 2.3 basic material on elliptical distributions is given. The reason for including elliptical distributions in this chapter is that many statistical procedures based on the normal distribution also hold for the much wider class of elliptical distributions. Hotelling's T^2-statistic may serve as an example.

The fourth section presents material about the Wishart distribution. Here basic results are given. However, several classical results are presented with new proofs. For example, when deriving the Wishart density we are using the characteristic function which is a straightforward way of derivation but usually is not utilized. Moreover, we are dealing with quite complicated moment relations. In particular, moments for the inverse Wishart matrix have been included, as well as some basic results on moments for multivariate beta distributed variables.

2.1 MOMENTS AND CUMULANTS

2.1.1 *Introduction*

There are many ways of introducing multivariate moments. In the univariate case one usually defines moments via expectation, $E[U]$, $E[U^2]$, $E[U^3]$, etc., and if we suppose, for example, that U has a density $f_U(u)$, the $k-$th moment $m_k[U]$ is given by the integral

$$m_k[U] = E[U^k] = \int_R u^k f_U(u)du.$$

Furthermore, the $k-$th order central moment $\overline{m}_k[U]$ is also given as an expectation:

$$\overline{m}_k[U] = E[(U - E[U])^k].$$

In particular, for $k = 2$, we note that from the last expression we obtain the variance of U:

$$D[U] = E[(U - E[U])^2].$$

Let us denote the variance by $D[U]$, i.e. dispersion. In the multivariate case $D[\mathbf{u}]$ denotes the dispersion matrix of a random vector $\mathbf{u}$. An important tool for deriving moments and cumulants is the characteristic function of a random variable. For a random variable U we denote the characteristic function by $\varphi_U(t)$ and define it as

$$\varphi_U(t) = E[e^{itU}], \tag{2.1.1}$$

where i is the imaginary unit. We will treat i as a constant. Furthermore, throughout this chapter we suppose that in (2.1.1) we can change the order of differentiation and taking expectation. Hence,

$$E[U^k] = \frac{1}{i^k}\frac{d^k}{dt^k}\varphi_U(t)\bigg|_{t=0}. \tag{2.1.2}$$

This relation will be generalized to cover random vectors as well as random matrices. The problem of deriving multivariate moments is illustrated by an example.

Example 2.1.1 Consider the random vector $\mathbf{x} = (X_1, X_2)'$. For the first order moments, $E[X_1]$ and $E[X_2]$ are of interest and it is natural to define the first moment as

$$E[\mathbf{x}] = (E[X_1], E[X_2])'$$

or

$$E[\mathbf{x}] = (E[X_1], E[X_2]).$$

Choosing between these two versions is a matter of taste. However, with the representation of the second moments we immediately run into problems. Now $E[X_1X_2]$, $E[X_1^2]$ and $E[X_2^2]$ are of interest. Hence, we have more expectations than the size of the original vector. Furthermore, it is not obvious how to present the three different moments. For example, a direct generalization of the univariate case in the form $E[\mathbf{x}^2]$ does not work, since $\mathbf{x}^2$ is not properly defined. One definition which seems to work fairly well is $E[\mathbf{x}^{\otimes 2}]$, where the Kroneckerian power of a vector is given by Definition 1.3.4. Let us compare this possibility with $E[\mathbf{x}\mathbf{x}']$. By definition of the Kronecker product it follows that

$$E[\mathbf{x}^{\otimes 2}] = E[(X_1^2, X_1X_2, X_2X_1, X_2^2)'], \tag{2.1.3}$$

whereas

$$E[\mathbf{x}\mathbf{x}'] = E\left[\begin{pmatrix} X_1^2 & X_1X_2 \\ X_2X_1 & X_2^2 \end{pmatrix}\right]. \tag{2.1.4}$$

A third choice could be

$$E[V^2(\mathbf{x}\mathbf{x}')] = E[(X_1^2, X_1X_2, X_2^2)'], \tag{2.1.5}$$

where the operator $V^2(\cdot)$ was introduced in §1.3.7. ∎

All these three definitions in Example 2.1.1 have their pros and cons. For example, the Kroneckerian definition, given in (2.1.3), can be easily extended to moments of arbitrary order, namely

$$E[\mathbf{x}^{\otimes k}], \qquad k = 1, 2, \ldots$$

One problem arising here is the recognition of the position of a certain mixed moment. If we are only interested in the elements $E[X_1^2]$, $E[X_2^2]$ and $E[X_1X_2]$ in $E[\mathbf{x}^{\otimes 2}]$, things are easy, but for higher order moments it is complicated to show where the moments of interest are situated in the vector $E[\mathbf{x}^{\otimes k}]$. Some authors prefer to use so-called coordinate free approaches (Eaton, 1983; Holmquist, 1985; Wong & Liu, 1994, 1995; Käärik & Tiit, 1995). For instance, moments can be considered as elements of a vector space, or treated with the help of index sets. From a mathematical point of view this is convenient but in applications we really need explicit representations of moments. One fact, influencing our choice, is that the shape of the second order central moment, i.e. the dispersion matrix $D[\mathbf{x}]$ of a $p-$vector $\mathbf{x}$, is defined as a $p \times p-$matrix:

$$D[\mathbf{x}] = E[(\mathbf{x} - E[\mathbf{x}])(\mathbf{x} - E[\mathbf{x}])'].$$

If one defines moments of a random vector it would be natural to have the second order central moment in a form which gives us the dispersion matrix. We see that if $E[\mathbf{x}] = \mathbf{0}$ the definition given by (2.1.4) is identical to the dispersion matrix, i.e. $E[\mathbf{x}\mathbf{x}'] = D[\mathbf{x}]$. This is advantageous, since many statistical methods are based on dispersion matrices. However, one drawback with the moment definitions given by (2.1.3) and (2.1.4) is that they comprise too many elements, since in both $E[X_1X_2]$ appears twice. The definition in (2.1.5) is designed to present the smallest number of all possible mixed moments.

2.1.2 *Basic statistical functions*

In this paragraph we shall introduce notation and notions of the basic functions to be used for random vectors and matrices. A *density function* of a $p-$vector $\mathbf{x}$ is denoted by $f_{\mathbf{x}}(\mathbf{x})$, where $\mathbf{x} \in \mathbb{R}^p$, and a density function of a random $p \times q-$matrix $\mathbf{X}$ is denoted by $f_{\mathbf{X}}(\mathbf{X})$, where $\mathbf{X} \in \mathbb{R}^{p\times q}$. Here we follow the classical notation used by Muirhead (1982), Fang & Zhang (1990), Anderson (2003), for example, which is somewhat confusing: the random matrix and the variable argument of the density are denoted by the same letter, but this difference can be easily understood from the context. Similarly, we denote by $F_{\mathbf{x}}(\mathbf{x})$ the *distribution function* of a random $p-$vector $\mathbf{x}$ and by $F_{\mathbf{X}}(\mathbf{X})$ the distribution function of a random matrix $\mathbf{X} : p \times q$.

The *characteristic function* of a random $p-$vector $\mathbf{x}$ is defined by the equality

$$\varphi_{\mathbf{x}}(\mathbf{t}) = E\left[e^{i\mathbf{t}'\mathbf{x}}\right], \qquad \mathbf{t} \in \mathbb{R}^p. \tag{2.1.6}$$

The characteristic function of a random matrix $\mathbf{X}$ is given by

$$\varphi_{\mathbf{X}}(\mathbf{T}) = E[e^{i\mathrm{tr}(\mathbf{T}'\mathbf{X})}], \tag{2.1.7}$$

where $\mathbf{T}$ is of the same size as $\mathbf{X}$. From property (1.3.32) of the vec−operator we get another representation of the characteristic function:

$$\varphi_{\mathbf{X}}(\mathbf{T}) = E[e^{i\mathrm{vec}'\mathbf{T}\mathrm{vec}\mathbf{X}}]. \tag{2.1.8}$$

While putting $\mathbf{t} = \mathrm{vec}\mathbf{T}$ and $\mathbf{x} = \mathrm{vec}\mathbf{X}$, we observe that the characteristic function of a matrix can be considered as the characteristic function of a vector, and so (2.1.7) and (2.1.8) can be reduced to (2.1.6). Things become more complicated for random matrices with repeated elements, such as symmetric matrices. When dealing with symmetric matrices, as well as with other structures, we have to first decide what is meant by the distribution of the matrix. In the subsequent we will only consider sets of non-repeated elements of a matrix. For example, consider the $p \times p$ symmetric matrix $\mathbf{X} = (X_{ij})$. Since $X_{ij} = X_{ji}$, we can take the lower triangle or the upper triangle of $\mathbf{X}$, i.e. the elements $X_{11}, \ldots, X_{pp}, X_{12}, \ldots, X_{1p}, X_{23}, \ldots, X_{2p}, \ldots, X_{(p-1)p}$. We may use that in the notation of §1.3.7, the vectorized upper triangle equals $V^2(\mathbf{X})$. From this we define the characteristic function of a symmetric $p \times p-$matrix $\mathbf{X}$ by the equality

$$\varphi_{\mathbf{X}}(\mathbf{T}) = E[e^{iV^{2\prime}(\mathbf{T})V^2(\mathbf{X})}]. \tag{2.1.9}$$

A general treatment of matrices with repeated elements is covered by results about patterned matrices (Definition 1.3.6). Let us assume that there are no repeated elements in a random patterned matrix $\mathbf{X}(K)$. The characteristic function of a patterned matrix $\mathbf{X}(K)$ is defined by

$$\varphi_{\mathbf{X}(K)}(\mathbf{T}(K)) = E\left[e^{i\mathrm{vec}'\mathbf{T}(K)\mathrm{vec}\mathbf{X}(K)}\right], \tag{2.1.10}$$

where $\mathbf{T}(K)$ is a real patterned matrix with pattern K.
We call the logarithm of the characteristic function the *cumulant function* of a $p-$vector $\mathbf{x}$

$$\psi_{\mathbf{x}}(\mathbf{t}) = \ln\varphi_{\mathbf{x}}(\mathbf{t}). \tag{2.1.11}$$

Analogously, the function

$$\psi_{\mathbf{X}}(\mathbf{T}) = \ln\varphi_{\mathbf{X}}(\mathbf{T}) \tag{2.1.12}$$

is the cumulant function of a $p \times q-$matrix $\mathbf{X}$, and

$$\psi_{\mathbf{X}(K)}(\mathbf{T}(K)) = \ln\varphi_{\mathbf{X}(K)}(\mathbf{T}(K)) \tag{2.1.13}$$

is the cumulant function of a patterned matrix $\mathbf{X}(K)$ with a pattern K.

2.1.3 *Moments and central moments*

It was indicated in (2.1.2) that univariate moments can be obtained by differentiating the characteristic function. This idea will now be extended to random vectors and matrices. However, it is first noted that throughout the text $E[\mathbf{X}] = (E[X_{ij}])$ for any matrix $\mathbf{X}$.

Definition 2.1.1. *Let the characteristic function $\varphi_{\mathbf{x}}(\mathbf{t})$ be k times differentiable. Then the $k-th$ moment of a p-vector $\mathbf{x}$ equals*

$$m_k[\mathbf{x}] = \frac{1}{i^k}\frac{d^k}{d\mathbf{t}^k}\varphi_{\mathbf{x}}(\mathbf{t})\Big|_{\mathbf{t}=\mathbf{0}}, \qquad \mathbf{t} \in \mathbb{R}^p, \tag{2.1.14}$$

where the $k-th$ order matrix derivative is defined by (1.4.41). ∎

Let us check, whether this definition gives us the second order moment of the form (2.1.4). We have to find the second order derivative

$$\frac{d^2\varphi_{\mathbf{x}}(\mathbf{t})}{d\mathbf{t}^2}$$

$$= E[\frac{d^2}{d\mathbf{t}^2}e^{i\mathbf{t}'\mathbf{x}}] \underset{(1.4.41)}{=} E[\frac{d}{d\mathbf{t}}(\frac{d}{d\mathbf{t}}e^{i\mathbf{t}'\mathbf{x}})] \underset{(1.4.14)}{=} E[\frac{d}{d\mathbf{t}}(i\mathbf{x}e^{i\mathbf{t}'\mathbf{x}})] \underset{(1.4.16)}{=} E[-\mathbf{x}e^{i\mathbf{t}'\mathbf{x}}\mathbf{x}']$$

and thus $m_2[\mathbf{x}] = E[\mathbf{x}\mathbf{x}']$, i.e. $m_2[\mathbf{x}]$ is of the same form as the moment expression given by (2.1.4). Central moments of a random vector are defined in a similar way.

Definition 2.1.2. *The $k-th$ central moment of a $p-$vector $\mathbf{x}$ is given by the equality*

$$\overline{m}_k[\mathbf{x}] = m_k[\mathbf{x} - E[\mathbf{x}]] = \frac{1}{i^k}\frac{d^k}{d\mathbf{t}^k}\varphi_{\mathbf{x}-E[\mathbf{x}]}(\mathbf{t})\Big|_{\mathbf{t}=\mathbf{0}}, \qquad \mathbf{t} \in \mathbb{R}^p, \tag{2.1.15}$$

where the $k-th$ order matrix derivative is defined by (1.4.41). ∎

From this definition it follows directly that when $k = 2$ we obtain from (2.1.15) the dispersion matrix $D[\mathbf{x}]$ of $\mathbf{x}$. It is convenient to obtain moments via differentiation, since we can derive them more or less mechanically. Approaches which do not involve differentiation often rely on combinatorial arguments. Above, we have extended a univariate approach to random vectors and now we will indicate how to extend this to matrices. Let $m_k[\mathbf{X}]$ denote the $k-$th moment of a random matrix $\mathbf{X}$.

Definition 2.1.3. *Let the characteristic function $\varphi_{\mathbf{X}}(\mathbf{T})$ be k times differentiable. Then the k–th moment of a random $p \times q$–matrix* $\mathbf{X}$ *equals*

$$m_k[\mathbf{X}] = \frac{1}{i^k} \frac{d^k}{d\mathbf{T}^k} \varphi_{\mathbf{X}}(\mathbf{T}) \bigg|_{\mathbf{T}=\mathbf{0}}, \qquad \mathbf{T} \in \mathbb{R}^{p\times q}, \tag{2.1.16}$$

where the k–th order matrix derivative is defined by (1.4.41). ∎

In a similar way we define central moments of a random matrix.

Definition 2.1.4. *The k–th central moment $\overline{m}_k[\mathbf{X}]$ of a random matrix* $\mathbf{X}$ *equals*

$$\overline{m}_k[\mathbf{X}] = m_k[\mathbf{X} - E[\mathbf{X}]] = \frac{1}{i^k} \frac{d^k}{d\mathbf{T}^k} \varphi_{\mathbf{X}-E[\mathbf{X}]}(\mathbf{T}) \bigg|_{\mathbf{T}=\mathbf{0}}, \qquad \mathbf{T} \in \mathbb{R}^{p\times q}, \tag{2.1.17}$$

where the k–th order matrix derivative is defined by (1.4.41). ∎

One should observe that

$$\varphi_{\mathbf{X}-E[\mathbf{X}]}(\mathbf{T}) = \varphi_{\mathbf{X}}(\mathbf{T}) g(\mathbf{T}), \tag{2.1.18}$$

where

$$g(\mathbf{T}) = e^{-i\mathrm{tr}(\mathbf{T}'E[\mathbf{X}])}.$$

For a random vector $\mathbf{x}$ we have

$$g(\mathbf{t}) = e^{-i\mathbf{t}'E[\mathbf{x}]}.$$

Now we get the central moments of a random vector and a random matrix in the following form

$$\overline{m}_k[\mathbf{x}] = \frac{1}{i^k} \frac{d^k}{d\mathbf{t}^k} \{\varphi_{\mathbf{x}}(\mathbf{t}) e^{-i\mathbf{t}'E[\mathbf{x}]}\} \bigg|_{\mathbf{t}=\mathbf{0}}, \tag{2.1.19}$$

$$\overline{m}_k[\mathbf{X}] = \frac{1}{i^k} \frac{d^k}{d\mathbf{T}^k} \{\varphi_{\mathbf{X}}(\mathbf{T}) e^{-i\mathrm{tr}(\mathbf{T}'E[\mathbf{X}])}\} \bigg|_{\mathbf{T}=\mathbf{0}}. \tag{2.1.20}$$

It is not obvious how to define the dispersion matrix $D[\mathbf{X}]$ for a random matrix $\mathbf{X}$. Here we adopt the definition

$$D[\mathbf{X}] = D[\mathrm{vec}\mathbf{X}], \tag{2.1.21}$$

which is the most common one. In the next, a theorem is presented which gives expressions for moments of a random vector as expectations, without the reference to the characteristic function.

Theorem 2.1.1. *For an arbitrary random vector* $\mathbf{x}$

(i) $$m_k[\mathbf{x}] = E[\mathbf{x}(\mathbf{x}')^{\otimes k-1}], \qquad k = 1, 2, \ldots; \tag{2.1.22}$$

(ii) $$\overline{m}_k[\mathbf{x}] = E[(\mathbf{x} - E[\mathbf{x}])((\mathbf{x} - E[\mathbf{x}])')^{\otimes k-1}], \qquad k = 1, 2, \ldots \tag{2.1.23}$$

PROOF: We are going to use the method of mathematical induction when differentiating the characteristic function $\varphi_{\mathbf{x}}(\mathbf{t})$ in (2.1.6). For $k = 1$ we have

$$\frac{de^{i\mathbf{t}'\mathbf{x}}}{d\mathbf{t}} \underset{(1.4.14)}{=} \frac{d\{i\mathbf{t}'\mathbf{x}\}}{d\mathbf{t}} \frac{de^{i\mathbf{t}'\mathbf{x}}}{d\{i\mathbf{t}'\mathbf{x}\}} \underset{(1.4.16)}{=} i\mathbf{x}e^{i\mathbf{t}'\mathbf{x}} \tag{2.1.24}$$

and

$$\left.\frac{d\varphi_{\mathbf{x}}(\mathbf{t})}{d\mathbf{t}}\right|_{\mathbf{t}=\mathbf{0}} = E[ie^{i\mathbf{t}'\mathbf{x}}\mathbf{x}]\Big|_{\mathbf{t}=\mathbf{0}} = im_1[\mathbf{x}].$$

Let us suppose that for $k = j - 1$ the equality

$$\frac{d^{j-1}\varphi_{\mathbf{x}}(\mathbf{t})}{d\mathbf{t}^{j-1}} = E[i^{j-1}\mathbf{x}e^{i\mathbf{t}'\mathbf{x}}(\mathbf{x}')^{\otimes(j-2)}]$$

holds, from where we get the statement of the theorem at $\mathbf{t} = \mathbf{0}$. We are going to prove that the statement is also valid for $k = j$.

$$\frac{d^j\varphi_{\mathbf{x}}(\mathbf{t})}{d\mathbf{t}^j} \underset{(1.4.16)}{=} i^{j-1}E[\frac{de^{i\mathbf{t}'\mathbf{x}}}{d\mathbf{t}}(\mathbf{x}')^{\otimes(j-1)}] \underset{(2.1.24)}{=} E[i^j\mathbf{x}e^{i\mathbf{t}'\mathbf{x}}(\mathbf{x}')^{\otimes(j-1)}].$$

At the point $\mathbf{t} = \mathbf{0}$ we obtain

$$\left.\frac{d^j\varphi_{\mathbf{x}}(\mathbf{t})}{d\mathbf{t}^j}\right|_{\mathbf{t}=\mathbf{0}} = i^jE[\mathbf{x}(\mathbf{x}')^{\otimes(j-1)}] = i^jm_j[\mathbf{x}].$$

Thus (2.1.22) is established. To complete the proof it remains to show that (2.1.23) is valid, but this follows immediately from the fact that the central moment of $\mathbf{x}$ is identical to the moment of $\mathbf{x} - E[\mathbf{x}]$. ∎

When defining moments we assume that we can change the order of differentiation and taking expectation when differentiating the characteristic function. From the proof of the next theorem we get a characterization of that assumption.

Corollary 2.1.1.1. *Let the vector* $\mathbf{x}^{\otimes k}$ *be absolutely integrable elementwise. Then all the moments* $m_j(\mathbf{x})$, $j \leq k$ *exist.*

PROOF: By assumption all mixed absolute moments of order k are finite: $E[|X_1^{i_1}X_2^{i_2} \times \cdots \times X_p^{i_p}|] < \infty$, $\sum_{j=1}^p i_j = k$, $i_j \in \{0, 1, \ldots, k\}$. Let us examine the modulus of the $gh-$th element of the expectation matrix

$$\left|\left(E[\frac{d^ke^{i\mathbf{t}'\mathbf{x}}}{d\mathbf{t}^k}]\right)_{gh}\right| = \left|i^k\left(E[\mathbf{x}(\mathbf{x}')^{\otimes(k-1)}e^{i\mathbf{t}'\mathbf{x}}]\right)_{gh}\right| \leq E[|\left(\mathbf{x}(\mathbf{x}')^{\otimes(k-1)}e^{i\mathbf{t}'\mathbf{x}}\right)_{gh}|]$$
$$= E[|\left(\mathbf{x}(\mathbf{x}')^{\otimes(k-1)}\right)_{gh}|].$$

By assumptions, we have the expectation from elements with finite modulus on the right hand side of our chain of relations. From calculus we know that this is a sufficient condition for changing the order of integration and differentiation. So the assumption for existence of moments of the k−th order is satisfied. The existence of all lower order moments follows from the existence of the k−th order moments (Feller, 1971; p. 136). ∎

Taking into account Definition 2.1.3, Definition 2.1.4 and (2.1.18), we also get from Theorem 2.1.1 expressions for moments of random matrices.

Corollary 2.1.1.2. *For an arbitrary random matrix* $\mathbf{X}$

(i) $$m_k[\mathbf{X}] = E[\mathrm{vec}\mathbf{X}(\mathrm{vec}'\mathbf{X})^{\otimes k-1}], \qquad k = 1, 2, \ldots; \tag{2.1.25}$$

(ii) $$\overline{m}_k[\mathbf{X}] = E[\mathrm{vec}(\mathbf{X} - E[\mathbf{X}])(\mathrm{vec}'(\mathbf{X} - E[\mathbf{X}]))^{\otimes k-1}], \quad k = 1, 2, \ldots \tag{2.1.26}$$

∎

The following theorem gives us a way of presenting the characteristic function as an expansion via moments. Let us first present the result for matrices.

Theorem 2.1.2. *Let the characteristic function* $\varphi_{\mathbf{X}}(\mathbf{T})$ *be* n *times differentiable. Then* $\varphi_{\mathbf{X}}(\mathbf{T})$ *can be presented as the following series expansion:*

$$\varphi_{\mathbf{X}}(\mathbf{T}) = 1 + \sum_{k=1}^{n} \frac{i^k}{k!}(\mathrm{vec}'\mathbf{T})^{\otimes k}\mathrm{vec}(m_k[\mathbf{X}])' + r_n, \tag{2.1.27}$$

where r_n *is the remainder term.*

PROOF: Corollary 1.4.8.2 gives us the Taylor expansion of the characteristic function at $\mathbf{T} = \mathbf{0}$ in the following form:

$$\varphi_{\mathbf{X}}(\mathbf{T}) = 1 + \sum_{k=1}^{n} \frac{1}{k!}(\mathrm{vec}'\mathbf{T})^{\otimes k} \mathrm{vec}\left(\frac{d^k\varphi_{\mathbf{X}}(\mathbf{T})}{d\mathbf{T}^k}\right)'\Bigg|_{\mathbf{T}=\mathbf{0}} + r_n.$$

From Definition 2.1.3 we get

$$\frac{d^k\varphi_{\mathbf{X}}(\mathbf{T})}{d\mathbf{T}^k}\Bigg|_{\mathbf{T}=\mathbf{0}} = i^k m_k[\mathbf{X}].$$

∎

Analogously we get a similar expansion for the characteristic function of a random vector.

Corollary 2.1.2.1. *Let the characteristic function* $\varphi_{\mathbf{x}}(\mathbf{t})$ *be* n *times differentiable. Then* $\varphi_{\mathbf{X}}(\mathbf{T})$ *can be presented as the following series expansion:*

$$\varphi_{\mathbf{x}}(\mathbf{t}) = 1 + \sum_{k=1}^{n} \frac{i^k}{k!}(\mathbf{t}')^{\otimes k}\mathrm{vec}(m_k[\mathbf{x}])' + r_n, \tag{2.1.28}$$

where r_n is the remainder term. ∎

From the above approach it follows that we can define moments simultaneously for several random variables, several random vectors or several random matrices, since all definitions fit with the definition of moments of matrices. Now we shall pay some attention to the *covariance* between two random matrices. Indeed, from the definition of the dispersion matrix given in (2.1.21) it follows that it is natural to define the covariance of two random matrices $\mathbf{X}$ and $\mathbf{Y}$ in the following way:

Definition 2.1.5. *The covariance of two arbitrary matrices* $\mathbf{X}$ *and* $\mathbf{Y}$ *is given by*

$$C[\mathbf{X},\mathbf{Y}] = E[(\mathrm{vec}\mathbf{X} - E[\mathrm{vec}\mathbf{X}])(\mathrm{vec}\mathbf{Y} - E[\mathrm{vec}\mathbf{Y}])'], \tag{2.1.29}$$

if the expectations exist. ∎

Furthermore, note that

$$D[\mathbf{X}:\mathbf{Y}] = \begin{pmatrix} D[\mathbf{X}] & C[\mathbf{X},\mathbf{Y}] \\ C[\mathbf{Y},\mathbf{X}] & D[\mathbf{Y}] \end{pmatrix} \tag{2.1.30}$$

and we will prove the following theorem, which is useful when considering singular covariance matrices.

Theorem 2.1.3. *For any random matrices* $\mathbf{X}$ *and* $\mathbf{Y}$

(i)
$$\mathcal{C}(C[\mathbf{X},\mathbf{Y}]) \subseteq \mathcal{C}(D[\mathbf{X}]);$$

(ii)
$$\mathcal{C}(C[\mathbf{Y},\mathbf{X}]) \subseteq \mathcal{C}(D[\mathbf{Y}]).$$

PROOF: Since $D[\mathbf{X}:\mathbf{Y}]$ by (2.1.30) is positive semidefinite, we can write

$$D[\mathbf{X}:\mathbf{Y}] = \boldsymbol{\tau}\boldsymbol{\tau}' = \begin{pmatrix} \boldsymbol{\tau}_1\boldsymbol{\tau}_1' & \boldsymbol{\tau}_1\boldsymbol{\tau}_2' \\ \boldsymbol{\tau}_2\boldsymbol{\tau}_1' & \boldsymbol{\tau}_2\boldsymbol{\tau}_2' \end{pmatrix}$$

for some $\boldsymbol{\tau} = (\boldsymbol{\tau}_1' : \boldsymbol{\tau}_2')'$, where the partition of $\boldsymbol{\tau}$ corresponds to the partition of $\mathbf{X}$ and $\mathbf{Y}$ in $(\mathbf{X}:\mathbf{Y})$. Now, because

$$\mathcal{C}(C[\mathbf{X},\mathbf{Y}]) = \mathcal{C}(\boldsymbol{\tau}_1\boldsymbol{\tau}_2') \subseteq \mathcal{C}(\boldsymbol{\tau}_1) = \mathcal{C}(\boldsymbol{\tau}_1\boldsymbol{\tau}_1') = \mathcal{C}(D[\mathbf{X}])$$

and

$$\mathcal{C}(C[\mathbf{Y},\mathbf{X}]) = \mathcal{C}(\boldsymbol{\tau}_2\boldsymbol{\tau}_1') \subseteq \mathcal{C}(\boldsymbol{\tau}_2) = \mathcal{C}(\boldsymbol{\tau}_2\boldsymbol{\tau}_2') = \mathcal{C}(D[\mathbf{Y}])$$

the theorem is established. ∎

If $\boldsymbol{\Sigma} = D[\mathbf{X}]$ is singular, there are some restrictions on the random matrix $\mathbf{X}$. Note that then there must exist a matrix $\boldsymbol{\Sigma}^o$ which spans the orthogonal complement $\mathcal{C}(\boldsymbol{\Sigma})^\perp$. Hence

$$D[\boldsymbol{\Sigma}^{o'}\mathrm{vec}\mathbf{X}] = \boldsymbol{\Sigma}^{o'}\boldsymbol{\Sigma}\boldsymbol{\Sigma}^o = \mathbf{0}$$

and therefore $\boldsymbol{\Sigma}^{o'}\mathrm{vec}\mathbf{X} = E[\boldsymbol{\Sigma}^{o'}\mathrm{vec}\mathbf{X}]$ with probability 1.

Next we study the dispersion matrix of $\mathbf{X}$ which is partitioned as

$$\mathbf{X} = \begin{pmatrix} \mathbf{X}_{11} & \mathbf{X}_{12} \\ \mathbf{X}_{21} & \mathbf{X}_{22} \end{pmatrix} \quad \begin{pmatrix} r \times s & r \times (n-s) \\ (p-r) \times s & (p-r) \times (n-s) \end{pmatrix}, \tag{2.1.31}$$

where the sizes of the blocks are given in the matrix on the right hand side.

Theorem 2.1.4. *For any random matrix* $\mathbf{X}$, *which is partitioned according to (2.1.31), the following statements hold:*

(i)

$$\begin{pmatrix}\mathbf{K}_{n,r} & \mathbf{0}\\ \mathbf{0} & \mathbf{K}_{n,p-r}\end{pmatrix} D[\mathbf{X}'] \begin{pmatrix}\mathbf{K}_{r,n} & \mathbf{0}\\ \mathbf{0} & \mathbf{K}_{p-r,n}\end{pmatrix}$$
$$= \begin{pmatrix} D[\mathbf{X}_{11}] & C[\mathbf{X}_{11},\mathbf{X}_{12}] & C[\mathbf{X}_{11},\mathbf{X}_{21}] & C[\mathbf{X}_{11},\mathbf{X}_{22}]\\ C[\mathbf{X}_{12},\mathbf{X}_{11}] & D[\mathbf{X}_{12}] & C[\mathbf{X}_{12},\mathbf{X}_{21}] & C[\mathbf{X}_{12},\mathbf{X}_{22}]\\ C[\mathbf{X}_{21},\mathbf{X}_{11}] & C[\mathbf{X}_{21},\mathbf{X}_{12}] & D[\mathbf{X}_{21}] & C[\mathbf{X}_{21},\mathbf{X}_{22}]\\ C[\mathbf{X}_{22},\mathbf{X}_{11}] & C[\mathbf{X}_{22},\mathbf{X}_{12}] & C[\mathbf{X}_{22},\mathbf{X}_{21}] & D[\mathbf{X}_{22}] \end{pmatrix};$$

(ii)

$$\begin{pmatrix}\mathbf{K}_{n,r} & \mathbf{0}\\ \mathbf{0} & \mathbf{K}_{n,p-r}\end{pmatrix} \mathbf{K}_{p,n} D[\mathbf{X}] \mathbf{K}_{n,p} \begin{pmatrix}\mathbf{K}_{r,n} & \mathbf{0}\\ \mathbf{0} & \mathbf{K}_{p-r,n}\end{pmatrix}$$
$$= \begin{pmatrix} D[\mathbf{X}_{11}] & C[\mathbf{X}_{11},\mathbf{X}_{12}] & C[\mathbf{X}_{11},\mathbf{X}_{21}] & C[\mathbf{X}_{11},\mathbf{X}_{22}]\\ C[\mathbf{X}_{12},\mathbf{X}_{11}] & D[\mathbf{X}_{12}] & C[\mathbf{X}_{12},\mathbf{X}_{21}] & C[\mathbf{X}_{12},\mathbf{X}_{22}]\\ C[\mathbf{X}_{21},\mathbf{X}_{11}] & C[\mathbf{X}_{21},\mathbf{X}_{12}] & D[\mathbf{X}_{21}] & C[\mathbf{X}_{21},\mathbf{X}_{22}]\\ C[\mathbf{X}_{22},\mathbf{X}_{11}] & C[\mathbf{X}_{22},\mathbf{X}_{12}] & C[\mathbf{X}_{22},\mathbf{X}_{21}] & D[\mathbf{X}_{22}] \end{pmatrix};$$

(iii) *Let* $\mathbf{X}_{\bullet 1} = (\mathbf{X}'_{11} : \mathbf{X}'_{21})'$. *Then*

$$(\mathbf{I}_{ps} : \mathbf{0}) D[\mathbf{X}] (\mathbf{I}_{ps} : \mathbf{0})' = D[\mathbf{X}_{\bullet 1}].$$

(iv) *Let* $\mathbf{X}_{1\bullet} = (\mathbf{X}_{11} : \mathbf{X}_{12})$. *Then*

$$(\mathbf{I}_n \otimes (\mathbf{I}_r : \mathbf{0})) D[\mathbf{X}] (\mathbf{I}_n \otimes (\mathbf{I}_r : \mathbf{0}))' = D[\mathbf{X}_{1\bullet}].$$

(v)

$$\begin{pmatrix}\mathbf{I}_s \otimes (\mathbf{I}_r : \mathbf{0}) & \mathbf{0}\\ \mathbf{0} & \mathbf{I}_{n-s} \otimes (\mathbf{0} : \mathbf{I}_{p-r})\end{pmatrix} D[\mathbf{X}] \begin{pmatrix}\mathbf{I}_s \otimes (\mathbf{I}_r : \mathbf{0})' & \mathbf{0}\\ \mathbf{0} & \mathbf{I}_{n-s} \otimes (\mathbf{0} : \mathbf{I}_{p-r})'\end{pmatrix}$$
$$= D[\begin{pmatrix}\mathrm{vec}\mathbf{X}_{11}\\ \mathrm{vec}\mathbf{X}_{22}\end{pmatrix}].$$

PROOF: All proofs of the statements are based mainly on the fundamental property of the commutation matrix given by (1.3.30). For (i) we note that

$$\begin{pmatrix}\mathbf{K}_{n,r} & \mathbf{0}\\ \mathbf{0} & \mathbf{K}_{n,p-r}\end{pmatrix} \begin{pmatrix} \mathrm{vec}\begin{pmatrix}\mathbf{X}'_{11}\\ \mathbf{X}'_{12}\end{pmatrix}\\ \mathrm{vec}\begin{pmatrix}\mathbf{X}'_{21}\\ \mathbf{X}'_{22}\end{pmatrix}\end{pmatrix}$$
$$= \begin{pmatrix}\mathrm{vec}(\mathbf{X}_{11} : \mathbf{X}_{12})\\ \mathrm{vec}(\mathbf{X}_{21} : \mathbf{X}_{22})\end{pmatrix} = (\mathrm{vec}'\mathbf{X}_{11} : \mathrm{vec}'\mathbf{X}_{12} : \mathrm{vec}'\mathbf{X}_{21} : \mathrm{vec}'\mathbf{X}_{22})'.$$

For (ii) we use that $\mathbf{K}_{p,n}\text{vec}\mathbf{X} = \text{vec}\mathbf{X}'$. Statement (iii) is established by the definition of the dispersion matrix and the fact that

$$\text{vec}\mathbf{X} = \begin{pmatrix} \text{vec}\begin{pmatrix} \mathbf{X}_{11} \\ \mathbf{X}_{21} \end{pmatrix} \\ \text{vec}\begin{pmatrix} \mathbf{X}_{12} \\ \mathbf{X}_{22} \end{pmatrix} \end{pmatrix}.$$

In order to verify (iv) it is noted that

$$(\mathbf{I}_n \otimes (\mathbf{I}_r : \mathbf{0}))\text{vec}\mathbf{X} \underset{(1.3.31)}{=} \text{vec}((\mathbf{I}_r : \mathbf{0})\mathbf{X}) = \text{vec}\mathbf{X}_{1\bullet},$$

and in order to verify (v)

$$\begin{pmatrix} \mathbf{I}_s \otimes (\mathbf{I}_r : \mathbf{0}) & \mathbf{0} \\ \mathbf{0} & \mathbf{I}_{n-s} \otimes (\mathbf{0} : \mathbf{I}_{p-r}) \end{pmatrix} \begin{pmatrix} \text{vec}\begin{pmatrix} \mathbf{X}_{11} \\ \mathbf{X}_{21} \end{pmatrix} \\ \text{vec}\begin{pmatrix} \mathbf{X}_{12} \\ \mathbf{X}_{22} \end{pmatrix} \end{pmatrix} = \begin{pmatrix} \text{vec}\mathbf{X}_{11} \\ \text{vec}\mathbf{X}_{22} \end{pmatrix}.$$

∎

One may observe that since $(\mathbf{I}_{pr} : \mathbf{0}) = (\mathbf{I}_r : \mathbf{0}) \otimes \mathbf{I}_p$, the statements (iii) and (iv) are more similar to each other than they seem at a first sight.

2.1.4 *Cumulants*

Most of what has been said about moments can be carried over to cumulants which sometimes are also called *semiinvariants*. Now we have, instead of the characteristic function, the cumulant function (2.1.11) of a random vector, or (2.1.12) in the case of a random matrix. As noted already in the previous paragraph, there are several ways to introduce moments and, additionally, there exist many natural possibilities to present multivariate moments. For cumulants we believe that the most natural way is to define them as derivatives of the cumulant function. As it has been seen before, the definition of multivariate moments depends on the definition of the matrix derivative. The same problem arises when representing multivariate cumulants. However, there will always be a link between a certain type of matrix derivative, a specific multivariate moment representation and a certain choice of multivariate cumulant. It is, as we have noted before, more or less a matter of taste which representations are to be preferred, since there are some advantages as well as some disadvantages with all of them.
Let $c_k[X]$, $c_k[\mathbf{x}]$ and $c_k[\mathbf{X}]$ denote the $k-$th cumulant of a random variable X, a random vector $\mathbf{x}$ and a random matrix $\mathbf{X}$, respectively.

Definition 2.1.6. *Let the cumulant function $\psi_{\mathbf{x}}(\mathbf{t})$ be k times differentiable at $\mathbf{t} = \mathbf{0}$. Then the $k-$th cumulant of a random vector $\mathbf{x}$ is given by*

$$c_k[\mathbf{x}] = \frac{1}{i^k}\frac{d^k}{d\mathbf{t}^k}\psi_{\mathbf{x}}(\mathbf{t})\Big|_{\mathbf{t}=\mathbf{0}}, \quad k = 1, 2, \ldots, \tag{2.1.32}$$

where the matrix derivative is given by (1.4.41) and the cumulant function is defined by (2.1.11). ∎

Definition 2.1.7. *Let the cumulant function $\psi_{\mathbf{X}}(\mathbf{T})$ be k times differentiable at $\mathbf{T} = \mathbf{0}$. Then the $k-$th cumulant of a random matrix $\mathbf{X}$ is defined by the equality*

$$c_k[\mathbf{X}] = \frac{1}{i^k}\frac{d^k}{d\mathbf{T}^k}\psi_{\mathbf{X}}(\mathbf{T})\bigg|_{\mathbf{T}=\mathbf{0}}, \qquad k = 1, 2, \ldots, \tag{2.1.33}$$

where the matrix derivative is given by (1.4.41) and $\psi_{\mathbf{X}}(\mathbf{T})$ is defined by (2.1.12). ∎

It was shown that moments could be obtained from the Taylor expansion of the characteristic function. Similarly, cumulants appear in the Taylor expansion of the cumulant function.

Theorem 2.1.5. *Let the cumulant function $\psi_{\mathbf{X}}(\mathbf{T})$ be n times differentiable. Then $\psi_{d\mathbf{X}}(\mathbf{T})$ can be presented as the following series expansion:*

$$\psi_{\mathbf{X}}(\mathbf{T}) = \sum_{k=1}^{n} \frac{i^k}{k!}(\mathrm{vec}'\mathbf{T})^{\otimes k}\mathrm{vec}(c_k[\mathbf{X}])' + r_n, \tag{2.1.34}$$

where r_n is the remainder term.

PROOF: The proof repeats the argumentation of establishing Theorem 2.1.2. From Corollary 1.4.8.2 we get the Taylor expansion of the cumulant function at $\mathbf{T} = \mathbf{0}$ in the following form:

$$\psi_{\mathbf{X}}(\mathbf{T}) = \sum_{k=1}^{n} \frac{1}{k!}(\mathrm{vec}'\mathbf{T})^{\otimes k}\ \mathrm{vec}\left(\frac{d^k\psi_{\mathbf{X}}(\mathbf{T})}{d\mathbf{T}^k}\right)'\bigg|_{\mathbf{T}=\mathbf{0}} + r_n.$$

From Definition 2.1.7. we have

$$\frac{d^k\psi_{\mathbf{X}}(\mathbf{T})}{d\mathbf{T}^k}\bigg|_{\mathbf{T}=\mathbf{0}} = i^k c_k[\mathbf{X}].$$

∎

If we consider a special case when $\mathbf{X}$ is a $p \times 1-$matrix, i.e. a $p-$vector, we get the following statement.

Corollary 2.1.5.1. *Let the cumulant function $\psi_{\mathbf{x}}(\mathbf{t})$ be n times differentiable. Then the cumulant function can be presented as the following series expansion:*

$$\psi_{\mathbf{x}}(\mathbf{t}) = \sum_{k=1}^{n} \frac{i^k}{k!}(\mathbf{t}')^{\otimes k}\mathrm{vec}(c_k[\mathbf{x}])' + r_n, \tag{2.1.35}$$

where r_n is the remainder term. ∎

2.1.5 *Moments and cumulants of patterned matrices*

In this paragraph we are going to introduce moments and cumulants of matrices which have some linear restrictions among their elements. For example, symmetric matrices, like the sample dispersion matrix or the sample correlation matrix, belong to this class of matrices. In general, let $\mathbf{X}(K)$ be a patterned matrix with a pattern K considered in §1.3.6.

Definition 2.1.8. *Let the characteristic function $\varphi_{\mathbf{X}(K)}(\mathbf{T}(K))$ be k times differentiable at $\mathbf{0}$. Then the $k-$th moment of $\mathbf{X}(K)$ equals*

$$m_k[\mathbf{X}(K)] = \frac{1}{i^k} \frac{d^k}{d\mathbf{T}(K)^k} \varphi_{\mathbf{X}(K)}(\mathbf{T}(K))\bigg|_{\mathbf{T}(K)=\mathbf{0}}, \qquad \mathbf{T} \in \mathbb{R}^{p\times q}, \qquad (2.1.36)$$

and the $k-$th central moment of $\mathbf{X}(K)$ is given by

$$\overline{m}_k[\mathbf{X}(K)] = m_k[\mathbf{X}(K) - E[\mathbf{X}(K)]] = \frac{1}{i^k} \frac{d^k}{d\mathbf{T}(K)^k} \varphi_{\mathbf{X}(K)-E[\mathbf{X}(K)]}(\mathbf{T}(K))\bigg|_{\mathbf{T}(K)=\mathbf{0}}, \qquad (2.1.37)$$

where the elements of $\mathbf{T}(K)$ are real and the matrix derivative is given by (1.4.43). ∎

In complete analogy with Corollary 2.1.1.2 in §2.1.3 we can present moments and central moments as expectations.

Theorem 2.1.6. *For an arbitrary random patterned matrix $\mathbf{X}(K)$*

(i) $$m_k[\mathbf{X}(K)] = E[\mathrm{vec}\mathbf{X}(K)(\mathrm{vec}'\mathbf{X}(K))^{\otimes k-1}], \quad k = 1, 2, \ldots,; \qquad (2.1.38)$$

(ii) $$\overline{m}_k[\mathbf{X}(K)] = E[\mathrm{vec}(\mathbf{X}(K) - E[\mathbf{X}(K)])(\mathrm{vec}'(\mathbf{X}(K) - E[\mathbf{X}(K)]))^{\otimes k-1}], \quad k = 1, 2, \ldots \qquad (2.1.39)$$

∎

The proof is omitted as it repeats the proof of Theorem 2.1.1 step by step if we change $\mathbf{x}$ to $\mathrm{vec}\mathbf{X}(K)$ and use the characteristic function (2.1.10) instead of (2.1.6). In the same way an analogue of Theorem 2.1.2 is also valid.

Theorem 2.1.7. *Let the characteristic function $\varphi_{\mathbf{X}(K)}(\mathbf{T}(K))$ be n times differentiable at $\mathbf{0}$. Then the characteristic function can be presented as the following series expansion:*

$$\varphi_{\mathbf{X}(K)}(\mathbf{T}(K)) = 1 + \sum_{k=1}^{n} \frac{i^k}{k!} (\mathrm{vec}'\mathbf{T}(K))^{\otimes k} \mathrm{vec}(m_k[\mathbf{X}(K)])' + r_n, \qquad (2.1.40)$$

where r_n is the remainder term. ∎

The cumulants of a patterned matrix are also defined similarly to the cumulants of an ordinary matrix.

Definition 2.1.9. *Let the cumulant function $\psi_{\mathbf{X}(K)}(\mathbf{T}(K))$ be k times differentiable at $\mathbf{0}$. Then the $k-$th cumulant of a random patterned matrix $\mathbf{X}(K)$ is defined by*

$$c_k[\mathbf{X}(K)] = \frac{1}{i^k} \frac{d^k}{d\mathbf{T}(K)^k} \psi_{\mathbf{X}(K)}(\mathbf{T}(K))\bigg|_{\mathbf{T}(K)=\mathbf{0}}, \qquad (2.1.41)$$

where the derivative is given by (1.4.43) and $\psi_{\mathbf{X}(K)}(\mathbf{T}(K))$ is given by (2.1.13). ∎

The next theorem presents the cumulant function through the cumulants of $\mathbf{X}(K)$. Again the proof copies straightforwardly the one of Theorem 2.1.5 and is therefore omitted.

Theorem 2.1.8. *Let the cumulant function $\psi_{\mathbf{X}(K)}(\mathbf{T}(K))$ be n times differentiable at $\mathbf{0}$. Then the cumulant function $\psi_{\mathbf{X}(K)}(\mathbf{T}(K))$ can be presented as the following series expansion:*

$$\psi_{\mathbf{X}(K)}(\mathbf{T}(K)) = \sum_{k=1}^{n} \frac{i^k}{k!} \mathrm{vec}'\mathbf{T}(K)^{\otimes k} \mathrm{vec}(c_k[\mathbf{X}(K)])' + r_n, \tag{2.1.42}$$

where r_n is the remainder term. ∎

The most important special case of patterned matrices is the class of symmetric matrices. The characteristic function of a symmetric matrix is given by (2.1.9). We shall immediately get the moments and cumulants of a symmetric matrix by differentiating the characteristic function (2.1.9) by using formulas (2.1.36), (2.1.37) and (2.1.41). In the following text symmetric matrices are differentiated very often. Therefore we are not going to indicate the pattern set K all the time, but write them as usual matrices, like $\mathbf{X}$, $\mathbf{Y}$, while keeping in mind that we are differentiating symmetric matrices

$$\frac{d\mathbf{Y}}{d\mathbf{X}} \equiv \frac{d(V^2(\mathbf{Y}))}{d(V^2(\mathbf{X}))}, \tag{2.1.43}$$

where $V^2(\bullet)$ is given by Definition 1.3.9.

2.1.6 *Minimal moments and cumulants*

The notions of minimal moments and cumulants are closely related to minimal derivatives defined in §1.4.8. We are going to introduce so-called minimal moments on the basis of the general vectorization operator $V^j(\bullet)$, given in Definition 1.3.9, and the product vectorization operator $R^j(\bullet)$, given in Definition 1.3.10, which were examined in §1.3.7. Let us use an example to explain the idea. Let $\mathbf{x}$ be a random $p-$vector and consider the fourth order moments of $\mathbf{x}$, which are often needed when multivariate distributions are approximated or when asymptotic distributions are to be found. As noted previously, there is one natural definition of the moments in a coordinate free setting. In order to do some calculations explicit moment expressions are needed. Unfortunately, as we have observed, the moments can not be presented in a unique way. On the basis of the direct product any of the expressions $E[x^{\otimes 4}]$ or $E[(\mathbf{x}\mathbf{x}')^{\otimes 2}]$ or $E[\mathbf{x}^{\otimes 3}\mathbf{x}']$, but also others, are possible. However, since $\mathbf{x}$ is $p-$dimensional, all these expressions consist of p^4 elements, whereas the number of different mixed moments (expectations of monomials) of fourth order equals $\binom{p+3}{4}$. The set which comprises one copy of all different mixed moments is called *minimal moment* of the random vector $\mathbf{x}$. Thus, for $p = 2$ we have 5 different mixed moments among 16 elements in $E[X^{\otimes 4}]$, i.e. $\approx 31\%$; for $p = 4$ the corresponding figures are 35 out of 256 ($\approx 14\%$); for $p = 10$ it is 715 of 10000 (7%). For large p there will be approximately 4% of different elements. Thus, in practice where, for example, $p > 50$ and we intend to use computers, it is really of advantage to use only those elements which are necessary for the calculations. Furthermore, the situation becomes much more drastic for higher order moments. The following presentation is based on papers by Kollo & von Rosen

(1995b, 1995c). We shall define minimal moments which collect expectations from all different monomials, i.e. $X_{i_1j_1}X_{i_2j_2}\times\cdots\times X_{i_pj_p}$. In our notation we shall add a letter "m" (from minimal) to the usual notation of moments and cumulants. So $m\overline{m}_k[\mathbf{x}]$ denotes the minimal $k-$th order central moment of $\mathbf{x}$, and $mc_k[\mathbf{X}]$ is the notation for the $k-$th minimal cumulant of $\mathbf{X}$, for example.

Definition 2.1.10. *The $k-$th minimal moment of a random $p-$vector $\mathbf{x}$ is given by*

$$mm_k[\mathbf{x}] = \frac{1}{i^k}\frac{\widehat{d}^k}{\widehat{d\mathbf{t}}^k}\varphi_{\mathbf{x}}(\mathbf{t})\bigg|_{\mathbf{t}=\mathbf{0}}, \tag{2.1.44}$$

and the $k-$th minimal central moment by

$$m\overline{m}_k[\mathbf{x}] = mm_k[\mathbf{x}-E[\mathbf{x}]] = \frac{1}{i^k}\frac{\widehat{d}^k}{\widehat{d\mathbf{t}}^k}\varphi_{\mathbf{x}-E[\mathbf{x}]}(\mathbf{t})\bigg|_{\mathbf{t}=\mathbf{0}}, \tag{2.1.45}$$

where $\varphi_{\mathbf{x}(\mathbf{t})}(\mathbf{t})$ is defined by (2.1.6) and the derivative is defined by (1.4.52) and (1.4.53). ∎

As the characteristic function of $\mathbf{X} : p\times q$ is the same as that of the $pq-$vector $\mathbf{x} = \text{vec}\mathbf{X}$, we can also write out formulas for moments of a random matrix:

$$mm_k[\mathbf{X}] = \frac{1}{i^k}\frac{\widehat{d}^k}{\widehat{d\mathbf{T}}^k}\varphi_{\mathbf{X}}(\mathbf{T})\bigg|_{\mathbf{T}=\mathbf{0}} \tag{2.1.46}$$

and

$$m\overline{m}_k[\mathbf{X}] = m\overline{m}_k[\mathbf{X}-E[\mathbf{X}]] = \frac{1}{i^k}\frac{\widehat{d}^k}{\widehat{d\mathbf{T}}^k}\varphi_{\mathbf{X}-E[\mathbf{X}]}(\mathbf{T})\bigg|_{\mathbf{T}=\mathbf{0}}. \tag{2.1.47}$$

We define the minimal cumulants in the same way.

Definition 2.1.11. *The $k-$th minimal cumulant of a random vector $\mathbf{x}$ is defined by*

$$mc_k[\mathbf{x}] = \frac{1}{i^k}\frac{\widehat{d}^k}{\widehat{d\mathbf{t}}^k}\psi_{\mathbf{x}}(\mathbf{t})\bigg|_{\mathbf{t}=\mathbf{0}}, \tag{2.1.48}$$

where the cumulant function $\psi_{\mathbf{x}}(\mathbf{t})$ is defined by (2.1.11) and the derivative by (1.4.52) and (1.4.53). ∎

The $k-$th minimal cumulant of $\mathbf{X} : p\times q$ is given by the equality

$$mc_k[\mathbf{X}] = \frac{1}{i^k}\frac{\widehat{d}^k}{\widehat{d\mathbf{T}}^k}\psi_{\mathbf{X}}(\mathbf{T})\bigg|_{\mathbf{T}=\mathbf{0}}, \tag{2.1.49}$$

where $\psi_{\mathbf{X}}(\mathbf{T})$ is defined by (2.1.12). Minimal moments can be expressed through the product vectorization operator $R^j(\mathbf{x})$.

Theorem 2.1.9. *For a random vector* $\mathbf{x}$

(i)
$$mm_k[\mathbf{x}] = E[R^k(\mathbf{x})];$$

(ii)
$$m\overline{m}_k[\mathbf{x}] = E[R^k(\mathbf{x} - E[\mathbf{x}])],$$

where $R^k(\mathbf{x})$ *is given by Definition 1.3.10.*

PROOF: In the proof we are going to explore the similarity in the structure of the k−th matrix derivative $\frac{\widehat{d}^k\mathbf{Y}}{\widehat{d}\mathbf{X}^k}$ and the vectorization operator $R^k(\bullet)$. Note that it is sufficient to prove statement (i). Statement (ii) follows immediately from (i) because, according to Definition 2.1.10, the central moment $m\overline{m}_k[\mathbf{x}]$ is the moment of the centered random vector $\mathbf{x} - E[\mathbf{x}]$. According to our definition of minimal derivatives, the considered derivative consists of all different partial derivatives of order k and each derivative appears only once. The derivatives are ordered according to the rule defined by the composition operator $Q^k(\bullet)$, given in (1.4.56). The $R^k(\bullet)$ operator organizes elements in the same way as the minimal differentiation operator because it is also a realization of the $Q^k(\bullet)$ operator. Therefore (i) holds. ∎

In Theorem 2.1.9 we considered minimal moments and cumulants of a random vector, but these notions are far more important for random matrices. As the characteristic function and the cumulant function of a random matrix $\mathbf{X}$ are defined via vectorized form vec$\mathbf{X}$ of the matrix, the results can be extended straightforwardly to random matrices and patterned matrices. Let us present the statement of Theorem 2.1.9 for matrices as a corollary.

Corollary 2.1.9.1. *For a random matrix* $\mathbf{X}$

(i)
$$mm_k[\mathbf{X}] = E[R^k(\mathbf{X})];$$

(ii)
$$m\overline{m}_k[\mathbf{X}] = E[R^k(\mathbf{X} - E[\mathbf{X}])],$$

where $R^k(\mathbf{X})$ *is given by Definition 1.3.10.* ∎

In the following we establish the connection between $mm_k[\mathbf{X}]$ and $E[\mathbf{X}^{\otimes k}]$ through formulas which show how $mm_k[\mathbf{X}]$ can be obtained from $E[\mathbf{X}^{\otimes k}]$ and vice versa. Of course, instead of $E[\mathbf{X}^{\otimes k}]$, we could also consider $E[\text{vec}\mathbf{X}(\text{vec}'\mathbf{X})^{\otimes k-1}]$.

Theorem 2.1.10. *Let*

$$\mathbf{T}(i_1, i_2, \ldots, i_j) = \mathbf{e}_{v_j}(\mathbf{e}'_{i_1} \otimes \mathbf{e}'_{i_2} \otimes \cdots \otimes \mathbf{e}'_{i_j}),$$

where

$$v_j = 1 + \sum_{r=2}^{j+1}(r, i_{r-1} - 1)$$

and (k, l) *is defined by (1.3.73). Then*

$$E[R^j(\mathbf{A})] = \sum_{I_j} \mathbf{T}(i_1, i_2, \ldots, i_j)\text{vec}(E[\mathbf{A}^{\otimes j}])$$

and

$$\mathrm{vec}(E[\mathbf{A}^{\otimes j}]) = \sum_{\tilde{I}_j} \mathbf{T}(i_1, i_2, \ldots, i_j)' E[R^j(\mathbf{A})],$$

where

$$I_j = \{(i_1, i_2, \ldots, i_j) : i_g = 1, 2, \ldots, n; 1 \leq u_{k-1} \leq (k, i_k), \quad k = 1, 2, \ldots, j\},$$

with u_s given in Theorem 1.3.15 and

$$\tilde{I}_j = \{(i_1, i_2, \ldots, i_j) : i_g = 1, 2, \ldots, n\}.$$

∎

Note that $\sum_{\tilde{I}_j} \mathbf{T}(i_1, i_2, \ldots, i_j)'$ is in fact a $g-$inverse of $\sum_{I_j} \mathbf{T}(i_1, i_2, \ldots, i_j)$.

2.1.7 *Relations between moments and cumulants*

We are going to present some general moment relations for random matrices. As we saw above, moments are obtained by differentiating the characteristic function, and cumulants are obtained by differentiating the cumulant function. Since the cumulant function is the logarithm of the characteristic function, there must be a straight connection between moments and cumulants. In the next theorem we write down the relations between moments and cumulants of low order. General relations of moments and cumulants have been given by Holmquist (1985a).

Theorem 2.1.11. *Let $\mathbf{x}$ be a random $p-$vector. Then*

(i) $$c_1[\mathbf{x}] = m_1[\mathbf{x}] = E[\mathbf{x}];$$

(ii) $$c_2[\mathbf{x}] = \overline{m}_2[\mathbf{x}] = D[\mathbf{x}]; \tag{2.1.50}$$

(iii) $$c_3[\mathbf{x}] = \overline{m}_3[\mathbf{x}], \tag{2.1.51}$$

where

$$\overline{m}_3[\mathbf{x}] = m_3[\mathbf{x}] - m_2[\mathbf{x}] \otimes E[\mathbf{x}]' - E[\mathbf{x}]' \otimes m_2[\mathbf{x}]$$
$$- E[\mathbf{x}]\mathrm{vec}' m_2[\mathbf{x}] + 2E[\mathbf{x}]E[\mathbf{x}]'^{\otimes 2}; \tag{2.1.52}$$

(iv) $$c_4[\mathbf{x}] = \overline{m}_4[\mathbf{x}] - \overline{m}_2[\mathbf{x}] \otimes \mathrm{vec}'\overline{m}_2[\mathbf{x}]$$
$$- (\mathrm{vec}'\overline{m}_2[\mathbf{x}] \otimes \overline{m}_2[\mathbf{x}])(\mathbf{I} + \mathbf{I}_p \otimes \mathbf{K}_{p,p}). \tag{2.1.53}$$

PROOF: Since $\psi_{\mathbf{x}}(\mathbf{t}) = \ln\varphi_{\mathbf{x}}(\mathbf{t})$ and $\varphi_{\mathbf{x}}(\mathbf{0}) = 1$, we obtain by Definition 2.1.1 and Definition 2.1.6 that

$$i\,c_1[\mathbf{x}] = \left.\frac{d\psi_{\mathbf{x}}(\mathbf{t})}{d\mathbf{t}}\right|_{\mathbf{t}=\mathbf{0}} = \left.\frac{d\varphi_{\mathbf{x}}(\mathbf{t})}{d\mathbf{t}}\frac{1}{\varphi_{\mathbf{x}}(\mathbf{t})}\right|_{\mathbf{t}=\mathbf{0}} = \left.\frac{d\varphi_{\mathbf{x}}(\mathbf{t})}{d\mathbf{t}}\right|_{\mathbf{t}=\mathbf{0}} = i\,m_1[\mathbf{x}]$$

and thus (i) is verified.

For (ii) and (iii) we copy the ideas of proving (i), but now we have to use additionally some algebra presented in Section 1.3 as well as properties of the matrix derivatives from Section 1.4. By the following calculations we get

$$
\begin{aligned}
-c_2[\mathbf{x}] &= \frac{d^2\psi_{\mathbf{x}}(\mathbf{t})}{d\mathbf{t}^2}\bigg|_{\mathbf{t}=\mathbf{0}} = \frac{d}{d\mathbf{t}}\{\frac{d\varphi_{\mathbf{x}}(\mathbf{t})}{d\mathbf{t}}\frac{1}{\varphi_{\mathbf{x}}(\mathbf{t})}\}\bigg|_{\mathbf{t}=\mathbf{0}} \\
&= \frac{d^2\varphi_{\mathbf{x}}(\mathbf{t})}{d\mathbf{t}^2}\frac{1}{\varphi_{\mathbf{x}}(\mathbf{t})} - \frac{d\varphi_{\mathbf{x}}(\mathbf{t})}{d\mathbf{t}}\frac{1}{\varphi_{\mathbf{x}}(\mathbf{t})^2}\left(\frac{d\varphi_{\mathbf{x}}(\mathbf{t})}{d\mathbf{t}}\right)'\bigg|_{\mathbf{t}=\mathbf{0}} \\
&= -\,m_2[\mathbf{x}] + m_1[\mathbf{x}]m_1[\mathbf{x}]' = -\overline{m}_2[\mathbf{x}]
\end{aligned}
$$

and hence (ii) is established. Moreover,

$$
\begin{aligned}
-i\,c_3[\mathbf{x}] &= \frac{d}{d\mathbf{t}}\{\frac{d^2\varphi_{\mathbf{x}}(\mathbf{t})}{d\mathbf{t}^2}\frac{1}{\varphi_{\mathbf{x}}(\mathbf{t})} - \frac{d\varphi_{\mathbf{x}}(\mathbf{t})}{d\mathbf{t}}\frac{1}{\varphi_{\mathbf{x}}(\mathbf{t})^2}\left(\frac{d\varphi_{\mathbf{x}}(\mathbf{t})}{d\mathbf{t}}\right)'\}\bigg|_{\mathbf{t}=\mathbf{0}} \\
&= \frac{d^3\varphi_{\mathbf{x}}(\mathbf{t})}{d\mathbf{t}^3}\frac{1}{\varphi_{\mathbf{x}}(\mathbf{t})} - \frac{d\varphi_{\mathbf{x}}(\mathbf{t})}{d\mathbf{t}}\frac{1}{\varphi_{\mathbf{x}}(\mathbf{t})^2}\mathrm{vec}'(\frac{d^2\varphi_{\mathbf{x}}(\mathbf{t})}{d\mathbf{t}^2}) \\
&\quad - \frac{1}{\varphi_{\mathbf{x}}(\mathbf{t})^2}\{(\frac{d\varphi_{\mathbf{x}}(\mathbf{t})}{d\mathbf{t}})'\otimes\frac{d^2\varphi_{\mathbf{x}}(\mathbf{t})}{d\mathbf{t}^2} + \frac{d^2\varphi_{\mathbf{x}}(\mathbf{t})}{d\mathbf{t}^2}\otimes(\frac{d\varphi_{\mathbf{x}}(\mathbf{t})}{d\mathbf{t}})'\} \\
&\quad + \frac{2}{\varphi_{\mathbf{x}}(\mathbf{t})^3}\frac{d\varphi_{\mathbf{x}}(\mathbf{t})}{d\mathbf{t}}((\frac{d\varphi_{\mathbf{x}}(\mathbf{t})}{d\mathbf{t}})^{\otimes 2})'\bigg|_{\mathbf{t}=\mathbf{0}} \\
&= -\,i\,m_3[\mathbf{x}] + i\,m_1[\mathbf{x}]\mathrm{vec}'(m_2[\mathbf{x}]) + i\,m_1[\mathbf{x}]'\otimes m_2[\mathbf{x}] + i\,m_2[\mathbf{x}]\otimes m_1[\mathbf{x}]' \\
&\quad - 2i\,m_1[\mathbf{x}](m_1[\mathbf{x}]')^{\otimes 2}.
\end{aligned}
$$

The sum on the right hand side equals $-i\overline{m}_3[\mathbf{x}]$. This follows immediately from (2.1.18) which gives the relation between the characteristic functions $\varphi_{\mathbf{x}}(\mathbf{t})$ and $\varphi_{\mathbf{x}-E[\mathbf{x}]}(\mathbf{t})$, i.e.

$$
\varphi_{\mathbf{x}}(\mathbf{t}) = \varphi_{\mathbf{x}-E[\mathbf{x}]}(\mathbf{t})\exp(i\mathbf{t}'E[\mathbf{x}]). \tag{2.1.54}
$$

Thus,

$$
\psi_{\mathbf{x}}(\mathbf{t}) = \psi_{\mathbf{x}-E[\mathbf{x}]}(\mathbf{t}) + i\,\mathbf{t}'E[\mathbf{x}].
$$

From this we can draw the important conclusion that starting from $k = 2$ the expressions presenting cumulants via moments also give the relations between cumulants and central moments. The last ones have a simpler form because the expectation of the centered random vector $\mathbf{x} - E[\mathbf{x}]$ equals zero. Hence, the expression of $-i\,c_3[\mathbf{x}]$ gives us directly (2.1.51), and the formula (2.1.52) has been proved at the same time.

It remains to show (2.1.53). Differentiating once more gives us

$$\begin{aligned}
&\frac{d}{d\mathbf{t}}\left\{\frac{d^3\varphi_{\mathbf{x}}(\mathbf{t})}{d\mathbf{t}^3}\frac{1}{\varphi_{\mathbf{x}}(\mathbf{t})} - \frac{d\varphi_{\mathbf{x}}(\mathbf{t})}{d\mathbf{t}}\frac{1}{\varphi_{\mathbf{x}}(\mathbf{t})^2}\mathrm{vec}'(\frac{d^2\varphi_{\mathbf{x}}(\mathbf{t})}{d\mathbf{t}^2})\right.\\
&\quad - \frac{1}{\varphi_{\mathbf{x}}(\mathbf{t})^2}((\frac{d\varphi_{\mathbf{x}}(\mathbf{t})}{d\mathbf{t}})'\otimes\frac{d^2\varphi_{\mathbf{x}}(\mathbf{t})}{d\mathbf{t}^2} + \frac{d^2\varphi_{\mathbf{x}}(\mathbf{t})}{d\mathbf{t}^2}\otimes(\frac{d\varphi_{\mathbf{x}}(\mathbf{t})}{d\mathbf{t}})')\\
&\quad \left.+ \frac{2}{\varphi_{\mathbf{x}}(\mathbf{t})^3}\frac{d\varphi_{\mathbf{x}}(\mathbf{t})}{d\mathbf{t}}((\frac{d\varphi_{\mathbf{x}}(\mathbf{t})}{d\mathbf{t}})^{\otimes 2})'\right\}\Bigg|_{\mathbf{t}=\mathbf{0}}\\
&= \left\{\frac{d^4\varphi_{\mathbf{x}}(\mathbf{t})}{d\mathbf{t}^4}\frac{1}{\varphi_{\mathbf{x}}(\mathbf{t})} - \frac{1}{\varphi_{\mathbf{x}}(\mathbf{t})^2}(\mathrm{vec}'\frac{d^2\varphi_{\mathbf{x}}(\mathbf{t})}{d\mathbf{t}^2}\otimes\frac{d^2\varphi_{\mathbf{x}}(\mathbf{t})}{d\mathbf{t}^2})\right.\\
&\quad - \frac{1}{\varphi_{\mathbf{x}}(\mathbf{t})^2}(\frac{d^2\varphi_{\mathbf{x}}(\mathbf{t})}{d\mathbf{t}^2}\otimes\mathrm{vec}'\frac{d^2\varphi_{\mathbf{x}}(\mathbf{t})}{d\mathbf{t}^2})\\
&\quad \left.- \frac{1}{\varphi_{\mathbf{x}}(\mathbf{t})^2}(\mathrm{vec}'\frac{d^2\varphi_{\mathbf{x}}(\mathbf{t})}{d\mathbf{t}^2}\otimes\frac{d^2\varphi_{\mathbf{x}}(\mathbf{t})}{d\mathbf{t}^2})(\mathbf{I}_p\otimes\mathbf{K}_{p,p})\right\} + \mathbf{R}(\mathbf{t})\Bigg|_{\mathbf{t}=\mathbf{0}},
\end{aligned}$$

where $\mathbf{R}(\mathbf{t})$ is the expression including functions of $\frac{d\varphi_{\mathbf{x}}(\mathbf{t})}{d\mathbf{t}}$, such that $\mathbf{R}(\mathbf{0}) = \mathbf{0}$ when $\mathbf{x}$ is a centered vector. Thus, when evaluating this expression at $\mathbf{t} = \mathbf{0}$ yields statement (iv) of the theorem. ∎

Similar statements to those in Theorem 2.1.11 will be given in the next corollary for random matrices.

Corollary 2.1.11.1. *Let $\mathbf{X} : p \times q$ be a random matrix. Then*

(i) $c_1[\mathbf{X}] = m_1[\mathbf{X}] = E[\mathrm{vec}\mathbf{X}]$;

(ii) $c_2[\mathbf{X}] = \overline{m}_2[\mathbf{X}] = D[\mathbf{X}]$; (2.1.55)

(iii) $c_3[\mathbf{X}] = \overline{m}_3[\mathbf{X}]$, (2.1.56)

where

$$\overline{m}_3[\mathbf{X}] = m_3[\mathbf{X}] - m_2[\mathbf{X}]\otimes m_1[\mathbf{X}]' - m_1[\mathbf{X}]'\otimes m_2[\mathbf{X}] - m_1[\mathbf{X}]\mathrm{vec}'m_2[\mathbf{X}] + 2m_1[\mathbf{X}]m_1[\mathbf{X}]'^{\otimes 2}; \quad (2.1.57)$$

(iv) $c_4[\mathbf{X}] = \overline{m}_4[\mathbf{X}] - \overline{m}_2[\mathbf{X}]\otimes\mathrm{vec}'\overline{m}_2[\mathbf{X}] - (\mathrm{vec}'\overline{m}_2[\mathbf{X}]\otimes\overline{m}_2[\mathbf{X}])(\mathbf{I} + \mathbf{I}_{pq}\otimes\mathbf{K}_{pq,pq})$, (2.1.58)

where the moments $m_k[\mathbf{X}]$ are given by (2.1.25) and the central moments $\overline{m}_k[\mathbf{X}]$ by (2.1.26) ∎

2.1.8 *Problems*

1. Show that $c_k[\mathbf{AXB}] = (\mathbf{B}' \otimes \mathbf{A})c_k[\mathbf{X}](\mathbf{B} \otimes \mathbf{A}')^{\otimes(k-1)}$ for a random matrix $\mathbf{X}$.
2. Find the expression of $\overline{m}_4[\mathbf{x}]$ through moments of $\mathbf{x}$.
3. Suppose that we know $C[\mathbf{X}_{i_1 j_1}, \mathbf{X}_{i_2 j_2}]$, $i_1, j_1, i_2, j_2 = 1, 2$. Express $D[\mathbf{X}]$ with the help of these quantities.
4. In Definition 2.1.1 use the matrix derivative given by (1.4.47) instead of $\dfrac{d\text{vec}'\mathbf{X}}{d\text{vec}\mathbf{T}}$. Express $c_1[\mathbf{X}]$ and $c_2[\mathbf{X}]$ as functions of $m_1[\mathbf{X}]$ and $m_2[\mathbf{X}]$.
5. Let $\mathbf{x}_1, \mathbf{x}_2, \ldots, \mathbf{x}_n$ be independent, with $E[\mathbf{x}_i] = \boldsymbol{\mu}$ and $D[\mathbf{x}_i] = \boldsymbol{\Sigma}$, and let

$$\mathbf{S} = \frac{1}{n-1}\sum_{i=1}^{n}(\mathbf{x}_i - \overline{\mathbf{x}})(\mathbf{x}_i - \overline{\mathbf{x}})'.$$

   Determine $E[\mathbf{S}]$ and $D[\mathbf{S}]$.
6. Let $\mathbf{x}_1, \mathbf{x}_2, \ldots, \mathbf{x}_n$ be independent with $E[\mathbf{x}_i] = \boldsymbol{\mu}$ and $D[\mathbf{x}_i] = \boldsymbol{\Sigma}$. Put $\mathbf{X} = (\mathbf{x}_1, \mathbf{x}_2, \ldots, \mathbf{x}_n)$ and let $\mathbf{S} = \mathbf{X}(\mathbf{I} - \mathbf{C}'(\mathbf{CC}')^-\mathbf{C})\mathbf{X}'$, where $\mathbf{C} : k \times n$, $k < n$. Determine $E[\mathbf{S}]$ and $D[\mathbf{S}]$.
7. Consider a 20−dimensional random vector. Find the number of non-repeated mixed moments of $6th$ order.
8. Let $\mathbf{z} = (Z_1, Z_2, Z_3)'$ be a random vector. Write out $mm_k(\mathbf{z})$ using Theorem 2.1.9.
9. The asymmetric Laplace distribution is defined by the characteristic function (Kotz, Kozubowski & Podgórski, 2001):

$$\varphi(\mathbf{t}) = \frac{1}{1 + \mathbf{t}'\boldsymbol{\mu} + \frac{1}{2}\mathbf{t}'\boldsymbol{\Sigma}\mathbf{t}},$$

   where $\boldsymbol{\mu} : p \times 1$ and $\boldsymbol{\Sigma} > 0 : p \times p$ are parameters. Find the mean and dispersion matrix of the distribution.
10. Find the dispersion matrix of a mixture distribution with density

$$f_{\mathbf{z}}(\mathbf{x}) = \gamma f_{N_p(\mathbf{0},\boldsymbol{\Sigma}_1)}(\mathbf{x}) + (1-\gamma) f_{N_p(\mathbf{0},\boldsymbol{\Sigma}_2)}(\mathbf{x}),$$

   where $\gamma \in (0, 1)$. Here $N_p(\mathbf{0}, \boldsymbol{\Sigma})$ denotes the multivariate normal distribution which is defined in the next section.

2.2 THE NORMAL DISTRIBUTION

2.2.1 *Introduction and definition*

One of the most fundamental distributions in statistics is the normal distribution. The univariate normal distribution has been known and used in statistics for about two hundred years. The multivariate normal distribution has also been in use for a long time. Usually, when it is referred to a multivariate normal distribution, it is a distribution of a vector. However, among specialists in multivariate analysis, a matrix version has been put forward which comprises the vector valued version. This must be quite obvious, since any matrix $\mathbf{A}$ may be considered in the vector form $\text{vec}\mathbf{A}$. The way of ordering the elements can have no effect on the distribution. It will be seen below that the results for the vector valued distribution can be generalized to the matrix case in a nice and constructive manner. On one hand, the matrix version is a "bilinear" extension of the vector version, and the multivariate structure is obtained from the covariance structure which will be presented as a Kronecker product of two dispersion matrices. However, on the other hand, the matrix normal distribution can always be obtained from the multivariate normal distribution by choosing a particular covariance structure.

Over the years many ways of defining the normal distribution have been presented. There are at least three different approaches for introducing the multivariate normal distribution. One is to utilize the density, provided that the density exists, another is via the characteristic function, and the third is by applying some characterization of the distribution. Our approach will rest on a characterization which stresses the connection between the normal distribution and linear (multilinear) transformations. Other characterizations can also be used.

To start the whole process of defining a matrix normal distribution, we begin with the definition of the univariate standard normal distribution which is defined via its density

$$f_U(u) = (2\pi)^{-1/2}e^{-\frac{1}{2}u^2}, \qquad -\infty < u < \infty \tag{2.2.1}$$

and denoted $U \sim N(0,1)$. It follows that $E[U] = 0$ and $D[U] = 1$. To define a univariate normal distribution with mean μ and variance $\sigma^2 > 0$ we observe that any variable X which has the same distribution as

$$\mu + \sigma U, \qquad \sigma > 0, \quad -\infty < \mu < \infty, \tag{2.2.2}$$

where the density of U is defined by (2.2.1), has a density

$$f_X(x) = (2\pi\sigma^2)^{-1/2}e^{-\frac{1}{2}\frac{(x-\mu)^2}{\sigma^2}}, \qquad -\infty < \mu, x < \infty, \quad \sigma > 0.$$

We say that $X \sim N(\mu, \sigma^2)$, and it is clear that we can use (2.2.2) as a definition for the normal distribution. One advantage of using (2.2.2) is illustrated in the following text. Consider kX, where k is a constant. From (2.2.2) it follows that kX has the same distribution as

$$k\mu + k\sigma U$$

and thus $kX \sim N(k\mu, (k\sigma)^2)$. Furthermore, (2.2.2) also holds in the singular case when $\sigma = 0$, because a random variable with zero variance equals its mean. From (2.2.2) it follows that X has the same distribution as μ and we can write $X = \mu$. Now, let $\mathbf{u} = (U_1, \ldots, U_p)'$ be a vector which consists of p independent identically distributed (i.i.d.) $N(0,1)$ elements. Due to independence, it follows from (2.2.1) that the density of $\mathbf{u}$ equals

$$f_{\mathbf{u}}(\mathbf{u}) = (2\pi)^{-\frac{1}{2}p} e^{-\frac{1}{2}\mathrm{tr}(\mathbf{u}\mathbf{u}')}, \tag{2.2.3}$$

and we say that $\mathbf{u} \sim N_p(\mathbf{0}, \mathbf{I})$. Note that $\mathrm{tr}(\mathbf{u}\mathbf{u}') = \mathbf{u}'\mathbf{u}$. The density in (2.2.3) serves as a definition of the standard multivariate normal density function. To obtain a general definition of a normal distribution for a vector valued variable we follow the scheme of the univariate case. Thus, let $\mathbf{x}$ be a p-dimensional vector with mean $E[\mathbf{x}] = \boldsymbol{\mu}$ and dispersion $D[\mathbf{x}] = \boldsymbol{\Sigma}$, where $\boldsymbol{\Sigma}$ is non-negative definite. From Theorem 1.1.3 it follows that $\boldsymbol{\Sigma} = \boldsymbol{\tau}\boldsymbol{\tau}'$ for some matrix $\boldsymbol{\tau}$, and if $\boldsymbol{\Sigma} > \mathbf{0}$, we can always choose $\boldsymbol{\tau}$ to be of full rank, $r(\boldsymbol{\tau}) = p$. Therefore, $\mathbf{x}$ is multivariate normally distributed, if $\mathbf{x}$ has the same distribution as

$$\boldsymbol{\mu} + \boldsymbol{\tau}\mathbf{u}, \tag{2.2.4}$$

where $\mathbf{u} \sim N_p(\mathbf{0}, \mathbf{I})$ and the distribution of $\mathbf{x}$ is denoted $\mathbf{x} \sim N_p(\boldsymbol{\mu}, \boldsymbol{\Sigma})$. If $\boldsymbol{\Sigma} > \mathbf{0}$, it follows from (2.2.4) and Theorem 1.4.14 by a substitution of variables that the density equals

$$f_{\mathbf{x}}(\mathbf{x}) = (2\pi)^{-\frac{1}{2}p} |\boldsymbol{\Sigma}|^{-1/2} e^{-\frac{1}{2}\mathrm{tr}\{\boldsymbol{\Sigma}^{-1}(\mathbf{x}-\boldsymbol{\mu})(\mathbf{x}-\boldsymbol{\mu})'\}}. \tag{2.2.5}$$

Now we turn to the main definition of this section, which introduces the matrix normal distribution.

Definition 2.2.1. *Let $\boldsymbol{\Sigma} = \boldsymbol{\tau}\boldsymbol{\tau}'$ and $\boldsymbol{\Psi} = \boldsymbol{\gamma}\boldsymbol{\gamma}'$, where $\boldsymbol{\tau} : p \times r$ and $\boldsymbol{\gamma} : n \times s$. A matrix $\mathbf{X} : p \times n$ is said to be matrix normally distributed with parameters $\boldsymbol{\mu}$, $\boldsymbol{\Sigma}$ and $\boldsymbol{\Psi}$, if it has the same distribution as*

$$\boldsymbol{\mu} + \boldsymbol{\tau}\mathbf{U}\boldsymbol{\gamma}', \tag{2.2.6}$$

where $\boldsymbol{\mu} : p \times n$ is non-random and $\mathbf{U} : r \times s$ consists of s i.i.d. $N_r(\mathbf{0}, \mathbf{I})$ vectors $\mathbf{U}_i$, $i = 1, 2, \ldots, s$. If $\mathbf{X} : p \times n$ is matrix normally distributed, this will be denoted $\mathbf{X} \sim N_{p,n}(\boldsymbol{\mu}, \boldsymbol{\Sigma}, \boldsymbol{\Psi})$. ∎

If $\boldsymbol{\Sigma}$ and $\boldsymbol{\Psi}$ are positive definite, then $\boldsymbol{\tau}$ and $\boldsymbol{\gamma}$ in (2.2.6) are both square and non-singular. In the subsequent we exclude the trivial cases $\boldsymbol{\Sigma} = \mathbf{0}$ $\boldsymbol{\Psi} = \mathbf{0}$.
Since vec$\mathbf{X}$ and $\mathbf{X}$ have the same distribution, it follows, by applying the vec-operator to (2.2.6), that $\mathbf{X}$ has the same distribution as

$$\mathrm{vec}\boldsymbol{\mu} + (\boldsymbol{\gamma} \otimes \boldsymbol{\tau})\mathrm{vec}\mathbf{U}.$$

Thus, from (2.2.4) it follows that $\mathbf{X} \sim N_{p,n}(\boldsymbol{\mu}, \boldsymbol{\Sigma}, \boldsymbol{\Psi})$ means the same as $\mathrm{vec}\mathbf{X} \sim N_{pn}(\mathrm{vec}\boldsymbol{\mu}, \boldsymbol{\Psi} \otimes \boldsymbol{\Sigma})$. Furthermore, since the expectation of $\mathbf{U}$ in (2.2.6) equals zero,

$E[\mathbf{X}] = \boldsymbol{\mu}$, and since by definition of the dispersion matrix $D[\mathbf{X}] = D[\text{vec}\mathbf{X}]$, we obtain that $D[\mathbf{X}] = \boldsymbol{\Psi} \otimes \boldsymbol{\Sigma}$.
For the interpretation of $D[\mathbf{X}] = \boldsymbol{\Psi} \otimes \boldsymbol{\Sigma}$ we note that $\boldsymbol{\Psi}$ describes the covariance between the columns of $\mathbf{X}$. These covariances will be the same for each row. On the other hand, $\boldsymbol{\Sigma}$ describes the covariances between the rows of $\mathbf{X}$ which will be the same for each column. Now, take into consideration the covariances between columns as well as the covariances between rows. Then $\boldsymbol{\Psi} \otimes \boldsymbol{\Sigma}$ tells us that the overall covariance consists of the products of these covariances, i.e.

$$\text{Cov}[x_{ij}, x_{kl}] = \sigma_{ik}\psi_{jl},$$

where $\mathbf{X} = (x_{ij})$, $\boldsymbol{\Sigma} = (\sigma_{ik})$ and $\boldsymbol{\Psi} = (\psi_{jl})$. Furthermore, let $\boldsymbol{\mu}_i$ denote the ith column of $\boldsymbol{\mu}$. Then, if $\boldsymbol{\Psi} = \mathbf{I}_n$, the columns of $\mathbf{X}$ are independently $N_p(\boldsymbol{\mu}_i, \boldsymbol{\Sigma})$ distributed. Moreover, if in addition $\boldsymbol{\Sigma} = \mathbf{I}_p$, then all elements of $\mathbf{X}$ are mutually independently distributed. If $\boldsymbol{\Psi} \otimes \boldsymbol{\Sigma}$ is positive definite, the density of $N_{p,n}(\boldsymbol{\mu}, \boldsymbol{\Sigma}, \boldsymbol{\Psi})$ is given by

$$f_{\mathbf{X}}(\mathbf{X}) = (2\pi)^{-\frac{1}{2}pn}|\boldsymbol{\Sigma}|^{-n/2}|\boldsymbol{\Psi}|^{-p/2}e^{-\frac{1}{2}\text{tr}\{\boldsymbol{\Sigma}^{-1}(\mathbf{X}-\boldsymbol{\mu})\boldsymbol{\Psi}^{-1}(\mathbf{X}-\boldsymbol{\mu})'\}}, \tag{2.2.7}$$

which can be obtained from (2.2.5) by using $\text{vec}\mathbf{X} \sim N_{pn}(\text{vec}\boldsymbol{\mu}, \boldsymbol{\Psi} \otimes \boldsymbol{\Sigma})$ and noting that

$$\text{vec}'\mathbf{X}(\boldsymbol{\Psi} \otimes \boldsymbol{\Sigma})^{-1}\text{vec}\mathbf{X} = \text{tr}(\boldsymbol{\Sigma}^{-1}\mathbf{X}\boldsymbol{\Psi}^{-1}\mathbf{X}')$$

and

$$|\boldsymbol{\Psi} \otimes \boldsymbol{\Sigma}| = |\boldsymbol{\Psi}|^p|\boldsymbol{\Sigma}|^n.$$

The first of these two equalities is obtained via Proposition 1.3.14 (iii) and the second is valid due to Proposition 1.3.12 (ix).

2.2.2 *Some properties of the matrix normal distribution*

In this paragraph some basic facts for matrix normally distributed matrices will be presented. From now on, in this paragraph, when partitioned matrices will be considered, the following notation and sizes of matrices will be used:

$$\mathbf{X} = \begin{pmatrix} \mathbf{X}_{11} & \mathbf{X}_{12} \\ \mathbf{X}_{21} & \mathbf{X}_{22} \end{pmatrix} \quad \boldsymbol{\mu} = \begin{pmatrix} \boldsymbol{\mu}_{11} & \boldsymbol{\mu}_{12} \\ \boldsymbol{\mu}_{21} & \boldsymbol{\mu}_{22} \end{pmatrix} \quad \begin{pmatrix} r \times s & r \times (n-s) \\ (p-r) \times s & (p-r) \times (n-s) \end{pmatrix}, \tag{2.2.8}$$

$$\mathbf{X}_{\bullet 1} = \begin{pmatrix} \mathbf{X}_{11} \\ \mathbf{X}_{21} \end{pmatrix} \quad \mathbf{X}_{\bullet 2} = \begin{pmatrix} \mathbf{X}_{12} \\ \mathbf{X}_{22} \end{pmatrix} \quad \mathbf{X}_{1\bullet} = (\mathbf{X}_{11} : \mathbf{X}_{12}) \quad \mathbf{X}_{2\bullet} = (\mathbf{X}_{21} : \mathbf{X}_{22}), \tag{2.2.9}$$

$$\boldsymbol{\mu}_{\bullet 1} = \begin{pmatrix} \boldsymbol{\mu}_{11} \\ \boldsymbol{\mu}_{21} \end{pmatrix} \quad \boldsymbol{\mu}_{\bullet 2} = \begin{pmatrix} \boldsymbol{\mu}_{12} \\ \boldsymbol{\mu}_{22} \end{pmatrix} \quad \boldsymbol{\mu}_{1\bullet} = (\boldsymbol{\mu}_{11} : \boldsymbol{\mu}_{12}) \quad \boldsymbol{\mu}_{2\bullet} = (\boldsymbol{\mu}_{21} : \boldsymbol{\mu}_{22}), \tag{2.2.10}$$

$$\boldsymbol{\Sigma} = \begin{pmatrix} \boldsymbol{\Sigma}_{11} & \boldsymbol{\Sigma}_{12} \\ \boldsymbol{\Sigma}_{21} & \boldsymbol{\Sigma}_{22} \end{pmatrix} \quad \begin{pmatrix} r \times r & r \times (p-r) \\ (p-r) \times r & (p-r) \times (p-r) \end{pmatrix}, \tag{2.2.11}$$

$$\boldsymbol{\Psi} = \begin{pmatrix} \boldsymbol{\Psi}_{11} & \boldsymbol{\Psi}_{12} \\ \boldsymbol{\Psi}_{21} & \boldsymbol{\Psi}_{22} \end{pmatrix} \quad \begin{pmatrix} s \times s & s \times (n-s) \\ (n-s) \times s & (n-s) \times (n-s) \end{pmatrix}. \tag{2.2.12}$$

The characteristic function and cumulant function for a matrix normal variable are given in the following theorem. As noted before, the characteristic function is fundamental in our approach when moments are derived. In the next paragraph the characteristic function will be utilized. The characteristic function could also have been used as a definition of the normal distribution. The function exists for both singular and non-singular covariance matrix.

Theorem 2.2.1. *Let* $\mathbf{X} \sim N_{p,n}(\boldsymbol{\mu}, \boldsymbol{\Sigma}, \boldsymbol{\Psi})$. *Then*

(i) *the characteristic function* $\varphi_{\mathbf{X}}(\mathbf{T})$ *is given by*

$$\varphi_{\mathbf{X}}(\mathbf{T}) = e^{i\mathrm{tr}(\mathbf{T}'\boldsymbol{\mu}) - \frac{1}{2}\mathrm{tr}(\boldsymbol{\Sigma}\mathbf{T}\boldsymbol{\Psi}\mathbf{T}')};$$

(ii) *the cumulant function* $\psi_{\mathbf{X}}(\mathbf{T})$ *is given by*

$$\psi_{\mathbf{X}}(\mathbf{T}) = i\mathrm{tr}(\mathbf{T}'\boldsymbol{\mu}) - \frac{1}{2}\mathrm{tr}(\boldsymbol{\Sigma}\mathbf{T}\boldsymbol{\Psi}\mathbf{T}').$$

PROOF: Let us again start with the univariate case. The characteristic function of $U \sim N(0,1)$ equals

$$\varphi_U(t) = e^{-\frac{1}{2}t^2}.$$

Therefore, because of independence of the elements of $\mathbf{U}$, given in Definition 2.2.1, the characteristic function of $\mathbf{U}$ equals (apply Proposition 1.1.4 (viii))

$$\varphi_{\mathbf{U}}(\mathbf{T}) = e^{-\frac{1}{2}\sum_{i,j} t_{ij}^2} = e^{-\frac{1}{2}\mathrm{tr}(\mathbf{T}\mathbf{T}')}. \tag{2.2.13}$$

By Definition 2.2.1, $\mathbf{X}$ has the same distribution as $\boldsymbol{\mu} + \boldsymbol{\tau}\mathbf{U}\boldsymbol{\gamma}'$ and thus

$$\begin{aligned}\varphi_{\mathbf{X}}(\mathbf{T}) &= E[e^{i\mathrm{tr}\{\mathbf{T}'(\boldsymbol{\mu}+\boldsymbol{\tau}\mathbf{U}\boldsymbol{\gamma}')\}}] = e^{i\mathrm{tr}(\mathbf{T}'\boldsymbol{\mu})}E[e^{i\mathrm{tr}(\boldsymbol{\gamma}'\mathbf{T}'\boldsymbol{\tau}\mathbf{U})}] \\ &\underset{(2.2.13)}{=} e^{i\mathrm{tr}(\mathbf{T}'\boldsymbol{\mu}) - \frac{1}{2}\mathrm{tr}(\boldsymbol{\tau}'\mathbf{T}\boldsymbol{\gamma}\boldsymbol{\gamma}'\mathbf{T}'\boldsymbol{\tau})} = e^{i\mathrm{tr}(\mathbf{T}'\boldsymbol{\mu}) - \frac{1}{2}\mathrm{tr}(\boldsymbol{\Sigma}\mathbf{T}\boldsymbol{\Psi}\mathbf{T}')}.\end{aligned}$$

By definition of the cumulant function given in (2.1.12), statement (ii) follows if we take the logarithm of the characteristic function in (i). ∎

The next theorem states the well-known property that normality is preserved under linear transformations. However, it also tells us that under bilinear transformations the property of being matrix normal is kept.

Theorem 2.2.2. *Let* $\mathbf{X} \sim N_{p,n}(\boldsymbol{\mu}, \boldsymbol{\Sigma}, \boldsymbol{\Psi})$. *For any* $\mathbf{A} : q \times p$ *and* $\mathbf{B} : m \times n$

$$\mathbf{A}\mathbf{X}\mathbf{B}' \sim N_{q,m}(\mathbf{A}\boldsymbol{\mu}\mathbf{B}', \mathbf{A}\boldsymbol{\Sigma}\mathbf{A}', \mathbf{B}\boldsymbol{\Psi}\mathbf{B}').$$

PROOF: From Definition 2.2.1 it follows that the matrix $\mathbf{X}$ has the same distribution as $\boldsymbol{\mu} + \boldsymbol{\tau}\mathbf{U}\boldsymbol{\gamma}'$, where $\boldsymbol{\Sigma} = \boldsymbol{\tau}\boldsymbol{\tau}'$ and $\boldsymbol{\Psi} = \boldsymbol{\gamma}\boldsymbol{\gamma}'$, and thus $\mathbf{A}\mathbf{X}\mathbf{B}'$ has the same distribution as $\mathbf{A}\boldsymbol{\mu}\mathbf{B}' + \mathbf{A}\boldsymbol{\tau}\mathbf{U}\boldsymbol{\gamma}'\mathbf{B}'$. Since $\mathbf{A}\boldsymbol{\tau}\boldsymbol{\tau}'\mathbf{A}' = \mathbf{A}\boldsymbol{\Sigma}\mathbf{A}'$ and $\mathbf{B}\boldsymbol{\gamma}\boldsymbol{\gamma}'\mathbf{B}' = \mathbf{B}\boldsymbol{\Psi}\mathbf{B}'$, the theorem is verified. ∎

Marginal distributions of a multivariate normal distribution are also normal. Here we present the results in terms of matrix normal variables.

Corollary 2.2.2.1. *Let* $\mathbf{X} \sim N_{p,n}(\boldsymbol{\mu}, \boldsymbol{\Sigma}, \boldsymbol{\Psi})$ *and* $\mathbf{X}$, $\boldsymbol{\mu}$, $\boldsymbol{\Sigma}$ *and* $\boldsymbol{\Psi}$ *are partitioned according to (2.2.8) – (2.2.12). Then*

(i) $\mathbf{X}_{11} \sim N_{r,s}(\boldsymbol{\mu}_{11}, \boldsymbol{\Sigma}_{11}, \boldsymbol{\Psi}_{11}), \qquad \mathbf{X}_{12} \sim N_{r,n-s}(\boldsymbol{\mu}_{12}, \boldsymbol{\Sigma}_{11}, \boldsymbol{\Psi}_{22});$

(ii) $\mathbf{X}_{21} \sim N_{p-r,s}(\boldsymbol{\mu}_{21}, \boldsymbol{\Sigma}_{22}, \boldsymbol{\Psi}_{11}), \qquad \mathbf{X}_{22} \sim N_{p-r,n-s}(\boldsymbol{\mu}_{22}, \boldsymbol{\Sigma}_{22}, \boldsymbol{\Psi}_{22});$

(iii) $\mathbf{X}_{\bullet 1} \sim N_{p,s}(\boldsymbol{\mu}_{\bullet 1}, \boldsymbol{\Sigma}, \boldsymbol{\Psi}_{11}), \qquad \mathbf{X}_{\bullet 2} \sim N_{p,n-s}(\boldsymbol{\mu}_{\bullet 2}, \boldsymbol{\Sigma}, \boldsymbol{\Psi}_{22});$
$\mathbf{X}_{1\bullet} \sim N_{r,n}(\boldsymbol{\mu}_{1\bullet}, \boldsymbol{\Sigma}_{11}, \boldsymbol{\Psi}), \qquad \mathbf{X}_{2\bullet} \sim N_{p-r,n}(\boldsymbol{\mu}_{2\bullet}, \boldsymbol{\Sigma}_{22}, \boldsymbol{\Psi}).$

PROOF: In order to obtain the distribution for $\mathbf{X}_{11}$, we choose the matrices $\mathbf{A} = (\mathbf{I}_r : \mathbf{0})$ and $\mathbf{B} = (\mathbf{I}_s : \mathbf{0})$ in Theorem 2.2.2. Other expressions can be verified in the same manner. ∎

Corollary 2.2.2.2. *Let* $\mathbf{X} \sim N_{p,n}(\mathbf{0}, \boldsymbol{\Sigma}, \mathbf{I})$ *and let* $\boldsymbol{\Gamma} : n \times n$ *be an orthogonal matrix which is independent of* $\mathbf{X}$. *Then* $\mathbf{X}$ *and* $\mathbf{X}\boldsymbol{\Gamma}$ *have the same distribution.* ∎

Theorem 2.2.3.

(i) *Let* $\mathbf{X}_j \sim N_{p_j,n}(\boldsymbol{\mu}_j, \boldsymbol{\Sigma}_j, \boldsymbol{\Psi})$, $j = 1, 2, \ldots, k$ *be mutually independently distributed. Let* $\mathbf{A}_j$, $j = 1, 2, \ldots, k$ *be of size* $q \times p_j$ *and* $\mathbf{B}$ *of size* $m \times n$. *Then*
$$\sum_{j=1}^{k} \mathbf{A}_j\mathbf{X}_j\mathbf{B}' \sim N_{q,m}(\sum_{j=1}^{k} \mathbf{A}_j\boldsymbol{\mu}_j\mathbf{B}', \sum_{j=1}^{k} \mathbf{A}_j\boldsymbol{\Sigma}_j\mathbf{A}_j', \mathbf{B}\boldsymbol{\Psi}\mathbf{B}').$$

(ii) *Let* $\mathbf{X}_j \sim N_{p,n_j}(\boldsymbol{\mu}_j, \boldsymbol{\Sigma}, \boldsymbol{\Psi}_j)$, $j = 1, 2, \ldots, k$ *be mutually independently distributed. Let* $\mathbf{B}_j$, $j = 1, 2, \ldots, k$ *be of size* $m \times n_j$ *and* $\mathbf{A}$ *of size* $q \times p$. *Then*
$$\sum_{j=1}^{k} \mathbf{A}\mathbf{X}_j\mathbf{B}_j' \sim N_{q,m}(\sum_{j=1}^{k} \mathbf{A}\boldsymbol{\mu}_j\mathbf{B}_j', \mathbf{A}\boldsymbol{\Sigma}\mathbf{A}', \sum_{j=1}^{k} \mathbf{B}_j\boldsymbol{\Psi}_j\mathbf{B}_j').$$

PROOF: We will just prove (i), because the result in (ii) follows by duality, i.e. if $\mathbf{X} \sim N_{p,n}(\boldsymbol{\mu}, \boldsymbol{\Sigma}, \boldsymbol{\Psi})$ then $\mathbf{X}' \sim N_{n,p}(\boldsymbol{\mu}', \boldsymbol{\Psi}, \boldsymbol{\Sigma})$. Let $\boldsymbol{\Sigma}_j = \boldsymbol{\tau}_j\boldsymbol{\tau}_j'$, $\boldsymbol{\tau}_j : p_j \times r$, and $\boldsymbol{\Psi} = \boldsymbol{\gamma}\boldsymbol{\gamma}'$, $\boldsymbol{\gamma} : n \times s$. Put $\mathbf{A} = (\mathbf{A}_1, \ldots, \mathbf{A}_k)$, $\mathbf{X} = (\mathbf{X}_1', \ldots, \mathbf{X}_k')'$, $\boldsymbol{\mu} = (\boldsymbol{\mu}_1', \ldots, \boldsymbol{\mu}_k')'$, $\mathbf{U} = (\mathbf{U}_1', \ldots, \mathbf{U}_k')'$, $\mathbf{U}_j : p_j \times s$. Since $\mathbf{X}_j$ has the same distribution as $\boldsymbol{\mu}_j + \boldsymbol{\tau}_j\mathbf{U}_j\boldsymbol{\gamma}'$, it follows, because of independence, that $\mathbf{X}$ has the same distribution as
$$\boldsymbol{\mu} + (\boldsymbol{\tau}_1, \ldots, \boldsymbol{\tau}_k)_{[d]}\mathbf{U}\boldsymbol{\gamma}'$$
and thus $\mathbf{A}\mathbf{X}\mathbf{B}'$ has the same distribution as
$$\mathbf{A}\boldsymbol{\mu}\mathbf{B}' + \mathbf{A}(\boldsymbol{\tau}_1, \ldots, \boldsymbol{\tau}_k)_{[d]}\mathbf{U}\boldsymbol{\gamma}'\mathbf{B}'.$$
Since
$$\mathbf{A}(\boldsymbol{\tau}_1, \ldots, \boldsymbol{\tau}_k)_{[d]}(\boldsymbol{\tau}_1, \ldots, \boldsymbol{\tau}_k)_{[d]}'\mathbf{A}' = \sum_{j=1}^{k} \mathbf{A}_j\boldsymbol{\Sigma}_j\mathbf{A}_j'$$

and $\mathbf{B}\gamma\gamma'\mathbf{B}' = \mathbf{B}\mathbf{\Psi}\mathbf{B}'$, (i) is verified. ∎

If $\mathbf{X}_1 \sim N_{p,n}(\boldsymbol{\mu}_1, \mathbf{\Sigma}_1, \mathbf{\Psi}_1)$ and $\mathbf{X}_2 \sim N_{p,n}(\boldsymbol{\mu}_2, \mathbf{\Sigma}_2, \mathbf{\Psi}_2)$, it is interesting to note that $\mathbf{X}_1 + \mathbf{X}_2$ is normally distributed but not matrix normally distributed. The reason for this is that in general there are no matrices $\mathbf{K}$ and $\mathbf{L}$ such that $(\mathbf{\Psi}_1 \otimes \mathbf{\Sigma}_1 + \mathbf{\Psi}_2 \otimes \mathbf{\Sigma}_2) = \mathbf{K} \otimes \mathbf{L}$ unless some restriction is put on $\mathbf{\Psi}_j, \mathbf{\Sigma}_j$, $j = 1, 2$. Therefore, one cannot combine (i) and (ii) in Theorem 2.2.3.

One of the most important properties of the normal distribution is that its dispersion matrix is connected to independence as well as to conditional independence. This fact is exploited in the next two theorems and in particular in Corollary 2.2.4.2. Furthermore, linear combinations of a normal vector $\mathbf{x}$, say $a'\mathbf{x}$ and $\mathbf{b}'\mathbf{x}$, are independent if and only if $\mathbf{a}$ and $\mathbf{b}$ are orthogonal. This is exploited in a more general case in the next theorem. Among others, this property is essential from a geometric point of view.

Theorem 2.2.4. *Let* $\mathbf{X} \sim N_{p,n}(\boldsymbol{\mu}, \mathbf{\Sigma}, \mathbf{\Psi})$, $\mathbf{Y} \sim N_{p,n}(\mathbf{0}, \mathbf{\Sigma}, \mathbf{\Psi})$ *and* $\mathbf{A}$, $\mathbf{B}$, $\mathbf{C}$, $\mathbf{D}$, $\mathbf{K}$ *and* $\mathbf{L}$ *are non-random matrices of proper sizes. Then*

(i) $\mathbf{AXK}$ *is independent of* $\mathbf{CXL}$ *for all constant matrices* $\mathbf{K}$ *and* $\mathbf{L}$ *if and only if* $\mathbf{A\Sigma C}' = \mathbf{0}$*;*

(ii) $\mathbf{KXB}'$ *is independent of* $\mathbf{LXD}'$ *for all constant matrices* $\mathbf{K}$ *and* $\mathbf{L}$ *if and only if* $\mathbf{B\Psi D}' = \mathbf{0}$*;*

(iii) $\mathbf{YAY}'$ *is independent of* $\mathbf{YBY}'$ *if and only if*

$$\mathbf{\Psi A \Psi B}'\mathbf{\Psi} = \mathbf{0}, \quad \mathbf{\Psi A}'\mathbf{\Psi B \Psi} = \mathbf{0},$$
$$\mathbf{\Psi A \Psi B \Psi} = \mathbf{0}, \quad \mathbf{\Psi A}'\mathbf{\Psi B}'\mathbf{\Psi} = \mathbf{0};$$

(iv) $\mathbf{YAY}'$ *is independent of* $\mathbf{YB}$ *if and only if*

$$\mathbf{B}'\mathbf{\Psi A}'\mathbf{\Psi} = \mathbf{0},$$
$$\mathbf{B}'\mathbf{\Psi A \Psi} = \mathbf{0}.$$

PROOF: We just prove (i), (iii) and (iv), since the proof of the second statement is identical to the proof of (i). Independence in (i) implies that $C[\mathbf{AXK}, \mathbf{CXL}] = \mathbf{0}$, which in turn is equivalent to

$$(\mathbf{K}' \otimes \mathbf{A})(\mathbf{\Psi} \otimes \mathbf{\Sigma})(\mathbf{L} \otimes \mathbf{C}') = \mathbf{0}.$$

Hence,

$$(\mathbf{K}'\mathbf{\Psi L}) \otimes (\mathbf{A\Sigma C}') = \mathbf{0}$$

and from Proposition 1.3.12 (xi) this holds if and only if $\mathbf{K}'\mathbf{\Psi L} = \mathbf{0}$ or $\mathbf{A\Sigma C}' = \mathbf{0}$. However, $\mathbf{K}$ and $\mathbf{L}$ are arbitrary and therefore $\mathbf{A\Sigma C}' = \mathbf{0}$ must hold.

For the converse it is noted that

$$\begin{pmatrix} \mathbf{A} \\ \mathbf{C} \end{pmatrix} \mathbf{X}(\mathbf{K} : \mathbf{L}) \sim N_{\bullet,\bullet}\{\begin{pmatrix} \mathbf{A} \\ \mathbf{C} \end{pmatrix} \boldsymbol{\mu}(\mathbf{K} : \mathbf{L}), \begin{pmatrix} \mathbf{A} \\ \mathbf{C} \end{pmatrix} \mathbf{\Sigma}(\mathbf{A}' : \mathbf{C}'), \begin{pmatrix} \mathbf{K}' \\ \mathbf{L}' \end{pmatrix} \mathbf{\Psi}(\mathbf{K} : \mathbf{L})\}.$$

However, by assumption,

$$\begin{pmatrix}\mathbf{A}\\\mathbf{C}\end{pmatrix}\boldsymbol{\Sigma}(\mathbf{A}':\mathbf{C}')=\begin{pmatrix}\mathbf{A}\boldsymbol{\Sigma}\mathbf{A}' & \mathbf{0}\\ \mathbf{0} & \mathbf{C}\boldsymbol{\Sigma}\mathbf{C}'\end{pmatrix}=\begin{pmatrix}\mathbf{A}\boldsymbol{\tau} & \mathbf{0}\\ \mathbf{0} & \mathbf{C}\boldsymbol{\tau}\end{pmatrix}\begin{pmatrix}\boldsymbol{\tau}'\mathbf{A}' & \mathbf{0}\\ \mathbf{0} & \boldsymbol{\tau}'\mathbf{C}'\end{pmatrix},$$

where $\boldsymbol{\Sigma}=\boldsymbol{\tau}\boldsymbol{\tau}'$. Thus $(\mathbf{A}':\mathbf{C}')'\mathbf{X}(\mathbf{K}:\mathbf{L})$ has the same distribution as

$$\begin{pmatrix}\mathbf{A}\\\mathbf{C}\end{pmatrix}\boldsymbol{\mu}(\mathbf{K}:\mathbf{L})+\begin{pmatrix}\mathbf{A}\boldsymbol{\tau} & \mathbf{0}\\ \mathbf{0} & \mathbf{C}\boldsymbol{\tau}\end{pmatrix}\mathbf{U}\boldsymbol{\gamma}'(\mathbf{K}:\mathbf{L}), \tag{2.2.14}$$

where $\boldsymbol{\Psi}=\boldsymbol{\gamma}\boldsymbol{\gamma}'$ and $\mathbf{U}\sim N_{\bullet,\bullet}(\mathbf{0},\mathbf{I},\mathbf{I})$. Let $\mathbf{U}=(\mathbf{U}_1':\mathbf{U}_2')'$ and the partition corresponds to the partition of the covariance matrix $\boldsymbol{\Sigma}$. From (2.2.14) it follows that $\mathbf{AX}(\mathbf{K}:\mathbf{L})$ and $\mathbf{CX}(\mathbf{K}:\mathbf{L})$ have the same distributions as

$$\mathbf{A}\boldsymbol{\mu}(\mathbf{K}:\mathbf{L})+\mathbf{A}\boldsymbol{\tau}\mathbf{U}_1\boldsymbol{\gamma}'(\mathbf{K}:\mathbf{L})$$

and

$$\mathbf{C}\boldsymbol{\mu}(\mathbf{K}:\mathbf{L})+\mathbf{C}\boldsymbol{\tau}\mathbf{U}_2\boldsymbol{\gamma}'(\mathbf{K}:\mathbf{L}),$$

respectively. Since $\mathbf{U}_1$ and $\mathbf{U}_2$ are independently distributed, $\mathbf{AX}(\mathbf{K}:\mathbf{L})$ and $\mathbf{CX}(\mathbf{K}:\mathbf{L})$ must also be independent. Hence, $\mathbf{AXK}$ and $\mathbf{CXL}$ are independent and (i) is established.

The proof of (iii) is based on Corollary 2.2.7.4 (iii). If $\mathbf{YAY}'$ is independent of $\mathbf{YBY}'$, it is also independent of $\mathbf{YB}'\mathbf{Y}'$. Furthermore, $\mathbf{YA}'\mathbf{Y}'$ is independent of $\mathbf{YBY}'$ and $\mathbf{YB}'\mathbf{Y}'$. We are going to show how independence between $\mathbf{YAY}'$ and $\mathbf{YB}'\mathbf{Y}'$ implies that $\boldsymbol{\Psi}\mathbf{A}\boldsymbol{\Psi}\mathbf{B}\boldsymbol{\Psi}=\mathbf{0}$ as well as the other relations of the statement. From Corollary 2.2.7.4 it follows that

$$\begin{aligned}
&E[\mathbf{Y}^{\otimes 8}](\mathrm{vec}\mathbf{A})^{\otimes 2}\otimes(\mathrm{vec}\mathbf{B})^{\otimes 2}=E[(\mathrm{vec}\mathbf{YAY}')^{\otimes 2}\otimes(\mathrm{vec}\mathbf{YBY}')^{\otimes 2}]\\
&=\sum_{i_1=2}^{8}\sum_{i_2=2}^{6}\sum_{i_3=2}^{4}(\mathbf{I}_p\otimes\mathbf{K}_{p^{i_1-2},p}\otimes\mathbf{I}_{p^{8-i_1}})(\mathbf{I}_{p^3}\otimes\mathbf{K}_{p^{i_2-2},p}\otimes\mathbf{I}_{p^{6-i_2}})\\
&\quad\times(\mathbf{I}_{p^5}\otimes\mathbf{K}_{p^{i_3-2},p}\otimes\mathbf{I}_{p^{4-i_3}})(\mathrm{vec}\boldsymbol{\Sigma})^{\otimes 4}(\mathrm{vec}'\boldsymbol{\Psi})^{\otimes 4}(\mathbf{I}_{n^5}\otimes\mathbf{K}_{n,n^{i_3-2}}\otimes\mathbf{I}_{n^{4-i_3}})\\
&\quad\times(\mathbf{I}_{n^3}\otimes\mathbf{K}_{n,n^{i_2-2}}\otimes\mathbf{I}_{n^{6-i_2}})(\mathbf{I}_n\otimes\mathbf{K}_{n,n^{i_1-2}}\otimes\mathbf{I}_{n^{8-i_1}})(\mathrm{vec}\mathbf{A})^{\otimes 2}\otimes(\mathrm{vec}\mathbf{B})^{\otimes 2}.
\end{aligned} \tag{2.2.15}$$

Moreover, due to independence we should have

$$E[(\mathrm{vec}(\mathbf{YAY}'))^{\otimes 2}\otimes(\mathrm{vec}(\mathbf{YBY}'))^{\otimes 2}]=E[(\mathrm{vec}(\mathbf{YAY}'))^{\otimes 2}]\otimes E[(\mathrm{vec}(\mathbf{YBY}'))^{\otimes 2}].$$

Since $\boldsymbol{\Sigma}$ is an arbitrary matrix, (2.2.15) and Corollary 2.2.7.4 (ii) imply that

$$\begin{aligned}
&(\mathbf{I}_p\otimes\mathbf{K}_{p^{i_1-2},p}\otimes\mathbf{I}_{p^{8-i_1}})(\mathbf{I}_{p^3}\otimes\mathbf{K}_{p^{i_2-2},p}\otimes\mathbf{I}_{p^{6-i_2}})(\mathbf{I}_{p^5}\otimes\mathbf{K}_{p^{i_3-2},p}\otimes\mathbf{I}_{p^{4-i_3}})\\
&\quad=(\mathbf{I}_p\otimes\mathbf{K}_{p^{i_1-2},p}\otimes\mathbf{I}_{p^{4-i_1}}\otimes\mathbf{I}_{p^4})(\mathbf{I}_{p^4}\otimes\mathbf{I}_p\otimes\mathbf{K}_{p^{i_3-2},p}\otimes\mathbf{I}_{p^{4-i_3}})
\end{aligned}$$

must be satisfied and this relation holds if $i_2=2$ and $i_1\leq 4$. Thus, if $i_1>4$ or $i_2\neq 2$, we have

$$\begin{aligned}
&(\mathrm{vec}'\boldsymbol{\Psi})^{\otimes 4}(\mathbf{I}_{n^5}\otimes\mathbf{K}_{n,n^{i_3-2}}\otimes\mathbf{I}_{n^{4-i_3}})(\mathbf{I}_{n^3}\otimes\mathbf{K}_{n,n^{i_2-2}}\otimes\mathbf{I}_{n^{6-i_2}})\\
&\qquad\times(\mathbf{I}_n\otimes\mathbf{K}_{n,n^{i_1-2}}\otimes\mathbf{I}_{n^{8-i_1}})(\mathrm{vec}\mathbf{A})^{\otimes 2}\otimes(\mathrm{vec}\mathbf{B})^{\otimes 2}=0.
\end{aligned}$$

In particular if $i_3=2$, $i_2=3$ and $i_1=7$

$$(\mathrm{vec}'\boldsymbol{\Psi})^{\otimes 4}(\mathbf{I}_{n^3}\otimes\mathbf{K}_{n,n}\otimes\mathbf{I}_{n^3})(\mathbf{I}_n\otimes\mathbf{K}_{n,n^5}\otimes\mathbf{I}_n)(\mathrm{vec}\mathbf{A})^{\otimes 2}\otimes(\mathrm{vec}\mathbf{B})^{\otimes 2}$$
$$=\mathrm{tr}(\mathbf{A}\boldsymbol{\Psi}\mathbf{B}\boldsymbol{\Psi}\mathbf{B}'\boldsymbol{\Psi}\mathbf{A}'\boldsymbol{\Psi})=0.$$

The equality $\mathrm{tr}(\mathbf{A}\boldsymbol{\Psi}\mathbf{B}\boldsymbol{\Psi}\mathbf{B}'\boldsymbol{\Psi}\mathbf{A}'\boldsymbol{\Psi})=0$ is equivalent to $\boldsymbol{\Psi}\mathbf{A}\boldsymbol{\Psi}\mathbf{B}\boldsymbol{\Psi}=\mathbf{0}$. Thus, one of the conditions of (iii) has been obtained. Other conditions follow by considering $E[\mathbf{Y}^{\otimes 8}]\mathrm{vec}\mathbf{A}\otimes\mathrm{vec}\mathbf{A}'\otimes(\mathrm{vec}\mathbf{B})^{\otimes 2}$, $E[\mathbf{Y}^{\otimes 8}](\mathrm{vec}\mathbf{A})^{\otimes 2}\otimes\mathrm{vec}\mathbf{B}\otimes\mathrm{vec}\mathbf{B}'$ and $E[\mathbf{Y}^{\otimes 8}]\mathrm{vec}\mathbf{A}\otimes\mathrm{vec}\mathbf{A}'\otimes\mathrm{vec}\mathbf{B}\otimes\mathrm{vec}\mathbf{B}'$.
On the other hand, $\mathbf{YAY}'=\mathbf{YA}\boldsymbol{\Psi}\boldsymbol{\Psi}^{-}\mathbf{Y}'=\mathbf{Y}\boldsymbol{\Psi}^{-}\boldsymbol{\Psi}\mathbf{AY}'$, and if the conditions in (iii) hold, according to (ii), $\mathbf{YA}\boldsymbol{\Psi}$ is independent of $\mathbf{YB}\boldsymbol{\Psi}$ and $\mathbf{YB}'\boldsymbol{\Psi}$, and $\mathbf{YA}'\boldsymbol{\Psi}$ is independent of $\mathbf{YB}\boldsymbol{\Psi}$ and $\mathbf{YB}'\boldsymbol{\Psi}$.
To prove (iv) we can use ideas similar to those in the proof of (iii). Suppose first that $\mathbf{YAY}'$ is independent of $\mathbf{YB}$. Thus, $\mathbf{YA}'\mathbf{Y}'$ is also independent of $\mathbf{YB}$. From Corollary 2.2.7.4 it follows that

$$E[\mathbf{Y}^{\otimes 6}](\mathrm{vec}\mathbf{A})^{\otimes 2}\otimes\mathbf{B}^{\otimes 2}=E[(\mathrm{vec}\mathbf{YAY}')^{\otimes 2}\otimes(\mathbf{YB})^{\otimes 2}]$$
$$=\sum_{i_1=2}^{6}\sum_{i_2=2}^{4}(\mathbf{I}_p\otimes\mathbf{K}_{p^{i_1-2},p}\otimes\mathbf{I}_{p^{6-i_1}})(\mathbf{I}_{p^3}\otimes\mathbf{K}_{p^{i_2-2},p}\otimes\mathbf{I}_{p^{4-i_2}})(\mathrm{vec}\boldsymbol{\Sigma})^{\otimes 3}(\mathrm{vec}'\boldsymbol{\Psi})^{\otimes 3}$$
$$\times(\mathbf{I}_{n^3}\otimes\mathbf{K}_{n,n^{i_2-2}}\otimes\mathbf{I}_{n^{4-i_2}})(\mathbf{I}_n\otimes\mathbf{K}_{n,n^{i_1-2}}\otimes\mathbf{I}_{n^{6-i_1}})(\mathrm{vec}\mathbf{A})^{\otimes 2}\otimes\mathbf{B}^{\otimes 2}.$$

Independence implies that

$$(\mathbf{I}_p\otimes\mathbf{K}_{p^{i_1-2},p}\otimes\mathbf{I}_{p^{6-i_1}})(\mathbf{I}_{p^3}\otimes\mathbf{K}_{p^{i_2-2},p}\otimes\mathbf{I}_{p^{4-i_2}})=(\mathbf{I}_p\otimes\mathbf{K}_{p^{i_1-2},p}\otimes\mathbf{I}_{p^{4-i_1}}\otimes\mathbf{I}_{p^2})$$

must hold, which in turn gives that if $i_1>4$,

$$(\mathrm{vec}'\boldsymbol{\Psi})^{\otimes 3}(\mathbf{I}_p\otimes\mathbf{K}_{p^{i_1-2},p}\otimes\mathbf{I}_{p^{4-i_1}}\otimes\mathbf{I}_{p^2})(\mathbf{I}_{p^4}\otimes\mathbf{I}_p\otimes\mathbf{K}_{p^{i_2-2},p}\otimes\mathbf{I}_{p^{4-i_2}})$$
$$\times(\mathrm{vec}\mathbf{A})^{\otimes 2}\otimes\mathbf{B}^{\otimes 2}=\mathbf{0}.$$

In particular, if $i_1=6$ and $i_2=3$, we obtain that

$$\mathrm{vec}(\mathbf{B}'\boldsymbol{\Psi}\mathbf{A}'\boldsymbol{\Psi}\mathbf{A}\boldsymbol{\Psi}\mathbf{B})=\mathbf{0},$$

which is equivalent to

$$\mathbf{B}'\boldsymbol{\Psi}\mathbf{A}'\boldsymbol{\Psi}=\mathbf{0}.$$

By symmetry it follows that $\mathbf{B}'\boldsymbol{\Psi}\mathbf{A}\boldsymbol{\Psi}=\mathbf{0}$ also is true. For the converse we rely on (ii) and notice that $\mathbf{B}'\boldsymbol{\Psi}\mathbf{A}'\boldsymbol{\Psi}=\mathbf{0}$ implies that $\mathbf{YB}$ and $\mathbf{YA}'\boldsymbol{\Psi}$ are independent, and $\mathbf{B}'\boldsymbol{\Psi}\mathbf{A}\boldsymbol{\Psi}=\mathbf{0}$ implies that $\mathbf{YB}$ and $\mathbf{YA}\boldsymbol{\Psi}$ are independent. Furthermore, note that $\mathbf{YAY}'=\mathbf{Y}\boldsymbol{\Psi}^{-}\boldsymbol{\Psi}\mathbf{A}\boldsymbol{\Psi}\boldsymbol{\Psi}^{-}\mathbf{Y}'=\mathbf{YA}\boldsymbol{\Psi}\boldsymbol{\Psi}^{-}\mathbf{Y}'=\mathbf{Y}\boldsymbol{\Psi}^{-}\boldsymbol{\Psi}\mathbf{AY}'$. Hence, $\mathbf{YB}$ and $\mathbf{YAY}'$ are also independent. ∎

One may notice that, in particular, the proofs of (iii) and (iv) utilize the first moments and not the characteristic function, which is usually the case when showing independence.
The proof of the theorem has induced the following corollaries.

Corollary 2.2.4.1. *Let* $\mathbf{X} \sim N_{p,n}(\boldsymbol{\mu}, \boldsymbol{\Sigma}, \boldsymbol{\Psi})$ *and* $\mathbf{X}$, $\boldsymbol{\mu}$, $\boldsymbol{\Sigma}$ *and* $\boldsymbol{\Psi}$ *be partitioned according to (2.2.8) – (2.2.12). Then*

(i) *if* $\boldsymbol{\Psi}_{22}^{-1}$ *is non-singular,* $\mathbf{X}_{\bullet 1} - \mathbf{X}_{\bullet 2}\boldsymbol{\Psi}_{22}^{-1}\boldsymbol{\Psi}_{21}$ *and* $\mathbf{X}_{\bullet 2}$ *are independent and normally distributed;*

(ii) *if* $\boldsymbol{\Sigma}_{22}^{-1}$ *is non-singular,* $\mathbf{X}_{1\bullet} - \boldsymbol{\Sigma}_{12}\boldsymbol{\Sigma}_{22}^{-1}\mathbf{X}_{2\bullet}$ *and* $\mathbf{X}_{2\bullet}$ *are independent and normally distributed.* ∎

Corollary 2.2.4.2. *Let* $\mathbf{X} \sim N_{p,n}(\boldsymbol{\mu}, \boldsymbol{\Sigma}, \boldsymbol{\Psi})$ *and* $\mathbf{X}$, $\boldsymbol{\mu}$, $\boldsymbol{\Sigma}$ *and* $\boldsymbol{\Psi}$ *be partitioned according to (2.2.8) – (2.2.12). Then*

(i) $\mathbf{X}_{1\bullet}$ *and* $\mathbf{X}_{2\bullet}$ *are independent if and only if* $\boldsymbol{\Sigma}_{12} = \mathbf{0}$*;*

(ii) $\mathbf{X}_{\bullet 1}$ *and* $\mathbf{X}_{\bullet 2}$ *are independent if and only if* $\boldsymbol{\Psi}_{12} = \mathbf{0}$*;*

(iii) $\mathbf{X}_{11}$ *and* $\mathbf{X}_{12}$, $\mathbf{X}_{22}$, $\mathbf{X}_{12}$ *are independent if and only if* $\boldsymbol{\Sigma}_{12} = \mathbf{0}$ *and* $\boldsymbol{\Psi}_{12} = \mathbf{0}$.

PROOF: For (i) choose $\mathbf{A} = (\mathbf{I} : \mathbf{0})$ and $\mathbf{C} = (\mathbf{0} : \mathbf{I})$ in Theorem 2.2.4 (i). For (ii) choose $\mathbf{B} = (\mathbf{I} : \mathbf{0})$ and $\mathbf{D} = (\mathbf{0} : \mathbf{I})$ in Theorem 2.2.4 (ii). For (iii) we combine (i) and (ii) of Theorem 2.2.4. In (i) we choose $\mathbf{A} = (\mathbf{I} : \mathbf{0})$, $\mathbf{K} = (\mathbf{I} : \mathbf{0})'$ and $\mathbf{C} = (\mathbf{0} : \mathbf{I})$ and in (ii) we choose $\mathbf{B} = (\mathbf{I} : \mathbf{0})$, $\mathbf{K} = (\mathbf{I} : \mathbf{0})$ and $\mathbf{D} = (\mathbf{0} : \mathbf{I})$. ∎

Now some results on conditioning in the matrix normal distribution are presented.

Theorem 2.2.5. *Let* $\mathbf{X} \sim N_{p,n}(\boldsymbol{\mu}, \boldsymbol{\Sigma}, \boldsymbol{\Psi})$ *and* $\mathbf{X}$, $\boldsymbol{\mu}$, $\boldsymbol{\Sigma}$ *and* $\boldsymbol{\Psi}$ *be partitioned according to (2.2.8) – (2.2.12). Put* $\boldsymbol{\Sigma}_{1\cdot 2} = \boldsymbol{\Sigma}_{11} - \boldsymbol{\Sigma}_{12}\boldsymbol{\Sigma}_{22}^{-1}\boldsymbol{\Sigma}_{21}$ *and* $\boldsymbol{\Psi}_{1\cdot 2} = \boldsymbol{\Psi}_{11} - \boldsymbol{\Psi}_{12}\boldsymbol{\Psi}_{22}^{-1}\boldsymbol{\Psi}_{21}$.

(i) *Suppose that* $\boldsymbol{\Psi}_{22}^{-1}$ *exists. Then*

$$\mathbf{X}_{\bullet 1}|\mathbf{X}_{\bullet 2} \sim N_{p,s}(\boldsymbol{\mu}_{\bullet 1} + (\mathbf{X}_{\bullet 2} - \boldsymbol{\mu}_{\bullet 2})\boldsymbol{\Psi}_{22}^{-1}\boldsymbol{\Psi}_{21}, \boldsymbol{\Sigma}, \boldsymbol{\Psi}_{1\cdot 2}).$$

(ii) *Suppose that* $\boldsymbol{\Sigma}_{22}^{-1}$ *exists. Then*

$$\mathbf{X}_{1\bullet}|\mathbf{X}_{2\bullet} \sim N_{r,n}(\boldsymbol{\mu}_{1\bullet} + \boldsymbol{\Sigma}_{12}\boldsymbol{\Sigma}_{22}^{-1}(\mathbf{X}_{2\bullet} - \boldsymbol{\mu}_{2\bullet}), \boldsymbol{\Sigma}_{1\cdot 2}, \boldsymbol{\Psi}).$$

PROOF: Let

$$\mathbf{H} = \begin{pmatrix} \mathbf{I} & -\boldsymbol{\Psi}_{12}\boldsymbol{\Psi}_{22}^{-1} \\ \mathbf{0} & \mathbf{I} \end{pmatrix}.$$

From Theorem 2.2.2 it follows that $\mathbf{X}\mathbf{H}' \sim N_{p,n}(\boldsymbol{\mu}\mathbf{H}', \boldsymbol{\Sigma}, \mathbf{H}\boldsymbol{\Psi}\mathbf{H}')$ and

$$\begin{aligned} \mathbf{X}\mathbf{H}' &= (\mathbf{X}_{\bullet 1} - \mathbf{X}_{\bullet 2}\boldsymbol{\Psi}_{22}^{-1}\boldsymbol{\Psi}_{21} : \mathbf{X}_{\bullet 2}), \\ \boldsymbol{\mu}\mathbf{H}' &= (\boldsymbol{\mu}_{\bullet 1} - \boldsymbol{\mu}_{\bullet 2}\boldsymbol{\Psi}_{22}^{-1}\boldsymbol{\Psi}_{21} : \boldsymbol{\mu}_{\bullet 2}). \end{aligned}$$

However, since

$$\mathbf{H}\boldsymbol{\Psi}\mathbf{H}' = \begin{pmatrix} \boldsymbol{\Psi}_{1\cdot 2} & \mathbf{0} \\ \mathbf{0} & \boldsymbol{\Psi}_{22} \end{pmatrix},$$

Corollary 2.2.4.1 (i) implies that $\mathbf{X}_{\bullet 1} - \mathbf{X}_{\bullet 2}\boldsymbol{\Psi}_{22}^{-1}\boldsymbol{\Psi}_{21}$ and $\mathbf{X}_{\bullet 2}$ are independently distributed. Hence,

$$\mathbf{X}_{\bullet 1} - \mathbf{X}_{\bullet 2}\boldsymbol{\Psi}_{22}^{-1}\boldsymbol{\Psi}_{21} \sim N_{p,s}(\boldsymbol{\mu}_{\bullet 1} - \boldsymbol{\mu}_{\bullet 2}\boldsymbol{\Psi}_{22}^{-1}\boldsymbol{\Psi}_{21}, \boldsymbol{\Sigma}, \boldsymbol{\Psi}_{1\cdot 2}),$$

and because of independence with $\mathbf{X}_{\bullet 2}$, the same expression when conditioning with respect to $\mathbf{X}_{\bullet 2}$ holds. Thus (i) is established. The relation (ii) can immediately be obtained by noting that $\mathbf{X}' \sim N_{n,p}(\boldsymbol{\mu}', \boldsymbol{\Psi}, \boldsymbol{\Sigma})$. ∎

It is straightforward to derive the conditional distribution of $\mathbf{X}_{\bullet 1}|\mathbf{X}_{12}$ or $\mathbf{X}_{\bullet 1}|\mathbf{X}_{22}$, etc. by using results for the vector valued normal distributions. It turns out that the conditional distributions are normal, of course, but they are not matrix normal distributions. An extension of Theorem 2.2.5 is given in the following corollary.

Corollary 2.2.5.1. *Let* $\mathbf{X} \sim N_{p,n}(\boldsymbol{\mu}, \boldsymbol{\Sigma}, \boldsymbol{\Psi})$ *and* $\mathbf{A} : q \times p$, $\mathbf{B} : m \times n$, $\mathbf{C} : r \times p$ *and* $\mathbf{D} : k \times n$. *Moreover, let* $\boldsymbol{\Sigma}_{1\cdot 2\mathbf{A}} = \mathbf{A}\boldsymbol{\Sigma}\mathbf{A}' - \mathbf{A}\boldsymbol{\Sigma}\mathbf{C}'(\mathbf{C}\boldsymbol{\Sigma}\mathbf{C}')^{-1}\mathbf{C}\boldsymbol{\Sigma}\mathbf{A}'$ *and* $\boldsymbol{\Psi}_{1\cdot 2\mathbf{B}} = \mathbf{B}\boldsymbol{\Psi}\mathbf{B}' - \mathbf{B}\boldsymbol{\Psi}\mathbf{D}'(\mathbf{D}\boldsymbol{\Psi}\mathbf{D}')^{-1}\mathbf{D}\boldsymbol{\Psi}\mathbf{B}'$.

(i) *If* $(\mathbf{C}\boldsymbol{\Sigma}\mathbf{C}')^{-1}$ *exists, then*

$$\mathbf{A}\mathbf{X}\mathbf{B}'|\mathbf{C}\mathbf{X}\mathbf{B}' \sim N_{q,m}(\mathbf{A}\boldsymbol{\mu}\mathbf{B}' + \mathbf{A}\boldsymbol{\Sigma}\mathbf{C}'(\mathbf{C}\boldsymbol{\Sigma}\mathbf{C}')^{-1}(\mathbf{C}\mathbf{X}\mathbf{B}' - \mathbf{C}\boldsymbol{\mu}\mathbf{B}'), \boldsymbol{\Sigma}_{1\cdot 2\mathbf{A}}, \mathbf{B}\boldsymbol{\Psi}\mathbf{B}').$$

(ii) *If* $(\mathbf{D}\boldsymbol{\Psi}^{-1}\mathbf{D}')^{-1}$ *exists, then*

$$\mathbf{A}\mathbf{X}\mathbf{B}'|\mathbf{A}\mathbf{X}\mathbf{D}' \sim N_{q,m}(\mathbf{A}\boldsymbol{\mu}\mathbf{B}' + (\mathbf{A}\mathbf{X}\mathbf{D}' - \mathbf{A}\boldsymbol{\mu}\mathbf{D}')(\mathbf{D}\boldsymbol{\Psi}\mathbf{D}')^{-1}\mathbf{D}\boldsymbol{\Psi}\mathbf{B}', \mathbf{A}\boldsymbol{\Sigma}\mathbf{A}', \boldsymbol{\Psi}_{1\cdot 2\mathbf{B}}).$$

∎

The next theorem gives us another generalization of Theorem 2.2.5. Here we do not suppose any full rank conditions.

Theorem 2.2.6. *Let* $\mathbf{X} \sim N_{p,n}(\boldsymbol{\mu}, \boldsymbol{\Sigma}, \boldsymbol{\Psi})$ *and the matrices* $\mathbf{X}$, $\boldsymbol{\mu}$, $\boldsymbol{\Sigma}$ *and* $\boldsymbol{\Psi}$ *be partitioned according to (2.2.8) – (2.2.12). Put* $\boldsymbol{\Sigma}_{1\cdot 2} = \boldsymbol{\Sigma}_{11} - \boldsymbol{\Sigma}_{12}\boldsymbol{\Sigma}_{22}^{-}\boldsymbol{\Sigma}_{21}$ *and* $\boldsymbol{\Psi}_{1\cdot 2} = \boldsymbol{\Psi}_{11} - \boldsymbol{\Psi}_{12}\boldsymbol{\Psi}_{22}^{-}\boldsymbol{\Psi}_{21}$. *Then*

(i)
$$\mathbf{X}_{\bullet 1}|\mathbf{X}_{\bullet 2} \sim N_{p,s}(\boldsymbol{\mu}_{\bullet 1} + (\mathbf{X}_{\bullet 2} - \boldsymbol{\mu}_{\bullet 2})\boldsymbol{\Psi}_{22}^{-}\boldsymbol{\Psi}_{21}, \boldsymbol{\Sigma}, \boldsymbol{\Psi}_{1\cdot 2});$$

(ii)
$$\mathbf{X}_{1\bullet}|\mathbf{X}_{2\bullet} \sim N_{r,n}(\boldsymbol{\mu}_{1\bullet} + \boldsymbol{\Sigma}_{12}\boldsymbol{\Sigma}_{22}^{-}(\mathbf{X}_{2\bullet} - \boldsymbol{\mu}_{2\bullet}), \boldsymbol{\Sigma}_{1\cdot 2}, \boldsymbol{\Psi}).$$

PROOF: The proof of Theorem 2.2.5 may be copied. For example, for (i) let

$$\mathbf{H} = \begin{pmatrix} \mathbf{I} & -\boldsymbol{\Psi}_{12}\boldsymbol{\Psi}_{22}^{-} \\ \mathbf{0} & \mathbf{I} \end{pmatrix}$$

and note that from Theorem 2.1.3 it follows that $C(\boldsymbol{\Psi}_{21}) \subseteq C(\boldsymbol{\Psi}_{22})$, which implies $\boldsymbol{\Psi}_{12}\boldsymbol{\Psi}_{22}^{-}\boldsymbol{\Psi}_{22} = \boldsymbol{\Psi}_{12}$. ∎

Let $\boldsymbol{\Sigma}_{22}^{o}$ and $\boldsymbol{\Psi}_{22}^{o}$ span the orthogonal complements to $C(\boldsymbol{\Sigma}_{22})$ and $C(\boldsymbol{\Psi}_{22})$, respectively. Then $D[\boldsymbol{\Sigma}_{22}^{o'}(\mathbf{X}_{2\bullet} - \boldsymbol{\mu}_{2\bullet})] = \mathbf{0}$ and $D[(\mathbf{X}_{\bullet 2} - \boldsymbol{\mu}_{\bullet 2})\boldsymbol{\Psi}_{22}^{o'}] = \mathbf{0}$. Thus, $C((\mathbf{X}_{\bullet 2} - \boldsymbol{\mu}_{\bullet 2})') \subseteq C(\boldsymbol{\Psi}_{22})$ and $C(\mathbf{X}_{2\bullet} - \boldsymbol{\mu}_{2\bullet}) \subseteq C(\boldsymbol{\Sigma}_{22})$, which imply that the relations in (i) and (ii) of Theorem 2.2.6 are invariant with respect to the choice of g-inverse, with probability 1.

2.2.3 *Moments of the matrix normal distribution*

Moments of the normal distribution are needed when approximating other distributions with the help of the normal distribution, for example. The order of moments needed in expansions depends on the number of used terms. Usually it is enough to have expressions of the first six moments, and many practical problems can be handled with the first three or four moments. In principle, moments of arbitrary order could be used but usually this will not improve the approximation. When applying the results, the moments in approximation formulas have to be estimated and therefore, when utilizing higher order moments, the performance of the approximation can get even worse than using an approximation which is based on the first three moments. Outliers, for example, can cause remarkable biases in the estimates of higher order moments and cumulants. Nevertheless, we shall briefly study moments of arbitrary order because sometimes the basic structure of moments is also valuable.

The next Theorem 2.2.7 gives the moments of a matrix normal variable up to the fourth order as well as moments of arbitrary order. The moments of arbitrary order are given in a recursive way. We could also have stated the moments non-recursively but these expressions are quite lengthy (e.g. see Holmquist, 1988; von Rosen, 1988b). One, maybe the most important part of the theorem lies in the proof of Lemma 2.2.1, given below. If we understand how the moments can be obtained, we can easily apply the ideas in other situations. In the lemma we adopt the convention that if $k < 0$,

$$\sum_{i=0}^{k} \mathbf{A}_i = \mathbf{0}$$

for arbitrary matrices $\mathbf{A}_i$.

Lemma 2.2.1. *Let $\mathbf{X} \sim N_{p,n}(\boldsymbol{\mu}, \boldsymbol{\Sigma}, \boldsymbol{\Psi})$ and*

$$\varphi_{\mathbf{X}}^{k}(\mathbf{T}) = \frac{d^k \varphi_{\mathbf{X}}(\mathbf{T})}{d\mathbf{T}^k},$$

where $\varphi_{\mathbf{X}}^{0}(\mathbf{T}) = \varphi_{\mathbf{X}}(\mathbf{T})$, and $\varphi_{\mathbf{X}}(\mathbf{T})$ is given in Theorem 2.2.1. Put

$$\mathbf{A}(\mathbf{T}) = i\mathrm{vec}'\boldsymbol{\mu} - \mathrm{vec}'(\mathbf{T})(\boldsymbol{\Psi} \otimes \boldsymbol{\Sigma})$$

and $\mathbf{A}^1(\mathbf{T}) = \frac{d\mathbf{A}(\mathbf{T})}{d\mathbf{T}} = -\boldsymbol{\Psi} \otimes \boldsymbol{\Sigma}$. Then, if $k > 1$,

$$\begin{aligned}\varphi_{\mathbf{X}}^{k}(\mathbf{T}) = {} & \mathbf{A}(\mathbf{T}) \otimes \varphi_{\mathbf{X}}^{k-1}(\mathbf{T}) + \mathbf{A}^1(\mathbf{T}) \otimes \mathrm{vec}'\varphi_{\mathbf{X}}^{k-2}(\mathbf{T}) \\ & + \sum_{i=0}^{k-3} (\mathrm{vec}'\mathbf{A}^1(\mathbf{T}) \otimes \varphi_{\mathbf{X}}^{k-2}(\mathbf{T}))(\mathbf{I}_{pn} \otimes \mathbf{K}_{pn,(pn)^{k-i-3}} \otimes \mathbf{I}_{(pn)^i}).\end{aligned}$$

Proof: Since $\mathbf{A}(\mathbf{T})$ is linear in $\mathbf{T}$, higher order derivatives of $\mathbf{A}(\mathbf{T})$ vanish. By differentiating $\varphi_{\mathbf{X}}(\mathbf{T})$ four times, we obtain with the help of §1.4.9

$$\varphi_{\mathbf{X}}^1(\mathbf{T}) \underset{\substack{(1.4.14)\\(1.4.28)}}{=} \mathbf{A}(\mathbf{T})'\varphi_{\mathbf{X}}(\mathbf{T}), \tag{2.2.16}$$

$$\varphi_{\mathbf{X}}^2(\mathbf{T}) \underset{(1.4.19)}{=} \mathbf{A}^1(\mathbf{T})\varphi_{\mathbf{X}}(\mathbf{T}) + \mathbf{A}(\mathbf{T}) \otimes \varphi_{\mathbf{X}}^1(\mathbf{T}), \tag{2.2.17}$$

$$\begin{aligned}\varphi_{\mathbf{X}}^3(\mathbf{T}) \underset{(1.4.23)}{=} & \operatorname{vec}'\mathbf{A}^1(\mathbf{T}) \otimes \varphi_{\mathbf{X}}^1(\mathbf{T}) + \mathbf{A}^1(\mathbf{T}) \otimes \operatorname{vec}'\varphi_{\mathbf{X}}^1(\mathbf{T}) \\ & + \mathbf{A}(\mathbf{T}) \otimes \varphi_{\mathbf{X}}^2(\mathbf{T}),\end{aligned} \tag{2.2.18}$$

$$\begin{aligned}\varphi_{\mathbf{X}}^4(\mathbf{T}) \underset{(1.4.23)}{=} & \mathbf{A}(\mathbf{T}) \otimes \varphi_{\mathbf{X}}^3(\mathbf{T}) + \mathbf{A}^1(\mathbf{T}) \otimes \operatorname{vec}'\varphi_{\mathbf{X}}^2(\mathbf{T}) \\ & + (\operatorname{vec}'\mathbf{A}^1(\mathbf{T}) \otimes \varphi_{\mathbf{X}}^2(\mathbf{T}))(\mathbf{I}_{pn} \otimes \mathbf{K}_{pn,pn}) + \operatorname{vec}'\mathbf{A}^1(\mathbf{T}) \otimes \varphi_{\mathbf{X}}^2(\mathbf{T}).\end{aligned} \tag{2.2.19}$$

Now suppose that the lemma holds for $k-1$. We are going to differentiate $\varphi_{\mathbf{X}}^{k-1}(\mathbf{T})$ and by assumption

$$\begin{aligned}\varphi_{\mathbf{X}}^k(\mathbf{T}) = \frac{d}{d\mathbf{T}}\varphi_{\mathbf{X}}^{k-1}(\mathbf{T}) = \frac{d}{d\mathbf{T}}\{\mathbf{A}(\mathbf{T}) \otimes \varphi_{\mathbf{X}}^{k-2}(\mathbf{T})\} + \frac{d}{d\mathbf{T}}\{\mathbf{A}^1(\mathbf{T}) \otimes \operatorname{vec}'\varphi_{\mathbf{X}}^{k-3}(\mathbf{T})\} \\ + \frac{d}{d\mathbf{T}}\{\sum_{i=0}^{k-4}(\operatorname{vec}'\mathbf{A}^1(\mathbf{T}) \otimes \varphi_{\mathbf{X}}^{k-3}(\mathbf{T}))(\mathbf{I}_{pn} \otimes \mathbf{K}_{pn,(pn)^{k-1-i-3}} \otimes \mathbf{I}_{(pn)^i})\}.\end{aligned} \tag{2.2.20}$$

Straightforward calculations yield

$$\frac{d}{d\mathbf{T}}\{\mathbf{A}(\mathbf{T}) \otimes \varphi_{\mathbf{X}}^{k-2}(\mathbf{T})\} \underset{(1.4.23)}{=} \mathbf{A}(\mathbf{T}) \otimes \varphi_{\mathbf{X}}^{k-1}(\mathbf{T}) + \mathbf{A}^1(\mathbf{T}) \otimes \operatorname{vec}'\varphi_{\mathbf{X}}^{k-2}(\mathbf{T}), \tag{2.2.21}$$

$$\frac{d}{d\mathbf{T}}\{\mathbf{A}^1(\mathbf{T}) \otimes \operatorname{vec}'\varphi_{\mathbf{X}}^{k-3}(\mathbf{T})\} \underset{(1.4.23)}{=} (\operatorname{vec}'\mathbf{A}^1(\mathbf{T}) \otimes \varphi_{\mathbf{X}}^{k-2}(\mathbf{T}))(\mathbf{I}_{pn} \otimes \mathbf{K}_{pn,(pn)^{k-3}}) \tag{2.2.22}$$

and

$$\begin{aligned}\frac{d}{d\mathbf{T}}\{(\operatorname{vec}'\mathbf{A}^1(\mathbf{T}) \otimes \varphi_{\mathbf{X}}^{k-3}(\mathbf{T}))(\mathbf{I}_{pn} \otimes \mathbf{K}_{pn,(pn)^{k-i-1-3}} \otimes \mathbf{I}_{(pn)^i})\} \\ \underset{(1.4.23)}{=} (\operatorname{vec}'\mathbf{A}^1(\mathbf{T}) \otimes \varphi_{\mathbf{X}}^{k-2}(\mathbf{T}))(\mathbf{I}_{pn} \otimes \mathbf{K}_{pn,(pn)^{k-i-1-3}} \otimes \mathbf{I}_{(pn)^{i+1}}).\end{aligned} \tag{2.2.23}$$

Summing over i in (2.2.23) and thereafter adding (2.2.22) yields

$$\begin{aligned}\sum_{i=0}^{k-4}(\operatorname{vec}'\mathbf{A}^1(\mathbf{T}) \otimes \varphi_{\mathbf{X}}^{k-2}(\mathbf{T}))(\mathbf{I}_{pn} \otimes \mathbf{K}_{pn,(pn)^{k-i-1-3}} \otimes \mathbf{I}_{(pn)^{i+1}}) \\ +(\operatorname{vec}'\mathbf{A}^1(\mathbf{T}) \otimes \varphi_{\mathbf{X}}^{k-2}(\mathbf{T}))(\mathbf{I}_{pn} \otimes \mathbf{K}_{pn,(pn)^{k-3}}).\end{aligned}$$

Reindexing, i.e. $i \to i-1$, implies that this expression is equivalent to

$$\begin{aligned}\sum_{i=1}^{k-3}(\operatorname{vec}'\mathbf{A}^1(\mathbf{T}) \otimes \varphi_{\mathbf{X}}^{k-2}(\mathbf{T}))(\mathbf{I}_{pn} \otimes \mathbf{K}_{pn,(pn)^{k-i-3}} \otimes \mathbf{I}_{(pn)^i}) \\ + (\operatorname{vec}'\mathbf{A}^1(\mathbf{T}) \otimes \varphi_{\mathbf{X}}^{k-2}(\mathbf{T}))(\mathbf{I}_{pn} \otimes \mathbf{K}_{pn,(pn)^{k-3}}) \\ = \sum_{i=0}^{k-3}(\operatorname{vec}'\mathbf{A}^1(\mathbf{T}) \otimes \varphi_{\mathbf{X}}^{k-2}(\mathbf{T}))(\mathbf{I}_{pn} \otimes \mathbf{K}_{pn,(pn)^{k-i-3}} \otimes \mathbf{I}_{(pn)^i}).\end{aligned} \tag{2.2.24}$$

Finally, summing the right hand side of (2.2.21) and (2.2.24), we obtain the statement of the lemma from (2.2.20). ∎

Theorem 2.2.7. *Let* $\mathbf{X} \sim N_{p,n}(\boldsymbol{\mu}, \boldsymbol{\Sigma}, \boldsymbol{\Psi})$. *Then*

(i) $m_2[\mathbf{X}] = (\boldsymbol{\Psi} \otimes \boldsymbol{\Sigma}) + \mathrm{vec}\boldsymbol{\mu}\mathrm{vec}'\boldsymbol{\mu};$

(ii)
$$m_3[\mathbf{X}] = \mathrm{vec}\boldsymbol{\mu}(\mathrm{vec}'\boldsymbol{\mu})^{\otimes 2} + \mathrm{vec}'\boldsymbol{\mu} \otimes \boldsymbol{\Psi} \otimes \boldsymbol{\Sigma} + \boldsymbol{\Psi} \otimes \boldsymbol{\Sigma} \otimes \mathrm{vec}'\boldsymbol{\mu} + \mathrm{vec}\boldsymbol{\mu}\mathrm{vec}'(\boldsymbol{\Psi} \otimes \boldsymbol{\Sigma});$$

(iii)
$$\begin{aligned} m_4[\mathbf{X}] = {} & \mathrm{vec}\boldsymbol{\mu}(\mathrm{vec}'\boldsymbol{\mu})^{\otimes 3} + (\mathrm{vec}'\boldsymbol{\mu})^{\otimes 2} \otimes \boldsymbol{\Psi} \otimes \boldsymbol{\Sigma} \\ & + \mathrm{vec}'\boldsymbol{\mu} \otimes \boldsymbol{\Psi} \otimes \boldsymbol{\Sigma} \otimes \mathrm{vec}'\boldsymbol{\mu} + \mathrm{vec}\boldsymbol{\mu}\mathrm{vec}'\boldsymbol{\mu} \otimes \mathrm{vec}'(\boldsymbol{\Psi} \otimes \boldsymbol{\Sigma}) \\ & + \boldsymbol{\Psi} \otimes \boldsymbol{\Sigma} \otimes \mathrm{vec}'(\boldsymbol{\Psi} \otimes \boldsymbol{\Sigma}) + \boldsymbol{\Psi} \otimes \boldsymbol{\Sigma} \otimes (\mathrm{vec}'\boldsymbol{\mu})^{\otimes 2} \\ & + \{\mathrm{vec}\boldsymbol{\mu}\mathrm{vec}'(\boldsymbol{\Psi} \otimes \boldsymbol{\Sigma}) \otimes \mathrm{vec}'\boldsymbol{\mu} + \mathrm{vec}'(\boldsymbol{\Psi} \otimes \boldsymbol{\Sigma}) \otimes \boldsymbol{\Psi} \otimes \boldsymbol{\Sigma}\} \\ & \times (\mathbf{I} + \mathbf{I}_{pn} \otimes \mathbf{K}_{pn,pn}); \end{aligned}$$

(iv)
$$\begin{aligned} m_k[\mathbf{X}] = {} & \mathrm{vec}'\boldsymbol{\mu} \otimes m_{k-1}[\mathbf{X}] + \boldsymbol{\Psi} \otimes \boldsymbol{\Sigma} \otimes \mathrm{vec}'(m_{k-2}[\mathbf{X}]) \\ & + \sum_{i=0}^{k-3} (\mathrm{vec}'(\boldsymbol{\Psi} \otimes \boldsymbol{\Sigma}) \otimes m_{k-2}[\mathbf{X}])(\mathbf{I}_{pn} \otimes \mathbf{K}_{pn,(pn)^{k-i-3}} \otimes \mathbf{I}_{(pn)^i}), \\ & \qquad k > 1. \end{aligned}$$

PROOF: The relations in (i), (ii) and (iii) follow immediately from (2.2.16) – (2.2.19) by setting $\mathbf{T} = \mathbf{0}$, since

$$\begin{aligned} \mathbf{A}(\mathbf{0}) = {} & i\mathrm{vec}'\boldsymbol{\mu}, \\ \mathbf{A}^1(\mathbf{0}) = {} & -\boldsymbol{\Psi} \otimes \boldsymbol{\Sigma}. \end{aligned}$$

The general case (iv) is obtained from Lemma 2.2.1, if we note that (apply Definition 2.1.3)

$$\begin{aligned} \varphi^k_{\mathbf{x}}(\mathbf{0}) &= i^k m_k[\mathbf{X}], \\ \mathbf{A}(\mathbf{0}) \otimes \varphi^{k-1}_{\mathbf{x}}(\mathbf{0}) &= i^k \mathrm{vec}'\boldsymbol{\mu} \otimes m_{k-1}[\mathbf{X}] \end{aligned}$$

and

$$\mathrm{vec}'\mathbf{A}^1(\mathbf{0}) \otimes \varphi^{k-2}_{\mathbf{x}}(\mathbf{0}) = i^k \mathrm{vec}'(\boldsymbol{\Psi} \otimes \boldsymbol{\Sigma}) \otimes m_{k-2}[\mathbf{X}],$$

where it has been utilized in the last equality that $i^k = -i^{k-2}$. ∎

Corollary 2.2.7.1. *Let* $\mathbf{Y} \sim N_{p,n}(\mathbf{0}, \boldsymbol{\Sigma}, \boldsymbol{\Psi})$. *Then* $m_k[\mathbf{Y}] = \mathbf{0}$, *if* k *is odd. When* k *is even, then*

(i) $m_2[\mathbf{Y}] = \boldsymbol{\Psi} \otimes \boldsymbol{\Sigma};$

(ii) $m_4[\mathbf{Y}] = \boldsymbol{\Psi} \otimes \boldsymbol{\Sigma} \otimes \mathrm{vec}'(\boldsymbol{\Psi} \otimes \boldsymbol{\Sigma})$

$$+ (\mathrm{vec}'(\boldsymbol{\Psi} \otimes \boldsymbol{\Sigma}) \otimes \boldsymbol{\Psi} \otimes \boldsymbol{\Sigma})(\mathbf{I} + \mathbf{I}_{pn} \otimes \mathbf{K}_{pn,pn});$$

(iii) $$m_k[\mathbf{Y}] = \boldsymbol{\Psi} \otimes \boldsymbol{\Sigma} \otimes \mathrm{vec}' m_{k-2}[\mathbf{Y}] + \sum_{i=0}^{k-3} (\mathrm{vec}'(\boldsymbol{\Psi} \otimes \boldsymbol{\Sigma}) \otimes m_{k-2}[\mathbf{Y}])(\mathbf{I}_{pn} \otimes \mathbf{K}_{pn,(pn)^{k-i-3}} \otimes \mathbf{I}_{(pn)^i}),$$
$$m_0[\mathbf{Y}] = 1, \qquad k = 2, 4, 6, \ldots;$$

(iv) $$\mathrm{vec}\, m_4[\mathbf{Y}] = (\mathbf{I} + \mathbf{I}_{pn} \otimes \mathbf{K}_{(pn)^2,pn} + \mathbf{I}_{pn} \otimes \mathbf{K}_{pn,pn} \otimes \mathbf{I}_{pn})(\mathrm{vec}(\boldsymbol{\Psi} \otimes \boldsymbol{\Sigma}))^{\otimes 2};$$

(v) $$\mathrm{vec}\, m_k[\mathbf{Y}] = \sum_{i=0}^{k-2} (\mathbf{I}_{pn} \otimes \mathbf{K}_{(pn)^{k-i-2},pn} \otimes \mathbf{I}_{(pn)^i})(\mathrm{vec}(\boldsymbol{\Psi} \otimes \boldsymbol{\Sigma}) \otimes \mathrm{vec}\, m_{k-2}[\mathbf{Y}]),$$
$$m_0[\mathbf{Y}] = 1, \qquad k = 2, 4, 6, \ldots$$

PROOF: Statements (i), (ii) and (iii) follow immediately from Theorem 2.2.7. By applying Proposition 1.3.14 (iv), one can see that (ii) implies (iv). In order to prove the last statement Proposition 1.3.14 (iv) is applied once again and after some calculations it follows from (iii) that

$$\begin{aligned}
\mathrm{vec}\, m_k[\mathbf{Y}] &= (\mathbf{I}_{pn} \otimes \mathbf{K}_{(pn)^{k-2},pn})\mathrm{vec}(\boldsymbol{\Psi} \otimes \boldsymbol{\Sigma}) \otimes \mathrm{vec}\, m_{k-2}[\mathbf{Y}] \\
&\quad + \sum_{i=0}^{k-3} (\mathbf{I}_{pn} \otimes \mathbf{K}_{(pn)^{k-i-3},pn} \otimes \mathbf{I}_{(pn)^i})\mathrm{vec}(\boldsymbol{\Psi} \otimes \boldsymbol{\Sigma}) \otimes \mathrm{vec}\, m_{k-2}[\mathbf{Y}] \\
&= \sum_{i=0}^{k-2} (\mathbf{I}_{pn} \otimes \mathbf{K}_{(pn)^{k-i-2},pn} \otimes \mathbf{I}_{(pn)^i})\mathrm{vec}(\boldsymbol{\Psi} \otimes \boldsymbol{\Sigma}) \otimes \mathrm{vec}\, m_{k-2}[\mathbf{Y}].
\end{aligned}$$

∎

In the next two corollaries we shall present the expressions of moments and central moments for a normally distributed random vector.

Corollary 2.2.7.2. *Let $\mathbf{x} \sim N_p(\boldsymbol{\mu}, \boldsymbol{\Sigma})$. Then moments of the p-vector are of the form:*

(i) $$m_2[\mathbf{x}] = \boldsymbol{\Sigma} + \boldsymbol{\mu}\boldsymbol{\mu}';$$

(ii) $$m_3[\mathbf{x}] = \boldsymbol{\mu}(\boldsymbol{\mu}')^{\otimes 2} + \boldsymbol{\mu}' \otimes \boldsymbol{\Sigma} + \boldsymbol{\Sigma} \otimes \boldsymbol{\mu}' + \boldsymbol{\mu}\mathrm{vec}'\boldsymbol{\Sigma};$$

(iii) $$\begin{aligned}
m_4[\mathbf{x}] &= \boldsymbol{\mu}(\boldsymbol{\mu}')^{\otimes 3} + (\boldsymbol{\mu}')^{\otimes 2} \otimes \boldsymbol{\Sigma} + \boldsymbol{\mu}' \otimes \boldsymbol{\Sigma} \otimes \boldsymbol{\mu}' + \boldsymbol{\mu}\boldsymbol{\mu}' \otimes \mathrm{vec}'\boldsymbol{\Sigma} \\
&\quad + \boldsymbol{\Sigma} \otimes \mathrm{vec}'\boldsymbol{\Sigma} + \boldsymbol{\Sigma} \otimes (\boldsymbol{\mu}')^{\otimes 2} \\
&\quad + \{\boldsymbol{\mu}\mathrm{vec}'\boldsymbol{\Sigma} \otimes \boldsymbol{\mu}' + \mathrm{vec}'\boldsymbol{\Sigma} \otimes \boldsymbol{\Sigma}\}(\mathbf{I}_{p^3} + \mathbf{I}_p \otimes \mathbf{K}_{p,p});
\end{aligned}$$

(iv) $$m_k[\mathbf{x}] = \boldsymbol{\mu}' \otimes m_{k-1}[\mathbf{x}] + \boldsymbol{\Sigma} \otimes \mathrm{vec}'(m_{k-2}[\mathbf{x}]) + \sum_{i=0}^{k-3} (\mathrm{vec}'\boldsymbol{\Sigma} \otimes m_{k-2}[\mathbf{x}])(\mathbf{I}_p \otimes \mathbf{K}_{p,p^{k-i-3}} \otimes \mathbf{I}_{p^i}), \qquad k > 1.$$

PROOF: All the statements follow directly from Theorem 2.2.7 if we take $\mathbf{\Psi} = 1$ and replace vec$\boldsymbol{\mu}$ by the expectation $\boldsymbol{\mu}$. ∎

The next results follow in the same way as those presented in Corollary 2.2.7.1.

Corollary 2.2.7.3. *Let* $\mathbf{x} \sim N_p(\boldsymbol{\mu}, \mathbf{\Sigma})$. *Then odd central moments of* $\mathbf{x}$ *equal zero and even central moments are given by the following equalities:*

(i) $\overline{m}_2[\mathbf{x}] = \mathbf{\Sigma};$

(ii) $\overline{m}_4[\mathbf{x}] = \mathbf{\Sigma} \otimes \mathrm{vec}'\mathbf{\Sigma} + (\mathrm{vec}'\mathbf{\Sigma} \otimes \mathbf{\Sigma})(\mathbf{I}_{p^3} + \mathbf{I}_p \otimes \mathbf{K}_{p,p});$

(iii)
$$\overline{m}_k[\mathbf{x}] = \mathbf{\Sigma} \otimes \mathrm{vec}'\overline{m}_{k-2}[\mathbf{x}] + \sum_{i=0}^{k-3}(\mathrm{vec}'\mathbf{\Sigma} \otimes \overline{m}_{k-2}[\mathbf{x}])(\mathbf{I}_p \otimes \mathbf{K}_{p,p^{k-i-3}} \otimes \mathbf{I}_{p^i}), \quad \overline{m}_0[\mathbf{x}] = 1, \quad k = 2, 4, 6, \ldots;$$

(iv) $\mathrm{vec}\,\overline{m}_4[\mathbf{x}] = (\mathbf{I}_{p^4} + \mathbf{I}_p \otimes \mathbf{K}_{p^2,p} + \mathbf{I}_p \otimes \mathbf{K}_{p,p} \otimes \mathbf{I}_p)(\mathrm{vec}\mathbf{\Sigma})^{\otimes 2};$

(v)
$$\mathrm{vec}\,\overline{m}_k[\mathbf{x}] = \sum_{i=0}^{k-2}(\mathbf{I}_p \otimes \mathbf{K}_{p^{k-i-2},p} \otimes \mathbf{I}_{p^i})(\mathrm{vec}\mathbf{\Sigma} \otimes \mathrm{vec}\,\overline{m}_{k-2}[\mathbf{x}]), \quad \overline{m}_0[\mathbf{x}] = 1, \quad k = 2, 4, 6, \ldots$$

∎

Since $\mathbf{I}_{pn} \otimes \mathbf{K}_{(pn)^{k-i-2},pn} \otimes \mathbf{I}_{(pn)^i}$ is a permutation operator, it is seen that the central moments in Corollary 2.2.7.1 (v) are generated with the help of permutations on tensors of the basis vectors of $\mathrm{vec}(\mathbf{\Psi} \otimes \mathbf{\Sigma})^{\otimes k}$. Note that not all permutations are used. However, one can show that all permutations $\mathbf{P}_i$ are used, which satisfy the condition that $\mathbf{P}_i\mathrm{vec}(\mathbf{\Psi} \otimes \mathbf{\Sigma})^{\otimes k}$ differs from $\mathbf{P}_j\mathrm{vec}(\mathbf{\Psi} \otimes \mathbf{\Sigma})^{\otimes k}$, if $\mathbf{P}_i \neq \mathbf{P}_j$.

Example 2.2.1. Consider $(\mathrm{vec}(\mathbf{\Psi} \otimes \mathbf{\Sigma}))^{\otimes 2}$, which can be written as a sum

$$\sum_{ijkl} a_{ij}a_{kl}\mathbf{e}_j^1 \otimes \mathbf{d}_i^1 \otimes \mathbf{e}_l^2 \otimes \mathbf{d}_k^2,$$

where $\mathrm{vec}(\mathbf{\Psi} \otimes \mathbf{\Sigma}) = \sum_{ij} a_{ij}\mathbf{e}_j^1 \otimes \mathbf{d}_i^1 = \sum_{kl} a_{kl}\mathbf{e}_l^2 \otimes \mathbf{d}_k^2$ and $\mathbf{e}_j^1$, $\mathbf{e}_j^2$, $\mathbf{d}_j^1$ and $\mathbf{d}_j^2$ are unit bases vectors. All possible permutations of the basis vectors are given by

$\mathbf{e}_j^1 \otimes \mathbf{d}_i^1 \otimes \mathbf{e}_l^2 \otimes \mathbf{d}_k^2$	$\mathbf{d}_i^1 \otimes \mathbf{e}_j^1 \otimes \mathbf{e}_l^2 \otimes \mathbf{d}_k^2$	$\mathbf{e}_l^2 \otimes \mathbf{d}_i^1 \otimes \mathbf{e}_j^1 \otimes \mathbf{d}_k^2$	$\mathbf{d}_k^2 \otimes \mathbf{e}_j^1 \otimes \mathbf{e}_l^2 \otimes \mathbf{d}_i^1$
$\mathbf{e}_j^1 \otimes \mathbf{d}_i^1 \otimes \mathbf{d}_k^2 \otimes \mathbf{e}_l^2$	$\mathbf{d}_i^1 \otimes \mathbf{e}_j^1 \otimes \mathbf{d}_k^2 \otimes \mathbf{e}_l^2$	$\mathbf{e}_l^2 \otimes \mathbf{d}_i^1 \otimes \mathbf{d}_k^2 \otimes \mathbf{e}_j^1$	$\mathbf{d}_k^2 \otimes \mathbf{e}_j^1 \otimes \mathbf{d}_i^1 \otimes \mathbf{e}_l^2$
$\mathbf{e}_j^1 \otimes \mathbf{e}_l^2 \otimes \mathbf{d}_k^2 \otimes \mathbf{d}_i^1$	$\mathbf{d}_i^1 \otimes \mathbf{e}_l^2 \otimes \mathbf{d}_k^2 \otimes \mathbf{e}_j^1$	$\mathbf{e}_l^2 \otimes \mathbf{e}_j^1 \otimes \mathbf{d}_k^2 \otimes \mathbf{d}_i^1$	$\mathbf{d}_k^2 \otimes \mathbf{e}_l^2 \otimes \mathbf{d}_i^1 \otimes \mathbf{e}_j^1$
$\mathbf{e}_j^1 \otimes \mathbf{e}_l^2 \otimes \mathbf{d}_i^1 \otimes \mathbf{d}_k^2$	$\mathbf{d}_i^1 \otimes \mathbf{e}_l^2 \otimes \mathbf{e}_j^1 \otimes \mathbf{d}_k^2$	$\mathbf{e}_l^2 \otimes \mathbf{e}_j^1 \otimes \mathbf{d}_i^1 \otimes \mathbf{d}_k^2$	$\mathbf{d}_k^2 \otimes \mathbf{e}_l^2 \otimes \mathbf{e}_j^1 \otimes \mathbf{d}_i^1$
$\mathbf{e}_j^1 \otimes \mathbf{d}_k^2 \otimes \mathbf{e}_l^2 \otimes \mathbf{d}_i^1$	$\mathbf{d}_i^1 \otimes \mathbf{d}_k^2 \otimes \mathbf{e}_l^2 \otimes \mathbf{e}_j^1$	$\mathbf{e}_l^2 \otimes \mathbf{d}_k^2 \otimes \mathbf{e}_j^1 \otimes \mathbf{d}_i^1$	$\mathbf{d}_k^2 \otimes \mathbf{d}_i^1 \otimes \mathbf{e}_l^2 \otimes \mathbf{e}_j^1$
$\mathbf{e}_j^1 \otimes \mathbf{d}_k^2 \otimes \mathbf{d}_i^1 \otimes \mathbf{e}_l^2$	$\mathbf{d}_i^1 \otimes \mathbf{d}_k^2 \otimes \mathbf{e}_j^1 \otimes \mathbf{e}_l^2$	$\mathbf{e}_l^2 \otimes \mathbf{d}_k^2 \otimes \mathbf{d}_i^1 \otimes \mathbf{e}_j^1$	$\mathbf{d}_k^2 \otimes \mathbf{d}_i^1 \otimes \mathbf{e}_j^1 \otimes \mathbf{e}_l^2$

Since $\mathbf{\Psi} \otimes \mathbf{\Sigma}$ is symmetric, some of the permutations are equal and we may reduce the number of them. For example, $\mathbf{e}_j^1 \otimes \mathbf{d}_i^1 \otimes \mathbf{e}_l^2 \otimes \mathbf{d}_k^2$ represents the same elements as $\mathbf{d}_i^1 \otimes \mathbf{e}_j^1 \otimes \mathbf{e}_l^2 \otimes \mathbf{d}_k^2$. Hence, the following permutations remain

$$\begin{array}{ll} \mathbf{e}_j^1 \otimes \mathbf{d}_i^1 \otimes \mathbf{e}_l^2 \otimes \mathbf{d}_k^2 & \mathbf{e}_l^2 \otimes \mathbf{e}_j^1 \otimes \mathbf{d}_k^2 \otimes \mathbf{d}_i^1 \\ \mathbf{e}_j^1 \otimes \mathbf{e}_l^2 \otimes \mathbf{d}_k^2 \otimes \mathbf{d}_i^1 & \mathbf{e}_l^2 \otimes \mathbf{e}_j^1 \otimes \mathbf{d}_i^1 \otimes \mathbf{d}_k^2 \\ \mathbf{e}_j^1 \otimes \mathbf{e}_l^2 \otimes \mathbf{d}_i^1 \otimes \mathbf{d}_k^2 & \mathbf{e}_l^2 \otimes \mathbf{d}_k^2 \otimes \mathbf{e}_j^1 \otimes \mathbf{d}_i^1 \end{array}.$$

However, $\mathbf{e}_j^1 \otimes \mathbf{e}_l^2 \otimes \mathbf{d}_k^2 \otimes \mathbf{d}_i^1$ represents the same element as $\mathbf{e}_l^2 \otimes \mathbf{e}_j^1 \otimes \mathbf{d}_i^1 \otimes \mathbf{d}_k^2$ and the final set of elements can be chosen as

$$\begin{array}{l} \mathbf{e}_j^1 \otimes \mathbf{d}_i^1 \otimes \mathbf{e}_l^2 \otimes \mathbf{d}_k^2 \\ \mathbf{e}_j^1 \otimes \mathbf{e}_l^2 \otimes \mathbf{d}_k^2 \otimes \mathbf{d}_i^1 \\ \mathbf{e}_j^1 \otimes \mathbf{e}_l^2 \otimes \mathbf{d}_i^1 \otimes \mathbf{d}_k^2 \end{array}.$$

The three permutations which correspond to this set of basis vectors and act on $(\mathrm{vec}(\mathbf{\Psi} \otimes \mathbf{\Sigma}))^{\otimes 2}$ are given by $\mathbf{I}$, $\mathbf{I}_{pn} \otimes \mathbf{K}_{(pn)^2,pn}$ and $\mathbf{I}_{pn} \otimes \mathbf{K}_{pn,pn} \otimes \mathbf{I}_{pn}$. These are the ones used in Corollary 2.2.7.1 (iv). ∎

Sometimes it is useful to rearrange the moments. For example, when proving Theorem 2.2.9, which is given below. In the next corollary the Kroneckerian power is used.

Corollary 2.2.7.4. *Let* $\mathbf{Y} \sim N_{p,n}(\mathbf{0}, \mathbf{\Sigma}, \mathbf{\Psi})$. *Then*

(i) $E[\mathbf{Y}^{\otimes 2}] = \mathrm{vec}\mathbf{\Sigma}\mathrm{vec}'\mathbf{\Psi}$;

(ii) $$E[\mathbf{Y}^{\otimes 4}] = (\mathrm{vec}\mathbf{\Sigma}\mathrm{vec}'\mathbf{\Psi})^{\otimes 2} + (\mathbf{I}_p \otimes \mathbf{K}_{p,p} \otimes \mathbf{I}_p)(\mathrm{vec}\mathbf{\Sigma}\mathrm{vec}'\mathbf{\Psi})^{\otimes 2}(\mathbf{I}_n \otimes \mathbf{K}_{n,n} \otimes \mathbf{I}_n)$$
$$+ (\mathbf{I}_p \otimes \mathbf{K}_{p^2,p})(\mathrm{vec}\mathbf{\Sigma}\mathrm{vec}'\mathbf{\Psi})^{\otimes 2}(\mathbf{I}_n \otimes \mathbf{K}_{n,n^2});$$

(iii) $$E[\mathbf{Y}^{\otimes k}] = \sum_{i=2}^{k}(\mathbf{I}_p \otimes \mathbf{K}_{p^{i-2},p} \otimes \mathbf{I}_{p^{k-i}})(\mathrm{vec}\mathbf{\Sigma}\mathrm{vec}'\mathbf{\Psi} \otimes E[\mathbf{Y}^{\otimes k-2}])$$
$$\times (\mathbf{I}_n \otimes \mathbf{K}_{n,n^{i-2}} \otimes \mathbf{I}_{n^{k-i}}), \qquad k = 2, 4, 6, \ldots$$

PROOF: It can be obtained from Corollary 2.2.7.1 (i) that

$$E[\mathrm{vec}\mathbf{Y} \otimes \mathrm{vec}\mathbf{Y}] = \mathrm{vec}(\mathbf{\Psi} \otimes \mathbf{\Sigma}).$$

Premultiplying this expression with $\mathbf{I}_n \otimes \mathbf{K}_{n,p} \otimes \mathbf{I}_p$ yields, according to Proposition 1.3.14 (iv),

$$\mathrm{vec}E[\mathbf{Y} \otimes \mathbf{Y}] = (\mathbf{I}_n \otimes \mathbf{K}_{n,p} \otimes \mathbf{I}_p)\mathrm{vec}(\mathbf{\Psi} \otimes \mathbf{\Sigma}) = \mathrm{vec}\mathbf{\Psi} \otimes \mathrm{vec}\mathbf{\Sigma} = \mathrm{vec}(\mathrm{vec}\mathbf{\Sigma}\mathrm{vec}'\mathbf{\Psi}).$$

Concerning (ii), the same idea as in the above given proof is applied. Corollary 2.2.7.1 (v) implies that if

$$\mathbf{P} = (\mathbf{I}_{n^3} \otimes \mathbf{K}_{n,p^3} \otimes \mathbf{I}_p)(\mathbf{I}_{n^2} \otimes \mathbf{K}_{n,p^2} \otimes \mathbf{I}_p \otimes \mathbf{I}_{pn})(\mathbf{I}_n \otimes \mathbf{K}_{n,p} \otimes \mathbf{I}_p \otimes \mathbf{I}_{(pn)^2}),$$

then

$$\mathrm{vec}E[\mathbf{Y}^{\otimes 4}] = \mathbf{P}E[(\mathrm{vec}\mathbf{Y})^{\otimes 4}] = \mathbf{P}\mathrm{vec}\, m_4[\mathbf{Y}],$$

which is equivalent to (ii).
In order to prove (iii) one can copy the above lines and apply an induction argument. Alternatively to the proof of this statement, as well as when proving (i) and (ii), one may apply the matrix derivative given by (1.4.2) instead of (1.4.4) when differentiating the characteristic function. ∎

Another way of obtaining the moments in Theorem 2.2.7 is to start with the moments given in Corollary 2.2.7.1 and then utilize that

$$m_k[\mathbf{X}] = E[\mathrm{vec}(\mathbf{Y}+\boldsymbol{\mu})\{\mathrm{vec}'(\mathbf{Y}+\boldsymbol{\mu})\}^{\otimes k-1}].$$

Now the idea is to expand $\{\mathrm{vec}(\mathbf{Y}+\boldsymbol{\mu})\}^{\otimes k-1}$. Thereafter certain permutation operators are applied to each term of the expansion so that the terms will take the form

$$\mathrm{vec}(\mathbf{Y}+\boldsymbol{\mu})((\mathrm{vec}'\mathbf{Y})^{\otimes j} \otimes (\mathrm{vec}'\boldsymbol{\mu})^{\otimes k-j}), \qquad j = 0, 1, \ldots, k.$$

An interesting feature of the normal distribution is that all cumulants are almost trivial to obtain. As the cumulant function in Theorem 2.2.1 consists of a linear term and a quadratic term in $\mathbf{T}$, the following theorem can be verified.

Theorem 2.2.8. *Let* $\mathbf{X} \sim N_{p,n}(\boldsymbol{\mu}, \boldsymbol{\Sigma}, \boldsymbol{\Psi})$. *Then*

(i) $$c_1[\mathbf{X}] = \mathrm{vec}\boldsymbol{\mu};$$

(ii) $$c_2[\mathbf{X}] = \boldsymbol{\Psi} \otimes \boldsymbol{\Sigma};$$

(iii) $$c_k[\mathbf{X}] = \mathbf{0}, \qquad k \geq 3.$$

∎

Quadratic forms play a key role in statistics. Now and then moments of quadratic forms are needed. In the next theorem some quadratic forms in matrix normal variables are considered. By studying the proof one understands how moments can be obtained when $\mathbf{X}$ is not normally distributed.

Theorem 2.2.9. *Let* $\mathbf{X} \sim N_{p,n}(\boldsymbol{\mu}, \boldsymbol{\Sigma}, \boldsymbol{\Psi})$. *Then*

(i) $E[\mathbf{XAX}'] = \mathrm{tr}(\boldsymbol{\Psi}\mathbf{A})\boldsymbol{\Sigma} + \boldsymbol{\mu}\mathbf{A}\boldsymbol{\mu}';$

(ii) $$E[\mathbf{XAX}' \otimes \mathbf{XBX}'] = \mathrm{tr}(\boldsymbol{\Psi}\mathbf{A})\mathrm{tr}(\boldsymbol{\Psi}\mathbf{B})\boldsymbol{\Sigma} \otimes \boldsymbol{\Sigma}$$
$$+ \mathrm{tr}(\boldsymbol{\Psi}\mathbf{A}\boldsymbol{\Psi}\mathbf{B}')\mathrm{vec}\boldsymbol{\Sigma}\mathrm{vec}'\boldsymbol{\Sigma} + \mathrm{tr}(\boldsymbol{\Psi}\mathbf{A}\boldsymbol{\Psi}\mathbf{B})\mathbf{K}_{p,p}(\boldsymbol{\Sigma} \otimes \boldsymbol{\Sigma})$$
$$+ \mathrm{tr}(\boldsymbol{\Psi}\mathbf{A})\boldsymbol{\Sigma} \otimes \boldsymbol{\mu}\mathbf{B}\boldsymbol{\mu}' + \mathrm{vec}(\boldsymbol{\mu}\mathbf{B}\boldsymbol{\Psi}\mathbf{A}'\boldsymbol{\mu}')\mathrm{vec}'\boldsymbol{\Sigma}$$

$$+\mathbf{K}_{p,p}\{\boldsymbol{\mu}\mathbf{B}\boldsymbol{\Psi}\mathbf{A}\boldsymbol{\mu}'\otimes\boldsymbol{\Sigma}+\boldsymbol{\Sigma}\otimes\boldsymbol{\mu}\mathbf{A}\boldsymbol{\Psi}\mathbf{B}\boldsymbol{\mu}'\}$$

$$+\mathrm{vec}\boldsymbol{\Sigma}\mathrm{vec}'(\boldsymbol{\mu}\mathbf{B}'\boldsymbol{\Psi}\mathbf{A}\boldsymbol{\mu}')+\mathrm{tr}(\boldsymbol{\Psi}\mathbf{B})\boldsymbol{\mu}\mathbf{A}\boldsymbol{\mu}'\otimes\boldsymbol{\Sigma}+\boldsymbol{\mu}\mathbf{A}\boldsymbol{\mu}'\otimes\boldsymbol{\mu}\mathbf{B}\boldsymbol{\mu}';$$

(iii) $$D[\mathbf{XAX}']=\mathrm{tr}(\boldsymbol{\Psi}\mathbf{A}\boldsymbol{\Psi}\mathbf{A}')\boldsymbol{\Sigma}\otimes\boldsymbol{\Sigma}+tr(\boldsymbol{\Psi}\mathbf{A}\boldsymbol{\Psi}\mathbf{A})\mathbf{K}_{p,p}(\boldsymbol{\Sigma}\otimes\boldsymbol{\Sigma})$$

$$+\boldsymbol{\Sigma}\otimes\boldsymbol{\mu}\mathbf{A}\boldsymbol{\Psi}\mathbf{A}'\boldsymbol{\mu}'+\boldsymbol{\mu}\mathbf{A}\boldsymbol{\Psi}\mathbf{A}\boldsymbol{\mu}'\otimes\boldsymbol{\Sigma}$$

$$+\mathbf{K}_{p,p}\{\boldsymbol{\mu}\mathbf{A}\boldsymbol{\Psi}\mathbf{A}\boldsymbol{\mu}'\otimes\boldsymbol{\Sigma}+\boldsymbol{\Sigma}\otimes\boldsymbol{\mu}\mathbf{A}'\boldsymbol{\Psi}\mathbf{A}\boldsymbol{\mu}'\}.$$

PROOF: In order to show (i) we are going to utilize Corollary 2.2.7.4. Observe that $\mathbf{X}$ and $\mathbf{Y}+\boldsymbol{\mu}$ have the same distribution when $\mathbf{Y}\sim N_{p,n}(\mathbf{0},\boldsymbol{\Sigma},\boldsymbol{\Psi})$. The odd moments of $\mathbf{Y}$ equal zero and

$$\begin{aligned}\mathrm{vec}E[\mathbf{XAX}']&=E[((\mathbf{Y}+\boldsymbol{\mu})\otimes(\mathbf{Y}+\boldsymbol{\mu}))\mathrm{vec}\mathbf{A}]=E[(\mathbf{Y}\otimes\mathbf{Y})\mathrm{vec}\mathbf{A}]+\mathrm{vec}(\boldsymbol{\mu}\mathbf{A}\boldsymbol{\mu}')\\&=\mathrm{vec}\boldsymbol{\Sigma}\mathrm{vec}'\boldsymbol{\Psi}\mathrm{vec}\mathbf{A}+\mathrm{vec}(\boldsymbol{\mu}\mathbf{A}\boldsymbol{\mu}')=\mathrm{tr}(\boldsymbol{\Psi}\mathbf{A})\mathrm{vec}\boldsymbol{\Sigma}+\mathrm{vec}(\boldsymbol{\mu}\mathbf{A}\boldsymbol{\mu}'),\end{aligned}$$

which establishes (i).
For (ii) it is noted that

$$\begin{aligned}&E[(\mathbf{XAX}')\otimes(\mathbf{XBX}')]\\&\quad=E[(\mathbf{YAY}')\otimes(\mathbf{YBY}')]+E[(\boldsymbol{\mu}\mathbf{A}\boldsymbol{\mu}')\otimes(\boldsymbol{\mu}\mathbf{B}\boldsymbol{\mu}')]\\&\qquad+E[(\mathbf{YA}\boldsymbol{\mu}')\otimes(\mathbf{YB}\boldsymbol{\mu}')]+E[(\mathbf{YA}\boldsymbol{\mu}')\otimes(\boldsymbol{\mu}\mathbf{BY}')]\\&\qquad+E[(\boldsymbol{\mu}\mathbf{AY}')\otimes(\mathbf{YB}\boldsymbol{\mu}')]+E[(\boldsymbol{\mu}\mathbf{AY}')\otimes(\boldsymbol{\mu}\mathbf{BY}')]\\&\qquad+E[(\boldsymbol{\mu}\mathbf{A}\boldsymbol{\mu}')\otimes(\mathbf{YBY}')]+E[(\mathbf{YAY}')\otimes(\boldsymbol{\mu}\mathbf{B}\boldsymbol{\mu}')].\end{aligned}\qquad(2.2.25)$$

We are going to consider the expressions in the right-hand side of (2.2.25) term by term. It follows by Proposition 1.3.14 (iii) and (iv) and Corollary 2.2.7.4 that

$$\begin{aligned}&\mathrm{vec}(E[(\mathbf{YAY}')\otimes(\mathbf{YBY}')])\\&\quad=(\mathbf{I}_p\otimes\mathbf{K}_{p,p}\otimes\mathbf{I}_p)E[\mathbf{Y}^{\otimes4}](\mathrm{vec}\mathbf{A}\otimes\mathrm{vec}\mathbf{B})\\&\quad=\mathrm{tr}(\boldsymbol{\Psi}\mathbf{A})\mathrm{tr}(\boldsymbol{\Psi}\mathbf{B})\mathrm{vec}(\boldsymbol{\Sigma}\otimes\boldsymbol{\Sigma})+\mathrm{tr}(\boldsymbol{\Psi}\mathbf{A}\boldsymbol{\Psi}\mathbf{B}')\mathrm{vec}(\mathrm{vec}\boldsymbol{\Sigma}\mathrm{vec}'\boldsymbol{\Sigma})\\&\qquad+\mathrm{tr}(\boldsymbol{\Psi}\mathbf{A}\boldsymbol{\Psi}\mathbf{B})\mathrm{vec}(\mathbf{K}_{p,p}(\boldsymbol{\Sigma}\otimes\boldsymbol{\Sigma})),\end{aligned}$$

which implies that

$$\begin{aligned}E[(\mathbf{YAY}')\otimes(\mathbf{YBY}')]&=\mathrm{tr}(\boldsymbol{\Psi}\mathbf{A})\mathrm{tr}(\boldsymbol{\Psi}\mathbf{B})\boldsymbol{\Sigma}\otimes\boldsymbol{\Sigma}\\&\quad+\mathrm{tr}(\boldsymbol{\Psi}\mathbf{A}\boldsymbol{\Psi}\mathbf{B}')\mathrm{vec}\boldsymbol{\Sigma}\mathrm{vec}'\boldsymbol{\Sigma}+\mathrm{tr}(\boldsymbol{\Psi}\mathbf{A}\boldsymbol{\Psi}\mathbf{B})\mathbf{K}_{p,p}(\boldsymbol{\Sigma}\otimes\boldsymbol{\Sigma}).\end{aligned}\qquad(2.2.26)$$

Some calculations give that

$$E[(\mathbf{YA}\boldsymbol{\mu}')\otimes(\mathbf{YB}\boldsymbol{\mu}')]=E[\mathbf{Y}^{\otimes2}](\mathbf{A}\boldsymbol{\mu}'\otimes\mathbf{B}\boldsymbol{\mu}')=\mathrm{vec}\boldsymbol{\Sigma}\mathrm{vec}'(\boldsymbol{\mu}\mathbf{B}'\boldsymbol{\Psi}\mathbf{A}\boldsymbol{\mu}')\qquad(2.2.27)$$

and

$$E[(\boldsymbol{\mu}\mathbf{A}\mathbf{Y}')\otimes(\boldsymbol{\mu}\mathbf{B}\mathbf{Y}')] = (\boldsymbol{\mu}\mathbf{A}\otimes\boldsymbol{\mu}\mathbf{B})E[(\mathbf{Y}')^{\otimes 2}] = \mathrm{vec}(\boldsymbol{\mu}\mathbf{B}\boldsymbol{\Psi}\mathbf{A}'\boldsymbol{\mu}')\mathrm{vec}'\boldsymbol{\Sigma}. \quad (2.2.28)$$

Furthermore,

$$E[\mathbf{Y}\otimes\mathbf{Y}'] = \mathbf{K}_{p,n}(\boldsymbol{\Psi}\otimes\boldsymbol{\Sigma}),$$

which implies that

$$\begin{aligned} E[(\mathbf{Y}\mathbf{A}\boldsymbol{\mu}')\otimes(\boldsymbol{\mu}\mathbf{B}\mathbf{Y}')] &= (\mathbf{I}_p\otimes\boldsymbol{\mu}\mathbf{B})E[\mathbf{Y}\otimes\mathbf{Y}'](\mathbf{A}\boldsymbol{\mu}'\otimes\mathbf{I}_p)\\ &= (\mathbf{I}_p\otimes\boldsymbol{\mu}\mathbf{B})\mathbf{K}_{p,n}(\boldsymbol{\Psi}\otimes\boldsymbol{\Sigma})(\mathbf{A}\boldsymbol{\mu}'\otimes\mathbf{I}_p) = \mathbf{K}_{p,p}(\boldsymbol{\mu}\mathbf{B}\boldsymbol{\Psi}\mathbf{A}\boldsymbol{\mu}'\otimes\boldsymbol{\Sigma}) \quad (2.2.29)\end{aligned}$$

and

$$\begin{aligned} E[(\boldsymbol{\mu}\mathbf{A}\mathbf{Y}')\otimes(\mathbf{Y}\mathbf{B}\boldsymbol{\mu}')] &= E[(\mathbf{Y}\mathbf{A}'\boldsymbol{\mu}')\otimes(\boldsymbol{\mu}\mathbf{B}'\mathbf{Y}')]'\\ &= (\boldsymbol{\mu}\mathbf{A}\boldsymbol{\Psi}\mathbf{B}\boldsymbol{\mu}'\otimes\boldsymbol{\Sigma})\mathbf{K}_{p,p} = \mathbf{K}_{p,p}(\boldsymbol{\Sigma}\otimes\boldsymbol{\mu}\mathbf{A}\boldsymbol{\Psi}\mathbf{B}\boldsymbol{\mu}'). \quad (2.2.30)\end{aligned}$$

Now, by applying (i),

$$E[(\boldsymbol{\mu}\mathbf{A}\boldsymbol{\mu}')\otimes(\mathbf{Y}\mathbf{B}\mathbf{Y}')] = \mathrm{tr}(\boldsymbol{\Psi}\mathbf{B})\boldsymbol{\mu}\mathbf{A}\boldsymbol{\mu}'\otimes\boldsymbol{\Sigma} \quad (2.2.31)$$

and

$$E[(\mathbf{Y}\mathbf{A}\mathbf{Y}')\otimes(\boldsymbol{\mu}\mathbf{B}\boldsymbol{\mu}')] = \mathrm{tr}(\boldsymbol{\Psi}\mathbf{A})\boldsymbol{\Sigma}\otimes\boldsymbol{\mu}\mathbf{B}\boldsymbol{\mu}'. \quad (2.2.32)$$

Hence, using (2.2.26) – (2.2.32) in (2.2.25) proves (ii).
Finally (iii) is going to be established. Since

$$D[\mathbf{X}\mathbf{A}\mathbf{X}'] = E[\mathrm{vec}(\mathbf{X}\mathbf{A}\mathbf{X}')\mathrm{vec}'(\mathbf{X}\mathbf{A}\mathbf{X}')] - E[\mathrm{vec}(\mathbf{X}\mathbf{A}\mathbf{X}')]E[\mathrm{vec}'(\mathbf{X}\mathbf{A}\mathbf{X}')]$$

and since by (i) $E[\mathrm{vec}(\mathbf{X}\mathbf{A}\mathbf{X}')]$ is known, we only need an expression for $E[\mathrm{vec}(\mathbf{X}\mathbf{A}\mathbf{X}')\mathrm{vec}'(\mathbf{X}\mathbf{A}\mathbf{X}')]$. From Proposition 1.3.14 (iv) it follows that

$$\begin{aligned}(\mathbf{I}_p\otimes\mathbf{K}_{p,p}\otimes\mathbf{I}_p)\mathrm{vec}(E[(\mathbf{X}\mathbf{A}\mathbf{X}')\otimes(\mathbf{X}\mathbf{A}\mathbf{X}')]) &= E[\mathrm{vec}(\mathbf{X}\mathbf{A}\mathbf{X}')\otimes\mathrm{vec}(\mathbf{X}\mathbf{A}\mathbf{X}')]\\ &= \mathrm{vec}(E[\mathrm{vec}(\mathbf{X}\mathbf{A}\mathbf{X}')\mathrm{vec}'(\mathbf{X}\mathbf{A}\mathbf{X}')]).\end{aligned}$$

Therefore, by utilizing (ii) we obtain

$$\begin{aligned}&\mathrm{vec}(E[\mathrm{vec}(\mathbf{X}\mathbf{A}\mathbf{X}')\mathrm{vec}'(\mathbf{X}\mathbf{A}\mathbf{X}')])\\ &\quad=(\mathrm{tr}(\boldsymbol{\Psi}\mathbf{A}))^2\mathrm{vec}(\mathrm{vec}\boldsymbol{\Sigma}\mathrm{vec}'\boldsymbol{\Sigma}) + \mathrm{tr}(\boldsymbol{\Psi}\mathbf{A}\boldsymbol{\Psi}\mathbf{A}')\mathrm{vec}(\boldsymbol{\Sigma}\otimes\boldsymbol{\Sigma})\\ &\qquad+ \mathrm{tr}(\boldsymbol{\Psi}\mathbf{A}\boldsymbol{\Psi}\mathbf{A})\mathrm{vec}(\mathbf{K}_{p,p}\boldsymbol{\Sigma}\otimes\boldsymbol{\Sigma}) + \mathrm{tr}(\boldsymbol{\Psi}\mathbf{A})\mathrm{vec}(\mathrm{vec}(\boldsymbol{\mu}\mathbf{A}\boldsymbol{\mu}')\mathrm{vec}'\boldsymbol{\Sigma})\\ &\qquad+ \mathrm{vec}(\boldsymbol{\Sigma}\otimes\boldsymbol{\mu}\mathbf{A}\boldsymbol{\Psi}\mathbf{A}'\boldsymbol{\mu}') + \mathrm{vec}(\mathbf{K}_{p,p}(\boldsymbol{\mu}\mathbf{A}\boldsymbol{\Psi}\mathbf{A}\boldsymbol{\mu}'\otimes\boldsymbol{\Sigma}))\\ &\qquad+ \mathrm{vec}(\mathbf{K}_{p,p}(\boldsymbol{\Sigma}\otimes\boldsymbol{\mu}\mathbf{A}'\boldsymbol{\Psi}\mathbf{A}'\boldsymbol{\mu}')) + \mathrm{vec}(\boldsymbol{\mu}\mathbf{A}'\boldsymbol{\Psi}\mathbf{A}\boldsymbol{\mu}'\otimes\boldsymbol{\Sigma})\\ &\qquad+ \mathrm{tr}(\boldsymbol{\Psi}\mathbf{A})\mathrm{vec}(\mathrm{vec}\boldsymbol{\Sigma}\mathrm{vec}'(\boldsymbol{\mu}\mathbf{A}\boldsymbol{\mu}')) + \mathrm{vec}(\mathrm{vec}(\boldsymbol{\mu}\mathbf{A}\boldsymbol{\mu}')\mathrm{vec}'(\boldsymbol{\mu}\mathbf{A}\boldsymbol{\mu}')).\end{aligned}$$

Hence,

$$
\begin{aligned}
&E[\mathrm{vec}(\mathbf{XAX}')\mathrm{vec}'(\mathbf{XAX}')] \\
&\quad=(\mathrm{tr}(\mathbf{\Psi A}))^2\mathrm{vec}\mathbf{\Sigma}\mathrm{vec}'\mathbf{\Sigma}+\mathrm{tr}(\mathbf{\Psi A\Psi A}')\mathbf{\Sigma}\otimes\mathbf{\Sigma} \\
&\qquad+\mathrm{tr}(\mathbf{\Psi A})\mathrm{tr}(\mathbf{\Psi A})\mathbf{K}_{p,p}(\mathbf{\Sigma}\otimes\mathbf{\Sigma})+\mathrm{tr}(\mathbf{\Psi A})\mathrm{vec}(\boldsymbol{\mu}\mathbf{A}\boldsymbol{\mu}')\mathrm{vec}'\mathbf{\Sigma} \\
&\qquad+\mathbf{\Sigma}\otimes\boldsymbol{\mu}\mathbf{A\Psi A}'\boldsymbol{\mu}'+\mathbf{K}_{p,p}((\boldsymbol{\mu}\mathbf{A\Psi A}\boldsymbol{\mu}')\otimes\mathbf{\Sigma}) \\
&\qquad+\mathbf{K}_{p,p}(\mathbf{\Sigma}\otimes\boldsymbol{\mu}\mathbf{A}'\mathbf{\Psi A}'\boldsymbol{\mu}')+\boldsymbol{\mu}\mathbf{A}'\mathbf{\Psi A}\boldsymbol{\mu}'\otimes\mathbf{\Sigma} \\
&\qquad+\mathrm{tr}(\mathbf{\Psi A})\mathrm{vec}\mathbf{\Sigma}\mathrm{vec}'(\boldsymbol{\mu}\mathbf{A}\boldsymbol{\mu}')+\mathrm{vec}(\boldsymbol{\mu}\mathbf{A}\boldsymbol{\mu}')\mathrm{vec}'(\boldsymbol{\mu}\mathbf{A}\boldsymbol{\mu}'). \qquad (2.2.33)
\end{aligned}
$$

Combining (2.2.33) with the expression for $E[\mathrm{vec}(\mathbf{XAX}')]$ in (i) establishes (iii). ∎

For some results on arbitrary moments on quadratic forms see Kang & Kim (1996).

2.2.4 *Hermite polynomials*

If we intend to present a multivariate density or distribution function through a multivariate normal distribution, there exist expansions where multivariate densities or distribution functions, multivariate cumulants and multivariate Hermite polynomials will appear in the formulas. Finding explicit expressions for the Hermite polynomials of low order will be the topic of this paragraph while in the next chapter we are going to apply these results.

In the univariate case the class of orthogonal polynomials known as Hermite polynomials can be defined in several ways. We shall use the definition which starts from the normal density. The polynomial $h_k(x)$ is a *Hermite polynomial of order* k if it satisfies the following equality:

$$\frac{d^k f_x(x)}{dx^k}=(-1)^k h_k(x)f_x(x), \qquad k=0,1,2,\ldots, \qquad (2.2.34)$$

where $f_x(x)$ is the density function of the standard normal distribution $N(0,1)$. Direct calculations give the first Hermite polynomials:

$$
\begin{aligned}
h_0(x)&=1,\\
h_1(x)&=x,\\
h_2(x)&=x^2-1,\\
h_3(x)&=x^3-3x.
\end{aligned}
$$

In the multivariate case we are going to use a multivariate normal distribution $N_p(\boldsymbol{\mu},\mathbf{\Sigma})$ which gives us a possibility to define multivariate Hermite polynomials depending on two parameters, namely the mean $\boldsymbol{\mu}$ and the dispersion matrix $\mathbf{\Sigma}$. A general coordinate-free treatment of multivariate Hermite polynomials has been given by Holmquist (1996), and in certain tensor notation they appear in the books by McCullagh (1987) and Barndorff-Nielsen & Cox (1989). A matrix representation was first given by Traat (1986) on the basis of the matrix derivative of MacRae (1974). In our notation multivariate Hermite polynomials were given by Kollo (1991), when $\boldsymbol{\mu}=\mathbf{0}$.

Definition 2.2.2. *The matrix $\mathbf{H}_k(\mathbf{x}, \boldsymbol{\mu}, \boldsymbol{\Sigma})$ is called multivariate Hermite polynomial of order k for the vector $\boldsymbol{\mu}$ and the matrix $\boldsymbol{\Sigma} > 0$, if it satisfies the equality:*

$$\frac{d^k f_{\mathbf{x}}(\mathbf{x})}{d\mathbf{x}^k} = (-1)^k \mathbf{H}_k(\mathbf{x}, \boldsymbol{\mu}, \boldsymbol{\Sigma}) f_{\mathbf{x}}(\mathbf{x}), \quad k = 0, 1, \ldots, \tag{2.2.35}$$

where $\dfrac{d^k}{d\mathbf{x}^k}$ is given by Definition 1.4.1 and Definition 1.4.3, and $f_{\mathbf{x}}(\mathbf{x})$ is the density function of the normal distribution $N_p(\boldsymbol{\mu}, \boldsymbol{\Sigma})$:

$$f_{\mathbf{x}}(\mathbf{x}) = (2\pi)^{-\frac{p}{2}} |\boldsymbol{\Sigma}|^{-\frac{1}{2}} \exp(-\frac{1}{2}(\mathbf{x} - \boldsymbol{\mu})'\boldsymbol{\Sigma}^{-1}(\mathbf{x} - \boldsymbol{\mu})).$$

∎

Usually Hermite polynomials are defined for a centered normal distribution. In the literature we find several generalizations of Hermite polynomials in different directions. Viskov (1991) considers Hermite polynomials as derivatives of an exponential function where the dispersion matrix $\boldsymbol{\Sigma}$ in the normal density in Definition 2.2.2 is replaced by an arbitrary non-singular square matrix. Chikuse (1992a, 1992b) develops a theory for Hermite polynomials with a symmetric (but not necessarily positive definite) matrix argument. For our purposes we get the necessary notion from Definition 2.2.2. The explicit formulas for the first three Hermite polynomials will be given in the next theorem.

Theorem 2.2.10. *Multivariate Hermite polynomials $\mathbf{H}_k(\mathbf{x}, \boldsymbol{\mu}, \boldsymbol{\Sigma})$, $k = 0, 1, 2, 3$, are of the form:*

(i) $H_0(\mathbf{x}, \boldsymbol{\mu}, \boldsymbol{\Sigma}) = 1;$

(ii) $$\mathbf{H}_1(\mathbf{x}, \boldsymbol{\mu}, \boldsymbol{\Sigma}) = \boldsymbol{\Sigma}^{-1}(\mathbf{x} - \boldsymbol{\mu}); \tag{2.2.36}$$

(iii) $$\mathbf{H}_2(\mathbf{x}, \boldsymbol{\mu}, \boldsymbol{\Sigma}) = \boldsymbol{\Sigma}^{-1}(\mathbf{x} - \boldsymbol{\mu})(\mathbf{x} - \boldsymbol{\mu})'\boldsymbol{\Sigma}^{-1} - \boldsymbol{\Sigma}^{-1}; \tag{2.2.37}$$

(iv) $$\begin{aligned}\mathbf{H}_3(\mathbf{x}, \boldsymbol{\mu}, \boldsymbol{\Sigma}) = {} & \boldsymbol{\Sigma}^{-1}(\mathbf{x} - \boldsymbol{\mu})((\mathbf{x} - \boldsymbol{\mu})'\boldsymbol{\Sigma}^{-1})^{\otimes 2} - \boldsymbol{\Sigma}^{-1}(\mathbf{x} - \boldsymbol{\mu})\mathrm{vec}'\boldsymbol{\Sigma}^{-1} \\ & - \{(\mathbf{x} - \boldsymbol{\mu})'\boldsymbol{\Sigma}^{-1}\} \otimes \boldsymbol{\Sigma}^{-1} - \boldsymbol{\Sigma}^{-1} \otimes \{(\mathbf{x} - \boldsymbol{\mu})'\boldsymbol{\Sigma}^{-1}\}.\end{aligned} \tag{2.2.38}$$

PROOF: For $k = 0$, the relation in Definition 2.2.2 turns into a trivial identity and we obtain that

$$H_0(\mathbf{x}, \boldsymbol{\mu}, \boldsymbol{\Sigma}) = 1.$$

For $k = 1$ the first order derivative equals

$$
\begin{aligned}
&\frac{df_{\mathbf{x}}(\mathbf{x})}{d\mathbf{x}} \\
&= (2\pi)^{-\frac{p}{2}}|\boldsymbol{\Sigma}|^{-\frac{1}{2}}\frac{d\exp(-\frac{1}{2}(\mathbf{x}-\boldsymbol{\mu})'\boldsymbol{\Sigma}^{-1}(\mathbf{x}-\boldsymbol{\mu}))}{d\mathbf{x}} \\
&\underset{(1.4.14)}{=} (2\pi)^{-\frac{p}{2}}|\boldsymbol{\Sigma}|^{-\frac{1}{2}}\exp(-\frac{1}{2}(\mathbf{x}-\boldsymbol{\mu})'\boldsymbol{\Sigma}^{-1}(\mathbf{x}-\boldsymbol{\mu}))\frac{d(-\frac{1}{2}(\mathbf{x}-\boldsymbol{\mu})'\boldsymbol{\Sigma}^{-1}(\mathbf{x}-\boldsymbol{\mu}))}{d\mathbf{x}} \\
&\underset{(1.4.19)}{=} -\frac{1}{2}f_{\mathbf{x}}(\mathbf{x})\{\frac{d(\mathbf{x}-\boldsymbol{\mu})'}{d\mathbf{x}}\boldsymbol{\Sigma}^{-1}(\mathbf{x}-\boldsymbol{\mu}) + \frac{d(\mathbf{x}-\boldsymbol{\mu})}{d\mathbf{x}}\boldsymbol{\Sigma}^{-1}(\mathbf{x}-\boldsymbol{\mu})\} \\
&\underset{(1.4.17)}{=} -\frac{1}{2}f_{\mathbf{x}}(\mathbf{x})\{\boldsymbol{\Sigma}^{-1}(\mathbf{x}-\boldsymbol{\mu}) + \boldsymbol{\Sigma}^{-1}(\mathbf{x}-\boldsymbol{\mu})\} = -f_{\mathbf{x}}(\mathbf{x})\boldsymbol{\Sigma}^{-1}(\mathbf{x}-\boldsymbol{\mu}).
\end{aligned}
$$

Thus, Definition 2.2.2 yields (ii). To prove (iii) we have to find the second order derivative, i.e.

$$
\begin{aligned}
\frac{d^2 f_{\mathbf{x}}(\mathbf{x})}{d\mathbf{x}^2} = \frac{d\frac{df}{d\mathbf{x}}}{d\mathbf{x}} &= -\frac{d(f_{\mathbf{x}}(\mathbf{x})\boldsymbol{\Sigma}^{-1}(\mathbf{x}-\boldsymbol{\mu}))}{d\mathbf{x}} \\
&= -\frac{d(\mathbf{x}-\boldsymbol{\mu})'}{d\mathbf{x}}\boldsymbol{\Sigma}^{-1}f_{\mathbf{x}}(\mathbf{x}) - \frac{df_{\mathbf{x}}(\mathbf{x})}{d\mathbf{x}}(\mathbf{x}-\boldsymbol{\mu})'\boldsymbol{\Sigma}^{-1} \\
&= -\boldsymbol{\Sigma}^{-1}f_{\mathbf{x}}(\mathbf{x}) + \boldsymbol{\Sigma}^{-1}(\mathbf{x}-\boldsymbol{\mu})(\mathbf{x}-\boldsymbol{\mu})'\boldsymbol{\Sigma}^{-1}f_{\mathbf{x}}(\mathbf{x}) \\
&= f_{\mathbf{x}}(\mathbf{x})(\boldsymbol{\Sigma}^{-1}(\mathbf{x}-\boldsymbol{\mu})(\mathbf{x}-\boldsymbol{\mu})'\boldsymbol{\Sigma}^{-1} - \boldsymbol{\Sigma}^{-1}).
\end{aligned}
$$

Thus, (iii) is verified.
To prove (iv), the function $f_{\mathbf{x}}(\mathbf{x})$ has to be differentiated once more and it follows that

$$
\begin{aligned}
\frac{d^3 f_{\mathbf{x}}(\mathbf{x})}{d\mathbf{x}^3} &= \frac{d\frac{d^2 f_{\mathbf{x}}(\mathbf{x})}{d\mathbf{x}^2}}{d\mathbf{x}} = \frac{d[f_{\mathbf{x}}(\mathbf{x})(\boldsymbol{\Sigma}^{-1}(\mathbf{x}-\boldsymbol{\mu})(\mathbf{x}-\boldsymbol{\mu})'\boldsymbol{\Sigma}^{-1})]}{d\mathbf{x}} - \frac{d(f_{\mathbf{x}}(\mathbf{x})\boldsymbol{\Sigma}^{-1})}{d\mathbf{x}} \\
&\underset{(1.3.31)}{=} \frac{d\{(\boldsymbol{\Sigma}^{-1}(\mathbf{x}-\boldsymbol{\mu})\otimes\boldsymbol{\Sigma}^{-1}(\mathbf{x}-\boldsymbol{\mu}))f_{\mathbf{x}}(\mathbf{x})\}}{d\mathbf{x}} - \frac{df_{\mathbf{x}}(\mathbf{x})\mathrm{vec}'\boldsymbol{\Sigma}^{-1}}{d\mathbf{x}} \\
&\underset{(1.4.23)}{=} f_{\mathbf{x}}(\mathbf{x})\frac{d(\boldsymbol{\Sigma}^{-1}(\mathbf{x}-\boldsymbol{\mu})\otimes\boldsymbol{\Sigma}^{-1}(\mathbf{x}-\boldsymbol{\mu}))}{d\mathbf{x}} \\
&\quad + \frac{df_{\mathbf{x}}(\mathbf{x})}{d\mathbf{x}}(\mathbf{x}-\boldsymbol{\mu})'\boldsymbol{\Sigma}^{-1}\otimes(\mathbf{x}-\boldsymbol{\mu})'\boldsymbol{\Sigma}^{-1} - \frac{df_{\mathbf{x}}(\mathbf{x})}{d\mathbf{x}}\mathrm{vec}'\boldsymbol{\Sigma}^{-1} \\
&\underset{(1.4.23)}{=} f_{\mathbf{x}}(\mathbf{x})\{(\mathbf{x}-\boldsymbol{\mu})'\boldsymbol{\Sigma}^{-1}\otimes\boldsymbol{\Sigma}^{-1} + \boldsymbol{\Sigma}^{-1}\otimes(\mathbf{x}-\boldsymbol{\mu})'\boldsymbol{\Sigma}^{-1}\} \\
&\quad - \boldsymbol{\Sigma}^{-1}(\mathbf{x}-\boldsymbol{\mu})((\mathbf{x}-\boldsymbol{\mu})'\boldsymbol{\Sigma}^{-1})^{\otimes 2} + \boldsymbol{\Sigma}^{-1}(\mathbf{x}-\boldsymbol{\mu})\mathrm{vec}'\boldsymbol{\Sigma}^{-1}.
\end{aligned}
$$

Hence, from Definition 2.2.2 the expression for $\mathbf{H}_3(\mathbf{x},\boldsymbol{\mu},\boldsymbol{\Sigma})$ is obtained. ∎

In most applications the centered multivariate normal distribution $N_p(\mathbf{0},\boldsymbol{\Sigma})$ is used. In this case Hermite polynomials have a slightly simpler form, and we denote them by $\mathbf{H}_k(\mathbf{x},\boldsymbol{\Sigma})$. The first three Hermite polynomials $\mathbf{H}_k(\mathbf{x},\boldsymbol{\Sigma})$ are given by the following corollary.

Corollary 2.2.10.1. *Multivariate Hermite polynomials* $\mathbf{H}_k(\mathbf{x}, \mathbf{\Sigma})$, $k = 0, 1, 2, 3$, *are of the form:*

(i) $H_0(\mathbf{x}, \mathbf{\Sigma}) = 1;$

(ii) $$\mathbf{H}_1(\mathbf{x}, \mathbf{\Sigma}) = \mathbf{\Sigma}^{-1}\mathbf{x}; \qquad (2.2.39)$$

(iii) $$\mathbf{H}_2(\mathbf{x}, \mathbf{\Sigma}) = \mathbf{\Sigma}^{-1}\mathbf{x}\mathbf{x}'\mathbf{\Sigma}^{-1} - \mathbf{\Sigma}^{-1}; \qquad (2.2.40)$$

(iv) $$\mathbf{H}_3(\mathbf{x}, \mathbf{\Sigma}) = \mathbf{\Sigma}^{-1}\mathbf{x}(\mathbf{x}'\mathbf{\Sigma}^{-1})^{\otimes 2} - \mathbf{\Sigma}^{-1}\mathbf{x}\mathrm{vec}'\mathbf{\Sigma}^{-1} - (\mathbf{x}'\mathbf{\Sigma}^{-1} \otimes \mathbf{\Sigma}^{-1}) - (\mathbf{\Sigma}^{-1} \otimes \mathbf{x}'\mathbf{\Sigma}^{-1}). \qquad (2.2.41)$$

∎

In the case $\boldsymbol{\mu} = \mathbf{0}$ and $\mathbf{\Sigma} = \mathbf{I}_p$ the formulas, which follow from Theorem 2.2.10, are multivariate versions of univariate Hermite polynomials.

Corollary 2.2.10.2. *Multivariate Hermite polynomials* $H_k(\mathbf{x}, \mathbf{I}_p)$, $k = 0, 1, 2, 3$, *equal:*

(i) $H_0(\mathbf{x}, \mathbf{I}_p) = 1;$

(ii) $\mathbf{H}_1(\mathbf{x}, \mathbf{I}_p) = \mathbf{x};$

(iii) $\mathbf{H}_2(\mathbf{x}, \mathbf{I}_p) = \mathbf{x}\mathbf{x}' - \mathbf{I}_p;$

(iv) $\mathbf{H}_3(\mathbf{x}, \mathbf{I}_p) = \mathbf{x}(\mathbf{x}')^{\otimes 2} - \mathbf{x}\mathrm{vec}'\mathbf{I}_p - (\mathbf{x}' \otimes \mathbf{I}_p) - (\mathbf{I}_p \otimes \mathbf{x}').$

∎

Up to now we have been thinking of Hermite polynomials as derivatives of the normal density. However, when the mean is zero, the characteristic function of the normal distribution has the same structure as that of the multivariate normal density function. Hence, one can immediately imagine that the moments of a multivariate normal distribution are connected to the Hermite polynomials, and the next theorem is a consequence of this fact.

Theorem 2.2.11. *Let* $\mathbf{x} \sim N_p(\mathbf{0}, \mathbf{\Sigma})$. *Then*

$$m_k[\mathbf{x}] = \frac{1}{i^k}(-1)^k\mathbf{H}_k(\mathbf{0}, \mathbf{\Sigma}^{-1}), \qquad k = 2, 4, 6, \ldots$$

∎

Furthermore, since recursive relations are given for the derivatives of the characteristic function in Lemma 2.2.1, we may follow up this result and present a recursive relation for the Hermite polynomials $\mathbf{H}_k(\mathbf{x}, \mathbf{\Sigma})$.

Theorem 2.2.12. *Let $\mathbf{H}_k(\mathbf{x},\boldsymbol{\Sigma})$ be given by Definition 2.2.2, when $\boldsymbol{\mu}=\mathbf{0}$. Then, if $k>1$,*

$$\mathbf{H}_k(\mathbf{x},\boldsymbol{\Sigma}) = \mathbf{x}'\boldsymbol{\Sigma}^{-1}\mathbf{H}_{k-1}(\mathbf{x},\boldsymbol{\Sigma}) - \boldsymbol{\Sigma}^{-1}\otimes \mathrm{vec}\mathbf{H}_{k-2}(\mathbf{x},\boldsymbol{\Sigma})$$
$$- (\mathrm{vec}'\boldsymbol{\Sigma}^{-1}\otimes \mathbf{H}_{k-2}(\mathbf{x},\boldsymbol{\Sigma}))\sum_{i=0}^{k-3}(\mathbf{I}_p\otimes \mathbf{K}_{p,p^{k-i-3}}\otimes \mathbf{I}_{p^i}).$$

∎

An interesting fact about Hermite polynomials is that they are orthogonal, i.e. $E[\mathrm{vec}\mathbf{H}_k(\mathbf{x},\boldsymbol{\Sigma})\mathrm{vec}'\mathbf{H}_l(\mathbf{x},\boldsymbol{\Sigma})]=\mathbf{0}$.

Theorem 2.2.13. *Let $\mathbf{H}_k(\mathbf{x},\boldsymbol{\Sigma})$ be given in Definition 2.2.2, where $\boldsymbol{\mu}=\mathbf{0}$. Then, if $k\neq l$,*

$$E[\mathrm{vec}\mathbf{H}_k(\mathbf{x},\boldsymbol{\Sigma})\mathrm{vec}'\mathbf{H}_l(\mathbf{x},\boldsymbol{\Sigma})]=\mathbf{0}.$$

PROOF: First, the correspondence between (2.2.35) and (1.4.67) is noted. Thereafter it is recognized that (1.4.68) holds because the exponential function converges to 0 faster than $\mathbf{H}_k(\mathbf{x},\boldsymbol{\Sigma})\to\pm\infty$, when any component in $\mathbf{x}\to\pm\infty$. Hence, (1.4.70) establishes the theorem. ∎

The Hermite polynomials, or equivalently the derivatives of the normal density function, may be useful when obtaining error bounds of expansions. This happens when derivatives of the normal density function appear in approximation formulas. With the help of the next theorem error bounds, independent of the argument $\mathbf{x}$ of the density function, can be found.

Theorem 2.2.14. *Let $\mathbf{Z}\sim N_{p,n}(\mathbf{0},\boldsymbol{\Sigma},\mathbf{I})$, $f_{\mathbf{Z}}(\mathbf{X})$ denote the corresponding density and $f^k_{\mathbf{Z}}(\mathbf{X})$ the $k-$th derivative of the density. Then, for any matrix $\mathbf{A}$ of proper size,*

(i) $|\mathrm{tr}(\mathbf{A}^{\otimes 2s}f^{2s}_{\mathbf{Z}}(\mathbf{X}))|\leq \mathrm{tr}(\mathbf{A}^{\otimes 2s}f^{2s}_{\mathbf{Z}}(\mathbf{0}))$;

(ii) $|\mathrm{tr}(\mathbf{A}^{\otimes 2s}\mathbf{H}_{2s}(\mathrm{vec}\mathbf{X},\boldsymbol{\Sigma})f_{\mathbf{Z}}(\mathbf{X}))|\leq (2\pi)^{-pn/2}|\boldsymbol{\Sigma}|^{-n/2}\mathrm{tr}(\mathbf{A}^{\otimes 2s}\mathbf{H}_{2s}(\mathrm{vec}\mathbf{X},\boldsymbol{\Sigma}))$.

PROOF: The statement in (ii) follows from (i) and Definition 2.2.2. In order to show (i) we make use of Corollary 3.2.1.L2, where the inverse Fourier transform is given and the derivatives of a density function are represented using the characteristic function.

Hence, from Corollary 3.2.1.L2 it follows that

$$\begin{aligned}
|\mathrm{tr}(\mathbf{A}^{\otimes 2s}f^{2s}_{\mathbf{Z}}(\mathbf{X}))| &= |\mathrm{vec}'(\mathbf{A}'^{\otimes 2s})\mathrm{vec}(f^{2s}_{\mathbf{Z}}(\mathbf{X}))|\\
&= |(2\pi)^{-pn}\int_{\mathbb{R}^{pn}}\varphi_{\mathbf{Z}}(\mathbf{T})\mathrm{vec}'(\mathbf{A}'^{\otimes 2s})(i\mathrm{vec}\mathbf{T})^{\otimes 2s}e^{-i\mathrm{tr}(\mathbf{T}'\mathbf{X})}d\mathbf{T}|\\
&\leq (2\pi)^{-pn}\int_{\mathbb{R}^{pn}}\varphi_{\mathbf{Z}}(\mathbf{T})|\mathrm{vec}'(\mathbf{A}'^{\otimes 2s})(i\mathrm{vec}\mathbf{T})^{\otimes 2s}|d\mathbf{T}\\
&= (2\pi)^{-pn}\int_{\mathbb{R}^{pn}}\varphi_{\mathbf{Z}}(\mathbf{T})(\mathrm{tr}(\mathbf{A}\mathbf{T}))^{2s}d\mathbf{T}\\
&= (2\pi)^{-pn}\int_{\mathbb{R}^{pn}}\varphi_{\mathbf{Z}}(\mathbf{T})\mathrm{vec}'(\mathbf{A}'^{\otimes 2s})(i\mathrm{vec}\mathbf{T})^{\otimes 2s}d\mathbf{T}\\
&= \mathrm{vec}'(\mathbf{A}'^{\otimes 2s})\mathrm{vec}(f^{2s}_{\mathbf{Z}}(\mathbf{0})) = \mathrm{tr}(\mathbf{A}^{\otimes 2s}f^{2s}_{\mathbf{Z}}(\mathbf{0})),
\end{aligned}$$

which establishes the theorem. ∎

2.2.5 *Multilinear normal distribution*

The matrix normal distribution can be regarded as a bilinear normal distribution. A straightforward step further leads to the multilinear normal distribution. Since algebra for treating the distribution is similar to the algebra applied in previous parts of this section, we consider the extension of the "bilinear" normal here. First, it is observed that a matrix normally distributed $\mathbf{X} : p \times n$, i.e. $\mathbf{X} \sim N_{p,n}(\boldsymbol{\mu}, \boldsymbol{\Sigma}, \boldsymbol{\Psi})$, can be written as

$$\sum_{ij} X_{ij}\mathbf{e}_i^1(\mathbf{e}_j^2)' = \sum_{ij} \mu_{ij}\mathbf{e}_i^1(\mathbf{e}_j^2)' + \sum_{ik}\sum_{nl}\sum_{mj} \tau_{ik}\gamma_{mj}U_{nl}\mathbf{e}_i^1(\mathbf{e}_k^1)'\mathbf{e}_n^1(\mathbf{e}_l^2)'\mathbf{e}_m^2(\mathbf{e}_j^2)',$$

where $\boldsymbol{\Sigma} = \boldsymbol{\tau}\boldsymbol{\tau}'$, $\boldsymbol{\Psi} = \boldsymbol{\gamma}\boldsymbol{\gamma}'$, $\mathbf{e}_i^1 : p \times 1$, $\mathbf{e}_j^2 : n \times 1$ are the unit basis vectors and $U_{nl} \sim N(0,1)$. This expression equals

$$\sum_{ij} X_{ij}\mathbf{e}_i^1(\mathbf{e}_j^2)' = \sum_{ij} \mu_{ij}\mathbf{e}_i^1(\mathbf{e}_j^2)' + \sum_{ij}\sum_{kl} \tau_{ik}\gamma_{mj}U_{km}\mathbf{e}_i^1(\mathbf{e}_j^2)'.$$

If the products of basis vectors are rearranged, i.e. $\mathbf{e}_i^1(\mathbf{e}_j^2)' \to \mathbf{e}_j^2 \otimes \mathbf{e}_i^1$, the vec-representation of the matrix normal distribution is obtained. The calculations and ideas above motivate the following extension of the matrix normal distribution.

Definition 2.2.3. *A matrix* $\mathbf{X}$ *is multilinear normal of order* k,

$$\mathbf{X} \sim N_{p_1,p_2,\ldots,p_k}(\boldsymbol{\mu}, \boldsymbol{\Sigma}_k, \boldsymbol{\Sigma}_1, \boldsymbol{\Sigma}_2, \ldots, \boldsymbol{\Sigma}_{k-1}),$$

if

$$\sum_{i_1,i_2,\ldots,i_k} X_{i_1\ldots i_k}\mathbf{e}_{i_1}^1 \otimes \mathbf{e}_{i_2}^2 \otimes \cdots \otimes \mathbf{e}_{i_k}^k = \sum_{i_1,i_2,\ldots,i_k} \mu_{i_1\ldots i_k}\mathbf{e}_{i_1}^1 \otimes \mathbf{e}_{i_2}^2 \otimes \cdots \otimes \mathbf{e}_{i_k}^k$$
$$+\sum_{i_1}^{p_1}\sum_{i_2}^{p_2}\cdots\sum_{i_k}^{p_k}\sum_{j_1}^{p_1}\sum_{j_2}^{p_2}\cdots\sum_{j_k}^{p_k} \tau_{i_1j_1}^1\tau_{i_2j_2}^2\cdots\tau_{i_kj_k}^k U_{j_1j_2\ldots j_k}\mathbf{e}_{i_1}^1 \otimes \mathbf{e}_{i_2}^2 \otimes \cdots \otimes \mathbf{e}_{i_k}^k,$$

where $\boldsymbol{\Sigma}_i = \boldsymbol{\tau}^i(\boldsymbol{\tau}^i)'$, $\mathbf{e}_{i_r}^r : p_r \times 1$ *and* $U_{j_1j_2\cdots j_k} \sim N(0,1)$. ∎

From Definition 2.2.3 it follows immediately, that if omitting the basis vectors, we have a multivariate normal distribution represented in a coordinate free language, namely

$$X_{i_1\ldots i_k} = \mu_{i_1\ldots i_k} + \sum_{j_1j_2\ldots j_k} \tau_{i_1j_1}^1\tau_{i_2j_2}^2\cdots\tau_{i_kj_k}^k U_{j_1j_2\ldots j_k}.$$

The next theorem spells out the connection between the matrix normal and the multilinear normal distributions.

Theorem 2.2.15. *Let* $\mathbf{X} \sim N_{p,n}(\boldsymbol{\mu}, \boldsymbol{\Sigma}_k, \boldsymbol{\Psi})$, *where*

$$\boldsymbol{\Psi} = \boldsymbol{\Sigma}_1 \otimes \boldsymbol{\Sigma}_2 \otimes \cdots \otimes \boldsymbol{\Sigma}_{k-1},$$

$\boldsymbol{\Sigma}_i : p_i \times p_i$, $i = 1, 2, \ldots, k-1$, *and* $\sum_{i=1}^{k-1} p_i = n$. *Then*

$$\mathbf{X} \sim N_{p,p_1,p_2,\ldots,p_{k-1}}(\boldsymbol{\mu}, \boldsymbol{\Sigma}_k, \boldsymbol{\Sigma}_1, \boldsymbol{\Sigma}_2, \ldots, \boldsymbol{\Sigma}_{k-1}).$$

PROOF: First, it is noted that the set $\{\mathbf{e}_j\}_{j=1}^n$, $\mathbf{e}_j : n \times 1$, is identical to

$$\{\mathbf{e}^1_{i_1} \otimes \mathbf{e}^2_{i_2} \otimes \cdots \otimes \mathbf{e}^{k-1}_{i_{k-1}}\}^{p_j}_{i_j=1,\, j=1,2,\ldots,k-1}, \tag{2.2.42}$$

where $\mathbf{e}^j_{i_j} : p_j \times 1$ and $n = \sum_{j=1}^{k-1} p_j$. Now, by assumption

$$\boldsymbol{\Psi} = \sum_{i_1,i_2,\ldots,i_{k-1}} \sum_{j_1,j_2,\ldots,j_{k-1}} \gamma^1_{i_1j_1}\gamma^2_{i_2j_2}\cdots\gamma^{k-1}_{i_{k-1}j_{k-1}} \mathbf{e}^1_{i_1}(\mathbf{e}^1_{j_1})' \otimes \mathbf{e}^2_{i_2}(\mathbf{e}^2_{j_2})' \otimes \cdots \otimes \mathbf{e}^{k-1}_{i_{k-1}}(\mathbf{e}^{k-1}_{j_{k-1}})'$$

and vec$\mathbf{X}$ can be written as

$$\sum_{i_k,i_v} X_{i_ki_v}\mathbf{e}_{i_v} \otimes \mathbf{e}^k_{i_k} = \sum_{i_k,i_v} \mu_{i_ki_v}\mathbf{e}_{i_v} \otimes \mathbf{e}^k_{i_k} + \sum_{i_k,i_v}\sum_{j_k,j_v} \tau^k_{i_kj_k}\gamma^v_{i_vj_v}U_{j_kj_v}\mathbf{e}_{i_v} \otimes \mathbf{e}^k_{i_k}, \tag{2.2.43}$$

where $\mathbf{e}_{i_v} : n \times 1$, $\mathbf{e}^k_{i_k} : p \times 1$. From (2.2.42) it follows that $\mathbf{e}_{i_v}$ may be replaced by elements from $\{\mathbf{e}^1_{i_1} \otimes \mathbf{e}^2_{i_2} \otimes \cdots \otimes \mathbf{e}^{k-1}_{i_{k-1}}\}$, and since

$$\boldsymbol{\Psi} = \boldsymbol{\gamma}\boldsymbol{\gamma}' = \boldsymbol{\gamma}_1\boldsymbol{\gamma}_1' \otimes \boldsymbol{\gamma}_2\boldsymbol{\gamma}_2' \otimes \cdots \otimes \boldsymbol{\gamma}_{k-1}\boldsymbol{\gamma}_{k-1}' = (\boldsymbol{\gamma}_1 \otimes \boldsymbol{\gamma}_2 \otimes \cdots \otimes \boldsymbol{\gamma}_{k-1})(\boldsymbol{\gamma}_1 \otimes \boldsymbol{\gamma}_2 \otimes \cdots \otimes \boldsymbol{\gamma}_{k-1})'$$

we may replace $\gamma^v_{i_vj_v}$ by $\gamma^v_{i_vj_v} = \gamma^1_{i_1j_1}\gamma^2_{i_2j_2}\cdots\gamma^{k-1}_{i_{k-1}j_{k-1}}$ in (2.2.43). Thus, (2.2.43) is equivalent to Definition 2.2.3. ∎

In Definition 2.2.3 we have an arbitrary $\boldsymbol{\mu}$. However, it is much more interesting to connect the mean structure with the dispersion structure $\boldsymbol{\Sigma}_1 \otimes \cdots \otimes \boldsymbol{\Sigma}_k$. For example, consider vec$\boldsymbol{\mu} = (\mathbf{F}_1 \otimes \cdots \otimes \mathbf{F}_k)\boldsymbol{\delta}$. Next it is supposed that this structure holds, and then it is said that $\mathbf{X}$ is multilinear normal with mean structure.

Definition 2.2.4. *The matrix* $\mathbf{X}$ *is multilinear normal of order* k *with mean structure if* $\mathbf{X} \sim N_{p_1,\ldots,p_k}(\boldsymbol{\mu}, \boldsymbol{\Sigma}_k, \boldsymbol{\Sigma}_1, \ldots, \boldsymbol{\Sigma}_{k-1})$ *and* vec$\boldsymbol{\mu} = (\mathbf{F}_1 \otimes \cdots \otimes \mathbf{F}_k)\boldsymbol{\delta}$, *where* $\mathbf{F}_i : p_i \times q_i$. *It will be denoted*

$$\mathbf{X} \sim N_{p_1\ldots p_k}(\mathbf{F}_1, \ldots, \mathbf{F}_k; \boldsymbol{\delta}, \boldsymbol{\Sigma}_k, \boldsymbol{\Sigma}_1, \ldots, \boldsymbol{\Sigma}_{k-1}).$$

∎

From Definition 2.2.4 it follows that vec$\mathbf{X}$ has the same distribution as

$$(\mathbf{F}_1 \otimes \cdots \otimes \mathbf{F}_k)\boldsymbol{\delta} + \{(\boldsymbol{\Sigma}_1)^{1/2} \otimes \cdots \otimes (\boldsymbol{\Sigma}_k)^{1/2}\}\mathbf{u},$$

where the elements of $\mathbf{u}$ are independent $N(0,1)$. For the results presented in Theorem 2.2.2 and Theorem 2.2.6 there exist analogous results for the multilinear normal distribution. Some of them are presented below. Let

$$\mathbf{X}_{\bullet r_l \bullet} = \sum_{i_1 \ldots i_{r-1} i_{r+1} \ldots i_k} \sum_{i_r=1}^{l} X_{i_1 \ldots i_k} \mathbf{e}^1_{i_1} \otimes \cdots \otimes \mathbf{e}^{r-1}_{i_{r-1}} \otimes \mathbf{e}^{r_l}_{i_r} \otimes \mathbf{e}^{r+1}_{i_{r+1}} \otimes \cdots \otimes \mathbf{e}^k_{i_k}, \tag{2.2.44}$$

$$\mathbf{X}_{\bullet r^l \bullet} = \sum_{i_1 \ldots i_{r-1} i_{r+1} \ldots i_k} \sum_{i_r=l+1}^{p_r} X_{i_1 \ldots i_k} \mathbf{e}^1_{i_1} \otimes \cdots \otimes \mathbf{e}^{r-1}_{i_{r-1}} \otimes \mathbf{e}^{r^l}_{i_r} \otimes \mathbf{e}^{r+1}_{i_{r+1}} \otimes \cdots \otimes \mathbf{e}^k_{i_k} \tag{2.2.45}$$

with $\mathbf{e}^{r_l}_{i_r} : l \times 1$ and $\mathbf{e}^{r^l}_{i_r} : (p_r - l) \times 1$. The special cases $r = k$ or $r = 1$ follow immediately, although a proper notation when $r = k$ would be $\mathbf{X}_{\bullet r_k}$, and $\mathbf{X}_{\bullet r^k}$. Furthermore, let

$$\boldsymbol{\mu}_{\bullet r_l \bullet} = \sum_{i_1 \ldots i_{r-1} i_{r+1} \ldots i_k} \sum_{i_r=1}^{l} \boldsymbol{\mu}_{i_1 \ldots i_k} e^1_{i_1} \otimes \cdots \otimes \mathbf{e}^{r-1}_{i_{r-1}} \otimes \mathbf{e}^{r_l}_{i_r} \otimes \mathbf{e}^{r+1}_{i_{r+1}} \otimes \cdots \otimes \mathbf{e}^k_{i_k}, \tag{2.2.46}$$

$$\boldsymbol{\mu}_{\bullet r^l \bullet} = \sum_{i_1 \ldots i_{r-1} i_{r+1} \ldots i_k} \sum_{i_r=l+1}^{p_r} \boldsymbol{\mu}_{i_1 \ldots i_k} \mathbf{e}^1_{i_1} \otimes \cdots \otimes \mathbf{e}^{r-1}_{i_{r-1}} \otimes \mathbf{e}^{r^l}_{i_r} \otimes \mathbf{e}^{r+1}_{i_{r+1}} \otimes \cdots \otimes \mathbf{e}^k_{i_k}, \tag{2.2.47}$$

$$\boldsymbol{\Sigma}_r = \begin{pmatrix} \boldsymbol{\Sigma}^r_{11} & \boldsymbol{\Sigma}^r_{12} \\ \boldsymbol{\Sigma}^r_{21} & \boldsymbol{\Sigma}^r_{22} \end{pmatrix} \quad \begin{pmatrix} l \times l & l \times (p_r - l) \\ (p_r - l) \times l & (p_r - l) \times (p_r - l) \end{pmatrix}$$

and

$$\mathbf{F}_r = \begin{pmatrix} \mathbf{F}^r_1 \\ \mathbf{F}^r_2 \end{pmatrix} \quad \begin{pmatrix} l \times 1 \\ (p_r - l) \times 1 \end{pmatrix}. \tag{2.2.48}$$

It follows from (2.2.44) and (2.2.45) that

$$\begin{pmatrix} \mathbf{X}_{\bullet r_l \bullet} \\ \mathbf{X}_{\bullet r^l \bullet} \end{pmatrix} = \begin{pmatrix} \mathbf{I}_{p_1 + \cdots + p_{r-1}} \otimes (\mathbf{I} : \mathbf{0}) \otimes \mathbf{I}_{p_{r+1} + \cdots + p_k} \\ \mathbf{I}_{p_1 + \cdots + p_{r-1}} \otimes (\mathbf{0} : \mathbf{I}) \otimes \mathbf{I}_{p_{r+1} + \cdots + p_k} \end{pmatrix} \mathrm{vec}\mathbf{X}.$$

Similarly,

$$\begin{pmatrix} \boldsymbol{\mu}_{\bullet r_l \bullet} \\ \boldsymbol{\mu}_{\bullet r^l \bullet} \end{pmatrix} = \begin{pmatrix} \mathbf{I}_{p_1 + \cdots + p_{r-1}} \otimes (\mathbf{I} : \mathbf{0}) \otimes \mathbf{I}_{p_{r+1} + \cdots + p_k} \\ \mathbf{I}_{p_1 + \cdots + p_{r-1}} \otimes (\mathbf{0} : \mathbf{I}) \otimes \mathbf{I}_{p_{r+1} + \cdots + p_k} \end{pmatrix} \mathrm{vec}\boldsymbol{\mu},$$

and thus

$$\boldsymbol{\mu}_{\bullet r_l \bullet} = (\mathbf{F}_1 \otimes \cdots \otimes \mathbf{F}_{r-1} \otimes \mathbf{F}^r_1 \otimes \mathbf{F}_{r+1} \otimes \cdots \otimes \mathbf{F}_k)\boldsymbol{\delta}, \tag{2.2.49}$$

$$\boldsymbol{\mu}_{\bullet r^l \bullet} = (\mathbf{F}_1 \otimes \cdots \otimes \mathbf{F}_{r-1} \otimes \mathbf{F}^r_2 \otimes \mathbf{F}_{r+1} \otimes \cdots \otimes \mathbf{F}_k)\boldsymbol{\delta}. \tag{2.2.50}$$

Now we give some results which correspond to Theorem 2.2.2, Theorem 2.2.4, Corollary 2.2.4.1, Theorem 2.2.5 and Theorem 2.2.6.

Theorem 2.2.16. *Let* $\mathbf{X} \sim N_{p_1,\ldots,p_k}(\mathbf{F}_1,\ldots,\mathbf{F}_k;\boldsymbol{\delta},\boldsymbol{\Sigma}_k,\boldsymbol{\Sigma}_1,\boldsymbol{\Sigma}_2,\ldots,\boldsymbol{\Sigma}_{k-1})$.

(i) *Let* $\mathbf{A}_i : q_i \times p_i, \quad i = 1,2,\ldots,k$. *Then*

$$\begin{aligned}(\mathbf{A}_1 \otimes \mathbf{A}_2 \otimes \cdots \otimes \mathbf{A}_k)\mathrm{vec}\mathbf{X} \\ \sim N_{q_1,\ldots,q_k}\Big(\mathbf{A}_1\mathbf{F}_1,\ldots,\mathbf{A}_k\mathbf{F}_k;\boldsymbol{\delta},\mathbf{A}_k\boldsymbol{\Sigma}_k\mathbf{A}_k',\mathbf{A}_1\boldsymbol{\Sigma}_1\mathbf{A}_1',\mathbf{A}_2\boldsymbol{\Sigma}_2\mathbf{A}_2',\ldots, \\ \mathbf{A}_{k-1}\boldsymbol{\Sigma}_{k-1}\mathbf{A}_{k-1}'\Big).\end{aligned}$$

(ii) $(\mathbf{A}_1 \otimes \mathbf{A}_2 \otimes \cdots \otimes \mathbf{A}_k)\mathrm{vec}\mathbf{X}$ *is independent of* $(\mathbf{B}_1 \otimes \mathbf{B}_2 \otimes \cdots \otimes \mathbf{B}_k)\mathrm{vec}\mathbf{X}$ *if for some* i, $i = 1,2,\ldots,k$, $\mathbf{A}_i\boldsymbol{\Sigma}_i\mathbf{B}_i' = \mathbf{0}$.

(iii) $\mathbf{X}_{\bullet r_l \bullet}$ *is independent of* $\mathbf{X}_{\bullet r^l \bullet}$ *if and only if* $\boldsymbol{\Sigma}_{12}^r = \mathbf{0}$.

(iv) $\mathbf{X}_{\bullet r_l \bullet}$ *is not independent of* $\mathbf{X}_{\bullet s^m \bullet}$, *if* $s \neq r$.

(v) *Let* $\boldsymbol{\mu}_{\bullet r_l \bullet}$ *and* $\boldsymbol{\mu}_{\bullet r^l \bullet}$ *be given by (2.2.49) and (2.2.50), respectively, and*

$$\boldsymbol{\Sigma}_{1\bullet 2}^r = \boldsymbol{\Sigma}_{11}^r - \boldsymbol{\Sigma}_{12}^r(\boldsymbol{\Sigma}_{22}^r)^{-1}\boldsymbol{\Sigma}_{21}.$$

Then $\mathbf{X}_{\bullet r_l \bullet}|\mathbf{X}_{\bullet r^l \bullet}$ *has the same distribution as*

$$\begin{aligned}\boldsymbol{\mu}_{\bullet r_l \bullet} + (\mathbf{I}_{p_1+\cdots+p_{r-1}} \otimes \boldsymbol{\Sigma}_{12}^r(\boldsymbol{\Sigma}_{22}^r)^{-1} \otimes \mathbf{I}_{p_{r+1}+\cdots+p_k})(\mathbf{X}_{\bullet r^l \bullet} - \boldsymbol{\mu}_{\bullet r^l \bullet}) \\ + (\boldsymbol{\Sigma}_1 \otimes \cdots \otimes \boldsymbol{\Sigma}_{r-1} \otimes \boldsymbol{\Sigma}_{1\bullet 2}^r \otimes \boldsymbol{\Sigma}_{r+1} \otimes \cdots \otimes \boldsymbol{\Sigma}_k)\mathrm{vec}\mathbf{U},\end{aligned}$$

where $\mathbf{U} \sim N_{p_1,\ldots,p_k}(\mathbf{0},\mathbf{I},\mathbf{I},\ldots,\mathbf{I})$.

PROOF: The statements in (i) and (ii) follow from Theorem 2.2.2 and Theorem 2.2.4, respectively. In the rest of the proof Proposition 1.3.12 will be frequently used. From Theorem 2.2.4 it follows that two normally distributed variables are independent if and only if they are uncorrelated. Therefore, we will study the equation

$$\begin{aligned}\mathbf{0} &= C[\mathbf{X}_{\bullet r_l \bullet},\mathbf{X}_{\bullet r^l \bullet}] \\ &= (\mathbf{I}_{p_1+\cdots+p_{r-1}} \otimes (\mathbf{I} : \mathbf{0}) \otimes \mathbf{I}_{p_{r+1}+\cdots+p_k})(\boldsymbol{\Sigma}_1 \otimes \cdots \otimes \boldsymbol{\Sigma}_k) \\ &\quad \times (\mathbf{I}_{p_1+\cdots+p_{r-1}} \otimes (\mathbf{0} : \mathbf{I})' \otimes \mathbf{I}_{p_{r+1}+\cdots+p_k}) \\ &= \boldsymbol{\Sigma}_1 \otimes \cdots \otimes \boldsymbol{\Sigma}_{r-1} \otimes \boldsymbol{\Sigma}_{12}^r \otimes \boldsymbol{\Sigma}_{r+1} \otimes \cdots \otimes \boldsymbol{\Sigma}_k. \end{aligned} \tag{2.2.51}$$

Since $\boldsymbol{\Sigma}_i$, $i = 1,2,\ldots k$ differ from zero, (2.2.51) holds if and only if $\boldsymbol{\Sigma}_{12}^r = \mathbf{0}$. The statement in (iv) can be proved by giving a simple example. For instance, for $\mathbf{X} \sim N_{p_1,p_2,p_3}(\mathbf{0},\boldsymbol{\Sigma}_3,\boldsymbol{\Sigma}_1,\boldsymbol{\Sigma}_2)$

$$\begin{aligned}C[(\mathbf{I} \otimes (\mathbf{I} :\mathbf{0}) \otimes \mathbf{I})\mathrm{vec}\mathbf{X}, ((\mathbf{0} : \mathbf{I}) \otimes \mathbf{I} \otimes \mathbf{I})'\mathrm{vec}\mathbf{X}] \\ = (\mathbf{I} \otimes (\mathbf{I} : \mathbf{0}) \otimes \mathbf{I})(\boldsymbol{\Sigma}_1 \otimes \boldsymbol{\Sigma}_2 \otimes \boldsymbol{\Sigma}_3)((\mathbf{0} : \mathbf{I}) \otimes \mathbf{I} \otimes \mathbf{I})' \\ = \left(\begin{pmatrix}\boldsymbol{\Sigma}_{12}^1 \\ \boldsymbol{\Sigma}_{22}^1\end{pmatrix} \otimes (\boldsymbol{\Sigma}_{11}^2 : \boldsymbol{\Sigma}_{12}^2) \otimes \boldsymbol{\Sigma}_3\right) \neq \mathbf{0}\end{aligned}$$

without any further assumptions on the dispersion matrices.

In order to show (v) we rely on Theorem 2.2.6. We know that the conditional distribution must be normal. Thus it is sufficient to investigate the conditional mean and the conditional dispersion matrix. For the mean we have

$$\begin{aligned}
\boldsymbol{\mu}_{\bullet r_l \bullet} &+ (\mathbf{I}_{p_1+\cdots+p_{r-1}} \otimes (\mathbf{I}:\mathbf{0}) \otimes \mathbf{I}_{p_{r+1}+\cdots+p_k})(\boldsymbol{\Sigma}_1 \otimes \cdots \otimes \boldsymbol{\Sigma}_k) \\
&\times (\mathbf{I}_{p_1+\cdots+p_{r-1}} \otimes (\mathbf{0}:\mathbf{I})' \otimes \mathbf{I}_{p_{r+1}+\cdots+p_k}) \\
&\times \Big\{(\mathbf{I}_{p_1+\cdots+p_{r-1}} \otimes (\mathbf{0}:\mathbf{I}) \otimes \mathbf{I}_{p_{r+1}+\cdots+p_k})(\boldsymbol{\Sigma}_1 \otimes \cdots \otimes \boldsymbol{\Sigma}_k) \\
&\qquad \times (\mathbf{I}_{p_1+\cdots+p_{r-1}} \otimes (\mathbf{0}:\mathbf{I})' \otimes \mathbf{I}_{p_{r+1}+\cdots+p_k})\Big\}^{-1} (\mathbf{X}_{\bullet r^l \bullet} - \boldsymbol{\mu}_{\bullet r^l \bullet}) \\
= \boldsymbol{\mu}_{\bullet r_l \bullet} &+ (\mathbf{I}_{p_1+\cdots+p_{r-1}} \otimes \boldsymbol{\Sigma}_{12}^r (\boldsymbol{\Sigma}_{22}^r)^{-1} \otimes \mathbf{I}_{p_{r+1}+\cdots+p_k})(\mathbf{X}_{\bullet r^l \bullet} - \boldsymbol{\mu}_{\bullet r^l \bullet})
\end{aligned}$$

and for the dispersion

$$\begin{aligned}
&(\mathbf{I}_{p_1+\cdots+p_{r-1}} \otimes (\mathbf{I}:\mathbf{0}) \otimes \mathbf{I}_{p_{r+1}+\cdots+p_k})(\boldsymbol{\Sigma}_1 \otimes \cdots \otimes \boldsymbol{\Sigma}_k) \\
&\quad \times (\mathbf{I}_{p_1+\cdots+p_{r-1}} \otimes (\mathbf{I}:\mathbf{0})' \otimes \mathbf{I}_{p_{r+1}+\cdots+p_k}) \\
&\quad - (\mathbf{I}_{p_1+\cdots+p_{r-1}} \otimes (\mathbf{I}:\mathbf{0}) \otimes \mathbf{I}_{p_{r+1}+\cdots+p_k})(\boldsymbol{\Sigma}_1 \otimes \cdots \otimes \boldsymbol{\Sigma}_k) \\
&\quad \times (\mathbf{I}_{p_1+\cdots+p_{r-1}} \otimes (\mathbf{0}:\mathbf{I})' \otimes \mathbf{I}_{p_{r+1}+\cdots+p_k}) \\
&\quad \times \Big\{(\mathbf{I}_{p_1+\cdots+p_{r-1}} \otimes (\mathbf{0}:\mathbf{I}) \otimes \mathbf{I}_{p_{r+1}+\cdots+p_k})(\boldsymbol{\Sigma}_1 \otimes \cdots \otimes \boldsymbol{\Sigma}_k) \\
&\qquad \times (\mathbf{I}_{p_1+\cdots+p_{r-1}} \otimes (\mathbf{0}:\mathbf{I})' \otimes \mathbf{I}_{p_{r+1}+\cdots+p_k})\Big\}^{-1} \\
&\quad \times (\mathbf{I}_{p_1+\cdots+p_{r-1}} \otimes (\mathbf{0}:\mathbf{I}) \otimes \mathbf{I}_{p_{r+1}+\cdots+p_k})(\boldsymbol{\Sigma}_1 \otimes \cdots \otimes \boldsymbol{\Sigma}_k) \\
&\quad \times (\mathbf{I}_{p_1+\cdots+p_{r-1}} \otimes (\mathbf{I}:\mathbf{0})' \otimes \mathbf{I}_{p_{r+1}+\cdots+p_k}) \\
&= \boldsymbol{\Sigma}_1 \otimes \cdots \otimes \boldsymbol{\Sigma}_{r-1} \otimes \boldsymbol{\Sigma}_{1\bullet 2}^r \otimes \boldsymbol{\Sigma}_{r+1} \otimes \cdots \otimes \boldsymbol{\Sigma}_k.
\end{aligned}$$

∎

2.2.6 *Problems*

1. Let $\mathbf{X}$ have the same distribution as $\boldsymbol{\mu} + \boldsymbol{\tau}\mathbf{U}\boldsymbol{\gamma}'$, where additionally it is supposed that $\mathbf{AUB} = \mathbf{0}$ for some matrices $\mathbf{A}$ and $\mathbf{B}$. Under what conditions on $\mathbf{A}$ and $\mathbf{B}$ the matrix $\mathbf{X}$ is still matrix normally distributed?
2. Let $\mathbf{X}_1 \sim N_{p,n}(\boldsymbol{\mu}_1, \boldsymbol{\Sigma}_1, \boldsymbol{\Psi}_1)$ and $\mathbf{X}_2 \sim N_{p,n}(\boldsymbol{\mu}_2, \boldsymbol{\Sigma}_2, \boldsymbol{\Psi}_2)$. Under what conditions on $\boldsymbol{\Sigma}_i$, $\boldsymbol{\Psi}_i$, $i = 1, 2$, the matrix $\mathbf{X}_1 + \mathbf{X}_2$ is matrix normally distributed?
3. Prove statement (ii) of Theorem 2.2.4.
4. Find an alternative proof of Theorem 2.2.5 via a factorization of the normal density function.
5. Let $\mathbf{X}_1 \sim N_{p,n}(\boldsymbol{\mu}_1, \boldsymbol{\Sigma}_1, \boldsymbol{\Psi}_1)$, $\mathbf{X}_2 \sim N_{p,n}(\boldsymbol{\mu}_2, \boldsymbol{\Sigma}_2, \boldsymbol{\Psi}_2)$ and $\mathbf{Z}$ have a mixture distribution
$$f_{\mathbf{Z}}(\mathbf{X}) = \gamma f_{\mathbf{X}_1}(\mathbf{X}) + (1-\gamma) f_{\mathbf{X}_2}(\mathbf{X}), \qquad 0 < \gamma < 1.$$
Express $D[\mathbf{Z}]$ with the help of $\boldsymbol{\Sigma}_i$ and $\boldsymbol{\Psi}_i$.
6. Give conditions on the matrices $\mathbf{A}$, $\mathbf{B}$, $\mathbf{C}$ and $\mathbf{D}$ such that $\mathbf{AXB}'|\mathbf{CXD}'$ is matrix normally distributed.

7. Define Hermite polynomials for the matrix normal distribution.
8. Is it true that if $\mathbf{x} : p \times 1$ has a symmetric distribution and $E[\mathbf{x}] = \mathbf{0}$,

$$c_k[\mathbf{x}] = \overline{m}_k[\mathbf{x}] - \overline{m}_2[\mathbf{x}] \otimes \mathrm{vec}'(\overline{m}_{k-2}[\mathbf{x}]) + \sum_{i=0}^{k-3} (\mathrm{vec}'(\overline{m}_{k-2}[\mathbf{x}]) \otimes \overline{m}_2[\mathbf{x}])(\mathbf{I}_p \otimes \mathbf{K}_{p,p^{k-i-3}} \otimes \mathbf{I}_{p^i})?$$

9. Derive $E[\mathbf{Y} \otimes \mathbf{Y} \otimes \mathbf{Y}' \otimes \mathbf{Y}']$ when $\mathbf{Y} \sim N_{p,n}(\boldsymbol{\mu}, \boldsymbol{\Sigma}, \boldsymbol{\Psi})$.
10. Let the $\mathbf{P}-$class of matrices be the class of orthogonal matrices $\mathbf{P}$ where $\mathbf{P}\mathbf{1}_p = \mathbf{1}_p$. Let $\mathbf{x} \sim N_p(\mathbf{0}, \boldsymbol{\Sigma})$ and $\mathbf{P}\mathbf{x}$ have the same distribution for all matrices $\mathbf{P}$ in the $\mathbf{P}-$class. Determine the necessary structure of $\boldsymbol{\Sigma}$.
    Let $\mathbf{P}_1$: $n_1 \times n_1$, $\mathbf{P}_2$: $n_2 \times n_2$, ..., $\mathbf{P}_s$: $n_s \times n_s$ be arbitrary $\mathbf{P}-$class matrices and let

$$(\mathbf{P}_1 \otimes \mathbf{P}_2 \otimes \cdots \otimes \mathbf{P}_s)\mathbf{x}$$

    have the same distribution for all choices of $\mathbf{P}_i$. What conditions do the size p and $\boldsymbol{\Sigma}$ have to satisfy?

2.3. ELLIPTICAL DISTRIBUTIONS

2.3.1 *Introduction, spherical distributions*

Classical multivariate analysis has been built up on the assumption of normality of the observations. Real data very seldom satisfy this assumption. One of the main tasks in developing multivariate analysis has been to generalize the assumption about normality of the model. Several robustness studies, asymptotic results and enlargement of the class of population distributions can be considered as developments in this direction. In particular, much attention has been paid to the elliptical distributions or elliptically contoured distributions. We prefer the shorter. These distributions have earned interest of many statisticians for several reasons. One is that the class contains the normal distributions. Another reason is that many results which are valid for the normal theory can be carried over to the elliptical models very easily. Especially the results in asymptotic theory of multivariate statistics will be similar to the case with the normal population, but at the same time a much wider class of distributions is covered. Besides the description of populations, the elliptical distributions have become important tools in robustness studies of multivariate analysis. As pointed out by Fang, Kotz & Ng (1990), the topic can be traced back to Maxwell (1860), but the modern study of the distributions starts from 1960s with the first description of the family by Kelker (1970).

The first books on the topic were written by Kariya & Sinha (1989), Fang & Zhang (1990) and Fang, Kotz & Ng (1990). Also in Muirhead (1982) an overview of the topic is given, and in Anderson (2003) the multivariate analysis is applied to elliptical distributions. Our presentation is mainly based on these references. Several classical results on elliptical distributions are presented in the following text without reproducing the proofs. The interested reader is referred to the literature in these cases.

In the class of elliptical distributions a *spherical* distribution has the same role as the standard multivariate normal distribution $N_p(\mathbf{0},\mathbf{I})$ in the family of multivariate normal distributions $N_p(\boldsymbol{\mu},\boldsymbol{\Sigma})$.

Definition 2.3.1. *A p-vector* $\mathbf{x}$ *is said to have a spherical distribution if* $\mathbf{x}$ *and* $\boldsymbol{\Gamma}'\mathbf{x}$ *have the same distribution for all orthogonal* $p\times p$*-matrices* $\boldsymbol{\Gamma}$. ▮

If $\mathbf{x}$ is a continuous random vector with a spherical distribution, then due to the equality $\boldsymbol{\Gamma}'\boldsymbol{\Gamma}=\mathbf{I}$, its density function must depend on the argument $\mathbf{x}$ through the value of $\mathbf{x}'\mathbf{x}$. Some examples will follow:

a) the normal distribution $N_p(\mathbf{0},\sigma^2\mathbf{I})$ with the density

$$f(\mathbf{x})=\frac{1}{(2\pi\sigma^2)^{\frac{p}{2}}}\exp(-\frac{1}{2\sigma^2}\mathbf{x}'\mathbf{x});$$

b) the mixture of two normal distributions, $N_p(\mathbf{0},\mathbf{I})$ and $N_p(\mathbf{0},\sigma^2\mathbf{I})$, i.e. the *ε-contaminated normal distribution*, with the density function

$$f(\mathbf{x})=(1-\varepsilon)\frac{1}{(2\pi)^{\frac{p}{2}}}\exp(-\frac{1}{2}\mathbf{x}'\mathbf{x})+\varepsilon\frac{1}{(2\pi\sigma^2)^{\frac{p}{2}}}\exp(-\frac{1}{2\sigma^2}\mathbf{x}'\mathbf{x}),\quad(0\leq\varepsilon\leq1); \tag{2.3.1}$$

c) the *multivariate* $\mathbf{t}$-*distribution* with n degrees of freedom and density

$$f(\mathbf{x}) = \frac{\Gamma\left(\frac{1}{2}(n+p)\right)}{\Gamma\left(\frac{1}{2}n\right) n\pi^{\frac{p}{2}}} \left(1 + \frac{1}{n}\mathbf{x}'\mathbf{x}\right)^{-\frac{n+p}{2}}. \tag{2.3.2}$$

In all these examples we see that the density remains the same if we replace $\mathbf{x}$ by $\mathbf{\Gamma}'\mathbf{x}$. The next theorem gives a characterization of a spherical distribution through its characteristic function.

Theorem 2.3.1. *A $p-$vector $\mathbf{x}$ has a spherical distribution if and only if its characteristic function $\varphi_{\mathbf{x}}(\mathbf{t})$ satisfies one of the following two equivalent conditions:*

(i) $\varphi_{\mathbf{x}}(\mathbf{\Gamma}'\mathbf{t}) = \varphi_{\mathbf{x}}(\mathbf{t})$ *for any orthogonal matrix* $\mathbf{\Gamma} : p \times p$;
(ii) *there exists a function $\phi(\cdot)$ of a scalar variable such that $\varphi_{\mathbf{x}}(\mathbf{t}) = \phi(\mathbf{t}'\mathbf{t})$.*

PROOF: For a square matrix $\mathbf{A}$ the characteristic function of $\mathbf{Ax}$ equals $\varphi_{\mathbf{x}}(\mathbf{A}'\mathbf{t})$. Thus (i) is equivalent to Definition 2.3.1. The condition (ii) implies (i), since

$$\varphi_{\mathbf{x}}(\mathbf{\Gamma}'\mathbf{t}) = \phi((\mathbf{\Gamma}'\mathbf{t})'(\mathbf{\Gamma}'\mathbf{t})) = \phi(\mathbf{t}'\mathbf{\Gamma}\mathbf{\Gamma}'\mathbf{t}) = \phi(\mathbf{t}'\mathbf{t}) = \varphi_{\mathbf{x}}(\mathbf{t}).$$

Conversely, (i) implies that $\varphi_{\mathbf{x}}(\mathbf{t})$ is invariant with respect to multiplication from left by an orthogonal matrix, but from the invariance properties of the orthogonal group $\mathcal{O}(p)$ (see Fang, Kotz & Ng, 1990, Section 1.3, for example) it follows that $\varphi_{\mathbf{x}}(\mathbf{t})$ must be a function of $\mathbf{t}'\mathbf{t}$. ∎

In the theory of spherical distributions an important role is played by the random $p-$vector $\mathbf{u}$, which is uniformly distributed on the unit sphere in $\mathbb{R}^p$. Fang & Zhang (1990) have shown that $\mathbf{u}$ is distributed according to a spherical distribution. When two random vectors $\mathbf{x}$ and $\mathbf{y}$ have the same distribution we shall use the notation

$$\mathbf{x} \stackrel{d}{=} \mathbf{y}.$$

Theorem 2.3.2. *Assume that the $p-$vector $\mathbf{x}$ is spherically distributed. Then $\mathbf{x}$ has the stochastic representation*

$$\mathbf{x} \stackrel{d}{=} R\mathbf{u}, \tag{2.3.3}$$

where $\mathbf{u}$ is uniformly distributed on the unit sphere, $R \sim F(x)$ is independent of $\mathbf{u}$, and $F(x)$ is a distribution function over $[0, \infty)$. ∎

The random variable R in (2.3.3) may be looked upon as a radius.
In the next theorem we shall give a characterization of the class of functions $\phi(\cdot)$, which appeared in Theorem 2.3.1. Denote

$$\Phi_p = \{\phi(\cdot) : \phi(\mathbf{t}'\mathbf{t}) \text{ is a characteristic function}\}.$$

Theorem 2.3.3. *A function $\phi(\cdot) \in \Phi_p$ if and only if*

$$\phi(x) = \int_0^\infty \Omega_p(xr^2)dF(r),$$

where $F(\cdot)$ is defined over $[0,\infty)$ and $\Omega_p(\mathbf{y}'\mathbf{y})$ is the characteristic function of $\mathbf{u}$*:*

$$\Omega_p(\mathbf{y}'\mathbf{y}) = \frac{1}{S_p}\int_{S:\mathbf{x}'\mathbf{x}=1} e^{i\mathbf{y}'\mathbf{x}}dS,$$

where S_p is the area of the unit sphere surface in $\mathbb{R}^p$. ∎

When characterizing elliptical distributions the multivariate *Dirichlet distribution* is useful. Therefore we will present a definition of the distribution. Consider random variables with the $\Gamma(\alpha)$-distribution with the density function

$$f_\Gamma(x) = \Gamma(\alpha)^{-1}x^{\alpha-1}e^{-x}, \quad x > 0.$$

Definition 2.3.2. *Let $X_1,\dots,X_{p+1}$ be independent random variables, where $X_i \sim \Gamma(\alpha_i)$ with $\alpha_i > 0$ and*

$$Y_j = \frac{X_j}{\sum_{i=1}^{p+1} X_i}, \qquad j = 1,2,\dots,p.$$

Then the distribution of $\mathbf{y} = (Y_1, Y_2,\dots,Y_p)'$ is called the Dirichlet distribution with parameters $\alpha_1,\dots\alpha_{p+1}$. ∎

The density and basic properties of the Dirichlet distribution can be found in Fang, Kotz & Ng (1990), §1.4, for example.
The following theorem represents three necessary and sufficient sets of conditions for a spherical distribution (Fang, Kotz & Ng, 1990).

Theorem 2.3.4. *Assume that $\mathbf{x}$ is a random $p-$vector. Then the following three statements are equivalent.*
(i) *The characteristic function of $\mathbf{x}$ has the form $\phi(\mathbf{t}'\mathbf{t})$.*
(ii) *$\mathbf{x}$ has the stochastic representation*

$$\mathbf{x} \stackrel{d}{=} R\mathbf{u},$$

where R is independent of the uniformly distributed $\mathbf{u}$.
(iii) *$\mathbf{x} \stackrel{d}{=} \mathbf{\Gamma}'\mathbf{x}$ for every $\mathbf{\Gamma} \in \mathcal{O}(p)$, where $\mathcal{O}(p)$ is the group of orthogonal $p \times p-$matrices.* ∎

The statement (ii) is specified in the next corollary.

Corollary 2.3.4.1. *Suppose $\mathbf{x} \stackrel{d}{=} R\mathbf{u}$, and $P(\mathbf{x} = \mathbf{0}) = 0$. Then*

$$||\mathbf{x}|| \stackrel{d}{=} R, \qquad \frac{\mathbf{x}}{||\mathbf{x}||} \stackrel{d}{=} \mathbf{u}$$

are independent, where $||\mathbf{x}||$ is the usual Euclidian norm $||\mathbf{x}|| = \sqrt{\sum_{i=1}^p x_i^2}$. ∎

In general, a random vector $\mathbf{x}$ with a spherical distribution does not necessarily have a density. However, if the density $f_{\mathbf{x}}(\mathbf{x})$ exists, then by using Lemma 3.2.1 we get from Theorem 2.3.1 that for some nonnegative function $g(\cdot)$ of a scalar variable the density must be of the form $g(\mathbf{x}'\mathbf{x})$.

2.3.2 *Elliptical distributions: definition and basic relations*
Now we shall give the main definition of the section.

Definition 2.3.3. *A random $p-$vector $\mathbf{x}$ is said to have an elliptical distribution with parameters $\boldsymbol{\mu} : p \times 1$ and $\mathbf{V} : p \times p$ if $\mathbf{x}$ has the same distribution as*

$$\boldsymbol{\mu} + \mathbf{A}\mathbf{y},$$

where $\mathbf{y}$ has a spherical distribution and $\mathbf{A} : p \times k$, $\mathbf{A}\mathbf{A}' = \mathbf{V}$ with rank $r(\mathbf{V}) = k$. ∎

We shall use the notation $\mathbf{x} \sim E_p(\boldsymbol{\mu}, \mathbf{V})$ when the distribution of $\mathbf{x} : p \times 1$ belongs to the elliptical family. The following theorem is valid.

Theorem 2.3.5. *Let $\mathbf{x} \sim E_p(\boldsymbol{\mu}, \mathbf{V})$ with $r(\mathbf{V}) = k$. Then the characteristic function $\varphi_{\mathbf{x}}(\mathbf{t})$ is of the form*

$$\varphi_{\mathbf{x}}(\mathbf{t}) = \exp(i\mathbf{t}'\boldsymbol{\mu})\phi(\mathbf{t}'\mathbf{V}\mathbf{t}) \tag{2.3.4}$$

for some function ϕ. The cumulant function equals

$$\psi_{\mathbf{x}}(\mathbf{t}) = i\mathbf{t}'\boldsymbol{\mu} + \ln\phi(\mathbf{t}'\mathbf{V}\mathbf{t}).$$

PROOF: By definition (2.1.6) of the characteristic function

$$\varphi_{\mathbf{x}}(\mathbf{t}) = E\left[\exp(i\mathbf{t}'(\boldsymbol{\mu} + \mathbf{A}\mathbf{y}))\right] = \exp(i\mathbf{t}'\boldsymbol{\mu})\varphi_{\mathbf{A}\mathbf{y}}(\mathbf{t}) = \exp(i\mathbf{t}'\boldsymbol{\mu})\varphi_{\mathbf{y}}(\mathbf{A}'\mathbf{t}),$$

where $\mathbf{A}$ is a $p \times k-$matrix. From Theorem 2.3.1 we get

$$\varphi_{\mathbf{x}}(\mathbf{t}) = \exp(i\mathbf{t}'\boldsymbol{\mu})\phi(\mathbf{t}'\mathbf{A}\mathbf{A}'\mathbf{t}) = \exp(i\mathbf{t}'\boldsymbol{\mu})\phi(\mathbf{t}'\mathbf{V}\mathbf{t})$$

for some function ϕ, which defines the characteristic function of the spherical distribution. The expression of the cumulant function is directly obtained by taking the logarithm in (2.3.4). ∎

Let us again give some examples.

a) The multivariate normal distribution $N_p(\boldsymbol{\mu}, \boldsymbol{\Sigma})$ belongs to the class of elliptical distributions, since if $\mathbf{x} \sim N_p(\boldsymbol{\mu}, \boldsymbol{\Sigma})$, the vector $\mathbf{x}$ can be represented as $\mathbf{x} = \boldsymbol{\mu} + \mathbf{A}\mathbf{y}$, where $\mathbf{y} \sim N_p(\mathbf{0}, \mathbf{I})$, and $\mathbf{A}\mathbf{A}' = \boldsymbol{\Sigma}$.

b) ε-contaminated distribution: When $\mathbf{y}$ is distributed according to (2.3.1), then $\mathbf{x} = \boldsymbol{\mu} + \mathbf{A}\mathbf{y}$ is elliptically distributed with $\mathbf{A}\mathbf{A}' = \boldsymbol{\Sigma}$.

c) Multivariate $\mathbf{t}$-distribution: We obtain a multivariate $\mathbf{t}$-distribution with parameters $\boldsymbol{\mu}$, $\boldsymbol{\Sigma} = \mathbf{A}\mathbf{A}'$ with n degrees of freedom with the same transformation $\mathbf{x} = \boldsymbol{\mu} + \mathbf{A}\mathbf{y}$, where $\mathbf{y}$ is spherically distributed with density (2.3.2). Then we write $\mathbf{x} \sim t_p(n, \boldsymbol{\mu}, \boldsymbol{\Sigma})$.

It follows from Theorem 2.3.2 that all marginal distributions of an elliptical distribution are elliptical. For example, partitioning $\mathbf{x}$, $\boldsymbol{\mu}$, and $\mathbf{V}$ as

$$\mathbf{x} = \begin{pmatrix} \mathbf{x}_1 \\ \mathbf{x}_2 \end{pmatrix}, \qquad \boldsymbol{\mu} = \begin{pmatrix} \boldsymbol{\mu}_1 \\ \boldsymbol{\mu}_2 \end{pmatrix}, \qquad \mathbf{V} = \begin{pmatrix} \mathbf{V}_{11} & \mathbf{V}_{12} \\ \mathbf{V}_{21} & \mathbf{V}_{22} \end{pmatrix},$$

where $\mathbf{x}_1$ and $\boldsymbol{\mu}_1$ are $k \times 1$ and $\mathbf{V}_{11}$ is $k \times k$, the characteristic function of $\mathbf{x}_1$ can be obtained from (2.3.4) by putting $\mathbf{t} = (\mathbf{t}_1' : \mathbf{0}')'$, where $\mathbf{t}_1$ is $k \times 1$:

$$\varphi_{\mathbf{x}_1}(\mathbf{t}_1) = \exp(i\mathbf{t}_1'\boldsymbol{\mu}_1)\phi(\mathbf{t}_1'\mathbf{V}_{11}\mathbf{t}_1),$$

which is the characteristic function of a random vector with an elliptical distribution $E_k(\boldsymbol{\mu}_1, \mathbf{V}_{11})$. So the well-known property (see Corollary 2.2.2.1) of the normal distribution that all the marginal distributions of a normal vector are normally distributed holds also in the case of elliptical distributions. At the same time we get a characterization of the multivariate normal distribution in the class of elliptical distributions.

Theorem 2.3.6. *Let $\mathbf{x} \sim E_p(\boldsymbol{\mu}, \mathbf{D})$, where $\mathbf{D}$ is diagonal. If $X_1, \ldots, X_p$ in $\mathbf{x}$ are independent, then $\mathbf{x}$ is normal.*

PROOF: Assume without loss of generality that $\boldsymbol{\mu} = \mathbf{0}$. Then the characteristic function of $\mathbf{x}$ has the form

$$\varphi_{\mathbf{x}}(\mathbf{t}) = \phi(\mathbf{t}'\mathbf{D}\mathbf{t}) = \phi(\sum_{i=1}^{p} t_i^2 d_{ii})$$

for some function ϕ. Since $X_1, \ldots, X_p$ are independent we get

$$\phi(\mathbf{t}'\mathbf{D}\mathbf{t}) = \prod_{i=1}^{p} \phi(t_i^2 d_{ii}).$$

Putting $u_i = t_i d_{ii}^{\frac{1}{2}}$ gives

$$\phi(\sum_{i=1}^{p} u_i^2) = \prod_{i=1}^{p} \phi(u_i^2).$$

The last equation is known as *Hamel's equation* and the only continuous solution of it is

$$\phi(z) = e^{kz}$$

for some constant k (e.g. see Feller, 1968, pp. 459–460). Hence the characteristic function has the form

$$\varphi(\mathbf{t}) = e^{k\mathbf{t}'\mathbf{D}\mathbf{t}}$$

and because it is a characteristic function, we must have $k \leq 0$, which implies that $\mathbf{x}$ is normally distributed. ∎

The conditional distribution will be examined in the following theorem.

Theorem 2.3.7. *If $\mathbf{x} \sim E_p(\boldsymbol{\mu}, \mathbf{V})$ and $\mathbf{x}$, $\boldsymbol{\mu}$ and $\mathbf{V}$ are partitioned as*

$$\mathbf{x} = \begin{pmatrix} \mathbf{x}_1 \\ \mathbf{x}_2 \end{pmatrix}, \qquad \boldsymbol{\mu} = \begin{pmatrix} \boldsymbol{\mu}_1 \\ \boldsymbol{\mu}_2 \end{pmatrix}, \qquad \mathbf{V} = \begin{pmatrix} \mathbf{V}_{11} & \mathbf{V}_{12} \\ \mathbf{V}_{21} & \mathbf{V}_{22} \end{pmatrix},$$

where $\mathbf{x}_1$ *and* $\boldsymbol{\mu}_1$ *are* $k-$*vectors and* $\mathbf{V}_{11}$ *is* $k \times k$, *then, provided* $E[\mathbf{x}_1|\mathbf{x}_2]$ *and* $D[\mathbf{x}_1|\mathbf{x}_2]$ *exist,*

$$E[\mathbf{x}_1|\mathbf{x}_2] = \boldsymbol{\mu}_1 + \mathbf{V}_{12}\mathbf{V}_{22}^{-1}(\mathbf{x}_2 - \boldsymbol{\mu}_2), \tag{2.3.5}$$

$$D[\mathbf{x}_1|\mathbf{x}_2] = h(\mathbf{x}_2)(\mathbf{V}_{11} - \mathbf{V}_{12}\mathbf{V}_{22}^{-1}\mathbf{V}_{21}) \tag{2.3.6}$$

for some function $h(\bullet)$. *Moreover, the conditional distribution of* $\mathbf{x}_1$ *given* $\mathbf{x}_2$ *is* $k-$*variate elliptical.* ∎

For the proof we refer to Muirhead (1982). The next theorem points out that the class of elliptical distributions is closed with respect to affine transformations which follows immediately from Definition 2.3.2.

Theorem 2.3.8. *Let* $\mathbf{x} \sim E_p(\boldsymbol{\mu}, \mathbf{V})$, $\mathbf{B} : m \times p$ *and* $\boldsymbol{\nu}$ *a* $m-$*vector. Then*

$$\boldsymbol{\nu} + \mathbf{B}\mathbf{x} \sim E_m(\boldsymbol{\nu} + \mathbf{B}\boldsymbol{\mu}, \mathbf{B}\mathbf{V}\mathbf{B}'). \tag{2.3.7}$$

∎

2.3.3 *Moments and cumulants*

Moments and cumulants of an elliptical distribution can be found by differentiating the characteristic function and the cumulant function, respectively. It is simpler to derive first the central moments and then go over to non-central moments than to start with the non-central moments directly. When the derivatives $\phi'(\mathbf{t}'\mathbf{V}\mathbf{t})$ and $\phi''(\mathbf{t}'\mathbf{V}\mathbf{t})$ will appear later on, the derivatives are always taken by the univariate argument t in $\phi(t)$.

Theorem 2.3.9. *Let* $\mathbf{x} \sim E_p(\boldsymbol{\mu}, \mathbf{V})$ *with the characteristic function (2.3.4). Then, if* $\phi(\bullet)$ *is as in (2.3.4),* $\phi(\bullet)'$ *and* $\phi(\bullet)''$ *denote the first and second derivative, respectively, and if* $\overline{m}_4[\mathbf{x}] < \infty$, *we have*

(i) $$E[\mathbf{x}] = \boldsymbol{\mu};$$

(ii) $$D[\mathbf{x}] = -2\phi'(0)\mathbf{V}; \tag{2.3.8}$$

(iii) $$\overline{m}_4[\mathbf{x}] = 4\phi''(0)\left[(\mathbf{V} \otimes \mathrm{vec}'\mathbf{V}) + (\mathrm{vec}'\mathbf{V} \otimes \mathbf{V})(\mathbf{I}_{p^3} + \mathbf{I}_p \otimes \mathbf{K}_{p,p})\right]. \tag{2.3.9}$$

All the odd central moments which exist equal zero.

PROOF: We have to differentiate the characteristic function four times.
(i): The first derivative equals

$$\frac{d\varphi_{\mathbf{x}}(\mathbf{t})}{d\mathbf{t}} = i\boldsymbol{\mu}\phi(\mathbf{t}'\mathbf{V}\mathbf{t})e^{i\mathbf{t}'\boldsymbol{\mu}} + 2\mathbf{V}\mathbf{t}\phi'(\mathbf{t}'\mathbf{V}\mathbf{t})e^{i\mathbf{t}'\boldsymbol{\mu}}.$$

By Definition 2.1.1 we have, since $\phi(0) = 1$,

$$E[\mathbf{x}] = \frac{1}{i}i\boldsymbol{\mu} = \boldsymbol{\mu}.$$

(ii): Observe that the central moments of $\mathbf{x}$ are the moments of $\mathbf{y} = \mathbf{x} - \boldsymbol{\mu}$ with the characteristic function

$$\varphi_{\mathbf{y}}(\mathbf{t}) = \phi(\mathbf{t}'\mathbf{V}\mathbf{t}).$$

So we have to find $\frac{d^2\phi(\mathbf{t}'\mathbf{V}\mathbf{t})}{d\mathbf{t}^2}$. From part (i) of the proof we have

$$\frac{d\phi(\mathbf{t}'\mathbf{V}\mathbf{t})}{d\mathbf{t}} = 2\mathbf{V}\mathbf{t}\phi'(\mathbf{t}'\mathbf{V}\mathbf{t})$$

and then

$$\frac{d^2\phi(\mathbf{t}'\mathbf{V}\mathbf{t})}{d\mathbf{t}^2} = \frac{d(2\mathbf{V}\mathbf{t}\phi'(\mathbf{t}'\mathbf{V}\mathbf{t}))}{d\mathbf{t}} = 2\mathbf{V}\phi'(\mathbf{t}'\mathbf{V}\mathbf{t}) + 4\mathbf{V}\mathbf{t}\phi''(\mathbf{t}'\mathbf{V}\mathbf{t})\mathbf{t}'\mathbf{V}. \qquad (2.3.10)$$

By (2.1.19),

$$D[\mathbf{x}] = -2\phi'(0)\mathbf{V},$$

and the second statement of the theorem is proved.
(iii): We have to differentiate the right hand side of (2.3.10) twice:

$$\begin{aligned}
&\frac{d^3\phi(\mathbf{t}'\mathbf{V}\mathbf{t})}{d\mathbf{t}^3}\\
&= \frac{d(2\mathbf{V}\phi'(\mathbf{t}'\mathbf{V}\mathbf{t}))}{d\mathbf{t}} + \frac{d(4\mathbf{V}\mathbf{t}\phi''(\mathbf{t}'\mathbf{V}\mathbf{t})\mathbf{t}'\mathbf{V})}{d\mathbf{t}}\\
&= 4\mathbf{V}\mathbf{t}\phi''(\mathbf{t}'\mathbf{V}\mathbf{t})\mathrm{vec}'\mathbf{V} + 4\frac{d(\mathbf{V}\mathbf{t})}{d\mathbf{t}}(\phi''(\mathbf{t}'\mathbf{V}\mathbf{t})\mathbf{t}'\mathbf{V}\otimes\mathbf{I}_p)\\
&\quad + 4\frac{d(\phi''(\mathbf{t}'\mathbf{V}\mathbf{t})\mathbf{t}'\mathbf{V})}{d\mathbf{t}}(\mathbf{I}_p\otimes\mathbf{t}'\mathbf{V})\\
&= 4\phi''(\mathbf{t}'\mathbf{V}\mathbf{t})\left\{\mathbf{V}\mathbf{t}\mathrm{vec}'\mathbf{V} + (\mathbf{t}'\mathbf{V}\otimes\mathbf{V}) + (\mathbf{V}\otimes\mathbf{t}'\mathbf{V})\right\} + 4\frac{d\phi''(\mathbf{t}'\mathbf{V}\mathbf{t})}{d\mathbf{t}}(\mathbf{t}'\mathbf{V}\otimes\mathbf{t}'\mathbf{V}).
\end{aligned}$$

In the next step we are interested in terms which after differentiating do not include $\mathbf{t}$ in positive power. Therefore, we can neglect the term outside the curly brackets in the last expression, and consider

$$\begin{aligned}
&\frac{d\left\{4\phi''(\mathbf{t}'\mathbf{V}\mathbf{t})\left\{\mathbf{V}\mathbf{t}\mathrm{vec}'\mathbf{V} + (\mathbf{t}'\mathbf{V}\otimes\mathbf{V}) + (\mathbf{V}\otimes\mathbf{t}'\mathbf{V})\right\}\right\}}{d\mathbf{t}}\\
&= 4\frac{d(\phi''(\mathbf{t}'\mathbf{V}\mathbf{t}))}{d\mathbf{t}}\otimes\mathrm{vec}'\left\{\mathbf{V}\mathbf{t}\mathrm{vec}'\mathbf{V} + (\mathbf{t}'\mathbf{V}\otimes\mathbf{V}) + (\mathbf{V}\otimes\mathbf{t}'\mathbf{V})\right\}\\
&\quad + 4\phi''(\mathbf{t}'\mathbf{V}\mathbf{t})\left\{\frac{d\mathbf{V}\mathbf{t}\mathrm{vec}'\mathbf{V}}{d\mathbf{t}} + \frac{d(\mathbf{t}'\mathbf{V}\otimes\mathbf{V})}{d\mathbf{t}} + \frac{d(\mathbf{V}\otimes\mathbf{t}'\mathbf{V})}{d\mathbf{t}}\right\}\\
&= 4\frac{d(\phi''(\mathbf{t}'\mathbf{V}\mathbf{t}))}{d\mathbf{t}}\otimes\mathrm{vec}'\left\{\mathbf{V}\mathbf{t}\mathrm{vec}'\mathbf{V} + (\mathbf{t}'\mathbf{V}\otimes\mathbf{V}) + (\mathbf{V}\otimes\mathbf{t}'\mathbf{V})\right\}\\
&\quad + 4\phi''(\mathbf{t}'\mathbf{V}\mathbf{t})\left\{(\mathbf{V}\otimes\mathrm{vec}'\mathbf{V})\mathbf{K}_{p,p^2} + (\mathbf{V}\otimes\mathrm{vec}'\mathbf{V}) + (\mathrm{vec}'\mathbf{V}\otimes\mathbf{V})(\mathbf{I}_p\otimes\mathbf{K}_{p,p})\right\}.
\end{aligned}$$

As the first term turns to zero at the point $\mathbf{t} = \mathbf{0}$, we get the following expression.

$$\overline{m}_4[\mathbf{x}] = 4\phi''(0)\left\{(\mathbf{V}\otimes\mathrm{vec}'\mathbf{V}) + (\mathrm{vec}'\mathbf{V}\otimes\mathbf{V})(\mathbf{I}_{p^3} + \mathbf{I}_p\otimes\mathbf{K}_{p,p})\right\},$$

which completes the proof. ∎

If we compare (2.3.8) and (2.3.9) with the expressions of the second and the fourth central moments of the normal distribution in Corollary 2.2.7.3, we remark that the only difference in the formulae concerns the multipliers $-2\phi'(0)$ and $4\phi''(0)$. As the expressions of the first moments of a normal vector are known already (Corollary 2.2.7.2), it is possible to write out the expressions for first moments of an elliptically distributed random vector.

Corollary 2.3.9.1. *Let* $\mathbf{x} \sim E_p(\boldsymbol{\mu}, \mathbf{V})$, *with the characteristic function (2.3.4). Then*

(i) $m_2[\mathbf{x}] = -2\phi'(0)\mathbf{V} + \boldsymbol{\mu}\boldsymbol{\mu}'$;

(ii) $m_3[\mathbf{x}] = \boldsymbol{\mu}(\boldsymbol{\mu}')^{\otimes 2} - 2\phi'(0)(\boldsymbol{\mu}' \otimes \mathbf{V} + \mathbf{V} \otimes \boldsymbol{\mu}' + \boldsymbol{\mu}\mathrm{vec}'\mathbf{V})$;

(iii)
$$m_4[\mathbf{x}] = \boldsymbol{\mu}(\boldsymbol{\mu}')^{\otimes^3} - 2\phi'(0)\boldsymbol{\mu}(\mathrm{vec}'\mathbf{V} \otimes \boldsymbol{\mu}')(\mathbf{I}_{p^3} + \mathbf{I}_p \otimes \mathbf{K}_{p,p})$$
$$- 2\phi'(0)\left\{(\boldsymbol{\mu}')^{\otimes^2} \otimes \mathbf{V} + \boldsymbol{\mu}' \otimes \mathbf{V} \otimes \boldsymbol{\mu}' + \boldsymbol{\mu}\boldsymbol{\mu}' \otimes \mathrm{vec}'\mathbf{V} + \mathbf{V} \otimes (\boldsymbol{\mu}')^{\otimes 2}\right\}$$
$$+ 4\phi''(0)\left\{\mathbf{V} \otimes \mathrm{vec}'\mathbf{V} + (\mathrm{vec}'\mathbf{V} \otimes \mathbf{V})(\mathbf{I}_{p^3} + \mathbf{I}_p \otimes \mathbf{K}_{p,p})\right\}.$$

∎

The next theorem gives us expressions of the first four cumulants of an elliptical distribution.

Theorem 2.3.10. *Let* $\mathbf{x} \sim E_p(\boldsymbol{\mu}, \mathbf{V})$, *with the characteristic function (2.3.4). Then, under the assumptions of Theorem 2.3.9,*

(i) $c_1[\mathbf{x}] = \boldsymbol{\mu}$;

(ii)
$$c_2[\mathbf{x}] = D[\mathbf{x}] = -2\phi'(0)\mathbf{V}; \tag{2.3.11}$$

(iii)
$$c_4[\mathbf{x}] = 4(\phi''(0) - (\phi'(0))^2)\{(\mathbf{V} \otimes \mathrm{vec}'\mathbf{V})$$
$$+ (\mathrm{vec}'\mathbf{V} \otimes \mathbf{V})(\mathbf{I}_{p^3} + \mathbf{I}_p \otimes \mathbf{K}_{p,p})\}. \tag{2.3.12}$$

(iv) *All the odd cumulants which exist equal zero.*

PROOF: To find the cumulants we have to differentiate the cumulant function

$$\psi_{\mathbf{x}}(\mathbf{t}) = \ln\varphi_{\mathbf{x}}(\mathbf{t}) = i\mathbf{t}'\boldsymbol{\mu} + \ln\phi(\mathbf{t}'\mathbf{V}\mathbf{t}).$$

(i): Observe that

$$\frac{d\psi_{\mathbf{x}}(\mathbf{t})}{d\mathbf{t}} = i\boldsymbol{\mu} + 2\mathbf{V}\mathbf{t}\frac{\phi'(\mathbf{t}'\mathbf{V}\mathbf{t})}{\phi(\mathbf{t}'\mathbf{V}\mathbf{t})},$$

from where, by (2.1.32), we get $c_1[\mathbf{x}] = \boldsymbol{\mu}$.

(ii): We get the second cumulant similarly to the variance in the proof of the previous theorem:

$$\begin{aligned}\frac{d^2\psi_{\mathbf{x}}(\mathbf{t})}{d\mathbf{t}^2} &= \frac{d}{d\mathbf{t}}\left(2\mathbf{V}\mathbf{t}\frac{\phi'(\mathbf{t}'\mathbf{V}\mathbf{t})}{\phi(\mathbf{t}'\mathbf{V}\mathbf{t})}\right)\\ &= 2\mathbf{V}\frac{\phi'(\mathbf{t}'\mathbf{V}\mathbf{t})}{\phi(\mathbf{t}'\mathbf{V}\mathbf{t})} + 4\mathbf{V}\mathbf{t}\frac{\phi''(\mathbf{t}'\mathbf{V}\mathbf{t})\phi(\mathbf{t}'\mathbf{V}\mathbf{t}) - (\phi'(\mathbf{t}'\mathbf{V}\mathbf{t}))^2}{(\phi(\mathbf{t}'\mathbf{V}\mathbf{t}))^2}\mathbf{t}'\mathbf{V},\end{aligned}$$

and at the point $\mathbf{t} = \mathbf{0}$ we obtain statement (ii).
(iii): Differentiating once again gives

$$\begin{aligned}\frac{d^3\psi_{\mathbf{x}}(\mathbf{t})}{d\mathbf{t}^3} &= 4\mathbf{V}\mathbf{t}\frac{\phi''(\mathbf{t}'\mathbf{V}\mathbf{t})\phi(\mathbf{t}'\mathbf{V}\mathbf{t}) - (\phi'(\mathbf{t}'\mathbf{V}\mathbf{t}))^2}{(\phi(\mathbf{t}'\mathbf{V}\mathbf{t}))^2}\mathrm{vec}'\mathbf{V}\\ &\quad + \frac{d}{d\mathbf{t}}\left[4\mathbf{V}\mathbf{t}\frac{\phi''(\mathbf{t}'\mathbf{V}\mathbf{t})\phi(\mathbf{t}'\mathbf{V}\mathbf{t}) - (\phi'(\mathbf{t}'\mathbf{V}\mathbf{t}))^2}{(\phi(\mathbf{t}'\mathbf{V}\mathbf{t}))^2}\mathbf{t}'\mathbf{V}\right].\end{aligned}$$

If we compare the expression on the right hand side of the last equality with the expression of the third derivative in the proof of Theorem 2.3.9, we see that the only difference is that $\phi''(\mathbf{t}'\mathbf{V}\mathbf{t})$ is changed to the ratio $\dfrac{\phi''(\mathbf{t}'\mathbf{V}\mathbf{t})\phi(\mathbf{t}'\mathbf{V}\mathbf{t}) - (\phi'(\mathbf{t}'\mathbf{V}\mathbf{t}))^2}{(\phi(\mathbf{t}'\mathbf{V}\mathbf{t}))^2}$. Since $\phi(0) = 1$, we get the final expression of the fourth cumulant when we put $\mathbf{t} = \mathbf{0}$. ∎

The fact that the second and fourth order moments and cumulants of elliptical and normal distributions differ only by a certain constant, which depends on the function $\phi(\cdot)$, is used in defining a kurtosis parameter κ. Following Muirhead (1982), we introduce it as a parameter

$$\kappa = \frac{\phi''(0) - (\phi'(0))^2}{(\phi'(0))^2}. \tag{2.3.13}$$

This means that any mixed fourth order cumulant for coordinates of an elliptically distributed vector $\mathbf{x} = (X_1, \ldots, X_p)'$ is determined by the covariances between the random variables and the parameter κ:

$$c_4[X_iX_jX_kX_l] = \kappa(\sigma_{ij}\sigma_{kl} + \sigma_{ik}\sigma_{jl} + \sigma_{il}\sigma_{jk}),$$

where $\sigma_{ij} = Cov(X_i, X_j)$.

2.3.4 *Density*

Although an elliptical distribution generally does not have a density, the most important examples all come from the class of continuous multivariate distributions with a density function. An elliptical distribution is defined via a spherical distribution which is invariant under orthogonal transformations. By the same argument as for the characteristic function we have that if the spherical distribution has a density, it must depend on an argument $\mathbf{x}$ via the product $\mathbf{x}'\mathbf{x}$ and be of the form $g(\mathbf{x}'\mathbf{x})$ for some non-negative function $g(\cdot)$. The density of a spherical distribution has been carefully examined in Fang & Zhang (1990). It is shown that $g(\mathbf{x}'\mathbf{x})$ is a density, if the following equality is satisfied:

$$1 = \int g(\mathbf{x}'\mathbf{x})d\mathbf{x} = \frac{\pi^{\frac{p}{2}}}{\Gamma(\frac{p}{2})}\int_0^\infty y^{\frac{p}{2}-1}g(y)dy. \tag{2.3.14}$$

The proof includes some results from function theory which we have not considered in this book. An interested reader can find the proof in Fang & Zhang (1990, pp. 59–60). Hence, a function $g(\cdot)$ defines the density of a spherical distribution if and only if

$$\int_0^\infty y^{\frac{p}{2}-1}g(y)dy < \infty. \tag{2.3.15}$$

By representation (2.3.3) a spherically distributed random vector can be presented as a product of a random variable R and a random vector $\mathbf{u}$ which is distributed uniformly on the unit sphere. The distribution of $\mathbf{u}$ is described by the Dirichlet distribution given in Definition 2.3.2, and so the existence of the density of a spherical distribution is determined by the existence of the density of the random variable R.

Theorem 2.3.11. *Let $\mathbf{x} \stackrel{d}{=} R\mathbf{u}$ be spherically distributed. Then $\mathbf{x}$ has a density $g(\cdot)$ if and only if R has a density $h(\cdot)$ and the two densities are related as follows:*

$$h(r) = \frac{2\pi^{\frac{p}{2}}}{\Gamma(\frac{p}{2})}r^{p-1}g(r^2). \tag{2.3.16}$$

PROOF: Assume that $\mathbf{x}$ has a density $g(\mathbf{x}'\mathbf{x})$. Let $f(\cdot)$ be any nonnegative Borel function. Then, using (2.3.14), we have

$$E[f(R)] = \int f(\{\mathbf{x}'\mathbf{x}\}^{\frac{1}{2}})g(\mathbf{x}'\mathbf{x})d\mathbf{x} = \frac{\pi^{\frac{p}{2}}}{\Gamma(\frac{p}{2})}\int_0^\infty f(y^{\frac{1}{2}})y^{\frac{p}{2}-1}g(y)dy.$$

Denote $r = y^{\frac{1}{2}}$, then

$$E[f(R)] = \frac{2\pi^{\frac{p}{2}}}{\Gamma(\frac{p}{2})}\int_0^\infty f(r)r^{p-1}g(r^2)dr.$$

The obtained equality shows that R has a density of the form (2.3.16). Conversely, when (2.3.16) is true, the statement follows immediately. ∎

Let us now consider an elliptically distributed vector $\mathbf{x} \sim E_p(\boldsymbol{\mu}, \mathbf{V})$. A necessary condition for existence of a density is that $r(\mathbf{V}) = p$. From Definition 2.3.3 and (2.3.3) we get the representation

$$\mathbf{x} = \boldsymbol{\mu} + R\mathbf{A}\mathbf{u},$$

where $\mathbf{V} = \mathbf{A}\mathbf{A}'$ and $\mathbf{A}$ is non-singular. Let us denote $\mathbf{y} = \mathbf{A}^{-1}(\mathbf{x} - \boldsymbol{\mu})$. Then $\mathbf{y}$ is spherically distributed and its characteristic function equals

$$\phi(\mathbf{t}'\mathbf{A}^{-1}\mathbf{V}(\mathbf{A}^{-1})'\mathbf{t}) = \phi(\mathbf{t}'\mathbf{t}).$$

We saw above that the density of $\mathbf{y}$ is of the form $g(\mathbf{y}'\mathbf{y})$, where $g(\cdot)$ satisfies (2.3.15). For $\mathbf{x} = \boldsymbol{\mu} + R\mathbf{A}\mathbf{u}$ the density $f_{\mathbf{x}}(\mathbf{x})$ is of the form:

$$f_{\mathbf{x}}(\mathbf{x}) = c_p|\mathbf{V}|^{-1/2}g((\mathbf{x}-\boldsymbol{\mu})'\mathbf{V}^{-1}(\mathbf{x}-\boldsymbol{\mu})). \qquad (2.3.17)$$

Every function $g(\cdot)$, which satisfies (2.3.15), can be considered as a function which defines a density of an elliptical distribution. The normalizing constant c_p equals

$$c_p = \frac{\Gamma(\frac{p}{2})}{2\pi^{\frac{p}{2}}\int_0^\infty r^{p-1}g(r^2)dr}.$$

So far we have only given few examples of representatives of the class of elliptical distributions. In fact, this class includes a variety of distributions which will be listed in Table 2.3.1. As all elliptical distributions are obtained as transformations of spherical ones, it is natural to list the main classes of spherical distributions. The next table, due to Jensen (1985), is given in Fang, Kotz & Ng (1990).

Table 2.3.1. Subclasses of p-dimensional spherical distributions (c denotes a normalizing constant).

Distribution	Density $f(\mathbf{x})$ or ch.f. $\varphi(\mathbf{t})$
Kotz type	$f(\mathbf{x}) = c\,(\mathbf{x}'\mathbf{x})^{N-1}\exp(-r(\mathbf{x}'\mathbf{x})^s),\ r,s>0,\ 2N+p>2$
multivariate normal	$f(\mathbf{x}) = c\exp(-\frac{1}{2}\mathbf{x}'\mathbf{x})$
Pearson type VII	$f(\mathbf{x}) = c\,(1+\frac{\mathbf{x}'\mathbf{x}}{s})^{-N},\ N>\frac{p}{2},\ s>0$
multivariate $\mathbf{t}$	$f(\mathbf{x}) = c\,(1+\frac{\mathbf{x}'\mathbf{x}}{s})^{-\frac{(p+m)}{2}},\ m\in\mathcal{N}$
multivariate Cauchy	$f(\mathbf{x}) = c\,(1+\frac{\mathbf{x}'\mathbf{x}}{s})^{-\frac{(p+1)}{2}},\ s>0$
Pearson type II	$f(\mathbf{x}) = c\,(1-\mathbf{x}'\mathbf{x})^m,\ m>0$
logistic	$f(\mathbf{x}) = c\exp(-\mathbf{x}'\mathbf{x})/\{1+\exp(-\mathbf{x}'\mathbf{x})\}^2$
multivariate Bessel	$f(\mathbf{x}) = c\left(\frac{\|\mathbf{x}\|}{\beta}\right)^a K_a\left(\frac{\|\mathbf{x}\|}{\beta}\right),\ a>-\frac{p}{2},\ \beta>0,\ K_a(\cdot)$ is a modified Bessel function of the 3rd kind
scale mixture	$f(\mathbf{x}) = c\int_0^\infty t^{-\frac{p}{2}}\exp(-\frac{\mathbf{x}'\mathbf{x}}{2t})dG(t)$
stable laws	$\varphi(\mathbf{t}) = \exp\{r(\mathbf{t}'\mathbf{t})^{\frac{a}{2}}\},\ 0<a\le 2,\ r<0$
multiuniform	$\varphi(\mathbf{t}) = {}_0F_1(\frac{p}{2};-\frac{1}{4}\|\mathbf{t}\|^2),\ {}_0F_1(\cdot)$ is a generalized hypergeometric function

For a modified Bessel function of the 3rd kind we refer to Kotz, Kozubowski & Podgórski (2001), and the definition of the generalized hypergeometric function can be found in Muirhead (1982, p. 20), for example.

2.3.5 *Elliptical matrix distributions*

Multivariate spherical and elliptical distributions have been used as population distributions as well as sample distributions. On matrix elliptical distributions we refer to Gupta & Varga (1993) and the volume of papers edited by Fang & Anderson (1990), but we shall follow a slightly different approach and notation in our presentation. There are many different ways to introduce the class of spherical matrix distributions in order to describe a random $p\times n$−matrix $\mathbf{X}$, where columns can be considered as observations from a p−dimensional population.

Definition 2.3.4. *Let* $\mathbf{X}$ *be a random* $p\times n-$*matrix. We call* $\mathbf{X}$ *matrix spherically distributed, if* $\mathrm{vec}\mathbf{X}$ *is spherically distributed.* ∎

Fang & Zhang (1990) say in this case that $\mathbf{X}$ is vector-spherically distributed. If we reformulate Theorem 2.3.4 for matrices, we get the following statement.

Theorem 2.3.12. *Assume that* $\mathbf{X}$ *is a random* $p \times n$*-matrix. Then the following three statements are equivalent.*

(i) *The characteristic function of* $\mathrm{vec}\mathbf{X}$ *has the form* $\phi(\mathrm{vec}'\mathbf{T}\mathrm{vec}\mathbf{T}) = \phi(\mathrm{tr}(\mathbf{T}'\mathbf{T}))$, *where* $\mathbf{T} \in \mathbb{R}^{p\times n}$.

(ii) $\mathbf{X}$ *has the stochastic representation*

$$\mathbf{X} \overset{d}{=} R\mathbf{U},$$

where $R \geq 0$ *is independent of* $\mathbf{U}$ *and* $\mathrm{vec}\mathbf{U}$ *is uniformly distributed on the unit sphere in* $\mathbb{R}^{pn}$.

(iii) $\mathrm{vec}\mathbf{X} \overset{d}{=} \mathbf{\Gamma}'\mathrm{vec}\mathbf{X}$ *for every* $\mathbf{\Gamma} \in \mathcal{O}(pn)$, *where* $\mathcal{O}(pn)$ *is the group of orthogonal* $pn \times pn-$*matrices.* ∎

If we consider continuous spherical distributions, all the results concerning existence and properties of the density of a spherical distribution can be directly converted to the spherical matrix distributions. So the density of $\mathbf{X}$ has the form $g(\mathrm{tr}(\mathbf{X}'\mathbf{X}))$ for some nonnegative function $g(\cdot)$. By Theorem 2.3.11, $\mathbf{X}$ has a density if and only if R has a density $h(\cdot)$, and these two density functions are connected by the following equality:

$$h(r) = \frac{2\pi^{\frac{np}{2}}}{\Gamma(\frac{pn}{2})} r^{np-1} g(r^2).$$

Definition 2.3.5. *Let* $\mathbf{X}$ *be an* $r \times s$ *spherically distributed random matrix, and* $\mathbf{V} = \boldsymbol{\tau}\boldsymbol{\tau}'$, $\mathbf{W} = \boldsymbol{\gamma}\boldsymbol{\gamma}'$ *be non-negative definite* $p\times p$ *and* $n\times n$ *matrices, respectively, where* $\boldsymbol{\tau} : p \times r$; $\boldsymbol{\gamma} : n \times s$. *A matrix* $\mathbf{Y} : p \times n$ *is said to be matrix elliptically distributed with parameters* $\boldsymbol{\mu}$, $\mathbf{V}$, *and* $\mathbf{W}$, $\mathbf{Y} \sim E_{p,n}(\boldsymbol{\mu}, \mathbf{V}, \mathbf{W})$, *if*

$$\mathbf{Y} \overset{d}{=} \boldsymbol{\mu} + \boldsymbol{\tau}\mathbf{X}\boldsymbol{\gamma}', \tag{2.3.18}$$

where $\boldsymbol{\mu} : p \times n$ *is a real matrix.* ∎

The next theorem gives us the general form of the characteristic function for an elliptical distribution.

Theorem 2.3.13. *Let* $\mathbf{Y} \sim E_{p,n}(\boldsymbol{\mu}, \mathbf{V}, \mathbf{W})$, *with* $\boldsymbol{\mu} : p \times n$, $\mathbf{V} = \boldsymbol{\tau}\boldsymbol{\tau}'$, *and* $\mathbf{W} = \boldsymbol{\gamma}\boldsymbol{\gamma}'$. *Then the characteristic function of* $\mathbf{Y}$ *is given by*

$$\varphi_{\mathbf{Y}}(\mathbf{T}) = e^{i\mathrm{vec}'\mathbf{T}\mathrm{vec}\boldsymbol{\mu}}\phi(\mathrm{vec}'\mathbf{T}(\mathbf{W} \otimes \mathbf{V})\mathrm{vec}\mathbf{T}) = e^{i\mathrm{tr}(\mathbf{T}'\boldsymbol{\mu})}\phi(\mathrm{tr}(\mathbf{T}'\mathbf{V}\mathbf{T}\mathbf{W})). \tag{2.3.19}$$

PROOF: By definition (2.1.7) of the characteristic function

$$\varphi_{\mathbf{Y}}(\mathbf{T}) = E[e^{i\mathrm{tr}(\mathbf{T}'\mathbf{Y})}] = E[e^{i\mathrm{vec}'\mathbf{T}\mathrm{vec}\mathbf{Y}}].$$

Definition 2.3.5 gives us

$$\varphi_{\mathbf{Y}}(\mathbf{T}) = e^{i\mathrm{vec}'\mathbf{T}\mathrm{vec}\boldsymbol{\mu}} E[e^{i\mathrm{tr}(\mathbf{T}'\boldsymbol{\tau}\mathbf{X}\boldsymbol{\gamma}')}] = e^{i\mathrm{vec}'\mathbf{T}\mathrm{vec}\boldsymbol{\mu}} E[e^{i\mathrm{tr}(\boldsymbol{\gamma}'\mathbf{T}'\boldsymbol{\tau}\mathbf{X})}]$$
$$= e^{i\mathrm{vec}'\mathbf{T}\mathrm{vec}\boldsymbol{\mu}} \varphi_{\mathbf{X}}(\boldsymbol{\tau}'\mathbf{T}\boldsymbol{\gamma}).$$

By Theorem 2.3.12,

$$\begin{aligned}\varphi_{\mathbf{Y}}(\mathbf{T}) &= e^{i\mathrm{vec}'\mathbf{T}\mathrm{vec}\boldsymbol{\mu}} \phi(\mathrm{vec}'(\boldsymbol{\tau}'\mathbf{T}\boldsymbol{\gamma})\mathrm{vec}(\boldsymbol{\tau}'\mathbf{T}\boldsymbol{\gamma}))\\ &= e^{i\mathrm{vec}'\mathbf{T}\mathrm{vec}\boldsymbol{\mu}} \phi(\mathrm{vec}'\mathbf{T}(\boldsymbol{\gamma}\otimes\boldsymbol{\tau})(\boldsymbol{\gamma}'\otimes\boldsymbol{\tau}')\mathrm{vec}\mathbf{T})\\ &= e^{i\mathrm{vec}'\mathbf{T}\mathrm{vec}\boldsymbol{\mu}} \phi(\mathrm{vec}'\mathbf{T}(\mathbf{W}\otimes\mathbf{V})\mathrm{vec}\mathbf{T}),\end{aligned}$$

and the first equality in (2.3.19) is proved. The second equality follows easily:

$$\varphi_{\mathbf{Y}}(\mathbf{T}) = e^{i\mathrm{tr}(\mathbf{T}'\boldsymbol{\mu})}\phi(\mathrm{tr}(\boldsymbol{\gamma}'\mathbf{T}'\boldsymbol{\tau}\boldsymbol{\tau}'\mathbf{T}\boldsymbol{\gamma})) = e^{i\mathrm{tr}(\mathbf{T}'\boldsymbol{\mu})}\phi(\mathrm{tr}(\mathbf{T}'\mathbf{V}\mathbf{T}\mathbf{W})).$$

∎

By definition of the characteristic function of a random matrix the distributions of $\mathbf{Y}$ and $\mathrm{vec}\mathbf{Y}$ are identical. Therefore

$$\mathrm{vec}\mathbf{Y} = \mathrm{vec}\boldsymbol{\mu} + (\boldsymbol{\gamma}\otimes\boldsymbol{\tau})\mathrm{vec}\mathbf{X}$$

has the same distribution as $\mathbf{Y}$. As $\mathbf{E}[\mathrm{vec}\mathbf{X}] = \mathbf{0}$, we have

$$\mathbf{E}[\mathrm{vec}\mathbf{Y}] = \mathrm{vec}\boldsymbol{\mu}.$$

The higher order moments will be derived in the next section.

Theorem 2.3.14. *Let $\mathbf{Y} \sim E_{p,n}(\boldsymbol{\mu}, \mathbf{V}, \mathbf{W})$, with $\mathbf{V} = \boldsymbol{\tau}\boldsymbol{\tau}'$ and $\mathbf{W} = \boldsymbol{\gamma}\boldsymbol{\gamma}'$. Then for any $\mathbf{A} : q\times p$ and $\mathbf{B} : m\times n$,*

$$\mathbf{AYB}' \sim E_{q,m}(\mathbf{A}\boldsymbol{\mu}\mathbf{B}', \mathbf{AVA}', \mathbf{BWB}').$$

PROOF: By Definition 2.3.5, $\mathbf{Y} \stackrel{d}{=} \boldsymbol{\mu} + \boldsymbol{\tau}\mathbf{X}\boldsymbol{\gamma}'$, where $\boldsymbol{\tau}\boldsymbol{\tau}' = \mathbf{V}$ and $\boldsymbol{\gamma}\boldsymbol{\gamma}' = \mathbf{W}$. Then

$$\mathbf{AYB}' \stackrel{d}{=} \mathbf{A}\boldsymbol{\mu}\mathbf{B}' + \mathbf{A}\boldsymbol{\tau}\mathbf{X}\boldsymbol{\gamma}'\mathbf{B}'.$$

As $\mathbf{A}\boldsymbol{\tau}\boldsymbol{\tau}'\mathbf{A}' = \mathbf{AVA}'$ and $\mathbf{B}\boldsymbol{\gamma}\boldsymbol{\gamma}'\mathbf{B}' = \mathbf{BWB}'$, the statement is proved. ∎

From §2.3.4 we know that an elliptical distribution does not necessarily have a density, but if the density exists we can find the general form of the density function. As the matrix elliptical distribution at the same time is an elliptical distribution of the $pn-$vector $\mathrm{vec}\mathbf{Y}$, the results for vectors are also valid for matrices.
So, if $\mathbf{X}$ is matrix spherically distributed, its density function must be of the form

$$f_{\mathbf{X}}(\mathbf{X}) = g(\mathrm{vec}'\mathbf{X}\mathrm{vec}\mathbf{X}) = g(\mathrm{tr}(\mathbf{X}'\mathbf{X})) = g(\mathrm{tr}(\mathbf{XX}')),$$

where $g(\cdot)$ is some non-negative function.

Note that $\mathbf{V}$ and $\mathbf{W}$ in $E_{p,n}(\boldsymbol{\mu}, \mathbf{V}, \mathbf{W})$ have to be positive definite, if the matrix elliptical distribution should have a density. For example, assume that $\mathbf{V} = \boldsymbol{\Gamma}'\mathbf{D}\boldsymbol{\Gamma}$ is singular, where $\boldsymbol{\Gamma}$ is an orthogonal matrix and $\mathbf{D}$ is diagonal. Singularity implies that at least one diagonal element is $\mathbf{D}$ equals 0. If $\mathbf{Y}$ has a density, then $\boldsymbol{\Gamma}\mathbf{Y}$ has also a density but for from Theorem 2.3.14 it follows that at least one component in $\boldsymbol{\Gamma}\mathbf{Y}$ is a constant which leads to a contradiction.
In the case of non-singular $\boldsymbol{\tau}$ and $\boldsymbol{\gamma}$, the equality (2.3.18) defines a one-to-one transformation between $\mathbf{X}$ and $\mathbf{Y}$ and therefore

$$f_{\mathbf{Y}}(\mathbf{Y}) = f_{\mathbf{X}}(\mathbf{X} = \mathbf{X}(\mathbf{Y}))|J(\mathbf{X} \to \mathbf{Y})|.$$

The Jacobian we get from the relation

$$\mathbf{X} = \boldsymbol{\tau}^{-1}(\mathbf{Y} - \boldsymbol{\mu})\boldsymbol{\gamma}'^{-1}$$

by differentiation (see also Theorem 1.4.14):

$$\begin{aligned}|J(\mathbf{X} \to \mathbf{Y})|_+ &= \left|\boldsymbol{\gamma}^{-1} \otimes \boldsymbol{\tau}^{-1}\right| = |\boldsymbol{\gamma}|^{-n}|\boldsymbol{\tau}|^{-p} \\ &= |\boldsymbol{\gamma}|^{-\frac{n}{2}}|\boldsymbol{\tau}|^{-\frac{p}{2}}|\boldsymbol{\gamma}'|^{-\frac{n}{2}}|\boldsymbol{\tau}'|^{-\frac{p}{2}} = |\mathbf{V}|^{-\frac{n}{2}}|\mathbf{W}|^{-\frac{p}{2}}.\end{aligned}$$

From here the expression for the density function follows:

$$\begin{aligned}f_{\mathbf{Y}}(\mathbf{Y}) &= |\mathbf{V}|^{-\frac{n}{2}}|\mathbf{W}|^{-\frac{p}{2}}g(\operatorname{tr}\{\boldsymbol{\tau}^{-1}(\mathbf{Y} - \boldsymbol{\mu})\boldsymbol{\gamma}'^{-1}\boldsymbol{\gamma}^{-1}(\mathbf{Y} - \boldsymbol{\mu})'\boldsymbol{\tau}'^{-1}\}) \\ &= |\mathbf{V}|^{-\frac{n}{2}}|\mathbf{W}|^{-\frac{p}{2}}g(\operatorname{tr}\{\mathbf{V}^{-1}(\mathbf{Y} - \boldsymbol{\mu})\mathbf{W}^{-1}(\mathbf{Y} - \boldsymbol{\mu})'\}).\end{aligned}$$

Thus the next theorem is established.

Theorem 2.3.15. *Let* $\mathbf{Y} \sim E_{p,n}(\boldsymbol{\mu}, \mathbf{V}, \mathbf{W})$, *with non-singular matrices* $\mathbf{V} = \boldsymbol{\tau}\boldsymbol{\tau}'$, $\mathbf{W} = \boldsymbol{\gamma}\boldsymbol{\gamma}'$, *and let* $\mathbf{Y}$ *have a density. Then*

$$f_{\mathbf{Y}}(\mathbf{Y}) = |\mathbf{V}|^{-\frac{n}{2}}|\mathbf{W}|^{-\frac{p}{2}}g(\operatorname{tr}\{\mathbf{V}^{-1}(\mathbf{Y} - \boldsymbol{\mu})\mathbf{W}^{-1}(\mathbf{Y} - \boldsymbol{\mu})'\}),$$

where $g(\cdot)$ *is some non-negative function.* ∎

2.3.6 *Moments and cumulants of matrix elliptical distributions*
In the previous paragraph we have shown that $E[\mathbf{Y}] = \boldsymbol{\mu}$, if $\mathbf{Y} \sim E_{p,n}(\boldsymbol{\mu}, \mathbf{V}, \mathbf{W})$. In principle, moments and cumulants of $\mathbf{Y}$ can be found by differentiation of the characteristic function and the cumulant function, respectively, but it appears that we can get the formulae for the first moments and cumulants of $\mathbf{Y}$ easily from known results about elliptical and matrix normal distributions.

Theorem 2.3.16. *Let* $\mathbf{Y} \sim E_{p,n}(\boldsymbol{\mu}, \mathbf{V}, \mathbf{W})$. *Then*

(i) $m_2[\mathbf{Y}] = -2\phi'(0)(\mathbf{W} \otimes \mathbf{V}) + \operatorname{vec}\boldsymbol{\mu}\operatorname{vec}'\boldsymbol{\mu}$;

(ii) $m_3[\mathbf{Y}] = \operatorname{vec}\boldsymbol{\mu}(\operatorname{vec}'\boldsymbol{\mu})^{\otimes 2}$

$$- 2\phi'(0)\{\mathbf{W} \otimes \mathbf{V} \otimes \mathrm{vec}'\boldsymbol{\mu} + \mathrm{vec}'\boldsymbol{\mu} \otimes \mathbf{W} \otimes \mathbf{V} + \mathrm{vec}\boldsymbol{\mu}\mathrm{vec}'(\mathbf{W} \otimes \mathbf{V})\};$$

(iii) $$m_4[\mathbf{Y}] = \mathrm{vec}\boldsymbol{\mu}(\mathrm{vec}'\boldsymbol{\mu})^{\otimes 3} - 2\phi'(0)\left\{(\mathrm{vec}'\boldsymbol{\mu})^{\otimes 2} \otimes \mathbf{W} \otimes \mathbf{V}\right.$$

$$+ \mathrm{vec}'\boldsymbol{\mu} \otimes \mathbf{W} \otimes \mathbf{V} \otimes \mathrm{vec}'\boldsymbol{\mu} + \mathrm{vec}\boldsymbol{\mu}\mathrm{vec}'\boldsymbol{\mu} \otimes \mathrm{vec}'(\mathbf{W} \otimes \mathbf{V})$$
$$+ \mathbf{W} \otimes \mathbf{V} \otimes (\mathrm{vec}'\boldsymbol{\mu})^{\otimes 2}$$

$$\left.+\mathrm{vec}\boldsymbol{\mu}\mathrm{vec}'(\mathbf{W} \otimes \mathbf{V}) \otimes \mathrm{vec}'\boldsymbol{\mu}(\mathbf{I}_{(pn)^3} + \mathbf{I}_{pn} \otimes \mathbf{K}_{pn,pn})\right\}$$

$$+ 4\phi''(0)\left\{\mathbf{W} \otimes \mathbf{V} \otimes \mathrm{vec}'(\mathbf{W} \otimes \mathbf{V})\right.$$

$$\left.+(\mathrm{vec}'(\mathbf{W} \otimes \mathbf{V}) \otimes \mathbf{W} \otimes \mathbf{V})(\mathbf{I}_{(pn)^3} + \mathbf{I}_{pn} \otimes \mathbf{K}_{pn,pn})\right\}.$$

PROOF: If we compare the characteristic functions of a matrix elliptical distribution (2.3.19) and elliptical distribution (2.3.4), we see that the only difference is that in (2.3.19) the vectors in (2.3.4) are replaced by vectorized matrices and the matrix $\mathbf{V}$ by the matrix $\mathbf{W} \otimes \mathbf{V}$. So the derivatives of $\varphi_{\mathbf{Y}}(\mathbf{T})$ have exactly the same structure as the derivatives of the characteristic function $\varphi_{\mathbf{y}}(\mathbf{t})$ of an elliptically distributed vector $\mathbf{y}$. At the same time, the matrix product in the trace expression is the same in (2.3.19) and in the characteristic function of the matrix normal distribution in Theorem 2.2.1. Therefore, we get the statement of our theorem directly from Theorem 2.2.7 when combining it with Corollary 2.3.9.1. ∎

The expressions of the first central moments of a matrix elliptical distribution we get from the theorem above if we take $\boldsymbol{\mu} = \mathbf{0}$.

Corollary 2.3.16.1. *Let $\mathbf{Y} \sim E_{p,n}(\boldsymbol{\mu}, \mathbf{V}, \mathbf{W})$. Then the first even central moments of $\mathbf{Y}$ are given by the following equalities, and all the odd central moments equal zero:*

(i) $$\overline{m}_2[\mathbf{Y}] = -2\phi'(0)(\mathbf{W} \otimes \mathbf{V});$$

(ii) $$\overline{m}_4[\mathbf{Y}] = 4\phi''(0)\left\{\mathbf{W} \otimes \mathbf{V} \otimes \mathrm{vec}'(\mathbf{W} \otimes \mathbf{V})\right.$$
$$\left.+ (\mathrm{vec}'(\mathbf{W} \otimes \mathbf{V}) \otimes \mathbf{W} \otimes \mathbf{V})(\mathbf{I}_{(pn)^3} + \mathbf{I}_{pn} \otimes \mathbf{K}_{pn,pn})\right\}.$$

∎

Cumulants of the matrix elliptical distribution can be found by differentiating the cumulant function

$$\psi_{\mathbf{Y}}(\mathbf{T}) = i\mathrm{tr}(\mathbf{T}'\boldsymbol{\mu}) + \ln(\phi(\mathrm{tr}\{\mathbf{T}'\mathbf{V}\mathbf{T}\mathbf{W}\})). \qquad (2.3.21)$$

Theorem 2.3.17. *Let $\mathbf{Y} \sim E_{p,n}(\boldsymbol{\mu}, \mathbf{V}, \mathbf{W})$. Then the first even cumulants of $\mathbf{Y}$ are given by the following equalities, and all the odd cumulants $c_k[\mathbf{Y}]$, $k = 3, 5, \ldots$, equal zero.*

(i) $$c_1[\mathbf{Y}] = \mathrm{vec}\boldsymbol{\mu};$$

(ii) $$c_2[\mathbf{Y}] = -2\phi'(0)(\mathbf{W} \otimes \mathbf{V});$$

(iii) $$c_4[\mathbf{Y}] = 4(\phi''(0) - (\phi'(0))^2)\Big\{\mathbf{W} \otimes \mathbf{V} \otimes \mathrm{vec}'(\mathbf{W} \otimes \mathbf{V}) + (\mathrm{vec}'(\mathbf{W} \otimes \mathbf{V}) \otimes \mathbf{W} \otimes \mathbf{V})(\mathbf{I}_{(pn)^3} + \mathbf{I}_{pn} \otimes \mathbf{K}_{pn,pn})\Big\}.$$

PROOF: Once again we shall make use of the similarity between the matrix normal and matrix elliptical distributions. If we compare the cumulant functions of the matrix elliptical distribution (2.3.21) and multivariate elliptical distribution in Theorem 2.3.5, we see that they coincide, if we change the vectors to vectorized matrices and the matrix $\mathbf{V}$ to $\mathbf{W} \otimes \mathbf{V}$. So the derivatives of $\psi_{\mathbf{Y}}(\mathbf{T})$ have the same structure as the derivatives of $\psi_{\mathbf{y}}(\mathbf{t})$ where $\mathbf{y}$ is elliptically distributed. The trace expression in formula (2.3.21) is exactly the same as in the case of the matrix normal distribution in Theorem 2.2.1. Therefore, we get the statements of the theorem by combining expressions from Theorem 2.3.10 and Theorem 2.2.8. ∎

2.3.7 *Problems*

1. Let $\mathbf{x}_1 \sim N_p(\mathbf{0}, \sigma_1^2\mathbf{I})$ and $\mathbf{x}_2 \sim N_p(\mathbf{0}, \sigma_2^2\mathbf{I})$. Is the mixture $\mathbf{z}$ with density
$$f_{\mathbf{z}}(\mathbf{x}) = \gamma f_{\mathbf{x}_1}(\mathbf{x}) + (1-\gamma) f_{\mathbf{x}_2}(\mathbf{x})$$
spherically distributed?
2. Construct an elliptical distribution which has no density.
3. Show that $||\mathbf{X}||$ and $\dfrac{\mathbf{X}}{||\mathbf{X}||}$ in Corollary 2.3.4.1 are independent.
4. Find the characteristic function of the Pearson II type distribution (see Table 2.3.1).
5. Find $m_2[\mathbf{x}]$ of the Pearson II type distribution.
6. A matrix generalization of the multivariate $\mathbf{t}-$distribution (see Table 2.3.1) is called the matrix $\mathbf{T}-$distribution. Derive the density of the matrix $\mathbf{T}-$distribution as well as its mean and dispersion matrix.
7. Consider the Kotz type elliptical distribution with $N = 1$ (Table 2.3.1). Derive its characteristic function.
8. Find $D[\mathbf{x}]$ and $c_4(\mathbf{x})$ for the Kotz type distribution with $N = 1$.
9. Consider a density of the form
$$f_{\mathbf{x}}(\mathbf{x}) = c(\mathbf{x} - \boldsymbol{\mu})'\boldsymbol{\Sigma}^{-1}(\mathbf{x} - \boldsymbol{\mu}) f_{N(\boldsymbol{\mu},\boldsymbol{\Sigma})}(\mathbf{x}),$$
where c is the normalizing constant and the normal density is defined by (2.2.5). Determine the constant c and find $E[\mathbf{x}]$ and $D[\mathbf{x}]$.
10. The symmetric Laplace distribution is defined by the characteristic function
$$\varphi(\mathbf{t}) = \frac{1}{1 + \frac{1}{2}\mathbf{t}'\boldsymbol{\Sigma}\mathbf{t}},$$
where $\boldsymbol{\Sigma} > 0 : p \times p$ is the parameter matrix. Find the kurtosis characteristic κ for the distribution.

2.4 THE WISHART DISTRIBUTION

2.4.1 *Definition and basic properties*

The matrix distribution, which is nowadays known as the Wishart distribution, was first derived by Wishart (1928). It is usually regarded as a multivariate extension of the chi-square distribution. An interesting extension of the Wishart distribution is given by Hassairi & Lajmi (2001) who study the so-called Riesz exponential family (see also Hassairi, Lajmi & Zine, 2004). There are many possibilities to define a Wishart distribution. We will adopt the following approach.

Definition 2.4.1. *The matrix* $\mathbf{W} : p \times p$ *is said to be Wishart distributed if and only if* $\mathbf{W} = \mathbf{X}\mathbf{X}'$ *for some matrix* $\mathbf{X}$, *where* $\mathbf{X} \sim N_{p,n}(\boldsymbol{\mu}, \boldsymbol{\Sigma}, \mathbf{I})$, $\boldsymbol{\Sigma} \geq 0$. *If* $\boldsymbol{\mu} = \mathbf{0}$, *we have a central Wishart distribution which will be denoted* $\mathbf{W} \sim W_p(\boldsymbol{\Sigma}, n)$, *and if* $\boldsymbol{\mu} \neq \mathbf{0}$ *we have a non-central Wishart distribution which will be denoted* $W_p(\boldsymbol{\Sigma}, n, \boldsymbol{\Delta})$, *where* $\boldsymbol{\Delta} = \boldsymbol{\mu}\boldsymbol{\mu}'$. ∎

The parameter $\boldsymbol{\Sigma}$ is usually supposed to be unknown, whereas the second parameter n, which stands for the degrees of freedom, is usually considered to be known. The third parameter $\boldsymbol{\Delta}$, which is used in the non-central Wishart distribution, is called the non-centrality parameter. Generally speaking, one should remember that if $\boldsymbol{\Delta} \neq \mathbf{0}$, things start to be complicated. The Wishart distribution belongs to the class of matrix distributions. However, it is somewhat misleading to speak about a matrix distribution since among the p^2 elements of $\mathbf{W}$ there are just $\frac{1}{2}p(p+1)$ non-repeated elements and often the distribution is given via these elements. Let $\mathbf{X} = (\mathbf{x}_1, \ldots, \mathbf{x}_n)$, where $\mathbf{x}_i \sim N_p(\boldsymbol{\mu}_i, \boldsymbol{\Sigma})$ and $\mathbf{x}_i$ is independent of $\mathbf{x}_j$, when $i \neq j$. From Definition 2.4.1 it follows that $\mathbf{W} = \sum_{i=1}^n \mathbf{x}_i\mathbf{x}_i'$ and that clearly spells out that $\mathbf{W}$ is a sum of n independently distributed matrices. If $p = 1$, $\boldsymbol{\mu} = 0$ and $\boldsymbol{\Sigma} = 1$, the Wishart matrix is identical to a central χ^2-variable with n degrees of freedom (see also Corollary 2.4.2.2) and its density is given by

$$f_{\chi^2}(x) = (2^{n/2}\Gamma(n/2))^{-1}x^{n/2-1}e^{-x/2}, \quad x > 0,$$

where $\Gamma(\cdot)$ is the Gamma function. The distribution of a random variable, which is central χ^2-distributed with n degrees of freedom, will be denoted $\chi^2(n)$. If $p = 1$, $\boldsymbol{\Sigma} = 1$ but $\mu \neq 0$, we have a non-central χ^2-distribution with n degrees of freedom and non-centrality parameter $\delta = \mu^2$. Its density can be written as an infinite series (see Srivastava & Khatri, 1979, p. 60, for example) but this will not be utilized in the subsequent. The distribution of a random variable, which is non-central χ^2-distributed with n degrees of freedom and non-centrality parameter δ, will be denoted $\chi^2(n, \delta)$.

The first result of this paragraph is a direct consequence of Definition 2.4.1.

Theorem 2.4.1.

(i) *Let* $\mathbf{W}_1 \sim \mathbf{W}_p(\boldsymbol{\Sigma}, n, \boldsymbol{\Delta}_1)$ *be independent of* $\mathbf{W}_2 \sim \mathbf{W}_p(\boldsymbol{\Sigma}, m, \boldsymbol{\Delta}_2)$. *Then*

$$\mathbf{W}_1 + \mathbf{W}_2 \sim W_p(\boldsymbol{\Sigma}, n + m, \boldsymbol{\Delta}_1 + \boldsymbol{\Delta}_2).$$

(ii) *Let* $\mathbf{X} \sim N_{p,n}(\boldsymbol{\mu}, \boldsymbol{\Sigma}, \boldsymbol{\Psi})$, *where* $C(\boldsymbol{\mu}') \subseteq C(\boldsymbol{\Psi})$. *Put* $\mathbf{W} = \mathbf{X}\boldsymbol{\Psi}^-\mathbf{X}'$. *Then*

$$\mathbf{W} \sim W_p(\boldsymbol{\Sigma}, r(\boldsymbol{\Psi}), \boldsymbol{\Delta}),$$

where $\boldsymbol{\Delta} = \boldsymbol{\mu}\boldsymbol{\Psi}^-\boldsymbol{\mu}'$.

PROOF: By Definition 2.4.1, $\mathbf{W}_1 = \mathbf{X}_1\mathbf{X}_1'$ for some $\mathbf{X}_1 \sim N_{p,n}(\boldsymbol{\mu}_1, \boldsymbol{\Sigma}, \mathbf{I})$, where $\boldsymbol{\Delta}_1 = \boldsymbol{\mu}_1\boldsymbol{\mu}_1'$, and $\mathbf{W}_2 = \mathbf{X}_2\mathbf{X}_2'$ for some $\mathbf{X}_2 \sim N_{p,m}(\boldsymbol{\mu}_2, \boldsymbol{\Sigma}, \mathbf{I})$, where $\boldsymbol{\Delta}_2 = \boldsymbol{\mu}_2\boldsymbol{\mu}_2'$. The result follows, since

$$\mathbf{X}_1 : \mathbf{X}_2 \sim N_{p,n+m}(\boldsymbol{\mu}_1 : \boldsymbol{\mu}_2, \boldsymbol{\Sigma}, \mathbf{I})$$

and $\mathbf{W}_1 + \mathbf{W}_2 = (\mathbf{X}_1 : \mathbf{X}_2)(\mathbf{X}_1 : \mathbf{X}_2)'$.
For (ii) it is noted that by assumption $C(\mathbf{X}') \subseteq C(\boldsymbol{\Psi})$ with probability 1. Therefore $\mathbf{W}$ does not depend on the choice of g-inverse $\boldsymbol{\Psi}^-$. From Corollary 1.2.39.1 it follows that we may let $\boldsymbol{\Psi} = \boldsymbol{\Gamma}_1\mathbf{D}\boldsymbol{\Gamma}_1'$, where $\mathbf{D}$ is diagonal with positive elements and $\boldsymbol{\Gamma}_1$ is semi-orthogonal. Then

$$\mathbf{X}\boldsymbol{\Gamma}_1\mathbf{D}^{-\frac{1}{2}} \sim N_{p,r(\boldsymbol{\Psi})}(\boldsymbol{\mu}\boldsymbol{\Gamma}_1\mathbf{D}^{-\frac{1}{2}}, \boldsymbol{\Sigma}, \mathbf{I})$$

and $\mathbf{X}\boldsymbol{\Gamma}_1\mathbf{D}^{-\frac{1}{2}}\mathbf{D}^{-\frac{1}{2}}\boldsymbol{\Gamma}'\mathbf{X}' = \mathbf{X}\boldsymbol{\Psi}^-\mathbf{X}'$. ∎

One of the fundamental properties is given in the next theorem. It corresponds to the fact that the normal distribution is closed under linear transformations.

Theorem 2.4.2. *Let* $\mathbf{W} \sim W_p(\boldsymbol{\Sigma}, n, \boldsymbol{\Delta})$ *and* $\mathbf{A} \in \mathbb{R}^{q\times p}$. *Then*

$$\mathbf{AWA}' \sim W_q(\mathbf{A}\boldsymbol{\Sigma}\mathbf{A}', n, \mathbf{A}\boldsymbol{\Delta}\mathbf{A}').$$

PROOF: The proof follows immediately, since according to Definition 2.4.1, there exists an $\mathbf{X}$ such that $\mathbf{W} = \mathbf{XX}'$. By Theorem 2.2.2, $\mathbf{AX} \sim N_{q,n}(\mathbf{A}\boldsymbol{\mu}, \mathbf{A}\boldsymbol{\Sigma}\mathbf{A}', \mathbf{I})$, and thus we have that $\mathbf{AX}(\mathbf{AX})'$ is Wishart distributed. ∎

Corollary 2.4.2.1. *Let* $\mathbf{W} \sim W_p(\mathbf{I}, n)$ *and let* $\boldsymbol{\Gamma} : p \times p$ *be an orthogonal matrix which is independent of* $\mathbf{W}$. *Then* $\mathbf{W}$ *and* $\boldsymbol{\Gamma}\mathbf{W}\boldsymbol{\Gamma}'$ *have the same distribution:* $\mathbf{W} \stackrel{d}{=} \boldsymbol{\Gamma}\mathbf{W}\boldsymbol{\Gamma}'$. ∎

By looking closer at the proof of the theorem we may state another corollary.

Corollary 2.4.2.2. *Let* W_{ii}, $i = 1, 2, \ldots, p$, *denote the diagonal elements of* $\mathbf{W} \sim W_p(k\mathbf{I}, n, \boldsymbol{\Delta})$. *Then*

(i) $\frac{1}{k}W_{ii} \sim \chi^2(n, \delta_{ii})$, *where* $\boldsymbol{\Delta} = (\delta_{ij})$ *and* W_{ii} *is independent of* W_{jj} *if* $i \neq j$;

(ii) $\frac{1}{k}\mathrm{tr}\mathbf{W} \sim \chi^2(pn, \mathrm{tr}\boldsymbol{\Delta})$.

PROOF: According to Definition 2.4.1, there exists a matrix $\mathbf{X} \sim N_{p,n}(\boldsymbol{\mu}, \mathbf{I}, \mathbf{I})$ such that $\frac{1}{k}\mathbf{W} = \mathbf{XX}'$ and $\boldsymbol{\Delta} = \boldsymbol{\mu}\boldsymbol{\mu}'$. Now

$$\tfrac{1}{k}W_{ii} = \mathbf{e}_i'\mathbf{XX}'\mathbf{e}_i,$$

where $\mathbf{e}_i$ is the i-th unit vector and

$$(\mathbf{e}_i : \mathbf{e}_j)'\mathbf{X} \sim N_{2,n}((\mathbf{e}_i : \mathbf{e}_j)'\boldsymbol{\mu}, \mathbf{I}_2, \mathbf{I}).$$

Since $\mathbf{e}_i'\boldsymbol{\mu}\boldsymbol{\mu}'\mathbf{e}_i = \delta_{ii}$, this shows independence as well as that $\frac{1}{k}W_{ii}$ is χ^2-distributed. In (ii) we are just summing p independent χ^2-variables. ∎

The next theorem gives a multivariate version of a χ^2-property.

Theorem 2.4.3. *Let* $\mathbf{X} \sim N_{p,n}(\boldsymbol{\mu}, \boldsymbol{\Sigma}, \mathbf{I})$ *and* $\mathbf{Q} : n \times n$ *be symmetric. Then* $\mathbf{XQX}'$ *is Wishart distributed if and only if* $\mathbf{Q}$ *is an idempotent matrix.*

PROOF: We will give a somewhat different proof from the usual one. For the standard proof see Rao (1973a, p. 186) or Srivastava & Khatri (1979, pp. 63-64), for example. For a generalization of the statement with a similar proof to the one given below see Mathew & Nordström (1997). Let $\mathbf{U} \sim N_{p,n}(\boldsymbol{\mu}, \mathbf{I}, \mathbf{I})$, and $\mathbf{UQU}'$ will be studied. Thus we are going to study a special case when $\boldsymbol{\Sigma} = \mathbf{I}$. However, the general case follows from Theorem 2.4.2, since if $\mathbf{UQU}'$ is Wishart distributed, $\boldsymbol{\Sigma}^{1/2}\mathbf{UQU}'\boldsymbol{\Sigma}^{1/2}$ is also Wishart distributed, where $\boldsymbol{\Sigma}^{1/2}$ is a symmetric square root. Definition 2.4.1 will be applied.

Suppose that $\mathbf{Q}$ is idempotent. Then from Corollary 1.2.39.1 and Proposition 1.2.3 (viii) the representation $\mathbf{Q} = \boldsymbol{\Gamma}\mathbf{D}\boldsymbol{\Gamma}'$ is obtained, where $\boldsymbol{\Gamma}$ is an orthogonal matrix and $\mathbf{D}$ is a diagonal matrix where the elements on the diagonal equal either 1 or 0. Hence,

$$\mathbf{UQU}' = \mathbf{U}\boldsymbol{\Gamma}\mathbf{D}\boldsymbol{\Gamma}'\mathbf{U}',$$

and since the covariance matrix of $\mathbf{U}$ equals the identity matrix, the matrix $\mathbf{U}\boldsymbol{\Gamma}$ is $N_{p,n}(\boldsymbol{\mu}\boldsymbol{\Gamma}, \mathbf{I}, \mathbf{I})$ distributed. Without loss of generality, suppose that

$$\mathbf{D} = \begin{pmatrix} \mathbf{I}_{r(\mathbf{Q})} & \mathbf{0} \\ \mathbf{0} & \mathbf{0} \end{pmatrix}.$$

Let us partition $\mathbf{U}\boldsymbol{\Gamma}$ as $\mathbf{U}\boldsymbol{\Gamma} = (\mathbf{U}\boldsymbol{\Gamma}_1 : \mathbf{U}\boldsymbol{\Gamma}_2)$, $(p \times r(\mathbf{Q}) : p \times (n - r(\mathbf{Q}))$ and, correspondingly, $\boldsymbol{\mu}\boldsymbol{\Gamma} = (\boldsymbol{\mu}\boldsymbol{\Gamma}_1 : \boldsymbol{\mu}\boldsymbol{\Gamma}_2)$. Thus, $\mathbf{UQU}'$ has the same distribution as $\mathbf{U}\boldsymbol{\Gamma}_1\boldsymbol{\Gamma}_1'\mathbf{U}'$ which by Definition 2.4.1 is Wishart distributed, i.e. $\mathbf{U}\boldsymbol{\Gamma}_1\boldsymbol{\Gamma}_1'\mathbf{U}' \sim W_p(\mathbf{I}, r(\mathbf{Q}), \boldsymbol{\Delta}_1)$, where $\boldsymbol{\Delta}_1 = \boldsymbol{\mu}\boldsymbol{\Gamma}_1\boldsymbol{\Gamma}_1'\boldsymbol{\mu}'$.

Now the converse will be shown, i.e. if $\mathbf{UQU}'$ is Wishart distributed, then $\mathbf{Q}$ must be idempotent. This is the tricky part. If $\mathbf{UQU}'$ is Wishart distributed there must exist a matrix $\mathbf{Z} \sim N_{p,m}(\boldsymbol{\mu}, \mathbf{I}, \mathbf{I})$ such that the distribution of $\mathbf{UQU}'$ is the same as that of $\mathbf{ZZ}'$. Once again the canonical decomposition $\mathbf{Q} = \boldsymbol{\Gamma}\mathbf{D}\boldsymbol{\Gamma}'$ is used where $\boldsymbol{\Gamma}$ is orthogonal and $\mathbf{D}$ is diagonal. Note that $m \leq n$, since otherwise there are more random variables in $\mathbf{Z}$ than in $\mathbf{U}$. If $m \leq n$, we use the partition $\mathbf{D} = (\mathbf{D}_1, \mathbf{D}_2)_{[d]}$, where $\mathbf{D}_1 : m \times m$. Similarly, partition $\mathbf{U}$ as $(\mathbf{U}_1 : \mathbf{U}_2)$, $(p \times m : p \times (n - m))$. Since $\mathbf{U}_1$ is independent of $\mathbf{U}_2$ and $\mathbf{Z}$, with a proper choice of $\boldsymbol{\mu}$ it has the same distribution as $\mathbf{U}_1$, and we note that $\mathbf{UQU}'$ has the same distribution as

$$\mathbf{Z}\mathbf{D}_1\mathbf{Z}' + \mathbf{U}_2\mathbf{D}_2\mathbf{U}_2', \tag{2.4.1}$$

which in turn, according to the assumption, should have the same distribution as $\mathbf{ZZ}'$. Unless $\mathbf{D}_2 = \mathbf{0}$ in (2.4.1), this is impossible. For example, if conditioning in (2.4.1) with respect to $\mathbf{Z}$ some randomness due to $\mathbf{U}_2$ is still left, whereas conditioning $\mathbf{ZZ}'$ with respect to $\mathbf{Z}$ leads to a constant. Hence, it has been shown that $\mathbf{D}_2$ must be zero. It remains to prove under which conditions $\mathbf{Z}\mathbf{D}_1\mathbf{Z}'$ has the same distribution as $\mathbf{ZZ}'$. Observe that if $\mathbf{Z}\mathbf{D}_1\mathbf{Z}'$ has the same distribution as $\mathbf{ZZ}'$, we must also have that $(\mathbf{Z} - \boldsymbol{\mu})\mathbf{D}_1(\mathbf{Z} - \boldsymbol{\mu})'$ has the same distribution as

$(\mathbf{Z}-\boldsymbol{\mu})(\mathbf{Z}-\boldsymbol{\mu})'$. We are going to use the first two moments of these expressions and obtain from Theorem 2.2.9

$$\begin{aligned} E[(\mathbf{Z}-\boldsymbol{\mu})\mathbf{D}_1(\mathbf{Z}-\boldsymbol{\mu})'] &= \mathrm{tr}(\mathbf{D}_1)\mathbf{I}_p, \\ E[(\mathbf{Z}-\boldsymbol{\mu})(\mathbf{Z}-\boldsymbol{\mu})'] &= m\mathbf{I}_p, \qquad (2.4.2)\\ D[(\mathbf{Z}-\boldsymbol{\mu})\mathbf{D}_1(\mathbf{Z}-\boldsymbol{\mu})'] &= \mathrm{tr}(\mathbf{D}_1\mathbf{D}_1)(\mathbf{I}_{p^2}+\mathbf{K}_{p,p}), \\ D[(\mathbf{Z}-\boldsymbol{\mu})(\mathbf{Z}-\boldsymbol{\mu})'] &= m(\mathbf{I}_{p^2}+\mathbf{K}_{p,p}). \qquad (2.4.3)\end{aligned}$$

From (2.4.2) and (2.4.3) it follows that if $(\mathbf{Z}-\boldsymbol{\mu})\mathbf{D}_1(\mathbf{Z}-\boldsymbol{\mu})'$ is Wishart distributed, then the following equation system must hold:

$$\begin{aligned} \mathrm{tr}\mathbf{D}_1 &= m, \\ \mathrm{tr}(\mathbf{D}_1\mathbf{D}_1) &= m, \end{aligned}$$

which is equivalent to

$$\begin{aligned} \mathrm{tr}(\mathbf{D}_1-\mathbf{I}_m) &= 0, \qquad (2.4.4)\\ \mathrm{tr}(\mathbf{D}_1\mathbf{D}_1-\mathbf{I}_m) &= 0. \qquad (2.4.5)\end{aligned}$$

However, (2.4.5) is equivalent to

$$0 = \mathrm{tr}(\mathbf{D}_1\mathbf{D}_1-\mathbf{I}) = \mathrm{tr}((\mathbf{D}_1-\mathbf{I})(\mathbf{D}_1-\mathbf{I})) + 2\mathrm{tr}(\mathbf{D}_1-\mathbf{I}).$$

By (2.4.4), $2\mathrm{tr}(\mathbf{D}_1-\mathbf{I})=0$, and since $\mathrm{tr}((\mathbf{D}_1-\mathbf{I})(\mathbf{D}_1-\mathbf{I}))\geq 0$, equality in (2.4.5) holds if and only if $\mathbf{D}_1=\mathbf{I}$. Thus, $\mathbf{ZDZ}'$ has the same distribution as $\mathbf{ZZ}'$, if and only if $\mathbf{D}_1=\mathbf{I}$ and $\mathbf{D}_2=\mathbf{0}$. If $\mathbf{D}_1=\mathbf{I}$ and $\mathbf{D}_2=\mathbf{0}$, the matrix $\mathbf{Q}=\boldsymbol{\Gamma}(\mathbf{D}_1,\mathbf{D}_2)_{[d]}\boldsymbol{\Gamma}'$ is idempotent. Hence, it has been shown that $\mathbf{Q}$ must be idempotent, if $\mathbf{UQU}'$ is Wishart distributed. ∎

Corollary 2.4.3.1. *Let $\mathbf{X}\sim N_{p,n}(\boldsymbol{\mu},\boldsymbol{\Sigma},\mathbf{I})$ and $\mathbf{Q}: n\times n$ be symmetric and idempotent, so that $\boldsymbol{\mu}\mathbf{Q}=\mathbf{0}$. Then $\mathbf{XQX}'\sim W_p(\boldsymbol{\Sigma},r(\mathbf{Q}))$.* ∎

A very useful result when considering properties of the Wishart matrix is the so-called *Bartlett decomposition.* The first proof of this result was given by Bartlett (1933).

Theorem 2.4.4. *Let $\mathbf{W}\sim W_p(\mathbf{I},n,\boldsymbol{\Delta})$, where $p\leq n$, $\boldsymbol{\Delta}=\boldsymbol{\mu}\boldsymbol{\mu}'$ and $\boldsymbol{\mu}: p\times n$.*

(i) *Then there exists a lower triangular matrix $\mathbf{T}$ with positive diagonal elements and all elements independent, $T_{ij}\sim N(0,1)$, $p\geq i>j\geq 1$, $T_{ii}^2\sim\chi^2(n-i+1)$, $i=1,2,\ldots,p$, and a matrix $\mathbf{U}\sim N_{p,n}(\mathbf{0},\mathbf{I},\mathbf{I})$ such that*

$$\mathbf{W}=\mathbf{TT}'+\boldsymbol{\mu}\mathbf{U}'+\mathbf{U}\boldsymbol{\mu}'+\boldsymbol{\Delta}.$$

If $\boldsymbol{\mu}=\mathbf{0}$, then $\mathbf{W}=\mathbf{TT}'$ which is the classical Bartlett decomposition.

(ii) *Let $\boldsymbol{\mu}=\binom{\boldsymbol{\mu}_1}{\mathbf{0}}$, where $\boldsymbol{\mu}_1: 1\times n$. Then there exists a lower triangular matrix $\mathbf{T}$ with positive diagonal elements, with all elements independently distributed,*

$T_{ij} \sim N(0,1)$, $p \geq i > j \geq 1$, $T_{11}^2 \sim \chi^2(n,\delta)$ *with* $\delta = \boldsymbol{\mu}_1\boldsymbol{\mu}_1'$ *and* $T_{ii}^2 \sim \chi^2(n-i+1)$, $i = 2,3,\ldots,p$, *such that* $\mathbf{W} = \mathbf{T}\mathbf{T}'$.

PROOF: Note that by definition $\mathbf{W} = \mathbf{X}\mathbf{X}'$ for some $\mathbf{X} \sim N_{p,n}(\boldsymbol{\mu},\mathbf{I},\mathbf{I})$. Then

$$\mathbf{W} = \mathbf{U}\mathbf{U}' + \boldsymbol{\mu}\mathbf{U}' + \mathbf{U}\boldsymbol{\mu}' + \boldsymbol{\mu}\boldsymbol{\mu}'.$$

Hence, we need to show that $\mathbf{U}\mathbf{U}'$ has the same distribution as $\mathbf{T}\mathbf{T}'$. With probability 1, the rank $r(\mathbf{U}\mathbf{U}') = p$ and thus the matrix $\mathbf{U}\mathbf{U}'$ is p.d. with probability 1. From Theorem 1.1.4 and its proof it follows that there exists a unique lower triangular matrix $\mathbf{T}$ such that $\mathbf{U}\mathbf{U}' = \mathbf{T}\mathbf{T}'$, with elements

$$\begin{aligned} T_{11} &= (U_{11}^2 + \mathbf{u}_{12}\mathbf{u}_{12}')^{1/2}, \\ \mathbf{t}_{21} &= (U_{11}^2 + \mathbf{u}_{12}\mathbf{u}_{12}')^{-1/2}(\mathbf{u}_{21}U_{11} + \mathbf{U}_{22}\mathbf{u}_{12}'), \end{aligned} \tag{2.4.6}$$

where

$$\mathbf{U} = \begin{pmatrix} U_{11} & \mathbf{u}_{12} \\ \mathbf{u}_{21} & \mathbf{U}_{22} \end{pmatrix}.$$

Furthermore,

$$\begin{aligned} \mathbf{T}_{22}\mathbf{T}_{22}' = {} & \mathbf{u}_{21}\mathbf{u}_{21}' + \mathbf{U}_{22}\mathbf{U}_{22}' \\ & - (\mathbf{u}_{21}U_{11} + \mathbf{U}_{22}\mathbf{u}_{12}')(U_{11}^2 + \mathbf{u}_{12}\mathbf{u}_{12}')^{-1}(U_{11}\mathbf{u}_{21}' + \mathbf{u}_{12}\mathbf{U}_{22}'). \end{aligned} \tag{2.4.7}$$

By assumption, the elements of $\mathbf{U}$ are independent and normally distributed: $U_{ij} \sim N(0,1)$, $i = 1,2,\ldots,p$, $j = 1,2,\ldots,n$. Since U_{11} and $\mathbf{u}_{12}$ are independently distributed, $T_{11}^2 \sim \chi^2(n)$. Moreover, consider the conditional distribution of $\mathbf{t}_{21}|U_{11},\mathbf{u}_{12}$ which by (2.4.6) and independence of the elements of $\mathbf{U}$ is normally distributed:

$$\mathbf{t}_{21}|U_{11},\mathbf{u}_{12} \sim N_{p-1}(\mathbf{0},\mathbf{I}). \tag{2.4.8}$$

However, by (2.4.8), $\mathbf{t}_{21}$ is independent of $(U_{11},\mathbf{u}_{12})$ and thus

$$\mathbf{t}_{21} \sim N_{p-1}(\mathbf{0},\mathbf{I}).$$

It remains to show that

$$\mathbf{T}_{22}\mathbf{T}_{22}' \sim W_{p-1}(\mathbf{I}, n-1). \tag{2.4.9}$$

We can restate the arguments above, but instead of n degrees of freedom we have now $(n-1)$. From (2.4.7) it follows that

$$\mathbf{T}_{22}\mathbf{T}_{22}' = (\mathbf{u}_{21} : \mathbf{U}_{22})\mathbf{P}(\mathbf{u}_{21} : \mathbf{U}_{22})',$$

where

$$\mathbf{P} = \mathbf{I} - (U_{11} : \mathbf{u}_{12})'(U_{11}^2 + \mathbf{u}_{12}\mathbf{u}_{12}')^{-1}(U_{11} : \mathbf{u}_{12}).$$

Since $\mathbf{P}$ is idempotent and $(\mathbf{u}_{21} : \mathbf{U}_{22})$ is independent of $(U_{11} : \mathbf{u}_{12})$, it follows from Theorem 2.4.3 that

$$\mathbf{T}_{22}\mathbf{T}_{22}'|U_{11} : \mathbf{u}_{12} \sim W_{p-1}(\mathbf{I}, r(\mathbf{P})).$$

Now, with probability 1, $r(\mathbf{P}) = \mathrm{tr}(\mathbf{P}) = n-1$ (see Proposition 1.1.4 (v)). Thus $\mathbf{T}_{22}\mathbf{T}'_{22}$ does not depend of $(U_{11} : \mathbf{u}_{12})$ and (2.4.9) is established.
The proof of (ii) is almost identical to the one given for (i). Instead of $\mathbf{U}$ we just have to consider $\mathbf{X} \sim N_{p,n}(\boldsymbol{\mu}, \mathbf{I}, \mathbf{I})$. Thus,

$$\begin{aligned} T_{11} &= (X_{11}^2 + \mathbf{x}_{12}\mathbf{x}'_{12})^{1/2}, \\ \mathbf{t}_{21} &= (X_{11}^2 + \mathbf{x}_{12}\mathbf{x}'_{12})^{-1/2}(\mathbf{x}_{21}X_{11} + \mathbf{X}_{22}\mathbf{x}'_{12}), \\ \mathbf{T}_{22}\mathbf{T}'_{22} &= \mathbf{x}_{21}\mathbf{x}'_{21} + \mathbf{X}_{22}\mathbf{X}'_{22} \\ &\quad - (\mathbf{x}_{21}X_{11} + \mathbf{X}_{22}\mathbf{x}'_{12})(X_{11}^2 + \mathbf{x}_{12}\mathbf{x}'_{12})^{-1}(X_{11}\mathbf{x}'_{21} + \mathbf{x}_{12}\mathbf{X}'_{22}). \end{aligned}$$

Then it follows that $T_{11}^2 \sim \chi^2(n, \delta)$ and the distribution of $\mathbf{t}_{21}|X_{11}, \mathbf{x}_{12}$ has to be considered. However,

$$E[\mathbf{t}_{21}|X_{11}, \mathbf{x}_{12}] = \mathbf{0},$$

because $\mathbf{x}_{21}$, $\mathbf{X}_{22}$ are independent of X_{11}, $\mathbf{x}_{21}$, $E[\mathbf{x}_{21}] = \mathbf{0}$ as well as $E[\mathbf{X}_{22}] = \mathbf{0}$. Furthermore,

$$D[\mathbf{t}_{21}|X_{11}, \mathbf{x}_{12}] = \mathbf{I}.$$

Thus, $\mathbf{t}_{21} \sim N_{p-1}(\mathbf{0}, \mathbf{I})$ and the rest of the proof follows from the proof of (i). ∎

Corollary 2.4.4.1. *Let $\mathbf{W} \sim W_p(\mathbf{I}, n)$, where $p \leq n$. Then there exists a lower triangular matrix $\mathbf{T}$, where all elements are independent, and the diagonal elements are positive, with $T_{ij} \sim N(0,1)$, $p \geq i > j \geq 1$, and $T_{ii}^2 \sim \chi^2(n-i+1)$ such that*

$$\mathbf{W} = \mathbf{T}\mathbf{T}'.$$

∎

Corollary 2.4.4.2. *Let $\mathbf{W} \sim W_p(k\mathbf{I}, n)$, $p \leq n$. Then*

$$V = \frac{|\mathbf{W}|}{(\frac{1}{p}\mathrm{tr}\mathbf{W})^p}$$

and $\mathrm{tr}\mathbf{W}$ are independently distributed.

PROOF: First it is noted that V does not change if $\mathbf{W}$ is replaced by $\frac{1}{k}\mathbf{W}$. Thus, we may assume that $k=1$. From Corollary 2.4.4.1 it follows that

$$V = \frac{\prod_{i=1}^p T_{ii}^2}{(\frac{1}{p}\sum_{i\geq j=1}^p T_{ij}^2)^p}$$

and $\mathrm{tr}\mathbf{W} = \sum_{i\geq j=1}^p T_{ij}^2$. It will be shown that the joint density of V and $\mathrm{tr}\mathbf{W}$ is a product of the marginal densities of V and $\mathrm{tr}\mathbf{W}$. Since $T_{ii}^2 \sim \chi^2(n-i+1)$, $T_{ij}^2 \sim \chi^2(1)$, $i > j$, and because of independence of the elements in $\mathbf{T}$, it follows that the joint density of $\{T_{ij}^2, i \geq j\}$ is given by

$$c\prod_{i=1}^p T_{ii}^{2(\frac{n-i+1}{2}-1)} \prod_{i>j=1}^{p-1} T_{ij}^{2(\frac{1}{2}-1)} e^{-\frac{1}{2}\sum_{i\geq j=1}^p T_{ij}^2}, \qquad (2.4.10)$$

where c is a normalizing constant. Make the following change of variables

$$Y = \frac{1}{p}\sum_{i\ge j=1}^{p} T_{ij}^2, \quad Z_{ij} = \frac{T_{ij}^2}{Y}, \; i \ge j.$$

Since $\sum_{i\ge j=1}^{p} Z_{ij} = p$ put

$$Z_{p,p-1} = p - \sum_{i=1}^{p} Z_{ii} - \sum_{i>j=1}^{p-2} Z_{ij}.$$

It will be shown that Y is independently distributed of $\{Z_{ij}\}$, which verifies the corollary since

$$V = \prod_{i=1}^{p} Z_{ii}.$$

Utilizing (2.4.10) gives the joint density of Z_{ii}, $i = 1,2,\ldots,p$, Z_{ij}, $i > j = 1,2,\ldots,p-2$, and Y:

$$c\prod_{i=1}^{p-1} Z_{ii}^{\frac{n-i+1}{2}-1}\prod_{i>j=1}^{p-2} Z_{ij}^{-\frac{1}{2}}\Big(p - \sum_{i=1}^{p} Z_{ii} - \sum_{i>j=1}^{p-2} Z_{ij}\Big)^{-\frac{1}{2}} Y^{a} e^{-\frac{p}{2}Y}|\mathbf{J}|_+,$$

where

$$a = \sum_{i=1}^{p}\big(\tfrac{n-i+1}{2} - 1\big) - \tfrac{1}{2}p(p-1) = \tfrac{1}{4}(2pn - 3p^2 - p)$$

and the Jacobian is defined as

$$|\mathbf{J}|_+ = |\mathbf{J}(T_{11}^2, T_{21}^2, \ldots, T_{p,p-1}^2, T_{pp}^2, \to Y, Z_{11}, Z_{21}, \ldots, Z_{p,p-2}, Z_{pp})|_+.$$

Thus, it remains to express the Jacobian. With the help of (1.1.6)

$$\left|\frac{d\,(T_{11}^2, T_{21}^2, \ldots, T_{p,p-1}^2, T_{pp}^2)}{d\,\mathrm{vec}(Y, Z_{11}, Z_{21}, \ldots, Z_{p,p-2}, Z_{pp})}\right|_+ = \left|\begin{array}{cccccc} Z_{11} & Z_{21} & Z_{22} & \ldots & Z_{p,p-1} & Z_{pp} \\ Y & 0 & 0 & \ldots & -Y & 0 \\ 0 & Y & 0 & \ldots & -Y & 0 \\ 0 & 0 & Y & \ldots & -Y & 0 \\ \vdots & \vdots & \vdots & \ddots & \vdots & \vdots \\ 0 & 0 & 0 & \ldots & -Y & Y \end{array}\right|_+$$

$$= |\sum_{i\ge j=1}^{p} Z_{ij}(-Y)Y^{\frac{1}{2}p(p+1)-2}|_+ = \sum_{i\ge j=1}^{p} Z_{ij}Y^{\frac{1}{2}p(p+1)-1} = pY^{\frac{1}{2}p(p+1)-1},$$

which implies that the density of $\{Z_{ij}\}$ and Y factors. ∎

The statistic V in Corollary 2.4.4.2 looks artificial but it arises when testing $\mathbf{\Sigma} = k\mathbf{I}$ (sphericity test) in a normal sample (e.g. see Muirhead (1982), §8.3.1). Moreover, it is interesting to note that we can create other functions besides V, and these are solely functions of $\{Z_{ij}\}$ and thus independent of $\mathrm{tr}\mathbf{W}$.

2.4.2 *Characteristic and density functions*

When considering the characteristic or density function of the Wishart matrix $\mathbf{W}$, we have to take into account that $\mathbf{W}$ is symmetric. There are many ways of doing this and one of them is not to take into account that the matrix is symmetric which, however, will not be applied here. Instead, when obtaining the characteristic function and density function of $\mathbf{W}$, we are going to obtain the characteristic function and density function of the elements of the upper triangular part of $\mathbf{W}$, i.e. W_{ij}, $i \leq j$. For a general reference on the topic see Olkin (2002).

Theorem 2.4.5. *Let* $\mathbf{W} \sim W_p(\mathbf{\Sigma}, n)$. *The characteristic function of* $\{W_{ij}, i \leq j\}$, *equals*

$$\varphi_{\mathbf{W}}(\mathbf{T}) = |\mathbf{I}_p - i\mathbf{M}(\mathbf{T})\mathbf{\Sigma}|^{-\frac{n}{2}},$$

where

$$\mathbf{M}(\mathbf{T}) = \sum_I t_{ij}(\mathbf{e}_i\mathbf{e}_j' + \mathbf{e}_j\mathbf{e}_i') = \mathbf{T} + \mathbf{T}_d, \tag{2.4.11}$$

$\mathbf{e}_i$ *is the* $i-th$ *column of* $\mathbf{I}_p$ *and* $I = \{i, j; 1 \leq i \leq j \leq p\}$.

PROOF: According to (2.1.9) we have to find $\varphi_{\mathbf{W}}(\mathbf{T}) = E[e^{iV^{2'}(\mathbf{W})V^2(\mathbf{T})}]$. First note that

$$\begin{aligned}\tfrac{1}{2}\mathrm{tr}(\mathbf{M}(\mathbf{T})\mathbf{W}) &= \tfrac{1}{2}\mathrm{tr}\{\sum_{i\leq j} t_{ij}(\mathbf{e}_j\mathbf{e}_i' + \mathbf{e}_i\mathbf{e}_j')\mathbf{W}\} = \tfrac{1}{2}\sum_{i\leq j} t_{ij}(\mathbf{e}_i'\mathbf{W}\mathbf{e}_j + \mathbf{e}_j'\mathbf{W}\mathbf{e}_i)\\ &= \tfrac{1}{2}\sum_{i\leq j} t_{ij}(W_{ij} + W_{ji}) = \sum_{i\leq j} t_{ij}W_{ij} = V^{2'}(\mathbf{T})V^2(\mathbf{W}).\end{aligned}$$

Then, by Corollary 1.2.39.1, there exist an orthogonal matrix $\mathbf{\Gamma}$ and a diagonal matrix $\mathbf{D}$ such that

$$\mathbf{\Sigma}^{1/2}\mathbf{M}(\mathbf{T})\mathbf{\Sigma}^{1/2} = \mathbf{\Gamma}\mathbf{D}\mathbf{\Gamma}',$$

where $\mathbf{\Sigma}^{1/2}$ denotes a symmetric square root. Let $\mathbf{V} \sim W_p(\mathbf{I}, n)$, which implies that the distribution $\mathbf{V}$ is rotational invariant and thus,

$$\varphi_{\mathbf{W}}(\mathbf{T}) = E[e^{\frac{i}{2}\mathrm{tr}(\mathbf{M}(\mathbf{T})\mathbf{W})}] = E[e^{\frac{i}{2}\mathrm{tr}(\mathbf{\Gamma}\mathbf{D}\mathbf{\Gamma}'\mathbf{V})}] = E[e^{\frac{i}{2}\mathrm{tr}(\mathbf{D}\mathbf{V})}] = E[e^{\frac{i}{2}\sum_k d_{kk}V_{kk}}].$$

Since, by Corollary 2.4.2.2, $V_{kk} \sim \chi^2(n)$, for $k = 1, 2, \ldots, p$, and mutually independent, we obtain that

$$\varphi_{\mathbf{W}}(\mathbf{T}) = \prod_{k=1}^{p}(1 - id_{kk})^{-n/2} = |\mathbf{I} - i\mathbf{D}|^{-n/2}.$$

Hence, it remains to express the characteristic function through the original matrix $\mathbf{\Sigma}$:

$$|\mathbf{I} - i\mathbf{D}| = |\mathbf{I} - i\mathbf{\Gamma}\mathbf{D}\mathbf{\Gamma}'| = |\mathbf{I} - i\mathbf{\Sigma}^{1/2}\mathbf{M}(\mathbf{T})\mathbf{\Sigma}^{1/2}| = |\mathbf{I} - i\mathbf{M}(\mathbf{T})\mathbf{\Sigma}|,$$

where in the last equality Proposition 1.1.1 (iii) has been used. ∎

Observe that in Theorem 2.4.5 the assumption $n \geq p$ is not required.

Theorem 2.4.6. *Let* $\mathbf{W} \sim W_p(\mathbf{\Sigma}, n)$. *If* $\mathbf{\Sigma} > 0$ *and* $n \geq p$, *then the matrix* $\mathbf{W}$ *has the density function*

$$f_{\mathbf{W}}(\mathbf{W}) = \begin{cases} \dfrac{1}{2^{\frac{pn}{2}}\Gamma_p(\frac{n}{2})|\mathbf{\Sigma}|^{\frac{n}{2}}}|\mathbf{W}|^{\frac{n-p-1}{2}}e^{-\frac{1}{2}\mathrm{tr}(\mathbf{\Sigma}^{-1}\mathbf{W})}, & \mathbf{W} > 0, \\ 0, & \textit{otherwise}, \end{cases} \tag{2.4.12}$$

where the multivariate gamma function $\Gamma_p(\frac{n}{2})$ *is given by*

$$\Gamma_p(\tfrac{n}{2}) = \pi^{\frac{p(p-1)}{4}}\prod_{i=1}^{p}\Gamma(\tfrac{1}{2}(n+1-i)). \tag{2.4.13}$$

PROOF: We are going to sketch an alternative proof to the standard one (e.g. see Anderson, 2003, pp. 252-255; Srivastava & Khatri, 1979, pp. 74-76 or Muirhead, 1982, pp. 85-86) where, however, all details for integrating complex variables will not be given. Relation (2.4.12) will be established for the special case $\mathbf{V} \sim W_p(\mathbf{I}, n)$. The general case is immediately obtained from Theorem 2.4.2. The idea is to use the Fourier transform which connects the characteristic and the density functions. Results related to the Fourier transform will be presented later in §3.2.2. From Theorem 2.4.5 and Corollary 3.2.1.L1 it follows that

$$f_{\mathbf{V}}(\mathbf{V}) = (2\pi)^{-\frac{1}{2}p(p+1)}\int_{\mathbb{R}^{p(p+1)/2}}|\mathbf{I} - i\mathbf{M}(\mathbf{T})|^{-n/2}e^{-\frac{i}{2}\mathrm{tr}(\mathbf{M}(\mathbf{T})\mathbf{V})}d\mathbf{T} \tag{2.4.14}$$

should be studied, where $\mathbf{M}(\mathbf{T})$ is given by (2.4.11) and $d\mathbf{T} = \prod_{i\leq j} dt_{ij}$. Put $\mathbf{Z} = (\mathbf{V}^{1/2})'(\mathbf{I} - i\mathbf{M}(\mathbf{T}))\mathbf{V}^{1/2}$. By Theorem 1.4.13 (i), the Jacobian for the transformation from $\mathbf{T}$ to $\mathbf{Z}$ equals $c|\mathbf{V}^{1/2}|^{-(p+1)} = c|\mathbf{V}|^{-\frac{1}{2}(p+1)}$ for some constant c. Thus, the right hand side of (2.4.14) can be written as

$$c\int|\mathbf{Z}|^{-n/2}e^{\frac{1}{2}\mathrm{tr}(\mathbf{Z})}d\mathbf{Z}|\mathbf{V}|^{\frac{1}{2}(n-p-1)}e^{-\frac{1}{2}\mathrm{tr}\mathbf{V}},$$

and therefore we know that the density equals

$$f_{\mathbf{V}}(\mathbf{V}) = c(p,n)|\mathbf{V}|^{\frac{1}{2}(n-p-1)}e^{-\frac{1}{2}\mathrm{tr}\mathbf{V}}, \tag{2.4.15}$$

where $c(p,n)$ is a normalizing constant. Let us find $c(p,n)$. From Theorem 1.4.18, and by integrating the standard univariate normal density function, it follows that

$$\begin{aligned}
1 =& c(p,n)\int|\mathbf{V}|^{\frac{1}{2}(n-p-1)}e^{-\frac{1}{2}\mathrm{tr}\mathbf{V}}d\mathbf{V} \\
=& c(p,n)\int 2^p\prod_{i=1}^{p}t_{ii}^{p-i+1}t_{ii}^{n-p-1}e^{-\frac{1}{2}t_{ii}^2}\prod_{i>j}e^{-\frac{1}{2}t_{ij}^2}d\mathbf{T} \\
=& c(p,n)2^p(2\pi)^{\frac{1}{4}p(p-1)}\prod_{i=1}^{p}\int t_{ii}^{n-i}e^{-\frac{1}{2}t_{ii}^2}dt_{ii} \\
=& c(p,n)2^p(2\pi)^{\frac{1}{4}p(p-1)}\prod_{i=1}^{p}2^{-1}\int v_i^{\frac{1}{2}(n-i-1)}e^{-\frac{1}{2}v_i}dv_i \\
=& c(p,n)(2\pi)^{\frac{1}{4}p(p-1)}\prod_{i=1}^{p}2^{\frac{1}{2}(n-i+1)}\Gamma(\tfrac{1}{2}(n-i+1)),
\end{aligned}$$

where in the last equality the expression for the χ^2-density has been used. Thus,

$$c(p,n)^{-1} = (\pi)^{\frac{1}{4}p(p-1)} 2^{\frac{1}{2}pn} \prod_{i=1}^{p} \Gamma(\tfrac{1}{2}(n-i+1)) \qquad (2.4.16)$$

and the theorem is established. ∎

Another straightforward way of deriving the density would be to combine Corollary 2.4.4.1 and Theorem 1.4.18.

Instead of Definition 2.4.1 we could have used either Theorem 2.4.5 or Theorem 2.4.6 as a definition of a Wishart distribution. It is a matter of taste which one to prefer. However, the various possibilities of defining the Wishart distribution are not completely equivalent. For example, the density in Theorem 2.4.6 is valid for all $n \geq p$ and not just for any positive integer n. Furthermore, from the characteristic function we could also get degenerated Wishart distributions, for example, when $n = 0$. A general discussion on the topic can be found in Faraut & Korányi (1994) (see also Casalis & Letac, 1996). Instead of the characteristic function we could have used the Laplace transform (see Herz, 1955; Muirhead, 1982, pp. 252-253).

From Theorem 2.4.6 a corollary is derived, where the density of $\mathbf{W}^{-1}$ is obtained when $\mathbf{W} \sim W_p(\mathbf{\Sigma}, n)$. The distribution of $\mathbf{W}^{-1}$ is called *inverted Wishart distribution.*

Corollary 2.4.6.1. *Let* $\mathbf{W} \sim W_p(\mathbf{\Sigma}, n)$, $p \leq n$ *and* $\mathbf{\Sigma} > 0$. *Then the density of* $\mathbf{V} = \mathbf{W}^{-1}$ *is given by*

$$f_{\mathbf{W}^{-1}}(\mathbf{V}) = (2^{\frac{1}{2}pn}\Gamma_p(\tfrac{n}{2}))^{-1}|\mathbf{\Sigma}|^{-n/2}|\mathbf{V}|^{-\frac{1}{2}(n+p+1)}e^{-\frac{1}{2}\mathrm{tr}(\mathbf{\Sigma}^{-1}\mathbf{V}^{-1})}, \qquad \mathbf{V} > 0.$$

PROOF: Relation (2.4.12) and Theorem 1.4.17 (ii) establish the equality. ∎

Another interesting consequence of the theorem is stated in the next corollary.

Corollary 2.4.6.2. *Let* $\mathbf{L} : p \times n$, $p \leq n$ *be semi-orthogonal, i.e.* $\mathbf{L}\mathbf{L}' = \mathbf{I}_p$, *and let the functionally independent elements of* $\mathbf{L} = (l_{kl})$ *be given by* $l_{12}, l_{13}, \ldots, l_{1n}, l_{23}, \ldots, l_{2n}, \ldots, l_{p(p+1)}, \ldots, l_{pn}$. *Denote these elements* $\mathbf{L}(K)$. *Then*

$$\int \prod_{i=1}^{p} |\mathbf{L}_i|_+ d\,\mathbf{L}(K) = c(p,n)(2\pi)^{\frac{1}{2}pn}2^p,$$

where $\mathbf{L}_i = (l_{kl})$, $k, l = 1, 2, \ldots, i$, *and* $c(p,n)^{-1} = 2^{\frac{1}{2}pn}\Gamma_p(\frac{n}{2})$.

PROOF: Suppose that $\mathbf{X} \sim N_{p,n}(\mathbf{0}, \mathbf{I}, \mathbf{I})$. Let $\mathbf{X} = \mathbf{T}\mathbf{L}$, where $\mathbf{T}$ is the lower triangular matrix with positive diagonal elements and $\mathbf{L}$ is the semi-orthogonal matrix mentioned above. According to Theorem 1.4.20, the joint density of $\mathbf{T}$ and $\mathbf{L}$ is given by

$$(2\pi)^{-\frac{1}{2}pn}e^{-\frac{1}{2}\mathrm{tr}\mathbf{T}\mathbf{T}'}\prod_{i=1}^{p} T_{ii}^{n-i}\prod_{i=1}^{p}|\mathbf{L}_i|_+.$$

Put $\mathbf{W} = \mathbf{TT}'$, and from Theorem 1.4.18 it follows that the joint density of $\mathbf{W}$ and $\mathbf{L}$ is given by

$$(2\pi)^{-\frac{1}{2}pn}e^{-\frac{1}{2}\mathrm{tr}\mathbf{W}}2^{-p}|\mathbf{W}|^{\frac{1}{2}(n-p-1)}\prod_{i=1}^{p}|\mathbf{L}_i|_+.$$

Thus, Theorem 2.4.6 implies that

$$(2\pi)^{-\frac{1}{2}pn}2^{-p}\int\prod_{i=1}^{p}|\mathbf{L}_i|_+d\,\mathbf{L}(K) = c(p,n).$$

∎

For the inverted Wishart distribution a result may be stated which is similar to the Bartlett decomposition given in Theorem 2.4.4.

Theorem 2.4.7. *Let $\mathbf{W} \sim W_p(\mathbf{I}, n)$. Then there exists a lower triangular matrix $\mathbf{T}$ with positive diagonal elements such that for $\mathbf{W}^{-1} = \mathbf{TT}'$, $\mathbf{T}^{-1} = (T^{ij})$, $T^{ij} \sim N(0,1)$, $p \geq i > j \geq 1$, $(T^{ii})^2 \sim \chi^2(n-p+1)$, $i = 1, 2, \ldots, p$, and all random variables T^{ij} are independently distributed.*

PROOF: Using Corollary 2.4.6.1 and Theorem 1.4.18, the density of $\mathbf{T}$ is given by

$$\begin{aligned} c|\mathbf{TT}'|^{-\frac{1}{2}(n+p+1)}&e^{-\frac{1}{2}\mathrm{tr}(\mathbf{TT}')^{-1}}\prod_{i=1}^{p}T_{ii}^{p-i+1} \\ &= c\prod_{i=1}^{p}T_{ii}^{-(n+i)}e^{-\frac{1}{2}\mathrm{tr}(\mathbf{TT}')^{-1}}, \end{aligned}$$

where c is the normalizing constant. Consider

$$(\mathbf{TT}')^{-1} = \begin{pmatrix} (T_{11}^{-1})^2 + \mathbf{v}'\mathbf{v} & -\mathbf{v}'\mathbf{T}_{22}^{-1} \\ -(\mathbf{T}_{22}')^{-1}\mathbf{v} & (\mathbf{T}_{22}')^{-1}\mathbf{T}_{22}^{-1} \end{pmatrix}, \quad \begin{pmatrix} 1\times 1 & 1\times(p-1) \\ (p-1)\times 1 & (p-1)\times(p-1) \end{pmatrix},$$

where

$$\mathbf{T} = \begin{pmatrix} T_{11} & \mathbf{0} \\ \mathbf{t}_{21} & \mathbf{T}_{22} \end{pmatrix}$$

and

$$\mathbf{v} = \mathbf{T}_{22}^{-1}\mathbf{t}_{21}T_{11}^{-1}.$$

Observe that

$$\mathbf{T}^{-1} = \begin{pmatrix} T_{11}^{-1} & \mathbf{0} \\ -\mathbf{v} & \mathbf{T}_{22}^{-1} \end{pmatrix}.$$

Now, Theorem 1.4.14 implies that

$$|\mathbf{J}(T_{11}, \mathbf{t}_{21}, \mathbf{T}_{22} \to T_{11}, \mathbf{v}, \mathbf{T}_{22})|_+ = T_{11}^{p-1}|\mathbf{T}_{22}|_+$$

and therefore the density equals

$$cT_{11}^{-(n-p+2)}e^{-\frac{1}{2}(T_{11}^{-1})^2}\prod_{i=2}^{p}T_{ii}^{-(n+i-1)}e^{-\frac{1}{2}\mathrm{tr}\{(\mathbf{T}_{22}\mathbf{T}_{22}')^{-1}\}}e^{-\frac{1}{2}\mathbf{v}'\mathbf{v}},$$

which means that T_{11}, $\mathbf{T}_{22}$ and $\mathbf{v}$ are independent, $(T_{11}^2)^{-1}\sim\chi^2(n-p+1)$ and the other elements in $\mathbf{T}^{-1}$ are standard normal. Then we may restart with $\mathbf{T}_{22}$ and by performing the same operations as above we see that T_{11} and T_{22} have the same distribution. Continuing in the same manner we obtain that $(T_{ii}^2)^{-1}\sim\chi^2(n-p+1)$. ∎

Observe that the above theorem does not follow immediately from the Bartlett decomposition since we have

$$\mathbf{W}^{-1}=\mathbf{T}\mathbf{T}'$$

and

$$\mathbf{W}=\widetilde{\mathbf{T}}\widetilde{\mathbf{T}}',$$

where $\mathbf{W}\sim W_p(\mathbf{I},n)$, $\mathbf{T}$ and $\widetilde{\mathbf{T}}$ are lower triangular matrices. However, $\mathbf{T}$ corresponds to $\widetilde{\mathbf{T}}'$ which is an upper triangular matrix, i.e. $\mathbf{T}\neq\widetilde{\mathbf{T}}'$.

2.4.3 *Multivariate beta distributions*

A random variable with *univariate beta distribution* has a density function which equals

$$f_\beta(x)=\begin{cases}\dfrac{\Gamma(\frac{1}{2}(m+n))}{\Gamma(\frac{1}{2}m)\Gamma(\frac{1}{2}n)}x^{\frac{1}{2}m-1}(1-x)^{\frac{1}{2}n-1} & 0<x<1,\quad m,n\geq 1,\\ 0 & \text{elsewhere.}\end{cases}\tag{2.4.17}$$

If a variable follows (2.4.17), it is denoted $\beta(m,n)$.

In this paragraph a number of different generalizations of the distribution given via (2.4.17) will be discussed. Two types of multivariate beta distributions will be introduced, which are both closely connected to the normal distribution and Wishart distribution. It is interesting to observe how the Jacobians of §1.4.12 will be utilized in the subsequent derivations of the multivariate beta densities.

Theorem 2.4.8. *Let* $\mathbf{W}_1\sim W_p(\mathbf{I},n)$, $p\leq n$, *and* $\mathbf{W}_2\sim W_p(\mathbf{I},m)$, $p\leq m$, *be independently distributed. Then*

$$\mathbf{F}=(\mathbf{W}_1+\mathbf{W}_2)^{-1/2}\mathbf{W}_2(\mathbf{W}_1+\mathbf{W}_2)^{-1/2}$$

has a density function given by

$$f_{\mathbf{F}}(\mathbf{F})=\begin{cases}\dfrac{c(p,n)c(p,m)}{c(p,n+m)}|\mathbf{F}|^{\frac{1}{2}(m-p-1)}|\mathbf{I}-\mathbf{F}|^{\frac{1}{2}(n-p-1)} & |\mathbf{I}-\mathbf{F}|>0,|\mathbf{F}|>0,\\ 0 & \text{otherwise,}\end{cases}\tag{2.4.18}$$

where

$$c(p,n) = (2^{\frac{pn}{2}}\Gamma_p(\tfrac{n}{2}))^{-1}$$

and $(\mathbf{W}_1+\mathbf{W}_2)^{-1/2}$ *is a symmetric square root.*

PROOF: From Theorem 2.4.6 it follows that the joint density of $\mathbf{W}_1$ and $\mathbf{W}_2$ is given by

$$c(p,n)c(p,m)|\mathbf{W}_1|^{\frac{1}{2}(n-p-1)}|\mathbf{W}_2|^{\frac{1}{2}(m-p-1)}e^{-\frac{1}{2}\mathrm{tr}(\mathbf{W}_1+\mathbf{W}_2)}. \tag{2.4.19}$$

Put $\mathbf{W}=\mathbf{W}_1+\mathbf{W}_2 \sim \mathbf{W}_p(\mathbf{I}, n+m)$. The Jacobian equals

$$|\mathbf{J}(\mathbf{W}_1,\mathbf{W}_2 \to \mathbf{W},\mathbf{W}_2)|_+ = |\mathbf{J}(\mathbf{W}-\mathbf{W}_2,\mathbf{W}_2 \to \mathbf{W},\mathbf{W}_2)|_+ = \begin{vmatrix} \mathbf{I} & \mathbf{0} \\ -\mathbf{I} & \mathbf{I} \end{vmatrix}_+ = 1$$

and thus the joint density of $\mathbf{W}$ and $\mathbf{W}_2$ can be written as

$$c(p,n)c(p,m)|\mathbf{W}-\mathbf{W}_2|^{\frac{1}{2}(n-p-1)}|\mathbf{W}_2|^{\frac{1}{2}(m-p-1)}e^{-\frac{1}{2}\mathrm{tr}\mathbf{W}}.$$

Now we are going to obtain the joint density of $\mathbf{F}$ and $\mathbf{W}$. The Jacobian is

$$\begin{aligned}|\mathbf{J}(\mathbf{W},\mathbf{W}_2 \to \mathbf{W},\mathbf{F})|_+ =&|\mathbf{J}(\mathbf{W},\mathbf{W}^{1/2}\mathbf{F}\mathbf{W}^{1/2} \to \mathbf{W},\mathbf{F})|_+ \\ =&|\mathbf{J}(\mathbf{W}^{1/2}\mathbf{F}\mathbf{W}^{1/2} \to \mathbf{F})|_+ = |\mathbf{W}|^{\frac{1}{2}(p+1)},\end{aligned}$$

where Theorem 1.4.13 (i) has been applied to obtain the last equality. Therefore, the joint density of $\mathbf{F}$ and $\mathbf{W}$ is given by

$$\begin{aligned}&c(p,n)c(p,m)|\mathbf{W}|^{\frac{n-p-1}{2}}|\mathbf{I}-\mathbf{F}|^{\frac{n-p-1}{2}}|\mathbf{F}|^{\frac{m-p-1}{2}}|\mathbf{W}|^{\frac{m-p-1}{2}}e^{-\frac{1}{2}\mathrm{tr}\mathbf{W}}|\mathbf{W}|^{\frac{1}{2}(p+1)} \\ &\quad=\frac{c(p,n)c(p,m)}{c(p,n+m)}|\mathbf{I}-\mathbf{F}|^{\frac{n-p-1}{2}}|\mathbf{F}|^{\frac{m-p-1}{2}}c(p,n+m)|\mathbf{W}|^{\frac{n+m-p-1}{2}}e^{-\frac{1}{2}\mathrm{tr}\mathbf{W}}.\end{aligned}$$

Integrating out $\mathbf{W}$ with the help of Theorem 2.4.6 establishes the theorem. ∎

REMARK: The assumption about a symmetric square root was used for notational convenience. From the proof it follows that we could have studied $\mathbf{W}=\mathbf{T}\mathbf{T}'$ and $\mathbf{F}=\mathbf{T}^{-1}\mathbf{W}_2(\mathbf{T}')^{-1}$, where $\mathbf{T}$ is a lower triangular matrix with positive diagonal elements.

Definition 2.4.2. *A random matrix which has a density given by (2.4.18) is said to have a multivariate beta distribution of type I. This will be denoted* $M\beta_I(p,m,n)$. ∎

Observe that other authors sometimes denote this distribution $M\beta_I(p,n,m)$.
As a consequence of the proof of Theorem 2.4.8 the next corollary can be stated.

Corollary 2.4.8.1. *The random matrices* $\mathbf{W}$ *and* $\mathbf{F}$ *in Theorem 2.4.8 are independently distributed.* ∎

Theorem 2.4.9. *Let* $\mathbf{W}_1 \sim W_p(\mathbf{I}, n)$, $p \leq n$ *and* $\mathbf{W}_2 \sim W_p(\mathbf{I}, m)$, $p \leq m$, *be independently distributed. Then*

$$\mathbf{Z} = \mathbf{W}_2^{-1/2}\mathbf{W}_1\mathbf{W}_2^{-1/2}$$

has the density function

$$f_{\mathbf{Z}}(\mathbf{Z}) = \begin{cases} \dfrac{c(p,n)c(p,m)}{c(p,n+m)}|\mathbf{Z}|^{\frac{1}{2}(n-p-1)}|\mathbf{I}+\mathbf{Z}|^{-\frac{1}{2}(n+m)} & |\mathbf{Z}| > 0, \\ 0 & \text{otherwise}, \end{cases} \qquad (2.4.20)$$

where $c(p, n)$ *is defined as in Theorem 2.4.8 and* $\mathbf{W}_2^{1/2}$ *is a symmetric square root.*

PROOF: As noted in the proof of Theorem 2.4.8, the joint density of $\mathbf{W}_1$ and $\mathbf{W}_2$ is given by (2.4.19). From Theorem 1.4.13 (i) it follows that

$$\begin{aligned} |\mathbf{J}(\mathbf{W}_2, \mathbf{W}_1 \to \mathbf{W}_2, \mathbf{Z})|_+ =& |\mathbf{J}(\mathbf{W}_2, \mathbf{W}_2^{1/2}\mathbf{Z}\mathbf{W}_2^{1/2} \to \mathbf{W}_2, \mathbf{Z})|_+ \\ =& |\mathbf{J}(\mathbf{W}_2^{1/2}\mathbf{Z}\mathbf{W}_2^{1/2} \to \mathbf{Z})|_+ = |\mathbf{W}_2|^{\frac{1}{2}(p+1)}. \end{aligned}$$

Hence, the joint density of $\mathbf{W}_2$ and $\mathbf{Z}$ equals

$$\begin{aligned} c(p,n)c(p,m)|\mathbf{W}_2|^{\frac{m-p-1}{2}}|\mathbf{Z}|^{\frac{n-p-1}{2}}|\mathbf{W}_2|^{\frac{n-p-1}{2}}|\mathbf{W}_2|^{\frac{1}{2}(p+1)}e^{-\frac{1}{2}\mathrm{tr}(\mathbf{W}_2(\mathbf{I}+\mathbf{Z}))} \\ = \frac{c(p,n)c(p,m)}{c(p,n+m)}|\mathbf{Z}|^{\frac{n-p-1}{2}}c(p,n+m)|\mathbf{W}_2|^{\frac{n+m-p-1}{2}}e^{-\frac{1}{2}\mathrm{tr}(\mathbf{W}_2(\mathbf{I}+\mathbf{Z}))} \end{aligned}$$

and integrating out $\mathbf{W}_2$, by utilizing Theorem 2.4.6, verifies the theorem. ∎

Definition 2.4.3. *A random matrix which has a density given by (2.4.20) is said to have a multivariate beta distribution of type II. This will be denoted* $M\beta_{II}(p, m, n)$. ∎

Sometimes the distribution in Definition 2.4.3 is denoted $M\beta_{II}(p, n, m)$. By putting

$$\mathbf{F}_0 = (\mathbf{I} + \mathbf{Z})^{-1}$$

it is observed that the density (2.4.18) may be directly obtained from the density (2.4.20). Just note that the Jacobian of the transformation $\mathbf{Z} \to \mathbf{F}_0$ equals

$$|\mathbf{J}(\mathbf{Z} \to \mathbf{F}_0)|_+ = |\mathbf{F}_0|^{-(p+1)},$$

and then straightforward calculations yield the result. Furthermore, $\mathbf{Z}$ has the same density as $\mathbf{Z}_0 = \mathbf{W}_1^{1/2}\mathbf{W}_2^{-1}\mathbf{W}_1^{1/2}$ and $\mathbf{F}$ has the same density as

$$\mathbf{F}_0 = (\mathbf{I} + \mathbf{Z})^{-1} = \mathbf{W}_2^{1/2}(\mathbf{W}_1 + \mathbf{W}_2)^{-1}\mathbf{W}_2^{1/2}.$$

Thus, for example, the moments for $\mathbf{F}$ may be obtained via the moments for $\mathbf{F}_0$, which will be used later.

Theorem 2.4.10. *Let* $\mathbf{W}_1 \sim W_p(\mathbf{I}, n)$, $n \geq p$, *and* $\mathbf{Y} \sim N_{p,m}(\mathbf{0}, \mathbf{I}, \mathbf{I})$, $m < p$, *be independent random matrices. Then*

$$\mathbf{G} = \mathbf{Y}'(\mathbf{W}_1 + \mathbf{Y}\mathbf{Y}')^{-1}\mathbf{Y}$$

has the multivariate density function

$$f_{\mathbf{G}}(\mathbf{G}) = \begin{cases} \dfrac{c(p,n)c(m,p)}{c(p,n+m)}|\mathbf{G}|^{\frac{p-m-1}{2}}|\mathbf{I} - \mathbf{G}|^{\frac{n-p-1}{2}} & |\mathbf{I} - \mathbf{G}| > 0, |\mathbf{G}| > 0, \\ 0 & \text{otherwise}, \end{cases}$$

where $c(p, n)$ *is defined in Theorem 2.4.8.*

PROOF: The joint density of $\mathbf{Y}$ and $\mathbf{W}_1$ is given by

$$c(p,n)(2\pi)^{-\frac{1}{2}pm}|\mathbf{W}_1|^{\frac{n-p-1}{2}}e^{-\frac{1}{2}\mathrm{tr}\mathbf{W}_1}e^{-\frac{1}{2}\mathrm{tr}\mathbf{Y}\mathbf{Y}'}.$$

Put $\mathbf{W} = \mathbf{W}_1 + \mathbf{Y}\mathbf{Y}'$ and, since $|\mathbf{J}(\mathbf{W}_1, \mathbf{Y} \to \mathbf{W}, \mathbf{Y})|_+ = 1$, the joint density of $\mathbf{W}$ and $\mathbf{Y}$ is given by

$$\begin{aligned} &c(p,n)(2\pi)^{-\frac{1}{2}pm}|\mathbf{W} - \mathbf{Y}\mathbf{Y}'|^{\frac{n-p-1}{2}}e^{-\frac{1}{2}\mathrm{tr}\mathbf{W}} \\ &\quad = c(p,n)(2\pi)^{-\frac{1}{2}pm}|\mathbf{W}|^{\frac{n-p-1}{2}}|\mathbf{I} - \mathbf{Y}'\mathbf{W}^{-1}\mathbf{Y}|^{\frac{n-p-1}{2}}e^{-\frac{1}{2}\mathrm{tr}\mathbf{W}}. \end{aligned}$$

Let $\mathbf{U} = \mathbf{W}^{-1/2}\mathbf{Y}$ and by Theorem 1.4.14 the joint density of $\mathbf{U}$ and $\mathbf{W}$ equals

$$c(p,n)(2\pi)^{-\frac{1}{2}pm}|\mathbf{I} - \mathbf{U}'\mathbf{U}|^{\frac{1}{2}(n-p-1)}|\mathbf{W}|^{\frac{1}{2}(n+m-p-1)}e^{-\frac{1}{2}\mathrm{tr}\mathbf{W}}.$$

Integrating out $\mathbf{W}$ yields

$$\frac{c(p,n)}{c(p,n+m)}(2\pi)^{-\frac{1}{2}pm}|\mathbf{I} - \mathbf{U}'\mathbf{U}|^{\frac{1}{2}(n-p-1)}.$$

Now, from Proposition 1.1.6 (ii) it follows that $\mathbf{U}' = \mathbf{T}\mathbf{L}$, where $\mathbf{T} : m \times m$ is a lower triangular matrix with positive diagonal elements and $\mathbf{L} : m \times p$ is semi-orthogonal. According to Theorem 1.4.20, the joint density of $\mathbf{T}$ and $\mathbf{L}$ equals

$$\frac{c(p,n)}{c(p,n+m)}(2\pi)^{-\frac{1}{2}pm}|\mathbf{I} - \mathbf{T}\mathbf{T}'|^{\frac{1}{2}(n-p-1)}\prod_{i=1}^{m} t_{ii}^{p-i}\prod_{i=1}^{m}|\mathbf{L}_i|_+,$$

where $\mathbf{L}_i = (l_{kl})$, $k, l = 1, 2, \ldots, i$. Theorem 1.4.18 implies that the joint density of $\mathbf{G} = \mathbf{T}\mathbf{T}'$ and $\mathbf{L}$ can be written as

$$\frac{c(p,n)}{c(p,n+m)}(2\pi)^{-\frac{1}{2}pm}2^{-m}|\mathbf{I} - \mathbf{G}|^{\frac{1}{2}(n-p-1)}|\mathbf{G}|^{\frac{1}{2}(p-m-1)}\prod_{i=1}^{m}|\mathbf{L}_i|_+.$$

Finally the theorem is established by Corollary 2.4.6.2 and integration of $\mathbf{L}$:

$$\frac{c(p,n)c(m,p)}{c(p,n+m)}|\mathbf{I} - \mathbf{G}|^{\frac{1}{2}(n-p-1)}|\mathbf{G}|^{\frac{1}{2}(p-m-1)}.$$

∎

Corollary 2.4.10.1. *The random matrices* $\mathbf{G}$ *and* $\mathbf{W}$ *in Theorem 2.4.10 are independently distributed.* ∎

The $M\beta_I(p,m,n)$ shares some properties with the Wishart distribution. To some extent, the similarities are remarkable. Here we give a result which corresponds to the Bartlett decomposition, i.e. Theorem 2.4.4 (i).

Theorem 2.4.11. *Let* $\mathbf{F} \sim M\beta_I(p,m,n)$ *and* $\mathbf{F} = \mathbf{T}\mathbf{T}'$ *where* $\mathbf{T}$ *is a lower triangular matrix with positive diagonal elements. Then* $T_{11}, T_{22}, \ldots, T_{pp}$ *are all independent and*

$$T_{ii}^2 \sim \beta(m+1-i, n+i-1), \quad i = 1, 2, \ldots, p.$$

PROOF: The density for $\mathbf{F}$ is given in Theorem 2.4.8 and combining this result with Theorem 1.4.18 yields

$$\frac{c(p,n)c(p,m)}{c(p,n+m)}|\mathbf{T}\mathbf{T}'|^{\frac{m-p-1}{2}}|\mathbf{I}-\mathbf{T}\mathbf{T}'|^{\frac{n-p-1}{2}} 2^p \prod_{i=1}^{p} T_{ii}^{p-i+1}$$
$$= 2^p \frac{c(p,n)c(p,m)}{c(p,n+m)} \prod_{i=1}^{p} T_{ii}^{m-i}|\mathbf{I}-\mathbf{T}\mathbf{T}'|^{\frac{n-p-1}{2}}.$$

First we will show that T_{11}^2 is beta distributed. Partition $\mathbf{T}$ as

$$\mathbf{T} = \begin{pmatrix} T_{11} & \mathbf{0} \\ \mathbf{t}_{21} & \mathbf{T}_{22} \end{pmatrix}.$$

Thus,

$$|\mathbf{I}-\mathbf{T}\mathbf{T}'| = (1-T_{11}^2)(1-\mathbf{v}'\mathbf{v})|\mathbf{I}-\mathbf{T}_{22}\mathbf{T}_{22}'|,$$

where

$$\mathbf{v} = (\mathbf{I}-\mathbf{T}_{22}\mathbf{T}_{22}')^{-1/2}\mathbf{t}_{21}(1-T_{11}^2)^{-1/2}.$$

We are going to make a change of variables, i.e. $T_{11}, \mathbf{t}_{21}, \mathbf{T}_{22} \to T_{11}, \mathbf{v}, \mathbf{T}_{22}$. The corresponding Jacobian equals

$$|\mathbf{J}(\mathbf{t}_{21}, T_{11}, \mathbf{T}_{22} \to \mathbf{v}, T_{11}, \mathbf{T}_{22})|_+ = |\mathbf{J}(\mathbf{t}_{21} \to \mathbf{v})|_+ = (1-T_{11}^2)^{\frac{p-1}{2}}|\mathbf{I}-\mathbf{T}_{22}\mathbf{T}_{22}'|^{\frac{1}{2}},$$

where the last equality was obtained by Theorem 1.4.14. Now the joint density of T_{11}, $\mathbf{v}$ and $\mathbf{T}_{22}$ can be written as

$$2^p \frac{c(p,n)c(p,m)}{c(p,n+m)} T_{11}^{m-1}(1-T_{11}^2)^{\frac{n-p-1}{2}} \prod_{i=2}^{p} T_{ii}^{m-i}$$
$$\times |\mathbf{I}-\mathbf{T}_{22}\mathbf{T}_{22}'|^{\frac{n-p-1}{2}}(1-\mathbf{v}'\mathbf{v})^{\frac{n-p-1}{2}}(1-T_{11}^2)^{\frac{p-1}{2}}|\mathbf{I}-\mathbf{T}_{22}\mathbf{T}_{22}'|^{\frac{1}{2}}$$
$$= 2^p \frac{c(p,n)c(p,m)}{c(p,n+m)}(T_{11}^2)^{\frac{m-1}{2}}(1-T_{11}^2)^{\frac{n}{2}-1}$$
$$\times \prod_{i=2}^{p} T_{ii}^{m-i}|\mathbf{I}-\mathbf{T}_{22}\mathbf{T}_{22}'|^{\frac{n-p}{2}}(1-\mathbf{v}'\mathbf{v})^{\frac{n-p-1}{2}}.$$

Hence, T_{11} is independent of both $\mathbf{T}_{22}$ and $\mathbf{v}$, and therefore T_{11}^2 follows a $\beta(m,n)$ distribution. In order to obtain the distribution for T_{22}, T_{33}, ...,T_{pp}, it is noted that $\mathbf{T}_{22}$ is independent of T_{11} and $\mathbf{v}$ and its density is proportional to

$$\prod_{i=2}^{p} T_{ii}^{m-i}|\mathbf{I}-\mathbf{T}_{22}\mathbf{T}'_{22}|^{\frac{n-p}{2}}.$$

Therefore, we have a density function which is of the same form as the one given in the beginning of the proof, and by repeating the arguments it follows that T_{22} is $\beta(m-1,n+1)$ distributed. The remaining details of the proof are straightforward to fill in. ∎

2.4.4 *Partitioned Wishart matrices*

We are going to consider some results for partitioned Wishart matrices. Let

$$\mathbf{W}=\begin{pmatrix}\mathbf{W}_{11} & \mathbf{W}_{12}\\ \mathbf{W}_{21} & \mathbf{W}_{22}\end{pmatrix}\quad\begin{pmatrix} r\times r & r\times(p-r)\\ (p-r)\times r & (p-r)\times(p-r)\end{pmatrix}, \tag{2.4.21}$$

where on the right-hand side the sizes of the matrices are indicated, and let

$$\boldsymbol{\Sigma}=\begin{pmatrix}\boldsymbol{\Sigma}_{11} & \boldsymbol{\Sigma}_{12}\\ \boldsymbol{\Sigma}_{21} & \boldsymbol{\Sigma}_{22}\end{pmatrix}\quad\begin{pmatrix} r\times r & r\times(p-r)\\ (p-r)\times r & (p-r)\times(p-r)\end{pmatrix}. \tag{2.4.22}$$

Theorem 2.4.12. *Let* $\mathbf{W}\sim W_p(\boldsymbol{\Sigma},n)$, *and let the matrices* $\mathbf{W}$ *and* $\boldsymbol{\Sigma}$ *be partitioned according to (2.4.21) and (2.4.22), respectively. Furthermore, put*

$$\mathbf{W}_{1\cdot 2}=\mathbf{W}_{11}-\mathbf{W}_{12}\mathbf{W}_{22}^{-1}\mathbf{W}_{21}$$

and

$$\boldsymbol{\Sigma}_{1\cdot 2}=\boldsymbol{\Sigma}_{11}-\boldsymbol{\Sigma}_{12}\boldsymbol{\Sigma}_{22}^{-1}\boldsymbol{\Sigma}_{21}.$$

Then the following statements are valid:

(i) $$\mathbf{W}_{1\cdot 2}\sim W_r(\boldsymbol{\Sigma}_{1\cdot 2},n-p+r);$$

(ii) $\mathbf{W}_{1\cdot 2}$ *and* $(\mathbf{W}_{12}:\mathbf{W}_{22})$ *are independently distributed;*

(iii) *for any square root* $\mathbf{W}_{22}^{-1/2}$

$$\mathbf{W}_{12}\mathbf{W}_{22}^{-1/2}-\boldsymbol{\Sigma}_{12}\boldsymbol{\Sigma}_{22}^{-1}\mathbf{W}_{22}^{1/2}\sim N_{r,p-r}(\mathbf{0},\boldsymbol{\Sigma}_{1\cdot 2},\mathbf{I});$$

(iv) $$\mathbf{W}_{12}|\mathbf{W}_{22}\sim N_{r,p-r}(\boldsymbol{\Sigma}_{12}\boldsymbol{\Sigma}_{22}^{-1}\mathbf{W}_{22},\boldsymbol{\Sigma}_{1\cdot 2},\mathbf{W}_{22}).$$

PROOF: By Definition 2.4.1, there exists a matrix $\mathbf{Z}'=(\mathbf{Z}'_1:\mathbf{Z}'_2)$, $(n\times r:n\times(p-r))$, such that $\mathbf{W}=\mathbf{Z}\mathbf{Z}'$, where $\mathbf{Z}\sim N_{p,n}(\mathbf{0},\boldsymbol{\Sigma},\mathbf{I})$. Then $\mathbf{W}_{1\cdot 2}$ has the same distribution as

$$\mathbf{Z}_1\mathbf{P}\mathbf{Z}'_1, \tag{2.4.23}$$

where

$$\mathbf{P} = \mathbf{I} - \mathbf{Z}_2'(\mathbf{Z}_2\mathbf{Z}_2')^{-1}\mathbf{Z}_2.$$

We are going to condition (2.4.23) with respect to $\mathbf{Z}_2$ and note that according to Theorem 2.2.5 (ii),

$$\mathbf{Z}_1 - \mathbf{\Sigma}_{12}\mathbf{\Sigma}_{22}^{-1}\mathbf{Z}_2|\mathbf{Z}_2 \sim N_{r,n}(\mathbf{0}, \mathbf{\Sigma}_{1\cdot 2}, \mathbf{I}), \tag{2.4.24}$$

which is independent of $\mathbf{Z}_2$. Conditionally on $\mathbf{Z}_2$, the matrix $\mathbf{P}$ is fixed and idempotent. Furthermore, $\mathbf{\Sigma}_{12}\mathbf{\Sigma}_{22}^{-1}\mathbf{Z}_2\mathbf{P} = \mathbf{0}$. It follows that $r(\mathbf{P}) = n - r(\mathbf{Z}_2) = n - p + r$, with probability 1. Thus, from Corollary 2.4.3.1 we obtain that

$$\mathbf{W}_{1\cdot 2}|\mathbf{Z}_2 \sim W_r(\mathbf{\Sigma}_{1\cdot 2}, n - p - r),$$

which is independent of $\mathbf{Z}_2$, and thus (i) is established. Observe that similar arguments were used in the proof of Theorem 2.4.4.

For (ii) we note that since $\mathbf{W}_{1\cdot 2}$ is independent of $\mathbf{Z}_2$, it must also be independent of $\mathbf{W}_{22} = \mathbf{Z}_2\mathbf{Z}_2'$. Furthermore, since $\mathbf{Z}_2\mathbf{P} = \mathbf{0}$, we have by Theorem 2.2.4 that given $\mathbf{Z}_2$, the matrices $\mathbf{W}_{12} = \mathbf{Z}_1\mathbf{Z}_2'$ and $\mathbf{W}_{1\cdot 2}$ are independently distributed. We are going to see that $\mathbf{W}_{12}$ and $\mathbf{W}_{1\cdot 2}$ are also unconditionally independent, which follows from the fact that $\mathbf{W}_{1\cdot 2}$ is independent of $\mathbf{Z}_2$:

$$f_{\mathbf{W}_{12},\mathbf{W}_{1\cdot 2}}(\mathbf{x}_1, \mathbf{x}_2) = \int f_{\mathbf{W}_{12},\mathbf{W}_{1\cdot 2},\mathbf{Z}_2}(\mathbf{x}_1, \mathbf{x}_2, \mathbf{x}_3)d\mathbf{x}_3$$

$$= \int f_{\mathbf{W}_{12},\mathbf{W}_{1\cdot 2}|\mathbf{Z}_2}(\mathbf{x}_1, \mathbf{x}_2)f_{\mathbf{Z}_2}(\mathbf{x}_3)d\mathbf{x}_3 = \int f_{\mathbf{W}_{12}|\mathbf{Z}_2}(\mathbf{x}_1)f_{\mathbf{W}_{1\cdot 2}|\mathbf{Z}_2}(\mathbf{x}_2)f_{\mathbf{Z}_2}(\mathbf{x}_3)d\mathbf{x}_3$$

$$= \int f_{\mathbf{W}_{12}|\mathbf{Z}_2}(\mathbf{x}_1)f_{\mathbf{W}_{1\cdot 2}}(\mathbf{x}_2)f_{\mathbf{Z}_2}(\mathbf{x}_3)d\mathbf{x}_3 = \int f_{\mathbf{W}_{1\cdot 2}}(\mathbf{x}_2)f_{\mathbf{W}_{12},\mathbf{Z}_2}(\mathbf{x}_1, \mathbf{x}_3)d\mathbf{x}_3$$

$$= f_{\mathbf{W}_{12}}(\mathbf{x}_1)f_{\mathbf{W}_{1\cdot 2}}(\mathbf{x}_2).$$

Now, (iii) and (iv) will follow from the next relation. The statement in (2.4.24) implies that

$$\mathbf{Z}_1\mathbf{Z}_2'(\mathbf{Z}_2\mathbf{Z}_2')^{-1/2} - \mathbf{\Sigma}_{12}\mathbf{\Sigma}_{22}^{-1}(\mathbf{Z}_2\mathbf{Z}_2')^{1/2}|\mathbf{Z}_2 \sim N_{r,p-r}(\mathbf{0}, \mathbf{\Sigma}_{1\cdot 2}, \mathbf{I})$$

holds, since

$$(\mathbf{Z}_2\mathbf{Z}_2')^{-1/2}\mathbf{Z}_2\mathbf{Z}_2'(\mathbf{Z}_2\mathbf{Z}_2')^{-1/2} = \mathbf{I}.$$

The expression is independent of $\mathbf{Z}_2$, as well as of $\mathbf{W}_{22}$, and has the same distribution as

$$\mathbf{W}_{12}\mathbf{W}_{22}^{-1/2} - \mathbf{\Sigma}_{12}\mathbf{\Sigma}_{22}^{-1}\mathbf{W}_{22}^{1/2} \sim N_{r,p-r}(\mathbf{0}, \mathbf{\Sigma}_{1\cdot 2}, \mathbf{I}).$$

▮

Corollary 2.4.12.1. *Let* $\mathbf{V} \sim W_p(\mathbf{I}, n)$ *and apply the same partition as in (2.4.21). Then*

$$\mathbf{V}_{12}\mathbf{V}_{22}^{-1/2} \sim N_{r,p-r}(\mathbf{0}, \mathbf{I}, \mathbf{I})$$

is independent of $\mathbf{V}_{22}$. ▮

There exist several interesting results connecting generalized inverse Gaussian distributions (GIG) and partitioned Wishart matrices. For example, $\mathbf{W}_{11}|\mathbf{W}_{12}$ is matrix GIG distributed (see Butler, 1998).

The next theorem gives some further properties for the inverted Wishart distribution.

Theorem 2.4.13. *Let* $\mathbf{W} \sim W_p(\mathbf{\Sigma}, n)$, $\mathbf{\Sigma} > 0$, *and* $\mathbf{A} \in \mathbb{R}^{p\times q}$. *Then*

(i) $\mathbf{A}(\mathbf{A}'\mathbf{W}^{-1}\mathbf{A})^{-}\mathbf{A}' \sim \mathbf{W}_p(\mathbf{A}(\mathbf{A}'\mathbf{\Sigma}^{-1}\mathbf{A})^{-}\mathbf{A}', n - p + r(\mathbf{A}))$;

(ii) $\mathbf{A}(\mathbf{A}'\mathbf{W}^{-1}\mathbf{A})^{-}\mathbf{A}'$ *and* $\mathbf{W} - \mathbf{A}(\mathbf{A}'\mathbf{W}^{-1}\mathbf{A})^{-}\mathbf{A}'$ *are independent;*

(iii) $\mathbf{A}(\mathbf{A}'\mathbf{W}^{-1}\mathbf{A})^{-}\mathbf{A}'$ *and* $\mathbf{I} - \mathbf{A}(\mathbf{A}'\mathbf{W}^{-1}\mathbf{A})^{-}\mathbf{A}'\mathbf{W}^{-1}$ *are independent.*

PROOF: Since $\mathbf{A}(\mathbf{A}'\mathbf{W}^{-1}\mathbf{A})^{-}\mathbf{A}'\mathbf{W}^{-1}$ is a projection operator (idempotent matrix), it follows, because of uniqueness of projection operators, that $\mathbf{A}$ can be replaced by any $\underline{\mathbf{A}} \in \mathbb{R}^{p\times r(\mathbf{A})}$, $C(\mathbf{A}) = C(\underline{\mathbf{A}})$ such that

$$\mathbf{A}(\mathbf{A}'\mathbf{W}^{-1}\mathbf{A})^{-}\mathbf{A}' = \underline{\mathbf{A}}(\underline{\mathbf{A}}'\mathbf{W}^{-1}\underline{\mathbf{A}})^{-1}\underline{\mathbf{A}}'.$$

It is often convenient to present a matrix in a canonical form. Applying Proposition 1.1.6 (ii) to $\underline{\mathbf{A}}'\mathbf{\Sigma}^{-1/2}$ we have

$$\underline{\mathbf{A}}' = \mathbf{T}(\mathbf{I}_{r(\mathbf{A})} : \mathbf{0})\mathbf{\Gamma}\mathbf{\Sigma}^{1/2} = \mathbf{T}\mathbf{\Gamma}_1\mathbf{\Sigma}^{1/2}, \tag{2.4.25}$$

where $\mathbf{T}$ is non-singular, $\mathbf{\Gamma}' = (\mathbf{\Gamma}_1' : \mathbf{\Gamma}_2')$ is orthogonal and $\mathbf{\Sigma}^{1/2}$ is a symmetric square root of $\mathbf{\Sigma}$. Furthermore, by Definition 2.4.1, $\mathbf{W} = \mathbf{Z}\mathbf{Z}'$, where $\mathbf{Z} \sim N_{p,n}(\mathbf{0}, \mathbf{\Sigma}, \mathbf{I})$. Put

$$\mathbf{V} = \mathbf{\Gamma}\mathbf{\Sigma}^{-1/2}\mathbf{W}\mathbf{\Sigma}^{-1/2}\mathbf{\Gamma}' \sim W_p(\mathbf{I}, n).$$

Now,

$$\begin{aligned}\mathbf{A}(\mathbf{A}'\mathbf{W}^{-1}\mathbf{A})^{-}\mathbf{A}' &= \mathbf{\Sigma}^{1/2}\mathbf{\Gamma}_1'((\mathbf{I} : \mathbf{0})\mathbf{V}^{-1}(\mathbf{I} : \mathbf{0})')^{-1}\mathbf{\Gamma}_1\mathbf{\Sigma}^{1/2} \\ &= \mathbf{\Sigma}^{1/2}\mathbf{\Gamma}_1'(\mathbf{V}^{11})^{-1}\mathbf{\Gamma}_1\mathbf{\Sigma}^{1/2},\end{aligned}$$

where $\mathbf{V}$ has been partitioned as $\mathbf{W}$ in (2.4.21). However, by Proposition 1.3.3 $(\mathbf{V}^{11})^{-1} = \mathbf{V}_{1\cdot 2}$, where $\mathbf{V}_{1\cdot 2} = \mathbf{V}_{11} - \mathbf{V}_{12}\mathbf{V}_{22}^{-1}\mathbf{V}_{21}$, and thus, by Theorem 2.4.12 (i),

$$\mathbf{\Sigma}^{1/2}\mathbf{\Gamma}_1'(\mathbf{V}^{11})^{-1}\mathbf{\Gamma}_1\mathbf{\Sigma}^{1/2} \sim W_p(\mathbf{\Sigma}^{1/2}\mathbf{\Gamma}_1'\mathbf{\Gamma}_1\mathbf{\Sigma}^{1/2}, n - p + r(\mathbf{A})).$$

When $\mathbf{\Sigma}^{1/2}\mathbf{\Gamma}_1'\mathbf{\Gamma}_1\mathbf{\Sigma}^{1/2}$ is expressed through the original matrices, we get

$$\mathbf{\Sigma}^{1/2}\mathbf{\Gamma}_1'\mathbf{\Gamma}_1\mathbf{\Sigma}^{1/2} = \underline{\mathbf{A}}(\underline{\mathbf{A}}'\mathbf{\Sigma}^{-1}\underline{\mathbf{A}})^{-1}\underline{\mathbf{A}}' = \mathbf{A}(\mathbf{A}'\mathbf{\Sigma}^{-1}\mathbf{A})^{-}\mathbf{A}'$$

and hence (i) is verified.

In order to show (ii) and (iii), the canonical representation given in (2.4.25) will be used again. Furthermore, by Definition 2.4.1, $\mathbf{V} = \mathbf{U}\mathbf{U}'$, where $\mathbf{U} \sim N_{p,n}(\mathbf{0}, \mathbf{I}, \mathbf{I})$. Let us partition $\mathbf{U}$ in correspondence with the partition of $\mathbf{V}$. Then

$$\begin{aligned}&\mathbf{W} - \mathbf{A}(\mathbf{A}'\mathbf{W}^{-1}\mathbf{A})^{-}\mathbf{A}' = \mathbf{\Sigma}^{1/2}\mathbf{\Gamma}'\mathbf{V}\mathbf{\Gamma}\mathbf{\Sigma}^{1/2} - \mathbf{\Sigma}^{1/2}\mathbf{\Gamma}'(\mathbf{I} : \mathbf{0})'(\mathbf{V}^{11})^{-1}(\mathbf{I} : \mathbf{0})\mathbf{\Gamma}\mathbf{\Sigma}^{1/2} \\ &= \mathbf{\Sigma}^{1/2}\mathbf{\Gamma}'\begin{pmatrix}\mathbf{V}_{12}\mathbf{V}_{22}^{-1}\mathbf{V}_{21} & \mathbf{V}_{12} \\ \mathbf{V}_{21} & \mathbf{V}_{22}\end{pmatrix}\mathbf{\Gamma}\mathbf{\Sigma}^{1/2} \\ &= \mathbf{\Sigma}^{1/2}\mathbf{\Gamma}'\begin{pmatrix}\mathbf{U}_1\mathbf{U}_2'(\mathbf{U}_2\mathbf{U}_2')^{-1}\mathbf{U}_2\mathbf{U}_1' & \mathbf{U}_1\mathbf{U}_2' \\ \mathbf{U}_2\mathbf{U}_1' & \mathbf{U}_2\mathbf{U}_2'\end{pmatrix}\mathbf{\Gamma}\mathbf{\Sigma}^{1/2}\end{aligned}$$

and

$$
\begin{aligned}
\mathbf{I}-\mathbf{A}(\mathbf{A}'\mathbf{W}^{-1}\mathbf{A})^{-}\mathbf{A}'\mathbf{W}^{-1} &= \mathbf{I}-\boldsymbol{\Sigma}^{1/2}\boldsymbol{\Gamma}_1'(\mathbf{I}:(\mathbf{V}^{11})^{-1}\mathbf{V}^{12})\boldsymbol{\Gamma}\boldsymbol{\Sigma}^{-1/2}\\
&= \mathbf{I}-\boldsymbol{\Sigma}^{1/2}\boldsymbol{\Gamma}_1'\boldsymbol{\Gamma}_1\boldsymbol{\Sigma}^{-1/2}+\boldsymbol{\Sigma}^{1/2}\boldsymbol{\Gamma}_1'\mathbf{V}_{12}\mathbf{V}_{22}^{-1}\boldsymbol{\Gamma}_2\boldsymbol{\Sigma}^{-1/2}\\
&= \mathbf{I}-\boldsymbol{\Sigma}^{1/2}\boldsymbol{\Gamma}_1'\boldsymbol{\Gamma}_1\boldsymbol{\Sigma}^{-1/2}+\boldsymbol{\Sigma}^{1/2}\boldsymbol{\Gamma}_1'\mathbf{U}_1\mathbf{U}_2'(\mathbf{U}_2\mathbf{U}_2')^{-1}\boldsymbol{\Gamma}_2\boldsymbol{\Sigma}^{-1/2}. \qquad (2.4.26)
\end{aligned}
$$

In (2.4.26) we used the equality $(\mathbf{V}^{11})^{-1}\mathbf{V}^{12} = -\mathbf{V}_{12}(\mathbf{V}_{22})^{-1}$ given in Proposition 1.3.4 (i). Now we are going to prove that

$$
\mathbf{U}_2,\ \mathbf{U}_2\mathbf{U}_1',\ \mathbf{U}_1\mathbf{U}_2'(\mathbf{U}_2\mathbf{U}_2')^{-1}\mathbf{U}_2\mathbf{U}_1',\ (\mathbf{U}_2\mathbf{U}_2')^{-1}\mathbf{U}_2\mathbf{U}_1' \qquad (2.4.27)
$$

and

$$
\mathbf{V}_{1\cdot 2} = \mathbf{U}_1(\mathbf{I}-\mathbf{U}_2'(\mathbf{U}_2\mathbf{U}_2')^{-1}\mathbf{U}_2)\mathbf{U}_1' \qquad (2.4.28)
$$

are independently distributed. However, from Theorem 2.2.4 it follows that, conditionally on $\mathbf{U}_2$, the matrices (2.4.27) and (2.4.28) are independent and since $\mathbf{V}_{1\cdot 2}$ in (2.4.28) is independent of $\mathbf{U}_2$, (ii) and (iii) are verified. Similar ideas were used in the proof of Theorem 2.4.12 (i). ∎

Corollary 2.4.13.1. *Let* $\mathbf{W} \sim W_p(\boldsymbol{\Sigma}, n)$, $\mathbf{A} : p \times q$ *and* $r(\mathbf{A}) = q$. *Then*

$$
(\mathbf{A}'\mathbf{W}^{-1}\mathbf{A})^{-1} \sim \mathbf{W}_p((\mathbf{A}'\boldsymbol{\Sigma}^{-1}\mathbf{A})^{-1}, n-p+q).
$$

PROOF: The statement follows from (i) of the theorem above and Theorem 2.4.2, since

$$
(\mathbf{A}'\mathbf{W}^{-1}\mathbf{A})^{-1} = (\mathbf{A}'\mathbf{A})^{-1}\mathbf{A}'\mathbf{A}(\mathbf{A}'\mathbf{W}^{-1}\mathbf{A})^{-1}\mathbf{A}'\mathbf{A}(\mathbf{A}'\mathbf{A})^{-1}.
$$

∎

2.4.5 *Moments of the Wishart distribution*

We are going to present some moment relations. They will be used in Sections 3.2 and 3.3 when we are going to approximate densities with the help of the Wishart density. In addition to the mean and dispersion we will consider inverse moments or, one can say, moments of the inverted Wishart distribution. The density of the inverted Wishart distribution under a somewhat different parameterization was introduced in Corollary 2.4.6.1. Often the formula

$$
f_{\mathbf{W}^{-1}}(\mathbf{W}) = \begin{cases} \dfrac{1}{2^{\frac{n-p-1}{2}}\boldsymbol{\Gamma}_p(\frac{n-p-1}{2})}|\mathbf{W}|^{-\frac{n}{2}}|\boldsymbol{\Sigma}|^{\frac{n-p-1}{2}}e^{-\frac{1}{2}tr(\boldsymbol{\Sigma}\mathbf{W}^{-1})} & \mathbf{W}>0\\ 0 & \text{otherwise,} \end{cases} \qquad (2.4.29)
$$

is used as a representation of the density function for the inverted Wishart distribution. It follows from the Wishart density, if we use the transformation $\mathbf{W} \to \mathbf{W}^{-1}$. The Jacobian of that transformation was given by Theorem 1.4.17 (ii), i.e.

$$
|\mathbf{J}(\mathbf{W} \to \mathbf{W}^{-1})|_+ = |\mathbf{W}|^{(p+1)}.
$$

The fact that a matrix $\mathbf{W} : p \times p$ is distributed according to the density (2.4.29) is usually denoted by $\mathbf{W} \sim W_p^{-1}(\mathbf{\Sigma}, n)$. If $\mathbf{W} \sim W_p(\mathbf{\Sigma}, n)$, then it follows that $\mathbf{W}^{-1} \sim W_p^{-1}(\mathbf{\Sigma}^{-1}, n+p+1)$. It is a bit unfortunate that the second parameter equals n for the Wishart distribution and $n+p+1$ for the inverted Wishart distribution. However, we are not going to use the density directly. Instead, we shall use some properties of the Wishart density which lead to the moments of the inverted Wishart distribution. We are also going to present some moment relations which involve both $\mathbf{W}^{-1}$ and $\mathbf{W}$.

Theorem 2.4.14. *Let $\mathbf{W} \sim W_p(\mathbf{\Sigma}, n)$. Then*

(i) $E[\mathbf{W}] = n\mathbf{\Sigma}$;

(ii) $D[\mathbf{W}] = n(\mathbf{I}_{p^2} + \mathbf{K}_{p,p})(\mathbf{\Sigma} \otimes \mathbf{\Sigma})$;

(iii) $E[\mathbf{W}^{-1}] = \frac{1}{n-p-1}\mathbf{\Sigma}^{-1}, \qquad n-p-1>0$;

(iv)
$$E[\mathbf{W}^{-1} \otimes \mathbf{W}^{-1}] = \frac{n-p-2}{(n-p)(n-p-1)(n-p-3)}\mathbf{\Sigma}^{-1} \otimes \mathbf{\Sigma}^{-1}$$
$$+ \frac{1}{(n-p)(n-p-1)(n-p-3)}(\mathrm{vec}\mathbf{\Sigma}^{-1}\mathrm{vec}'\mathbf{\Sigma}^{-1} + \mathbf{K}_{p,p}(\mathbf{\Sigma}^{-1} \otimes \mathbf{\Sigma}^{-1})),$$
$$n-p-3>0;$$

(v)
$$E[\mathrm{vec}\mathbf{W}^{-1}\mathrm{vec}'\mathbf{W}] = \frac{n}{n-p-1}\mathrm{vec}\mathbf{\Sigma}^{-1}\mathrm{vec}'\mathbf{\Sigma} - \frac{1}{n-p-1}(\mathbf{I} + \mathbf{K}_{p,p}),$$
$$n-p-1>0;$$

(vi)
$$E[\mathbf{W}^{-1}\mathbf{W}^{-1}] = \frac{1}{(n-p)(n-p-3)}\mathbf{\Sigma}^{-1}\mathbf{\Sigma}^{-1} + \frac{1}{(n-p)(n-p-1)(n-p-3)}\mathbf{\Sigma}^{-1}\mathrm{tr}\mathbf{\Sigma}^{-1},$$
$$n-p-3>0;$$

(vii) $E[\mathrm{tr}(\mathbf{W}^{-1})\mathbf{W}] = \frac{1}{n-p-1}(n\mathbf{\Sigma}\mathrm{tr}\mathbf{\Sigma}^{-1} - 2\mathbf{I}), \qquad n-p-1>0$;

(viii)
$$E[\mathrm{tr}(\mathbf{W}^{-1})\mathbf{W}^{-1}] = \frac{2}{(n-p)(n-p-1)(n-p-3)}\mathbf{\Sigma}^{-1}\mathbf{\Sigma}^{-1}$$
$$+ \frac{n-p-2}{(n-p)(n-p-1)(n-p-3)}\mathbf{\Sigma}^{-1}\mathrm{tr}\mathbf{\Sigma}^{-1}, \qquad n-p-3>0.$$

PROOF: The statements in (i) and (ii) immediately follow from Theorem 2.2.9 (i) and (iii). Now we shall consider the inverse Wishart moments given by the statements (iii) – (viii) of the theorem. The proof is based on the ideas similar to those used when considering the multivariate integration by parts formula in §1.4.11. However, in the proof we will modify our differentiation operator somewhat. Let $\mathbf{Y} \in \mathbb{R}^{q \times r}$ and $\mathbf{X} \in \mathbb{R}^{p \times n}$. Instead of using Definition 1.4.1 or the equivalent formula

$$\frac{d\mathbf{Y}}{d\mathbf{X}} = \sum_I \frac{\partial y_{ij}}{\partial x_{kl}}(\mathbf{f}_l \otimes \mathbf{g}_k)(\mathbf{e}_j \otimes \mathbf{d}_i)',$$

where $I = \{i,j,k,l; 1 \le i \le q, 1 \le j \le r, 1 \le k \le p, 1 \le l \le n\}$, and $\mathbf{d}_i$, $\mathbf{e}_j$, $\mathbf{g}_k$ and $\mathbf{f}_l$ are the i–th, j–th, k–th and l–th column of $\mathbf{I}_q$, $\mathbf{I}_r$, $\mathbf{I}_p$ and $\mathbf{I}_n$, respectively, we will use a derivative analogous to (1.4.48):

$$\frac{d\mathbf{Y}}{d\mathbf{X}} = \sum_I \frac{\partial y_{ij}}{\partial x_{kl}} (\mathbf{f}_l \otimes \mathbf{g}_k)\epsilon_{kl}(\mathbf{e}_j \otimes \mathbf{d}_i)', \quad \epsilon_{kl} = \begin{cases} 1 & k = l, \\ \frac{1}{2} & k \neq l. \end{cases}$$

All rules for matrix derivatives, especially those in Table 1.4.2 in §1.4.9, hold except $\frac{d\mathbf{X}}{d\mathbf{X}}$, which equals

$$\frac{d\mathbf{X}}{d\mathbf{X}} = \frac{1}{2}(\mathbf{I} + \mathbf{K}_{p,p}) \tag{2.4.30}$$

for symmetric $\mathbf{X} \in \mathbb{R}^{p\times p}$. Let $f_{\mathbf{W}}(\mathbf{W})$ denote the Wishart density given in Theorem 2.4.6, $\int_{\mathbf{W}>0}$ means an ordinary multiple integral with integration performed over the subset of $\mathbb{R}^{1/2p(p+1)}$, where $\mathbf{W}$ is positive definite and $d\mathbf{W}$ denotes the Lebesque measure $\prod_{k\leq l} dW_{kl}$ in $\mathbb{R}^{1/2p(p+1)}$. At first, let us verify that

$$\int_{\mathbf{W}>0} \frac{df_{\mathbf{W}}(\mathbf{W})}{d\mathbf{W}} d\mathbf{W} = \mathbf{0}. \tag{2.4.31}$$

The statement is true, if we can show that

$$\int_{\mathbf{W}>0} \frac{df_{\mathbf{W}}(\mathbf{W})}{dW_{ij}} d\mathbf{W} = 0, \qquad i,j = 1,2,\ldots,p, \tag{2.4.32}$$

holds. The relation in (2.4.32) will be proven for the two special cases: $(i,j) = (p,p)$ and $(i,j) = (p-1,p)$. Then the general formula follows by symmetry. We are going to integrate over a subset of $\mathbb{R}^{1/2p(p+1)}$, where $\mathbf{W} > 0$, and one representation of this subset is given by the principal minors of $\mathbf{W}$, i.e.

$$W_{11} > 0, \begin{vmatrix} W_{11} & W_{12} \\ W_{21} & W_{22} \end{vmatrix} > 0, \ldots, |\mathbf{W}| > 0. \tag{2.4.33}$$

Observe that $|\mathbf{W}| > 0$ is the only relation in (2.4.33) which leads to some restrictions on W_{pp} and W_{p-1p}. Furthermore, by using the definition of a determinant, the condition $|\mathbf{W}| > 0$ in (2.4.33) can be replaced by

$$0 \leq \psi_1(W_{11}, W_{12}, \ldots, W_{p-1p}) < W_{pp} < \infty \tag{2.4.34}$$

or

$$\psi_2(W_{11}, W_{12}, \ldots, W_{p-2p}, W_{pp}) < W_{p-1p} < \psi_3(W_{11}, W_{12}, \ldots, W_{p-2p}, W_{pp}), \tag{2.4.35}$$

where $\psi_1(\bullet)$, $\psi_2(\bullet)$ and $\psi_3(\bullet)$ are continuous functions. Hence, integration of W_{pp} and W_{p-1p} in (2.4.32) can be performed over the intervals given by (2.4.34) and (2.4.35), respectively. Note that if in these intervals $W_{pp} \to \psi_1(\bullet)$, $W_{p-1p} \to \psi_2(\bullet)$ or $W_{p-1p} \to \psi_3(\bullet)$, then $|\mathbf{W}| \to 0$. Thus,

$$\begin{aligned} \int_{\mathbf{W}>0} \frac{df_{\mathbf{W}}(\mathbf{W})}{dW_{pp}} d\mathbf{W} &= \int\int \cdots \int_{\psi_1}^{\infty} \frac{df_{\mathbf{W}}(\mathbf{W})}{dW_{pp}} dW_{pp} \ldots dW_{12} dW_{11} \\ &= \int\int \cdots \int \lim_{W_{pp}\to\infty} f_{\mathbf{W}}(\mathbf{W}) - \lim_{W_{pp}\to\psi_1} f_{\mathbf{W}}(\mathbf{W}) \ldots dW_{12} dW_{11} = 0, \end{aligned}$$

since $\lim_{W_{pp}\to\psi_1} f_{\mathbf{W}}(\mathbf{W}) = 0$ and $\lim_{W_{pp}\to\infty} f_{\mathbf{W}}(\mathbf{W}) = 0$. These two statements hold since for the first limit $\lim_{W_{pp}\to\psi_1} |\mathbf{W}| = 0$ and for the second one $\lim_{W_{pp}\to\infty} |\mathbf{W}|\exp(-1/2\mathrm{tr}(\boldsymbol{\Sigma}^{-1}\mathbf{W})) = 0$, which follows from the definition of a determinant. Furthermore, when considering W_{p-1p}, another order of integration yields

$$\int_{\mathbf{W}>0} \frac{df_{\mathbf{W}}(\mathbf{W})}{dW_{p-1p}} d\mathbf{W} = \int\int \cdots \int_{\psi_2}^{\psi_3} \frac{df_{\mathbf{W}}(\mathbf{W})}{dW_{p-1p}} dW_{p-1p} \ldots dW_{12}dW_{11}$$
$$= \int \cdots \int \lim_{W_{p-1p}\to\psi_3} f_{\mathbf{W}}(\mathbf{W}) - \lim_{W_{p-1p}\to\psi_2} f_{\mathbf{W}}(\mathbf{W}) dW_{p-2,p} \ldots dW_{12}dW_{11} = 0,$$

since $\lim_{W_{p-1p}\to\psi_3} f_{\mathbf{W}}(\mathbf{W}) = 0$ and $\lim_{W_{p-1p}\to\psi_2} f_{\mathbf{W}}(\mathbf{W}) = 0$. Hence, (2.4.31) as well as (2.4.32) are established. By definition of the Wishart density it follows that (2.4.31) is equivalent to

$$c\int_{\mathbf{W}>0} \frac{d}{d\mathbf{W}}\{|\mathbf{W}|^{\frac{1}{2}(n-p-1)}|\boldsymbol{\Sigma}^{-1}|^{\frac{1}{2}n}\exp(-1/2\mathrm{tr}(\boldsymbol{\Sigma}^{-1}\mathbf{W}))\}d\mathbf{W} = \mathbf{0}, \qquad (2.4.36)$$

where c is the normalization constant. However, after applying Proposition 1.4.10, it follows that (2.4.36), in turn, is equivalent to

$$\int_{\mathbf{W}>0} \{\tfrac{1}{2}(n-p-1)\mathrm{vec}\mathbf{W}^{-1}f_{\mathbf{W}}(\mathbf{W}) - \tfrac{1}{2}\mathrm{vec}\boldsymbol{\Sigma}^{-1}f_{\mathbf{W}}(\mathbf{W})\}d\mathbf{W} = \mathbf{0}.$$

Thus, $E[\mathbf{W}^{-1}] = \frac{1}{n-p-1}\boldsymbol{\Sigma}^{-1}$.
For the verification of (iv) we need to establish that

$$\int_{\mathbf{W}>0} \frac{d}{d\mathbf{W}}\mathbf{W}^{-1}f_{\mathbf{W}}(\mathbf{W})d\mathbf{W} = \mathbf{0}. \qquad (2.4.37)$$

By copying the proof of (2.4.31), the relation in (2.4.37) follows if it can be shown that

$$\lim_{W_{pp}\to\infty} \mathbf{W}^{-1}f_{\mathbf{W}}(\mathbf{W}) = \mathbf{0}, \qquad \lim_{W_{pp}\to\psi_1} \mathbf{W}^{-1}f_{\mathbf{W}}(\mathbf{W}) = \mathbf{0}, \qquad (2.4.38)$$
$$\lim_{W_{p-1p}\to\psi_3} \mathbf{W}^{-1}f_{\mathbf{W}}(\mathbf{W}) = \mathbf{0}, \qquad \lim_{W_{p-1p}\to\psi_2} \mathbf{W}^{-1}f_{\mathbf{W}}(\mathbf{W}) = \mathbf{0}. \qquad (2.4.39)$$

Now $\lim_{W_{pp}\to\infty} \mathbf{W}^{-1}f_{\mathbf{W}}(\mathbf{W}) = \mathbf{0}$, since $\lim_{W_{pp}\to\infty} f_{\mathbf{W}}(\mathbf{W}) = \mathbf{0}$ and the elements of $\mathbf{W}^{-1}$ are finite, as $W_{pp} \to \infty$. For other statements in (2.4.38) and (2.4.39) we will apply that

$$\mathbf{W}^{-1}|\mathbf{W}| = \mathrm{adj}(\mathbf{W}), \qquad (2.4.40)$$

where adj denotes the adjoint matrix of $\mathbf{W}$, i.e. the elements of $\mathbf{W}$ are replaced by their cofactors (see §1.1.2) and then transposed. The point is that the adjoint matrix exists whether or not $|\mathbf{W}| = 0$. Therefore, it follows that

$$\lim_{|\mathbf{W}|\to 0} \mathbf{W}^{-1}f_{\mathbf{W}}(\mathbf{W}) = c\lim_{|\mathbf{W}|\to 0} \mathbf{W}^{-1}|\mathbf{W}|^{1/2(n-p-1)}|\boldsymbol{\Sigma}^{-1}|^{1/2n}\exp(-1/2\mathrm{tr}(\boldsymbol{\Sigma}^{-1}\mathbf{W}))$$
$$= c\lim_{|\mathbf{W}|\to 0} \mathrm{adj}(\mathbf{W})|\mathbf{W}|^{1/2(n-p-3)}|\boldsymbol{\Sigma}^{-1}|^{1/2n}\exp(-1/2\mathrm{tr}(\boldsymbol{\Sigma}^{-1}\mathbf{W})) = \mathbf{0} \qquad (2.4.41)$$

holds. Thus, since $W_{pp} \to \psi_1$, $W_{p-1p} \to \psi_3$ and $W_{p-1p} \to \psi_2$ together imply that $|\mathbf{W}| \to 0$, and from (2.4.41) it follows that (2.4.38) and (2.4.39) are true which establishes (2.4.37).

Now (2.4.37) will be examined. From (2.4.30) and Table 1.4.1 it follows that

$$\begin{aligned}\frac{d}{d\mathbf{W}}\mathbf{W}^{-1}f_{\mathbf{W}}(\mathbf{W}) &= f_{\mathbf{W}}(\mathbf{W})\frac{d\mathbf{W}^{-1}}{d\mathbf{W}} + \text{vec}'\mathbf{W}^{-1}\frac{df_{\mathbf{W}}(\mathbf{W})}{d\mathbf{W}}\\ &= -\tfrac{1}{2}(\mathbf{I}+\mathbf{K}_{p,p})(\mathbf{W}^{-1}\otimes\mathbf{W}^{-1})f_{\mathbf{W}}(\mathbf{W})\\ &\quad + \text{vec}\mathbf{W}^{-1}\{\tfrac{1}{2}(n-p-1)\text{vec}'\mathbf{W}^{-1} - \tfrac{1}{2}\text{vec}'\mathbf{\Sigma}^{-1}\}f_{\mathbf{W}}(\mathbf{W}),\end{aligned}$$

which, utilizing (iii), after integration gives

$$\begin{aligned}-E[\mathbf{W}^{-1}\otimes\mathbf{W}^{-1}] - \mathbf{K}_{p,p}E[\mathbf{W}^{-1}\otimes\mathbf{W}^{-1}] + (n-p-1)E[\text{vec}\mathbf{W}^{-1}\text{vec}'\mathbf{W}^{-1}]&\\ = \tfrac{1}{n-p-1}\text{vec}\mathbf{\Sigma}^{-1}\text{vec}'\mathbf{\Sigma}^{-1}.&\end{aligned} \quad (2.4.42)$$

Let

$$\begin{aligned}\mathbf{T} &=(E[\mathbf{W}^{-1}\otimes\mathbf{W}^{-1}] : \mathbf{K}_{p,p}E[\mathbf{W}^{-1}\otimes\mathbf{W}^{-1}] : E[\text{vec}\mathbf{W}^{-1}\text{vec}'\mathbf{W}^{-1}])',\\ \mathbf{M} &=(\text{vec}\mathbf{\Sigma}^{-1}\text{vec}'\mathbf{\Sigma}^{-1} : \mathbf{K}_{p,p}(\mathbf{\Sigma}^{-1}\otimes\mathbf{\Sigma}^{-1}) : \mathbf{\Sigma}^{-1}\otimes\mathbf{\Sigma}^{-1})'.\end{aligned}$$

Applying Proposition 1.3.14 (vi) and premultiplying (2.4.42) by $\mathbf{K}_{p,p}$ yields the following matrix equation:

$$\mathbf{QT} = \tfrac{1}{n-p-1}\mathbf{M}, \quad (2.4.43)$$

where

$$\mathbf{Q} = \begin{pmatrix} -\mathbf{I}_{p^2} & -\mathbf{I}_{p^2} & (n-p-1)\mathbf{I}_{p^2}\\ -\mathbf{I}_{p^2} & (n-p-1)\mathbf{I}_{p^2} & -\mathbf{I}_{p^2}\\ (n-p-1)\mathbf{I}_{p^2} & -\mathbf{I}_{p^2} & -\mathbf{I}_{p^2}\end{pmatrix}.$$

Thus,

$$\mathbf{T} = \tfrac{1}{n-p-1}\mathbf{Q}^{-1}\mathbf{M}$$

and since

$$\mathbf{Q}^{-1} = \frac{1}{(n-p)(n-p-3)}\begin{pmatrix} \mathbf{I}_{p^2} & \mathbf{I}_{p^2} & (n-p-2)\mathbf{I}_{p^2}\\ \mathbf{I}_{p^2} & (n-p-2)\mathbf{I}_{p^2} & \mathbf{I}_{p^2}\\ (n-p-2)\mathbf{I}_{p^2} & \mathbf{I}_{p^2} & \mathbf{I}_{p^2}\end{pmatrix},$$

the relation in (iv) can be obtained with the help of some calculations, i.e.

$$E[\mathbf{W}^{-1}\otimes\mathbf{W}^{-1}] = \frac{1}{(n-p)(n-p-1)(n-p-3)}(\mathbf{I}_{p^2} : \mathbf{I}_{p^2} : (n-p-2)\mathbf{I}_{p^2})\mathbf{M}.$$

For (v) we differentiate both sides in (iii) with respect to $\mathbf{\Sigma}^{-1}$ and get

$$E[n\,\text{vec}\mathbf{W}^{-1}\text{vec}'\mathbf{\Sigma} - \text{vec}\mathbf{W}^{-1}\text{vec}'\mathbf{W}] = \tfrac{1}{n-p-1}(\mathbf{I}_{p^2} + \mathbf{K}_{p,p}).$$

The statement in (vi) is obtained from (iv), because

$$\text{vec}E[\mathbf{W}^{-1}\mathbf{W}^{-1}] = E[\mathbf{W}^{-1}\otimes\mathbf{W}^{-1}]\text{vec}\mathbf{I},$$

$$\mathbf{K}_{p,p}(\boldsymbol{\Sigma}^{-1}\otimes\boldsymbol{\Sigma}^{-1})\mathrm{vec}\mathbf{I} = \mathrm{vec}(\boldsymbol{\Sigma}^{-1}\boldsymbol{\Sigma}^{-1})$$

and

$$\mathrm{vec}\boldsymbol{\Sigma}^{-1}\mathrm{vec}'\boldsymbol{\Sigma}^{-1}\mathrm{vec}\mathbf{I} = \mathrm{vec}\boldsymbol{\Sigma}^{-1}\mathrm{tr}\boldsymbol{\Sigma}^{-1}.$$

To get (vii) we multiply (v) by $\mathrm{vec}\mathbf{I}$ and obtain the statement.
For (viii) it will be utilized that

$$\sum_{m=1}^{p}(\mathbf{e}'_m\otimes\mathbf{I})\mathbf{W}^{-1}\otimes\mathbf{W}^{-1}(\mathbf{e}_m\otimes\mathbf{I}) = \mathrm{tr}(\mathbf{W}^{-1})\mathbf{W}^{-1}.$$

Since

$$\sum_{m=1}^{p}(\mathbf{e}'_m\otimes\mathbf{I})\boldsymbol{\Sigma}^{-1}\otimes\boldsymbol{\Sigma}^{-1}(\mathbf{e}_m\otimes\mathbf{I}) = \mathrm{tr}(\boldsymbol{\Sigma}^{-1})\boldsymbol{\Sigma}^{-1},$$

$$\sum_{m=1}^{p}(\mathbf{e}_m\otimes\mathbf{I})\mathrm{vec}\boldsymbol{\Sigma}^{-1}\mathrm{vec}'\boldsymbol{\Sigma}^{-1}(\mathbf{e}'_m\otimes\mathbf{I}) = \sum_{m=1}^{p}\boldsymbol{\Sigma}^{-1}\mathbf{e}_m\mathbf{e}'_m\boldsymbol{\Sigma}^{-1} = \boldsymbol{\Sigma}^{-1}\boldsymbol{\Sigma}^{-1}$$

and

$$\begin{aligned}\sum_{m=1}^{p}(\mathbf{e}'_m\otimes\mathbf{I})\mathbf{K}_{p,p}(\boldsymbol{\Sigma}^{-1}\otimes\boldsymbol{\Sigma}^{-1})(\mathbf{e}'_m\otimes\mathbf{I}) &= \sum_{m=1}^{p}(\mathbf{I}\otimes\mathbf{e}'_m)(\boldsymbol{\Sigma}^{-1}\otimes\boldsymbol{\Sigma}^{-1})(\mathbf{e}'_m\otimes\mathbf{I})\\ &= \sum_{m=1}^{p}\boldsymbol{\Sigma}^{-1}\mathbf{e}_m\mathbf{e}'_m\boldsymbol{\Sigma}^{-1} = \boldsymbol{\Sigma}^{-1}\boldsymbol{\Sigma}^{-1},\end{aligned}$$

(viii) is verified. ∎

Note that from the proof of (iv) we could easily have obtained $E[\mathrm{vec}\mathbf{W}^{-1}\mathrm{vec}'\mathbf{W}^{-1}]$ which then would have given $D[\mathbf{W}^{-1}]$, i.e.

$$\begin{aligned}D[\mathbf{W}^{-1}] &= \tfrac{2}{(n-p)(n-p-1)^2(n-p-3)}\mathrm{vec}\boldsymbol{\Sigma}^{-1}\mathrm{vec}'\boldsymbol{\Sigma}^{-1}\\ &+ \tfrac{1}{(n-p)(n-p-1)(n-p-3)}(\mathbf{I}+\mathbf{K}_{p,p})(\boldsymbol{\Sigma}^{-1}\otimes\boldsymbol{\Sigma}^{-1}),\quad n-p-3>0. \qquad (2.4.44)\end{aligned}$$

Furthermore, by copying the above proof we can extend (2.4.37) and prove that

$$\int_{\mathbf{W}>0}\frac{d}{d\mathbf{W}}\mathrm{vec}\mathbf{W}^{-1}\mathrm{vec}'(\mathbf{W}^{-1})^{\otimes r-1}f_{\mathbf{W}}(\mathbf{W})d\mathbf{W} = \mathbf{0},\quad n-p-1-2r>0. \quad (2.4.45)$$

Since

$$\frac{df_{\mathbf{W}}(\mathbf{W})}{d\mathbf{W}} = \tfrac{1}{2}(n-p-1)\mathrm{vec}\mathbf{W}^{-1}f_{\mathbf{W}}(\mathbf{W}) - \tfrac{1}{2}\mathrm{vec}\boldsymbol{\Sigma}^{-1}f_{\mathbf{W}}(\mathbf{W}),$$

we obtain from (2.4.45) that

$$\begin{aligned}&2E[\frac{d}{d\mathbf{W}}\mathrm{vec}\mathbf{W}^{-1}\mathrm{vec}'(\mathbf{W}^{-1})^{\otimes r-1}] + (n-p-1)E[\mathrm{vec}\mathbf{W}^{-1}\mathrm{vec}'(\mathbf{W}^{-1})^{\otimes r}]\\ &= E[\mathrm{vec}\boldsymbol{\Sigma}^{-1}\mathrm{vec}'(\mathbf{W}^{-1})^{\otimes r}],\qquad n-p-1-2r>0. \qquad (2.4.46)\end{aligned}$$

It follows from (2.4.46) that in order to obtain $E[(\mathbf{W}^{-1})^{\otimes r+1}]$ one has to use the sum

$$2E[\text{vec}(\frac{d}{d\mathbf{W}}(\mathbf{W}^{-1})^{\otimes r})] + (n-p-1)E[\text{vec}((\mathbf{W}^{-1})^{\otimes r+1})].$$

Here one stacks algebraic problems when an explicit solution is of interest. On the other hand the problem is straightforward, although complicated. For example, we can mention difficulties which occurred when (2.4.46) was used in order to obtain the moments of the third order (von Rosen, 1988a).

Let us consider the basic structure of the first two moments of symmetric matrices which are *rotationally invariant*, i.e. $\mathbf{\Gamma U \Gamma}'$ has the the same distribution as $\mathbf{U}$ for every orthogonal matrix $\mathbf{\Gamma}$. We are interested in $E[\mathbf{U}]$ and of course

$$E[\text{vec}\mathbf{U}] = \text{vec}\mathbf{A}$$

for some symmetric $\mathbf{A}$. There are many ways to show that $\mathbf{A} = c\mathbf{I}$ for some constant c if $\mathbf{U}$ is rotationally invariant. For example, $\mathbf{A} = \mathbf{HDH}'$ where $\mathbf{H}$ is orthogonal and $\mathbf{D}$ diagonal implies that $E[\text{vec}\mathbf{U}] = \text{vec}\mathbf{D}$ and therefore $\mathbf{\Gamma D \Gamma}' = \mathbf{D}$ for all $\mathbf{\Gamma}$. By using different $\mathbf{\Gamma}$ we observe that $\mathbf{D}$ must be proportional to $\mathbf{I}$. Thus

$$E[\mathbf{U}] = c\mathbf{I}$$

and $c = E[U_{11}]$.

When proceeding with the second order moments, we consider

$$\text{vec}E[\mathbf{U}\otimes\mathbf{U}] = E[\sum_{ijkl} U_{ij}U_{kl}\mathbf{e}_j\otimes\mathbf{e}_l\otimes\mathbf{e}_i\otimes\mathbf{e}_k].$$

Because $\mathbf{U}$ is symmetric and obviously $E[U_{ij}U_{kl}] = E[U_{kl}U_{ij}]$, the expectation $\text{vec}E[\mathbf{U}\otimes\mathbf{U}]$ should be the same when interchanging the pairs of indices (i,j) and (k,l) as well as when interchanging either i and j or k and l. Thus we may write that for some elements a_{ij} and constants d_1, d_2 and d_3,

$$\text{vec}E[\mathbf{U}\otimes\mathbf{U}] = \sum_{ijkl}(d_1a_{ij}a_{kl} + d_2a_{ik}a_{jl} + d_2a_{il}a_{jk} + d_3)\mathbf{e}_j\otimes\mathbf{e}_l\otimes\mathbf{e}_i\otimes\mathbf{e}_k,$$

where $a_{mn} = a_{nm}$. Thus, for $\mathbf{A} = (a_{ij})$,

$$\text{vec}E[\mathbf{U}\otimes\mathbf{U}] = d_1\text{vec}(\mathbf{A}\otimes\mathbf{A}) + d_2\text{vec}\mathbf{A}\otimes\text{vec}\mathbf{A} + d_2\text{vec}(\mathbf{K}_{p,p}(\mathbf{A}\otimes\mathbf{A})) + d_3\text{vec}\mathbf{I}.$$

Now we will study the consequence of $\mathbf{U}$ being "rotationally invariant". Since $\mathbf{A} = \mathbf{\Gamma D \Gamma}'$ for an orthogonal matrix $\mathbf{\Gamma}$ and a diagonal matrix $\mathbf{D}$,

$$\begin{aligned}\text{vec}E[\mathbf{U}\otimes\mathbf{U}] &= (\mathbf{\Gamma}\otimes\mathbf{\Gamma}\otimes\mathbf{\Gamma}\otimes\mathbf{\Gamma})\text{vec}E[\mathbf{U}\otimes\mathbf{U}]\\ &= d_1\text{vec}(\mathbf{D}\otimes\mathbf{D}) + d_2\text{vec}\mathbf{D}\otimes\text{vec}\mathbf{D} + d_2\text{vec}(\mathbf{K}_{p,p}(\mathbf{D}\otimes\mathbf{D})) + d_3\text{vec}\mathbf{I}.\end{aligned}$$

Furthermore, by premultiplying with properly chosen orthogonal matrices, we may conclude that $\mathbf{D} = c\mathbf{I}$. Thus, if $c_1 = cd_1 + d_3$ and $c_2 = cd_2$,

$$E[\mathbf{U}\otimes\mathbf{U}] = c_1\mathbf{I} + c_2\text{vec}\mathbf{I}\text{vec}'\mathbf{I} + c_2\mathbf{K}_{p,p}.$$

Moreover,

$$\begin{aligned}E[U_{11}U_{22}] &= c_1,\\ E[U_{12}^2] &= c_2.\end{aligned}$$

The results are summarized in the next lemma.

Lemma 2.4.1. *Let* $\mathbf{U} : p \times p$ *be symmetric and rotationally invariant. Then*

(i) $E[\mathbf{U}] = c\mathbf{I}, \qquad c = E[U_{11}];$

(ii) $E[\mathbf{U} \otimes \mathbf{U}] = c_1\mathbf{I} + c_2\mathrm{vec}\mathbf{I}\mathrm{vec}'\mathbf{I} + c_2\mathbf{K}_{p,p}, \quad c_1 = E[U_{11}U_{22}],\ c_2 = E[U_{12}^2];$

(iii) $D[\mathbf{U}] = c_2\mathbf{I} + (c_1 - c^2)\mathrm{vec}\mathbf{I}\mathrm{vec}'\mathbf{I} + c_2\mathbf{K}_{p,p}.$

PROOF: The relations in (i) and (ii) were motivated before the lemma. The statement in (iii) follows from (i) and (ii) when Proposition 1.3.14 (vi) is applied. ∎

Observe that the first two moments of the Wishart and inverted Wishart distribution follow the structure of the moments given in the lemma.
In §2.4.3 two versions of the multivariate beta distribution were introduced. Now, the mean and the dispersion matrix for both are given.

Theorem 2.4.15. *Let* $\mathbf{Z} \sim M\beta_{II}(p, m, n)$ *and* $\mathbf{F} \sim M\beta_I(p, m, n)$. *Then*

(i) $E[\mathbf{Z}] = \frac{n}{m-p-1}\mathbf{I}, \qquad m - p - 1 > 0;$

(ii) $D[\mathbf{Z}] = \frac{2n(m-p-2)+n^2}{(m-p)(m-p-1)(m-p-3)}(\mathbf{I} + \mathbf{K}_{p,p}) + \frac{2n(m-p-1)+2n^2}{(m-p)(m-p-1)^2(m-p-3)}\mathrm{vec}\mathbf{I}\mathrm{vec}'\mathbf{I},$

$m - p - 3 > 0;$

(iii) $E[\mathbf{F}] = \frac{m}{n+m}\mathbf{I};$

(iv) $D[\mathbf{F}] = (c_3 - \frac{m^2}{(n+m)^2})\mathrm{vec}\mathbf{I}\mathrm{vec}'\mathbf{I} + c_4(\mathbf{I} + \mathbf{K}_{p,p}),$
where

$$c_4 = \frac{n-p-1}{(n+m-1)(n+m+2)}\{(n-p-2+\tfrac{1}{n+m})c_2 - (1 + \tfrac{n-p-1}{n+m})c_1\},$$

$$c_3 = \frac{n-p-1}{n+m}((n-p-2)c_2 - c_1) - (n+m+1)c_4,$$

$$c_1 = \frac{n^2(m-p-2)+2n}{(m-p)(m-p-1)(m-p-3)}, \qquad c_2 = \frac{n(m-p-2)+n^2+n}{(m-p)(m-p-1)(m-p-3)},$$

$m - p - 3 > 0.$

PROOF: By definition, $\mathbf{Z} = \mathbf{W}_2^{-1/2}\mathbf{W}_1\mathbf{W}_2^{-1/2}$, where $\mathbf{W}_1 \sim W_p(\mathbf{I}, n)$ and $\mathbf{W}_2 \sim W_p(\mathbf{I}, m)$ are independent. Hence, by Theorem 2.4.14,

$$E[\mathbf{Z}] = E[\mathbf{W}_2^{-1/2}\mathbf{W}_1\mathbf{W}_2^{-1/2}] = E[\mathbf{W}_2^{-1/2}E[\mathbf{W}_1]\mathbf{W}_2^{-1/2}] = nE[\mathbf{W}_2^{-1}] = \frac{n}{m-p-1}\mathbf{I}$$

and (i) is established. For (ii) we get

$$\begin{aligned} D[\mathbf{Z}] &= E[D[\mathbf{Z}|\mathbf{W}_2]] + D[E[\mathbf{Z}|\mathbf{W}_2]] \\ &= E[n(\mathbf{I} + \mathbf{K}_{p,p})(\mathbf{W}_2^{-1} \otimes \mathbf{W}_2^{-1})] + D[n\mathbf{W}_2^{-1}] \end{aligned}$$

and then, by Theorem 2.4.14 (iv) and (2.4.44), the statement follows.

In (iii) and (iv) it will be utilized that $(\mathbf{I}+\mathbf{Z})^{-1}$ has the same distribution as $\mathbf{F}=(\mathbf{W}_1+\mathbf{W}_2)^{-1/2}\mathbf{W}_2(\mathbf{W}_1+\mathbf{W}_2)^{-1/2}$. A technique similar to the one which was used when moments for the inverted Wishart matrix were derived will be applied. From (2.4.20) it follows that

$$\mathbf{0}=\int \frac{d}{d\mathbf{Z}}c|\mathbf{Z}|^{\frac{1}{2}(n-p-1)}|\mathbf{I}+\mathbf{Z}|^{-\frac{1}{2}(n+m)}d\mathbf{Z},$$

where c is the normalizing constant and the derivative is the same as the one which was used in the proof of Theorem 2.4.14. In particular, the projection in (2.4.30) holds, i.e.

$$\frac{d\mathbf{Z}}{d\mathbf{Z}}=\frac{1}{2}(\mathbf{I}+\mathbf{K}_{p,p}).$$

By differentiation we obtain

$$\mathbf{0}=\tfrac{1}{2}(n-p-1)\frac{d\mathbf{Z}}{d\mathbf{Z}}E[\mathrm{vec}\mathbf{Z}^{-1}]-\tfrac{1}{2}(n+m)\frac{d\mathbf{Z}}{d\mathbf{Z}}E[\mathrm{vec}(\mathbf{I}+\mathbf{Z})^{-1}],$$

which is equivalent to

$$E[\mathrm{vec}(\mathbf{I}+\mathbf{Z})^{-1}]=\tfrac{n-p-1}{n+m}E[\mathrm{vec}\mathbf{Z}^{-1}].$$

However, by definition of $\mathbf{Z}$,

$$E[\mathbf{Z}^{-1}]=E[\mathbf{W}_2^{1/2}\mathbf{W}_1^{-1}\mathbf{W}_2^{1/2}]=\tfrac{m}{n-p-1}\mathbf{I}$$

and (iii) is established.

Turning to (iv), it is first observed that

$$\begin{aligned}\mathbf{0}&=\int \frac{d}{d\mathbf{Z}}\mathbf{Z}^{-1}c|\mathbf{Z}|^{\frac{1}{2}(n-p-1)}|\mathbf{I}+\mathbf{Z}|^{-\frac{1}{2}(n+m)}d\mathbf{Z}\\&=-\frac{d\mathbf{Z}}{d\mathbf{Z}}E[\mathbf{Z}^{-1}\otimes\mathbf{Z}^{-1}]+\tfrac{1}{2}(n-p-1)\frac{d\mathbf{Z}}{d\mathbf{Z}}E[\mathrm{vec}\mathbf{Z}^{-1}\mathrm{vec}'\mathbf{Z}^{-1}]\\&\quad-\tfrac{1}{2}(n+m)\frac{d\mathbf{Z}}{d\mathbf{Z}}E[\mathrm{vec}(\mathbf{I}+\mathbf{Z})^{-1}\mathrm{vec}'\mathbf{Z}^{-1}].\end{aligned}$$

Moreover, instead of considering $E[\frac{d}{d\mathbf{Z}}\mathbf{Z}^{-1}]$, we can study $E[\frac{d}{d\mathbf{Z}}(\mathbf{I}+\mathbf{Z})^{-1}]$ and obtain

$$\begin{aligned}\frac{d\mathbf{Z}}{d\mathbf{Z}}E[(\mathbf{I}+\mathbf{Z})^{-1}\otimes(\mathbf{I}+\mathbf{Z})^{-1}]&=\tfrac{1}{2}(n-p-1)E[\mathrm{vec}\mathbf{Z}^{-1}\mathrm{vec}'(\mathbf{I}+\mathbf{Z})^{-1}]\\&\quad-\tfrac{1}{2}(n+m)E[\mathrm{vec}(\mathbf{I}+\mathbf{Z})^{-1}\mathrm{vec}'(\mathbf{I}+\mathbf{Z})^{-1}].\end{aligned}$$

By transposing this expression it follows, since

$$(\mathbf{I}+\mathbf{K}_{p,p})E[\mathbf{Z}^{-1}\otimes\mathbf{Z}^{-1}]=E[\mathbf{Z}^{-1}\otimes\mathbf{Z}^{-1}](\mathbf{I}+\mathbf{K}_{p,p}),$$

that

$$E[\text{vec}(\mathbf{I}+\mathbf{Z})^{-1}\text{vec}'\mathbf{Z}^{-1}] = E[\text{vec}\mathbf{Z}^{-1}\text{vec}'(\mathbf{I}+\mathbf{Z})^{-1}].$$

Thus, using the relations given above,

$$(\mathbf{I}+\mathbf{K}_{p,p})E[(\mathbf{I}+\mathbf{Z})^{-1}\otimes(\mathbf{I}+\mathbf{Z})^{-1}] = -\tfrac{n-p-1}{n+m}(\mathbf{I}+\mathbf{K}_{p,p})E[\mathbf{Z}^{-1}\otimes\mathbf{Z}^{-1}]$$
$$+\tfrac{(n-p-1)^2}{n+m}E[\text{vec}\mathbf{Z}^{-1}\text{vec}\mathbf{Z}^{-1}] - (n+m)E[\text{vec}(\mathbf{I}+\mathbf{Z})^{-1}\text{vec}'(\mathbf{I}+\mathbf{Z})^{-1}]. \quad (2.4.47)$$

Now Proposition 1.3.14 (vi) implies that

$$\begin{aligned} &E[\text{vec}(\mathbf{I}+\mathbf{Z})^{-1}\text{vec}'(\mathbf{I}+\mathbf{Z})^{-1}] + \mathbf{K}_{p,p}E[(\mathbf{I}+\mathbf{Z})^{-1}\otimes(\mathbf{I}+\mathbf{Z})^{-1}] \\ &\quad= -\tfrac{n-p-1}{n+m}(E[\text{vec}\mathbf{Z}^{-1}\text{vec}'\mathbf{Z}^{-1}] + \mathbf{K}_{p,p})E[\mathbf{Z}^{-1}\otimes\mathbf{Z}^{-1}]) \\ &\quad\quad+\tfrac{(n-p-1)^2}{n+m}E[\mathbf{Z}^{-1}\otimes\mathbf{Z}^{-1}] - (n+m)E[\text{vec}(\mathbf{I}+\mathbf{Z})^{-1}\text{vec}'(\mathbf{I}+\mathbf{Z})^{-1}] \end{aligned}$$

and by multiplying this equation by $\mathbf{K}_{p,p}$ we obtain a third set of equations. Put

$$\begin{aligned}\mathbf{E} =\Big(&E[\text{vec}(\mathbf{I}+\mathbf{Z})^{-1}\text{vec}'(\mathbf{I}+\mathbf{Z})^{-1}], E[(\mathbf{I}+\mathbf{Z})^{-1}\otimes(\mathbf{I}+\mathbf{Z})^{-1}], \\ &\mathbf{K}_{p,p}E[(\mathbf{I}+\mathbf{Z})^{-1}\otimes(\mathbf{I}+\mathbf{Z})^{-1}]\Big)'\end{aligned}$$

and

$$\mathbf{M} =(E[\text{vec}\mathbf{Z}^{-1}\text{vec}'\mathbf{Z}^{-1}], E[\mathbf{Z}^{-1}\otimes\mathbf{Z}^{-1}], \mathbf{K}_{p,p}E[\mathbf{Z}^{-1}\otimes\mathbf{Z}^{-1}])'.$$

Then, from the above relations it follows that

$$\begin{pmatrix} (n+m)\mathbf{I}_{p^2} & \mathbf{I}_{p^2} & \mathbf{I}_{p^2} \\ \mathbf{I}_{p^2} & (n+m)\mathbf{I}_{p^2} & \mathbf{I}_{p^2} \\ \mathbf{I}_{p^2} & \mathbf{I}_{p^2} & (n+m)\mathbf{I}_{p^2} \end{pmatrix}\mathbf{E}$$
$$= -\frac{n-p-1}{n+m}\begin{pmatrix} (n-p-1)\mathbf{I}_{p^2} & -\mathbf{I}_{p^2} & -\mathbf{I}_{p^2} \\ -\mathbf{I}_{p^2} & (n-p-1)\mathbf{I}_{p^2} & -\mathbf{I}_{p^2} \\ -\mathbf{I}_{p^2} & -\mathbf{I}_{p^2} & (n-p-1)\mathbf{I}_{p^2} \end{pmatrix}\mathbf{M}.$$

After $E[\mathbf{Z}^{-1}\otimes\mathbf{Z}^{-1}]$ is obtained, $\mathbf{M}$ can be determined and then we have a linear equation system in $\mathbf{E}$ with a unique solution. Some calculations yield

$$E[\mathbf{Z}^{-1}\otimes\mathbf{Z}^{-1}] = c_1\mathbf{I} + c_2\text{vec}\mathbf{I}\text{vec}'\mathbf{I} + c_2\mathbf{K}_{p,p},$$

where c_1 and c_2 are given in statement (iv). Via Proposition 1.3.14 (vi) $\mathbf{M}$ is found. This approach is similar to the one used for obtaining moments of the inverted Wishart distribution.

However, instead of following this route to the end, which consists of routine work, we are going to utilize Lemma 2.4.1, which implies that for constants c_3 and c_4

$$\begin{aligned} E[(\mathbf{I}+\mathbf{Z})^{-1}\otimes(\mathbf{I}+\mathbf{Z})^{-1}] &=c_3\mathbf{I} + c_4\text{vec}\mathbf{I}\text{vec}'\mathbf{I} + c_4\mathbf{K}_{p,p}, \\ E[\text{vec}(\mathbf{I}+\mathbf{Z})^{-1}\text{vec}'(\mathbf{I}+\mathbf{Z})^{-1}] &=c_3\text{vec}\mathbf{I}\text{vec}'\mathbf{I} + c_4(\mathbf{I}+\mathbf{K}_{p,p}). \end{aligned}$$

From (2.4.47) it follows that

$$
\begin{aligned}
(n+m)c_4(\mathbf{I}+\mathbf{K}_{p,p}) + (n+m)c_3\mathrm{vec}\mathbf{I}\mathrm{vec}'\mathbf{I} + (c_3+c_4)(\mathbf{I}+\mathbf{K}_{p,p}) + 2c_4\mathrm{vec}\mathbf{I}\mathrm{vec}'\mathbf{I} \\
= -\tfrac{n-p-1}{n+m}\{(c_1+c_2)(\mathbf{I}+\mathbf{K}_{p,p}) + 2c_2\mathrm{vec}\mathbf{I}\mathrm{vec}'\mathbf{I}\} \\
+ \tfrac{(n-p-1)^2}{n+m}\{c_1\mathrm{vec}\mathbf{I}\mathrm{vec}'\mathbf{I} + c_2(\mathbf{I}+\mathbf{K}_{p,p})\},
\end{aligned}
$$

which is identical to

$$
\begin{aligned}
&((n+m+1)c_4+c_3)(\mathbf{I}+\mathbf{K}_{p,p}) + ((n+m)c_3+2c_4)\mathrm{vec}\mathbf{I}\mathrm{vec}'\mathbf{I} \\
&= \{\tfrac{(n-p-1)^2}{n+m}c_2 - \tfrac{n-p-1}{n+m}(c_1+c_2)\}(\mathbf{I}+\mathbf{K}_{p,p}) + \{\tfrac{(n-p-1)^2}{n+m}c_1 - \tfrac{n-p-1}{n+m}2c_2\}\mathrm{vec}\mathbf{I}\mathrm{vec}'\mathbf{I}.
\end{aligned}
$$

Vaguely speaking, $\mathbf{I}$, $\mathbf{K}_{p,p}$ and $\mathrm{vec}\mathbf{I}\mathrm{vec}'\mathbf{I}$ act in some sense as elements of a basis and therefore the constants c_4 and c_3 are determined by the following equations:

$$
\begin{aligned}
(n+m+1)c_4 + c_3 &= \tfrac{n-p-1}{n+m}((n-p-2)c_2 - c_1), \\
(n+m)c_3 + c_4 &= \tfrac{n-p-1}{n+m}((n-p-1)c_1 - 2c_2).
\end{aligned}
$$

Thus,

$$
E[\mathrm{vec}(\mathbf{I}+\mathbf{Z})^{-1}\mathrm{vec}'(\mathbf{I}+\mathbf{Z})^{-1}] = E[\mathrm{vec}\mathbf{F}\mathrm{vec}'\mathbf{F}]
$$

is obtained and now, with the help of (iii), the dispersion in (iv) is verified. ∎

2.4.6 *Cumulants of the Wishart matrix*

Later, when approximating with the help of the Wishart distribution, cumulants of low order of $\mathbf{W} \sim W_p(\mathbf{\Sigma}, n)$ are needed. In §2.1.4 cumulants were defined as derivatives of the cumulant function at the point zero. Since $\mathbf{W}$ is symmetric, we apply Definition 2.1.9 to the Wishart matrix, which gives the following result.

Theorem 2.4.16. *The first four cumulants of the Wishart matrix* $\mathbf{W} \sim W_p(\mathbf{\Sigma}, n)$ *are given by the equalities*

(i) $c_1[\mathbf{W}] = n\mathbf{G}_p\mathrm{vec}\mathbf{\Sigma} = nV^2(\mathbf{\Sigma})$,
where $V^2(\bullet)$ *is given by Definition 1.3.9;*

(ii) $c_2[\mathbf{W}] = n\mathbf{J}_p(\mathbf{\Sigma}\otimes\mathbf{\Sigma})\mathbf{G}'_p$;

(iii) $c_3[\mathbf{W}] = n\mathbf{J}_p(\mathbf{\Sigma}\otimes\mathbf{\Sigma}\otimes\mathrm{vec}'\mathbf{\Sigma})(\mathbf{I}_{p^4}+\mathbf{K}_{p^2,p^2})(\mathbf{I}_p\otimes\mathbf{K}_{p,p}\otimes\mathbf{I}_p)(\mathbf{G}'_p\otimes\mathbf{J}'_p)$;

(iv)
$$
\begin{aligned}
c_4[\mathbf{W}] = {}& n\mathbf{J}_p(\mathbf{\Sigma}\otimes\mathbf{\Sigma}\otimes(\mathrm{vec}'\mathbf{\Sigma})^{\otimes 2})(\mathbf{I}_{p^3}\otimes\mathbf{K}_{p,p^2})(\mathbf{I}+\mathbf{K}_{p,p}\otimes\mathbf{I}_{p^2}\otimes\mathbf{K}_{p,p}) \\
&\times(\mathbf{I}+\mathbf{K}_{p,p}\otimes\mathbf{K}_{p,p}\otimes\mathbf{I}_{p^2})(\mathbf{G}'_p\otimes\mathbf{J}'_p\otimes\mathbf{J}'_p) \\
&+ n\mathbf{J}_p(\mathbf{\Sigma}\otimes\mathbf{\Sigma}\otimes\mathrm{vec}'(\mathbf{\Sigma}\otimes\mathbf{\Sigma}))(\mathbf{K}_{p^2,p}\otimes\mathbf{I}_p^3)(\mathbf{I}+\mathbf{K}_{p,p}\otimes\mathbf{I}_{p^2}\otimes\mathbf{K}_{p,p}) \\
&\times(\mathbf{I}+\mathbf{K}_{p,p}\otimes\mathbf{K}_{p,p}\otimes\mathbf{I}_{p^2})(\mathbf{G}'_p\otimes\mathbf{J}'_p\otimes\mathbf{J}'_p),
\end{aligned}
$$

where $\mathbf{G}_p$ is given by (1.3.49) and (1.3.50) (see also $\mathbf{T}(u)$ in Proposition 1.3.22 which is identical to $\mathbf{G}_p$), and

$$\mathbf{J}_p = \mathbf{G}_p(\mathbf{I} + \mathbf{K}_{p,p}).$$

PROOF: By using the definitions of $\mathbf{G}_p$ and $\mathbf{J}_p$, (i) and (ii) follow immediately from Theorem 2.1.11 and Theorem 2.4.14. In order to show (iii) we have to perform some calculations. We start by differentiating the cumulant function three times. Indeed, when deriving (iii), we will prove (i) and (ii) in another way. The characteristic function was given in Theorem 2.4.5 and thus the cumulant function equals

$$\psi_{\mathbf{W}}(\mathbf{T}) = -\frac{n}{2} ln|\mathbf{I}_p - i\mathbf{M}(\mathbf{T})\mathbf{\Sigma}|, \tag{2.4.48}$$

where $\mathbf{M}(\mathbf{T})$ is given by (2.4.11). The matrix

$$\mathbf{J}_p = \frac{d\mathbf{M}(\mathbf{T})}{dV^2(\mathbf{T})} = \mathbf{G}_p(\mathbf{I}_{p^2} + \mathbf{K}_{p,p}) \tag{2.4.49}$$

will be used several times when establishing the statements. The last equality in (2.4.49) follows, because ($\mathbf{f}_{k(k,l)}$ and $\mathbf{e}_i$ are the same as in §1.3.6)

$$\begin{aligned}
\frac{d\mathbf{M}(\mathbf{T})}{dV^2(\mathbf{T})} =& \sum_{\substack{k\leq l\\ i\leq j}} \frac{dt_{ij}}{dt_{kl}} \mathbf{f}_{k(k,l)}(\mathbf{e}_i \otimes \mathbf{e}_j + \mathbf{e}_j \otimes \mathbf{e}_i)' \\
=& \sum_{i\leq j} \mathbf{f}_{k(i,j)}(\mathbf{e}_i \otimes \mathbf{e}_j + \mathbf{e}_j \otimes \mathbf{e}_i)' = \mathbf{G}_p \sum_{i\leq j} (\mathbf{e}_j \otimes \mathbf{e}_i)(\mathbf{e}_i \otimes \mathbf{e}_j + \mathbf{e}_j \otimes \mathbf{e}_i)' \\
=& \mathbf{G}_p \sum_{i,j} (\mathbf{e}_j \otimes \mathbf{e}_i)(\mathbf{e}_i \otimes \mathbf{e}_j + \mathbf{e}_j \otimes \mathbf{e}_i)' = \mathbf{G}_p(\mathbf{I}_{p^2} + \mathbf{K}_{p,p}).
\end{aligned}$$

Now

$$\begin{aligned}
\frac{d\psi_{\mathbf{W}}(\mathbf{T})}{dV^2(\mathbf{T})} \underset{(1.4.14)}{=} & -\frac{n}{2}\frac{d|\mathbf{I}_p - i\mathbf{M}(\mathbf{T})\mathbf{\Sigma}|}{dV^2(\mathbf{T})}\frac{d(\ln|\mathbf{I}_p - i\mathbf{M}(\mathbf{T})\mathbf{\Sigma}|)}{d|\mathbf{I}_p - i\mathbf{M}(\mathbf{T})\mathbf{\Sigma}|} \\
\underset{(1.4.14)}{=} & -\frac{n}{2}\frac{d(\mathbf{I}_p - i\mathbf{M}(\mathbf{T})\mathbf{\Sigma})}{dV^2(\mathbf{T})}\frac{d|\mathbf{I}_p - i\mathbf{M}(\mathbf{T})\mathbf{\Sigma}|}{d(\mathbf{I}_p - i\mathbf{M}(\mathbf{T})\mathbf{\Sigma})}\frac{1}{|\mathbf{I}_p - i\mathbf{M}(\mathbf{T})\mathbf{\Sigma}|} \\
= & \, i\frac{n}{2}\frac{d\mathbf{M}(\mathbf{T})}{dV^2(\mathbf{T})}(\mathbf{\Sigma} \otimes \mathbf{I}_p)\mathrm{vec}\{(\mathbf{I}_p - i\mathbf{\Sigma}\mathbf{M}(\mathbf{T}))^{-1}\}.
\end{aligned}$$

Thus,

$$c_1[\mathbf{W}] = i^{-1}\frac{d\psi_{\mathbf{W}}(\mathbf{T})}{dV^2(\mathbf{T})}\bigg|_{\mathbf{T}=\mathbf{0}} = \frac{n}{2}\mathbf{J}_p\mathrm{vec}\mathbf{\Sigma} = n\mathbf{G}_p\mathrm{vec}\mathbf{\Sigma} = nV^2(\mathbf{\Sigma}).$$

Before going further with the proof of (ii), (iii) and (iv), we shall express the matrix $(\mathbf{I}_p - i\mathbf{\Sigma}\mathbf{M}(\mathbf{T}))^{-1}$ with the help of the series expansion

$$(\mathbf{I} - \mathbf{A})^{-1} = \mathbf{I} + \mathbf{A} + \mathbf{A}^2 + \cdots = \sum_{k=0}^{\infty} \mathbf{A}^k$$

and obtain

$$(\mathbf{I}_p - i\boldsymbol{\Sigma}\mathbf{M}(\mathbf{T}))^{-1} = \sum_{k=0}^{\infty} i^k(\boldsymbol{\Sigma}\mathbf{M}(\mathbf{T}))^k.$$

For $\mathbf{T}$ close to zero, the series converges. Thus, the first derivative takes the form

$$\begin{aligned}\frac{d\psi_{\mathbf{W}}(\mathbf{T})}{dV^2(\mathbf{T})} &= i\frac{n}{2}\mathbf{J}_p(\boldsymbol{\Sigma}\otimes\mathbf{I}_p)\mathrm{vec}\{\sum_{k=0}^{\infty} i^k(\boldsymbol{\Sigma}\mathbf{M}(\mathbf{T}))^k\}\\ &= n\sum_{k=0}^{\infty} i^{k+1}V^2((\boldsymbol{\Sigma}\mathbf{M}(\mathbf{T}))^k\boldsymbol{\Sigma}).\end{aligned} \tag{2.4.50}$$

Because $\mathbf{M}(\mathbf{T})$ is a linear function in $\mathbf{T}$, it follows from (2.4.50) and the definition of cumulants that

$$c_k[\mathbf{W}] = n\frac{d^{k-1}V^2((\boldsymbol{\Sigma}\mathbf{M}(\mathbf{T}))^{k-1}\boldsymbol{\Sigma})}{dV^2(\mathbf{T})^{k-1}}, \qquad k = 2, 3, \ldots \tag{2.4.51}$$

Now equality (2.4.51) is studied when $k = 2$. Straightforward calculations yield

$$c_2[\mathbf{W}] = n\frac{dV^2(\boldsymbol{\Sigma}\mathbf{M}(\mathbf{T})\boldsymbol{\Sigma})}{dV^2(\mathbf{T})} = n\frac{d\mathbf{M}(\mathbf{T})}{dV^2(\mathbf{T})}(\boldsymbol{\Sigma}\otimes\boldsymbol{\Sigma})\mathbf{G}'_p = n\mathbf{J}_p(\boldsymbol{\Sigma}\otimes\boldsymbol{\Sigma})\mathbf{G}'_p.$$

Thus, the second order cumulant is obtained which could also have been presented via Theorem 2.4.14 (ii).

For the third order cumulant the product in (2.4.51) has to be differentiated twice. Hence,

$$\begin{aligned}c_3[\mathbf{W}] =& n\frac{d}{dV^2(\mathbf{T})}\frac{dV^2(\boldsymbol{\Sigma}\mathbf{M}(\mathbf{T})\boldsymbol{\Sigma}\mathbf{M}(\mathbf{T})\boldsymbol{\Sigma})}{dV^2(\mathbf{T})} = n\frac{d}{dV^2(\mathbf{T})}\frac{d(\boldsymbol{\Sigma}\mathbf{M}(\mathbf{T})\boldsymbol{\Sigma}\mathbf{M}(\mathbf{T})\boldsymbol{\Sigma})}{dV^2(\mathbf{T})}\mathbf{G}'_p\\ =& n\frac{d}{dV^2(\mathbf{T})}\frac{d\mathbf{M}(\mathbf{T})}{dV^2(\mathbf{T})}\{(\boldsymbol{\Sigma}\mathbf{M}(\mathbf{T})\boldsymbol{\Sigma}\otimes\boldsymbol{\Sigma}) + (\boldsymbol{\Sigma}\otimes\boldsymbol{\Sigma}\mathbf{M}(\mathbf{T})\boldsymbol{\Sigma})\}\mathbf{G}'_p\\ =& n\frac{d}{dV^2(\mathbf{T})}\{\mathbf{J}_p((\boldsymbol{\Sigma}\mathbf{M}(\mathbf{T})\boldsymbol{\Sigma}\otimes\boldsymbol{\Sigma}) + (\boldsymbol{\Sigma}\otimes\boldsymbol{\Sigma}\mathbf{M}(\mathbf{T})\boldsymbol{\Sigma}))\mathbf{G}'_p\}\\ =& n(\frac{d(\boldsymbol{\Sigma}\mathbf{M}(\mathbf{T})\boldsymbol{\Sigma}\otimes\boldsymbol{\Sigma})}{dV^2(\mathbf{T})} + \frac{d(\boldsymbol{\Sigma}\otimes\boldsymbol{\Sigma}\mathbf{M}(\mathbf{T})\boldsymbol{\Sigma})}{dV^2(\mathbf{T})})(\mathbf{G}'_p\otimes\mathbf{J}'_p)\\ =& n(\frac{d(\boldsymbol{\Sigma}\mathbf{M}(\mathbf{T})\boldsymbol{\Sigma})}{dV^2(\mathbf{T})}\otimes\mathrm{vec}'\boldsymbol{\Sigma})(\mathbf{I}+\mathbf{K}_{p^2,p^2})(\mathbf{I}_p\otimes\mathbf{K}_{p,p}\otimes\mathbf{I}_p)(\mathbf{G}'_p\otimes\mathbf{J}'_p)\\ =& n\mathbf{J}_p(\boldsymbol{\Sigma}\otimes\boldsymbol{\Sigma}\otimes\mathrm{vec}'\boldsymbol{\Sigma})(\mathbf{I}+\mathbf{K}_{p^2,p^2})(\mathbf{I}_p\otimes\mathbf{K}_{p,p}\otimes\mathbf{I}_p)(\mathbf{G}'_p\otimes\mathbf{J}'_p).\end{aligned} \tag{2.4.52}$$

Finally, we consider the fourth order cumulants. The third order derivative of the expression in (2.4.51) gives $c_4[\mathbf{W}]$ and we note that

$$\begin{aligned}c_4[\mathbf{W}] =& n\frac{d^3\{V^2((\boldsymbol{\Sigma}\mathbf{M}(\mathbf{T}))^3\boldsymbol{\Sigma})\}}{dV^2(\mathbf{T})^3} = n\frac{d^2}{V^2(\mathbf{T})^2}\left\{\frac{d\{(\boldsymbol{\Sigma}\mathbf{M}(\mathbf{T}))^3\boldsymbol{\Sigma}\}}{dV^2(\mathbf{T})}\mathbf{G}'_p\right\}\\ =& n\frac{d^2}{V^2(\mathbf{T})}\left\{\frac{d\{(\boldsymbol{\Sigma}\mathbf{M}(\mathbf{T}))^2\boldsymbol{\Sigma}\otimes\boldsymbol{\Sigma} + (\boldsymbol{\Sigma}\mathbf{M}(\mathbf{T})\boldsymbol{\Sigma})^{\otimes 2} + \boldsymbol{\Sigma}\otimes(\boldsymbol{\Sigma}\mathbf{M}(\mathbf{T}))^2\boldsymbol{\Sigma}\}}{dV^2(\mathbf{T})}\mathbf{G}'_p\right\}\\ =& n\frac{d^2}{V^2(\mathbf{T})}\left\{\frac{d\{(\boldsymbol{\Sigma}\mathbf{M}(\mathbf{T}))^2\boldsymbol{\Sigma}\otimes\boldsymbol{\Sigma}\}}{dV^2(\mathbf{T})}(\mathbf{I}+\mathbf{K}_{p,p}\otimes\mathbf{K}_{p,p})(\mathbf{G}'_p\otimes\mathbf{J}'_p)\right\}\\ &+ n\frac{d^2}{V^2(\mathbf{T})}\left\{\frac{d\{(\boldsymbol{\Sigma}\mathbf{M}(\mathbf{T})\boldsymbol{\Sigma})^{\otimes 2}\}}{dV^2(\mathbf{T})}(\mathbf{G}'_p\otimes\mathbf{J}'_p)\right\},\end{aligned} \tag{2.4.53}$$

since

$$\mathrm{vec}'(\mathbf{\Sigma}\otimes(\mathbf{\Sigma M(T)})^2\mathbf{\Sigma})(\mathbf{K}_{p,p}\otimes\mathbf{K}_{p,p})=\mathrm{vec}'((\mathbf{\Sigma M(T)})^2\mathbf{\Sigma}\otimes\mathbf{\Sigma}).$$

When proceeding with the calculations, we obtain

$$\begin{aligned}c_4[\mathbf{W}]=&n\frac{d}{dV^2(\mathbf{T})}\Big\{(\frac{d((\mathbf{\Sigma M(T)})^2\mathbf{\Sigma})}{dV^2(\mathbf{T})}\otimes\mathrm{vec}'\mathbf{\Sigma})(\mathbf{I}+\mathbf{K}_{p,p}\otimes\mathbf{K}_{p,p})(\mathbf{G}_p'\otimes\mathbf{J}_p')\Big\}\\&+n\frac{d}{dV^2(\mathbf{T})}\Big\{(\frac{d(\mathbf{\Sigma M(T)\Sigma})}{dV^2(\mathbf{T})}\otimes\mathrm{vec}'(\mathbf{\Sigma}M(\mathbf{T})\mathbf{\Sigma}))(\mathbf{I}+\mathbf{K}_{p^2,p^2})\\&\times(\mathbf{I}\otimes\mathbf{K}_{p,p}\otimes\mathbf{I})(\mathbf{G}_p'\otimes\mathbf{J}_p')\Big\},\end{aligned}$$

which is equivalent to

$$\begin{aligned}&c_4[\mathbf{W}]\\&=n\frac{d}{dV^2(\mathbf{T})}\Big\{(\frac{d\mathbf{M(T)}}{dV^2(\mathbf{T})}(\mathbf{\Sigma M(T)\Sigma}\otimes\mathbf{\Sigma}+\mathbf{\Sigma}\otimes\mathbf{\Sigma M(T)\Sigma})\otimes\mathrm{vec}'\mathbf{\Sigma})\\&\quad\times(\mathbf{I}+\mathbf{K}_{p,p}\otimes\mathbf{K}_{p,p})(\mathbf{G}_p'\otimes\mathbf{J}_p')\Big\}\\&\quad+n\frac{d}{dV^2(\mathbf{T})}\Big\{(\frac{d\mathbf{M(T)}}{dV^2(\mathbf{T})}(\mathbf{\Sigma}\otimes\mathbf{\Sigma})\otimes\mathrm{vec}'(\mathbf{\Sigma M(T)\Sigma}))(\mathbf{I}+\mathbf{K}_{p^2,p^2})\\&\quad\times(\mathbf{I}\otimes\mathbf{K}_{p,p}\otimes\mathbf{I})(\mathbf{G}_p'\otimes\mathbf{J}_p')\Big\}\\&=n\frac{d((\mathbf{\Sigma M(T)\Sigma}\otimes\mathbf{\Sigma}+\mathbf{\Sigma}\otimes\mathbf{\Sigma M(T)\Sigma})\otimes\mathrm{vec}'\mathbf{\Sigma})}{dV^2(\mathbf{T})}\\&\quad\times\{(\mathbf{I}+\mathbf{K}_{p,p}\otimes\mathbf{K}_{p,p})(\mathbf{G}_p'\otimes\mathbf{J}_p')\otimes\mathbf{J}_p'\}\\&\quad+n\frac{d\mathbf{\Sigma}\otimes\mathbf{\Sigma}\otimes\mathrm{vec}'(\mathbf{\Sigma M(T)\Sigma})}{dV^2(\mathbf{T})}\{(\mathbf{I}+\mathbf{K}_{p^2,p^2})(\mathbf{I}\otimes\mathbf{K}_{p,p}\otimes\mathbf{I})(\mathbf{G}_p'\otimes\mathbf{J}_p')\otimes\mathbf{J}_p'\}.\end{aligned}\tag{2.4.54}$$

Now

$$\begin{aligned}&\mathrm{vec}'(\mathbf{\Sigma M(T)\Sigma}\otimes\mathbf{\Sigma}\otimes\mathrm{vec}'\mathbf{\Sigma}+\mathbf{\Sigma}\otimes\mathbf{\Sigma M(T)\Sigma}\otimes\mathrm{vec}'\mathbf{\Sigma})\\&\quad=\mathrm{vec}'(\mathbf{\Sigma M(T)\Sigma}\otimes\mathbf{\Sigma}\otimes\mathrm{vec}'\mathbf{\Sigma})(\mathbf{I}+\mathbf{K}_{p,p}\otimes\mathbf{I}_{p^2}\otimes\mathbf{K}_{p,p})\end{aligned}\tag{2.4.55}$$

and

$$\begin{aligned}\frac{d\mathbf{\Sigma M(T)\Sigma}\otimes\mathbf{\Sigma}\otimes\mathrm{vec}'\mathbf{\Sigma}}{dV^2(\mathbf{T})}=&\mathbf{J}_p(\mathbf{\Sigma}\otimes\mathbf{\Sigma})\otimes\mathrm{vec}'(\mathbf{\Sigma}\otimes\mathrm{vec}'\mathbf{\Sigma})\\=&\mathbf{J}_p(\mathbf{\Sigma}\otimes\mathbf{\Sigma}\otimes(\mathrm{vec}'\mathbf{\Sigma})^{\otimes2}(\mathbf{I}_{p^3}\otimes\mathbf{K}_{p,p^2}).\end{aligned}\tag{2.4.56}$$

Thus, from (2.4.55) and (2.4.56) it follows that the first term in (2.4.54) equals

$$\begin{aligned}&n\mathbf{J}_p(\mathbf{\Sigma}\otimes\mathbf{\Sigma}\otimes(\mathrm{vec}'\mathbf{\Sigma})^{\otimes2})(\mathbf{I}_{p^3}\otimes\mathbf{K}_{p,p^2})(\mathbf{I}+\mathbf{K}_{p,p}\otimes\mathbf{I}_{p^2}\otimes\mathbf{K}_{p,p})\\&\qquad\times(\mathbf{I}+\mathbf{K}_{p,p}\otimes\mathbf{K}_{p,p}\otimes\mathbf{I}_{p^2})(\mathbf{G}_p'\otimes\mathbf{J}_p'\otimes\mathbf{J}_p').\end{aligned}\tag{2.4.57}$$

For the second term we get

$$\begin{aligned}\frac{d(\boldsymbol{\Sigma}\otimes\boldsymbol{\Sigma}\otimes \mathrm{vec}'(\boldsymbol{\Sigma}\mathbf{M}(\mathbf{T})\boldsymbol{\Sigma}))}{dV^2(\mathbf{T})} &= \left(\frac{d(\boldsymbol{\Sigma}\mathbf{M}(\mathbf{T})\boldsymbol{\Sigma})}{dV^2(\mathbf{T})}\otimes\boldsymbol{\Sigma}\otimes\boldsymbol{\Sigma}\right)(\mathbf{K}_{p^2,p}\otimes\mathbf{I}_p^3)\\ &= \frac{d(\boldsymbol{\Sigma}\mathbf{M}(\mathbf{T})\boldsymbol{\Sigma}\otimes\boldsymbol{\Sigma}\otimes\boldsymbol{\Sigma})}{dV^2(\mathbf{T})}(\mathbf{K}_{p^2,p}\otimes\mathbf{I}_p^3)\\ &\underset{(1.4.24)}{=} \mathbf{J}_p(\boldsymbol{\Sigma}\otimes\boldsymbol{\Sigma}\otimes\mathrm{vec}'(\boldsymbol{\Sigma}\otimes\boldsymbol{\Sigma}))(\mathbf{K}_{p^2,p}\otimes\mathbf{I}_p^3).\end{aligned} \quad (2.4.58)$$

Hence, (2.4.58) gives an expression for the second term in (2.4.54), i.e.

$$\begin{aligned}&n\mathbf{J}_p(\boldsymbol{\Sigma}\otimes\boldsymbol{\Sigma}\otimes\mathrm{vec}'(\boldsymbol{\Sigma}\otimes\boldsymbol{\Sigma}))(\mathbf{K}_{p^2,p}\otimes\mathbf{I}_p^3)(\mathbf{I}+\mathbf{K}_{p,p}\otimes\mathbf{I}_{p^2}\otimes\mathbf{K}_{p,p})\\ &\qquad(\mathbf{I}+\mathbf{K}_{p,p}\otimes\mathbf{K}_{p,p}\otimes\mathbf{I}_{p^2})(\mathbf{G}_p'\otimes\mathbf{J}_p'\otimes\mathbf{J}_p').\end{aligned} \quad (2.4.59)$$

Summing the expressions (2.4.57) and (2.4.59) establishes $c_4[\mathbf{W}]$. ∎

2.4.7 *Derivatives of the Wishart density*

When approximating densities with the Wishart density we need, besides the first cumulants, also the first derivatives of the Wishart density. In this paragraph we will use our standard version of matrix derivative which differs somewhat from the one which was used in §2.4.5. The reason is that now the derivative should be used in a Fourier transform, while in the previous paragraph the derivative was used as an artificial operator. Let $f_{\mathbf{W}}(\mathbf{W})$ be the density function of the Wishart matrix.

Theorem 2.4.17. *Let* $\mathbf{W}\sim W_p(\boldsymbol{\Sigma},n)$. *Then*

$$\frac{d^k f_{\mathbf{W}}(\mathbf{W})}{dV^2(\mathbf{W})^k} = (-1)^k\mathbf{L}_k(\mathbf{W},\boldsymbol{\Sigma})f_{\mathbf{W}}(\mathbf{W}), \quad k=0,1,2,\ldots, \quad (2.4.60)$$

where $V^2(\mathbf{W})$ *is defined in (1.3.75). For* $k=0,1,2,3$ *the matrices* $L_k(\mathbf{W},\boldsymbol{\Sigma})$ *are of the form*

$$L_0(\mathbf{W},\boldsymbol{\Sigma}) = 1,$$

$$\mathbf{L}_1(\mathbf{W},\boldsymbol{\Sigma}) = -\tfrac{1}{2}\mathbf{G}_p\mathbf{H}_p\mathrm{vec}(s\mathbf{W}^{-1}-\boldsymbol{\Sigma}^{-1}), \quad (2.4.61)$$

$$\begin{aligned}\mathbf{L}_2(\mathbf{W},\boldsymbol{\Sigma}) = -\tfrac{1}{2}\mathbf{G}_p\mathbf{H}_p\{&s(\mathbf{W}^{-1}\otimes\mathbf{W}^{-1})\\ &-\tfrac{1}{2}\mathrm{vec}(s\mathbf{W}^{-1}-\boldsymbol{\Sigma}^{-1})\mathrm{vec}'(s\mathbf{W}^{-1}-\boldsymbol{\Sigma}^{-1})\}\mathbf{H}_p\mathbf{G}_p',\end{aligned} \quad (2.4.62)$$

$$\begin{aligned}\mathbf{L}_3(\mathbf{W},\boldsymbol{\Sigma}) &= -\tfrac{1}{2}\mathbf{G}_p\mathbf{H}_p\\ &\Big\{s(\mathbf{W}^{-1}\otimes\mathbf{W}^{-1}\otimes\mathrm{vec}'\mathbf{W}^{-1}+\mathrm{vec}'\mathbf{W}^{-1}\otimes\mathbf{W}^{-1}\otimes\mathbf{W}^{-1})(\mathbf{I}_p\otimes\mathbf{K}_{p,p}\otimes\mathbf{I}_p)\\ &-\tfrac{s}{2}(\mathbf{W}^{-1}\otimes\mathbf{W}^{-1})\{(\mathbf{I}_{p^2}\otimes\mathrm{vec}'(s\mathbf{W}^{-1}-\boldsymbol{\Sigma}^{-1}))+(\mathrm{vec}'(s\mathbf{W}^{-1}-\boldsymbol{\Sigma}^{-1})\otimes\mathbf{I}_{p^2})\}\\ &-\tfrac{s}{2}\mathrm{vec}(s\mathbf{W}^{-1}-\boldsymbol{\Sigma}^{-1})\mathrm{vec}'(\mathbf{W}^{-1}\otimes\mathbf{W}^{-1})\\ &+\tfrac{1}{4}\mathrm{vec}(s\mathbf{W}^{-1}-\boldsymbol{\Sigma}^{-1})(\mathrm{vec}'(s\mathbf{W}^{-1}-\boldsymbol{\Sigma}^{-1})\otimes\mathrm{vec}'(s\mathbf{W}^{-1}-\boldsymbol{\Sigma}^{-1}))\Big\}(\mathbf{H}_p\mathbf{G}_p')^{\otimes 2},\end{aligned} \quad (2.4.63)$$

where $s = n - p - 1$, $\mathbf{K}_{p,p}$ is the commutation matrix given by (1.3.11), $\mathbf{G}_p$ is defined by (1.3.49) and (1.3.50), and $\mathbf{H}_p = \mathbf{I} + \mathbf{K}_{p,p} - (\mathbf{K}_{p,p})_d$.

PROOF: Consider the Wishart density (2.4.12), which is of the form

$$f_{\mathbf{W}}(\mathbf{W}) = c|\mathbf{W}|^{\frac{s}{2}} e^{-\frac{1}{2}\mathrm{tr}(\boldsymbol{\Sigma}^{-1}\mathbf{W})}, \quad \mathbf{W} > 0,$$

where

$$c = \frac{1}{2^{\frac{pn}{2}}\boldsymbol{\Gamma}_p(\frac{n}{2})|\boldsymbol{\Sigma}|^{\frac{n}{2}}}.$$

To obtain (2.4.61), (2.4.62) and (2.4.63) we must differentiate the Wishart density three times. However, before carrying out these calculations it is noted that by (1.4.30) and (1.4.28)

$$\begin{aligned}\frac{d|\mathbf{W}|^a}{dV^2(\mathbf{W})} =& a\mathbf{G}_p\mathbf{H}_p\mathrm{vec}\mathbf{W}^{-1}|\mathbf{W}|^a,\\ \frac{de^{a\mathrm{tr}(\mathbf{W}\boldsymbol{\Sigma}^{-1})}}{dV^2(\mathbf{W})} =& a\mathbf{G}_p\mathbf{H}_p\mathrm{vec}\boldsymbol{\Sigma}^{-1}e^{a\mathrm{tr}(\mathbf{W}\boldsymbol{\Sigma}^{-1})}\end{aligned}$$

and therefore (2.4.60) must hold for some function $\mathbf{L}_k(\mathbf{W},\boldsymbol{\Sigma})$.
Observe that

$$\frac{d^0 f_{\mathbf{W}}(\mathbf{W})}{dV^2(\mathbf{W})^0} = f_{\mathbf{W}}(\mathbf{W}) = (-1)^0 \times 1 \times f_{\mathbf{W}}(\mathbf{W})$$

implies that

$$L_0(\mathbf{W},\boldsymbol{\Sigma}) = 1.$$

The vector $\mathbf{L}_1(\mathbf{W},\boldsymbol{\Sigma})$ will be obtained from the first order derivative. It is observed that

$$\begin{aligned}\frac{df_{\mathbf{W}}(\mathbf{W})}{dV^2(\mathbf{W})} =& c\frac{d|\mathbf{W}|^{\frac{s}{2}}}{dV^2(\mathbf{W})}e^{-\frac{1}{2}\mathrm{tr}(\boldsymbol{\Sigma}^{-1}\mathbf{W})} + c|\mathbf{W}|^{\frac{s}{2}}\frac{de^{-\frac{1}{2}\mathrm{tr}(\boldsymbol{\Sigma}^{-1}\mathbf{W})}}{dV^2(\mathbf{W})}\\ =& \tfrac{s}{2}\mathbf{G}_p\mathbf{H}_p\mathrm{vec}\mathbf{W}^{-1}f_{\mathbf{W}}(\mathbf{W}) - \tfrac{1}{2}\mathbf{G}_p\mathbf{H}_p\mathrm{vec}\boldsymbol{\Sigma}^{-1}f_{\mathbf{W}}(\mathbf{W})\\ =& \tfrac{1}{2}\mathbf{G}_p\mathbf{H}_p(s\mathrm{vec}\mathbf{W}^{-1} - \mathrm{vec}\boldsymbol{\Sigma}^{-1})f_{\mathbf{W}}(\mathbf{W}).\end{aligned}$$

Thus, $\mathbf{L}_1(\mathbf{W},\boldsymbol{\Sigma})$ is given by (2.4.61).
In order to find $\mathbf{L}_2(\mathbf{W},\boldsymbol{\Sigma})$, we have to take the second order derivative of the Wishart density function and obtain

$$\begin{aligned}\frac{d^2 f_{\mathbf{W}}(\mathbf{W})}{dV^2(\mathbf{W})^2} =& \frac{d\{-\mathbf{L}_1(\mathbf{W},\boldsymbol{\Sigma})f_{\mathbf{W}}(\mathbf{W})\}}{dV^2(\mathbf{W})}\\ =& \frac{d(-\mathbf{L}_1(\mathbf{W},\boldsymbol{\Sigma}))}{dV^2(\mathbf{W})}f_{\mathbf{W}}(\mathbf{W}) - \frac{df_{\mathbf{W}}(\mathbf{W})}{dV^2(\mathbf{W})}\mathbf{L}_1'(\mathbf{W},\boldsymbol{\Sigma})\\ =& \frac{d\mathbf{W}^{-1}}{dV^2(\mathbf{W})}\mathbf{H}_p\mathbf{G}_p'\tfrac{s}{2}f_{\mathbf{W}}(\mathbf{W}) + \mathbf{L}_1(\mathbf{W},\boldsymbol{\Sigma})\mathbf{L}_1'(\mathbf{W},\boldsymbol{\Sigma})f_{\mathbf{W}}(\mathbf{W}),\end{aligned} \quad (2.4.64)$$

which implies that $\mathbf{L}_2(\mathbf{W},\boldsymbol{\Sigma})$ is given by (2.4.62).
For the third derivative we have the following chain of equalities:

$$
\begin{aligned}
\frac{d^3 f_{\mathbf{W}}(\mathbf{W})}{dV^2(\mathbf{W})^3} &= \frac{d}{dV^2(\mathbf{W})}\frac{d^2 f_{\mathbf{W}}(\mathbf{W})}{dV^2(\mathbf{W})^2} \\
&\underset{(2.4.64)}{=} -\frac{s}{2}\frac{d(\mathbf{W}^{-1}\otimes\mathbf{W}^{-1})}{dV^2(\mathbf{W})}(\mathbf{H}_p\mathbf{G}_p'\otimes\mathbf{H}_p\mathbf{G}_p')f_{\mathbf{W}}(\mathbf{W}) \\
&\quad -\frac{s}{2}\frac{df_{\mathbf{W}}(\mathbf{W})}{dV^2(\mathbf{W})}\mathrm{vec}'(\mathbf{W}^{-1}\otimes\mathbf{W}^{-1})(\mathbf{H}_p\mathbf{G}_p'\otimes\mathbf{H}_p\mathbf{G}_p') \\
&\quad +\frac{d\{\mathbf{L}_1(\mathbf{W},\boldsymbol{\Sigma})f_{\mathbf{W}}(\mathbf{W})\}}{dV^2(\mathbf{W})}(\mathbf{I}_{p^2}\otimes\mathbf{L}_1'(\mathbf{W},\boldsymbol{\Sigma})) \\
&\quad +\frac{d\mathbf{L}_1(\mathbf{W},\boldsymbol{\Sigma})}{dV^2(\mathbf{W})}(\mathbf{L}_1(\mathbf{W},\boldsymbol{\Sigma})\otimes\mathbf{I}_{p^2})f_{\mathbf{W}}(\mathbf{W}) \\
&= \frac{1}{2}\mathbf{G}_p\mathbf{H}_p\Big\{s(\mathbf{W}^{-1}\otimes\mathbf{W}^{-1}\otimes\mathrm{vec}'\mathbf{W}^{-1}+\mathrm{vec}'\mathbf{W}^{-1}\otimes\mathbf{W}^{-1}\otimes\mathbf{W}^{-1}) \\
&\quad \times(\mathbf{I}_p\otimes\mathbf{K}_{p,p}\otimes\mathbf{I}_p) \\
&\quad -\frac{1}{2}\mathrm{vec}(s\mathbf{W}^{-1}-\boldsymbol{\Sigma}^{-1})\mathrm{vec}'(\mathbf{W}^{-1}\otimes\mathbf{W}^{-1})\Big\}(\mathbf{H}_p\mathbf{G}_p'\otimes\mathbf{H}_p\mathbf{G}_p')f_{\mathbf{W}}(\mathbf{W}) \\
&\quad -\frac{s}{4}\mathbf{G}_p\mathbf{H}_p(\mathbf{W}^{-1}\otimes\mathbf{W}^{-1})(\mathbf{I}_{p^2}\otimes\mathrm{vec}'(s\mathbf{W}^{-1}-\boldsymbol{\Sigma}^{-1}))f_{\mathbf{W}}(\mathbf{W}) \\
&\quad -\frac{1}{8}\mathbf{G}_p\mathbf{H}_p\mathrm{vec}(s\mathbf{W}^{-1}-\boldsymbol{\Sigma}^{-1})(\mathrm{vec}'(s\mathbf{W}^{-1}-\boldsymbol{\Sigma}^{-1})\otimes\mathrm{vec}'(s\mathbf{W}^{-1}-\boldsymbol{\Sigma}^{-1})) \\
&\quad \times(\mathbf{H}_p\mathbf{G}_p'\otimes\mathbf{H}_p\mathbf{G}_p')f_{\mathbf{W}}(\mathbf{W}) \\
&\quad -\frac{s}{4}\mathbf{G}_p\mathbf{H}_p(\mathbf{W}^{-1}\otimes\mathbf{W}^{-1})(\mathrm{vec}'(s\mathbf{W}^{-1}-\boldsymbol{\Sigma}^{-1})\otimes\mathbf{I}_{p^2}) \\
&\quad \times(\mathbf{H}_p\mathbf{G}_p'\otimes\mathbf{H}_p\mathbf{G}_p')f_{\mathbf{W}}(\mathbf{W}),
\end{aligned}
$$

which is identical to (2.4.63). ∎

2.4.8 *Centered Wishart distribution*

Let $\mathbf{W}\sim W_p(\boldsymbol{\Sigma},n)$. We are going to use the Wishart density as an approximating density in Sections 3.2 and 3.3. One complication with the Wishart approximation is that the derivatives of the density of a Wishart distributed matrix, or more precisely $\mathbf{L}_k(\mathbf{W},\boldsymbol{\Sigma})$ given by (2.4.60), increase with n. When considering $\mathbf{L}_k(\mathbf{W},\boldsymbol{\Sigma})$, we first point out that $\mathbf{W}$ in $\mathbf{L}_k(\mathbf{W},\boldsymbol{\Sigma})$ can be any $\mathbf{W}$ which is positive definite. On the other hand, from a theoretical point of view we must always have $\mathbf{W}$ of the form: $\mathbf{W}=\sum_{i=1}^n\mathbf{x}_i\mathbf{x}_i'$ and then, under some conditions, it can be shown that the Wishart density is $O(n^{-p/2})$. The result follows by an application of Stirling's formula to $\boldsymbol{\Gamma}_p(n/2)$ in (2.4.12). Moreover, when $n\to\infty$, $1/n\mathbf{W}-\boldsymbol{\Sigma}\to\mathbf{0}$ in probability. Hence, asymptotic properties indicate that the derivatives of the Wishart density, although they depend on n, are fairly stable. So, from an asymptotic point of view, the Wishart density can be used. However, when approximating densities, we are often interested in tail probabilities and in many cases it will not be realistic to suppose, when approximating with the Wishart distribution, that $\mathbf{W} - n\mathbf{\Sigma}$ is small. Generally, we see that $\mathbf{L}_k(\mathbf{W}, \mathbf{\Sigma})$ is an increasing function in n. However, it seems wise to adjust the density so that the derivatives decrease with n. Indeed, some authors have not observed this negative property of the Wishart distribution. In order to overcome this problem we propose a translation of the Wishart matrix, so that a centered version is used, i.e.

$$\mathbf{V} = \mathbf{W} - n\mathbf{\Sigma}.$$

From Theorem 2.4.6 it follows that the matrix $\mathbf{V}$ has the density function

$$f_{\mathbf{V}}(\mathbf{V}) = \begin{cases} \dfrac{1}{2^{\frac{pn}{2}}\mathbf{\Gamma}_p(\frac{n}{2})|\mathbf{\Sigma}|^{\frac{n}{2}}}|\mathbf{V}+n\mathbf{\Sigma}|^{\frac{n-p-1}{2}}e^{-\frac{1}{2}tr(\mathbf{\Sigma}^{-1}(\mathbf{V}+n\mathbf{\Sigma}))}, & \mathbf{V}+n\mathbf{\Sigma} > 0, \\ 0, & \text{otherwise.} \end{cases} \tag{2.4.65}$$

If a symmetric matrix $\mathbf{V}$ has the density (2.4.65), we say that $\mathbf{V}$ follows a centered Wishart distribution. It is important to observe that since we have performed a translation with the mean, the first moment of $\mathbf{V}$ equals zero and the cumulants of higher order are identical to the corresponding cumulants of the Wishart distribution.

In order to use the density of the centered Wishart distribution when approximating an unknown distribution, we need the first derivatives of the density functions. The derivatives of $f_{\mathbf{V}}(\mathbf{X})$ can be easily obtained by simple transformations of $\mathbf{L}_i(\mathbf{X}, \mathbf{\Sigma})$, if we take into account the expressions of the densities (2.4.12) and (2.4.65). Analogously to Theorem 2.4.17 we have

Lemma 2.4.2. *Let $\mathbf{V} = \mathbf{W} - n\mathbf{\Sigma}$, where $\mathbf{W} \sim W_p(\mathbf{\Sigma}, n)$, let $\mathbf{G}_p$ and $\mathbf{H}_p$ be as in Theorem 2.4.17. Then*

$$f_{\mathbf{V}}^{(k)}(\mathbf{V}) = \frac{d^k f_{\mathbf{V}}(\mathbf{V})}{dV^2\mathbf{V}^k} = (-1)^k \mathbf{L}_k^\star(\mathbf{V}, \mathbf{\Sigma}) f_{\mathbf{V}}(\mathbf{V}), \tag{2.4.66}$$

where

$$\mathbf{L}_k^\star(\mathbf{V}, \mathbf{\Sigma}) = \mathbf{L}_k(\mathbf{V} + n\mathbf{\Sigma}, \mathbf{\Sigma}), \quad k = 0, 1, 2, \ldots$$

and $\mathbf{L}_k(\mathbf{W}, \mathbf{\Sigma})$ are given in Theorem 2.4.17.

For $n \gg p$

$$\mathbf{L}_1^\star(\mathbf{V}, \mathbf{\Sigma}) \approx \tfrac{1}{2n}\mathbf{G}_p\mathbf{H}_p\mathrm{vec}(\mathbf{B}_1), \tag{2.4.67}$$
$$\mathbf{L}_2^\star(\mathbf{V}, \mathbf{\Sigma}) \approx -\tfrac{1}{2n}\mathbf{G}_p\mathbf{H}_p\mathbf{B}_2\mathbf{H}_p\mathbf{G}_p' - \tfrac{1}{4n^2}\mathbf{G}_p\mathbf{H}_p\mathrm{vec}(\mathbf{B}_1)\mathrm{vec}'(\mathbf{B}_1)\mathbf{H}_p\mathbf{G}_p', \tag{2.4.68}$$

where

$$\begin{aligned} \mathbf{B}_1 &= \mathbf{\Sigma}^{-1}\mathbf{V}^{\frac{1}{2}}(\tfrac{1}{n}\mathbf{V}^{\frac{1}{2}}\mathbf{\Sigma}^{-1}\mathbf{V}^{\frac{1}{2}} + \mathbf{I}_p)^{-1}\mathbf{V}^{\frac{1}{2}}\mathbf{\Sigma}^{-1}, \\ \mathbf{B}_2 &= (\mathbf{V}/n + \mathbf{\Sigma})^{-1} \otimes (\mathbf{V}/n + \mathbf{\Sigma})^{-1}. \end{aligned}$$

For $k = 3, 4, \ldots$ the matrix $L_k^\star(\mathbf{V}, \mathbf{\Sigma})$ is of order $n^{-(k-1)}$.

PROOF: The relation in (2.4.66) follows directly from (2.4.60) in Theorem 2.4.17 if we replace $\mathbf{W}$ with the expression $\mathbf{V} + n\mathbf{\Sigma}$, since $\mathbf{W} = \mathbf{V} + n\mathbf{\Sigma}$. For $\mathbf{L}_1^\star(\mathbf{V}, \mathbf{\Sigma})$ we have

$$\begin{aligned}\mathbf{L}_1^\star(\mathbf{V}, \mathbf{\Sigma}) &= -\frac{1}{2}\mathbf{G}_p\mathbf{H}_p\mathrm{vec}\{(n-p-1)(\mathbf{V}+n\mathbf{\Sigma})^{-1} - \mathbf{\Sigma}^{-1}\}\\ &= -\frac{1}{2}\mathbf{G}_p\mathbf{H}_p\mathrm{vec}\{\tfrac{n-p-1}{n}(\mathbf{V}/n+\mathbf{\Sigma})^{-1} - \mathbf{\Sigma}^{-1}\}.\end{aligned}$$

If $n \gg p$, we have

$$\mathbf{L}_1^\star(\mathbf{V}, \mathbf{\Sigma}) \approx -\frac{1}{2}\mathbf{G}_p\mathbf{H}_p\mathrm{vec}\{(\mathbf{V}/n+\mathbf{\Sigma})^{-1} - \mathbf{\Sigma}^{-1}\},$$

and using Proposition 1.3.6 and that $\mathbf{V}$ is p.d., we get

$$\mathbf{L}_1^\star(\mathbf{V}, \mathbf{\Sigma}) \approx \tfrac{1}{2n}\mathbf{G}_p\mathbf{H}_p\mathrm{vec}\{\mathbf{\Sigma}^{-1}\mathbf{V}^{\frac{1}{2}}(\frac{1}{n}\mathbf{V}^{\frac{1}{2}}\mathbf{\Sigma}^{-1}\mathbf{V}^{\frac{1}{2}} + \mathbf{I}_p)^{-1}\mathbf{V}^{\frac{1}{2}}\mathbf{\Sigma}^{-1}\}.$$

Hence, (2.4.67) has been proved.
For $k = 2$ we obtain in a similar way $(s = n-p-1)$

$$\begin{aligned}\mathbf{L}_2^\star(\mathbf{V}, \mathbf{\Sigma}) &= -\frac{1}{2}\mathbf{G}_p\mathbf{H}_p\Big\{s(\mathbf{V}+n\mathbf{\Sigma})^{-1}\otimes(\mathbf{V}+n\mathbf{\Sigma})^{-1}\\ &\quad -\frac{1}{2}\mathrm{vec}(s(\mathbf{V}+n\mathbf{\Sigma})^{-1} - \mathbf{\Sigma}^{-1})\mathrm{vec}'(s(\mathbf{V}+n\mathbf{\Sigma})^{-1} - \mathbf{\Sigma}^{-1})\Big\}\mathbf{H}_p\mathbf{G}_p'\\ &\approx -\frac{1}{2n}\mathbf{G}_p\mathbf{H}_p(\mathbf{V}/n+\mathbf{\Sigma})^{-1}\otimes(\mathbf{V}/n+\mathbf{\Sigma})^{-1}\mathbf{H}_p\mathbf{G}_p\\ &\quad -\frac{1}{4n^2}\mathbf{G}_p\mathbf{H}_p(\mathrm{vec}\mathbf{B}_1\mathrm{vec}'\mathbf{B}_1)\mathbf{H}_p\mathbf{G}_p'.\end{aligned}$$

Thus (2.4.68) is established.
To complete the proof we remark that from (2.4.63) we have, with the help of (2.4.67) and (2.4.68), that $\mathbf{L}_3^\star(\mathbf{V}, \mathbf{\Sigma})$ is of order n^{-2}. From the recursive definition of the matrix derivative the last statement of the lemma is established. ∎

Previously we noted that the order of magnitude of $f_{\mathbf{W}}(\mathbf{W})$ is $O(n^{-1/2})$ when it is supposed that $\mathbf{W} = \sum_{i=1}^n \mathbf{x}_i\mathbf{x}_i'$. Since we have translated the Wishart distribution with $n\mathbf{\Sigma}$, it follows that the order of magnitude is also $O(n^{-1/2})$.
Another property which will be used later is given in the next theorem.

Lemma 2.4.3. *Let $\mathbf{W} \sim W_p(\mathbf{\Sigma}, n)$ and put $\mathbf{V} = \mathbf{W} - n\mathbf{\Sigma}$, where*

$$\mathbf{V} = \begin{pmatrix}\mathbf{V}_{11} & \mathbf{V}_{12}\\ \mathbf{V}_{21} & \mathbf{V}_{22}\end{pmatrix} \qquad \begin{pmatrix} q\times q & q\times(p-q)\\ (p-q)\times q & (p-q)\times(p-q)\end{pmatrix}.$$

Partition $\mathbf{\Sigma}$ in the same way as $\mathbf{V}$ and let

$$\mathbf{W}_{1\cdot 2} = \mathbf{V}_{11} + n\mathbf{\Sigma}_{11} - (\mathbf{V}_{12}+n\mathbf{\Sigma}_{12})(\mathbf{V}_{22}+n\mathbf{\Sigma}_{22})^{-1}(\mathbf{V}_{21}+n\mathbf{\Sigma}_{21}).$$

Then

$$\begin{aligned}\mathbf{W}_{1\cdot 2} &\sim W_q(\boldsymbol{\Sigma}_{1\cdot 2}, n-p-q),\\ \mathbf{V}_{12}|\mathbf{V}_{22} &\sim N_{q,p-q}(\boldsymbol{\Sigma}_{12}\boldsymbol{\Sigma}_{22}^{-1}\mathbf{V}_{22}, \boldsymbol{\Sigma}_{1\cdot 2}, \mathbf{V}_{22}+n\boldsymbol{\Sigma}_{22}),\\ \mathbf{V}_{22}+n\boldsymbol{\Sigma}_{22} = \mathbf{W}_{22} &\sim W_{p-q}(\boldsymbol{\Sigma}_{22}, n)\end{aligned}$$

and $\mathbf{W}_{1\cdot 2}$ *is independent of* $\mathbf{V}_{12}, \mathbf{V}_{22}$.

PROOF: The proof follows from a non-centered version of this lemma which was given as Theorem 2.4.12, since the Jacobian from $\mathbf{W}_{1\cdot 2}, \mathbf{W}_{12}, \mathbf{W}_{22}$ to $\mathbf{W}_{1\cdot 2}, \mathbf{V}_{12}$, $\mathbf{V}_{22}$ equals one. ∎

2.4.9 *Problems*

1. Let $\mathbf{W} \sim W_p(\boldsymbol{\Sigma}, n)$ and $\mathbf{A}$: $p \times q$. Prove that
$$E[\mathbf{A}(\mathbf{A}'\mathbf{W}\mathbf{A})^-\mathbf{A}'\mathbf{W}] = \mathbf{A}(\mathbf{A}'\boldsymbol{\Sigma}\mathbf{A})^-\mathbf{A}'\boldsymbol{\Sigma}$$
and
$$E[\mathbf{A}(\mathbf{A}'\mathbf{W}^{-1}\mathbf{A})^-\mathbf{A}'\mathbf{W}^{-1}] = \mathbf{A}(\mathbf{A}'\boldsymbol{\Sigma}^{-1}\mathbf{A})^-\mathbf{A}'\boldsymbol{\Sigma}^{-1}.$$
2. Let $\mathbf{V} \sim W_p(\mathbf{I}, n)$ and $\mathbf{A}$: $p \times q$. Show that, if $n-p-1>0$,
$$\begin{aligned}E[\mathbf{V}^{-1}\mathbf{A}(\mathbf{A}'\mathbf{V}^{-1}\mathbf{A})^-\mathbf{A}'\mathbf{V}^{-1}] = &\frac{r(\mathbf{A})}{(n-p-1)(n-p+r(\mathbf{A})-1)}\mathbf{I}\\ &+\frac{1}{n-p+r(\mathbf{A})-1}\mathbf{A}(\mathbf{A}'\mathbf{A})^-\mathbf{A}'.\end{aligned}$$
3. Let $\mathbf{V} \sim W_p(\mathbf{I}, n)$ and $\mathbf{A}$: $p \times q$. Prove that, if $n-p+r(A)-1>0$,
$$\begin{aligned}&E[\mathbf{V}^{-1}\mathbf{A}(\mathbf{A}'\mathbf{V}^{-1}\mathbf{A})^-\mathbf{A}'\mathbf{V}^{-1}\mathbf{V}^{-1}\mathbf{A}(\mathbf{A}'\mathbf{V}^{-1}\mathbf{A})^-\mathbf{A}'\mathbf{V}^{-1}]\\ &=\frac{n-1}{n-p+r(\mathbf{A})-1}\mathbf{A}(\mathbf{A}'\mathbf{A})^-\mathbf{A}'.\end{aligned}$$
4. Let $\mathbf{W} \sim W_p(\boldsymbol{\Sigma}, n)$. Show that $E[\mathbf{L}_k(\mathbf{W}, \boldsymbol{\Sigma})] = \mathbf{0}$.
5. Let $\mathbf{W} \sim W_p(\boldsymbol{\Sigma}, n)$. Show that the cumulant function
$$\psi_{\mathrm{tr}\mathbf{W}}(t) = -\tfrac{n}{2}\ln|(\mathbf{I}-2i\,t\boldsymbol{\Sigma})|.$$
6. Find $c_i[\mathrm{tr}\mathbf{W}]$, $i=1,2,3$, using the cumulant function in Problem 5.
7. Derive the Wishart density, under proper assumptions, by starting with Corollary 2.4.4.1 and thereafter applying Theorem 1.4.18.
8. Differentiate the Wishart density with respect to $\boldsymbol{\Sigma}$ and obtain the first two moments of the Wishart distribution by integration.
9. Derive the density of the eigenvalues of $\mathbf{W} \sim W_p(\mathbf{I}, n)$.
10. Derive the density of the eigenvalues of $\mathbf{G}$ in Theorem 2.4.10.
11. Calculate the mean and variance of V in Corollary 2.4.4.2.

CHAPTER III

Distribution Expansions

In statistical estimation theory one of the main problems is the approximation of the distribution of a specific statistic. Even if it is assumed that the observations follow a normal distribution the exact distributions of the statistics of interest are seldom known or they are complicated to obtain and to use. The approximation of the distributions of the eigenvalues and eigenvectors of the sample covariance matrix may serve as an example. While for a normal population the distribution of eigenvalues is known (James, 1960), the distribution of eigenvectors has not yet been described in a convenient manner. Moreover, the distribution of eigenvalues can only be used via approximations, because the density of the exact distribution is expressed as an infinite sum of terms, including expressions of complicated polynomials. Furthermore, the treatment of data via an assumption of normality of the population is too restrictive in many cases. Often the existence of the first few moments is the only assumption which can be made. This implies that distribution free or asymptotic methods will be valuable. For example, approaches based on the asymptotic normal distribution or approaches relying on the chi-square distribution are both important. In this chapter we are going to examine different approximations and expansions. All of them stem from Taylor expansions of important functions in mathematical statistics, such as the characteristic function and the cumulant function as well as some others. The first section treats asymptotic distributions. Here we shall also analyze the Taylor expansion of a random vector. Then we shall deal with multivariate normal approximations. This leads us to the well-known Edgeworth expansions. In the third section we shall give approximation formulas for the density of a random matrix via the density of another random matrix of the same size, such as a matrix normally distributed matrix. The final section, which is a direct extension of Section 3.3, presents an approach to multivariate approximations of densities of random variables of different sizes. Throughout distribution expansions of several well-known statistics from multivariate analysis will be considered as applications.

3.1 ASYMPTOTIC NORMALITY

3.1.1 *Taylor series of a random vector*

In the following we need the notions of convergence in distribution and in probability. In notation we follow Billingsley (1999). Let $\{\mathbf{x}_n\}$ be a sequence of random p-vectors. We shall denote *convergence in distribution* or *weak convergence* of $\{\mathbf{x}_n\}$, when $n \to \infty$, as

$$\mathbf{x}_n \xrightarrow{\mathcal{D}} P_{\mathbf{x}}$$

or

$$\mathbf{x}_n \xrightarrow{\mathcal{D}} \mathbf{x},$$

depending on which notation is more convenient to use. The sequence $\{\mathbf{x}_n\}$ converges in distribution if and only if $F_{\mathbf{x}_n}(\mathbf{y}) \longrightarrow F_{\mathbf{x}}(\mathbf{y})$ at any point $\mathbf{y}$, where the limiting distribution function $F_{\mathbf{x}}(\mathbf{y})$ is continuous. The sequence of random matrices $\{\mathbf{X}_n\}$ converges in distribution to $\mathbf{X}$ if

$$\mathrm{vec}\mathbf{X}_n \xrightarrow{\mathcal{D}} \mathrm{vec}\mathbf{X}.$$

Convergence in probability of the sequence $\{\mathbf{x}_n\}$, when $n \to \infty$, is denoted by

$$\mathbf{x}_n \xrightarrow{\mathcal{P}} \mathbf{x}.$$

This means, with $n \to \infty$, that

$$P\{\omega : \rho(\mathbf{x}_n(\omega), \mathbf{x}(\omega)) > \varepsilon\} \to 0$$

for any $\varepsilon > 0$, where $\rho(\cdot, \cdot)$ is the Euclidean distance in $\mathbb{R}^p$. The convergence in probability of random matrices is traced to random vectors:

$$\mathbf{X}_n \xrightarrow{\mathcal{P}} \mathbf{X},$$

when

$$\mathrm{vec}\mathbf{X}_n \xrightarrow{\mathcal{P}} \mathrm{vec}\mathbf{X}.$$

Let $\{\varepsilon_i\}$ be a sequence of positive numbers and $\{X_i\}$ be a sequence of random variables. Then, following Rao (1973a, pp. 151–152), we write $X_i = o_P(\varepsilon_i)$ if

$$\frac{X_i}{\varepsilon_i} \xrightarrow{\mathcal{P}} 0.$$

For a sequence of random vectors $\{\mathbf{x}_i\}$ we use similar notation: $\mathbf{x}_i = o_P(\varepsilon_i)$, if

$$\frac{\mathbf{x}_i}{\varepsilon_i} \xrightarrow{\mathcal{P}} \mathbf{0}.$$

Let $\{\mathbf{X}_i\}$ be a sequence of random $p \times q$-matrices. Then we write $\mathbf{X}_i = o_P(\varepsilon_i)$, if

$$\mathrm{vec}\mathbf{X}_i = o_P(\varepsilon_i).$$

So, a sequence of matrices is considered via the corresponding notion for random vectors. Similarly, we introduce the concept of $O_P(\cdot)$ in the notation given above (Rao, 1973a, pp. 151–152). We say $X_i = O_P(\varepsilon_i)$ or $\frac{X_i}{\varepsilon_i}$ is *bounded in probability*, if for each δ there exist m_δ and n_δ such that $P(\frac{X_i}{\varepsilon_i} > m_\delta) < \delta$ for $i > n_\delta$. We say that a p-vector $\mathbf{x}_n = O_P(\varepsilon_n)$, if the coordinates $(\mathbf{x}_n)_i = O_P(\varepsilon_n)$, $i = 1, \ldots, p$. A $p \times q$-matrix $\mathbf{X}_n = O_P(\varepsilon_n)$, if $\mathrm{vec}\mathbf{X}_n = O_P(\varepsilon_n)$.

Lemma 3.1.1. *Let $\{\mathbf{X}_n\}$ and $\{\mathbf{Y}_n\}$ be sequences of random matrices with $\mathbf{X}_n \xrightarrow{\mathcal{D}} \mathbf{X}$ and $\mathbf{Y}_n \xrightarrow{\mathcal{P}} \mathbf{0}$. Then, provided the operations are well defined,*

$$\mathbf{X}_n \otimes \mathbf{Y}_n \xrightarrow{\mathcal{P}} \mathbf{0},$$
$$\mathrm{vec}\mathbf{X}_n \mathrm{vec}'\mathbf{Y}_n \xrightarrow{\mathcal{P}} \mathbf{0},$$
$$\mathbf{X}_n + \mathbf{Y}_n \xrightarrow{\mathcal{D}} \mathbf{X}.$$

PROOF: The proof repeats the argument in Rao (1973a, pp. 122–123) for random variables, if we change the expressions $|x|$, $|x-y|$ to the Euclidean distances $\rho(\mathbf{x},\mathbf{0})$ and $\rho(\mathbf{x},\mathbf{y})$, respectively. ∎

Observe that $\mathbf{X}_n$ and $\mathbf{Y}_n$ do not have to be independent. Thus, if for example $\mathbf{X}_n \xrightarrow{\mathcal{D}} \mathbf{X}$ and $\mathbf{g}(\mathbf{X}_n) \xrightarrow{\mathcal{P}} \mathbf{0}$, the asymptotic distribution of $\mathbf{X}_n + \mathbf{g}(\mathbf{X}_n)$ is also $P_{\mathbf{X}}$. In applications the following corollary of Lemma 3.1.1 is useful.

Corollary 3.1.1.1L. *Let $\{\mathbf{X}_n\}$, $\{\mathbf{Y}_n\}$ and $\{\mathbf{Z}_n\}$ be sequences of random matrices, with $\mathbf{X}_n \xrightarrow{\mathcal{D}} \mathbf{X}$, $\mathbf{Y}_n = o_P(\varepsilon_n)$ and $\mathbf{Z}_n = o_P(\varepsilon_n)$, and $\{\varepsilon_n\}$ be a sequence of positive real numbers. Then*

$$\mathbf{X}_n \otimes \mathbf{Y}_n = o_P(\varepsilon_n),$$
$$\mathrm{vec}\mathbf{X}_n \mathrm{vec}'\mathbf{Y}_n = o_P(\varepsilon_n),$$
$$\mathbf{Z}_n \otimes \mathbf{Y}_n = o_P(\varepsilon_n^2).$$

If the sequence $\{\varepsilon_n\}$ is bounded, then

$$\mathbf{X}_n + \mathbf{Y}_n \xrightarrow{\mathcal{D}} \mathbf{X}.$$

PROOF: By assumptions

$$\mathrm{vec}\frac{\mathbf{Y}_n}{\varepsilon_n} \xrightarrow{\mathcal{P}} \mathbf{0}$$

and

$$\mathrm{vec}\frac{\mathbf{Z}_n}{\varepsilon_n} \xrightarrow{\mathcal{P}} \mathbf{0}.$$

The first three statements of the corollary follow now directly from Lemma 3.1.1. If $\{\varepsilon_n\}$ is bounded, $\mathbf{Y}_n \xrightarrow{\mathcal{P}} \mathbf{0}$, and the last statement follows immediately from the lemma. ∎

Corollary 3.1.1.2L. *If $\sqrt{n}(\mathbf{X}_n - \mathbf{A}) \xrightarrow{\mathcal{D}} \mathbf{X}$ and $\mathbf{X}_n - \mathbf{A} = o_P(\varepsilon_n)$ for some constant matrix $\mathbf{A}$, then*

$$n^{\frac{1}{2}(k-1)}(\mathbf{X}_n - \mathbf{A})^{\otimes k} = o_P(\varepsilon_n), \qquad k = 2, 3, \ldots.$$

PROOF: Write

$$n^{\frac{1}{2}(k-1)}(\mathbf{X}_n - \mathbf{A})^{\otimes k} = \left(\sqrt{n}(\mathbf{X}_n - \mathbf{A})\right)^{\otimes k-1} \otimes (\mathbf{X}_n - \mathbf{A})$$

and then Corollary 3.1.1.1L shows the result, since $(\sqrt{n}(\mathbf{X}_n - \mathbf{A}))^{\otimes k-1}$ converges in distribution to $\mathbf{X}^{\otimes(k-1)}$. ∎

Expansions of different multivariate functions of random vectors, which will be applied in the subsequent, are often based on relations given below as Theorem 3.1.1, Theorem 3.1.2 and Theorem 3.1.3.

Theorem 3.1.1. *Let $\{\mathbf{x}_n\}$ and $\{\varepsilon_n\}$ be sequences of random $p-$vectors and positive numbers, respectively, and let $\mathbf{x}_n - \mathbf{a} = o_P(\varepsilon_n)$, where $\varepsilon_n \to 0$ if $n \to \infty$. If the function $\mathbf{g}(\mathbf{x})$ from $\mathbb{R}^p$ to $\mathbb{R}^q$ can be expanded into the Taylor series (1.4.59) at the point $\mathbf{a}$:*

$$\mathbf{g}(\mathbf{x}) = \mathbf{g}(\mathbf{a}) + \sum_{k=1}^{m} \frac{1}{k!}((\mathbf{x}-\mathbf{a})^{\otimes k-1} \otimes \mathbf{I}_q)' \left(\frac{d^k\mathbf{g}(\mathbf{x})}{d\mathbf{x}^k}\right)'\Bigg|_{\mathbf{x}=\mathbf{a}} (\mathbf{x}-\mathbf{a}) + o(\rho^m(\mathbf{x},\mathbf{a})),$$

then

$$\mathbf{g}(\mathbf{x}_n) - \mathbf{g}(\mathbf{a}) - \sum_{k=1}^{m} \frac{1}{k!}((\mathbf{x}_n-\mathbf{a})^{\otimes k-1} \otimes \mathbf{I}_q)' \left(\frac{d^k\mathbf{g}(\mathbf{x}_n)}{d\mathbf{x}_n^k}\right)'\Bigg|_{\mathbf{x}_n=\mathbf{a}} (\mathbf{x}_n-\mathbf{a}) = o_P(\varepsilon_n^m),$$

where the matrix derivative is given in (1.4.41).

PROOF: Denote

$$\mathbf{r}_m(\mathbf{x}_n,\mathbf{a}) = \mathbf{g}(\mathbf{x}_n) - \mathbf{g}(\mathbf{a}) - \sum_{k=1}^{m} \frac{1}{k!}((\mathbf{x}_n-\mathbf{a})^{\otimes k-1} \otimes \mathbf{I}_q)' \left(\frac{d^k\mathbf{g}(\mathbf{x}_n)}{d\mathbf{x}_n^k}\right)'\Bigg|_{\mathbf{x}_n=\mathbf{a}} (\mathbf{x}_n-\mathbf{a}).$$

We have to show that, if $n \to \infty$,

$$\frac{\mathbf{r}_m(\mathbf{x}_n,\mathbf{a})}{\varepsilon_n^m} \xrightarrow{\mathcal{P}} \mathbf{0},$$

i.e.

$$P\left\{\omega : \frac{\rho(\mathbf{r}_m(\mathbf{x}_n(\omega),\mathbf{a}),\mathbf{0})}{\varepsilon_n^m} > \eta\right\} \longrightarrow 0$$

for any $\eta > 0$. Now

$$\begin{aligned} P&\left\{\omega : \frac{\rho(\mathbf{r}_m(\mathbf{x}_n(\omega),\mathbf{a}),\mathbf{0})}{\varepsilon_n^m} > \eta\right\} \\ &= P\left\{\omega : \frac{\rho(\mathbf{r}_m(\mathbf{x}_n(\omega),\mathbf{a}),\mathbf{0})}{\varepsilon_n^m} > \eta,\ \frac{\rho(\mathbf{x}_n(\omega),\mathbf{a})}{\varepsilon_n} > \nu\right\} \\ &\quad + P\left\{\omega : \frac{\rho(\mathbf{r}_m(\mathbf{x}_n(\omega),\mathbf{a}),\mathbf{0})}{\varepsilon_n^m} > \eta,\ \frac{\rho(\mathbf{x}_n(\omega),\mathbf{a})}{\varepsilon_n} \leq \nu\right\}. \end{aligned}$$

By assumption, when $n \to \infty$,

$$\frac{\mathbf{x}_n - \mathbf{a}}{\varepsilon_n} \xrightarrow{\mathcal{P}} \mathbf{0},$$

and therefore the first term tends to zero. Moreover,

$$P\left\{\omega : \frac{\rho(\mathbf{r}_m(\mathbf{x}_n(\omega),\mathbf{a}),\mathbf{0})}{\varepsilon_n^m} > \eta,\ \frac{\rho(\mathbf{x}_n(\omega),\mathbf{a})}{\varepsilon_n} \le \nu\right\}$$
$$\le P\left\{\omega : \frac{\rho(\mathbf{r}_m(\mathbf{x}_n(\omega),\mathbf{a}),\mathbf{0})}{\rho^m(\mathbf{x}_n(\omega),\mathbf{a})} > \frac{\eta}{\nu^m},\ \frac{\rho(\mathbf{x}_n(\omega),\mathbf{a})}{\varepsilon_n} \le \nu\right\}.$$

By our assumptions, the Taylor expansion of order m is valid for $\mathbf{g}(\mathbf{x})$, and therefore it follows from the definition of $o(\rho^m(\mathbf{x},\mathbf{a}))$ that

$$\frac{\rho(\mathbf{r}_m(\mathbf{x}_n(\omega),\mathbf{a}),\mathbf{0})}{\rho^m(\mathbf{x}_n(\omega),\mathbf{a})} \longrightarrow 0,$$

if $\mathbf{x}_n(\omega) \to \mathbf{a}$. Now choose ν to be small and increase n so that

$$P\left\{\omega : \frac{\rho(\mathbf{r}_m(\mathbf{x}_n),\mathbf{a})}{\varepsilon_n} \le \nu\right\}$$

becomes small and then

$$P\left\{\omega : \frac{\rho(\mathbf{r}_m(\mathbf{x}_n,\mathbf{a}),\mathbf{0})}{\rho^m(\mathbf{x}_n(\omega),\mathbf{a})} > \frac{\eta}{\nu^m},\ \rho(\mathbf{x}_n(\omega),\mathbf{a}) \le \varepsilon_n\nu\right\} \longrightarrow 0.$$

∎

Corollary 3.1.1.1. *Let $\{\mathbf{X}_n(K)\}$ and $\{\varepsilon_n\}$ be sequences of random $p\times q$ pattern matrices and positive numbers, respectively, and let $\mathbf{X}_n(K) - \mathbf{A}(K) = o_P(\varepsilon_n)$, where $\varepsilon_n \to 0$, if $n \to \infty$. If the function $\mathbf{G}(\mathbf{X}(K))$ can be expanded into the Taylor series (1.4.59) at the point $\mathbf{G}(\mathbf{A}(K))$, then*

$$\operatorname{vec}\{\mathbf{G}(\mathbf{X}_n(K)) - \mathbf{G}(\mathbf{A}(K))\} - \sum_{k=1}^{m}\frac{1}{k!}(\operatorname{vec}(\mathbf{X}_n(K) - \mathbf{A}(K))^{\otimes k-1}\otimes \mathbf{I}_l)'$$
$$\times \left(\frac{d^k\mathbf{G}(\mathbf{X}_n(K))}{d\mathbf{X}_n(K)^k}\right)'\Bigg|_{\mathbf{X}_n(K)=\mathbf{A}(K)} \operatorname{vec}(\mathbf{X}_n(K) - \mathbf{A}(K)) = o_P(\varepsilon_n^m),$$

where the matrix derivative is given in (1.4.43) and l equals the size of $\operatorname{vec}\mathbf{X}_n(K)$.

∎

Theorem 3.1.2. *Let $\{\mathbf{x}_n\}$ and $\{\varepsilon_n\}$ be sequences of random $p-$vectors and positive numbers, respectively. Let $\frac{\mathbf{x}_n - \mathbf{a}}{\varepsilon_n} \xrightarrow{\mathcal{D}} \mathbf{Z}$, for some $\mathbf{Z}$, and $\mathbf{x}_n - \mathbf{a} \xrightarrow{\mathcal{P}} \mathbf{0}$. If the function $\mathbf{g}(\mathbf{x})$ from $\mathbb{R}^p$ to $\mathbb{R}^q$ can be expanded into the Taylor series (1.4.59) at the point $\mathbf{a}$:*

$$\mathbf{g}(\mathbf{x}) = \mathbf{g}(\mathbf{a}) + \sum_{k=1}^{m}\frac{1}{k!}((\mathbf{x}-\mathbf{a})^{\otimes k-1}\otimes \mathbf{I}_q)'\left(\frac{d^k\mathbf{g}(\mathbf{x})}{d\mathbf{x}^k}\right)'\Bigg|_{\mathbf{x}=\mathbf{a}}(\mathbf{x}-\mathbf{a}) + \mathbf{r}_m(\mathbf{x},\mathbf{a}),$$

where

$$\mathbf{r}_m(\mathbf{x},\mathbf{a}) = \frac{1}{(m+1)!}((\mathbf{x}-\mathbf{a})^{\otimes m}\otimes\mathbf{I}_q)'\left(\frac{d^{m+1}\mathbf{g}(\mathbf{x})}{d\mathbf{x}^{m+1}}\right)'\Bigg|_{\mathbf{x}=\boldsymbol{\theta}}(\mathbf{x}-\mathbf{a}),$$

$\boldsymbol{\theta}$ is some element in a neighborhood of $\mathbf{a}$ and the derivative in (1.4.41) is used, then

$$\mathbf{g}(\mathbf{x}_n) = \mathbf{g}(\mathbf{a}) + \sum_{k=1}^{m}\frac{1}{k!}((\mathbf{x}_n-\mathbf{a})^{\otimes k-1}\otimes\mathbf{I}_q)'\left(\frac{d^k\mathbf{g}(\mathbf{x})}{d\mathbf{x}^k}\right)'\Bigg|_{\mathbf{x}=\mathbf{a}}(\mathbf{x}_n-\mathbf{a}) + o_P(\varepsilon_n^m).$$

PROOF: From the assumption it follows that $(\mathbf{x}_n-\mathbf{a})^{\otimes 2} = o_P(\varepsilon_n)$. Moreover, since $\mathbf{x}_n \xrightarrow{\mathcal{P}} \mathbf{a}$, it is enough to consider

$$P\left\{\omega : \frac{\mathbf{r}_m(\mathbf{x}_n(\omega,\mathbf{a})}{\varepsilon_n^m} > \eta,\ \rho(\mathbf{x}_n(\omega),\mathbf{a}) \le \nu\right\}.$$

Since we assume differentiability in Taylor expansions, the derivative in $\mathbf{r}_m(\mathbf{x},\mathbf{a})$ is continuous and bounded when $\rho(\mathbf{x}_n(\omega),\mathbf{a}) \le \nu$. Furthermore, since $\frac{\mathbf{x}_n-\mathbf{a}}{\varepsilon_n} \xrightarrow{\mathcal{D}} \mathbf{Z}$, the error term $\frac{\mathbf{r}_m(\mathbf{x}_n,\mathbf{a})}{\varepsilon_n^m}$ converges to $\mathbf{Y}(\mathbf{x}_n-\mathbf{a})$, for some random matrix $\mathbf{Y}$, which in turn, according to Corollary 3.1.1.1L, converges in probability to $\mathbf{0}$, since $\mathbf{x}_n \xrightarrow{\mathcal{P}} \mathbf{a}$. ∎

Corollary 3.1.2.1. *Let $\{\mathbf{X}_n(K)\}$ and $\{\varepsilon_n\}$ be sequences of random pattern $p\times q$−matrices and positive numbers, respectively. Let $\frac{\mathbf{X}_n(K)-\mathbf{A}(K)}{\varepsilon_n} \xrightarrow{\mathcal{D}} \mathbf{Z}(K)$ for some $\mathbf{Z}(K)$ and constant matrix $\mathbf{A} : p\times q$, and suppose that $\mathbf{X}_n(K)-\mathbf{A}(K) \xrightarrow{\mathcal{P}} \mathbf{0}$. If the function $\mathbf{G}(\mathbf{X}(K))$ from $\mathbb{R}^{p\times q}$ to $\mathbb{R}^{r\times s}$ can be expanded into Taylor series (1.4.59) at the point $\mathbf{A}(K)$,*

$$\operatorname{vec}\{\mathbf{G}(\mathbf{X}(K))-\mathbf{G}(\mathbf{A}(K))\} - \sum_{k=1}^{m}\frac{1}{k!}(\operatorname{vec}(\mathbf{X}(K)-\mathbf{A}(K))^{\otimes k-1}\otimes\mathbf{I}_l)'$$
$$\times\left(\frac{d^k\mathbf{G}(\mathbf{X}(K))}{d\mathbf{X}(K)^k}\right)'\Bigg|_{\mathbf{X}(K)=\mathbf{A}(K)}\operatorname{vec}(\mathbf{X}(K)-\mathbf{A}(K)) = r_m(\mathbf{X}_n(K),\mathbf{A}(K)),$$

where the matrix derivative in (1.4.43) is used,

$$\mathbf{r}_m(\mathbf{X}(K),\mathbf{A}(K)) = \frac{1}{(m+1)!}(\operatorname{vec}(\mathbf{X}(K)-\mathbf{A}(K))^{\otimes m}\otimes\mathbf{I}_l)'$$
$$\times\left(\frac{d^{m+1}\mathbf{G}(\mathbf{X}(K))}{d\mathbf{X}(K)^{m+1}}\right)'\Bigg|_{\mathbf{X}(K)=\boldsymbol{\theta}}\operatorname{vec}(\mathbf{X}(K)-\mathbf{A}(K)),$$

l is the size of $\operatorname{vec}\mathbf{X}_n(K)$ and $\boldsymbol{\theta}$ is some element in a neighborhood of $\mathbf{A}(K)$, then in the Taylor expansion of $\mathbf{G}(\mathbf{X}_n(K))$

$$\mathbf{r}_m(\mathbf{X}_n(K),\mathbf{A}(K)) = o_P(\varepsilon_n^m).$$

∎

3.1.2 *Asymptotic normality of functions of random vectors*

We are going to consider the simplest possible distribution approximations via Taylor expansion and the normal distribution. Let $\mathbf{X} = (\mathbf{x}_1, \ldots, \mathbf{x}_n)$ be a sample of size n and consider a *statistic* $\mathbf{T} = \mathbf{T}(\mathbf{X})$. To point out the dependence of $\mathbf{T}(\mathbf{X})$ on n, we write $\mathbf{T}_n$ or $\mathbf{T}(n)$ and obtain a sequence $\{\mathbf{T}_n\}$, when $n \to \infty$. Let

$$\mathbf{T}_n \xrightarrow{\mathcal{D}} N_p(\cdot, \cdot)$$

mean that the statistic $\mathbf{T}_n$ is asymptotically normally distributed. Asymptotic normality is one of the most fundamental properties of statistics, which gives us a simple way of finding approximate interval estimates and a possibility of testing hypotheses approximately for a wide class of underlying distributions. A theoretical basis for these results is the classical central limit theorem which will be considered here in one of its simplest forms:

$$\frac{1}{\sqrt{n}} \sum_{i=1}^{n} (\mathbf{x}_i - \boldsymbol{\mu}) \xrightarrow{\mathcal{D}} N_p(\mathbf{0}, \boldsymbol{\Sigma}),$$

when $n \to \infty$, where $\{\mathbf{x}_i\}$ is a sequence of i.i.d. random vectors with $E[\mathbf{x}_i] = \boldsymbol{\mu}$ and $D[\mathbf{x}_i] = \boldsymbol{\Sigma}$. One of the most important results for applications states that all smooth functions of an asymptotic normal vector are asymptotically normally distributed. Following Anderson (2003, pp. 132–133), let us present this result as a mathematical statement.

Theorem 3.1.3. *Assume that for* $\{\mathbf{x}_n\}$

$$\sqrt{n}(\mathbf{x}_n - \mathbf{a}) \xrightarrow{\mathcal{D}} N_p(\mathbf{0}, \boldsymbol{\Sigma}), \tag{3.1.1}$$

and

$$\mathbf{x}_n \xrightarrow{\mathcal{P}} \mathbf{a},$$

when $n \to \infty$. *Let the function* $\mathbf{g}(\mathbf{x}) : \mathbb{R}^p \longrightarrow \mathbb{R}^q$ *have continuous partial derivatives in a neighborhood of* $\mathbf{a}$. *Then, if* $n \to \infty$,

$$\sqrt{n}\,(\mathbf{g}(\mathbf{x}_n) - \mathbf{g}(\mathbf{a})) \xrightarrow{\mathcal{D}} N_q(\mathbf{0}, \boldsymbol{\xi}'\boldsymbol{\Sigma}\boldsymbol{\xi}), \tag{3.1.2}$$

where

$$\boldsymbol{\xi} = \left.\frac{d\mathbf{g}(\mathbf{x})}{d\mathbf{x}}\right|_{\mathbf{x}=\mathbf{a}} \neq \mathbf{0}$$

is the matrix derivative given by Definition 1.4.1.

PROOF: To prove the convergence (3.1.2) it is sufficient to show that the characteristic function, $\varphi_{g(\mathbf{x}_n)}(\mathbf{t})$, converges to the characteristic function of the limit distribution, which should be continuous at $\mathbf{t} = \mathbf{0}$. By assumptions we have

$$\lim_{n\to\infty} \varphi_{\sqrt{n}(\mathbf{x}_n - \mathbf{a})}(\mathbf{t}) = e^{-\frac{1}{2}\mathbf{t}'\boldsymbol{\Sigma}\mathbf{t}}, \quad \mathbf{t} \in \mathbb{R}^p.$$

From Theorem 3.1.2

$$\lim_{n\to\infty}\varphi_{\sqrt{n}\{g(\mathbf{x}_n)-g(\mathbf{a})\}}(\mathbf{t}) = \lim_{n\to\infty}\varphi_{\left\{\left(\frac{dg(\mathbf{x})}{d\mathbf{x}}\Big|_{\mathbf{x}=\mathbf{a}}\right)'\sqrt{n}(\mathbf{x}_n-\mathbf{a})+o_P(n^{-\frac{1}{2}})\right\}}(\mathbf{t})$$
$$= \lim_{n\to\infty}\varphi_{\sqrt{n}(\mathbf{x}_n-\mathbf{a})}\left(\frac{dg(\mathbf{x})}{d\mathbf{x}}\Big|_{\mathbf{x}=\mathbf{a}}\mathbf{t}\right)$$
$$= \exp\left\{-\frac{1}{2}\mathbf{t}'\left(\frac{dg(\mathbf{x})}{d\mathbf{x}}\Big|_{\mathbf{x}=\mathbf{a}}\right)'\boldsymbol{\Sigma}\left(\frac{dg(\mathbf{x})}{d\mathbf{x}}\Big|_{\mathbf{x}=\mathbf{a}}\right)\mathbf{t}\right\}$$
$$= \varphi_{N(\mathbf{0},\boldsymbol{\xi}'\boldsymbol{\Sigma}\boldsymbol{\xi})}(\mathbf{t}),$$

where $\boldsymbol{\xi} = \frac{dg(\mathbf{x})}{d\mathbf{x}}\Big|_{\mathbf{x}=\mathbf{a}}$. ∎

Corollary 3.1.3.1. *Suppose that Theorem 3.1.3 holds for* $\mathbf{g}(\mathbf{x}) : \mathbb{R}^p \longrightarrow \mathbb{R}^q$ *and there exists a function* $\mathbf{g}_0(\mathbf{x}) : \mathbb{R}^p \longrightarrow \mathbb{R}^q$ *which satisfies the assumptions of Theorem 3.1.3 such that*

$$\boldsymbol{\xi} = \frac{d\mathbf{g}_0(\mathbf{x})}{d\mathbf{x}}\Big|_{\mathbf{x}=\mathbf{a}} = \frac{d\mathbf{g}(\mathbf{x})}{d\mathbf{x}}\Big|_{\mathbf{x}=\mathbf{a}}.$$

Then

$$\sqrt{n}\,(\mathbf{g}_0(\mathbf{x}_n) - \mathbf{g}_0(\mathbf{a})) \xrightarrow{\mathcal{D}} N_q(\mathbf{0}, \boldsymbol{\xi}'\boldsymbol{\Sigma}\boldsymbol{\xi}).$$

∎

The implication of the corollary is important, since now, when deriving asymptotic distributions, one can always change the original function $\mathbf{g}(\mathbf{x})$ to another $\mathbf{g}_0(\mathbf{x})$ as long as the difference between the functions is of order $o_P(n^{-1/2})$ and the partial derivatives are continuous in a neighborhood of $\mathbf{a}$. Thus we may replace $\mathbf{g}(\mathbf{x})$ by $\mathbf{g}_0(\mathbf{x})$ and it may simplify calculations.
The fact that the derivative $\boldsymbol{\xi}$ is not equal to zero is essential. When $\boldsymbol{\xi} = \mathbf{0}$, the term including the first derivative in the Taylor expansion of $\mathbf{g}(\mathbf{x}_n)$ in the proof of Theorem 3.1.3 vanishes. Asymptotic behavior of the function is determined by the first non-zero term of the expansion. Thus, in the case $\boldsymbol{\xi} = \mathbf{0}$, the asymptotic distribution is not normal.
Very often procedures of multivariate statistical analysis are based upon statistics as functions of the sample mean

$$\overline{\mathbf{x}} = \frac{1}{n}\sum_{i=1}^{n}\mathbf{x}_i$$

and the sample dispersion matrix

$$\mathbf{S} = \frac{1}{n-1}\sum_{i=1}^{n}(\mathbf{x}_i - \overline{\mathbf{x}})(\mathbf{x}_i - \overline{\mathbf{x}})', \tag{3.1.3}$$

which are sufficient statistics under various distribution assumptions. As examples of such functions we have the determinant $|\mathbf{S}|$ of $\mathbf{S}$, which is called the generalized variance, the sample correlation matrix $\mathbf{R} = \mathbf{S}_d^{-\frac{1}{2}}\mathbf{S}\mathbf{S}_d^{-\frac{1}{2}}$, Hotelling's T^2-statistic $\overline{\mathbf{x}}'\mathbf{S}^{-1}\overline{\mathbf{x}}$, eigenvalues and eigenvectors of $\mathbf{S}$ and $\mathbf{R}$, etc.
The list can be made longer and this suggests the idea that we should prove asymptotic normality for $\overline{\mathbf{x}}(n)$ and $\mathbf{S}(n)$ because then, by Theorem 3.1.3, the asymptotic normality will hold for all the above mentioned statistics.

Theorem 3.1.4. *Let $\mathbf{x}_1, \mathbf{x}_2, \ldots, \mathbf{x}_n$ be an i.i.d. sample of size n from a $p-$dimensional population with $E[\mathbf{x}_i] = \boldsymbol{\mu}$, $D[\mathbf{x}_i] = \boldsymbol{\Sigma}$, $m_4[\mathbf{x}_i] < \infty$, and let $\mathbf{S}$ be given by (3.1.3). Then, if $n \to \infty$,*

(i) $$\overline{\mathbf{x}} \xrightarrow{\mathcal{P}} \boldsymbol{\mu};$$

(ii) $$\mathbf{S} \xrightarrow{\mathcal{P}} \boldsymbol{\Sigma};$$

(iii) $$\sqrt{n}(\overline{\mathbf{x}} - \boldsymbol{\mu}) \xrightarrow{\mathcal{D}} N_p(\mathbf{0}, \boldsymbol{\Sigma});$$

(iv) $$\sqrt{n}\mathrm{vec}(\mathbf{S} - \boldsymbol{\Sigma}) \xrightarrow{\mathcal{D}} N_{p^2}(\mathbf{0}, \boldsymbol{\Pi}), \tag{3.1.4}$$

where $\boldsymbol{\Pi} : p^2 \times p^2$ consists of the fourth and second order central moments:

$$\begin{aligned}\boldsymbol{\Pi} &= D[(\mathbf{x}_i - \boldsymbol{\mu})(\mathbf{x}_i - \boldsymbol{\mu})'] \\ &= E[(\mathbf{x}_i - \boldsymbol{\mu}) \otimes (\mathbf{x}_i - \boldsymbol{\mu})' \otimes (\mathbf{x}_i - \boldsymbol{\mu}) \otimes (\mathbf{x}_i - \boldsymbol{\mu})'] - \mathrm{vec}\boldsymbol{\Sigma}\mathrm{vec}'\boldsymbol{\Sigma}; \end{aligned} \tag{3.1.5}$$

(v) $$\sqrt{n}V^2(\mathbf{S} - \boldsymbol{\Sigma}) \xrightarrow{\mathcal{D}} N_{p(p+1)/2}(\mathbf{0}, \mathbf{G}_p\boldsymbol{\Pi}\mathbf{G}_p'), \tag{3.1.6}$$

where $V^2(\bullet)$ and $\mathbf{G}_p$ are given by Definition 1.3.9 and (1.3.49), respectively.

PROOF: The convergence in (i) and (iii) follows directly from the law of large numbers and the central limit theorem, respectively. For example, in (iii)

$$\sqrt{n}(\overline{\mathbf{x}} - \boldsymbol{\mu}) = \sqrt{n}(\frac{1}{n}\sum_{i=1}^{n}(\mathbf{x}_i - \boldsymbol{\mu})) = \frac{1}{\sqrt{n}}\sum_{i=1}^{n}(\mathbf{x}_i - \boldsymbol{\mu}) \xrightarrow{\mathcal{D}} N_p(\mathbf{0}, \boldsymbol{\Sigma}).$$

For (ii) $E[\mathbf{S} - \boldsymbol{\Sigma}] = \mathbf{0}$ and $D[\mathbf{S}] \to \mathbf{0}$ hold. Thus the statement is established. When proving (3.1.4) we prove first the convergence

$$\sqrt{n}\mathrm{vec}(\mathbf{S}^* - \boldsymbol{\Sigma}) \xrightarrow{\mathcal{D}} N_{p^2}(\mathbf{0}, \boldsymbol{\Pi}),$$

where $\mathbf{S}^* = \frac{n-1}{n}\mathbf{S}$. It follows that

$$\begin{aligned}\sqrt{n}\mathrm{vec}(\mathbf{S}^* - \boldsymbol{\Sigma}) &= \sqrt{n}\mathrm{vec}(\frac{1}{n}\sum_{i=1}^{n}\mathbf{x}_i\mathbf{x}_i' - \boldsymbol{\Sigma} - \overline{\mathbf{x}}\overline{\mathbf{x}}') \\ &= \sqrt{n}\mathrm{vec}(\frac{1}{n}\sum_{i=1}^{n}(\mathbf{x}_i - \boldsymbol{\mu})(\mathbf{x}_i - \boldsymbol{\mu})' - \boldsymbol{\Sigma}) - \sqrt{n}\mathrm{vec}((\overline{\mathbf{x}} - \boldsymbol{\mu})(\overline{\mathbf{x}} - \boldsymbol{\mu})'). \end{aligned} \tag{3.1.7}$$

Now,

$$E[(\mathbf{x}_i - \boldsymbol{\mu})(\mathbf{x}_i - \boldsymbol{\mu})'] = \boldsymbol{\Sigma}$$

and

$$\begin{aligned}
&D[(\mathbf{x}_i - \boldsymbol{\mu})(\mathbf{x}_i - \boldsymbol{\mu})'] \\
&\quad = E[\mathrm{vec}\,((\mathbf{x}_i - \boldsymbol{\mu})(\mathbf{x}_i - \boldsymbol{\mu})')\,\mathrm{vec}'\,((\mathbf{x}_i - \boldsymbol{\mu})(\mathbf{x}_i - \boldsymbol{\mu})')] \\
&\qquad - E[\mathrm{vec}\,((\mathbf{x}_i - \boldsymbol{\mu})(\mathbf{x}_i - \boldsymbol{\mu})')]E[\mathrm{vec}'\,((\mathbf{x}_i - \boldsymbol{\mu})(\mathbf{x}_i - \boldsymbol{\mu})')] \\
&\quad \underset{\substack{(1.3.31)\\(1.3.14)}}{=} E[((\mathbf{x}_i - \boldsymbol{\mu}) \otimes (\mathbf{x}_i - \boldsymbol{\mu}))\,((\mathbf{x}_i - \boldsymbol{\mu})' \otimes (\mathbf{x}_i - \boldsymbol{\mu})')] - \mathrm{vec}\boldsymbol{\Sigma}\mathrm{vec}'\boldsymbol{\Sigma} \\
&\quad \underset{\substack{(1.3.14)\\(1.3.16)}}{=} E[(\mathbf{x}_i - \boldsymbol{\mu}) \otimes (\mathbf{x}_i - \boldsymbol{\mu})' \otimes (\mathbf{x}_i - \boldsymbol{\mu}) \otimes (\mathbf{x}_i - \boldsymbol{\mu})'] - \mathrm{vec}\boldsymbol{\Sigma}\mathrm{vec}'\boldsymbol{\Sigma} = \Pi.
\end{aligned}$$

Thus, by the central limit theorem,

$$\sqrt{n}\mathrm{vec}\left(\frac{1}{n}\sum_{i=1}^{n}(\mathbf{x}_i - \boldsymbol{\mu})(\mathbf{x}_i - \boldsymbol{\mu})' - \boldsymbol{\Sigma}\right) \quad \xrightarrow{\mathcal{D}} \quad N_{p^2}(\mathbf{0}, \boldsymbol{\Pi}).$$

Finally it is observed that if $n \to \infty$,

$$\sqrt{n}\mathrm{vec}((\overline{\mathbf{x}} - \boldsymbol{\mu})(\overline{\mathbf{x}} - \boldsymbol{\mu})') \quad \xrightarrow{\mathcal{P}} \quad \mathbf{0},$$

since $E[\overline{\mathbf{x}}] = \boldsymbol{\mu}$ and $D[\overline{\mathbf{x}}] = \frac{1}{n}\boldsymbol{\Sigma}$ imply that $n^{\frac{1}{4}}(\overline{\mathbf{x}} - \boldsymbol{\mu})\xrightarrow{\mathcal{P}}\mathbf{0}$. Hence,

$$\sqrt{n}\mathrm{vec}(\mathbf{S}^* - \boldsymbol{\Sigma}) \quad \xrightarrow{\mathcal{D}} \quad N_{p^2}(\mathbf{0}, \boldsymbol{\Pi})$$

and the statement in (3.1.4) is obtained, since obviously $\mathbf{S}^*\xrightarrow{\mathcal{P}}\mathbf{S}$. The relation in (v) follows immediately from $V^2(\mathbf{S}) = \mathbf{G}_p\mathrm{vec}(\mathbf{S})$. ∎

When $\{\mathbf{x}_i\}$ is normally distributed, convergence (3.1.4) takes a simpler form.

Corollary 3.1.4.1. *Let* $\mathbf{x}_i \sim N_p(\boldsymbol{\mu}, \boldsymbol{\Sigma})$. *Then*

(i)
$$\sqrt{n}\mathrm{vec}(\mathbf{S} - \boldsymbol{\Sigma}) \quad \xrightarrow{\mathcal{D}} \quad N_{p^2}(\mathbf{0}, \boldsymbol{\Pi}^N),$$

where

$$\boldsymbol{\Pi}^N = (\mathbf{I}_{p^2} + \mathbf{K}_{p,p})(\boldsymbol{\Sigma} \otimes \boldsymbol{\Sigma});$$

(ii)
$$\sqrt{n}V^2(\mathbf{S} - \boldsymbol{\Sigma}) \quad \xrightarrow{\mathcal{D}} \quad N_{p(p+1)/2}(\mathbf{0}, \mathbf{G}_p\boldsymbol{\Pi}^N\mathbf{G}_p').$$

PROOF: Since $(n-1)\mathbf{S} \sim W_p(\boldsymbol{\Sigma}, n-1)$ holds, $\boldsymbol{\Pi}^N$ follows from Theorem 2.4.14 (ii). ∎

The next corollary gives us the asymptotic distribution of $\mathbf{S}$ in the case of an elliptically distributed population.

Corollary 3.1.4.2. *Let* $\mathbf{x}_i \sim E_p(\boldsymbol{\mu}, \Upsilon)$ *and* $D[\mathbf{x}_i] = \boldsymbol{\Sigma}$. *Then*

(i)
$$\sqrt{n}\text{vec}(\mathbf{S} - \boldsymbol{\Sigma}) \xrightarrow{\mathcal{D}} N_{p^2}(\mathbf{0}, \boldsymbol{\Pi}^E),$$

where
$$\boldsymbol{\Pi}^E = (1+\kappa)(\mathbf{I}_{p^2} + \mathbf{K}_{p,p})(\boldsymbol{\Sigma} \otimes \boldsymbol{\Sigma}) + \kappa\text{vec}\boldsymbol{\Sigma}\text{vec}'\boldsymbol{\Sigma},$$

with κ *being the kurtosis parameter defined by (2.3.13);*

(ii)
$$\sqrt{n}V^2(\mathbf{S} - \boldsymbol{\Sigma}) \xrightarrow{\mathcal{D}} N_{p(p+1)/2}(\mathbf{0}, \mathbf{G}_p\boldsymbol{\Pi}^E\mathbf{G}_p').$$

∎

3.1.3 *Asymptotic distribution of statistics with functionally dependent arguments*
From the previous paragraph we know that asymptotic normality can be established for smooth functions $\mathbf{g}(\mathbf{x}_n)$ of an asymptotically normal variable $\mathbf{x}_n$. The expression of the asymptotic variance matrix in (3.1.2) includes the matrix derivative $\frac{d\mathbf{g}(\mathbf{x})}{d\mathbf{x}}$ at a certain fixed point $\mathbf{a}$. How should one find the derivative, when some elements of $\mathbf{x}_n$ are functionally dependent (off-diagonal elements of the sample dispersion $\mathbf{S}$ are symmetric, for example) or functionally dependent and constant (off-diagonal elements of the sample correlation matrix $\mathbf{R}$ are symmetric, whereas the diagonal ones are constant)? The problem can be solved using the notion of patterned matrices of §1.3.6, where we consider non-repeated and non-constant elements in the patterned matrix only. Let $\{\mathbf{X}_n\}$ be a sequence of $p \times q$-matrices converging in distribution when $n \to \infty$:

$$\sqrt{n}\text{vec}(\mathbf{X}_n - \mathbf{A}) \xrightarrow{\mathcal{D}} N_{pq}(\mathbf{0}, \boldsymbol{\Sigma}_\mathbf{X}), \tag{3.1.8}$$

where $\mathbf{A} : p \times q$ is a constant matrix. Furthermore, suppose that $\mathbf{X}_n \xrightarrow{\mathcal{P}} \mathbf{A}$. If a certain coordinate, say the i−th coordinate of vec$\mathbf{X}_n$ is constant, then in $\boldsymbol{\Sigma}_\mathbf{X}$ in (3.1.8), the i−th row and column consist of zeros. Let $\mathbf{X}_n(K)$ denote the patterned matrix obtained from $\mathbf{X}_n$ by leaving out the constant and repeated elements. Then obviously for $\mathbf{X}_n(K)$ the convergence in distribution also takes place:

$$\sqrt{n}\text{vec}(\mathbf{X}_n(K) - \mathbf{A}(K)) \xrightarrow{\mathcal{D}} N_\bullet(\mathbf{0}, \boldsymbol{\Sigma}_{\mathbf{X}(K)}), \tag{3.1.9}$$

where $\bullet$ in the index stands for the number of elements in vec$(\mathbf{X}_n(K))$. Let $\mathbf{G} : r \times s$ be a matrix where the elements are certain functions of the elements of $\mathbf{X}_n$. By Theorem 3.1.3

$$\sqrt{n}\text{vec}\{\mathbf{G}(\mathbf{T}^+(K)\text{vec}\mathbf{X}_n(K)) - \mathbf{G}(\mathbf{T}^+(K)\text{vec}\mathbf{A}(K))\} \xrightarrow{\mathcal{D}} N_\bullet(\mathbf{0}, \boldsymbol{\Sigma}_\mathbf{G}),$$

where

$$\boldsymbol{\Sigma}_\mathbf{G} = \boldsymbol{\xi}'\boldsymbol{\Sigma}_{\mathbf{X}(K)}\boldsymbol{\xi}, \tag{3.1.10}$$

and

$$\begin{aligned}\boldsymbol{\xi} &= \frac{d\mathbf{G}(\mathbf{T}^{+}(K)\mathrm{vec}\mathbf{X}_n(K))}{d\mathbf{X}_n(K)}\bigg|_{\mathbf{X}_n(K)=\mathbf{A}(K)} \\ &= \frac{d\mathbf{T}^{+}(K)\mathrm{vec}\mathbf{X}(K)}{d\mathbf{X}(K)}\frac{d\mathbf{G}(\mathbf{Z})}{d\mathbf{Z}}\bigg|_{\mathrm{vec}\mathbf{Z}=\mathbf{T}^{+}(K)\mathrm{vec}\mathbf{A}(K)} \\ &= (\mathbf{T}^{+}(K))'\frac{d\mathbf{G}(\mathbf{Z})}{d\mathbf{Z}}\bigg|_{\mathrm{vec}\mathbf{Z}=\mathbf{T}^{+}(K)\mathrm{vec}\mathbf{A}(K)},\end{aligned}$$

where $\mathbf{T}^{+}(K)$ is defined in (1.3.60). The expression in (3.1.10) of the asymptotic dispersion matrix $\boldsymbol{\Sigma}_{\mathbf{G}}$ may be somewhat complicated to use. However, by §1.3.6 (see also Kollo, 1994), it follows that

$$\begin{aligned}\boldsymbol{\Sigma}_{\mathbf{G}} &= \left(\frac{d\mathbf{G}(\mathbf{Z})}{d\mathbf{Z}}\bigg|_{\mathrm{vec}\mathbf{Z}=\mathbf{T}^{+}(K)\mathrm{vec}\mathbf{A}(K)}\right)'\mathbf{T}^{+}(K) \\ &\quad\times\boldsymbol{\Sigma}_{\mathbf{X}(K)}(\mathbf{T}^{+}(K))'\frac{d\mathbf{G}(\mathbf{Z})}{d\mathbf{Z}}\bigg|_{\mathrm{vec}\mathbf{Z}=\mathbf{T}^{+}(K)\mathrm{vec}\mathbf{A}(K)} \\ &= \left(\frac{d\mathbf{G}(\mathbf{Z})}{d\mathbf{Z}}\bigg|_{\mathrm{vec}\mathbf{Z}=\mathbf{T}^{+}(K)\mathrm{vec}\mathbf{A}(K)}\right)'\boldsymbol{\Sigma}_{\mathbf{X}}\frac{d\mathbf{G}(\mathbf{Z})}{d\mathbf{Z}}\bigg|_{\mathrm{vec}\mathbf{Z}=\mathbf{T}^{+}(K)\mathrm{vec}\mathbf{A}(K)},\end{aligned}$$

since

$$\mathbf{T}^{+}(K)\boldsymbol{\Sigma}_{\mathbf{X}(K)}(\mathbf{T}^{+}(K))' = \boldsymbol{\Sigma}_{\mathbf{X}}.$$

Therefore the following theorem can be stated.

Theorem 3.1.5. *Let $\{\mathbf{X}_n\}$ be a sequence of $p\times q-$matrices such that*

$$\sqrt{n}\mathrm{vec}(\mathbf{X}_n-\mathbf{A}) \xrightarrow{\mathcal{D}} N_{pq}(\mathbf{0},\boldsymbol{\Sigma}_{\mathbf{X}})$$

and

$$\mathbf{X}_n \xrightarrow{\mathcal{P}} \mathbf{A},$$

where $\mathbf{A}: p\times q$ is a constant matrix. Let $\mathbf{X}_n(K)$ be the patterned matrix obtained from $\mathbf{X}_n$ by excluding the constant and repeated elements. Let $\mathbf{G}: r\times s$ be a matrix with the elements being functions of $\mathbf{X}_n$. Then

$$\sqrt{n}\mathrm{vec}(\mathbf{G}(\mathbf{X}_n(K))-\mathbf{G}(\mathbf{A}(K))) \xrightarrow{\mathcal{D}} N_{rs}(\mathbf{0},\boldsymbol{\Sigma}_{\mathbf{G}}),$$

with

$$\boldsymbol{\Sigma}_{\mathbf{G}} = \left(\frac{d\mathbf{G}(\mathbf{Z})}{d\mathbf{Z}}\bigg|_{\mathbf{Z}=\mathbf{A}}\right)'\boldsymbol{\Sigma}_{\mathbf{X}}\frac{d\mathbf{G}(\mathbf{Z})}{d\mathbf{Z}}\bigg|_{\mathbf{Z}=\mathbf{A}}, \tag{3.1.11}$$

where in $\frac{d\mathbf{G}(\mathbf{Z})}{d\mathbf{Z}}$ the matrix $\mathbf{Z}$ is regarded as unstructured. ∎

REMARK: In (3.1.9) vec(•) is the operation given in §1.3.6, whereas vec(•) in Theorem 3.1.5 follows the standard definition given in §1.3.4, what can be understood from the context.

Note that Theorem 3.1.5 implies that when obtaining asymptotic distributions we do not have to care about repetition of elements, since the result of the theorem is in agreement with

$$\sqrt{n}\mathrm{vec}(\mathbf{G}(\mathbf{X}_n) - \mathbf{G}(\mathbf{A})) \xrightarrow{\mathcal{D}} N_{rs}\left(\mathbf{0}, \left(\frac{d\mathbf{G}(\mathbf{Z})}{d\mathbf{Z}}\bigg|_{\mathbf{Z}=\mathbf{A}}\right)' \mathbf{\Sigma}_{\mathbf{X}} \frac{d\mathbf{G}(\mathbf{Z})}{d\mathbf{Z}}\bigg|_{\mathbf{Z}=\mathbf{A}}\right).$$

3.1.4 *Asymptotic distribution of the sample correlation matrix*

The correlation matrix of a random $p-$vector $\mathbf{x}$ is defined through its dispersion matrix $\mathbf{\Sigma}$:

$$\mathbf{\Omega} = \mathbf{\Sigma}_d^{-\frac{1}{2}}\mathbf{\Sigma}\mathbf{\Sigma}_d^{-\frac{1}{2}}, \tag{3.1.12}$$

where the diagonal matrix $\mathbf{\Sigma}_d$ is defined by (1.1.4). The corresponding sample correlation matrix $\mathbf{R}$ from a sample of size n is a function of the sample dispersion matrix (3.1.3):

$$\mathbf{R} = \mathbf{S}_d^{-\frac{1}{2}}\mathbf{S}\mathbf{S}_d^{-\frac{1}{2}}.$$

In the case of a normal population the asymptotic normal distribution of a non-diagonal element of $\sqrt{n}\mathrm{vec}(\mathbf{R}-\mathbf{\Omega})$ was derived by Girshik (1939). For the general case when the existence of the fourth order moments is assumed, the asymptotic normal distribution in matrix form can be found in Kollo (1984) or Neudecker & Wesselman (1990); see also Magnus (1988, Chapter 10), Nel (1985).

Theorem 3.1.6. *Let $\mathbf{x}_1, \mathbf{x}_2, \ldots, \mathbf{x}_n$ be a sample of size n from a $p-$dimensional population, with $E[\mathbf{x}_i] = \boldsymbol{\mu}$, $D[\mathbf{x}_i] = \mathbf{\Sigma}$ and $m_4[\mathbf{x}_i] < \infty$. Then, if $n \to \infty$,*

$$\sqrt{n}\mathrm{vec}(\mathbf{R} - \mathbf{\Omega}) \xrightarrow{\mathcal{D}} N_{p^2}(\mathbf{0}, \mathbf{\Psi}_{\mathbf{R}}),$$

where $\mathbf{\Psi}_{\mathbf{R}} = \Xi'_{\mathbf{R}}\mathbf{\Pi}\Xi_{\mathbf{R}}$, $\mathbf{\Pi}$ is given by (3.1.5) and the matrix derivative is

$$\Xi_{\mathbf{R}} = \mathbf{\Sigma}_d^{-\frac{1}{2}} \otimes \mathbf{\Sigma}_d^{-\frac{1}{2}} - \tfrac{1}{2}(\mathbf{K}_{p,p})_d(\mathbf{I}_p \otimes \mathbf{\Sigma}_d^{-1}\mathbf{\Omega} + \mathbf{\Sigma}_d^{-1}\mathbf{\Omega} \otimes \mathbf{I}_p). \tag{3.1.13}$$

REMARK: The asymptotic dispersion matrix $\mathbf{\Psi}_{\mathbf{R}}$ is singular with $r(\mathbf{\Psi}_{\mathbf{R}}) = \frac{p(p-1)}{2}$.

PROOF: The matrix $\mathbf{R}$ is a function of $p(p+1)/2$ different elements in $\mathbf{S}$. From Theorem 3.1.5 it follows that we should find $\frac{d\mathbf{R}}{d\mathbf{S}}$. The idea of the proof is first to approximate $\mathbf{R}$ and thereafter calculate the derivative of the approximated $\mathbf{R}$. Such a procedure simplifies the calculations. The correctness of the procedure follows from Corollary 3.1.3.1.

Let us find the first terms of the Taylor expansion of $\mathbf{R}$ through the matrices $\mathbf{S}$ and $\mathbf{\Sigma}$. Observe that

$$\mathbf{R} = (\mathbf{\Sigma} + (\mathbf{S}-\mathbf{\Sigma}))_d^{-\frac{1}{2}}(\mathbf{\Sigma} + (\mathbf{S}-\mathbf{\Sigma}))(\mathbf{\Sigma} + (\mathbf{S}-\mathbf{\Sigma}))_d^{-\frac{1}{2}}. \tag{3.1.14}$$

Since

$$(\mathbf{\Sigma} + (\mathbf{S}-\mathbf{\Sigma}))_d^{-\frac{1}{2}} = (\mathbf{\Sigma}_d + (\mathbf{S}-\mathbf{\Sigma})_d)^{-\frac{1}{2}}$$

and Theorem 3.1.4 (i) and (iv) hold, it follows by the Taylor expansion given in Theorem 3.1.2, that

$$(\boldsymbol{\Sigma}+(\mathbf{S}-\boldsymbol{\Sigma}))_d^{-\frac{1}{2}}=\boldsymbol{\Sigma}_d^{-\frac{1}{2}}-\tfrac{1}{2}\boldsymbol{\Sigma}_d^{-\frac{3}{2}}(\mathbf{S}-\boldsymbol{\Sigma})_d+o_P(n^{-\frac{1}{2}}).$$

Note that we can calculate the derivatives in the Taylor expansion of the diagonal matrix elementwise. Equality (3.1.14) turns now into the relation

$$\mathbf{R}=\boldsymbol{\Omega}+\boldsymbol{\Sigma}_d^{-\frac{1}{2}}(\mathbf{S}-\boldsymbol{\Sigma})\boldsymbol{\Sigma}_d^{-\frac{1}{2}}-\tfrac{1}{2}\left(\boldsymbol{\Omega}\boldsymbol{\Sigma}_d^{-1}(\mathbf{S}-\boldsymbol{\Sigma})_d+(\mathbf{S}-\boldsymbol{\Sigma})_d\boldsymbol{\Sigma}_d^{-1}\boldsymbol{\Omega}\right)+o_P(n^{-\frac{1}{2}}).$$

Thus, since according to Theorem 3.1.5 the matrix $\mathbf{S}$ could be treated as unstructured,

$$\begin{aligned}\frac{d\mathbf{R}}{d\mathbf{S}}&=\frac{d\mathbf{S}}{d\mathbf{S}}(\boldsymbol{\Sigma}_d^{-\frac{1}{2}}\otimes\boldsymbol{\Sigma}_d^{-\frac{1}{2}})-\tfrac{1}{2}\frac{d\mathbf{S}_d}{d\mathbf{S}}\left(\mathbf{I}\otimes\boldsymbol{\Sigma}_d^{-1}\boldsymbol{\Omega}+\boldsymbol{\Sigma}_d^{-1}\boldsymbol{\Omega}\otimes\mathbf{I}\right)+o_P(n^{-\frac{1}{2}})\\&=\boldsymbol{\Sigma}_d^{-\frac{1}{2}}\otimes\boldsymbol{\Sigma}_d^{-\frac{1}{2}}-\tfrac{1}{2}(\mathbf{K}_{p,p})_d(\mathbf{I}\otimes\boldsymbol{\Sigma}_d^{-1}\boldsymbol{\Omega}+\boldsymbol{\Sigma}_d^{-1}\boldsymbol{\Omega}\otimes\mathbf{I})+o_P(n^{-\frac{1}{2}}),\end{aligned}$$

where $\dfrac{d\mathbf{S}_d}{d\mathbf{S}}$ is given by (1.4.37). ∎

Corollary 3.1.6.1. *Let* $\mathbf{x}_i\sim N_p(\boldsymbol{\mu},\boldsymbol{\Sigma})$. *If* $n\to\infty$, *then*

$$\sqrt{n}\mathrm{vec}(\mathbf{R}-\boldsymbol{\Omega})\quad\xrightarrow{\mathcal{D}}\quad N_{p^2}(\mathbf{0},\boldsymbol{\Psi}_R^N),$$

where

$$\boldsymbol{\Psi}_R^N=\mathbf{A}_1-\mathbf{A}_2-\mathbf{A}_2'+\mathbf{A}_3, \tag{3.1.15}$$

with

$$\mathbf{A}_1=(\mathbf{I}+\mathbf{K}_{p,p})(\boldsymbol{\Omega}\otimes\boldsymbol{\Omega}), \tag{3.1.16}$$
$$\mathbf{A}_2=(\boldsymbol{\Omega}\otimes\mathbf{I}_p+\mathbf{I}_p\otimes\boldsymbol{\Omega})(\mathbf{K}_{p,p})_d(\boldsymbol{\Omega}\otimes\boldsymbol{\Omega}), \tag{3.1.17}$$
$$\mathbf{A}_3=\frac{1}{2}(\boldsymbol{\Omega}\otimes\mathbf{I}_p+\mathbf{I}_p\otimes\boldsymbol{\Omega})(\mathbf{K}_{p,p})_d(\boldsymbol{\Omega}\otimes\boldsymbol{\Omega})(\mathbf{K}_{p,p})_d(\boldsymbol{\Omega}\otimes\mathbf{I}_p+\mathbf{I}_p\otimes\boldsymbol{\Omega}). \tag{3.1.18}$$

PROOF: From Theorem 3.1.6 and Corollary 3.1.4.1 we have

$$\boldsymbol{\Psi}_R^N=\Xi_{\mathbf{R}}'(\mathbf{I}+\mathbf{K}_{p,p})(\boldsymbol{\Sigma}\otimes\boldsymbol{\Sigma})\Xi_{\mathbf{R}},$$

where $\Xi_{\mathbf{R}}$ is given in (3.1.13). Thus,

$$\begin{aligned}\boldsymbol{\Psi}_R^N=&\{\boldsymbol{\Sigma}_d^{-\frac{1}{2}}\otimes\boldsymbol{\Sigma}_d^{-\frac{1}{2}}-\frac{1}{2}\left(\boldsymbol{\Omega}\boldsymbol{\Sigma}_d^{-1}\otimes\mathbf{I}_p+\mathbf{I}_p\otimes\boldsymbol{\Omega}\boldsymbol{\Sigma}_d^{-1}\right)(\mathbf{K}_{p,p})_d\}(\mathbf{I}+\mathbf{K}_{p,p})\\&\times(\boldsymbol{\Sigma}\otimes\boldsymbol{\Sigma})\{\boldsymbol{\Sigma}_d^{-\frac{1}{2}}\otimes\boldsymbol{\Sigma}_d^{-\frac{1}{2}}-\frac{1}{2}(\mathbf{K}_{p,p})_d\left(\boldsymbol{\Sigma}_d^{-1}\boldsymbol{\Omega}\otimes\mathbf{I}_p+\mathbf{I}_p\otimes\boldsymbol{\Sigma}_d^{-1}\boldsymbol{\Omega}\right)\}\\=&\mathbf{A}_1-\mathbf{A}_2-\mathbf{A}_2'+\mathbf{A}_3,\end{aligned}$$

where

$$
\begin{aligned}
\mathbf{A}_1 &= (\boldsymbol{\Sigma}_d^{-\frac{1}{2}} \otimes \boldsymbol{\Sigma}_d^{-\frac{1}{2}})(\mathbf{I} + \mathbf{K}_{p,p})(\boldsymbol{\Sigma} \otimes \boldsymbol{\Sigma})(\boldsymbol{\Sigma}_d^{-\frac{1}{2}} \otimes \boldsymbol{\Sigma}_d^{-\frac{1}{2}}),\\
\mathbf{A}_2 &= \frac{1}{2}(\boldsymbol{\Omega}\boldsymbol{\Sigma}_d^{-1} \otimes \mathbf{I}_p + \mathbf{I}_p \otimes \boldsymbol{\Omega}\boldsymbol{\Sigma}_d^{-1})(\mathbf{K}_{p,p})_d(\boldsymbol{\Sigma} \otimes \boldsymbol{\Sigma})(\mathbf{I} + \mathbf{K}_{p,p})(\boldsymbol{\Sigma}_d^{-\frac{1}{2}} \otimes \boldsymbol{\Sigma}_d^{-\frac{1}{2}}),\\
\mathbf{A}_3 &= \frac{1}{4}(\boldsymbol{\Omega}\boldsymbol{\Sigma}_d^{-1} \otimes \mathbf{I}_p + \mathbf{I}_p \otimes \boldsymbol{\Omega}\boldsymbol{\Sigma}_d^{-1})(\mathbf{K}_{p,p})_d(\mathbf{I} + \mathbf{K}_{p,p})(\boldsymbol{\Sigma} \otimes \boldsymbol{\Sigma})(\mathbf{K}_{p,p})_d\\
&\quad \times (\boldsymbol{\Sigma}_d^{-1}\boldsymbol{\Omega} \otimes \mathbf{I}_p + \mathbf{I}_p \otimes \boldsymbol{\Sigma}_d^{-1}\boldsymbol{\Omega}).
\end{aligned}
$$

Let us show that $\mathbf{A}_i$, $i = 1, 2, 3$, are identical to (3.1.16), (3.1.17) and (3.1.18). For $\mathbf{A}_1$ we get

$$
\begin{aligned}
\mathbf{A}_1 &\underset{(1.3.15)}{=} (\mathbf{I} + \mathbf{K}_{p,p})(\boldsymbol{\Sigma}_d^{-\frac{1}{2}} \otimes \boldsymbol{\Sigma}_d^{-\frac{1}{2}})(\boldsymbol{\Sigma} \otimes \boldsymbol{\Sigma})(\boldsymbol{\Sigma}_d^{-\frac{1}{2}} \otimes \boldsymbol{\Sigma}_d^{-\frac{1}{2}})\\
&\underset{\substack{(1.3.14)\\(3.1.12)}}{=} (\mathbf{I} + \mathbf{K}_{p,p})(\boldsymbol{\Omega} \otimes \boldsymbol{\Omega}).
\end{aligned}
$$

To establish (3.1.17), it is observed that

$$
\begin{aligned}
\mathbf{A}_2 &= (\boldsymbol{\Omega}\boldsymbol{\Sigma}_d^{-1} \otimes \mathbf{I}_p + \mathbf{I}_p \otimes \boldsymbol{\Omega}\boldsymbol{\Sigma}_d^{-1})(\mathbf{K}_{p,p})_d(\boldsymbol{\Sigma} \otimes \boldsymbol{\Sigma})(\boldsymbol{\Sigma}_d^{-\frac{1}{2}} \otimes \boldsymbol{\Sigma}_d^{-\frac{1}{2}})\\
&= (\boldsymbol{\Omega} \otimes \mathbf{I}_p + \mathbf{I}_p \otimes \boldsymbol{\Omega})(\mathbf{I} \otimes \boldsymbol{\Sigma}_d^{-1})(\mathbf{K}_{p,p})_d(\boldsymbol{\Sigma} \otimes \boldsymbol{\Sigma})(\boldsymbol{\Sigma}_d^{-\frac{1}{2}} \otimes \boldsymbol{\Sigma}_d^{-\frac{1}{2}})\\
&= (\boldsymbol{\Omega} \otimes \mathbf{I}_p + \mathbf{I}_p \otimes \boldsymbol{\Omega})(\mathbf{K}_{p,p})_d(\boldsymbol{\Omega} \otimes \boldsymbol{\Omega}),
\end{aligned}
$$

since $(\mathbf{K}_{p,p})_d(\mathbf{I} \otimes \boldsymbol{\Sigma}_d^{-1}) = (\mathbf{K}_{p,p})_d(\boldsymbol{\Sigma}_d^{-1} \otimes \mathbf{I}) = (\boldsymbol{\Sigma}_d^{-\frac{1}{2}} \otimes \boldsymbol{\Sigma}_d^{-\frac{1}{2}})(\mathbf{K}_{p,p})_d$ as well as $\mathbf{K}_{p,p}(\mathbf{K}_{p,p})_d = (\mathbf{K}_{p,p})_d$.
By similar calculations

$$
\begin{aligned}
\mathbf{A}_3 &= \frac{1}{2}\,(\boldsymbol{\Omega} \otimes \mathbf{I}_p + \mathbf{I}_p \otimes \boldsymbol{\Omega})\,(\mathbf{I} \otimes \boldsymbol{\Sigma}_d^{-1})(\mathbf{K}_{p,p})_d(\boldsymbol{\Sigma} \otimes \boldsymbol{\Sigma})(\mathbf{K}_{p,p})_d\\
&\qquad \times (\mathbf{I} \otimes \boldsymbol{\Sigma}_d^{-1})\,(\boldsymbol{\Omega} \otimes \mathbf{I}_p + \mathbf{I}_p \otimes \boldsymbol{\Omega})\\
&= \frac{1}{2}(\boldsymbol{\Omega} \otimes \mathbf{I}_p + \mathbf{I}_p \otimes \boldsymbol{\Omega})(\mathbf{K}_{p,p})_d(\boldsymbol{\Omega} \otimes \boldsymbol{\Omega})(\mathbf{K}_{p,p})_d(\boldsymbol{\Omega} \otimes \mathbf{I}_p + \mathbf{I}_p \otimes \boldsymbol{\Omega}).
\end{aligned}
$$

∎

In the case of elliptically distributed vectors the formula of the asymptotic dispersion matrix of $\mathbf{R}$ has a similar construction as in the normal population case, and the proof directly repeats the steps of proving Corollary 3.1.6.1. Therefore we shall present the result in the next corollary without a proof, which we leave to the interested reader as an exercise.

Corollary 3.1.6.2. *Let $\mathbf{x}_i \sim E_p(\boldsymbol{\mu}, \mathbf{V})$ and $D[\mathbf{x}_i] = \boldsymbol{\Sigma}$. If $n \to \infty$, then*

$$
\sqrt{n}\mathrm{vec}(\mathbf{R} - \boldsymbol{\Omega}) \quad \xrightarrow{\mathcal{D}} \quad N_p(\mathbf{0}, \Psi_R^E),
$$

where

$$
\Psi_R^E = \mathbf{B}_1 - \mathbf{B}_2 - \mathbf{B}_2' + \mathbf{B}_3 \tag{3.1.19}
$$

and

$$\mathbf{B}_1 = (1+\kappa)(\mathbf{I}+\mathbf{K}_{p,p})(\boldsymbol{\Omega}\otimes\boldsymbol{\Omega}) + \kappa\text{vec}\boldsymbol{\Omega}\text{vec}'\boldsymbol{\Omega}, \tag{3.1.20}$$

$$\mathbf{B}_2 = (\boldsymbol{\Omega}\otimes\mathbf{I}_p + \mathbf{I}_p\otimes\boldsymbol{\Omega})(\mathbf{K}_{p,p})_d\{(1+\kappa)(\boldsymbol{\Omega}\otimes\boldsymbol{\Omega}) + \frac{\kappa}{2}\text{vec}\boldsymbol{\Omega}\text{vec}'\boldsymbol{\Omega}\}, \tag{3.1.21}$$

$$\mathbf{B}_3 = \frac{1}{2}(\boldsymbol{\Omega}\otimes\mathbf{I}_p + \mathbf{I}_p\otimes\boldsymbol{\Omega})(\mathbf{K}_{p,p})_d\{(1+\kappa)(\boldsymbol{\Omega}\otimes\boldsymbol{\Omega}) + \frac{\kappa}{2}\text{vec}\boldsymbol{\Omega}\text{vec}'\boldsymbol{\Omega}\} \times (\mathbf{K}_{p,p})_d(\boldsymbol{\Omega}\otimes\mathbf{I}_p + \mathbf{I}_p\otimes\boldsymbol{\Omega}), \tag{3.1.22}$$

where the kurtosis parameter κ is defined by (2.3.13). ∎

3.1.5 *Asymptotics of eigenvalues and eigenvectors of a symmetric matrix*

Dealing with eigenvectors of a matrix one has to keep in mind that there are two possibilities for vector orientation in a linear space. If one wants to determine a matrix of eigenvectors uniquely one has to fix the direction of the vectors. To do so, the two most commonly used assumptions are:

(i) The diagonal elements of the matrix of eigenvectors are non-negative;
(ii) The first coordinate in each eigenvector is non-negative.

In the following we shall adopt assumption (i). There exist two different normalizations of eigenvectors of a symmetric matrix which are of interest in statistical applications, namely, the unit-length eigenvectors forming an orthogonal matrix and the eigenvalue normed eigenvectors. Historically the latter have been a basis for principal component analysis, although nowadays the first are mainly used in this kind of analysis. It seems that the length of a vector cannot be a reason for a separate study of these eigenvectors, but it appears that for asymptotic distribution theory this difference is essential. While the asymptotic distribution of the unit-length eigenvectors of $\mathbf{S}$ is singular, the eigenvalue-normed eigenvectors of $\mathbf{S}$ have a non-singular asymptotic distribution. Therefore, we shall discuss both normalizations in the sequel in some detail.

Consider a symmetric matrix $\boldsymbol{\Sigma}$ of order p with eigenvalues $\lambda_1 > \lambda_2 > \cdots > \lambda_p > 0$ and associated eigenvectors $\boldsymbol{\gamma}_i$ of length $\sqrt{\lambda_i}$, with $\gamma_{ii} > 0$, $i = 1, \ldots, p$. Thus,

$$\boldsymbol{\Sigma\Gamma} = \boldsymbol{\Gamma\Lambda}, \tag{3.1.23}$$

$$\boldsymbol{\Gamma}'\boldsymbol{\Gamma} = \boldsymbol{\Lambda}, \tag{3.1.24}$$

where $\boldsymbol{\Gamma} = (\boldsymbol{\gamma}_1, \ldots, \boldsymbol{\gamma}_p)$ and $\boldsymbol{\Lambda}$ is a diagonal matrix consisting of p (distinct) eigenvalues of $\boldsymbol{\Sigma}$. Observe that this means that $\boldsymbol{\Sigma} = \boldsymbol{\Gamma\Gamma}'$. In parallel we consider the set of unit-length eigenvectors $\boldsymbol{\psi}_i$, $i = 1, \ldots, p$, corresponding to the eigenvalues λ_i, where $\psi_{ii} > 0$. Then

$$\boldsymbol{\Sigma\Psi} = \boldsymbol{\Psi\Lambda}, \tag{3.1.25}$$

$$\boldsymbol{\Psi}'\boldsymbol{\Psi} = \mathbf{I}_p, \tag{3.1.26}$$

where $\boldsymbol{\Psi} = (\boldsymbol{\psi}_1, \ldots, \boldsymbol{\psi}_p)$.

Now, consider a symmetric random matrix $\mathbf{V}(n)$ (n again denotes the sample size) with eigenvalues $d(n)_i$, these are the estimators of the aforementioned $\boldsymbol{\Sigma}$ and

λ_i, respectively. Let $\mathbf{H}(n)$ consist of the associated eigenvalue-normed orthogonal eigenvectors $\mathbf{h}(n)_i$. We shall omit n in the subsequent development whenever that is convenient. Hence,

$$\mathbf{VH} = \mathbf{HD}, \tag{3.1.27}$$
$$\mathbf{H}'\mathbf{H} = \mathbf{D}. \tag{3.1.28}$$

Similarly, $\mathbf{P}(n)$ consists of unit-length eigenvectors $\mathbf{p}_i(n)$ and we have

$$\mathbf{VP} = \mathbf{PD}, \tag{3.1.29}$$
$$\mathbf{P}'\mathbf{P} = \mathbf{I}_p. \tag{3.1.30}$$

In the next lemma, let the matrix derivative be defined by

$$\frac{d\,\mathbf{A}}{d_\bullet \mathbf{B}} = \frac{d\,\mathrm{vec}'\mathbf{A}}{d\,V^2(\mathbf{B})}, \tag{3.1.31}$$

where $\mathbf{B}$ is supposed to be symmetric. Moreover, by using patterned matrices in §1.3.6 we may obtain derivatives for certain structures of $\mathbf{A}$ and $\mathbf{B}$ from $d\,\mathrm{vec}\mathbf{A}$, for instance when $\mathbf{A}$ is a diagonal matrix.

Lemma 3.1.3. *Let* $\mathbf{ZG} = \mathbf{GL}$, $\mathbf{G}'\mathbf{G} = \mathbf{I}_p$, *where* $\mathbf{Z} : p \times p$ *is symmetric,* $\mathbf{G} : p \times p$ *is orthogonal and* $\mathbf{L} : p \times p$ *is diagonal with different non-zero diagonal elements. Then*

(i) $$\frac{d\,\mathbf{L}}{d_\bullet \mathbf{Z}} = \frac{d\,\mathbf{Z}}{d_\bullet \mathbf{Z}}(\mathbf{G} \otimes \mathbf{G})(\mathbf{K}_{p,p})_d;$$

(ii) $$\frac{d\,\mathbf{G}}{d_\bullet \mathbf{Z}} = -\frac{d\,\mathbf{Z}}{d_\bullet \mathbf{Z}}(\mathbf{G} \otimes \mathbf{G})(\mathbf{I} - (\mathbf{K}_{p,p})_d)(\mathbf{I} \otimes \mathbf{L} - \mathbf{L} \otimes \mathbf{I})^-(\mathbf{I} \otimes \mathbf{G}').$$

PROOF: The identity $\mathbf{G}'\mathbf{G} = \mathbf{I}$ implies that

$$\mathbf{0} = \frac{d\,\mathbf{G}'}{d_\bullet \mathbf{Z}}(\mathbf{G} \otimes \mathbf{I}) + \frac{d\,\mathbf{G}}{d_\bullet \mathbf{Z}}(\mathbf{I} \otimes \mathbf{G}) = \frac{d\,\mathbf{G}}{d_\bullet \mathbf{Z}}(\mathbf{I} \otimes \mathbf{G})(\mathbf{I} + \mathbf{K}_{p,p}). \tag{3.1.32}$$

Thus, since $(\mathbf{I} + \mathbf{K}_{p,p})(\mathbf{K}_{p,p})_d = 2(\mathbf{K}_{p,p})_d$, equation (3.1.32) establishes

$$\frac{d\,\mathbf{G}}{d_\bullet \mathbf{Z}}(\mathbf{I} \otimes \mathbf{G})(\mathbf{K}_{p,p})_d = \mathbf{0}. \tag{3.1.33}$$

The relation $\mathbf{ZG} = \mathbf{GL}$ leads to the equality

$$\frac{d\,\mathbf{Z}}{d_\bullet \mathbf{Z}}(\mathbf{G} \otimes \mathbf{I}) + \frac{d\,\mathbf{G}}{d_\bullet \mathbf{Z}}(\mathbf{I} \otimes \mathbf{Z}) = \frac{d\,\mathbf{G}}{d_\bullet \mathbf{Z}}(\mathbf{L} \otimes \mathbf{I}) + \frac{d\,\mathbf{L}}{d_\bullet \mathbf{Z}}(\mathbf{I} \otimes \mathbf{G}'). \tag{3.1.34}$$

Postmultiplying (3.1.34) by $(\mathbf{I} \otimes \mathbf{G})$ yields

$$\frac{d\,\mathbf{Z}}{d_\bullet \mathbf{Z}}(\mathbf{G} \otimes \mathbf{G}) + \frac{d\,\mathbf{G}}{d_\bullet \mathbf{Z}}(\mathbf{I} \otimes \mathbf{G})(\mathbf{I} \otimes \mathbf{L} - \mathbf{L} \otimes \mathbf{I}) = \frac{d\,\mathbf{L}}{d_\bullet \mathbf{Z}}. \tag{3.1.35}$$

Now observe that since diag$\mathbf{L}$ consists of different non-zero elements,

$$\mathcal{C}(\mathbf{I}\otimes\mathbf{L}-\mathbf{L}\otimes\mathbf{I})=\mathcal{C}(\mathbf{I}-(\mathbf{K}_{p,p})_d). \tag{3.1.36}$$

To show (3.1.36), note that $\mathbf{I}-(\mathbf{K}_{p,p})_d$ is a projection and

$$(\mathbf{K}_{p,p})_d(\mathbf{I}\otimes\mathbf{L}-\mathbf{L}\otimes\mathbf{I})=\mathbf{0},$$

which means that

$$\mathcal{C}(\mathbf{I}\otimes\mathbf{L}-\mathbf{L}\otimes\mathbf{I})\subseteq\mathcal{C}((\mathbf{K}_{p,p})_d)^{\perp}=\mathcal{C}(\mathbf{I}-(\mathbf{K}_{p,p})_d).$$

However, $r(\mathbf{I}\otimes\mathbf{L}-\mathbf{L}\otimes\mathbf{I})=p^2-p$, and from Proposition 1.3.20 it follows that in $\mathbf{B}(d_p)$ we have p basis vectors, which means that $r((\mathbf{K}_{p,p})_d)=p$. Thus, the rank $r(\mathbf{I}-(\mathbf{K}_{p,p})_d)=p^2-p$ and (3.1.36) is verified.
Postmultiplying (3.1.35) by $(\mathbf{K}_{p,p})_d$ and $\mathbf{I}-(\mathbf{K}_{p,p})_d$ gives us the equations

$$\begin{cases}\dfrac{d\mathbf{Z}}{d_\bullet\mathbf{Z}}(\mathbf{G}\otimes\mathbf{G})(\mathbf{K}_{p,p})_d=\dfrac{d\mathbf{L}}{d_\bullet\mathbf{Z}}(\mathbf{K}_{p,p})_d, & (3.1.37)\\ \dfrac{d\mathbf{Z}}{d_\bullet\mathbf{Z}}(\mathbf{G}\otimes\mathbf{G})(\mathbf{I}-(\mathbf{K}_{p,p})_d)+\dfrac{d\mathbf{G}}{d_\bullet\mathbf{Z}}(\mathbf{I}\otimes\mathbf{G})(\mathbf{I}\otimes\mathbf{L}-\mathbf{L}\otimes\mathbf{I})(\mathbf{I}-(\mathbf{K}_{p,p})_d) & \\ \quad=\dfrac{d\mathbf{L}}{d_\bullet\mathbf{Z}}(\mathbf{I}-(\mathbf{K}_{p,p})_d). & (3.1.38)\end{cases}$$

Since $\mathbf{L}$ is diagonal,

$$\frac{d\mathbf{L}}{d_\bullet\mathbf{Z}}(\mathbf{K}_{p,p})_d=\frac{d\mathbf{L}}{d_\bullet\mathbf{Z}},$$

and then from (3.1.37) it follows that (i) is established. Moreover, using this result in (3.1.38) together with (3.1.36) and the fact that $\mathbf{I}-(\mathbf{K}_{p,p})_d$ is a projector, yields

$$\frac{d\mathbf{Z}}{d_\bullet\mathbf{Z}}(\mathbf{G}\otimes\mathbf{G})(\mathbf{I}-(\mathbf{K}_{p,p})_d)+\frac{d\mathbf{G}}{d_\bullet\mathbf{Z}}(\mathbf{I}\otimes\mathbf{G})(\mathbf{I}\otimes\mathbf{L}-\mathbf{L}\otimes\mathbf{I})=\mathbf{0}.$$

This is a linear equation in $\dfrac{d\mathbf{G}}{d_\bullet\mathbf{Z}}(\mathbf{I}\otimes\mathbf{G})$, which according to §1.3.5, equals

$$\frac{d\mathbf{G}}{d_\bullet\mathbf{Z}}(\mathbf{I}\otimes\mathbf{G})=-\frac{d\mathbf{Z}}{d_\bullet\mathbf{Z}}(\mathbf{G}\otimes\mathbf{G})(\mathbf{I}-(\mathbf{K}_{p,p})_d)(\mathbf{I}\otimes\mathbf{L}-\mathbf{L}\otimes\mathbf{I})^-+\mathbf{Q}(\mathbf{K}_{p,p})_d, \tag{3.1.39}$$

where $\mathbf{Q}$ is an arbitrary matrix. However, we are going to show that $\mathbf{Q}(\mathbf{K}_{p,p})_d=\mathbf{0}$. If $\mathbf{Q}=\mathbf{0}$, postmultiplying (3.1.39) by $\mathbf{I}+\mathbf{K}_{p,p}$ yields, according to (3.1.32),

$$\frac{d\mathbf{Z}}{d_\bullet\mathbf{Z}}(\mathbf{G}\otimes\mathbf{G})(\mathbf{I}-(\mathbf{K}_{p,p})_d)(\mathbf{I}\otimes\mathbf{L}-\mathbf{L}\otimes\mathbf{I})^-(\mathbf{I}+\mathbf{K}_{p,p})=\mathbf{0}.$$

Hence, for arbitrary $\mathbf{Q}$ in (3.1.39)

$$\mathbf{0}=\mathbf{Q}(\mathbf{K}_{p,p})_d(\mathbf{I}+\mathbf{K}_{p,p})=2\mathbf{Q}(\mathbf{K}_{p,p})_d$$

and the lemma is verified. ∎

REMARK: The lemma also holds if the eigenvectors are not distinct, since

$$\mathcal{C}(\mathbf{I}\otimes\mathbf{L}-\mathbf{L}\otimes\mathbf{I})\subseteq\mathcal{C}(\mathbf{I}-(\mathbf{K}_{p,p})_d)$$

is always true.
For the eigenvalue-standardized eigenvectors the next results can be established.

Lemma 3.1.4. *Let* $\mathbf{ZF} = \mathbf{FL}$, $\mathbf{F}'\mathbf{F} = \mathbf{L}$, *where* $\mathbf{Z} : p \times p$ *is symmetric,* $\mathbf{F} : p \times p$ *is non-singular and* $\mathbf{L} : p \times p$ *is diagonal with different non-zero diagonal elements. Then*

(i) $$\frac{d\mathbf{L}}{d_\bullet\mathbf{Z}} = \frac{d\mathbf{Z}}{d_\bullet\mathbf{Z}}(\mathbf{F} \otimes \mathbf{FL}^{-1})(\mathbf{K}_{p,p})_d;$$

(ii) $$\frac{d\mathbf{F}}{d_\bullet\mathbf{Z}} = \frac{d\mathbf{Z}}{d_\bullet\mathbf{Z}}(\mathbf{F} \otimes \mathbf{F})(\mathbf{L} \otimes \mathbf{I} - \mathbf{I} \otimes \mathbf{L} + 2(\mathbf{I} \otimes \mathbf{L})(\mathbf{K}_{p,p})_d)^{-1}(\mathbf{I} \otimes \mathbf{L}^{-1}\mathbf{F}').$$

PROOF: From the relation $\mathbf{ZF} = \mathbf{FL}$ it follows that

$$\frac{d\mathbf{Z}}{d_\bullet\mathbf{Z}}(\mathbf{F} \otimes \mathbf{I}) + \frac{d\mathbf{F}}{d_\bullet\mathbf{Z}}(\mathbf{I} \otimes \mathbf{Z}) = \frac{d\mathbf{F}}{d_\bullet\mathbf{Z}}(\mathbf{L} \otimes \mathbf{I}) + \frac{d\mathbf{L}}{d_\bullet\mathbf{Z}}(\mathbf{I} \otimes \mathbf{F}'),$$

and by multiplying both sides by $\mathbf{I} \otimes \mathbf{F}$ we obtain

$$\frac{d\mathbf{Z}}{d_\bullet\mathbf{Z}}(\mathbf{F} \otimes \mathbf{F}) + \frac{d\mathbf{F}}{d_\bullet\mathbf{Z}}(\mathbf{I} \otimes \mathbf{F})(\mathbf{I} \otimes \mathbf{L} - \mathbf{L} \otimes \mathbf{I}) = \frac{d\mathbf{L}}{d_\bullet\mathbf{Z}}(\mathbf{I} \otimes \mathbf{L}). \tag{3.1.40}$$

Postmultiplying (3.1.40) by $(\mathbf{K}_{p,p})_d$ implies that

$$\frac{d\mathbf{Z}}{d_\bullet\mathbf{Z}}(\mathbf{F} \otimes \mathbf{F})(\mathbf{K}_{p,p})_d = \frac{d\mathbf{L}}{d_\bullet\mathbf{Z}}(\mathbf{I} \otimes \mathbf{L}),$$

since (3.1.36) holds, and

$$\frac{d\mathbf{L}}{d_\bullet\mathbf{Z}}(\mathbf{I} \otimes \mathbf{L})(\mathbf{K}_{p,p})_d = \frac{d\mathbf{L}}{d_\bullet\mathbf{Z}}(\mathbf{K}_{p,p})_d(\mathbf{I} \otimes \mathbf{L}) = \frac{d\mathbf{L}}{d_\bullet\mathbf{Z}}(\mathbf{I} \otimes \mathbf{L}).$$

This gives us

$$\frac{d\mathbf{L}}{d_\bullet\mathbf{Z}} = \frac{d\mathbf{Z}}{d_\bullet\mathbf{Z}}(\mathbf{F} \otimes \mathbf{F})(\mathbf{K}_{p,p})_d(\mathbf{I} \otimes \mathbf{L}^{-1})$$

and (i) is established.

For (ii) it is first noted that $\mathbf{F}'\mathbf{F} = \mathbf{L}$ implies

$$\frac{d\mathbf{F}'\mathbf{F}}{d_\bullet\mathbf{Z}} = \frac{d\mathbf{F}}{d_\bullet\mathbf{Z}}(\mathbf{I} \otimes \mathbf{F})(\mathbf{I} + \mathbf{K}_{p,p}) = \frac{d\mathbf{L}}{d_\bullet\mathbf{Z}}. \tag{3.1.41}$$

Moreover, postmultiplying (3.1.41) by $(\mathbf{K}_{p,p})_d$ yields

$$\frac{d\mathbf{L}}{d_\bullet\mathbf{Z}} = 2\frac{d\mathbf{F}}{d_\bullet\mathbf{Z}}(\mathbf{I} \otimes \mathbf{F})(\mathbf{K}_{p,p})_d,$$

since $(\mathbf{I} + \mathbf{K}_{p,p})(\mathbf{K}_{p,p})_d = 2(\mathbf{K}_{p,p})_d$ and

$$\frac{d\mathbf{L}}{d_\bullet\mathbf{Z}}(\mathbf{K}_{p,p})_d = \frac{d\mathbf{L}}{d_\bullet\mathbf{Z}}.$$

Hence, (3.1.40) is equivalent to

$$\frac{d\mathbf{F}}{d_\bullet\mathbf{Z}}(\mathbf{I}\otimes\mathbf{F})(\mathbf{L}\otimes\mathbf{I}-\mathbf{I}\otimes\mathbf{L}+2(\mathbf{I}\otimes\mathbf{L})(\mathbf{K}_{p,p})_d)=\frac{d\mathbf{Z}}{d_\bullet\mathbf{Z}}(\mathbf{F}\otimes\mathbf{F})$$

and since $\mathbf{L}\otimes\mathbf{I}-\mathbf{I}\otimes\mathbf{L}+2(\mathbf{I}\otimes\mathbf{L})(\mathbf{K}_{p,p})_d$ is non-singular, (ii) is established. ∎

In (3.1.23)–(3.1.26) relations between eigenvalues and eigenvectors were given for a positive definite matrix $\mathbf{\Sigma}$. Since we are going to estimate $\mathbf{\Sigma}$ by $\mathbf{V}$ when $\mathbf{V}$ is close to $\mathbf{\Sigma}$, it is important to understand whether there exists some function behind the eigenvectors and eigenvalues which is continuous in its arguments. Furthermore, for approximations performed later, differentiability is an important issue. The solution to this problem is given by the *implicit function theorem* (see Rudin, 1976, pp. 223–228, for example). For instance, consider the case when the eigenvectors are normed according to (3.1.24). Let

$$\widetilde{\mathbf{F}}(\mathbf{Z},\mathbf{F},\mathbf{L})=\mathbf{Z}\mathbf{F}-\mathbf{F}\mathbf{L}.$$

At the point $(\mathbf{\Sigma},\mathbf{\Gamma},\mathbf{\Lambda})$ the function $\widetilde{\mathbf{F}}(\mathbf{Z},\mathbf{F},\mathbf{L})=\mathbf{0}$, and it is assumed that the derivatives of this function with respect to $\mathbf{F},\mathbf{L}$ differ from zero. Then, according to the implicit function theorem, there exists a neighborhood U of $(\mathbf{\Sigma},\mathbf{\Gamma},\mathbf{\Lambda})$ such that

$$\widetilde{\mathbf{F}}(\mathbf{Z},\mathbf{F},\mathbf{L})=\mathbf{0},\qquad \mathbf{Z},\mathbf{F},\mathbf{L}\in U,$$

and in a neighborhood of $\mathbf{\Sigma}$ there exists a continuous and differentiable function

$$\mathrm{f}:\mathbf{Z}\to\mathbf{F},\mathbf{L}.$$

Hence, if $\mathbf{V}$ converges to $\mathbf{\Sigma}$ and in $\mathbf{\Sigma}$ there is no specific structure other than symmetry, we can find asymptotic expressions for eigenvalues and eigenvectors of $\mathbf{V}$ via Taylor expansions, where the functions are differentiated with respect to $\mathbf{V}$ and then evaluated at the point $\mathbf{V}=\mathbf{\Sigma}$. Similarly, the discussion above can be applied to the unit-length eigenvectors.

Another way of formulating the consequences of the implicit functions theorem is to note that for all positive definite matrices $\mathbf{Z}$, in some neighborhood $U(\mathbf{\Sigma})\subset\mathbb{R}^{p\times p}$ of $\mathbf{\Sigma}$, there exist vector functions $\mathbf{f}_i(\mathbf{Z})$, $\mathbf{g}_i(\mathbf{Z})$ and scalar functions $l_i(\mathbf{Z})>0$ such that

$$\mathbf{f}_i(\mathbf{Z})=\mathbf{f}_i,\quad \mathbf{g}_i(\mathbf{\Sigma})=\mathbf{g}_i,\quad l_i(\mathbf{\Sigma})=l_i,\quad i=1,\dots,p,$$

and

$$\mathbf{Z}\mathbf{F}=\mathbf{F}\mathbf{L}, \tag{3.1.42}$$

$$\mathbf{F}'\mathbf{F}=\mathbf{L}, \tag{3.1.43}$$

where $\mathbf{F}=(\mathbf{f}_1,\dots,\mathbf{f}_p)$ and $\mathbf{L}=(l_1,\dots,l_p)_d$,

$$\mathbf{Z}\mathbf{G}=\mathbf{G}\mathbf{L}, \tag{3.1.44}$$

$$\mathbf{G}'\mathbf{G}=\mathbf{I}_p, \tag{3.1.45}$$

with $\mathbf{G} = (\mathbf{g}_1, \ldots, \mathbf{g}_p)$. In our applications it is supposed that within $U(\boldsymbol{\Sigma})$ the functions $\mathbf{f}_i$, $\mathbf{g}_i$ and l_i are differentiable a sufficient number of times.
Now the preparation for treating eigenvalues and eigenvectors is finished. Suppose that, when the sample size $n \to \infty$, the following convergence relations hold

$$\mathbf{V}(n) \xrightarrow{\mathcal{P}} \boldsymbol{\Sigma}, \qquad \sqrt{n}\mathrm{vec}(\mathbf{V}(n) - \boldsymbol{\Sigma}) \xrightarrow{\mathcal{D}} N_{p^2}(\mathbf{0}, \boldsymbol{\Sigma}_{\mathbf{V}}). \tag{3.1.46}$$

Since $\mathbf{V}$ is symmetric, $\boldsymbol{\Sigma}_{\mathbf{V}}$ is singular. From Theorem 3.1.5, Lemma 3.1.3 and Lemma 3.1.4 we get the following two theorems.

Theorem 3.1.7. *Let $\mathbf{H}(n)$ and $\mathbf{D}(n)$ be defined by (3.1.27) and (3.1.28). Let $\boldsymbol{\Gamma}$ and $\boldsymbol{\Lambda}$ be defined by (3.1.23) and (3.1.24). Suppose that (3.1.46) holds. Put*

$$\mathbf{N} = \boldsymbol{\Lambda} \otimes \mathbf{I} - \mathbf{I} \otimes \boldsymbol{\Lambda} + 2(\mathbf{I} \otimes \boldsymbol{\Lambda})(\mathbf{K}_{p,p})_d. \tag{3.1.47}$$

Then, if $n \to \infty$,

(i)
$$\sqrt{n}\mathrm{vec}\big(\mathbf{D}(n) - \boldsymbol{\Lambda}\big) \xrightarrow{\mathcal{D}} N_{p^2}(\mathbf{0}, \boldsymbol{\Sigma}_{\boldsymbol{\Lambda}}),$$

where

$$\boldsymbol{\Sigma}_{\boldsymbol{\Lambda}} = (\mathbf{K}_{p,p})_d(\boldsymbol{\Gamma}' \otimes \boldsymbol{\Lambda}^{-1}\boldsymbol{\Gamma}')\boldsymbol{\Sigma}_{\mathbf{V}}(\boldsymbol{\Gamma} \otimes \boldsymbol{\Gamma}\boldsymbol{\Lambda}^{-1})(\mathbf{K}_{p,p})_d;$$

(ii)
$$\sqrt{n}\mathrm{vec}\big(\mathbf{H}_{(n)} - \boldsymbol{\Gamma}\big) \xrightarrow{\mathcal{D}} N_{p^2}(\mathbf{0}, \boldsymbol{\Sigma}_{\boldsymbol{\Gamma}}),$$

where

$$\boldsymbol{\Sigma}_{\boldsymbol{\Gamma}} = (\mathbf{I} \otimes \boldsymbol{\Gamma}\boldsymbol{\Lambda}^{-1})\mathbf{N}^{-1}(\boldsymbol{\Gamma}' \otimes \boldsymbol{\Gamma}')\boldsymbol{\Sigma}_{\mathbf{V}}(\boldsymbol{\Gamma} \otimes \boldsymbol{\Gamma})\mathbf{N}^{-1}(\mathbf{I} \otimes \boldsymbol{\Lambda}^{-1}\boldsymbol{\Gamma}'),$$

with

$$\mathbf{N}^{-1} = (\boldsymbol{\Lambda} \otimes \mathbf{I} - \mathbf{I} \otimes \boldsymbol{\Lambda})^{+} + \tfrac{1}{2}(\mathbf{I} \otimes \boldsymbol{\Lambda}^{-1})(\mathbf{K}_{p,p})_d;$$

(iii)
$$\sqrt{n}\big(\mathbf{h}_i(n) - \boldsymbol{\gamma}_i\big) \xrightarrow{D} N_p(\mathbf{0}, [\boldsymbol{\Sigma}_{\boldsymbol{\Gamma}_{ii}}]), \qquad i = 1, \ldots, p,$$

where

$$[\boldsymbol{\Sigma}_{\boldsymbol{\Gamma}_{ii}}] = \big(\boldsymbol{\gamma}_i' \otimes \boldsymbol{\Gamma}\boldsymbol{\Lambda}^{-1}[\mathbf{N}_{ii}^{-1}]\boldsymbol{\Gamma}'\big)\boldsymbol{\Sigma}_{\mathbf{V}}\big(\boldsymbol{\gamma}_i \otimes \boldsymbol{\Gamma}[\mathbf{N}_{ii}^{-1}]\boldsymbol{\Lambda}^{-1}\boldsymbol{\Gamma}'\big),$$

with

$$\mathbf{N}_{ii}^{-1} = (\boldsymbol{\Lambda} - \lambda\mathbf{I})^{+} + \tfrac{1}{2}\boldsymbol{\Lambda}^{-1}\mathbf{e}_i\mathbf{e}_i',$$

and $\mathbf{e}_i$ is the $i-th$ unit basis vector;

(iv) *the asymptotic covariance between $\sqrt{n}\mathbf{h}_i(n)$ and $\sqrt{n}\mathbf{h}_j(n)$ equals*

$$[\boldsymbol{\Sigma}_{\boldsymbol{\Gamma}_{ij}}] = \big(\boldsymbol{\gamma}_i' \otimes \boldsymbol{\Gamma}\boldsymbol{\Lambda}^{-1}[\mathbf{N}_{ii}^{-1}]\boldsymbol{\Gamma}'\big)\boldsymbol{\Sigma}_{\mathbf{V}}\big(\boldsymbol{\gamma}_j \otimes \boldsymbol{\Gamma}[\mathbf{N}_{jj}^{-1}]\boldsymbol{\Lambda}^{-1}\boldsymbol{\Gamma}'\big), \quad i \neq j;\ i,j = 1, 2, \ldots, p.$$

∎

Theorem 3.1.8. *Let* $\mathbf{P}(n)$ *and* $\mathbf{D}(n)$ *be defined by (3.1.29) and (3.1.30). Let* $\boldsymbol{\Psi}$ *and* $\boldsymbol{\Lambda}$ *be defined by (3.1.25) and (3.1.26). Suppose that (3.1.46) holds with an asymptotic dispersion matrix* $\boldsymbol{\Sigma}_{\mathbf{V}}$. *Then, if* $n \to \infty$,

(i)
$$\sqrt{n}\mathrm{vec}\big(\mathbf{D}(n) - \boldsymbol{\Lambda}\big) \xrightarrow{\mathcal{D}} N_{p^2}(\mathbf{0}, \boldsymbol{\Sigma}_{\boldsymbol{\Lambda}}),$$

where
$$\boldsymbol{\Sigma}_{\boldsymbol{\Lambda}} = (\mathbf{K}_{p,p})_d(\boldsymbol{\Psi}' \otimes \boldsymbol{\Psi}')\boldsymbol{\Sigma}_{\mathbf{V}}(\boldsymbol{\Psi} \otimes \boldsymbol{\Psi})(\mathbf{K}_{p,p})_d;$$

(ii)
$$\sqrt{n}\mathrm{vec}\big(\mathbf{P}(n) - \boldsymbol{\Psi}\big) \xrightarrow{\mathcal{D}} N_{p^2}(\mathbf{0}, \boldsymbol{\Sigma}_{\boldsymbol{\Psi}}),$$

where
$$\begin{aligned}\boldsymbol{\Sigma}_{\boldsymbol{\Psi}} = &(\mathbf{I} \otimes \boldsymbol{\Psi})(\mathbf{I} \otimes \boldsymbol{\Lambda} - \boldsymbol{\Lambda} \otimes \mathbf{I})^{-}(\mathbf{I} - (\mathbf{K}_{p,p})_d)(\boldsymbol{\Psi}' \otimes \boldsymbol{\Psi}')\boldsymbol{\Sigma}_{\mathbf{V}} \\ &\times (\boldsymbol{\Psi} \otimes \boldsymbol{\Psi})(\mathbf{I} - (\mathbf{K}_{p,p})_d)(\mathbf{I} \otimes \boldsymbol{\Lambda} - \boldsymbol{\Lambda} \otimes \mathbf{I})^{-}(\mathbf{I} \otimes \boldsymbol{\Psi}');\end{aligned}$$

(iii)
$$\sqrt{n}\big(\mathbf{p}_i(n) - \boldsymbol{\psi}_i\big) \xrightarrow{D} N_p(\mathbf{0}, [\boldsymbol{\Sigma}_{\boldsymbol{\Psi}_{ii}}]), \quad i = 1, 2, \ldots, p,$$

with
$$[\boldsymbol{\Sigma}_{\boldsymbol{\Psi}_{ii}}] = \big(\boldsymbol{\psi}_i' \otimes \boldsymbol{\Psi}(\boldsymbol{\Lambda} - \lambda_i\mathbf{I})^{+}\boldsymbol{\Psi}'\big)\boldsymbol{\Sigma}_{\mathbf{V}}\big(\boldsymbol{\psi}_i \otimes \boldsymbol{\Psi}(\boldsymbol{\Lambda} - \lambda_i\mathbf{I})^{+}\boldsymbol{\Psi}'\big);$$

(iv) *The asymptotic covariance between* $\sqrt{n}\mathbf{p}_i(n)$ *and* $\sqrt{n}\mathbf{p}_j(n)$ *equals*

$$[\boldsymbol{\Sigma}_{\boldsymbol{\Psi}_{ij}}] = \big(\boldsymbol{\psi}_i' \otimes \boldsymbol{\Psi}(\boldsymbol{\Lambda} - \lambda_i\mathbf{I})^{+}\boldsymbol{\Psi}'\boldsymbol{\Sigma}_{\mathbf{V}}(\boldsymbol{\psi}_j \otimes \boldsymbol{\Psi}(\boldsymbol{\Lambda} - \lambda_j\mathbf{I})^{+}\boldsymbol{\Psi}'\big),$$
$$i \neq j; \; i, j = 1, 2, \ldots, p.$$

REMARK: If we use a reflexive g-inverse $(\mathbf{I} \otimes \boldsymbol{\Lambda} - \boldsymbol{\Lambda} \otimes \mathbf{I})^{-}$ in the theorem, then

$$(\mathbf{I} - (\mathbf{K}_{p,p})_d)(\mathbf{I} \otimes \boldsymbol{\Lambda} - \boldsymbol{\Lambda} \otimes \mathbf{I})^{-} = (\mathbf{I} \otimes \boldsymbol{\Lambda} - \boldsymbol{\Lambda} \otimes \mathbf{I})^{-}.$$

Observe that Theorem 3.1.8 (i) corresponds to Theorem 3.1.7 (i), since

$$\begin{aligned}&(\mathbf{K}_{p,p})_d(\mathbf{I} \otimes \boldsymbol{\Lambda}^{-1})(\boldsymbol{\Gamma}' \otimes \boldsymbol{\Gamma}') \\ &\quad = (\mathbf{K}_{p,p})_d(\boldsymbol{\Lambda}^{-1/2} \otimes \boldsymbol{\Lambda}^{-1/2})(\boldsymbol{\Gamma}' \otimes \boldsymbol{\Gamma}')(\mathbf{K}_{p,p})_d(\boldsymbol{\Lambda}^{-1/2}\boldsymbol{\Gamma}' \otimes \boldsymbol{\Lambda}^{-1/2}\boldsymbol{\Gamma}') \\ &\quad = (\mathbf{K}_{p,p})_d(\boldsymbol{\Psi} \otimes \boldsymbol{\Psi}).\end{aligned}$$

3.1.6 *Asymptotic normality of eigenvalues and eigenvectors of* $\mathbf{S}$

Asymptotic distribution of eigenvalues and eigenvectors of the sample dispersion matrix

$$\mathbf{S} = \frac{1}{n-1}\sum_{i=1}^{n}(\mathbf{x}_i - \overline{\mathbf{x}})(\mathbf{x}_i - \overline{\mathbf{x}})'$$

has been examined through several decades. The first paper dates back to Girshick (1939), who derived expressions for the variances and covariances of the asymptotic normal distributions of the eigenvalues and the coordinates of eigenvalue normed eigenvectors, assuming that the eigenvalues λ_i of the population dispersion matrix are all different and $\mathbf{x}_i \sim N_p(\boldsymbol{\mu}, \boldsymbol{\Sigma})$. Anderson (1963) considered the case $\lambda_1 \geq \lambda_2 \geq \ldots \geq \lambda_p$ and described the asymptotic distribution of the eigenvalues and the coordinates of the unit-length eigenvectors. Waternaux (1976, 1984) generalized the results of Girshick from the normal population to the case when only the fourth order population moments are assumed to exist. Fujikoshi (1980) found asymptotic expansions of the distribution functions of the eigenvalues of $\mathbf{S}$ up to the term of order n^{-1} under the assumptions of Waternaux. Fang & Krishnaiah (1982) presented a general asymptotic theory for functions of eigenvalues of $\mathbf{S}$ in the case of multiple eigenvalues, i.e. all eigenvalues do not have to be different. For unit-length eigenvectors the results of Anderson (1963) were generalized to the case of finite fourth order moments by Davis (1977). As in the case of multiple eigenvalues, the eigenvectors are not uniquely determined. Therefore, in this case it seems reasonable to use eigenprojectors. We are going to present the convergence results for eigenvalues and eigenvectors of $\mathbf{S}$, assuming that all eigenvalues are different. When applying Theorem 3.1.3 we get directly from Theorem 3.1.8 the asymptotic distribution of eigenvalues and unit-length eigenvectors of the sample variance matrix $\mathbf{S}$.

Theorem 3.1.9. *Let the dispersion matrix* $\boldsymbol{\Sigma}$ *have eigenvalues* $\lambda_1 > \ldots > \lambda_p > 0$, $\boldsymbol{\Lambda} = (\lambda_1, \ldots, \lambda_p)_d$ *and associated unit-length eigenvectors* $\boldsymbol{\psi}_i$ *with* $\psi_{ii} > 0$, $i = 1, \ldots, p$. *The latter are assembled into the matrix* $\boldsymbol{\Psi}$, *where* $\boldsymbol{\Psi}'\boldsymbol{\Psi} = \mathbf{I}_p$. *Let the sample dispersion matrix* $\mathbf{S}(n)$ *have* p *eigenvalues* $d_i(n)$, $\mathbf{D}(n) = (d_1(n), \ldots, d_p(n))_d$, *where* n *is the sample size. Furthermore, let* $\mathbf{P}(n)$ *consist of the associated orthonormal eigenvectors* $\mathbf{p}_i(n)$. *Put*

$$\mathbf{M}_4 = E[(\mathbf{x}_i - \boldsymbol{\mu}) \otimes (\mathbf{x}_i - \boldsymbol{\mu})' \otimes (\mathbf{x}_i - \boldsymbol{\mu}) \otimes (\mathbf{x}_i - \boldsymbol{\mu})'] < \infty.$$

Then, when $n \to \infty$,

(i)
$$\sqrt{n}\mathrm{vec}(\mathbf{D}(n) - \boldsymbol{\Lambda}) \xrightarrow{\mathcal{D}} N_{p^2}(\mathbf{0}, \boldsymbol{\Sigma}_{\boldsymbol{\Lambda}}),$$

where

$$\boldsymbol{\Sigma}_{\boldsymbol{\Lambda}} = (\mathbf{K}_{p,p})_d(\boldsymbol{\Psi}' \otimes \boldsymbol{\Psi}')\mathbf{M}_4(\boldsymbol{\Psi} \otimes \boldsymbol{\Psi})(\mathbf{K}_{p,p})_d - \mathrm{vec}\boldsymbol{\Lambda}\mathrm{vec}'\boldsymbol{\Lambda};$$

(ii)
$$\sqrt{n}\mathrm{vec}(\mathbf{P}(n) - \boldsymbol{\Psi}) \xrightarrow{\mathcal{D}} N_{p^2}(\mathbf{0}, \boldsymbol{\Sigma}_{\boldsymbol{\Psi}}),$$

where

$$\begin{aligned}\boldsymbol{\Sigma}_{\boldsymbol{\Psi}} =& (\mathbf{I} \otimes \boldsymbol{\Psi})(\mathbf{I} \otimes \boldsymbol{\Lambda} - \boldsymbol{\Lambda} \otimes \mathbf{I})^-(\mathbf{I} - (\mathbf{K}_{p,p})_d)(\boldsymbol{\Psi}' \otimes \boldsymbol{\Psi}')\mathbf{M}_4 \\ & \times (\boldsymbol{\Psi} \otimes \boldsymbol{\Psi})(\mathbf{I} - (\mathbf{K}_{p,p})_d)(\mathbf{I} \otimes \boldsymbol{\Lambda} - \boldsymbol{\Lambda} \otimes \mathbf{I})^-(\mathbf{I} \otimes \boldsymbol{\Psi}').\end{aligned}$$

PROOF: In (ii) we have used that

$$(\mathbf{I} - (\mathbf{K}_{p,p})_d)(\boldsymbol{\Psi}' \otimes \boldsymbol{\Psi}')\mathrm{vec}\boldsymbol{\Sigma} = \mathbf{0}.$$

∎

When it is supposed that we have an underlying normal distribution, i.e. $\mathbf{S}$ is Wishart distributed, the theorem can be simplified via Corollary 3.1.4.1.

Corollary 3.1.9.1. *Let* $\mathbf{x}_i \sim N_p(\boldsymbol{\mu}, \boldsymbol{\Sigma})$, $i = 1, 2, \ldots, n$. *Then, if* $n \to \infty$,

(i)
$$\sqrt{n}\text{vec}(\mathbf{D}(n) - \boldsymbol{\Lambda}) \xrightarrow{\mathcal{D}} N_{p^2}(\mathbf{0}, \boldsymbol{\Sigma}_{\boldsymbol{\Lambda}}^N),$$

where
$$\boldsymbol{\Sigma}_{\boldsymbol{\Lambda}}^N = 2(\mathbf{K}_{p,p})_d(\boldsymbol{\Lambda} \otimes \boldsymbol{\Lambda})(\mathbf{K}_{p,p})_d;$$

(ii)
$$\sqrt{n}\text{vec}(\mathbf{P}(n) - \boldsymbol{\Psi}) \xrightarrow{\mathcal{D}} N_{p^2}(\mathbf{0}, \boldsymbol{\Sigma}_{\boldsymbol{\Psi}}^N),$$

where
$$\begin{aligned}\boldsymbol{\Sigma}_{\boldsymbol{\Psi}}^N =& (\mathbf{I} \otimes \boldsymbol{\Psi})(\mathbf{I} \otimes \boldsymbol{\Lambda} - \boldsymbol{\Lambda} \otimes \mathbf{I})^-(\mathbf{I} - (\mathbf{K}_{p,p})_d) \\ &\times (\mathbf{I} + \mathbf{K}_{p,p})(\boldsymbol{\Lambda} \otimes \boldsymbol{\Lambda})(\mathbf{I} - (\mathbf{K}_{p,p})_d)(\mathbf{I} \otimes \boldsymbol{\Lambda} - \boldsymbol{\Lambda} \otimes \mathbf{I})^-(\mathbf{I} \otimes \boldsymbol{\Psi}').\end{aligned}$$

∎

REMARK: If $(\mathbf{I} \otimes \boldsymbol{\Lambda} - \boldsymbol{\Lambda} \otimes \mathbf{I})^-$ is reflexive,

$$\boldsymbol{\Sigma}_{\boldsymbol{\Psi}}^N = (\mathbf{I} \otimes \boldsymbol{\Psi})(\mathbf{I} \otimes \boldsymbol{\Lambda} - \boldsymbol{\Lambda} \otimes \mathbf{I})^-(\mathbf{I} + \mathbf{K}_{p,p})(\boldsymbol{\Lambda} \otimes \boldsymbol{\Lambda})(\mathbf{I} \otimes \boldsymbol{\Lambda} - \boldsymbol{\Lambda} \otimes \mathbf{I})^-(\mathbf{I} \otimes \boldsymbol{\Psi}').$$

In the case of an elliptical population the results are only slightly more complicated than in the normal case.

Corollary 3.1.9.2. *Let* $\mathbf{x}_i \sim E_p(\boldsymbol{\mu}, \boldsymbol{\Upsilon})$, $D[\mathbf{x}_i] = \boldsymbol{\Sigma}$, $i = 1, 2, \ldots, n$, *and the kurtosis parameter be denoted by* κ. *Then, if* $n \to \infty$,

(i)
$$\sqrt{n}\text{vec}(\mathbf{D}(n) - \boldsymbol{\Lambda}) \xrightarrow{\mathcal{D}} N_{p^2}(\mathbf{0}, \boldsymbol{\Sigma}_{\boldsymbol{\Lambda}}^E),$$

where
$$\boldsymbol{\Sigma}_{\boldsymbol{\Lambda}}^E = 2(1+\kappa)(\mathbf{K}_{p,p})_d(\boldsymbol{\Lambda} \otimes \boldsymbol{\Lambda})(\mathbf{K}_{p,p})_d + \kappa\text{vec}\boldsymbol{\Lambda}\text{vec}'\boldsymbol{\Lambda};$$

(ii)
$$\sqrt{n}\text{vec}(\mathbf{P}(n) - \boldsymbol{\Psi}) \xrightarrow{\mathcal{D}} N_{p^2}(\mathbf{0}, \boldsymbol{\Sigma}_{\boldsymbol{\Psi}}^E),$$

where
$$\begin{aligned}\boldsymbol{\Sigma}_{\boldsymbol{\Psi}}^E =& (1+\kappa)(\mathbf{I} \otimes \boldsymbol{\Psi})(\mathbf{I} \otimes \boldsymbol{\Lambda} - \boldsymbol{\Lambda} \otimes \mathbf{I})^-(\mathbf{I} - (\mathbf{K}_{p,p})_d)(\mathbf{I} + \mathbf{K}_{p,p})(\boldsymbol{\Lambda} \otimes \boldsymbol{\Lambda}) \\ &\times (\mathbf{I} - (\mathbf{K}_{p,p})_d)(\mathbf{I} \otimes \boldsymbol{\Lambda} - \boldsymbol{\Lambda} \otimes \mathbf{I})^-(\mathbf{I} \otimes \boldsymbol{\Psi}').\end{aligned}$$

∎

Applying Theorem 3.1.7 to the matrix $\mathbf{S}$ gives us the following result.

Theorem 3.1.10. *Let the dispersion matrix* $\mathbf{\Sigma}$ *have eigenvalues* $\lambda_1 > \ldots > \lambda_p > 0$, $\mathbf{\Lambda} = (\lambda_1, \ldots, \lambda_p)_d$, *and associated eigenvectors* $\boldsymbol{\gamma}_i$ *be of length* $\sqrt{\lambda_i}$, *with* $\gamma_{ii} > 0$, $i = 1, \ldots, p$. *The latter are collected into the matrix* $\mathbf{\Gamma}$, *where* $\mathbf{\Gamma}'\mathbf{\Gamma} = \mathbf{\Lambda}$. *Let the sample variance matrix* $\mathbf{S}(n)$ *have* p *eigenvalues* $d_i(n)$, $\mathbf{D}(n) = (d_1(n), \ldots, d_p(n))_d$, *where* n *is the sample size. Furthermore, let* $\mathbf{H}(n)$ *consist of the associated eigenvalue-normed orthogonal eigenvectors* $\mathbf{h}_i(n)$. *Put*

$$\mathbf{M}_4 = E[(\mathbf{x}_i - \boldsymbol{\mu}) \otimes (\mathbf{x}_i - \boldsymbol{\mu})' \otimes (\mathbf{x}_i - \boldsymbol{\mu}) \otimes (\mathbf{x}_i - \boldsymbol{\mu})'] < \infty.$$

Then, if $n \to \infty$,

(i)
$$\sqrt{n}\mathrm{vec}(\mathbf{D}(n) - \mathbf{\Lambda}) \xrightarrow{\mathcal{D}} N_{p^2}(\mathbf{0}, \mathbf{\Sigma}_{\mathbf{\Lambda}}),$$

where

$$\mathbf{\Sigma}_{\mathbf{\Lambda}} = (\mathbf{K}_{p,p})_d(\mathbf{\Gamma}' \otimes \mathbf{\Lambda}^{-1}\mathbf{\Gamma}')\mathbf{M}_4(\mathbf{\Gamma} \otimes \mathbf{\Gamma}\mathbf{\Lambda}^{-1})(\mathbf{K}_{p,p})_d - \mathrm{vec}\mathbf{\Lambda}\mathrm{vec}'\mathbf{\Lambda};$$

(ii)
$$\sqrt{n}\mathrm{vec}(\mathbf{H}(n) - \mathbf{\Gamma}) \xrightarrow{D} N_{p^2}(\mathbf{0}, \mathbf{\Sigma}_{\mathbf{\Gamma}}),$$

where

$$\begin{aligned}\mathbf{\Sigma}_{\mathbf{\Gamma}} =& (\mathbf{I} \otimes \mathbf{\Gamma}\mathbf{\Lambda}^{-1})\mathbf{N}^{-1}(\mathbf{\Gamma}' \otimes \mathbf{\Gamma}')\mathbf{M}_4(\mathbf{\Gamma} \otimes \mathbf{\Gamma})\mathbf{N}^{-1}(\mathbf{I} \otimes \mathbf{\Lambda}^{-1}\mathbf{\Gamma}') \\ &- (\mathbf{I} \otimes \mathbf{\Gamma})\mathbf{N}^{-1}\mathrm{vec}\mathbf{\Lambda}\mathrm{vec}'\mathbf{\Lambda}\mathbf{N}^{-1}(\mathbf{I} \otimes \mathbf{\Gamma}'),\end{aligned}$$

with $\mathbf{N}$ *given in by (3.1.47).*

PROOF: In (ii) we have used that

$$(\mathbf{I} \otimes \mathbf{\Gamma}\mathbf{\Lambda}^{-1})\mathbf{N}^{-1}(\mathbf{\Gamma}' \otimes \mathbf{\Gamma}')\mathrm{vec}\mathbf{\Sigma} = (\mathbf{I} \otimes \mathbf{\Gamma})\mathbf{N}^{-1}\mathrm{vec}\mathbf{\Lambda}.$$

∎

In the following corollary we apply Theorem 3.1.10 in the case of an elliptical distribution with the help of Corollary 3.1.4.2.

Corollary 3.1.10.1. *Let* $\mathbf{x}_i \sim E_p(\boldsymbol{\mu}, \boldsymbol{\Upsilon})$, $D[\mathbf{x}_i] = \mathbf{\Sigma}$, $i = 1, 2, \ldots, n$ *and the kurtosis parameter be denoted by* κ. *Then, if* $n \to \infty$,

(i)
$$\sqrt{n}\mathrm{vec}(\mathbf{D}(n) - \mathbf{\Lambda}) \xrightarrow{\mathcal{D}} N_{p^2}(\mathbf{0}, \mathbf{\Sigma}_{\mathbf{\Lambda}}^E),$$

where

$$\mathbf{\Sigma}_{\mathbf{\Lambda}}^E = 2(1 + \kappa)(\mathbf{K}_{p,p})_d(\mathbf{\Lambda} \otimes \mathbf{\Lambda}) + \kappa\mathrm{vec}\mathbf{\Lambda}\mathrm{vec}'\mathbf{\Lambda};$$

(ii)
$$\sqrt{n}\mathrm{vec}(\mathbf{H}(n) - \mathbf{\Gamma}) \xrightarrow{\mathcal{D}} N_{p^2}(\mathbf{0}, \mathbf{\Sigma}_{\mathbf{\Gamma}}^E),$$

where

$$\begin{aligned}\mathbf{\Sigma}_{\mathbf{\Gamma}}^E =& (1 + \kappa)(\mathbf{I} \otimes \mathbf{\Gamma})\mathbf{N}^{-1}(\mathbf{\Lambda}^2 \otimes \mathbf{I})\mathbf{N}^{-1}(\mathbf{I} \otimes \mathbf{\Gamma}') \\ &+ (1 + \kappa)(\mathbf{I} \otimes \mathbf{\Gamma})\mathbf{N}^{-1}\mathbf{K}_{p,p}(\mathbf{\Lambda} \otimes \mathbf{\Lambda})\mathbf{N}^{-1}(\mathbf{I} \otimes \mathbf{\Gamma}') \\ &+ \kappa(\mathbf{I} \otimes \mathbf{\Gamma})\mathbf{N}^{-1}\mathrm{vec}\mathbf{\Lambda}\mathrm{vec}'\mathbf{\Lambda}\mathbf{N}^{-1}(\mathbf{I} \otimes \mathbf{\Gamma}')\end{aligned}$$

and $\mathbf{N}$ *is given by (3.1.47).*

PROOF: When proving (i) it has been noted that

$$\begin{aligned}&(\mathbf{K}_{p,p})_d(\mathbf{\Gamma}' \otimes \mathbf{\Lambda}^{-1}\mathbf{\Gamma}')(\mathbf{I}_{p^2} + \mathbf{K}_{p,p})(\mathbf{\Sigma} \otimes \mathbf{\Sigma})(\mathbf{\Gamma} \otimes \mathbf{\Gamma}\mathbf{\Lambda}^{-1})(\mathbf{K}_{p,p})_d \\ &\quad = (\mathbf{K}_{p,p})_d(\mathbf{\Lambda}^2 \otimes \mathbf{I}) + (\mathbf{K}_{p,p})_d(\mathbf{\Lambda} \otimes \mathbf{\Lambda}) = 2(\mathbf{K}_{p,p})_d(\mathbf{\Lambda} \otimes \mathbf{\Lambda}).\end{aligned}$$

∎

Moreover, by putting $\kappa = 0$, from Corollary 3.1.10.1 the corresponding relations for a normally distributed population will follow.

Corollary 3.1.10.2. *Let* $\mathbf{x}_i \sim N_p(\boldsymbol{\mu}, \boldsymbol{\Sigma})$, $i = 1, 2, \ldots, n$. *Then, if* $n \to \infty$,

(i)
$$\sqrt{n}\operatorname{vec}(\mathbf{D}(n) - \boldsymbol{\Lambda}) \xrightarrow{\mathcal{D}} N_{p^2}(\mathbf{0}, \boldsymbol{\Sigma}_{\boldsymbol{\Lambda}}^N),$$

where
$$\boldsymbol{\Sigma}_{\boldsymbol{\Lambda}}^N = 2(\mathbf{K}_{p,p})_d(\boldsymbol{\Lambda} \otimes \boldsymbol{\Lambda});$$

(ii)
$$\sqrt{n}\operatorname{vec}(\mathbf{H}(n) - \boldsymbol{\Gamma}) \xrightarrow{\mathcal{D}} N_{p^2}(\mathbf{0}, \boldsymbol{\Sigma}_{\boldsymbol{\Gamma}}^N),$$

where
$$\boldsymbol{\Sigma}_{\boldsymbol{\Gamma}}^N = (\mathbf{I} \otimes \boldsymbol{\Lambda})\mathbf{N}^{-1}(\boldsymbol{\Lambda}^2 \otimes \mathbf{I})\mathbf{N}^{-1}(\mathbf{I} \otimes \boldsymbol{\Gamma}') + (\mathbf{I} \otimes \boldsymbol{\Lambda})\mathbf{N}^{-1}\mathbf{K}_{p,p}(\boldsymbol{\Lambda} \otimes \boldsymbol{\Lambda})\mathbf{N}^{-1}(\mathbf{I} \otimes \boldsymbol{\Gamma}')$$

and $\mathbf{N}$ *is given by (3.1.47).* ∎

3.1.7 *Asymptotic normality of eigenvalues and eigenvectors of* R

Asymptotic distributions of eigenvalues and eigenvectors of the sample correlation matrix
$$\mathbf{R} = \mathbf{S}_d^{-\frac{1}{2}}\mathbf{S}\mathbf{S}_d^{-\frac{1}{2}}$$
are much more complicated than the distributions of the same functions of the sample dispersion matrix. When studying asymptotic behavior of the eigenfunctions of **R** the pioneering paper by Girshick (1939) should again be mentioned. He obtained the asymptotic normal distribution of the eigenvalues of **R** for the case of a normal population $N_p(\boldsymbol{\mu}, \boldsymbol{\Sigma})$. Kollo (1977) presented his results in a matrix form for an enlarged class of population distributions. Konishi (1979) gave asymptotic expansion of the distribution function of an eigenvalue of **R** for $N_p(\boldsymbol{\mu}, \boldsymbol{\Sigma})$, assuming that eigenvalues of the theoretical correlation matrix can be multiple. Fang & Krishnaiah (1982) generalized his results and only assumed the existence of the fourth order moments of the population distribution. The asymptotic distribution of the eigenvectors of **R** was derived by Kollo (1977), for a class of population distributions which includes the normal distributions, assuming that the theoretical eigenvalues are not multiple. Konishi (1979) gave asymptotic expansions of the coordinates of eigenvectors of **R** for a normal population when the eigenvalues of the theoretical correlation matrix can be multiple. For the general case, when existence of the fourth order moments is assumed, the asymptotic distributions of eigenvalues and eigenvectors of **R** follow from the general asymptotic distribution formula for a symmetric matrix, which was given by Kollo & Neudecker (1993), where also the normal and elliptic populations were considered as special cases (see also Schott, 1997a).

The general scheme of getting asymptotic distributions of eigenvalues and eigenvectors of **R** is the same as in the case of the dispersion matrix but the formulas become more complicated because the correlation matrix is a function of the dispersion matrix as it can be seen from (3.1.12). Therefore we shall leave out details in proofs and only present the general lines in the following text.

Let us first consider eigenvalue-normed eigenvectors. In the general case, when we assume that the fourth order population moments are finite, we get, by applying Theorem 3.1.5, the asymptotic distribution directly from Theorem 3.1.6 and Theorem 3.1.7.

Theorem 3.1.11. *Let the population correlation matrix $\mathbf{\Omega}$, which is defined by (3.1.12), have eigenvalues $\lambda_1 > \ldots > \lambda_p > 0$, $\mathbf{\Lambda} = (\lambda_1, \ldots, \lambda_p)_d$, and let the associated eigenvectors $\boldsymbol{\gamma}_i$ be of length $\sqrt{\lambda_i}$, with $\gamma_{ii} > 0$, $i = 1, \ldots, p$. The latter are collected into the matrix $\mathbf{\Gamma}$, where $\mathbf{\Gamma}'\mathbf{\Gamma} = \mathbf{\Lambda}$. Let the sample correlation matrix $\mathbf{R}(n)$ have eigenvalues $d_i(n)$, $\mathbf{D}(n) = (d_1(n), \ldots, d_p(n))_d$, where n is the sample size. Furthermore, let $\mathbf{H}(n) : p \times p$ consist of the associated eigenvalue-normed orthogonal eigenvectors $\mathbf{h}_i(n)$. Then, if $n \to \infty$,*

(i)
$$\sqrt{n}\mathrm{vec}(\mathbf{D}(n) - \mathbf{\Lambda}) \xrightarrow{\mathcal{D}} N_{p^2}(\mathbf{0}, \mathbf{\Sigma}_\Lambda),$$
where
$$\mathbf{\Sigma}_{\mathbf{\Lambda}} = (\mathbf{K}_{p,p})_d(\mathbf{\Gamma}' \otimes \mathbf{\Lambda}^{-1}\mathbf{\Gamma}')\mathbf{\Xi}_R\mathbf{\Pi}\mathbf{\Xi}'_R(\mathbf{\Gamma} \otimes \mathbf{\Gamma}\mathbf{\Lambda}^{-1})(\mathbf{K}_{p,p})_d$$
and the matrices $\mathbf{\Pi}$ and $\mathbf{\Xi}_R$ are given by (3.1.5) and (3.1.13), respectively;

(ii)
$$\sqrt{n}\mathrm{vec}(\mathbf{H}(n) - \mathbf{\Gamma}) \xrightarrow{D} N_{p^2}(\mathbf{0}, \mathbf{\Sigma}_{\mathbf{\Gamma}}),$$
where
$$\mathbf{\Sigma}_{\mathbf{\Gamma}} = (\mathbf{I} \otimes \mathbf{\Gamma}\mathbf{\Lambda}^{-1})\mathbf{N}^{-1}(\mathbf{\Gamma}' \otimes \mathbf{\Gamma}')\mathbf{\Xi}_R\mathbf{\Pi}\mathbf{\Xi}'_R(\mathbf{\Gamma} \otimes \mathbf{\Gamma})\mathbf{N}^{-1}(\mathbf{I} \otimes \mathbf{\Lambda}^{-1}\mathbf{\Gamma}'),$$
with $\mathbf{N}$ given in (3.1.47), and $\mathbf{\Pi}$ and $\mathbf{\Xi}_R$ as in (i). ∎

As before we shall apply the theorem to the elliptical distribution.

Corollary 3.1.11.1. *Let $\mathbf{x}_i \sim E_p(\boldsymbol{\mu}, \mathbf{\Upsilon})$, $D[\mathbf{x}_i] = \mathbf{\Sigma}$, $i = 1, 2, \ldots, n$, and the kurtosis parameter be denoted by κ. Then, if $n \to \infty$,*

(i)
$$\sqrt{n}\mathrm{vec}(\mathbf{D}(n) - \mathbf{\Lambda}) \xrightarrow{\mathcal{D}} N_{p^2}(\mathbf{0}, \mathbf{\Sigma}^E_{\mathbf{\Lambda}}),$$
where
$$\mathbf{\Sigma}^E_{\mathbf{\Lambda}} = (\mathbf{K}_{p,p})_d(\mathbf{\Gamma}' \otimes \mathbf{\Lambda}^{-1}\mathbf{\Gamma}')\mathbf{\Xi}_R\mathbf{\Pi}^E\mathbf{\Xi}'_R(\mathbf{\Gamma} \otimes \mathbf{\Gamma}\mathbf{\Lambda}^{-1})(\mathbf{K}_{p,p})_d$$
and $\mathbf{\Pi}^E$ is defined in Corollary 3.1.4.2 and $\mathbf{\Xi}_R$ is given by (3.1.13);

(ii)
$$\sqrt{n}\mathrm{vec}(\mathbf{H}(n) - \mathbf{\Gamma}) \xrightarrow{D} N_{p^2}(\mathbf{0}, \mathbf{\Sigma}^E_{\mathbf{\Gamma}}),$$
where
$$\mathbf{\Sigma}^E_{\mathbf{\Gamma}} = (\mathbf{I} \otimes \mathbf{\Gamma}\mathbf{\Lambda}^{-1})\mathbf{N}^{-1}(\mathbf{\Gamma}' \otimes \mathbf{\Gamma}')\mathbf{\Xi}_R\mathbf{\Pi}^E\mathbf{\Xi}'_R(\mathbf{\Gamma} \otimes \mathbf{\Gamma})\mathbf{N}^{-1}(\mathbf{I} \otimes \mathbf{\Lambda}^{-1}\mathbf{\Gamma}'),$$
with $\mathbf{N}$ given in (3.1.47), and $\mathbf{\Pi}^E$ and $\mathbf{\Xi}_R$ as in (i). ∎

Corollary 3.1.11.2. *Let $\mathbf{x}_i \sim N_p(\boldsymbol{\mu}, \mathbf{\Sigma})$, $i = 1, 2, \ldots, n$. Then, if $n \to \infty$,*

(i)
$$\sqrt{n}\mathrm{vec}(\mathbf{D}(n) - \mathbf{\Lambda}) \xrightarrow{\mathcal{D}} N_{p^2}(\mathbf{0}, \mathbf{\Sigma}^N_{\mathbf{\Lambda}}),$$
where
$$\mathbf{\Sigma}^N_{\mathbf{\Lambda}} = (\mathbf{K}_{p,p})_d(\mathbf{\Gamma}' \otimes \mathbf{\Lambda}^{-1}\mathbf{\Gamma}')\mathbf{\Xi}_R\mathbf{\Pi}^N\mathbf{\Xi}'_R(\mathbf{\Gamma} \otimes \mathbf{\Gamma}\mathbf{\Lambda}^{-1})(\mathbf{K}_{p,p})_d$$
and $\mathbf{\Pi}^N$ and $\mathbf{\Xi}_R$ are given in Corollary 3.1.4.1 and (3.1.13), respectively;

(ii)
$$\sqrt{n}\mathrm{vec}(\mathbf{H}(n) - \mathbf{\Gamma}) \xrightarrow{D} N_{p^2}(\mathbf{0}, \mathbf{\Sigma}^N_{\mathbf{\Gamma}}),$$
where
$$\mathbf{\Sigma}^N_{\mathbf{\Gamma}} = (\mathbf{I} \otimes \mathbf{\Gamma}\mathbf{\Lambda}^{-1})\mathbf{N}^{-1}(\mathbf{\Gamma}' \otimes \mathbf{\Gamma}')\mathbf{\Xi}_R\mathbf{\Pi}^N\mathbf{\Xi}'_R(\mathbf{\Gamma} \otimes \mathbf{\Gamma})\mathbf{N}^{-1}(\mathbf{I} \otimes \mathbf{\Lambda}^{-1}\mathbf{\Gamma}'),$$
with $\mathbf{N}$ given by (3.1.47), and $\mathbf{\Pi}^N$ and $\mathbf{\Xi}_R$ as in (i). ∎

Now we present the asymptotic distributions of the eigenvalues and corresponding unit-length eigenvectors, which directly can be obtained from Theorem 3.1.8 by applying Theorem 3.1.6.

Theorem 3.1.12. *Let the population correlation matrix* $\mathbf{\Omega}$ *have eigenvalues* $\lambda_1 > \ldots > \lambda_p > 0$, $\mathbf{\Lambda} = (\lambda_1 \ldots \lambda_p)_d$ *and associated unit-length eigenvectors* $\boldsymbol{\psi}_i$ *with* $\psi_{ii} > 0, i = 1, \ldots, p$. *The latter are assembled into the matrix* $\mathbf{\Psi}$, *where* $\mathbf{\Psi}'\mathbf{\Psi} = \mathbf{I}_p$. *Let the sample correlation matrix* $\mathbf{R}(n)$ *have eigenvalues* $d_i(n)$, *where* n *is the sample size. Furthermore, let* $\mathbf{P}(n)$ *consist of associated orthonormal eigenvectors* $\mathbf{p}_i(n)$. *Then, if* $n \to \infty$,

(i)
$$\sqrt{n}\mathrm{vec}(\mathbf{D}(n) - \mathbf{\Lambda}) \xrightarrow{\mathcal{D}} N_{p^2}(\mathbf{0}, \mathbf{\Sigma}_{\mathbf{\Lambda}}),$$

where

$$\mathbf{\Sigma}_{\mathbf{\Lambda}} = (\mathbf{K}_{p,p})_d(\mathbf{\Psi}' \otimes \mathbf{\Psi}')\mathbf{\Xi}_R\mathbf{\Pi}\mathbf{\Xi}_R'(\mathbf{\Psi} \otimes \mathbf{\Psi})(\mathbf{K}_{p,p})_d$$

and $\mathbf{\Pi}$ *and* $\mathbf{\Xi}_R$ *are defined by (3.1.5) and (3.1.13), respectively;*

(ii)
$$\sqrt{n}\mathrm{vec}(\mathbf{P}(n) - \mathbf{\Psi}) \xrightarrow{\mathcal{D}} N_{p^2}(\mathbf{0}, \mathbf{\Sigma}_{\mathbf{\Psi}}),$$

where

$$\begin{aligned}\mathbf{\Sigma}_{\mathbf{\Psi}} =& (\mathbf{I} \otimes \mathbf{\Psi})(\mathbf{I} \otimes \mathbf{\Lambda} - \mathbf{\Lambda} \otimes \mathbf{I})^-(\mathbf{I} - (\mathbf{K}_{p,p})_d)(\mathbf{\Psi}' \otimes \mathbf{\Psi}')\mathbf{\Xi}_R\mathbf{\Pi}\mathbf{\Xi}_R' \\ & \times (\mathbf{\Psi} \otimes \mathbf{\Psi})(\mathbf{I} - (\mathbf{K}_{p,p})_d)(\mathbf{I} \otimes \mathbf{\Lambda} - \mathbf{\Lambda} \otimes \mathbf{I})^-(\mathbf{I} \otimes \mathbf{\Psi}')\end{aligned}$$

and the matrices $\mathbf{\Pi}$ *and* $\mathbf{\Xi}_R$ *are as in (i).* ∎

The theorem is going to be applied to normal and elliptical populations.

Corollary 3.1.12.1. *Let* $\mathbf{x}_i \sim E_p(\boldsymbol{\mu}, \mathbf{\Upsilon})$, $D[\mathbf{x}_i] = \mathbf{\Sigma}$, $i = 1, 2, \ldots, n$, *and the kurtosis parameter be denoted by* κ. *Then, if* $n \to \infty$,

(i)
$$\sqrt{n}\mathrm{vec}(\mathbf{D}(n) - \mathbf{\Lambda}) \xrightarrow{\mathcal{D}} N_{p^2}(\mathbf{0}, \mathbf{\Sigma}_{\mathbf{\Lambda}}^E),$$

where

$$\mathbf{\Sigma}_{\mathbf{\Lambda}}^E = (\mathbf{K}_{p,p})_d(\mathbf{\Psi}' \otimes \mathbf{\Psi}')\mathbf{\Xi}_R\mathbf{\Pi}^E\mathbf{\Xi}_R'(\mathbf{\Psi} \otimes \mathbf{\Psi})(\mathbf{K}_{p,p})_d$$

and the matrices $\mathbf{\Pi}^E$ *and* $\mathbf{\Xi}_R$ *are given by Corollary 3.1.4.2 and (3.1.13), respectively;*

(ii)
$$\sqrt{n}\mathrm{vec}(\mathbf{P}(n) - \mathbf{\Psi}) \xrightarrow{\mathcal{D}} N_{p^2}(\mathbf{0}, \mathbf{\Sigma}_{\mathbf{\Psi}}^E),$$

where

$$\begin{aligned}\mathbf{\Sigma}_{\mathbf{\Psi}}^E =& (\mathbf{I} \otimes \mathbf{\Psi})(\mathbf{I} \otimes \mathbf{\Lambda} - \mathbf{\Lambda} \otimes \mathbf{I})^-(\mathbf{I} - (\mathbf{K}_{p,p})_d)(\mathbf{\Psi}' \otimes \mathbf{\Psi}')\mathbf{\Xi}_R\mathbf{\Pi}^E\mathbf{\Xi}_R' \\ & \times (\mathbf{\Psi} \otimes \mathbf{\Psi})(\mathbf{I} - (\mathbf{K}_{p,p})_d)(\mathbf{I} \otimes \mathbf{\Lambda} - \mathbf{\Lambda} \otimes \mathbf{I})^-(\mathbf{I} \otimes \mathbf{\Psi}')\end{aligned}$$

and the matrices $\mathbf{\Pi}^E$ *and* $\mathbf{\Xi}_R$ *are as in (i).*

Corollary 3.1.12.2. *Let* $\mathbf{x}_i \sim N_p(\boldsymbol{\mu}, \boldsymbol{\Sigma})$. *Then, if* $n \to \infty$,

(i)
$$\sqrt{n}\text{vec}(\mathbf{D}(n) - \boldsymbol{\Lambda}) \xrightarrow{\mathcal{D}} N_{p^2}(\mathbf{0}, \boldsymbol{\Sigma}_{\boldsymbol{\Lambda}}^N),$$

where

$$\boldsymbol{\Sigma}_{\boldsymbol{\Lambda}}^N = (\mathbf{K}_{p,p})_d(\boldsymbol{\Psi}' \otimes \boldsymbol{\Psi}')\boldsymbol{\Xi}_R\boldsymbol{\Pi}^N\boldsymbol{\Xi}_R'(\boldsymbol{\Psi} \otimes \boldsymbol{\Psi})(\mathbf{K}_{p,p})_d$$

and the matrices $\boldsymbol{\Pi}^N$ *and* $\boldsymbol{\Xi}_R$ *are given by Corollary 3.1.4.1 and (3.1.13), respectively;*

(ii)
$$\sqrt{n}\text{vec}(\mathbf{P}(n) - \boldsymbol{\Psi}) \xrightarrow{\mathcal{D}} N_{p^2}(\mathbf{0}, \boldsymbol{\Sigma}_{\boldsymbol{\Psi}}^N),$$

where

$$\begin{aligned}\boldsymbol{\Sigma}_{\boldsymbol{\Psi}}^N =&(\mathbf{I} \otimes \boldsymbol{\Psi})(\mathbf{I} \otimes \boldsymbol{\Lambda} - \boldsymbol{\Lambda} \otimes \mathbf{I})^-(\mathbf{I} - (\mathbf{K}_{p,p})_d)(\boldsymbol{\Psi}' \otimes \boldsymbol{\Psi}')\boldsymbol{\Xi}_R\boldsymbol{\Pi}^N\boldsymbol{\Xi}_R' \\ &\times (\boldsymbol{\Psi} \otimes \boldsymbol{\Psi})(\mathbf{I} - (\mathbf{K}_{p,p})_d)(\mathbf{I} \otimes \boldsymbol{\Lambda} - \boldsymbol{\Lambda} \otimes \mathbf{I})^-(\mathbf{I} \otimes \boldsymbol{\Psi}')\end{aligned}$$

and the matrices $\boldsymbol{\Pi}^N$ *and* $\boldsymbol{\Xi}_R$ *are as in (i).* ∎

3.1.8 *Asymptotic distribution of eigenprojectors of* **S** *and* **R**

Let us come back to the notations of §3.1.5 and consider a symmetric matrix $\boldsymbol{\Sigma}$ with eigenvalues λ_i and corresponding unit-length eigenvectors $\boldsymbol{\psi}_i$. Their sample estimators are denoted by $\mathbf{V}$, d_i and $\mathbf{p}_i$, respectively. It makes sense to talk about the distribution of an eigenvector $\mathbf{p}_i$ which corresponds to the eigenvalue d_i, if λ_i is not multiple. When λ_i is a multiple characteristic root of $\boldsymbol{\Sigma}$, then the eigenvectors corresponding to the eigenvalue λ_i form a subspace which can be characterized by an eigenprojector. Eigenprojectors were briefly considered in §1.2.8. Their distributions are needed in the following testing problem (see Tyler (1983) and Schott (1997a), for instance). Let $\lambda_1 \geq \lambda_2 \geq \ldots \geq \lambda_p > 0$ and let $\mathbf{A} : p \times r$ be a real matrix of rank r. Assume that for fixed i and m, $\lambda_{i-1} \neq \lambda_i$; $\lambda_{i+m-1} \neq \lambda_{i+m}$ and when $r < m$, we consider the hypothesis

H_0: the columns of $\mathbf{A}$ belong to the subspace spanned by the eigenvectors of $\boldsymbol{\Sigma}$ which correspond to the eigenvalues $\lambda_i, \ldots, \lambda_{i+m-1}$.

This type of testing problem arises when we use principal components and we want to omit the components belonging to the subspace spanned by the eigenvectors corresponding to small eigenvalues. Let there be k distinct eigenvalues among the p eigenvalues of $\boldsymbol{\Sigma}$. From now on denote the distinct ones by $\lambda_1, \ldots, \lambda_k$ and their multiplicities by $m_1, \ldots, m_k$. The eigenprojector corresponding to λ_i will be denoted by $\mathbf{P}_{\lambda_i}$ or simply $\mathbf{P}_i$. The projection $\mathbf{P}_i$ is constructed with the help of orthogonal unit-length eigenvectors $\boldsymbol{\psi}_{i_1}, \ldots, \boldsymbol{\psi}_{i_m}$, which all are connected to λ_i,

$$\mathbf{P}_{\lambda_i} = \sum_{j=1}^{m_i} \boldsymbol{\psi}_{i_j}\boldsymbol{\psi}_{i_j}'.$$

Let an arbitrary subset of $\lambda_1, \ldots, \lambda_k$ be denoted by w and the eigenprojector corresponding to the eigenvalues from w by $\mathbf{P}_w$. Suppose that all eigenvalues of $\mathbf{V}$ are ordered so that $d_{i_j} > d_{i'_{j'}}$ if $i < i'$, or if $i = i'$ and $j < j'$. Then a natural estimator of $\mathbf{P}_w$ is

$$\widehat{\mathbf{P}}_w = \sum_{\substack{i \\ \lambda_i \in w}} \sum_{j=1}^{m_i} \mathbf{p}_{i_j} \mathbf{p}'_{i_j}.$$

The asymptotic distribution of eigenprojectors of the sample dispersion matrix was obtained by Tyler (1981) under the assumption of normality or ellipticity of the population. Kollo (1984) found the asymptotic distributions of the eigenprojectors of the sample dispersion and correlation matrices when the existence of the fourth order population moments was assumed (see also Schott, 1999; Kollo, 2000). In the next lemma an approximation for $\widehat{\mathbf{P}}_w$ is found. For alternative proofs see Kato (1972), Tyler (1981) and Watson (1983, Appendix B).

Lemma 3.1.5. *Suppose that for a consistent estimator* $\mathbf{V}$ *of* $\boldsymbol{\Sigma}$, *i.e.* $\mathbf{V} \xrightarrow{\mathcal{P}} \boldsymbol{\Sigma}$, *when* $n \to \infty$, *the convergence*

$$\sqrt{n}\mathrm{vec}(\mathbf{V} - \boldsymbol{\Sigma}) \quad \xrightarrow{\mathcal{D}} \quad N_{p^2}(\mathbf{0}, \boldsymbol{\Sigma}_{\boldsymbol{\Sigma}}) \tag{3.1.48}$$

takes place. Then

$$\widehat{\mathbf{P}}_w = \mathbf{P}_w - \sum_{\substack{i \\ \lambda_i \in w}} \left(\mathbf{P}_{\lambda_i} \boldsymbol{\Sigma}(\mathbf{V} - \boldsymbol{\Sigma})(\boldsymbol{\Sigma} - \lambda_i \mathbf{I}_p)^+ + (\boldsymbol{\Sigma} - \lambda_i \mathbf{I}_p)^+ (\mathbf{V} - \boldsymbol{\Sigma}) \mathbf{P}_{\lambda_i} \right) + o_P(n^{-\frac{1}{2}}). \tag{3.1.49}$$

PROOF: Since $\mathbf{V} \xrightarrow{\mathcal{P}} \boldsymbol{\Sigma}$, the subspace $\mathcal{C}(\mathbf{p}_{i_1}, \mathbf{p}_{i_2}, \ldots, \mathbf{p}_{i_{m_i}})$ must be close to the subspace $\mathcal{C}(\boldsymbol{\psi}_{i_1}, \boldsymbol{\psi}_{i_2}, \ldots, \boldsymbol{\psi}_{i_{m_i}})$, when $n \to \infty$, i.e. $\mathbf{p}_{i_1}, \mathbf{p}_{i_2}, \ldots, \mathbf{p}_{i_{m_i}}$ must be close to $(\boldsymbol{\psi}_{i_1}, \boldsymbol{\psi}_{i_2}, \ldots, \boldsymbol{\psi}_{i_{m_i}})\mathbf{Q}$, where $\mathbf{Q}$ is an orthogonal matrix, as $\mathbf{k}_{i_j}$ and $\boldsymbol{\psi}_{i_j}$ are of unit length and orthogonal. It follows from the proof that the choice of $\mathbf{Q}$ is immaterial. The idea of the proof is to approximate $\mathbf{p}_{i_1}, \mathbf{p}_{i_2}, \ldots, \mathbf{p}_{i_{m_i}}$ via a Taylor expansion and then to construct $\widehat{\mathbf{P}}_w$. From Corollary 3.1.2.1 it follows that

$$\begin{aligned}
&\mathrm{vec}(\mathbf{p}_{i_1}, \ldots, \mathbf{p}_{i_{m_i}}) \\
&= \mathrm{vec}((\boldsymbol{\psi}_{i_1}, \ldots, \boldsymbol{\psi}_{i_{m_i}})\mathbf{Q}) + \left(\frac{d\, \mathbf{p}_{i_1}, \ldots, \mathbf{p}_{i_{m_i}}}{d\mathbf{V}(K)} \right)' \Bigg|_{\mathbf{V}(K) = \boldsymbol{\Sigma}(K)} \mathrm{vec}(\mathbf{V}(K) - \boldsymbol{\Sigma}(K)) \\
&\qquad + o_P(n^{-\frac{1}{2}}),
\end{aligned}$$

where $\mathbf{V}(K)$ and $\boldsymbol{\Sigma}(K)$ stand for the $\frac{1}{2}p(p+1)$ different elements of $\mathbf{V}$ and $\boldsymbol{\Sigma}$, respectively. In order to determine the derivative, Lemma 3.1.3 will be used. As

$\mathbf{p}_{i_1}, \ldots, \mathbf{p}_{i_{m_i}} = \mathbf{P}(\mathbf{e}_{i_1} : \mathbf{e}_{i_2} : \ldots : \mathbf{e}_{i_{m_i}})$, where $\mathbf{e}_\bullet$ are unit basis vectors, Lemma 3.1.3 (ii) yields

$$\frac{d\mathbf{P}}{d\mathbf{V}(K)}\{(\mathbf{e}_{i_1} : \ldots : \mathbf{e}_{i_{m_i}}) \otimes \mathbf{I}\} = \frac{d(\mathbf{p}_{i_1} : \ldots : \mathbf{p}_{i_{m_i}})}{d\mathbf{V}(K)}$$
$$= -\frac{d\mathbf{V}}{d\mathbf{V}(K)}(\mathbf{P} \otimes \mathbf{P})(\mathbf{I} \otimes \mathbf{D} - \mathbf{D} \otimes \mathbf{I})^+\{(\mathbf{e}_{i_1} : \ldots : \mathbf{e}_{i_{m_i}}) \otimes \mathbf{P}'\}.$$

Now

$$(\mathbf{P} \otimes \mathbf{P})(\mathbf{I} \otimes \mathbf{D} - \mathbf{D} \otimes \mathbf{I})^+\{(\mathbf{e}_{i_1} : \ldots : \mathbf{e}_{i_{m_i}}) \otimes \mathbf{P}'\}$$
$$=(\mathbf{P} \otimes \mathbf{I})(\mathbf{I} \otimes \mathbf{V} - \mathbf{D} \otimes \mathbf{I})^+\{(\mathbf{e}_{i_1} : \ldots : \mathbf{e}_{i_{m_i}}) \otimes \mathbf{I}\}$$
$$=\{(\mathbf{p}_{i_1} : \ldots : \mathbf{p}_{i_{m_i}}) \otimes \mathbf{I}\}(\mathbf{V} - d_{i_1}\mathbf{I}, \mathbf{V} - d_{i_2}\mathbf{I}, \ldots, \mathbf{V} - d_{i_{m_i}}\mathbf{I})^+_{[d]}.$$

Moreover,

$$\frac{d\mathbf{V}}{d\mathbf{V}(K)} = (\mathbf{T}^+(s))',$$

where $\mathbf{T}^+(s)$ is given in Proposition 1.3.18. Thus,

$$\left(\frac{d\,\mathbf{p}_{i_1} : \ldots : \mathbf{p}_{i_{m_i}}}{d\mathbf{V}(K)}\right)'\Bigg|_{\mathbf{V}(K)=\mathbf{\Sigma}(K)} \mathrm{vec}(\mathbf{V}(K) - \mathbf{\Sigma}(K))$$
$$= -(\mathbf{I} \otimes (\mathbf{\Sigma} - \lambda_i\mathbf{I})^+)\{\mathbf{Q}'(\boldsymbol{\psi}_{i_1} : \ldots : \boldsymbol{\psi}_{i_{m_i}})' \otimes \mathbf{I}\}\mathbf{T}^+(s)\mathrm{vec}(\mathbf{V}(K) - \mathbf{\Sigma}(K))$$
$$= -(\mathbf{I} \otimes (\mathbf{\Sigma} - \lambda_i\mathbf{I})^+)\mathrm{vec}\{(\mathbf{V} - \mathbf{\Sigma})(\boldsymbol{\psi}_{i_1} : \ldots : \boldsymbol{\psi}_{i_{m_i}})\mathbf{Q}\},$$

since $\mathbf{T}^+(s)\mathrm{vec}(\mathbf{V}(K) - \mathbf{\Sigma}(K)) = \mathrm{vec}(\mathbf{V} - \mathbf{\Sigma})$ and

$$\mathrm{vec}(\mathbf{p}_{i_1}, \ldots, \mathbf{p}_{i_{m_i}})$$
$$= \mathrm{vec}\{(\boldsymbol{\psi}_{i_1}, \ldots, \boldsymbol{\psi}_{i_{m_i}})\mathbf{Q}\} - (\mathbf{I} \otimes (\mathbf{\Sigma} - \lambda_i\mathbf{I})^+)\mathrm{vec}\{(\mathbf{V} - \mathbf{\Sigma})(\boldsymbol{\psi}_{i_1}, \ldots, \boldsymbol{\psi}_{i_{m_i}})\mathbf{Q}\}$$
$$+o_P(n^{-\frac{1}{2}}).$$

Next it will be utilized that

$$\widehat{\mathbf{P}}_{\lambda_i} = \sum_{j=1}^{m_i} \mathbf{p}_{i_j}\mathbf{p}'_{i_j} = \sum_{j=1}^{m_i}(\mathbf{d}'_j \otimes \mathbf{I})\mathrm{vec}(\mathbf{p}_{i_1}, \ldots, \mathbf{p}_{i_{m_i}})\mathrm{vec}'(\mathbf{p}_{i_1}, \ldots, \mathbf{p}_{i_{m_i}})(\mathbf{d}_j \otimes \mathbf{I}),$$

where $\mathbf{d}_j$ is the $j-$th unit vector. This implies that

$$\widehat{\mathbf{P}}_{\lambda_i} = \sum_{j=1}^{m_i}(\boldsymbol{\psi}_{i_1}, \ldots, \boldsymbol{\psi}_{i_{m_i}})\mathbf{Q}\mathbf{d}_j\mathbf{d}'_j\mathbf{Q}'(\boldsymbol{\psi}_{i_1}, \ldots, \boldsymbol{\psi}_{i_{m_i}})'$$
$$- \sum_{j=1}^{m_i}(\mathbf{\Sigma} - \lambda_i\mathbf{I})^+(\mathbf{V} - \mathbf{\Sigma})(\boldsymbol{\psi}_{i_1}, \ldots, \boldsymbol{\psi}_{i_{m_i}})\mathbf{Q}\mathbf{d}_j\mathbf{d}'_j\mathbf{Q}'(\boldsymbol{\psi}_{i_1}, \ldots, \boldsymbol{\psi}_{i_{m_i}})'$$
$$- \sum_{j=1}^{m_i}(\boldsymbol{\psi}_{i_1}, \ldots, \boldsymbol{\psi}_{i_{m_i}})\mathbf{Q}\mathbf{d}_j\mathbf{d}'_j\mathbf{Q}'(\mathbf{V} - \mathbf{\Sigma})(\mathbf{\Sigma} - \lambda_i\mathbf{I})^+ + o_P(n^{-1/2})$$
$$=\mathbf{P}_{\lambda_i} - (\mathbf{\Sigma} - \lambda_i\mathbf{I})^+(\mathbf{V} - \mathbf{\Sigma})\mathbf{P}_{\lambda_i} - \mathbf{P}_{\lambda_i}(\mathbf{V} - \mathbf{\Sigma})(\mathbf{\Sigma} - \lambda_i\mathbf{I})^+ + o_P(n^{-1/2}),$$

since $\sum_{j=1}^{m_i} \mathbf{Q}\mathbf{d}_j\mathbf{d}_j'\mathbf{Q}' = \mathbf{I}$. Hence, the lemma is established, because

$$\mathbf{P}_w = \sum_{\lambda_i \in w} \mathbf{P}_{\lambda_i}.$$

∎

The asymptotic distribution of the eigenprojector $\widehat{\mathbf{P}}_w$ is given by the following theorem.

Theorem 3.1.13. *Let the convergence (3.1.48) take place for the matrix* $\mathbf{V}$*. Then for the eigenprojector* $\mathbf{P}_w$ *which corresponds to the eigenvalues* $\lambda_i \in w$ *of* $\mathbf{\Sigma}$*,*

$$\sqrt{n}\mathrm{vec}(\widehat{\mathbf{P}}_w - \mathbf{P}_w) \xrightarrow{\mathcal{D}} N_{p^2}(\mathbf{0}, (\mathbf{\Xi}_{P_w})'\mathbf{\Sigma}_{\mathbf{\Sigma}}\mathbf{\Xi}_{P_w}), \tag{3.1.50}$$

where $\mathbf{\Sigma}_{\mathbf{\Sigma}}$ *is the asymptotic dispersion matrix of* $\mathbf{V}$ *in (3.1.48) and*

$$\mathbf{\Xi}_{P_w} = \sum_{\substack{i \\ \lambda_i \in w}} \sum_{\substack{j \\ \lambda_j \notin w}} \frac{1}{\lambda_i - \lambda_j} (\mathbf{P}_i \otimes \mathbf{P}_j + \mathbf{P}_j \otimes \mathbf{P}_i). \tag{3.1.51}$$

PROOF: From Theorem 3.1.3 it follows that the statement of the theorem holds, if instead of $\mathbf{\Xi}_{P_w}$

$$\boldsymbol{\xi}_{P_w} = \left. \frac{d\widehat{\mathbf{P}}_w}{d\mathbf{V}(n)} \right|_{\mathbf{V}(n)=\mathbf{\Sigma}}$$

is used. Let us prove that this derivative can be approximated by (3.1.51) with an error $o_P(n^{-1/2})$. Differentiating the main term in (3.1.49) yields

$$\frac{d\widehat{\mathbf{P}}_w}{d\mathbf{V}} = - \sum_{\substack{i \\ \lambda_i \in w}} \left((\mathbf{\Sigma} - \lambda_i \mathbf{I}_p)^+ \otimes \mathbf{P}_{\lambda_i} + \mathbf{P}_{\lambda_i} \otimes (\mathbf{\Sigma} - \lambda_i \mathbf{I}_p)^+\right) + o_P(n^{-1/2}).$$

As noted in Theorem 1.2.30, the Moore-Penrose inverse can be presented through eigenvalues and eigenprojectors:

$$(\mathbf{\Sigma} - \lambda_i \mathbf{I}_p)^+ = (\sum_j \lambda_j \mathbf{P}_j - \lambda_i \sum_j \mathbf{P}_j)^+ = (\sum_{i \neq j} (\lambda_j - \lambda_i)\mathbf{P}_j)^+ = \sum_{i \neq j} \frac{1}{\lambda_j - \lambda_i}\mathbf{P}_j.$$

The last equality follows from the uniqueness of the Moore-Penrose inverse and the fact that the matrix $\sum_{i \neq j} \frac{1}{\lambda_j - \lambda_i}\mathbf{P}_j$ satisfies the defining equalities (1.1.16) – (1.1.19) of the Moore-Penrose inverse. Then

$$\begin{aligned} \frac{d\widehat{\mathbf{P}}_w}{d\mathbf{V}} &= - \sum_{\substack{i \\ \lambda_i \in w}} \sum_{i \neq j} \frac{1}{\lambda_j - \lambda_i} (\mathbf{P}_i \otimes \mathbf{P}_j + \mathbf{P}_j \otimes \mathbf{P}_i) + o_P(n^{-1/2}) \\ &= \sum_{\substack{i \\ \lambda_i \in w}} \sum_{\substack{j \\ \lambda_j \notin w}} \frac{1}{\lambda_i - \lambda_j} (\mathbf{P}_i \otimes \mathbf{P}_j + \mathbf{P}_j \otimes \mathbf{P}_i) + o_P(n^{-1/2}). \end{aligned}$$

∎

From the theorem we immediately get the convergence results for eigenprojectors of the sample dispersion matrix and the correlation matrix.

Corollary 3.1.13.1. *Let* $(\mathbf{x}_1, \ldots, \mathbf{x}_n)$ *be a sample of size* n *from a* p*-dimensional population with* $D[\mathbf{x}_i] = \mathbf{\Sigma}$ *and* $m_4[\mathbf{x}_i] < \infty$. *Then the sample estimator* $\widehat{\mathbf{P}}_w$ *of the eigenprojector* $\mathbf{P}_w$, *corresponding to the eigenvalues* $\lambda_i \in w$ *of the dispersion matrix* $\mathbf{\Sigma}$, *converges in distribution to the normal law, if* $n \to \infty$:

$$\sqrt{n}\mathrm{vec}(\widehat{\mathbf{P}}_w - \mathbf{P}_w) \xrightarrow{\mathcal{D}} N_{p^2}(\mathbf{0}, (\mathbf{\Xi}_{P_w})'\mathbf{\Pi}\mathbf{\Xi}_{P_w}),$$

where the matrices $\mathbf{\Xi}_{P_w}$ *and* $\mathbf{\Pi}$ *are defined by (3.1.51) and (3.1.5), respectively.* ▮

Corollary 3.1.13.2. *Let* $(\mathbf{x}_1, \ldots, \mathbf{x}_n)$ *be a sample of size* n *from a* p*-dimensional population with* $D[\mathbf{x}_i] = \mathbf{\Sigma}$ *and* $m_4[\mathbf{x}_i] < \infty$. *Then the sample estimator of the eigenprojector* $\mathbf{P}_w$, *which corresponds to the eigenvalues* $\lambda_i \in w$ *of the theoretical correlation matrix* $\mathbf{\Omega}$, *converges in distribution to the normal law, if* $n \to \infty$:

$$\sqrt{n}\mathrm{vec}(\widehat{\mathbf{P}}_w - \mathbf{P}_w) \xrightarrow{\mathcal{D}} N_{p^2}(\mathbf{0}, (\mathbf{\Xi}_{P_w})'(\mathbf{\Xi}_R)'\mathbf{\Pi}\mathbf{\Xi}_R\mathbf{\Xi}_{P_w}),$$

where the matrices $\mathbf{\Xi}_{P_w}$, $\mathbf{\Xi}_R$ *and* $\mathbf{\Pi}$ *are defined by (3.1.51), (3.1.13) and (3.1.5), respectively.*

PROOF: The corollary is established by copying the statements of Lemma 3.1.5 and Theorem 3.1.13, when instead of $\mathbf{V}$ the correlation matrix $\mathbf{R}$ is used. ▮

3.1.9 *Asymptotic normality of the MANOVA matrix*

In this and the next paragraph we shall consider statistics which are functions of two multivariate arguments. The same technique as before can be applied but now we have to deal with partitioned matrices and covariance matrices with block structures. Let

$$\mathbf{S}_j = \frac{1}{n-1}\sum_{i=1}^{n}(\mathbf{x}_{ij} - \overline{\mathbf{x}}_j)(\mathbf{x}_{ij} - \overline{\mathbf{x}}_j)', \qquad j = 1, 2, \tag{3.1.52}$$

be the sample dispersion matrices from independent samples which both are of size n, and let $\mathbf{\Sigma}_1$, $\mathbf{\Sigma}_2$ be the corresponding population dispersion matrices. In 1970s a series of papers appeared on asymptotic distributions of different functions of the MANOVA matrix

$$\mathbf{T} = \mathbf{S}_1\mathbf{S}_2^{-1}.$$

It was assumed that $\mathbf{x}_{ij}$ are normally distributed, i.e. $\mathbf{S}_1$ and $\mathbf{S}_2$ are Wishart distributed (see Chang, 1973; Hayakawa, 1973; Krishnaiah & Chattopadhyay, 1975; Sugiura, 1976; Constantine & Muirhead, 1976; Fujikoshi, 1977; Khatri & Srivastava, 1978; for example). Typical functions of interest are the determinant, the trace, the eigenvalues and the eigenvectors of the matrices, which all play an important role in hypothesis testing in multivariate analysis. Observe that these functions of $\mathbf{T}$ may be considered as functions of $\mathbf{Z}$ in §2.4.2, following a multivariate beta type II distribution. In this paragraph we are going to obtain the dispersion matrix of the asymptotic normal law of the $\mathbf{T}$-matrix in the case of finite fourth order moments following Kollo (1990). As an application, the determinant of the matrix will be considered.

The notation

$$\mathbf{z} = \begin{pmatrix} \mathrm{vec}\mathbf{S}_1 \\ \mathrm{vec}\mathbf{S}_2 \end{pmatrix}, \qquad \boldsymbol{\sigma}_0 = \begin{pmatrix} \mathrm{vec}\boldsymbol{\Sigma}_1 \\ \mathrm{vec}\boldsymbol{\Sigma}_2 \end{pmatrix}$$

will be used. Moreover, let $\boldsymbol{\Pi}_1$ and $\boldsymbol{\Pi}_2$ denote the asymptotic dispersion matrices of $\sqrt{n}\mathrm{vec}(\mathbf{S}_1 - \boldsymbol{\Sigma}_1)$ and $\sqrt{n}\mathrm{vec}(\mathbf{S}_2 - \boldsymbol{\Sigma}_2)$, respectively. The asymptotic distribution of the **T**-matrix follows from Theorem 3.1.3.

Theorem 3.1.14. *Let the p-vectors $\mathbf{x}_{ij}$ in (3.1.52) satisfy $D[\mathbf{x}_{ij}] = \boldsymbol{\Sigma}_j$ and $m_4[\mathbf{x}_{ij}] < \infty$, $i = 1, 2, \ldots, n$, $j = 1, 2$. Then, if $n \to \infty$,*

$$\sqrt{n}\mathrm{vec}(\mathbf{S}_1\mathbf{S}_2^{-1} - \boldsymbol{\Sigma}_1\boldsymbol{\Sigma}_2^{-1}) \xrightarrow{\mathcal{D}} N_{p^2}(\mathbf{0}, \boldsymbol{\Psi}),$$

where

$$\boldsymbol{\Psi} = (\boldsymbol{\Sigma}_2^{-1} \otimes \mathbf{I}_p)\boldsymbol{\Pi}_1(\boldsymbol{\Sigma}_2^{-1} \otimes \mathbf{I}_p) + (\boldsymbol{\Sigma}_2^{-1} \otimes \boldsymbol{\Sigma}_1\boldsymbol{\Sigma}_2^{-1})\boldsymbol{\Pi}_2(\boldsymbol{\Sigma}_2^{-1} \otimes \boldsymbol{\Sigma}_2^{-1}\boldsymbol{\Sigma}_1), \quad (3.1.53)$$

and $\boldsymbol{\Pi}_1$ and $\boldsymbol{\Pi}_2$ are defined by (3.1.5).

PROOF: Applying Theorem 3.1.3 gives us

$$\boldsymbol{\Psi} = \left(\frac{d(\mathbf{S}_1\mathbf{S}_2^{-1})}{d\mathbf{z}}\bigg|_{\mathbf{z}=\boldsymbol{\sigma}_0} \right)' \boldsymbol{\Sigma}_{\mathbf{z}} \frac{d(\mathbf{S}_1\mathbf{S}_2^{-1})}{d\mathbf{z}}\bigg|_{\mathbf{z}=\boldsymbol{\sigma}_0},$$

where $\boldsymbol{\Sigma}_{\mathbf{z}}$ denotes the asymptotic dispersion matrix of $\mathbf{z}$, i.e.

$$\sqrt{n}(\mathbf{z} - \boldsymbol{\sigma}_0) \xrightarrow{\mathcal{D}} N_{2p^2}(\mathbf{0}, \boldsymbol{\Sigma}_{\mathbf{z}}).$$

From Theorem 3.1.4 and independence of the samples it follows that

$$\boldsymbol{\Sigma}_{\mathbf{z}} = \begin{pmatrix} \boldsymbol{\Pi}_1 & \mathbf{0} \\ \mathbf{0} & \boldsymbol{\Pi}_2 \end{pmatrix}.$$

Let us find the derivative $\dfrac{d(\mathbf{S}_1\mathbf{S}_2^{-1})}{d\mathbf{z}}$. According to Theorem 3.1.5, we are going to treat $\mathbf{S}_1$ and $\mathbf{S}_2$ as unstructured matrices.

$$\begin{aligned} \frac{d(\mathbf{S}_1\mathbf{S}_2^{-1})}{d\mathbf{z}} &\underset{(1.4.19)}{=} \frac{d\mathbf{S}_1}{d\mathbf{z}}(\mathbf{S}^{-1} \otimes \mathbf{I}_p) + \frac{d\mathbf{S}_2^{-1}}{d\mathbf{z}}(\mathbf{I}_p \otimes \mathbf{S}_1) \\ &\underset{(1.4.21)}{=} \begin{pmatrix} \mathbf{I} \\ \mathbf{0} \end{pmatrix} (\mathbf{S}_2^{-1} \otimes \mathbf{I}_p) - \begin{pmatrix} \mathbf{0} \\ (\mathbf{S}_2^{-1} \otimes \mathbf{S}_2^{-1}) \end{pmatrix} (\mathbf{I}_p \otimes \mathbf{S}_1). \quad (3.1.54) \end{aligned}$$

Thus,

$$\begin{aligned} \boldsymbol{\Psi} &= (\boldsymbol{\Sigma}_2^{-1} \otimes \mathbf{I}_p : -\boldsymbol{\Sigma}_2^{-1} \otimes \boldsymbol{\Sigma}_1\boldsymbol{\Sigma}_2^{-1}) \begin{pmatrix} \boldsymbol{\Pi}_1 & \mathbf{0} \\ \mathbf{0} & \boldsymbol{\Pi}_2 \end{pmatrix} \begin{pmatrix} (\boldsymbol{\Sigma}_2^{-1} \otimes \mathbf{I}_p) \\ -(\boldsymbol{\Sigma}_2^{-1} \otimes \boldsymbol{\Sigma}_2^{-1}\boldsymbol{\Sigma}_1) \end{pmatrix} \\ &= (\boldsymbol{\Sigma}_2^{-1} \otimes \mathbf{I}_p)\boldsymbol{\Pi}_1(\boldsymbol{\Sigma}_2^{-1} \otimes \mathbf{I}_p) + (\boldsymbol{\Sigma}_2^{-1} \otimes \boldsymbol{\Sigma}_1\boldsymbol{\Sigma}_2^{-1})\boldsymbol{\Pi}_2(\boldsymbol{\Sigma}_2^{-1} \otimes \boldsymbol{\Sigma}_2^{-1}\boldsymbol{\Sigma}_1). \end{aligned}$$

∎

Corollary 3.1.14.1. *Let* $\mathbf{x}_{1j} \sim N_p(\boldsymbol{\mu}_1, \boldsymbol{\Sigma}_1)$ *and* $\mathbf{x}_{2j} \sim N_p(\boldsymbol{\mu}_2, \boldsymbol{\Sigma}_2)$, $j = 1, 2, \ldots, n$. *If* $n \to \infty$, *then*

$$\sqrt{n}\mathrm{vec}(\mathbf{S}_1\mathbf{S}_2^{-1} - \boldsymbol{\Sigma}_1\boldsymbol{\Sigma}_2^{-1}) \quad \xrightarrow{\mathcal{D}} \quad N_{p^2}(\mathbf{0}, \boldsymbol{\Psi}^N),$$

where

$$\boldsymbol{\Psi}^N = \boldsymbol{\Sigma}_2^{-1}\boldsymbol{\Sigma}_1\boldsymbol{\Sigma}_2^{-1} \otimes \boldsymbol{\Sigma}_1 + \boldsymbol{\Sigma}_2^{-1} \otimes \boldsymbol{\Sigma}_1\boldsymbol{\Sigma}_2^{-1}\boldsymbol{\Sigma}_1 + 2\mathbf{K}_{p,p}(\boldsymbol{\Sigma}_1\boldsymbol{\Sigma}_2^{-1} \otimes \boldsymbol{\Sigma}_2^{-1}\boldsymbol{\Sigma}_1). \quad (3.1.55)$$

In the special case, when $\boldsymbol{\Sigma}_1 = \boldsymbol{\Sigma}_2 = \boldsymbol{\Sigma}$, *we have*

$$\boldsymbol{\Psi}^N = 2\boldsymbol{\Sigma}^{-1} \otimes \boldsymbol{\Sigma} + 2\mathbf{K}_{p,p}.$$

∎

The statement of the theorem has a simple form in the case of an elliptical distribution.

Corollary 3.1.14.2. *Let* $\mathbf{x}_{1j} \sim E_p(\boldsymbol{\mu}_1, \boldsymbol{\Upsilon}_1)$, $D[\mathbf{x}_{1j}] = \boldsymbol{\Sigma}_1$ *and* $\mathbf{x}_{2j} \sim E_p(\boldsymbol{\mu}_2, \boldsymbol{\Upsilon}_2)$, $D[\mathbf{x}_{2j}] = \boldsymbol{\Sigma}_2$, $j = 1, 2, \ldots, n$, *with kurtosis parameters* κ_1 *and* κ_2, *respectively. If* $n \to \infty$, *then*

$$\sqrt{n}\mathrm{vec}(\mathbf{S}_1\mathbf{S}_2^{-1} - \boldsymbol{\Sigma}_1\boldsymbol{\Sigma}_2^{-1}) \quad \xrightarrow{\mathcal{D}} \quad N_{p^2}(\mathbf{0}, \boldsymbol{\Psi}^E),$$

where

$$\boldsymbol{\Psi}^E = (\boldsymbol{\Sigma}_2^{-1} \otimes \mathbf{I}_p)\boldsymbol{\Pi}_1^E(\boldsymbol{\Sigma}_2^{-1} \otimes \mathbf{I}_p) + (\boldsymbol{\Sigma}_2^{-1} \otimes \boldsymbol{\Sigma}_1\boldsymbol{\Sigma}_2^{-1})\boldsymbol{\Pi}_2^E(\boldsymbol{\Sigma}_2^{-1} \otimes \boldsymbol{\Sigma}_2^{-1}\boldsymbol{\Sigma}_1), \quad (3.1.56)$$

and $\boldsymbol{\Pi}_1^E$ *and* $\boldsymbol{\Pi}_2^E$ *are given in Corollary 3.1.4.2. In the special case, when* $\boldsymbol{\Sigma}_1 = \boldsymbol{\Sigma}_2 = \boldsymbol{\Sigma}$ *and* $\kappa_1 = \kappa_2$,

$$\boldsymbol{\Psi}^E = 2(1 + \kappa)(\boldsymbol{\Sigma}^{-1} \otimes \boldsymbol{\Sigma} + \mathbf{K}_{p,p}) + 2\kappa\mathrm{vec}\mathbf{I}\mathrm{vec}'\mathbf{I}.$$

∎

Theorem 3.1.14 makes it possible to find the asymptotic distributions of functions of the $\mathbf{T}$ matrix in a convenient way. As an example, let us consider the determinant of $\mathbf{T}$. By Theorem 3.1.3 the following convergence holds:

$$\sqrt{n}(|\mathbf{S}_1\mathbf{S}_2^{-1}| - |\boldsymbol{\Sigma}_1\boldsymbol{\Sigma}_2^{-1}|) \quad \xrightarrow{\mathcal{D}} \quad N(0, \beta), \quad (3.1.57)$$

where

$$\beta = \left(\left. \frac{d|\mathbf{S}_1\mathbf{S}_2^{-1}|}{d\mathbf{z}} \right|_{\mathbf{z}=\boldsymbol{\sigma}_0} \right)' \boldsymbol{\Sigma}_{\mathbf{z}} \left. \frac{d|\mathbf{S}_1\mathbf{S}_2^{-1}|}{d\mathbf{z}} \right|_{\mathbf{z}=\boldsymbol{\sigma}_0}$$

and $\boldsymbol{\Psi}$ is given by (3.1.53). Relation (1.4.30) yields

$$\frac{d|\mathbf{S}_1\mathbf{S}_2^{-1}|}{d\mathbf{z}} = |\mathbf{S}_1||\mathbf{S}_2^{-1}|\frac{d\mathbf{S}_1\mathbf{S}_2^{-1}}{d\mathbf{z}}\mathrm{vec}(\mathbf{S}_1^{-1}\mathbf{S}_2)$$

and $\dfrac{d\mathbf{S}_1\mathbf{S}_2^{-1}}{d\mathbf{z}}$ was obtained in (3.1.54).

3.1.10 *Asymptotics of Hotelling T^2-statistic*

In this paragraph the asymptotic behavior of the Hotelling T^2-statistic will be examined. This statistic is a function which depends on both the sample mean and the sample dispersion matrix. At the same time, it is a good introduction to the next section, where we shall deal with asymptotic expansions. It appears that different types of asymptotic distributions are valid for T^2. Its asymptotic behavior depends on the parameters of the distribution. Different distributions can be analyzed from the same point of view by using two terms in a Taylor expansion of the statistic.

Suppose that we have $p-$variate random vectors with finite first moments: $E[\mathbf{x}_i] = \boldsymbol{\mu}$, $D[\mathbf{x}_i] = \boldsymbol{\Sigma}$ and $m_4[\mathbf{x}_i] < \infty$. The statistic $\overline{\mathbf{x}}'\mathbf{S}^{-1}\overline{\mathbf{x}}$ is called *Hotelling T^2-statistic*, where, as usual, $\overline{\mathbf{x}}$ is the sample mean and $\mathbf{S}$ is the sample dispersion matrix. The asymptotic distribution of the vector $\sqrt{n}(\mathbf{z}-\boldsymbol{\sigma}_0)$ with

$$\mathbf{z} = \begin{pmatrix} \overline{\mathbf{x}} \\ \mathrm{vec}\mathbf{S} \end{pmatrix}, \qquad \boldsymbol{\sigma}_0 = \begin{pmatrix} \boldsymbol{\mu} \\ \mathrm{vec}\boldsymbol{\Sigma} \end{pmatrix} \tag{3.1.58}$$

is given by

$$\sqrt{n}(\mathbf{z}-\boldsymbol{\sigma}_0) \quad \xrightarrow{\mathcal{D}} \quad N_{p+p^2}(\mathbf{0}, \boldsymbol{\Sigma}_{\mathbf{z}}),$$

where

$$\boldsymbol{\Sigma}_{\mathbf{z}} = \begin{pmatrix} \boldsymbol{\Sigma} & \mathbf{M}_3' \\ \mathbf{M}_3 & \boldsymbol{\Pi} \end{pmatrix}, \tag{3.1.59}$$

the matrix $\boldsymbol{\Pi}$ is defined by (3.1.5) and

$$\mathbf{M}_3 = E[(\mathbf{x}-\boldsymbol{\mu}) \otimes (\mathbf{x}-\boldsymbol{\mu})' \otimes (\mathbf{x}-\boldsymbol{\mu})]. \tag{3.1.60}$$

Using these notations, we formulate the following asymptotic result for T^2.

Theorem 3.1.15. *Let $\mathbf{x}_1, \ldots, \mathbf{x}_n$ be a sample of size n from a p-dimensional population with the first moments $E[\mathbf{x}_i] = \boldsymbol{\mu} \neq \mathbf{0}$, $D[\mathbf{x}_i] = \boldsymbol{\Sigma}_i$ and $m_4[\mathbf{x}_i] < \infty$. Then, if $n \to \infty$,*

$$\sqrt{n}(\overline{\mathbf{x}}'\mathbf{S}^{-1}\overline{\mathbf{x}} - \boldsymbol{\mu}'\boldsymbol{\Sigma}^{-1}\boldsymbol{\mu}) \quad \xrightarrow{\mathcal{D}} \quad N(0,\tau),$$

where

$$\tau = \boldsymbol{\xi}'\boldsymbol{\Sigma}_{\mathbf{z}}\boldsymbol{\xi},$$

$\boldsymbol{\Sigma}_{\mathbf{z}}$ is given by (3.1.59) and

$$\boldsymbol{\xi} = \begin{pmatrix} 2\boldsymbol{\Sigma}^{-1}\boldsymbol{\mu} \\ -\boldsymbol{\Sigma}^{-1}\boldsymbol{\mu} \otimes \boldsymbol{\Sigma}^{-1}\boldsymbol{\mu} \end{pmatrix}. \tag{3.1.61}$$

If the distribution of $\mathbf{x}_i$ is symmetric, then

$$\tau = 4\boldsymbol{\mu}'\boldsymbol{\Sigma}^{-1}\boldsymbol{\mu} + (\boldsymbol{\mu}'\boldsymbol{\Sigma}^{-1} \otimes \boldsymbol{\mu}'\boldsymbol{\Sigma}^{-1})\boldsymbol{\Pi}(\boldsymbol{\Sigma}^{-1}\boldsymbol{\mu} \otimes \boldsymbol{\Sigma}^{-1}\boldsymbol{\mu}). \tag{3.1.62}$$

PROOF: From Theorem 3.1.3,

$$\tau = \left(\frac{d(\overline{\mathbf{x}}'\mathbf{S}^{-1}\overline{\mathbf{x}})}{d\mathbf{z}}\bigg|_{\mathbf{z}=\boldsymbol{\sigma}_0} \right)' \boldsymbol{\Sigma}_{\mathbf{z}} \frac{d(\overline{\mathbf{x}}'\mathbf{S}^{-1}\overline{\mathbf{x}})}{d\mathbf{z}}\bigg|_{\mathbf{z}=\boldsymbol{\sigma}_0},$$

where $\mathbf{z}$ and $\boldsymbol{\sigma}_0$ are given by (3.1.58) and $\boldsymbol{\Sigma}_{\mathbf{z}}$ by (3.1.59). Let us find the derivative

$$\begin{aligned} \frac{d(\overline{\mathbf{x}}'\mathbf{S}^{-1}\overline{\mathbf{x}})}{d\mathbf{z}} &\underset{(1.4.19)}{=} \frac{d\overline{\mathbf{x}}'}{d\mathbf{z}}\mathbf{S}^{-1}\overline{\mathbf{x}} + \frac{d(\mathbf{S}^{-1}\overline{\mathbf{x}})}{d\mathbf{z}}\overline{\mathbf{x}} \\ &= 2\begin{pmatrix} \mathbf{S}^{-1}\overline{\mathbf{x}} \\ \mathbf{0} \end{pmatrix} - \begin{pmatrix} \mathbf{0} \\ \mathbf{S}^{-1}\overline{\mathbf{x}} \otimes \mathbf{S}^{-1}\overline{\mathbf{x}} \end{pmatrix}, \end{aligned}$$

where, as before when dealing with asymptotics, we differentiate as if $\mathbf{S}$ is non-symmetric. At the point $\mathbf{z} = \boldsymbol{\sigma}_0$, the matrix $\boldsymbol{\xi}$ in (3.1.61) is obtained, which also gives the main statement of the theorem. It remains to consider the case when the population distribution is symmetric, i.e. $\mathbf{M}_3 = \mathbf{0}$. Multiplying the matrices in the expression of τ under this condition yields (3.1.62). ∎

Corollary 3.1.15.1. *Let $\mathbf{x}_i \sim N_p(\boldsymbol{\mu}, \boldsymbol{\Sigma})$, $i = 1, 2, \ldots, n$, and $\boldsymbol{\mu} \neq \mathbf{0}$. If $n \to \infty$, then*

$$\sqrt{n}(\overline{\mathbf{x}}'\mathbf{S}^{-1}\overline{\mathbf{x}} - \boldsymbol{\mu}'\boldsymbol{\Sigma}^{-1}\boldsymbol{\mu}') \xrightarrow{\mathcal{D}} N(0, \tau^N),$$

where

$$\tau^N = 4\boldsymbol{\mu}'\boldsymbol{\Sigma}^{-1}\boldsymbol{\mu} + 2(\boldsymbol{\mu}'\boldsymbol{\Sigma}^{-1}\boldsymbol{\mu})^2.$$

∎

Corollary 3.1.15.2. *Let $\mathbf{x}_i \sim E_p(\boldsymbol{\mu}, \boldsymbol{\Upsilon})$, $i = 1, 2, \ldots, n$, $D[\mathbf{x}_i] = \boldsymbol{\Sigma}$, with kurtosis parameter κ and $\boldsymbol{\mu} \neq \mathbf{0}$. If $n \to \infty$, then*

$$\sqrt{n}(\overline{\mathbf{x}}'\mathbf{S}^{-1}\overline{\mathbf{x}} - \boldsymbol{\mu}'\boldsymbol{\Sigma}^{-1}\boldsymbol{\mu}') \xrightarrow{\mathcal{D}} N(0, \tau^E),$$

where

$$\tau^E = 4\boldsymbol{\mu}'\boldsymbol{\Sigma}^{-1}\boldsymbol{\mu} + (2 + 3\kappa)(\boldsymbol{\mu}'\boldsymbol{\Sigma}^{-1}\boldsymbol{\mu})^2.$$

∎

As noted before, the asymptotic behavior of the Hotelling T^2-statistic is an interesting object to study. When $\boldsymbol{\mu} \neq \mathbf{0}$, we get the asymptotic normal distribution as the limiting distribution, while for $\boldsymbol{\mu} = \mathbf{0}$ the asymptotic distribution is a chi-square distribution. Let us consider this case in more detail. From (3.1.61) it follows that if $\boldsymbol{\mu} = \mathbf{0}$, the first derivative $\boldsymbol{\xi} = \mathbf{0}$. This means that the second term in the Taylor series of the statistic $\overline{\mathbf{x}}'\mathbf{S}^{-1}\overline{\mathbf{x}}$ equals zero and its asymptotic behavior is determined by the next term in the expansion. From Theorem 3.1.1 it follows that we have to find the second order derivative at the point $\boldsymbol{\sigma}_0$. In the proof of Theorem 3.1.15 the first derivative was derived. From here

$$\frac{d^2(\overline{\mathbf{x}}'\mathbf{S}^{-1}\overline{\mathbf{x}})}{d\mathbf{z}^2} = 2\frac{d\mathbf{S}^{-1}\overline{\mathbf{x}}}{d\mathbf{z}}(\mathbf{I} : \mathbf{0}) - \frac{d\mathbf{S}^{-1}\overline{\mathbf{x}} \otimes \mathbf{S}^{-1}\overline{\mathbf{x}}}{d\mathbf{z}}(\mathbf{0} : \mathbf{I}).$$

However, evaluating the derivative at the point $\mathbf{z} = \boldsymbol{\sigma}_0$ with $\boldsymbol{\mu} = \mathbf{0}$ yields

$$\left.\frac{d^2(\overline{\mathbf{x}}'\mathbf{S}^{-1}\overline{\mathbf{x}})}{d\mathbf{z}^2}\right|_{\mathbf{z}=\boldsymbol{\sigma}_0} = 2\begin{pmatrix} \boldsymbol{\Sigma}^{-1} & \mathbf{0} \\ \mathbf{0} & \mathbf{0} \end{pmatrix}.$$

From Theorem 3.1.1 we get that the first non-zero and non-constant term in the Taylor expansion of $\overline{\mathbf{x}}'\mathbf{S}^{-1}\overline{\mathbf{x}}$ is $\overline{\mathbf{x}}'\boldsymbol{\Sigma}^{-1}\overline{\mathbf{x}}$. As all the following terms will be $o_P(n^{-1})$, the asymptotic distribution of the T^2-statistic is determined by the asymptotic distribution of $\overline{\mathbf{x}}'\boldsymbol{\Sigma}^{-1}\overline{\mathbf{x}}$. If $\boldsymbol{\mu} = \mathbf{0}$, then we know that

$$n\overline{\mathbf{x}}'\mathbf{S}^{-1}\overline{\mathbf{x}} \xrightarrow{\mathcal{D}} \chi^2(p),$$

if $n \to \infty$ (see Moore, 1977, for instance). Here $\chi^2(p)$ denotes the chi-square distribution with p degrees of freedom. Hence, the next theorem can be formulated.

Theorem 3.1.16. *Let $\mathbf{x}_1, \dots, \mathbf{x}_n$ be a sample of the size n, $\mathbf{E}[\mathbf{x}_i] = \boldsymbol{\mu} = \mathbf{0}$ and $D[\mathbf{x}_i] = \boldsymbol{\Sigma}$. If $n \to \infty$, then*

$$n\overline{\mathbf{x}}'\mathbf{S}^{-1}\overline{\mathbf{x}} \xrightarrow{\mathcal{D}} \chi^2(p).$$

∎

Example 3.1.1. To illustrate the convergence of the T^2-statistic let us consider the following example. It is based on the normal distribution $N_3(\boldsymbol{\mu}, \boldsymbol{\Sigma})$, where $\boldsymbol{\mu} = a\mathbf{e}$, $\mathbf{e} = (1, 1, 1)'$, a is a constant which takes different values, and

$$\boldsymbol{\Sigma} = \begin{pmatrix} 1.0 & 0.1 & 0.2 \\ 0.1 & 1.1 & 0.3 \\ 0.2 & 0.3 & 1.2 \end{pmatrix}.$$

A simulation experiment was carried out by the following scheme. The empirical and the asymptotic normal distributions of the T^2-statistic were compared for different sample sizes when $\boldsymbol{\mu} \to \mathbf{0}$, and the parameter a was varied within the range $0.1 - 1$. In the tables the number of replications k was 300. The tendencies in the tables given below were the same when k was larger (≥ 1000). Let

$$Y_n = \sqrt{n}(\overline{\mathbf{x}}'\mathbf{S}^{-1}\overline{\mathbf{x}} - \boldsymbol{\mu}'\boldsymbol{\Sigma}^{-1}\boldsymbol{\mu}')$$

and its simulated values y_i, $i = 1, \dots, k$. Let $\overline{y} = \frac{1}{k}\sum_{i=1}^k y_i$ and the estimated value of the asymptotic variance of Y_n be $\overline{\tau}^N$. From the tables we see how the asymptotic variance changes when the parameter a tends to zero. The variance and its estimator $\overline{\tau}^N$ are presented. To examine the goodness-of-fit between empirical and asymptotic normal distributions the standard chi-square test for good-ness-of-fit was used with 13 degrees of freedom.

Table 3.1.1. Goodness-of-fit of empirical and asymptotic normal distributions of the T^2-statistic, with $n = 200$ and 300 replicates.

a	1	0.7	0.5	0.3	0.2	0.15	0.1
$\overline{y}$	0.817	0.345	0.493	0.248	0.199	0.261	0.247
$\overline{\tau}^N$	18.86	5.301	3.022	0.858	0.400	0.291	0.139
τ^N	16.489	5.999	2.561	0.802	0.340	0.188	0.082
$\chi^2(13)$	29.56	21.69	44.12	31.10	95.33	92.88	82.57

The critical value of the chi-square statistic is 22.36 at the significance level 0.05. The results of the simulations indicate that the speed of convergence to the asymptotic normal law is low and for $n = 200$, in one case only, we do not reject the null-hypothesis. At the same time, we see that the chi-square coefficient starts to grow drastically, when $a \leq 0.2$.

Table 3.1.2. Goodness-of-fit of empirical and asymptotic normal distributions of the T^2-statistic with $n = 1000$ and 300 replicates.

a	1	0.5	0.2	0.17	0.15	0.13	0.1
$\overline{y}$	0.258	0.202	0.115	0.068	0.083	0.095	−0.016
$\overline{\tau}^N$	16.682	2.467	0.317	0.248	0.197	0.117	0.097
τ^N	16.489	2.561	0.340	0.243	0.188	0.140	0.082
$\chi^2(13)$	16.87	12.51	15.67	16.52	23.47	29.03	82.58

We see that if the sample size is as large as $n = 1000$, the fit between the asymptotic normal distribution and the empirical one is good for larger values of a, while the chi-square coefficient starts to grow from $a = 0.15$ and becomes remarkably high when $a = 0.1$. It is interesting to note that the Hotelling T^2-statistic remains biased even in the case of a sample size of $n = 1000$ and the bias has a tendency to grow with the parameter a.

Now we will us examine the convergence of Hotelling's T^2-statistic to the chi-square distribution. Let $Y_n = n\overline{\mathbf{x}}'\mathbf{S}^{-1}\overline{\mathbf{x}}$, $\overline{y}$ be as in the previous tables and denote $\widehat{D[Y_n]}$ for the estimated value of the variance. From the convergence results above we know that Y_n converges to the chi-square distribution $Y \sim \chi^2(3)$ with $E[Y] = 3$ and $D[Y] = 6$, when $\boldsymbol{\mu} = \mathbf{0}$.

Table 3.1.3. Goodness-of-fit of empirical and asymptotic chi-square distributions of the T^2-statistic based on 400 replicates.

n	100	200	500	1000
$\overline{y}$	2.73	3.09	2.96	3.11
$\widehat{D[Y_n]}$	4.93	8.56	6.71	6.67
$\chi^2(13)$	14.00	11.26	16,67	8.41

The convergence of the T^2-statistic to the chi-square distribution is much faster than to the normal distribution. The asymptotic variance starts to be stable from $n = 500$. In the next table we show the convergence results to the chi-square distribution, when the parameter a is growing.

Table 3.1.4. Goodness-of-fit of the empirical and asymptotic chi-square distributions of the T^2-statistic with $n = 200$ and 300 replicates.

a	0	0.01	0.02	0.03	0.04	0.05	0.1
$\overline{y}$	3.09	3.09	3.38	3.30	3.72	4.11	7.38
$\widehat{D[Y_n]}$	8.56	5.92	7.67	6.66	7.10	12.62	23.96
$\chi^2(13)$	11.275	13.97	19.43	21.20	38.73	60.17	1107.0

We see that if the value of a is very small, the convergence to the chi-square distribution still holds, but starting from the value $a = 0.04$ it breaks down in our example. From the experiment we can conclude that there is a certain area of values of the mean vector $\boldsymbol{\mu}$ where we can describe the asymptotic behavior of the T^2-statistic neither by the asymptotic normal nor by the asymptotic chi-square distribution. ∎

3.1.11 *Problems*

1. Show that for $\mathbf{G}$ and $\mathbf{Z}$ in Lemma 3.1.3
$$\frac{d\mathbf{G}}{d_\bullet\mathbf{Z}} = -\frac{d\mathbf{Z}}{d_\bullet\mathbf{Z}}(\mathbf{G}\otimes\mathbf{G})\mathbf{P}_2(\mathbf{I}\otimes(\mathbf{ZG})^{-1}),$$
where
$$\mathbf{P}_2 = \tfrac{1}{2}(\mathbf{I}+\mathbf{K}_{p,p}) - (\mathbf{K}_{p,p})_d.$$
2. Show that in Theorem 3.1.7
$$\mathbf{N}^{-1} = (\mathbf{\Lambda}\otimes\mathbf{I}-\mathbf{I}\otimes\mathbf{\Lambda})^+ + \tfrac{1}{2}(\mathbf{I}\otimes\mathbf{\Lambda}^{-1})(\mathbf{K}_{p,p})_d.$$
3. Prove Corollary 3.1.6.2.
4. Show that if $\mathbf{x}_{ij} \sim N_p(\boldsymbol{\mu},\mathbf{\Sigma})$, $i = 1,2,\ldots,n$, $j = 1,2$, then the asymptotic variance in (3.1.57) equals
$$\beta = 4p.$$
5. Let $\mathbf{x}_{ij} \sim E_p(\boldsymbol{\mu},\mathbf{V})$, $i = 1,2,\ldots,n$, $j = 1,2$. Show that in (3.1.57)
$$\beta = 4(1+\kappa)p + 2\kappa p^2,$$
where κ is the kurtosis parameter.
6. Let $\mathbf{x}_i \sim N_p(\boldsymbol{\mu},\mathbf{\Sigma})$, $i = 1,\ldots,n$. Find the asymptotic dispersion matrix for the vector $\sqrt{n}\mathrm{diag}(\mathbf{D}(n)-\mathbf{\Lambda})$. Follow the notation of Theorem 3.1.10.
7. Find an explicit expression of the asymptotic dispersion matrix of the eigenprojector in Corollary 3.1.13.1 when $\mathbf{x}_i \sim N_p(\boldsymbol{\mu},\mathbf{\Sigma})$.
8. Find the asymptotic dispersion matrix for the i–th eigenvector $\mathbf{h}_i(n)$ of the sample dispersion matrix $\mathbf{S}$ under the assumptions of Corollary 3.1.10.2.
9. Let the population be normal, i.e. $\mathbf{x} \sim N_p(\boldsymbol{\mu},\mathbf{\Sigma})$. Find the asymptotic normal law for the inverse sample dispersion matrix $\mathbf{S}^{-1}$.
10. Let the population be elliptical, i.e. $\mathbf{x} \sim E_p(\boldsymbol{\mu},\mathbf{V})$. Find the asymptotic normal law for the inverse sample dispersion matrix $\mathbf{S}^{-1}$.

3.2. MULTIVARIATE FORMAL DENSITY EXPANSIONS IN $\mathbb{R}^p$

3.2.1 *Introduction*

In §3.1.10 we saw that the asymptotic distribution of Hotelling's T^2-statistic was normal if its first derivative was non-zero at the point where the function was expanded into a Taylor series. At the same time, if the first derivative equals zero then a chi-square distribution describes the behavior of the T^2-statistic. In short: when the population expectation $\boldsymbol{\mu} = \mathbf{0}$, we get an asymptotic chi-square distribution as the limit distribution for T^2, and when $\boldsymbol{\mu} \neq \mathbf{0}$, an asymptotic normal distribution is obtained. Certainly, when $\boldsymbol{\mu} \to \mathbf{0}$, the convergence to the normal distribution becomes slower. For small values of $\boldsymbol{\mu}$ the asymptotic chi-square distribution is not valid. This was clearly seen from the small simulation experiment given in Example 3.1.1 in §3.1.10. In this case it would be natural to use both terms which were considered when characterizing the asymptotic distribution of T^2. Such a situation is not exceptional. Indeed it occurs quite often that the convergence of a test statistic depends on the parameters to be tested. From this point of view, almost the only possibility to characterize the distribution of interest is to use several terms from the Taylor expansion. Another point is, of course, that by using more terms in the approximation of the distribution, one hopes to obtain better quality of the approximation which, however, may not be true, if some quantity in the expansion has to be estimated. In this case some new errors are introduced, which can be relatively large if higher order moments are involved.

In the following we are going to consider density approximations. It is often fairly straightforward to obtain formulae for distribution functions from the relations between the densities which will be presented in the subsequent.

3.2.2 *General relation between two densities in* $\mathbb{R}^p$

In statistical approximation theory the most common tool for approximating the density or the distribution function of some statistic is the Edgeworth expansion or related expansions, such as tilted Edgeworth (e.g. see Barndorff-Nielsen & Cox, 1989). In such a case a distribution is approximated by the standard normal distribution, and the derivatives of its density function and the first cumulants of the statistic are involved. However, for approximating a skewed random variable it is natural to use some skewed distribution. This idea was used by Hall (1983) to approximate a sum of independent random variables with chi-square distribution and it is exploited in insurance mathematics for approximating claim distributions via the Γ-distribution (see Gerber 1979, for instance).

Similar ideas can also be applied in the multivariate case. For different multivariate statistics, Edgeworth expansions have been derived on the basis of the multivariate normal distribution, $N_p(\mathbf{0}, \boldsymbol{\Sigma})$ (e.g. see Traat, 1986; Skovgaard, 1986; McCullagh, 1987; Barndorff-Nielsen & Cox, 1989), but in many cases it seems more natural to use multivariate approximations via some skewed distribution such as non-symmetric mixtures or the Wishart distribution. A majority of the test statistics in multivariate analysis is based on functions of quadratic forms. Therefore, it is reasonable to believe, at least when the statistics are based on normal samples, that we can expect good approximations for these statistics using the Wishart density. Let $\mathbf{x}$ and $\mathbf{y}$ be two $p-$vectors with densities $f_{\mathbf{x}}(\mathbf{x})$ and $f_{\mathbf{y}}(\mathbf{x})$, characteristic functions $\varphi_{\mathbf{x}}(\mathbf{t})$, $\varphi_{\mathbf{y}}(\mathbf{t})$ and cumulant functions $\psi_{\mathbf{x}}(\mathbf{t})$ and $\psi_{\mathbf{y}}(\mathbf{t})$. Our aim is to present the more complicated density function, say $f_{\mathbf{y}}(\mathbf{x})$, through the simpler one, $f_{\mathbf{x}}(\mathbf{x})$. In the univariate case, the problem in this setup was examined by Cornish & Fisher (1937), who obtained the principal solution to the problem and used it in the case when $X \sim N(0,1)$. Finney (1963) generalized the idea for the multivariate case and gave a general expression of the relation between two densities. In his paper Finney applied the idea in the univariate case, presenting one density through the other. From later presentations we mention McCullagh (1987) and Barndorff-Nielsen & Cox (1989), who briefly considered generalized formal Edgeworth expansions in tensor notation. When comparing our approach with the coordinate free tensor approach, it is a matter of taste which one to prefer. The tensor notation approach, as used by McCullagh (1987), gives compact expressions. However, these can sometimes be difficult to apply in real calculations. Before going over to expansions, remember that the characteristic function of a continuous random $p-$vector $\mathbf{y}$ can be considered as the Fourier transform of the density function:

$$\varphi_{\mathbf{y}}(\mathbf{t}) = \int_{\mathbb{R}^p} e^{i\mathbf{t}'\mathbf{x}} f_{\mathbf{y}}(\mathbf{x}) d\mathbf{x}.$$

To establish our results we need some properties of the inverse Fourier transforms. An interested reader is referred to Esséen (1945). The next lemma gives the basic relation which connects the characteristic function with the derivatives of the density function.

Lemma 3.2.1. *Assume that*

$$\lim_{|x_{i_k}|\to\infty} \frac{\partial^{k-1} f_{\mathbf{y}}(\mathbf{x})}{\partial x_{i_1} \dots \partial x_{i_{k-1}}} = 0.$$

Then the k-th derivative of the density $f_{\mathbf{y}}(\mathbf{x})$ is connected with the characteristic function $\varphi_{\mathbf{y}}(\mathbf{t})$ of a $p-$vector $\mathbf{y}$ by the following relation:

$$(i\mathbf{t})^{\otimes k}\varphi_{\mathbf{y}}(\mathbf{t}) = (-1)^k \int_{\mathbb{R}^p} \mathbf{e}^{i\mathbf{t}'\mathbf{x}} \mathrm{vec}\frac{d^k f_{\mathbf{y}}(\mathbf{x})}{d\mathbf{x}^k} d\mathbf{x}, \qquad k = 0, 1, 2, \dots, \tag{3.2.1}$$

where i is the imaginary unit and $\frac{d^k f_{\mathbf{y}}(\mathbf{x})}{d\mathbf{x}^k}$ is the matrix derivative defined by (1.4.7).

PROOF: The relation in (3.2.1) will be proved by induction. When $k = 0$, the equality (3.2.1) is identical to the definition of the characteristic function. For $k = 1$ we get the statement by assumption, taking into account that $|\mathbf{e}^{i\mathbf{t}'\mathbf{x}}| \leq 1$ and Corollary 1.4.9.1:

$$-\int_{\mathbb{R}^p} \frac{d f_{\mathbf{y}}(\mathbf{x})}{d\mathbf{x}} e^{i\mathbf{t}'\mathbf{x}} d\mathbf{x} = \int_{\mathbb{R}^p} \frac{d e^{i\mathbf{t}'\mathbf{x}}}{d\mathbf{x}} f_{\mathbf{y}}(\mathbf{x}) d\mathbf{x} = i\mathbf{t} \int_{\mathbb{R}^p} e^{i\mathbf{t}'\mathbf{x}} f_{\mathbf{y}}(\mathbf{x}) d\mathbf{x} = i\mathbf{t}\varphi_{\mathbf{y}}(\mathbf{t}).$$

Suppose that the relation (3.2.1) holds for $k = s - 1$. By assumption,

$$e^{i\mathbf{t}'\mathbf{x}} \frac{d^{s-1} f_{\mathbf{y}}(\mathbf{x})}{d\,\mathbf{x}^{s-1}} \bigg|_{\mathbf{x} \in \partial \mathbb{R}^p} = \mathbf{0},$$

where ∂ denotes the boundary. Thus, once again applying Corollary 1.4.9.1 we get

$$- \int_{\mathbb{R}^p} \frac{d^s f_{\mathbf{y}}(\mathbf{x})}{d\,\mathbf{x}^s} e^{i\mathbf{t}'\mathbf{x}} d\mathbf{x} = \int_{\mathbb{R}^p} \frac{d\, e^{i\mathbf{t}'\mathbf{x}}}{d\,\mathbf{x}} \mathrm{vec}' \frac{d^{s-1} f_{\mathbf{y}}(\mathbf{x})}{d\,\mathbf{x}^{s-1}} d\mathbf{x}$$
$$= i\mathbf{t} \int_{\mathbb{R}^p} e^{i\mathbf{t}'\mathbf{x}} \mathrm{vec}' \frac{d^{s-1} f_{\mathbf{y}}(\mathbf{x})}{d\,\mathbf{x}^{s-1}} d\mathbf{x} = (-1)^{s-1} i\mathbf{t}(i\mathbf{t}')^{\otimes s-1} \varphi_{\mathbf{y}}(\mathbf{t}),$$

which means that the statement holds for $k = s$. ∎

In the subsequent text we are using the notation $\mathbf{f}^k_{\mathbf{Y}}(\mathbf{X})$, $k = 0, 1, 2, \ldots$, instead of $\frac{d^k f_{\mathbf{Y}}(\mathbf{X})}{d\,\mathbf{X}^k}$, where $f^0_{\mathbf{Y}}(\mathbf{X}) = f_{\mathbf{Y}}(\mathbf{X})$. A formula for the inverse Fourier transform is often needed. This transform is given in the next corollary.

Corollary 3.2.1.L1. *Let* $\mathbf{y}$ *be a random* $p-$*vector and* $\mathbf{a}$ *an arbitrary constant* p^k-*vector. Then*

$$(-1)^k \mathbf{a}' \mathrm{vec}\, \mathbf{f}^k_{\mathbf{y}}(\mathbf{x}) = (2\pi)^{-p} \int_{\mathbb{R}^p} \varphi_{\mathbf{y}}(\mathbf{t}) \mathbf{a}'(i\mathbf{t})^{\otimes k} \mathbf{e}^{-i\mathbf{t}'\mathbf{x}} d\mathbf{t}, \qquad k = 0, 1, 2, \ldots, \quad (3.2.3)$$

where i *is the imaginary unit.*

PROOF: By Lemma 3.2.1, the vector $(i\mathbf{t})^{\otimes k} \varphi_{\mathbf{y}}(\mathbf{t})$ is the Fourier transform of the vectorized derivative $\dfrac{d^k f_{\mathbf{y}}(\mathbf{x})}{d\mathbf{x}^k}$. Then the vector of derivatives is obtained from its inverse Fourier transform, i.e.

$$(-1)^k \mathrm{vec} \frac{d^k f_{\mathbf{y}}(\mathbf{x})}{d\mathbf{x}^k} = (2\pi)^{-p} \int_{\mathbb{R}^p} \varphi_{\mathbf{y}}(\mathbf{t})(i\mathbf{t})^{\otimes k} \mathbf{e}^{-i\mathbf{t}'\mathbf{x}} d\mathbf{t}.$$

If multiplying the left-hand side by a p^k-vector $\mathbf{a}'$, the necessary result is obtained. ∎

Taking into account that Taylor expansions of matrices are realized through their vector representations, we get immediately a result for random matrices.

Corollary 3.2.1.L2. *Let* $\mathbf{Y}$ *be a random* $p \times q-$*matrix and* $\mathbf{a}$ *an arbitrary constant* $(pq)^k-$*vector. Then*

$$(-1)^k \mathbf{a}' \mathrm{vec}\, \mathbf{f}^k_{\mathbf{Y}}(\mathbf{X}) = (2\pi)^{-pq} \int_{\mathbb{R}^{pq}} \varphi_{\mathbf{Y}}(\mathbf{T}) \mathbf{a}'(i\mathrm{vec}\mathbf{T})^{\otimes k} \mathbf{e}^{-i\mathrm{vec}'\mathbf{T}\mathrm{vec}\mathbf{X}} d\mathrm{vec}\mathbf{T},$$
$$k = 0, 1, 2, \ldots,$$

where i *is the imaginary unit and* $\mathbf{T}, \mathbf{X} \in \mathbb{R}^{p \times q}$. ∎

Now we are can present the main result of the paragraph.

Theorem 3.2.1. *If* $\mathbf{y}$ *and* $\mathbf{x}$ *are two random* $p-$*vectors, the density* $f_{\mathbf{y}}(\mathbf{x})$ *can be presented through the density* $f_{\mathbf{x}}(\mathbf{x})$ *by the following formal expansion:*

$$\begin{aligned} f_{\mathbf{y}}(\mathbf{x}) = f_{\mathbf{x}}(\mathbf{x}) &- (E[\mathbf{y}] - E[\mathbf{x}])'\mathbf{f}^1_{\mathbf{x}}(\mathbf{x}) \\ &+ \tfrac{1}{2}\mathrm{vec}'\{D[\mathbf{y}] - D[\mathbf{x}] + (E[\mathbf{y}] - E[\mathbf{x}])(E[\mathbf{y}] - E[\mathbf{x}])'\}\mathrm{vec}\mathbf{f}^2_{\mathbf{x}}(\mathbf{x}) \\ &- \tfrac{1}{6}\{\mathrm{vec}'(c_3[\mathbf{y}] - c_3[\mathbf{x}]) + 3\mathrm{vec}'(D[\mathbf{y}] - D[\mathbf{x}]) \otimes (E[\mathbf{y}] - E[\mathbf{x}])' \\ &+ (E[\mathbf{y}] - E[\mathbf{x}])'^{\otimes 3}\}\mathrm{vec}\mathbf{f}^3_{\mathbf{x}}(\mathbf{x}) + \cdots . \end{aligned} \tag{3.2.4}$$

PROOF: Using the expansion (2.1.35) of the cumulant function we have

$$\psi_{\mathbf{y}}(\mathbf{t}) - \psi_{\mathbf{x}}(\mathbf{t}) = \sum_{k=1}^{\infty} \tfrac{i^k}{k!}\mathbf{t}'(c_k[\mathbf{y}] - c_k[\mathbf{x}])\mathbf{t}^{\otimes k-1},$$

and thus

$$\varphi_{\mathbf{y}}(\mathbf{t}) = \varphi_{\mathbf{x}}(\mathbf{t}) \prod_{k=1}^{\infty} \exp\{\tfrac{1}{k!}i\mathbf{t}'(c_k[\mathbf{y}] - c_k[\mathbf{x}])(i\mathbf{t})^{\otimes k-1}\}.$$

By using a series expansion of the exponential function we obtain, after ordering the terms according to i^k, the following equality:

$$\begin{aligned} \varphi_{\mathbf{y}}(\mathbf{t}) = \varphi_{\mathbf{x}}(\mathbf{t})\Big\{&1 + i(c_1[\mathbf{y}] - c_1[\mathbf{x}])'\mathbf{t} \\ &+ \tfrac{i^2}{2}\mathbf{t}'\{c_2[\mathbf{y}] - c_2[\mathbf{x}] + (c_1[\mathbf{y}] - c_1[\mathbf{x}])(c_1[\mathbf{y}] - c_1[\mathbf{x}])'\}\mathbf{t} \\ &+ \tfrac{i^3}{6}\Big(\mathbf{t}'\{(c_3[\mathbf{y}] - c_3[\mathbf{x}]) \\ &\quad + (c_1[\mathbf{y}] - c_1[\mathbf{x}])' \otimes (c_1[\mathbf{y}] - c_1[\mathbf{x}])(c_1[\mathbf{y}] - c_1[\mathbf{x}])'\}\mathbf{t}^{\otimes 2} \\ &\quad + 3(c_1[\mathbf{y}] - c_1[\mathbf{x}])'\mathbf{t}\mathbf{t}'(c_2[\mathbf{y}] - c_2[\mathbf{x}])\mathbf{t}\Big) + \cdots\Big\}. \end{aligned}$$

Applying the equality (1.3.31) repeatedly we obtain, when using the vec-operator,

$$\begin{aligned} \varphi_{\mathbf{y}}(\mathbf{t}) = \varphi_{\mathbf{x}}(\mathbf{t})\Big\{&1 + i(c_1[\mathbf{y}] - c_1[\mathbf{x}])'\mathbf{t} \\ &+ \tfrac{i^2}{2}\mathrm{vec}'\{c_2[\mathbf{y}] - c_2[\mathbf{x}] + (c_1[\mathbf{y}] - c_1[\mathbf{x}])(c_1[\mathbf{y}] - c_1[\mathbf{x}])'\}\mathbf{t}^{\otimes 2} \\ &+ \tfrac{i^3}{6}\Big(\mathrm{vec}'(c_3[\mathbf{y}] - c_3[\mathbf{x}]) + 3\mathrm{vec}'(c_2[\mathbf{y}] - c_2[\mathbf{x}]) \otimes (c_1[\mathbf{y}] - c_1[\mathbf{x}])' \\ &\quad + (c_1[\mathbf{y}] - c_1[\mathbf{x}])'^{\otimes 3}\Big)\mathbf{t}^{\otimes 3} + \cdots\Big\}. \end{aligned}$$

This equality can be inverted by applying the inverse Fourier transform given in Corollary 3.2.1.L1. Then the characteristic functions turn into density functions and, taking into account that $c_1[\bullet] = E[\bullet]$ and $c_2[\bullet] = D[\bullet]$, the theorem is established. ∎

By applying the theorem, a similar result can be stated for random matrices.

Corollary 3.2.1.1. *If* $\mathbf{Y}$ *and* $\mathbf{X}$ *are two random* $p\times q-$*matrices, the density* $f_{\mathbf{Y}}(\mathbf{X})$ *can be presented through the density* $f_{\mathbf{X}}(\mathbf{X})$ *as the following formal expansion:*

$$\begin{aligned}f_{\mathbf{Y}}(\mathbf{X}) = f_{\mathbf{X}}(\mathbf{X}) &- \mathrm{vec}'(E[\mathbf{Y}]-E[\mathbf{X}])\mathbf{f}^1_{\mathbf{X}}(\mathbf{X})\\ &+\tfrac{1}{2}\mathrm{vec}'\{D[\mathbf{Y}]-D[\mathbf{X}]+\mathrm{vec}(E[\mathbf{Y}]-E[\mathbf{X}])\mathrm{vec}'(E[\mathbf{Y}]-E[\mathbf{X}])\}\mathrm{vec}\mathbf{f}^2_{\mathbf{X}}(\mathbf{X})\\ &-\tfrac{1}{6}\Big(\mathrm{vec}'(c_3[\mathbf{Y}]-c_3[\mathbf{X}])+3\mathrm{vec}'(D[\mathbf{Y}]-D[\mathbf{X}])\otimes\mathrm{vec}'(E[\mathbf{Y}]-E[\mathbf{X}])\\ &\quad+\mathrm{vec}'(E[\mathbf{Y}]-E[\mathbf{X}])^{\otimes 3}\Big)\mathrm{vec}\mathbf{f}^3_{\mathbf{X}}(\mathbf{X})+\cdots.\end{aligned}$$

∎

3.2.3 *Multivariate Edgeworth type expansions*

Consider the univariate case. When the known distribution of X is the standard normal distribution, from (3.2.4) one gets a classical Edgeworth type expansion where the density of the standardized random variable Y is presented as a series expansion through its cumulants and derivatives of the normal density, expressed via Hermite polynomials, as in §2.2.4. In the multivariate case, we have the vector $\mathbf{x}\sim N_p(\mathbf{0},\mathbf{\Sigma})$ in the role of $X\sim N(0,1)$. Now, the multivariate Hermite polynomials $\mathbf{H}_i(\mathbf{x},\mathbf{\Sigma})$, defined by (2.2.39) – (2.2.41), will be of use. However, in the multivariate case we can not divide by the standard deviation and therefore, we shall present the unknown density of $\mathbf{y}$ through the normal density $N_p(\mathbf{0},\mathbf{\Sigma})$ given by (2.2.5). When $\mathbf{x}\sim N_p(\mathbf{0},\mathbf{\Sigma})$ in the expansion (3.2.4), we say that we have a multivariate Edgeworth type expansion for the density $f_{\mathbf{y}}(\mathbf{x})$.

Theorem 3.2.2. *Let* $\mathbf{y}$ *be a random* $p-$*vector with finite first four moments, then the density* $f_{\mathbf{y}}(\mathbf{x})$ *can be presented through the density* $f_N(\mathbf{x})$ *of the distribution* $N_p(\mathbf{0},\mathbf{\Sigma})$ *by the following formal Edgeworth type expansion:*

$$\begin{aligned}f_{\mathbf{y}}(\mathbf{x}) = f_N(\mathbf{x})\Big\{&1+E[\mathbf{y}]'\mathrm{vec}\mathbf{H}_1(\mathbf{x},\mathbf{\Sigma})\\ &+\tfrac{1}{2}\mathrm{vec}'\{D[\mathbf{y}]-\mathbf{\Sigma}+E[\mathbf{y}](E[\mathbf{y}])'\}\mathrm{vec}\mathbf{H}_2(\mathbf{x},\mathbf{\Sigma})\\ &+\tfrac{1}{6}\,\{\mathrm{vec}'c_3[\mathbf{y}]+3\mathrm{vec}'(D[\mathbf{y}]-\mathbf{\Sigma})\otimes(E[\mathbf{y}])'\\ &\quad+(E[\mathbf{y}])'^{\otimes 3}\}\,\mathrm{vec}\mathbf{H}_3(\mathbf{x},\mathbf{\Sigma})+\cdots\Big\},\end{aligned}\tag{3.2.5}$$

where the multivariate Hermite polynomials $\mathbf{H}_i(\mathbf{x},\mathbf{\Sigma})$ *are given by the relations in (2.2.39) – (2.2.41).* ∎

As cumulants of random matrices are defined through their vector representations, a formal expansion for random matrices can be given in a similar way.

Corollary 3.2.2.1. *If* $\mathbf{Y}$ *is a random* $p\times q-$*matrix with finite first four moments, then the density* $f_{\mathbf{Y}}(\mathbf{X})$ *can be presented through the density* $f_N(\mathbf{X})$ *of the* *distribution $N_{pq}(\mathbf{0},\mathbf{\Sigma})$ by the following formal matrix Edgeworth type expansion:*

$$f_{\mathbf{Y}}(\mathbf{X}) = f_N(\mathbf{X})\Big\{1 + E[\mathrm{vec}\mathbf{Y}]'\mathrm{vec}\mathbf{H}_1(\mathrm{vec}\mathbf{X},\mathbf{\Sigma})$$
$$+ \tfrac{1}{2}\mathrm{vec}'\{D[\mathrm{vec}\mathbf{Y}] - \mathbf{\Sigma} + E[\mathrm{vec}\mathbf{Y}]E[\mathrm{vec}\mathbf{Y}]'\}\mathrm{vec}\mathbf{H}_2(\mathrm{vec}\mathbf{X},\mathbf{\Sigma})$$
$$+ \tfrac{1}{6}\Big(\mathrm{vec}'(c_3[\mathbf{Y}] + 3\mathrm{vec}'(D[\mathrm{vec}\mathbf{Y}] - \mathbf{\Sigma}) \otimes E[\mathrm{vec}'\mathbf{Y}]$$
$$+ E[\mathrm{vec}'\mathbf{Y}]^{\otimes 3}\Big)\mathrm{vec}\mathbf{H}_3(\mathrm{vec}\mathbf{X},\mathbf{\Sigma}) + \cdots\Big\}. \tag{3.2.6}$$

∎

The importance of Edgeworth type expansions is based on the fact that asymptotic normality holds for a very wide class of statistics. Better approximation may be obtained if a centered version of the statistic of interest is considered. From the applications' point of view the main interest is focused on statistics $\mathbf{T}$ which are functions of sample moments, especially the sample mean and the sample dispersion matrix. If we consider the formal expansion (3.2.5) for a $p-$dimensional $\mathbf{T}$, the cumulants of $\mathbf{T}$ depend on the sample size. Let us assume that the cumulants $c_i[\mathbf{T}]$ depend on the sample size n in the following way:

$$c_1[\mathbf{T}] = n^{-\frac{1}{2}}\mathbf{\Gamma}_1(\mathbf{T}) + o(n^{-1}), \tag{3.2.7}$$
$$c_2[\mathbf{T}] = \mathbf{K}_2(\mathbf{T}) + n^{-1}\mathbf{\Gamma}_2(\mathbf{T}) + o(n^{-1}), \tag{3.2.8}$$
$$c_3[\mathbf{T}] = n^{-\frac{1}{2}}\mathbf{\Gamma}_3(\mathbf{T}) + o(n^{-1}), \tag{3.2.9}$$
$$c_j[\mathbf{T}] = o(n^{-1}),\ j \geq 4, \tag{3.2.10}$$

where $\mathbf{K}_2(\mathbf{T})$ and $\mathbf{\Gamma}_i(\mathbf{T})$ depend on the underlying distribution but not on n. This choice of cumulants guarantees that centered sample moments and their functions multiplied to $\sqrt{n}$ can be used. For example, cumulants of the statistic $\sqrt{n}(g(\overline{\mathbf{x}}) - g(\mu))$ will satisfy (3.2.7) – (3.2.10) for smooth functions $g(\bullet)$. In the univariate case the Edgeworth expansion of the density of a statistic T, with the cumulants satisfying (3.2.7) – (3.2.10), is of the form

$$f_T(x) = f_{N(0,k_2(T))}(x)\{1 + n^{-\frac{1}{2}}\{\gamma_1(T)h_1(x) + \frac{1}{6}\gamma_3(T)h_3(x)\} + o(n^{-\frac{1}{2}})\}, \tag{3.2.11}$$

where the Hermite polynomials $h_k(x)$ are defined in §2.2.4. The form of (3.2.11) can be carried over to the multivariate case.

Corollary 3.2.2.2. *Let $\mathbf{T}(n)$ be a $p-$dimensional statistic with cumulants satisfying (3.2.7) - (3.2.10). Then, for $\mathbf{T}(n)$, the following Edgeworth type expansion is valid:*

$$f_{\mathbf{T}(n)}(\mathbf{x}) = f_{N_p(\mathbf{0},\mathbf{K}_2(\mathbf{T}))}(\mathbf{x})\{1 + n^{-\frac{1}{2}}((\mathbf{\Gamma}_1(\mathbf{T}))'\mathrm{vec}\mathbf{H}_1(\mathbf{x},\mathbf{\Sigma})$$
$$+ \tfrac{1}{6}\mathrm{vec}'(c_3[\mathbf{y}])\mathrm{vec}\mathbf{H}_3(\mathbf{x},\mathbf{\Sigma})) + o(n^{-\frac{1}{2}})\}. \tag{3.2.12}$$

∎

If $\mathbf{T}(n)$ is a random matrix, a similar statement follows from Corollary 3.2.2.1.

Corollary 3.2.2.3. *Let $\mathbf{T}(n) : p \times q$ be a statistic with cumulants satisfying (3.2.7) – (3.2.10). Then, for $\mathbf{T}(n)$ the following formal Edgeworth type expansion is valid:*

$$
\begin{aligned}
f_{\mathbf{T}(n)}(\mathbf{X}) = f_{N_{pq}(\mathbf{0},\mathbf{K}_2(\mathbf{T}))}(\mathbf{X})\big\{1 + n^{-\frac{1}{2}}\{(\boldsymbol{\Gamma}_1(\mathbf{T}))'\mathrm{vec}\mathbf{H}_1(\mathrm{vec}\mathbf{X}, \boldsymbol{\Sigma}) \\
+ \tfrac{1}{6}\mathrm{vec}'\big(c_3[\mathbf{T}]\big)\mathrm{vec}\mathbf{H}_3(\mathrm{vec}\mathbf{X}, \boldsymbol{\Sigma})\} + o(n^{-\frac{1}{2}})\big\}.
\end{aligned}
\qquad (3.2.13)
$$

∎

3.2.4 *Wishart expansions*

For different multivariate statistics, Edgeworth type expansions have been obtained on the basis of the multivariate normal distribution $N_p(\mathbf{0}, \boldsymbol{\Sigma})$, but in many cases it seems more natural to use multivariate approximations via the Wishart distribution or some other multivariate skewed distribution. As noted in the beginning of this section, many test statistics in multivariate analysis are based on functions of quadratic forms. Therefore, if some properties of a quadratic form are transmitted to the test statistic under consideration we may hope that the Wishart density, which often is the density of the quadratic form, will be appropriate to use. The starting point for our study in this paragraph is Theorem 3.2.1 and its Corollary 3.2.1.1. In Section 2.4 the Wishart distribution was examined. We are going to use the centered Wishart distribution, which was considered in §2.4.8. The reason for using it is that the derivatives of its density are decreasing as functions of the degrees of freedom. This is not the case with the ordinary Wishart distribution.

As the Wishart matrix is symmetric, we shall deal with cumulants of symmetric matrices in the following text. In Section 2.1 we agreed not to point out symmetricity in the notation, and we wrote $f_{\mathbf{X}}(\mathbf{X})$ and $\varphi_{\mathbf{X}}(\mathbf{T})$ for the density and characteristic functions of a symmetric $\mathbf{X} : p \times p$, while remembering that we use the $\frac{1}{2}p(p+1)$ elements of the upper triangles of $\mathbf{X}$ and $\mathbf{T}$. The formal expansion through the centered Wishart distribution is given in the next theorem (Kollo & von Rosen, 1995a).

Theorem 3.2.3. *Let $\mathbf{W}$, $\mathbf{Y}$ and $\mathbf{V}$ be $p \times p$ random symmetric matrices with $\mathbf{W} \sim W_p(\boldsymbol{\Sigma}, n)$ and $\mathbf{V} = \mathbf{W} - n\boldsymbol{\Sigma}$. Then, for the density $f_{\mathbf{Y}}(\mathbf{X})$, the following formal expansion holds:*

$$
\begin{aligned}
f_{\mathbf{Y}}(\mathbf{X}) = f_{\mathbf{V}}(\mathbf{X})\Big\{1 + E[V^2(\mathbf{Y})]'\mathbf{L}_1^\star(\mathbf{X}, \boldsymbol{\Sigma}) \\
+ \tfrac{1}{2}\mathrm{vec}'(D[V^2(\mathbf{Y})] - D[V^2(\mathbf{V})] + E[V^2(\mathbf{Y})]E[V^2(\mathbf{Y})]')\mathrm{vec}\mathbf{L}_2^\star(\mathbf{X}, \boldsymbol{\Sigma}) \\
+ \tfrac{1}{6}\Big(\mathrm{vec}'(c_3[V^2(\mathbf{Y})] - c_3[V^2(\mathbf{V})])' \\
+ 3\mathrm{vec}'(D[V^2(\mathbf{Y})] - D[V^2(\mathbf{V})]) \otimes E[V^2(\mathbf{Y})]' + E[V^2(\mathbf{Y})]'^{\otimes 3}\Big) \\
\times \mathrm{vec}\mathbf{L}_3^\star(\mathbf{X}, \boldsymbol{\Sigma}) + \cdots\Big\}, \qquad \mathbf{X} > 0,
\end{aligned}
\qquad (3.2.14)
$$

where $V^2(\bullet)$ is given in Definition 1.3.9 and $\mathbf{L}_i^\star(\mathbf{X},\boldsymbol{\Sigma})$ are defined in Lemma 2.4.2 by (2.4.66).

PROOF: We get the statement of the theorem directly from Corollary 3.2.1.1, if we take into account that by Lemma 2.4.2 the derivative of the centered Wishart density equals

$$f^k{}_{\mathbf{V}}(\mathbf{V}) = (-1)^k \mathbf{L}^\star{}_k(\mathbf{V},\boldsymbol{\Sigma}) f_{\mathbf{V}}(\mathbf{V}).$$

∎

Theorem 3.2.3 will be used in the following, when considering an approximation of the density of the sample dispersion matrix $\mathbf{S}$ with the density of a centered Wishart distribution. An attempt to approximate the distribution of the sample dispersion matrix with the Wishart distribution was probably first made by Tan (1980), but he did not present general explicit expressions. Only in the two-dimensional case formulae were derived.

Corollary 3.2.3.1. *Let $\mathbf{Z} = (\mathbf{z}_1,\ldots,\mathbf{z}_n)$ be a sample of size n from a p-dimensional population with $E[\mathbf{z}_i] = \boldsymbol{\mu}$, $D[\mathbf{z}_i] = \boldsymbol{\Sigma}$ and $\overline{m}_k[\mathbf{z}_i] < \infty$, $k = 3,4,\ldots$ and let $\mathbf{S}$ denote the sample dispersion matrix given by (3.1.3). Then the density function $f_{\mathbf{S}^\star}(\mathbf{X})$ of $\mathbf{S}^\star = n(\mathbf{S}-\boldsymbol{\Sigma})$ has the following representation through the centered Wishart density $f_{\mathbf{V}}(\mathbf{X})$, where $\mathbf{V} = \mathbf{W} - n\boldsymbol{\Sigma}$, $\mathbf{W} \sim W_p(\boldsymbol{\Sigma}, n)$ and $n \geq p$:*

$$f_{\mathbf{S}^\star}(\mathbf{X}) = f_{\mathbf{V}}(\mathbf{X})\Big\{1 - \tfrac{1}{4}\mathrm{vec}'\big\{\mathbf{G}_p\{\overline{m}_4[\mathbf{z}_i] - \mathrm{vec}\boldsymbol{\Sigma}\mathrm{vec}'\boldsymbol{\Sigma} - (\mathbf{I}_{p^2} + \mathbf{K}_{p,p})(\boldsymbol{\Sigma}\otimes\boldsymbol{\Sigma})\}\mathbf{G}_p'\big\}$$
$$\times\, \mathrm{vec}\big\{\mathbf{G}_p\mathbf{H}_p\{(\mathbf{X}/n+\boldsymbol{\Sigma})^{-1}\otimes(\mathbf{X}/n+\boldsymbol{\Sigma})^{-1}\}\mathbf{H}_p\mathbf{G}_p'\big\} + O(\tfrac{1}{n})\Big\}, \quad \mathbf{X}>0, \tag{3.2.15}$$

where $\mathbf{G}_p$ is defined by (1.3.49) – (1.3.50), and $\mathbf{H}_p = \mathbf{I} + \mathbf{K}_{p,p} - (\mathbf{K}_{p,p})_d$ is from Corollary 1.4.3.1.

PROOF: To obtain (3.2.15), we have to insert the expressions of $\mathbf{L}_k^\star(\mathbf{X},\boldsymbol{\Sigma})$ and cumulants $c_k[V^2(\mathbf{S}^\star)]$ and $c_k[V^2(\mathbf{V})]$ in (3.2.14) and examine the result. Note that $c_1[V^2(\mathbf{S}^\star)] = \mathbf{0}$ and in (3.2.14) all terms including $E[V^2(\mathbf{Y})]$ vanish. By Kollo & Neudecker (1993, Appendix 1),

$$\begin{aligned} D[\sqrt{n}\mathbf{S}] &= \overline{m}_4[\mathbf{z}_i] - \mathrm{vec}\boldsymbol{\Sigma}\mathrm{vec}'\boldsymbol{\Sigma} + \tfrac{1}{n-1}(\mathbf{I}_{p^2}+\mathbf{K}_{p,p})(\boldsymbol{\Sigma}\otimes\boldsymbol{\Sigma}) \\ &= \overline{m}_4[\mathbf{z}_i] - vec\boldsymbol{\Sigma}\mathrm{vec}'\boldsymbol{\Sigma} + O(\tfrac{1}{n}). \end{aligned}$$

Hence, by the definition of $\mathbf{G}_p$, we have

$$D[V^2(\mathbf{S}^\star)] = n\mathbf{G}_p(\overline{m}_4[\mathbf{z}_i] - \mathrm{vec}\boldsymbol{\Sigma}\mathrm{vec}'\boldsymbol{\Sigma})\mathbf{G}_p' + O(1). \tag{3.2.16}$$

In Lemma 2.4.2 we have shown that $\mathbf{L}_2^\star(\mathbf{X},\boldsymbol{\Sigma})$ is of order n^{-1}, and therefore in the approximation we can neglect the second term in $\mathbf{L}_2^\star(\mathbf{X},\boldsymbol{\Sigma})$. Thus, from (3.2.16), (2.4.62) and Theorem 2.4.16 (ii) it follows that

$$\tfrac{1}{2}\mathrm{vec}'(D[V^2(\mathbf{S}^\star)] - D[V^2(\mathbf{V})])\mathrm{vec}\mathbf{L}_2^\star(\mathbf{X},\boldsymbol{\Sigma}) = -\tfrac{1}{4}\mathrm{vec}'\Big(\mathbf{G}_p\{\overline{m}_4[\mathbf{z}_i] - \mathrm{vec}\boldsymbol{\Sigma}\mathrm{vec}'\boldsymbol{\Sigma}$$
$$- (\mathbf{I}_{p^2}+\mathbf{K}_{p,p})(\boldsymbol{\Sigma}\otimes\boldsymbol{\Sigma})\}\mathbf{G}_p'\Big)\mathrm{vec}\big(\mathbf{G}_p\mathbf{H}_p((\mathbf{X}/n+\boldsymbol{\Sigma})^{-1}\otimes(\mathbf{X}/n+\boldsymbol{\Sigma})^{-1})\mathbf{H}_p\mathbf{G}_p'\big)$$
$$+O(\tfrac{1}{n}).$$

To complete the proof we have to show that in (3.2.14) the remaining terms are $O(\frac{1}{n})$. Let us first show that the term including $\mathbf{L}_3^\star(\mathbf{X},\boldsymbol{\Sigma})$ in (3.2.14) is of order n^{-1}. From Lemma 2.4.2 we have that $\mathbf{L}_3^\star(\mathbf{X},\boldsymbol{\Sigma})$ is of order n^{-2}. From Theorem 2.4.16 (iii) it follows that the cumulant $c_3[V^2(\mathbf{V})]$ is of order n. Traat (1984) has found a matrix $\mathbf{M}_3$, independent of n, such that

$$c_3[\mathbf{S}] = n^{-2}\mathbf{M}_3 + O(n^{-3}).$$

Therefore,

$$c_3[V^2(\mathbf{S}^\star)] = n\mathbf{K}_3 + O(1),$$

where the matrix $\mathbf{K}_3$ is independent of n. Thus, the difference of the third order cumulants is of order n, and multiplying it with $\mathrm{vec}\mathbf{L}_3^\star(\mathbf{X},\boldsymbol{\Sigma})$ gives us that the product is $O(n^{-1})$. All other terms in (3.2.14) are scalar products of vectors which dimension do not depend on n. Thus, when examining the order of these terms, it is sufficient to consider products of $\mathbf{L}_k^\star(\mathbf{X},\boldsymbol{\Sigma})$ and differences of cumulants $c_k[V^2(\mathbf{S}^\star)] - c_k[V^2(\mathbf{V})]$. Remember that it was shown in Lemma 2.4.2 that $\mathbf{L}_k^\star(\mathbf{X},\boldsymbol{\Sigma})$, $k \geq 4$ is of order n^{-k+1}. Furthermore, from Theorem 2.4.16 and properties of sample cumulants of k-statistics (Kendall & Stuart, 1958, Chapter 12), it follows that the differences $c_k[V^2(\mathbf{S}^\star)] - c_k[V^2(\mathbf{V})]$, $k \geq 2$, are of order n. Then, from the construction of the formal expansion (3.2.14), we have that for $k = 2p$, $p = 2, 3, \ldots$, the term including $\mathbf{L}_k^\star(\mathbf{X},\boldsymbol{\Sigma})$ is of order n^{-p+1}. The main term of the cumulant differences is the term where the second order cumulants have been multiplied p times. Hence, the $\mathbf{L}_4^\star(\mathbf{X},\boldsymbol{\Sigma})-$term, i.e. the expression including $\mathbf{L}_4^\star(\mathbf{X},\boldsymbol{\Sigma})$ and the product of $D[V^2(\mathbf{S}^\star)] - D[V^2(\mathbf{V})]$ with itself, is $O(n^{-1})$, the $\mathbf{L}_6^\star(\mathbf{X},\boldsymbol{\Sigma})$-term is $O(n^{-2})$, etc.

For $k = 2p+1$, $p = 2, 3, \ldots$, the order of the $\mathbf{L}_k^\star(\mathbf{X},\boldsymbol{\Sigma})-$term is determined by the product of $\mathbf{L}_k^\star(\mathbf{X},\boldsymbol{\Sigma})$, the $(p-1)$ products of the differences of the second order cumulants and a difference of the third order cumulants. So the order of the $\mathbf{L}_k^\star(\mathbf{X},\boldsymbol{\Sigma})-$term ($k = 2p+1$) is $n^{-2p} \times n^{p-1} \times n = n^{-p}$. Thus, the $\mathbf{L}_5^\star(\mathbf{X},\boldsymbol{\Sigma})$-term is $O(n^{-2})$, the $\mathbf{L}_7^\star(\mathbf{X},\boldsymbol{\Sigma}-)$term is $O(n^{-3})$ and so on. The presented arguments complete the proof. ∎

Our second application concerns the non-central Wishart distribution. It turns out that Theorem 3.2.3 gives a very convenient way to describe the non-central Wishart density. The approximation of the non-central Wishart distribution by the Wishart distributions has, among others, previously been considered by Steyn & Roux (1972) and Tan (1979). Both Steyn & Roux (1972) and Tan (1979) perturbed the covariance matrix in the Wishart distribution so that moments of the Wishart distribution and the non-central Wishart distribution became close to each other. Moreover, Tan (1979) based his approximation on Finney's (1963) approach, but never explicitly calculated the derivatives of the density. It was not considered that the density is dependent on n. Although our approach is a matrix version of Finney's, there is a fundamental difference with the approach by Steyn & Roux (1972) and Tan (1979). Instead of perturbing the covariance matrix, we use the idea of centering the non-central Wishart distribution. Indeed, as shown below, this will also simplify the calculations because we are now able to describe the difference between the cumulants in a convenient way, instead of treating

the cumulants of the Wishart distribution and non-central Wishart distribution separately.
Let $\mathbf{Y} \sim W_p(\boldsymbol{\Sigma}, n, \boldsymbol{\mu})$, i.e. $\mathbf{Y}$ follows the non-central Wishart distribution with a non-centrality parameter $\boldsymbol{\mu}$. If $\boldsymbol{\Sigma} > 0$, the matrix $\mathbf{Y}$ has the characteristic function (see Muirhead, 1982)

$$\varphi_{\mathbf{Y}}(\mathbf{T}) = \varphi_{\mathbf{W}}(\mathbf{T}) e^{-\frac{1}{2} tr(\boldsymbol{\Sigma}^{-1}\boldsymbol{\mu}\boldsymbol{\mu}')} e^{\frac{1}{2}\mathrm{tr}(\boldsymbol{\Sigma}^{-1}\boldsymbol{\mu}\boldsymbol{\mu}'(\mathbf{I}_p - i\mathbf{M}(\mathbf{T})\boldsymbol{\Sigma})^{-1})}, \tag{3.2.17}$$

where $\mathbf{M}(\mathbf{T})$ and $\varphi_{\mathbf{W}}(\mathbf{T})$, the characteristic function of $\mathbf{W} \sim W_p(\boldsymbol{\Sigma}, n)$, are given in Theorem 2.4.5.
We shall again consider the centered versions of $\mathbf{Y}$ and $\mathbf{W}$, where $\mathbf{W} \sim W_p(\boldsymbol{\Sigma}, n)$. Let $\mathbf{Z} = \mathbf{Y} - n\boldsymbol{\Sigma} - \boldsymbol{\mu}\boldsymbol{\mu}'$ and $\mathbf{V} = \mathbf{W} - n\boldsymbol{\Sigma}$. Since we are interested in the differences $c_k[\mathbf{Z}] - c_k[\mathbf{V}]$, $k = 1, 2, 3, \ldots$, we take the logarithm on both sides in (3.2.17) and obtain the difference of the cumulant functions

$$\begin{aligned}\psi_{\mathbf{Z}}(\mathbf{T}) - &\psi_{\mathbf{V}}(\mathbf{T}) \\ &= -\tfrac{1}{2}\mathrm{tr}(\boldsymbol{\Sigma}^{-1}\boldsymbol{\mu}\boldsymbol{\mu}') - i\tfrac{1}{2}\mathrm{tr}\{\mathbf{M}(\mathbf{T})\boldsymbol{\mu}\boldsymbol{\mu}'\} + \tfrac{1}{2}\mathrm{tr}\{\boldsymbol{\Sigma}^{-1}\boldsymbol{\mu}\boldsymbol{\mu}'(\mathbf{I} - i\mathbf{M}(\mathbf{T})\boldsymbol{\Sigma})^{-1}\}.\end{aligned}$$

After expanding the matrix $(\mathbf{I} - i\mathbf{M}(\mathbf{T})\boldsymbol{\Sigma})^{-1}$ (Kato, 1972, for example), we have

$$\psi_{\mathbf{Z}}(\mathbf{T}) - \psi_{\mathbf{V}}(\mathbf{T}) = \tfrac{1}{2}\sum_{j=2}^{\infty} i^j \mathrm{tr}\{\boldsymbol{\Sigma}^{-1}\boldsymbol{\mu}\boldsymbol{\mu}'(\mathbf{M}(\mathbf{T})\boldsymbol{\Sigma})^j\}. \tag{3.2.18}$$

From (3.2.18) it follows that $c_1[\mathbf{Z}] - c_1[\mathbf{V}] = \mathbf{0}$, which, of course, must be true, because $E[\mathbf{Z}] = E[\mathbf{V}] = 0$. In order to obtain the difference of the second order cumulants, we have to differentiate (3.2.18) and we obtain

$$c_2[V^2(\mathbf{Z})] - c_2[V^2(\mathbf{V})] = \tfrac{1}{2}\frac{d^2\mathrm{tr}(\boldsymbol{\mu}\boldsymbol{\mu}'\mathbf{M}(\mathbf{T})\boldsymbol{\Sigma}\mathbf{M}(\mathbf{T}))}{d\,V^2(\mathbf{T})^2} = \mathbf{J}_p(\boldsymbol{\mu}\boldsymbol{\mu}' \otimes \boldsymbol{\Sigma} + \boldsymbol{\Sigma} \otimes \boldsymbol{\mu}\boldsymbol{\mu}')\mathbf{G}_p', \tag{3.2.19}$$

where $\mathbf{J}_p$ and $\mathbf{G}_p$ are as in Theorem 2.4.16. Moreover,

$$\begin{aligned}c_3[V^2(\mathbf{Z})] &- c_3[V^2(\mathbf{V})] \\ &= \mathbf{J}_p\Big\{\boldsymbol{\Sigma} \otimes \boldsymbol{\Sigma} \otimes \mathrm{vec}'(\boldsymbol{\mu}\boldsymbol{\mu}') + \mathrm{vec}'(\boldsymbol{\mu}\boldsymbol{\mu}') \otimes \boldsymbol{\Sigma} \otimes \boldsymbol{\Sigma} + \mathrm{vec}'\boldsymbol{\Sigma} \otimes \boldsymbol{\mu}\boldsymbol{\mu}' \otimes \boldsymbol{\Sigma} \\ &\qquad + \boldsymbol{\mu}\boldsymbol{\mu}' \otimes \boldsymbol{\Sigma} \otimes \mathrm{vec}'\boldsymbol{\Sigma} + \boldsymbol{\Sigma} \otimes \boldsymbol{\mu}\boldsymbol{\mu}' \otimes \mathrm{vec}'\boldsymbol{\Sigma} \\ &\qquad + \mathrm{vec}'\boldsymbol{\Sigma} \otimes \boldsymbol{\Sigma} \otimes \boldsymbol{\mu}\boldsymbol{\mu}')(\mathbf{I}_p \otimes \mathbf{K}_{p,p} \otimes \mathbf{I}_p)\Big\}\mathbf{J}_p'\mathbf{G}_p'.\end{aligned} \tag{3.2.20}$$

Hence, the next theorem is established.

Theorem 3.2.4. *Let* $\mathbf{Z} = \mathbf{Y} - n\boldsymbol{\Sigma} - \boldsymbol{\mu}\boldsymbol{\mu}'$, *where* $\mathbf{Y} \sim W_p(\boldsymbol{\Sigma}, n, \boldsymbol{\mu})$, *and* $\mathbf{V} = \mathbf{W} - n\boldsymbol{\Sigma}$, *where* $\mathbf{W} \sim W_p(\boldsymbol{\Sigma}, n)$. *Then*

$$\begin{aligned}f_{\mathbf{Z}}(\mathbf{X}) = f_{\mathbf{V}}(\mathbf{X})\Big\{1 &+ \frac{1}{2}\mathrm{vec}'(\boldsymbol{\mu}\boldsymbol{\mu}' \otimes \boldsymbol{\Sigma} + \boldsymbol{\Sigma} \otimes \boldsymbol{\mu}\boldsymbol{\mu}')(\mathbf{G}_p' \otimes \mathbf{J}_p')\mathrm{vec}\mathbf{L}_2^\star(\mathbf{X}, \boldsymbol{\Sigma}) \\ &+ \tfrac{1}{6}\mathrm{vec}'(c_3[V^2(\mathbf{Z})] - c_3[V^2(\mathbf{V})])\mathrm{vec}\mathbf{L}_3^\star(\mathbf{X}, \boldsymbol{\Sigma}) + o(n^{-2})\Big\}, \quad \mathbf{X} > 0,\end{aligned} \tag{3.2.21}$$

where $\mathbf{L}_k^\star(\mathbf{X},\boldsymbol{\Sigma})$, $k=2,3$, *are given in Lemma 2.4.2 and* $(c_3[V^2(\mathbf{Z})]-c_3[V^2(\mathbf{V})])$ *is determined by (3.2.20).*

PROOF: The proof follows from (3.2.14) if we replace the difference of the second order cumulants by (3.2.19), taking into account that, by Lemma 2.4.2, $\mathbf{L}_k^\star(\mathbf{X},\boldsymbol{\Sigma})$ is of order n^{-k+1}, $k\geq 3$, and note that the differences of the cumulants $c_k[V^2(\mathbf{Z})]-c_k[V^2(\mathbf{V})]$ do not depend on n. ∎

We get an approximation of order n^{-1} from Theorem 3.2.4 in the following way.

Corollary 3.2.4.1.

$$f_{\mathbf{Z}}(\mathbf{X})=f_{\mathbf{V}}(\mathbf{X})\Big\{1-\frac{1}{4n}\mathrm{vec}'(\boldsymbol{\mu}\boldsymbol{\mu}'\otimes\boldsymbol{\Sigma}+\boldsymbol{\Sigma}\otimes\boldsymbol{\mu}\boldsymbol{\mu}')$$
$$\times(\mathbf{G}_p'\mathbf{G}_p\mathbf{H}_p'\otimes\mathbf{J}_p'\mathbf{G}_p\mathbf{H}_p')\mathrm{vec}((\mathbf{X}/n+\boldsymbol{\Sigma})^{-1}\otimes(\mathbf{X}/n+\boldsymbol{\Sigma})^{-1})+o(\tfrac{1}{n})\Big\},\quad \mathbf{X}>0,$$

where, as previously, $\mathbf{H}_p=\mathbf{I}+\mathbf{K}_{p,p}-(\mathbf{K}_{p,p})_d$.

PROOF: The statement follows from (3.2.21) if we omit the $\mathbf{L}_3^\star(\mathbf{X},\boldsymbol{\Sigma})$-term, which is of order n^{-2}, and then use the n^{-1} term in $\mathbf{L}_2^\star(\mathbf{X},\boldsymbol{\Sigma})$ in (2.4.68). ∎

3.2.5 *Problems*

1. Let $\mathbf{S}$ be Wishart distributed. Show that assumptions (3.2.7) and (3.2.8) are satisfied for $\mathbf{S}^{-1}$.
2. Let $\mathbf{Z}\sim E(\boldsymbol{\mu},\mathbf{V},\phi)$. Find the approximation (3.2.15) for the elliptical population.
3. Find a matrix $\mathbf{M}_3$ such that

$$c_3[\mathbf{S}]=n^{-2}\mathbf{M}_3+O(n^{-3}),$$

   where $\mathbf{M}_3$ does not depend on n.
4. Establish (3.2.19).
5. Establish (3.2.20).
6. Present the expansion (3.2.21) in such a way that the terms of order $\frac{1}{n}$ and $\frac{1}{n^2}$ are presented separately.
7. Write out a formal density expansion for a $p-$vector $\mathbf{y}$ through the mixture $f(\mathbf{x})$ of two normal distributions

$$f(\mathbf{x})=\gamma f_{N(\mathbf{0},\boldsymbol{\Sigma}_1)}(\mathbf{x})+(1-\gamma)f_{N(\mathbf{0},\boldsymbol{\Sigma}_2)}(\mathbf{x}).$$

8. Find a Wishart expansion for the sample dispersion matrix when the population has a symmetric Laplace distribution with the characteristic function

$$\varphi(\mathbf{t})=\frac{1}{1+\frac{1}{2}\mathbf{t}'\boldsymbol{\Sigma}\mathbf{t}},$$

   where $\boldsymbol{\Sigma}$ is the dispersion matrix.
9. Find a Wishart approximation for $\mathbf{S}^{-1}$ using two terms.

10. The *multivariate skew normal distribution* $SN_p(\mathbf{\Sigma}, \boldsymbol{\alpha})$ has a density function

$$f(\mathbf{x}) = 2f_{N(\mathbf{0},\mathbf{\Sigma})}(\mathbf{x})\Phi(\boldsymbol{\alpha}'\mathbf{x}),$$

where $\Phi(x)$ is the distribution function of $N(0,1)$, $\boldsymbol{\alpha}$ is a $p-$vector and $\mathbf{\Sigma}$: $p \times p$ is positive definite. The moment generating function of $SN_p(\mathbf{\Sigma}, \boldsymbol{\alpha})$ is of the form

$$M(\mathbf{t}) = 2e^{\frac{1}{2}\mathbf{t}'\mathbf{\Sigma}\mathbf{t}}\Phi\left(\frac{\boldsymbol{\alpha}'\mathbf{\Sigma}\mathbf{t}}{\sqrt{1+\boldsymbol{\alpha}'\mathbf{\Sigma}\boldsymbol{\alpha}}}\right)$$

(see Azzalini & Dalla Valle, 1996; Azzalini & Capitanio, 1999). Find a formal density expansion for a $p-$vector $\mathbf{y}$ through $SN_p(\mathbf{\Sigma}, \boldsymbol{\alpha})$ which includes the first and second order cumulants.

3.3 GENERAL MULTIVARIATE EXPANSIONS

3.3.1 *General relation between two densities*

In the previous section we obtained approximation formulae for density functions of $p-$dimensional distributions via densities of the same size. For multivariate analysis this situation is somewhat restrictive. In this section we are going to consider the approximation of the distribution of a $p-$dimensional random vector via a $r-$dimensional distribution when $p \leq r$. There are many situations when it would be natural to approximate a distribution of a multivariate statistic with a distribution of higher dimension. Let us give some examples. The sample correlation matrix $\mathbf{R}$ is a function of the sample dispersion matrix $\mathbf{S}$. If the population is normally distributed, the matrix $\mathbf{S}$ is Wishart distributed. Therefore, it can be of interest to approximate the distribution of $\mathbf{R}$ by the Wishart distribution. However, there are $\frac{1}{2}p(p+1)$ different random variables in $\mathbf{S}$, whereas there are $\frac{1}{2}p(p-1)$ different variables in $\mathbf{R}$. We have a similar situation when approximating the density of an eigenvector of $\mathbf{S}$ with the Wishart distribution, or if one wants to approximate the distribution of the generalized variance $|\mathbf{S}|$ with the multivariate normal population distribution, for example. Very little has been written on approximation of distributions in the case of different dimensionalities. Kolassa (1994) examines different dimensionalities when approximating a conditional distribution, and we can also refer to Skovgaard (1987). The following presentation is based on the paper by Kollo & von Rosen (1998). A general relation between density functions will be obtained with the help of the next lemma, which gives a representation of a $p-$dimensional density through an integral over the $r-$dimensional Euclidean space. For the presentation we use the following notation: let $\mathbf{y}$ be a random $p-$vector and $\mathbf{t}_1$, $\mathbf{t}$ real $p-$ and $r-$vectors, respectively, with $p \leq r$. Let $\mathbf{P} : p \times r$ be a real matrix of rank $r(\mathbf{P}) = p$. Consider in $\mathbb{R}^r$ the following one-to-one transformation:

$$\mathbf{t} \longrightarrow \begin{pmatrix} \mathbf{t}_1 \\ \mathbf{z} \end{pmatrix},$$

so that $\mathbf{t}_1 = \mathbf{P}\mathbf{t}$, $\quad \mathbf{z} = (\mathbf{P}')^{o'}\mathbf{A}\mathbf{t}$, where $\mathbf{A} : r \times r$ is a positive definite matrix and $(\mathbf{P}')^o : r \times (r-p)$ is any full rank matrix which, as previously, spans the orthogonal complement of the column space of $\mathbf{P}'$, i.e.

$$\begin{pmatrix} \mathbf{t}_1 \\ \mathbf{z} \end{pmatrix} = \begin{pmatrix} \mathbf{P} \\ (\mathbf{P}')^{o'}\mathbf{A} \end{pmatrix} \mathbf{t}. \tag{3.3.1}$$

In the given notation the following lemma is valid.

Lemma 3.3.1. *Let* $\mathbf{y}$ *be a random* $p-$*vector with density* $f_{\mathbf{y}}(\mathbf{x})$ *and matrices* $\mathbf{P}$, $(\mathbf{P}')^o$ *given in (3.3.1). Then*

$$f_{\mathbf{y}}(\mathbf{x}_0) = |\mathbf{A}|^{\frac{1}{2}} |\mathbf{P}\mathbf{A}^{-1}\mathbf{P}'|^{\frac{1}{2}} \frac{1}{(2\pi)^{\frac{1}{2}(r+p)}}$$
$$\times \int_{R^r} \exp(-i(\mathbf{P}\mathbf{t})'\mathbf{x}_0)\varphi_{\mathbf{y}}(\mathbf{P}\mathbf{t})\exp\{-\tfrac{1}{2}\mathbf{t}'\mathbf{A}(\mathbf{P}')^o((\mathbf{P}')^{o'}\mathbf{A}(\mathbf{P}')^o)^{-1}(\mathbf{P}')^{o'}\mathbf{A}\mathbf{t}\}d\mathbf{t}. \tag{3.3.2}$$

PROOF: Denote the right hand side of (3.3.2) by J. Our aim is to get an approximation of the density $f_{\mathbf{y}}(\mathbf{x})$ of the p–dimensional $\mathbf{y}$. According to (3.3.1), a change of variables in J will be carried out. For the Jacobian of the transformation, the following determinant is calculated:

$$\left|\begin{matrix}\mathbf{P}\\ (\mathbf{P}')^{o'}\mathbf{A}\end{matrix}\right| = \left|\begin{pmatrix}\mathbf{P}\\ (\mathbf{P}')^{o'}\mathbf{A}\end{pmatrix}(\mathbf{P}' : \mathbf{A}(\mathbf{P}')^o)\right|^{\frac{1}{2}} = \left|\begin{pmatrix}\mathbf{P}\mathbf{A}^{-1}\\ (\mathbf{P}')^{o'}\end{pmatrix}\mathbf{A}\,(\mathbf{P}' : \mathbf{A}(\mathbf{P}')^o)\right|^{\frac{1}{2}}$$
$$= |\mathbf{A}|^{\frac{1}{2}}|\mathbf{P}\mathbf{A}^{-1}\mathbf{P}'|^{\frac{1}{2}}|(\mathbf{P}')^{o'}\mathbf{A}(\mathbf{P}')^o|^{\frac{1}{2}}.$$

This means that the Jacobian of the transformation (3.3.1) equals

$$|\mathbf{A}|^{-\frac{1}{2}}|\mathbf{P}\mathbf{A}^{-1}\mathbf{P}'|^{-\frac{1}{2}}|(\mathbf{P}')^{o'}\mathbf{A}(\mathbf{P}')^o|^{-\frac{1}{2}}$$

and we obtain

$$J = \frac{1}{(2\pi)^p}\int_{R^p}\exp(-i\mathbf{t}_1{}'\mathbf{x}_0)\varphi_{\mathbf{y}}(\mathbf{t}_1)d\mathbf{t}_1$$
$$\times\frac{1}{(2\pi)^{\frac{1}{2}(r-p)}}\int_{R^{r-p}}|(\mathbf{P}')^{o'}\mathbf{A}(\mathbf{P}')^o|^{-\frac{1}{2}}\exp(-\tfrac{1}{2}\mathbf{z}'((\mathbf{P}')^{o'}\mathbf{A}(\mathbf{P}')^o)^{-1}\mathbf{z})d\mathbf{z}.$$

The last integral equals 1, since it is an integral over a multivariate normal density, i.e. $\mathbf{z}\sim N_{r-p}(\mathbf{0},(\mathbf{P}')^{o'}\mathbf{A}(\mathbf{P}')^o)$. Thus, by the inversion formula given in Corollary 3.2.1.L1,

$$J = \frac{1}{(2\pi)^p}\int_{\mathbb{R}^p}\exp(-i\mathbf{t}_1'\mathbf{x}_0)\varphi_{\mathbf{y}}(\mathbf{t}_1)d\mathbf{t}_1 = f_{\mathbf{y}}(\mathbf{x}_0)$$

and the statement is established. ∎

If $\mathbf{x}$ and $\mathbf{y}$ are vectors of the same dimensionality, our starting point for getting a relation between the two densities in the previous paragraph was the equality

$$\varphi_{\mathbf{y}}(\mathbf{t}) = \frac{\varphi_{\mathbf{y}}(\mathbf{t})}{\varphi_{\mathbf{x}}(\mathbf{t})}\varphi_{\mathbf{x}}(\mathbf{t}),$$

where it is supposed that $\varphi_{\mathbf{x}}(\mathbf{t})\neq 0$. However, since in this paragraph the dimensions of the two distributions are different, we have to modify this identity. Therefore, instead of the trivial equality consider the more complicated one

$$|\mathbf{A}|^{\frac{1}{2}}|\mathbf{P}\mathbf{A}^{-1}\mathbf{P}'|^{\frac{1}{2}}(2\pi)^{-\frac{1}{2}(r+p)}\exp(-i\mathbf{t}_1'\mathbf{y}_0)\varphi_{\mathbf{y}}(\mathbf{t}_1)$$
$$\times\exp\{-\tfrac{1}{2}\mathbf{t}'\mathbf{A}(\mathbf{P}')^o((\mathbf{P}')^{o'}\mathbf{A}(\mathbf{P}')^o)^{-1}(\mathbf{P}')^{o'}\mathbf{A}\mathbf{t}\}$$
$$= |\mathbf{A}|^{\frac{1}{2}}|\mathbf{P}\mathbf{A}^{-1}\mathbf{P}'|^{\frac{1}{2}}(2\pi)^{-r}k(\mathbf{t}_1,\mathbf{t})\exp(-i\mathbf{t}'\mathbf{x}_0)\varphi_{\mathbf{x}}(t), \qquad (3.3.3)$$

where

$$k(\mathbf{t}_1,\mathbf{t}) = (2\pi)^{\frac{1}{2}(r-p)}\frac{\varphi_{\mathbf{y}}(\mathbf{t}_1)}{\varphi_{\mathbf{x}}(\mathbf{t})}$$
$$\times\exp\{i\mathbf{t}'\mathbf{x}_0 - i\mathbf{t}_1'\mathbf{y}_0 - \tfrac{1}{2}\mathbf{t}'\mathbf{A}(\mathbf{P}')^o((\mathbf{P}')^{o'}\mathbf{A}(\mathbf{P}')^o)^{-1}(\mathbf{P}')^{o'}\mathbf{A}\mathbf{t}\}. \quad (3.3.4)$$

The approximations in this paragraph arise from the approximation of the right hand side of (3.3.4), i.e. $k(\mathbf{t}_1, \mathbf{t})$. If $\mathbf{x}$ and $\mathbf{y}$ are of the same size we may choose $\mathbf{t}_1 = \mathbf{t}$ and $\mathbf{x}_0 = \mathbf{y}_0$, which is not necessarily the best choice. In this case, $k(\mathbf{t}_1, \mathbf{t})$ reduces to $\dfrac{\varphi_{\mathbf{y}}(\mathbf{t})}{\varphi_{\mathbf{x}}(\mathbf{t})}$, which was presented as the product

$$\frac{\varphi_{\mathbf{y}}(\mathbf{t})}{\varphi_{\mathbf{x}}(\mathbf{t})} = \prod_{k=1}^{\infty} \exp\{\tfrac{i^k}{k!}\mathbf{t}'(c_k[\mathbf{y}] - c_k[\mathbf{x}])\mathbf{t}^{\otimes k-1}\}, \tag{3.3.5}$$

when proving Theorem 3.2.1. In the general case, i.e. when $\mathbf{t}_1 = \mathbf{P}\mathbf{t}$ holds for some $\mathbf{P}$, we perform a Taylor expansion of (3.3.4) and from now on $k(\mathbf{t}_1, \mathbf{t}) = k(\mathbf{P}\mathbf{t}, \mathbf{t})$. The Taylor expansion of $k(\mathbf{P}\mathbf{t}, \mathbf{t}) : \mathbb{R}^r \to \mathbb{R}$, is given by Corollary 1.4.8.1 and equals

$$k(\mathbf{P}\mathbf{t}, \mathbf{t}) = k(\mathbf{0}, \mathbf{0}) + \sum_{j=1}^{m} \tfrac{1}{j!}\mathbf{t}'\mathbf{k}^j(\mathbf{0}, \mathbf{0})\mathbf{t}^{\otimes j-1} + r_m(\mathbf{t}), \tag{3.3.6}$$

where the derivative

$$\mathbf{k}^j(\mathbf{P}\mathbf{t}, \mathbf{t}) = \frac{d^j k(\mathbf{P}\mathbf{t}, \mathbf{t})}{d\mathbf{t}^j}$$

is given by (1.4.7) and (1.4.41). The remainder term $r_m(\mathbf{t})$ equals

$$r_m(\mathbf{t}) = \tfrac{1}{(m+1)!}\mathbf{t}'\mathbf{k}^{m+1}(\mathbf{P}(\boldsymbol{\theta} \circ \mathbf{t}), \boldsymbol{\theta} \circ \mathbf{t})\mathbf{t}^{\otimes m}, \tag{3.3.7}$$

where $\boldsymbol{\theta} \circ \mathbf{t}$ is the Hadamard product of $\boldsymbol{\theta}$ and $\mathbf{t}$, and $\boldsymbol{\theta}$ is an $r-$vector with elements between 0 and 1. Using expression (3.3.6), the relation in (3.3.3) may be rewritten. The next lemma is a reformulation of the equality (3.3.3) via the expansion (3.3.6).

Lemma 3.3.2. *Let* $\mathbf{k}(\mathbf{t}_1, \mathbf{t}) = \mathbf{k}(\mathbf{P}\mathbf{t}, \mathbf{t})$ *be given by (3.3.4), where* $\mathbf{t}$ *is an* $r-$*vector and* $\mathbf{P}$: $p \times r$. *Then*

$$\begin{aligned}
&|\mathbf{A}|^{\frac{1}{2}}|\mathbf{P}\mathbf{A}^{-1}\mathbf{P}'|^{\frac{1}{2}}(2\pi)^{-\frac{1}{2}(r+p)}\\
&\quad\times \exp(-i(\mathbf{P}\mathbf{t})'\mathbf{y}_0)\varphi_{\mathbf{y}}(\mathbf{P}\mathbf{t})\exp\{-\tfrac{1}{2}\mathbf{t}'\mathbf{A}(\mathbf{P}')^o((\mathbf{P}')^{o'}\mathbf{A}(\mathbf{P}')^o)^{-1}(\mathbf{P}')^{o'}\mathbf{A}\mathbf{t}\}\\
&= |\mathbf{A}|^{\frac{1}{2}}|\mathbf{P}\mathbf{A}^{-1}\mathbf{P}'|^{\frac{1}{2}}\left\{(2\pi)^{\frac{1}{2}(r-p)} + \sum_{j=1}^{m}\frac{1}{j!}\mathbf{t}'\mathbf{k}^j(\mathbf{0}, \mathbf{0})\mathbf{t}^{\otimes j-1} + r_m(\mathbf{t})\right\}\\
&\quad\times (2\pi)^{-r}\exp(-i\mathbf{t}'\mathbf{x}_0)\varphi_{\mathbf{x}}(\mathbf{t}),
\end{aligned}$$

where $r_m(\mathbf{t})$ *is given by (3.3.7),* $\mathbf{A} : r \times r$ *is positive definite and* $(\mathbf{P}')^o : r \times (r-p)$ *is introduced in (3.3.1).* ∎

Put

$$\mathbf{h}_j(\mathbf{t}) = i^{-j}\mathbf{k}^j(\mathbf{P}\mathbf{t}, \mathbf{t}) \tag{3.3.8}$$

and note that $\mathbf{h}_j(\mathbf{0})$ is real. Now we are able to write out a formal density expansion of general form.

Theorem 3.3.1. *Let* $\mathbf{y}$ *and* $\mathbf{x}$ *be random* $p-$*vector and* $r-$*vector, respectively,* $p \leq r$, $\mathbf{P} : p \times r$ *be a real matrix of rank* $r(\mathbf{P}) = p$, *and* $\mathbf{A} : r \times r$ *positive definite. Then*

$$f_{\mathbf{y}}(\mathbf{y}_0) = |\mathbf{A}|^{\frac{1}{2}}|\mathbf{PA}^{-1}\mathbf{P}'|^{\frac{1}{2}}$$
$$\left\{(2\pi)^{\frac{1}{2}(r-p)} f_{\mathbf{x}}(\mathbf{x}_0) + \sum_{k=1}^{m}(-1)^k \frac{1}{k!}\text{vec}'\mathbf{h}_k(\mathbf{0})\text{vec}\mathbf{f}_{\mathbf{x}}^k(\mathbf{x}_0) + r_m^*\right\}$$

and

$$r_m^* = (2\pi)^{-r}\int_{R^r} r_m(\mathbf{t})\exp(-i\mathbf{t}'\mathbf{x}_0)\varphi_{\mathbf{x}}(\mathbf{t})d\mathbf{t},$$

where $r_m(\mathbf{t})$ *is defined by (3.3.7) and* $\mathbf{h}_k(\mathbf{t})$ *is given by (3.3.8). When* $|\mathbf{t}'\mathbf{h}_{m+1}(\mathbf{v})\mathbf{t}^{\otimes m}| \leq |\mathbf{c}'_{m+1}\mathbf{t}^{\otimes m+1}|$ *for some constant* $r^{m+1}-$*vector* $\mathbf{c}_{m+1}$, $\mathbf{v} \in \mathcal{D}$, *where* $\mathcal{D}$ *is a neighborhood of* $\mathbf{0}$, *then*

$$|r_m^*| \leq \frac{1}{(m+1)!}(2\pi)^{-r}\int_{R^r}|\mathbf{c}'_{m+1}\mathbf{t}^{\otimes m+1}|\varphi_{\mathbf{x}}(\mathbf{t})d\mathbf{t}.$$

PROOF: Using the basic property of the vec-operator (1.3.31), i.e. $\text{vec}(\mathbf{ABC}) = (\mathbf{C}' \otimes \mathbf{A})\text{vec}\mathbf{B}$, we get

$$\mathbf{t}'\mathbf{h}_k(\mathbf{0})\mathbf{t}^{\otimes k-1} = \text{vec}'(\mathbf{h}_k(\mathbf{0}))\mathbf{t}^{\otimes k}.$$

Then it follows from Corollary 3.2.1.L1 that

$$(2\pi)^{-r}\int_{R^r}(i\mathbf{t}')\mathbf{h}_k(\mathbf{0})(i\mathbf{t})^{\otimes k-1}\exp(-i\mathbf{t}'\mathbf{x}_0)\varphi_{\mathbf{x}}(\mathbf{t})d\mathbf{t} = (-1)^k\text{vec}'\mathbf{h}_k(\mathbf{0})\text{vec}\mathbf{f}_{\mathbf{x}}^k(\mathbf{x}_0)$$

and with the help of Theorem 3.2.1, Lemma 3.3.1 and Lemma 3.3.2 the general relation between $f_{\mathbf{y}}(\mathbf{y}_0)$ and $f_{\mathbf{x}}(\mathbf{x}_0)$ is established. The upper bound of the error term r_m^* follows from the fact that the error bound of the density approximation is given by

$$(2\pi)^{-r}|\int_{R^r} r_m(\mathbf{t})e^{-i\mathbf{t}'\mathbf{x}_0}\varphi_{\mathbf{x}}(\mathbf{t})d\mathbf{t}| \leq (2\pi)^{-r}\int_{R^r}|r_m(\mathbf{t})|\varphi_{\mathbf{x}}(\mathbf{t})d\mathbf{t}$$

and we get the statement by assumption, Lemma 3.3.1 and Lemma 3.3.2. ∎

We have an important application of the expression for the error bound when $\mathbf{x} \sim N_{nr}(\mathbf{0}, \mathbf{I}_n \otimes \mathbf{\Sigma})$. Then $\varphi_{\mathbf{x}}(\mathbf{t}) = \exp(-\frac{1}{2}\mathbf{t}'(\mathbf{I}_n \otimes \mathbf{\Sigma})\mathbf{t})$. If $m+1$ is even, we see that the integral in the error bound in Theorem 3.3.1 can be immediately obtained by using moment relations for multivariate normally distributed variables (see §2.2.2). Indeed, the problem of obtaining error bounds depends on finding a constant vector $\mathbf{c}_{m+1}$, which has to be considered separately for every case.

In the next corollary we give a representation of the density via cumulants of the distributions.

Corollary 3.3.1.1. *Let* $\mathbf{y}$ *be a* $p-$*vector,* $\mathbf{x}$ *an* $r-$*vector,* $\mathbf{P} : p \times r$ *of rank* $r(\mathbf{P}) = p$, $\mathbf{A} : r \times r$ *positive definite and* $(\mathbf{P}')^o : r \times (r-p)$ *of rank* $r((\mathbf{P}')^o) = r - p$. *Then*

$$f_{\mathbf{y}}(\mathbf{y}_0) = |\mathbf{A}|^{\frac{1}{2}}|\mathbf{P}\mathbf{A}^{-1}\mathbf{P}'|^{\frac{1}{2}}(2\pi)^{\frac{1}{2}(r-p)}\Big\{f_{\mathbf{x}}(\mathbf{x}_0) - (\mathbf{M}_0 + \mathbf{M}_1)'\mathrm{vec}\mathbf{f}^1_{\mathbf{x}}(\mathbf{x}_0)$$
$$+ \tfrac{1}{2}((\mathbf{M}_0 + \mathbf{M}_1)'^{\otimes 2} + \mathrm{vec}'\mathbf{M}_2)\mathrm{vec}\mathbf{f}^2_{\mathbf{x}}(\mathbf{x}_0)$$
$$- \tfrac{1}{6}\{\mathrm{vec}'\mathbf{M}_3 + (\mathbf{M}_0 + \mathbf{M}_1)'^{\otimes 3} + \mathrm{vec}'\mathbf{M}_2 \otimes (\mathbf{M}_0 + \mathbf{M}_1)'(\mathbf{I}_{r^3} + \mathbf{I}_r \otimes \mathbf{K}_{r,r} + \mathbf{K}_{r^2,r})\}$$
$$\times \mathrm{vec}\mathbf{f}^3_{\mathbf{x}}(\mathbf{x}_0) + r_3^*\Big\},$$

where

$$\mathbf{M}_0 = \mathbf{x}_0 - \mathbf{P}'\mathbf{y}_0, \tag{3.3.9}$$
$$\mathbf{M}_1 = \mathbf{P}'c_1[\mathbf{y}] - c_1[\mathbf{x}] = \mathbf{P}'E[\mathbf{y}] - E[\mathbf{x}], \tag{3.3.10}$$
$$\mathbf{M}_2 = \mathbf{P}'c_2[\mathbf{y}]\mathbf{P} - c_2[\mathbf{x}] + \mathbf{Q} = \mathbf{P}'D[\mathbf{y}]\mathbf{P} - D[\mathbf{x}] + \mathbf{Q}, \tag{3.3.11}$$
$$\mathbf{M}_3 = \mathbf{P}'c_3[\mathbf{y}]\mathbf{P}^{\otimes 2} - c_3[\mathbf{x}], \tag{3.3.12}$$
$$\mathbf{Q} = \mathbf{A}(\mathbf{P}')^o((\mathbf{P}')^{o'}\mathbf{A}(\mathbf{P}')^o)^{-1}(\mathbf{P}')^{o'}\mathbf{A}. \tag{3.3.13}$$

PROOF: For the functions $\mathbf{h}_k(\mathbf{t})$ given by (3.3.4) and (3.3.8) we have

$$\mathbf{h}_1(\mathbf{t}) = i^{-1}\mathbf{k}^1(\mathbf{P}\mathbf{t},\mathbf{t}) = i^{-1}(i\mathbf{x}_0 - i\mathbf{P}'\mathbf{y}_0 - \mathbf{Q}\mathbf{t} + \frac{d\ln\varphi_{\mathbf{P}'\mathbf{y}}(\mathbf{t})}{d\mathbf{t}} - \frac{d\ln\varphi_{\mathbf{x}}(\mathbf{t})}{d\mathbf{t}})k(\mathbf{P}\mathbf{t},\mathbf{t}),$$
$$\mathbf{h}_2(\mathbf{t}) \underset{(1.4.19)}{=} i^{-2}\mathbf{k}^2(\mathbf{P}\mathbf{t},\mathbf{t}) = i^{-2}\Big\{(\frac{d^2\ln\varphi_{\mathbf{P}'\mathbf{y}}(\mathbf{t})}{d\mathbf{t}^2} - \frac{d^2\ln\varphi_{\mathbf{x}}(\mathbf{t})}{d\mathbf{t}^2} - \mathbf{Q})k(\mathbf{P}\mathbf{t},\mathbf{t})$$
$$+ \mathbf{k}^1(\mathbf{P}\mathbf{t},\mathbf{t})\{i\mathbf{x}_0 - i\mathbf{P}'\mathbf{y}_0 - \mathbf{Q}\mathbf{t} + \frac{d\ln\varphi_{\mathbf{P}'\mathbf{y}}(\mathbf{t})}{d\mathbf{t}} - \frac{d\ln\varphi_{\mathbf{x}}(\mathbf{t})}{d\mathbf{t}}\}'\Big\}$$

and

$$\mathbf{h}_3(\mathbf{t}) \underset{\substack{(1.4.19)\\(1.4.23)}}{=} i^{-3}\mathbf{k}^3(\mathbf{P}\mathbf{t},\mathbf{t}) = i^{-3}\Big\{(\frac{d^3\ln\varphi_{\mathbf{P}'\mathbf{y}}(\mathbf{t})}{d\mathbf{t}^3} - \frac{d^3\ln\varphi_{\mathbf{x}}(\mathbf{t})}{d\mathbf{t}^3})k(\mathbf{P}\mathbf{t},\mathbf{t})$$
$$+ \mathbf{k}^1(\mathbf{P}\mathbf{t},\mathbf{t})\mathrm{vec}'(\frac{d^2\ln\varphi_{\mathbf{P}'\mathbf{y}}(\mathbf{t})}{d\mathbf{t}^2} - \frac{d^2\ln\varphi_{\mathbf{x}}(\mathbf{t})}{d\mathbf{t}^2} - \mathbf{Q})$$
$$+ (\frac{d^2\ln\varphi_{\mathbf{P}'\mathbf{y}}(\mathbf{t})}{d\mathbf{t}^2} - \frac{d^2\ln\varphi_{\mathbf{x}}(\mathbf{t})}{d\mathbf{t}^2} - \mathbf{Q}) \otimes \big(\mathbf{k}^1(\mathbf{P}\mathbf{t},\mathbf{t})\big)'$$
$$+ (i\mathbf{x}_0 - i\mathbf{P}'\mathbf{y}_0 - \mathbf{Q}\mathbf{t} + \frac{d\ln\varphi_{\mathbf{P}'\mathbf{y}}(\mathbf{t})}{d\mathbf{t}} - \frac{d\ln\varphi_{\mathbf{x}}(\mathbf{t})}{d\mathbf{t}})' \otimes \mathbf{k}^2(\mathbf{P}\mathbf{t},\mathbf{t})\Big\}.$$

Thus, according to Definition 2.1.6,

$$\mathbf{h}_1(\mathbf{0}) = (\mathbf{x}_0 - \mathbf{P}'\mathbf{y}_0 + \mathbf{P}'c_1[\mathbf{y}] - c_1[\mathbf{x}])k(\mathbf{0},\mathbf{0}), \tag{3.3.14}$$
$$\mathbf{h}_2(\mathbf{0}) = (\mathbf{P}'c_2[\mathbf{y}]\mathbf{P} - c_2[\mathbf{x}] + \mathbf{Q})k(\mathbf{0},\mathbf{0}) + \mathbf{h}_1(\mathbf{0})(\mathbf{x}_0 - \mathbf{P}'\mathbf{y}_0 + \mathbf{P}'c_1[\mathbf{y}] - c_1[\mathbf{x}])' \tag{3.3.15}$$

and

$$
\begin{aligned}
\mathbf{h}_3(\mathbf{0}) =&(\mathbf{P}'c_3[\mathbf{y}]\mathbf{P}^{\otimes 2} - c_3[\mathbf{x}])k(\mathbf{0},\mathbf{0}) + \mathbf{h}_1(\mathbf{0})\mathrm{vec}'(\mathbf{P}'c_2[\mathbf{y}]\mathbf{P} - c_2[\mathbf{x}] + \mathbf{Q}) \\
&+ (\mathbf{P}'c_2[\mathbf{y}]\mathbf{P} - c_2[\mathbf{x}] + \mathbf{Q}) \otimes \left(\mathbf{h}_1(\mathbf{0})\right)' \\
&+ (\mathbf{x}_0 - \mathbf{P}'\mathbf{y}_0 + \mathbf{P}'c_1[\mathbf{y}] - c_1[\mathbf{x}]) \otimes \mathbf{h}_2(\mathbf{0}).
\end{aligned} \tag{3.3.16}
$$

Now $k(\mathbf{0},\mathbf{0}) = (2\pi)^{\frac{1}{2}(r-p)}$ and, using (3.3.9) – (3.3.12),

$$
\begin{aligned}
\mathbf{h}_1(\mathbf{0}) =&(\mathbf{M}_0 + \mathbf{M}_1)(2\pi)^{\frac{1}{2}(r-p)}, \\
\mathbf{h}_2(\mathbf{0}) =&\{\mathbf{M}_2 + (\mathbf{M}_0 + \mathbf{M}_1)(\mathbf{M}_0 + \mathbf{M}_1)'\}(2\pi)^{\frac{1}{2}(r-p)}, \\
\mathbf{h}_3(\mathbf{0}) =&\big\{\mathbf{M}_3 + (\mathbf{M}_0 + \mathbf{M}_1)\mathrm{vec}'\mathbf{M}_2 + \mathbf{M}_2 \otimes (\mathbf{M}_0 + \mathbf{M}_1)' + (\mathbf{M}_0 + \mathbf{M}_1)' \otimes \mathbf{M}_2 \\
&+ (\mathbf{M}_0 + \mathbf{M}_1)' \otimes (\mathbf{M}_0 + \mathbf{M}_1)(\mathbf{M}_0 + \mathbf{M}_1)'\big\}(2\pi)^{\frac{1}{2}(r-p)},
\end{aligned}
$$

the statement of the corollary follows from Theorem 3.3.1. ∎

We end this paragraph by giving an example where a well-known problem of error estimation is treated through our approach.

Example 3.3.1. Consider

$$
\mathbf{y} = \mathbf{x} - \mathbf{u},
$$

where $\mathbf{x}$ and $\mathbf{u}$ are independent (Fujikoshi, 1985, 1987). It will be shown that if we know the distribution of $\mathbf{y}$ and $\mathbf{x}$, which implies that we know the moments of $\mathbf{u}$, then it is possible to approximate the density of $\mathbf{y}$ with an upper error bound. In this case we may choose $\mathbf{A} = \mathbf{I}$, $\mathbf{x}_0 = \mathbf{y}_0$ and $\mathbf{P} = \mathbf{I}$. Moreover, using Definition 2.1.1 of $m_k[\mathbf{u}]$, the expression in Corollary 2.1.2.1, and applying (3.3.4) and (3.3.6) gives us

$$
k(\mathbf{t},\mathbf{t}) = \varphi_{-\mathbf{u}}(\mathbf{t}) = 1 + \sum_{k=1}^{m} \tfrac{(-i)^k}{k!}\mathbf{t}'m_k[\mathbf{u}]\mathbf{t}^{\otimes k-1} + r_m(\mathbf{t}),
$$

where a representation of the error term similar to (3.3.7) can be given as

$$
r_m(\mathbf{t}) = \tfrac{1}{(m+1)!}\mathbf{t}'\mathbf{k}^{m+1}(\boldsymbol{\theta} \circ \mathbf{t}, \boldsymbol{\theta} \circ \mathbf{t})\mathbf{t}^{\otimes m}.
$$

Thus, from Theorem 3.3.1 it follows that

$$
f_{\mathbf{y}}(\mathbf{x}_0) = f_{\mathbf{x}}(\mathbf{x}_0) + \sum_{k=1}^{m} \tfrac{1}{k!}\mathrm{vec}'E[\mathbf{u}^{\otimes k}]\mathrm{vec}\mathbf{f}^k_{\mathbf{x}}(\mathbf{x}_0) + r^*_m
$$

and

$$
r^*_m = (2\pi)^{-r}\int_{R^r} \tfrac{1}{(m+1)!}\mathrm{vec}'(\mathbf{k}^{m+1}(\boldsymbol{\theta} \circ \mathbf{t}, \boldsymbol{\theta} \circ \mathbf{t}))\mathbf{t}^{\otimes m+1}\exp(-i\mathbf{t}'\mathbf{x}_0)\varphi_{\mathbf{x}}(\mathbf{t})d\mathbf{t}.
$$

In order to obtain an error bound, we note that

$$
|r^*_m| \leq (2\pi)^{-r}\tfrac{1}{(m+1)!}\int_{R^r} |\mathrm{vec}'(\mathbf{k}^{m+1}(\boldsymbol{\theta} \circ \mathbf{t}, \boldsymbol{\theta} \circ \mathbf{t}))\mathbf{t}^{\otimes m+1}||\varphi_{\mathbf{x}}(\mathbf{t})|d\mathbf{t},
$$

and since $k(\mathbf{t},\mathbf{t}) = \varphi_{-\mathbf{u}}(\mathbf{t})$, it follows by Definition 2.1.1 that

$$|\mathrm{vec}'(k^{m+1}(\boldsymbol{\theta}\circ\mathbf{t},\boldsymbol{\theta}\circ\mathbf{t}))\mathbf{t}^{\otimes m+1}| = |\mathrm{vec}'(E[\mathbf{u}(\mathbf{u}')^{\otimes m}e^{-i\mathbf{u}'(\mathbf{t}\circ\boldsymbol{\theta})}])(-i)^{m+1}\mathbf{t}^{\otimes m+1}|$$
$$= |E[(\mathbf{u}'\mathbf{t})^{\otimes m+1}e^{-i\mathbf{u}'(\mathbf{t}\circ\boldsymbol{\theta})}](-i)^{m+1}| \leq E[|(\mathbf{u}'\mathbf{t})^{\otimes m+1}|]. \tag{3.3.17}$$

If $m+1$ is even, the power $(\mathbf{u}'\mathbf{t})^{\otimes m+1}$ is obviously positive, which implies that the absolute moment can be changed into the ordinary one given in (3.3.17). Thus, if $m+1$ is even, the right hand side of (3.3.17) equals $E[(\mathbf{u}')^{\otimes m+1}]\mathbf{t}^{\otimes m+1}$ and if it is additionally supposed that $\varphi_{\mathbf{x}}(\mathbf{t})$ is real valued, then

$$|r_m^*| \leq (2\pi)^{-r}\tfrac{1}{(m+1)!}E[(\mathbf{u}')^{\otimes m+1}]\int_{R^r}\mathbf{t}^{\otimes m+1}\varphi_{\mathbf{x}}(\mathbf{t})d\mathbf{t}, \tag{3.3.18}$$

which is sometimes easy to calculate. For example, if $\mathbf{x}$ is normally distributed with mean zero, the integral is obtained from moments of normally distributed vectors. As an application of the example we may obtain multivariate extensions of some of Fujikoshi's (1987) results. ∎

3.3.2 *Normal expansions of densities of different dimensions*

In this paragraph a formal density approximation is examined in the case when $\mathbf{x}$ in Theorem 3.3.1 is normally distributed. In statistical applications $\mathbf{y}$ is typically a statistic which is asymptotically normal, and we would like to present its density through a multivariate normal density. Assume that the cumulants of $\mathbf{y}$ depend on the sample size n in the same way as in (3.2.7) – (3.2.10). It means that the first and third order cumulants are $O(n^{-\frac{1}{2}})$ and the variance of $\mathbf{y}$ is $O(1)$. It is important to observe that from Corollary 3.3.1.1 we get a normal approximation for the density of $\mathbf{y}$ only in the case when the term $\mathbf{M}_2 = \mathbf{0}$. Otherwise the products of the second order moments will appear in all even moments of higher order and these terms will also be of order $O(1)$. So far, we have not made any attempts to specify the unknown parameters $\mathbf{x}_0$, $\mathbf{P}$, $(\mathbf{P}')^o$ and $\mathbf{A}$ in the formal expansions. The point $\mathbf{x}_0$ and the matrices can be chosen in many ways and this certainly depends on what type of approximation we are interested in. A normal approximation can be effective for one set of parameters, while for a Wishart approximation we may need a different choice. In the following text we shall only describe some possible choices. First, we may always choose $\mathbf{x}_0$ as a function of $\mathbf{P}$, so that

$$\mathbf{x}_0 = \mathbf{P}'\mathbf{y}_0 - \mathbf{P}'E[\mathbf{y}] + E[\mathbf{x}], \tag{3.3.19}$$

which implies

$$\mathbf{h}_1(\mathbf{0}) = \mathbf{0},$$

and in Corollary 3.3.1.1

$$\mathbf{M}_0 + \mathbf{M}_1 = \mathbf{0}.$$

Thus, Corollary 3.3.1.1 yields the following formal expansion:

$$f_{\mathbf{y}}(\mathbf{y}_0)$$
$$= |\mathbf{A}|^{\frac{1}{2}}|\mathbf{P}\mathbf{A}^{-1}\mathbf{P}'|^{\frac{1}{2}}(2\pi)^{\frac{1}{2}(r-p)}\Big\{f_{\mathbf{x}}(\mathbf{x}_0) + \tfrac{1}{2}\mathrm{vec}'(\mathbf{P}'D[\mathbf{y}]\mathbf{P} - D[\mathbf{x}] + \mathbf{Q})\mathrm{vec}\mathbf{f}_{\mathbf{x}}^2(\mathbf{x}_0)$$
$$- \tfrac{1}{6}\mathrm{vec}'(\mathbf{P}'c_3[\mathbf{y}]\mathbf{P}^{\otimes 2} - c_3[\mathbf{x}])\mathrm{vec}\mathbf{f}_{\mathbf{x}}^3(\mathbf{x}_0) + r_3^*\Big\}, \tag{3.3.20}$$

where $\mathbf{P}$ comes from (3.3.1) and $\mathbf{Q}$ is given by (3.3.13). However, this does not mean that the point $\mathbf{x}_0$ has always to be chosen according to (3.3.19). The choice $\mathbf{x}_0 = \mathbf{P}'\mathbf{y}_0$, for example, would also be natural, but first we shall examine the case (3.3.19) in some detail. When approximating $f_{\mathbf{y}}(\mathbf{y})$ via $f_{\mathbf{x}}(\mathbf{x})$, one seeks for a relation where the terms are of diminishing order, and it is desirable to make the terms following the first as small as possible. Expansion (3.3.20) suggests the idea of finding $\mathbf{P}$ and $\mathbf{A}$ in such a way that the term $\mathbf{M}_2$ will be "small". Let us assume that the dispersion matrices $D[\mathbf{x}]$ and $D[\mathbf{y}]$ are non-singular and the eigenvalues of $D[\mathbf{x}]$ are all different. The last assumption guarantees that the eigenvectors of $D[\mathbf{x}]$ will be orthogonal which we shall use later. However, we could also study a general eigenvector system and obtain, after orthogonalization, an orthogonal system of vectors in $\mathbb{R}^r$. Suppose first that there exist matrices $\mathbf{P}$ and $\mathbf{A}$ such that $\mathbf{M}_2 = \mathbf{0}$, where $\mathbf{A}$ is positive definite. In the following we shall make use of the identity given in Corollary 1.2.25.1:

$$\mathbf{A} = \mathbf{P}'(\mathbf{P}\mathbf{A}^{-1}\mathbf{P}')^{-1}\mathbf{P} + \mathbf{A}(\mathbf{P}')^{o}((\mathbf{P}')^{o'}\mathbf{A}(\mathbf{P}')^{o})^{-1}(\mathbf{P}')^{o'}\mathbf{A}.$$

With help of the identity we get from (3.3.13) the condition

$$\mathbf{M}_2 = \mathbf{P}'D[\mathbf{y}]\mathbf{P} - D[\mathbf{x}] + \mathbf{A} - \mathbf{P}'(\mathbf{P}\mathbf{A}^{-1}\mathbf{P}')^{-1}\mathbf{P} = \mathbf{0}.$$

Here, taking

$$\mathbf{A} = D[\mathbf{x}] \tag{3.3.21}$$

yields

$$\mathbf{P}'D[\mathbf{y}]\mathbf{P} - \mathbf{P}'(\mathbf{P}\mathbf{A}^{-1}\mathbf{P}')^{-1}\mathbf{P} = \mathbf{0},$$

from where it follows that $\mathbf{P}$ must satisfy the equation

$$(D[\mathbf{y}])^{-1} = \mathbf{P}(D[\mathbf{x}])^{-1}\mathbf{P}'. \tag{3.3.22}$$

Then $\mathbf{P}$ will be a solution to (3.3.22), if

$$\mathbf{P} = (D[\mathbf{y}])^{-\frac{1}{2}}\mathbf{V}', \tag{3.3.23}$$

where $\mathbf{V} : r \times p$ is a matrix with columns being p eigenvalue-normed eigenvectors of $D[\mathbf{x}]$: $\mathbf{v}_i'\mathbf{v}_i = \lambda_i$, where λ_i is an eigenvalue of $D[\mathbf{x}]$. Let us present the expansion when $\mathbf{x}_0$, $\mathbf{A}$ and $\mathbf{P}$ are chosen as above.

Theorem 3.3.2. *Let $D[\mathbf{x}]$ be a non-singular matrix with different eigenvalues λ_i, $i = 1, \ldots, r$. Let $D[\mathbf{y}]$ be non-singular and $D[\mathbf{y}]^{\frac{1}{2}}$ any square root of $D[\mathbf{y}]$. Then*

$$f_{\mathbf{y}}(\mathbf{y}_0) = |D[\mathbf{x}]|^{\frac{1}{2}}|D[\mathbf{y}]|^{-\frac{1}{2}}(2\pi)^{\frac{1}{2}(r-p)} \times \left\{ f_{\mathbf{x}}(\mathbf{x}_0) - \tfrac{1}{6}\mathrm{vec}'\left(\mathbf{P}'c_3[\mathbf{y}]\mathbf{P}^{\otimes 2} - c_3[\mathbf{x}]\right)\mathrm{vec}\mathbf{f}^3_{\mathbf{x}}(\mathbf{x}_0) + r_3^* \right\}, \tag{3.3.24}$$

where $\mathbf{P}$ is defined by (3.3.23) and $\mathbf{x}_0$ is determined by (3.3.19).

PROOF: The equality (3.3.24) follows directly from the expression (3.3.20) after applying (3.3.21), (3.3.23) and taking into account that $\mathbf{M}_2 = \mathbf{0}$. ∎

With the choice of $\mathbf{x}_0$, $\mathbf{P}$ and $\mathbf{A}$ as in Theorem 3.3.2, we are able to omit the first two terms in the expansion in Corollary 3.3.1.1. If one has in mind a normal approximation, then it is necessary, as noted above, that the second term vanishes in the expansion. However, one could also think about the classical Edgeworth expansion, where the term including the first derivative of the density is present. This type of formal expansion is obtained from Corollary 3.3.1.1 with a different choice of $\mathbf{x}_0$. Take $\mathbf{x}_0 = \mathbf{P}'\mathbf{y}_0$, which implies that $\mathbf{M}_0 = \mathbf{0}$, and applying the same arguments as before we reach the following expansion.

Theorem 3.3.3. *Let $D[\mathbf{x}]$ be a non-singular matrix with different eigenvalues λ_i, $i = 1, \ldots, r$, and let $D[\mathbf{y}]$ be non-singular. Then, if $E[\mathbf{x}] = \mathbf{0}$,*

$$f_{\mathbf{y}}(\mathbf{y}_0) = |D[\mathbf{x}]|^{\frac{1}{2}} |m_2[\mathbf{y}]|^{-\frac{1}{2}} (2\pi)^{\frac{1}{2}(r-p)} \Big\{ f_{\mathbf{x}}(\mathbf{x}_0) - E[\mathbf{y}]'\mathbf{P}\mathbf{f}^1_{\mathbf{x}}(\mathbf{x}_0)$$
$$- \tfrac{1}{6}\mathrm{vec}'\left(\mathbf{P}'(c_3[\mathbf{y}]\mathbf{P}^{\otimes 2} - c_3[\mathbf{x}] - 2\mathbf{M}_1^{\otimes 3}\right)\mathrm{vec}\mathbf{f}^3_{\mathbf{x}}(\mathbf{x}_0) + r_3^* \Big\}, \quad (3.3.25)$$

where $\mathbf{x}_0 = \mathbf{P}'\mathbf{y}_0$,

$$\mathbf{P} = (m_2[\mathbf{y}])^{-\frac{1}{2}}\mathbf{V}'$$

and $\mathbf{V}$ is a matrix with columns being p eigenvalue-normed eigenvectors of $D[\mathbf{x}]$.

PROOF: Now, $\mathbf{M}_0 = \mathbf{0}$ and $\mathbf{M}_1 = \mathbf{P}'E[\mathbf{y}]$ in Corollary 3.3.1.1. Moreover, the sum $(\mathbf{M}_1')^{\otimes 2} + \mathrm{vec}'\mathbf{M}_2$ should be put equal to $\mathbf{0}$. If we choose $\mathbf{A} = D[\mathbf{x}]$, we obtain, as in the previous theorem, the equation

$$(m_2[\mathbf{y}])^{-1} = \mathbf{P}D[\mathbf{x}]\mathbf{P}',$$

and the rest follows from the proof of Theorem 3.3.2. ∎

In both Theorem 3.3.2 and Theorem 3.3.3 a matrix $\mathbf{P}$ has been chosen to be a function of p eigenvectors. However, it has not been said which eigenvectors should be used. To answer this question we suggest the following idea.

The density $f_{\mathbf{x}}(\mathbf{x}_0)$ should be close to $f_{\mathbf{y}}(\mathbf{y}_0)$. To achieve this, let us consider the moments of respective distributions and make them close to each other. By centering we may always get that both $E[\mathbf{x}]$ and $E[\mathbf{y}]$ are zero. Therefore, as the next step, $\mathbf{P}$ is chosen so that the second moments will become close in some sense, i.e. the difference

$$\mathbf{P}'D[\mathbf{y}]\mathbf{P} - D[\mathbf{x}]$$

will be studied. One way of doing this is to minimize the norm

$$\mathrm{tr}\{(\mathbf{P}'D[\mathbf{y}]\mathbf{P} - D[\mathbf{x}])(\mathbf{P}'D[\mathbf{y}]\mathbf{P} - D[\mathbf{x}])'\} \quad (3.3.26)$$

with respect to $\mathbf{P}$. The obtained result is stated in the next lemma.

Lemma 3.3.3. *Expression (3.3.26) is minimized, when*

$$\mathbf{P} = (D[\mathbf{y}])^{-\frac{1}{2}}\mathbf{V}',$$

where $\mathbf{V} : r \times p$ *is a matrix with columns being eigenvalue-normed eigenvectors, which correspond to the* p *largest eigenvalues of* $D[\mathbf{x}]$.

PROOF: Let

$$\mathbf{Q} = \mathbf{P}'D[\mathbf{y}]\mathbf{P} - D[\mathbf{x}]$$

and then we are going to solve the least squares problem $\frac{d\,\mathrm{tr}\mathbf{Q}\mathbf{Q}'}{d\mathbf{P}} = \mathbf{0}$. Thus, we have to find the derivative

$$\begin{aligned}\frac{d\,\mathrm{tr}\mathbf{Q}\mathbf{Q}'}{d\mathbf{P}} &\underset{(1.4.28)}{=} \frac{d\,\mathbf{Q}\mathbf{Q}'}{d\mathbf{P}}\mathrm{vec}\mathbf{I} = \frac{d\,\mathbf{Q}}{d\mathbf{P}}\mathrm{vec}\mathbf{Q} + \frac{d\,\mathbf{Q}}{d\mathbf{P}}\mathbf{K}_{p,r}\mathrm{vec}\mathbf{Q} = 2\frac{d\,\mathbf{Q}}{d\mathbf{P}}\mathrm{vec}\mathbf{Q}\\ &= 2\{\frac{d\mathbf{P}'}{d\mathbf{P}}(D[\mathbf{y}]\mathbf{P}\otimes\mathbf{I}) + \frac{d\mathbf{P}}{d\mathbf{P}}(\mathbf{I}\otimes D[\mathbf{y}]\mathbf{P})\}\mathrm{vec}\mathbf{Q}\\ &= 4\mathrm{vec}\{D[\mathbf{y}]\mathbf{P}(\mathbf{P}'D[\mathbf{y}]\mathbf{P} - D[\mathbf{x}])\}.\end{aligned}$$

The minimum has to satisfy the equation

$$\mathrm{vec}\{D[\mathbf{y}]\mathbf{P}(\mathbf{P}'D[\mathbf{y}]\mathbf{P} - D[\mathbf{x}])\} = \mathbf{0}. \tag{3.3.27}$$

Now we make the important observation that $\mathbf{P}$ in (3.3.23) satisfies (3.3.27). Let $(\mathbf{V} : \mathbf{V}^0) : r\times r$ be a matrix of the eigenvalue-normed eigenvectors of $D[\mathbf{x}]$. Observe that $D[\mathbf{x}] = (\mathbf{V} : \mathbf{V}^0)(\mathbf{V} : \mathbf{V}^0)'$. Thus, for $\mathbf{P}$ in (3.3.23),

$$D[\mathbf{y}]\mathbf{P}(\mathbf{P}'D[\mathbf{y}]\mathbf{P} - D[\mathbf{x}]) = D[\mathbf{y}]^{\frac{1}{2}}\mathbf{V}'(\mathbf{V}\mathbf{V}' - \mathbf{V}\mathbf{V}' - \mathbf{V}^0(\mathbf{V}^0)') = \mathbf{0},$$

since $\mathbf{V}'\mathbf{V}^0 = \mathbf{0}$. Furthermore,

$$\begin{aligned}\mathrm{tr}\mathbf{Q}\mathbf{Q}' &= \mathrm{tr}\{(\mathbf{V}\mathbf{V}' - \mathbf{V}\mathbf{V}' - \mathbf{V}^0(\mathbf{V}^0)')(\mathbf{V}\mathbf{V}' - \mathbf{V}\mathbf{V}' - \mathbf{V}^0(\mathbf{V}^0)')'\}\\ &= \mathrm{tr}(\mathbf{V}^0(\mathbf{V}^0)'\mathbf{V}^0(\mathbf{V}^0)') = \sum_{i=1}^{r-p}(\lambda_i^0)^2,\end{aligned}$$

where λ_i^0, $i = 1, 2, \ldots, r-p$, are the eigenvalues of $D[\mathbf{x}]$ with corresponding eigenvector $\mathbf{v}_i^o$, $\mathbf{V}^0 = (\mathbf{v}_1^o, \ldots, \mathbf{v}_{r-p}^o)$. This sum attains its minimum, if λ_i^0 are the $r-p$ smallest eigenvalues of $D[\mathbf{x}]$. Thus the norm is minimized if $\mathbf{P}$ is defined as stated. ∎

In Lemma 3.3.3 we used the norm given by (3.3.26). If it is assumed that $E[\mathbf{x}] = \mathbf{0}$, as in Theorem 3.3.3, it is more natural to use $\mathbf{P}'m_2[\mathbf{y}]\mathbf{P} - m_2[\mathbf{x}]$ and consider the norm

$$\mathrm{tr}\{(\mathbf{P}'m_2[\mathbf{y}]\mathbf{P} - m_2[\mathbf{x}])(\mathbf{P}'m_2[\mathbf{y}]\mathbf{P} - m_2[\mathbf{x}])'\}. \tag{3.3.28}$$

By copying the proof of the previous lemma we immediately get the next statement.

Lemma 3.3.4. *The expression in (3.3.28) is minimized when*

$$\mathbf{P} = (m_2[\mathbf{y}])^{-\frac{1}{2}}\mathbf{V}',$$

where $\mathbf{V} : r \times p$ *is a matrix with columns being eigenvalue-normed eigenvectors which correspond to the* p *largest eigenvalues of* $m_2[\mathbf{x}]$. ∎

On the basis of Theorems 3.3.2 and 3.3.3 let us formulate two results for normal approximations.

Theorem 3.3.4. *Let* $\mathbf{x} \sim N_r(\boldsymbol{\mu}, \boldsymbol{\Sigma})$, *where* $\boldsymbol{\Sigma} > 0$, λ_i, $i = 1, \dots, p$, *be the eigenvalues of* $\boldsymbol{\Sigma}$, *and* $\mathbf{V}$ *the matrix of corresponding eigenvalue-normed eigenvectors. Let* $D[\mathbf{y}]$ *be non-singular and* $D[\mathbf{y}]^{\frac{1}{2}}$ *any square root of* $D[\mathbf{y}]$. *Then*

$$f_{\mathbf{y}}(\mathbf{y}_0) = |\Sigma|^{\frac{1}{2}}|D[\mathbf{y}]|^{-\frac{1}{2}}(2\pi)^{\frac{1}{2}(r-p)} f_{N(\boldsymbol{\mu},\boldsymbol{\Sigma})}(\mathbf{x}_0)$$
$$\times \left\{1 + \tfrac{1}{6}\mathrm{vec}'\left(\mathbf{P}'c_3[\mathbf{y}]\mathbf{P}^{\otimes 2}\right)\mathrm{vec}\mathbf{H}_3(\mathbf{x}_0, \boldsymbol{\mu}, \boldsymbol{\Sigma}) + r_3^*\right\},$$

where $\mathbf{x}_0$ *and* $\mathbf{P}$ *are defined by (3.3.19) and (3.3.23), respectively, and the multivariate Hermite polynomial* $\mathbf{H}_3(\mathbf{x}_0, \boldsymbol{\mu}, \boldsymbol{\Sigma})$ *is given by (2.2.38). The expansion is optimal in the sense of the norm* $||\mathbf{Q}|| = \sqrt{\mathrm{tr}(\mathbf{QQ}')}$, *if* λ_i, $i = 1, \dots, p$, *comprise the* p *largest different eigenvalues of* $\boldsymbol{\Sigma}$.

PROOF: The expansion is obtained from Theorem 3.3.2 if we apply the definition of multivariate Hermite polynomials (2.2.35) and take into account that $c_i[\mathbf{x}] = \mathbf{0}$, $i \geq 3$, for the normal distribution. The optimality follows from Lemma 3.3.3. ∎

Theorem 3.3.5. *Let* $\mathbf{x} \sim N_r(\mathbf{0}, \boldsymbol{\Sigma})$, *where* $\boldsymbol{\Sigma} > 0$, λ_i, $i = 1, \dots, p$, *be the eigenvalues of* Σ, *and* $\mathbf{V}$ *the matrix of corresponding eigenvalue-normed eigenvectors. Let* $m_2[\mathbf{y}]$ *be non-singular and* $m_2[\mathbf{y}]^{\frac{1}{2}}$ *any square root of* $m_2[\mathbf{y}]$. *Then*

$$f_{\mathbf{y}}(\mathbf{y}_0) = |\boldsymbol{\Sigma}|^{\frac{1}{2}}|m_2[\mathbf{y}]|^{-\frac{1}{2}}(2\pi)^{\frac{1}{2}(r-p)} f_{N(\mathbf{0},\boldsymbol{\Sigma})}(\mathbf{x}_0)$$
$$\times \left\{1 + (E[\mathbf{y}])'\mathbf{P}\mathbf{H}_1(\mathbf{x}_0, \boldsymbol{\Sigma}) + \tfrac{1}{6}\mathrm{vec}'\left(\mathbf{P}'c_3[\mathbf{y}]\mathbf{P}^{\otimes 2} - 2\mathbf{M}_1^{\otimes 3}\right)\mathrm{vec}\mathbf{H}_3(\mathbf{x}_0, \boldsymbol{\Sigma}) + r_3^*\right\},$$

where $\mathbf{x}_0$ *and* $\mathbf{P}$ *are defined as in Theorem 3.3.3 and the multivariate Hermite polynomials* $\mathbf{H}_i(\mathbf{x}_0, \boldsymbol{\Sigma})$, $i = 1, 3$, *are defined by (2.2.39) and (2.2.41). The expansion is optimal in the sense of the norm* $||\mathbf{Q}|| = \sqrt{\mathrm{tr}(\mathbf{QQ}')}$, *if* λ_i, $i = 1, \dots, p$ *comprise the* p *largest different eigenvalues of* $\boldsymbol{\Sigma}$.

PROOF: The expansion is a straightforward consequence of Theorem 3.3.3 if we apply the definition of multivariate Hermite polynomials (2.2.35) for the case $\boldsymbol{\mu} = \mathbf{0}$ and take into account that $c_i[\mathbf{x}] = \mathbf{0}$, $i \geq 3$, for the normal distribution. The optimality follows from Lemma 3.3.4. ∎

Let us now consider some examples. Remark that if we take the sample mean $\overline{\mathbf{x}}$ of $\mathbf{x}_1, \mathbf{x}_2, \dots, \mathbf{x}_n$, where $E[\mathbf{x}_i] = \mathbf{0}$ and $D[\mathbf{x}_i] = \boldsymbol{\Sigma}$, we get from Theorem 3.3.4 the usual multivariate Edgeworth type expansion. Taking $\mathbf{y} = \sqrt{n}\overline{\mathbf{x}}$, the cumulants of $\mathbf{y}$ are

$$E[\mathbf{y}] = \mathbf{0}, \quad D[\mathbf{y}] = \boldsymbol{\Sigma}, \quad c_3[\mathbf{y}] = \frac{1}{\sqrt{n}}c_3[\mathbf{x}_i], \quad c_k[\mathbf{y}] = o(\tfrac{1}{\sqrt{n}}), \quad k \geq 4.$$

If we choose $\mathbf{P} = \mathbf{I}$ and $\mathbf{y}_0 = \mathbf{x}_0$, it follows from Theorem 3.3.4 that

$$f_{\mathbf{y}}(\mathbf{x}_0) = f_{\mathbf{x}}(\mathbf{x}_0)\{1 + \frac{1}{6\sqrt{n}}\mathrm{vec}'c_3[\mathbf{y}_i]\mathrm{vec}\mathbf{H}_3(\mathbf{x}_0, \boldsymbol{\Sigma}) + r_3^*\}.$$

The following terms in the expansion are diminishing in powers of $n^{-\frac{1}{2}}$, and the first term in r_3^* is of order $\frac{1}{n}$.

Example 3.3.2. We are going to approximate the distribution of the trace of the sample covariance matrix $\mathbf{S}$, where $\mathbf{S} \sim W_p(\frac{1}{n}\boldsymbol{\Sigma}, n)$. Let $\mathbf{x}_1, \mathbf{x}_2, \dots, \mathbf{x}_{n+1}$ be a sample of size $n+1$ from a $p-$dimensional normal population: $\mathbf{x}_i \sim N_p(\boldsymbol{\mu}, \boldsymbol{\Sigma})$. For a Wishart distributed matrix $\mathbf{W} \sim W_p(\boldsymbol{\Sigma}, n)$ the first cumulants of $\mathrm{tr}\mathbf{W}$ can be found by differentiating the logarithm of the characteristic function

$$\psi_{\mathrm{tr}\mathbf{W}}(t) = -\tfrac{n}{2}\ln|\mathbf{I} - 2i\,t\boldsymbol{\Sigma}|.$$

Thus, differentiating $\psi_{\mathrm{tr}\mathbf{W}}(t)$ gives us

$$c_k[\mathrm{tr}\mathbf{W}] = n2^{k-1}(k-1)\mathrm{tr}(\boldsymbol{\Sigma}^k), \qquad k = 1, 2, \dots, \tag{3.3.29}$$

(see also Problems 5 and 6 in §2.4.9). Let us take

$$Y = \sqrt{n}\mathrm{tr}(\mathbf{S} - \boldsymbol{\Sigma}).$$

Because $\mathbf{S} \sim W_p(\frac{1}{n}\boldsymbol{\Sigma}, n)$, from the general expression (3.3.29) of the cumulants of $\mathrm{tr}\mathbf{W}$ we have:

$$\begin{aligned} c_1[Y] &= \mathbf{0}, \\ c_2[Y] &= 2\mathrm{tr}\boldsymbol{\Sigma}^2, \\ c_3[Y] &= 8\frac{1}{\sqrt{n}}\mathrm{tr}\boldsymbol{\Sigma}^3, \\ c_k[Y] &= o(\frac{1}{\sqrt{n}}), \qquad k \geq 4. \end{aligned}$$

Let us approximate Y by $\mathbf{x} \sim N_p(\mathbf{0}, \boldsymbol{\Sigma})$. Theorem 3.3.4 gives the following expansion:

$$\begin{aligned} f_Y(y_0) = &(2\pi)^{\frac{1}{2}(p-1)}|\boldsymbol{\Sigma}|^{\frac{1}{2}}(2\mathrm{tr}\boldsymbol{\Sigma}^2)^{-\frac{1}{2}} f_{\mathbf{x}}(\mathbf{x}_0) \\ &\times \left\{1 + \frac{\sqrt{2}}{3\sqrt{n}}\frac{\mathrm{tr}(\boldsymbol{\Sigma}^3)}{(\mathrm{tr}\boldsymbol{\Sigma}^2)^{\frac{3}{2}}}\mathbf{v}_1'^{\otimes 3}\mathrm{vec}\mathbf{H}_3(\mathbf{x}_0, \boldsymbol{\Sigma}) + r_3^*\right\}, \end{aligned}$$

where $\mathbf{H}_3(\mathbf{x}_0, \boldsymbol{\Sigma})$ is defined by (2.2.41), and

$$\mathbf{x}_0 = \{2\mathrm{tr}(\boldsymbol{\Sigma}^2)\}^{-\frac{1}{2}}\mathbf{v}_1 y_0,$$

where $\mathbf{v}_1$ is the eigenvector of $\boldsymbol{\Sigma}$ with length $\sqrt{\lambda_1}$ corresponding to the largest eigenvalue λ_1 of $\boldsymbol{\Sigma}$. The terms in the remainder are of diminishing order in powers of $n^{-\frac{1}{2}}$, with the first term in r_3^* being of order $\frac{1}{n}$. ∎

Example 3.3.3. Let us present the density of an eigenvalue of the sample variance matrix through the multivariate normal density, $\mathbf{x} \sim N_p(\mathbf{0}, \boldsymbol{\Sigma})$. From Siotani, Hayakawa & Fujikoshi (1985) we get the first cumulants for $\sqrt{n}(d_i - \lambda_i)$, where

$\lambda_i \neq 0$ is an eigenvalue of the population variance matrix $\mathbf{\Sigma}$: $p \times p$, and d_i is an eigenvalue of the sample variance matrix $\mathbf{S}$:

$$\begin{aligned} c_1(\sqrt{n}(d_i - \lambda_i)) =& a_i n^{-\frac{1}{2}} + O(n^{-\frac{3}{2}}), \\ c_2(\sqrt{n}(d_i - \lambda_i)) =& 2\lambda_i^2 + 2b_i n^{-1} + O(n^{-2}), \\ c_3(\sqrt{n}(d_i - \lambda_i)) =& 8\lambda_i^3 n^{-\frac{1}{2}} + O(n^{-\frac{3}{2}}), \\ c_k(\sqrt{n}(d_i - \lambda_i)) =& o(n^{-\frac{1}{2}}), \qquad k \geq 4, \end{aligned}$$

where

$$a_i = \sum_{k \neq i}^{p} \frac{\lambda_i \lambda_k}{\lambda_i - \lambda_k}, \qquad b_i = \sum_{k \neq i}^{p} \frac{\lambda_i^2 \lambda_k^2}{(\lambda_i - \lambda_k)^2}.$$

Since the moments cannot be calculated exactly, this example differs somewhat from the previous one. We obtain the following expansion from Theorem 3.3.4:

$$\begin{aligned} f_{\sqrt{n}(d_i - \lambda_i)}(y_0) =& (2\pi)^{\frac{1}{2}(p-1)} |\mathbf{\Sigma}|^{\frac{1}{2}} (2\lambda_i^2 + \frac{1}{n} 2b_i)^{-\frac{1}{2}} f_{\mathbf{x}}(\mathbf{x}_0) \\ & \times \{1 + \frac{\sqrt{2}}{3\sqrt{n}} \lambda_i^3 (\lambda_i^2 + \frac{1}{n} b_i)^{-\frac{3}{2}} \mathbf{v}_1^{\prime \otimes 3} \mathrm{vec}\mathbf{H}_3(\mathbf{x}_0, \mathbf{\Sigma}) + r_3^*\}, \qquad (3.3.30) \end{aligned}$$

where $\mathbf{v}_1$ is the eigenvector corresponding to λ_1, of length $\sqrt{\lambda_1}$, where λ_1 is the largest eigenvalue of $\mathbf{\Sigma}$. The point $\mathbf{x}_0$ should be chosen according to (3.3.19), and thus

$$\mathbf{x}_0 = (2\lambda_i^2 + \frac{2b_i}{n})^{-\frac{1}{2}} \mathbf{v}_1' y_0 - \frac{a_i}{\sqrt{n}}.$$

The first term in the remainder is again of order $\frac{1}{n}$. We get a somewhat different expansion if we apply Theorem 3.3.5. Then

$$\begin{aligned} f_{\sqrt{n}(d_i - \lambda_i)}(y_0) =& (2\pi)^{\frac{1}{2}(p-1)} |\mathbf{\Sigma}|^{\frac{1}{2}} (2\lambda_i^2 + \frac{1}{n}(a_i^2 + 2b_i))^{-\frac{1}{2}} f_{\mathbf{x}}(\mathbf{x}_0) \\ & \times \{1 + \frac{1}{\sqrt{n}} \Big(a_i (2\lambda_i^2 + \frac{1}{n}(a_i^2 + 2b_i))^{-\frac{1}{2}} \mathbf{v}_1' \mathrm{vec}\mathbf{H}_1(\mathbf{x}_0, \mathbf{\Sigma}) \\ & + \frac{\sqrt{2}}{3} \lambda_i^3 (\lambda_i^2 + \frac{1}{n} b_i)^{-\frac{3}{2}} \mathbf{v}_1^{\prime \otimes 3} \mathrm{vec}\mathbf{H}_3(\mathbf{x}_0, \mathbf{\Sigma}) \Big) + r_3^*\}, \end{aligned}$$

where

$$\mathbf{x}_0 = (2\lambda_i^2 + \frac{1}{n}(a_i^2 + 2b_i))^{-\frac{1}{2}} \mathbf{v}_1' y_0$$

and $\mathbf{v}_1$ is defined as above, since the assumption $E[\mathbf{x}] = \mathbf{0}$ implies $\mathbf{\Sigma} = m_2[\mathbf{x}]$. The first term in the remainder is again of order $\frac{1}{n}$. ∎

3.3.3. *Wishart expansions for different dimensional densities*

In paragraph 3.2.4 we saw that in Wishart expansions, unlike in normal approximations, the second and third order cumulants are both included into the approximations. So, it is desirable to have the second order cumulants included into the following expansions. This can be obtained by a choice of point $\mathbf{x}_0$ and the matrices $\mathbf{P}$, $\mathbf{P}^o$ and $\mathbf{A}$ different from the one for the normal approximation. Let us again fix the point $\mathbf{x}_0$ by (3.3.19), but let us have a different choice for $\mathbf{A}$. Instead of (3.3.21), take

$$\mathbf{A} = \mathbf{I}.$$

Then (3.3.13) turns into the equality

$$\mathbf{Q} = (\mathbf{P}')^o((\mathbf{P}')^{o'}(\mathbf{P}')^o)^{-1}(\mathbf{P}')^{o'} = \mathbf{I} - \mathbf{P}'(\mathbf{P}\mathbf{P}')^{-1}\mathbf{P}. \tag{3.3.31}$$

We are going to make the term

$$\mathbf{M}_2 = \mathbf{P}'D[\mathbf{y}]\mathbf{P} - D[\mathbf{x}] + \mathbf{Q}$$

as small as possible in the sense of the norm

$$\| \mathbf{M}_2 \| = \{\mathrm{tr}(\mathbf{M}_2\mathbf{M}_2')\}^{\frac{1}{2}}. \tag{3.3.32}$$

From Lemma 3.3.2 and (3.3.15) it follows that if we are going to choose $\mathbf{P}$ such that $\mathbf{t}'\mathbf{h}^2(0)\mathbf{t}$ is small, we have to consider $\mathbf{t}'\mathbf{M}_2\mathbf{t}$. By the Cauchy-Schwarz inequality, this expression is always smaller than

$$\mathbf{t}'\mathbf{t}\{\mathrm{tr}(\mathbf{M}_2\mathbf{M}_2')\}^{1/2},$$

which motivates the norm (3.3.32). With the choice $\mathbf{A} = \mathbf{I}$, the term in the density function which includes $\mathbf{M}_2$ will not vanish. The norm (3.3.32) is a rather complicated expression in $\mathbf{P}$. As an alternative to applying numerical methods to minimize the norm, we can find an explicit expression of $\mathbf{P}$ which makes (3.3.32) reasonably small. In a straightforward way it can be shown that

$$\mathrm{tr}(\mathbf{M}_2\mathbf{M}_2') = \mathrm{tr}(\mathbf{M}_2\mathbf{P}'(\mathbf{P}\mathbf{P}')^{-1}\mathbf{P}\mathbf{M}_2) + \mathrm{tr}(\mathbf{M}_2(\mathbf{I} - \mathbf{P}'(\mathbf{P}\mathbf{P}')^{-1}\mathbf{P})\mathbf{M}_2). \tag{3.3.33}$$

Since we are not able to find explicitly an expression for $\mathbf{P}$ minimizing (3.3.33), the strategy will be to minimize $\mathrm{tr}(\mathbf{M}_2\mathbf{A}\mathbf{P}'(\mathbf{P}\mathbf{P}')^{-1}\mathbf{P}\mathbf{M}_2)$ and thereafter make

$$\mathrm{tr}(\mathbf{M}_2(\mathbf{I} - \mathbf{P}'(\mathbf{P}\mathbf{P}')^{-1}\mathbf{P})\mathbf{M}_2)$$

as small as possible.

Let $\mathbf{U} : r \times p$ and $\mathbf{U}^o : r \times (r-p)$ be matrices of full rank such that $\mathbf{U}'\mathbf{U}^o = \mathbf{0}$ and

$$\begin{aligned} D[\mathbf{x}] =& \mathbf{U}\mathbf{U}' + \mathbf{U}^o\mathbf{U}^{o'}, \\ (\mathbf{U} : \mathbf{U}^o)'(\mathbf{U} : \mathbf{U}^o) =& \mathbf{\Lambda} = \begin{pmatrix} \mathbf{\Lambda}_1 & \mathbf{0} \\ \mathbf{0} & \mathbf{\Lambda}_0 \end{pmatrix}, \end{aligned}$$

where $\mathbf{\Lambda}$ is diagonal. The last equalities define $\mathbf{U}$ and $\mathbf{U}^o$ as the matrices consisting of eigenvectors of $D[\mathbf{x}]$:

$$D[\mathbf{x}](\mathbf{U} : \mathbf{U}^o) = (\mathbf{U}\mathbf{\Lambda}_1 : \mathbf{U}^o\mathbf{\Lambda}_0).$$

Let us take

$$\mathbf{P} = D[\mathbf{y}]^{-1/2}\mathbf{U}'. \tag{3.3.34}$$

Then, inserting (3.3.34) into (3.3.11), we obtain

$$\mathbf{M}_2 = \mathbf{U}\mathbf{U}' - \mathbf{U}\mathbf{U}' - \mathbf{U}^o\mathbf{U}^{o'} + \mathbf{Q} = \mathbf{Q} - \mathbf{U}^o\mathbf{U}^{o'},$$

which implies that $\mathbf{M}_2$ is orthogonal to $\mathbf{P}$, which allows us to conclude that

$$\mathrm{tr}(\mathbf{M}_2\mathbf{P}'(\mathbf{P}\mathbf{P}')^{-1}\mathbf{P}\mathbf{M}_2) = 0.$$

The choice (3.3.34) for $\mathbf{P}$ gives us the lower bound of $\mathrm{tr}(\mathbf{M}_2\mathbf{P}'(\mathbf{P}\mathbf{P}')^{-1}\mathbf{P}\mathbf{M}_2)$, and next we examine

$$\mathrm{tr}\{\mathbf{M}_2(\mathbf{I} - \mathbf{P}'(\mathbf{P}\mathbf{P}')^{-1}\mathbf{P})\mathbf{M}_2\}.$$

Notice that from (3.3.31) $\mathbf{Q}$ is idempotent. Since $\mathbf{P}\mathbf{Q} = \mathbf{0}$, we obtain

$$\begin{aligned}
\mathrm{tr}\{\mathbf{M}_2(\mathbf{I} - \mathbf{P}'(\mathbf{P}\mathbf{P}')^{-1}\mathbf{P})\mathbf{M}_2\} &= \mathrm{tr}\{(\mathbf{I} - D[\mathbf{x}])(\mathbf{I} - \mathbf{P}'(\mathbf{P}\mathbf{P}')^{-1}\mathbf{P})(\mathbf{I} - D[\mathbf{x}])\} \\
&= \mathrm{tr}\{\mathbf{U}^o(\mathbf{U}^{o'}\mathbf{U}^o)^{-1}\mathbf{U}^{o'} - 2\mathbf{U}^o\mathbf{U}^{o'} + \mathbf{U}^o\mathbf{U}^{o'}\mathbf{U}^o\mathbf{U}^{o'}\} \\
&= \mathrm{tr}(\mathbf{I}_{r-p} - 2\mathbf{\Lambda}_0 + \mathbf{\Lambda}_0^2) \\
&= \mathrm{tr}(\mathbf{\Lambda}_0 - \mathbf{I}_{r-p})^2 = \sum_{i=1}^{r-p}(1-\lambda_i^o)^2,
\end{aligned}$$

where λ_i^o are diagonal elements of $\mathbf{\Lambda}_0$. The trace is small when the eigenvalues λ_i^o are close to 1. Thus, a reasonable choice of $\mathbf{P}$ will be based on p eigenvectors, which correspond to the p eigenvalues λ_i of $D[\mathbf{x}]$ with the largest absolute values $|\lambda_i - 1|$.

Noting that the choice of $D[\mathbf{y}]^{\frac{1}{2}}$ is immaterial, we formulate the following result.

Theorem 3.3.6. *Let $D[\mathbf{x}]$ be a non-singular matrix with different eigenvalues λ_i, $D[\mathbf{y}]$ non-singular, and $\mathbf{u}_i$, $i = 1, 2, \ldots p$, be the eigenvectors of $D[\mathbf{x}]$ corresponding to λ_i, of length $\sqrt{\lambda_i}$. Put $\mu_i = |\lambda_i - 1|$ and let $\mu_{(i)}$ denote the diminishingly ordered values of μ_i with $\mu_{(i)} = |\lambda_{(i)} - 1|$. Let $\mathbf{U} = (\mathbf{u}_{(1)}, \ldots, \mathbf{u}_{(p)})$, $\mathbf{\Lambda}_{(1)} = (\lambda_{(1)}, \ldots, \lambda_{(p)})_d$ and $D[\mathbf{y}]^{\frac{1}{2}}$ denote any square root of $D[\mathbf{y}]$. Then the matrix*

$$\mathbf{P} = D[\mathbf{y}]^{-\frac{1}{2}}\mathbf{U}'$$

minimizes $\mathbf{M}_2$ in (3.3.11), in the sense of the norm (3.3.32), and the following expansion of the density $f_{\mathbf{y}}(\mathbf{y}_0)$ holds:

$$\begin{aligned}
f_{\mathbf{y}}(\mathbf{y}_0) = (2\pi)^{\frac{1}{2}(r-p)}|D[\mathbf{y}]|^{-\frac{1}{2}}\prod_{k=1}^{p}\sqrt{\lambda_{(k)}}\Big\{ & f_{\mathbf{x}}(\mathbf{x}_0) \\
& + \tfrac{1}{2}\mathrm{vec}'(\mathbf{I} + \mathbf{U}(\mathbf{I}_p - \mathbf{\Lambda}_{(1)}^{-1})\mathbf{U}' - D[\mathbf{x}])\mathrm{vec} f_{\mathbf{x}}^2(\mathbf{x}_0) \\
& - \tfrac{1}{6}\mathrm{vec}'\left(\mathbf{P}'c_3[\mathbf{y}]\mathbf{P}^{\otimes 2} - c_3[\mathbf{x}]\right)\mathrm{vec} f_y^3(y_0) + r_3^*\Big\},
\end{aligned}$$

where $\mathbf{x}_0$ is determined by (3.3.19). ∎

Let $\mathbf{V} = \mathbf{W} - n\mathbf{\Sigma}$, where $\mathbf{W} \sim W(\mathbf{\Sigma}, n)$ (for the centered Wishart distribution see §2.4.8). Then the following formal Wishart approximation of the density $f_{\mathbf{y}}(\mathbf{y}_0)$ can be written out.

Theorem 3.3.7. *Let* $\mathbf{y}$ *be a random* $q-$*vector and* $D[\mathbf{y}]$ *non-singular, and let eigenvalues of* $D[V^2(\mathbf{V})]$ *be denoted by* λ_i *and the corresponding eigenvalue-normed eigenvectors by* $\mathbf{u}_i$, $i = 1, \ldots, \frac{1}{2}p(p+1)$. *Put* $\mu_i = |\lambda_i - 1|$ *and let* $\mu_{(i)}$ *be the diminishingly ordered values of* μ_i *with* $\mu_{(i)} = |\lambda_{(i)} - 1|$. *Let* $\mathbf{U}$ *consist of eigenvectors* $\mathbf{u}_{(i)}$, $i = 1, 2, \ldots, q$, *corresponding to* $\lambda_{(i)}$, $\mathbf{\Lambda}_{(1)} = (\lambda_{(1)}, \ldots, \lambda_{(q)})_d$, *and* $D[\mathbf{y}]^{\frac{1}{2}}$ *denote any square root of* $D[\mathbf{y}]$. *Then the matrix*

$$\mathbf{P} = D[\mathbf{y}]^{-\frac{1}{2}}\mathbf{U}'$$

minimizes the matrix $\mathbf{M}_2$ *in (3.3.11), in the sense of the norm (3.3.32), and the following formal Wishart expansion of the density* $f_{\mathbf{y}}(\mathbf{y}_0)$ *holds:*

$$f_{\mathbf{y}}(\mathbf{y}_0) = (2\pi)^{\frac{1}{2}(\frac{1}{2}p(p+1)-q)}|D[\mathbf{y}]|^{-\frac{1}{2}} \prod_{k=1}^{q} \sqrt{\lambda_{(k)}} f_{\mathbf{V}}(\mathbf{V}_0)$$
$$\times \Big\{ 1 + \tfrac{1}{2}\mathrm{vec}'(\mathbf{I} + \mathbf{U}(\mathbf{I}_q - \mathbf{\Lambda}_{(1)}^{-1})\mathbf{U}' - c_2[V^2(\mathbf{V})])\mathrm{vec}\mathbf{L}_2^*(\mathbf{V}_0, \mathbf{\Sigma})$$
$$+ \tfrac{1}{6}\mathrm{vec}'\left(\mathbf{P}'c_3[\mathbf{y}]\mathbf{P}^{\otimes 2} - c_3[V^2(\mathbf{V})]\right)\mathrm{vec}\mathbf{L}_3^*(\mathbf{V}_0, \mathbf{\Sigma}) + r_3^* \Big\}, \qquad (3.3.35)$$

where the symmetric $\mathbf{V}_0$ *is given via its vectorized upper triangle* $V^2(\mathbf{V}_0) = \mathbf{P}'(\mathbf{y}_0 - E[\mathbf{y}])$, $\mathbf{L}_i^*(\mathbf{V}_0, \mathbf{\Sigma})$, $i = 2, 3$, *are defined by (2.4.66), (2.4.62), (2.4.63), and* $c_i[V^2(\mathbf{V})]$, $i = 2, 3$, *are given in Theorem 2.4.16.*

PROOF: The expansion follows directly from Theorem 3.3.6, if we replace $f_{\mathbf{x}}^k(\mathbf{x}_0)$ according to (2.4.66) and take into account the expressions of cumulants of the Wishart distribution in Theorem 2.4.16. ∎

REMARK: Direct calculations show that in the second term of the expansion the product $\mathrm{vec}'c_2[V^2(\mathbf{V})]\mathrm{vec}\mathbf{L}_2^*(\mathbf{V}_0, \mathbf{\Sigma})$ is $O(\frac{1}{n})$, and the main term of the product $\mathrm{vec}'c_3[V^2(\mathbf{V})]\mathrm{vec}\mathbf{L}_3^*(\mathbf{V}_0, \mathbf{\Sigma})$ is $O(\frac{1}{n^2})$. It gives a hint that the Wishart distribution itself can be a reasonable approximation to the density of $\mathbf{y}$, and that it makes sense to examine in more detail the approximation where the third cumulant term has been neglected.

As the real order of terms can be estimated only in applications where the statistic $\mathbf{y}$ is also specified, we shall finish this paragraph with Wishart approximations of the same statistics as in the end of the previous paragraph.

Example 3.3.4. Consider the approximation of $\mathrm{tr}\mathbf{S}$ by the Wishart distribution, where $\mathbf{S}$ is the sample dispersion matrix. Take

$$Y = n\mathrm{tr}(\mathbf{S} - \mathbf{\Sigma}).$$

From (3.3.29) we get the expressions of the first cumulants:

$$\begin{aligned} c_1[Y] &= 0, \\ c_2[Y] &= 2n\mathrm{tr}\mathbf{\Sigma}^2, \\ c_3[Y] &= 8n\mathrm{tr}\mathbf{\Sigma}^3, \\ c_k[Y] &= O(n), \qquad k \geq 4. \end{aligned}$$

The expansion (3.3.35) takes the form

$$f_Y(y_0) = (2\pi)^{\frac{1}{2}(\frac{1}{2}p(p+1)-1)}(2n\mathrm{tr}\boldsymbol{\Sigma}^2)^{-\frac{1}{2}}\lambda_{(1)}f_{\mathbf{V}}(\mathbf{V}_0)$$
$$\times\Big\{1+\tfrac{1}{2}\mathrm{vec}'(\mathbf{I}+\mathbf{u}_{(1)}(1-\lambda_{(1)}^{-1})\mathbf{u}'_{(1)}-\mathbf{c}_2[V^2(\mathbf{V})])\mathrm{vec}\mathbf{L}_2^*(\mathbf{V}_0,\boldsymbol{\Sigma})$$
$$+\tfrac{1}{6}\mathrm{vec}'\left(\frac{2\sqrt{2}\mathrm{tr}\boldsymbol{\Sigma}^3}{\sqrt{n}\sqrt{\mathrm{tr}\boldsymbol{\Sigma}^2}^3}\mathbf{u}_{(1)}\mathbf{u}'^{\otimes 2}_{(1)}-c_3[V^2(\mathbf{V})]\right)\mathrm{vec}\mathbf{L}_3^*(\mathbf{V}_0,\boldsymbol{\Sigma})+r_3^*\Big\},$$

where $\lambda_{(1)}$ is the eigenvalue of $D[V^2(\mathbf{V})]$ with the corresponding eigenvector $\mathbf{u}_{(1)}$ of length $\sqrt{\lambda_{(1)}}$. The index (1) in $\lambda_{(1)}$, $\mathbf{V}_0$, $\mathbf{L}_i^*(\mathbf{V}_0,\boldsymbol{\Sigma})$ and $c_i[V^2(\mathbf{V})]$ are defined in Theorem 3.3.7.

At first it seems not an easy task to estimate the order of terms in this expansion, as both, the cumulants $c_i[V^2(\mathbf{V})]$ and derivatives $\mathbf{L}_k^*(\mathbf{V}_0,\boldsymbol{\Sigma})$ depend on n. However, if $n \gg p$, we have proved in Lemma 2.4.2 that $\mathbf{L}_k^*(\mathbf{V}_0,\boldsymbol{\Sigma}) \sim O(n^{-(k-1)})$, $k = 2,3,\ldots$, while $c_k[V^2(\mathbf{V})] \sim O(n)$. From here it follows that the $\mathbf{L}_2^*(\mathbf{V}_0,\boldsymbol{\Sigma})-$term is $O(1)$, $\mathbf{L}_3^*(\mathbf{V}_0,\boldsymbol{\Sigma})-$term is $O(\frac{1}{n})$ and the first term in R_3^* is $O(\frac{1}{n^2})$. ∎

Example 3.3.5. We are going to consider a Wishart approximation of the eigenvalues of the sample dispersion matrix $\mathbf{S}$. Denote

$$Y_i = n(d_i - \delta_i),$$

where d_i and δ_i are the $i-$th eigenvalues of $\mathbf{S}$ and $\boldsymbol{\Sigma}$, respectively. We get the expressions of the first cumulants of Y_i from Siotani, Hayakawa & Fujikoshi (1985), for example:

$$\begin{aligned}
c_1(Y_i) =& a_i + O(n^{-1}),\\
c_2(Y_i) =& 2n\delta_i^2 + 2b_i + O(n^{-1}),\\
c_3(Y_i) =& 8n\delta_i^3 + O(n^{-1}),\\
c_k(Y_i) =& O(n), \qquad k \geq 4,
\end{aligned}$$

where

$$a_i = \sum_{k\neq i}^{p}\frac{\delta_i\delta_k}{\delta_i-\delta_k}, \qquad b_i = \sum_{k\neq i}^{p}\frac{\delta_i^2\delta_k^2}{(\delta_i-\delta_k)^2}.$$

From (3.3.35) we get the following expansions:

$$f_{Y_i}(y_0) = (2\pi)^{\frac{1}{2}(\frac{1}{2}p(p+1)-1)}\frac{\lambda_{(1)}}{\sqrt{2n}\delta_i}f_{\mathbf{V}}(\mathbf{V}_0)$$
$$\times\Big\{1+\tfrac{1}{2}\mathrm{vec}'(\mathbf{I}+\mathbf{u}_{(1)}(1-\lambda_{(1)}^{-1})\mathbf{u}'_{(1)}-c_2[V^2(\mathbf{V})])\mathrm{vec}\mathbf{L}_2^*(\mathbf{V}_0,\boldsymbol{\Sigma})$$
$$+\tfrac{1}{6}\mathrm{vec}'\left(\frac{2\sqrt{2}}{\sqrt{n}}\mathbf{u}_{(1)}\mathbf{u}'^{\otimes 2}_{(1)}-c_3[V^2(\mathbf{V})]\right)\mathrm{vec}\mathbf{L}_3^*(\mathbf{V}_0,\boldsymbol{\Sigma})+r_3^*\Big\}.$$

Here again $\lambda_{(1)}$ is the eigenvalue of $D[V^2(\mathbf{V})]$ with the corresponding eigenvector $\mathbf{u}_{(1)}$ of length $\sqrt{\lambda_{(1)}}$. The index (1) in $\lambda_{(1)}$, $\mathbf{V}_0$, $\mathbf{L}_i^*(\mathbf{V}_0, \mathbf{\Sigma})$ and $c_i[V^2(\mathbf{V})]$ are given in Theorem 3.3.7.

Repeating the same argument as in the end of the previous example, we get that the $\mathbf{L}_2^*(\mathbf{V}_0, \mathbf{\Sigma})$−term is $O(1)$, $\mathbf{L}_3^*(\mathbf{V}_0, \mathbf{\Sigma})$−term is $O(\frac{1}{n})$ and the first term in r_3^* is $O(\frac{1}{n^2})$. ∎

3.3.4 *Density expansions of* $\mathbf{R}$

We are going to finish the chapter with a non-trivial example of using general multivariate approximations. In §3.1.4. we derived the asymptotic normal distribution of the sample correlation matrix $\mathbf{R}$. The exact distribution of the sample correlation matrix is complicated. Even the density of the sample correlation coefficient r in the classical normal case is represented as an infinite sum of complicated terms (Fisher, 1915). The normalizing Fisher z−transformation, which is commonly used for the correlation coefficient, cannot be applied to the matrix $\mathbf{R}$. Expansions of the distribution of the sample correlation coefficient and some univariate functions of $\mathbf{R}$ have been studied for many years. Here we refer to Konishi (1979) and Fang & Krishnaiah (1982) as the basic references on the topic (see also Boik, 2003). Following Kollo & Ruul (2003), in this paragraph we are going to present multivariate normal and Wishart approximations for the density of $\mathbf{R}$. To apply Theorems 3.3.4 – 3.3.7, we need the first three cumulants of $\mathbf{R}$. The cumulants can be obtained by differentiating the cumulant function. Unfortunately, the characteristic function and the cumulant function of $\mathbf{R}$ are not available in the literature. In the next theorem we present an expansion of the characteristic function.

Theorem 3.3.8. *The first order approximation of the characteristic function of the sample correlation matrix* $\mathbf{R}$ *is of the form:*

$$\varphi_{\mathbf{R}}(\mathbf{T}) = e^{i\mathrm{vec}'\mathbf{T}\mathrm{vec}\mathbf{\Omega}}\left\{1 - \frac{1}{2}\mathrm{vec}'\mathbf{T}\mathbf{D}_1\overline{m}_2[\mathrm{vec}\mathbf{S}]\mathbf{D}_1'\mathrm{vec}\mathbf{T}\right.$$
$$\left. + \frac{i}{2}\mathrm{vec}'\mathbf{T}\{\mathrm{vec}'(\overline{m}_2[\mathrm{vec}\mathbf{S}]) \otimes \mathbf{I}_{p^2}\}\mathrm{vec}\mathbf{D}_2 + o\left(\frac{1}{n}\right)\right\}, \qquad (3.3.36)$$

where $\mathbf{D}_1$ *is given by (3.1.13),*

$$\overline{m}_2(\mathrm{vec}\mathbf{S}) = \frac{1}{n}\left\{E[(\mathbf{X}-\boldsymbol{\mu})\otimes(\mathbf{X}-\boldsymbol{\mu})'\otimes(\mathbf{X}-\boldsymbol{\mu})\otimes(\mathbf{X}-\boldsymbol{\mu})']\right.$$
$$\left. - \mathrm{vec}\mathbf{\Sigma}\mathrm{vec}'\mathbf{\Sigma} + \frac{1}{n-1}(\mathbf{I}_{p^2} + \mathbf{K}_{p,p})(\mathbf{\Sigma}\otimes\mathbf{\Sigma})\right\} \qquad (3.3.37)$$

and

$$\mathbf{D}_2 = -\frac{1}{2}(\mathbf{I}_p \otimes \mathbf{K}_{p,p} \otimes \mathbf{I}_p)\left(\mathbf{I}_{p^2} \otimes \mathrm{vec}\boldsymbol{\Sigma}_d^{-\frac{1}{2}} + \mathrm{vec}\boldsymbol{\Sigma}_d^{-\frac{1}{2}} \otimes I_{p^2}\right)(\mathbf{I}_p \otimes \boldsymbol{\Sigma}_d^{-\frac{3}{2}})(\mathbf{K}_{p,p})_d$$
$$-\frac{1}{2}\left(\mathbf{I}_{p^2} \otimes (\mathbf{K}_{p,p})_d\right)(\mathbf{I}_p \otimes \mathbf{K}_{p,p} \otimes \mathbf{I}_p)\left(\mathbf{I}_{p^2} \otimes \mathrm{vec}\mathbf{I}_p + \mathrm{vec}\mathbf{I}_p \otimes \mathbf{I}_{p^2}\right)$$
$$\times\left\{\boldsymbol{\Sigma}_d^{-\frac{1}{2}} \otimes \boldsymbol{\Sigma}_d^{-\frac{3}{2}} - \frac{1}{2}\left(\mathbf{I}_p \otimes \boldsymbol{\Sigma}_d^{-\frac{3}{2}}\boldsymbol{\Sigma}\boldsymbol{\Sigma}_d^{-\frac{3}{2}} + 3(\boldsymbol{\Sigma}_d^{-\frac{1}{2}}\boldsymbol{\Sigma}\boldsymbol{\Sigma}_d^{-\frac{3}{2}} \otimes \boldsymbol{\Sigma}_d^{-1})\right)(\mathbf{K}_{p,p})_d\right\}. \tag{3.3.38}$$

PROOF: The starting point for the subsequent derivation is the Taylor expansion of $\mathrm{vec}\mathbf{R}$

$$\mathrm{vec}\mathbf{R} = \mathrm{vec}\boldsymbol{\Omega} + \left(\frac{d\mathbf{R}}{d\mathbf{S}}\right)'\Bigg|_{\mathbf{S}=\boldsymbol{\Sigma}} \mathrm{vec}(\mathbf{S}-\boldsymbol{\Sigma})$$
$$+\frac{1}{2}\{\mathrm{vec}(\mathbf{S}-\boldsymbol{\Sigma}) \otimes \mathbf{I}_{p^2}\}'\left(\frac{d^2\mathbf{R}}{d\mathbf{S}^2}\right)'\Bigg|_{\mathbf{S}=\boldsymbol{\Sigma}} \mathrm{vec}(\mathbf{S}-\boldsymbol{\Sigma}) + \mathbf{r}_2.$$

Denote

$$\mathrm{vec}\mathbf{R}^* = \mathrm{vec}\mathbf{R} - \mathbf{r}_2, \qquad \mathbf{D}_1 = \left(\frac{d\mathbf{R}}{d\mathbf{S}}\right)'\Bigg|_{\mathbf{S}=\boldsymbol{\Sigma}}, \qquad \mathbf{D}_2 = \left(\frac{d^2\mathbf{R}}{d\mathbf{S}^2}\right)'\Bigg|_{\mathbf{S}=\boldsymbol{\Sigma}}.$$

In this notation the characteristic function of $\mathrm{vec}\mathbf{R}^*$ has the following presentation:

$$\varphi_{\mathbf{R}^*}(\mathbf{T}) = e^{i\mathrm{vec}'\mathbf{T}\mathrm{vec}\boldsymbol{\Omega}}E\left[e^{i\mathrm{vec}'\mathbf{T}\mathbf{A}}e^{i\mathrm{vec}'\mathbf{T}\mathbf{B}}\right], \tag{3.3.39}$$

where

$$\mathbf{A} = \mathbf{D}_1\mathrm{vec}(\mathbf{S}-\boldsymbol{\Sigma}),$$
$$\mathbf{B} = \frac{1}{2}\{\mathrm{vec}(\mathbf{S}-\boldsymbol{\Sigma}) \otimes \mathbf{I}_{p^2}\}'\mathbf{D}_2\mathrm{vec}(\mathbf{S}-\boldsymbol{\Sigma}).$$

From the convergence (3.1.4) it follows that $\mathbf{A} \sim O_P(\frac{1}{\sqrt{n}})$ and $\mathbf{B} \sim O_P(\frac{1}{n})$. Expanding the exponential functions into Taylor series and taking expectation gives

$$\varphi_{\mathbf{R}^*}(\mathbf{T}) = e^{i\mathrm{vec}'\mathbf{T}\mathrm{vec}\boldsymbol{\Omega}}\left\{1 - \frac{1}{2}\mathrm{vec}'\mathbf{T}\mathbf{D}_1\overline{m}_2(\mathrm{vec}\mathbf{S})\mathbf{D}_1'\mathrm{vec}\mathbf{T}\right.$$
$$\left.+\frac{i}{2}\mathrm{vec}'\mathbf{T}\,E\left[\left(\mathrm{vec}(\mathbf{S}-\boldsymbol{\Sigma}) \otimes \mathbf{I}_{p^2}\right)'\mathbf{D}_2\mathrm{vec}(\mathbf{S}-\boldsymbol{\Sigma})\right] + o_P\left(\frac{1}{n}\right)\right\}.$$

The expansion (3.3.36) in the theorem follows after using properties of the vec-operator and Kronecker product. Now we only have to find the expressions of $\overline{m}_2(\mathrm{vec}\mathbf{S})$ and $\mathbf{D}_2$. The first one, $\overline{m}_2(\mathrm{vec}\mathbf{S})$, is well known and can be found in matrix form in Kollo & Neudecker (1993, Appendix I), for instance. To obtain $\mathbf{D}_2$, we have to differentiate **R** by **S** twice. The equality (3.3.38) follows after tedious calculations which we leave as an exercise to the reader (Problem 5 in §3.3.5). The proof is completed by noting that the approximation of the characteristic function of $\mathbf{R}^*$ is also an approximation of the characteristic function of **R**. ∎

From (3.3.36) we get an approximation of the cumulant function of **R**:

$$\psi_{\mathbf{R}}(\mathbf{T}) = \ln\varphi_{\mathbf{R}}(\mathbf{T}) = i\mathrm{vec}'\mathbf{T}\mathrm{vec}\boldsymbol{\Omega} + \ln\left\{1 - \frac{1}{2}\mathrm{vec}'\mathbf{T}\mathbf{D}_1\overline{m}_2[\mathrm{vec}\mathbf{S}]\mathbf{D}_1'\mathrm{vec}\mathbf{T}\right.$$
$$\left. + \frac{i}{2}\mathrm{vec}'\mathbf{T}\left\{\mathrm{vec}'(\overline{m}_2[\mathrm{vec}\mathbf{S}]) \otimes \mathbf{I}_{p^2}\right\}\mathrm{vec}\mathbf{D}_2\right\} + o\left(\frac{1}{n}\right). \tag{3.3.40}$$

Denote

$$\mathbf{M} = \mathbf{D}_1\overline{m}_2[\mathrm{vec}\mathbf{S}]\mathbf{D}_1', \tag{3.3.41}$$
$$\mathbf{N} = \left(\mathrm{vec}'\{\overline{m}_2[\mathrm{vec}\mathbf{S}]\} \otimes \mathbf{I}_{p^2}\right)\mathrm{vec}\mathbf{D}_2. \tag{3.3.42}$$

In this notation the main terms of the first cumulants of **R** will be presented in the next

Theorem 3.3.9. *The main terms of the first three cumulants $c_i^*[\mathbf{R}]$ of the sample correlation matrix* **R** *are of the form*

$$c_1^*[\mathbf{R}] = vec\boldsymbol{\Omega} + \frac{1}{2}\mathbf{N}, \tag{3.3.43}$$
$$c_2^*[\mathbf{R}] = \frac{1}{2}\{\mathbf{L}(\mathrm{vec}\mathbf{M} \otimes \mathbf{I}_{p^2}) - \frac{1}{2}\mathbf{N}\mathbf{N}'\}, \tag{3.3.44}$$
$$c_3^*[\mathbf{R}] = \frac{1}{4}\left\{\left(\frac{1}{2}\mathbf{N}\mathbf{N}' - \mathbf{L}(\mathrm{vec}\mathbf{M} \otimes \mathbf{I}_{p^2})\right)(\mathbf{N}' \otimes \mathbf{I}_{p^2})(\mathbf{I}_{p^4} + \mathbf{K}_{p^2,p^2})\right.$$
$$\left. - \mathbf{N}\mathrm{vec}'\mathbf{L}(\mathrm{vec}\mathbf{M} \otimes \mathbf{I}_{p^2})\right\}, \tag{3.3.45}$$

where $\boldsymbol{\Omega}$, **M** *and* **N** *are defined by (3.1.12), (3.3.41) and (3.3.42), respectively, and*

$$\mathbf{L} = (\mathbf{I}_{p^2} \otimes \mathrm{vec}'\mathbf{I}_{p^2})(\mathbf{I}_{p^6} + \mathbf{K}_{p^2,p^4}). \tag{3.3.46}$$

PROOF: In order to get the expressions of the cumulants, we have to differentiate the cumulant function (3.3.40) three times. To simplify the derivations, put

$$\psi_{\mathbf{R}}^*(\mathbf{T}) = i\mathrm{vec}'\mathbf{T}\mathrm{vec}\boldsymbol{\Omega} + \ln\left\{1 - \frac{1}{2}\mathrm{vec}'\mathbf{T}\mathbf{D}_1\overline{m}_2[\mathrm{vec}\mathbf{S}]\mathbf{D}_1'\mathrm{vec}\mathbf{T}\right.$$
$$\left. + \frac{i}{2}\mathrm{vec}'\mathbf{T}\left\{\mathrm{vec}'(\overline{m}_2[\mathrm{vec}\mathbf{S}]) \otimes \mathbf{I}_{p^2}\right\}\mathrm{vec}\mathbf{D}_2\right\}.$$

The first derivative equals

$$\frac{d\psi_{\mathbf{R}}^*(\mathbf{T})}{d\mathbf{T}} = i\mathrm{vec}\boldsymbol{\Omega} + \frac{1}{1 - \frac{1}{2}\mathrm{vec}'\mathbf{T}(\mathbf{M}\mathrm{vec}\mathbf{T} - i\mathbf{N})}$$
$$\times \frac{1}{2}\left\{i\mathbf{N} - (\mathbf{I}_{p^2} \otimes \mathrm{vec}'\mathbf{T} + \mathrm{vec}'\mathbf{T} \otimes \mathbf{I}_{p^2})\mathrm{vec}\mathbf{M}\right\},$$

from where the equality (3.3.42) follows by (2.1.33).
The second derivative is given by

$$\frac{d^2\psi^*_{\mathbf{R}}(\mathbf{T})}{d\mathbf{T}^2} = \frac{i}{2}\frac{d(\mathbf{H}_2\mathbf{H}_1)}{d\mathbf{T}} = \frac{i}{2}\Big(\frac{d\mathbf{H}_1}{d\mathbf{T}}\mathbf{H}_2' + \frac{d\mathbf{H}_2}{d\mathbf{T}}\mathbf{H}_1\Big), \tag{3.3.47}$$

where

$$\mathbf{H}_1 = \big(1 - \frac{1}{2}\mathrm{vec}'\mathbf{T}\mathbf{M}\mathrm{vec}\mathbf{T} + \frac{i}{2}\mathrm{vec}'\mathbf{T}\mathbf{N}\big)^{-1},$$
$$\mathbf{H}_2 = \mathbf{N} + i(\mathbf{I}_{p^2} \otimes \mathrm{vec}'\mathbf{T} + \mathrm{vec}'\mathbf{T} \otimes \mathbf{I}_{p^2})\mathrm{vec}\mathbf{M}.$$

The two derivatives which appear in (3.3.47) are of the form:

$$\begin{aligned}
\frac{d\mathbf{H}_1}{d\mathbf{T}} &= \frac{1}{2}\frac{d(i\mathrm{vec}'\mathbf{T}\mathbf{N} - \mathrm{vec}'\mathbf{T}\mathbf{M}\mathrm{vec}\mathbf{T})}{d\mathbf{T}}\frac{d\mathbf{H}_1}{d\left(1 - \frac{1}{2}\mathrm{vec}'\mathbf{T}\mathbf{M}\mathrm{vec}\mathbf{T} + \frac{i}{2}\mathrm{vec}'\mathbf{T}\mathbf{N}\right)}\\
&= \frac{1}{2}\left[(\mathbf{I}_{p^2} \otimes \mathrm{vec}'\mathbf{T})(\mathbf{I}_{p^4} + \mathbf{K}_{p^2,p^2})\mathrm{vec}\mathbf{M} - i\mathbf{N}\right]\mathbf{H}_1^2,\\
\frac{d\mathbf{H}_2}{d\mathbf{T}} &= i(\mathbf{I}_{p^2} \otimes \mathrm{vec}'\mathbf{I}_{p^2})(\mathbf{I}_{p^6} + \mathbf{K}_{p^2,p^4})(\mathrm{vec}\mathbf{M} \otimes \mathbf{I}_{p^2}).
\end{aligned}$$

After replacing the obtained derivatives into (3.3.47), at $\mathbf{T} = \mathbf{0}$, we get the main term of the second cumulant

$$c_2^*[\mathbf{R}] = \frac{1}{2}\Big(\mathbf{L}(\mathrm{vec}\mathbf{M} \otimes \mathbf{I}_{p^2}) - \frac{1}{2}\mathbf{N}\mathbf{N}'\Big),$$

where $\mathbf{L}$ is given in (3.3.46).
We get the third cumulant after differentiating the right-hand side of (3.3.47). Denote

$$\frac{d\mathbf{H}_2}{d\mathbf{T}} = i\mathbf{L}_1 = i\mathbf{L}(\mathrm{vec}\mathbf{M} \otimes \mathbf{I}_{p^2}).$$

Then

$$\begin{aligned}
c_3^*[\mathbf{R}] &= \frac{1}{i^3}\frac{d}{d\mathbf{T}}\Big\{\frac{i^2}{2}\big(-\frac{1}{2}\mathbf{H}_2\mathbf{H}_1^2\mathbf{H}_2' + \mathbf{L}_1\mathbf{H}_1\big)\Big\}\Big|_{\mathbf{T}=\mathbf{0}}\\
&= \frac{1}{2i}\left\{-\frac{1}{2}\frac{d\mathbf{H}_2\mathbf{H}_1(\mathbf{H}_2\mathbf{H}_1)'}{d\mathbf{T}} + \frac{d\mathbf{H}_1\mathrm{vec}'\mathbf{L}_1}{d\mathbf{T}}\right\}\Big|_{\mathbf{T}=\mathbf{0}}\\
&= \frac{1}{2}\left\{\frac{i}{2}\frac{d\mathbf{H}_2\mathbf{H}_1}{d\mathbf{T}}(\mathbf{H}_1\mathbf{H}_2' \otimes \mathbf{I}_{p^2} + \mathbf{I}_{p^2} \otimes \mathbf{H}_1\mathbf{H}_2') - \frac{d\mathbf{H}_1}{d\mathbf{T}}i\mathrm{vec}'\mathbf{L}_1\right\}\Big|_{\mathbf{T}=\mathbf{0}}\\
&= \frac{1}{4}\left\{\big(\frac{1}{2}\mathbf{N}\mathbf{N}' - \mathbf{L}_1\big)\left(\mathbf{N}' \otimes \mathbf{I}_{p^2} + \mathbf{I}_{p^2} \otimes \mathbf{N}'\right) - \mathbf{N}\mathrm{vec}'\mathbf{L}_1\right\}.
\end{aligned}$$

∎

Now we shall switch over to the density expansions of $\mathbf{R}$. We are going to find approximations of the density function of the $\frac{1}{2}p(p-1)$–vector

$$\mathbf{z} = \sqrt{n}\mathbf{T}(c_p)\mathrm{vec}(\mathbf{R} - \mathbf{\Omega}),$$

where $\mathbf{T}(c_p)$ is the $\frac{1}{2}p(p-1)\times p^2$–matrix defined in Proposition 1.3.21, which eliminates constants and repeated elements from $\mathrm{vec}\mathbf{R}$ and $\mathrm{vec}\boldsymbol{\Omega}$. It is most essential to construct expansions on the basis of the asymptotic distribution of $\mathbf{R}$. Let us denote the asymptotic dispersion matrices of $\mathbf{z}$ and $\sqrt{n}\mathrm{vec}(\mathbf{R}-\boldsymbol{\Omega})$ by $\boldsymbol{\Sigma}_{\mathbf{z}}$ and $\boldsymbol{\Sigma}_{\mathbf{R}}$. The matrix $\boldsymbol{\Sigma}_{\mathbf{R}}$ was derived in Theorem 3.1.6:

$$\boldsymbol{\Sigma}_{\mathbf{R}} = \mathbf{D}_1\boldsymbol{\Pi}\mathbf{D}_1',$$

where $\boldsymbol{\Pi}$ and $\mathbf{D}_1$ are given by (3.1.5) and (3.1.13), respectively. In the important special case of the normal population, $\mathbf{x}\sim N_p(\boldsymbol{\mu},\boldsymbol{\Sigma})$, the asymptotic dispersion matrix $\boldsymbol{\Pi}_N$ is given in Corollary 3.1.4.1. Theorem 3.1.4 also specifies another matrix of interest:

$$\boldsymbol{\Sigma}_{\mathbf{z}} = \mathbf{T}(c_p)\boldsymbol{\Sigma}_{\mathbf{R}}\mathbf{T}(c_p)'.$$

Using (3.3.37) and the expressions (3.3.43) – (3.3.45) of $c_i^*[\mathbf{R}]$, $i=1,2,3$, it follows that

$$c_1^*[\mathbf{z}] = \frac{1}{\sqrt{n}}\boldsymbol{\Gamma}_1 + o(n^{-\frac{1}{2}}),$$
$$c_2^*[\mathbf{z}] = \boldsymbol{\Xi} + \frac{1}{n}\boldsymbol{\Gamma}_2 + o(n^{-1}),$$
$$c_3^*[\mathbf{z}] = \frac{1}{\sqrt{n}}\boldsymbol{\Gamma}_3 + o(n^{-\frac{1}{2}}),$$

where

$$\boldsymbol{\Gamma}_1 = \frac{1}{2}\mathbf{T}(c_p)(\mathrm{vec}'\boldsymbol{\Pi}\otimes\mathbf{I}_{p^2})\mathrm{vec}\mathbf{D}_2, \tag{3.3.48}$$

$$\boldsymbol{\Xi} = \frac{1}{2}\mathbf{T}(c_p)\mathbf{L}\{\mathrm{vec}(\mathbf{D}_1\mathrm{vec}\boldsymbol{\Pi}\mathbf{D}_1')\otimes\mathbf{I}_{p^2}\}\mathbf{T}(c_p)', \tag{3.3.49}$$

$$\boldsymbol{\Gamma}_2 = -\frac{1}{4}\mathbf{T}(c_p)(\mathrm{vec}'\boldsymbol{\Pi}\otimes\mathbf{I}_{p^2})\mathrm{vec}\mathbf{D}_2\mathrm{vec}'\mathbf{D}_2(\mathrm{vec}\boldsymbol{\Pi}\otimes\mathbf{I}_{p^2})\mathbf{T}(c_p)', \tag{3.3.50}$$

$$\begin{aligned}\boldsymbol{\Gamma}_3 = \frac{1}{4}\mathbf{T}(c_p)\Big\{&(\mathrm{vec}'\boldsymbol{\Pi}\otimes\mathbf{I}_{p^2})\mathrm{vec}\mathbf{D}_2\mathrm{vec}'\mathbf{D}_2(\mathrm{vec}\boldsymbol{\Pi}\otimes\mathbf{I}_{p^2}) \\ &- \mathbf{L}(\mathrm{vec}(\mathbf{D}_1\mathrm{vec}\boldsymbol{\Pi}\mathbf{D}_1')\otimes\mathbf{I}_{p^2})(+vec'\mathbf{D}_2(\mathrm{vec}\boldsymbol{\Pi}\otimes\mathbf{I}_{p^2})\otimes\mathbf{I}_{p^2})(\mathbf{I}_{p^4}+\mathbf{K}_{p^2,p^2}) \\ &- (\mathrm{vec}'\boldsymbol{\Pi}\otimes\mathbf{I}_{p^2})\mathrm{vec}\mathbf{D}_2\mathrm{vec}'\{\mathbf{L}(\mathrm{vec}(\mathbf{D}_1\mathrm{vec}\boldsymbol{\Pi}\mathbf{D}_1')\otimes\mathbf{I}_{p^2})\}\Big\}\mathbf{T}(c_p)'^{\otimes 2}.\end{aligned} \tag{3.3.51}$$

When approximating with the normal distribution, we are going to use the same dimensions and will consider the normal distribution $N_r(\mathbf{0},\boldsymbol{\Xi})$ as the approximating distribution, where $r=\frac{1}{2}p(p-1)$. This guarantees the equality of the two variance matrices of interest up to an order of n^{-1}.

Theorem 3.3.10. *Let $\mathbf{X}=(\mathbf{x}_1,\ldots,\mathbf{x}_n)$ be a sample of size n from a $p-$dimensional population with $E[\mathbf{x}]=\boldsymbol{\mu}$, $D[\mathbf{x}]=\boldsymbol{\Sigma}$, $m_4[\mathbf{x}]<\infty$, and let $\boldsymbol{\Omega}$ and $\mathbf{R}$ be the population and the sample correlation matrices, respectively. Then, for the density function of the $r-$vector $\mathbf{z}=\sqrt{n}\mathbf{T}(c_p)\mathrm{vec}(\mathbf{R}-\boldsymbol{\Omega})$, $r=\frac{1}{2}p(p-1)$, the following formal expansions are valid:*

(i) $$f_{\mathbf{z}}(\mathbf{y}_0) = f_{N_r(\mathbf{0},\boldsymbol{\Xi})}(\mathbf{x}_0)\Big\{1+\frac{1}{6\sqrt{n}}\mathrm{vec}'(\mathbf{V}\boldsymbol{\Xi}^{-\frac{1}{2}}\boldsymbol{\Gamma}_3(\boldsymbol{\Xi}^{-\frac{1}{2}}\mathbf{V}')^{\otimes 2})$$

$$\times \ \mathrm{vec}\mathbf{H}_3(\mathbf{x}_0, \Xi) + o(\frac{1}{\sqrt{n}})\Big\},$$

where $\mathbf{x}_0$ *is determined by (3.3.19) and (3.3.23),* $\mathbf{V} = (\mathbf{v}_1, \ldots, \mathbf{v}_{\frac{1}{2}p(p-1)})$ *is the matrix of eigenvalue-normed eigenvectors* $\mathbf{v}_i$ *of* Ξ*, which correspond to the* r *largest eigenvalues of* Ξ*;*

$$\text{(ii)}\ \ f_{\mathbf{z}}(\mathbf{y}_0) = f_{N_r(\mathbf{0},\Xi)}(\mathbf{x}_0)\Big\{1 + \frac{1}{\sqrt{n}}\boldsymbol{\Gamma}_1'(m_2[\mathbf{z}])^{-\frac{1}{2}}\mathbf{U}'\mathbf{H}_1(\mathbf{x}_0, \Xi)$$
$$+ \frac{1}{6\sqrt{n}}\mathrm{vec}'\Big\{\mathbf{U}'(m_2[\mathbf{z}])^{-\frac{1}{2}}\boldsymbol{\Gamma}_3((m_2[\mathbf{z}])^{-\frac{1}{2}}\mathbf{U}')^{\otimes 2}\Big\}\mathrm{vec}\mathbf{H}_3(\mathbf{x}_0, \Xi)$$
$$+o(\frac{1}{\sqrt{n}})\Big\},$$

where $\mathbf{x}_0 = \mathbf{U}(m_2[\mathbf{y}])^{-\frac{1}{2}}\mathbf{y}_0$ *and* $\mathbf{U} = (\mathbf{u}_1, \ldots, \mathbf{u}_{\frac{1}{2}p(p-1)})$ *is the matrix of eigenvalue-normed eigenvectors* $\mathbf{u}_i$ *of the moment matrix* $m_2[\mathbf{z}]$*, which correspond to the* r *largest eigenvalues of* $m_2[\mathbf{z}]$*. The Hermite polynomials* $\mathbf{H}_i(\mathbf{x}_0, \Xi)$ *are defined by (2.2.39) and (2.2.41), and* $\boldsymbol{\Gamma}_1$ *and* $\boldsymbol{\Gamma}_3$ *are given by (3.3.48) and (3.3.51), respectively.*

PROOF: The expansion (i) directly follows from Theorem 3.3.4, and the expansion (ii) is a straightforward conclusion from Theorem 3.3.5, if we include the expressions of the cumulants of **R** into the expansions and take into account the remainder terms of the cumulants. ∎

It is also of interest to calculate $f_{\mathbf{z}}(\cdot)$ and $f_{N_r(\mathbf{0},\Xi)}(\cdot)$ at the same points, when $\mathbf{z}$ has the same dimension as the approximating normal distribution. This will be realized in the next corollary.

Corollary 3.3.10.1. *In the notation of Theorem 3.3.10, the following density expansion holds:*

$$f_{\mathbf{z}}(\mathbf{y}_0) = f_{N_r(\mathbf{0},\Xi)}(\mathbf{x}_0)\Big\{1 + \frac{1}{\sqrt{n}}\boldsymbol{\Gamma}_1'\mathbf{H}_1(\mathbf{x}_0, \Xi)$$
$$+\frac{1}{6\sqrt{n}}\mathrm{vec}'\boldsymbol{\Gamma}_3\mathrm{vec}\mathbf{H}_3(\mathbf{x}_0, \Xi) + o(\frac{1}{\sqrt{n}})\Big\}.$$

PROOF: The expansion follows from Corollary 3.3.1.1, when $\mathbf{P} = \mathbf{I}$ and if we take into account that $D[\mathbf{z}] - \Xi = o\left(\frac{1}{n}\right)$. ∎

For the Wishart approximations we shall consider

$$\mathbf{y} = n\mathbf{T}(c_p)\mathrm{vec}(\mathbf{R} - \boldsymbol{\Omega}). \tag{3.3.52}$$

Now the main terms of the first three cumulants of $\mathbf{y}$ are obtained from (3.3.43) – (3.3.45) and (3.3.48) – (3.3.51):

$$c_1^*[\mathbf{y}] = \boldsymbol{\Gamma}_1, \tag{3.3.53}$$
$$c_2^*[\mathbf{y}] = n\Xi - \boldsymbol{\Gamma}_2, \tag{3.3.54}$$
$$c_3^*[\mathbf{y}] = n\boldsymbol{\Gamma}_3. \tag{3.3.55}$$

The next theorem gives us expansions of the sample correlation matrix through the Wishart distribution. In the approximation we shall use the $\frac{1}{2}p(p+1)-$vector $V^2(\mathbf{V})$, where $\mathbf{V} = \mathbf{W} - n\boldsymbol{\Sigma}$, $\mathbf{W} \sim W_p(\boldsymbol{\Sigma}, n)$.

Theorem 3.3.11. *Let the assumptions of Theorem 3.3.10 about the sample and population distribution hold, and let* $\mathbf{y} = n(\mathbf{r} - \boldsymbol{\omega})$, *where* $\mathbf{r}$ *and* $\boldsymbol{\omega}$ *are the vectors of upper triangles of* $\mathbf{R}$ *and* $\boldsymbol{\Omega}$ *without the main diagonal. Then, for the density function of the* $\frac{1}{2}p(p-1)-$*vector* $\mathbf{y}$, *the following expansions are valid:*

$$\text{(i)}\quad f_{\mathbf{y}}(\mathbf{y}_0) = (2\pi)^{\frac{p}{2}}|c_2[V^2(\mathbf{V})]|^{\frac{1}{2}}|c_2[\mathbf{y}]|^{-\frac{1}{2}}f_{\mathbf{V}}(\mathbf{V}_0)$$
$$\times \Big\{1 + \frac{1}{6}\mathrm{vec}'\{\mathbf{U}(c_2^*[\mathbf{y}])^{-\frac{1}{2}}c_3^*[\mathbf{y}]((c_2^*[\mathbf{y}])^{-\frac{1}{2}}\mathbf{U}')^{\otimes 2} - c_3^*[V^2(\mathbf{V})]\}$$
$$\times\, \mathrm{vec}\mathbf{L}_3^*(\mathbf{V}_0, \boldsymbol{\Sigma}) + r_3^*\Big\},$$

where $\mathbf{V}_0$ *is given by its vectorized upper triangle* $V^2(\mathbf{V}_0) = \mathbf{U}(c_2^*[\mathbf{y}])^{-\frac{1}{2}}(\mathbf{y}_0 - E[\mathbf{y}])$ *and* $\mathbf{U} = (\mathbf{u}_1, \ldots, \mathbf{u}_{\frac{1}{2}p(p-1)})$ *is the* $\big(\frac{1}{2}p(p+1)\big)\times\big(\frac{1}{2}p(p-1)\big)$ *−matrix of eigenvalue-normed eigenvectors* $\mathbf{u}_i$ *corresponding to the* $\frac{1}{2}p(p-1)$ *largest eigenvalues of* $c_2[V^2(\mathbf{V})]$;

$$\text{(ii)}\quad f_{\mathbf{y}}(\mathbf{y}_0) = (2\pi)^{\frac{p}{2}}|c_2[V^2(\mathbf{V})]|^{\frac{1}{2}}|c_2^*[\mathbf{y}]|^{-\frac{1}{2}}f_{\mathbf{V}}(\mathbf{V}_0)\Big\{1 + \boldsymbol{\Gamma}_1'\mathbf{U}(m_2^*[\mathbf{y}])^{-\frac{1}{2}}\mathbf{L}_1^*(\mathbf{V}_0, \boldsymbol{\Sigma})$$
$$+ \frac{1}{6}\mathrm{vec}'\{\mathbf{U}(m_2^*[\mathbf{y}])^{-\frac{1}{2}}c_3^*[\mathbf{y}]((m_2^*[\mathbf{y}])^{-\frac{1}{2}}\mathbf{U}')^{\otimes 2} - c_3[V^2(\mathbf{V})]\}$$
$$\times\, \mathrm{vec}\mathbf{L}_3^*(\mathbf{V}_0, \boldsymbol{\Sigma}) + r_3^*\Big\},$$

where $\mathbf{V}_0$ *is given by the vectorized upper triangle* $V^2(\mathbf{V}_0) = \mathbf{U}(c_2^*[\mathbf{y}])^{-\frac{1}{2}}(\mathbf{y}_0 - E[\mathbf{y}])$ *and* $\mathbf{U} = (\mathbf{u}_1, \ldots, \mathbf{u}_{\frac{1}{2}p(p-1)})$ *is the matrix of eigenvalue-normed eigenvectors* $\mathbf{u}_i$ *corresponding to the* $\frac{1}{2}p(p-1)$ *largest eigenvalues of* $m_2[V^2(\mathbf{V})]$. *In (i) and (ii) the matrices* $\mathbf{L}_i^*(\mathbf{V}_0, \boldsymbol{\Sigma})$ *are defined by (2.4.66), (2.4.61) and (2.4.63),* $c_i^*[\mathbf{y}]$ *are given in (3.3.53) – (3.3.55).*
If $n \gg p$, *then* $r_3^* = o(n^{-1})$ *in (i) and (ii).* ∎

PROOF: The terms in the expansions (i) and (ii) directly follow from Theorems 3.3.2 and 3.3.3, respectively, if we replace derivatives $f_{\mathbf{V}}^k(\mathbf{V}_0, \boldsymbol{\Sigma})$ by $\mathbf{L}_k^*(\mathbf{V}_0, \boldsymbol{\Sigma})$ according to (2.4.66). Now, we have to show that the remainder terms are $o(n^{-1})$. In Lemma 2.4.2 it was established that when $p \ll n$,

$$\mathbf{L}_1^*(\mathbf{V}_0, \boldsymbol{\Sigma}) \sim O(n^{-1}),$$
$$\mathbf{L}_k^*(\mathbf{V}_0, \boldsymbol{\Sigma}) \sim O(n^{-(k-1)}), \qquad k \geq 2.$$

From (3.3.53) – (3.3.55), combined with (3.3.43) – (3.3.45) and (3.3.41), we get

$$c_1^*[\mathbf{y}] \sim O(1),$$
$$c_k^*[\mathbf{y}] \sim O(n), \qquad k \geq 2,$$

and as $c_k[V^2(\mathbf{V})] \sim O(n)$, $k \geq 2$, the remainder terms in (i) and (ii) are $o(n^{-1})$. ∎

If we compare the expansions in Theorems 3.3.10 and 3.3.11, we can conclude that Theorem 3.3.11 gives us higher order accuracy, but also more complicated formulae are involved in this case. The simplest Wishart approximation with the error term $O(n^{-1})$ will be obtained, if we use just the Wishart distribution itself with the multiplier of $f_{\mathbf{V}}(\mathbf{V}_0)$ in the expansions. Another possibility to get a Wishart approximation for $\mathbf{R}$ stems from Theorem 3.3.7. This will be presented in the next theorem.

Theorem 3.3.12. *Let $\mathbf{y}$ be the random $\frac{1}{2}p(p-1)-$vector defined in (3.3.52), $D[\mathbf{y}]$ non-singular, and let eigenvalues of $D[V^2(\mathbf{V})]$ be denoted by λ_i and the corresponding eigenvalue-normed eigenvectors by $\mathbf{u}_i$, $i = 1,\ldots,\frac{1}{2}p(p+1)$. Put $\mu_i = |\lambda_i - 1|$ and let $\mu_{(i)}$ be the diminishingly ordered values of μ_i with $\mu_{(i)} = |\lambda_{(i)} - 1|$. Let $\mathbf{U}$ consist of eigenvectors $\mathbf{u}_{(i)}$, $i = 1, 2, \ldots, q$, corresponding to $\lambda_{(i)}$, $\mathbf{\Lambda}_{(1)} = (\lambda_{(1)},\ldots,\lambda_{(q)})_d$, and $D[\mathbf{y}]^{\frac{1}{2}}$ denote any square root of $D[\mathbf{y}]$. Then the matrix*

$$\mathbf{P} = D[\mathbf{y}]^{-\frac{1}{2}}\mathbf{U}'$$

minimizes the matrix $\mathbf{M}_2$ in (3.3.11) in the sense of the norm (3.3.32) and the following formal Wishart expansion of the density $f_{\mathbf{y}}(\mathbf{y}_0)$ holds:

$$f_{\mathbf{y}}(\mathbf{y}_0) = (2\pi)^{\frac{p}{2}}|D[\mathbf{y}]|^{-\frac{1}{2}} \prod_{k=1}^{\frac{1}{2}p(p-1)} \sqrt{\lambda_{(k)}} f_{\mathbf{V}}(\mathbf{V}_0)$$

$$\times \Big\{1 + \tfrac{1}{2}\mathrm{vec}'(\mathbf{I}_{\frac{1}{2}p(p+1)} + \mathbf{U}(\mathbf{I}_{\frac{1}{2}p(p-1)} - \mathbf{\Lambda}_{(1)}^{-1})\mathbf{U}' - c_2[V^2(\mathbf{V})])\mathrm{vec}\mathbf{L}_2^*(\mathbf{V}_0,\mathbf{\Sigma})$$

$$+ \tfrac{1}{6}\mathrm{vec}'\Big(\mathbf{U}(D[\mathbf{y}])^{-\frac{1}{2}}c_3[\mathbf{y}]((D[\mathbf{y}])^{-\frac{1}{2}}\mathbf{U}')^{\otimes 2} - c_3[V^2(\mathbf{V})]\Big)\,\mathrm{vec}\mathbf{L}_3^*(\mathbf{V}_0,\mathbf{\Sigma}) + r_3^*\Big\},$$

where $\mathbf{V}_0$ is given by its vectorized upper triangle $V^2(\mathbf{V}_0) = \mathbf{P}'(\mathbf{y}_0 - E[\mathbf{y}])$, $\mathbf{L}_i^*(\mathbf{V}_0,\mathbf{\Sigma})$, $i = 2,3$ are defined by (2.4.66), (2.4.62), (2.4.63), and $c_i[V^2(\mathbf{V})]$, $i = 2,3$, are given in Theorem 2.4.16. If $n \gg p$, then $r_3^* = o(n^{-1})$.

PROOF: The expansion stems from Theorem 3.3.7, if we take $q = \frac{1}{2}p(p-1)$. The remainder term is $o(n^{-1})$ by the same argument as in the proof of Theorem 3.3.11. ∎

From Theorem 3.3.12 we get an approximation of order $O(n^{-1})$, if we do not use the $\mathbf{L}_3^*(\mathbf{V}_0,\mathbf{\Sigma})-$term.

3.3.5 *Problems*

1. Verify the identity

$$\mathrm{vec}'\mathbf{M}_2 \otimes (\mathbf{M}_0 + \mathbf{M}_1)' = \mathrm{vec}'(\mathbf{M}_2 \otimes (\mathbf{M}_0 + \mathbf{M}_1)),$$

   which was used in the proof of Corollary 3.3.1.1.
2. Show that

$$\mathbf{P}(D[\mathbf{y}])^{-1/2}\mathbf{V}$$

   in (3.2.23) is a solution to (3.2.22).
3. Prove Lemma 3.3.4.

4. Show that equality (3.3.33) is valid.
5. Find the second order derivative (3.3.38).
6. Find the first two terms in the expansion of the characteristic function of $\mathbf{R}^{-1}$.
7. Let $\mathbf{S}$ be the sample dispersion matrix for a normal population $N_p(\boldsymbol{\mu}, \boldsymbol{\Sigma})$. Construct a formal density expansion for $\sqrt{n}\mathrm{tr}(\mathbf{S} - \boldsymbol{\Sigma})$ through the skew normal distribution $SN_p(\boldsymbol{\Sigma}, \boldsymbol{\alpha})$ (see Problem 10 in §3.2.5).
8. Show that if $n \gg p$, then $c_1^*[\mathbf{y}] \sim O(1)$ and $c_2^*[\mathbf{y}] \sim O(n)$.
9. Write out the $\frac{1}{n}$–term in Theorem 3.3.12, if $n \gg p$.
10. Construct the density expansion for the sample correlation matrix $\mathbf{R}$ based on the skew normal distribution $SN_p(\boldsymbol{\Sigma}, \boldsymbol{\alpha})$ (see Problem 10 in §3.2.5).

CHAPTER IV

Multivariate Linear Models

Linear models play a key role in statistics. If exact inference is not possible then at least a linear approximate approach can often be carried out. We are going to focus on multivariate linear models. Throughout this chapter various results presented in earlier chapters will be utilized. The Growth Curve model by Potthoff & Roy (1964) will serve as a starting point. Although the Growth Curve model is a linear model, the maximum likelihood estimator of its mean parameters is a non-linear random expression which causes many difficulties when considering these estimators. The first section presents various multivariate linear models as well as maximum likelihood estimators of the parameters in the models. Since the estimators are non-linear stochastic expressions, their distributions have to be approximated and we are going to rely on the results from Chapter 3. However, as it is known from that chapter, one needs detailed knowledge about moments. It turns out that even if we do not know the distributions of the estimators, exact moment expressions can often be obtained, or at least it is possible to approximate the moments very accurately. We base our moment expressions on the moments of the matrix normal distribution, the Wishart distribution and the inverted Wishart distribution. Necessary relations were obtained in Chapter 2 and in this chapter these results are applied. Section 4.2 comprises some of the models considered in Section 4.1 and the goal is to obtain moments of the estimators. These moments will be used in Section 4.3 when finding approximations of the distributions.

4.1 THE GROWTH CURVE MODEL AND ITS EXTENSIONS

4.1.1 *Introduction*

In this section multivariate linear models are introduced. Particularly, the Growth Curve model with some extensions is studied. In the subsequent $L(\mathbf{B},\mathbf{\Sigma})$ denotes the likelihood function with parameters $\mathbf{B}$ and $\mathbf{\Sigma}$. Moreover, in order to shorten matrix expressions we will write $(\mathbf{X})()'$ instead of $(\mathbf{X})(\mathbf{X})'$. See (4.1.3) below, for example.

Suppose that we have an observation vector $\mathbf{x}_i \in \mathbb{R}^p$, which follows the linear model

$$\mathbf{x}_i = \boldsymbol{\mu} + \mathbf{\Sigma}^{1/2}\mathbf{e}_i,$$

where $\boldsymbol{\mu} \in \mathbb{R}^p$ and $\mathbf{\Sigma} \in \mathbb{R}^{p\times p}$ are unknown parameters, $\mathbf{\Sigma}^{1/2}$ is a symmetric square root of the positive definite matrix $\mathbf{\Sigma}$, and $\mathbf{e}_i \sim N_p(\mathbf{0},\mathbf{I})$. Our crucial assumption is that $\boldsymbol{\mu} \in C(\mathbf{A})$, i.e. $\boldsymbol{\mu} = \mathbf{A}\boldsymbol{\beta}$. Here $\boldsymbol{\beta}$ is unknown, whereas $\mathbf{A} : p \times q$ is known. Hence we have a linear model. Note that if $\mathbf{A} \in \mathbb{R}^{p\times q}$ spans the whole space, i.e. $r(\mathbf{A}) = p$, there are no restrictions on $\boldsymbol{\mu}$, or in other words, there is no non-trivial linear model for $\boldsymbol{\mu}$.

The observations $\mathbf{x}_i$ may be regarded as repeated measurements on some individual or as a short time series. Up to now the model assumption about $\boldsymbol{\mu}$ is a pure within-individuals model assumption. However, in order to estimate the parameters, more than one individual is needed. In many situations it is also natural to have a between-individuals model. For example, if there are more than one treatment group among the individuals. Let $\mathbf{X} \in \mathbb{R}^{p\times n}$ be a matrix where each column corresponds to one individual. Then, instead of $\boldsymbol{\beta}$ in $\boldsymbol{\mu} = \mathbf{A}\boldsymbol{\beta}$, n parameter vectors $\boldsymbol{\beta}_1, \ldots, \boldsymbol{\beta}_n$ exist, and under a between-individuals linear model assumption we have

$$\boldsymbol{\beta}_1, \ldots, \boldsymbol{\beta}_n = \mathbf{BC},$$

where $\mathbf{C} \in \mathbb{R}^{k\times n}$ is a known between-individuals design matrix and $\mathbf{B} \in \mathbb{R}^{q\times k}$ is an unknown parameter matrix. Another type of model is considered when instead of $\boldsymbol{\mu} = \mathbf{A}(\boldsymbol{\beta}_1, \boldsymbol{\beta}_2, \ldots, \boldsymbol{\beta}_n)$ we put some rank condition on $\boldsymbol{\mu}$, i.e. instead of supposing that $\boldsymbol{\mu} \in \mathcal{C}(\mathbf{A})$ it is only supposed that $r(\boldsymbol{\mu})$ is known. By Proposition 1.1.6 (i) it follows for this model that $\boldsymbol{\mu} = \mathbf{A}(\boldsymbol{\beta}_1, \boldsymbol{\beta}_2, \ldots, \boldsymbol{\beta}_n)$, but this time both $\mathbf{A}$ and $\boldsymbol{\beta}_i$ are unknown. In the subsequent, all matrices will be supposed to be real.

Definition 4.1.1 GROWTH CURVE MODEL. *Let $\mathbf{X} : p \times n$, $\mathbf{A} : p \times q$, $q \leq p$, $\mathbf{B} : q \times k$, $\mathbf{C} : k \times n$, $r(\mathbf{C}) + p \leq n$ and $\boldsymbol{\Sigma} : p \times p$ be p.d. Then*

$$\mathbf{X} = \mathbf{ABC} + \boldsymbol{\Sigma}^{1/2}\mathbf{E} \tag{4.1.1}$$

defines the Growth Curve model, where $\mathbf{E} \sim N_{p,n}(\mathbf{0}, \mathbf{I}_p, \mathbf{I}_n)$, $\mathbf{A}$ and $\mathbf{C}$ are known matrices, and $\mathbf{B}$ and $\boldsymbol{\Sigma}$ are unknown parameter matrices. ∎

Observe that the columns of $\mathbf{X}$ are independent normally distributed $p-$vectors with an unknown dispersion matrix $\boldsymbol{\Sigma}$, and expectation of $\mathbf{X}$ equals $E[\mathbf{X}] = \mathbf{ABC}$. The model in (4.1.1) which has a long history is usually called Growth Curve model but other names are also used. Before Potthoff & Roy (1964) introduced the model, growth curve problems in a similar set-up had been considered by Rao (1958), and others. For general reviews of the model we refer to Woolson & Leeper (1980), von Rosen (1991) and Srivastava & von Rosen (1999). A recent elementary textbook on growth curves which is mainly based on the model (4.1.1) has been written by Kshirsagar & Smith (1995). Note that if $\mathbf{A} = \mathbf{I}$, the model is an ordinary multivariate analysis of variance (MANOVA) model treated in most texts on multivariate analysis. Moreover, the *between-individuals design matrix* $\mathbf{C}$ is precisely the same design matrix as used in the theory of univariate linear models which includes univariate analysis of variance and regression analysis. The matrix $\mathbf{A}$ is often called a *within-individuals design matrix*.

The mean structure in (4.1.1) is bilinear, contrary to the MANOVA model, which is linear. So, from the point of view of bilinearity the Growth Curve model is the first fundamental multivariate linear model, whereas the MANOVA model is a univariate model. Mathematics also supports this classification since in MANOVA, as well as in univariate linear models, linear spaces are the proper objects to consider, whereas for the Growth Curve model decomposition of tensor spaces is shown to be a natural tool to use. Before going over to the technical details a simple artificial example is presented which illustrates the Growth Curve model. Later on we shall return to that example.

Example 4.1.1. Let there be 3 treatment groups of animals, with n_j animals in the $j-$th group, and each group is subjected to different treatment conditions. The aim is to investigate the weight increase of the animals in the groups. All animals have been measured at the same p time points (say t_r, $r = 1, 2, \ldots, p$). This is a necessary condition for applying the Growth Curve model. It is assumed that the measurements of an animal are multivariate normally distributed with a dispersion matrix $\boldsymbol{\Sigma}$, and that the measurements of different animals are independent.

We will consider a case where the expected growth curve, and the mean of the distribution for each treatment group, is supposed to be polynomial in time of degree $q - 1$. Hence, the mean μ_j of the $j-$th treatment group at time t is

$$\mu_j = \beta_{1j} + \beta_{2j}t + \cdots + \beta_{qj}t^{q-1}, \qquad j = 1, 2, 3,$$

where β_{ij} are unknown parameters.

Furthermore, data form a random matrix $\mathbf{X} : p \times n$ where $n = n_1 + n_2 + n_3$. Each animal is represented by a column in $\mathbf{X}$. If the first n_1 columns of $\mathbf{X}$ represent group one, the following n_2 columns group two, and so on, we get the between-individuals design matrix $\mathbf{C} : 3 \times n$:

$$\mathbf{C} = \begin{pmatrix} 1 & 1 & \ldots & 1 & 0 & 0 & \ldots & 0 & 0 & 0 & \ldots & 0 \\ 0 & 0 & \ldots & 0 & 1 & 1 & \ldots & 1 & 0 & 0 & \ldots & 0 \\ 0 & 0 & \ldots & 0 & 0 & 0 & \ldots & 0 & 1 & 1 & \ldots & 1 \end{pmatrix}$$

and by the polynomial mean structure we have that the within-individuals design matrix $\mathbf{A} : p \times q$ equals

$$\mathbf{A} = \begin{pmatrix} 1 & t_1 & \ldots & t_1^{q-1} \\ 1 & t_2 & \ldots & t_2^{q-1} \\ \vdots & \vdots & \ddots & \vdots \\ 1 & t_p & \ldots & t_p^{q-1} \end{pmatrix}.$$

Hence, the expectation of the data matrix $\mathbf{X}$ can be presented as

$$E[\mathbf{X}] = \mathbf{ABC},$$

where $\mathbf{B} : q \times 3$ is a matrix of the parameters $\boldsymbol{\beta}_{ij}$, $i = 1, 2, \ldots q$, and $j = 1, 2, 3$. Note that the columns of $\mathbf{B}$ describe the different groups and the rows the expected growth. ∎

Instead of (4.1.1) we may equivalently consider the identity

$$\mathrm{vec}\mathbf{X} = (\mathbf{C}' \otimes \mathbf{A})\mathrm{vec}\mathbf{B} + (\mathbf{I} \otimes \boldsymbol{\Sigma}^{1/2})\mathrm{vec}\mathbf{E},$$

which is a special case of the ordinary multivariate regression model

$$\mathbf{x} = \mathbf{T}\boldsymbol{\beta} + \mathbf{e}.$$

However, it is not fruitful to consider the model in this form because the interesting part lies in the connection between the tensor spaces generated by $\mathbf{C}' \otimes \mathbf{A}$ and $\mathbf{I} \otimes \boldsymbol{\Sigma}$.
There are several important extensions of the Growth Curve model. We are going to focus on extensions which comprise more general mean structures than $E[\mathbf{X}] = \mathbf{ABC}$ in the Growth Curve model. In Section 4.1 we are going to find maximum likelihood estimators when

$$E[\mathbf{X}] = \boldsymbol{\mu}\mathbf{C}, \quad r(\boldsymbol{\mu}) = q,$$

as well as when

$$E[\mathbf{X}] = \sum_{i=1}^{m} \mathbf{A}_i\mathbf{B}_i\mathbf{C}_i, \quad C(\mathbf{C}'_m) \subseteq C(\mathbf{C}'_{m-1}) \subseteq \ldots \subseteq C(\mathbf{C}'_1).$$

We mainly discuss the special case $m = 3$. Furthermore, we will consider

$$E[\mathbf{X}] = \sum_{i=1}^{m} \mathbf{A}_i\mathbf{B}_i\mathbf{C}_i + \mathbf{B}_{m+1}\mathbf{C}_{m+1}, \quad C(\mathbf{C}'_m) \subseteq C(\mathbf{C}'_{m-1}) \subseteq \ldots \subseteq C(\mathbf{C}'_1),$$

which is a multivariate covariance model. All these models are treated in §4.1.4. In §4.1.6 we study various restrictions, which can be put on $\mathbf{B}$ in $E[\mathbf{X}] = \mathbf{ABC}$, i.e.

$$E[\mathbf{X}] = \mathbf{ABC}, \qquad \mathbf{GBH} = \mathbf{0},$$

and

$$E[\mathbf{X}] = \mathbf{ABC}, \qquad \mathbf{G}_1\mathbf{BH}_1 = \mathbf{0}, \quad \mathbf{G}_2\mathbf{BH}_2 = \mathbf{0},$$

with various subspace conditions on $\mathbf{G}_i$, $i = 1, 2$, and $\mathbf{H}_i$, $i = 1, 2$, which are supposed to be known. Also, the system of equations $\mathbf{G}_i\mathbf{BH}_i = \mathbf{0}$, $i = 1, 2, \ldots, s$, will be discussed briefly. A different type of restriction is given by

$$E[\mathbf{X}] = \mathbf{ABC}, \qquad \mathbf{G}_1\boldsymbol{\Theta}\mathbf{H}_1 + \mathbf{G}_2\mathbf{BH}_2 = \mathbf{0},$$

where $\boldsymbol{\Theta}$ is also an unknown parameter.
Furthermore, in §4.1.3 we give a brief treatment of the case when the dispersion matrix is supposed to be singular. We will not go into details because calculations are complicated. Only some results for the Growth Curve model with a singular $\boldsymbol{\Sigma}$ will be presented. The fact is that in the above given extensions it could have been additionally assumed that $\boldsymbol{\Sigma}$ is singular.
Finally it is noted that in §4.1.5 we examine the conditions that give us unique estimators. This is an important issue, but unfortunately the results usually rely on rather tedious calculations.

4.1.2 *Maximum likelihood estimators*

Several approaches of finding maximum likelihood estimators will be presented in this section. We point out that we are mainly interested in the mean structure and we suppose that we have no information about any structure in $\mathbf{\Sigma}$. Therefore, solely an arbitrary $\mathbf{\Sigma}$ is considered. We will usually assume that $\mathbf{\Sigma}$ is p.d. However, §4.1.3 is devoted to the case when $\mathbf{\Sigma}$ is not of full rank. The reason for presenting different approaches to the same problem is that it is useful to have some knowledge about various techniques. The approaches discussed show several aspects which can be useful in other situations.

In the next lemma a useful inequality is presented which will be referred to several times. Among others, it has been used by Watson (1964), Srivastava & Khatri (1979) and Anderson (2003).

Lemma 4.1.1. *Let $\mathbf{\Sigma}$ and $\mathbf{S}$ be positive definite matrices of size $p \times p$. Then*

$$|\mathbf{\Sigma}|^{-\frac{1}{2}n}e^{-\frac{1}{2}\mathrm{tr}(\mathbf{\Sigma}^{-1}\mathbf{S})} \leq |\tfrac{1}{n}\mathbf{S}|^{-\frac{1}{2}n}e^{-\frac{1}{2}np},$$

where equality holds if and only if $\mathbf{\Sigma} = \frac{1}{n}\mathbf{S}$.

PROOF: It follows from Corollary 1.2.42.1 that there exist matrices $\mathbf{H}$ and $\mathbf{D}$ such that

$$\mathbf{\Sigma} = \mathbf{H}\mathbf{D}^{-1}\mathbf{H}', \qquad \mathbf{S} = \mathbf{H}\mathbf{H}',$$

where $\mathbf{H}$ is non-singular and $\mathbf{D} = (d_1, d_2, \ldots, d_p)_d$. Thus,

$$\begin{aligned}|\mathbf{\Sigma}|^{-\frac{1}{2}n}e^{-\frac{1}{2}\mathrm{tr}(\mathbf{\Sigma}^{-1}\mathbf{S})} &= |\mathbf{H}\mathbf{D}^{-1}\mathbf{H}'|^{-\frac{1}{2}n}e^{-\frac{1}{2}\mathrm{tr}\mathbf{D}} = |\mathbf{H}\mathbf{H}'|^{-\frac{1}{2}n}\prod_{i=1}^{p} d_i^{\frac{n}{2}}e^{-\frac{1}{2}d_i} \\ &\leq |\mathbf{H}\mathbf{H}'|^{-\frac{1}{2}n}n^{\frac{pn}{2}}e^{-\frac{1}{2}pn} = |\tfrac{1}{n}\mathbf{S}|^{-\frac{1}{2}n}e^{-\frac{1}{2}np}\end{aligned}$$

and equality holds if and only if $d_i = n$ which means that $n\mathbf{\Sigma} = \mathbf{S}$. ∎

Now we start with the first method of obtaining maximum likelihood estimators in the Growth Curve model. The approach is based on a direct use of the likelihood function. From the density of the matrix normal distribution, given in (2.2.7), we obtain that the likelihood function becomes

$$L(\mathbf{B}, \mathbf{\Sigma}) = (2\pi)^{-\frac{1}{2}pn}|\mathbf{\Sigma}|^{-\frac{1}{2}n}e^{-\frac{1}{2}\mathrm{tr}\{\mathbf{\Sigma}^{-1}(\mathbf{X}-\mathbf{ABC})(\mathbf{X}-\mathbf{ABC})'\}}. \qquad (4.1.2)$$

Remember that we sometimes write $(\mathbf{X} - \mathbf{ABC})()'$ instead of $(\mathbf{X} - \mathbf{ABC})(\mathbf{X} - \mathbf{ABC})'$. Lemma 4.1.1 yields

$$|\mathbf{\Sigma}|^{-\frac{1}{2}n}e^{-\frac{1}{2}\mathrm{tr}\{\mathbf{\Sigma}^{-1}(\mathbf{X}-\mathbf{ABC})()'\}} \leq |\tfrac{1}{n}(\mathbf{X} - \mathbf{ABC})()'|^{-\frac{1}{2}n}e^{-\frac{1}{2}np}, \qquad (4.1.3)$$

and equality holds if and only if $n\mathbf{\Sigma} = (\mathbf{X} - \mathbf{ABC})()'$. Observe that $\mathbf{\Sigma}$ has been estimated as a function of the mean and now the mean is estimated. In many models one often starts with an estimate of the mean and thereafter searches for an estimate of the dispersion (covariance) matrix. The aim is achieved if a lower bound of

$$|(\mathbf{X} - \mathbf{ABC})(\mathbf{X} - \mathbf{ABC})'|,$$

which is independent of $\mathbf{B}$, is found and if that bound can be obtained for at least one specific choice of $\mathbf{B}$.
In order to find a lower bound two ideas are used. First we split the product $(\mathbf{X} - \mathbf{ABC})(\mathbf{X} - \mathbf{ABC})'$ into two parts, i.e.

$$(\mathbf{X} - \mathbf{ABC})(\mathbf{X} - \mathbf{ABC})' = \mathbf{S} + \mathbf{VV}', \tag{4.1.4}$$

where

$$\mathbf{S} = \mathbf{X}(\mathbf{I} - \mathbf{C}'(\mathbf{CC}')^{-}\mathbf{C})\mathbf{X}' \tag{4.1.5}$$

and

$$\mathbf{V} = \mathbf{XC}'(\mathbf{CC}')^{-}\mathbf{C} - \mathbf{ABC}. \tag{4.1.6}$$

Note that $\mathbf{S}$ does not depend on the parameter $\mathbf{B}$. Furthermore, if $\mathbf{S}$ is re-scaled, i.e. $1/(n - r(\mathbf{C}))\mathbf{S}$, the matrix is often called sample dispersion matrix. However, under the Growth Curve model it is not appropriate to call the scaled $\mathbf{S}$ the sample dispersion matrix because there exist alternatives which are more relevant to use in this role, as it will be shown below. Proposition 1.1.1 (ii) and (iii) imply

$$|\mathbf{S} + \mathbf{VV}'| = |\mathbf{S}||\mathbf{I} + \mathbf{S}^{-1}\mathbf{VV}'| = |\mathbf{S}||\mathbf{I} + \mathbf{V}'\mathbf{S}^{-1}\mathbf{V}|.$$

Here it is assumed that $\mathbf{S}^{-1}$ exists, which is true with probability 1 (see Corollary 4.1.2.1L in §4.1.3).
The second idea is based on Corollary 1.2.25.1 which yields

$$\mathbf{S}^{-1} = \mathbf{S}^{-1}\mathbf{A}(\mathbf{A}'\mathbf{S}^{-1}\mathbf{A})^{-}\mathbf{A}'\mathbf{S}^{-1} + \mathbf{A}^{o}(\mathbf{A}^{o'}\mathbf{SA}^{o})^{-}\mathbf{A}^{o'}.$$

Therefore,

$$\begin{aligned}
|(\mathbf{X}{-}\mathbf{ABC})()'| &= |\mathbf{S}||\mathbf{I} + \mathbf{V}'\mathbf{S}^{-1}\mathbf{V}| \\
&= |\mathbf{S}||\mathbf{I} + \mathbf{V}'\mathbf{S}^{-1}\mathbf{A}(\mathbf{A}'\mathbf{S}^{-1}\mathbf{A})^{-}\mathbf{A}'\mathbf{S}^{-1}\mathbf{V} + \mathbf{V}'\mathbf{A}^{o}(\mathbf{A}^{o'}\mathbf{SA}^{o})^{-}\mathbf{A}^{o'}\mathbf{V}| \\
&\geq |\mathbf{S}||\mathbf{I} + \mathbf{V}'\mathbf{A}^{o}(\mathbf{A}^{o'}\mathbf{SA}^{o})^{-}\mathbf{A}^{o'}\mathbf{V}| \text{ (Proposition 1.1.5 (viii))},
\end{aligned}$$

which is independent of $\mathbf{B}$, since $\mathbf{A}^{o'}\mathbf{V} = \mathbf{A}^{o'}\mathbf{XC}'(\mathbf{CC}')^{-}\mathbf{C}$ is independent of $\mathbf{B}$. Equality holds if and only if

$$\mathbf{V}'\mathbf{S}^{-1}\mathbf{A}(\mathbf{A}'\mathbf{S}^{-1}\mathbf{A})^{-}\mathbf{A}'\mathbf{S}^{-1}\mathbf{V} = \mathbf{0},$$

which is equivalent to

$$\mathbf{V}'\mathbf{S}^{-1}\mathbf{A}(\mathbf{A}'\mathbf{S}^{-1}\mathbf{A})^{-} = \mathbf{0}.$$

This is a linear equation in $\mathbf{B}$, and from Theorem 1.3.4 it follows that the general solution is given by

$$\widehat{\mathbf{B}} = (\mathbf{A}'\mathbf{S}^{-1}\mathbf{A})^{-}\mathbf{A}'\mathbf{S}^{-1}\mathbf{XC}'(\mathbf{CC}')^{-} + (\mathbf{A}')^{o}\mathbf{Z}_1 + \mathbf{A}'\mathbf{Z}_2\mathbf{C}^{o'}, \tag{4.1.7}$$

where $\mathbf{Z}_1$ and $\mathbf{Z}_2$ are arbitrary matrices. If $\mathbf{A}$ and $\mathbf{C}$ are of full rank, i.e. $r(\mathbf{A}) = q$ and $r(\mathbf{C}) = k$, a unique solution exists:

$$\widehat{\mathbf{B}} = (\mathbf{A}'\mathbf{S}^{-1}\mathbf{A})^{-1}\mathbf{A}'\mathbf{S}^{-1}\mathbf{XC}'(\mathbf{CC}')^{-1}.$$

Furthermore, the maximum likelihood estimator of $\boldsymbol{\Sigma}$ is given by

$$n\widehat{\boldsymbol{\Sigma}} = (\mathbf{X} - \mathbf{A}\widehat{\mathbf{B}}\mathbf{C})(\mathbf{X} - \mathbf{A}\widehat{\mathbf{B}}\mathbf{C})' = \mathbf{S} + \widehat{\mathbf{V}}\widehat{\mathbf{V}}', \tag{4.1.8}$$

where $\widehat{\mathbf{V}}$ is the matrix $\mathbf{V}$ given by (4.1.6) with $\mathbf{B}$ replaced by $\widehat{\mathbf{B}}$, i.e.

$$\widehat{\mathbf{V}} = \mathbf{X}\mathbf{C}'(\mathbf{C}\mathbf{C}')^{-}\mathbf{C} - \mathbf{A}\widehat{\mathbf{B}}\mathbf{C}.$$

From (4.1.7) and Proposition 1.2.2 (x) it follows that

$$\mathbf{A}\widehat{\mathbf{B}}\mathbf{C} = \mathbf{A}(\mathbf{A}'\mathbf{S}^{-1}\mathbf{A})^{-}\mathbf{A}'\mathbf{S}^{-1}\mathbf{X}\mathbf{C}'(\mathbf{C}\mathbf{C}')^{-}\mathbf{C}$$

is always unique, i.e. the expression does not depend on the choices of g-inverses, and therefore $\widehat{\boldsymbol{\Sigma}}$ is also uniquely estimated. The conditions, under which the parameters in the mean structure or linear combinations of them can be uniquely estimated, will be studied in §4.1.5 in some detail. The above given results are summarized in the next theorem.

Theorem 4.1.1. *For the Growth Curve model (4.1.1) the maximum likelihood estimator of* $\mathbf{B}$ *is given by (4.1.7) and the unique estimator of* $\boldsymbol{\Sigma}$ *by (4.1.8).* ∎

The approach above can be modified in order to get more precise information about the covariance structure. By utilizing Corollary 1.2.39.1, the matrix normal density in the likelihood function can be written in the following way

$$L(\mathbf{B},\boldsymbol{\Sigma}) = c|\mathbf{D}|^{-\frac{n}{2}}e^{-\frac{1}{2}\mathrm{tr}\{\mathbf{D}^{-1}\boldsymbol{\Gamma}'(\mathbf{X}-\mathbf{A}\mathbf{B}\mathbf{C})(\mathbf{X}-\mathbf{A}\mathbf{B}\mathbf{C})'\boldsymbol{\Gamma}\}}, \quad c = (2\pi)^{-\frac{1}{2}pn}, \tag{4.1.9}$$

where $\mathbf{D}$ is the diagonal matrix which consists of the p positive eigenvalues of $\boldsymbol{\Sigma}$, $\boldsymbol{\Sigma}^{-1} = \boldsymbol{\Gamma}\mathbf{D}^{-1}\boldsymbol{\Gamma}'$, $\boldsymbol{\Gamma} : p \times p$, where $\boldsymbol{\Gamma}'\boldsymbol{\Gamma} = \mathbf{I}_p$.

As before we are going to maximize (4.1.9) with respect to $\mathbf{B}$ and $\boldsymbol{\Sigma}$, i.e. $\mathbf{B}$, $\mathbf{D}$ and $\boldsymbol{\Gamma}$. Since $\mathbf{D}$ is diagonal, (4.1.9) can be rewritten as

$$L(\mathbf{B},\boldsymbol{\Sigma}) = c|\mathbf{D}|^{-\frac{n}{2}}e^{-\frac{1}{2}\mathrm{tr}\{\mathbf{D}^{-1}\{\boldsymbol{\Gamma}'(\mathbf{X}-\mathbf{A}\mathbf{B}\mathbf{C})(\mathbf{X}-\mathbf{A}\mathbf{B}\mathbf{C})'\boldsymbol{\Gamma}\}_d\}}.$$

One can see that

$$L(\mathbf{B},\boldsymbol{\Sigma}) \leq c|\tfrac{1}{n}\{\boldsymbol{\Gamma}'(\mathbf{X} - \mathbf{A}\mathbf{B}\mathbf{C})(\mathbf{X} - \mathbf{A}\mathbf{B}\mathbf{C})'\boldsymbol{\Gamma}\}_d|^{-\frac{n}{2}}e^{-\frac{1}{2}pn}, \tag{4.1.10}$$

where equality holds if and only if

$$n\mathbf{D} = \{\boldsymbol{\Gamma}'(\mathbf{X} - \mathbf{A}\mathbf{B}\mathbf{C})(\mathbf{X} - \mathbf{A}\mathbf{B}\mathbf{C})'\boldsymbol{\Gamma}\}_d.$$

From (4.1.10) it follows that we have to examine the determinant, and first we observe that

$$|\{\boldsymbol{\Gamma}'(\mathbf{X} - \mathbf{A}\mathbf{B}\mathbf{C})(\mathbf{X} - \mathbf{A}\mathbf{B}\mathbf{C})'\boldsymbol{\Gamma}\}_d| \geq |\boldsymbol{\Gamma}'(\mathbf{X} - \mathbf{A}\mathbf{B}\mathbf{C})(\mathbf{X} - \mathbf{A}\mathbf{B}\mathbf{C})'\boldsymbol{\Gamma}|,$$

since $\mathbf{\Gamma}$ is non-singular. Equality holds if $\mathbf{\Gamma}$ is a matrix of unit length eigenvectors of $(\mathbf{X}-\mathbf{ABC})(\mathbf{X}-\mathbf{ABC})'$. The inequality is a special case of Hadamard's inequality (see Problem 14 in §1.1.7). This approach will be used in the next paragraph when considering models with a singular dispersion matrix.

Although two ways of finding the maximum likelihood estimators have been presented, we will continue with two more alternative approaches. The third approach is based on differentiation of the likelihood function. The obtained likelihood equations are then solved. In the fourth approach, which is completely different, the model is rewritten in a reparameterized form.

When differentiating the likelihood function in (4.1.2) with respect to $\mathbf{B}$, it follows, by Proposition 1.4.9 (iii), that the likelihood equation is

$$\mathbf{A}'\mathbf{\Sigma}^{-1}(\mathbf{X}-\mathbf{ABC})\mathbf{C}' = \mathbf{0}. \tag{4.1.11}$$

Moreover, (4.1.2) will be differentiated with respect to $(\mathbf{\Sigma}(K))^{-1}$, i.e. the $\frac{1}{2}p(p+1)$ different elements of $\mathbf{\Sigma}^{-1}$. Hence, from (4.1.2) it follows that the derivatives

$$\frac{d|\mathbf{\Sigma}^{-1}|^{1/2}}{d\,\mathbf{\Sigma}(K)^{-1}}, \qquad \frac{d\,\mathrm{tr}\{\mathbf{\Sigma}^{-1}(\mathbf{X}-\mathbf{ABC})(\mathbf{X}-\mathbf{ABC})'\}}{d\,\mathbf{\Sigma}(K)^{-1}}$$

have to be found. Using the notation of §1.3.6, the equality (1.3.32) yields

$$\begin{aligned}\frac{d\,\mathrm{tr}\{\mathbf{\Sigma}^{-1}(\mathbf{X}-\mathbf{ABC})(\mathbf{X}-\mathbf{ABC})'\}}{d\,\mathbf{\Sigma}(K)^{-1}} &= \frac{d\,\mathbf{\Sigma}^{-1}}{d\,\mathbf{\Sigma}(K)^{-1}}\mathrm{vec}\{(\mathbf{X}-\mathbf{ABC})(\mathbf{X}-\mathbf{ABC})'\}\\ &= (\mathbf{T}^{+}(s))'\mathrm{vec}\{(\mathbf{X}-\mathbf{ABC})(\mathbf{X}-\mathbf{ABC})'\}.\end{aligned} \tag{4.1.12}$$

From the proof of (1.4.31) it follows that

$$\begin{aligned}\frac{d|\mathbf{\Sigma}^{-1}|^{\frac{1}{2}n}}{d\,\mathbf{\Sigma}(K)^{-1}} &= \frac{d|\mathbf{\Sigma}^{-1}|^{1/2}}{d\mathbf{\Sigma}(K)^{-1}}\frac{d|\mathbf{\Sigma}^{-1}|^{\frac{1}{2}n}}{d|\mathbf{\Sigma}^{-1}|^{1/2}} = n|\mathbf{\Sigma}^{-1}|^{\frac{1}{2}(n-1)}\frac{d|\mathbf{\Sigma}^{-1}|^{1/2}}{d\,\mathbf{\Sigma}^{-1}(K)}\\ &= n|\mathbf{\Sigma}^{-1}|^{\frac{1}{2}(n-1)}\frac{1}{2}\frac{d\,\mathbf{\Sigma}^{-1}}{d\,\mathbf{\Sigma}^{-1}(K)}|\mathbf{\Sigma}^{-1}|^{\frac{1}{2}}\mathrm{vec}\mathbf{\Sigma}\\ &= \frac{n}{2}|\mathbf{\Sigma}^{-1}|^{\frac{n}{2}}(\mathbf{T}^{+}(s))'\mathrm{vec}\mathbf{\Sigma}.\end{aligned} \tag{4.1.13}$$

Thus, differentiating the likelihood function (4.1.2), with the use of (4.1.12) and (4.1.13), leads to the relation

$$\begin{aligned}&\frac{d\,L(\mathbf{B},\mathbf{\Sigma})}{d\,\mathbf{\Sigma}(K)^{-1}}\\ &= \Big\{\frac{n}{2}(\mathbf{T}^{+}(s))'\mathrm{vec}\mathbf{\Sigma} - \frac{1}{2}(\mathbf{T}^{+}(s))'\mathrm{vec}((\mathbf{X}-\mathbf{ABC})(\mathbf{X}-\mathbf{ABC})')\Big\}L(\mathbf{B},\mathbf{\Sigma}).\end{aligned} \tag{4.1.14}$$

Since for any symmetric $\mathbf{W}$ the equation $(\mathbf{T}^{+}(s))'\mathrm{vec}\mathbf{W} = \mathbf{0}$ is equivalent to $\mathrm{vec}\mathbf{W} = \mathbf{0}$, we obtain from (4.1.14) the following equality.

$$n\mathbf{\Sigma} = (\mathbf{X}-\mathbf{ABC})(\mathbf{X}-\mathbf{ABC})'. \tag{4.1.15}$$

Observe that we have differentiated with respect to the concentration matrix $\mathbf{\Sigma}^{-1}$, which is, in some sense, a more natural parameter than $\mathbf{\Sigma}$ as the log-likelihood function, besides the determinant, is a linear function in $\mathbf{\Sigma}^{-1}$. Now (4.1.11) and (4.1.15) are equivalent to

$$\mathbf{A}'\mathbf{\Sigma}^{-1}\mathbf{V}\mathbf{C}' = \mathbf{0}, \tag{4.1.16}$$

$$n\mathbf{\Sigma} = \mathbf{S} + \mathbf{V}\mathbf{V}', \tag{4.1.17}$$

where $\mathbf{S}$ and $\mathbf{V}$ are given by (4.1.5) and (4.1.6), respectively. From Proposition 1.3.6 it follows that instead of (4.1.16),

$$\mathbf{A}'\mathbf{S}^{-1}\mathbf{V}\mathbf{M}\mathbf{C}' = \mathbf{0} \tag{4.1.18}$$

holds, where

$$\mathbf{M} = (\mathbf{V}'\mathbf{S}^{-1}\mathbf{V} + \mathbf{I})^{-1}.$$

Thus, (4.1.18) is independent of $\mathbf{\Sigma}$, but we have now a system of non-linear equations. However,

$$\mathbf{V}\mathbf{M}\mathbf{C}' = (\mathbf{X}\mathbf{C}'(\mathbf{C}\mathbf{C}')^{-} - \mathbf{A}\mathbf{B})\mathbf{C}\mathbf{M}\mathbf{C}',$$

and since $\mathcal{C}(\mathbf{C}\mathbf{M}\mathbf{C}') = \mathcal{C}(\mathbf{C})$ it follows that (4.1.18) is equivalent to

$$\mathbf{A}'\mathbf{S}^{-1}\mathbf{V} = \mathbf{0},$$

which is a linear equation in $\mathbf{B}$. Hence, by applying Theorem 1.3.4, an explicit solution for $\mathbf{B}$ is obtained via (4.1.17) that leads to the maximum likelihood estimator of $\mathbf{\Sigma}$. The solutions were given in (4.1.7) and (4.1.8). However, one should observe that a solution of the likelihood equations does not necessarily lead to the maximum likelihood estimate, because local maxima can appear or maximum can be obtained on the boundary of the parameter space. Therefore, after solving the likelihood equations, one has to do some additional work in order to guarantee that the likelihood estimators have been obtained. From this point of view it is advantageous to work with the likelihood function directly. On the other hand, it may be simpler to solve the likelihood equations and the obtained results can then be directing guidelines for further studies.

The likelihood approach is always invariant under non-singular transformations. In the fourth approach a transformation is applied, so that the within-individual model is presented in a canonical form. Without loss of generality it is assumed that $r(\mathbf{A}) = q$ and $r(\mathbf{C}) = k$. These assumptions can be made because otherwise it follows from Proposition 1.1.6 that $\mathbf{A} = \mathbf{A}_1\mathbf{A}_2$ and $\mathbf{C}' = \mathbf{C}_1'\mathbf{C}_2'$ where $\mathbf{A}_1 : p \times s$, $r(\mathbf{A}_1) = s$, $\mathbf{A}_2 : s \times q$, $r(\mathbf{A}_2) = s$, $\mathbf{C}_1' : n \times t$, $r(\mathbf{C}_1) = t$ and $\mathbf{C}_2' : t \times k$, $r(\mathbf{C}_2) = t$. Then the mean of the Growth Curve model equals

$$E[\mathbf{X}] = \mathbf{A}_1\mathbf{A}_2\mathbf{B}\mathbf{C}_2\mathbf{C}_1 = \mathbf{A}_1\mathbf{\Theta}\mathbf{C}_1,$$

where $\mathbf{A}_1$ and $\mathbf{C}_1$ are of full rank. This reparameterization is, of course, not one-to-one. However, $\mathbf{B}$ can never be uniquely estimated, whereas $\mathbf{\Theta}$ is always estimated uniquely. Therefore we use $\mathbf{\Theta}$ and obtain uniquely estimated linear combinations of $\mathbf{B}$. The idea is to construct a non-singular transformation matrix which is independent of the parameters. Let

$$\mathbf{Q}' = (\mathbf{A}(\mathbf{A}'\mathbf{A})^{-1} : \mathbf{A}^o), \tag{4.1.19}$$

where $\mathbf{Q} \in \mathbb{R}^{p\times p}$ and $\mathbf{A}^o \in \mathbb{R}^{p\times(p-q)}$, which, as previously, satisfies $\mathcal{C}(\mathbf{A}^o) = \mathcal{C}(\mathbf{A})^\perp$. Hence, $\mathbf{Q}$ is non-singular and a new transformed model equals

$$\mathbf{QX} = \begin{pmatrix} \mathbf{I} \\ \mathbf{0} \end{pmatrix} \mathbf{BC} + \mathbf{Q}\mathbf{\Sigma}^{1/2}\mathbf{E}.$$

The corresponding likelihood function is given by

$$L(\mathbf{B}, \mathbf{\Sigma}) =(2\pi)^{-\frac{1}{2}pn}|\mathbf{Q}\mathbf{\Sigma}\mathbf{Q}'|^{-\frac{1}{2}n}\exp\{-\tfrac{1}{2}\mathrm{tr}\{(\mathbf{Q}\mathbf{\Sigma}\mathbf{Q}')^{-1}(\mathbf{QX} - \begin{pmatrix} \mathbf{I} \\ \mathbf{0} \end{pmatrix} \mathbf{BC})()'\}\}. \tag{4.1.20}$$

Define $\mathbf{Y}$ and $\mathbf{\Psi}$ through the following relations:

$$\begin{aligned} \mathbf{QX} =& \mathbf{Y} = (\mathbf{Y}_1' : \mathbf{Y}_2')', \\ \mathbf{Q}\mathbf{\Sigma}\mathbf{Q}' =& \begin{pmatrix} \mathbf{\Psi}_{11} & \mathbf{\Psi}_{12} \\ \mathbf{\Psi}_{21} & \mathbf{\Psi}_{22} \end{pmatrix} \end{aligned}$$

and

$$\mathbf{\Psi}_{1\cdot2} = \mathbf{\Psi}_{11} - \mathbf{\Psi}_{12}\mathbf{\Psi}_{22}^{-1}\mathbf{\Psi}_{21}.$$

By Proposition 1.3.2 and Proposition 1.3.3, the equality (4.1.20) is identical to

$$L(\mathbf{B}, \mathbf{\Sigma}) = (2\pi)^{-\frac{1}{2}pn}|\mathbf{\Psi}_{22}|^{-\frac{1}{2}n}|\mathbf{\Psi}_{1\cdot2}|^{-\frac{1}{2}n}e^{-\frac{1}{2}\mathrm{tr}\{\mathbf{\Psi}_{1\cdot2}^{-1}(\mathbf{Y}_1 - \mathbf{BC} - \mathbf{\Theta}\mathbf{Y}_2)()' + \mathbf{\Psi}_{22}^{-1}\mathbf{Y}_2\mathbf{Y}_2'\}}, \tag{4.1.21}$$

where $\mathbf{\Theta} = \mathbf{\Psi}_{12}\mathbf{\Psi}_{22}^{-1}$. However, there is an one-to-one correspondence between the parameter space $(\mathbf{\Sigma}, \mathbf{B})$ and the space $(\mathbf{\Psi}_{1\cdot2}, \mathbf{\Psi}_{22}, \mathbf{\Theta}, \mathbf{B})$. Thus, the parameters $(\mathbf{\Psi}_{1\cdot2}, \mathbf{\Psi}_{22}, \mathbf{\Theta}, \mathbf{B})$ may be considered, and from the likelihood function (4.1.21) it follows that the maximum likelihood estimators are obtained via ordinary multivariate regression analysis. One may view this approach as a conditioning procedure, i.e.

$$L_{\mathbf{Y}}(\mathbf{B}, \mathbf{\Psi}) = L_{\mathbf{Y}_2}(\mathbf{\Psi}_{22})L_{\mathbf{Y}_1|\mathbf{Y}_2}(\mathbf{B}, \mathbf{\Theta}, \mathbf{\Psi}_{1\cdot2})$$

in usual notation. Thus,

$$\begin{aligned} n\widehat{\mathbf{\Psi}}_{22} =& \mathbf{Y}_2\mathbf{Y}_2', \\ (\widehat{\mathbf{B}}, \widehat{\mathbf{\Theta}}) =& \mathbf{Y}_1(\mathbf{C}' : \mathbf{Y}_2') \begin{pmatrix} \mathbf{CC}' & \mathbf{CY}_2' \\ \mathbf{Y}_2\mathbf{C}' & \mathbf{Y}_2\mathbf{Y}_2' \end{pmatrix}^{-1}, \\ n\widehat{\mathbf{\Psi}}_{1\cdot2} =& (\mathbf{Y}_1 - \widehat{\mathbf{B}}\mathbf{C} - \widehat{\mathbf{\Theta}}\mathbf{Y}_2)(\mathbf{Y}_1 - \widehat{\mathbf{B}}\mathbf{C} - \widehat{\mathbf{\Theta}}\mathbf{Y}_2)'. \end{aligned}$$

It remains to convert these results into expressions which comprise the original matrices. Put

$$\mathbf{P}_{\mathbf{C}'} = \mathbf{C}'(\mathbf{CC}')^{-1}\mathbf{C}$$

and

$$
\begin{aligned}
\mathbf{N} =&\mathbf{Y}_1(\mathbf{I}-\mathbf{P}_{\mathbf{C}'})\mathbf{Y}_1' - \mathbf{Y}_1(\mathbf{I}-\mathbf{P}_{\mathbf{C}'})\mathbf{Y}_2'\{\mathbf{Y}_2(\mathbf{I}-\mathbf{P}_{\mathbf{C}'})\mathbf{Y}_2'\}^{-1}\mathbf{Y}_2(\mathbf{I}-\mathbf{P}_{\mathbf{C}'})\mathbf{Y}_1'\\
=&\left((\mathbf{I}:\mathbf{0})\left\{\begin{pmatrix}\mathbf{Y}_1\\ \mathbf{Y}_2\end{pmatrix}(\mathbf{I}-\mathbf{P}_{\mathbf{C}'})(\mathbf{Y}_1':\mathbf{Y}_2')\right\}^{-1}\begin{pmatrix}\mathbf{I}\\ \mathbf{0}\end{pmatrix}\right)^{-1}\\
=&(\mathbf{A}'\mathbf{Q}'(\mathbf{Q}\mathbf{S}\mathbf{Q}')^{-1}\mathbf{Q}\mathbf{A})^{-1} = (\mathbf{A}'\mathbf{S}^{-1}\mathbf{A})^{-1},
\end{aligned}
$$

where $\mathbf{Q}$ is defined in (4.1.19). From Proposition 1.3.3, Proposition 1.3.4 and Proposition 1.3.6 it follows that

$$
\begin{aligned}
\widehat{\mathbf{B}} =&\mathbf{Y}_1\mathbf{C}'\{\mathbf{C}\mathbf{C}' - \mathbf{C}\mathbf{Y}_2'(\mathbf{Y}_2\mathbf{Y}_2')^{-1}\mathbf{Y}_2\mathbf{C}'\}^{-1}\\
&\qquad - \mathbf{Y}_1\mathbf{Y}_2'\{\mathbf{Y}_2(\mathbf{I}-\mathbf{P}_{\mathbf{C}'})\mathbf{Y}_2'\}^{-1}\mathbf{Y}_2\mathbf{C}'(\mathbf{C}\mathbf{C}')^{-1}\\
=&\mathbf{Y}_1\mathbf{C}'(\mathbf{C}\mathbf{C}')^{-1} - \mathbf{Y}_1(\mathbf{I}-\mathbf{P}_{\mathbf{C}'})\mathbf{Y}_2'(\mathbf{Y}_2(\mathbf{I}-\mathbf{P}_{\mathbf{C}'})\mathbf{Y}_2')^{-1}\mathbf{Y}_2\mathbf{C}'(\mathbf{C}\mathbf{C}')^{-1}\\
=&\mathbf{N}\big(\mathbf{N}^{-1} : -\mathbf{N}^{-1}\mathbf{Y}_1(\mathbf{I}-\mathbf{P}_{\mathbf{C}'})\mathbf{Y}_2'\{\mathbf{Y}_2(\mathbf{I}-\mathbf{P}_{\mathbf{C}'})\mathbf{Y}_2'\}^{-1}\big)\begin{pmatrix}\mathbf{Y}_1\\ \mathbf{Y}_2\end{pmatrix}\mathbf{C}'(\mathbf{C}\mathbf{C}')^{-1}\\
=&(\mathbf{A}'\mathbf{S}^{-1}\mathbf{A})^{-1}\mathbf{A}'\mathbf{Q}'(\mathbf{Q}\mathbf{S}\mathbf{Q}')^{-1}\mathbf{Q}\mathbf{X}\mathbf{C}'(\mathbf{C}\mathbf{C}')^{-1}\\
=&(\mathbf{A}'\mathbf{S}^{-1}\mathbf{A})^{-1}\mathbf{A}'\mathbf{S}^{-1}\mathbf{X}\mathbf{C}'(\mathbf{C}\mathbf{C}')^{-1},
\end{aligned}
$$

since $\mathbf{Q}$ is non-singular and $\mathbf{A}'\mathbf{Q}' = (\mathbf{I} : \mathbf{0})$. Hence, the maximum likelihood estimator of $\mathbf{B}$ in the Growth Curve model has been obtained. Furthermore,

$$
\widehat{\boldsymbol{\Theta}} = \mathbf{Y}_1(\mathbf{I}-\mathbf{P}_{\mathbf{C}'})\mathbf{Y}_2'\{\mathbf{Y}_2(\mathbf{I}-\mathbf{P}_{\mathbf{C}'})\mathbf{Y}_2'\}^{-1}
$$

and

$$
\begin{aligned}
\mathbf{Y}_1-&(\widehat{\mathbf{B}} : \widehat{\boldsymbol{\Theta}})(\mathbf{C}' : \mathbf{Y}_2')'\\
&= \mathbf{Y}_1(\mathbf{I}-\mathbf{P}_{\mathbf{C}'}) - \mathbf{Y}_1(\mathbf{I}-\mathbf{P}_{\mathbf{C}'})\mathbf{Y}_2'\{\mathbf{Y}_2(\mathbf{I}-\mathbf{P}_{\mathbf{C}'})\mathbf{Y}_2'\}^{-1}\mathbf{Y}_2(\mathbf{I}-\mathbf{P}_{\mathbf{C}'}),
\end{aligned}
$$

which yields

$$
\begin{aligned}
n\widehat{\boldsymbol{\Psi}}_{1\cdot 2} =&\mathbf{Y}_1(\mathbf{I}-\mathbf{P}_{\mathbf{C}'})\mathbf{Y}_1' - \mathbf{Y}_1(\mathbf{I}-\mathbf{P}_{\mathbf{C}'})\mathbf{Y}_2'\{\mathbf{Y}_2(\mathbf{I}-\mathbf{P}_{\mathbf{C}'})\mathbf{Y}_2'\}^{-1}\mathbf{Y}_2(\mathbf{I}-\mathbf{P}_{\mathbf{C}'})\mathbf{Y}_1',\\
n\widehat{\boldsymbol{\Psi}}_{12} =&\mathbf{Y}_1(\mathbf{I}-\mathbf{P}_{\mathbf{C}'})\mathbf{Y}_2'\{\mathbf{Y}_2(\mathbf{I}-\mathbf{P}_{\mathbf{C}'})\mathbf{Y}_2'\}^{-1}\mathbf{Y}_2\mathbf{Y}_2'.
\end{aligned}
$$

Since $\boldsymbol{\Psi}_{11} = \boldsymbol{\Psi}_{1\cdot 2} + \boldsymbol{\Psi}_{12}\boldsymbol{\Psi}_{22}^{-1}\boldsymbol{\Psi}_{21}$, we have

$$
\begin{aligned}
n\widehat{\boldsymbol{\Psi}}_{11} =&\mathbf{Y}_1(\mathbf{I}-\mathbf{P}_{\mathbf{C}'})\mathbf{Y}_1' + \mathbf{Y}_1(\mathbf{I}-\mathbf{P}_{\mathbf{C}'})\mathbf{Y}_2'\{\mathbf{Y}_2(\mathbf{I}-\mathbf{P}_{\mathbf{C}'})\mathbf{Y}_2'\}^{-1}\mathbf{Y}_2\mathbf{P}_{\mathbf{C}'}\\
&\times \mathbf{Y}_2'\{\mathbf{Y}_2(\mathbf{I}-\mathbf{P}_{\mathbf{C}'})\mathbf{Y}_2'\}^{-1}\mathbf{Y}_2(\mathbf{I}-\mathbf{P}_{\mathbf{C}'})\mathbf{Y}_1'.
\end{aligned}
$$

Thus,

$$
\begin{aligned}
n\begin{pmatrix}\widehat{\boldsymbol{\Psi}}_{11} & \widehat{\boldsymbol{\Psi}}_{12}\\ \widehat{\boldsymbol{\Psi}}_{21} & \widehat{\boldsymbol{\Psi}}_{22}\end{pmatrix} =& \begin{pmatrix}\mathbf{Y}_1\\ \mathbf{Y}_2\end{pmatrix}\Big(\mathbf{I}-\mathbf{P}_{\mathbf{C}'} + (\mathbf{I}-\mathbf{P}_{\mathbf{C}'})\mathbf{Y}_2'\{\mathbf{Y}_2(\mathbf{I}-\mathbf{P}_{\mathbf{C}'})\mathbf{Y}_2'\}^{-1}\\
&\times \mathbf{Y}_2\mathbf{P}_{\mathbf{C}'}\mathbf{Y}_2'\{\mathbf{Y}_2(\mathbf{I}-\mathbf{P}_{\mathbf{C}'})\mathbf{Y}_2'\}^{-1}\mathbf{Y}_2(\mathbf{I}-\mathbf{P}_{\mathbf{C}'})\Big)(\mathbf{Y}_1' : \mathbf{Y}_2').
\end{aligned}
$$

By definition of $\mathbf{Q}$ and $\mathbf{Y}_2$,

$$\mathbf{Y}_2 = (\mathbf{0} : \mathbf{I})\mathbf{Q}\mathbf{X} = \mathbf{A}^{o'}\mathbf{X},$$

which implies that

$$\begin{aligned} n\begin{pmatrix} \widehat{\mathbf{\Psi}}_{11} & \widehat{\mathbf{\Psi}}_{12} \\ \widehat{\mathbf{\Psi}}_{21} & \widehat{\mathbf{\Psi}}_{22} \end{pmatrix} & \\ = n\mathbf{Q}\mathbf{X}\big(\mathbf{I} - \mathbf{P}_{\mathbf{C}'} &+ (\mathbf{I} - \mathbf{P}_{\mathbf{C}'})\mathbf{X}'\mathbf{A}^o\{\mathbf{A}^{o'}\mathbf{X}(\mathbf{I} - \mathbf{P}_{\mathbf{C}'})\mathbf{X}'\mathbf{A}^o\}^{-1} \\ &\times \mathbf{A}^{o'}\mathbf{X}\mathbf{P}_{\mathbf{C}'}\mathbf{X}'\mathbf{A}^o\{\mathbf{A}^{o'}\mathbf{X}(\mathbf{I} - \mathbf{P}_{\mathbf{C}'})\mathbf{X}'\mathbf{A}^o\}^{-1}\mathbf{A}^{o'}\mathbf{X}(\mathbf{I} - \mathbf{P}_{\mathbf{C}'})\big)\mathbf{X}'\mathbf{Q}' \\ = n\mathbf{Q}\mathbf{S}\mathbf{Q}' &+ \mathbf{Q}\mathbf{S}\mathbf{A}^o(\mathbf{A}^{o'}\mathbf{S}\mathbf{A}^o)^{-1}\mathbf{A}^{o'}\mathbf{X}\mathbf{P}_{\mathbf{C}'}\mathbf{X}'\mathbf{A}^o(\mathbf{A}^{o'}\mathbf{S}\mathbf{A}^o)^{-1}\mathbf{A}^{o'}\mathbf{S}\mathbf{Q}' \\ = \mathbf{Q}(\mathbf{S} &+ \widehat{\mathbf{V}}\widehat{\mathbf{V}}')\mathbf{Q}' = n\mathbf{Q}\widehat{\mathbf{\Sigma}}\mathbf{Q}', \end{aligned}$$

where $\mathbf{S}$, $\mathbf{V}$ and $\widehat{\mathbf{\Sigma}}$ are given by (4.1.5), (4.1.6) and (4.1.8), respectively, and $\widehat{\mathbf{V}}$ is as in (4.1.8). Since $\mathbf{Q}$ is non-singular, again the maximum likelihood estimator of $\mathbf{\Sigma}$ has been derived.

A modified version of the last approach is obtained when we utilize Proposition 1.1.6 (ii), which implies that if $\mathbf{A} : p \times q$ is of full rank, i.e. $r(\mathbf{A}) = q$,

$$\mathbf{A}' = \mathbf{T}(\mathbf{I}_q : \mathbf{0})\mathbf{\Gamma},$$

where $\mathbf{T}$ is non-singular and $\mathbf{\Gamma}$ is orthogonal. Then, instead of (4.1.20), it can be established that

$$\begin{aligned} L(\mathbf{B}, \mathbf{\Sigma}) &= (2\pi)^{-\frac{1}{2}pn}|\mathbf{\Sigma}|^{-\frac{1}{2}n}e^{-\frac{1}{2}\mathrm{tr}\{\mathbf{\Sigma}^{-1}(\mathbf{X} - \mathbf{\Gamma}'(\mathbf{I}_q:\mathbf{0})'\mathbf{T}'\mathbf{B}\mathbf{C})()'\}} \\ &= (2\pi)^{-\frac{1}{2}pn}|\mathbf{\Gamma}\mathbf{\Sigma}\mathbf{\Gamma}'|^{-\frac{1}{2}n}e^{-\frac{1}{2}\mathrm{tr}\{\mathbf{\Gamma}\mathbf{\Sigma}^{-1}\mathbf{\Gamma}'(\mathbf{\Gamma}\mathbf{X} - (\mathbf{I}_q:\mathbf{0})'\mathbf{T}'\mathbf{B}\mathbf{C})()'\}}. \end{aligned}$$

Put $\mathbf{\Theta} = \mathbf{T}'\mathbf{B}$ and the likelihood function has the same structure as in (4.1.20).

4.1.3 *The Growth Curve model with a singular dispersion matrix*

Univariate linear models with correlated observations and singular dispersion matrix have been extensively studied: Mitra & Rao (1968), Khatri (1968), Zyskind & Martin (1969), Rao (1973b), Alalouf (1978), Pollock (1979), Feuerverger & Fraser (1980) and Nordström (1985) are some examples. Here the dispersion matrix is supposed to be known. However, very few results have been presented for multivariate linear models with an unknown dispersion matrix. In this paragraph the Growth Curve model with a singular dispersion matrix is studied. For some alternative approaches see Wong & Cheng (2001) and Srivastava & von Rosen (2002). In order to handle the Growth Curve model with a singular dispersion matrix, we start by performing an one-to-one transformation. It should, however, be noted that the transformation consists of unknown parameters and can not be given explicitly at this stage. This means also that we cannot guarantee that we do not lose some information when estimating the parameters, even if the transformation is one-to-one.

Suppose $r(\boldsymbol{\Sigma}) = r < p$. From (4.1.1) it follows that we should consider

$$\begin{pmatrix} \boldsymbol{\Gamma}' \\ \boldsymbol{\Gamma}^{o'} \end{pmatrix} \mathbf{X} = \begin{pmatrix} \boldsymbol{\Gamma}' \\ \boldsymbol{\Gamma}^{o'} \end{pmatrix} \mathbf{ABC} + \begin{pmatrix} \boldsymbol{\Gamma}'\boldsymbol{\Sigma}^{\frac{1}{2}}\mathbf{E} \\ \mathbf{0} \end{pmatrix}, \tag{4.1.22}$$

where $(\boldsymbol{\Gamma} : \boldsymbol{\Gamma}^o)$ spans the whole space, $\boldsymbol{\Gamma}'\boldsymbol{\Gamma}^o = \mathbf{0}$, $\boldsymbol{\Gamma}^{o'}\boldsymbol{\Sigma} = \mathbf{0}$, $\boldsymbol{\Gamma}'\boldsymbol{\Gamma} = \mathbf{I}_r$, $\boldsymbol{\Gamma} : p \times r$, $\boldsymbol{\Sigma} = \boldsymbol{\Gamma}\boldsymbol{\Lambda}\boldsymbol{\Gamma}'$, where $\boldsymbol{\Lambda}$: $r \times r$ is a diagonal matrix of positive eigenvalues of $\boldsymbol{\Sigma}$ and $\boldsymbol{\Gamma}^o : p \times (p - r)$. It follows that with probability 1

$$\boldsymbol{\Gamma}^{o'}\mathbf{X} = \boldsymbol{\Gamma}^{o'}\mathbf{ABC}. \tag{4.1.23}$$

Thus, by treating (4.1.23) as a linear equation in $\mathbf{B}$, it follows from Theorem 1.3.4 that, with probability 1,

$$\mathbf{B} = (\boldsymbol{\Gamma}^{o'}\mathbf{A})^{-}\boldsymbol{\Gamma}^{o'}\mathbf{XC}^{-} + (\mathbf{A}'\boldsymbol{\Gamma}^o)^o\boldsymbol{\Theta} + \mathbf{A}'\boldsymbol{\Gamma}^o\boldsymbol{\Theta}_2\mathbf{C}^{o'}, \tag{4.1.24}$$

where $\boldsymbol{\Theta}$ and $\boldsymbol{\Theta}_2$ are interpreted as new parameters. Inserting (4.1.24) into (4.1.22) gives

$$\boldsymbol{\Gamma}'\mathbf{X} = \boldsymbol{\Gamma}'\mathbf{A}(\boldsymbol{\Gamma}^{o'}\mathbf{A})^{-}\boldsymbol{\Gamma}^{o'}\mathbf{X} + \boldsymbol{\Gamma}'\mathbf{A}(\mathbf{A}'\boldsymbol{\Gamma}^o)^o\boldsymbol{\Theta}\mathbf{C} + \boldsymbol{\Gamma}'\boldsymbol{\Sigma}^{1/2}\mathbf{E}. \tag{4.1.25}$$

Since by Theorem 1.2.26 (ii) $\mathcal{C}(\mathbf{A}(\mathbf{A}'\boldsymbol{\Gamma}^o)^o) = \mathcal{C}(\boldsymbol{\Gamma}(\boldsymbol{\Gamma}'\mathbf{A}^o)^o)$, and $\boldsymbol{\Gamma}'\boldsymbol{\Gamma} = \mathbf{I}$ we may consider $(\boldsymbol{\Gamma}'\mathbf{A}^o)^o\boldsymbol{\Theta}_1$ instead of $\boldsymbol{\Gamma}'\mathbf{A}(\mathbf{A}'\boldsymbol{\Gamma}^o)^o\boldsymbol{\Theta}$. Furthermore, let

$$\mathbf{F} = (\boldsymbol{\Gamma}'\mathbf{A}^o)^o, \quad \mathbf{Z} = (\mathbf{I} - \mathbf{A}(\boldsymbol{\Gamma}^{o'}\mathbf{A})^{-}\boldsymbol{\Gamma}^{o'})\mathbf{X}.$$

Assume that $\mathcal{C}(\boldsymbol{\Gamma}) \cap \mathcal{C}(\mathbf{A}) \neq \{\mathbf{0}\}$, which implies that $\mathbf{F}$ differs from $\mathbf{0}$. The case $\mathbf{F} = \mathbf{0}$ will be treated later. Hence, utilizing a reparameterization, the model in (4.1.1) can be written as

$$\boldsymbol{\Gamma}'\mathbf{Z} = \mathbf{F}\boldsymbol{\Theta}_1\mathbf{C} + \boldsymbol{\Gamma}'\boldsymbol{\Sigma}^{1/2}\mathbf{E} \tag{4.1.26}$$

with the corresponding likelihood function

$$L(\boldsymbol{\Gamma}, \boldsymbol{\Lambda}, \boldsymbol{\Theta}_1) = c|\boldsymbol{\Gamma}'\boldsymbol{\Sigma}\boldsymbol{\Gamma}|^{-\frac{1}{2}n} e^{\mathrm{tr}\{-\frac{1}{2}(\boldsymbol{\Gamma}'\boldsymbol{\Sigma}\boldsymbol{\Gamma})^{-1}(\boldsymbol{\Gamma}'\mathbf{Z}-\mathbf{F}\boldsymbol{\Theta}_1\mathbf{C})(\boldsymbol{\Gamma}'\mathbf{Z}-\mathbf{F}\boldsymbol{\Theta}_1\mathbf{C})'\}} \tag{4.1.27}$$

where $c = (2\pi)^{-\frac{1}{2}rn}$.

First we will consider the parameters which build up the dispersion matrix $\boldsymbol{\Sigma}$. Since $\boldsymbol{\Lambda} = \boldsymbol{\Gamma}'\boldsymbol{\Sigma}\boldsymbol{\Gamma}$ is diagonal, (4.1.27) is identical to

$$L(\boldsymbol{\Gamma}, \boldsymbol{\Lambda}, \boldsymbol{\Theta}_1) = c|\boldsymbol{\Lambda}|^{-\frac{1}{2}n} e^{\mathrm{tr}\{-\frac{1}{2}\boldsymbol{\Lambda}^{-1}\{(\boldsymbol{\Gamma}'\mathbf{Z}-\mathbf{F}\boldsymbol{\Theta}_1\mathbf{C})(\boldsymbol{\Gamma}'\mathbf{Z}-\mathbf{F}\boldsymbol{\Theta}_1\mathbf{C})'\}_d\}}. \tag{4.1.28}$$

According to the proof of Lemma 4.1.1, the likelihood function in (4.1.28) satisfies

$$L(\boldsymbol{\Gamma}, \boldsymbol{\Lambda}, \boldsymbol{\Theta}_1) \leq c|\frac{1}{n}\{(\boldsymbol{\Gamma}'\mathbf{Z} - \mathbf{F}\boldsymbol{\Theta}_1\mathbf{C})(\boldsymbol{\Gamma}'\mathbf{Z} - \mathbf{F}\boldsymbol{\Theta}_1\mathbf{C})'\}_d|^{-n/2} e^{-\frac{1}{2}rn} \tag{4.1.29}$$

and equality holds if and only if

$$n\mathbf{\Lambda} = \{(\mathbf{\Gamma}'\mathbf{Z} - \mathbf{F}\mathbf{\Theta}_1\mathbf{C})(\mathbf{\Gamma}'\mathbf{Z} - \mathbf{F}\mathbf{\Theta}_1\mathbf{C})'\}_d.$$

It remains to find estimators of $\mathbf{\Gamma}$ and $\mathbf{\Theta}_1$. These will be obtained by deriving a lower bound of $|\{(\mathbf{\Gamma}'\mathbf{Z} - \mathbf{F}\mathbf{\Theta}_1\mathbf{C})(\mathbf{\Gamma}'\mathbf{Z} - \mathbf{F}\mathbf{\Theta}_1\mathbf{C})'\}_d|$ which is independent of the parameters $\mathbf{\Gamma}$ and $\mathbf{\Theta}_1$. By using the definition of $\mathbf{F}$ we note that for the diagonal matrix in (4.1.29)

$$\begin{aligned}
&|\{\mathbf{\Gamma}'(\mathbf{Z} - \mathbf{\Gamma}(\mathbf{\Gamma}'\mathbf{A}^o)^o\mathbf{\Theta}_1\mathbf{C})(\mathbf{Z} - \mathbf{\Gamma}(\mathbf{\Gamma}'\mathbf{A}^o)^o\mathbf{\Theta}_1\mathbf{C})'\mathbf{\Gamma}\}_d| \\
&\quad \geq |\mathbf{\Gamma}'(\mathbf{Z} - \mathbf{\Gamma}(\mathbf{\Gamma}'\mathbf{A}^o)^o\mathbf{\Theta}_1\mathbf{C})(\mathbf{Z} - \mathbf{\Gamma}(\mathbf{\Gamma}'\mathbf{A}^o)^o\mathbf{\Theta}_1\mathbf{C})'\mathbf{\Gamma}| = |\mathbf{\Gamma}'\mathbf{S}_\mathbf{Z}\mathbf{\Gamma} + \mathbf{V}\mathbf{V}'|,
\end{aligned} \tag{4.1.30}$$

where

$$\begin{aligned}
\mathbf{S}_\mathbf{Z} &= \mathbf{Z}(\mathbf{I} - \mathbf{C}'(\mathbf{C}\mathbf{C}')^-\mathbf{C})\mathbf{Z}', \\
\mathbf{V} &= \mathbf{\Gamma}'\mathbf{Z}\mathbf{C}'(\mathbf{C}\mathbf{C}')^-\mathbf{C} - \mathbf{\Gamma}(\mathbf{\Gamma}'\mathbf{A}^o)^o\mathbf{\Theta}_1\mathbf{C}.
\end{aligned} \tag{4.1.31}$$

Equality holds in (4.1.30), if $\mathbf{\Gamma}$ is a matrix of eigenvectors of the product

$$(\mathbf{Z} - \mathbf{\Gamma}(\mathbf{\Gamma}'\mathbf{A}^o)^o\mathbf{\Theta}_1\mathbf{C})(\mathbf{Z} - \mathbf{\Gamma}(\mathbf{\Gamma}'\mathbf{A}^o)^o\mathbf{\Theta}_1\mathbf{C})'.$$

From the calculations presented via (4.1.4) – (4.1.7) it follows that (4.1.30) is minimized, if

$$\begin{aligned}
&(\mathbf{\Gamma}'\mathbf{A}^o)^o\widehat{\mathbf{\Theta}}_1\mathbf{C} \\
&\quad = (\mathbf{\Gamma}'\mathbf{A}^o)^o\{(\mathbf{\Gamma}'\mathbf{A}^o)^{o'}(\mathbf{\Gamma}'\mathbf{S}_\mathbf{Z}\mathbf{\Gamma})^{-1}(\mathbf{\Gamma}'\mathbf{A}^o)^o\}^-(\mathbf{\Gamma}'\mathbf{A}^o)^{o'}(\mathbf{\Gamma}'\mathbf{S}_\mathbf{Z}\mathbf{\Gamma})^{-1}\mathbf{\Gamma}'\mathbf{Z}\mathbf{C}'(\mathbf{C}\mathbf{C}')^-\mathbf{C} \\
&\quad = \{\mathbf{I} - \mathbf{\Gamma}'\mathbf{S}_\mathbf{Z}\mathbf{\Gamma}\mathbf{\Gamma}'\mathbf{A}^o(\mathbf{A}^{o'}\mathbf{\Gamma}\mathbf{\Gamma}'\mathbf{S}_\mathbf{Z}\mathbf{\Gamma}\mathbf{\Gamma}'\mathbf{A}^o)^-\mathbf{A}^{o'}\mathbf{\Gamma}\}\mathbf{\Gamma}'\mathbf{Z}\mathbf{C}'(\mathbf{C}\mathbf{C}')^-\mathbf{C},
\end{aligned} \tag{4.1.32}$$

where Corollary 1.2.25.1 has been used for the last equality. Since the linear functions in $\mathbf{X}$, $(\mathbf{I} - \mathbf{A}(\mathbf{\Gamma}^{o'}\mathbf{A})^-\mathbf{\Gamma}^{o'})\mathbf{X}(\mathbf{I} - \mathbf{C}'(\mathbf{C}\mathbf{C}')^-\mathbf{C})$ and $\mathbf{X}(\mathbf{I} - \mathbf{C}'(\mathbf{C}\mathbf{C}')^-\mathbf{C})$ have the same mean and dispersion they also have the same distribution, because $\mathbf{X}$ is normally distributed. This implies that $\mathbf{S}_\mathbf{Z}$ and

$$\mathbf{S} = \mathbf{X}(\mathbf{I} - \mathbf{C}'(\mathbf{C}\mathbf{C}')^-\mathbf{C})\mathbf{X}'$$

have the same distribution, which is a crucial observation since $\mathbf{S}$ is independent of any unknown parameter. Thus, in (4.1.32), the matrix $\mathbf{S}_\mathbf{Z}$ can be replaced by $\mathbf{S}$ and

$$\begin{aligned}
&(\mathbf{\Gamma}'\mathbf{A}^o)^o\widehat{\mathbf{\Theta}}_1\mathbf{C} \\
&= \mathbf{\Gamma}'\mathbf{Z}\mathbf{C}'(\mathbf{C}\mathbf{C}')^-\mathbf{C} - \mathbf{\Gamma}'\mathbf{S}\mathbf{\Gamma}\mathbf{\Gamma}'\mathbf{A}^o(\mathbf{A}^{o'}\mathbf{\Gamma}\mathbf{\Gamma}'\mathbf{S}\mathbf{\Gamma}\mathbf{\Gamma}'\mathbf{A}^o)^-\mathbf{A}^{o'}\mathbf{\Gamma}\mathbf{\Gamma}'\mathbf{Z}\mathbf{C}'(\mathbf{C}\mathbf{C}')^-\mathbf{C}
\end{aligned} \tag{4.1.33}$$

is obtained.

In the next lemma a key result for inference problems in the Growth Curve model with a singular dispersion matrix is presented.

Lemma 4.1.2. *Let* $\mathbf{S} \sim W_p(\boldsymbol{\Sigma}, n)$. *Then, with probability 1,* $\mathcal{C}(\mathbf{S}) \subseteq \mathcal{C}(\boldsymbol{\Sigma})$ *and if* $n \geq r = r(\boldsymbol{\Sigma})$, $\mathcal{C}(\mathbf{S}) = \mathcal{C}(\boldsymbol{\Sigma})$.

PROOF: Since $\mathbf{S}$ is Wishart distributed, $\mathbf{S} = \mathbf{ZZ}'$, where $\mathbf{Z} \sim N_{p,n}(\mathbf{0}, \boldsymbol{\Sigma}, \mathbf{I}_n)$. Furthermore, $\mathbf{Z} = \Psi\mathbf{U}$, where $\boldsymbol{\Sigma} = \Psi\Psi'$, $\Psi : p \times r$ and the elements in $\mathbf{U} : r \times n$ are independent $N(0,1)$. Thus, $\mathcal{C}(\mathbf{S}) = \mathcal{C}(\mathbf{Z}) \subseteq \mathcal{C}(\Psi) = \mathcal{C}(\boldsymbol{\Sigma})$. If $n \geq r$, $r(\mathbf{U}) = r$, with probability 1, then equality holds. ∎

Corollary 4.1.2.1L. *If* $\boldsymbol{\Sigma}$ *is p.d. and* $n \geq p$, *then* $\mathbf{S}$ *is p.d. with probability 1.* ∎

Let $\mathbf{S} = \mathbf{HLH}'$, where $\mathbf{H} : p \times r$ is semiorthogonal and $\mathbf{L} : r \times r$ is a diagonal matrix with the non-zero eigenvalues of $\mathbf{S}$ on the diagonal. From Lemma 4.1.2 it follows that

$$\boldsymbol{\Gamma} = \mathbf{HQ} \tag{4.1.34}$$

for some matrix $\mathbf{Q}$. Because both $\boldsymbol{\Gamma}$ and $\mathbf{H}$ are semiorthogonal, $\mathbf{Q}$ must be orthogonal. Hence, $\boldsymbol{\Gamma}\boldsymbol{\Gamma}' = \mathbf{HH}'$ and (with probability 1)

$$\mathbf{S}\boldsymbol{\Gamma}\boldsymbol{\Gamma}' = \mathbf{S}. \tag{4.1.35}$$

Instead of (4.1.33) we may thus write

$$(\boldsymbol{\Gamma}'\mathbf{A}^o)^o\widehat{\boldsymbol{\Theta}}_1\mathbf{C} = \boldsymbol{\Gamma}'(\mathbf{I} - \mathbf{SA}^o(\mathbf{A}^{o'}\mathbf{SA}^o)^-\mathbf{A}^{o'}\mathbf{HH}')\mathbf{ZC}'(\mathbf{CC}')^-\mathbf{C}.$$

We also have

$$\mathbf{A}^{o'}\boldsymbol{\Gamma}\mathbf{V} = \mathbf{A}^{o'}\mathbf{HH}'(\mathbf{I} - \mathbf{A}(\mathbf{H}^{o'}\mathbf{A})^-\mathbf{H}^{o'})\mathbf{XC}'(\mathbf{CC}')^-\mathbf{C},$$

and therefore

$$\begin{aligned}
&|\boldsymbol{\Gamma}'\mathbf{S_Z}\boldsymbol{\Gamma} + \mathbf{VV}'| \\
&\quad \geq |\boldsymbol{\Gamma}'\mathbf{S_Z}\boldsymbol{\Gamma}||\mathbf{I} + \mathbf{C}'(\mathbf{CC}')^-\mathbf{CX}'(\mathbf{I} - \mathbf{A}(\mathbf{H}^{o'}\mathbf{A})^-\mathbf{H}^{o'})'\mathbf{HH}'\mathbf{A}^o(\mathbf{A}^{o'}\mathbf{SA}^o)^-\mathbf{A}^{o'}\mathbf{H} \\
&\qquad \times \mathbf{H}'(\mathbf{I} - \mathbf{A}(\mathbf{H}^{o'}\mathbf{A})^-\mathbf{H}^{o'})\mathbf{XC}'(\mathbf{CC}')^-\mathbf{C}|.
\end{aligned} \tag{4.1.36}$$

With probability 1 it follows that

$$|\boldsymbol{\Gamma}'\mathbf{S_Z}\boldsymbol{\Gamma}| = |\boldsymbol{\Gamma}'\mathbf{S}\boldsymbol{\Gamma}| = |\mathbf{Q}'\mathbf{H}'\mathbf{SHQ}| = |\mathbf{H}'\mathbf{SH}| = |\mathbf{L}|.$$

Furthermore, we can always choose $\mathbf{H}^o$ so that $\mathbf{H}^o\mathbf{H}^{o'} + \mathbf{HH}' = \mathbf{I}$. Then, after some calculations, it follows that with probability 1

$$\mathbf{A}^{o'}\mathbf{HH}'\mathbf{Z} = \mathbf{A}^{o'}\mathbf{HH}'(\mathbf{I} - \mathbf{A}(\mathbf{H}^{o'}\mathbf{A})^-\mathbf{H}^{o'})\mathbf{X} = \mathbf{A}^{o'}\mathbf{X}.$$

Thus,

$$(\boldsymbol{\Gamma}'\mathbf{A}^o)^o\widehat{\boldsymbol{\Theta}}_1\mathbf{C} = \boldsymbol{\Gamma}'\mathbf{ZC}'(\mathbf{CC}')^-\mathbf{C} - \boldsymbol{\Gamma}'\mathbf{SA}^o(\mathbf{A}^{o'}\mathbf{SA}^o)^-\mathbf{A}^{o'}\mathbf{XC}'(\mathbf{CC}')^-\mathbf{C} \tag{4.1.37}$$

and the lower bound in (4.1.36) equals

$$|\mathbf{L}||\mathbf{I} + \mathbf{C}'(\mathbf{CC}')^-\mathbf{CX}'\mathbf{A}^o(\mathbf{A}^{o'}\mathbf{SA}^o)^-\mathbf{A}^{o'}\mathbf{XC}'(\mathbf{CC}')^-\mathbf{C}|,$$

which is free of parameters. Finally, from (4.1.37) it follows that

$$\begin{aligned}
&\mathbf{\Gamma}'(\mathbf{Z}-\mathbf{\Gamma}(\mathbf{\Gamma}'\mathbf{A}^o)^o\widehat{\mathbf{\Theta}}_1\mathbf{C})\\
&\quad=\mathbf{\Gamma}'\{\mathbf{Z}(\mathbf{I}-\mathbf{C}'(\mathbf{CC}')^-\mathbf{C})+\mathbf{SA}^o(\mathbf{A}^{o'}\mathbf{SA}^o)^-\mathbf{A}^{o'}\mathbf{XC}'(\mathbf{CC}')^-\mathbf{C}\}\\
&\quad=\mathbf{\Gamma}'\{\mathbf{X}(\mathbf{I}-\mathbf{C}'(\mathbf{CC}')^-\mathbf{C})+\mathbf{SA}^o(\mathbf{A}^{o'}\mathbf{SA}^o)^-\mathbf{A}^{o'}\mathbf{XC}'(\mathbf{CC}')^-\mathbf{C}\}
\end{aligned}$$

holds with probability 1. Thus, using (4.1.29) we see that an estimator of $\mathbf{\Gamma}$ is given by the eigenvectors, which correspond to the $r = r(\mathbf{\Sigma})$ largest eigenvalues of

$$\mathbf{T}=\mathbf{S}+\mathbf{SA}^o(\mathbf{A}^{o'}\mathbf{SA}^o)^-\mathbf{A}^{o'}\mathbf{XC}'(\mathbf{CC}')^-\mathbf{CX}'\mathbf{A}^o(\mathbf{A}^{o'}\mathbf{SA}^o)^-\mathbf{A}^{o'}\mathbf{S}. \tag{4.1.38}$$

This means that we have found estimators of $\mathbf{\Lambda}$, $\mathbf{\Gamma}$, $\mathbf{\Gamma}^o$ and $\mathbf{\Sigma}$. It remains to find the estimator of $\mathbf{B}$. Since $\mathbf{A}(\mathbf{A}'\mathbf{\Gamma}^o)^o\widehat{\mathbf{\Theta}}\mathbf{C}=\mathbf{\Gamma}(\mathbf{\Gamma}'\mathbf{A}^o)^o\widehat{\mathbf{\Theta}}_1\mathbf{C}$, it follows from (4.1.23) and (4.1.31) that, with probability 1,

$$\begin{aligned}
\mathbf{A}\widehat{\mathbf{B}}\mathbf{C}=&\mathbf{A}(\mathbf{\Gamma}^{o'}\mathbf{A})^-\mathbf{\Gamma}^{o'}\mathbf{XC}^-\mathbf{C}\\
&+\mathbf{\Gamma}\{\mathbf{I}-\mathbf{\Gamma}'\mathbf{S}\mathbf{\Gamma}\mathbf{\Gamma}'\mathbf{A}^o(\mathbf{A}^{o'}\mathbf{\Gamma}\mathbf{\Gamma}'\mathbf{S}\mathbf{\Gamma}\mathbf{\Gamma}'\mathbf{A}^o)^-\mathbf{A}^{o'}\mathbf{\Gamma}\}\mathbf{\Gamma}'\mathbf{ZC}'(\mathbf{CC}')^-\mathbf{C}.
\end{aligned}$$

By (4.1.34) and (4.1.22), and since $\mathbf{\Gamma}^{o'}\mathbf{Z}=\mathbf{0}$,

$$\begin{aligned}
\mathbf{A}\widehat{\mathbf{B}}\mathbf{C}=&\mathbf{A}(\mathbf{\Gamma}^{o'}\mathbf{A})^-\mathbf{\Gamma}^{o'}\mathbf{X}+(\mathbf{I}-\mathbf{SA}^o(\mathbf{A}^{o'}\mathbf{SA}^o)^-\mathbf{A}^{o'})\mathbf{\Gamma}\mathbf{\Gamma}'\mathbf{ZC}'(\mathbf{CC}')^-\mathbf{C}\\
=&\mathbf{A}(\mathbf{\Gamma}^{o'}\mathbf{A})^-\mathbf{\Gamma}^{o'}\mathbf{X}+\mathbf{ZC}'(\mathbf{CC}')^-\mathbf{C}-\mathbf{SA}^o(\mathbf{A}^{o'}\mathbf{SA}^o)^-\mathbf{A}^{o'}\mathbf{ZC}'(\mathbf{CC}')^-\mathbf{C}\\
=&\mathbf{XC}'(\mathbf{CC}')^-\mathbf{C}+\mathbf{A}(\mathbf{\Gamma}^{o'}\mathbf{A})^-\mathbf{\Gamma}^{o'}\mathbf{X}(\mathbf{I}-\mathbf{C}'(\mathbf{CC}')^-\mathbf{C})\\
&-\mathbf{SA}^o(\mathbf{A}^{o'}\mathbf{SA}^o)^-\mathbf{A}^{o'}\mathbf{XC}'(\mathbf{CC}')^-\mathbf{C}\\
=&(\mathbf{I}-\mathbf{SA}^o(\mathbf{A}^{o'}\mathbf{SA}^o)^-\mathbf{A}^{o'})\mathbf{XC}'(\mathbf{CC}')^-\mathbf{C}.
\end{aligned}$$

If $r(\mathbf{C})=k$, $r(\mathbf{A})=q$, i.e. both design matrices are of full rank, then

$$\widehat{\mathbf{B}}=(\mathbf{A}'\mathbf{A})^{-1}\mathbf{A}'(\mathbf{I}-\mathbf{SA}^o(\mathbf{A}^{o'}\mathbf{SA}^o)^-\mathbf{A}^{o'})\mathbf{XC}'(\mathbf{CC}')^{-1}.$$

Theorem 4.1.2. *Let $r=r(\mathbf{\Sigma})$ and $\mathbf{M}: p\times r$ be a matrix of eigenvectors which correspond to the r largest eigenvalues d_i of $\mathbf{T}$, given in (4.1.38), and let $\mathbf{D}$ be diagonal with diagonal elements d_i. If $n-r(\mathbf{X})\geq r$, and $\mathcal{C}(\mathbf{A})\cap\mathcal{C}(\mathbf{\Sigma})\neq\{\mathbf{0}\}$, then estimators of the parameters in (4.1.1) are given by*

$$\begin{aligned}
\widehat{\mathbf{\Sigma}}=&\tfrac{1}{n}\mathbf{MDM}',\\
\mathbf{A}\widehat{\mathbf{B}}\mathbf{C}=&(\mathbf{I}-\mathbf{SA}^o(\mathbf{A}^{o'}\mathbf{SA}^o)^-\mathbf{A}^{o'})\mathbf{XC}'(\mathbf{CC}')^-\mathbf{C}.
\end{aligned}$$

∎

According to Corollary 1.2.25.1, the maximum likelihood estimators of $\mathbf{\Sigma}$ and $\mathbf{ABC}$ in the full rank case $r(\mathbf{\Sigma})=p$ are given by

$$\begin{aligned}
n\widehat{\mathbf{\Sigma}}=&\mathbf{T},\\
\mathbf{A}\widehat{\mathbf{B}}\mathbf{C}=&\mathbf{A}(\mathbf{A}'\mathbf{S}^{-1}\mathbf{A})^-\mathbf{A}'\mathbf{S}^{-1}\mathbf{XC}'(\mathbf{CC}')^-\mathbf{C},
\end{aligned}$$

and hence $n\widehat{\mathbf{\Sigma}}$ is identical to (4.1.8) and $\mathbf{A}\widehat{\mathbf{B}}\mathbf{C}$ is of the same form as the expression obtained from (4.1.7) when multiplying by $\mathbf{A}$ and $\mathbf{C}$.
In Theorem 4.1.2 we assumed that $\mathcal{C}(\mathbf{A})\cap\mathcal{C}(\mathbf{\Sigma})\neq\{\mathbf{0}\}$. Note that $\mathcal{C}(\mathbf{A})\cap\mathcal{C}(\mathbf{\Sigma})=\{\mathbf{0}\}$ is equivalent to $\mathbf{F}=\mathbf{0}$ in (4.1.26). As $(\mathbf{\Gamma}^{o'}\mathbf{A})^-\mathbf{\Gamma}^{o'}=(\widetilde{\mathbf{\Gamma}}^{o'}\mathbf{A})^-\widetilde{\mathbf{\Gamma}}^{o'}$ for all full rank matrices $\widetilde{\mathbf{\Gamma}}^o$ such that $\mathcal{C}(\mathbf{\Gamma}^o)=\mathcal{C}(\widetilde{\mathbf{\Gamma}}^o)$, we have established the next theorem.

Theorem 4.1.3. *Let $r = r(\boldsymbol{\Sigma})$ and $\mathbf{M} : p \times r$ be a matrix of eigenvectors which correspond to the r non-zero eigenvalues d_i of $\widehat{\mathbf{Z}}\widehat{\mathbf{Z}}'$, where*

$$\widehat{\mathbf{Z}} = (\mathbf{I} - \mathbf{A}(\widetilde{\boldsymbol{\Gamma}}^{o'}\mathbf{A})^{-}\widetilde{\boldsymbol{\Gamma}}^{o'})\mathbf{X},$$

$\widetilde{\boldsymbol{\Gamma}}^{o} : p \times (p-r)$ is any matrix which generates $\mathcal{C}(\mathbf{S})^{\perp}$, and let $\mathbf{D}$ be diagonal with diagonal elements d_i. If $\mathcal{C}(\mathbf{A}) \cap \mathcal{C}(\boldsymbol{\Sigma}) = \{\mathbf{0}\}$, then estimators of the parameters in (4.1.1) are given by

$$\begin{aligned}\widehat{\boldsymbol{\Sigma}} =& \tfrac{1}{n}\mathbf{MDM}'\\ \mathbf{A}\widehat{\mathbf{B}}\mathbf{C} =& \mathbf{A}(\widetilde{\boldsymbol{\Gamma}}^{o'}\mathbf{A})^{-}\widetilde{\boldsymbol{\Gamma}}^{o'}\mathbf{X}.\end{aligned}$$

If $r(\mathbf{C}) = k$, then

$$\mathbf{A}\widehat{\mathbf{B}} = \mathbf{A}(\widetilde{\boldsymbol{\Gamma}}^{o'}\mathbf{A})^{-}\widetilde{\boldsymbol{\Gamma}}^{o'}\mathbf{X}\mathbf{C}'(\mathbf{C}\mathbf{C}')^{-1}.$$

If $r(\mathbf{C}) = k$ and $r(\mathbf{A}) = q$, then

$$\widehat{\mathbf{B}} = (\widetilde{\boldsymbol{\Gamma}}^{o'}\mathbf{A})^{-}\widetilde{\boldsymbol{\Gamma}}^{o'}\mathbf{X}\mathbf{C}'(\mathbf{C}\mathbf{C}')^{-1}.$$

∎

In the next theorem we are going to show that if $\mathcal{C}(\mathbf{A}) \cap \mathcal{C}(\boldsymbol{\Sigma}) = \{\mathbf{0}\}$ holds, the expressions of $\mathbf{A}\widehat{\mathbf{B}}\mathbf{C}$ in Theorem 4.1.2 and Theorem 4.1.3 are identical. This means that we can always use the representation

$$\mathbf{A}\widehat{\mathbf{B}}\mathbf{C} = (\mathbf{I} - \mathbf{S}\mathbf{A}^{o}(\mathbf{A}^{o'}\mathbf{S}\mathbf{A}^{o})^{-}\mathbf{A}^{o'})\mathbf{X}\mathbf{C}'(\mathbf{C}\mathbf{C}')^{-}\mathbf{C},$$

but the interpretation of the estimator depends heavily on the conditions $\mathcal{C}(\mathbf{A}) \cap \mathcal{C}(\boldsymbol{\Sigma}) = \{\mathbf{0}\}$ and $\mathcal{C}(\mathbf{A}) \cap \mathcal{C}(\boldsymbol{\Sigma}) \neq \{\mathbf{0}\}$.

Theorem 4.1.4. *If $\mathcal{C}(\mathbf{A}) \cap \mathcal{C}(\boldsymbol{\Sigma}) = \mathbf{0}$ holds, then $\mathbf{A}\widehat{\mathbf{B}}\mathbf{C}$ and $\widehat{\boldsymbol{\Sigma}}$ in Theorem 4.1.2 equal $\mathbf{A}\widehat{\mathbf{B}}\mathbf{C}$ and $\widehat{\boldsymbol{\Sigma}}$ in Theorem 4.1.3, with probability 1.*

PROOF: Since $\mathbf{X} \in \mathcal{C}(\mathbf{A} : \boldsymbol{\Sigma})$, we have $\mathbf{X} = \mathbf{A}\mathbf{Q}_1 + \boldsymbol{\Gamma}\mathbf{Q}_2$ for some matrices $\mathbf{Q}_1$ and $\mathbf{Q}_2$. Then

$$\begin{aligned}\mathbf{A}(\widetilde{\boldsymbol{\Gamma}}^{o'}\mathbf{A})^{-}\widetilde{\boldsymbol{\Gamma}}^{o'}\mathbf{X}\mathbf{C}'(\mathbf{C}\mathbf{C}')^{-}\mathbf{C} &= \mathbf{A}(\widetilde{\boldsymbol{\Gamma}}^{o'}\mathbf{A})^{-}\widetilde{\boldsymbol{\Gamma}}^{o'}\mathbf{A}\mathbf{Q}_1\mathbf{C}'(\mathbf{C}\mathbf{C}')^{-}\mathbf{C}\\ &= \mathbf{A}\mathbf{Q}_1\mathbf{C}'(\mathbf{C}\mathbf{C}')^{-}\mathbf{C},\end{aligned}$$

where for the last equality to hold it is observed that according to Theorem 1.2.12 $\mathcal{C}(\mathbf{A}) \cap \mathcal{C}(\boldsymbol{\Sigma}) = \{\mathbf{0}\}$ implies $\mathcal{C}(\mathbf{A}') = \mathcal{C}(\mathbf{A}'\boldsymbol{\Gamma}^{o})$. Furthermore, let $\mathbf{H}$ be defined as before, i.e. $\mathbf{S} = \mathbf{HLH}'$, and since $\mathcal{C}(\boldsymbol{\Sigma}) = \mathcal{C}(\mathbf{S})$ implies that for some $\mathbf{Q}_3$, $\mathbf{X} = \mathbf{A}\mathbf{Q}_1 + \mathbf{S}\mathbf{Q}_3$ with probability 1, and thus

$$\begin{aligned}&(\mathbf{I} - \mathbf{S}\mathbf{A}^{o}(\mathbf{A}^{o'}\mathbf{S}\mathbf{A}^{o})^{-}\mathbf{A}^{o'})\mathbf{X}\mathbf{C}'(\mathbf{C}\mathbf{C}')^{-}\mathbf{C}\\ &\quad= \mathbf{X}\mathbf{C}'(\mathbf{C}\mathbf{C}')^{-}\mathbf{C} - \mathbf{S}\mathbf{A}^{o}(\mathbf{A}^{o'}\mathbf{S}\mathbf{A}^{o})^{-}\mathbf{A}^{o'}\mathbf{S}\mathbf{Q}_3\mathbf{C}'(\mathbf{C}\mathbf{C}')^{-}\mathbf{C}\\ &\quad= \mathbf{X}\mathbf{C}'(\mathbf{C}\mathbf{C}')^{-}\mathbf{C} - \Big(\mathbf{S} - \mathbf{H}(\mathbf{H}'\mathbf{A}^{o})^{o}\{(\mathbf{H}'\mathbf{A}^{o})^{o'}\mathbf{H}'\mathbf{S}^{+}\mathbf{H}(\mathbf{H}'\mathbf{A}^{o})^{o}\}^{-}\\ &\qquad\quad \times (\mathbf{H}'\mathbf{A}^{o})^{o'}\mathbf{H}'\Big)\mathbf{Q}_3\mathbf{C}'(\mathbf{C}\mathbf{C}')^{-}\mathbf{C}\\ &\quad= \mathbf{X}\mathbf{C}'(\mathbf{C}\mathbf{C}')^{-}\mathbf{C} - \mathbf{S}\mathbf{Q}_3\mathbf{C}'(\mathbf{C}\mathbf{C}')^{-}\mathbf{C} = \mathbf{A}\mathbf{Q}_1\mathbf{C}'(\mathbf{C}\mathbf{C}')^{-}\mathbf{C},\end{aligned}$$

where $\mathbf{S}^{+} = \mathbf{HL}^{+}\mathbf{H}'$, with $\mathbf{L}$ being the diagonal matrix formed by the eigenvalues of $\mathbf{S}$, and the last equality is true because by assumption $\mathcal{C}(\mathbf{H}(\mathbf{H}'\mathbf{A}^o)) = \mathcal{C}(\mathbf{H}) \cap \mathcal{C}(\mathbf{A}) = \{\mathbf{0}\}$. Moreover,

$$\begin{aligned}\mathbf{SA}^o(\mathbf{A}^{o'}\mathbf{SA}^o)^{-}\mathbf{A}^{o'}\mathbf{XC}'(\mathbf{CC}')^{-}\mathbf{C} =& \mathbf{XC}'(\mathbf{CC}')^{-}\mathbf{C} - \mathbf{A}\widehat{\mathbf{B}}\mathbf{C} \\ =& (\mathbf{I} - \mathbf{A}(\widetilde{\mathbf{\Gamma}}^{o'}\mathbf{A})^{-}\widetilde{\mathbf{\Gamma}}^{o'})\mathbf{XC}'(\mathbf{CC}')^{-}\mathbf{C},\end{aligned}$$

with probability 1, and since $\mathbf{S}_{\mathbf{Z}} = \mathbf{S}$ also holds with probability 1, the expressions of $\widehat{\mathbf{\Sigma}}$ in both theorems are identical. ∎

4.1.4 *Extensions of the Growth Curve model*

In our first extension we are going to consider the model with a rank restriction on the parameters describing the mean structure in an ordinary MANOVA model. This is an extension of the Growth Curve model due to the following definition.

Definition 4.1.2 GROWTH CURVE MODEL WITH RANK RESTRICTION. *Let* $\mathbf{X} : p \times q$, $\boldsymbol{\mu} : p \times k$, $r(\boldsymbol{\mu}) = q < p$, $\mathbf{C} : k \times n$, $r(\mathbf{C}) + p \leq n$ *and* $\mathbf{\Sigma} : p \times p$ *be p.d. Then*

$$\mathbf{X} = \boldsymbol{\mu}\mathbf{C} + \mathbf{\Sigma}^{1/2}\mathbf{E} \tag{4.1.39}$$

is called a Growth Curve model with rank restriction, where $\mathbf{E} \sim N_{p,n}(\mathbf{0}, \mathbf{I}_p, \mathbf{I}_n)$, $\mathbf{C}$ *is a known matrix, and* $\boldsymbol{\mu}$, $\mathbf{\Sigma}$ *are unknown parameter matrices.* ∎

By Proposition 1.1.6 (i),

$$\boldsymbol{\mu} = \mathbf{AB},$$

where $\mathbf{A} : p \times q$ and $\mathbf{B} : q \times k$. If $\mathbf{A}$ is known we are back in the ordinary Growth Curve model which was treated in §4.1.2. In this paragraph it will be assumed that both $\mathbf{A}$ and $\mathbf{B}$ are unknown. The model is usually called a regression model with rank restriction or reduced rank regression model and has a long history. Seminal work was performed by Fisher (1939), Anderson (1951), (2002) and Rao (1973a). For results on the Growth Curve model, and for an overview on multivariate reduced rank regression, we refer to Reinsel & Velu (1998, 2003). In order to estimate the parameters we start from the likelihood function, as before, and initially note, as in §4.41.2, that

$$\begin{aligned}L(\mathbf{A}, \mathbf{B}, \mathbf{\Sigma}) =& (2\pi)^{-\frac{1}{2}pn}|\mathbf{\Sigma}|^{-\frac{1}{2}n}e^{-\frac{1}{2}\mathrm{tr}\{\mathbf{\Sigma}^{-1}(\mathbf{X}-\mathbf{ABC})(\mathbf{X}-\mathbf{ABC})'\}} \\ \leq& (2\pi)^{-\frac{1}{2}pn}|\tfrac{1}{n}(\mathbf{X} - \mathbf{ABC})(\mathbf{X} - \mathbf{ABC})'|^{-\frac{1}{2}n}e^{-\frac{1}{2}pn}.\end{aligned}$$

Recall $\mathbf{S}$ and $\mathbf{V}$, which were defined by (4.1.5) and (4.1.6), respectively. Then

$$\begin{aligned}|(\mathbf{X}-\mathbf{ABC})(\mathbf{X} - \mathbf{ABC})'| &= |\mathbf{S}||\mathbf{I} + \mathbf{V}'\mathbf{S}^{-1}\mathbf{V}| \\ &\geq |\mathbf{S}||\mathbf{I} + \mathbf{C}'(\mathbf{CC}')^{-}\mathbf{CX}'\mathbf{A}^o(\mathbf{A}^{o'}\mathbf{SA}^o)^{-1}\mathbf{A}^{o'}\mathbf{XC}'(\mathbf{CC}')^{-}\mathbf{C}|.\end{aligned} \tag{4.1.40}$$

Put

$$\mathbf{F} = (\mathbf{A}^{o'}\mathbf{SA}^o)^{-1/2}\mathbf{A}^{o'}\mathbf{S}^{1/2}, \tag{4.1.41}$$

where $\mathbf{M}^{1/2}$ denotes a symmetric square root of $\mathbf{M}$. Since $\mathbf{FF}' = \mathbf{I}$, the right hand side of (4.1.40) can be written as

$$|\mathbf{S}||\mathbf{F}(\mathbf{I} + \mathbf{S}^{-1/2}\mathbf{XC}'(\mathbf{CC}')^{-}\mathbf{CX}'\mathbf{S}^{-1/2})\mathbf{F}'|. \tag{4.1.42}$$

However, from Proposition 1.2.3 (xviii) it follows that

$$|\mathbf{F}(\mathbf{I} + \mathbf{S}^{-1/2}\mathbf{XC}'(\mathbf{CC}')^{-}\mathbf{CX}'\mathbf{S}^{-1/2})\mathbf{F}'| \geq \prod_{i=q+1}^{p} v_{(i)},$$

where $v_{(i)}$ is the i–th largest eigenvalue (suppose $v_{(q)} > v_{(q+1)}$) of

$$\mathbf{I} + \mathbf{S}^{-1/2}\mathbf{XC}'(\mathbf{CC}')^{-}\mathbf{CX}'\mathbf{S}^{-1/2} = \mathbf{S}^{-1/2}\mathbf{XX}'\mathbf{S}^{-1/2}.$$

The minimum value of (4.1.42) is attained when $\mathbf{F}'$ is estimated by the eigenvectors corresponding to the $p-q$ smallest eigenvalues of $\mathbf{S}^{-1/2}\mathbf{XX}'\mathbf{S}^{-1/2}$. It remains to find a matrix $\mathbf{A}^o$ or, equivalently, a matrix $\mathbf{A}$ which satisfies (4.1.41). However, since $\widehat{\mathbf{F}}\widehat{\mathbf{F}}' = \mathbf{I}$, it follows that

$$\widehat{\mathbf{A}}^o = \mathbf{S}^{-1/2}\widehat{\mathbf{F}}'.$$

Furthermore,

$$\widehat{\mathbf{A}} = \mathbf{S}^{1/2}(\widehat{\mathbf{F}}')^o.$$

As $(\widehat{\mathbf{F}}')^o$ we may choose the eigenvectors which correspond to the q largest eigenvalues of $\mathbf{S}^{-1/2}\mathbf{XX}'\mathbf{S}^{-1/2}$. Thus, from the treatment of the ordinary Growth Curve model it is observed that

$$\begin{aligned}\widehat{\mathbf{B}}\mathbf{C} =&\{(\widehat{\mathbf{F}}')^{o'}(\widehat{\mathbf{F}}')^o\}^{-1}(\widehat{\mathbf{F}}')^{o'}\mathbf{S}^{-1/2}\mathbf{XC}'(\mathbf{CC}')^{-}\mathbf{C}\\ =&(\widehat{\mathbf{F}}')^{o'}\mathbf{S}^{-1/2}\mathbf{XC}'(\mathbf{CC}')^{-}\mathbf{C}.\end{aligned}$$

Theorem 4.1.5. *For the model in (4.1.39), where $r(\boldsymbol{\mu}) = q < p$, the maximum likelihood estimators are given by*

$$\begin{aligned}n\widehat{\boldsymbol{\Sigma}} =&(\mathbf{X} - \widehat{\boldsymbol{\mu}}\mathbf{C})(\mathbf{X} - \widehat{\boldsymbol{\mu}}\mathbf{C})',\\ \widehat{\boldsymbol{\mu}} =&(\widehat{\mathbf{F}}')^o\widehat{\mathbf{B}},\\ \widehat{\mathbf{B}}\mathbf{C} =&(\widehat{\mathbf{F}}')^{o'}\mathbf{S}^{-1/2}\mathbf{XC}'(\mathbf{CC}')^{-}\mathbf{C}.\end{aligned}$$

∎

A different type of extension of the Growth Curve model is the following one. In the subsequent the model will be utilized several times.

Definition 4.1.3 EXTENDED GROWTH CURVE MODEL. *Let* $\mathbf{X} : p\times n$, $\mathbf{A}_i : p\times q_i$, $q_i \leq p$, $\mathbf{B}_i : q_i \times k_i$, $\mathbf{C}_i : k_i \times n$, $r(\mathbf{C}_1)+p \leq n$, $i = 1, 2, \dots, m$, $\mathcal{C}(\mathbf{C}'_i) \subseteq \mathcal{C}(\mathbf{C}'_{i-1})$, $i = 2, 3, \dots, m$ *and* $\boldsymbol{\Sigma} : p \times p$ *be p.d. Then*

$$\mathbf{X} = \sum_{i=1}^{m} \mathbf{A}_i\mathbf{B}_i\mathbf{C}_i + \boldsymbol{\Sigma}^{1/2}\mathbf{E} \tag{4.1.43}$$

is called an Extended Growth Curve model, where $\mathbf{E} \sim N_{p,n}(\mathbf{0}, \mathbf{I}_p, \mathbf{I}_n)$, $\mathbf{A}_i$ *and* $\mathbf{C}_i$ *are known matrices, and* $\mathbf{B}_i$ *and* $\boldsymbol{\Sigma}$ *are unknown parameter matrices.* ∎

The only difference with the Growth Curve model (4.1.1) is the presence of a more general mean structure. If $m = 1$, the model is identical to the Growth Curve model. Sometimes the model has been called *sum of profiles model* (Verbyla & Venables, 1988). An alternative name is *multivariate linear normal model with mean* $\sum_{i=1}^{m} \mathbf{A}_i\mathbf{B}_i\mathbf{C}_i$, which can be abbreviated MLNM$(\sum_{i=1}^{m} \mathbf{A}_i\mathbf{B}_i\mathbf{C}_i)$. Note that the Growth Curve model may be called a MLNM$(\mathbf{ABC})$. Furthermore, instead of $\mathcal{C}(\mathbf{C}'_s) \subseteq \mathcal{C}(\mathbf{C}'_{s-1})$, we may assume that $\mathcal{C}(\mathbf{A}_s) \subseteq \mathcal{C}(\mathbf{A}_{s-1})$ holds.

In the subsequent presentation we shall focus on the special case $m = 3$. The general case will now and then be considered but results will usually be stated without complete proofs. The reader who grasps the course of derivation when $m = 3$ will also succeed in the treatment of the general case. As it often happens in multivariate analysis the problems hang much on convenient notation. In particular, when working with the MLNM$(\sum_{i=1}^{m} \mathbf{A}_i\mathbf{B}_i\mathbf{C}_i)$, the notation problem is important.

Before starting with the technical derivation, Example 4.1.1 is continued in order to shed some light on the extension MLNM$(\sum_{i=1}^{3} \mathbf{A}_i\mathbf{B}_i\mathbf{C}_i)$ and, in particular, to illuminate the nested subspace condition $\mathcal{C}(\mathbf{C}'_3) \subseteq \mathcal{C}(\mathbf{C}'_2) \subseteq \mathcal{C}(\mathbf{C}'_1)$.

Example 4.1.2 (Example 4.1.1 continued). Suppose that for the three groups in Example 4.1.1 there exist three different responses

$$\begin{aligned}
&\beta_{11}+\beta_{21}t + \cdots + \beta_{(q-2)1}t^{q-3},\\
&\beta_{12}+\beta_{22}t + \cdots + \beta_{(q-2)2}t^{q-3} + \beta_{(q-1)2}t^{q-2},\\
&\beta_{13}+\beta_{23}t + \cdots + \beta_{(q-2)3}t^{q-3} + \beta_{(q-1)3}t^{q-2} + \beta_{q3}t^{q-1}.
\end{aligned}$$

In order to describe these different responses, consider the model

$$E[\mathbf{X}] = \mathbf{A}_1\mathbf{B}_1\mathbf{C}_1 + \mathbf{A}_2\mathbf{B}_2\mathbf{C}_2 + \mathbf{A}_3\mathbf{B}_3\mathbf{C}_3,$$

where the matrices in $E[\mathbf{X}]$ are defined as below:

$$\mathbf{A}_1 = \begin{pmatrix} 1 & t_1 & \dots & t_1^{q-3} \\ 1 & t_2 & \dots & t_2^{q-3} \\ \vdots & \vdots & \ddots & \vdots \\ 1 & t_p & \dots & t_p^{q-3} \end{pmatrix}, \quad \mathbf{A}_2 = \begin{pmatrix} t_1^{q-2} \\ t_2^{q-2} \\ \vdots \\ t_p^{q-2} \end{pmatrix}, \quad \mathbf{A}_3 = \begin{pmatrix} t_1^{q-1} \\ t_2^{q-1} \\ \vdots \\ t_p^{q-1} \end{pmatrix},$$

$$\mathbf{B}_1 = \begin{pmatrix} \beta_{11} & \beta_{12} & \beta_{13} \\ \beta_{21} & \beta_{22} & \beta_{23} \\ \vdots & \vdots & \vdots \\ \beta_{(q-2)1} & \beta_{(q-2)2} & \beta_{(q-2)3} \end{pmatrix},$$

$$\mathbf{B}_2 = (\, \beta_{(q-1)1} \quad \beta_{(q-1)2} \quad \beta_{(q-1)3} \,), \quad \mathbf{B}_3 = (\, \beta_{q1} \quad \beta_{q2} \quad \beta_{q3} \,),$$

$$\mathbf{C}_1 = \begin{pmatrix} 1 & 1 & \dots & 1 & 0 & 0 & \dots & 0 & 0 & 0 & \dots & 0 \\ 0 & 0 & \dots & 0 & 1 & 1 & \dots & 1 & 0 & 0 & \dots & 0 \\ 0 & 0 & \dots & 0 & 0 & 0 & \dots & 0 & 1 & 1 & \dots & 1 \end{pmatrix},$$

$$\mathbf{C}_2 = \begin{pmatrix} 0 & 0 & \dots & 0 & 0 & 0 & \dots & 0 & 0 & 0 & \dots & 0 \\ 0 & 0 & \dots & 0 & 1 & 1 & \dots & 1 & 0 & 0 & \dots & 0 \\ 0 & 0 & \dots & 0 & 0 & 0 & \dots & 0 & 1 & 1 & \dots & 1 \end{pmatrix},$$

$$\mathbf{C}_3 = \begin{pmatrix} 0 & 0 & \dots & 0 & 0 & 0 & \dots & 0 & 0 & 0 & \dots & 0 \\ 0 & 0 & \dots & 0 & 0 & 0 & \dots & 0 & 0 & 0 & \dots & 0 \\ 0 & 0 & \dots & 0 & 0 & 0 & \dots & 0 & 1 & 1 & \dots & 1 \end{pmatrix}.$$

Beside those rows which consist of zeros, the rows of $\mathbf{C}_3$ and $\mathbf{C}_2$ are represented in $\mathbf{C}_2$ and $\mathbf{C}_1$, respectively. Hence, $\mathcal{C}(\mathbf{C}_3') \subseteq \mathcal{C}(\mathbf{C}_2') \subseteq \mathcal{C}(\mathbf{C}_1')$. Furthermore, there is no way of modelling the mean structure with $E[\mathbf{X}] = \mathbf{ABC}$ for some $\mathbf{A}$ and $\mathbf{C}$ and arbitrary elements in $\mathbf{B}$. Thus, if different treatment groups have different polynomial responses, this case cannot be treated with a Growth Curve model (4.1.1). ∎

The likelihood equations for the MLNM($\sum_{i=1}^{3} \mathbf{A}_i\mathbf{B}_i\mathbf{C}_i$) are:

$$\begin{aligned} &\mathbf{A}_1'\mathbf{\Sigma}^{-1}(\mathbf{X} - \mathbf{A}_1\mathbf{B}_1\mathbf{C}_1 - \mathbf{A}_2\mathbf{B}_2\mathbf{C}_2 - \mathbf{A}_3\mathbf{B}_3\mathbf{C}_3)\mathbf{C}_1' = \mathbf{0}, \\ &\mathbf{A}_2'\mathbf{\Sigma}^{-1}(\mathbf{X} - \mathbf{A}_1\mathbf{B}_1\mathbf{C}_1 - \mathbf{A}_2\mathbf{B}_2\mathbf{C}_2 - \mathbf{A}_3\mathbf{B}_3\mathbf{C}_3)\mathbf{C}_2' = \mathbf{0}, \\ &\mathbf{A}_3'\mathbf{\Sigma}^{-1}(\mathbf{X} - \mathbf{A}_1\mathbf{B}_1\mathbf{C}_1 - \mathbf{A}_2\mathbf{B}_2\mathbf{C}_2 - \mathbf{A}_3\mathbf{B}_3\mathbf{C}_3)\mathbf{C}_3' = \mathbf{0}, \\ &n\mathbf{\Sigma} = (\mathbf{X} - \mathbf{A}_1\mathbf{B}_1\mathbf{C}_1 - \mathbf{A}_2\mathbf{B}_2\mathbf{C}_2 - \mathbf{A}_3\mathbf{B}_3\mathbf{C}_3)()'. \end{aligned}$$

However, these equations are not going to be solved. Instead we are going to work directly with the likelihood function, as has been done in §4.1.2 and when proving Theorem 4.1.5. As noted before, one advantage of this approach is that the global maximum is obtained, whereas when solving the likelihood equations we have to be sure that it is not a local maximum that has been found. For the MLNM($\sum_{i=1}^{m} \mathbf{A}_i\mathbf{B}_i\mathbf{C}_i$) this is not a trivial task.

The likelihood function equals

$$\begin{aligned} &L(\mathbf{B}_1,\mathbf{B}_2,\mathbf{B}_3,\mathbf{\Sigma}) \\ &\quad =(2\pi)^{-\frac{1}{2}pn}|\mathbf{\Sigma}|^{-\frac{1}{2}n}e^{-\frac{1}{2}\mathrm{tr}\{\mathbf{\Sigma}^{-1}(\mathbf{X}-\mathbf{A}_1\mathbf{B}_1\mathbf{C}_1-\mathbf{A}_2\mathbf{B}_2\mathbf{C}_2-\mathbf{A}_3\mathbf{B}_3\mathbf{C}_3)()'\}}. \end{aligned}$$

By Lemma 4.1.1, the likelihood function satisfies the inequality

$$\begin{aligned}&L(\mathbf{B}_1,\mathbf{B}_2,\mathbf{B}_3,\boldsymbol{\Sigma})\\&\leq(2\pi)^{-\frac{1}{2}np}|\frac{1}{n}(\mathbf{X}-\mathbf{A}_1\mathbf{B}_1\mathbf{C}_1-\mathbf{A}_2\mathbf{B}_2\mathbf{C}_2-\mathbf{A}_3\mathbf{B}_3\mathbf{C}_3)()'|^{-\frac{1}{2}n}e^{-\frac{1}{2}np},\end{aligned}$$

where equality holds if and only if

$$n\boldsymbol{\Sigma}=(\mathbf{X}-\mathbf{A}_1\mathbf{B}_1\mathbf{C}_1-\mathbf{A}_2\mathbf{B}_2\mathbf{C}_2-\mathbf{A}_3\mathbf{B}_3\mathbf{C}_3)()'. \quad (4.1.44)$$

Therefore, the determinant

$$|(\mathbf{X}-\mathbf{A}_1\mathbf{B}_1\mathbf{C}_1-\mathbf{A}_2\mathbf{B}_2\mathbf{C}_2-\mathbf{A}_3\mathbf{B}_3\mathbf{C}_3)()'| \quad (4.1.45)$$

is going to be minimized with respect to the parameters $\mathbf{B}_1$, $\mathbf{B}_2$ and $\mathbf{B}_3$. Since $\mathcal{C}(\mathbf{C}_3')\subseteq\mathcal{C}(\mathbf{C}_2')\subseteq\mathcal{C}(\mathbf{C}_1')$ implies $\mathbf{C}_1'(\mathbf{C}_1\mathbf{C}_1')^-\mathbf{C}_1(\mathbf{C}_3':\mathbf{C}_2')=(\mathbf{C}_3':\mathbf{C}_2')$ we can present (4.1.45) in the following form:

$$|\mathbf{S}_1+\mathbf{V}_1\mathbf{V}_1'|, \quad (4.1.46)$$

where

$$\mathbf{S}_1=\mathbf{X}(\mathbf{I}-\mathbf{C}_1'(\mathbf{C}_1\mathbf{C}_1')^-\mathbf{C}_1)\mathbf{X}' \quad (4.1.47)$$

and

$$\mathbf{V}_1=\mathbf{X}\mathbf{C}_1'(\mathbf{C}_1\mathbf{C}_1')^-\mathbf{C}_1-\mathbf{A}_1\mathbf{B}_1\mathbf{C}_1-\mathbf{A}_2\mathbf{B}_2\mathbf{C}_2-\mathbf{A}_3\mathbf{B}_3\mathbf{C}_3.$$

Note that $\mathbf{S}_1$ is identical to $\mathbf{S}$ given by (4.1.5), and the matrices in (4.1.46) and (4.1.4) have a similar structure. Thus, the approach used for the Growth Curve model may be copied. For any pair of matrices $\mathbf{S}$ and $\mathbf{A}$, where $\mathbf{S}$ is p.d., we will use the notation

$$\mathbf{P}_{\mathbf{A},\mathbf{S}}=\mathbf{A}(\mathbf{A}'\mathbf{S}^{-1}\mathbf{A})^-\mathbf{A}'\mathbf{S}^{-1},$$

which will shorten the subsequent matrix expressions. Observe that $\mathbf{P}_{\mathbf{A},\mathbf{S}}$ is a projector and Corollary 1.2.25.1 yields

$$\mathbf{P}_{\mathbf{A},\mathbf{S}}=\mathbf{I}-\mathbf{P}'_{\mathbf{A}^o,\mathbf{S}^{-1}}=\mathbf{I}-\mathbf{S}\mathbf{A}^o(\mathbf{A}^{o\prime}\mathbf{S}\mathbf{A}^o)^-\mathbf{A}^{o\prime}.$$

Moreover, $\mathbf{P}^o_{\mathbf{A},\mathbf{S}}=\mathbf{I}-\mathbf{P}'_{\mathbf{A},\mathbf{S}}=\mathbf{P}_{\mathbf{A}^o,\mathbf{S}^{-1}}$. If $\mathbf{S}=\mathbf{I}$, instead of $\mathbf{P}_{\mathbf{A},\mathbf{I}}$ the notation $\mathbf{P}_{\mathbf{A}}$ will be used.
We start by modifying (4.1.46), using Corollary 1.2.25.1 and Proposition 1.1.1 (iii):

$$\begin{aligned}|\mathbf{S}_1+\mathbf{V}_1\mathbf{V}_1'|=&|\mathbf{S}_1||\mathbf{I}+\mathbf{S}_1^{-1}\mathbf{V}_1\mathbf{V}_1'|=|\mathbf{S}_1||\mathbf{I}+\mathbf{V}_1'\mathbf{S}_1^{-1}\mathbf{V}_1|\\=&|\mathbf{S}_1||\mathbf{I}+\mathbf{V}_1'\mathbf{P}'_{\mathbf{A}_1,\mathbf{S}_1}\mathbf{S}_1^{-1}\mathbf{P}_{\mathbf{A}_1,\mathbf{S}_1}\mathbf{V}_1+\mathbf{V}_1'\mathbf{P}_{\mathbf{A}_1^o,\mathbf{S}_1^{-1}}\mathbf{S}_1^{-1}\mathbf{P}'_{\mathbf{A}_1^o,\mathbf{S}_1^{-1}}\mathbf{V}_1|\\\geq&|\mathbf{S}_1||\mathbf{I}+\mathbf{V}_1'\mathbf{P}_{\mathbf{A}_1^o,\mathbf{S}_1^{-1}}\mathbf{S}_1^{-1}\mathbf{P}'_{\mathbf{A}_1^o,\mathbf{S}_1^{-1}}\mathbf{V}_1|\\=&|\mathbf{S}_1||\mathbf{I}+\mathbf{W}_1'\mathbf{P}_{\mathbf{A}_1^o,\mathbf{S}_1^{-1}}\mathbf{S}_1^{-1}\mathbf{P}'_{\mathbf{A}_1^o,\mathbf{S}_1^{-1}}\mathbf{W}_1|,\end{aligned} \quad (4.1.48)$$

where

$$\mathbf{W}_1=\mathbf{X}\mathbf{C}_1'(\mathbf{C}_1\mathbf{C}_1')^-\mathbf{C}_1-\mathbf{A}_2\mathbf{B}_2\mathbf{C}_2-\mathbf{A}_3\mathbf{B}_3\mathbf{C}_3. \quad (4.1.49)$$

The point is that (4.1.49) is functionally independent of $\mathbf{B}_1$. It will be supposed that $\mathbf{S}_1$ is non-singular, which is known to be true with probability 1 (Corollary 4.1.2.1L). Moreover, it is observed that equality in (4.1.48) holds if and only if $\mathbf{P}_{\mathbf{A}_1,\mathbf{S}_1}\mathbf{V}_1 = \mathbf{0}$, which is equivalent to

$$\mathbf{A}_1'\mathbf{S}_1^{-1}\mathbf{V}_1 = \mathbf{0}. \tag{4.1.50}$$

Later this system of equations will be solved. By Proposition 1.1.1 (ii) it follows that the right hand side of (4.1.48) is identical to

$$|\mathbf{S}_1||\mathbf{I} + \mathbf{S}_1^{-1}\mathbf{P}'_{\mathbf{A}_1^o,\mathbf{S}_1^{-1}}\mathbf{W}_1\mathbf{W}_1'\mathbf{P}_{\mathbf{A}_1^o,\mathbf{S}_1^{-1}}| = |\mathbf{S}_1 + \mathbf{P}'_{\mathbf{A}_1^o,\mathbf{S}_1^{-1}}\mathbf{W}_1\mathbf{W}_1'\mathbf{P}_{\mathbf{A}_1^o,\mathbf{S}_1^{-1}}|,$$

which can be written as

$$|\mathbf{S}_1 + \mathbf{T}_1\mathbf{W}_1\mathbf{W}_1'\mathbf{T}_1'|, \tag{4.1.51}$$

where

$$\mathbf{T}_1 = \mathbf{I} - \mathbf{P}_{\mathbf{A}_1,\mathbf{S}_1}. \tag{4.1.52}$$

The next calculations are based on the similarity between (4.1.46) and (4.1.51). Let

$$\begin{aligned}\mathbf{S}_2 =&\mathbf{S}_1 + \mathbf{T}_1\mathbf{X}\mathbf{P}_{\mathbf{C}_1'}(\mathbf{I} - \mathbf{C}_2'(\mathbf{C}_2\mathbf{C}_2')^-\mathbf{C}_2)\mathbf{P}_{\mathbf{C}_1'}\mathbf{X}'\mathbf{T}_1', \qquad (4.1.53)\\ \mathbf{V}_2 =&\mathbf{T}_1\mathbf{X}\mathbf{C}_2'(\mathbf{C}_2\mathbf{C}_2')^-\mathbf{C}_2 - \mathbf{T}_1\mathbf{A}_2\mathbf{B}_2\mathbf{C}_2 - \mathbf{T}_1\mathbf{A}_3\mathbf{B}_3\mathbf{C}_3.\end{aligned}$$

Since $\mathbf{S}_1$ is positive definite with probability 1, it follows that $\mathbf{S}_2$ is non-singular with probability 1. Using these definitions, it is obvious that

$$|\mathbf{S}_1 + \mathbf{T}_1\mathbf{W}_1\mathbf{W}_1'\mathbf{T}_1'| = |\mathbf{S}_2 + \mathbf{V}_2\mathbf{V}_2'|, \tag{4.1.54}$$

since $C(\mathbf{C}_3') \subseteq C(\mathbf{C}_2')$. Moreover, by copying the derivation of (4.1.48), it follows that from (4.1.54) we get

$$\begin{aligned}&|\mathbf{S}_2||\mathbf{I} + \mathbf{V}_2\mathbf{S}_2^{-1}\mathbf{V}_2'|\\ &=|\mathbf{S}_2||\mathbf{I} + \mathbf{V}_2'\mathbf{P}'_{\mathbf{T}_1\mathbf{A}_2,\mathbf{S}_2}\mathbf{S}_2^{-1}\mathbf{P}_{\mathbf{T}_1\mathbf{A}_2,\mathbf{S}_2}\mathbf{V}_2 + \mathbf{V}_2'\mathbf{P}_{(\mathbf{T}_1\mathbf{A}_2)^o,\mathbf{S}_2^{-1}}\mathbf{S}_2^{-1}\mathbf{P}'_{(\mathbf{T}_1\mathbf{A}_2)^o,\mathbf{S}_2^{-1}}\mathbf{V}_2|\\ &\geq|\mathbf{S}_2||\mathbf{I} + \mathbf{W}_2'\mathbf{P}_{(\mathbf{T}_1\mathbf{A}_2)^o,\mathbf{S}_2^{-1}}\mathbf{S}_2^{-1}\mathbf{P}'_{(\mathbf{T}_1\mathbf{A}_2)^o,\mathbf{S}_2^{-1}}\mathbf{W}_2|\\ &=|\mathbf{S}_2||\mathbf{I} + \mathbf{S}_2^{-1}\mathbf{P}'_{(\mathbf{T}_1\mathbf{A}_2)^o,\mathbf{S}_2^{-1}}\mathbf{W}_2\mathbf{W}_2'\mathbf{P}_{(\mathbf{T}_1\mathbf{A}_2)^o,\mathbf{S}_2^{-1}}|\\ &=|\mathbf{S}_2 + \mathbf{P}'_{(\mathbf{T}_1\mathbf{A}_2)^o,\mathbf{S}_2^{-1}}\mathbf{W}_2\mathbf{W}_2'\mathbf{P}_{(\mathbf{T}_1\mathbf{A}_2)^o,\mathbf{S}_2^{-1}}|\\ &=|\mathbf{S}_2 + \mathbf{P}_3\mathbf{W}_2\mathbf{W}_2'\mathbf{P}_3'|, \qquad (4.1.55)\end{aligned}$$

where

$$\begin{aligned}\mathbf{W}_2 =&\mathbf{X}\mathbf{C}_2'(\mathbf{C}_2\mathbf{C}_2')^-\mathbf{C}_2 - \mathbf{A}_3\mathbf{B}_3\mathbf{C}_3,\\ \mathbf{P}_3 =&\mathbf{T}_2\mathbf{T}_1, \qquad (4.1.56)\\ \mathbf{T}_2 =&\mathbf{I} - \mathbf{P}_{\mathbf{T}_1\mathbf{A}_2,\mathbf{S}_2}. \qquad (4.1.57)\end{aligned}$$

Equality in (4.1.55) holds if and only if

$$\mathbf{A}_2'\mathbf{T}_1'\mathbf{S}_2^{-1}\mathbf{V}_2 = \mathbf{0}. \tag{4.1.58}$$

Moreover, since $\mathbf{W}_2$ does not include $\mathbf{B}_1$ and $\mathbf{B}_2$, by (4.1.55) we are back to the relations used when finding estimators for the ordinary Growth Curve model. Thus, by continuing the calculations from (4.1.55) in a straightforward manner we get

$$|\mathbf{S}_2 + \mathbf{P}_3\mathbf{W}_2\mathbf{W}_2'\mathbf{P}_3'| = |\mathbf{S}_3 + \mathbf{V}_3\mathbf{V}_3'|, \tag{4.1.59}$$

where

$$\begin{aligned}\mathbf{S}_3 =& \mathbf{S}_2 + \mathbf{P}_3\mathbf{X}\mathbf{P}_{\mathbf{C}_2'}(\mathbf{I} - \mathbf{P}_{\mathbf{C}_3'})\mathbf{P}_{\mathbf{C}_2'}\mathbf{X}'\mathbf{P}_3', \\ \mathbf{V}_3 =& \mathbf{P}_3\mathbf{X}\mathbf{C}_3'(\mathbf{C}_3\mathbf{C}_3')^-\mathbf{C}_3 - \mathbf{P}_3\mathbf{A}_3\mathbf{B}_3\mathbf{C}_3.\end{aligned} \tag{4.1.60}$$

The matrix $\mathbf{S}_3$ is non-singular with probability 1. Furthermore, from (4.1.59) a chain of calculations can be started:

$$\begin{aligned}&|\mathbf{S}_3||\mathbf{I} + \mathbf{V}_3\mathbf{S}_3^{-1}\mathbf{V}_3'| \\ &=|\mathbf{S}_3||\mathbf{I} + \mathbf{V}_3'\mathbf{P}_{\mathbf{P}_3\mathbf{A}_3,\mathbf{S}_3}'\mathbf{S}_3^{-1}\mathbf{P}_{\mathbf{P}_3\mathbf{A}_3,\mathbf{S}_3}\mathbf{V}_3 + \mathbf{V}_3'\mathbf{P}_{(\mathbf{P}_3\mathbf{A}_3)^o,\mathbf{S}_3^{-1}}\mathbf{S}_3^{-1}\mathbf{P}_{(\mathbf{P}_3\mathbf{A}_3)^o,\mathbf{S}_3^{-1}}'\mathbf{V}_3| \\ &\geq|\mathbf{S}_3||\mathbf{I} + \mathbf{W}_3'\mathbf{P}_{(\mathbf{P}_3\mathbf{A}_3)^o,\mathbf{S}_3^{-1}}\mathbf{S}_3^{-1}\mathbf{P}_{(\mathbf{P}_3\mathbf{A}_3)^o,\mathbf{S}_3^{-1}}'\mathbf{W}_3| \\ &=|\mathbf{S}_3||\mathbf{I} + \mathbf{S}_3^{-1}\mathbf{P}_{(\mathbf{P}_3\mathbf{A}_3)^o,\mathbf{S}_3^{-1}}'\mathbf{W}_3\mathbf{W}_3'\mathbf{P}_{(\mathbf{P}_3\mathbf{A}_3)^o,\mathbf{S}_3^{-1}}| \\ &=|\mathbf{S}_3 + \mathbf{P}_{(\mathbf{P}_3\mathbf{A}_3)^o,\mathbf{S}_3^{-1}}'\mathbf{W}_3\mathbf{W}_3'\mathbf{P}_{(\mathbf{P}_3\mathbf{A}_3)^o,\mathbf{S}_3^{-1}}| \\ &=|\mathbf{S}_3 + \mathbf{P}_4\mathbf{W}_3\mathbf{W}_3'\mathbf{P}_4'|,\end{aligned} \tag{4.1.61}$$

where

$$\begin{aligned}\mathbf{W}_3 =& \mathbf{X}\mathbf{C}_3'(\mathbf{C}_3\mathbf{C}_3')^-\mathbf{C}_3, \\ \mathbf{P}_4 =& \mathbf{T}_3\mathbf{T}_2\mathbf{T}_1, \\ \mathbf{T}_3 =& \mathbf{I} - \mathbf{P}_{\mathbf{P}_3\mathbf{A}_3,\mathbf{S}_3}.\end{aligned}$$

Note that $\mathbf{S}_3^{-1}$ exists with probability 1, and equality in (4.1.61) holds if and only if

$$\mathbf{A}_3'\mathbf{P}_3'\mathbf{S}_3^{-1}\mathbf{V}_3 = \mathbf{0}. \tag{4.1.62}$$

Furthermore, it is observed that the right hand side of (4.1.61) does not include any parameter. Thus, if we can find values of $\mathbf{B}_1$, $\mathbf{B}_2$ and $\mathbf{B}_3$, i.e. parameter estimators, which satisfy (4.1.50), (4.1.58) and (4.1.62), the maximum likelihood estimators of the parameters have been found, when $\mathbf{X} \sim N_{p,n}(\sum_{i=1}^3 \mathbf{A}_i\mathbf{B}_i\mathbf{C}_i, \mathbf{\Sigma}, \mathbf{I}_n)$.

Theorem 4.1.6. *Let* $\mathbf{X} \sim N_{p,n}(\sum_{i=1}^3 \mathbf{A}_i\mathbf{B}_i\mathbf{C}_i, \mathbf{\Sigma}, \mathbf{I}_n)$ $n \geq p + r(\mathbf{C}_1)$, *where* $\mathbf{B}_i$, $i = 1, 2, 3$, *and* $\mathbf{\Sigma}$ *are unknown parameters. Representations of their maximum* *likelihood estimators are given by*

$$\begin{aligned}
\widehat{\mathbf{B}}_3 =&(\mathbf{A}_3'\mathbf{P}_3'\mathbf{S}_3^{-1}\mathbf{P}_3\mathbf{A}_3)^-\mathbf{A}_3'\mathbf{P}_3'\mathbf{S}_3^{-1}\mathbf{X}\mathbf{C}_3'(\mathbf{C}_3\mathbf{C}_3')^-\\
&+(\mathbf{A}_3'\mathbf{P}_3')^o\mathbf{Z}_{31}+\mathbf{A}_3'\mathbf{P}_3'\mathbf{Z}_{32}\mathbf{C}_3^{o'},\\
\widehat{\mathbf{B}}_2 =&(\mathbf{A}_2'\mathbf{T}_1'\mathbf{S}_2^{-1}\mathbf{T}_1\mathbf{A}_2)^-\mathbf{A}_2'\mathbf{T}_1'\mathbf{S}_2^{-1}(\mathbf{X}-\mathbf{A}_3\widehat{\mathbf{B}}_3\mathbf{C}_3)\mathbf{C}_2'(\mathbf{C}_2\mathbf{C}_2')^-\\
&+(\mathbf{A}_2'\mathbf{T}_1')^o\mathbf{Z}_{21}+\mathbf{A}_2'\mathbf{T}_1'\mathbf{Z}_{22}\mathbf{C}_2^{o'},\\
\widehat{\mathbf{B}}_1 =&(\mathbf{A}_1'\mathbf{S}_1^{-1}\mathbf{A}_1)^-\mathbf{A}_1'\mathbf{S}_1^{-1}(\mathbf{X}-\mathbf{A}_2\widehat{\mathbf{B}}_2\mathbf{C}_2-\mathbf{A}_3\widehat{\mathbf{B}}_3\mathbf{C}_3)\mathbf{C}_1'(\mathbf{C}_1\mathbf{C}_1')^-\\
&+(\mathbf{A}_1')^o\mathbf{Z}_{11}+\mathbf{A}_1'\mathbf{Z}_{12}\mathbf{C}_1^{o'},\\
n\widehat{\mathbf{\Sigma}} =&(\mathbf{X}-\mathbf{A}_1\widehat{\mathbf{B}}_1\mathbf{C}_1-\mathbf{A}_2\widehat{\mathbf{B}}_2\mathbf{C}_2-\mathbf{A}_3\widehat{\mathbf{B}}_3\mathbf{C}_3)()',
\end{aligned}$$

where $\mathbf{Z}_{ij}$, $i=1,2,3$, $j=1,2$, *are arbitrary matrices, and* $\mathbf{P}_3$, $\mathbf{S}_3$ $\mathbf{T}_1$, $\mathbf{S}_2$ *and* $\mathbf{S}_1$ *are defined by (4.1.56), (4.1.60), (4.1.52), (4.1.53) and (4.1.47), respectively.*

PROOF: The equations (4.1.50), (4.1.58) and (4.1.62) are going to be solved. The last equation is a linear equation in $\mathbf{B}_3$ and using Theorem 1.3.4 the estimator $\widehat{\mathbf{B}}_3$ is obtained. Inserting this solution into (4.1.58) implies that we have a linear equation in $\mathbf{B}_2$ and via Theorem 1.3.4 the estimator $\widehat{\mathbf{B}}_2$ is found. Now, inserting both $\widehat{\mathbf{B}}_3$ and $\widehat{\mathbf{B}}_2$ into (4.1.50) and then solving the equation gives $\widehat{\mathbf{B}}_1$. Finally, plugging $\widehat{\mathbf{B}}_i$, $i=1,2,3$, into (4.1.44) establishes the theorem. ∎

It has already been mentioned that the results of Theorem 4.1.6 can be extended so that maximum likelihood estimators can be found for the parameters in the MLNM($\sum_{i=1}^m \mathbf{A}_i\mathbf{B}_i\mathbf{C}_i$) (von Rosen, 1989). These estimators are presented in the next theorem.

Theorem 4.1.7. *Let*

$$\begin{aligned}
\mathbf{P}_r =&\mathbf{T}_{r-1}\mathbf{T}_{r-2}\times\cdots\times\mathbf{T}_0, \quad \mathbf{T}_0=\mathbf{I}, \quad r=1,2,\ldots,m+1,\\
\mathbf{T}_i =&\mathbf{I}-\mathbf{P}_i\mathbf{A}_i(\mathbf{A}_i'\mathbf{P}_i'\mathbf{S}_i^{-1}\mathbf{P}_i\mathbf{A}_i)^-\mathbf{A}_i'\mathbf{P}_i'\mathbf{S}_i^{-1}, \quad i=1,2,\ldots,m,\\
\mathbf{S}_i =&\sum_{j=1}^{i}\mathbf{K}_j, \qquad i=1,2,\ldots,m,\\
\mathbf{K}_j =&\mathbf{P}_j\mathbf{X}\mathbf{P}_{\mathbf{C}_{j-1}'}(\mathbf{I}-\mathbf{P}_{\mathbf{C}_j'})\mathbf{P}_{\mathbf{C}_{j-1}'}\mathbf{X}'\mathbf{P}_j', \quad \mathbf{C}_0=\mathbf{I},\\
\mathbf{P}_{\mathbf{C}_{j-1}'} =&\mathbf{C}_{j-1}'(\mathbf{C}_{j-1}\mathbf{C}_{j-1}')^-\mathbf{C}_{j-1}.
\end{aligned}$$

Assume that $\mathbf{S}_1$ *is p.d., then representations of the maximum likelihood estimators for the* MLNM($\sum_{i=1}^m \mathbf{A}_i\mathbf{B}_i\mathbf{C}_i$) *are given by*

$$\begin{aligned}
\widehat{\mathbf{B}}_r =&(\mathbf{A}_r'\mathbf{P}_r'\mathbf{S}_r^{-1}\mathbf{P}_r\mathbf{A}_r)^-\mathbf{A}_r'\mathbf{P}_r'\mathbf{S}_r^{-1}(\mathbf{X}-\sum_{i=r+1}^{m}\mathbf{A}_i\widehat{\mathbf{B}}_i\mathbf{C}_i)\mathbf{C}_r'(\mathbf{C}_r\mathbf{C}_r')^-\\
&+(\mathbf{A}_r'\mathbf{P}_r')^o\mathbf{Z}_{r1}+\mathbf{A}_r'\mathbf{P}_r'\mathbf{Z}_{r2}\mathbf{C}_r^{o'}, \qquad r=1,2,\ldots,m,\\
n\widehat{\mathbf{\Sigma}} =&(\mathbf{X}-\sum_{i=1}^{m}\mathbf{A}_i\widehat{\mathbf{B}}_i\mathbf{C}_i)(\mathbf{X}-\sum_{i=1}^{m}\mathbf{A}_i\widehat{\mathbf{B}}_i\mathbf{C}_i)'\\
=&\mathbf{S}_m+\mathbf{P}_{m+1}\mathbf{X}\mathbf{C}_m'(\mathbf{C}_m\mathbf{C}_m')^-\mathbf{C}_m\mathbf{X}'\mathbf{P}_{m+1}',
\end{aligned}$$

where the matrices $\mathbf{Z}_{rj}$ *are arbitrary. Here* $\sum_{i=m+1}^{m} \mathbf{A}_i \widehat{\mathbf{B}}_i \mathbf{C}_i = \mathbf{0}$.

PROOF: One can show that

$$
\begin{aligned}
&|\mathbf{S}_{r-1} + \mathbf{P}_r(\mathbf{X} - \sum_{i=r}^{m} \mathbf{A}_i\mathbf{B}_i\mathbf{C}_i)\mathbf{C}'_{r-1}(\mathbf{C}_{r-1}\mathbf{C}'_{r-1})^{-}\mathbf{C}_{r-1}(\mathbf{X} - \sum_{i=r}^{m} \mathbf{A}_i\mathbf{B}_i\mathbf{C}_i)\mathbf{P}'_r| \\
&\quad \geq |\mathbf{S}_r + \mathbf{P}_{r+1}(\mathbf{X} - \sum_{i=r+1}^{m} \mathbf{A}_i\mathbf{B}_i\mathbf{C}_i)\mathbf{C}'_r(\mathbf{C}_r\mathbf{C}'_r)^{-}\mathbf{C}_r(\mathbf{X} - \sum_{i=r+1}^{m} \mathbf{A}_i\mathbf{B}_i\mathbf{C}_i)\mathbf{P}'_{r+1}| \\
&\quad \geq |n\widehat{\mathbf{\Sigma}}|, \qquad r = 2, 3, \ldots, m.
\end{aligned}
$$

Therefore, we get an explicit expression of the upper bound of the likelihood function as in the proof of Theorem 4.1.6. ∎

In order to shed some light on the expressions in Theorem 4.1.7 the next lemma presents an important algebraic fact about $\mathbf{P}_r$ and $\mathbf{S}_r$.

Lemma 4.1.3. *Let* $\mathbf{P}_r$ *and* $\mathbf{S}_r$ *be as in Theorem 4.1.7, and*

$$
\begin{aligned}
\mathbf{G}_{r+1} &= \mathbf{G}_r(\mathbf{G}'_r\mathbf{A}_{r+1})^o, \quad \mathbf{G}_0 = \mathbf{I}, \\
\mathbf{W}_r &= \mathbf{X}(\mathbf{I} - \mathbf{C}'_r(\mathbf{C}_r\mathbf{C}'_r)^{-}\mathbf{C}_r)\mathbf{X}' \sim W_p(\mathbf{\Sigma}, n - r(\mathbf{C}_r)),
\end{aligned}
$$

where $(\mathbf{G}'_r\mathbf{A}_{r+1})^o$ *is chosen so that* $\mathbf{G}_{r+1}$ *is of full rank. Then*

(i) $\quad \mathbf{P}'_r\mathbf{S}_r^{-1}\mathbf{P}_r = \mathbf{P}'_r\mathbf{S}_r^{-1}, \qquad \mathbf{P}_r\mathbf{P}_r = \mathbf{P}_r, \qquad r = 1, 2, \ldots, m;$

(ii) $\quad \mathbf{P}'_r\mathbf{S}_r^{-1}\mathbf{P}_r = \mathbf{G}_{r-1}(\mathbf{G}'_{r-1}\mathbf{W}_r\mathbf{G}_{r-1})^{-1}\mathbf{G}'_{r-1}, \qquad r = 1, 2, \ldots, m.$

PROOF: Both (i) and (ii) will be proved via induction. Note that (i) is true for $r = 1, 2$. Suppose that $\mathbf{P}'_{r-1}\mathbf{S}_{r-1}^{-1}\mathbf{P}_{r-1} = \mathbf{P}'_{r-1}\mathbf{S}_{r-1}^{-1}$ and $\mathbf{P}_{r-1}\mathbf{P}_{r-1} = \mathbf{P}_{r-1}$. By applying Proposition 1.3.6 to $\mathbf{S}_r^{-1} = (\mathbf{S}_{r-1} + \mathbf{K}_r)^{-1}$ we get that it is sufficient to show that $\mathbf{P}'_r\mathbf{S}_{r-1}^{-1}\mathbf{P}_r = \mathbf{P}'_r\mathbf{S}_{r-1}^{-1} = \mathbf{S}_{r-1}^{-1}\mathbf{P}_r$. However,

$$
\mathbf{P}'_r\mathbf{S}_{r-1}^{-1}\mathbf{P}_r = \mathbf{P}'_{r-1}\mathbf{S}_{r-1}^{-1}\mathbf{P}_r = \mathbf{S}_{r-1}^{-1}\mathbf{P}_{r-1}\mathbf{P}_r = \mathbf{S}_{r-1}^{-1}\mathbf{P}_r
$$

and

$$
\mathbf{P}_r\mathbf{P}_r = \mathbf{T}_{r-1}\mathbf{P}_{r-1}\mathbf{P}_r = \mathbf{T}_{r-1}\mathbf{P}_r = \mathbf{P}_r.
$$

Concerning (ii), it is noted that the case $r = 1$ is trivial. For $r = 2$, if

$$
\mathbf{F}_j = \mathbf{P}_{\mathbf{C}'_{j-1}}(\mathbf{I} - \mathbf{P}_{\mathbf{C}'_j}),
$$

where as previously $\mathbf{P}_{\mathbf{C}'_j} = \mathbf{C}'_j(\mathbf{C}_j\mathbf{C}'_j)^{-}\mathbf{C}_j$, we obtain by using Proposition 1.3.6 that

$$
\begin{aligned}
\mathbf{P}'_2\mathbf{S}_2^{-1}\mathbf{P}_2 &= \mathbf{T}'_1\mathbf{S}_2^{-1}\mathbf{T}_1 \\
&= \mathbf{T}'_1\mathbf{S}_1^{-1}\mathbf{T}_1 - \mathbf{T}'_1\mathbf{S}_1^{-1}\mathbf{T}_1\mathbf{X}\mathbf{F}_2 \\
&\qquad \times \{\mathbf{I} + \mathbf{F}'_2\mathbf{X}'\mathbf{T}'_1\mathbf{S}_1^{-1}\mathbf{T}_1\mathbf{X}\mathbf{F}_2\}^{-1}\mathbf{F}'_2\mathbf{X}'\mathbf{T}'_1\mathbf{S}_1^{-1}\mathbf{T}_1 \\
&= \mathbf{G}_1(\mathbf{G}'_1\mathbf{W}_2\mathbf{G}_1)^{-1}\mathbf{G}'_1.
\end{aligned}
$$

In the next suppose that (ii) holds for $r-1$. The following relation will be used in the subsequent:

$$
\begin{aligned}
&\mathbf{P}_r'\mathbf{S}_{r-1}^{-1}\mathbf{P}_r\\
=&\mathbf{P}_{r-1}'\{\mathbf{S}_{r-1}^{-1}-\mathbf{S}_{r-1}^{-1}\mathbf{P}_{r-1}\mathbf{A}_{r-1}(\mathbf{A}_{r-1}'\mathbf{P}_{r-1}'\mathbf{S}_{r-1}^{-1}\mathbf{P}_{r-1}\mathbf{A}_{r-1})^{-}\mathbf{A}_{r-1}'\mathbf{P}_{r-1}'\mathbf{S}_{r-1}^{-1}\}\mathbf{P}_{r-1}\\
=&\mathbf{G}_{r-1}(\mathbf{G}_{r-1}'\mathbf{W}_r\mathbf{G}_{r-1})^{-1}\mathbf{G}_{r-1}',
\end{aligned}
$$

which can be verified via Proposition 1.3.6. Thus,

$$
\begin{aligned}
&\mathbf{P}_r'\mathbf{S}_r^{-1}\mathbf{P}_r\\
&=\mathbf{P}_r'\{\mathbf{S}_{r-1}^{-1}-\mathbf{S}_{r-1}^{-1}\mathbf{P}_r\mathbf{X}\mathbf{F}_r(\mathbf{F}_r'\mathbf{X}'\mathbf{P}_r'\mathbf{S}_r^{-1}\mathbf{P}_r\mathbf{X}\mathbf{F}_r+\mathbf{I})^{-1}\mathbf{F}_r'\mathbf{X}'\mathbf{P}_r'\mathbf{S}_r^{-1}\}\mathbf{P}_r\\
&=\mathbf{G}_{r-1}\{(\mathbf{G}_{r-1}'\mathbf{W}_{r-1}\mathbf{G}_{r-1})^{-1}\\
&\qquad-(\mathbf{G}_{r-1}'\mathbf{W}_{r-1}\mathbf{G}_{r-1})^{-1}\mathbf{G}_{r-1}'\mathbf{X}\mathbf{F}_r\mathbf{F}_r'\mathbf{X}'\mathbf{G}_{r-1}(\mathbf{G}_{r-1}'\mathbf{W}_{r-1}\mathbf{G}_{r-1})^{-1}\}\mathbf{G}_{r-1}'\\
&=\mathbf{G}_{r-1}(\mathbf{G}_{r-1}'\mathbf{W}_{r-1}\mathbf{G}_{r-1})^{-1}\mathbf{G}_{r-1}'
\end{aligned}
$$

and (ii) is established. ∎

Besides the algebraic identities, the lemma tells us that according to (i), $\mathbf{P}_r$ is idempotent. Hence, $\mathbf{P}_r$ is a projector and thus gives us a possibility to interpret the results geometrically. Furthermore, note that in (ii)

$$C(\mathbf{G}_r)=C(\mathbf{A}_1:\mathbf{A}_2:\ldots:\mathbf{A}_r)^{\perp}$$

and that the distribution of $\mathbf{P}_r'\mathbf{S}_r^{-1}\mathbf{P}_r$ follows an Inverted Wishart distribution. Moreover, $\mathbf{G}_r$ consists of p rows and m_r columns, where

$$
\begin{aligned}
m_r=&p-r(\mathbf{G}_{r-1}'\mathbf{A}_r)\\
=&p-r(\mathbf{A}_1:\mathbf{A}_2:\ldots:\mathbf{A}_r)+r(\mathbf{A}_1:\mathbf{A}_2:\ldots:\mathbf{A}_{r-1}),\quad r>1, &(4.1.63)\\
m_1=&p-r(\mathbf{A}_1),\quad m_0=p. &(4.1.64)
\end{aligned}
$$

Now a lemma is presented which is very important since it gives the basis for rewriting the general MLNM($\sum_{i=1}^m\mathbf{A}_i\mathbf{B}_i\mathbf{C}_i$) in a canonical form. In the next section this lemma will be utilized when deriving the dispersion matrices of the maximum likelihood estimators. The existence of the matrices in the lemma follows from Proposition 1.1.6.

Lemma 4.1.4. *Let the non-singular matrix $\mathbf{H}_r: m_r\times m_r$, where m_r is given in (4.1.63) and (4.1.64), and the orthogonal matrix*

$$\mathbf{\Gamma}^r=((\mathbf{\Gamma}_1^r)':(\mathbf{\Gamma}_2^r)')',\qquad \mathbf{\Gamma}^r:m_{r-1}\times m_{r-1},\ \mathbf{\Gamma}_1^r:m_r\times m_{r-1},$$

be defined by

$$
\begin{aligned}
&\mathbf{G}_1=\mathbf{A}_1^{o'}=\mathbf{H}_1(\mathbf{I}_{p-r(\mathbf{A}_1)}:\mathbf{0})\mathbf{\Gamma}^1\mathbf{\Sigma}^{-1/2}=\mathbf{H}_1\mathbf{\Gamma}_1^1\mathbf{\Sigma}^{-1/2},\\
&(\mathbf{G}_{r-1}'\mathbf{A}_r)^{o'}\mathbf{H}_{r-1}=\mathbf{H}_r(\mathbf{I}_{m_r}:\mathbf{0})\mathbf{\Gamma}^r=\mathbf{H}_r\mathbf{\Gamma}_1^r,\quad \mathbf{\Gamma}^0=\mathbf{I},\ \mathbf{H}_0=\mathbf{I},
\end{aligned}
$$

where $\mathbf{G}_s$ *is defined in Lemma 4.1.3. Furthermore, let*

$$\begin{aligned}
\mathbf{V}^r =&\boldsymbol{\Gamma}^r\boldsymbol{\Gamma}_1^{r-1} \times \cdots \times \boldsymbol{\Gamma}_1^1\boldsymbol{\Sigma}^{-1/2}\mathbf{W}_r\boldsymbol{\Sigma}^{-1/2}(\boldsymbol{\Gamma}_1^1)' \\
&\times \cdots \times (\boldsymbol{\Gamma}_1^{r-1})'(\boldsymbol{\Gamma}^r)' \sim W_{m_{r-1}}(\mathbf{I}, n - r(\mathbf{C}_r)),\\
\underline{\mathbf{V}}^r =&\boldsymbol{\Gamma}^{r-1}\boldsymbol{\Gamma}_1^{r-2} \times \cdots \times \boldsymbol{\Gamma}_1^1\boldsymbol{\Sigma}^{-1/2}\mathbf{W}_r\boldsymbol{\Sigma}^{-1/2}(\boldsymbol{\Gamma}_1^1)' \times \cdots \times (\boldsymbol{\Gamma}_1^{r-2})'(\boldsymbol{\Gamma}_1^{r-1})', \; r > 1,\\
\underline{\mathbf{V}}^1 =&\boldsymbol{\Sigma}^{-1/2}\mathbf{W}_1\boldsymbol{\Sigma}^{-1/2},\\
\mathbf{M}_r =&\mathbf{W}_r\mathbf{G}_r(\mathbf{G}_r'\mathbf{W}_r\mathbf{G}_r)^{-1}\mathbf{G}_r'\\
\mathbf{F}_r =&\mathbf{C}_{r-1}'(\mathbf{C}_{r-1}\mathbf{C}_{r-1}')^{-}\mathbf{C}_{r-1}(\mathbf{I} - \mathbf{C}_r'(\mathbf{C}_r\mathbf{C}_r')^{-}\mathbf{C}_r),\\
\mathbf{Z}_{r,s} =&\boldsymbol{\Gamma}_1^{r-1}\boldsymbol{\Gamma}_1^{r-2} \times \cdots \times \boldsymbol{\Gamma}_1^1\boldsymbol{\Sigma}^{-1/2}\mathbf{M}_r\mathbf{M}_{r+1} \times \cdots \times \mathbf{M}_s, \; s \geq r \geq 2,\\
\mathbf{Z}_{1,s} =&\mathbf{M}_1\mathbf{M}_2 \times \cdots \times \mathbf{M}_s,\\
\mathbf{N}_r' =&(\mathbf{I} : (\mathbf{V}_{11}^r)^{-1}\mathbf{V}_{12}^r),
\end{aligned}$$

where $\mathbf{V}_{11}^r$ *and* $\mathbf{V}_{12}^r$ *refer to a standard partition of* $\mathbf{V}^r$. *Then*

(i) $\mathbf{G}_r' = \mathbf{H}_r\boldsymbol{\Gamma}_1^r\boldsymbol{\Gamma}_1^{r-1} \times \cdots \times \boldsymbol{\Gamma}_1^1\boldsymbol{\Sigma}^{-1/2}$;

(ii) $\mathbf{P}_r = \mathbf{Z}_{1,r-1}, \quad \mathbf{Z}_{1,0} = \mathbf{I}$; *(*$\mathbf{P}_r$ *is as in Theorem 4.1.7)*

(iii) $\boldsymbol{\Gamma}_1^{r-1}\boldsymbol{\Gamma}_1^{r-2} \times \cdots \times \boldsymbol{\Gamma}_1^1\boldsymbol{\Sigma}^{-1/2}\mathbf{M}_r = (\boldsymbol{\Gamma}^r)'\mathbf{N}_r\boldsymbol{\Gamma}_1^r\boldsymbol{\Gamma}_1^{r-1} \times \cdots \times \boldsymbol{\Gamma}_1^1\boldsymbol{\Sigma}^{-1/2}$;

(iv) $$\begin{aligned}\underline{\mathbf{V}}_{11}^r = \mathbf{V}_{11}^{r-1} &+ \boldsymbol{\Gamma}_1^{r-1}\boldsymbol{\Gamma}_1^{r-2} \times \cdots \times \boldsymbol{\Gamma}_1^1\boldsymbol{\Sigma}^{-1/2}\mathbf{X}\mathbf{F}_r\\ &\times \mathbf{F}_r'\mathbf{X}'\boldsymbol{\Sigma}^{-1/2}(\boldsymbol{\Gamma}_1^1)' \times \cdots \times (\boldsymbol{\Gamma}_1^{r-2})'(\boldsymbol{\Gamma}_1^{r-1})';\end{aligned}$$

(v) $$\begin{aligned}&\boldsymbol{\Sigma}^{1/2}(\boldsymbol{\Gamma}_1^1)' \times \cdots \times (\boldsymbol{\Gamma}_1^{r-1})'(\boldsymbol{\Gamma}_1^r)'\boldsymbol{\Gamma}_1^r\boldsymbol{\Gamma}_1^{r-1} \times \cdots \times \boldsymbol{\Gamma}_1^1\boldsymbol{\Sigma}^{1/2}\\ &\quad = \boldsymbol{\Sigma}\mathbf{G}_r(\mathbf{G}_r'\boldsymbol{\Sigma}\mathbf{G}_r)^{-1}\mathbf{G}_r'\boldsymbol{\Sigma};\end{aligned}$$

(vi) $$\begin{aligned}&\boldsymbol{\Sigma}^{1/2}(\boldsymbol{\Gamma}_1^1)' \times \cdots \times (\boldsymbol{\Gamma}_1^{r-1})'(\boldsymbol{\Gamma}_2^r)'\boldsymbol{\Gamma}_2^r\boldsymbol{\Gamma}_1^{r-1} \times \cdots \times \boldsymbol{\Gamma}_1^1\boldsymbol{\Sigma}^{1/2}\\ &\quad = \boldsymbol{\Sigma}\mathbf{G}_{r-1}(\mathbf{G}_{r-1}'\boldsymbol{\Sigma}\mathbf{G}_{r-1})^{-1}\mathbf{G}_{r-1}'\boldsymbol{\Sigma} - \boldsymbol{\Sigma}\mathbf{G}_r(\mathbf{G}_r'\boldsymbol{\Sigma}\mathbf{G}_r)^{-1}\mathbf{G}_r'\boldsymbol{\Sigma}.\end{aligned}$$

PROOF: By definition of $\mathbf{G}_r$, given in Lemma 4.1.3, and the assumptions we obtain

$$\begin{aligned}
\mathbf{G}_r' =&(\mathbf{G}_{r-1}'\mathbf{A}_r)^{o'}\mathbf{G}_{r-1}' = \mathbf{H}_r\boldsymbol{\Gamma}_1^r\mathbf{H}_{r-1}^{-1}\mathbf{G}_{r-1}' = \mathbf{H}_r\boldsymbol{\Gamma}_1^r\boldsymbol{\Gamma}_1^{r-1}\mathbf{H}_{r-2}^{-1}\mathbf{G}_{r-2}'\\
=&\ldots = \mathbf{H}_r\boldsymbol{\Gamma}_1^r\boldsymbol{\Gamma}_1^{r-1} \times \cdots \times \boldsymbol{\Gamma}_1^1\boldsymbol{\Sigma}^{-1/2}
\end{aligned}$$

and thus (i) has been established.
Concerning (ii), first observe that by Corollary 1.2.25.1

$$\mathbf{P}_2 = \mathbf{I} - \mathbf{A}_1(\mathbf{A}_1'\mathbf{S}_1^{-1}\mathbf{A}_1)^{-}\mathbf{A}_1'\mathbf{S}_1^{-1} = \mathbf{W}_1\mathbf{G}_1(\mathbf{G}_1'\mathbf{W}_1\mathbf{G}_1)^{-1}\mathbf{G}_1',$$

which by definition is identical to $\mathbf{M}_1$. Then we assume that the statement is true for $2, \ldots, r-1$, i.e. $\mathbf{P}_{r-1} = \mathbf{Z}_{1,r-2}$. Hence, the definition of $\mathbf{P}_r$, given in Theorem 4.1.7 and Lemma 4.1.3 (ii), yields

$$\begin{aligned}
\mathbf{P}_r =&\mathbf{T}_{r-1}\mathbf{P}_{r-1} = \mathbf{P}_{r-2}\mathbf{W}_{r-2}\mathbf{G}_{r-2}(\mathbf{G}_{r-2}'\mathbf{W}_{r-2}\mathbf{G}_{r-2})^{-1}\\
&\times \Big\{\mathbf{I} - \mathbf{G}_{r-2}'\mathbf{A}_{r-1}\{\mathbf{A}_{r-1}'\mathbf{G}_{r-2}(\mathbf{G}_{r-2}'\mathbf{W}_{r-1}\mathbf{G}_{r-2})^{-1}\mathbf{G}_{r-2}'\mathbf{A}_{r-1}\}^{-}\\
&\quad \times \mathbf{A}_{r-1}'\mathbf{G}_{r-2}(\mathbf{G}_{r-2}'\mathbf{W}_{r-1}\mathbf{G}_{r-2})^{-1}\Big\}\mathbf{G}_{r-2}'\\
=&\mathbf{P}_{r-2}\mathbf{Q}_{r-2}\mathbf{W}_{r-1}\mathbf{G}_{r-2}\{(\mathbf{G}_{r-2}'\mathbf{A}_{r-1})^{o'}\mathbf{G}_{r-2}'\mathbf{W}_{r-1}\mathbf{G}_{r-2}(\mathbf{G}_{r-2}'\mathbf{A}_{r-1})^{o}\}^{-1}\\
&\times (\mathbf{G}_{r-2}'\mathbf{A}_{r-1})^{o'}\mathbf{G}_{r-2}'\\
=&\mathbf{P}_{r-2}\mathbf{Q}_{r-2}\mathbf{Q}_{r-1} = \mathbf{Z}_{1,r-1}.
\end{aligned}$$

The statement in (iii) follows immediately from (i) and the definition of $\mathbf{V}^r$. Furthermore, (iv) is true since

$$\mathbf{W}_r = \mathbf{W}_{r-1} + \mathbf{X}\mathbf{F}_r\mathbf{F}_r'\mathbf{X}',$$

and (v) and (vi) are direct consequences of (i). ∎

The estimators $\widehat{\mathbf{B}}_i$ in Theorem 4.1.7 are given by a recursive formula. In order to present the expressions in a non-recursive way one has to impose some kind of uniqueness conditions to the model, such as $\sum_{i=r+1}^m \mathbf{A}_i\widehat{\mathbf{B}}_i\mathbf{C}_i$ to be unique. Otherwise expressions given in a non-recursive way are rather hard to interpret. However, without any further assumptions, $\mathbf{P}_r\sum_{i=r}^m \mathbf{A}_i\widehat{\mathbf{B}}_i\mathbf{C}_i$ is always unique. The next theorem gives its expression in a non-recursive form.

Theorem 4.1.8. *For the estimators $\widehat{\mathbf{B}}_i$, given in Theorem 4.1.7,*

$$\mathbf{P}_r\sum_{i=r}^m \mathbf{A}_i\widehat{\mathbf{B}}_i\mathbf{C}_i = \sum_{i=r}^m (\mathbf{I} - \mathbf{T}_i)\mathbf{X}\mathbf{C}_i'(\mathbf{C}_i\mathbf{C}_i')^-\mathbf{C}_i.$$

PROOF: In Lemma 4.1.3 the relation $\mathbf{P}_r'\mathbf{S}_r^{-1}\mathbf{P}_r = \mathbf{P}_r'\mathbf{S}_r^{-1}$ was verified. Thus, it follows that $(\mathbf{I} - \mathbf{T}_r) = (\mathbf{I} - \mathbf{T}_r)\mathbf{P}_r$. Then

$$\begin{aligned}
&\mathbf{P}_r\sum_{i=r}^m \mathbf{A}_i\widehat{\mathbf{B}}_i\mathbf{C}_i \\
&= (\mathbf{I} - \mathbf{T}_r)\mathbf{X}\mathbf{C}_r'(\mathbf{C}_r\mathbf{C}_r')^-\mathbf{C}_r - (\mathbf{I} - \mathbf{T}_r)\sum_{i=r+1}^m \mathbf{A}_i\widehat{\mathbf{B}}_i\mathbf{C}_i + \mathbf{P}_r\sum_{i=r+1}^m \mathbf{A}_i\widehat{\mathbf{B}}_i\mathbf{C}_i \\
&= (\mathbf{I} - \mathbf{T}_r)\mathbf{X}\mathbf{C}_r'(\mathbf{C}_r\mathbf{C}_r')^-\mathbf{C}_r + \mathbf{T}_r\mathbf{P}_r\sum_{i=r+1}^m \mathbf{A}_i\widehat{\mathbf{B}}_i\mathbf{C}_i \\
&= (\mathbf{I} - \mathbf{T}_r)\mathbf{X}\mathbf{C}_r'(\mathbf{C}_r\mathbf{C}_r')^-\mathbf{C}_r + \mathbf{P}_{r+1}\sum_{i=r+1}^m \mathbf{A}_i\widehat{\mathbf{B}}_i\mathbf{C}_i,
\end{aligned}$$

which establishes the theorem. ∎

A useful application of the theorem lies in the estimation of the mean structure in the MLNM($\sum_{i=1}^m \mathbf{A}_i\mathbf{B}_i\mathbf{C}_i$).

Corollary 4.1.8.1. $\widehat{E[\mathbf{X}]} = \sum_{i=1}^m \mathbf{A}_i\widehat{\mathbf{B}}_i\mathbf{C}_i = \sum_{i=1}^m (\mathbf{I} - \mathbf{T}_i)\mathbf{X}\mathbf{C}_i'(\mathbf{C}_i\mathbf{C}_i')^-\mathbf{C}_i$. ∎

Furthermore, Theorem 4.1.8 also shows that we do not have to worry about uniqueness properties of the maximum likelihood estimator of $\mathbf{\Sigma}$.

Corollary 4.1.8.2. *In the* MLNM($\sum_{i=1}^m \mathbf{A}_i\mathbf{B}_i\mathbf{C}_i$), *the maximum likelihood estimator of $\mathbf{\Sigma}$ is always unique.* ∎

In the Growth Curve model a canonical version of the model was used in the fourth approach of estimating the parameters. Correspondingly, there exists a canonical version of the MLNM($\sum_{i=1}^m \mathbf{A}_i\mathbf{B}_i\mathbf{C}_i$). The canonical version of the MLNM($\sum_{i=1}^m \mathbf{A}_i\mathbf{B}_i\mathbf{C}_i$) can be written as

$$\mathbf{X} = \mathbf{\Theta} + \mathbf{\Sigma}^{1/2}\mathbf{E}$$

where $\mathbf{E} \sim N_{p,n}(\mathbf{0}, \mathbf{I}_p, \mathbf{I}_n)$ and

$$\mathbf{\Theta} = \begin{pmatrix} \theta_{11} & \theta_{12} & \theta_{13} & \dots & \theta_{1k-1} & \theta_{1k} \\ \theta_{21} & \theta_{22} & \theta_{23} & \dots & \theta_{2k-1} & 0 \\ \vdots & \vdots & \vdots & \ddots & \vdots & \vdots \\ \theta_{q-21} & \theta_{q-22} & \theta_{q-23} & \dots & 0 & 0 \\ \theta_{q-11} & \theta_{q-12} & 0 & \dots & 0 & 0 \\ \theta_{q1} & 0 & 0 & \dots & 0 & 0 \end{pmatrix}. \tag{4.1.65}$$

The Growth Curve model can be identified with

$$\mathbf{\Theta} = \begin{pmatrix} \mathbf{\Theta}_{11} & \mathbf{\Theta}_{12} \\ \mathbf{\Theta}_{21} & \mathbf{0} \end{pmatrix}.$$

Note that the zeros in (4.1.65) form a *stairs structure*. In order to find explicit estimators, this type of structure is really needed. For example, in the ordinary MLNM($\sum_{i=1}^m \mathbf{A}_i\mathbf{B}_i\mathbf{C}_i$) the structure corresponds to the condition $C(\mathbf{C}'_m) \subseteq C(\mathbf{C}'_{m-1}) \subseteq \ldots \subseteq C(\mathbf{C}'_1)$. If the stairs structure does not exist, we have to suppose a suitable structure for $\mathbf{\Sigma}$, or, more precisely, there should be zeros on proper places in $\mathbf{\Sigma}^{-1}$ so that in the trace function of the likelihood function we get something similar to the stairs structure. Note that an off-diagonal element of $\mathbf{\Sigma}^{-1}$, which equals 0 corresponds to a conditional independence assumption. An interesting mathematical treatment of conditional independence, missing values and extended growth curve models has been presented by Andersson & Perlman (1993, 1998).

The canonical version could have been used in order to obtain the estimators in Theorem 4.1.8, although it is more straightforward to use the original matrices.

Now we switch over to another extension of the Growth Curve model. The extension is designed for handling covariates. Let us introduce the model via continuing with Example 4.1.1. A formal definition is given after the example .

Example 4.1.3 (Example 4.1.1 continued). Assume that there exist some background variables which influence growth and are regarded as non-random variables similarly to univariate or multivariate covariance analysis (e.g. see Srivastava, 2002). It is assumed, analogously to covariance analysis, that the expectation of $\mathbf{X}$ is of the form

$$E[\mathbf{X}] = \mathbf{ABC} + \mathbf{B}_2\mathbf{C}_2,$$

where $\mathbf{A}$, $\mathbf{B}$ and $\mathbf{C}$ are as in the Growth Curve model (4.1.1), $\mathbf{C}_2$ is a known matrix taking the values of the concomitant variables (covariables), and $\mathbf{B}_2$ is a matrix of parameters. The model will be referred to as the MLNM($\mathbf{ABC}+\mathbf{B}_2\mathbf{C}_2$). Note that the MLNM($\mathbf{ABC} + \mathbf{B}_2\mathbf{C}_2$) is a special case of the MLNM($\sum_{i=1}^2 \mathbf{A}_i\mathbf{B}_i\mathbf{C}_i$), since if $\mathbf{A}_1 = \mathbf{A}$, $\mathbf{A}_2 = \mathbf{A}^o$, $\mathbf{C}_1 = (\mathbf{C}' : \mathbf{C}'_2)'$ are chosen and $\mathbf{C}_2$ is unrestricted, where obviously $C(\mathbf{C}_2') \subseteq C(\mathbf{C}_1')$ holds. On the other hand, any MLNM($\sum_{i=1}^{2} \mathbf{A}_i\mathbf{B}_i\mathbf{C}_i$) can be presented as a MLNM($\mathbf{ABC}+\mathbf{B}_2\mathbf{C}_2$) after some manipulations. Moreover, if we choose in the MLNM($\sum_{i=1}^{3} \mathbf{A}_i\mathbf{B}_i\mathbf{C}_i$)

$$\mathbf{C}_1 = (\mathbf{C}_1' : \mathbf{C}_3')', \quad \mathbf{C}_2 = (\mathbf{C}_2' : \mathbf{C}_3')', \quad \mathbf{A}_3 = (\mathbf{A}_1 : \mathbf{A}_2)^o,$$

where $\mathbf{A}_1$, $\mathbf{A}_2$ and $\mathbf{C}_3$ are arbitrary, we see that the MLNM($\mathbf{A}_1\mathbf{B}_1\mathbf{C}_1+\mathbf{A}_2\mathbf{B}_2\mathbf{C}_2+\mathbf{B}_3\mathbf{C}_3$) is a special case of a MLNM($\sum_{i=1}^{3} \mathbf{A}_i\mathbf{B}_i\mathbf{C}_i$), if $C(\mathbf{C}_2') \subseteq C(\mathbf{C}_1')$ holds. The reason for presenting a separate treatment of the covariance model is that in this case it is easy to obtain expressions through the original matrices.
Let the birth weight be measured in Example 4.1.1 , and suppose that birth weight influences growth. Additionally, suppose that the influence from birth weight on growth differs between the three treatment groups. Then $\mathbf{C}_2$ is defined as

$$\mathbf{C}_2 = \begin{pmatrix} w_{11} & w_{12} & \dots & w_{1n_1} & 0 & 0 & \dots & 0 & 0 & 0 & \dots & 0 \\ 0 & 0 & \dots & 0 & w_{21} & w_{22} & \dots & w_{2n_2} & 0 & 0 & \dots & 0 \\ 0 & 0 & \dots & 0 & 0 & 0 & \dots & 0 & w_{31} & w_{32} & \dots & w_{3n_3} \end{pmatrix}, \tag{4.1.66}$$

where w_{ij} is the birth weight of the $j-$th animal in the $i-th$ group. On the other hand, if the influence of birth weight on growth is the same in each group, then

$$\mathbf{C}_2 = (\, w_{11} \quad w_{12} \quad \dots \quad w_{1n_1} \quad w_{21} \quad w_{22} \quad \dots \quad w_{2n_2} \quad w_{31} \quad w_{32} \quad \dots \quad w_{3n_3}\,). \tag{4.1.67}$$

Finally, note that in the model building a new set of parameters has been used for each time point. Thus, when (4.1.66) holds, $3p$ parameters are needed to describe the covariates, and in (4.1.67) only p parameters are needed. ∎
The model presented in Example 4.1.3 will now be explicitly defined.

Definition 4.1.4. *Let* $\mathbf{X} : p\times n$, $\mathbf{A} : p\times q$, $q \leq p$, $\mathbf{B} : q\times k$, $\mathbf{C} : k\times n$, $r(\mathbf{C})+p \leq n$, $\mathbf{B}_2 : p \times k_2$, $\mathbf{C}_2 : k_2 \times n$ *and* $\mathbf{\Sigma} : p \times p$ *be p.d. Then*

$$\mathbf{X} = \mathbf{ABC} + \mathbf{B}_2\mathbf{C}_2 + \mathbf{\Sigma}^{1/2}\mathbf{E}$$

is called MLNM($\mathbf{ABC} + \mathbf{B}_2\mathbf{C}_2$), *where* $\mathbf{E} \sim N_{p,n}(\mathbf{0}, \mathbf{I}_p, \mathbf{I}_n)$, $\mathbf{A}$ *and* $\mathbf{C}$ *are known matrices, and* $\mathbf{B}$, $\mathbf{B}_2$ *and* $\mathbf{\Sigma}$ *are unknown parameter matrices.* ∎

Maximum likelihood estimators of the parameters in the MLNM($\mathbf{ABC} + \mathbf{B}_2\mathbf{C}_2$) are going to be obtained. This time the maximum likelihood estimators are derived by solving the likelihood equations. From the likelihood equations for the more general MLNM($\sum_{i=1}^{3} \mathbf{A}_i\mathbf{B}_i\mathbf{C}_i$) it follows that the likelihood equations for the MLNM($\mathbf{ABC} + \mathbf{B}_2\mathbf{C}_2$) equal

$$\mathbf{A}'\mathbf{\Sigma}^{-1}(\mathbf{X} - \mathbf{ABC} - \mathbf{B}_2\mathbf{C}_2)\mathbf{C}' = \mathbf{0}, \tag{4.1.68}$$
$$\mathbf{\Sigma}^{-1}(\mathbf{X} - \mathbf{ABC} - \mathbf{B}_2\mathbf{C}_2)\mathbf{C}_2' = \mathbf{0}, \tag{4.1.69}$$
$$n\mathbf{\Sigma} =(\mathbf{X} - \mathbf{ABC} - \mathbf{B}_2\mathbf{C}_2)(\mathbf{X} - \mathbf{ABC} - \mathbf{B}_2\mathbf{C}_2)'. \tag{4.1.70}$$

In order to solve these equations we start with the simplest one, i.e. (4.1.69), and note that since $\mathbf{\Sigma}$ is p.d., (4.1.69) is equivalent to the linear equation in $\mathbf{B}_2$:

$$(\mathbf{X} - \mathbf{ABC} - \mathbf{B}_2\mathbf{C}_2)\mathbf{C}_2' = \mathbf{0}.$$

By applying Theorem 1.3.4, a solution for $\mathbf{B}_2$ is obtained as a function of $\mathbf{B}$, i.e.

$$\mathbf{B}_2 = (\mathbf{X} - \mathbf{ABC})\mathbf{C}_2'(\mathbf{C}_2\mathbf{C}_2')^- + \mathbf{Z}\mathbf{C}_2^{o'}, \tag{4.1.71}$$

where $\mathbf{Z}$ is an arbitrary matrix of proper size. Thus, it remains to find $\mathbf{B}$ and $\mathbf{\Sigma}$. From (4.1.71) it follows that $\mathbf{B}_2\mathbf{C}_2$ is always unique and equals

$$\mathbf{B}_2\mathbf{C}_2 = (\mathbf{X} - \mathbf{ABC})\mathbf{C}_2'(\mathbf{C}_2\mathbf{C}_2')^-\mathbf{C}_2. \tag{4.1.72}$$

Inserting (4.1.72) into (4.1.68) and (4.1.70) leads to a new system of equations

$$\mathbf{A}'\mathbf{\Sigma}^{-1}(\mathbf{X} - \mathbf{ABC})(\mathbf{I} - \mathbf{C}_2'(\mathbf{C}_2\mathbf{C}_2')^-\mathbf{C}_2)\mathbf{C}' = \mathbf{0}, \tag{4.1.73}$$

$$n\mathbf{\Sigma} = \{(\mathbf{X} - \mathbf{ABC})(\mathbf{I} - \mathbf{C}_2'(\mathbf{C}_2\mathbf{C}_2')^-\mathbf{C}_2)\}\{\}'. \tag{4.1.74}$$

Put

$$\mathbf{Y} = \mathbf{X}(\mathbf{I} - \mathbf{C}_2'(\mathbf{C}_2\mathbf{C}_2')^-\mathbf{C}_2), \tag{4.1.75}$$

$$\mathbf{H} = \mathbf{C}(\mathbf{I} - \mathbf{C}_2'(\mathbf{C}_2\mathbf{C}_2')^-\mathbf{C}_2). \tag{4.1.76}$$

Note that $\mathbf{I} - \mathbf{C}_2'(\mathbf{C}_2\mathbf{C}_2')^-\mathbf{C}_2$ is idempotent, and thus (4.1.73) and (4.1.74) are equivalent to

$$\mathbf{A}'\mathbf{\Sigma}^{-1}(\mathbf{Y} - \mathbf{ABH})\mathbf{H}' = \mathbf{0},$$
$$n\mathbf{\Sigma} = (\mathbf{Y} - \mathbf{ABH})(\mathbf{Y} - \mathbf{ABH})'.$$

These equations are identical to the equations for the ordinary Growth Curve model, i.e. (4.1.11) and (4.1.15). Thus, via Theorem 4.1.1, the next theorem is established.

Theorem 4.1.9. *Let* $\mathbf{Y}$ *and* $\mathbf{H}$ *be defined by (4.1.75) and (4.1.76), respectively. For the model given in Definition 4.1.4, representations of the maximum likelihood estimators are given by*

$$\widehat{\mathbf{B}} = (\mathbf{A}'\mathbf{S}_1^{-1}\mathbf{A})^-\mathbf{A}'\mathbf{S}_1^{-1}\mathbf{Y}\mathbf{H}'(\mathbf{H}\mathbf{H}')^- + (\mathbf{A}')^o\mathbf{Z}_1 + \mathbf{A}'\mathbf{Z}_2\mathbf{H}^{o'},$$
$$\widehat{\mathbf{B}}_2 = (\mathbf{X} - \mathbf{A}\widehat{\mathbf{B}}\mathbf{C})\mathbf{C}_2'(\mathbf{C}_2\mathbf{C}_2')^- + \mathbf{Z}_3\mathbf{C}_2^{o'},$$
$$n\widehat{\mathbf{\Sigma}} = (\mathbf{Y} - \mathbf{A}\widehat{\mathbf{B}}\mathbf{H})(\mathbf{Y} - \mathbf{A}\widehat{\mathbf{B}}\mathbf{H})',$$

where $\mathbf{Z}_i$, $i = 1, 2, 3$, *are arbitrary matrices and*

$$\mathbf{S}_1 = \mathbf{Y}(\mathbf{I} - \mathbf{H}'(\mathbf{H}\mathbf{H}')^-\mathbf{H})\mathbf{Y}'.$$

∎

By using Proposition 1.3.3 and the fact that $(\mathbf{I} - \mathbf{C}_2'(\mathbf{C}_2\mathbf{C}_2')^-\mathbf{C}_2)$ is idempotent, it can be shown that $\mathbf{S}_1$ in Theorem 4.1.9 can be presented in a different form. Let as before, $\mathbf{P}_{\mathbf{C}'} = \mathbf{C}'(\mathbf{C}\mathbf{C}')^-\mathbf{C}$. Then

$$\begin{aligned}
\mathbf{S}_1 =& \mathbf{Y}(\mathbf{I} - \mathbf{H}'(\mathbf{H}\mathbf{H}')^-\mathbf{H})\mathbf{Y}' \\
=& \mathbf{X}(\mathbf{I} - \mathbf{P}_{\mathbf{C}_2'})\{\mathbf{I} - (\mathbf{I} - \mathbf{P}_{\mathbf{C}_2'})\mathbf{C}_1'(\mathbf{C}_1(\mathbf{I} - \mathbf{P}_{\mathbf{C}_2'})\mathbf{C}_1')^-\mathbf{C}_1(\mathbf{I} - \mathbf{P}_{\mathbf{C}_2'})\}(\mathbf{I} - \mathbf{P}_{\mathbf{C}_2'})\mathbf{X}' \\
=& \mathbf{X}\{\mathbf{I} - \mathbf{P}_{\mathbf{C}_2'} - (\mathbf{I} - \mathbf{P}_{\mathbf{C}_2'})\mathbf{C}_1'(\mathbf{C}_1(\mathbf{I} - \mathbf{P}_{\mathbf{C}_2'})\mathbf{C}_1')^-\mathbf{C}_1(\mathbf{I} - \mathbf{P}_{\mathbf{C}_2'})\}\mathbf{X}' \\
=& \mathbf{X}(\mathbf{I} - \mathbf{P}_{\mathbf{C}_2':\mathbf{C}_1'})\mathbf{X}'.
\end{aligned}$$

It is noted without a proof that the MLNM($\mathbf{ABC}+\mathbf{B}_2\mathbf{C}_2$) can be extended to a sum of profiles model with covariates, i.e. a MLNM ($\sum_{i=1}^m \mathbf{A}_i\mathbf{B}_i\mathbf{C}_i+\mathbf{B}_{m+1}\mathbf{C}_{m+1}$). To present a formal proof one can copy the ideas from the proof of Theorem 4.1.9, where first the parameters for the covariates are expressed and thereafter one relies on expressions for the MLNM ($\sum_{i=1}^m \mathbf{A}_i\mathbf{B}_i\mathbf{C}_i$). The result is presented in the next theorem.

Theorem 4.1.10. *Let*

$$\begin{aligned}
\mathbf{H}_i =&\mathbf{C}_i(\mathbf{I}-\mathbf{C}'_{m+1}(\mathbf{C}_{m+1}\mathbf{C}'_{m+1})^-\mathbf{C}_{m+1}), \quad r=1,2,\ldots,m;\\
\mathbf{P}_r =&\mathbf{T}_{r-1}\mathbf{T}_{r-2}\times\cdots\times\mathbf{T}_0, \quad \mathbf{T}_0=\mathbf{I}, \quad r=1,2,\ldots,m+1;\\
\mathbf{T}_i =&\mathbf{I}-\mathbf{P}_i\mathbf{A}_i(\mathbf{A}'_i\mathbf{P}'_i\mathbf{S}_i^{-1}\mathbf{P}_i\mathbf{A}_i)^-\mathbf{A}'_i\mathbf{P}'_i\mathbf{S}_i^{-1}, \quad i=1,2,\ldots,m;\\
\mathbf{S}_i =&\sum_{j=1}^{i}\mathbf{K}_j, \qquad i=1,2,\ldots,m;\\
\mathbf{K}_j =&\mathbf{P}_j\mathbf{X}\mathbf{P}_{\mathbf{H}'_{j-1}}(\mathbf{I}-\mathbf{P}_{\mathbf{H}'_j})\mathbf{P}_{\mathbf{H}'_{j-1}}\mathbf{X}'\mathbf{P}'_j, \quad j=2,3,\ldots,m;\\
\mathbf{K}_1 =&\mathbf{X}(\mathbf{I}-\mathbf{P}_{\mathbf{C}'_{m+1}})\mathbf{X}';\\
\mathbf{P}_{\mathbf{H}'_j} =&\mathbf{H}'_j(\mathbf{H}_j\mathbf{H}'_j)^-\mathbf{H}_j.
\end{aligned}$$

Assume that $\mathbf{S}_1$ *is p.d. and* $\mathcal{C}(\mathbf{H}'_m)\subseteq\mathcal{C}(\mathbf{H}'_{m-1})\subseteq\ldots\subseteq\mathcal{C}(\mathbf{H}'_1)$, *then representations of the maximum likelihood estimators for the* MLNM($\sum_{i=1}^m \mathbf{A}_i\mathbf{B}_i\mathbf{C}_i+\mathbf{B}_{m+1}\mathbf{C}_{m+1}$) *are given by*

$$\begin{aligned}
\widehat{\mathbf{B}}_r =&(\mathbf{A}'_r\mathbf{P}'_r\mathbf{S}_r^{-1}\mathbf{P}_r\mathbf{A}_r)^-\mathbf{A}'_r\mathbf{P}'_r\mathbf{S}_r^{-1}(\mathbf{X}-\sum_{i=r+1}^{m}\mathbf{A}_i\widehat{\mathbf{B}}_i\mathbf{H}_i)\mathbf{H}'_r(\mathbf{H}_r\mathbf{H}'_r)^-\\
&+(\mathbf{A}'_r\mathbf{P}'_r)^o\mathbf{Z}_{r1}+\mathbf{A}'_r\mathbf{P}'_r\mathbf{Z}_{r2}\mathbf{H}_r^{o'}, \qquad r=1,2,\ldots,m;\\
\widehat{\mathbf{B}}_{m+1} =&(\mathbf{X}-\sum_{i=1}^{m}\mathbf{A}_i\widehat{\mathbf{B}}_i\mathbf{C}_i)\mathbf{C}'_{m+1}(\mathbf{C}_{m+1}\mathbf{C}'_{m+1})^-+\mathbf{Z}_{m+1}\mathbf{C}_{m+1}^{o'};\\
n\widehat{\mathbf{\Sigma}} =&(\mathbf{X}-\sum_{i=1}^{m}\mathbf{A}_i\widehat{\mathbf{B}}_i\mathbf{C}_i-\widehat{\mathbf{B}}_{m+1})(\mathbf{X}-\sum_{i=1}^{m}\mathbf{A}_i\widehat{\mathbf{B}}_i\mathbf{C}_i-\widehat{\mathbf{B}}_{m+1})'\\
=&\mathbf{S}_m+\mathbf{P}_{m+1}\mathbf{X}\mathbf{H}'_m(\mathbf{H}_m\mathbf{H}'_m)^-\mathbf{H}_m\mathbf{X}'\mathbf{P}'_{m+1},
\end{aligned}$$

where the $\mathbf{Z}-$*matrices are arbitrary.* ∎

We remark that the condition $\mathcal{C}(\mathbf{C}'_m)\subseteq\mathcal{C}(\mathbf{C}'_{m-1})\subseteq\ldots\subseteq\mathcal{C}(\mathbf{C}'_1)$ used in the MLNM($\sum_{i=1}^m \mathbf{A}_i\mathbf{B}_i\mathbf{C}_i$) implies $\mathcal{C}(\mathbf{H}'_m)\subseteq\mathcal{C}(\mathbf{H}'_{m-1})\subseteq\ldots\subseteq\mathcal{C}(\mathbf{H}'_1)$.

It is interesting to note that when studying multivariate linear models, i.e. the MLNM($\sum_{i=1}^m \mathbf{A}_i\mathbf{B}_i\mathbf{C}_i$) or the MLNM($\sum_{i=1}^m \mathbf{A}_i\mathbf{B}_i\mathbf{C}_i+\mathbf{B}_{m+1}\mathbf{C}_{m+1}$), we rely heavily on g-inverses. If g-inverses are not used, we have problems when estimating the parameters. The problem does not rise in univariate linear models, since reparametrizations can be performed straightforwardly. This remark also applies to the MANOVA model, i.e. the MLNM($\mathbf{BC}$). For the MLNM($\sum_{i=1}^m \mathbf{A}_i\mathbf{B}_i\mathbf{C}_i$) the reparametrizations can be worked out, although it will be difficult to interpret the parameters.

4.1.5 *When are the maximum likelihood estimators unique?*
In Theorem 4.1.6, Theorem 4.1.7, Theorem 4.1.9 and Theorem 4.1.10 the parameters describing the mean are not uniquely estimated. Now, conditions will be given on the design matrices so that unique maximum likelihood estimators are obtained. This is important because in practice the parameters represent certain effects and to have the estimators of the effects interpretable they must be unique. This paragraph involves technical calculations which are often based on theorems and propositions of Chapter 1. However, apparently there is a need to explain results from a geometrical point of view. Projection operators are used in several places. For the reader who is not interested in details, we refer to the results of Theorem 4.1.11, Theorem 4.1.12 and Theorem 4.1.13.
Let us start with the Growth Curve model. In Theorem 4.1.1 the maximum likelihood estimators for the Growth Curve model were given. It has already been shown in Theorem 4.1.1 that $\widehat{\boldsymbol{\Sigma}}$ is always unique. According to Theorem 4.1.1 the estimator of $\mathbf{B}$ equals

$$\widehat{\mathbf{B}} = (\mathbf{A}'\mathbf{S}^{-1}\mathbf{A})^{-}\mathbf{A}'\mathbf{S}^{-1}\mathbf{X}\mathbf{C}'(\mathbf{C}\mathbf{C}')^{-} + (\mathbf{A}')^{o}\mathbf{Z}_1 + \mathbf{A}'\mathbf{Z}_2\mathbf{C}^{o'}.$$

Since $\mathbf{Z}_i$, $i = 1, 2$, are arbitrary matrices, it turns out that $\widehat{\mathbf{B}}$ is uniquely estimated if and only if $(\mathbf{A}')^{o} = \mathbf{0}$ and $\mathbf{C}^{o'} = \mathbf{0}$. Thus, $\mathbf{A}'$ and $\mathbf{C}$ span the whole spaces which is equivalent to the requirements $r(\mathbf{A}) = q$ and $r(\mathbf{C}) = k$. In this case the g-inverses $(\mathbf{A}'\mathbf{S}^{-1}\mathbf{A})^{-}$ and $(\mathbf{C}\mathbf{C}')^{-}$ become inverse matrices. Moreover, let $\mathbf{K}$ and $\mathbf{L}$ be non-random known matrices of proper sizes. Then $\mathbf{K}\widehat{\mathbf{B}}\mathbf{L}$ is unique if and only if $\mathbf{K}(\mathbf{A}')^{o} = \mathbf{0}$ and $\mathbf{C}^{o'}\mathbf{L} = \mathbf{0}$. These conditions are equivalent to the inclusions $\mathcal{C}(\mathbf{K}') \subseteq \mathcal{C}(\mathbf{A}')$ and $\mathcal{C}(\mathbf{L}) \subseteq \mathcal{C}(\mathbf{C})$.

Theorem 4.1.11. *The maximum likelihood estimator $\widehat{\mathbf{B}}$, given by (4.1.7), is unique if and only if*

$$r(\mathbf{A}) = q, \quad r(\mathbf{C}) = k.$$

If $\widehat{\mathbf{B}}$ is unique, then

$$\widehat{\mathbf{B}} = (\mathbf{A}'\mathbf{S}^{-1}\mathbf{A})^{-1}\mathbf{A}'\mathbf{S}^{-1}\mathbf{X}\mathbf{C}'(\mathbf{C}\mathbf{C}')^{-1}.$$

The linear combination $\mathbf{K}\widehat{\mathbf{B}}\mathbf{L}$ is unique if and only if $\mathcal{C}(\mathbf{K}') \subseteq \mathcal{C}(\mathbf{A}')$ and $\mathcal{C}(\mathbf{L}) \subseteq \mathcal{C}(\mathbf{C})$ hold. ∎

Now we turn to the MLNM($\sum_{i=1}^{3}\mathbf{A}_i\mathbf{B}_i\mathbf{C}_i$). In Theorem 4.1.6 it was stated that the maximum likelihood estimators of the mean parameters $\widehat{\mathbf{B}}_i$ equal

$$\begin{aligned}\widehat{\mathbf{B}}_3 =&(\mathbf{A}_3'\mathbf{P}_3'\mathbf{S}_3^{-1}\mathbf{P}_3\mathbf{A}_3)^{-}\mathbf{A}_3'\mathbf{P}_3'\mathbf{S}_3^{-1}\mathbf{X}\mathbf{C}_3'(\mathbf{C}_3\mathbf{C}_3')^{-}\\ &+ (\mathbf{A}_3'\mathbf{P}_3')^{o}\mathbf{Z}_{31} + \mathbf{A}_3'\mathbf{P}_3'\mathbf{Z}_{32}\mathbf{C}_3^{o'}, &(4.1.77)\\ \widehat{\mathbf{B}}_2 =&(\mathbf{A}_2'\mathbf{T}_1'\mathbf{S}_2^{-1}\mathbf{T}_1\mathbf{A}_2)^{-}\mathbf{A}_2'\mathbf{T}_1'\mathbf{S}_2^{-1}(\mathbf{X} - \mathbf{A}_3\widehat{\mathbf{B}}_3\mathbf{C}_3)\mathbf{C}_2'(\mathbf{C}_2\mathbf{C}_2')^{-}\\ &+ (\mathbf{A}_2'\mathbf{T}_1')^{o}\mathbf{Z}_{21} + \mathbf{A}_2'\mathbf{T}_1'\mathbf{Z}_{22}\mathbf{C}_2^{o'}, &(4.1.78)\\ \widehat{\mathbf{B}}_1 =&(\mathbf{A}_1'\mathbf{S}_1^{-1}\mathbf{A}_1)^{-}\mathbf{A}_1'\mathbf{S}_1^{-1}(\mathbf{X} - \mathbf{A}_2\widehat{\mathbf{B}}_2\mathbf{C}_2 - \mathbf{A}_3\widehat{\mathbf{B}}_3\mathbf{C}_3)\mathbf{C}_1'(\mathbf{C}_1\mathbf{C}_1')^{-}\\ &+ (\mathbf{A}_1')^{o}\mathbf{Z}_{11} + \mathbf{A}_1'\mathbf{Z}_{12}\mathbf{C}_1^{o'}. &(4.1.79)\end{aligned}$$

These estimators are now studied in more detail. Since $\widehat{\mathbf{B}}_2$ and $\widehat{\mathbf{B}}_1$ are functions of $\widehat{\mathbf{B}}_3$, we start with $\widehat{\mathbf{B}}_3$. It follows immediately that $\widehat{\mathbf{B}}_3$ is unique if and only if

$$\begin{aligned} r(\mathbf{C}_3) =& k_3, \\ r(\mathbf{A}_3'\mathbf{P}_3') =& r(\mathbf{A}_3') = q_3. \end{aligned} \tag{4.1.80}$$

The matrix $\mathbf{P}_3 = \mathbf{T}_2\mathbf{T}_1$ given by (4.1.56) is a function of the observations through $\mathbf{S}_2$. However, in Lemma 4.1.3, it was shown that $\mathbf{P}_3$ is idempotent. Thus it is a projector on a certain space which is independent of the observations, since $\mathbf{S}_2$ acts as an estimator of the inner product matrix and has nothing to do with the space where $\mathbf{A}_3$ is projected on. Thus the condition (4.1.80) should be independent of the observations. Next we prove this in detail and present (4.1.80) in an alternative way.

Since $\mathcal{C}(\mathbf{A}_3'\mathbf{P}_3') = \mathcal{C}(\mathbf{A}_3')$ is equivalent to $r(\mathbf{A}_3'\mathbf{P}_3') = r(\mathbf{A}_3')$, it follows from Theorem 1.2.12 that the equality $r(\mathbf{A}_3'\mathbf{P}_3') = r(\mathbf{A}_3')$ in (4.1.80) is equivalent to

$$\mathcal{C}(\mathbf{A}_3) \cap \mathcal{C}(\mathbf{P}_3)^\perp = \{\mathbf{0}\}. \tag{4.1.81}$$

Hence, we have to determine $\mathcal{C}(\mathbf{P}_3)$. From Proposition 1.2.2 (ix), (4.1.57) and (4.1.52) it follows that $(\mathbf{A}_1 : \mathbf{A}_2)'\mathbf{T}_1'\mathbf{T}_2' = \mathbf{0}$, $\mathbf{P}_3$ as well as $\mathbf{T}_1$ and $\mathbf{T}_2$ are idempotent, and $\mathbf{T}_1'$ spans the orthogonal complement to $\mathcal{C}(\mathbf{A}_1)$. Then, it follows from Proposition 1.1.4 (v) and Theorem 1.2.17 that

$$\begin{aligned} r(\mathbf{P}_3) =& \mathrm{tr}(\mathbf{P}_3) = \mathrm{tr}(\mathbf{T}_1) - \mathrm{tr}(\mathbf{T}_1\mathbf{A}_2(\mathbf{A}_2'\mathbf{T}_1'\mathbf{S}_2^{-1}\mathbf{T}_1\mathbf{A}_2)^-\mathbf{A}_2'\mathbf{T}_1'\mathbf{S}_2^{-1}\mathbf{T}_1) \\ =& p - r(\mathbf{A}_1) - r(\mathbf{T}_1\mathbf{A}_2) = p - r(\mathbf{A}_1 : \mathbf{A}_2) = r((\mathbf{A}_1 : \mathbf{A}_2)^o). \end{aligned}$$

Thus, (4.1.81) is equivalent to

$$\mathcal{C}(\mathbf{A}_3) \cap \mathcal{C}(\mathbf{A}_1 : \mathbf{A}_2) = \{\mathbf{0}\},$$

which is independent of the observations. Moreover, $\mathbf{K}\widehat{\mathbf{B}}_3\mathbf{L}$ is unique if and only if

$$\mathcal{C}(\mathbf{K}') \subseteq \mathcal{C}(\mathbf{A}_3'\mathbf{P}_3') = \mathcal{C}(\mathbf{A}_3'(\mathbf{A}_1 : \mathbf{A}_2)^o), \tag{4.1.82}$$

$$\mathcal{C}(\mathbf{L}) \subseteq \mathcal{C}(\mathbf{C}_3). \tag{4.1.83}$$

Now we continue with discussing uniqueness properties of the maximum likelihood estimator of $\mathbf{B}_2$ given by (4.1.78). By applying considerations similar to those used when deriving uniqueness properties of $\widehat{\mathbf{B}}_3$ as well as by an application of (4.1.82) and (4.1.83), we have that $\widehat{\mathbf{B}}_2$ is unique if and only if

$$r(\mathbf{A}_2'\mathbf{T}_1') = r(\mathbf{A}_2'), \tag{4.1.84}$$

$$r(\mathbf{C}_2) = k_2, \qquad r(\mathbf{A}_2) = q_2,$$

$$\mathcal{C}(\mathbf{C}_3\mathbf{C}_2'(\mathbf{C}_2\mathbf{C}_2')^{-1}) \subseteq \mathcal{C}(\mathbf{C}_3), \tag{4.1.85}$$

$$\mathcal{C}(\mathbf{A}_3'\mathbf{T}_1'\mathbf{S}_2^{-1}\mathbf{T}_1\mathbf{A}_2) \subseteq \mathcal{C}(\mathbf{A}_3'\mathbf{T}_1'\mathbf{T}_2'). \tag{4.1.86}$$

Note that (4.1.85) and (4.1.86) refer to the uniqueness of linear functions of $\widehat{\mathbf{B}}_3$. Inclusion (4.1.85) is obviously true. Since $\mathcal{C}(\mathbf{T}_1') = \mathcal{C}(\mathbf{A}_1)^\perp$, by Theorem 1.2.12, (4.1.84) is equivalent to

$$\mathcal{C}(\mathbf{A}_1) \cap \mathcal{C}(\mathbf{A}_2) = \{\mathbf{0}\}. \tag{4.1.87}$$

Thus, it remains to interpret (4.1.86). Proposition 1.2.2 (ix) states that (4.1.86) is equivalent to

$$\begin{aligned}&\mathcal{C}(\mathbf{A}_3'\mathbf{T}_1'\mathbf{T}_2') \supseteq \mathcal{C}(\mathbf{A}_3'\mathbf{T}_1'\mathbf{S}_2^{-1}\mathbf{T}_1\mathbf{A}_2)\\ &\quad = \mathcal{C}(\mathbf{A}_3'\mathbf{T}_1'\mathbf{S}_2^{-1}\mathbf{T}_1\mathbf{A}_2(\mathbf{A}_2'\mathbf{T}_1'\mathbf{S}_2^{-1}\mathbf{T}_1\mathbf{A}_2)^-\mathbf{A}_2'\mathbf{T}_1') = \mathcal{C}(\mathbf{A}_3'\mathbf{T}_1'(\mathbf{I}-\mathbf{T}_2')).\end{aligned} \tag{4.1.88}$$

By using Proposition 1.2.2 (ii) it follows that (4.1.88) is equivalent to

$$\mathcal{C}(\mathbf{A}_3'\mathbf{T}_1') \subseteq \mathcal{C}(\mathbf{A}_3'\mathbf{T}_1'\mathbf{T}_2'). \tag{4.1.89}$$

Let

$$\mathbf{P}_{\mathbf{A}_1^o} = \mathbf{I} - \mathbf{A}_1(\mathbf{A}_1'\mathbf{A}_1)^-\mathbf{A}_1', \tag{4.1.90}$$

$$\mathbf{P}_{(\mathbf{A}_1:\mathbf{A}_2)_1^o} = \mathbf{P}_{\mathbf{A}_1^o} - \mathbf{P}_{\mathbf{A}_1^o}\mathbf{A}_2(\mathbf{A}_2'\mathbf{P}_{\mathbf{A}_1^o}\mathbf{A}_2)^-\mathbf{A}_2'\mathbf{P}_{\mathbf{A}_1^o} \tag{4.1.91}$$

determine $\mathbf{A}_1^o$ and $(\mathbf{A}_1 : \mathbf{A}_2)^o$, respectively. Observe that these matrices are projectors. Since $\mathcal{C}(\mathbf{T}_1') = \mathcal{C}(\mathbf{P}_{\mathbf{A}_1^o})$ and $\mathcal{C}(\mathbf{T}_1'\mathbf{T}_2') = \mathcal{C}(\mathbf{P}_{(\mathbf{A}_1:\mathbf{A}_2)_1^o})$, by Proposition 1.2.2 (iii), the inclusion (4.1.89) is equivalent to

$$\mathcal{C}(\mathbf{A}_3'\mathbf{P}_{\mathbf{A}_1^o}) \subseteq \mathcal{C}(\mathbf{A}_3'\mathbf{P}_{(\mathbf{A}_1:\mathbf{A}_2)_1^o}). \tag{4.1.92}$$

Moreover, (4.1.87) and Theorem 1.2.12 yield $\mathcal{C}(\mathbf{A}_2') = \mathcal{C}(\mathbf{A}_2'\mathbf{P}_{\mathbf{A}_1^o})$. Thus, from Proposition 1.2.2 (ix) it follows that the matrices $\mathbf{A}_2(\mathbf{A}_2'\mathbf{P}_{\mathbf{A}_1^o}\mathbf{A}_2)^-\mathbf{A}_2'\mathbf{P}_{\mathbf{A}_1^o}$ and $\mathbf{I} - \mathbf{A}_2(\mathbf{A}_2'\mathbf{P}_{\mathbf{A}_1^o}\mathbf{A}_2)^-\mathbf{A}_2'\mathbf{P}_{\mathbf{A}_1^o}$ are idempotent matrices which implies the identity $\mathcal{C}(\mathbf{I} - \mathbf{A}_2(\mathbf{A}_2'\mathbf{P}_{\mathbf{A}_1^o}\mathbf{A}_2)^-\mathbf{A}_2'\mathbf{P}_{\mathbf{A}_1^o}) = \mathcal{C}(\mathbf{P}_{\mathbf{A}_1^o}\mathbf{A}_2)^\perp$. Hence, by applying Theorem 1.2.12 and (4.1.91), the inclusion (4.1.92) can be written as

$$\mathcal{C}(\mathbf{P}_{\mathbf{A}_1^o}\mathbf{A}_3) \cap \mathcal{C}(\mathbf{P}_{\mathbf{A}_1^o}\mathbf{A}_2) = \{\mathbf{0}\}.$$

Thus, since $\mathbf{P}_{\mathbf{A}_1^o}$ is a projector, it follows from Theorem 1.2.16 that instead of (4.1.86), the equality

$$\mathcal{C}(\mathbf{A}_1)^\perp \cap \mathcal{C}(\mathbf{A}_1 : \mathbf{A}_2) \cap \mathcal{C}(\mathbf{A}_1 : \mathbf{A}_3) = \{\mathbf{0}\}$$

can be used.

Now we consider linear combinations of elements in $\widehat{\mathbf{B}}_2$. Necessary and sufficient conditions for uniqueness of $\mathbf{K}\widehat{\mathbf{B}}_2\mathbf{L}$ are given by

$$\mathcal{C}(\mathbf{L}) \subseteq \mathcal{C}(\mathbf{C}_2),$$

$$\mathcal{C}(\mathbf{K}') \subseteq \mathcal{C}(\mathbf{A}_2'\mathbf{T}_1'), \tag{4.1.93}$$

$$\mathcal{C}(\mathbf{A}_3'\mathbf{T}_1'\mathbf{S}_2^{-1}\mathbf{T}_1\mathbf{A}_2(\mathbf{A}_2'\mathbf{T}_1'\mathbf{S}_2^{-1}\mathbf{T}_1\mathbf{A}_2)^-\mathbf{K}') \subseteq \mathcal{C}(\mathbf{A}_3'\mathbf{T}_1'\mathbf{T}_2'). \tag{4.1.94}$$

With the help of Proposition 1.2.2 (iii) we get that (4.1.93) is equivalent to

$$\mathcal{C}(\mathbf{K}') \subseteq \mathcal{C}(\mathbf{A}_2'\mathbf{A}_1^o), \tag{4.1.95}$$

since $\mathcal{C}(\mathbf{T}_1') = \mathcal{C}(\mathbf{A}_1^o)$. Moreover, by utilizing Proposition 1.2.2 (i), (4.1.95) implies that

$$\mathbf{K}' = \mathbf{A}_2'\mathbf{A}_1^o\mathbf{Q}_1 = \mathbf{A}_2'\mathbf{T}_1'\mathbf{Q}_2\mathbf{Q}_1, \tag{4.1.96}$$

for some matrices $\mathbf{Q}_1$ and $\mathbf{Q}_2$, where $\mathbf{A}_1^o = \mathbf{T}_1'\mathbf{Q}_2$. Thus, the left hand side of (4.1.94) is equivalent to

$$\begin{aligned}\mathcal{C}(\mathbf{A}_3'\mathbf{T}_1'\mathbf{S}_2^{-1}\mathbf{T}_1\mathbf{A}_2(\mathbf{A}_2'\mathbf{T}_1'S_2^{-1}\mathbf{T}_1\mathbf{A}_2)^-\mathbf{A}_2'\mathbf{T}_1'\mathbf{Q}_2\mathbf{Q}_1)\\ = \mathcal{C}(\mathbf{A}_3'\mathbf{T}_1'(\mathbf{I} - \mathbf{T}_2')\mathbf{Q}_2\mathbf{Q}_1),\end{aligned}$$

where $\mathbf{T}_2$ is defined by (4.1.57). Therefore, Proposition 1.2.2 (ii) implies that (4.1.94) is equivalent to

$$\mathcal{C}(\mathbf{A}_3'\mathbf{T}_1'\mathbf{Q}_2\mathbf{Q}_1) \subseteq \mathcal{C}(\mathbf{A}_3'\mathbf{T}_1'\mathbf{T}_2'),$$

which is identical to

$$\mathcal{C}(\mathbf{A}_3'\mathbf{A}_1^o\mathbf{Q}_1) \subseteq \mathcal{C}(\mathbf{A}_3'(\mathbf{A}_1 : \mathbf{A}_2)^o), \tag{4.1.97}$$

since $\mathbf{T}_1'\mathbf{Q}_2 = \mathbf{A}_1^o$ and $\mathcal{C}(\mathbf{T}_1'\mathbf{T}_2') = \mathcal{C}(\mathbf{A}_1 : \mathbf{A}_1)^\perp$. It is noted that by (4.1.97) we get a linear equation in $\mathbf{Q}_1$:

$$(\mathbf{A}_3'(\mathbf{A}_1 : \mathbf{A}_2)^o)^{o'}\mathbf{A}_3'\mathbf{A}_1^o\mathbf{Q}_1 = \mathbf{0}. \tag{4.1.98}$$

In Theorem 1.3.4 a general representation of all solutions is given, and using (4.1.96) the set of all possible matrices $\mathbf{K}$ is obtained, i.e.

$$\mathbf{K}' = \mathbf{A}_2'\mathbf{A}_1^o(\mathbf{A}_1^{o'}\mathbf{A}_3\{\mathbf{A}_3'(\mathbf{A}_1 : \mathbf{A}_2)^o\}^o)^o\mathbf{Z},$$

where $\mathbf{Z}$ is an arbitrary matrix. Thus, the set of all possible matrices $\mathbf{K}$ is identified by the subspace

$$\mathcal{C}(\mathbf{A}_2'\mathbf{A}_1^o(\mathbf{A}_1^{o'}\mathbf{A}_3\{\mathbf{A}_3'(\mathbf{A}_1 : \mathbf{A}_2)^o\}^o)^o). \tag{4.1.99}$$

Since

$$\begin{aligned}&\mathcal{C}(\mathbf{A}_1^o(\mathbf{A}_1^{o'}\mathbf{A}_3\{\mathbf{A}_3'(\mathbf{A}_1 : \mathbf{A}_2)^o\}^o)^o)\\ &\quad = \mathcal{C}(\mathbf{A}_1)^\perp \cap \{\mathcal{C}(\mathbf{A}_3) \cap \mathcal{C}(\mathbf{A}_1 : \mathbf{A}_2)\}^\perp = \mathcal{C}(\mathbf{A}_1 : \mathbf{A}_3)^\perp + \mathcal{C}(\mathbf{A}_1 : \mathbf{A}_2)^\perp,\end{aligned}$$

the subspace (4.1.99) has a simpler representation:

$$\mathcal{C}(\mathbf{A}_2'(\mathbf{A}_1 : \mathbf{A}_3)^o).$$

Moreover, note that if $(\mathbf{A}_3'(\mathbf{A}_1 : \mathbf{A}_2)^o)^{o'}\mathbf{A}_3'\mathbf{A}_1^o = \mathbf{0}$, relation (4.1.98), as well as (4.1.94), are insignificant and therefore $\mathbf{K}$ has to satisfy (4.1.95) in order to

determine $\mathbf{K}\widehat{\mathbf{B}}_2\mathbf{L}$ uniquely. However, if $(\mathbf{A}_3'(\mathbf{A}_1 : \mathbf{A}_2)^o)^{o'}\mathbf{A}_3'\mathbf{A}_1^o = \mathbf{0}$ holds, (4.1.99) equals the right hand side of (4.1.95). Thus (4.1.99) always applies.
In the next, $\widehat{\mathbf{B}}_1$ given by (4.1.79) will be examined. First, necessary conditions for uniqueness of $\widehat{\mathbf{B}}_1$ are obtained. Afterwards it is shown that these conditions are also sufficient.
From (4.1.79) it follows that

$$r(\mathbf{A}_1) = q_1, \qquad r(\mathbf{C}_1) = k_1 \tag{4.1.100}$$

must hold. However, if $\widehat{\mathbf{B}}_1$ is unique, it is necessary that

$$(\mathbf{A}_1'\mathbf{S}_1^{-1}\mathbf{A}_1)^{-1}(\mathbf{A}_1'\mathbf{S}_1^{-1}(\mathbf{A}_2\widehat{\mathbf{B}}_2\mathbf{C}_2 + \mathbf{A}_3\widehat{\mathbf{B}}_3\mathbf{C}_3)\mathbf{C}_1'(\mathbf{C}_1\mathbf{C}_1')^{-1} \tag{4.1.101}$$

is unique. The expression (4.1.101) consists of linear functions of $\widehat{\mathbf{B}}_2$ and $\widehat{\mathbf{B}}_3$. In order to guarantee that the linear functions of $\widehat{\mathbf{B}}_2$, given by (4.1.78), are unique, it is necessary to cancel the matrix expressions which involve the $\mathbf{Z}$ matrices. A necessary condition for this is

$$\mathcal{C}(\mathbf{A}_2'\mathbf{S}_1^{-1}\mathbf{A}_1(\mathbf{A}_1'\mathbf{S}_1^{-1}\mathbf{A}_1)^{-1}) \subseteq \mathcal{C}(\mathbf{A}_2'\mathbf{T}_1'). \tag{4.1.102}$$

However, (4.1.78) also includes matrix expressions which are linear functions of $\widehat{\mathbf{B}}_3$, and therefore it follows from (4.1.78) and (4.1.101) that

$$\begin{aligned} &-(\mathbf{A}_1'\mathbf{S}_1^{-1}\mathbf{A}_1)^{-1}\mathbf{A}_1'\mathbf{S}_1^{-1}\mathbf{A}_2(\mathbf{A}_2'\mathbf{T}_1'\mathbf{S}_2^{-1}\mathbf{T}_1\mathbf{A}_2)^{-}\mathbf{A}_2'\mathbf{T}_1'\mathbf{S}_2^{-1}\mathbf{T}_1\mathbf{A}_3\widehat{\mathbf{B}}_3\mathbf{C}_3 \\ &\quad \times \mathbf{C}_2'(\mathbf{C}_2\mathbf{C}_2')^{-}\mathbf{C}_2\mathbf{C}_1'(\mathbf{C}_1\mathbf{C}_1')^{-1} + (\mathbf{A}_1'\mathbf{S}_1^{-1}\mathbf{A}_1)^{-1}\mathbf{A}_1'\mathbf{S}_1^{-1}\mathbf{A}_3\widehat{\mathbf{B}}_3\mathbf{C}_3\mathbf{C}_1'(\mathbf{C}_1\mathbf{C}_1')^{-1} \end{aligned}$$

has to be considered. This expression equals

$$\begin{aligned} &(\mathbf{A}_1'\mathbf{S}_1^{-1}\mathbf{A}_1)^{-1}\mathbf{A}_1'\mathbf{S}_1^{-1} \\ &\quad \times (\mathbf{I} - \mathbf{A}_2(\mathbf{A}_2'\mathbf{T}_1'\mathbf{S}_2^{-1}\mathbf{T}_1\mathbf{A}_2)^{-}\mathbf{A}_2'\mathbf{T}_1'\mathbf{S}_2^{-1}\mathbf{T}_1)\mathbf{A}_3\widehat{\mathbf{B}}_3\mathbf{C}_3\mathbf{C}_1'(\mathbf{C}_1\mathbf{C}_1')^{-1}, \end{aligned} \tag{4.1.103}$$

since by assumption, $\mathcal{C}(\mathbf{C}_3') \subseteq \mathcal{C}(\mathbf{C}_2') \subseteq \mathcal{C}(\mathbf{C}_1')$. Put

$$\mathbf{P} = \mathbf{I} - \mathbf{A}_2(\mathbf{A}_2'\mathbf{T}_1'\mathbf{S}_2^{-1}\mathbf{T}_1\mathbf{A}_2)^{-}\mathbf{A}_2'\mathbf{T}_1'\mathbf{S}_2^{-1}\mathbf{T}_1. \tag{4.1.104}$$

Thus, from (4.1.77) it follows that (4.1.103) is unique if and only if

$$\mathcal{C}(\mathbf{A}_3'\mathbf{P}'\mathbf{S}_1^{-1}\mathbf{A}_1(\mathbf{A}_1'\mathbf{S}_1^{-1}\mathbf{A}_1)^{-1}) \subseteq \mathcal{C}(\mathbf{A}_3'\mathbf{T}_1'\mathbf{T}_2'). \tag{4.1.105}$$

Hence, if $\widehat{\mathbf{B}}_1$ is unique, it is necessary that (4.1.100), (4.1.102) and (4.1.105) hold and now we are going to see that (4.1.102) and (4.1.105) can be simplified by using the ideas already applied above. By definition of $\mathbf{T}_1$, given by (4.1.52), inclusion (4.1.102) is equivalent to

$$\mathcal{C}(\mathbf{A}_2'(\mathbf{I} - \mathbf{T}_1')) \subseteq \mathcal{C}(\mathbf{A}_2'\mathbf{T}_1'),$$

which by Proposition 1.2.2 (ii) is equivalent to

$$\mathcal{C}(\mathbf{A}_2') \subseteq \mathcal{C}(\mathbf{A}_2'\mathbf{T}_1'), \tag{4.1.106}$$

and by Theorem 1.2.12, inclusion (4.1.106) holds if and only if

$$\mathcal{C}(\mathbf{A}_2) \cap \mathcal{C}(\mathbf{A}_1) = \{\mathbf{0}\}. \tag{4.1.107}$$

Before starting to examine (4.1.105) note first that from (4.1.104) and definition of $\mathbf{T}_2$, given by (4.1.57), it follows that

$$\mathbf{T}_1'\mathbf{T}_2' = \mathbf{P}'\mathbf{T}_1'.$$

Since

$$\mathcal{C}(\mathbf{A}_3'\mathbf{P}'\mathbf{S}_1^{-1}\mathbf{A}_1(\mathbf{A}_1'\mathbf{S}_1^{-1}\mathbf{A}_1)^{-1}) = \mathcal{C}(\mathbf{A}_3'\mathbf{P}'\mathbf{S}_1^{-1}\mathbf{A}_1(\mathbf{A}_1'\mathbf{S}_1^{-1}\mathbf{A}_1)^{-1}\mathbf{A}_1'),$$

the inclusion (4.1.105) is identical to

$$\mathcal{C}(\mathbf{A}_3'\mathbf{P}'(\mathbf{I} - \mathbf{T}_1')) \subseteq \mathcal{C}(\mathbf{A}_3'\mathbf{P}'\mathbf{T}_1').$$

By Proposition 1.2.2 (ii) we have

$$\mathcal{C}(\mathbf{A}_3'\mathbf{P}') = \mathcal{C}(\mathbf{A}_3'\mathbf{P}'\mathbf{T}_1') = \mathcal{C}(\mathbf{A}_3'\mathbf{T}_1'\mathbf{T}_2'). \tag{4.1.108}$$

However, from the definition of $\mathbf{P}$, given by (4.1.104) and (4.1.106), it follows that $\mathbf{P}$ is idempotent, $r(\mathbf{P}) = p - r(\mathbf{T}_1\mathbf{A}_2) = p - r(\mathbf{A}_2)$ and $\mathbf{A}_2'\mathbf{P}' = \mathbf{0}$. Hence, $\mathbf{P}'$ spans the orthogonal complement to $\mathcal{C}(\mathbf{A}_2)$. Analogously to (4.1.90) and (4.1.91), the projectors

$$\begin{aligned}\mathbf{P}_{\mathbf{A}_2^o} &= \mathbf{I} - \mathbf{A}_2(\mathbf{A}_2'\mathbf{A}_2)^-\mathbf{A}_2',\\ \mathbf{P}_{(\mathbf{A}_1:\mathbf{A}_2)_2^o} &= \mathbf{P}_{\mathbf{A}_2^o} - \mathbf{P}_{\mathbf{A}_2^o}\mathbf{A}_1(\mathbf{A}_1'\mathbf{P}_{\mathbf{A}_2^o}\mathbf{A}_1)^-\mathbf{A}_1'\mathbf{P}_{\mathbf{A}_2^o}\end{aligned} \tag{4.1.109}$$

define one choice of $\mathbf{A}_2^o$ and $(\mathbf{A}_1 : \mathbf{A}_2)^o$, respectively. Since $\mathcal{C}(\mathbf{P}') = \mathcal{C}(\mathbf{P}_{\mathbf{A}_2^o})$ and $\mathcal{C}(\mathbf{T}_1'\mathbf{T}_2') = \mathcal{C}(\mathbf{P}_{(\mathbf{A}_1:\mathbf{A}_2)_2^o})$, by Proposition 1.2.2 (iii), the equality (4.1.108) is equivalent to

$$\mathcal{C}(\mathbf{A}_3'\mathbf{P}_{\mathbf{A}_2^o}) = \mathcal{C}(\mathbf{A}_3'\mathbf{P}_{(\mathbf{A}_1:\mathbf{A}_2)_2^o}).$$

Thus, since $\mathcal{C}(\mathbf{I} - \mathbf{A}_1(\mathbf{A}_1'\mathbf{P}_{\mathbf{A}_2^o}\mathbf{A}_1)^-\mathbf{A}_1'\mathbf{P}_{\mathbf{A}_2^o}) = \mathcal{C}(\mathbf{P}_{\mathbf{A}_2^o}\mathbf{A}_1)^\perp$, Theorem 1.2.12 and (4.1.109) imply that (4.1.108) is equivalent to

$$\mathcal{C}(\mathbf{P}_{\mathbf{A}_2^o}\mathbf{A}_3) \cap \mathcal{C}(\mathbf{P}_{\mathbf{A}_2^o}\mathbf{A}_1) = \{\mathbf{0}\}. \tag{4.1.110}$$

However, using properties of a projector it follows from Theorem 1.2.16 that (4.1.110) is equivalent to

$$\mathcal{C}(\mathbf{A}_2)^\perp \cap \{\mathcal{C}(\mathbf{A}_2) + \mathcal{C}(\mathbf{A}_3)\} \cap \{\mathcal{C}(\mathbf{A}_2) + \mathcal{C}(\mathbf{A}_1)\} = \{\mathbf{0}\}. \tag{4.1.111}$$

Hence, if $\widehat{\mathbf{B}}_1$ given by (4.1.79) is unique, then (4.1.100), (4.1.107) and (4.1.111) hold.

Now we show briefly that the conditions given by (4.1.100), (4.1.107) and (4.1.111) are also sufficient. If (4.1.107) holds, it follows that $\mathbf{P}'$ spans the orthogonal complement to $C(\mathbf{A}_2)$, and in this case it has already been shown that (4.1.111) is equivalent to (4.1.105), which means that if (4.1.111) holds, (4.1.103) is unique. Now, the uniqueness of (4.1.103) together with (4.1.107) imply that (4.1.101) is unique and hence, if (4.1.100) is also satisfied, $\widehat{\mathbf{B}}_1$ is unique.

Finally, the uniqueness of $\mathbf{K}\widehat{\mathbf{B}}_1\mathbf{L}$ will be considered and here the discussion for $\widehat{\mathbf{B}}_1$ will be utilized. Thus, when $\mathbf{K}\widehat{\mathbf{B}}_1\mathbf{L}$ is unique, in correspondence with (4.1.100), (4.1.102) and (4.1.105),

$$C(\mathbf{L}) \subseteq C(\mathbf{C}_1),$$

$$C(\mathbf{K}') \subseteq C(\mathbf{A}_1'), \tag{4.1.112}$$

$$C(\mathbf{A}_2'\mathbf{S}_1^{-1}\mathbf{A}_1(\mathbf{A}_1'\mathbf{S}_1^{-1}\mathbf{A}_1)^-\mathbf{K}') \subseteq C(\mathbf{A}_2'\mathbf{T}_1'), \tag{4.1.113}$$

$$C(\mathbf{A}_3'\mathbf{P}'\mathbf{S}_1^{-1}\mathbf{A}_1(\mathbf{A}_1'\mathbf{S}_1^{-1}\mathbf{A}_1)^-\mathbf{K}') \subseteq C(\mathbf{A}_3'\mathbf{T}_1'\mathbf{T}_2') \tag{4.1.114}$$

hold. We are not going to summarize (4.1.112) – (4.1.114) into one relation as it was done with (4.1.93) and (4.1.94), since such a relation would be rather complicated. Instead it will be seen that there exist inclusions which are equivalent to (4.1.113) and (4.1.114) and which do not depend on the observations in the same way as (4.1.95) and (4.1.99) were equivalent to (4.1.93) and (4.1.94).

From (4.1.112) and Proposition 1.2.2 (i) it follows that $\mathbf{K}' = \mathbf{A}_1'\mathbf{Q}_1$ for some matrix $\mathbf{Q}_1$, and thus (4.1.113) is equivalent to

$$C(\mathbf{A}_2'\mathbf{S}_1^{-1}\mathbf{A}_1(\mathbf{A}_1'\mathbf{S}_1^{-1}\mathbf{A}_1)^-\mathbf{A}_1'\mathbf{Q}_1) \subseteq C(\mathbf{A}_2'\mathbf{T}_1'),$$

which can be written as

$$C(\mathbf{A}_2'(\mathbf{I} - \mathbf{T}_1')\mathbf{Q}_1) \subseteq C(\mathbf{A}_2'\mathbf{T}_1'). \tag{4.1.115}$$

By Proposition 1.2.2 (ii) it follows that (4.1.115) holds if and only if

$$C(\mathbf{A}_2'\mathbf{Q}_1) \subseteq C(\mathbf{A}_2'\mathbf{T}_1'). \tag{4.1.116}$$

The projectors $\mathbf{P}_{\mathbf{A}_1^o}$ and $\mathbf{P}_{(\mathbf{A}_1:\mathbf{A}_2)_1^o}$ appearing in the following discussion are given by (4.1.90) and (4.1.91), respectively. Since $C(\mathbf{A}_2'\mathbf{T}_1') = C(\mathbf{A}_2'\mathbf{P}_{\mathbf{A}_1^o})$ holds, using Proposition 1.2.2 (ii) and inclusion (4.1.115), we get

$$\mathbf{A}_2'\mathbf{Q}_1 = \mathbf{A}_2'\mathbf{P}_{\mathbf{A}_1^o}\mathbf{Q}_3 = \mathbf{A}_2'\mathbf{T}_1'\mathbf{Q}_2\mathbf{Q}_3, \quad \mathbf{P}_{\mathbf{A}_1^o} = \mathbf{T}_1'\mathbf{Q}_2, \tag{4.1.117}$$

for some matrices $\mathbf{Q}_2$ and $\mathbf{Q}_3$.

We leave (4.1.116) and (4.1.113) for a while and proceed with (4.1.114). Inclusion (4.1.114) is equivalent to

$$C(\mathbf{A}_3'\mathbf{P}'(\mathbf{I} - \mathbf{T}_1')\mathbf{Q}_1) \subseteq C(\mathbf{A}_3'\mathbf{T}_1'\mathbf{T}_2') = C(\mathbf{A}_3'\mathbf{P}'\mathbf{T}_1'),$$

which by Proposition 1.2.2 (ii) holds if and only if

$$\mathcal{C}(\mathbf{A}_3'\mathbf{P}'\mathbf{Q}_1) \subseteq \mathcal{C}(\mathbf{A}_3'\mathbf{T}_1'\mathbf{T}_2'). \tag{4.1.118}$$

When expanding the left-hand side of (4.1.118), by using (4.1.117) and the definition of $\mathbf{P}$ in (4.1.104), we obtain the equalities

$$\begin{aligned} &\mathcal{C}(\mathbf{A}_3'(\mathbf{I}-\mathbf{T}_1'\mathbf{S}_2^{-1}\mathbf{T}_1\mathbf{A}_2(\mathbf{A}_2'\mathbf{T}_1'\mathbf{S}_2^{-1}\mathbf{T}_1\mathbf{A}_2)^-\mathbf{A}_2')\mathbf{Q}_1) \\ &\quad= \mathcal{C}(\mathbf{A}_3'\mathbf{Q}_1 - \mathbf{A}_3'\mathbf{T}_1'\mathbf{S}_2^{-1}\mathbf{T}_1\mathbf{A}_2(\mathbf{A}_2'\mathbf{T}_1'\mathbf{S}_2^{-1}\mathbf{T}_1\mathbf{A}_2)^-\mathbf{A}_2'\mathbf{T}_1'\mathbf{Q}_2\mathbf{Q}_3) \\ &\quad= \mathcal{C}(\mathbf{A}_3'\mathbf{Q}_1 - \mathbf{A}_3'(\mathbf{T}_1' - \mathbf{T}_1'\mathbf{T}_2')\mathbf{Q}_2\mathbf{Q}_3). \end{aligned}$$

Hence, (4.1.118) can be written as

$$\mathcal{C}(\mathbf{A}_3'\mathbf{Q}_1 - \mathbf{A}_3'(\mathbf{T}_1' - \mathbf{T}_1'\mathbf{T}_2')\mathbf{Q}_2\mathbf{Q}_3) \subseteq \mathcal{C}(\mathbf{A}_3'\mathbf{T}_1'\mathbf{T}_2'),$$

which by Proposition 1.2.2 (ii) is identical to

$$\mathcal{C}(\mathbf{A}_3'\mathbf{Q}_1 - \mathbf{A}_3'\mathbf{T}_1'\mathbf{Q}_2\mathbf{Q}_3) \subseteq \mathcal{C}(\mathbf{A}_3'\mathbf{T}_1'\mathbf{T}_2').$$

By applying (4.1.117) and $\mathcal{C}(\mathbf{T}_1'\mathbf{T}_2') = \mathcal{C}(\mathbf{A}_1 : \mathbf{A}_2)$ it follows that

$$\mathcal{C}(\mathbf{A}_3'\mathbf{Q}_1 - \mathbf{A}_3'\mathbf{P}_{\mathbf{A}_1^o}\mathbf{Q}_3) \subseteq \mathcal{C}(\mathbf{A}_3'\mathbf{T}_1'\mathbf{T}_2') = \mathcal{C}(\mathbf{A}_3'(\mathbf{A}_1 : \mathbf{A}_2)^o). \tag{4.1.119}$$

Since $\mathcal{C}((\mathbf{A}_1 : \mathbf{A}_2)^o) = \mathcal{C}(\mathbf{P}_{(\mathbf{A}_1:\mathbf{A}_2)_1^o})$, the relation in (4.1.119) holds if and only if

$$\mathcal{C}(\mathbf{A}_3'\mathbf{Q}_1 - \mathbf{A}_3'(\mathbf{P}_{\mathbf{A}_1^o} - \mathbf{P}_{(\mathbf{A}_1:\mathbf{A}_2)_1^o})\mathbf{Q}_3) \subseteq \mathcal{C}(\mathbf{A}_3'(\mathbf{A}_1 : \mathbf{A}_2)^o), \tag{4.1.120}$$

and by utilizing (4.1.91), the inclusion (4.1.120) can be written

$$\mathcal{C}(\mathbf{A}_3'\mathbf{Q}_1 - \mathbf{A}_3'\mathbf{P}_{\mathbf{A}_1^o}\mathbf{A}_2(\mathbf{A}_2'\mathbf{P}_{\mathbf{A}_1^o}\mathbf{A}_2)^-\mathbf{A}_2'\mathbf{P}_{\mathbf{A}_1^o}\mathbf{Q}_3) \subseteq \mathcal{C}(\mathbf{A}_3'(\mathbf{A}_1 : \mathbf{A}_2)^o). \tag{4.1.121}$$

However, applying (4.1.117) to (4.1.121) yields

$$\mathcal{C}(\mathbf{A}_3'(\mathbf{I} - \mathbf{P}_{\mathbf{A}_1^o}\mathbf{A}_2(\mathbf{A}_2'\mathbf{P}_{\mathbf{A}_1^o}\mathbf{A}_2)^-\mathbf{A}_2')\mathbf{Q}_1) \subseteq \mathcal{C}(\mathbf{A}_3'(\mathbf{A}_1 : \mathbf{A}_2')^o),$$

and by definition of $\mathbf{P}_{(\mathbf{A}_1:\mathbf{A}_2)_1^o}$ in (4.1.91) we get that (4.1.121) is equivalent to

$$\mathcal{C}(\mathbf{A}_3'(\mathbf{I}-\mathbf{P}_{\mathbf{A}_1^o}\mathbf{A}_2(\mathbf{A}_2'\mathbf{P}_{\mathbf{A}_1^o}\mathbf{A}_2)^-\mathbf{A}_2')(\mathbf{I}-\mathbf{P}_{\mathbf{A}_1^o})\mathbf{Q}_1) \subseteq \mathcal{C}(\mathbf{A}_3'(\mathbf{A}_1 : \mathbf{A}_2')^o), \tag{4.1.122}$$

since $\mathcal{C}(\mathbf{A}_3'(\mathbf{A}_1 : \mathbf{A}_2)^o) = \mathcal{C}(\mathbf{A}_3'\mathbf{P}_{(\mathbf{A}_1:\mathbf{A}_2)_1^o})$. Thus, by definition of $\mathbf{P}_{\mathbf{A}_1^o}$, from (4.1.122) and the equality $\mathbf{K}' = \mathbf{A}_1'\mathbf{Q}_1'$, it follows that (4.1.114) is equivalent to

$$\mathcal{C}(\mathbf{A}_3'(\mathbf{I} - \mathbf{P}_{\mathbf{A}_1^o}\mathbf{A}_2(\mathbf{A}_2'\mathbf{P}_{\mathbf{A}_1^o}\mathbf{A}_2)^-\mathbf{A}_2')\mathbf{A}_1(\mathbf{A}_1'\mathbf{A}_1)^-\mathbf{K}') \subseteq \mathcal{C}(\mathbf{A}_3'(\mathbf{A}_1 : \mathbf{A}_2')^o). \tag{4.1.123}$$

Returning to (4.1.116) we get by applying Proposition 1.2.2 (ii) that (4.1.116) is equivalent to

$$\mathcal{C}(\mathbf{A}_2'(\mathbf{I} - \mathbf{P}_{\mathbf{A}_1^o})\mathbf{Q}_1) \subseteq \mathcal{C}(\mathbf{A}_2'\mathbf{A}_1^o),$$

since $\mathcal{C}(\mathbf{A}_2'\mathbf{T}_1') = \mathcal{C}(\mathbf{A}_2'\mathbf{A}_1^o) = \mathcal{C}(\mathbf{A}_2'\mathbf{P}_{\mathbf{A}_1^o})$. By definition of $\mathbf{P}_{\mathbf{A}_1^o}$ and $\mathbf{Q}_1$, the obtained inclusion is identical to

$$\mathcal{C}(\mathbf{A}_2'\mathbf{A}_1(\mathbf{A}_1'\mathbf{A}_1)^-\mathbf{K}') \subseteq \mathcal{C}(\mathbf{A}_2'\mathbf{A}_1^o). \tag{4.1.124}$$

Hence, (4.1.113) is equivalent to (4.1.124), and by (4.1.123) and (4.1.124) alternatives to (4.1.114) and (4.1.113) have been found which do not depend on the observations.

The next theorem summarizes some of the results given above.

Theorem 4.1.12. *Let $\widehat{\mathbf{B}}_i$, $i = 1,2,3$, be given in Theorem 4.1.6 and let $\mathbf{K}\widehat{\mathbf{B}}_i\mathbf{L}$, $i = 1,2,3$, be linear combinations of $\widehat{\mathbf{B}}_i$, where $\mathbf{K}$ and $\mathbf{L}$ are known non-random matrices of proper sizes. Then the following statements hold:*

(i) $\widehat{\mathbf{B}}_3$ *is unique if and only if*

$$r(\mathbf{A}_3) = q_3, \quad r(\mathbf{C}_3) = k_3$$

and

$$\mathcal{C}(\mathbf{A}_3) \cap \mathcal{C}(\mathbf{A}_1 : \mathbf{A}_2) = \{\mathbf{0}\};$$

(ii) $\mathbf{K}\widehat{\mathbf{B}}_3\mathbf{L}$ *is unique if and only if*

$$\mathcal{C}(\mathbf{L}) \subseteq \mathcal{C}(\mathbf{C}_3)$$

and

$$\mathcal{C}(\mathbf{K}') \subseteq \mathcal{C}(\mathbf{A}_3'(\mathbf{A}_1 : \mathbf{A}_2)^o);$$

(iii) $\widehat{\mathbf{B}}_2$ *is unique if and only if*

$$r(\mathbf{A}_2) = q_2, \quad r(\mathbf{C}_2) = k_2,$$
$$\mathcal{C}(\mathbf{A}_1) \cap \mathcal{C}(\mathbf{A}_2) = \{\mathbf{0}\}$$

and

$$\mathcal{C}(\mathbf{A}_1)^{\perp} \cap \mathcal{C}(\mathbf{A}_1 : \mathbf{A}_2) \cap \mathcal{C}(\mathbf{A}_1 : \mathbf{A}_3) = \{\mathbf{0}\};$$

(iv) $\mathbf{K}\widehat{\mathbf{B}}_2\mathbf{L}$ *is unique if and only if*

$$\mathcal{C}(\mathbf{L}) \subseteq \mathcal{C}(\mathbf{C}_2)$$

and

$$\mathcal{C}(\mathbf{K}') \subseteq \mathcal{C}(\mathbf{A}_2'(\mathbf{A}_1 : \mathbf{A}_3)^o);$$

(v) $\widehat{\mathbf{B}}_1$ *is unique if and only if*

$$r(\mathbf{A}_1) = q_1, \quad r(\mathbf{C}_1) = k_1,$$
$$\mathcal{C}(\mathbf{A}_1) \cap \mathcal{C}(\mathbf{A}_2) = \{\mathbf{0}\}$$

and

$$\mathcal{C}(\mathbf{A}_2)^{\perp} \cap \mathcal{C}(\mathbf{A}_1 : \mathbf{A}_2) \cap \mathcal{C}(\mathbf{A}_2 : \mathbf{A}_3) = \{\mathbf{0}\};$$

(vi) $\mathbf{K}\widehat{\mathbf{B}}_1\mathbf{L}$ *is unique if and only if*

$$\mathcal{C}(\mathbf{L}) \subseteq \mathcal{C}(\mathbf{C}_1),$$
$$\mathcal{C}(\mathbf{K}') \subseteq \mathcal{C}(\mathbf{A}_1'),$$
$$\mathcal{C}(\mathbf{A}_3'(\mathbf{I} - \mathbf{P}_{\mathbf{A}_1^o}\mathbf{A}_2(\mathbf{A}_2'\mathbf{P}_{\mathbf{A}_1^o}\mathbf{A}_2)^-\mathbf{A}_2')\mathbf{A}_1(\mathbf{A}_1'\mathbf{A}_1)^-\mathbf{K}') \subseteq \mathcal{C}(\mathbf{A}_3'(\mathbf{A}_1 : \mathbf{A}_2')^o),$$

where $\mathbf{P}_{\mathbf{A}_1^o}$ is defined in (4.1.90) and

$$\mathcal{C}(\mathbf{A}_2'\mathbf{A}_1(\mathbf{A}_1'\mathbf{A}_1)^-\mathbf{K}') \subseteq \mathcal{C}(\mathbf{A}_2'\mathbf{A}_1^o).$$

∎

Theorem 4.1.12 can be extended to cover the general MLNM($\sum_{i=1}^m \mathbf{A}_i\mathbf{B}_i\mathbf{C}_i$). The next theorem presents the results. A proof can be found in von Rosen (1993).

Theorem 4.1.13. *Let*

$$\begin{aligned}
\mathbf{A}_{s,r} =&(\mathbf{A}_1 : \mathbf{A}_2 : \ldots : \mathbf{A}_{r-1} : \mathbf{A}_{r+1} : \ldots : \mathbf{A}_{r+s+1}), \quad s = 2,3,\ldots,m-r, r>1;\\
\mathbf{A}_{s,r} =&(\mathbf{A}_2 : \mathbf{A}_3 : \ldots : \mathbf{A}_s), \quad s = 2,3,\ldots,m-r, r=1;\\
\mathbf{A}_{s,r} =&(\mathbf{A}_1 : \mathbf{A}_2 : \ldots : \mathbf{A}_{r-1}), \quad s = 1 \leq m-r, r>1;\\
\mathbf{A}_{s,r} =&\mathbf{0}, \quad s=1, r=1.
\end{aligned}$$

Then $\widehat{\mathbf{B}}_r$ *given in Theorem 4.1.7 is unique if and only if*

$$\begin{aligned}
&r(\mathbf{A}_r) = m_r, \quad r(\mathbf{C}_r) = k_r;\\
&C(\mathbf{A}_r) \cap C(\mathbf{A}_1 : \mathbf{A}_2 : \ldots : \mathbf{A}_{r-1}) = \{\mathbf{0}\}, \quad r > 1;\\
&C(\mathbf{A}_{s,r})^{\perp} \cap \{C(\mathbf{A}_{s,r}) + C(\mathbf{A}_{r+s})\} \cap \{C(\mathbf{A}_{s,r}) + C(\mathbf{A}_r)\} = \{\mathbf{0}\},\\
&\qquad\qquad\qquad\qquad\qquad\qquad\qquad\qquad\qquad s = 1,2,\ldots,m-r.
\end{aligned}$$

∎

4.1.6 *Restrictions on* **B** *in the Growth Curve model*

When treating restrictions on the mean parameters in the Growth Curve model, we mainly rely on previous results. Note that if it is possible to present a theory for estimating parameters under restrictions, one can immediately find test statistics for testing hypothesis. Of course, the problem of finding the distribution of the test statistics still remains. For many statistics, which can be obtained from the forthcoming results, the distributions are unknown and have to be approximated. Let us proceed with Example 4.1.1 again.

Example 4.1.4 (Example 4.1.1 continued). In the case when the three different treatment groups had different polynomial responses, the model had the mean

$$\mathbf{E}[\mathbf{X}] = \mathbf{A}_1\mathbf{B}_1\mathbf{C}_1 + \mathbf{A}_2\mathbf{B}_2\mathbf{C}_2 + \mathbf{A}_3\mathbf{B}_3\mathbf{C}_3. \tag{4.1.125}$$

The different responses were explicitly given by

$$\begin{aligned}
&\beta_{11}+\beta_{21}t + \cdots + \beta_{(q-2)1}t^{q-3},\\
&\beta_{12}+\beta_{22}t + \cdots + \beta_{(q-2)2}t^{q-3} + \beta_{(q-1)2}t^{q-2},\\
&\beta_{13}+\beta_{23}t + \cdots + \beta_{(q-2)3}t^{q-3} + \beta_{(q-1)3}t^{q-2} + \beta_{q3}t^{q-1}.
\end{aligned}$$

Equivalently, this can be written in the form

$$\begin{aligned}
&\beta_{11}+\beta_{21}t + \cdots + \beta_{(q-2)1}t^{q-3} + \beta_{(q-1)1}t^{q-2} + \beta_{q1}t^{q-1},\\
&\beta_{12}+\beta_{22}t + \cdots + \beta_{(q-2)2}t^{q-3} + \beta_{(q-1)2}t^{q-2} + \beta_{q2}t^{q-1},\\
&\beta_{13}+\beta_{23}t + \cdots + \beta_{(q-2)3}t^{q-3} + \beta_{(q-1)3}t^{q-2} + \beta_{q3}t^{q-1},
\end{aligned}$$

with

$$\beta_{(q-1)1} = 0, \quad \beta_{q1} = 0, \quad \beta_{q2} = 0. \tag{4.1.126}$$

In matrix notation the restrictions in (4.1.126) can be presented by

$$\mathbf{G}_1\mathbf{B}\mathbf{H}_1 = \mathbf{0}, \tag{4.1.127}$$
$$\mathbf{G}_2\mathbf{B}\mathbf{H}_2 = \mathbf{0}, \tag{4.1.128}$$

where $\mathbf{B} = (b_{ij})$: $q \times 3$,

$$\mathbf{G}_1 = \begin{pmatrix} 0 & 0 & \cdots & 1 & 0 \end{pmatrix}, \quad \mathbf{H}_1' = \begin{pmatrix} 1 & 0 & 0 \end{pmatrix},$$
$$\mathbf{G}_2 = \begin{pmatrix} 0 & 0 & \cdots & 0 & 1 \end{pmatrix}, \quad \mathbf{H}_2' = \begin{pmatrix} 1 & 0 & 0 \\ 0 & 1 & 0 \end{pmatrix}.$$

Hence, (4.1.125) is equivalent to $E[\mathbf{X}] = \mathbf{ABC}$ if (4.1.127) and (4.1.128) hold. In general, it follows that any model in the class of the MLNM($\sum_{i=1}^{m} \mathbf{A}_i\mathbf{B}_i\mathbf{C}_i$) can be written as a MLNM(**ABC**), with restrictions on **B**. ∎

Consider the Growth Curve model given in Definition 4.1.1. Various types of restrictions on the parameter **B** are going to be treated:

(i) $\mathbf{B} = \mathbf{F} + \mathbf{G}\boldsymbol{\Theta}\mathbf{H}$, (4.1.129)
where $\mathbf{F} : q \times k$, $\mathbf{G} : q \times s$, $\mathbf{H} : t \times k$ are known matrices and $\boldsymbol{\Theta} : s \times t$ is a matrix of unknown parameters;

(ii) $\mathbf{GB} = \mathbf{0}$, $\mathbf{BH} = \mathbf{0}$, (4.1.130)
where $\mathbf{G} : s \times q$ and $\mathbf{H} : k \times t$ are known matrices;

(iii) $\mathbf{GBH} = \mathbf{0}$, (4.1.131)
where $\mathbf{G} : s \times q$, $\mathbf{H} : k \times t$ are known matrices;

(iv) $\mathbf{G}_1\mathbf{B}\mathbf{H}_1 = \mathbf{0}$, $\mathbf{G}_2\mathbf{B}\mathbf{H}_2 = \mathbf{0}$, (4.1.132)
where $\mathbf{G}_i : s_i \times q$, $\mathbf{H}_i : k \times t_i$, $i = 1, 2$, are known matrices.

In (ii), (iii) and (iv) it is worth observing that the homogeneous restrictions can be replaced by non-homogenous restrictions, i.e. in (ii) $\mathbf{GB} = \mathbf{F}_1$, $\mathbf{BH} = \mathbf{F}_2$, in (iii) $\mathbf{GBH} = \mathbf{F}$ and in (iv) $\mathbf{G}_1\mathbf{B}\mathbf{H}_1 = \mathbf{F}_1$, $\mathbf{G}_2\mathbf{B}\mathbf{H}_2 = \mathbf{F}_2$ where some additional conditions on the **F**−matrices must be imposed. However, it is left to the reader to carry out the treatments for the non-homogeneous restrictions. Furthermore, in order to obtain explicit estimators under the restrictions given in (iv), some conditions have to be put on the known matrices.

The restrictions in (ii) and (iii) are special cases of (iv), but since (ii) is very easy to handle and (iii) is frequently used in multivariate analysis of variance, we are going to treat them separately. When deriving the maximum likelihood estimators $\widehat{\mathbf{B}}$ and $\widehat{\boldsymbol{\Sigma}}$ in (i), (ii), (iii) and (iv) it becomes clear how to obtain estimators when more general constraints exist. Other restrictions than those given in (4.1.129) – (4.1.132) will also be treated (see (4.1.166) and (4.1.169) given below).

The main idea of this paragraph is to identify each of the four types of different restrictions with either another unrestricted Growth Curve model or a MLNM($\sum_{i=1}^{m} \mathbf{A}_i\mathbf{B}_i\mathbf{C}_i$). It will be shown that (i) and (ii) can be formulated via the ordinary Growth Curve model, whereas (iii) and (iv) utilize the more general MLNM($\sum_{i=1}^{m} \mathbf{A}_i\mathbf{B}_i\mathbf{C}_i$). These results are obtained by interpreting the restrictions given by (4.1.130), (4.1.131) or (4.1.132) as a system of linear equations in the parameter matrix **B**. Thereafter the system is solved.

Starting with (i) in (4.1.129), it is seen that we have an ordinary Growth Curve model with a translated data matrix $\mathbf{X}$ with a mean structure which has been changed.

Theorem 4.1.14. *For the* MLNM($\mathbf{ABC}$) *with restriction* (i) *on* $\mathbf{B}$, *given by (4.1.129), the maximum likelihood estimators equal*

$$\begin{aligned}\widehat{\mathbf{B}} =&\mathbf{F} + \mathbf{G}(\mathbf{G}'\mathbf{A}'\mathbf{S}^{-1}\mathbf{A}\mathbf{G})^{-}\mathbf{G}'\mathbf{A}'\mathbf{S}^{-1}(\mathbf{X} - \mathbf{AFC})\mathbf{C}'\mathbf{H}'(\mathbf{HCC}'\mathbf{H}')^{-}\\ &+ (\mathbf{G}'\mathbf{A}')^{o}\mathbf{Z}_1 + \mathbf{G}'\mathbf{A}'\mathbf{Z}_2(\mathbf{HC})^{o'},\end{aligned}$$

where

$$\mathbf{S} = (\mathbf{X} - \mathbf{AFC})(\mathbf{I} - \mathbf{C}'\mathbf{H}'(\mathbf{HCC}'\mathbf{H}')^{-}\mathbf{HC})(\mathbf{X} - \mathbf{AFC})'$$

is assumed to be p.d., $\mathbf{Z}_i$, $i = 1, 2$, *are arbitrary matrices and*

$$n\widehat{\boldsymbol{\Sigma}} = (\mathbf{X} - \mathbf{A}\widehat{\mathbf{B}}\mathbf{C})(\mathbf{X} - \mathbf{A}\widehat{\mathbf{B}}\mathbf{C})'.$$

The estimator $\widehat{\mathbf{B}}$ *is unique if and only if*

$$\mathcal{C}(\mathbf{G}) \cap \mathcal{C}(\mathbf{A}')^{\perp} = \{\mathbf{0}\}, \tag{4.1.133}$$
$$\mathcal{C}(\mathbf{H}') \cap \mathcal{C}(\mathbf{C})^{\perp} = \{\mathbf{0}\}. \tag{4.1.134}$$

For known matrices $\mathbf{K}$ *and* $\mathbf{L}$ *the linear combination* $\mathbf{K}\widehat{\mathbf{B}}\mathbf{L}$ *is unique if and only if*

$$\mathcal{C}(\mathbf{G}'\mathbf{K}') \subseteq \mathcal{C}(\mathbf{G}'\mathbf{A}'), \tag{4.1.135}$$
$$\mathcal{C}(\mathbf{HL}) \subseteq \mathcal{C}(\mathbf{HC}). \tag{4.1.136}$$

Note that the conditions (4.1.135) and (4.1.136) hold if $\mathcal{C}(\mathbf{K}') \subseteq \mathcal{C}(\mathbf{A}')$ and $\mathcal{C}(\mathbf{L}) \subseteq \mathcal{C}(\mathbf{C})$ which, however, are not necessary conditions.

PROOF: Let

$$\begin{aligned}\mathbf{Y} =&\mathbf{X} - \mathbf{AFC},\\ \mathbf{M} =&\mathbf{AG},\\ \mathbf{N} =&\mathbf{HC}.\end{aligned}$$

Hence, when inserting (4.1.129) into the mean of the MLNM($\mathbf{ABC}$), the following Growth Curve model is obtained:

$$\mathbf{Y} = \mathbf{M}\boldsymbol{\Theta}\mathbf{N} + \boldsymbol{\Sigma}^{1/2}\mathbf{E},$$

where, as previously, $\boldsymbol{\Sigma}^{1/2}$ is a symmetric square root of $\boldsymbol{\Sigma}$ and $\mathbf{E} \sim N_{p,n}(\mathbf{0}, \mathbf{I}_p, \mathbf{I}_n)$. Thus, via Theorem 4.1.1, it follows from (4.1.7) and (4.1.8) that

$$\widehat{\boldsymbol{\Theta}} = (\mathbf{M}'\mathbf{S}^{-1}\mathbf{M})^{-}\mathbf{M}'\mathbf{S}^{-1}\mathbf{Y}\mathbf{N}'(\mathbf{NN}')^{-} + (\mathbf{M}')^{o}\mathbf{Z}_1 + \mathbf{M}'\mathbf{Z}_2\mathbf{N}^{o'},$$

where $\mathbf{Z}_1$ and $\mathbf{Z}_2$ are arbitrary matrices,

$$\mathbf{S} = \mathbf{Y}(\mathbf{I} - \mathbf{N}'(\mathbf{N}\mathbf{N}')^{-}\mathbf{N})\mathbf{Y}',$$

$$\mathbf{M}\widehat{\boldsymbol{\Theta}}\mathbf{N} = \mathbf{M}(\mathbf{M}'\mathbf{S}^{-1}\mathbf{M})^{-}\mathbf{M}'\mathbf{S}^{-1}\mathbf{Y}\mathbf{N}'(\mathbf{N}\mathbf{N}')^{-}\mathbf{N} \tag{4.1.137}$$

and

$$n\widehat{\boldsymbol{\Sigma}} = (\mathbf{Y} - \mathbf{M}\widehat{\boldsymbol{\Theta}}\mathbf{N})(\mathbf{Y} - \mathbf{M}\widehat{\boldsymbol{\Theta}}\mathbf{N})'. \tag{4.1.138}$$

According to Proposition 1.2.2 (x) the expression in (4.1.137) is invariant with respect to any choice of g-inverse, because the expression does not depend on the arbitrary matrices $\mathbf{Z}_i$, $i = 1, 2$. Thus, $n\widehat{\boldsymbol{\Sigma}}$ given by (4.1.138) is invariant with respect to g-inverses, and hence $n\widehat{\boldsymbol{\Sigma}}$ is unique. By (4.1.129) it follows that

$$\widehat{\mathbf{B}} = \mathbf{F} + \mathbf{G}\widehat{\boldsymbol{\Theta}}\mathbf{H}.$$

The maximum likelihood estimator $\widehat{\mathbf{B}}$ is unique if and only if

$$\mathcal{C}(\mathbf{G}') \subseteq \mathcal{C}(\mathbf{G}'\mathbf{A}'),$$
$$\mathcal{C}(\mathbf{H}) \subseteq \mathcal{C}(\mathbf{H}\mathbf{C})$$

hold, which by Theorem 1.2.12 are equivalent to the conditions (4.1.133) and (4.1.134). For $\mathbf{K}\widehat{\mathbf{B}}\mathbf{L}$ we also immediately obtain the conditions given by (4.1.135) and (4.1.136). ∎

By applying Corollary 1.3.6.2 we obtain for restrictions (ii), given by (4.1.130), that all possible matrices $\mathbf{B}$ satisfying (4.1.130) are given by

$$\mathbf{B} = (\mathbf{G}')^{o}\boldsymbol{\Theta}\mathbf{H}^{o'}, \tag{4.1.139}$$

where $\boldsymbol{\Theta}$ is arbitrary, i.e. $\boldsymbol{\Theta}$ is a matrix of new parameters. From (4.1.139) it follows that the restrictions in (ii) can be formulated by restrictions (i), and the next corollary is immediately established.

Corollary 4.1.14.1. *For a* MLNM($\mathbf{ABC}$) *with restrictions* (ii) *on* $\mathbf{B}$, *given by (4.1.130), the maximum likelihood estimators equal*

$$\begin{aligned}\widehat{\mathbf{B}} =& (\mathbf{G}')^{o}\{(\mathbf{G}')^{o'}\mathbf{A}'\mathbf{S}^{-1}\mathbf{A}(\mathbf{G}')^{o}\}^{-}(\mathbf{G}')^{o'}\mathbf{A}'\mathbf{S}^{-1}\mathbf{X}\mathbf{C}'\mathbf{H}^{o}(\mathbf{H}^{o'}\mathbf{C}\mathbf{C}'\mathbf{H}^{o})^{-}\\ &+ ((\mathbf{G}')^{o'}\mathbf{A}')^{o}\mathbf{Z}_1 + \mathbf{G}'\mathbf{A}'\mathbf{Z}_2(\mathbf{H}^{o'}\mathbf{C})^{o'},\end{aligned}$$

where

$$\mathbf{S} = \mathbf{X}(\mathbf{I} - \mathbf{C}'\mathbf{H}^{o}(\mathbf{H}^{o'}\mathbf{C}\mathbf{C}'\mathbf{H}^{o})^{-}\mathbf{H}^{o'}\mathbf{C})\mathbf{X}'$$

is assumed to be p.d., $\mathbf{Z}_i$, $i = 1, 2$, *are arbitrary matrices and*

$$n\widehat{\boldsymbol{\Sigma}} = (\mathbf{X} - \mathbf{A}\widehat{\mathbf{B}}\mathbf{C})(\mathbf{X} - \mathbf{A}\widehat{\mathbf{B}}\mathbf{C})'.$$

The estimator $\widehat{\mathbf{B}}$ *is unique if and only if*

$$\mathcal{C}(\mathbf{G}')^{\perp} \cap \mathcal{C}(\mathbf{A}')^{\perp} = \{\mathbf{0}\}, \tag{4.1.140}$$
$$\mathcal{C}(\mathbf{H}^{o}) \cap \mathcal{C}(\mathbf{C})^{\perp} = \{\mathbf{0}\}. \tag{4.1.141}$$

For known matrices, $\mathbf{K}$ *and* $\mathbf{L}$*, the linear combination* $\mathbf{K}\widehat{\mathbf{B}}\mathbf{L}$ *is unique, if and only if*

$$\mathcal{C}((\mathbf{G}')^{o'}\mathbf{K}') \subseteq \mathcal{C}((\mathbf{G}')^{o'}\mathbf{A}'), \tag{4.1.142}$$

$$\mathcal{C}(\mathbf{H}^{o'}\mathbf{L}) \subseteq \mathcal{C}(\mathbf{H}^{o'}\mathbf{C}). \tag{4.1.143}$$

∎

If $\mathcal{C}((\mathbf{G}')^{o}) \subseteq \mathcal{C}(\mathbf{A}')$ and $\mathcal{C}(\mathbf{H}^{o}) \subseteq \mathcal{C}(\mathbf{C})$, we obtain from (4.1.140) and (4.1.141) that $\widehat{\mathbf{B}}$ is always unique and these conditions are automatically satisfied if $\mathbf{A}$ and $\mathbf{C}$ are of full rank. Furthermore, if $\mathcal{C}(\mathbf{K}') \subseteq \mathcal{C}(\mathbf{A}')$ and $\mathcal{C}(\mathbf{L}) \subseteq \mathcal{C}(\mathbf{C})$, the inclusions (4.1.142) and (4.1.143) are always satisfied.
Now let us consider the restrictions (iii) given by (4.1.131). Since (4.1.131) forms a homogeneous equation system, it follows from Theorem 1.3.4 that $\mathbf{B}$ has to satisfy

$$\mathbf{B} = (\mathbf{G}')^{o}\boldsymbol{\Theta}_1 + \mathbf{G}'\boldsymbol{\Theta}_2\mathbf{H}^{o'}, \tag{4.1.144}$$

where $\boldsymbol{\Theta}_1$ and $\boldsymbol{\Theta}_2$ are new parameter matrices. Inserting (4.1.144) into $E[\mathbf{X}] = \mathbf{ABC}$ yields the mean

$$E[\mathbf{X}] = \mathbf{A}(\mathbf{G}')^{o}\boldsymbol{\Theta}_1\mathbf{C} + \mathbf{A}\mathbf{G}'\boldsymbol{\Theta}_2\mathbf{H}^{o'}\mathbf{C}.$$

Since $\mathcal{C}(\mathbf{C}'\mathbf{H}^{o}) \subseteq \mathcal{C}(\mathbf{C}')$ always holds we have a MLNM($\sum_{i=1}^{2}\mathbf{A}_i\mathbf{B}_i\mathbf{C}_i$). Thus, we have found that the MLNM($\mathbf{ABC}$) with restrictions (iii) can be reformulated as a MLNM($\sum_{i=1}^{2}\mathbf{A}_i\mathbf{B}_i\mathbf{C}_i$).

Theorem 4.1.15. *For the* MLNM($\mathbf{ABC}$) *with restrictions* (iii) *on* $\mathbf{B}$*, given by (4.1.131), the maximum likelihood estimators equal*

$$\widehat{\mathbf{B}} = (\mathbf{G}')^{o}\widehat{\boldsymbol{\Theta}}_1 + \mathbf{G}'\widehat{\boldsymbol{\Theta}}_2\mathbf{H}^{o'},$$

where

$$\begin{aligned}\widehat{\boldsymbol{\Theta}}_2 =&(\mathbf{G}\mathbf{A}'\mathbf{T}_1'\mathbf{S}_2^{-1}\mathbf{T}_1\mathbf{A}\mathbf{G}')^{-}\mathbf{G}\mathbf{A}'\mathbf{T}_1'\mathbf{S}_2^{-1}\mathbf{T}_1\mathbf{X}\mathbf{C}'\mathbf{H}^{o}(\mathbf{H}^{o'}\mathbf{C}\mathbf{C}'\mathbf{H}^{o})^{-}\\ &+(\mathbf{G}\mathbf{A}'\mathbf{T}_1')^{o}\mathbf{Z}_{11} + \mathbf{G}\mathbf{A}'\mathbf{T}_1'\mathbf{Z}_{12}(\mathbf{H}^{o'}\mathbf{C})^{o'},\end{aligned}$$

with

$$\begin{aligned}\mathbf{T}_1 =&\mathbf{I} - \mathbf{A}(\mathbf{G}')^{o}\{(\mathbf{G}')^{o'}\mathbf{A}'\mathbf{S}_1^{-1}\mathbf{A}(\mathbf{G}')^{o}\}^{-}(\mathbf{G}')^{o'}\mathbf{A}'\mathbf{S}_1^{-1},\\ \mathbf{S}_1 =&\mathbf{X}(\mathbf{I} - \mathbf{C}'(\mathbf{C}\mathbf{C}')^{-}\mathbf{C})\mathbf{X}',\\ \mathbf{S}_2 =&\mathbf{S}_1 + \mathbf{T}_1\mathbf{X}\mathbf{C}'(\mathbf{C}\mathbf{C}')^{-}\mathbf{C}\{\mathbf{I} - \mathbf{C}'\mathbf{H}^{o}(\mathbf{H}^{o'}\mathbf{C}\mathbf{C}'\mathbf{H}^{o})^{-}\mathbf{H}^{o'}\mathbf{C}\}\mathbf{C}'(\mathbf{C}\mathbf{C}')^{-}\mathbf{C}\mathbf{X}'\mathbf{T}_1',\end{aligned}$$

$$\begin{aligned}\widehat{\boldsymbol{\Theta}}_1 =&\{(\mathbf{G}')^{o'}\mathbf{A}'\mathbf{S}_1^{-1}\mathbf{A}(\mathbf{G}')^{o}\}^{-}(\mathbf{G}')^{o'}\mathbf{A}'\mathbf{S}_1^{-1}(\mathbf{X} - \mathbf{A}\mathbf{G}'\widehat{\boldsymbol{\Theta}}_2\mathbf{H}^{o'}\mathbf{C})\mathbf{C}'(\mathbf{C}\mathbf{C}')^{-}\\ &+((\mathbf{G}')^{o'}\mathbf{A}')^{o}\mathbf{Z}_{21} + (\mathbf{G}')^{o'}\mathbf{A}'\mathbf{Z}_{22}\mathbf{C}^{o'},\end{aligned}$$

where $\mathbf{S}_1$ *is assumed to be p.d.,* $\mathbf{Z}_{ij}$ *are arbitrary matrices, and*

$$n\widehat{\boldsymbol{\Sigma}} = (\mathbf{X} - \mathbf{A}\widehat{\mathbf{B}}\mathbf{C})(\mathbf{X} - \mathbf{A}\widehat{\mathbf{B}}\mathbf{C})'.$$

The estimator $\widehat{\mathbf{B}}$ *is unique if and only if*

$$\begin{aligned}&r(\mathbf{C}) = k,\\ &\mathcal{C}(\mathbf{G}')^{\perp} \cap \mathcal{C}(\mathbf{A}')^{\perp} = \{\mathbf{0}\},\\ &\mathcal{C}(\mathbf{G}') \cap \{\mathcal{C}(\mathbf{G}')^{\perp} + \mathcal{C}(\mathbf{A}')^{\perp}\} = \{\mathbf{0}\}.\end{aligned}$$

For known matrices, $\mathbf{K}$ *and* $\mathbf{L}$*, the linear combinations* $\mathbf{K}\widehat{\mathbf{B}}\mathbf{L}$ *are unique, if and only if*

$$\begin{aligned}\mathcal{C}(\mathbf{L}) &\subseteq \mathcal{C}(\mathbf{C}),\\ \mathcal{C}((\mathbf{G}')^{o'}\mathbf{K}') &\subseteq \mathcal{C}((\mathbf{G}')^{o'}\mathbf{A}'),\\ \mathcal{C}(\mathbf{H}^{o'}\mathbf{L}) &\subseteq \mathcal{C}(\mathbf{H}^{o'}\mathbf{C}),\\ \mathcal{C}(\mathbf{G}\mathbf{K}') &\subseteq \mathcal{C}(\mathbf{G}\mathbf{P}_1'\mathbf{A}'),\end{aligned}$$

where

$$\mathbf{P}_1 = \mathbf{I} - (\mathbf{G}')^{o}\{(\mathbf{G}')^{o'}\mathbf{A}'\mathbf{A}(\mathbf{G}')^{o}\}^{-}(\mathbf{G}')^{o'}\mathbf{A}'\mathbf{A}.$$

PROOF: The estimators $\widehat{\boldsymbol{\Theta}}_1$ and $\widehat{\mathbf{B}}$ follow from Theorem 4.1.7, with

$$\mathbf{A}_1 = \mathbf{A}(\mathbf{G}')^{o}, \quad \mathbf{C}_1 = \mathbf{C}, \quad \mathbf{A}_2 = \mathbf{A}\mathbf{G}', \quad \mathbf{C}_2 = \mathbf{H}^{o'}\mathbf{C}, \quad \mathbf{B}_1 = \boldsymbol{\Theta}_1, \quad \mathbf{B}_2 = \boldsymbol{\Theta}_2.$$

To prove the uniqueness conditions for $\widehat{\mathbf{B}}$ as well as $\mathbf{K}\widehat{\mathbf{B}}\mathbf{L}$, one has to copy the ideas and technique of §4.1.5 where uniqueness conditions for the estimators in the MLNM($\sum_{i=1}^{3}\mathbf{A}_i\mathbf{B}_i\mathbf{C}_i$) were discussed. ∎

In the next restrictions (iv), given by (4.1.132), will be examined. The reader will immediately recognize the ideas from previous discussions. From Theorem 1.3.6 it follows that a common solution of the equations in $\mathbf{B}$, given by (4.1.132), equals

$$\begin{aligned}\mathbf{B} =&(\mathbf{G}_1' : \mathbf{G}_2')^{o}\boldsymbol{\Theta}_1 + (\mathbf{G}_2' : (\mathbf{G}_1' : \mathbf{G}_2')^{o})^{o}\boldsymbol{\Theta}_2\mathbf{H}_1^{o'} + (\mathbf{G}_1' : (\mathbf{G}_1' : \mathbf{G}_2')^{o})^{o}\boldsymbol{\Theta}_3\mathbf{H}_2^{o'}\\ &+ ((\mathbf{G}_1')^{o} : (\mathbf{G}_2')^{o})^{o}\boldsymbol{\Theta}_4(\mathbf{H}_1 : \mathbf{H}_2)^{o'},\end{aligned} \tag{4.1.145}$$

where $\boldsymbol{\Theta}_1$, $\boldsymbol{\Theta}_2$, $\boldsymbol{\Theta}_3$ and $\boldsymbol{\Theta}_4$ are arbitrary matrices interpreted as unknown parameters.

For the MLNM($\mathbf{ABC}$) with restrictions (iv) the mean equals $E[\mathbf{X}] = \mathbf{ABC}$, where $\mathbf{B}$ is given by (4.1.145). However, when $\mathbf{B}$ is given by (4.1.145), the model does not necessarily belong to the MLNM($\sum_{i=1}^{m}\mathbf{A}_i\mathbf{B}_i\mathbf{C}_i$) without further assumptions, since $\mathcal{C}(\mathbf{C}'\mathbf{H}_2^{o}) \subseteq \mathcal{C}(\mathbf{C}'\mathbf{H}_1^{o})$ or $\mathcal{C}(\mathbf{C}'\mathbf{H}_1^{o}) \subseteq \mathcal{C}(\mathbf{C}'\mathbf{H}_2^{o})$ may not hold. In a canonical formulation this means that we do not have a stairs structure for the mean. Instead, the mean structure can be identified with

$$\boldsymbol{\Theta} = \begin{pmatrix}\mathbf{0} & \boldsymbol{\Theta}_{12}\\ \boldsymbol{\Theta}_{21} & \mathbf{0}\end{pmatrix}.$$

Therefore, in order to get explicit estimators, some additional conditions have to be imposed. In fact, there are several different alternatives which could be of interest.
From (4.1.145) it follows that in order to utilize results for the general model MLNM($\sum_{i=1}^m \mathbf{A}_i\mathbf{B}_i\mathbf{C}_i$), it is natural to consider the conditions given below:

$$\mathbf{A}(\mathbf{G}_2' : (\mathbf{G}_1' : \mathbf{G}_2')^o)^o = \mathbf{0}, \quad (4.1.146)$$
$$\mathbf{A}(\mathbf{G}_1' : (\mathbf{G}_1' : \mathbf{G}_2')^o)^o = \mathbf{0},$$
$$\mathcal{C}(\mathbf{C}'\mathbf{H}_2^o) \subseteq \mathcal{C}(\mathbf{C}'\mathbf{H}_1^o), \quad (4.1.147)$$
$$\mathcal{C}(\mathbf{C}'\mathbf{H}_1^o) \subseteq \mathcal{C}(\mathbf{C}'\mathbf{H}_2^o).$$

By symmetry it follows that only (4.1.146) and (4.1.147) have to be investigated, and from a point of view of applications it is useful to make a finer division of these conditions, namely

a) $$(\mathbf{G}_2' : (\mathbf{G}_1' : \mathbf{G}_2')^o)^o = \mathbf{0}, \quad (4.1.148)$$

b) $$\mathbf{A}(\mathbf{G}_2' : (\mathbf{G}_1' : \mathbf{G}_2')^o)^o = \mathbf{0}, \quad (\mathbf{G}_2' : (\mathbf{G}_1' : \mathbf{G}_2')^o)^o \neq \mathbf{0}, \quad (4.1.149)$$

c) $$\mathcal{C}(\mathbf{H}_2^o) \subseteq \mathcal{C}(\mathbf{H}_1^o), \quad (4.1.150)$$

d) $$\mathcal{C}(\mathbf{C}'\mathbf{H}_1^o) \subseteq \mathcal{C}(\mathbf{C}'\mathbf{H}_2^o), \quad \mathcal{C}(\mathbf{H}_2^o) \nsubseteq \mathcal{C}(\mathbf{H}_1^o). \quad (4.1.151)$$

In the subsequent, a brief discussion of the various alternatives, i.e. (4.1.148) – (4.1.151), will be given. Unfortunately, several of the calculations are straightforward but quite tedious and therefore they will not be included. In particular, this applies to case d), given by (4.1.151). Because there are so many uniqueness conditions corresponding to the different alternatives given in (4.1.148) – (4.1.151), we will not summarize them in any theorem. Instead some of them will appear in the text presented below.
We start by investigating the consequences of (4.1.148). If (4.1.148) holds, then $\mathbf{G}_2' : (\mathbf{G}_1' : \mathbf{G}_2')^o$ spans the entire space, which in turn implies that

$$\mathcal{C}(\mathbf{G}_2')^{\perp} = \mathcal{C}(\mathbf{G}_1' : \mathbf{G}_2)^{\perp}, \quad (4.1.152)$$

since $\mathbf{G}_2(\mathbf{G}_1' : \mathbf{G}_2')^o = 0$. Moreover, (4.1.152) is equivalent to

$$\mathcal{C}(\mathbf{G}_1') \subseteq \mathcal{C}(\mathbf{G}_2'),$$

implying

$$\mathcal{C}(((\mathbf{G}_1')^o : (\mathbf{G}_2')^o)^o) = \mathcal{C}(\mathbf{G}_1').$$

Hence, in the case a), instead of (4.1.145), consider

$$\mathbf{B} = (\mathbf{G}_2')^o\mathbf{\Theta}_1 + (\mathbf{G}_1' : (\mathbf{G}_2')^o)^o\mathbf{\Theta}_3\mathbf{H}_2^{o'} + \mathbf{G}_1'\mathbf{\Theta}_4(\mathbf{H}_1 : \mathbf{H}_2)^{o'}. \quad (4.1.153)$$

Maximum likelihood estimators of $\mathbf{\Theta}_1$, $\mathbf{\Theta}_3$, $\mathbf{\Theta}_4$, as well as $\mathbf{B}$ and $\mathbf{\Sigma}$, are now available from Theorem 4.1.6. Uniqueness conditions for $\widehat{\mathbf{\Theta}}_1$, $\widehat{\mathbf{\Theta}}_3$ and $\widehat{\mathbf{\Theta}}_4$ are given in Theorem 4.1.12. However, we are mainly interested in uniqueness conditions for $\widehat{\mathbf{B}}$ or $\mathbf{K}\widehat{\mathbf{B}}\mathbf{L}$. It can be shown that $\widehat{\mathbf{B}}$ is unique if and only if

$$\begin{aligned} &r(\mathbf{C}) = k, \\ &\mathcal{C}(\mathbf{G}_2')^{\perp} \cap \mathcal{C}(\mathbf{A}')^{\perp} = \{\mathbf{0}\}, && (4.1.154)\\ &\mathcal{C}(\mathbf{G}_1')^{\perp} \cap \mathcal{C}(\mathbf{G}_2') \cap \{\mathcal{C}(\mathbf{A}')^{\perp} + \mathcal{C}(\mathbf{G}_2')^{\perp}\} = \{\mathbf{0}\}, && (4.1.155)\\ &\mathcal{C}(\mathbf{G}_1') \cap \{\mathcal{C}(\mathbf{A}')^{\perp} + \mathcal{C}(\mathbf{G}_1')^{\perp}\} = \{\mathbf{0}\}. && (4.1.156) \end{aligned}$$

If $\mathbf{A}$ is of full rank, i.e. $r(\mathbf{A}) = q$, then (4.1.154), (4.1.155) and (4.1.156) are all satisfied. Hence, $\widehat{\mathbf{B}}$ is unique if $\mathbf{A}$ and $\mathbf{C}$ are of full rank, and this holds for all choices of $\mathbf{G}_1$, $\mathbf{G}_2$, $\mathbf{H}_1$ and $\mathbf{H}_2$, as long as $\mathcal{C}(\mathbf{G}_1') \subseteq \mathcal{C}(\mathbf{G}_2')$.
Below, necessary and sufficient conditions for uniqueness of $\mathbf{K}\widehat{\mathbf{B}}\mathbf{L}$ are given:

$$\begin{aligned} &\mathcal{C}(\mathbf{L}) \subseteq \mathcal{C}(\mathbf{C}), \\ &\mathcal{C}((\mathbf{G}_2')^{o'}\mathbf{K}') \subseteq \mathcal{C}((\mathbf{G}_2')^{o'}\mathbf{A}'), \\ &\mathcal{C}(\{\mathbf{G}_1' : (\mathbf{G}_2')^{o}\}^{o'}\mathbf{R}_1'\mathbf{K}') \subseteq \mathcal{C}(\{\mathbf{G}_1' : (\mathbf{G}_2')^{o}\}^{o'}\mathbf{A}'(\mathbf{A}(\mathbf{G}_2')^{o})^{o}), \\ &\mathcal{C}(\mathbf{G}_1\mathbf{R}_2'\mathbf{R}_1'\mathbf{K}') \subseteq \mathcal{C}(\mathbf{G}_1\mathbf{A}'\{\mathbf{A}(\mathbf{G}_1')^{o}\}^{o}), \end{aligned}$$

where

$$\mathbf{R}_1 = \mathbf{I} - (\mathbf{G}_1' : \mathbf{G}_2')^{o}\{(\mathbf{G}_1' : \mathbf{G}_2')^{o'}\mathbf{A}'\mathbf{A}(\mathbf{G}_1' : \mathbf{G}_2')^{o}\}^{-}(\mathbf{G}_1' : \mathbf{G}_2')^{o'}\mathbf{A}'\mathbf{A}, \qquad (4.1.157)$$

$$\begin{aligned}\mathbf{R}_2 = \mathbf{I} - &\{\mathbf{G}_1' : (\mathbf{G}_1' : \mathbf{G}_2')^{o}\}^{o}\big(\{\mathbf{G}_1' : (\mathbf{G}_1' : \mathbf{G}_2')^{o}\}^{o'}\mathbf{R}_1'\mathbf{A}'\mathbf{A}\mathbf{R}_1\{\mathbf{G}_1' : (\mathbf{G}_1' : \mathbf{G}_2')^{o}\}^{o}\big)^{-} \\ &\times \{\mathbf{G}_1' : (\mathbf{G}_1' : \mathbf{G}_2')^{o}\}^{o'}\mathbf{R}_1'\mathbf{A}'\mathbf{A}\mathbf{R}_1. \qquad (4.1.158)\end{aligned}$$

If $\mathcal{C}(\mathbf{K}') \subseteq \mathcal{C}(\mathbf{A}')$ and $\mathcal{C}(\mathbf{L}) \subseteq \mathcal{C}(\mathbf{C})$, it follows that $\mathbf{K}\widehat{\mathbf{B}}\mathbf{L}$ is unique.
Let us now briefly consider case b), given by (4.1.149). There are no principal differences between a) and b). In both cases, instead of (4.1.145), the mean structure can be written as

$$\begin{aligned}\mathbf{B} = &(\mathbf{G}_1' : \mathbf{G}_2')^{o}\boldsymbol{\Theta}_1 + \{\mathbf{G}_1' : (\mathbf{G}_1' : \mathbf{G}_2')^{o}\}^{o}\boldsymbol{\Theta}_3\mathbf{H}_2^{o'} \\ &+ \{(\mathbf{G}_1')^{o} : (\mathbf{G}_2')^{o}\}^{o}\boldsymbol{\Theta}_4(\mathbf{H}_1 : \mathbf{H}_2)^{o'}. \qquad (4.1.159)\end{aligned}$$

The difference compared with (4.1.153) is that in a) $\mathcal{C}(\mathbf{G}_1') \subseteq \mathcal{C}(\mathbf{G}_2')$ holds, which simplifies (4.1.159) somewhat. Now it follows that $\widehat{\mathbf{B}}$ is unique if and only if

$$\begin{aligned} &r(\mathbf{C}) = k, && (4.1.160)\\ &\mathcal{C}(\mathbf{G}_1' : \mathbf{G}_2')^{\perp} \cap \mathcal{C}(\mathbf{A}')^{\perp} = \{\mathbf{0}\}, && (4.1.161)\\ &\mathcal{C}(\mathbf{G}_1')^{\perp} \cap \mathcal{C}(\mathbf{G}_1 : \mathbf{G}_2') \cap \{\mathcal{C}(\mathbf{A}')^{\perp} + \mathcal{C}(\mathbf{G}_2')^{\perp}\} = \{\mathbf{0}\}, && (4.1.162)\\ &\mathcal{C}(\mathbf{G}_1') \cap \mathcal{C}(\mathbf{G}_2') \cap \{\mathcal{C}(\mathbf{A}')^{\perp} + \mathcal{C}(\mathbf{G}_1')^{\perp}\} = \{\mathbf{0}\}, && (4.1.163) \end{aligned}$$

and $\mathbf{K}\widehat{\mathbf{B}}\mathbf{L}$ is unique if and only if

$$\begin{aligned} &\mathcal{C}(\mathbf{L}) \subseteq \mathcal{C}(\mathbf{C}), \\ &\mathcal{C}((\mathbf{G}_1' : \mathbf{G}_2')^{o'}\mathbf{K}') \subseteq \mathcal{C}((\mathbf{G}_1 : \mathbf{G}_2')^{o'}\mathbf{A}'), \\ &\mathcal{C}(\{\mathbf{G}_1' : (\mathbf{G}_1' : \mathbf{G}_2')^{o}\}^{o'}\mathbf{R}_1'\mathbf{K}') \subseteq \mathcal{C}(\{\mathbf{G}_1' : (\mathbf{G}_1' : \mathbf{G}_2')^{o}\}^{o'}\mathbf{A}'(\mathbf{A}(\mathbf{G}_2')^{o})^{o}), \\ &\mathcal{C}(\{(\mathbf{G}_1')^{o} : (\mathbf{G}_2')^{o}\}^{o'}\mathbf{R}_2'\mathbf{R}_1'\mathbf{K}') \subseteq \mathcal{C}(\{(\mathbf{G}_1')^{o} : (\mathbf{G}_2')^{o}\}^{o'}\mathbf{A}'(\mathbf{A}(\mathbf{G}_1')^{o})^{o}), \end{aligned}$$

where $\mathbf{R}_1$ and $\mathbf{R}_2$ are given by (4.1.157) and (4.1.158), respectively.
Finally, we turn to case *c*). If (4.1.150) holds, then
$C(\mathbf{H}_1 : \mathbf{H}_2)^\perp = C(\mathbf{H}_2)^\perp$ and $C(\mathbf{H}_1) \subseteq C(\mathbf{H}_2)$. Hence, instead of $\widehat{\mathbf{B}}$ given by (4.1.145), we observe that

$$\begin{aligned}\mathbf{B} =&(\mathbf{G}_1' : \mathbf{G}_2')^o\mathbf{\Theta}_1 + \{\mathbf{G}_2' : (\mathbf{G}_1' : \mathbf{G}_2')^o\}^o\mathbf{\Theta}_2\mathbf{H}_1^{o'}\\ &+ \{(\mathbf{G}_1' : (\mathbf{G}_1' : \mathbf{G}_2')^o)^o : ((\mathbf{G}_1')^o : (\mathbf{G}_2')^o)^o\}\mathbf{\Theta}_3\mathbf{H}_2^{o'}.\end{aligned}$$

Since $C(\mathbf{C}'\mathbf{H}_2^o) \subseteq C(\mathbf{C}'\mathbf{H}_1^o) \subseteq C(\mathbf{C}')$, we get as before, that if restrictions (iv) given by (4.1.132) together with (4.1.150) hold, one may equivalently consider a MLNM($\sum_{i=1}^3 \mathbf{A}_i\mathbf{B}_i\mathbf{C}_i$). Hence, via Theorem 4.1.6, estimators are available for $\mathbf{\Theta}_1$, $\mathbf{\Theta}_2$, $\mathbf{\Theta}_3$ as well as for $\mathbf{B}$ and $\mathbf{\Sigma}$. Furthermore, $\widehat{\mathbf{B}}$ is unique if and only if

$$\begin{aligned}&r(\mathbf{C}) = k,\\ &C(\mathbf{G}_1' : \mathbf{G}_2')^\perp \cap C(\mathbf{A}')^\perp = \{\mathbf{0}\},\\ &C(\mathbf{G}_2')^\perp \cap C(\mathbf{G}_1' : \mathbf{G}_2') \cap \{C(\mathbf{A}')^\perp + C(\mathbf{G}_1' : \mathbf{G}_2')^\perp\} = \{\mathbf{0}\}, \qquad (4.1.164)\\ &C(\mathbf{G}_2') \cap \{C(\mathbf{A}')^\perp + C(\mathbf{G}_2')^\perp\} = \{\mathbf{0}\}. \qquad (4.1.165)\end{aligned}$$

In comparison with case *b*), where the corresponding conditions for uniqueness were given by (4.1.160) – (4.1.163), it is seen that the differences between *b*) and *c*) lie in the difference between (4.1.162) and (4.1.164), and the difference between (4.1.163) and (4.1.165). Furthermore, if $\mathbf{A}$ and $\mathbf{C}$ are of full rank, $\widehat{\mathbf{B}}$ is uniquely estimated.
Moreover, linear combinations of $\widehat{\mathbf{B}}$, i.e. $\mathbf{K}\widehat{\mathbf{B}}\mathbf{L}$, are unique if and only if

$$\begin{aligned}&C(\mathbf{L}) \subseteq C(\mathbf{C}),\\ &C((\mathbf{G}_1' : \mathbf{G}_2')^{o'}\mathbf{K}') \subseteq C((\mathbf{G}_1' : \mathbf{G}_2')^{o'}\mathbf{A}'),\\ &C(\{\mathbf{G}_2' : (\mathbf{G}_1' : \mathbf{G}_2')^o\}^{o'}\mathbf{R}_1'\mathbf{K}')\\ &\qquad\qquad \subseteq C(\{\mathbf{G}_2' : (\mathbf{G}_1' : \mathbf{G}_2')^o\}^{o'}\mathbf{A}'\{\mathbf{A}(\mathbf{G}_1' : \mathbf{G}_2')^o\}^o),\\ &C(\mathbf{G}_2'\mathbf{R}_2'\mathbf{R}_1'\mathbf{K}') \subseteq C(\mathbf{G}_2\mathbf{A}'(\mathbf{A}(\mathbf{G}_2')^o)^o),\end{aligned}$$

where $\mathbf{R}_1$ and $\mathbf{R}_2$ are given by (4.1.157) and (4.1.158), respectively.
For case *d*) it is just noted that since

$$C(\mathbf{C}'(\mathbf{H}_1 : \mathbf{H}_2)^o) \subseteq C(\mathbf{C}'\mathbf{H}_2^o) \subseteq C(\mathbf{C}'\mathbf{H}_1^o) \subseteq C(\mathbf{C}'),$$

the mean structure given by (4.1.145) implies a mean structure in the general MLNM($\sum_{i=1}^4 \mathbf{A}_i\mathbf{B}_i\mathbf{C}_i$). This model has not been considered previously in any detail and therefore no relations are presented for d).
Now more general restrictions on $\mathbf{B}$ in the MLNM($\mathbf{ABC}$) than restrictions (iv) given by (4.1.132) will be considered:

$$\mathbf{G}_i\mathbf{B}\mathbf{H}_i = \mathbf{0}, \qquad i = 1, 2, \dots, s. \qquad (4.1.166)$$

We have to be aware of two problems related to the restrictions in (4.1.166). When considering (4.1.166) as a linear homogeneous system of equations, the first problem is to find suitable representations of solutions. For $i > 3$, it is not known how to present the general solution in an interpretable way. The special case $i = 3$ has been solved by von Rosen (1993). Thus, some restrictions have to be put on $\mathbf{G}_i$ and $\mathbf{H}_i$ in $\mathbf{G}_i\mathbf{B}\mathbf{H}_i = \mathbf{0}$, $i = 1, 2, \ldots, s$. The second problem is how to stay within the class of MLNM($\sum_{i=1}^m \mathbf{A}_i\mathbf{B}_i\mathbf{C}_i$), i.e. a model where the nested subspace condition $C(\mathbf{C}'_m) \subseteq C(\mathbf{C}'_{m-1}) \subseteq \ldots \subseteq C(\mathbf{C}'_1)$ has to be fulfilled. There exist many possibilities. As an example, one may suppose that

$$C(\mathbf{H}_s) \subseteq C(\mathbf{H}_{s-1}) \subseteq \ldots \subseteq C(\mathbf{H}_1) \tag{4.1.167}$$

or

$$C(\mathbf{G}'_s) \subseteq C(\mathbf{G}'_{s-1}) \subseteq \ldots \subseteq C(\mathbf{G}'_1). \tag{4.1.168}$$

For these cases the solutions are easy to obtain and the model immediately belongs to the class of MLNM($\sum_{i=1}^m \mathbf{A}_i\mathbf{B}_i\mathbf{C}_i$).

Theorem 4.1.16. *Suppose that for the* MLNM($\mathbf{ABC}$) *the parameter matrix* $\mathbf{B}$ *satisfies the restrictions given by (4.1.166). Then the following statements are valid.*

(i) *If (4.1.167) holds, then an equivalent model belonging to the* MLNM($\sum_{i=1}^m \mathbf{A}_i\mathbf{B}_i\mathbf{C}_i$) *is given by a multivariate linear model with mean*

$$E[\mathbf{X}] = \mathbf{A}\mathbf{N}_s^o\mathbf{\Theta}_1\mathbf{C} + \mathbf{A}\sum_{i=2}^{s}(\mathbf{N}_{i-1} : \mathbf{N}_i^o)^o\mathbf{\Theta}_i\mathbf{H}_i^{o'}\mathbf{C} + \mathbf{A}\mathbf{G}'_1\mathbf{\Theta}_{s+1}\mathbf{H}_1^{o'}\mathbf{C},$$

where $\mathbf{\Theta}_i$, $i = 1, 2, \ldots, s$, *are new parameters and*

$$\mathbf{N}_i = (\mathbf{G}'_1 : \mathbf{G}'_2 : \ldots : \mathbf{G}'_i).$$

(ii) *If (4.1.168) holds, then an equivalent model belonging to the* MLNM($\sum_{i=1}^m \mathbf{A}_i\mathbf{B}_i\mathbf{C}_i$) *is given by a multivariate linear model with mean*

$$E[\mathbf{X}] = \mathbf{A}\mathbf{G}'_s\mathbf{\Theta}_1\mathbf{M}_s^{o'}\mathbf{C} + \mathbf{A}\sum_{i=1}^{s-1}((\mathbf{G}'_i)^o : \mathbf{G}'_{i+1})^o\mathbf{\Theta}_{i+1}\mathbf{M}_i^{o'}\mathbf{C} + \mathbf{A}(\mathbf{G}'_1)^o\mathbf{\Theta}_{s+1}\mathbf{C},$$

where $\mathbf{\Theta}_i$, $i = 1, 2, \ldots, s$, *are new parameters and*

$$\mathbf{M}_i = (\mathbf{H}_1 : \mathbf{H}_2 : \ldots : \mathbf{H}_i).$$

PROOF: Both statements are immediately obtained by solving (4.1.166) with respect to $\mathbf{B}$ with the help of Theorem 1.3.9. ∎

Hence, one can find explicit estimators and discuss properties of these estimators for fairly general multivariate linear models. However, we are not going to use this

model anymore in this text. Instead, we will consider another type of restriction on $\mathbf{B}$ in the MLNM($\mathbf{ABC}$).
Suppose that for the MLNM($\mathbf{ABC}$) the following restriction holds:

$$\mathbf{G}_1\boldsymbol{\Theta}\mathbf{H}_1 + \mathbf{G}_2\mathbf{B}\mathbf{H}_2 = \mathbf{0}, \tag{4.1.169}$$

where, as previously, $\mathbf{G}_i$ and $\mathbf{H}_i$, $i = 1, 2$, are known matrices, and $\mathbf{B}$ and $\boldsymbol{\Theta}$ are unknown. Let us return to Example 4.1.1 in order to examine an application of the restrictions in (4.1.169).
Example 4.1.5 (Example 4.1.1 continued). It has been stated previously that the mean structure of the example equals

$$\begin{aligned}
&\beta_{11}+\beta_{21}t + \cdots + \beta_{(q-2)1}t^{q-3} + \beta_{(q-1)1}t^{q-2} + \beta_{q1}t^{q-1},\\
&\beta_{12}+\beta_{22}t + \cdots + \beta_{(q-2)2}t^{q-3} + \beta_{(q-1)2}t^{q-2} + \beta_{q2}t^{q-1},\\
&\beta_{13}+\beta_{23}t + \cdots + \beta_{(q-2)3}t^{q-3} + \beta_{(q-1)3}t^{q-2} + \beta_{q3}t^{q-1}.
\end{aligned}$$

Furthermore, restrictions have been put on $\mathbf{B}$ by setting various elements in $\mathbf{B}$ equal to zero. For example, in (4.1.126)

$$\beta_{(q-1)1} = 0, \quad \beta_{q1} = 0, \quad \beta_{q2} = 0$$

was considered in Example 4.1.4, which implies a MLNM($\sum_{i=1}^{3} \mathbf{A}_i\mathbf{B}_i\mathbf{C}_i$).

Instead of these conditions on $\mathbf{B}$, assume now that for two treatment groups the intercepts, i.e. β_{12} and β_{13}, are both proportional to the same unknown constant. This could be realistic in many situations. For instance, when there is one factor which influences the treatment groups but depends on the treatment conditions, we may have a difference between the influence in each group. Mathematically, this may be expressed as $(\beta_{12} : \beta_{13}) = \Theta(f_1 : f_2)$ or $\beta_{12} = \beta_{13}f$, where f_1, f_2 and f are known constants, or equivalently as $\mathbf{G}\mathbf{B}\mathbf{H}_1 = \boldsymbol{\Theta}\mathbf{H}_2$, for some $\mathbf{G}$ and $\mathbf{H}_i$, $i = 1, 2$, where $\mathbf{B}$ and $\boldsymbol{\Theta}$ are unknown. Hence, we have linear restrictions which differ from those previously discussed. ∎

Theorem 4.1.17. *The* MLNM($\mathbf{ABC}$) *with restriction* $\mathbf{G}_1\boldsymbol{\Theta}\mathbf{H}_1 + \mathbf{G}_2\mathbf{B}\mathbf{H}_2 = \mathbf{0}$, *where* $\mathbf{G}_i$, $\mathbf{H}_i$, $i = 1, 2$, *are known and* $\boldsymbol{\Theta}$, $\mathbf{B}$ *unknown, can equivalently be described by a multivariate linear model with mean*

$$E[\mathbf{X}] = \mathbf{A}(\mathbf{G}_2')^o\boldsymbol{\Theta}_1\mathbf{C} + \mathbf{A}\{\mathbf{G}_2'\mathbf{G}_1^o : (\mathbf{G}_2')^o\}^o\boldsymbol{\Theta}_2(\mathbf{H}_2(\mathbf{H}_1')^o)^{o'}\mathbf{C} + \mathbf{A}\mathbf{G}_2'\mathbf{G}_1^o\boldsymbol{\Theta}_3\mathbf{H}_2^{o'}\mathbf{C},$$

which belongs to the MLNM($\sum_{i=1}^{3} \mathbf{A}_i\mathbf{B}_i\mathbf{C}_i$). *The maximum likelihood estimators of the mean parameters are given by*

$$\widehat{\mathbf{B}} =(\mathbf{G}_2')^o\widehat{\boldsymbol{\Theta}}_1 + \{\mathbf{G}_2'\mathbf{G}_1^o : (\mathbf{G}_2')^o\}^o\widehat{\boldsymbol{\Theta}}_2(\mathbf{H}_2(\mathbf{H}_1')^o)^{o'} + \mathbf{G}_2'\mathbf{G}_1^o\widehat{\boldsymbol{\Theta}}_3\mathbf{H}_2^{o'}, \tag{4.1.170}$$
$$\widehat{\boldsymbol{\Theta}} = -\mathbf{G}_1^-\mathbf{G}_2\{\mathbf{G}_2'\mathbf{G}_1^o : (\mathbf{G}_2')^o\}^o\widehat{\boldsymbol{\Theta}}_2(\mathbf{H}_2(\mathbf{H}_1')^o)^{o'} + (\mathbf{G}_1')^o\mathbf{Z}_1 + \mathbf{G}_1'\mathbf{Z}_2\mathbf{H}_1^{o'}, \tag{4.1.171}$$

where the estimators $\widehat{\boldsymbol{\Theta}}_1$, $\widehat{\boldsymbol{\Theta}}_2$ *and* $\widehat{\boldsymbol{\Theta}}_3$ *follow from Theorem 4.1.6, and* $\mathbf{Z}_i$, $i = 1, 2$, *are arbitrary matrices of proper size. Furthermore,*

(i) $\widehat{\mathbf{\Theta}}_3$ *is unique if and only if*

$$C(\mathbf{AG}_2'\mathbf{G}_1^o) \cap C(\mathbf{A}(\mathbf{G}_2'\mathbf{G}_1^o)^o) = \{\mathbf{0}\},$$

and both $\mathbf{AG}_2'\mathbf{G}_1^o$ *and* $\mathbf{C}'\mathbf{H}_2^o$ *are of full rank;*

(ii) $\widehat{\mathbf{\Theta}}_2$ *is unique if and only if*

$$C(\mathbf{A}(\mathbf{G}_2')^o) \cap C(\mathbf{A}\{\mathbf{G}_2'\mathbf{G}_1^o : (\mathbf{G}_2')^o\}^o) = \{\mathbf{0}\},$$
$$C(\mathbf{A}(\mathbf{G}_2')^o)^\perp \cap C(\mathbf{A}(\mathbf{G}_2'\mathbf{G}_1^o)^o) \cap C(\mathbf{A}\{\mathbf{G}_2'\mathbf{G}_1^o : (\mathbf{G}_2')^o\}) = \{\mathbf{0}\},$$

and both $\mathbf{A}((\mathbf{G}_2')^o : \mathbf{G}_2'\mathbf{G}_1^o)^o$ *and* $\mathbf{C}'(\mathbf{H}_2(\mathbf{H}_1')^o)^o$ *are of full rank;*

(iii) $\widehat{\mathbf{\Theta}}_1$ *is unique if and only if*

$$C(\mathbf{A}(\mathbf{G}_2')^o) \cap C(\mathbf{A}\{\mathbf{G}_2'\mathbf{G}_1^o : (\mathbf{G}_2')^o\}^o) = \{\mathbf{0}\},$$
$$C(\mathbf{A}\{\mathbf{G}_2'\mathbf{G}_1^o : (\mathbf{G}_2')^o\}^o)^\perp \cap C(\mathbf{A}(\mathbf{G}_2'\mathbf{G}_1^o)^o) \cap C(\mathbf{AG}_2') = \{\mathbf{0}\},$$

$r(\mathbf{C}) = k$, *and* $\mathbf{A}(\mathbf{G}_2')^o$ *is of full rank;*

(iv) *put* $\mathbf{M} = \mathbf{G}_2'\mathbf{G}_1^o : (\mathbf{G}_2')^o$. *Then* $\widehat{\mathbf{\Theta}}$ *is unique if and only if*

$$C(\mathbf{G}_2'\mathbf{G}_1) \subseteq C(\mathbf{M}) + C(\mathbf{A}') \cap C(\mathbf{M})^\perp,$$

and both $\mathbf{G}_1$ *and* $\mathbf{H}_1$ *are of full rank;*

(v) $\widehat{\mathbf{B}}$ *is unique if and only if*

$$C(\mathbf{G}_2')^\perp \cap C(\mathbf{A}')^\perp = \{\mathbf{0}\},$$
$$C(\mathbf{G}_2\{\mathbf{G}_2'\mathbf{G}_1^o : (\mathbf{G}_2')^o\}^o) \cap C(\mathbf{G}_2(\mathbf{A}')^o) = \{\mathbf{0}\},$$
$$C(\mathbf{G}_2\mathbf{G}_1^o) \subseteq C(\mathbf{A}'),$$
$$r(\mathbf{C}) = k.$$

PROOF: By Theorem 1.3.8 it follows that the general solution to

$$\mathbf{G}_1\mathbf{\Theta}\mathbf{H}_1 + \mathbf{G}_2\mathbf{B}\mathbf{H}_2 = \mathbf{0}$$

is given by (4.1.170) and (4.1.171). For (i) – (iv) of the theorem we can rely on Theorem 4.1.12, whereas for (v) some further calculations have to be carried out which, however, will be omitted here. ∎

Another case, where the restriction $\mathbf{G}_1\mathbf{B}_1\mathbf{H}_1 + \mathbf{G}_2\mathbf{B}_2\mathbf{H}_2 = \{\mathbf{0}\}$ is useful, is when studying the MLNM($\sum_{i=1}^m \mathbf{A}_i\mathbf{B}_i\mathbf{C}_i$), e.g. a MLNM($\mathbf{A}_1\mathbf{B}_1\mathbf{C}_1 + \mathbf{A}_2\mathbf{B}_2\mathbf{C}_2$) where $\mathbf{G}_1\mathbf{B}_1\mathbf{H}_1 + \mathbf{G}_2\mathbf{B}_2\mathbf{H}_2 = \{\mathbf{0}\}$ holds. However, for this case, besides the inclusion $C(\mathbf{C}_2') \subseteq C(\mathbf{C}_1')$, we have to impose some further conditions so that explicit estimators can be obtained. This can be done in a manner similar to deriving the results given above.

4.1.7 *Problems*

1. Give a detailed proof of Theorem 4.1.17.
2. Let $\mathbf{X} = \mathbf{ABC} + \mathbf{\Sigma}^{1/2}\mathbf{EW}^{1/2}$, where $\mathbf{X}$, $\mathbf{A}$, $\mathbf{B}$, $\mathbf{C}$ and $\mathbf{\Sigma}$ are as in the Growth Curve model, $\mathbf{W}$ is known and positive definite and $\mathbf{E} \sim N(\mathbf{0}, \mathbf{I}_p, \mathbf{I}_n)$. Find the maximum likelihood estimators of the parameters.
3. Derive the maximum likelihood estimators of the parameters in the MLNM($\sum_{i=1}^{2} \mathbf{A}_i\mathbf{B}_i\mathbf{C}_i$).
4. Show that $\mathbf{X} = \sum_{i=1}^{m} \mathbf{A}_i\mathbf{B}_i\mathbf{C}_i + \mathbf{\Sigma}^{1/2}\mathbf{E}$, where $C(\mathbf{A}_m) \subseteq C(\mathbf{A}_{m-1}) \subseteq \ldots \subseteq C(\mathbf{A}_1)$ is a MLNM($\sum_{i=1}^{m} \mathbf{A}_i\mathbf{B}_i\mathbf{C}_i$).
5. Use the canonical form of the MLNM($\sum_{i=1}^{m} \mathbf{A}_i\mathbf{B}_i\mathbf{C}_i$) defined via (4.1.65). By performing a reparametrization show that the likelihood function can be factorized similarly to (4.1.21).
6. If in the MLNM($\sum_{i=1}^{3} \mathbf{A}_i\mathbf{B}_i\mathbf{C}_i$) the condition $C(\mathbf{A}_2) \subseteq C(\mathbf{A}_3)$ holds, then indicate the changes it causes in the statements of Theorem 4.1.12. What happens if we assume $C(\mathbf{A}_1) \subseteq C(\mathbf{A}_2)$ instead of $C(\mathbf{A}_2) \subseteq C(\mathbf{A}_3)$?
7. The *Complex normal distribution* of $\mathbf{X}_1 + i\,\mathbf{X}_2$ is expressed through the following normal distribution:

$$\begin{pmatrix} \mathbf{X}_1 \\ \mathbf{X}_2 \end{pmatrix} \sim N_{2p,n}\left(\begin{pmatrix} \boldsymbol{\mu}_1 \\ \boldsymbol{\mu}_2 \end{pmatrix}, \begin{pmatrix} \mathbf{\Sigma}_{11} & \mathbf{\Sigma}_{12} \\ \mathbf{\Sigma}_{21} & \mathbf{\Sigma}_{11} \end{pmatrix}, \mathbf{I}_n \right),$$

   where $\mathbf{X}_i : p \times n$, $\boldsymbol{\mu}_i : p \times n$, $\mathbf{\Sigma}_{11} > 0$ and $\mathbf{\Sigma}_{12} = -\mathbf{\Sigma}_{21}'$. Estimate $\boldsymbol{\mu}_1$, $\boldsymbol{\mu}_2$, $\mathbf{\Sigma}_{11}$ and $\mathbf{\Sigma}_{12}$.
8. The *Quaternion normal distribution* of $\mathbf{X}_1 + i\,\mathbf{X}_2 + j\,\mathbf{X}_3 + k\,\mathbf{X}_4$ is expressed through the following normal distribution:

$$\begin{pmatrix} \mathbf{X}_1 \\ \mathbf{X}_2 \\ \mathbf{X}_3 \\ \mathbf{X}_4 \end{pmatrix} \sim N_{4p,n}\left(\begin{pmatrix} \boldsymbol{\mu}_1 \\ \boldsymbol{\mu}_2 \\ \boldsymbol{\mu}_3 \\ \boldsymbol{\mu}_4 \end{pmatrix}, \begin{pmatrix} \mathbf{\Sigma}_{11} & \mathbf{\Sigma}_{12} & \mathbf{\Sigma}_{13} & \mathbf{\Sigma}_{14} \\ -\mathbf{\Sigma}_{12} & \mathbf{\Sigma}_{11} & -\mathbf{\Sigma}_{14} & \mathbf{\Sigma}_{13} \\ -\mathbf{\Sigma}_{13} & \mathbf{\Sigma}_{14} & \mathbf{\Sigma}_{11} & -\mathbf{\Sigma}_{12} \\ -\mathbf{\Sigma}_{14} & -\mathbf{\Sigma}_{13} & \mathbf{\Sigma}_{12} & \mathbf{\Sigma}_{11} \end{pmatrix}, \mathbf{I}_n \right),$$

   where $\mathbf{X}_i : p \times n$, $\boldsymbol{\mu}_i : p \times n$ and $\mathbf{\Sigma}_{11} > 0$. Estimate $\boldsymbol{\mu}_i$, $i = 1, 2, 3, 4$, and $\mathbf{\Sigma}_{11}$, $\mathbf{\Sigma}_{12}$, $\mathbf{\Sigma}_{13}$ and $\mathbf{\Sigma}_{14}$. For quaternions see §1.3.1.
9. *Monotone missing structure* (see Wu & Perlman, 2000; Srivastava, 2002). Let $\mathbf{x} \sim N_p(\boldsymbol{\mu}, \mathbf{\Sigma})$, $\mathbf{\Sigma} > 0$, where $\mathbf{x}' = (X_1, X_2, \ldots, X_p)$. Suppose that we have n_1 observations on $\mathbf{x}$, n_2 observations on $\mathbf{x}_{p-1} = (X_1, X_2, \ldots, X_{p-1})$, n_3 observations on $\mathbf{x}_{p-2} = (X_1, X_2, \ldots, X_{p-2})$ and so on until n_k observations on $\mathbf{x}_{p-k+1} = (X_1, X_2, \ldots, X_{p-k+1})$. Estimate $\boldsymbol{\mu}$ and $\mathbf{\Sigma}$.
10. In §4.1.3 many statements hold with probability 1. Let $\widetilde{\mathbf{X}}$ and $\widetilde{\mathbf{S}}$ denote the observations of $\mathbf{X}$ and $\mathbf{S}$, respectively, and assume that

$$\widetilde{\mathbf{X}} \in C(\mathbf{A}, \mathbf{\Sigma}) \quad \textit{and}\ C(\widetilde{\mathbf{S}}) = C(\mathbf{\Sigma}).$$

   Prove in the above assumptions the analogues of the statements in §4.1.3, which hold with probability 1. Remark that now the condition "with probability 1" is not needed in the obtained results.

4.2. MOMENTS AND MULTIVARIATE LINEAR MODELS

The object of this section is to discuss some of the multivariate linear models from Section 4.1, namely the Growth Curve model, given in Definition 4.1.1, and the $\mathrm{MLNM}(\sum_{i=1}^{m}\mathbf{A}_i\mathbf{B}_i\mathbf{C}_i)$, given in Definition 4.1.3. In this section we are going to find the first moments of the maximum likelihood estimators of the parameter matrices. These results are needed for approximating the distributions of the estimators. As previously, we emphasize that readers have to work with the details by themselves. Most of the ideas are straightforward but it is difficult to achieve any results without the techniques presented.

4.2.1 *Moments of the mean estimator of the Growth Curve model*

Let us dissect the maximum likelihood estimator of $\mathbf{B}$ in the Growth Curve model, i.e. the $\mathrm{MLNM}(\mathbf{ABC})$. In Theorem 4.1.11 the maximum likelihood estimator was given by

$$\widehat{\mathbf{B}} = (\mathbf{A}'\mathbf{S}^{-1}\mathbf{A})^{-1}\mathbf{A}'\mathbf{S}^{-1}\mathbf{X}\mathbf{C}'(\mathbf{C}\mathbf{C}')^{-1}, \tag{4.2.1}$$

where it was assumed that $r(\mathbf{A}) = q$ and $r(\mathbf{C}) = k$, in order to have a unique estimator. The uniqueness is needed as we want to discuss properties of $\widehat{\mathbf{B}}$. This would be meaningless if the estimator is not unique. In the non-unique case, $\widehat{\mathbf{B}}$ is given by (4.1.7) and equals

$$\widehat{\mathbf{B}} = (\mathbf{A}'\mathbf{S}^{-1}\mathbf{A})^{-}\mathbf{A}'\mathbf{S}^{-1}\mathbf{X}\mathbf{C}'(\mathbf{C}\mathbf{C}')^{-} + (\mathbf{A}')^{o}\mathbf{Z}_1 + \mathbf{A}'\mathbf{Z}_2\mathbf{C}^{o'}, \tag{4.2.2}$$

where

$$\mathbf{S} = \mathbf{X}(\mathbf{I} - \mathbf{C}'(\mathbf{C}\mathbf{C}')^{-}\mathbf{C})\mathbf{X}'. \tag{4.2.3}$$

When considering $\widehat{\mathbf{B}}$ in (4.2.2), we have to treat the estimator separately for each choice of $\mathbf{Z}_i$, $i = 1, 2$. If $\mathbf{Z}_i$ is non-random we just have a translation of $\widehat{\mathbf{B}}$, and as it will be seen later, we have a biased estimator. If $\mathbf{Z}_i$ is random, everything is more complicated.

As an alternative to the assumption of uniqueness of $\widehat{\mathbf{B}}$, one may consider a linear combination $\mathbf{K}\widehat{\mathbf{B}}\mathbf{L}$ where $\mathbf{K}$ and $\mathbf{L}$ are known matrices. In this paragraph

$$\mathbf{A}\widehat{\mathbf{B}}\mathbf{C} = \mathbf{A}(\mathbf{A}'\mathbf{S}^{-1}\mathbf{A})^{-}\mathbf{A}'\mathbf{S}^{-1}\mathbf{X}\mathbf{C}'(\mathbf{C}\mathbf{C}')^{-}\mathbf{C} \tag{4.2.4}$$

is going to be treated, which according to Theorem 4.1.11 is unique. However, since Theorem 4.1.11 states that $\mathbf{K}\widehat{\mathbf{B}}\mathbf{L}$ is unique if and only if $\mathcal{C}(\mathbf{K}') \subseteq \mathcal{C}(\mathbf{A}')$, $\mathcal{C}(\mathbf{L}) \subseteq \mathcal{C}(\mathbf{C})$, and by Proposition 1.2.2 (i) these conditions are equivalent to $\mathbf{K}' = \mathbf{A}'\mathbf{Q}_1$ and $\mathbf{L} = \mathbf{C}\mathbf{Q}_2$ for some matrices $\mathbf{Q}_1$ and $\mathbf{Q}_2$, we obtain that $\mathbf{K}\widehat{\mathbf{B}}\mathbf{L} = \mathbf{Q}_1'\mathbf{A}\widehat{\mathbf{B}}\mathbf{C}\mathbf{Q}_2$ and thus it will be sufficient to consider $\mathbf{A}\widehat{\mathbf{B}}\mathbf{C}$. Note that if $\mathbf{A}$ and $\mathbf{C}$ are both of full rank, the expression in (4.2.4) can be pre- and post-multiplied by $(\mathbf{A}'\mathbf{A})^{-1}\mathbf{A}'$ and $\mathbf{C}'(\mathbf{C}\mathbf{C}')^{-1}$, respectively, which gives $\widehat{\mathbf{B}}$ in (4.2.1).

From (4.2.1) it follows that $\widehat{\mathbf{B}}$ is a non-linear estimator. It consists of two random parts, namely

$$(\mathbf{A}'\mathbf{S}^{-1}\mathbf{A})^{-1}\mathbf{A}'\mathbf{S}^{-1} \tag{4.2.5}$$

and

$$\mathbf{XC}'(\mathbf{CC}')^{-1}. \tag{4.2.6}$$

In (4.2.1) the matrix $\mathbf{S}$, given by (4.2.3), is random, so the expression in (4.2.5) as well as $\widehat{\mathbf{B}}$ are quite complicated non-linear random expressions.

Gleser & Olkin (1970) were the first to derive the distributions for $\widehat{\mathbf{B}}$ (under full rank assumptions) and $\widehat{\mathbf{\Sigma}}$ given in Theorem 4.1.2. This was performed through the canonical reformulation of the model presented in Section 4.1.2. Kabe (1975) presented an alternative approach when working directly with the original matrices. Kenward (1986) expressed the density of $\widehat{\mathbf{B}}$ with the help of hypergeometric functions. It is well to notice that these results are all quite complicated and they are difficult to apply without suitable approximations. Fujikoshi (1985, 1987) derived asymptotic expansions with explicit error bounds for the density of $\widehat{\mathbf{B}}$.

Since the distribution of $\widehat{\mathbf{B}}$ is not available in a simple form, one strategy is to compare $\widehat{\mathbf{B}}$ with some other statistic whose distribution is easier to utilize. Alternatives to (4.2.1) are found in the class of estimators proposed by Potthoff & Roy (1964);

$$\widehat{\mathbf{B}}_{\mathbf{G}} = (\mathbf{A}'\mathbf{G}^{-1}\mathbf{A})^{-1}\mathbf{A}'\mathbf{G}^{-1}\mathbf{XC}'(\mathbf{CC}')^{-1}, \tag{4.2.7}$$

where, for simplicity, $\mathbf{G}$ is supposed to be a non-random positive definite matrix. One choice is $\mathbf{G} = \mathbf{I}$. According to Theorem 2.2.2, the distribution of $\widehat{\mathbf{B}}_{\mathbf{G}}$ is matrix normal. Therefore, it can be of value to compare moments of $\widehat{\mathbf{B}}$ with the corresponding moments of $\widehat{\mathbf{B}}_{\mathbf{G}}$ in order to understand how the distribution of $\widehat{\mathbf{B}}$ differs from the normal one. Furthermore, it is tempting to use a conditional approach for inference problems concerning $\widehat{\mathbf{B}}$, i.e. conditioning on $\mathbf{S}$, since the distribution of $\mathbf{S}$ does not involve the parameter $\mathbf{B}$. Hence, it is of interest to study how an omission of the variation in $\mathbf{S}$ affects the moments of $\widehat{\mathbf{B}}$.

Theorem 4.2.1. *Let $\widehat{\mathbf{B}}$ be given by (4.2.1) and $\mathbf{A}\widehat{\mathbf{B}}\mathbf{C}$ by (4.2.4). Then the following statements hold:*

(i)
$$E[\widehat{\mathbf{B}}] = \mathbf{B};$$

(ii) *if $n - k - p + q - 1 > 0$, then*

$$D[\widehat{\mathbf{B}}] = \frac{n-k-1}{n-k-p+q-1}(\mathbf{CC}')^{-1} \otimes (\mathbf{A}'\mathbf{\Sigma}^{-1}\mathbf{A})^{-1};$$

(iii)
$$E[\mathbf{A}\widehat{\mathbf{B}}\mathbf{C}] = \mathbf{ABC};$$

(iv) *if $n - r(\mathbf{C}) - p + r(\mathbf{A}) - 1 > 0$, then*

$$D[\mathbf{A}\widehat{\mathbf{B}}\mathbf{C}] = \frac{n-r(\mathbf{C})-1}{n-r(\mathbf{C})-p+r(\mathbf{A})-1}\mathbf{C}'(\mathbf{CC}')^{-}\mathbf{C} \otimes \mathbf{A}(\mathbf{A}'\mathbf{\Sigma}^{-1}\mathbf{A})^{-}\mathbf{A}'.$$

PROOF: Let us first verify (iii). Since, by Theorem 2.2.4 (iii), $\mathbf{S}$ and $\mathbf{XC}'$ are independent,

$$E[\mathbf{A}\widehat{\mathbf{B}}\mathbf{C}] = E[\mathbf{A}(\mathbf{A}'\mathbf{S}^{-1}\mathbf{A})^{-}\mathbf{A}'\mathbf{S}^{-1}]E[\mathbf{XC}'(\mathbf{CC}')^{-}\mathbf{C}]. \tag{4.2.8}$$

However, since $E[\mathbf{X}] = \mathbf{ABC}$ implies $E[\mathbf{XC}'(\mathbf{CC}')^-\mathbf{C}] = \mathbf{ABC}$, the expression in (4.2.8) is equivalent to

$$\begin{aligned} E[\mathbf{A}\widehat{\mathbf{B}}\mathbf{C}] =& E[\mathbf{A}(\mathbf{A}'\mathbf{S}^{-1}\mathbf{A})^-\mathbf{A}'\mathbf{S}^{-1}]\mathbf{ABC} \\ =& E[\mathbf{A}(\mathbf{A}'\mathbf{S}^{-1}\mathbf{A})^-\mathbf{A}'\mathbf{S}^{-1}\mathbf{A}]\mathbf{BC} = \mathbf{ABC}, \end{aligned}$$

where in the last equality Proposition 1.2.2 (ix) has been used. Similarly (i) can be verified.

Now we start to consider

$$D[\mathbf{A}\widehat{\mathbf{B}}\mathbf{C}] = E[\mathrm{vec}(\mathbf{A}(\widehat{\mathbf{B}} - \mathbf{B})\mathbf{C})\mathrm{vec}'(\mathbf{A}(\widehat{\mathbf{B}} - \mathbf{B})\mathbf{C})] \tag{4.2.9}$$

and note that

$$\mathbf{A}(\widehat{\mathbf{B}} - \mathbf{B})\mathbf{C} = \mathbf{A}(\mathbf{A}'\mathbf{S}^{-1}\mathbf{A})^-\mathbf{A}'\mathbf{S}^{-1}(\mathbf{X} - \mathbf{ABC})\mathbf{C}'(\mathbf{CC}')^-\mathbf{C}. \tag{4.2.10}$$

It will be utilized that

$$\mathbf{A}(\mathbf{A}'\mathbf{S}^{-1}\mathbf{A})^-\mathbf{A}'\mathbf{S}^{-1} = \underline{\mathbf{A}}(\underline{\mathbf{A}}'\mathbf{S}^{-1}\underline{\mathbf{A}})^{-1}\underline{\mathbf{A}}'\mathbf{S}^{-1}, \tag{4.2.11}$$

where $\underline{\mathbf{A}}$ is any matrix of full rank such that $\mathcal{C}(\mathbf{A}) = \mathcal{C}(\underline{\mathbf{A}})$ which follows from the uniqueness property of projectors given in Proposition 1.2.1 (vi).

Put

$$\mathbf{Y} = (\mathbf{X} - \mathbf{ABC})\mathbf{C}'(\mathbf{CC}')^-\mathbf{C},$$

which by Theorem 2.2.4 (iv) is independent of $\mathbf{S}$, and the dispersion of $\mathbf{Y}$ equals

$$D[\mathbf{Y}] = \mathbf{C}'(\mathbf{CC}')^-\mathbf{C} \otimes \mathbf{\Sigma}. \tag{4.2.12}$$

Thus, from (4.2.9) and (4.2.11) it follows that

$$\begin{aligned} &D[\mathbf{A}\widehat{\mathbf{B}}\mathbf{C}] \\ &\quad = E[(\mathbf{I} \otimes \underline{\mathbf{A}}(\underline{\mathbf{A}}'\mathbf{S}^{-1}\underline{\mathbf{A}})^{-1}\underline{\mathbf{A}}'\mathbf{S}^{-1})E[\mathrm{vec}\mathbf{Y}\mathrm{vec}'\mathbf{Y}](\mathbf{I} \otimes \mathbf{S}^{-1}\underline{\mathbf{A}}(\underline{\mathbf{A}}'\mathbf{S}^{-1}\underline{\mathbf{A}})^{-1}\underline{\mathbf{A}}')] \\ &\quad = E[(\mathbf{I} \otimes \underline{\mathbf{A}}(\underline{\mathbf{A}}'\mathbf{S}^{-1}\underline{\mathbf{A}})^{-1}\underline{\mathbf{A}}'\mathbf{S}^{-1})D[\mathbf{Y}](\mathbf{I} \otimes \mathbf{S}^{-1}\underline{\mathbf{A}}(\underline{\mathbf{A}}'\mathbf{S}^{-1}\underline{\mathbf{A}})^{-1}\underline{\mathbf{A}}')] \\ &\quad = \mathbf{C}'(\mathbf{CC}')^-\mathbf{C} \otimes E[\underline{\mathbf{A}}(\underline{\mathbf{A}}'\mathbf{S}^{-1}\underline{\mathbf{A}})^{-1}\underline{\mathbf{A}}'\mathbf{S}^{-1}\mathbf{\Sigma}\mathbf{S}^{-1}\underline{\mathbf{A}}(\underline{\mathbf{A}}'\mathbf{S}^{-1}\underline{\mathbf{A}})^{-1}\underline{\mathbf{A}}']. \end{aligned} \tag{4.2.13}$$

When proceeding (see also Problem 3 in §2.4.9), we will make use of a canonical representation of $\underline{\mathbf{A}}'\mathbf{\Sigma}^{-1/2}$, where $\mathbf{\Sigma}^{-1/2}$ is a symmetric square root of $\mathbf{\Sigma}^{-1}$. Proposition 1.1.6 implies that there exist a non-singular matrix $\mathbf{H}$ and an orthogonal matrix $\mathbf{\Gamma}$ such that

$$\underline{\mathbf{A}}'\mathbf{\Sigma}^{-1/2} = \mathbf{H}(\mathbf{I}_{r(\mathbf{A})} : \mathbf{0})\mathbf{\Gamma} = \mathbf{H}\mathbf{\Gamma}_1, \tag{4.2.14}$$

where $\mathbf{\Gamma}' = (\mathbf{\Gamma}_1' : \mathbf{\Gamma}_2')$, $(p \times r(\mathbf{A}) : p \times (p - r(\mathbf{A}))$. Let

$$\mathbf{V} = \mathbf{\Sigma}^{-1/2}\mathbf{S}\mathbf{\Sigma}^{-1/2}, \tag{4.2.15}$$

and from Theorem 2.4.2 we have $\mathbf{V} \sim W_p(\mathbf{I}, n - r(\mathbf{C}))$. Furthermore, the matrices $\mathbf{V}$ and $\mathbf{V}^{-1}$ will be partitioned:

$$\mathbf{V} = \begin{pmatrix} \mathbf{V}_{11} & \mathbf{V}_{12} \\ \mathbf{V}_{21} & \mathbf{V}_{22} \end{pmatrix}, \qquad \begin{matrix} r(\mathbf{A}) \times r(\mathbf{A}) & r(\mathbf{A}) \times (p - r(\mathbf{A})) \\ (p - r(\mathbf{A})) \times r(\mathbf{A}) & (p - r(\mathbf{A})) \times (p - r(\mathbf{A})) \end{matrix}, \quad (4.2.16)$$

$$\mathbf{V}^{-1} = \begin{pmatrix} \mathbf{V}^{11} & \mathbf{V}^{12} \\ \mathbf{V}^{21} & \mathbf{V}^{22} \end{pmatrix}, \qquad \begin{matrix} r(\mathbf{A}) \times r(\mathbf{A}) & r(\mathbf{A}) \times (p - r(\mathbf{A})) \\ (p - r(\mathbf{A})) \times r(\mathbf{A}) & (p - r(\mathbf{A})) \times (p - r(\mathbf{A})) \end{matrix}. \quad (4.2.17)$$

Thus, from (4.2.14), (4.2.15) and (4.2.17) it follows that (once again compare with Problem 3 in §2.4.9)

$$\begin{aligned} &E[\underline{\mathbf{A}}(\underline{\mathbf{A}}'\mathbf{S}^{-1}\underline{\mathbf{A}})^{-1}\underline{\mathbf{A}}'\mathbf{S}^{-1}\boldsymbol{\Sigma}\mathbf{S}^{-1}\underline{\mathbf{A}}(\underline{\mathbf{A}}'\mathbf{S}^{-1}\underline{\mathbf{A}})^{-1}\underline{\mathbf{A}}'] \\ &\quad = E[\boldsymbol{\Sigma}^{1/2}\boldsymbol{\Gamma}_1'(\mathbf{V}^{11})^{-1}(\mathbf{V}^{11} : \mathbf{V}^{12})(\mathbf{V}^{11} : \mathbf{V}^{12})'(\mathbf{V}^{11})^{-1}\boldsymbol{\Gamma}_1\boldsymbol{\Sigma}^{1/2}] \\ &\quad = E[\boldsymbol{\Sigma}^{1/2}\boldsymbol{\Gamma}_1'(\mathbf{I} : (\mathbf{V}^{11})^{-1}\mathbf{V}^{12})(\mathbf{I} : (\mathbf{V}^{11})^{-1}\mathbf{V}^{12})'\boldsymbol{\Gamma}_1\boldsymbol{\Sigma}^{1/2}] \\ &\quad = E[\boldsymbol{\Sigma}^{1/2}\boldsymbol{\Gamma}_1'\{\mathbf{I} + (\mathbf{V}^{11})^{-1}\mathbf{V}^{12}\mathbf{V}^{21}(\mathbf{V}^{11})^{-1}\}\boldsymbol{\Gamma}_1\boldsymbol{\Sigma}^{1/2}]. \end{aligned} \quad (4.2.18)$$

From Proposition 1.3.4 (i) we utilize that $(\mathbf{V}^{11})^{-1}\mathbf{V}^{12} = -\mathbf{V}_{12}\mathbf{V}_{22}^{-1}$. Thus, the last line of (4.2.18) is identical to

$$\boldsymbol{\Sigma}^{1/2}\boldsymbol{\Gamma}_1'\boldsymbol{\Gamma}_1\boldsymbol{\Sigma}^{1/2} + \boldsymbol{\Sigma}^{1/2}\boldsymbol{\Gamma}_1'\mathbf{V}_{12}\mathbf{V}_{22}^{-1}\mathbf{V}_{22}^{-1}\mathbf{V}_{21}\boldsymbol{\Gamma}_1\boldsymbol{\Sigma}^{1/2}. \quad (4.2.19)$$

Next we proceed by focusing on $\mathbf{V}_{12}\mathbf{V}_{22}^{-1}\mathbf{V}_{22}^{-1}\mathbf{V}_{21}$. Since $\mathbf{V} \sim W_p(\mathbf{I}, n - r(\mathbf{C}))$, there exists, according to Definition 2.4.1, a matrix $\mathbf{U} \sim N_{p,n-r(\mathbf{C})}(\mathbf{0}, \mathbf{I}, \mathbf{I})$ such that $\mathbf{V} = \mathbf{U}\mathbf{U}'$. Partition $\mathbf{U} = (\mathbf{U}_1' : \mathbf{U}_2')'$ so that $\mathbf{V}_{21} = \mathbf{U}_2\mathbf{U}_1'$ and $\mathbf{V}_{11} = \mathbf{U}_1\mathbf{U}_1'$. Then

$$E[\mathbf{V}_{12}\mathbf{V}_{22}^{-1}\mathbf{V}_{22}^{-1}\mathbf{V}_{21}] = E[\mathbf{U}_1\mathbf{U}_2'(\mathbf{U}_2\mathbf{U}_2')^{-1}(\mathbf{U}_2\mathbf{U}_2')^{-1}\mathbf{U}_2\mathbf{U}_1']. \quad (4.2.20)$$

By Theorem 2.2.9 (i) and independence of $\mathbf{U}_1$ and $\mathbf{U}_2$, this is equivalent to

$$\begin{aligned} E[\mathbf{V}_{12}\mathbf{V}_{22}^{-1}\mathbf{V}_{22}^{-1}\mathbf{V}_{21}] &= E[\mathrm{tr}\{\mathbf{U}_2'(\mathbf{U}_2\mathbf{U}_2')^{-1}(\mathbf{U}_2\mathbf{U}_2')^{-1}\mathbf{U}_2\}]\mathbf{I} \\ &= E[\mathrm{tr}(\mathbf{U}_2\mathbf{U}_2')^{-1}]\mathbf{I}. \end{aligned} \quad (4.2.21)$$

Furthermore, since $\mathbf{U}_2\mathbf{U}_2' \sim \mathbf{W}_{p-r(\mathbf{A})}(\mathbf{I}, n - r(\mathbf{C}))$, it follows from Theorem 2.4.14 (iii) that (4.2.21) equals

$$E[\mathbf{V}_{12}\mathbf{V}_{22}^{-1}\mathbf{V}_{22}^{-1}\mathbf{V}_{21}] = \frac{p - r(\mathbf{A})}{n - r(\mathbf{C}) - p + r(\mathbf{A}) - 1}\mathbf{I}. \quad (4.2.22)$$

Alternatively, we could have used Theorem 2.4.12 (iii), which yields

$$\mathbf{U}_1\mathbf{U}_2'(\mathbf{U}_2\mathbf{U}_2')^{-\frac{1}{2}} \sim N_{r(\mathbf{A}),p-r(\mathbf{A})}(\mathbf{0}, \mathbf{I}, \mathbf{I}),$$

which in turn is independent of $\mathbf{U}_2\mathbf{U}_2'$. Finally, it is noted that since $\boldsymbol{\Gamma}_1\boldsymbol{\Gamma}_1' = \mathbf{I}$ and $\mathbf{H}$ is non-singular,

$$\boldsymbol{\Sigma}^{1/2}\boldsymbol{\Gamma}_1'\boldsymbol{\Gamma}_1\boldsymbol{\Sigma}^{1/2} = \mathbf{A}(\mathbf{A}'\boldsymbol{\Sigma}^{-1}\mathbf{A})^{-}\mathbf{A}', \quad (4.2.23)$$

and by combining (4.2.13), (4.2.18), (4.2.22) and (4.2.23), statement (iv) of the theorem is established.

Under full rank assumptions, (ii) follows immediately from (iv). ∎

Next we are going to consider moments of higher order and we have chosen to use the representation

$$E[(\mathbf{A}(\widehat{\mathbf{B}}-\mathbf{B})\mathbf{C})^{\otimes k}].$$

Theorem 4.2.2. *Let $\mathbf{A}\widehat{\mathbf{B}}\mathbf{C}$ be given by (4.2.4). Put*

$$\begin{aligned}\mathbf{v}(\mathbf{A}) &= \mathrm{vec}(\mathbf{A}(\mathbf{A}'\boldsymbol{\Sigma}^{-1}\mathbf{A})^{-}\mathbf{A}'),\\ \mathbf{v}(\mathbf{C}') &= \mathrm{vec}(\mathbf{C}'(\mathbf{C}\mathbf{C}')^{-}\mathbf{C}).\end{aligned}$$

In this notation the following statements hold:

(i) $E[(\mathbf{A}(\widehat{\mathbf{B}}-\mathbf{B})\mathbf{C})^{\otimes r}] = 0,$ *for odd r;*

(ii) *if $n-r(\mathbf{C})-p+r(\mathbf{A})-1>0$, then*

$$E[(\mathbf{A}(\widehat{\mathbf{B}}-\mathbf{B})\mathbf{C})^{\otimes 2}] = \frac{p-r(\mathbf{A})}{n-r(\mathbf{C})-p+r(\mathbf{A})-1}\mathbf{v}(\mathbf{A})\mathbf{v}'(\mathbf{C}');$$

(iii) *if $n-r(\mathbf{C})-p+r(\mathbf{A})-3>0$, then*

$$\begin{aligned}E[(\mathbf{A}(\widehat{\mathbf{B}}-\mathbf{B})\mathbf{C})^{\otimes 4}] =& (1+2c_1)\{\mathbf{v}(\mathbf{A})\mathbf{v}'(\mathbf{C}')\}^{\otimes 2}\\ &+ (1+2c_1)(\mathbf{I}_p\otimes\mathbf{K}_{p,p}\otimes\mathbf{I}_p)\{\mathbf{v}(\mathbf{A})\mathbf{v}'(\mathbf{C}')\}^{\otimes 2}(\mathbf{I}_n\otimes\mathbf{K}_{n,n}\otimes\mathbf{I}_n)\\ &+ (1+2c_1)\mathbf{K}_{p,p^3}\{\mathbf{v}(\mathbf{A})\mathbf{v}'(\mathbf{C}')\}^{\otimes 2}\mathbf{K}_{n^3,n}\\ &+ (c_2\mathbf{I}+c_3\{(\mathbf{I}_p\otimes\mathbf{K}_{p,p}\otimes\mathbf{I}_p)+\mathbf{K}_{p,p^3}\})\{\mathbf{v}(\mathbf{A})\mathbf{v}'(\mathbf{C}')\}^{\otimes 2},\end{aligned}$$

where

$$\begin{aligned}c_1 =& \tfrac{p-r(\mathbf{A})}{n-r(\mathbf{C})-p+r(\mathbf{A})-1},\\ c_2 =& \tfrac{2(p-r(\mathbf{A}))(n-r(\mathbf{C})-p+r(\mathbf{A})-1)+\{2+(n-r(\mathbf{C})-p+r(\mathbf{A}))(n-r(C)-p+r(\mathbf{A})-3)\}(p-r(\mathbf{A}))^2}{(n-r(\mathbf{C})-p+r(\mathbf{A}))(n-r(\mathbf{C})-p+r(A)-1)^2(n-r(\mathbf{C})-p+r(\mathbf{A})-3)},\\ c_3 =& \tfrac{p-r(\mathbf{A})}{(n-r(\mathbf{C})-p+r(\mathbf{A}))(n-r(\mathbf{C})-p+r(\mathbf{A})-3)}\\ &+ \tfrac{(p-r(\mathbf{A}))^2}{(n-r(\mathbf{C})-p+r(\mathbf{A}))(n-r(\mathbf{C})-p+r(\mathbf{A})-1)(n-r(\mathbf{C})-p+r(\mathbf{A})-3)};\end{aligned}$$

(iv) *if $n-r(\mathbf{C})-p+r(\mathbf{A})-2r+1>0$, then*

$$E[(\mathbf{A}\widehat{\mathbf{B}}\mathbf{C})^{\otimes 2r}] = O(n^{-r}).$$

PROOF: First of all we note that due to independence between $\mathbf{S}$ and $\mathbf{X}\mathbf{C}'$,

$$\begin{aligned}E[(\mathbf{A}(\widehat{\mathbf{B}}-\mathbf{B})\mathbf{C})^{\otimes r}] =& E[(\mathbf{A}(\mathbf{A}'\mathbf{S}^{-1}\mathbf{A})^{-}\mathbf{A}'\mathbf{S}^{-1})^{\otimes r}]\\ &\times E[(\mathbf{X}-\mathbf{A}\mathbf{B}\mathbf{C})^{\otimes r}](\mathbf{C}'(\mathbf{C}\mathbf{C}')^{-}\mathbf{C})^{\otimes r}, \quad (4.2.24)\end{aligned}$$

and since $E[(\mathbf{X}-\mathbf{A}\mathbf{B}\mathbf{C})^{\otimes r}] = \mathbf{0}$, for odd r, (i) is established.

The result in (ii) follows from Theorem 4.2.1 (iv) and Proposition 1.3.14 (vi). Now (iii) is considered. In Corollary 2.2.7.4 (ii) the fourth order moments of a matrix normally distributed variable were given, and applying this result implies that (4.2.24), for $r = 4$, equals

$$
\begin{aligned}
E[(\mathbf{A}(\widehat{\mathbf{B}}-\mathbf{B})\mathbf{C})^{\otimes 4}] &= E[(\mathbf{A}(\mathbf{A}'\mathbf{S}^{-1}\mathbf{A})^{-}\mathbf{A}'\mathbf{S}^{-1})^{\otimes 4}] \\
&\times \Big\{(\mathrm{vec}\boldsymbol{\Sigma}\mathrm{vec}'\mathbf{I})^{\otimes 2} + (\mathbf{I}_p\otimes\mathbf{K}_{p,p}\otimes\mathbf{I}_p)(\mathrm{vec}\boldsymbol{\Sigma}\mathrm{vec}'\mathbf{I})^{\otimes 2}(\mathbf{I}_n\otimes\mathbf{K}_{n,n}\otimes\mathbf{I}_n) \\
&\quad + \mathbf{K}_{p,p^3}(\mathrm{vec}\boldsymbol{\Sigma}\mathrm{vec}'\mathbf{I})^{\otimes 2}\mathbf{K}_{n^3,n}\Big\}(\mathbf{C}'(\mathbf{C}\mathbf{C}')^{-}\mathbf{C})^{\otimes 4}. \qquad (4.2.25)
\end{aligned}
$$

Put

$$
\begin{aligned}
\mathbf{K}^{1,i} &= \mathbf{I}_i, \\
\mathbf{K}^{2,i} &= \mathbf{I}_i\otimes\mathbf{K}_{i,i}\otimes\mathbf{I}_i, \qquad (4.2.26)\\
\mathbf{K}^{3,i} &= \mathbf{K}_{i,i^3},
\end{aligned}
$$

where the size of the matrices is indicated by i, which according to the applications may equal p, n, $r(\mathbf{A})$, $p-r(\mathbf{A})$, $r(\mathbf{C})$ or $n-r(\mathbf{C})$. Moreover, $\mathbf{K}^{j,i'}$ denotes the transpose of $\mathbf{K}^{j,i}$.
Proposition 1.3.12 (viii) implies that for $j = 1, 2, 3$,

$$
(\mathbf{A}(\mathbf{A}'\mathbf{S}^{-1}\mathbf{A})^{-}\mathbf{A}'\mathbf{S}^{-1})^{\otimes 4}\mathbf{K}^{j,p} = \mathbf{K}^{j,p}(\mathbf{A}(\mathbf{A}'\mathbf{S}^{-1}\mathbf{A})^{-}\mathbf{A}'\mathbf{S}^{-1})^{\otimes 4} \qquad (4.2.27)
$$

and

$$
\mathbf{K}^{j,n'}(\mathbf{C}'(\mathbf{C}\mathbf{C}')^{-}\mathbf{C})^{\otimes 4} = (\mathbf{C}'(\mathbf{C}\mathbf{C}')^{-}\mathbf{C})^{\otimes 4}\mathbf{K}^{j,n'}. \qquad (4.2.28)
$$

Hence, (4.2.25) equals

$$
\begin{aligned}
&E[(\mathbf{A}(\widehat{\mathbf{B}}-\mathbf{B})\mathbf{C})^{\otimes 4}] \\
&= \sum_{j=1}^{3}\mathbf{K}^{j,p}\{E[\mathrm{vec}(\boldsymbol{\Gamma}_1'(\mathbf{I}+\mathbf{V}_{12}\mathbf{V}_{22}^{-1}\mathbf{V}_{22}^{-1}\mathbf{V}_{21})\boldsymbol{\Gamma}_1)^{\otimes 2}]\mathbf{v}'(\mathbf{C}')^{\otimes 2}\}\mathbf{K}^{j,n'}, \qquad (4.2.29)
\end{aligned}
$$

where $\boldsymbol{\Gamma}$ and $\mathbf{V}$ are defined in (4.2.14) and (4.2.15), respectively. Expanding (4.2.29) gives

$$
\begin{aligned}
E[(\mathbf{A}(\widehat{\mathbf{B}}-\mathbf{B})\mathbf{C})^{\otimes 4}] &= \sum_{j=1}^{3}\mathbf{K}^{j,p}\Big\{\mathrm{vec}(\boldsymbol{\Gamma}_1'\boldsymbol{\Gamma}_1)^{\otimes 2} \\
&+ \mathrm{vec}(\boldsymbol{\Gamma}_1'\boldsymbol{\Gamma}_1)\otimes E[\mathrm{vec}(\boldsymbol{\Gamma}_1'\mathbf{V}_{12}\mathbf{V}_{22}^{-1}\mathbf{V}_{22}^{-1}\mathbf{V}_{21}\boldsymbol{\Gamma}_1)] \\
&+ E[\mathrm{vec}(\boldsymbol{\Gamma}_1'\mathbf{V}_{12}\mathbf{V}_{22}^{-1}\mathbf{V}_{22}^{-1}\mathbf{V}_{21}\boldsymbol{\Gamma}_1)]\otimes\mathrm{vec}(\boldsymbol{\Gamma}_1\boldsymbol{\Gamma}_1') \\
&+ E[\mathrm{vec}(\boldsymbol{\Gamma}_1'\mathbf{V}_{12}\mathbf{V}_{22}^{-1}\mathbf{V}_{22}^{-1}\mathbf{V}_{21}\boldsymbol{\Gamma}_1)^{\otimes 2}]\Big\}\mathbf{v}'(\mathbf{C}')^{\otimes 2}\mathbf{K}^{j,n'}. \qquad (4.2.30)
\end{aligned}
$$

According to (4.2.22),

$$
E[\boldsymbol{\Gamma}_1'\mathbf{V}_{12}\mathbf{V}_{22}^{-1}\mathbf{V}_{22}^{-1}\mathbf{V}_{21}\boldsymbol{\Gamma}_1] = c_1\boldsymbol{\Gamma}_1'\boldsymbol{\Gamma}_1,
$$

and now we will study

$$E[\mathrm{vec}(\mathbf{\Gamma}_1'\mathbf{V}_{12}\mathbf{V}_{22}^{-1}\mathbf{V}_{22}^{-1}\mathbf{V}_{21}\mathbf{\Gamma}_1)^{\otimes 2}]. \tag{4.2.31}$$

By using the same matrix $\mathbf{U} = (\mathbf{U}_1' : \mathbf{U}_2')'$ as in (4.2.20), where $\mathbf{U}_1$ and $\mathbf{U}_2$ in particular are independent, it is seen that (4.2.31) can be written as

$$\begin{aligned}&E[\mathrm{vec}(\mathbf{\Gamma}_1'\mathbf{V}_{12}\mathbf{V}_{22}^{-1}\mathbf{V}_{22}^{-1}\mathbf{V}_{21}\mathbf{\Gamma}_1)^{\otimes 2}]\\ &\quad= (\mathbf{\Gamma}_1')^{\otimes 4}E[\mathbf{U}_1^{\otimes 4}]E[\mathrm{vec}(\mathbf{U}_2'(\mathbf{U}_2\mathbf{U}_2')^{-1}(\mathbf{U}_2\mathbf{U}_2')^{-1}\mathbf{U}_2)^{\otimes 2}]. \end{aligned}\tag{4.2.32}$$

From Corollary 2.2.7.4 (ii) it follows that (4.2.32) equals

$$\begin{aligned}&(\mathbf{\Gamma}_1')^{\otimes 4}\sum_{j=1}^{3}\mathbf{K}^{j,r(\mathbf{A})}(\mathrm{vec}(\mathbf{I}_{r(\mathbf{A})})^{\otimes 2}(\mathrm{vec}'\mathbf{I}_{n-r(\mathbf{C})})^{\otimes 2}\mathbf{K}^{j,(n-r(\mathbf{C}))'}\\ &\qquad\times E[(\mathrm{vec}(\mathbf{U}_2'(\mathbf{U}_2\mathbf{U}_2')^{-1}(\mathbf{U}_2\mathbf{U}_2')^{-1}\mathbf{U}_2)^{\otimes 2}]. \end{aligned}\tag{4.2.33}$$

However, Proposition 1.3.12 (viii) implies that

$$(\mathbf{\Gamma}_1')^{\otimes 4}\mathbf{K}^{j,r(\mathbf{A})} = \mathbf{K}^{j,p}(\mathbf{\Gamma}_1')^{\otimes 4} \tag{4.2.34}$$

holds, Proposition 1.3.14 (iii) and (4.2.26) give us that

$$\begin{aligned}&(\mathrm{vec}'\mathbf{I}_{n-r(\mathbf{C})})^{\otimes 2}\mathbf{K}^{1,(n-r(\mathbf{C}))'}E[\mathrm{vec}(\mathbf{U}_2'(\mathbf{U}_2\mathbf{U}_2')^{-1}(\mathbf{U}_2\mathbf{U}_2')^{-1}\mathbf{U}_2)^{\otimes 2}]\\ &\quad= E[\{\mathrm{tr}(\mathbf{U}_2'(\mathbf{U}_2\mathbf{U}_2')^{-1}(\mathbf{U}_2\mathbf{U}_2')^{-1}\mathbf{U}_2)\}^2] = E[\{\mathrm{tr}(\mathbf{U}_2\mathbf{U}_2')^{-1}\}^2] \end{aligned}\tag{4.2.35}$$

holds, as well as that for $j = 2, 3$,

$$\begin{aligned}&(\mathrm{vec}'\mathbf{I}_{n-r(\mathbf{C})})^{\otimes 2}\mathbf{K}^{j,(n-r(\mathbf{C}))'}E[\mathrm{vec}(\mathbf{U}_2'(\mathbf{U}_2\mathbf{U}_2')^{-1}(\mathbf{U}_2\mathbf{U}_2')^{-1}\mathbf{U}_2)^{\otimes 2}]\\ &\quad= E[\mathrm{tr}(\mathbf{U}_2'(\mathbf{U}_2\mathbf{U}_2')^{-1}\mathbf{U}_2\mathbf{U}_2'(\mathbf{U}_2\mathbf{U}_2')^{-1}\mathbf{U}_2)]\\ &\quad= E[\mathrm{tr}\{(\mathbf{U}_2\mathbf{U}_2')^{-1}(\mathbf{U}_2\mathbf{U}_2')^{-1}\}] \end{aligned}\tag{4.2.36}$$

holds. Hence, from (4.2.34) – (4.2.36) it follows that (4.2.33) is equivalent to

$$\begin{aligned}&(\mathbf{\Gamma}_1')^{\otimes 4}(\mathrm{vec}\mathbf{I}_{r(\mathbf{A})})^{\otimes 2}E[\{\mathrm{tr}(\mathbf{U}_2\mathbf{U}_2')^{-1}\}^2]\\ &\qquad+\mathbf{K}^{2,p}(\mathbf{\Gamma}_1')^{\otimes 4}(\mathrm{vec}\mathbf{I}_{r(\mathbf{A})})^{\otimes 2}E[\mathrm{tr}\{(\mathbf{U}_2\mathbf{U}_2')^{-1}(\mathbf{U}_2\mathbf{U}_2')^{-1}\}]\\ &\qquad+\mathbf{K}^{3,p}(\mathbf{\Gamma}_1')^{\otimes 4}(\mathrm{vec}\mathbf{I}_{r(\mathbf{A})})^{2\otimes 2}E[\mathrm{tr}\{(\mathbf{U}_2\mathbf{U}_2')^{-1}(\mathbf{U}_2\mathbf{U}_2')^{-1}\}]\\ &\qquad= (\mathrm{vec}(\mathbf{\Gamma}_1'\mathbf{\Gamma}_1))^{\otimes 2}E[\{\mathrm{tr}(\mathbf{U}_2\mathbf{U}_2')^{-1}\}^2]\\ &\qquad+(\mathbf{K}^{2,p}+\mathbf{K}^{3,p})(\mathrm{vec}(\mathbf{\Gamma}_1'\mathbf{\Gamma}_1))^{\otimes 2}E[\mathrm{tr}\{(\mathbf{U}_2\mathbf{U}_2')^{-1}(\mathbf{U}_2\mathbf{U}_2')^{-1}\}]. \end{aligned}\tag{4.2.37}$$

Now, according to Theorem 2.4.14 (viii) and (vi), since $n-r(\mathbf{C})-p+r(\mathbf{A})-3>0$,

$$\begin{aligned}c_2 &= E[\{\mathrm{tr}(\mathbf{U}_2\mathbf{U}_2')^{-1}\}^2],\\ c_3 &= E[\mathrm{tr}\{(\mathbf{U}_2\mathbf{U}_2')^{-1}(\mathbf{U}_2\mathbf{U}_2')^{-1}\}],\end{aligned}$$

which establish the statement
In order to prove (iv), it is first observed that

$$E[(\mathbf{A}\widehat{\mathbf{B}}\mathbf{C})^{\otimes 2r}] = E[(\mathbf{A}(\mathbf{A}'\mathbf{S}^{-1}\mathbf{A})^{-}\mathbf{A}'\mathbf{S}^{-1})^{\otimes 2r}]E[(\mathbf{X}\mathbf{C}'(\mathbf{C}\mathbf{C}')^{-}\mathbf{C})^{\otimes 2r}].$$

Since we are just interested in the order of magnitude and no explicit expressions of $E[(\mathbf{A}\widehat{\mathbf{B}}\mathbf{C})^{\otimes 2r}]$, it follows from Theorem 2.2.7 (iv) that from now on it is sufficient to consider

$$E[\mathrm{vec}(\mathbf{A}(\mathbf{A}'\mathbf{S}^{-1}\mathbf{A})^{-}\mathbf{A}'\mathbf{S}^{-1}\boldsymbol{\Sigma}\mathbf{S}^{-1}\mathbf{A}(\mathbf{A}'\mathbf{S}^{-1}\mathbf{A})^{-}\mathbf{A}')^{\otimes r}].$$

This expression can be presented in a canonical form as in (4.2.31):

$$E[(\mathrm{vec}(\boldsymbol{\Gamma}_1'\mathbf{V}_{12}\mathbf{V}_{22}^{-1}\mathbf{V}_{22}^{-1}\mathbf{V}_{21}\boldsymbol{\Gamma}))^{\otimes r}] = E[(\boldsymbol{\Gamma}_1'\mathbf{V}_{12}\mathbf{V}_{22}^{-1/2})^{\otimes 2r}]E[(\mathrm{vec}\mathbf{V}_{22}^{-1})^{\otimes r}],$$

where the equality follows from Theorem 2.4.12 (iii). Now Theorem 2.4.14 (iii) and (2.4.46) give

$$E[(\mathrm{vec}\mathbf{V}_{22}^{-1})^{\otimes r}] = O(n^{-r}).$$

Since Theorem 2.4.12 (iii) states that the matrix $\mathbf{V}_{12}\mathbf{V}_{22}^{-1/2}$ is normally distributed with moments independent of n, the theorem is established. ∎

4.2.2 $E[\widehat{\boldsymbol{\Sigma}}]$ *and* $D[\widehat{\boldsymbol{\Sigma}}]$ *for the Growth Curve model*

In this section we are going to find the first and second moments for $\widehat{\boldsymbol{\Sigma}}$ in the Growth Curve model. The estimator $\widehat{\boldsymbol{\Sigma}}$ is given in (4.1.8) and equals

$$n\widehat{\boldsymbol{\Sigma}} = \mathbf{S} + (\mathbf{X}\mathbf{C}'(\mathbf{C}\mathbf{C}')^{-}\mathbf{C} - \mathbf{A}\widehat{\mathbf{B}}\mathbf{C})(\mathbf{X}\mathbf{C}'(\mathbf{C}\mathbf{C}')^{-}\mathbf{C} - \mathbf{A}\widehat{\mathbf{B}}\mathbf{C})', \qquad (4.2.38)$$

where $\mathbf{S}$ and $\mathbf{A}\widehat{\mathbf{B}}\mathbf{C}$ are given by (4.2.3) and (4.2.4), respectively, and equal

$$\mathbf{S} = \mathbf{X}(\mathbf{I} - \mathbf{C}'(\mathbf{C}\mathbf{C}')^{-}\mathbf{C})\mathbf{X}', \qquad (4.2.39)$$

$$\mathbf{A}\widehat{\mathbf{B}}\mathbf{C} = \mathbf{A}(\mathbf{A}'\mathbf{S}^{-1}\mathbf{A})^{-}\mathbf{A}'\mathbf{S}^{-1}\mathbf{X}\mathbf{C}'(\mathbf{C}\mathbf{C}')^{-}\mathbf{C}. \qquad (4.2.40)$$

Inserting (4.2.40) into (4.2.38) and then applying Corollary 1.2.25.1 yields that instead of (4.2.38) we may consider

$$n\widehat{\boldsymbol{\Sigma}} = \mathbf{S} + \mathbf{S}\mathbf{A}^{o}(\mathbf{A}^{o\prime}\mathbf{S}\mathbf{A}^{o})^{-}\mathbf{A}^{o\prime}\mathbf{X}\mathbf{C}'(\mathbf{C}\mathbf{C}')^{-}\mathbf{C}\mathbf{X}'\mathbf{A}^{o}(\mathbf{A}^{o\prime}\mathbf{S}\mathbf{A}^{o})^{-}\mathbf{A}^{o\prime}\mathbf{S}, \qquad (4.2.41)$$

where $\mathbf{A}^{o}$ has been chosen to be of full column rank, i.e. $\mathbf{A}^{o} : p \times (p - r(\mathbf{A}))$. There is no principal advantage of using (4.2.41) instead of (4.2.38), only the subsequent presentation will be somewhat shorter.

Theorem 4.2.3. *Let* $\widehat{\boldsymbol{\Sigma}}$ *be as in (4.2.38).*
(i) *If* $n - r(\mathbf{C}) - p + r(\mathbf{A}) - 1 > 0$, *then*

$$E[\widehat{\boldsymbol{\Sigma}}] = \boldsymbol{\Sigma} - r(\mathbf{C})\frac{1}{n}\frac{n - r(\mathbf{C}) - 2(p - r(\mathbf{A})) - 1}{n - r(\mathbf{C}) - p + r(\mathbf{A}) - 1}\mathbf{A}(\mathbf{A}'\boldsymbol{\Sigma}^{-1}\mathbf{A})^{-}\mathbf{A}'.$$

(ii) *If* $n - r(\mathbf{C}) - p + r(\mathbf{A}) - 3 > 0$, *then*

$$
\begin{aligned}
D[\widehat{\boldsymbol{\Sigma}}] = & d_1(\mathbf{I} + \mathbf{K}_{p,p})\{(\mathbf{A}(\mathbf{A}'\boldsymbol{\Sigma}^{-1}\mathbf{A})^{-}\mathbf{A}') \otimes (\mathbf{A}(\mathbf{A}'\boldsymbol{\Sigma}^{-1}\mathbf{A})^{-}\mathbf{A}')\} \\
& + d_2(\mathbf{I} + \mathbf{K}_{p,p})\{(\mathbf{A}(\mathbf{A}'\boldsymbol{\Sigma}^{-1}\mathbf{A})^{-}\mathbf{A}') \otimes (\boldsymbol{\Sigma} - \mathbf{A}(\mathbf{A}'\boldsymbol{\Sigma}^{-1}\mathbf{A})^{-}\mathbf{A}')\} \\
& + d_2(\mathbf{I} + \mathbf{K}_{p,p})\{(\boldsymbol{\Sigma} - \mathbf{A}(\mathbf{A}'\boldsymbol{\Sigma}^{-1}\mathbf{A})^{-}\mathbf{A}') \otimes (\mathbf{A}(\mathbf{A}'\boldsymbol{\Sigma}^{-1}\mathbf{A})^{-}\mathbf{A}')\} \\
& + \frac{1}{n}(\mathbf{I} + \mathbf{K}_{p,p})\{(\boldsymbol{\Sigma} - \mathbf{A}(\mathbf{A}'\boldsymbol{\Sigma}^{-1}\mathbf{A})^{-}\mathbf{A}') \otimes (\boldsymbol{\Sigma} - \mathbf{A}(\mathbf{A}'\boldsymbol{\Sigma}^{-1}\mathbf{A})^{-}\mathbf{A}')\} \\
& + d_3\mathrm{vec}(\mathbf{A}(\mathbf{A}'\boldsymbol{\Sigma}^{-1}\mathbf{A})^{-}\mathbf{A}')\mathrm{vec}'(\mathbf{A}(\mathbf{A}'\boldsymbol{\Sigma}^{-1}\mathbf{A})^{-}\mathbf{A}'),
\end{aligned}
$$

where

$$
d_1 = \tfrac{n-r(\mathbf{C})}{n^2} + 2r(\mathbf{C})\tfrac{p-r(\mathbf{A})}{n^2(n-r(\mathbf{C})-p+r(\mathbf{A})-1)} + r(\mathbf{C})\tfrac{2c_1+c_2+c_3}{n^2} + r(\mathbf{C})^2\tfrac{c_3}{n^2},
$$

with c_1, c_2 *and* c_3 *as in Theorem 4.2.2, and*

$$
\begin{aligned}
d_2 = & \tfrac{n-p+r(\mathbf{A})-1}{n(n-r(\mathbf{C})-p+r(\mathbf{A})-1)}, \\
d_3 = & \tfrac{2r(\mathbf{C})(n-r(\mathbf{C})-1)(n-p+r(\mathbf{A})-1)(p-r(\mathbf{A}))}{n^2(n-r(\mathbf{C})-p+r(\mathbf{A}))(n-r(\mathbf{C})-p+r(\mathbf{A})-1)^2(n-r(\mathbf{C})-p+r(\mathbf{A})-3)}.
\end{aligned}
$$

PROOF: We already know that $\mathbf{S} \sim W_p(\boldsymbol{\Sigma}, n - r(\mathbf{C}))$. From Corollary 2.4.3.1 and Theorem 2.4.2 it follows that

$$
\mathbf{A}^{o'}\mathbf{X}\mathbf{C}'(\mathbf{C}\mathbf{C}')^{-}\mathbf{C}\mathbf{X}'\mathbf{A}^{o} \sim W_{p-r(\mathbf{A})}(\mathbf{A}^{o'}\boldsymbol{\Sigma}\mathbf{A}^{o}, r(\mathbf{C})). \tag{4.2.42}
$$

Furthermore, Theorem 2.2.4 (iii) states that $\mathbf{S}$ and $\mathbf{A}^{o'}\mathbf{X}\mathbf{C}'(\mathbf{C}\mathbf{C}')^{-}\mathbf{C}\mathbf{X}'\mathbf{A}^{o}$ are independently distributed. Now, utilizing the first moments of a Wishart matrix, given in Theorem 2.4.14 (i), relation (4.2.41) yields

$$
\begin{aligned}
E[n\widehat{\boldsymbol{\Sigma}}] = & E[\mathbf{S}] + E[\mathbf{S}\mathbf{A}^{o}(\mathbf{A}^{o'}\mathbf{S}\mathbf{A}^{o})^{-}\mathbf{A}^{o'}\mathbf{X}\mathbf{C}'(\mathbf{C}\mathbf{C}')^{-}\mathbf{C}\mathbf{X}'\mathbf{A}^{o}(\mathbf{A}^{o'}\mathbf{S}\mathbf{A}^{o})^{-}\mathbf{A}^{o'}\mathbf{S}] \\
= & E[\mathbf{S}] + E[\mathbf{S}\mathbf{A}^{o}(\mathbf{A}^{o'}\mathbf{S}\mathbf{A}^{o})^{-}E[\mathbf{A}^{o'}\mathbf{X}\mathbf{C}'(\mathbf{C}\mathbf{C}')^{-}\mathbf{C}\mathbf{X}'\mathbf{A}^{o}](\mathbf{A}^{o'}\mathbf{S}\mathbf{A}^{o})^{-}\mathbf{A}^{o'}\mathbf{S}] \\
= & (n - r(\mathbf{C}))\boldsymbol{\Sigma} + r(\mathbf{C})E[\mathbf{S}\mathbf{A}^{o}(\mathbf{A}^{o'}\mathbf{S}\mathbf{A}^{o})^{-}\mathbf{A}^{o'}\boldsymbol{\Sigma}\mathbf{A}^{o}(\mathbf{A}^{o'}\mathbf{S}\mathbf{A}^{o})^{-}\mathbf{A}^{o'}\mathbf{S}].
\end{aligned} \tag{4.2.43}
$$

In the following the same technique as in §4.2.1 will be used. Let $\boldsymbol{\Sigma}^{1/2}$ be a symmetric square root of $\boldsymbol{\Sigma}$. From Proposition 1.1.6 (ii) it follows that there exist a non-singular matrix $\mathbf{H}$ and an orthogonal matrix $\boldsymbol{\Gamma} = (\boldsymbol{\Gamma}_1' : \boldsymbol{\Gamma}_2')'$, where $\boldsymbol{\Gamma}_1 : (p - r(\mathbf{A})) \times p$ and $\boldsymbol{\Gamma}_2 : r(\mathbf{A}) \times p$, such that

$$
\mathbf{A}^{o'}\boldsymbol{\Sigma}^{-1/2} = \mathbf{H}(\mathbf{I}_{p-r(\mathbf{A})} : \mathbf{0})\boldsymbol{\Gamma}
$$

holds, which is identical to

$$
\mathbf{A}^{o'} = \mathbf{H}(\mathbf{I}_{p-r(\mathbf{A})} : \mathbf{0})\boldsymbol{\Gamma}\boldsymbol{\Sigma}^{-1/2} = \mathbf{H}\boldsymbol{\Gamma}_1\boldsymbol{\Sigma}^{-1/2}. \tag{4.2.44}
$$

Utilizing (4.2.44) gives

$$
\begin{aligned}
&\mathbf{SA}^o(\mathbf{A}^{o'}\mathbf{SA}^o)^-\mathbf{A}^{o'}\boldsymbol{\Sigma}\mathbf{A}^o(\mathbf{A}^{o'}\mathbf{SA}^o)^-\mathbf{A}^{o'}\mathbf{S}\\
&\quad=\mathbf{S}\boldsymbol{\Sigma}^{-1/2}\boldsymbol{\Gamma}_1'(\boldsymbol{\Gamma}_1\boldsymbol{\Sigma}^{-1/2}\mathbf{S}\boldsymbol{\Sigma}^{-1/2}\boldsymbol{\Gamma}_1')^{-1}\boldsymbol{\Gamma}_1\boldsymbol{\Sigma}^{-1/2}\boldsymbol{\Sigma}\\
&\qquad\times\boldsymbol{\Sigma}^{-1/2}\boldsymbol{\Gamma}_1'(\boldsymbol{\Gamma}_1\boldsymbol{\Sigma}^{-1/2}\mathbf{S}\boldsymbol{\Sigma}^{-1/2}\boldsymbol{\Gamma}_1')^{-1}\boldsymbol{\Gamma}_1\boldsymbol{\Sigma}^{-1/2}\mathbf{S}\\
&\quad=\mathbf{S}\boldsymbol{\Sigma}^{-1/2}\boldsymbol{\Gamma}_1'(\boldsymbol{\Gamma}_1\boldsymbol{\Sigma}^{-1/2}\mathbf{S}\boldsymbol{\Sigma}^{-1/2}\boldsymbol{\Gamma}_1')^{-1}(\boldsymbol{\Gamma}_1\boldsymbol{\Sigma}^{-1/2}\mathbf{S}\boldsymbol{\Sigma}^{-1/2}\boldsymbol{\Gamma}_1')^{-1}\boldsymbol{\Gamma}_1\boldsymbol{\Sigma}^{-1/2}\mathbf{S},
\end{aligned}
\tag{4.2.45}
$$

where it has been used that orthogonality of $\boldsymbol{\Gamma}$ implies $\boldsymbol{\Gamma}_1\boldsymbol{\Gamma}_1'=\mathbf{I}$, which in turn gives $\boldsymbol{\Gamma}_1\boldsymbol{\Sigma}^{-1/2}\boldsymbol{\Sigma}\boldsymbol{\Sigma}^{-1/2}\boldsymbol{\Gamma}_1'=\mathbf{I}$. In order to simplify (4.2.45), the matrix

$$
\mathbf{V}=\boldsymbol{\Gamma}\boldsymbol{\Sigma}^{-1/2}\mathbf{S}\boldsymbol{\Sigma}^{-1/2}\boldsymbol{\Gamma}', \tag{4.2.46}
$$

already defined in the previous paragraph, will be used. This time the matrix is partitioned as

$$
\mathbf{V}=\begin{pmatrix}\mathbf{V}_{11} & \mathbf{V}_{12}\\ \mathbf{V}_{21} & \mathbf{V}_{22}\end{pmatrix},\qquad \begin{matrix}(p-r(\mathbf{A}))\times(p-r(\mathbf{A})) & (p-r(\mathbf{A}))\times r(\mathbf{A})\\ r(\mathbf{A})\times(p-r(\mathbf{A})) & r(\mathbf{A})\times r(\mathbf{A})\end{matrix}. \tag{4.2.47}
$$

After performing some calculations, we get

$$
\begin{aligned}
&\mathbf{SA}^o(\mathbf{A}^{o'}\mathbf{SA}^o)^-\mathbf{A}^{o'}\boldsymbol{\Sigma}\mathbf{A}^o(\mathbf{A}^{o'}\mathbf{SA}^o)^-\mathbf{A}^{o'}\mathbf{S}\\
&\quad=\boldsymbol{\Sigma}^{-1/2}\boldsymbol{\Gamma}'\begin{pmatrix}\mathbf{I} & \mathbf{V}_{11}^{-1}\mathbf{V}_{12}\\ \mathbf{V}_{21}\mathbf{V}_{11}^{-1} & \mathbf{V}_{21}\mathbf{V}_{11}^{-1}\mathbf{V}_{11}^{-1}\mathbf{V}_{12}\end{pmatrix}\boldsymbol{\Gamma}\boldsymbol{\Sigma}^{-1/2}.
\end{aligned}
\tag{4.2.48}
$$

Therefore, when returning to (4.2.43),

$$
E[n\widehat{\boldsymbol{\Sigma}}]=E[\mathbf{S}]+r(\mathbf{C})\boldsymbol{\Sigma}^{-1/2}\boldsymbol{\Gamma}'E\begin{bmatrix}\mathbf{I} & \mathbf{V}_{11}^{-1}\mathbf{V}_{12}\\ \mathbf{V}_{21}\mathbf{V}_{11}^{-1} & \mathbf{V}_{21}\mathbf{V}_{11}^{-1}\mathbf{V}_{11}^{-1}\mathbf{V}_{12}\end{bmatrix}\boldsymbol{\Gamma}\boldsymbol{\Sigma}^{-1/2}. \tag{4.2.49}
$$

Since expectation of a matrix is defined by the expectation of its elements, we treat the submatrices on the right hand side of (4.2.49) separately. The same technique as in the previous section will be used.
Since $\mathbf{V}\sim W_p(\mathbf{I},n-r(\mathbf{C}))$, there exists a matrix $\mathbf{U}\sim N_{p,n-r(\mathbf{C})}(\mathbf{0},\mathbf{I},\mathbf{I})$ such that $\mathbf{V}=\mathbf{UU}'$. Furthermore, $\mathbf{U}$ is partitioned in correspondence with the partition of $\mathbf{V}$, i.e.

$$
\mathbf{U}'=(\mathbf{U}_1':\mathbf{U}_2'),\qquad (n-r(\mathbf{C}))\times(p-r(\mathbf{A})):(n-r(\mathbf{C}))\times r(\mathbf{A}).
$$

Note that since $D[\mathbf{U}]=\mathbf{I}$, the matrices $\mathbf{U}_1$ and $\mathbf{U}_2$ are independently distributed. Hence,

$$
E[\mathbf{V}_{21}\mathbf{V}_{11}^{-1}]=E[\mathbf{U}_2\mathbf{U}_1'(\mathbf{U}_1\mathbf{U}_1')^{-1}]=E[\mathbf{U}_2]E[\mathbf{U}_1'(\mathbf{U}_1\mathbf{U}_1')^{-1}]=\mathbf{0}, \tag{4.2.50}
$$

since $E[\mathbf{U}_2]=\mathbf{0}$. By symmetry it follows that $E[\mathbf{V}_{11}^{-1}\mathbf{V}_{12}]=\mathbf{0}$. In the next lines we copy the derivation of (4.2.22). Thus, by Theorem 2.2.9 (i) and independence of $\mathbf{U}_1$ and $\mathbf{U}_2$,

$$
\begin{aligned}
E[\mathbf{V}_{21}\mathbf{V}_{11}^{-1}\mathbf{V}_{11}^{-1}\mathbf{V}_{12}]&=E[\mathbf{U}_2\mathbf{U}_1'(\mathbf{U}_1\mathbf{U}_1')^{-1}(\mathbf{U}_1\mathbf{U}_1')^{-1}\mathbf{U}_1\mathbf{U}_2']\\
&=E[\mathrm{tr}(\mathbf{U}_1'(\mathbf{U}_1\mathbf{U}_1')^{-1}(\mathbf{U}_1\mathbf{U}_1')^{-1}\mathbf{U}_1)]\mathbf{I}=E[\mathrm{tr}(\mathbf{U}_1\mathbf{U}_1')^{-1}]\mathbf{I}\\
&=\frac{p-r(\mathbf{A})}{n-r(\mathbf{C})-p+r(\mathbf{A})-1}\mathbf{I}_{r(\mathbf{A})}=c_1\mathbf{I}_{r(\mathbf{A})},
\end{aligned}
\tag{4.2.51}
$$

where Theorem 2.4.14 (iii) has been applied for the last line to

$$\mathbf{U}_1\mathbf{U}_1' \sim W_{p-r(\mathbf{A})}(\mathbf{I}, n - r(\mathbf{C})),$$

and c_1 is the same constant as in Theorem 4.2.2. Thus, using the Wishartness of $\mathbf{S}$, as well as applying the relations (4.2.50) and (4.2.51) we see that

$$\begin{aligned} E[n\widehat{\boldsymbol{\Sigma}}] =&(n - r(\mathbf{C}))\boldsymbol{\Sigma} + r(\mathbf{C})\boldsymbol{\Sigma}^{-1/2}\boldsymbol{\Gamma}'\begin{pmatrix}\mathbf{I} & \mathbf{0}\\ \mathbf{0} & c_1\mathbf{I}_{r(\mathbf{A})}\end{pmatrix}\boldsymbol{\Gamma}\boldsymbol{\Sigma}^{-1/2}\\ =&(n - r(\mathbf{C}))\boldsymbol{\Sigma} + r(\mathbf{C})\boldsymbol{\Sigma}^{-1/2}\boldsymbol{\Gamma}_1'\boldsymbol{\Gamma}_1\boldsymbol{\Sigma}^{-1/2} + r(\mathbf{C})c_1\boldsymbol{\Sigma}^{-1/2}\boldsymbol{\Gamma}_2\boldsymbol{\Gamma}_2'\boldsymbol{\Sigma}^{-1/2}. \end{aligned} \quad (4.2.52)$$

In order to verify (i) of the theorem, the products $\boldsymbol{\Sigma}^{-1/2}\boldsymbol{\Gamma}_i'\boldsymbol{\Gamma}_i\boldsymbol{\Sigma}^{-1/2}$, $i = 1, 2$, have to be expressed in the original matrices. By definition of orthogonality, we obtain that

$$\mathbf{I} = \boldsymbol{\Gamma}'\boldsymbol{\Gamma} = \boldsymbol{\Gamma}_1'\boldsymbol{\Gamma}_1 + \boldsymbol{\Gamma}_2'\boldsymbol{\Gamma}_2 \quad (4.2.53)$$

holds, and (4.2.44) implies

$$\mathbf{H}^{-1}\mathbf{A}^{o'} = \boldsymbol{\Gamma}_1\boldsymbol{\Sigma}^{-1/2}. \quad (4.2.54)$$

Furthermore,

$$\mathbf{A}^{o'}\boldsymbol{\Sigma}\mathbf{A}^{o} = \mathbf{A}^{o'}\boldsymbol{\Sigma}^{1/2}\boldsymbol{\Sigma}^{1/2}\mathbf{A}^{o} = \mathbf{H}(\mathbf{I}:\mathbf{0})\boldsymbol{\Gamma}\boldsymbol{\Gamma}'(\mathbf{I}:\mathbf{0})'\mathbf{H}' = \mathbf{H}\mathbf{H}'. \quad (4.2.55)$$

Thus, utilizing (4.2.54) and (4.2.55) gives

$$\boldsymbol{\Sigma}^{1/2}\boldsymbol{\Gamma}_1'\boldsymbol{\Gamma}_1\boldsymbol{\Sigma}^{1/2} = \boldsymbol{\Sigma}\mathbf{A}^{o}(\mathbf{H}^{-1})'\mathbf{H}^{-1}\mathbf{A}^{o'}\boldsymbol{\Sigma} = \boldsymbol{\Sigma}\mathbf{A}^{o}(\mathbf{A}^{o'}\boldsymbol{\Sigma}\mathbf{A}^{o})^{-1}\mathbf{A}^{o'}\boldsymbol{\Sigma}. \quad (4.2.56)$$

However, Corollary 1.2.25.1 states that (4.2.56) is equivalent to

$$\boldsymbol{\Sigma}^{1/2}\boldsymbol{\Gamma}_1'\boldsymbol{\Gamma}_1\boldsymbol{\Sigma}^{1/2} = \boldsymbol{\Sigma} - \mathbf{A}(\mathbf{A}'\boldsymbol{\Sigma}^{-1}\mathbf{A})^{-}\mathbf{A}', \quad (4.2.57)$$

and then (4.2.53) gives

$$\boldsymbol{\Sigma}^{1/2}\boldsymbol{\Gamma}_2'\boldsymbol{\Gamma}_2\boldsymbol{\Sigma}^{1/2} = \mathbf{A}(\mathbf{A}'\boldsymbol{\Sigma}^{-1}\mathbf{A})^{-}\mathbf{A}'. \quad (4.2.58)$$

Thus, the expectations given by (4.2.52) can be expressed in the original matrices, and from (4.2.57) and (4.2.58) it follows that (4.2.52) can be written

$$E[n\widehat{\boldsymbol{\Sigma}}] = (n - r(\mathbf{C}))\boldsymbol{\Sigma} + r(\mathbf{C})(\boldsymbol{\Sigma} - \mathbf{A}(\mathbf{A}'\boldsymbol{\Sigma}^{-1}\mathbf{A})^{-}\mathbf{A}') + r(\mathbf{C})c_1\mathbf{A}(\mathbf{A}'\boldsymbol{\Sigma}^{-1}\mathbf{A})^{-}\mathbf{A}',$$

which is identical to (i) of the theorem.
In the next the dispersion matrix of $\widehat{\boldsymbol{\Sigma}}$ will be studied. The expression for $\widehat{\boldsymbol{\Sigma}}$ given by (4.2.41) is taken as a starting point. Consider

$$\mathbf{W} = \mathbf{A}^{o'}\mathbf{X}\mathbf{C}'(\mathbf{C}\mathbf{C}')^{-}\mathbf{X}'\mathbf{A}^{o} \sim W_{p-r(\mathbf{A})}(\mathbf{A}^{o'}\boldsymbol{\Sigma}\mathbf{A}^{o}, r(\mathbf{C})). \quad (4.2.59)$$

As before, $\mathbf{S}$ and $\mathbf{W}$ are independently distributed. Note that for an arbitrary random matrix $\mathbf{Q}$ and arbitrary non-random matrices $\mathbf{P}_1$ and $\mathbf{P}_2$ such that the product $\mathbf{P}_1\mathbf{Q}\mathbf{P}_2$ is well defined, the dispersion matrix $D[\mathbf{P}_1\mathbf{Q}\mathbf{P}_2]$ equals

$$D[\mathbf{P}_1\mathbf{Q}\mathbf{P}_2] = D[\mathrm{vec}(\mathbf{P}_1\mathbf{Q}\mathbf{P}_2)] = (\mathbf{P}_2' \otimes \mathbf{P}_1)D[\mathbf{Q}](\mathbf{P}_2 \otimes \mathbf{P}_1'). \qquad (4.2.60)$$

Now, from (4.2.41) it follows that

$$\begin{aligned} D[n\widehat{\mathbf{\Sigma}}] =& E[(\mathbf{S}\mathbf{A}^o(\mathbf{A}^{o'}\mathbf{S}\mathbf{A}^o)^{-1})^{\otimes 2}D[\mathbf{W}]((\mathbf{A}^{o'}\mathbf{S}\mathbf{A}^o)^{-1}\mathbf{A}^{o'}\mathbf{S})^{\otimes 2}] \\ &+ D[\mathbf{S} + \mathbf{S}\mathbf{A}^o(\mathbf{A}^{o'}\mathbf{S}\mathbf{A}^o)^{-1}E[\mathbf{W}](\mathbf{A}^{o'}\mathbf{S}\mathbf{A}^o)^{-1}\mathbf{A}^{o'}\mathbf{S}], \qquad (4.2.61) \end{aligned}$$

which is established after lengthy matrix calculations by utilizing (4.2.60) and the independence between $\mathbf{S}$ and $\mathbf{W}$. Alternatively we could apply a conditional approach:

$$D[n\widehat{\mathbf{\Sigma}}] = E[D[n\widehat{\mathbf{\Sigma}}|\mathbf{S}]] + D[E[n\widehat{\mathbf{\Sigma}}|\mathbf{S}]].$$

However, in order to utilize the conditional formula, one has to show that the moments exist and this is not always straightforward. For example, for some vectors $\mathbf{p}$ and $\mathbf{q}$, $E[\mathbf{p}] = E[E[\mathbf{p}|\mathbf{q}]]$ may not hold since the elements in $E[\mathbf{p}|\mathbf{q}]$ may all be zeros, while some elements in $E[\mathbf{p}]$ are infinite. Therefore, we prefer the more tedious matrix derivation. From Theorem 2.4.14 (ii) it follows that

$$D[\mathbf{W}] = r(\mathbf{C})(\mathbf{I} + \mathbf{K}_{r,r})((\mathbf{A}^{o'}\mathbf{\Sigma}\mathbf{A}^o) \otimes (\mathbf{A}^{o'}\mathbf{\Sigma}\mathbf{A}^o)), \qquad (4.2.62)$$

where $r = p - r(\mathbf{A})$. Furthermore, Theorem 2.4.14 (i) states that

$$E[\mathbf{W}] = r(\mathbf{C})\mathbf{A}^{o'}\mathbf{\Sigma}\mathbf{A}^o. \qquad (4.2.63)$$

By means of (4.2.62) and (4.2.63), equality (4.2.61) is expressed in the following form:

$$\begin{aligned} D[n\widehat{\mathbf{\Sigma}}] =& r(\mathbf{C})(\mathbf{I} + \mathbf{K}_{p,p})E[(\mathbf{S}\mathbf{A}^o(\mathbf{A}^{o'}\mathbf{S}\mathbf{A}^o)^{-1}\mathbf{A}^{o'}\mathbf{\Sigma}\mathbf{A}^o(\mathbf{A}^{o'}\mathbf{S}\mathbf{A}^o)^{-1}\mathbf{A}^{o'}\mathbf{S})^{\otimes 2}] \\ &+ D[\mathbf{S} + r(\mathbf{C})\mathbf{S}\mathbf{A}^o(\mathbf{A}^{o'}\mathbf{S}\mathbf{A}^o)^{-1}\mathbf{A}^{o'}\mathbf{\Sigma}\mathbf{A}^o(\mathbf{A}^{o'}\mathbf{S}\mathbf{A}^o)^{-1}\mathbf{A}^{o'}\mathbf{S}]. \qquad (4.2.64) \end{aligned}$$

The two expressions on the right hand side of (4.2.64) will be treated separately, and we start from the second term. As before the canonical representation

$$\mathbf{A}^{o'}\mathbf{\Sigma}^{1/2} = \mathbf{H}(\mathbf{I}_{p-r(\mathbf{A})} : \mathbf{0})\mathbf{\Gamma}$$

is used, where $\mathbf{H}$ is non-singular and $\mathbf{\Gamma}$ is an orthogonal matrix. Furthermore, the matrix $\mathbf{V}$ defined by (4.2.46), as well as the partition of $\mathbf{V}$ given by (4.2.47), will be utilized. From (4.2.46) it follows that

$$\mathbf{S} = \mathbf{\Sigma}^{1/2}\mathbf{\Gamma}'\mathbf{V}\mathbf{\Gamma}\mathbf{\Sigma}^{1/2}.$$

Hence, applying (4.2.48) to the second term in (4.2.64) yields

$$\begin{aligned} D[\mathbf{S}+&r(\mathbf{C})\mathbf{S}\mathbf{A}^o(\mathbf{A}^{o'}\mathbf{S}\mathbf{A}^o)^{-1}\mathbf{A}^{o'}\mathbf{\Sigma}\mathbf{A}^o(\mathbf{A}^{o'}\mathbf{S}\mathbf{A}^o)^{-1}\mathbf{A}^{o'}\mathbf{S}] \\ &= D[\mathbf{\Sigma}^{1/2}\mathbf{\Gamma}'\{\mathbf{V} + r(\mathbf{C})\begin{pmatrix} \mathbf{I} & \mathbf{V}_{11}^{-1}\mathbf{V}_{12} \\ \mathbf{V}_{21}\mathbf{V}_{11}^{-1} & \mathbf{V}_{21}\mathbf{V}_{11}^{-1}\mathbf{V}_{11}^{-1}\mathbf{V}_{12} \end{pmatrix}\}\mathbf{\Gamma}\mathbf{\Sigma}^{1/2}]. \qquad (4.2.65) \end{aligned}$$

If $\mathbf{V}$ is represented through $\mathbf{U}$ as in (4.2.50), it follows that the right hand side of (4.2.65) can be written in the form

$$(\boldsymbol{\Sigma}^{1/2}\boldsymbol{\Gamma}')^{\otimes 2}D[(\mathbf{U}_1':\mathbf{U}_2')'\mathbf{G}(\mathbf{U}_1':\mathbf{U}_2')](\boldsymbol{\Sigma}^{1/2}\boldsymbol{\Gamma}')^{\otimes 2}, \tag{4.2.66}$$

where

$$\mathbf{G}=\mathbf{I}_{n-r(\mathbf{C})}+r(\mathbf{C})\mathbf{U}_1'(\mathbf{U}_1\mathbf{U}_1')^{-1}(\mathbf{U}_1\mathbf{U}_1')^{-1}\mathbf{U}_1. \tag{4.2.67}$$

Note that $\mathbf{G}$ is a symmetric matrix which is a function of $\mathbf{U}_1$ solely. Since $\mathbf{U}_1$ and $\mathbf{U}_2$ are independently distributed, it would be possible to use this fact in order to simplify (4.2.66). However, it is more straightforward to condition with respect to $\mathbf{U}_1$. This approach leads to expressions which are relatively easy to handle. Furthermore, note that since $\mathbf{U}_1$ is independent of $\mathbf{U}_2$, the conditional moments always exist. Hence, conditioning (4.2.66) with respect to $\mathbf{U}_1$ gives

$$\begin{aligned}(\boldsymbol{\Sigma}^{1/2}\boldsymbol{\Gamma}')^{\otimes 2}\Big\{&E[D[(\mathbf{U}_1':\mathbf{U}_2')'\mathbf{G}(\mathbf{U}_1':\mathbf{U}_2')|\mathbf{U}_1]]\\&+D[E[(\mathbf{U}_1':\mathbf{U}_2')'\mathbf{G}(\mathbf{U}_1':\mathbf{U}_2')|\mathbf{U}_1]]\Big\}(\boldsymbol{\Sigma}^{1/2}\boldsymbol{\Gamma}')^{\otimes 2}.\end{aligned} \tag{4.2.68}$$

Utilizing

$$(\mathbf{U}_1':\mathbf{U}_2')'|\mathbf{U}_1\sim N_{p,n-r(\mathbf{C})}(\begin{pmatrix}\mathbf{U}_1\\\mathbf{0}\end{pmatrix},\begin{pmatrix}\mathbf{0}&\mathbf{0}\\\mathbf{0}&\mathbf{I}_{r(\mathbf{A})}\end{pmatrix},\mathbf{I}) \tag{4.2.69}$$

we get, by applying Theorem 2.2.7 (i) and (ii), that (4.2.68) equals

$$\begin{aligned}&(\mathbf{I}+\mathbf{K}_{p,p})(\boldsymbol{\Sigma}^{1/2}\boldsymbol{\Gamma}_2'\boldsymbol{\Gamma}_2\boldsymbol{\Sigma}^{1/2})^{\otimes 2}E[\mathrm{tr}(\mathbf{G}\mathbf{G})]\\&\quad+\boldsymbol{\Sigma}^{1/2}\boldsymbol{\Gamma}_1'E[\mathbf{U}_1\mathbf{G}\mathbf{G}'\mathbf{U}_1']\boldsymbol{\Gamma}_1\boldsymbol{\Sigma}^{1/2}\otimes\boldsymbol{\Sigma}^{1/2}\boldsymbol{\Gamma}_2'\boldsymbol{\Gamma}_2\boldsymbol{\Sigma}^{1/2}\\&\quad+\boldsymbol{\Sigma}^{1/2}\boldsymbol{\Gamma}_2'\boldsymbol{\Gamma}_2\boldsymbol{\Sigma}^{1/2}\otimes\boldsymbol{\Sigma}^{1/2}\boldsymbol{\Gamma}_1'E[\mathbf{U}_1\mathbf{G}\mathbf{G}'\mathbf{U}_1']\boldsymbol{\Gamma}_1\boldsymbol{\Sigma}^{1/2}\\&\quad+(\boldsymbol{\Sigma}^{1/2}\boldsymbol{\Gamma}')^{\otimes 2}D[E[\mathbf{U}\mathbf{G}\mathbf{U}'|\mathbf{U}_1]](\boldsymbol{\Gamma}\boldsymbol{\Sigma}^{1/2})^{\otimes 2}.\end{aligned} \tag{4.2.70}$$

The terms in (4.2.70) will be studied in some detail. It is observed that

$$\begin{aligned}E[\mathrm{tr}(\mathbf{G}\mathbf{G})]=&(n-r(\mathbf{C})+2r(\mathbf{C}))E[\mathrm{tr}(\mathbf{U}_1\mathbf{U}_1')^{-1}]\\&+r(\mathbf{C})^2E[\mathrm{tr}\{(\mathbf{U}_1\mathbf{U}_1')^{-1}(\mathbf{U}_1\mathbf{U}_1')^{-1}\}],\end{aligned} \tag{4.2.71}$$

$$E[\mathbf{U}_1\mathbf{G}\mathbf{G}\mathbf{U}_1']=E[\mathbf{U}_1\mathbf{U}_1']+2r(\mathbf{C})\mathbf{I}_{p-r(\mathbf{A})}+r(\mathbf{C})^2E[(\mathbf{U}_1\mathbf{U}_1')^{-1}] \tag{4.2.72}$$

and

$$\begin{aligned}&(\boldsymbol{\Sigma}^{1/2}\boldsymbol{\Gamma}')^{\otimes 2}D[E[\mathbf{U}\mathbf{G}\mathbf{U}'|\mathbf{U}_1]](\boldsymbol{\Gamma}\boldsymbol{\Sigma}^{1/2})^{\otimes 2}\\&\quad=r(\mathbf{C})^2D[\mathrm{tr}(\mathbf{U}_1\mathbf{U}_1')^{-1}]\mathrm{vec}(\boldsymbol{\Sigma}^{1/2}\boldsymbol{\Gamma}_2'\boldsymbol{\Gamma}_2\boldsymbol{\Sigma}^{1/2})\mathrm{vec}'(\boldsymbol{\Sigma}^{1/2}\boldsymbol{\Gamma}_2'\boldsymbol{\Gamma}_2\boldsymbol{\Sigma}^{1/2})\\&\qquad+(\boldsymbol{\Sigma}^{1/2}\boldsymbol{\Gamma}_1')^{\otimes 2}D[\mathbf{U}_1\mathbf{U}_1'](\boldsymbol{\Gamma}_1\boldsymbol{\Sigma}^{1/2})^{\otimes 2}\\&\qquad+r(\mathbf{C})(\boldsymbol{\Sigma}^{1/2}\boldsymbol{\Gamma}_1')^{\otimes 2}C[\mathbf{U}_1\mathbf{U}_1',\mathrm{tr}(\mathbf{U}_1\mathbf{U}_1')^{-1}]\mathrm{vec}'(\boldsymbol{\Sigma}^{1/2}\boldsymbol{\Gamma}_2'\boldsymbol{\Gamma}_2\boldsymbol{\Sigma}^{1/2})\\&\qquad+r(\mathbf{C})\mathrm{vec}(\boldsymbol{\Sigma}^{1/2}\boldsymbol{\Gamma}_2'\boldsymbol{\Gamma}_2\boldsymbol{\Sigma}^{1/2})C[\mathrm{tr}(\mathbf{U}_1\mathbf{U}_1')^{-1},\mathbf{U}_1\mathbf{U}_1'](\boldsymbol{\Gamma}_1\boldsymbol{\Sigma}^{1/2})^{\otimes 2}.\end{aligned} \tag{4.2.73}$$

It follows from Theorem 2.4.14 (iii) and (vi) that (4.2.71) equals

$$d_{21} \equiv E[\mathrm{tr}(\mathbf{GG})] = n - r(\mathbf{C}) + 2r(\mathbf{C})\frac{p-r(\mathbf{A})}{n-r(\mathbf{C})-p+r(\mathbf{A})-1} + r(\mathbf{C})^2\frac{(p-r(\mathbf{A}))(n-r(\mathbf{C})-1)}{(n-r(\mathbf{C})-p+r(\mathbf{A}))(n-r(\mathbf{C})-p+r(\mathbf{A})-1)(n-r(\mathbf{C})-p+r(\mathbf{A})-3)}. \quad (4.2.74)$$

Moreover, the moments in (4.2.72) are calculated:

$$d_{31}\mathbf{I}_{p-r(\mathbf{A})} \equiv E[\mathbf{U}_1\mathbf{GGU}_1'] = \{n - r(\mathbf{C}) + 2r(\mathbf{C}) + r(\mathbf{C})^2\frac{1}{n-r(\mathbf{C})-p+r(\mathbf{A})-1}\}\mathbf{I}_{p-r(\mathbf{A})}, \quad (4.2.75)$$

and by Theorem 2.4.14 (vii) the covariance matrix in (4.2.73) is identical to

$$\begin{aligned} d_{51}\mathrm{vec}(\mathbf{I}_{p-r(\mathbf{A})}) &\equiv C[\mathbf{U}_1\mathbf{U}_1', \mathrm{tr}(\mathbf{U}_1\mathbf{U}_1')^{-1}] \\ &= E[\mathbf{U}_1\mathbf{U}_1'\mathrm{tr}(\mathbf{U}_1\mathbf{U}_1')^{-1}] - E[\mathbf{U}_1\mathbf{U}_1']E[\mathrm{tr}(\mathbf{U}_1\mathbf{U}_1')^{-1}] \\ &= \frac{(n-r(\mathbf{C}))(p-r(\mathbf{A}))-2}{n-r(\mathbf{C})-p+r(\mathbf{A})-1}\mathrm{vec}(\mathbf{I}_{p-r(\mathbf{A})}) - \frac{(n-r(\mathbf{C}))(p-r(\mathbf{A}))}{n-r(\mathbf{C})-p+r(\mathbf{A})-1}\mathrm{vec}(\mathbf{I}_{p-r(\mathbf{A})}) \\ &= -\frac{2}{n-r(\mathbf{C})-p+r(\mathbf{A})-1}\mathrm{vec}(\mathbf{I}_{p-r(\mathbf{A})}). \end{aligned} \quad (4.2.76)$$

Thus, by combining (4.2.65) – (4.2.76) we find that the second term in (4.2.64) equals

$$\begin{aligned} &D[\mathbf{S} + r(\mathbf{C})\mathbf{SA}^o(\mathbf{A}^{o\prime}\mathbf{SA}^o)^{-1}\mathbf{A}^{o\prime}\mathbf{\Sigma A}^o(\mathbf{A}^{o\prime}\mathbf{SA}^o)^{-1}\mathbf{A}^{o\prime}\mathbf{S}] \\ &= d_{21}(\mathbf{I} + \mathbf{K}_{p,p})(\mathbf{\Sigma}^{1/2}\mathbf{\Gamma}_2'\mathbf{\Gamma}_2\mathbf{\Sigma}^{1/2})^{\otimes 2} \\ &\quad + d_{31}\mathbf{\Sigma}^{1/2}\mathbf{\Gamma}_1'\mathbf{\Gamma}_1\mathbf{\Sigma}^{1/2} \otimes \mathbf{\Sigma}^{1/2}\mathbf{\Gamma}_2'\mathbf{\Gamma}_2\mathbf{\Sigma}^{1/2} \\ &\quad + d_{31}\mathbf{\Sigma}^{1/2}\mathbf{\Gamma}_2'\mathbf{\Gamma}_2\mathbf{\Sigma}^{1/2} \otimes \mathbf{\Sigma}^{1/2}\mathbf{\Gamma}_1'\mathbf{\Gamma}_1\mathbf{\Sigma}^{1/2} \\ &\quad + d_{41}\mathrm{vec}(\mathbf{\Sigma}^{1/2}\mathbf{\Gamma}_2'\mathbf{\Gamma}_2\mathbf{\Sigma}^{1/2})\mathrm{vec}'(\mathbf{\Sigma}^{1/2}\mathbf{\Gamma}_2'\mathbf{\Gamma}_2\mathbf{\Sigma}^{1/2}) \\ &\quad + (n - r(\mathbf{C}))(\mathbf{I} + \mathbf{K}_{p,p})(\mathbf{\Sigma}^{1/2}\mathbf{\Gamma}_1'\mathbf{\Gamma}_1\mathbf{\Sigma}^{1/2})^{\otimes 2} \\ &\quad + d_{51}\mathrm{vec}(\mathbf{\Sigma}^{1/2}\mathbf{\Gamma}_1'\mathbf{\Gamma}_1\mathbf{\Sigma}^{1/2})\mathrm{vec}'(\mathbf{\Sigma}^{1/2}\mathbf{\Gamma}_2'\mathbf{\Gamma}_2\mathbf{\Sigma}^{1/2}) \\ &\quad + d_{51}\mathrm{vec}(\mathbf{\Sigma}^{1/2}\mathbf{\Gamma}_2'\mathbf{\Gamma}_2\mathbf{\Sigma}^{1/2})\mathrm{vec}'(\mathbf{\Sigma}^{1/2}\mathbf{\Gamma}_1'\mathbf{\Gamma}_1\mathbf{\Sigma}^{1/2}). \end{aligned} \quad (4.2.77)$$

Now (4.2.77) is expressed in the original matrices. In (4.2.57) and (4.2.58) it was stated that

$$\begin{aligned} \mathbf{\Sigma}^{1/2}\mathbf{\Gamma}_2'\mathbf{\Gamma}_2\mathbf{\Sigma}^{1/2} &= \mathbf{A}(\mathbf{A}'\mathbf{\Sigma}^{-1}\mathbf{A})^{-}\mathbf{A}', \\ \mathbf{\Sigma}^{1/2}\mathbf{\Gamma}_1'\mathbf{\Gamma}_1\mathbf{\Sigma}^{1/2} &= \mathbf{\Sigma} - \mathbf{A}(\mathbf{A}'\mathbf{\Sigma}^{-1}\mathbf{A})^{-}\mathbf{A}'. \end{aligned}$$

Hence, (4.2.77) is identical to

$$\begin{aligned} &D[\mathbf{S} + r(\mathbf{C})\mathbf{SA}^o(\mathbf{A}^{o\prime}\mathbf{SA}^o)^{-1}\mathbf{A}^{o\prime}\mathbf{\Sigma A}^o(\mathbf{A}^{o\prime}\mathbf{SA}^o)^{-1}\mathbf{A}^{o\prime}\mathbf{S}] \\ &= d_{21}(\mathbf{I} + \mathbf{K}_{p,p})(\mathbf{A}(\mathbf{A}'\mathbf{\Sigma}^{-1}\mathbf{A})^{-}\mathbf{A}')^{\otimes 2} \\ &\quad + d_{31}(\mathbf{\Sigma} - \mathbf{A}(\mathbf{A}'\mathbf{\Sigma}^{-1}\mathbf{A})^{-}\mathbf{A}') \otimes \mathbf{A}(\mathbf{A}'\mathbf{\Sigma}^{-1}\mathbf{A})^{-}\mathbf{A}' \\ &\quad + d_{31}\mathbf{A}(\mathbf{A}'\mathbf{\Sigma}^{-1}\mathbf{A})^{-}\mathbf{A}' \otimes (\mathbf{\Sigma} - \mathbf{A}(\mathbf{A}'\mathbf{\Sigma}^{-1}\mathbf{A})^{-}\mathbf{A}') \\ &\quad + d_{41}\mathrm{vec}(\mathbf{A}(\mathbf{A}'\mathbf{\Sigma}^{-1}\mathbf{A})^{-}\mathbf{A}')\mathrm{vec}'(\mathbf{A}(\mathbf{A}'\mathbf{\Sigma}^{-1}\mathbf{A})^{-}\mathbf{A}') \\ &\quad + (n - r(\mathbf{C}))(\mathbf{I} + \mathbf{K}_{p,p})(\mathbf{\Sigma} - \mathbf{A}(\mathbf{A}'\mathbf{\Sigma}^{-1}\mathbf{A})^{-}\mathbf{A}')^{\otimes 2} \\ &\quad + d_{51}\mathrm{vec}(\mathbf{\Sigma} - \mathbf{A}(\mathbf{A}'\mathbf{\Sigma}^{-1}\mathbf{A})^{-}\mathbf{A}')\mathrm{vec}'(\mathbf{A}(\mathbf{A}'\mathbf{\Sigma}^{-1}\mathbf{A})^{-}\mathbf{A}') \\ &\quad + d_{51}\mathrm{vec}(\mathbf{A}(\mathbf{A}'\mathbf{\Sigma}^{-1}\mathbf{A})^{-}\mathbf{A}')\mathrm{vec}'(\mathbf{\Sigma} - \mathbf{A}(\mathbf{A}'\mathbf{\Sigma}^{-1}\mathbf{A})^{-}\mathbf{A}') \end{aligned} \quad (4.2.78)$$

and the second expression in (4.2.64) has been determined. Proceeding with the first one in (4.2.64), the term can be rewritten in the following way:

$$
\begin{aligned}
&r(\mathbf{C})(\mathbf{I}+\mathbf{K}_{p,p})E[(\mathbf{SA}^o(\mathbf{A}^{o'}\mathbf{SA}^o)^{-1}\mathbf{A}^{o'}\boldsymbol{\Sigma}\mathbf{A}^o(\mathbf{A}^{o'}\mathbf{SA}^o)^{-1}\mathbf{A}^{o'}\mathbf{S})^{\otimes 2}]\\
&\quad=r(\mathbf{C})(\mathbf{I}+\mathbf{K}_{p,p})(\boldsymbol{\Sigma}^{1/2}\boldsymbol{\Gamma}')^{\otimes 2}\\
&\qquad\times E[E[(\mathbf{U}\mathbf{U}_1'(\mathbf{U}_1\mathbf{U}_1')^{-1}(\mathbf{U}_1\mathbf{U}_1')^{-1}\mathbf{U}_1\mathbf{U}')^{\otimes 2}|\mathbf{U}_1]](\boldsymbol{\Gamma}\boldsymbol{\Sigma}^{1/2})^{\otimes 2},
\end{aligned}
$$

which by (4.2.69) and Theorem 2.2.9 (ii) equals

$$
\begin{aligned}
r(\mathbf{C})(\mathbf{I}+\mathbf{K}_{p,p})\Big\{&(\boldsymbol{\Sigma}^{1/2}\boldsymbol{\Gamma}_1'\boldsymbol{\Gamma}_1\boldsymbol{\Sigma}^{1/2})^{\otimes 2}+(\boldsymbol{\Sigma}^{1/2}\boldsymbol{\Gamma}_2'\boldsymbol{\Gamma}_2\boldsymbol{\Sigma}^{1/2})^{\otimes 2}E[(\mathrm{tr}(\mathbf{U}_1\mathbf{U}_1')^{-1})^2]\\
&+\mathrm{vec}(\boldsymbol{\Sigma}^{1/2}\boldsymbol{\Gamma}_2'\boldsymbol{\Gamma}_2\boldsymbol{\Sigma}^{1/2})\mathrm{vec}'(\boldsymbol{\Sigma}^{1/2}\boldsymbol{\Gamma}_2'\boldsymbol{\Gamma}_2\boldsymbol{\Sigma}^{1/2})E[\mathrm{tr}\{(\mathbf{U}_1\mathbf{U}_1')^{-1}(\mathbf{U}_1\mathbf{U}_1')^{-1}\}]\\
&+(\boldsymbol{\Sigma}^{1/2}\boldsymbol{\Gamma}_2'\boldsymbol{\Gamma}_2\boldsymbol{\Sigma}^{1/2})^{\otimes 2}E[\mathrm{tr}\{(\mathbf{U}_1\mathbf{U}_1')^{-1}(\mathbf{U}_1\mathbf{U}_1')^{-1}\}]\\
&+(\boldsymbol{\Sigma}^{1/2}\boldsymbol{\Gamma}_2'\boldsymbol{\Gamma}_2\boldsymbol{\Sigma}^{1/2})\otimes(\boldsymbol{\Sigma}^{1/2}\boldsymbol{\Gamma}_1'\boldsymbol{\Gamma}_1\boldsymbol{\Sigma}^{1/2})E[\mathrm{tr}(\mathbf{U}_1\mathbf{U}_1')^{-1}]\\
&+(\boldsymbol{\Sigma}^{1/2}\boldsymbol{\Gamma}_1'\boldsymbol{\Gamma}_1\boldsymbol{\Sigma}^{1/2})\otimes(\boldsymbol{\Sigma}^{1/2}\boldsymbol{\Gamma}_2'\boldsymbol{\Gamma}_2\boldsymbol{\Sigma}^{1/2})E[\mathrm{tr}(\mathbf{U}_1\mathbf{U}_1')^{-1}]\\
&+\mathrm{vec}(E[\boldsymbol{\Sigma}^{1/2}\boldsymbol{\Gamma}_1'(\mathbf{U}_1\mathbf{U}_1')^{-1}\boldsymbol{\Gamma}_1\boldsymbol{\Sigma}^{1/2}])\mathrm{vec}'(\boldsymbol{\Sigma}^{1/2}\boldsymbol{\Gamma}_2'\boldsymbol{\Gamma}_2\boldsymbol{\Sigma}^{1/2})\\
&+\mathrm{vec}(\boldsymbol{\Sigma}^{1/2}\boldsymbol{\Gamma}_2'\boldsymbol{\Gamma}_2\boldsymbol{\Sigma}^{1/2})\mathrm{vec}'(E[\boldsymbol{\Sigma}^{1/2}\boldsymbol{\Gamma}_1'(\mathbf{U}_1\mathbf{U}_1')^{-1}\boldsymbol{\Gamma}_1\boldsymbol{\Sigma}^{1/2}])\\
&+E[\boldsymbol{\Sigma}^{1/2}\boldsymbol{\Gamma}_1'(\mathbf{U}_1\mathbf{U}_1')^{-1}\boldsymbol{\Gamma}_1\boldsymbol{\Sigma}^{1/2}]\otimes\boldsymbol{\Sigma}^{1/2}\boldsymbol{\Gamma}_2'\boldsymbol{\Gamma}_2\boldsymbol{\Sigma}^{1/2}\\
&+\boldsymbol{\Sigma}^{1/2}\boldsymbol{\Gamma}_2'\boldsymbol{\Gamma}_2\boldsymbol{\Sigma}^{1/2}\otimes E[\boldsymbol{\Sigma}^{1/2}\boldsymbol{\Gamma}_1'(\mathbf{U}_1\mathbf{U}_1')^{-1}\boldsymbol{\Gamma}_1\boldsymbol{\Sigma}^{1/2}]\Big\}.
\end{aligned}
\qquad (4.2.79)
$$

In (4.2.79) it was used that

$$
(\mathbf{I}+\mathbf{K}_{p,p})(\boldsymbol{\Sigma}^{1/2}\boldsymbol{\Gamma}')^{\otimes 2}\mathbf{K}_{p,p}=(\mathbf{I}+\mathbf{K}_{p,p})(\boldsymbol{\Sigma}^{1/2}\boldsymbol{\Gamma}')^{\otimes 2}.
$$

Furthermore,

$$
(\mathbf{I}+\mathbf{K}_{p,p})\mathrm{vec}(E[\boldsymbol{\Sigma}^{1/2}\boldsymbol{\Gamma}_1'(\mathbf{U}_1\mathbf{U}_1')^{-1}\boldsymbol{\Gamma}_1\boldsymbol{\Sigma}^{1/2}])=2\mathrm{vec}(E[\boldsymbol{\Sigma}^{1/2}\boldsymbol{\Gamma}_1'(\mathbf{U}_1\mathbf{U}_1')^{-1}\boldsymbol{\Gamma}_1\boldsymbol{\Sigma}^{1/2}])
$$

and

$$
(\mathbf{I}+\mathbf{K}_{p,p})\mathrm{vec}(\boldsymbol{\Sigma}^{1/2}\boldsymbol{\Gamma}_2'\boldsymbol{\Gamma}_2\boldsymbol{\Sigma}^{1/2})=2\mathrm{vec}(\boldsymbol{\Sigma}^{1/2}\boldsymbol{\Gamma}_2'\boldsymbol{\Gamma}_2\boldsymbol{\Sigma}^{1/2}).
$$

Finally, applying (4.2.70) and (4.2.79) yields (ii) of the theorem, since

$$
\begin{aligned}
n^2d_2&=d_{21}+r(\mathbf{C})E[(\mathrm{tr}(\mathbf{U}_1\mathbf{U}_1')^{-1})^2]+r(\mathbf{C})E[\mathrm{tr}\{(\mathbf{U}_1\mathbf{U}_1')^{-1}(\mathbf{U}_1\mathbf{U}_1')^{-1}\}],\\
n^2d_3&=d_{31}+r(\mathbf{C})(1+1/\{p-r(\mathbf{A})\})E[\mathrm{tr}(\mathbf{U}_1\mathbf{U}_1')^{-1}],\\
n^2d_4&=d_{41}+2r(\mathbf{C})E[\mathrm{tr}\{(\mathbf{U}_1\mathbf{U}_1')^{-1}(\mathbf{U}_1\mathbf{U}_1')^{-1}\}],\\
d_{51}&+2r(\mathbf{C})/(n-r(\mathbf{C})-p+r(\mathbf{A})-1)=0.
\end{aligned}
$$

∎

A consequence of Theorem 4.2.3 (i) is that $\widehat{\boldsymbol{\Sigma}}$ is a biased estimator of $\boldsymbol{\Sigma}$. Usually, when working with linear models, it is possible to correct the bias through multiplication by a properly chosen constant. This is impossible in the Growth Curve model since $E[\widehat{\boldsymbol{\Sigma}}]$ is a function of the within-individuals design matrix $\mathbf{A}$. On the other hand, unbiased estimators exist and, for example, one is given by

$$\frac{1}{n-r(\mathbf{C})}\mathbf{S}. \tag{4.2.80}$$

The statistic (4.2.80) is often used as an estimator of $\boldsymbol{\Sigma}$ in multivariate analysis, in principal component analysis, canonical correlation analysis and different versions of factor analysis, for example. However, if a mean value is structured like in the Growth Curve model, i.e. $E[\mathbf{X}] = \mathbf{ABC}$, (4.2.80) is to some extent an unnatural estimator since $\boldsymbol{\Sigma}$ is the quadratic variation around the model $\mathbf{ABc}_i$, where $\mathbf{c}_i$ is the ith column of $\mathbf{C}$. The statistic (4.2.80) describes only the variation around the sample mean. The sample mean in turn varies around the estimated model, but this is not utilized when applying (4.2.80).
The use of (4.2.80) in the Growth Curve model is solely based on its unbiasedness. However, when the number of columns n in $\mathbf{C}$ is small, this argument must be taken with caution, since $\mathbf{S}$ is Wishart distributed and the Wishart distribution is a non-symmetric distribution.
In the standard MANOVA model, i.e. when $\mathbf{A} = \mathbf{I}$ in the Growth Curve model, the statistic (4.2.80) is used as an unbiased estimator of $\boldsymbol{\Sigma}$. However, it is known that (4.2.80) is inadmissible with respect to many reasonable loss functions. Furthermore, using a conditional argument in the MANOVA model, the likelihood approach leads to (4.2.80). Hence, besides unbiasedness, (4.2.80) has also other desirable qualities. Therefore, we will require that any unbiased estimator of the dispersion matrix in the Growth Curve model reduces to (4.2.80), if $\mathbf{A} = \mathbf{I}$.
The estimator $\widehat{\boldsymbol{\Sigma}}$, given by (4.2.38), combines the deviation between the observations and the sample mean, as well as the deviation between the sample mean and the estimated model. Hence, the maximum likelihood estimator uses two sources of information, whereas (4.2.80) uses only one. Intuitively the maximum likelihood estimator should be preferable, although one must remember that everything depends on the number of observations and the choice of design matrix $\mathbf{A}$.
As previously mentioned, the main drawback of $\widehat{\boldsymbol{\Sigma}}$ is that its distribution is not available in a simple form, and it seems difficult to master this problem. Furthermore, it follows by Theorem 4.2.3 (i) that $\widehat{\boldsymbol{\Sigma}}$ underestimates $\boldsymbol{\Sigma}$ on average. This can be overcome, and the next theorem presents an unbiased estimator, which is solely a function of $\widehat{\boldsymbol{\Sigma}}$ and reduces to (4.2.80), if $\mathbf{A} = \mathbf{I}$.

Theorem 4.2.4. *Let $\widehat{\boldsymbol{\Sigma}}$ be given by (4.2.38) and*

$$e_1 = r(\mathbf{C})\frac{n-r(\mathbf{C})-2p+2r(\mathbf{A})-1}{(n-r(\mathbf{C})-p+r(\mathbf{A})-1)(n-r(\mathbf{C})-p+r(\mathbf{A}))}.$$

Then $\widehat{\boldsymbol{\Sigma}} + e_1\mathbf{A}(\mathbf{A}'\widehat{\boldsymbol{\Sigma}}^{-1}\mathbf{A})^{-}\mathbf{A}'$ is an unbiased estimator of $\boldsymbol{\Sigma}$.

PROOF: Observe that $\mathbf{A}'\widehat{\boldsymbol{\Sigma}}^{-1} = n\mathbf{A}'\mathbf{S}^{-1}$, and according to Theorem 2.4.13 (i) and Theorem 2.4.14 (i),

$$E[\mathbf{A}(\mathbf{A}'\mathbf{S}^{-1}\mathbf{A})^{-}\mathbf{A}'] = (n - r(\mathbf{C}) - p + r(\mathbf{A}))\mathbf{A}(\mathbf{A}'\boldsymbol{\Sigma}^{-1}\mathbf{A})^{-}\mathbf{A}'.$$

Hence, from Theorem 4.2.3 it follows that

$$E[\widehat{\boldsymbol{\Sigma}} + e_1\mathbf{A}(\mathbf{A}'\widehat{\boldsymbol{\Sigma}}^{-1}\mathbf{A})^-\mathbf{A}']$$
$$=\boldsymbol{\Sigma} + \Big\{-r(\mathbf{C})\tfrac{1}{n}\tfrac{n-r(\mathbf{C})-2p+2r(\mathbf{A})-1}{n-r(\mathbf{C})-p+r(\mathbf{A})-1}$$
$$+r(\mathbf{C})\tfrac{n-r(\mathbf{C})-2p+2r(\mathbf{A})-1}{(n-r(\mathbf{C})-p+r(\mathbf{A})-1)(n-r(\mathbf{C})-p+r(\mathbf{A}))}\tfrac{1}{n}(n-r(\mathbf{C})-p+r(\mathbf{A}))\Big\}$$
$$\times\,\mathbf{A}(\mathbf{A}'\boldsymbol{\Sigma}^{-1}\mathbf{A})^-\mathbf{A}' = \boldsymbol{\Sigma}.$$

∎

Since there exist two unbiased estimators of $\boldsymbol{\Sigma}$, i.e. the one in Theorem 4.2.4 and another in (4.2.80), we have to decide which of them should be used. Intuitively, the estimator of Theorem 4.2.4 would be better since it uses two sources of variation. However, if the mean model $E[\mathbf{X}] = \mathbf{ABC}$ does not fit the data, (4.2.80) would be more natural to use. One way to compare these estimators is to study their dispersion matrices. It follows from Theorem 2.4.14 (ii) that

$$D[\frac{1}{n-r(\mathbf{C})}\mathbf{S}] = \frac{1}{n-r(\mathbf{C})}(\mathbf{I}+\mathbf{K}_{p,p})\boldsymbol{\Sigma}\otimes\boldsymbol{\Sigma}. \tag{4.2.81}$$

Moreover, since by Theorem 2.4.13 (iii),

$$\mathbf{I} - \mathbf{A}(\mathbf{A}'\mathbf{S}^{-1}\mathbf{A})^-\mathbf{A}'\mathbf{S}^{-1} = \mathbf{S}\mathbf{A}^o(\mathbf{A}^{o'}\mathbf{S}\mathbf{A}^o)^-\mathbf{A}^{o'}$$

is independent of $\mathbf{A}(\mathbf{A}'\mathbf{S}^{-1}\mathbf{A})^-\mathbf{A}'$, and by Theorem 2.2.4 (iii), $\mathbf{XC}'(\mathbf{CC}')^-\mathbf{CX}'$ is independent of $\mathbf{S}$:

$$D[\widehat{\boldsymbol{\Sigma}} + \frac{1}{n}e_1\mathbf{A}(\mathbf{A}'\mathbf{S}^{-1}\mathbf{A})^-\mathbf{A}'] = D[\widehat{\boldsymbol{\Sigma}}] + D[\frac{1}{n}e_1\mathbf{A}(\mathbf{A}'\mathbf{S}^{-1}\mathbf{A})^-\mathbf{A}']$$
$$+\,C[\frac{1}{n}\mathbf{S}, \frac{1}{n}e_1\mathbf{A}(\mathbf{A}'\mathbf{S}^{-1}\mathbf{A})^-\mathbf{A}'] + C[\frac{1}{n}e_1\mathbf{A}(\mathbf{A}'\mathbf{S}^{-1}\mathbf{A})^-\mathbf{A}', \frac{1}{n}\mathbf{S}].$$

However, Theorem 2.4.13 (iii) states that the matrices $\mathbf{A}(\mathbf{A}'\mathbf{S}^{-1}\mathbf{A})^-\mathbf{A}'$ and $\mathbf{S} - \mathbf{A}(\mathbf{A}'\mathbf{S}^{-1}\mathbf{A})^-\mathbf{A}'$ are independently distributed and thus the covariance is given by

$$C[\frac{1}{n}\mathbf{S}, \frac{1}{n}e_1\mathbf{A}(\mathbf{A}'\mathbf{S}^{-1}\mathbf{A})^-\mathbf{A}'] = C[\frac{1}{n}\mathbf{S} - \frac{1}{n}\mathbf{A}(\mathbf{A}'\mathbf{S}^{-1}\mathbf{A})^-\mathbf{A}', \frac{1}{n}e_1\mathbf{A}(\mathbf{A}'\mathbf{S}^{-1}\mathbf{A})^-\mathbf{A}']$$
$$+\,C[\frac{1}{n}\mathbf{A}(\mathbf{A}'\mathbf{S}^{-1}\mathbf{A})^-\mathbf{A}', \frac{1}{n}e_1\mathbf{A}(\mathbf{A}'\mathbf{S}^{-1}\mathbf{A})^-\mathbf{A}'] = \frac{1}{n^2}e_1D[\mathbf{A}(\mathbf{A}'\mathbf{S}^{-1}\mathbf{A})^-\mathbf{A}'].$$

Consequently,

$$D[\widehat{\boldsymbol{\Sigma}} + \frac{1}{n}e_1\mathbf{A}(\mathbf{A}'\mathbf{S}^{-1}\mathbf{A})^-\mathbf{A}'] = D[\widehat{\boldsymbol{\Sigma}}] + \frac{e_1^2+2e_1}{n^2}D[\mathbf{A}(\mathbf{A}'\mathbf{S}^{-1}\mathbf{A})^-\mathbf{A}']$$
$$= D[\widehat{\boldsymbol{\Sigma}}] + \frac{e_1^2+2e_1}{n^2}(n-r(\mathbf{C})-p+r(\mathbf{A}))(\mathbf{I}+\mathbf{K}_{p,p})(\mathbf{A}(\mathbf{A}'\boldsymbol{\Sigma}^{-1}\mathbf{A})^-\mathbf{A}')^{\otimes 2}, \tag{4.2.82}$$

where Theorem 2.4.14 (ii) has been used in the last equality. Hence, (4.2.81) should be compared to (4.2.82). Unfortunately, in order to illuminate the difference between (4.2.81) and (4.2.82), it seems difficult to discuss it without involving a particular matrix $\mathbf{A}$.

4.2.3 *Moments of estimators for the* MLNM($\mathbf{ABC}+\mathbf{B}_2\mathbf{C}_2$)

For the MLNM($\mathbf{ABC}+\mathbf{B}_2\mathbf{C}_2$), the estimators were presented in Theorem 4.1.9. Let us recall the estimators and the definitions of the involved matrices:

$$\mathbf{K}\widehat{\mathbf{B}}\mathbf{L}_1 = \mathbf{K}(\mathbf{A}'\mathbf{S}_1^{-1}\mathbf{A})^-\mathbf{A}'\mathbf{S}_1^{-1}\mathbf{Y}\mathbf{H}'(\mathbf{H}\mathbf{H}')^-\mathbf{L}_1, \tag{4.2.83}$$

where $C(\mathbf{K}') \subseteq C(\mathbf{A}')$, $C(\mathbf{L}_1) \subseteq C(\mathbf{H})$,

$$\begin{aligned}\mathbf{Y} &= \mathbf{X}(\mathbf{I}-\mathbf{C}_2'(\mathbf{C}_2\mathbf{C}_2')^-\mathbf{C}_2),\\ \mathbf{H} &= \mathbf{C}(\mathbf{I}-\mathbf{C}_2'(\mathbf{C}_2\mathbf{C}_2')^-\mathbf{C}_2),\\ \mathbf{S}_1 &= \mathbf{Y}(\mathbf{I}-\mathbf{H}'(\mathbf{H}\mathbf{H}')^-\mathbf{H})\mathbf{Y}',\end{aligned}$$

and

$$\widehat{\mathbf{B}}_2\mathbf{L}_2 = (\mathbf{X}-\mathbf{A}\widehat{\mathbf{B}}\mathbf{C})\mathbf{C}_2'(\mathbf{C}_2\mathbf{C}_2')^-\mathbf{L}_2, \tag{4.2.84}$$

where $C(\mathbf{L}_2) \subseteq C(\mathbf{C}_2(\mathbf{C}_1')^o)$, and

$$\begin{aligned}n\widehat{\boldsymbol\Sigma} &= (\mathbf{X}-\mathbf{A}\widehat{\mathbf{B}}\mathbf{C}-\widehat{\mathbf{B}}_2\mathbf{C}_2)(\mathbf{X}-\mathbf{A}\widehat{\mathbf{B}}_2\mathbf{C}-\widehat{\mathbf{B}}\mathbf{C}_2)\\ &= \mathbf{S}_1+\mathbf{S}_1\mathbf{A}^o(\mathbf{A}^{o'}\mathbf{S}_1\mathbf{A}^o)^-\mathbf{A}^{o'}\mathbf{X}\mathbf{H}'(\mathbf{H}\mathbf{H}')^-\mathbf{H}\mathbf{X}'\mathbf{A}^o(\mathbf{A}^{o'}\mathbf{S}_1\mathbf{A}^o)^-\mathbf{A}^{o'}\mathbf{S}_1.\end{aligned} \tag{4.2.85}$$

Observe that the estimators in (4.2.83) and (4.2.85) have a similar structure as the estimators for the Growth Curve model given by (4.1.7) and (4.1.8). Therefore, the techniques developed in the previous paragraph can be applied. The only difference is that in (4.2.83) and (4.2.85)

$$\begin{aligned}\mathbf{S}_1 &\sim \mathbf{W}_p(\boldsymbol\Sigma, n-r(\mathbf{C}':\mathbf{C}_2')),\\ \mathbf{A}^{o'}\mathbf{X}\mathbf{H}'(\mathbf{H}\mathbf{H}')^-\mathbf{H}\mathbf{X}'\mathbf{A}^o &\sim W_{p-r(\mathbf{A})}(\mathbf{A}^{o'}\boldsymbol\Sigma\mathbf{A}^o, r(\mathbf{C}':\mathbf{C}_2')-r(\mathbf{C}_2')),\end{aligned}$$

whereas in (4.1.7) and (4.1.8), $\mathbf{S}\sim W_p(\boldsymbol\Sigma, n-r(\mathbf{C}))$ and $\mathbf{A}^{o'}\mathbf{X}\mathbf{C}'(\mathbf{C}\mathbf{C}')^-\mathbf{C}\mathbf{X}'\mathbf{A}^o \sim W_{p-r(\mathbf{A})}(\mathbf{A}^{o'}\boldsymbol\Sigma\mathbf{A}^o, r(\mathbf{C}))$, i.e. there is a difference in the number of degrees of freedom. Furthermore, it follows from (4.2.84) that $\widehat{\mathbf{B}}_2\mathbf{L}_2$ is a linear function of $\mathbf{A}\widehat{\mathbf{B}}\mathbf{C}$. Therefore, the moments for $\widehat{\mathbf{B}}_2\mathbf{L}_2$ are fairly easy to obtain from the moments of $\mathbf{A}\widehat{\mathbf{B}}\mathbf{C}$. In the next theorem we list the first and second order moments.

Theorem 4.2.5. *For the* MLNM($\mathbf{ABC}+\mathbf{B}_2\mathbf{C}_2$), *let* $\mathbf{K}\widehat{\mathbf{B}}\mathbf{L}_1$, $\widehat{\mathbf{B}}_2\mathbf{L}_2$ *and* $\widehat{\boldsymbol\Sigma}$ *be given by (4.2.83), (4.2.84) and (4.2.85), respectively. Put*

$$\begin{aligned}c_1 &= \frac{n-r(\mathbf{C}':\mathbf{C}_2')-1}{n-r(\mathbf{C}':\mathbf{C}_2')-p+r(\mathbf{A})-1},\\ c_2 &= \frac{n}{n-r(\mathbf{C}':\mathbf{C}_2')-p+r(\mathbf{A})}c_1,\\ c_3 &= \frac{1}{n}\frac{n-r(\mathbf{C}':\mathbf{C}_2')-2(r(\mathbf{A})-p)-1}{n-r(\mathbf{C}':\mathbf{C}_2')-p+r(\mathbf{A})-1}r(\mathbf{C}':\mathbf{C}_2'),\\ c_4 &= \frac{n}{n-r(\mathbf{C}':\mathbf{C}_2')-p+r(\mathbf{A})}c_3.\end{aligned}$$

Then,

(i) $E[\mathbf{K}\widehat{\mathbf{B}}\mathbf{L}_1] = \mathbf{KBL}_1$;

(ii) $D[\mathbf{K}\widehat{\mathbf{B}}\mathbf{L}_1] = c_1\mathbf{L}_1'(\mathbf{HH}')^-\mathbf{L}_1 \otimes \mathbf{K}(\mathbf{A}'\mathbf{\Sigma}^{-1}\mathbf{A})^-\mathbf{K}'$;

(iii) $\widehat{D[\mathbf{K}\widehat{\mathbf{B}}\mathbf{L}_1]} = c_2\mathbf{L}_1'(\mathbf{HH}')^-\mathbf{L}_1 \otimes \mathbf{K}(\mathbf{A}'\widehat{\mathbf{\Sigma}}^{-1}\mathbf{A})^-\mathbf{K}'$

is an unbiased estimator of $D[\mathbf{K}\widehat{\mathbf{B}}\mathbf{L}_1]$;

(iv) $E[\widehat{\mathbf{B}}_2\mathbf{L}_2] = \mathbf{B}_2\mathbf{L}_2$;

(v) $$D[\widehat{\mathbf{B}}_2\mathbf{L}_2] = \mathbf{L}_2'(\mathbf{C}_2\mathbf{C}_2')^-\mathbf{L}_2 \otimes \mathbf{\Sigma} + c_1\mathbf{L}_2'(\mathbf{C}_2\mathbf{C}_2')^-\mathbf{C}_2\mathbf{C}'(\mathbf{HH}')^-\mathbf{CC}_2'(\mathbf{C}_2\mathbf{C}_2')^-\mathbf{L}_2 \otimes \mathbf{A}(\mathbf{A}'\mathbf{\Sigma}^{-1}\mathbf{A})^-\mathbf{A}';$$

(vi) $E[n\widehat{\mathbf{\Sigma}}] = \mathbf{\Sigma} - c_3\mathbf{A}(\mathbf{A}'\mathbf{\Sigma}^{-1}\mathbf{A})^-\mathbf{A}'$;

(vii) $$\widehat{D[\widehat{\mathbf{B}}_2\mathbf{L}_2]} = \tfrac{n}{n-r(\mathbf{C}_2)}\mathbf{L}_2'(\mathbf{C}_2\mathbf{C}_2')^-\mathbf{L}_2 \otimes \widehat{\mathbf{\Sigma}} + c_4\mathbf{L}_2'(\mathbf{C}_2\mathbf{C}_2')^-\mathbf{L}_2 \otimes \mathbf{A}(\mathbf{A}'\widehat{\mathbf{\Sigma}}^{-1}\mathbf{A})^-\mathbf{A}' + c_2\mathbf{L}_2'(\mathbf{C}_2\mathbf{C}_2')^-\mathbf{C}_2\mathbf{C}'(\mathbf{HH}')^-\mathbf{CC}_2'(\mathbf{C}_2\mathbf{C}_2')^-\mathbf{L}_2 \otimes \mathbf{A}(\mathbf{A}'\widehat{\mathbf{\Sigma}}^{-1}\mathbf{A})^-\mathbf{A}'$$

is an unbiased estimator of $D[\widehat{\mathbf{B}}_2\mathbf{L}_2]$;

(viii) $$\frac{1}{n-r(\mathbf{C}:\mathbf{C}_2')}\mathbf{S}_1 \quad \textit{and} \quad \widehat{\mathbf{\Sigma}} + c_4\mathbf{A}(\mathbf{A}'\widehat{\mathbf{\Sigma}}\mathbf{A})^-\mathbf{A}'$$

are both unbiased estimators of $\mathbf{\Sigma}$.

PROOF: By comparing Theorem 4.1.1 and Theorem 4.1.9 it follows that (i), (ii), (iii), (vi) and (viii) are established via Theorem 4.2.1 (iii) and (iv), Theorem 4.2.3 (i) and Theorem 4.2.4.
Moreover,

$$\begin{aligned} E[\widehat{\mathbf{B}}_2\mathbf{L}_2] &= E[(\mathbf{X} - \mathbf{A}\widehat{\mathbf{B}}\mathbf{C})]\mathbf{C}_2'(\mathbf{C}_2\mathbf{C}_2')^-\mathbf{L}_2 \\ &= (\mathbf{ABC} + \mathbf{B}_2\mathbf{C}_2 - \mathbf{ABC})\mathbf{C}_2'(\mathbf{C}_2\mathbf{C}_2')^-\mathbf{L}_2 = \mathbf{B}_2\mathbf{L}_2, \end{aligned}$$

and (iv) is verified.
Concerning (v), observe that $\mathbf{XC}_2'$ is independent of $\mathbf{S}_1$. Therefore, we will condition on $\mathbf{S}_1$ and obtain

$$D[\widehat{\mathbf{B}}_2\mathbf{L}_2] = E[D[\widehat{\mathbf{B}}_2\mathbf{L}_2|\mathbf{S}_1]] + D[E[\widehat{\mathbf{B}}_2\mathbf{L}_2|\mathbf{S}_1]] = E[D[\widehat{\mathbf{B}}_2\mathbf{L}_2|\mathbf{S}_1]],$$

since

$$\begin{aligned} E[\widehat{\mathbf{B}}_2\mathbf{L}_2|\mathbf{S}_1] &= \mathbf{ABCC}_2'(\mathbf{C}_2\mathbf{C}_2')^-\mathbf{L}_2 + \mathbf{B}_2\mathbf{L} - E[\mathbf{A}\widehat{\mathbf{B}}\mathbf{CC}_2'(\mathbf{C}_2\mathbf{C}_2')^-\mathbf{L}_2|\mathbf{S}_1] \\ &= \mathbf{ABCC}_2'(\mathbf{C}_2\mathbf{C}_2')^-\mathbf{L}_2 + \mathbf{B}_2\mathbf{L} - \mathbf{A}(\mathbf{A}'\mathbf{S}_1^{-1}\mathbf{A})^-\mathbf{A}'\mathbf{S}_1^{-1}\mathbf{ABCC}_2'(\mathbf{C}_2\mathbf{C}_2')^-\mathbf{L}_2 \\ &= \mathbf{B}_2\mathbf{L}_2. \end{aligned}$$

Now

$$E[D[\widehat{\mathbf{B}}_2\mathbf{L}_2|\mathbf{S}_1]] = D[\mathbf{X}\mathbf{C}_2'(\mathbf{C}_2\mathbf{C}_2')^-\mathbf{L}_2] + E[D[\mathbf{A}\widehat{\mathbf{B}}\mathbf{C}\mathbf{C}_2'(\mathbf{C}_2\mathbf{C}_2)^-\mathbf{L}_2|\mathbf{S}_1]]$$
$$= \mathbf{L}_2'(\mathbf{C}_2\mathbf{C}_2')^-\mathbf{L}_2 \otimes \mathbf{\Sigma} + D[\mathbf{A}\widehat{\mathbf{B}}\mathbf{C}\mathbf{C}_2'(\mathbf{C}_2\mathbf{C}_2')^-\mathbf{L}_2],$$

where in the last equality it has been used that

$$D[\mathbf{A}\widehat{\mathbf{B}}\mathbf{C}\mathbf{C}_2'(\mathbf{C}_2\mathbf{C}_2')^-\mathbf{L}_2] = E[D[\mathbf{A}\widehat{\mathbf{B}}\mathbf{C}\mathbf{C}_2'(\mathbf{C}_2\mathbf{C}_2')^-\mathbf{L}_2|\mathbf{S}_1]].$$

Finally, for proving (vii), we observe that

$$E[\mathbf{L}_2'(\mathbf{C}_2\mathbf{C}_2')^-\mathbf{L}_2 \otimes (\widehat{\mathbf{\Sigma}} + c_4\mathbf{A}(\mathbf{A}'\widehat{\mathbf{\Sigma}}^{-1}\mathbf{A})^-\mathbf{A}')] = \mathbf{L}_2'(\mathbf{C}_2\mathbf{C}_2')^-\mathbf{L}_2 \otimes \mathbf{\Sigma}$$

and

$$E[c_2\mathbf{A}(\mathbf{A}'\widehat{\mathbf{\Sigma}}^{-1}\mathbf{A})^-\mathbf{A}'] = \mathbf{A}(\mathbf{A}'\mathbf{\Sigma}^{-1}\mathbf{A})^-\mathbf{A}'.$$

∎

4.2.4 *Moments of estimators for the* MLNM($\sum_{i=1}^3 \mathbf{A}_i\mathbf{B}_i\mathbf{C}_i$)

Already for the ordinary Growth Curve model, i.e. MLNM($\mathbf{ABC}$), distributions of the maximum likelihood estimators are difficult to find. In Theorem 4.1.6 estimators for the MLNM($\sum_{i=1}^3 \mathbf{A}_i\mathbf{B}_i\mathbf{C}_i$) were given, and one can see that the expressions are stochastically much more complicated than the estimators for the MLNM($\mathbf{ABC}$). Therefore, approximations are needed and then at least the first and second order moments of estimators have to be obtained. Before studying $\widehat{\mathbf{B}}_i$, $i = 1,2,3$, the estimated mean structure $\widehat{E[\mathbf{X}]} = \sum_{i=1}^m \mathbf{A}_i\widehat{\mathbf{B}}_i\mathbf{C}_i$ and $\widehat{\mathbf{\Sigma}}$ in the general model MLNM($\sum_{i=1}^m \mathbf{A}_i\mathbf{B}_i\mathbf{C}_i$) will be treated. Thereafter we look into some details of calculating $D[\widehat{\mathbf{B}}_i]$, $i = 1,2,3$. The scheme of derivation and calculations is very similar to the ones presented for obtaining $D[\widehat{E[\mathbf{X}]}]$ and $E[\widehat{\mathbf{\Sigma}}]$ for the MLNM($\mathbf{ABC}$).

First of all we are going to show that in the MLNM($\sum_{i=1}^m \mathbf{A}_i\mathbf{B}_i\mathbf{C}_i$), under the uniqueness conditions presented in Theorem 4.1.13, the maximum likelihood estimators of $\mathbf{B}_i$ are unbiased, and then it follows that $\widehat{E[\mathbf{X}]} = \sum_{i=1}^m \mathbf{A}_i\widehat{\mathbf{B}}_i\mathbf{C}_i$ is also unbiased. In Theorem 4.1.7, the estimators $\widehat{\mathbf{B}}_r$, $r = 1,2,\ldots,m$, were presented. Since $\mathcal{C}(\mathbf{C}_j') \subseteq \mathcal{C}(\mathbf{C}_k')$, for $j \geq k$, it follows from Lemma 4.1.3 that $\mathbf{P}_r'\mathbf{S}_r^{-1}$ is independent of $\mathbf{XC}'_r$. Hence,

$$
\begin{aligned}
E[\widehat{\mathbf{B}}_r] &= E[(\mathbf{A}'_r\mathbf{P}'_r\mathbf{S}_r^{-1}\mathbf{P}_r\mathbf{A}_r)^{-1}\mathbf{A}'_r\mathbf{P}'_r\mathbf{S}_r^{-1}(E[\mathbf{X}] - \sum_{i=r+1}^{m}\mathbf{A}_i\widehat{\mathbf{B}}_i\mathbf{C}_i)]\mathbf{C}'_r(\mathbf{C}_r\mathbf{C}'_r)^{-1}\\
&= \mathbf{B}_r - E\Big[(\mathbf{A}'_r\mathbf{P}'_r\mathbf{S}_r^{-1}\mathbf{P}_r\mathbf{A}_r)^{-1}\mathbf{A}'_r\mathbf{P}'_r\mathbf{S}_r^{-1}\\
&\qquad \times\Big\{\sum_{i=r+1}^{m}\mathbf{A}_i\widehat{\mathbf{B}}_i\mathbf{C}_i - \sum_{i=r+1}^{m}\mathbf{A}_i\mathbf{B}_i\mathbf{C}_i\Big\}\Big]\mathbf{C}'_r(\mathbf{C}_r\mathbf{C}'_r)^{-1}\\
&= \mathbf{B}_r - E\Big[(\mathbf{A}'_r\mathbf{P}'_r\mathbf{S}_r^{-1}\mathbf{P}_r\mathbf{A}_r)^{-1}\mathbf{A}'_r\mathbf{P}'_r\mathbf{S}_r^{-1}\\
&\qquad \times\Big\{(\mathbf{I} - \mathbf{T}_{r+1})E[\mathbf{XC}'_{r+1}(\mathbf{C}_{r+1}\mathbf{C}'_{r+1})^{-}\mathbf{C}_{r+1}]\\
&\qquad\quad + \mathbf{T}_{r+1}\sum_{i=r+2}^{m}\mathbf{A}_i\widehat{\mathbf{B}}_i\mathbf{C}_i - \sum_{i=r+1}^{m}\mathbf{A}_i\mathbf{B}_i\mathbf{C}_i\Big\}\Big]\mathbf{C}'_r(\mathbf{C}_r\mathbf{C}'_r)^{-1}\\
&= \mathbf{B}_r - E\Big[(\mathbf{A}'_r\mathbf{P}'_r\mathbf{S}_r^{-1}\mathbf{P}_r\mathbf{A}_r)^{-1}\mathbf{A}'_r\mathbf{P}'_r\mathbf{S}_r^{-1}\mathbf{T}_{r+1}\\
&\qquad \times\Big\{\sum_{i=r+2}^{m}\mathbf{A}_i\widehat{\mathbf{B}}_i\mathbf{C}_i - \sum_{i=r+2}^{m}\mathbf{A}_i\mathbf{B}_i\mathbf{C}_i\Big\}\Big]\mathbf{C}'_r(\mathbf{C}_r\mathbf{C}'_r)^{-1}\\
&= \mathbf{B}_r - E\Big[(\mathbf{A}'_r\mathbf{P}'_r\mathbf{S}_r^{-1}\mathbf{P}_r\mathbf{A}_r)^{-1}\mathbf{A}'_r\mathbf{P}'_r\mathbf{S}_r^{-1}\mathbf{T}_{r+1}\\
&\qquad \times\Big\{(\mathbf{I} - \mathbf{T}_{r+2})E[\mathbf{XC}'_{r+2}(\mathbf{C}_{r+2}\mathbf{C}'_{r+2})^{-}\mathbf{C}_{r+2}]\\
&\quad + \sum_{i=r+2}^{m}\mathbf{A}_i\widehat{\mathbf{B}}_i\mathbf{C}_i - \mathbf{T}_{r+2}\sum_{i=r+2}^{m}\mathbf{A}_i\widehat{\mathbf{B}}_i\mathbf{C}_i - \sum_{i=r+2}^{m}\mathbf{A}_i\mathbf{B}_i\mathbf{C}_i\Big\}\Big]\mathbf{C}'_r(\mathbf{C}_r\mathbf{C}'_r)^{-1}\\
&= \mathbf{B}_r - E\Big[(\mathbf{A}'_r\mathbf{P}'_r\mathbf{S}_r^{-1}\mathbf{P}_r\mathbf{A}_r)^{-1}\mathbf{A}'_r\mathbf{P}'_r\mathbf{S}_r^{-1}\mathbf{T}_{r+1}\mathbf{T}_{r+2}\\
&\qquad\qquad \times\Big\{\sum_{i=r+3}^{m}\mathbf{A}_i\widehat{\mathbf{B}}_i\mathbf{C}_i - \sum_{i=r+3}^{m}\mathbf{A}_i\mathbf{B}_i\mathbf{C}_i\Big\}\Big]\mathbf{C}'_r(\mathbf{C}_r\mathbf{C}'_r)^{-1}\\
&= \ldots = \mathbf{B}_r.
\end{aligned}
$$

These calculations establish the following theorem.

Theorem 4.2.6. *The estimator $\widehat{\mathbf{B}}_i$ in Theorem 4.1.7 is an unbiased estimator under the uniqueness conditions in Theorem 4.1.13.* ∎

Most of the results in this paragraph will rest on Lemma 4.1.3, Lemma 4.1.4 and the following result.

Lemma 4.2.1. *Let $\mathbf{V}_{11}^r$ and $\underline{\mathbf{V}}_{11}^r$ be as in Lemma 4.1.4 (iv), and let $h(\bullet)$ be any measurable function of $(\underline{\mathbf{V}}_{11}^r)^{-1}$. Then*

$$E[(\mathbf{V}_{11}^{r-1})^{-1}h((\underline{\mathbf{V}}_{11}^r)^{-1})] = c_{r-1}E[(\underline{\mathbf{V}}_{11}^r)^{-1}h((\underline{\mathbf{V}}_{11}^r)^{-1})],$$

where

$$c_{r-1} = \frac{n - r(\mathbf{C}_r) - m_{r-1} - 1}{n - r(\mathbf{C}_{r-1}) - m_{r-1} - 1}, \tag{4.2.86}$$

$$m_r = p - r(\mathbf{A}_1 : \mathbf{A}_2 : \ldots : \mathbf{A}_r) + r(\mathbf{A}_1 : \mathbf{A}_2 : \ldots : \mathbf{A}_{r-1}). \tag{4.2.87}$$

PROOF: First Lemma 4.1.4 (iv), Definition 2.4.2 and Corollary 2.4.8.1 will be utilized. It follows that there exists a unique lower triangular matrix $\mathbf{T}$ such that

$$\begin{aligned}\underline{\mathbf{V}}_{11}^{r} &= \mathbf{T}\mathbf{T}',\\ \mathbf{V}_{11}^{r-1} &= \mathbf{T}\mathbf{F}\mathbf{T}',\end{aligned}$$

where $\mathbf{F} \sim M\beta_I(m_{r-1}, n - r(\mathbf{C}_{r-1}), r(\mathbf{C}_{r-1}) - r(\mathbf{C}_r))$ is independent of $\mathbf{T}$. The lemma is verified, if $E[\mathbf{F}^{-1}]$ is shown to equal to $c_{r-1}\mathbf{I}$. Since $\mathbf{F}^{-1} = \mathbf{I} + \mathbf{Z}$, where $\mathbf{Z} \sim M\beta_{II}(m_{r-1}, n - r(\mathbf{C}_{r-1}), r(\mathbf{C}_{r-1}) - r(\mathbf{C}_r))$, Theorem 2.4.15 (i) gives

$$E[\mathbf{F}^{-1}] = c_{r-1}\mathbf{I}.$$

∎

In Corollary 4.1.8.1, $\widehat{E[\mathbf{X}]} = \sum_{i=1}^{m} \mathbf{A}_i\widehat{\mathbf{B}}_i\mathbf{C}_i$ was presented as a sum of m dependent random variables, where each variable is of the same type as $\widehat{E[\mathbf{X}]}$ in the MLNM($\mathbf{ABC}$). This will be utilized now. We will show how to derive the dispersion matrix $D[\widehat{E[\mathbf{X}]}]$, where $\widehat{E[\mathbf{X}]}$ is given by Corollary 4.1.8.1. In the same way as unbiasedness of $\widehat{\mathbf{B}}_i$ was verified, it can be shown that $\sum_{i=1}^{m} \mathbf{A}_i\widehat{\mathbf{B}}_i\mathbf{C}_i$ is unbiased. The difference is that now we do not have to rely on uniqueness conditions, because $\sum_{i=1}^{m} \mathbf{A}_i\widehat{\mathbf{B}}_i\mathbf{C}_i$ is always unique. Thus, $E[\widehat{E[\mathbf{X}]}] = E[\mathbf{X}]$ and

$$D[\widehat{E[\mathbf{X}]}] = D[\widehat{E[\mathbf{X}]} - E[\mathbf{X}]] = D[\sum_{i=1}^{m}(\mathbf{I} - \mathbf{T}_i)(\mathbf{X} - E[\mathbf{X}])\mathbf{C}_i'(\mathbf{C}_i\mathbf{C}_i')^-\mathbf{C}_i].$$

Hence,

$$D[(\mathbf{I} - \mathbf{T}_r)(\mathbf{X} - E[\mathbf{X}])\mathbf{C}_r'(\mathbf{C}_r\mathbf{C}_r')^-\mathbf{C}_r]$$

and the covariance matrix

$$C[(\mathbf{I} - \mathbf{T}_r)(\mathbf{X} - E[\mathbf{X}])\mathbf{C}_r'(\mathbf{C}_r\mathbf{C}_r')^-\mathbf{C}_r, (\mathbf{I} - \mathbf{T}_s)(\mathbf{X} - E[\mathbf{X}])\mathbf{C}_s'(\mathbf{C}_s\mathbf{C}_s')^-\mathbf{C}_s]$$

will be considered for arbitrary r and s. By summing up all necessary dispersion and covariance matrices, $D[\widehat{E[\mathbf{X}]}]$ is obtained. The derivation, however, is left to the reader.

Theorem 4.2.7. *Let c_r and m_r be defined by (4.2.86) and (4.2.87), respectively. Put*

$$d_{i,r} = c_i c_{i+1} \times \cdots \times c_{r-1} e_r,$$

where

$$e_r = \frac{(p-m_r)(n-r(\mathbf{C}_r)-1)}{(n-r(\mathbf{C}_r)-m_{r-1}-1)(n-r(\mathbf{C}_r)-p+m_r-1)}.$$

Furthermore, define

$$f_r = \frac{n-r(\mathbf{C}_r)-1}{n-r(\mathbf{C}_r)-m_r-1}$$

and

$$\begin{aligned}\mathbf{K}_i =&\boldsymbol{\Sigma}\mathbf{G}_{i-1}(\mathbf{G}'_{i-1}\boldsymbol{\Sigma}\mathbf{G}_{i-1})^{-1}\mathbf{G}'_{i-1}\boldsymbol{\Sigma} - \boldsymbol{\Sigma}\mathbf{G}_i(\mathbf{G}'_i\boldsymbol{\Sigma}\mathbf{G}_i)^{-1}\mathbf{G}'_i\boldsymbol{\Sigma},\\ \mathbf{L}_i =&\boldsymbol{\Sigma}\mathbf{G}_{i-1}(\mathbf{G}'_{i-1}\boldsymbol{\Sigma}\mathbf{G}_{i-1})^{-1}\mathbf{G}'_{i-1}\mathbf{A}_i(\mathbf{A}'_i\mathbf{G}_{i-1}(\mathbf{G}'_{i-1}\boldsymbol{\Sigma}\mathbf{G}_{i-1})^{-1}\mathbf{G}'_{i-1}\mathbf{A}_i)^{-}\\ &\times \mathbf{A}'_i\mathbf{G}_{i-1}(\mathbf{G}'_{i-1}\boldsymbol{\Sigma}\mathbf{G}_{i-1})^{-1}\mathbf{G}'_{i-1}\boldsymbol{\Sigma}.\end{aligned}$$

Assume that e_r and f_r are finite and positive. Then, for the $\mathrm{MLNM}(\sum_{i=1}^m \mathbf{A}_i\mathbf{B}_i\mathbf{C}_i)$

$$D[(\mathbf{I}-\mathbf{T}_r)(\mathbf{X}-E[\mathbf{X}])\mathbf{C}'_r(\mathbf{C}_r\mathbf{C}'_r)^{-}\mathbf{C}_r] = \mathbf{C}'_r(\mathbf{C}_r\mathbf{C}'_r)^{-}\mathbf{C}_r \otimes (\sum_{i=1}^{r-1} d_{i,r}\mathbf{K}_i + f_r\mathbf{L}_r)$$

in the notation of Theorem 4.1.7 and Lemma 4.1.3.

PROOF: Let

$$\begin{aligned}&\mathbf{R}_{r-1}\\ &= \mathbf{A}_r(\mathbf{A}'_r\mathbf{G}_{r-1}(\mathbf{G}'_{r-1}\mathbf{W}_r\mathbf{G}_{r-1})^{-1}\mathbf{G}'_{r-1}\mathbf{A}'_r)^{-}\mathbf{A}'_r\mathbf{G}_{r-1}(\mathbf{G}'_{r-1}\mathbf{W}_r\mathbf{G}_{r-1})^{-1}\mathbf{G}'_{r-1}.\end{aligned} \tag{4.2.88}$$

Since

$$\begin{aligned}&D[(\mathbf{I}-\mathbf{T}_r)(\mathbf{X}-E[\mathbf{X}])\mathbf{C}'_r(\mathbf{C}_r\mathbf{C}'_r)^{-}\mathbf{C}_r]\\ &\quad = \mathbf{C}'_r(\mathbf{C}_r\mathbf{C}'_r)^{-}\mathbf{C}_r \otimes E[(\mathbf{I}-\mathbf{T}_r)\boldsymbol{\Sigma}(\mathbf{I}-\mathbf{T}_r)'],\end{aligned} \tag{4.2.89}$$

$E[(\mathbf{I}-\mathbf{T}_r)\boldsymbol{\Sigma}(\mathbf{I}-\mathbf{T}_r)']$ has to be considered. When utilizing the results and notation of Lemma 4.1.4 and Lemma 4.2.1, it follows that

$$\begin{aligned}&E[(\mathbf{I}-\mathbf{T}_r)\boldsymbol{\Sigma}(\mathbf{I}-\mathbf{T}_r)']\\ &\quad =E[\mathbf{Z}_{1,r-1}\mathbf{R}_{r-1}\boldsymbol{\Sigma}\mathbf{R}'_{r-1}\mathbf{Z}'_{1,r-1}]\\ &\quad =\boldsymbol{\Sigma}^{1/2}(\boldsymbol{\Gamma}_1)'E[\mathbf{N}_1\mathbf{Z}_{2,r-1}\mathbf{R}_{r-1}\boldsymbol{\Sigma}\mathbf{R}'_{r-1}\mathbf{Z}'_{2,r-1}\mathbf{N}'_1]\boldsymbol{\Gamma}_1\boldsymbol{\Sigma}^{1/2}\\ &\quad =\boldsymbol{\Sigma}^{1/2}(\boldsymbol{\Gamma}_1^1)'E[\mathbf{Z}_{2,r-1}\mathbf{R}_{r-1}\boldsymbol{\Sigma}\mathbf{R}'_{r-1}\mathbf{Z}'_{2,r-1}]\boldsymbol{\Gamma}_1^1\boldsymbol{\Sigma}^{1/2}\\ &\qquad + E[\mathrm{tr}\{(\mathbf{V}_{11}^1)^{-1}\mathbf{Z}_{2,r-1}\mathbf{R}_{r-1}\boldsymbol{\Sigma}\mathbf{R}'_{r-1}\mathbf{Z}'_{2,r-1}\}]\boldsymbol{\Sigma}^{1/2}(\boldsymbol{\Gamma}_2^1)'\boldsymbol{\Gamma}_2^1\boldsymbol{\Sigma}^{1/2}.\end{aligned} \tag{4.2.90}$$

For the last equality in (4.2.90) it has been utilized that $\mathbf{V}_{12}^1 = \mathbf{U}_1\mathbf{U}'_2$ for some normally distributed $\mathbf{U}' = (\mathbf{U}'_1 : \mathbf{U}'_2)$ where $\mathbf{U}_2$ is uncorrelated with $\mathbf{U}_1$, $\mathbf{Z}_{2,r-1}$ and $\mathbf{R}_{r-1}$. It is worth observing that (4.2.90) is a recurrence relation. Now the expectation of the trace function on the right-hand side of (4.2.90) will be calculated. Note that $\mathbf{R}_{r-1}$ is a function of $\mathbf{G}'_{r-1}\mathbf{W}_r\mathbf{G}_{r-1}$:

$$\mathbf{R}_{r-1} = f(\mathbf{G}'_{r-1}\mathbf{W}_r\mathbf{G}_{r-1}),$$

and with obvious notation,

$$\mathbf{Z}_{2,r-1} = f(\mathbf{G}'_1\mathbf{W}_2\mathbf{G}_1, \mathbf{G}'_2\mathbf{W}_3\mathbf{G}_2, \ldots, \mathbf{G}'_{r-1}\mathbf{W}_r\mathbf{G}_{r-1}).$$

However,

$$\mathbf{G}'_i\mathbf{W}_{i+1}\mathbf{G}_i = \mathbf{G}_i\mathbf{W}_2\mathbf{G}_i + \mathbf{G}'_i\mathbf{X}(\mathbf{P}_{\mathbf{C}'_2} - \mathbf{P}_{\mathbf{C}'_{i+1}})\mathbf{X}'\mathbf{G}_i.$$

Thus, according to Lemma 4.1.4, both $\mathbf{R}_{r-1}$ and $\mathbf{Z}_{2,r-1}$ are functions of $\underline{\mathbf{V}}^2_{11}$ and random quantities independent of $\mathbf{V}^1_{11}$ and $\underline{\mathbf{V}}^2_{11}$. This means that we can apply Lemma 4.2.1 and obtain

$$\begin{aligned}
&E[\mathrm{tr}\{(\mathbf{V}^1_{11})^{-1}\mathbf{Z}_{2,r-1}\mathbf{R}_{r-1}\boldsymbol{\Sigma}\mathbf{R}'_{r-1}\mathbf{Z}'_{2,r-1}\}] \\
&\quad = c_1E[\mathrm{tr}\{(\underline{\mathbf{V}}^2_{11})^{-1}\underline{\mathbf{V}}^2_{11}(\boldsymbol{\Gamma}^2_1)'(\mathbf{V}^2_{11})^{-1}\mathbf{Z}_{3,r-1}\mathbf{R}_{r-1}\boldsymbol{\Sigma}\mathbf{R}'_{r-1}\mathbf{Z}'_{3,r-1}(\mathbf{V}^2_{11})^{-1}\boldsymbol{\Gamma}^2_1\underline{\mathbf{V}}^2_{11}\}] \\
&\quad = c_1E[\mathrm{tr}\{(\mathbf{V}^2_{11})^{-1}\mathbf{Z}_{3,r-1}\mathbf{R}_{r-1}\boldsymbol{\Sigma}\mathbf{R}'_{r-1}\mathbf{Z}'_{3,r-1}\}], \qquad (4.2.91)
\end{aligned}$$

which is a recurrence relation. Therefore, when continuing in the same manner,

$$\begin{aligned}
&E[\mathrm{tr}\{(\mathbf{V}^1_{11})^{-1}\mathbf{Z}_{2,r-1}\mathbf{R}_{r-1}\boldsymbol{\Sigma}\mathbf{R}'_{r-1}\mathbf{Z}'_{2,r-1}\}] \\
&\quad = c_1c_2 \times \cdots \times c_{r-1}E[\mathrm{tr}\{(\underline{\mathbf{V}}^r_{11})^{-1}\mathbf{E}_r(\mathbf{E}'_r(\underline{\mathbf{V}}^r_{11})^{-1}\mathbf{E}_r)^-\mathbf{E}'_r(\underline{\mathbf{V}}^r_{11})^{-1}\}] = d_{1,r}, \qquad (4.2.92)
\end{aligned}$$

where

$$\mathbf{E}_r = \boldsymbol{\Gamma}^{r-1}_1\boldsymbol{\Gamma}^{r-2}_1 \times \cdots \times \boldsymbol{\Gamma}^1_1\boldsymbol{\Sigma}^{-1/2}\mathbf{A}_r. \qquad (4.2.93)$$

For the last equality in (4.2.92) observe that we have to calculate $E[\mathrm{tr}\{(\underline{\mathbf{V}}^r_{11})^{-1}\} - \mathrm{tr}\{(\mathbf{V}^r_{11})^{-1}\}]$ (see also Problem 2 in §2.4.9). Finally, we return to (4.2.90) and then, utilizing that it is a recurrence relation together with Lemma 4.1.4 (v) and (vi) and Lemma 4.2.1, we get

$$\begin{aligned}
&E[(\mathbf{I}-\mathbf{T}_r)\boldsymbol{\Sigma}(\mathbf{I}-\mathbf{T}_r)'] \\
&\quad = \sum_{i=1}^{r-1} d_{i,r}\mathbf{K}_i + \boldsymbol{\Sigma}^{1/2}(\boldsymbol{\Gamma}^1_1)' \times \cdots \times (\boldsymbol{\Gamma}^{r-2}_1)'(\boldsymbol{\Gamma}^{r-1}_1)'E\big[\mathbf{E}_r(\mathbf{E}'_r(\underline{\mathbf{V}}^r_{11})^{-1}\mathbf{E}_r)^- \\
&\qquad \times \mathbf{E}'_r(\underline{\mathbf{V}}^r_{11})^{-1}(\underline{\mathbf{V}}^r_{11})^{-1}\mathbf{E}_r(\mathbf{E}'_r(\underline{\mathbf{V}}^r_{11})^{-1}\mathbf{E}_r)^-\mathbf{E}'_r\big]\boldsymbol{\Gamma}^{r-1}_1\boldsymbol{\Gamma}^{r-2}_1 \times \cdots \times \boldsymbol{\Gamma}^1_1\boldsymbol{\Sigma}^{1/2} \\
&\quad = \sum_{i=1}^{r-1} d_{i,r}\mathbf{K}_i + f_r\mathbf{L}_r. \qquad (4.2.94)
\end{aligned}$$

∎

Theorem 4.2.8. *Let $r < s$ and let $\mathbf{K}_i$ and $\mathbf{L}_i$ be defined as in Theorem 4.2.7. Then, using the notation of Theorem 4.1.7 for the* MLNM$(\sum_{i=1}^m \mathbf{A}_i\mathbf{B}_i\mathbf{C}_i)$, *when e_r is finite and positive,*

$$C[(\mathbf{I}-\mathbf{T}_r)(\mathbf{X}-E[\mathbf{X}])\mathbf{C}_r'(\mathbf{C}_r\mathbf{C}_r')^-\mathbf{C}_r, (\mathbf{I}-\mathbf{T}_s)(\mathbf{X}-E[\mathbf{X}])\mathbf{C}_s'(\mathbf{C}_s\mathbf{C}_s')^-\mathbf{C}_s]$$
$$= \begin{cases} -\mathbf{C}_s'(\mathbf{C}_s\mathbf{C}_s')^-\mathbf{C}_s \otimes d_{r,s}\mathbf{K}_r, & \text{if } r > 1; \\ \mathbf{C}_s'(\mathbf{C}_s\mathbf{C}_s')^-\mathbf{C}_s \otimes (d_{1,s}\mathbf{K}_1 + \mathbf{L}_s), & \text{if } r = 1. \end{cases}$$

PROOF: In order to simplify the presentation we once again use the notation $\mathbf{P}_{\mathbf{C}_j'} = \mathbf{C}_j'(\mathbf{C}_j\mathbf{C}_j')^-\mathbf{C}_j$. The first observation is that $(\mathbf{X}-E[\mathbf{X}])\mathbf{P}_{\mathbf{C}_r'}$ can be written

$$(\mathbf{X}-E[\mathbf{X}])(\mathbf{P}_{\mathbf{C}_s'} + \mathbf{P}_{\mathbf{C}_r'} - \mathbf{P}_{\mathbf{C}_s'}),$$

and notice that $(\mathbf{X}-E[\mathbf{X}])\mathbf{P}_{\mathbf{C}_s'}$ is independent of $\mathbf{I}-\mathbf{T}_r$, $\mathbf{I}-\mathbf{T}_s$ and $(\mathbf{X}-E[\mathbf{X}])(\mathbf{P}_{\mathbf{C}_r'} - \mathbf{P}_{\mathbf{C}_s'})$. Hence,

$$\begin{aligned} C[(\mathbf{I}-&\mathbf{T}_r)(\mathbf{X}-E[\mathbf{X}])\mathbf{P}_{\mathbf{C}_r'}, (\mathbf{I}-\mathbf{T}_s)(\mathbf{X}-E[\mathbf{X}])\mathbf{P}_{\mathbf{C}_s'}] \\ &= \mathbf{P}_{\mathbf{C}_s'} \otimes E[(\mathbf{I}-\mathbf{T}_r)\mathbf{\Sigma}(\mathbf{I}-\mathbf{T}_s')]. \end{aligned}$$

From now on we will discuss $E[(\mathbf{I}-\mathbf{T}_r)\mathbf{\Sigma}(\mathbf{I}-\mathbf{T}_s')]$ in a manner similar to the case when $r = s$, and it turns out that when $r < s$, $r > 1$, there exists a recurrence relation which is somewhat easier to handle than when $r = s$. Now we can show that

$$\begin{aligned} E[(\mathbf{I}-&\mathbf{T}_r)\mathbf{\Sigma}(\mathbf{I}-\mathbf{T}_s')] \\ &= E[\mathbf{Z}_{1,r-1}\mathbf{R}_{r-1}\mathbf{\Sigma}\mathbf{R}_{s-1}'\mathbf{Z}_{1,s-1}'] \\ &= \mathbf{\Sigma}^{1/2}(\mathbf{\Gamma}_1^1)'E[\mathbf{Z}_{2,r-1}\mathbf{R}_{r-1}\mathbf{\Sigma}\mathbf{R}_{s-1}'\mathbf{Z}_{2,s-1}']\mathbf{\Gamma}_1^1\mathbf{\Sigma}^{1/2} \\ &= \ldots = \mathbf{\Sigma}^{1/2}(\mathbf{\Gamma}_1^1)' \times \cdots \times (\mathbf{\Gamma}_1^{r-2})'(\mathbf{\Gamma}_1^{r-1})'E\big[\mathbf{\Gamma}_1^{r-1}\mathbf{\Gamma}_1^{r-2} \\ &\quad \times \cdots \times \mathbf{\Gamma}_1^1\mathbf{\Sigma}^{-1/2}\mathbf{R}_{r-1}\mathbf{\Sigma}\mathbf{R}_{s-1}'\mathbf{Z}_{r,s-1}'\big]\mathbf{\Gamma}_1^{r-1}\mathbf{\Gamma}_1^{r-2} \times \cdots \times \mathbf{\Gamma}_1^1\mathbf{\Sigma}^{1/2}, \end{aligned} \tag{4.2.95}$$

and this is true if

$$E[\mathrm{tr}\{(\mathbf{V}_{11}^k)^{-1}\mathbf{Z}_{k+1,r-1}\mathbf{R}_{r-1}\mathbf{\Sigma}\mathbf{R}_{s-1}'\mathbf{Z}_{k+1,s-1}'\}] = 0, \quad k+1 < r < s. \tag{4.2.96}$$

To verify (4.2.96), some calculations are performed, and the expectation on the left hand side equals:

$$\begin{aligned} c_kc_{k+1} \times \cdots \times c_{r-2}&E[\mathrm{tr}\{(\mathbf{V}_{11}^{r-1})^{-1}\mathbf{\Gamma}_1^{r-1}\mathbf{\Gamma}_1^{r-2} \\ &\times \cdots \times \mathbf{\Gamma}_1^1\mathbf{\Sigma}^{-1/2}(\mathbf{I}-\mathbf{M}_r)\mathbf{\Sigma}\mathbf{R}_{s-1}'\mathbf{Z}_{r,s-1}'\}] \\ =c_kc_{k+1} \times \cdots \times c_{r-1}&E[\mathrm{tr}\{(\mathbf{V}_{11}^{r})^{-1}\mathbf{\Gamma}_1^{r}\mathbf{\Gamma}_1^{r-1} \\ &\times \cdots \times \mathbf{\Gamma}_1^1\mathbf{\Sigma}^{-1/2}(\mathbf{I}-\mathbf{I})\mathbf{\Sigma}\mathbf{R}_{s-1}'\mathbf{Z}_{r+1,s-1}'\}] = 0. \end{aligned}$$

When continuing, we obtain that the right hand side of (4.2.95) equals

$$\begin{aligned} \mathbf{\Sigma}^{1/2}(\mathbf{\Gamma}_1^1)' \times \cdots \times &(\mathbf{\Gamma}_1^{r-2})'(\mathbf{\Gamma}_1^{r-1})'E[\mathbf{\Gamma}_1^{r-1}\mathbf{\Gamma}_1^{r-2} \\ &\times \cdots \times \mathbf{\Gamma}_1^1\mathbf{\Sigma}^{-1/2}(\mathbf{I}-\mathbf{M}_r)\mathbf{\Sigma}\mathbf{R}_{s-1}'\mathbf{Z}_{r,s-1}']\mathbf{\Gamma}_1^{r-1}\mathbf{\Gamma}_1^{r-2} \times \cdots \times \mathbf{\Gamma}_1^1\mathbf{\Sigma}^{1/2}. \end{aligned}$$

After some additional calculations, this expression can be shown to be identical to

$$
\begin{aligned}
-E[&\mathrm{tr}\{(\mathbf{V}_{11}^r)^{-1}\boldsymbol{\Gamma}_1^r\boldsymbol{\Gamma}_1^{r-1}\times\cdots\times\boldsymbol{\Gamma}_1^1\boldsymbol{\Sigma}^{-1/2}\boldsymbol{\Sigma}\mathbf{R}_{s-1}'\mathbf{Z}_{r+1,s-1}'\}]\boldsymbol{\Sigma}^{1/2}(\boldsymbol{\Gamma}_1^1)'\\
&\times\cdots\times(\boldsymbol{\Gamma}_1^{r-1})'(\boldsymbol{\Gamma}_2^r)'\boldsymbol{\Gamma}_2^r\boldsymbol{\Gamma}_1^{r-1}\times\cdots\times\boldsymbol{\Gamma}_1^1\boldsymbol{\Sigma}^{1/2}\\
=&-c_rc_{r+1}\times\cdots\times c_{s-1}E[\mathrm{tr}\{(\underline{\mathbf{V}}_{11}^s)^{-1}\mathbf{E}_s(\mathbf{E}_s'(\underline{\mathbf{V}}_{11}^s)^{-1}\mathbf{E}_s)^-\mathbf{E}_s'(\underline{\mathbf{V}}_{11}^s)^{-1}\}]\\
&\times\boldsymbol{\Sigma}^{1/2}(\boldsymbol{\Gamma}_1^1)'\times\cdots\times(\boldsymbol{\Gamma}_1^{r-1})'(\boldsymbol{\Gamma}_2^r)'\boldsymbol{\Gamma}_2^r\boldsymbol{\Gamma}_1^{r-1}\times\cdots\times\boldsymbol{\Gamma}_1^1\boldsymbol{\Sigma}^{1/2}\\
=&-d_{r,s}\mathbf{K}_r,
\end{aligned}
$$

where $\mathbf{E}_s$ is given in (4.2.93).
It remains to examine the case $r=1$:

$$
\begin{aligned}
E[(\mathbf{I}-\mathbf{T}_1)\boldsymbol{\Sigma}(\mathbf{I}-\mathbf{T}_s)] &= E[\mathbf{M}_1\boldsymbol{\Sigma}\mathbf{R}_{s-1}'\mathbf{Z}_{1,s-1}']\\
&=\boldsymbol{\Sigma}^{1/2}(\boldsymbol{\Gamma}_1^1)'E[\boldsymbol{\Gamma}_1^1\boldsymbol{\Sigma}^{-1/2}\boldsymbol{\Sigma}\mathbf{R}_{s-1}'\mathbf{Z}_{2,s-1}']\boldsymbol{\Gamma}_1^1\boldsymbol{\Sigma}^{1/2}\\
&\quad+E[\mathrm{tr}\{(\mathbf{V}_{11}^1)^{-1}\boldsymbol{\Gamma}_1^1\boldsymbol{\Sigma}^{-1/2}\boldsymbol{\Sigma}\mathbf{R}_{s-1}'\mathbf{Z}_{2,s-1}'\}]\boldsymbol{\Sigma}^{1/2}(\boldsymbol{\Gamma}_2^1)'\boldsymbol{\Gamma}_2^1\boldsymbol{\Sigma}^{1/2}\\
&=\boldsymbol{\Sigma}^{1/2}(\boldsymbol{\Gamma}_1^1)'\boldsymbol{\Gamma}_1^1\boldsymbol{\Sigma}^{-1/2}\boldsymbol{\Sigma}E\big[\boldsymbol{\Sigma}^{-1/2}(\boldsymbol{\Gamma}_1^1)'\\
&\quad\times\cdots\times(\boldsymbol{\Gamma}_1^{s-1})'(\underline{\mathbf{V}}_{11}^{s-1})^{-1}\mathbf{E}_s(\mathbf{E}_s'(\underline{\mathbf{V}}_{11}^{s-1})^{-1}\mathbf{E}_s)^-\mathbf{E}_s'\big]\boldsymbol{\Gamma}_1^{s-1}\\
&\quad\times\cdots\times\boldsymbol{\Gamma}_1^1\boldsymbol{\Sigma}^{1/2}+d_{1,s}\mathbf{K}_1=\mathbf{L}_s+d_{1,s}\mathbf{K}_1,
\end{aligned}
$$

where the last identity follows from Problem 1 in §2.4.9. ∎

Now the expectation of $\widehat{\boldsymbol{\Sigma}}$ will be considered, where $\widehat{\boldsymbol{\Sigma}}$ is given in Theorem 4.1.7. The necessary calculations are similar but easier to carry out than those presented above, because the matrix $\mathbf{R}_i$ is not included in the expressions given below.

Theorem 4.2.9. *Let $\mathbf{K}_i$, $i=1,2,\ldots,m$, be defined in Theorem 4.2.7, let c_r and m_j be defined in Lemma 4.2.1, $\mathbf{C}_0=\mathbf{I}_n$, and put*

$$
\begin{aligned}
g_{i,j}&=c_ic_{i+1}\times\cdots\times c_{j-1}m_j/(n-r(\mathbf{C}_j)-m_j-1), \qquad i<j,\\
g_{j,j}&=m_j/(n-r(\mathbf{C}_j)-m_j-1)
\end{aligned}
$$

which are supposed to be finite and positive. Then, for $\widehat{\boldsymbol{\Sigma}}$ given in Theorem 4.1.7,

$$
\begin{aligned}
E[n\widehat{\boldsymbol{\Sigma}}]&=\sum_{j=1}^{m}(r(\mathbf{C}_{j-1})-r(\mathbf{C}_j))\\
&\quad\times\left(\sum_{i=1}^{j-1}g_{i,j-1}\mathbf{K}_i+\boldsymbol{\Sigma}\mathbf{G}_{j-1}(\mathbf{G}_{j-1}'\boldsymbol{\Sigma}\mathbf{G}_{j-1})^{-1}\mathbf{G}_{j-1}'\boldsymbol{\Sigma}\right)\\
&\quad+r(\mathbf{C}_m)\left(\sum_{i=1}^{m}g_{i,m}\mathbf{K}_i+\boldsymbol{\Sigma}\mathbf{G}_{r-1}(\mathbf{G}_{r-1}'\boldsymbol{\Sigma}\mathbf{G}_{r-1})^{-1}\mathbf{G}_{r-1}'\boldsymbol{\Sigma}\right),
\end{aligned}
$$

where $\sum_{i=1}^{0}g_{i,j-1}\mathbf{K}_i=\mathbf{0}$.

PROOF: From Theorem 4.1.7 it follows that the expectation of $\mathbf{P}_j\mathbf{X}\mathbf{F}_j\mathbf{F}_j'\mathbf{X}'\mathbf{P}_j'$, $j=1,2,\ldots,m$, and $\mathbf{P}_{m+1}\mathbf{X}\mathbf{P}_{\mathbf{C}_m'}\mathbf{X}'\mathbf{P}_{m+1}'$ are needed, where $\mathbf{F}_j=\mathbf{P}_{\mathbf{C}_{j-1}'}(\mathbf{I}-\mathbf{P}_{\mathbf{C}_j'})$.

Since $\mathbf{XF}_j\mathbf{F}_j'\mathbf{X}'$ as well as $\mathbf{XP}_{\mathbf{C}_m'}\mathbf{X}'$ are independent of $\mathbf{P}_j$ and $\mathbf{P}_{m+1}$, respectively, and

$$E[\mathbf{XF}_j\mathbf{F}_j'\mathbf{X}'] =(r(\mathbf{C}_{j-1}) - r(\mathbf{C}_j))\mathbf{\Sigma},$$
$$E[\mathbf{XP}_{\mathbf{C}_m'}\mathbf{X}'] =r(\mathbf{C}_m)\mathbf{\Sigma},$$

the theorem is verified, since the following lemma is valid. ∎

Lemma 4.2.2. *In the notation of Theorem 4.2.9*

$$E[\mathbf{P}_r\mathbf{\Sigma}\mathbf{P}_r'] = \sum_{i=1}^{r-1} g_{i,r-1}\mathbf{K}_i + \mathbf{\Sigma}\mathbf{G}_{r-1}(\mathbf{G}_{r-1}'\mathbf{\Sigma}\mathbf{G}_{r-1})^{-1}\mathbf{G}_{r-1}'\mathbf{\Sigma}, \quad r = 1,2,\ldots,m+1,$$

where $\sum_{i=1}^{0} g_{i,r-1}\mathbf{K}_i = \mathbf{0}$.

PROOF: From Lemma 4.1.4 it follows that

$$\mathbf{P}_r\mathbf{\Sigma}\mathbf{P}_r' = \mathbf{Z}_{1,r-1}\mathbf{\Sigma}\mathbf{Z}_{1,r-1}'$$

and then the calculations, started at (4.2.90), may be repeated with the exception that we do not have to take care of $\mathbf{R}_{r-1}$ in (4.2.90). ∎

As shown in Theorem 4.2.9, the estimator $\widehat{\mathbf{\Sigma}}$ is not an unbiased estimator. This is by no means unexpected because this has already been observed when $m = 1$ in the MLNM($\sum_{i=1}^{m} \mathbf{A}_i\mathbf{B}_i\mathbf{C}_i$), i.e. the Growth Curve model. The next lemma will serve as a basis for obtaining unbiased estimators of $\mathbf{\Sigma}$.

Lemma 4.2.3. *Let*

$$k_{i,r} = n(g_{i+1,r-1} - g_{i,r-1})/(n - r(\mathbf{C}_i) - m_i),$$

where $g_{i,r-1}$, $i = 1,2,\ldots,r-1$, *is defined in Theorem 4.2.9,* $g_{r,r-1} = 1$, *and suppose that* $k_{i,r}$ *are positive and finite. Then*

$$E[\mathbf{P}_r\mathbf{\Sigma}\mathbf{P}_r'] + E[\sum_{i=1}^{r-1} k_{i,r}\mathbf{A}_i(\mathbf{A}_i'\mathbf{P}_i'\widehat{\mathbf{\Sigma}}^{-1}\mathbf{P}_i\mathbf{A}_i)^{-}\mathbf{A}_i'] = \mathbf{\Sigma},$$

where $\sum_{i=1}^{0} k_{i,r}\mathbf{A}_i(\mathbf{A}_i'\mathbf{P}_i'\widehat{\mathbf{\Sigma}}^{-1}\mathbf{P}_i\mathbf{A}_i)^{-}\mathbf{A}_i' = \mathbf{0}$.

PROOF: First we are going to show that

$$\mathbf{A}_i'\mathbf{P}_i'\mathbf{S}_i^{-1}\mathbf{P}_{r+1} = \mathbf{0}, \quad r = i, i+1, \ldots, m, \tag{4.2.97}$$

holds. If $r = i$, we have

$$\mathbf{A}_i'\mathbf{P}_i'\mathbf{S}_i^{-1}\mathbf{P}_{i+1} = \mathbf{A}_i'\mathbf{P}_i'\mathbf{S}_i^{-1}\mathbf{T}_i\mathbf{P}_i = \mathbf{0},$$

since $\mathbf{A}_i'\mathbf{P}_i'\mathbf{S}_i^{-1}\mathbf{T}_i = \mathbf{0}$. Then we may consider

$$\mathbf{A}_i'\mathbf{P}_i'\mathbf{S}_i^{-1}\mathbf{P}_{i+2} = \mathbf{A}_i'\mathbf{P}_i'\mathbf{S}_i^{-1}\mathbf{T}_{i+1}\mathbf{P}_{i+1} = \mathbf{A}_i'\mathbf{P}_i'\mathbf{S}_i^{-1}\mathbf{P}_{i+1} = \mathbf{0}.$$

Continuing in the same manner we establish (4.2.97). Thus, from Theorem 4.1.7 it follows that

$$\mathbf{A}_i'\mathbf{P}_i'\widehat{\boldsymbol{\Sigma}}^{-1} = n\mathbf{A}_i'\mathbf{P}_i'\mathbf{S}_i^{-1},$$

and with the help of Lemma 4.1.3 and some calculations it implies that

$$\begin{aligned} &E[k_{ir}\mathbf{A}_i(\mathbf{A}_i'\mathbf{P}_i'\widehat{\boldsymbol{\Sigma}}^{-1}\mathbf{P}_i\mathbf{A}_i)^-\mathbf{A}_i'] \\ &\qquad = (g_{i+1,r-1} - g_{i,r-1})(\boldsymbol{\Sigma} - \boldsymbol{\Sigma}\mathbf{G}_i(\mathbf{G}_i'\boldsymbol{\Sigma}\mathbf{G}_i)^{-1}\mathbf{G}_i'\boldsymbol{\Sigma}). \end{aligned}$$

Summing over i establishes the lemma. ∎

By combining Theorem 4.2.9, Lemma 4.2.2 and Lemma 4.2.3 the next theorem is verified.

Theorem 4.2.10. *For $\widehat{\boldsymbol{\Sigma}}$ given in Theorem 4.1.7, the expression*

$$\begin{aligned} &\widehat{\boldsymbol{\Sigma}} + \frac{1}{n}\sum_{j=1}^{m}(r(\mathbf{C}_{j-1}) - r(\mathbf{C}_j))\sum_{i=1}^{j-1}k_{i,j}\mathbf{A}_i(\mathbf{A}_i'\mathbf{P}_i'\widehat{\boldsymbol{\Sigma}}^{-1}\mathbf{P}_i\mathbf{A}_i)^-\mathbf{A}_i' \\ &\quad + \frac{1}{n}r(\mathbf{C}_m)\sum_{i=1}^{m}k_{i,m+1}\mathbf{A}_i(\mathbf{A}_i'\mathbf{P}_i'\widehat{\boldsymbol{\Sigma}}^{-1}\mathbf{P}_i\mathbf{A}_i)^-\mathbf{A}_i', \quad \mathbf{C}_0 = \mathbf{I}_n, \end{aligned}$$

is an unbiased estimator of $\boldsymbol{\Sigma}$, where $0 < k_{i,j} < \infty$ are defined in Lemma 4.2.3 and $\sum_{i=1}^{0}k_{i,j}\mathbf{A}_i(\mathbf{A}_i'\mathbf{P}_i'\widehat{\boldsymbol{\Sigma}}^{-1}\mathbf{P}_i\mathbf{A}_i)^-\mathbf{A}_i' = \mathbf{0}$. ∎

Now we start to derive the moments of $\widehat{\mathbf{B}}_1$, $\widehat{\mathbf{B}}_2$, $\widehat{\mathbf{B}}_3$ given in Theorem 4.1.6. Fortunately, we can rely on notation and ideas given in the previous discussion about $\widehat{\boldsymbol{\Sigma}}$ and $\widehat{E[\mathbf{X}]}$. However, some modifications have to be performed and since the topic is complicated we once again go into details. Throughout, the estimators are assumed to be unique and it is of some interest to see how and when the uniqueness conditions in Theorem 4.1.12 have to be utilized. In Theorem 4.2.6 it was observed that $\widehat{\mathbf{B}}_1$, $\widehat{\mathbf{B}}_2$ and $\widehat{\mathbf{B}}_3$ are unbiased under uniqueness conditions. When considering the dispersion matrices of $\widehat{\mathbf{B}}_1$, $\widehat{\mathbf{B}}_2$ and $\widehat{\mathbf{B}}_3$, we start with $\widehat{\mathbf{B}}_3$.

For $\widehat{\mathbf{B}}_3$, it follows from Theorem 4.1.6 and Lemma 4.1.3 that

$$\widehat{\mathbf{B}}_3 = (\mathbf{A}_3'\mathbf{G}_2(\mathbf{G}_2'\mathbf{W}_3\mathbf{G}_2)^{-1}\mathbf{G}_2'\mathbf{A}_3)^{-1}\mathbf{A}_3'\mathbf{G}_2(\mathbf{G}_2'\mathbf{W}_3\mathbf{G}_2)^{-1}\mathbf{G}_2'\mathbf{X}\mathbf{C}_3'(\mathbf{C}_3\mathbf{C}_3')^{-1}, \tag{4.2.98}$$

where $\mathbf{G}_2$ and $\mathbf{W}_3$ are defined in Lemma 4.1.3. Thus, we see from (4.2.98) that $\widehat{\mathbf{B}}_3$ has the same structure as $\widehat{\mathbf{B}}$ in the MLNM($\mathbf{ABC}$) and therefore, by utilizing Theorem 4.2.1 (ii), if $n - k_3 - m_2 + q_3 - 1 > 0$,

$$D[\widehat{\mathbf{B}}_3] = \frac{n - k_3 - 1}{n - k_3 - m_2 + q_3 - 1}(\mathbf{C}_3\mathbf{C}_3')^{-1} \otimes (\mathbf{A}_3'\mathbf{G}_2(\mathbf{G}_2'\boldsymbol{\Sigma}\mathbf{G}_2)^{-1}\mathbf{G}_2'\mathbf{A}_3)^{-1}.$$

It is more complicated to obtain the dispersion matrix for $\widehat{\mathbf{B}}_2$. The reason is that $\widehat{\mathbf{B}}_2$ is a function of $\widehat{\mathbf{B}}_1$ which has to be taken into account. From Lemma 4.1.3 (ii) it follows that

$$\begin{aligned} \mathbf{T}_1'\mathbf{S}_2^{-1}\mathbf{T}_1 &= \mathbf{G}_1(\mathbf{G}_1'\mathbf{W}_2\mathbf{G}_1)^{-1}\mathbf{G}_1', \\ \mathbf{P}_3'\mathbf{S}_3^{-1}\mathbf{P}_3 &= \mathbf{G}_2(\mathbf{G}_2'\mathbf{W}_3\mathbf{G}_2)^{-1}\mathbf{G}_2'. \end{aligned}$$

Because of the assumption of uniqueness, we can instead of $\widehat{\mathbf{B}}_2$ study the linear combinations

$$\mathbf{G}_1'\mathbf{A}_2\widehat{\mathbf{B}}_2\mathbf{C}_2$$

since

$$(\mathbf{A}_2'\mathbf{G}_1\mathbf{G}_1'\mathbf{A}_2)^{-1}\mathbf{A}_2'\mathbf{G}_1\mathbf{G}_1'\mathbf{A}_2\widehat{\mathbf{B}}_2\mathbf{C}_2\mathbf{C}_2'(\mathbf{C}_2\mathbf{C}_2')^{-1} = \widehat{\mathbf{B}}_2.$$

Furthermore, using the notation of Theorem 4.1.6 and Lemma 4.1.3, we have shown that $\widehat{\mathbf{B}}_2$ is an unbiased estimator and therefore, when obtaining the dispersion matrix of $\widehat{\mathbf{B}}_2$, we consider

$$\mathbf{G}_1'\mathbf{A}_2(\widehat{\mathbf{B}}_2 - \mathbf{B}_2)\mathbf{C}_2. \tag{4.2.99}$$

From here the technical derivations start. After some transformations the expression in (4.2.99) can be written as follows:

$$\begin{aligned}
&\mathbf{G}_1'\mathbf{A}_2(\widehat{\mathbf{B}}_2 - \mathbf{B}_2)\mathbf{C}_2\\
&=\mathbf{G}_1'\mathbf{A}_2(\mathbf{A}_2'\mathbf{G}_1(\mathbf{G}_1'\mathbf{W}_2\mathbf{G}_1)^{-1}\mathbf{G}_1'\mathbf{A}_2)^{-1}\mathbf{A}_2'\mathbf{G}_1(\mathbf{G}_1'\mathbf{W}_2\mathbf{G}_1)^{-1}\mathbf{G}_1'\\
&\qquad \times (\mathbf{X} - E[\mathbf{X}] - \mathbf{A}_3(\widehat{\mathbf{B}}_3 - \mathbf{B}_3)\mathbf{C}_3)\mathbf{C}_2'(\mathbf{C}_2\mathbf{C}_2')^{-1}\mathbf{C}_2\\
&=\mathbf{G}_1'\mathbf{R}_1\{(\mathbf{X} - E[\mathbf{X}])(\mathbf{P}_{\mathbf{C}_2'} - \mathbf{P}_{\mathbf{C}_3'}) + (\mathbf{X} - E[\mathbf{X}])\mathbf{P}_{\mathbf{C}_3'} - \mathbf{R}_2(\mathbf{X} - E[\mathbf{X}])\mathbf{P}_{\mathbf{C}_3'}\}\\
&=\mathbf{G}_1'\mathbf{R}_1(\mathbf{X} - E[\mathbf{X}])(\mathbf{P}_{\mathbf{C}_2'} - \mathbf{P}_{\mathbf{C}_3'}) + \mathbf{G}_1'\mathbf{R}_1(\mathbf{I} - \mathbf{R}_2)(\mathbf{X} - E[\mathbf{X}])\mathbf{P}_{\mathbf{C}_3'},
\end{aligned}$$

where $\mathbf{R}_1$ and $\mathbf{R}_2$ are as in (4.2.88), and

$$\mathbf{P}_{\mathbf{C}_i'} = \mathbf{C}_i'(\mathbf{C}_i\mathbf{C}_i')^{-}\mathbf{C}_i.$$

The basic idea behind the decomposition of $\mathbf{G}_1'\mathbf{A}_2(\widehat{\mathbf{B}}_2 - \mathbf{B}_2)\mathbf{C}_2$ is that the components are partly independent: $(\mathbf{X}-E[\mathbf{X}])\mathbf{P}_{\mathbf{C}_3'}$ is independent of $(\mathbf{X}-E[\mathbf{X}])(\mathbf{P}_{\mathbf{C}_2'} - \mathbf{P}_{\mathbf{C}_3'})$, $\mathbf{R}_1$ and $\mathbf{R}_2$, since (see Theorem 2.2.4 (ii)) $(\mathbf{P}_{\mathbf{C}_2'} - \mathbf{P}_{\mathbf{C}_3'})\mathbf{P}_{\mathbf{C}_3'} = \mathbf{0}$, $(\mathbf{I} - \mathbf{P}_{\mathbf{C}_2'})(\mathbf{P}_{\mathbf{C}_2'} - \mathbf{P}_{\mathbf{C}_3'}) = \mathbf{0}$ and $(\mathbf{I} - \mathbf{P}_{\mathbf{C}_2'})\mathbf{P}_{\mathbf{C}_3'} = \mathbf{0}$. Thus,

$$C[\mathbf{G}_1'\mathbf{R}_1(\mathbf{X} - E[\mathbf{X}])(\mathbf{P}_{\mathbf{C}_2'} - \mathbf{P}_{\mathbf{C}_3'}), \mathbf{G}_1'\mathbf{R}_1(\mathbf{I} - \mathbf{R}_2)(\mathbf{X} - E[\mathbf{X}])\mathbf{P}_{\mathbf{C}_3'}] = \mathbf{0},$$

and therefore

$$\begin{aligned}
&D[\mathbf{G}_1'\mathbf{A}_2(\widehat{\mathbf{B}}_2 - \mathbf{B}_2)\mathbf{C}_2]\\
&\quad = D[\mathbf{G}_1'\mathbf{R}_1(\mathbf{X} - E[\mathbf{X}])(\mathbf{P}_{\mathbf{C}_2'} - \mathbf{P}_{\mathbf{C}_3'})] + D[\mathbf{G}_1'\mathbf{R}_1(\mathbf{I} - \mathbf{R}_2)(\mathbf{X} - E[\mathbf{X}])\mathbf{P}_{\mathbf{C}_3'}].
\end{aligned} \tag{4.2.100}$$

The two expressions on the right hand side of (4.2.100) will be treated separately. The first term equals

$$D[\mathbf{G}_1'\mathbf{R}_1(\mathbf{X} - E[\mathbf{X}])(\mathbf{P}_{\mathbf{C}_2'} - \mathbf{P}_{\mathbf{C}_3'})] = (\mathbf{P}_{\mathbf{C}_2'} - \mathbf{P}_{\mathbf{C}_3'}) \otimes E[\mathbf{G}_1'\mathbf{R}_1\boldsymbol{\Sigma}\mathbf{R}_1'\mathbf{G}_1],$$

and the rest follows from the treatment of the MLNM(**ABC**); e.g. see (4.2.13). Thus,

$$\begin{aligned}
&D[\mathbf{G}_1'\mathbf{R}_1(\mathbf{X} - E[\mathbf{X}])(\mathbf{P}_{\mathbf{C}_2'} - \mathbf{P}_{\mathbf{C}_3'})] = (\mathbf{P}_{\mathbf{C}_2'} - \mathbf{P}_{\mathbf{C}_3'})\\
&\quad \otimes \frac{n - r(\mathbf{C}_2) - 1}{n - r(\mathbf{C}_1) - m_1 + r(\mathbf{G}_1'\mathbf{A}_2) - 1}\mathbf{G}_1'\mathbf{A}_2(\mathbf{A}_2'\mathbf{G}_1(\mathbf{G}_1'\boldsymbol{\Sigma}\mathbf{G}_1)^{-1}\mathbf{G}_1'\mathbf{A}_2)^{-1}\mathbf{A}_2'\mathbf{G}_1.
\end{aligned} \tag{4.2.101}$$

The second expression in (4.2.100) is more complicated to deal with. In the subsequent, we are going to make use of the notation $\mathbf{\Gamma}_1^r$, $\mathbf{\Gamma}_2^r$, $\mathbf{H}_r$, $\mathbf{W}_r$, $\mathbf{M}_r$, $\mathbf{N}_r$, $\mathbf{V}^r$ and $\underline{\mathbf{V}}^r$, which were all defined in Lemma 4.1.3 and Lemma 4.1.4. Observe first that

$$\begin{aligned} &D[\mathbf{G}_1'\mathbf{R}_1(\mathbf{I}-\mathbf{R}_2)(\mathbf{X}-E[\mathbf{X}])\mathbf{P}_{\mathbf{C}_3'}] \\ &\quad = \mathbf{P}_{\mathbf{C}_3'} \otimes E[\mathbf{G}_1'\mathbf{R}_1(\mathbf{I}-\mathbf{R}_2)\mathbf{\Sigma}(\mathbf{I}-\mathbf{R}_2')\mathbf{R}_1'\mathbf{G}_1] \end{aligned}$$

and

$$\begin{aligned} \mathbf{G}_1'\mathbf{R}_1 &= \mathbf{G}_1' - \mathbf{H}_1\mathbf{\Gamma}_1'\mathbf{\Sigma}^{-1/2}\mathbf{M}_2 \\ &= \mathbf{H}_1(\mathbf{\Gamma}_2^2)'\mathbf{\Gamma}_2^2\mathbf{\Gamma}_1^1\mathbf{\Sigma}^{-1/2} - \mathbf{H}_1(\mathbf{\Gamma}_2^2)'\mathbf{V}_{21}^2(\mathbf{V}_{11}^2)^{-1}\mathbf{\Gamma}_1^2\mathbf{\Gamma}_1^1\mathbf{\Sigma}^{-1/2}. \end{aligned}$$

Then, since $\mathbf{V}_{21}^2 = \mathbf{U}_2\mathbf{U}_1'$ for some normally distributed $\mathbf{U}' = (\mathbf{U}_1' : \mathbf{U}_2')$, where $\mathbf{U}_1$ and $\mathbf{U}_2$ are independent and $\mathbf{U}_2$ is also independent of $\mathbf{R}_2$ and $\mathbf{V}_{11}^2$,

$$\begin{aligned} &E[\mathbf{G}_1'\mathbf{R}_1(\mathbf{I}-\mathbf{R}_2)\mathbf{\Sigma}(\mathbf{I}-\mathbf{R}_2')\mathbf{R}_1'\mathbf{G}_1] \\ &\quad = E[\mathbf{H}_1(\mathbf{\Gamma}_2^2)'\mathbf{\Gamma}_2^2\mathbf{\Gamma}_1^1\mathbf{\Sigma}^{-1/2}(\mathbf{I}-\mathbf{R}_2)\mathbf{\Sigma}(\mathbf{I}-\mathbf{R}_2')\mathbf{\Sigma}^{-1/2}(\mathbf{\Gamma}_1^1)'(\mathbf{\Gamma}_2^2)'\mathbf{\Gamma}_2^2\mathbf{H}_1'] \\ &\qquad + E\Big[\mathbf{H}_1(\mathbf{\Gamma}_2^2)'\mathbf{V}_{21}^2(\mathbf{V}_{11}^2)^{-1}\mathbf{\Gamma}_1^2\mathbf{\Gamma}_1^1\mathbf{\Sigma}^{-1/2}(\mathbf{I}-\mathbf{R}_2)\mathbf{\Sigma}(\mathbf{I}-\mathbf{R}_2') \\ &\qquad\qquad \times \mathbf{\Sigma}^{-1/2}(\mathbf{\Gamma}_1^1)'(\mathbf{\Gamma}_1^2)'(\mathbf{V}_{11}^2)^{-1}\mathbf{V}_{12}^2\mathbf{\Gamma}_2^2\mathbf{H}_1'\Big]. \end{aligned} \tag{4.2.102}$$

From now on the two terms on the right-hand side of (4.2.102) will be considered separately. First it is noted that from the uniqueness conditions in Theorem 4.1.12 it follows that

$$\mathbf{\Gamma}_1'\mathbf{\Sigma}^{-1/2}\mathbf{A}_3 = \mathbf{D}\mathbf{\Gamma}_1^2\mathbf{\Gamma}_1^1\mathbf{\Sigma}^{-1/2}\mathbf{A}_3$$

for some matrix $\mathbf{D}$. Thus,

$$\begin{aligned} \mathbf{\Gamma}_1^1\mathbf{\Sigma}^{-1/2}(\mathbf{I}-\mathbf{R}_2) &= \mathbf{\Gamma}_1^1\mathbf{\Sigma}^{-1/2} - \mathbf{D}\mathbf{H}_2^{-1}(\mathbf{I}-\mathbf{G}_2'\mathbf{M}_3)\mathbf{G}_2' \\ &= \mathbf{\Gamma}_1^1\mathbf{\Sigma}^{-1/2} - \mathbf{D}(\mathbf{\Gamma}_2^3)'\mathbf{\Gamma}_2^3\mathbf{\Gamma}_1^2\mathbf{\Gamma}_1^1\mathbf{\Sigma}^{-1/2} + \mathbf{D}(\mathbf{\Gamma}_2^3)'\mathbf{V}_{21}^3(\mathbf{V}_{11}^3)^{-1}\mathbf{\Gamma}_1^3\mathbf{\Gamma}_1^2\mathbf{\Gamma}_1^1\mathbf{\Sigma}^{-1/2}. \end{aligned}$$

Next it will be utilized that

$$E[\mathbf{V}_{21}^3(\mathbf{V}_{11}^3)^{-1}] = \mathbf{0}$$

and

$$\mathbf{H}_1(\mathbf{\Gamma}_2^2)'\mathbf{\Gamma}_2^2\mathbf{D}(\mathbf{\Gamma}_2^3)'\mathbf{\Gamma}_2^3\mathbf{\Gamma}_1^2(\mathbf{\Gamma}_2^2)'\mathbf{\Gamma}_2^2\mathbf{H}_1' = \mathbf{0},$$

since $\mathbf{\Gamma}_1^2(\mathbf{\Gamma}_2^2)' = \mathbf{0}$. Thus,

$$\begin{aligned} &E[\mathbf{H}_1(\mathbf{\Gamma}_2^2)'\mathbf{\Gamma}_2^2\mathbf{\Gamma}_1^1\mathbf{\Sigma}^{-1/2}(\mathbf{I}-\mathbf{R}_2)\mathbf{\Sigma}(\mathbf{I}-\mathbf{R}_2')\mathbf{\Sigma}^{-1/2}(\mathbf{\Gamma}_1^1)'(\mathbf{\Gamma}_2^1)'\mathbf{\Gamma}_2^2\mathbf{H}_1'] \\ &\quad = \mathbf{H}_1(\mathbf{\Gamma}_2^2)'\mathbf{\Gamma}_2^2\mathbf{H}_1' + \mathbf{H}_1(\mathbf{\Gamma}_2^2)'\mathbf{\Gamma}_2^2\mathbf{D}(\mathbf{\Gamma}_2^3)'\mathbf{\Gamma}_2^3\mathbf{D}'(\mathbf{\Gamma}_2^2)'\mathbf{\Gamma}_2^2\mathbf{H}_1' \\ &\qquad + E[\mathbf{H}_1(\mathbf{\Gamma}_2^2)'\mathbf{\Gamma}_2^2\mathbf{D}(\mathbf{\Gamma}_2^3)'\mathbf{V}_{21}^3(\mathbf{V}_{11}^3)^{-1}(\mathbf{V}_{11}^3)^{-1}\mathbf{V}_{12}^3\mathbf{\Gamma}_2^3\mathbf{D}'(\mathbf{\Gamma}_2^2)'\mathbf{\Gamma}_2^2\mathbf{H}_1']. \end{aligned}$$

Since $\mathbf{V}_{11}^3 \sim W_{m_3}(\mathbf{I}_{m_3}, n - r(\mathbf{C}_3))$, it follows, similarly to (4.2.22), that

$$E[\mathbf{H}_1(\mathbf{\Gamma}_2^2)'\mathbf{\Gamma}_2^2\mathbf{D}(\mathbf{\Gamma}_2^3)'\mathbf{V}_{21}^3(\mathbf{V}_{11}^3)^{-1}(\mathbf{V}_{11}^3)^{-1}\mathbf{V}_{12}^3\mathbf{\Gamma}_2^3\mathbf{D}'(\mathbf{\Gamma}_2^2)'\mathbf{\Gamma}_2^2\mathbf{H}_1'] \\ = \frac{m_3}{n - r(\mathbf{C}_3) - m_3 - 1}\mathbf{H}_1(\mathbf{\Gamma}_2^2)'\mathbf{\Gamma}_2^2\mathbf{D}(\mathbf{\Gamma}_2^3)'\mathbf{\Gamma}_2^3\mathbf{D}'(\mathbf{\Gamma}_2^2)'\mathbf{\Gamma}_2^2\mathbf{H}_1'.$$

Therefore it remains to express $\mathbf{H}_1(\mathbf{\Gamma}_2^2)'\mathbf{\Gamma}_2^2\mathbf{H}_1'$ and

$$\mathbf{H}_1(\mathbf{\Gamma}_2^2)'\mathbf{\Gamma}_2^2\mathbf{D}(\mathbf{\Gamma}_2^3)'\mathbf{\Gamma}_2^3\mathbf{D}'(\mathbf{\Gamma}_2^2)'\mathbf{\Gamma}_2^2\mathbf{H}_1'$$

in the original matrices. Since

$$(\mathbf{\Gamma}_2^3)'\mathbf{\Gamma}_2^3 = \mathbf{\Gamma}_1^2\mathbf{\Gamma}_1^1\mathbf{\Sigma}^{-1/2}\mathbf{A}_3(\mathbf{A}_3'\mathbf{G}_2(\mathbf{G}_2'\mathbf{\Sigma}\mathbf{G}_2)^{-1}\mathbf{G}_2'\mathbf{A}_3')^{-}\mathbf{A}_3'\mathbf{\Sigma}^{-1/2}(\mathbf{\Gamma}_1^1)'(\mathbf{\Gamma}_1^2)', \tag{4.2.103}$$

$$(\mathbf{\Gamma}_2^2)'\mathbf{\Gamma}_2^2 = \mathbf{\Gamma}_1^1\mathbf{\Sigma}^{-1/2}\mathbf{A}_2(\mathbf{A}_2'\mathbf{G}_1(\mathbf{G}_1'\mathbf{\Sigma}\mathbf{G}_1)^{-1}\mathbf{G}_1'\mathbf{A}_2)^{-}\mathbf{A}_2'\mathbf{\Sigma}^{-1/2}(\mathbf{\Gamma}_1^1)', \tag{4.2.104}$$

$$(\mathbf{\Gamma}_1^1)'\mathbf{\Gamma}_1^1 = \mathbf{\Sigma}^{1/2}\mathbf{G}_1(\mathbf{G}_1'\mathbf{\Sigma}\mathbf{G}_1)^{-1}\mathbf{G}_1'\mathbf{\Sigma}^{1/2}, \tag{4.2.105}$$

we get from (4.2.104)

$$\mathbf{H}_1(\mathbf{\Gamma}_2^2)'\mathbf{\Gamma}_2^2\mathbf{H}_1' = \mathbf{G}_1\mathbf{A}_2(\mathbf{A}_2'\mathbf{G}_1(\mathbf{G}_1'\mathbf{\Sigma}\mathbf{G}_1)^{-1}\mathbf{G}_1'\mathbf{A}_2)^{-}\mathbf{A}_2'\mathbf{G}_1, \tag{4.2.106}$$

and from (4.2.103)

$$\mathbf{D}(\mathbf{\Gamma}_2^3)'\mathbf{\Gamma}_2^3\mathbf{D}' = \mathbf{\Gamma}_1^1\mathbf{\Sigma}^{-1/2}\mathbf{A}_3(\mathbf{A}_3'\mathbf{G}_2(\mathbf{G}_2'\mathbf{\Sigma}\mathbf{G}_2)^{-1}\mathbf{G}_2'\mathbf{A}_3)^{-}\mathbf{A}_3'\mathbf{\Sigma}^{-1/2}(\mathbf{\Gamma}_1^1)'.$$

Then, using (4.2.105), the first expression on the right hand side of (4.2.102) equals

$$E[\mathbf{H}_1(\mathbf{\Gamma}_2^2)'\mathbf{\Gamma}_2^2\mathbf{\Gamma}_1^1\mathbf{\Sigma}^{-1/2}(\mathbf{I} - \mathbf{R}_2)\mathbf{\Sigma}(\mathbf{I} - \mathbf{R}_2')\mathbf{\Sigma}^{-1/2}(\mathbf{\Gamma}_1^1)'(\mathbf{\Gamma}_2^2)'\mathbf{\Gamma}_2^2\mathbf{H}_1'] \\ = \frac{n - r(\mathbf{C}_3) - 1}{n - r(\mathbf{C}_3) - m_3 - 1}\mathbf{H}(\mathbf{G}_1'\mathbf{\Sigma}\mathbf{G}_1)^{-1}\mathbf{G}_1'\mathbf{A}_3 \\ \times (\mathbf{A}_3'\mathbf{G}_2(\mathbf{G}_2'\mathbf{\Sigma}\mathbf{G}_2)^{-1}\mathbf{G}_2'\mathbf{A}_3)^{-}\mathbf{A}_3'\mathbf{G}_1(\mathbf{G}_1'\mathbf{\Sigma}\mathbf{G}_1)^{-1}\mathbf{H}', \tag{4.2.107}$$

where

$$\mathbf{H} = \mathbf{G}_1'\mathbf{A}_2(\mathbf{A}_2'\mathbf{G}_1(\mathbf{G}_1'\mathbf{\Sigma}\mathbf{G}_1)^{-1}\mathbf{G}_1'\mathbf{A}_2)^{-}\mathbf{A}_2'\mathbf{G}_1. \tag{4.2.108}$$

Now we will study the second term on the right hand side of (4.2.102). Since $\mathbf{V}^2 = \mathbf{U}\mathbf{U}'$, where $\mathbf{U} \sim N_{p,n-r(\mathbf{C}_2)}(\mathbf{0}, \mathbf{I}, \mathbf{I})$ and $\mathbf{V}_{12} = \mathbf{U}_1\mathbf{U}_2'$, $\mathbf{V}_{11} = \mathbf{U}_1\mathbf{U}_1'$ with $\mathbf{U}_1$ and $\mathbf{U}_2$ being independent, it follows from Theorem 2.2.9 (i) that

$$E\Big[\mathbf{H}_1(\mathbf{\Gamma}_2^2)'\mathbf{V}_{21}^2(\mathbf{V}_{11}^2)^{-1}\mathbf{\Gamma}_1^2\mathbf{\Gamma}_1^1\mathbf{\Sigma}^{-1/2}(\mathbf{I} - \mathbf{R}_2)\mathbf{\Sigma}(\mathbf{I} - \mathbf{R}_2')\mathbf{\Sigma}^{-1/2}(\mathbf{\Gamma}_1^1)'(\mathbf{\Gamma}_1^2)' \\ \times (\mathbf{V}_{11}^2)^{-1}\mathbf{V}_{12}^2\mathbf{\Gamma}_2^2\mathbf{H}_1'\Big] \\ = \mathbf{H}_1(\mathbf{\Gamma}_2^2)'\mathbf{\Gamma}_2^2\mathbf{H}_1'E[\mathrm{tr}\{(\mathbf{V}_{11}^2)^{-1}\mathbf{\Gamma}_1^2\mathbf{\Gamma}_1^1\mathbf{\Sigma}^{-1/2}(\mathbf{I} - \mathbf{R}_2\mathbf{\Sigma}(\mathbf{I} - \mathbf{R}_2')\mathbf{\Sigma}^{-1/2}(\mathbf{\Gamma}_1^1)'(\mathbf{\Gamma}_1^2)'\}]. \tag{4.2.109}$$

It remains to obtain the expectation of the trace function in (4.2.109). Once again Lemma 4.2.1 will be of utmost importance. The lemma can be applied, because $\mathbf{M}_3$ is a function of $\underline{\mathbf{V}}^3_{11}$ and

$$\begin{aligned}\boldsymbol{\Gamma}^2_1\boldsymbol{\Gamma}^1_1\boldsymbol{\Sigma}^{-1/2}(\mathbf{I}-\mathbf{R}_2) &= \boldsymbol{\Gamma}^2_1\boldsymbol{\Gamma}^1_1\boldsymbol{\Sigma}^{-1/2}\mathbf{M}_3\\ &= \underline{\mathbf{V}}^3_{11}(\boldsymbol{\Gamma}^3_1)'(\mathbf{V}^3_{11})^{-1}\boldsymbol{\Gamma}^3_1\boldsymbol{\Gamma}^2_1\boldsymbol{\Gamma}^1_1\boldsymbol{\Sigma}^{-1/2}.\end{aligned}$$

Since $\mathbf{V}^3_{11} \sim W_{m_3}(\mathbf{I}, n-r(\mathbf{C}_3))$, this yields

$$\begin{aligned}&E[\mathrm{tr}\{(\mathbf{V}^2_{11})^{-1}\boldsymbol{\Gamma}^2_1\boldsymbol{\Gamma}^1_1\boldsymbol{\Sigma}^{-1/2}(\mathbf{I}-\mathbf{R}_2)\boldsymbol{\Sigma}(\mathbf{I}-\mathbf{R}'_2)\boldsymbol{\Sigma}^{-1/2}(\boldsymbol{\Gamma}^1_1)'(\boldsymbol{\Gamma}^2_1)'\}]\\ &\quad = c_2E[\mathrm{tr}\{(\underline{\mathbf{V}}^3_{11})^{-1}\underline{\mathbf{V}}^3_{11}(\boldsymbol{\Gamma}^3_1)'(\mathbf{V}^3_{11})^{-1}(\mathbf{V}^3_{11})^{-1}\boldsymbol{\Gamma}^3_1\underline{\mathbf{V}}^3_{11}\}]\\ &\quad = c_2E[\mathrm{tr}\{(\mathbf{V}^3_{11})^{-1}\}] = c_2\frac{m_3}{n-r(\mathbf{C}_3)-m_3-1}.\end{aligned} \qquad (4.2.110)$$

Hence, by using (4.2.101), (4.2.102), (4.2.107) – (4.2.110), $D[\widehat{\mathbf{B}}_2]$ has been found. Concerning $D[\widehat{\mathbf{B}}_1]$ it should be noticed that the treatment of $D[\widehat{\mathbf{B}}_1]$ is similar to the treatment of $D[\widehat{\mathbf{B}}_2]$, although some additional argument is needed. For uniqueness of $\widehat{\mathbf{B}}_1$ it follows from Theorem 4.1.12 (v) that $\mathbf{A}_1$ and $\mathbf{C}_1$ must be of full rank. Therefore, instead of $\widehat{\mathbf{B}}_1$ the linear combinations $\mathbf{A}_1(\widehat{\mathbf{B}}_1-\mathbf{B}_1)\mathbf{C}_1$ will be studied. Observe that by the unbiasedness result in Theorem 4.2.6, $E[\mathbf{A}_1(\widehat{\mathbf{B}}_1-\mathbf{B}_1)\mathbf{C}_1] = \mathbf{0}$. First, we decompose $\mathbf{A}_1(\widehat{\mathbf{B}}_1-\mathbf{B}_1)\mathbf{C}_1$, as when treating (4.2.99):

$$\begin{aligned}&\mathbf{A}_1(\widehat{\mathbf{B}}_1-\mathbf{B}_1)\mathbf{C}_1\\ &\quad = \mathbf{R}_0(\mathbf{X}-E[\mathbf{X}])(\mathbf{P}_{\mathbf{C}'_1}-\mathbf{P}_{\mathbf{C}'_2}) + \mathbf{R}_0(\mathbf{I}-\mathbf{R}_1)(\mathbf{X}-E[\mathbf{X}])(\mathbf{P}_{\mathbf{C}'_2}-\mathbf{P}_{\mathbf{C}'_3})\\ &\qquad + \mathbf{R}_0(\mathbf{I}-\mathbf{R}_1)(\mathbf{I}-\mathbf{R}_2)(\mathbf{X}-E[\mathbf{X}])\mathbf{P}_{\mathbf{C}'_3},\end{aligned}$$

where $\mathbf{R}_i$ is given in (4.2.88). Since $(\mathbf{X}-E[\mathbf{X}])\mathbf{P}_{\mathbf{C}'_3}$ is independent (see Theorem 2.2.4) of $\mathbf{R}_0$, $\mathbf{R}_1$, $\mathbf{R}_2$, $(\mathbf{X}-E[\mathbf{X}])(\mathbf{P}_{\mathbf{C}'_2}-\mathbf{P}_{\mathbf{C}'_3})$, and $(\mathbf{X}-E[\mathbf{X}])(\mathbf{P}_{\mathbf{C}'_2}-\mathbf{P}_{\mathbf{C}'_3})$ is independent of $\mathbf{R}_0$, $\mathbf{R}_1$, $(\mathbf{X}-E[\mathbf{X}])(\mathbf{P}_{\mathbf{C}'_1}-\mathbf{P}_{\mathbf{C}'_2})$ we obtain a basic decomposition of the dispersion matrix:

$$\begin{aligned}&D[\mathbf{A}_1(\widehat{\mathbf{B}}_1-\mathbf{B}_1)\mathbf{C}_1]\\ &\quad = D[\mathbf{R}_0(\mathbf{X}-E[\mathbf{X}])(\mathbf{P}_{\mathbf{C}'_1}-\mathbf{P}_{\mathbf{C}'_2})] + D[\mathbf{R}_0(\mathbf{I}-\mathbf{R}_1)(\mathbf{X}-E[\mathbf{X}])(\mathbf{P}_{\mathbf{C}'_2}-\mathbf{P}_{\mathbf{C}'_3})]\\ &\qquad + D[\mathbf{R}_0(\mathbf{I}-\mathbf{R}_1)(\mathbf{I}-\mathbf{R}_2)(\mathbf{X}-E[\mathbf{X}])\mathbf{P}_{\mathbf{C}'_3}].\end{aligned} \qquad (4.2.111)$$

The terms on the right hand side of (4.2.111) will be treated one by one. After some manipulations and using the independence between $\mathbf{R}_0$ and $(\mathbf{X}-E[\mathbf{X}])(\mathbf{P}_{\mathbf{C}'_1}-\mathbf{P}_{\mathbf{C}'_2})$,

$$D[\mathbf{R}_0(\mathbf{X}-E[\mathbf{X}])(\mathbf{P}_{\mathbf{C}'_1}-\mathbf{P}_{\mathbf{C}'_2})] = (\mathbf{P}_{\mathbf{C}'_1}-\mathbf{P}_{\mathbf{C}'_2})\otimes E[\mathbf{R}_0\boldsymbol{\Sigma}\mathbf{R}'_0]. \qquad (4.2.112)$$

Now we can rely on the MLNM($\mathbf{ABC}$), i.e. (4.2.18) and (4.2.22), and get that

$$E[\mathbf{R}_0\boldsymbol{\Sigma}\mathbf{R}'_0] = \frac{n-k_1-1}{n-k_1-p+q_1-1}\mathbf{A}_1(\mathbf{A}'_1\boldsymbol{\Sigma}^{-1}\mathbf{A}_1)^{-1}\mathbf{A}'_1. \qquad (4.2.113)$$

For the second expression it is noted that $(\mathbf{X}-E[\mathbf{X}])(\mathbf{P}_{\mathbf{C}_2'}-\mathbf{P}_{\mathbf{C}_3'})$ is independent of $\mathbf{R}_0$ and $\mathbf{R}_1$. Thus,

$$\begin{aligned} &D[\mathbf{R}_0(\mathbf{I}-\mathbf{R}_1)(\mathbf{X}-E[\mathbf{X}])(\mathbf{P}_{\mathbf{C}_2'}-\mathbf{P}_{\mathbf{C}_3'})] \\ &\qquad =(\mathbf{P}_{\mathbf{C}_2'}-\mathbf{P}_{\mathbf{C}_3'})\otimes E[\mathbf{R}_0(\mathbf{I}-\mathbf{R}_1)\boldsymbol{\Sigma}(\mathbf{I}-\mathbf{R}_1')\mathbf{R}_0']. \qquad (4.2.114)\end{aligned}$$

In order to find an explicit expression of (4.2.114) we copy the approach of finding the expectation in (4.2.102). Let us start by rewriting $\mathbf{R}_0$ in a canonical form, i.e.

$$\begin{aligned}\mathbf{R}_0 &= \mathbf{I}-\mathbf{M}_1=\mathbf{I}-\boldsymbol{\Sigma}^{1/2}(\boldsymbol{\Gamma}^1)'\mathbf{N}_1\boldsymbol{\Gamma}_1^1\boldsymbol{\Sigma}^{-1/2} \\ &= \mathbf{I}-\boldsymbol{\Sigma}^{1/2}(\boldsymbol{\Gamma}_1^1)'\boldsymbol{\Gamma}_1^1\boldsymbol{\Sigma}^{-1/2}-\boldsymbol{\Sigma}^{1/2}(\boldsymbol{\Gamma}_2^1)'\mathbf{V}_{21}^1(\mathbf{V}_{11}^1)^{-1}\boldsymbol{\Gamma}_1^1\boldsymbol{\Sigma}^{-1/2} \\ &= \boldsymbol{\Sigma}^{1/2}(\boldsymbol{\Gamma}_2^1)'\boldsymbol{\Gamma}_2^1\boldsymbol{\Sigma}^{-1/2}-\boldsymbol{\Sigma}^{1/2}(\boldsymbol{\Gamma}_2^1)'\mathbf{V}_{21}^1(\mathbf{V}_{11}^1)^{-1}\boldsymbol{\Gamma}_1^1\boldsymbol{\Sigma}^{-1/2}.\end{aligned}$$

Since $\mathbf{V}_{21}^1$ is uncorrelated with $\mathbf{V}_{11}^1$ and $\mathbf{R}_1$,

$$\begin{aligned}&E[\mathbf{R}_0(\mathbf{I}-\mathbf{R}_1)\boldsymbol{\Sigma}(\mathbf{I}-\mathbf{R}_1')\mathbf{R}_0'] \\ &\quad = E[\boldsymbol{\Sigma}^{1/2}(\boldsymbol{\Gamma}_2^1)'\boldsymbol{\Gamma}_2^1\boldsymbol{\Sigma}^{-1/2}(\mathbf{I}-\mathbf{R}_1)\boldsymbol{\Sigma}(\mathbf{I}-\mathbf{R}_1')\boldsymbol{\Sigma}^{-1/2}(\boldsymbol{\Gamma}_2^1)'\boldsymbol{\Gamma}_2^1\boldsymbol{\Sigma}^{1/2}] \\ &\quad\; + E\Big[\boldsymbol{\Sigma}^{1/2}(\boldsymbol{\Gamma}_2^1)'\mathbf{V}_{21}^1(\mathbf{V}_{11}^1)^{-1}\boldsymbol{\Gamma}_1^1\boldsymbol{\Sigma}^{-1/2}(\mathbf{I}-\mathbf{R}_1)\boldsymbol{\Sigma}(\mathbf{I}-\mathbf{R}_1')\boldsymbol{\Sigma}^{-1/2}(\boldsymbol{\Gamma}_1^1)' \\ &\qquad \times(\mathbf{V}_{11}^1)^{-1}\mathbf{V}_{12}^1\boldsymbol{\Gamma}_2^1\boldsymbol{\Sigma}^{1/2}\Big]. \qquad (4.2.115)\end{aligned}$$

The estimator $\widehat{\mathbf{B}}_1$ is unique if (4.1.106) holds. This implies that we have $\mathcal{C}(\mathbf{A}_2')=\mathcal{C}(\mathbf{A}_2'\mathbf{G}_1)$, and then

$$\mathbf{A}_2=\mathbf{D}_1\mathbf{G}_1'\mathbf{A}_2, \qquad (4.2.116)$$

for some matrix $\mathbf{D}_1$. Now we will consider the first expression on the right hand side of (4.2.115). From (4.2.88) it follows that

$$\begin{aligned}\mathbf{R}_1 &= \mathbf{D}_1\mathbf{G}_1'\mathbf{A}_2(\mathbf{A}_2'\mathbf{G}_1(\mathbf{G}_1'\mathbf{W}_2\mathbf{G}_1)^{-1}\mathbf{G}_1'\mathbf{A}_2)^{-}\mathbf{A}_2'\mathbf{G}_1(\mathbf{G}_1'\mathbf{W}_2\mathbf{G}_1)^{-1}\mathbf{G}_1' \\ &= \mathbf{D}_1\mathbf{G}_1'(\mathbf{I}-\mathbf{M}_2)=\mathbf{D}_1\mathbf{G}_1'-\mathbf{D}_1\mathbf{H}_1(\boldsymbol{\Gamma}^2)'\mathbf{N}_2\boldsymbol{\Gamma}_1^2\boldsymbol{\Gamma}_1^1\boldsymbol{\Sigma}^{-1/2} \\ &= \mathbf{D}_1\mathbf{G}_1'-\mathbf{D}_1\mathbf{H}_1(\boldsymbol{\Gamma}_1^2)'\boldsymbol{\Gamma}_1^2\boldsymbol{\Gamma}_1^1\boldsymbol{\Sigma}^{-1/2}-\mathbf{D}_1\mathbf{H}_1(\boldsymbol{\Gamma}_2^2)'\mathbf{V}_{21}^2(\mathbf{V}_{11}^2)^{-1}\boldsymbol{\Gamma}_1^2\boldsymbol{\Gamma}_1^1\boldsymbol{\Sigma}^{-1/2} \\ &= \mathbf{D}_1\mathbf{H}_1(\boldsymbol{\Gamma}_2^2)'\boldsymbol{\Gamma}_2^2\boldsymbol{\Gamma}_1^1\boldsymbol{\Sigma}^{-1/2}-\mathbf{D}_1\mathbf{H}_1(\boldsymbol{\Gamma}_2^2)'\mathbf{V}_{21}^2(\mathbf{V}_{11}^2)^{-1}\boldsymbol{\Gamma}_1^2\boldsymbol{\Gamma}_1^1\boldsymbol{\Sigma}^{-1/2}.\end{aligned}$$

Then, since $E[\mathbf{V}_{21}^2]=\mathbf{0}$ and $\mathbf{V}_{21}^2$ is uncorrelated with $\mathbf{V}_{11}^2$,

$$\begin{aligned}&E[(\mathbf{I}-\mathbf{R}_1)\boldsymbol{\Sigma}(\mathbf{I}-\mathbf{R}_1)'] \\ &\quad =(\mathbf{I}-\mathbf{D}_1\mathbf{H}_1(\boldsymbol{\Gamma}_2^2)'\boldsymbol{\Gamma}_2^2\boldsymbol{\Gamma}_1^1\boldsymbol{\Sigma}^{-1/2}))\boldsymbol{\Sigma}(\mathbf{I}-\mathbf{D}_1\mathbf{H}_1(\boldsymbol{\Gamma}_2^2)'\boldsymbol{\Gamma}_2^2\boldsymbol{\Gamma}_1^1\boldsymbol{\Sigma}^{-1/2})' \\ &\qquad + E[\mathbf{D}_1\mathbf{H}_1(\boldsymbol{\Gamma}_2^2)'\mathbf{V}_{21}^2(\mathbf{V}_{11}^2)^{-1}(\mathbf{V}_{11}^2)^{-1}\mathbf{V}_{12}^2\boldsymbol{\Gamma}_2^2\mathbf{H}_1'\mathbf{D}_1'],\end{aligned}$$

and since $\boldsymbol{\Gamma}_2^1(\boldsymbol{\Gamma}_1^1)'=\mathbf{0}$,

$$\begin{aligned}&E[\boldsymbol{\Sigma}^{1/2}(\boldsymbol{\Gamma}_2^1)'\boldsymbol{\Gamma}_2^1\boldsymbol{\Sigma}^{-1/2}(\mathbf{I}-\mathbf{R}_1)\boldsymbol{\Sigma}(\mathbf{I}-\mathbf{R}_1)'\boldsymbol{\Sigma}^{-1/2}(\boldsymbol{\Gamma}_2^1)'\boldsymbol{\Gamma}_2^1\boldsymbol{\Sigma}^{1/2}] \\ &\; =\boldsymbol{\Sigma}^{1/2}(\boldsymbol{\Gamma}_2^1)'\boldsymbol{\Gamma}_2^1\boldsymbol{\Sigma}^{1/2}+\boldsymbol{\Sigma}^{1/2}(\boldsymbol{\Gamma}_2^1)'\boldsymbol{\Gamma}_2^1\boldsymbol{\Sigma}^{-1/2}\mathbf{D}_1\mathbf{H}_1(\boldsymbol{\Gamma}_2^2)'\boldsymbol{\Gamma}_2^2\mathbf{H}_1'\mathbf{D}_1'\boldsymbol{\Sigma}^{-1/2}(\boldsymbol{\Gamma}_2^1)'\boldsymbol{\Gamma}_2^1\boldsymbol{\Sigma}^{1/2} \\ &\quad +\boldsymbol{\Sigma}^{1/2}(\boldsymbol{\Gamma}_2^1)'\boldsymbol{\Gamma}_2^1\boldsymbol{\Sigma}^{-1/2}\mathbf{D}_1\mathbf{H}_1(\boldsymbol{\Gamma}_2^2)'\boldsymbol{\Gamma}_2^2\mathbf{H}_1'\mathbf{D}_1'\boldsymbol{\Sigma}^{-1/2}(\boldsymbol{\Gamma}_2^1)'\boldsymbol{\Gamma}_2^1\boldsymbol{\Sigma}^{1/2}E[\mathrm{tr}\{(\mathbf{V}_{11}^2)^{-1}\}].\end{aligned}$$

Here we use the expression of $(\mathbf{\Gamma}_2^2)'\mathbf{\Gamma}_2^2$ given in (4.2.104),

$$\mathbf{\Sigma}^{1/2}(\mathbf{\Gamma}_2^1)'\mathbf{\Gamma}_2^1\mathbf{\Sigma}^{1/2} = \mathbf{A}_1(\mathbf{A}_1'\mathbf{\Sigma}^{-1}\mathbf{A}_1)^{-1}\mathbf{A}_1', \tag{4.2.117}$$

$$\mathbf{D}_1\mathbf{H}_1\mathbf{\Gamma}_1^1\mathbf{\Sigma}^{-1/2}\mathbf{A}_2 = \mathbf{A}_2, \tag{4.2.118}$$

and

$$E[\mathrm{tr}\{(\mathbf{V}_{11}^2)^{-1}\}] = \frac{m_2}{n - r(\mathbf{C}_2) - m_2 - 1}. \tag{4.2.119}$$

Thus, $E[\mathbf{\Sigma}^{1/2}(\mathbf{\Gamma}_2^1)'\mathbf{\Gamma}_2^1\mathbf{\Sigma}^{-1/2}(\mathbf{I} - \mathbf{R}_1)\mathbf{\Sigma}(\mathbf{I} - \mathbf{R}_1)'\mathbf{\Sigma}^{-1/2}(\mathbf{\Gamma}_2^1)'\mathbf{\Gamma}_2^1\mathbf{\Sigma}^{1/2}]$ equals

$$\begin{aligned}
&\mathbf{A}_1(\mathbf{A}_1'\mathbf{\Sigma}^{-1}\mathbf{A}_1)^{-1}\mathbf{A}_1' \\
&\quad + \frac{n - r(\mathbf{C}_2) - 1}{n - r(\mathbf{C}_2) - m_2 - 1}\mathbf{A}_1(\mathbf{A}_1'\mathbf{\Sigma}^{-1}\mathbf{A}_1)^{-1}\mathbf{A}_1'\mathbf{\Sigma}^{-1}\mathbf{A}_2 \\
&\quad \times (\mathbf{A}_2'\mathbf{G}_1(\mathbf{G}_1'\mathbf{\Sigma}\mathbf{G}_1)^{-1}\mathbf{G}_1'\mathbf{A}_2)^{-}\mathbf{A}_2'\mathbf{\Sigma}^{-1}\mathbf{A}_1(\mathbf{A}_1'\mathbf{\Sigma}^{-1}\mathbf{A}_1)^{-1}\mathbf{A}_1'.
\end{aligned} \tag{4.2.120}$$

Since $\mathbf{R}_1$ is a function of $\underline{\mathbf{V}}_{11}^2$, we obtain from Lemma 4.2.1 that the second expression on the right hand side of (4.1.115) equals

$$\begin{aligned}
&\mathbf{\Sigma}^{1/2}(\mathbf{\Gamma}_2^1)'\mathbf{\Gamma}_2^1\mathbf{\Sigma}^{1/2}E[\mathrm{tr}\{(\mathbf{V}_{11}^1)^{-1}\mathbf{\Gamma}_1^1\mathbf{\Sigma}^{-1/2}(\mathbf{I} - \mathbf{R}_1)\mathbf{\Sigma}(\mathbf{I} - \mathbf{R}_1')\mathbf{\Sigma}^{-1/2}(\mathbf{\Gamma}_1^1)'\}] \\
&\quad = \mathbf{\Sigma}^{1/2}(\mathbf{\Gamma}_2^1)'\mathbf{\Gamma}_2^1\mathbf{\Sigma}^{1/2}c_1E[\mathrm{tr}\{(\underline{\mathbf{V}}_{11}^2)^{-1}\mathbf{\Gamma}_1^1\mathbf{\Sigma}^{-1/2}(\mathbf{I} - \mathbf{R}_1)\mathbf{\Sigma}(\mathbf{I} - \mathbf{R}_1')\mathbf{\Sigma}^{-1/2}(\mathbf{\Gamma}_1^1)'\}].
\end{aligned} \tag{4.2.121}$$

Moreover,

$$\begin{aligned}
\mathbf{\Gamma}_1^1\mathbf{\Sigma}^{-1/2}(\mathbf{I} - \mathbf{R}_1) &= \mathbf{H}_1^{-1}\mathbf{G}_1'(\mathbf{I} - \mathbf{R}_1) \\
&= \mathbf{H}_1^{-1}\mathbf{G}_1' - \mathbf{H}_1^{-1}\mathbf{G}_1'\mathbf{A}_2(\mathbf{A}_2'\mathbf{G}_1(\mathbf{G}_1'\mathbf{W}_2\mathbf{G}_1)^{-1}\mathbf{G}_1'\mathbf{A}_2)^{-}\mathbf{A}_2'\mathbf{G}_1(\mathbf{G}_1'\mathbf{W}_2\mathbf{G}_1)^{-1}\mathbf{G}_1' \\
&= \mathbf{H}_1^{-1}\mathbf{G}_1'\mathbf{W}_2\mathbf{G}_2(\mathbf{G}_2'\mathbf{W}_2\mathbf{G}_2)^{-}\mathbf{G}_2' = \underline{\mathbf{V}}_{11}^2(\mathbf{\Gamma}_1^2)'(\mathbf{V}_{11}^2)^{-1}\mathbf{\Gamma}_1^2\mathbf{\Gamma}_1^1\mathbf{\Sigma}^{-1/2},
\end{aligned}$$

and therefore

$$\begin{aligned}
&E[\mathrm{tr}\{(\underline{\mathbf{V}}_{11}^2)^{-1}\mathbf{\Gamma}_1^1\mathbf{\Sigma}^{-1/2}(\mathbf{I} - \mathbf{R}_1)\mathbf{\Sigma}(\mathbf{I} - \mathbf{R}_1')\mathbf{\Sigma}^{-1/2}(\mathbf{\Gamma}_1^1)'\}] \\
&\quad = E[\mathrm{tr}\{(\underline{\mathbf{V}}_{11}^2)^{-1}\underline{\mathbf{V}}_{11}^2(\mathbf{\Gamma}_1^2)'(\mathbf{V}_{11}^2)^{-1}(\mathbf{V}_{11}^2)^{-1}\mathbf{\Gamma}_1^2\underline{\mathbf{V}}_{11}^2\}] \\
&\quad = E[\mathrm{tr}\{(\mathbf{V}_{11}^2)^{-1}\}] = \frac{m_2}{n - r(\mathbf{C}_2) - m_2 - 1},
\end{aligned} \tag{4.2.122}$$

where (4.2.119) has been used. Hence, applying (4.2.122) and (4.2.117) yields

$$\begin{aligned}
&E[\mathbf{\Sigma}^{1/2}(\mathbf{\Gamma}_2^1)'\mathbf{V}_{21}^1(\mathbf{V}_{11}^1)^{-1}\mathbf{\Gamma}_1^1\mathbf{\Sigma}^{-1/2}(\mathbf{I} - \mathbf{R}_1)\mathbf{\Sigma} \\
&\qquad \times (\mathbf{I} - \mathbf{R}_1')\mathbf{\Sigma}^{-1/2}(\mathbf{\Gamma}_1^1)'(\mathbf{V}_{11}^1)^{-1}\mathbf{V}_{12}^1\mathbf{\Gamma}_2^1\mathbf{\Sigma}^{1/2}] \\
&\quad = c_1\frac{m_2}{n - r(\mathbf{C}_2) - m_2 - 1}\mathbf{\Sigma}^{1/2}(\mathbf{\Gamma}_2^1)'\mathbf{\Gamma}_2^1\mathbf{\Sigma}^{1/2} \\
&\quad = c_1\frac{m_2}{n - r(\mathbf{C}_2) - m_2 - 1}\mathbf{A}_1(\mathbf{A}_1'\mathbf{\Sigma}^{-1}\mathbf{A}_1)^{-1}\mathbf{A}_1'.
\end{aligned} \tag{4.2.123}$$

Thus, using (4.2.120) and (4.2.121) we may state that the expectation in (4.2.115) equals

$$\begin{aligned}
&E[\mathbf{R}_0(\mathbf{I}-\mathbf{R}_1)\boldsymbol{\Sigma}(\mathbf{I}-\mathbf{R}_1')\mathbf{R}_0'] \\
&\quad=(1+c_1\frac{m_2}{n-r(\mathbf{C}_2)-m_2-1})\mathbf{A}_1(\mathbf{A}_1'\boldsymbol{\Sigma}^{-1}\mathbf{A}_1)^{-1}\mathbf{A}_1' \\
&\qquad+\frac{n-r(\mathbf{C}_2)-1}{n-r(\mathbf{C}_2)-m_2-1}\mathbf{A}_1(\mathbf{A}_1'\boldsymbol{\Sigma}^{-1}\mathbf{A}_1)^{-1}\mathbf{A}_1'\boldsymbol{\Sigma}^{-1}\mathbf{A}_2 \\
&\qquad\times(\mathbf{A}_2'\mathbf{G}_1(\mathbf{G}_1'\boldsymbol{\Sigma}\mathbf{G}_1)^{-1}\mathbf{G}_1'\mathbf{A}_2)^{-}\mathbf{A}_2'\boldsymbol{\Sigma}^{-1}\mathbf{A}_1(\mathbf{A}_1'\boldsymbol{\Sigma}^{-1}\mathbf{A}_1)^{-1}\mathbf{A}_1'.
\end{aligned}$$

Now we will consider the third term in (4.2.111),

$$\begin{aligned}
&D[\mathbf{R}_0(\mathbf{I}-\mathbf{R}_1)(\mathbf{I}-\mathbf{R}_2)(\mathbf{X}-E[\mathbf{X}])\mathbf{P}_{\mathbf{C}_3'}] \\
&\qquad=\mathbf{P}_{\mathbf{C}_3'}\otimes E[\mathbf{R}_0(\mathbf{I}-\mathbf{R}_1)(\mathbf{I}-\mathbf{R}_2)\boldsymbol{\Sigma}(\mathbf{I}-\mathbf{R}_2')(\mathbf{I}-\mathbf{R}_1')\mathbf{R}_0'].
\end{aligned}$$

The expectation equals

$$\begin{aligned}
&E[\boldsymbol{\Sigma}^{1/2}(\boldsymbol{\Gamma}_2^1)'\boldsymbol{\Gamma}_2^1\boldsymbol{\Sigma}^{-1/2}(\mathbf{I}-\mathbf{R}_1)(\mathbf{I}-\mathbf{R}_2)\boldsymbol{\Sigma}(\mathbf{I}-\mathbf{R}_2')(\mathbf{I}-\mathbf{R}_1')\boldsymbol{\Sigma}^{-1/2}(\boldsymbol{\Gamma}_2^1)'\boldsymbol{\Gamma}_2^1\boldsymbol{\Sigma}^{1/2}] \\
&\quad+E\Big[\boldsymbol{\Sigma}^{1/2}(\boldsymbol{\Gamma}_2^1)'\mathbf{V}_{21}^1(\mathbf{V}_{11}^1)^{-1}\boldsymbol{\Gamma}_1^1\boldsymbol{\Sigma}^{-1/2}(\mathbf{I}-\mathbf{R}_1)(\mathbf{I}-\mathbf{R}_2)\boldsymbol{\Sigma} \\
&\qquad\times(\mathbf{I}-\mathbf{R}_2')(\mathbf{I}-\mathbf{R}_1')\boldsymbol{\Sigma}^{-1/2}(\boldsymbol{\Gamma}_1^1)'(\mathbf{V}_{11}^1)^{-1}\mathbf{V}_{12}^1\boldsymbol{\Gamma}_2^1\boldsymbol{\Sigma}^{1/2}\Big]. \qquad (4.2.124)
\end{aligned}$$

Both expectations in (4.2.124) will be explored further. Using (4.2.118) we get

$$\begin{aligned}
&E[(\mathbf{I}-\mathbf{R}_1)(\mathbf{I}-\mathbf{R}_2)\boldsymbol{\Sigma}(\mathbf{I}-\mathbf{R}_2')(\mathbf{I}-\mathbf{R}_1')] \\
&\quad=(\mathbf{I}-\mathbf{D}_1\mathbf{H}_1(\boldsymbol{\Gamma}_2^2)'\boldsymbol{\Gamma}_2^2\boldsymbol{\Gamma}_1^1\boldsymbol{\Sigma}^{-1/2})E[(\mathbf{I}-\mathbf{R}_2)\boldsymbol{\Sigma}(\mathbf{I}-\mathbf{R}_2')] \\
&\qquad\times(\mathbf{I}-\boldsymbol{\Sigma}^{-1/2}(\boldsymbol{\Gamma}_1^1)'(\boldsymbol{\Gamma}_2^2)'\boldsymbol{\Gamma}_2^2\mathbf{H}_1'\mathbf{D}_1') \\
&\quad+\mathbf{D}_1\mathbf{H}_1(\boldsymbol{\Gamma}_2^2)'E\Big[\mathbf{V}_{21}^2(\mathbf{V}_{11}^2)^{-1}\boldsymbol{\Gamma}_1^2\boldsymbol{\Gamma}_1^1\boldsymbol{\Sigma}^{-1/2}(\mathbf{I}-\mathbf{R}_2)\boldsymbol{\Sigma} \\
&\qquad\times(\mathbf{I}-\mathbf{R}_2')\boldsymbol{\Sigma}^{-1/2}(\boldsymbol{\Gamma}_1^1)'(\boldsymbol{\Gamma}_1^2)'(\mathbf{V}_{11}^2)^{-1}\mathbf{V}_{12}^2\Big]\boldsymbol{\Gamma}_2^2\mathbf{H}_1'\mathbf{D}_1'. \qquad (4.2.125)
\end{aligned}$$

Observe that from (4.2.104) and (4.2.118)

$$\begin{aligned}
&\mathbf{D}_1\mathbf{H}_1(\boldsymbol{\Gamma}_2^2)'\boldsymbol{\Gamma}_2^2\boldsymbol{\Gamma}_1^1\boldsymbol{\Sigma}^{-1/2} \\
&\quad=\mathbf{D}_1\mathbf{G}_1'\mathbf{A}_2(\mathbf{A}_2'\mathbf{G}_1(\mathbf{G}_1'\boldsymbol{\Sigma}\mathbf{G}_1)^{-1}\mathbf{G}_1'\mathbf{A}_2)^{-}\mathbf{A}_2'\mathbf{G}_1(\mathbf{G}_1'\boldsymbol{\Sigma}\mathbf{G}_1)^{-1}\mathbf{G}_1' \\
&\quad=\mathbf{A}_2(\mathbf{A}_2'\mathbf{G}_1(\mathbf{G}_1'\boldsymbol{\Sigma}\mathbf{G}_1)^{-1}\mathbf{G}_1'\mathbf{A}_2)^{-}\mathbf{A}_2'\mathbf{G}_1(\mathbf{G}_1'\boldsymbol{\Sigma}\mathbf{G}_1)^{-1}\mathbf{G}_1'.
\end{aligned}$$

Since $\mathcal{C}(\mathbf{A}_2')=\mathcal{C}(\mathbf{A}_2'\mathbf{G}_1)$, we have that

$$\mathcal{C}(\mathbf{I}-\boldsymbol{\Sigma}^{-1/2}(\boldsymbol{\Gamma}_1^1)'(\boldsymbol{\Gamma}_2^2)'\boldsymbol{\Gamma}_2^2\mathbf{H}_1'\mathbf{D}_1')=\mathcal{C}(\mathbf{A}_2')^{\perp}.$$

Moreover, by (4.1.106), since $\mathcal{C}(\mathbf{P}')$ also equals $\mathcal{C}(\mathbf{A}_2')^{\perp}$,

$$\mathcal{C}(\mathbf{A}_3'(\mathbf{I}-\boldsymbol{\Sigma}^{-1/2}(\boldsymbol{\Gamma}_1^1)'(\boldsymbol{\Gamma}_2^2)'\boldsymbol{\Gamma}_2^2\mathbf{H}_1'\mathbf{D}_1'))=\mathcal{C}(\mathbf{A}_3'\mathbf{G}_2).$$

Thus,

$$(\mathbf{I} - \mathbf{D}_1\mathbf{H}_1(\mathbf{\Gamma}_2^2)'\mathbf{\Gamma}_2^2\mathbf{\Gamma}_1^1\mathbf{\Sigma}^{-1/2})\mathbf{A}_3 = \mathbf{D}_2\mathbf{G}_2'\mathbf{A}_3,$$

for some matrix $\mathbf{D}_2$. Therefore,

$$\begin{aligned}
&(\mathbf{I}-\mathbf{D}_1\mathbf{H}_1(\mathbf{\Gamma}_2^2)'\mathbf{\Gamma}_2^2\mathbf{\Gamma}_1^1\mathbf{\Sigma}^{-1/2})\mathbf{R}_2 \\
&\quad = \mathbf{D}_2\mathbf{G}_2' - \mathbf{D}_2\mathbf{G}_2'\mathbf{M}_3 = \mathbf{D}_2\mathbf{G}_2' - \mathbf{D}_2\mathbf{H}_2(\mathbf{\Gamma}^3)'\mathbf{N}_3\mathbf{\Gamma}_1^3\mathbf{\Gamma}_1^2\mathbf{\Gamma}_1^1\mathbf{\Sigma}^{-1/2} \\
&\quad = \mathbf{D}_2\mathbf{H}_2(\mathbf{\Gamma}_2^3)'\mathbf{\Gamma}_2^3\mathbf{\Gamma}_1^2\mathbf{\Gamma}_1^1\mathbf{\Sigma}^{-1/2} - \mathbf{D}_2\mathbf{H}_2(\mathbf{\Gamma}_2^3)'\mathbf{V}_{21}^3(\mathbf{V}_{11}^3)^{-1}\mathbf{\Gamma}_1^3\mathbf{\Gamma}_1^2\mathbf{\Gamma}_1^1\mathbf{\Sigma}^{-1/2}.
\end{aligned}$$

Now, by copying calculations from previous parts, we obtain

$$\begin{aligned}
&E[\mathbf{\Sigma}^{1/2}(\mathbf{\Gamma}_2^1)'\mathbf{\Gamma}_2^1\mathbf{\Sigma}^{-1/2}(\mathbf{I}-\mathbf{R}_1)(\mathbf{I}-\mathbf{R}_2)\mathbf{\Sigma}(\mathbf{I}-\mathbf{R}_2')(\mathbf{I}-\mathbf{R}_1')\mathbf{\Sigma}^{-1/2}(\mathbf{\Gamma}_2^1)'\mathbf{\Gamma}_2^1\mathbf{\Sigma}^{1/2}] \\
&= \mathbf{\Sigma}^{1/2}(\mathbf{\Gamma}_2^1)'\mathbf{\Gamma}_2^1\mathbf{\Sigma}^{1/2} + \mathbf{\Sigma}^{1/2}(\mathbf{\Gamma}_2^1)'\mathbf{\Gamma}_2^1\mathbf{\Sigma}^{-1/2}\mathbf{D}_1\mathbf{H}_1(\mathbf{\Gamma}_2^2)'\mathbf{\Gamma}_2^2\mathbf{H}_1'\mathbf{D}_1'\mathbf{\Sigma}^{-1/2}(\mathbf{\Gamma}_2^1)'\mathbf{\Gamma}_2^1\mathbf{\Sigma}^{1/2} \\
&\quad + \mathbf{\Sigma}^{1/2}(\mathbf{\Gamma}_2^1)'\mathbf{\Gamma}_2^1\mathbf{\Sigma}^{-1/2}\mathbf{D}_2\mathbf{H}_2(\mathbf{\Gamma}_2^3)'\mathbf{\Gamma}_2^3\mathbf{H}_2'\mathbf{D}_2'\mathbf{\Sigma}^{-1/2}(\mathbf{\Gamma}_2^1)'\mathbf{\Gamma}_2^1\mathbf{\Sigma}^{1/2} \\
&\quad + \mathbf{\Sigma}^{1/2}(\mathbf{\Gamma}_2^1)'\mathbf{\Gamma}_2^1\mathbf{\Sigma}^{-1/2}\mathbf{D}_2\mathbf{H}_2(\mathbf{\Gamma}_2^3)'\mathbf{\Gamma}_2^3\mathbf{H}_2'\mathbf{D}_2'\mathbf{\Sigma}^{-1/2}(\mathbf{\Gamma}_2^1)'\mathbf{\Gamma}_2^1\mathbf{\Sigma}^{1/2}E[\mathrm{tr}\{(\mathbf{V}_{11}^3)^{-1}\}].
\end{aligned}$$

In (4.2.117) and (4.2.104) $(\mathbf{\Gamma}_2^1)'\mathbf{\Gamma}_2^1$ and $(\mathbf{\Gamma}_2^2)'\mathbf{\Gamma}_2^2$ were expressed in the original matrices, respectively, and now $(\mathbf{\Gamma}_2^3)'\mathbf{\Gamma}_2^3$ is presented:

$$(\mathbf{\Gamma}_2^3)'\mathbf{\Gamma}_2^3 = \mathbf{\Gamma}_1^2\mathbf{\Gamma}_1^1\mathbf{\Sigma}^{-1/2}\mathbf{A}_3(\mathbf{A}_3'\mathbf{G}_2(\mathbf{G}_2'\mathbf{\Sigma}\mathbf{G}_2)^{-1}\mathbf{G}_2'\mathbf{A}_3)^{-}\mathbf{A}_3'\mathbf{\Sigma}^{-1/2}(\mathbf{\Gamma}_1^1)'(\mathbf{\Gamma}_1^2)'. \tag{4.2.126}$$

Furthermore,

$$\begin{aligned}
\mathbf{D}_1\mathbf{H}_1(\mathbf{\Gamma}_2^2)'\mathbf{\Gamma}_2^2\mathbf{H}_1'\mathbf{D}_1' &= \mathbf{A}_2(\mathbf{A}_2'\mathbf{G}_1(\mathbf{G}_1'\mathbf{\Sigma}\mathbf{G}_1)^{-1}\mathbf{G}_1'\mathbf{A}_2)^{-}\mathbf{A}_2', \\
\mathbf{D}_2\mathbf{H}_2(\mathbf{\Gamma}_2^3)'\mathbf{\Gamma}_2^3\mathbf{H}_2'\mathbf{D}_2' &= \mathbf{D}_2\mathbf{G}_2'\mathbf{A}_3(\mathbf{A}_3'\mathbf{G}_2(\mathbf{G}_2'\mathbf{\Sigma}\mathbf{G}_2)^{-1}\mathbf{G}_2'\mathbf{A}_3)^{-}\mathbf{A}_3'\mathbf{G}_2\mathbf{D}_2' \\
&= \mathbf{H}\mathbf{A}_3(\mathbf{A}_3'\mathbf{G}_2(\mathbf{G}_2'\mathbf{\Sigma}\mathbf{G}_2)^{-1}\mathbf{G}_2'\mathbf{A}_3)^{-}\mathbf{A}_3'\mathbf{H}',
\end{aligned}$$

where

$$\mathbf{H} = \mathbf{I} - \mathbf{A}_2(\mathbf{A}_2'\mathbf{G}_1(\mathbf{G}_1'\mathbf{\Sigma}\mathbf{G}_1)^{-1}\mathbf{G}_1'\mathbf{A}_2)^{-}\mathbf{A}_2'\mathbf{G}_1(\mathbf{G}_1'\mathbf{\Sigma}\mathbf{G}_1)^{-1}\mathbf{G}_1',$$

and

$$E[\mathrm{tr}\{(\mathbf{V}_{11}^3)^{-1}\}] = \frac{m_3}{n - r(\mathbf{C}_3) - m_3 - 1}.$$

Using all these lengthy expressions we may write out

$$E[\mathbf{\Sigma}^{1/2}(\mathbf{\Gamma}_2^1)'\mathbf{\Gamma}_2^1\mathbf{\Sigma}^{-1/2}(\mathbf{I}-\mathbf{R}_1)(\mathbf{I}-\mathbf{R}_2)\mathbf{\Sigma}(\mathbf{I}-\mathbf{R}_2')(\mathbf{I}-\mathbf{R}_1')\mathbf{\Sigma}^{-1/2}(\mathbf{\Gamma}_2^1)'\mathbf{\Gamma}_2^1\mathbf{\Sigma}^{1/2}]$$

as

$$\begin{aligned}
&\mathbf{A}_1(\mathbf{A}_1'\mathbf{\Sigma}^{-1}\mathbf{A}_1)^{-1}\mathbf{A}_1' \\
&\quad + \mathbf{A}_1(\mathbf{A}_1'\mathbf{\Sigma}^{-1}\mathbf{A}_1)^{-1}\mathbf{A}_1'\mathbf{\Sigma}^{-1}\mathbf{A}_2(\mathbf{A}_2'\mathbf{G}_1(\mathbf{G}_1'\mathbf{\Sigma}\mathbf{G}_1)^{-1}\mathbf{G}_1'\mathbf{A}_2)^{-} \\
&\qquad \times \mathbf{A}_2'\mathbf{\Sigma}^{-1}\mathbf{A}_1(\mathbf{A}_1'\mathbf{\Sigma}^{-1}\mathbf{A}_1)^{-1}\mathbf{A}_1' \\
&\quad + \tfrac{n-r(\mathbf{C}_3)-1}{n-r(\mathbf{C}_3)-m_3-1}\mathbf{A}_1(\mathbf{A}_1'\mathbf{\Sigma}^{-1}\mathbf{A}_1)^{-1}\mathbf{A}_1'\mathbf{\Sigma}^{-1}\mathbf{H}\mathbf{A}_3(\mathbf{A}_3'\mathbf{G}_2(\mathbf{G}_2'\mathbf{\Sigma}\mathbf{G}_2)^{-1}\mathbf{G}_2'\mathbf{A}_3)^{-} \\
&\qquad \times \mathbf{A}_3'\mathbf{H}'\mathbf{\Sigma}^{-1}\mathbf{A}_1(\mathbf{A}_1'\mathbf{\Sigma}^{-1}\mathbf{A}_1)^{-1}\mathbf{A}_1'.
\end{aligned}$$

Finally, to obtain the expected value of the second expression in (4.2.124), it is noted that $\mathbf{V}_{21}^1$ is uncorrelated with $\mathbf{V}_{11}^1$, $\mathbf{R}_1$ and $\mathbf{R}_2$, and that $\mathbf{R}_2$ is a function of $\underline{\mathbf{V}}_{11}^2$ and independent of $\mathbf{V}_{21}^1$:

$$
\begin{aligned}
E\Big[&\boldsymbol{\Sigma}^{1/2}(\boldsymbol{\Gamma}_2^1)'\mathbf{V}_{21}^1(\mathbf{V}_{11}^1)^{-1}\boldsymbol{\Gamma}_1^1\boldsymbol{\Sigma}^{-1/2}(\mathbf{I}-\mathbf{R}_1)(\mathbf{I}-\mathbf{R}_2)\boldsymbol{\Sigma} \\
&\times(\mathbf{I}-\mathbf{R}_2')(\mathbf{I}-\mathbf{R}_1')\boldsymbol{\Sigma}^{-1/2}(\boldsymbol{\Gamma}_1^1)'(\mathbf{V}_{11}^1)^{-1}\mathbf{V}_{12}^1\boldsymbol{\Gamma}_2^1\boldsymbol{\Sigma}^{1/2}\Big] \\
&=\boldsymbol{\Sigma}^{1/2}(\boldsymbol{\Gamma}_2^1)'\boldsymbol{\Gamma}_2^1\boldsymbol{\Sigma}^{1/2} \\
&\quad\times E[\mathrm{tr}\{(\mathbf{V}_{11}^1)^{-1}\boldsymbol{\Gamma}_1^1\boldsymbol{\Sigma}^{-1/2}(\mathbf{I}-\mathbf{R}_1)(\mathbf{I}-\mathbf{R}_2)\boldsymbol{\Sigma}(\mathbf{I}-\mathbf{R}_2')(\mathbf{I}-\mathbf{R}_1')\boldsymbol{\Sigma}^{-1/2}(\boldsymbol{\Gamma}_1^1)'\}] \\
&=\boldsymbol{\Sigma}^{1/2}(\boldsymbol{\Gamma}_2^1)'\boldsymbol{\Gamma}_2^1\boldsymbol{\Sigma}^{1/2}c_1E\Big[\mathrm{tr}\{(\underline{\mathbf{V}}_{11}^2)^{-1}\boldsymbol{\Gamma}_1^1\boldsymbol{\Sigma}^{-1/2}(\mathbf{I}-\mathbf{R}_1)(\mathbf{I}-\mathbf{R}_2)\boldsymbol{\Sigma} \\
&\quad\times(\mathbf{I}-\mathbf{R}_2')(\mathbf{I}-\mathbf{R}_1')\boldsymbol{\Sigma}^{-1/2}(\boldsymbol{\Gamma}_1^1)'\}\Big]. \qquad (4.2.127)
\end{aligned}
$$

Moreover,

$$
\boldsymbol{\Gamma}_1^1\boldsymbol{\Sigma}^{-1/2}(\mathbf{I}-\mathbf{R}_1)=\underline{\mathbf{V}}_{11}^2(\boldsymbol{\Gamma}_1^2)'(\mathbf{V}_{11}^2)^{-1}\boldsymbol{\Gamma}_1^2\boldsymbol{\Gamma}_1^1\boldsymbol{\Sigma}^{-1/2}, \qquad (4.2.128)
$$

which implies that the right hand side of (4.2.127) equals

$$
\boldsymbol{\Sigma}^{1/2}(\boldsymbol{\Gamma}_2^1)'\boldsymbol{\Gamma}_2^1\boldsymbol{\Sigma}^{1/2}c_1E[\mathrm{tr}\{(\mathbf{V}_{11}^2)^{-1}\boldsymbol{\Gamma}_1^2\boldsymbol{\Gamma}_1^1\boldsymbol{\Sigma}^{-1/2}(\mathbf{I}-\mathbf{R}_2)\boldsymbol{\Sigma}(\mathbf{I}-\mathbf{R}_2')\boldsymbol{\Sigma}^{-1/2}(\boldsymbol{\Gamma}_1^1)'(\boldsymbol{\Gamma}_1^2)'\}]. \qquad (4.2.129)
$$

Hence, (4.2.110) yields that (4.2.129) can be written as

$$
\mathbf{A}_1(\mathbf{A}_1'\boldsymbol{\Sigma}^{-1}\mathbf{A}_1)^{-1}\mathbf{A}_1'c_1c_2\frac{m_3}{n-r(\mathbf{C}_3)-m_3-1},
$$

since $\boldsymbol{\Sigma}^{1/2}(\boldsymbol{\Gamma}_2^1)'\boldsymbol{\Gamma}_2^1\boldsymbol{\Sigma}^{1/2}=\mathbf{A}_1(\mathbf{A}_1'\boldsymbol{\Sigma}^{-1}\mathbf{A}_1)^{-1}\mathbf{A}_1'$. The paragraph is ended by summarizing all the above calculations and stating $\mathbf{D}[\widehat{\mathbf{B}}_i]$, $i=1,2,3$, explicitly.

Theorem 4.2.11. *Let $\widehat{\mathbf{B}}_i$, $i=1,2,3$ be given in Theorem 4.1.6 and suppose that for each $\widehat{\mathbf{B}}_i$ the uniqueness conditions in Theorem 4.1.12 are satisfied. Let the matrix $\mathbf{G}_i$ be defined in Lemma 4.1.3, and c_i and m_i in Lemma 4.2.1. Then, if the dispersion matrices are supposed to exist,*

(i) $D[\widehat{\mathbf{B}}_3]=\frac{n-k_3-1}{n-k_3-m_2+q_3-1}(\mathbf{C}_3\mathbf{C}_3')^{-1}\otimes(\mathbf{A}_3'\mathbf{G}_2(\mathbf{G}_2'\boldsymbol{\Sigma}\mathbf{G}_2)^{-1}\mathbf{G}_2'\mathbf{A}_3)^{-1}$;

(ii) $D[\widehat{\mathbf{B}}_2]$

$$
\begin{aligned}
&=D[(\mathbf{A}_2'\mathbf{G}_1\mathbf{G}_1'\mathbf{A}_2)^{-1}\mathbf{A}_2'\mathbf{G}_1\mathbf{G}_1'\mathbf{R}_1(\mathbf{X}-E[\mathbf{X}])(\mathbf{P}_{\mathbf{C}_2'}-\mathbf{P}_{\mathbf{C}_3'})\mathbf{C}_2'(\mathbf{C}_2\mathbf{C}_2')^{-1}] \\
&\quad+D[(\mathbf{A}_2'\mathbf{G}_1\mathbf{G}_1'\mathbf{A}_2)^{-1}\mathbf{A}_2'\mathbf{G}_1\mathbf{G}_1'\mathbf{R}_1(\mathbf{I}-\mathbf{R}_2)(\mathbf{X}-E[\mathbf{X}])\mathbf{P}_{\mathbf{C}_3'}\mathbf{C}_2'(\mathbf{C}_2\mathbf{C}_2')^{-1}],
\end{aligned}
$$

where

$$
\begin{aligned}
&D[(\mathbf{A}_2'\mathbf{G}_1\mathbf{G}_1'\mathbf{A}_2)^{-1}\mathbf{A}_2'\mathbf{G}_1\mathbf{G}_1'\mathbf{R}_1(\mathbf{X}-E[\mathbf{X}])(\mathbf{P}_{\mathbf{C}_2'}-\mathbf{P}_{\mathbf{C}_3'})\mathbf{C}_2'(\mathbf{C}_2\mathbf{C}_2')^{-1}] \\
&\quad=(\mathbf{C}_2\mathbf{C}_2')^{-1}\mathbf{C}_2(\mathbf{P}_{\mathbf{C}_2'}-\mathbf{P}_{\mathbf{C}_3'})\mathbf{C}_2'(\mathbf{C}_2\mathbf{C}_2')^{-1} \\
&\quad\quad\otimes\frac{n-k_2-1}{n-r(\mathbf{C}_1)-m_1+q_2-1}(\mathbf{A}_2'\mathbf{G}_1(\mathbf{G}_1\boldsymbol{\Sigma}\mathbf{G}_1)^{-1}\mathbf{G}_1'\mathbf{A}_2)^{-1},
\end{aligned}
$$

and

$$
\begin{aligned}
&D[(\mathbf{A}_2'\mathbf{G}_1\mathbf{G}_1'\mathbf{A}_2)^{-1}\mathbf{A}_2'\mathbf{G}_1\mathbf{G}_1'\mathbf{R}_1(\mathbf{I}-\mathbf{R}_2)(\mathbf{X}-E[\mathbf{X}])\mathbf{P}_{\mathbf{C}_3'}\mathbf{C}_2'(\mathbf{C}_2\mathbf{C}_2')^{-1}]\\
&\quad=(\mathbf{C}_2\mathbf{C}_2')^{-1}\mathbf{C}_2\mathbf{P}_{\mathbf{C}_3'}\mathbf{C}_2'(\mathbf{C}_2\mathbf{C}_2')^{-1}\\
&\qquad\otimes\Big\{\tfrac{n-r(\mathbf{C}_3)-1}{n-r(\mathbf{C}_3)-m_3-1}\mathbf{P}^1\mathbf{A}_3(\mathbf{A}_3'\mathbf{G}_2(\mathbf{G}_2'\boldsymbol{\Sigma}\mathbf{G}_2)^{-1}\mathbf{G}_2'\mathbf{A}_3)^-\mathbf{A}_3'(\mathbf{P}^1)'\\
&\qquad\quad+(1+c_2\tfrac{m_3}{n-r(\mathbf{C}_3)-m_3-1})(\mathbf{A}_2'\mathbf{G}_1(\mathbf{G}_1'\boldsymbol{\Sigma}\mathbf{G}_1)^{-1}\mathbf{G}_1'\mathbf{A}_2)^{-1}\Big\},
\end{aligned}
$$

where

$$
\mathbf{P}^1=(\mathbf{A}_2'\mathbf{G}_1(\mathbf{G}_1'\boldsymbol{\Sigma}\mathbf{G}_1)^{-1}\mathbf{G}_1'\mathbf{A}_2)^{-1}\mathbf{A}_2'\mathbf{G}_1(\mathbf{G}_1'\boldsymbol{\Sigma}\mathbf{G}_1)^{-1}\mathbf{G}_1';
$$

(iii)
$$
\begin{aligned}
&D[\widehat{\mathbf{B}}_1]\\
&=D[(\mathbf{A}_1'\mathbf{A}_1)^{-1}\mathbf{A}_1\mathbf{R}_0(\mathbf{X}-E[\mathbf{X}])(\mathbf{P}_{\mathbf{C}_1'}-\mathbf{P}_{\mathbf{C}_2'})\mathbf{C}_1'(\mathbf{C}_1\mathbf{C}_1')^{-1}]\\
&\quad+D[(\mathbf{A}_1'\mathbf{A}_1)^{-1}\mathbf{A}_1\mathbf{R}_0(\mathbf{I}-\mathbf{R}_1)(\mathbf{X}-E[\mathbf{X}])(\mathbf{P}_{\mathbf{C}_2'}-\mathbf{P}_{\mathbf{C}_3'})\mathbf{C}_1'(\mathbf{C}_1\mathbf{C}_1')^{-1}]\\
&\quad+D[(\mathbf{A}_1'\mathbf{A}_1)^{-1}\mathbf{A}_1\mathbf{R}_0(\mathbf{I}-\mathbf{R}_1)(\mathbf{I}-\mathbf{R}_2)(\mathbf{X}-E[\mathbf{X}])\mathbf{P}_{\mathbf{C}_3'}\mathbf{C}_1'(\mathbf{C}_1\mathbf{C}_1')^{-1}],
\end{aligned}
$$

where

$$
\begin{aligned}
&D[(\mathbf{A}_1'\mathbf{A}_1)^{-1}\mathbf{A}_1\mathbf{R}_0(\mathbf{X}-E[\mathbf{X}])(\mathbf{P}_{\mathbf{C}_1'}-\mathbf{P}_{\mathbf{C}_2'})\mathbf{C}_1'(\mathbf{C}_1\mathbf{C}_1')^{-1}]\\
&\quad=(\mathbf{C}_1\mathbf{C}_1')^{-1}\mathbf{C}_1(\mathbf{P}_{\mathbf{C}_1'}-\mathbf{P}_{\mathbf{C}_2'})\mathbf{C}_1'(\mathbf{C}_1\mathbf{C}_1')^{-1}\otimes\tfrac{n-k_1-1}{n-k_1-p+q_1-1}(\mathbf{A}_1'\boldsymbol{\Sigma}^{-1}\mathbf{A}_1)^{-1},\\
&D[(\mathbf{A}_1'\mathbf{A}_1)^{-1}\mathbf{A}_1\mathbf{R}_0(\mathbf{I}-\mathbf{R}_1)(\mathbf{X}-E[\mathbf{X}])(\mathbf{P}_{\mathbf{C}_2'}-\mathbf{P}_{\mathbf{C}_3'})\mathbf{C}_1'(\mathbf{C}_1\mathbf{C}_1')^{-1}]\\
&\quad=(\mathbf{C}_1\mathbf{C}_1')^{-1}\mathbf{C}_1(\mathbf{P}_{\mathbf{C}_2'}-\mathbf{P}_{\mathbf{C}_3'})\mathbf{C}_1'(\mathbf{C}_1\mathbf{C}_1')^{-1}\\
&\qquad\otimes\Big\{(1+c_1\tfrac{m_2}{n-r(\mathbf{C}_2)-m_2-1})(\mathbf{A}_1'\boldsymbol{\Sigma}^{-1}\mathbf{A}_1)^{-1}\\
&\qquad\quad+\tfrac{n-r(\mathbf{C}_2)-1}{n-r(\mathbf{C}_2)-m_2-1}\mathbf{P}^2\mathbf{A}_2(\mathbf{A}_2'\mathbf{G}_1(\mathbf{G}_1'\boldsymbol{\Sigma}\mathbf{G}_1)^{-1}\mathbf{G}_1'\mathbf{A}_2)^-\mathbf{A}_2'(\mathbf{P}^2)'\Big\}.
\end{aligned}
$$

Here

$$
\mathbf{P}^2=(\mathbf{A}_1'\boldsymbol{\Sigma}^{-1}\mathbf{A}_1)^{-1}\mathbf{A}_1'\boldsymbol{\Sigma}^{-1}
$$

and

$$
\begin{aligned}
&D[(\mathbf{A}_1'\mathbf{A}_1)^{-1}\mathbf{A}_1\mathbf{R}_0(\mathbf{I}-\mathbf{R}_1)(\mathbf{I}-\mathbf{R}_2)(\mathbf{X}-E[\mathbf{X}])\mathbf{P}_{\mathbf{C}_3'}\mathbf{C}_1'(\mathbf{C}_1\mathbf{C}_1')^{-1}]\\
&\quad=(\mathbf{C}_1\mathbf{C}_1')^{-1}\mathbf{C}_1\mathbf{P}_{\mathbf{C}_3'}\mathbf{C}_1'(\mathbf{C}_1\mathbf{C}_1')^{-1}\\
&\qquad\otimes\Big\{(1+c_1c_2\tfrac{m_3}{n-r(\mathbf{C}_3)-m_3-1})(\mathbf{A}_1'\boldsymbol{\Sigma}^{-1}\mathbf{A}_1)^{-1}\\
&\qquad\quad+\mathbf{P}^2\mathbf{A}_2(\mathbf{A}_2'\mathbf{G}_1(\mathbf{G}_1'\boldsymbol{\Sigma}\mathbf{G}_1)^{-1}\mathbf{G}_1'\mathbf{A}_2)^-\mathbf{A}_2'(\mathbf{P}^2)'\\
&\qquad\quad+\tfrac{n-r(\mathbf{C}_3)-1}{n-r(\mathbf{C}_3)-m_3-1}\mathbf{P}^3\mathbf{A}_3(\mathbf{A}_3'\mathbf{G}_2(\mathbf{G}_2'\boldsymbol{\Sigma}\mathbf{G}_2)^{-1}\mathbf{G}_2'\mathbf{A}_3)^-\mathbf{A}_3'(\mathbf{P}^2)'\Big\},
\end{aligned}
$$

where

$$\mathbf{P}^3 = \mathbf{P}^2(\mathbf{I} - \mathbf{A}_2\mathbf{P}^1).$$

∎

4.2.5 *Problems*

1. Compare the moments of $\widehat{\mathbf{B}}_{\mathbf{G}}$ in (4.2.7) with the corresponding moments in Theorem 4.2.1.
2. Compare (4.2.81) and (4.2.82), if

$$\mathbf{A} = \begin{pmatrix} 1 & 1 & 1 \\ 1 & 6 & 36 \\ 1 & 12 & 144 \\ 1 & 48 & 2304 \end{pmatrix}, \quad \mathbf{\Sigma} = \begin{pmatrix} 4.4 & 3.0 & -0.3 & -0.5 \\ 3.0 & 25.4 & -6.3 & -14.4 \\ -0.3 & -6.3 & 4.1 & 3.2 \\ -0.5 & -14.4 & 3.2 & 31.1 \end{pmatrix}, \quad r(\mathbf{C}) = 2.$$

3. Construct 5 different unbiased estimators of $\mathbf{\Sigma}$ in the MLNM($\sum_{i=1}^m \mathbf{A}_i\mathbf{B}_i\mathbf{C}_i$), $m \geq 5$.
4. Find $D[\widehat{E[\mathbf{X}]}]$ for the MLNM($\sum_{i=1}^2 \mathbf{A}_i\mathbf{B}_i\mathbf{C}_i$).
5. Show that $\widehat{\mathbf{\Sigma}} \xrightarrow{\mathcal{P}} \mathbf{\Sigma}$ in the MLNM($\sum_{i=1}^m \mathbf{A}_i\mathbf{B}_i\mathbf{C}_i$).
7. Show that $\widehat{E[\mathbf{X}]} = \sum_{i=1}^m \mathbf{A}_i\widehat{\mathbf{B}}_i\mathbf{C}_i$ is unbiased.
8. Find an unbiased estimator of $D[\mathbf{K}\widehat{\mathbf{B}}\mathbf{L}]$ in the MLNM($\mathbf{ABC}$) when $\mathcal{C}(\mathbf{K}') \subseteq \mathcal{C}(\mathbf{A}')$ and $\mathcal{C}(\mathbf{L}) \subseteq \mathcal{C}(\mathbf{C})$.
9. Three natural residuals for the MLNM($\mathbf{ABC}$) are given by:

$$\begin{aligned}
\mathbf{R}_1 &= \mathbf{S}\mathbf{A}^o(\mathbf{A}^{o'}\mathbf{S}\mathbf{A}^o)^-\mathbf{A}^{o'}\mathbf{X}(\mathbf{I} - \mathbf{C}'(\mathbf{C}\mathbf{C}')^-\mathbf{C}),\\
\mathbf{R}_2 &= \mathbf{A}(\mathbf{A}'\mathbf{S}^{-1}\mathbf{A})^-\mathbf{A}'\mathbf{S}^{-1}\mathbf{X}(\mathbf{I} - \mathbf{C}'(\mathbf{C}\mathbf{C}')^-\mathbf{C}),\\
\mathbf{R}_3 &= \mathbf{S}\mathbf{A}^o(\mathbf{A}^{o'}\mathbf{S}\mathbf{A}^o)^-\mathbf{A}^{o'}\mathbf{X}\mathbf{C}'(\mathbf{C}\mathbf{C}')^-\mathbf{C}.
\end{aligned}$$

   Derive $D[\mathbf{R}_i]$, $i = 1, 2, 3$.
10. Are the residuals in Problem 9 uncorrelated? Are they independent?

4.3. APPROXIMATIONS IN MULTIVARIATE LINEAR MODELS

4.3.1 *Introduction*

In this section we are going to approximate distributions of several estimators obtained in Section 4.1. In some cases, the distributions of the estimators are available in the literature (see Gleser & Olkin, 1970; Kabe, 1975; Kenward, 1986). However, often the distributions are given as integral expressions. Therefore, we will focus on approximations of these distributions in this section. The results of Section 4.2 will be of utmost importance. In order to utilize these results, we will rely on Section 3.2 and Section 3.3. The maximum likelihood estimators of the parameters in the MLNM($\mathbf{ABC}$), the MLNM($\mathbf{ABC}+\mathbf{B}_2\mathbf{C}_2$) and the MLNM($\sum_{i=3}^{3}\mathbf{A}_i\mathbf{B}_i\mathbf{C}_i$) will be considered. From Theorem 4.1.1 we have in the Growth Curve model

$$\widehat{\mathbf{B}} = (\mathbf{A}'\mathbf{S}^{-1}\mathbf{A})^{-1}\mathbf{A}'\mathbf{S}^{-1}\mathbf{X}\mathbf{C}'(\mathbf{C}\mathbf{C}')^{-1}, \tag{4.3.1}$$

where it is assumed that $\mathbf{A}$ and $\mathbf{C}$ are of full rank, i.e. $r(\mathbf{A}) = q$ and $r(\mathbf{C}) = k$,

$$\mathbf{S} = \mathbf{X}(\mathbf{I} - \mathbf{C}'(\mathbf{C}\mathbf{C}')^{-1}\mathbf{C})\mathbf{X}',$$

and

$$\begin{aligned} n\widehat{\mathbf{\Sigma}} &= (\mathbf{X} - \mathbf{A}\widehat{\mathbf{B}}\mathbf{C})(\mathbf{X} - \mathbf{A}\widehat{\mathbf{B}}\mathbf{C})' \\ &= \mathbf{S} + (\mathbf{X}\mathbf{C}'(\mathbf{C}\mathbf{C}')^{-1}\mathbf{C} - \mathbf{A}\widehat{\mathbf{B}}\mathbf{C})(\mathbf{X}\mathbf{C}'(\mathbf{C}\mathbf{C}')^{-1}\mathbf{C} - \mathbf{A}\widehat{\mathbf{B}}\mathbf{C})'. \end{aligned} \tag{4.3.2}$$

According to Theorem 4.1.9 and the full rank conditions, the maximum likelihood estimators in the MLNM($\mathbf{ABC}+\mathbf{B}_2\mathbf{C}_2$) are given by

$$\begin{aligned} \widehat{\mathbf{B}} &= (\mathbf{A}'\mathbf{S}_1^{-1}\mathbf{A})^{-1}\mathbf{A}'\mathbf{S}_1^{-1}\mathbf{Y}\mathbf{H}'(\mathbf{H}\mathbf{H}')^{-1}, \\ \widehat{\mathbf{B}}_2 &= (\mathbf{X} - \mathbf{A}\widehat{\mathbf{B}}\mathbf{C})\mathbf{C}_2'(\mathbf{C}_2\mathbf{C}_2')^{-1}, \\ n\widehat{\mathbf{\Sigma}} &= (\mathbf{Y} - \mathbf{A}\widehat{\mathbf{B}}\mathbf{H})(\mathbf{Y} - \mathbf{A}\widehat{\mathbf{B}}\mathbf{H})', \end{aligned}$$

where

$$\mathbf{S}_1 = \mathbf{Y}(\mathbf{I} - \mathbf{H}'(\mathbf{H}\mathbf{H}')^{-1}\mathbf{H})\mathbf{Y}',$$

and

$$\mathbf{Y} = \mathbf{X}(\mathbf{I} - \mathbf{C}_2'(\mathbf{C}_2\mathbf{C}_2')^{-1}\mathbf{C}_2), \tag{4.3.3}$$

$$\mathbf{H} = \mathbf{C}(\mathbf{I} - \mathbf{C}_2'(\mathbf{C}_2\mathbf{C}_2')^{-1}\mathbf{C}_2). \tag{4.3.4}$$

Theorem 4.1.6 gives the maximum likelihood estimators of the parameters in the MLNM($\sum_{i=1}^{3}\mathbf{A}_i\mathbf{B}_i\mathbf{C}_i$):

$$\widehat{\mathbf{B}}_3 = (\mathbf{A}_3'\mathbf{P}_3'\mathbf{S}_3^{-1}\mathbf{P}_3\mathbf{A}_3)^{-1}\mathbf{A}_3'\mathbf{P}_3'\mathbf{S}_3^{-1}\mathbf{X}\mathbf{C}_3'(\mathbf{C}_3\mathbf{C}_3')^{-1}, \tag{4.3.5}$$

$$\widehat{\mathbf{B}}_2 = (\mathbf{A}_2'\mathbf{P}_2'\mathbf{S}_2^{-1}\mathbf{P}_2\mathbf{A}_2)^{-1}\mathbf{A}_2'\mathbf{P}_2'\mathbf{S}_2^{-1}(\mathbf{X} - \mathbf{A}_3\widehat{\mathbf{B}}_3\mathbf{C}_3)\mathbf{C}_2'(\mathbf{C}_2\mathbf{C}_2')^{-1}, \tag{4.3.6}$$

$$\widehat{\mathbf{B}}_1 = (\mathbf{A}_1'\mathbf{S}_1^{-1}\mathbf{A}_1)^{-1}\mathbf{A}_1'\mathbf{S}_1^{-1}(\mathbf{X} - \mathbf{A}_2\widehat{\mathbf{B}}_2\mathbf{C}_2 - \mathbf{A}_3\widehat{\mathbf{B}}_3\mathbf{C}_3)\mathbf{C}_1'(\mathbf{C}_1\mathbf{C}_1')^{-1}, \tag{4.3.7}$$

$$\begin{aligned} n\widehat{\mathbf{\Sigma}} &= (\mathbf{X} - \mathbf{A}_1\widehat{\mathbf{B}}_1\mathbf{C}_1 - \mathbf{A}_2\widehat{\mathbf{B}}_2\mathbf{C}_2 - \mathbf{A}_3\widehat{\mathbf{B}}_3\mathbf{C}_3)()' \\ &= \mathbf{S}_3 + \mathbf{P}_4\mathbf{X}\mathbf{C}_3'(\mathbf{C}_3\mathbf{C}_3')\mathbf{C}_3\mathbf{X}'\mathbf{P}_4', \end{aligned} \tag{4.3.8}$$

where it is assumed that $\widehat{\mathbf{B}}_i$, $i = 1, 2, 3$, are unique and $\mathbf{P}_i$, $i = 2, 3, 4$, as well as $\mathbf{S}_i$, $i = 1, 2, 3$, are the same as in Theorem 4.1.7 and Theorem 4.1.6. Observe that we always assume the inverses in (4.3.5) – (4.3.7) to exist.

Throughout this section $f_{\mathbf{X}}(\mathbf{X}_0)$ will denote the density of $\mathbf{X}$ evaluated at the point $\mathbf{X}_0$. Furthermore, as in Chapter 3, $\mathbf{f}^k_{\mathbf{X}}(\mathbf{X}_0)$ denotes the $k-$th derivative of $f_{\mathbf{X}}(\mathbf{X})$ evaluated at the point $\mathbf{X}_0$.

4.3.2 *Approximation of the density of* $\widehat{\mathbf{B}}$ *in the Growth Curve model*

There are several strategies which could be followed when approximating the distribution of $\widehat{\mathbf{B}}$, given in (4.3.1). The first is to use a distribution obtained from asymptotic considerations. For example, if $\widehat{\mathbf{B}}$ converges to a normal distribution, the normal distribution would be appropriate to use. Another strategy is to use a distribution which is easy to compute, and to perform some kind of corrections afterwards. The third approach is to mimic some properties of the distribution of $\widehat{\mathbf{B}}$ in the approximating distribution. For instance, we may use a long tailed distribution if $\widehat{\mathbf{B}}$ has a long tail.

The starting point of this paragraph is a convergence result which is very similar to Theorem 3.1.4 (ii), i.e. for $\mathbf{S}$ in (4.3.2),

$$\frac{1}{n-k}\mathbf{S} \xrightarrow{\mathcal{P}} \mathbf{\Sigma}.$$

Therefore it is natural to approximate $\widehat{\mathbf{B}}$ by

$$\mathbf{B}_N = (\mathbf{A}'\mathbf{\Sigma}^{-1}\mathbf{A})^{-1}\mathbf{A}'\mathbf{\Sigma}^{-1}\mathbf{X}\mathbf{C}'(\mathbf{C}\mathbf{C}')^{-1}. \tag{4.3.9}$$

Since (4.3.9) is a linear function in $\mathbf{X}$ the distribution of $\widehat{\mathbf{B}}$ will be approximated by a normal distribution $N_{q,k}(\mathbf{B}, (\mathbf{A}'\mathbf{\Sigma}^{-1}\mathbf{A})^{-1}, (\mathbf{C}\mathbf{C}')^{-1})$ (see Theorem 2.2.2). When comparing $\widehat{\mathbf{B}}$ and $\mathbf{B}_N$, it is observed that due to statements (i) and (ii) of Theorem 4.2.1 and Corollary 2.2.7.1 (i),

$$E[\widehat{\mathbf{B}}] = E[\mathbf{B}_N] = \mathbf{B}$$

and

$$D[\widehat{\mathbf{B}}] - D[\mathbf{B}_N] = \frac{p-q}{n-k-p+q-1}(\mathbf{C}\mathbf{C}')^{-1} \otimes (\mathbf{A}'\mathbf{\Sigma}^{-1}\mathbf{A})^{-1},$$

which is positive definite. Thus $\mathbf{B}_N$ underestimates the variation in $\widehat{\mathbf{B}}$, which can be expected because the random matrix $\mathbf{S}$ in $\widehat{\mathbf{B}}$ has been replaced by the non-random matrix $\mathbf{\Sigma}$.

Next we are going to apply Example 3.3.1 which in turn relies on Theorem 3.3.1. We are going to achieve a better approximation than given by a normal distribution. In fact, the normal distribution will be used with a correction term. The point is that according to Example 3.3.1 it is possible to find error bounds for the obtained density approximation. Observe that

$$\begin{aligned}\widehat{\mathbf{B}} &= (\mathbf{A}'\mathbf{S}^{-1}\mathbf{A})^{-1}\mathbf{A}'\mathbf{S}^{-1}\mathbf{X}\mathbf{C}'(\mathbf{C}\mathbf{C}')^{-1}\\ &= (\mathbf{A}'\mathbf{\Sigma}^{-1}\mathbf{A})^{-1}\mathbf{A}'\mathbf{\Sigma}^{-1}\mathbf{X}\mathbf{C}'(\mathbf{C}\mathbf{C}')^{-1}\\ &\quad + (\mathbf{A}'\mathbf{S}^{-1}\mathbf{A})^{-1}\mathbf{A}'\mathbf{S}^{-1}(\mathbf{I} - \mathbf{A}(\mathbf{A}'\mathbf{\Sigma}^{-1}\mathbf{A})^{-1}\mathbf{A}'\mathbf{\Sigma}^{-1})\mathbf{X}\mathbf{C}'(\mathbf{C}\mathbf{C}')^{-1}.\end{aligned} \tag{4.3.10}$$

Theorem 2.2.4 (i) yields that

$$(\mathbf{A}'\boldsymbol{\Sigma}^{-1}\mathbf{A})^{-1}\mathbf{A}'\boldsymbol{\Sigma}^{-1}\mathbf{X}\mathbf{C}'(\mathbf{C}\mathbf{C}')^{-1}$$

and

$$(\mathbf{I}-\mathbf{A}(\mathbf{A}'\boldsymbol{\Sigma}^{-1}\mathbf{A})^{-1}\mathbf{A}'\boldsymbol{\Sigma}^{-1})\mathbf{X}\mathbf{C}'(\mathbf{C}\mathbf{C}')^{-1}$$

are independently distributed. Moreover, by Theorem 2.2.4 (iv) the sums of squares matrix $\mathbf{S}$ and "mean" $\mathbf{X}\mathbf{C}'(\mathbf{C}\mathbf{C}')^{-1}$ are independent. Hence, similarly to $\mathbf{u}$ in Example 3.3.1, we may take

$$\mathbf{U}=(\mathbf{A}'\mathbf{S}^{-1}\mathbf{A})^{-1}\mathbf{A}'\mathbf{S}^{-1}(\mathbf{I}-\mathbf{A}(\mathbf{A}'\boldsymbol{\Sigma}^{-1}\mathbf{A})^{-1}\mathbf{A}'\boldsymbol{\Sigma}^{-1})\mathbf{X}\mathbf{C}'(\mathbf{C}\mathbf{C}')^{-1}, \qquad (4.3.11)$$

which is independent of $\mathbf{B}_N$. Since $E[(\mathrm{vec}\mathbf{U})^{\otimes k}]=\mathbf{0}$ for odd k, it follows from Example 3.3.1 that the decomposition in (4.3.10) implies the next theorem. Furthermore, besides that $\mathbf{U}$ in $\widehat{\mathbf{B}}=\mathbf{B}_N+\mathbf{U}$ is independent of $\mathbf{B}_N$, we also have that $E[\mathbf{U}]=\mathbf{0}$. Thus it is very reasonable to approximate $\widehat{\mathbf{B}}$ by $\mathbf{B}_N$.

Theorem 4.3.1. *Let $\widehat{\mathbf{B}}$, $\mathbf{B}_N$ and $\mathbf{U}$ be given by (4.3.1), (4.3.9) and (4.3.11), respectively. Then an Edgeworth-type expansion of the density of $\widehat{\mathbf{B}}$ equals*

$$f_{\widehat{\mathbf{B}}}(\mathbf{B}_0)=f_{\mathbf{B}_E}(\mathbf{B}_0)+\ldots,$$

where

$$f_{\mathbf{B}_E}(\mathbf{B}_0)=\left\{1+\tfrac{1}{2}s\left(\mathrm{tr}\{\mathbf{A}'\boldsymbol{\Sigma}^{-1}\mathbf{A}(\mathbf{B}_0-\mathbf{B})\mathbf{C}\mathbf{C}'(\mathbf{B}_0-\mathbf{B})'\}-kq\right)\right\}f_{\mathbf{B}_N}(\mathbf{B}_0), \qquad (4.3.12)$$

$$s=\frac{p-q}{n-k-p+q-1}.$$

An upper error bound of the density approximation (4.3.12) is given by

$$\begin{aligned}&|f_{\widehat{\mathbf{B}}}(\mathbf{B}_0)-f_{\mathbf{B}_E}(\mathbf{B}_0)|\\&\quad\leq(2\pi)^{-\frac{1}{2}qk}\frac{1}{4!}E[(\mathrm{vec}'\mathbf{U})^{\otimes 4}]|\mathbf{A}'\boldsymbol{\Sigma}^{-1}\mathbf{A}|^{k/2}|\mathbf{C}\mathbf{C}'|^{q/2}\mathrm{vec}(\overline{m}_4[\mathbf{B}_N]). \qquad (4.3.13)\end{aligned}$$

where

$$\begin{aligned}E[(\mathrm{vec}\mathbf{U})^{\otimes 4}]=&\ \mathrm{vec}\,\overline{m}_4[\widehat{\mathbf{B}}]-(\mathbf{I}+\mathbf{I}_{qk}\otimes\mathbf{K}_{qk,qk}\otimes\mathbf{I}_{qk}+\mathbf{I}_{qk}\otimes\mathbf{K}_{(qk)^2,qk})\\&\times(\mathrm{vec}\,D[\mathbf{B}_N]\otimes E[(\mathrm{vec}\mathbf{U})^{\otimes 2}]+E[(\mathrm{vec}\mathbf{U})^{\otimes 2}]\otimes\mathrm{vec}\,D[\mathbf{B}_N])\\&-\mathrm{vec}\,\overline{m}_4[\mathbf{B}_N].\end{aligned}$$

Here $E[(\mathrm{vec}\mathbf{U})^{\otimes 2}]=s(\mathbf{C}\mathbf{C}')^{-1}\otimes(\mathbf{A}'\boldsymbol{\Sigma}^{-1}\mathbf{A})^{-1}$,

$$\begin{aligned}&\mathrm{vec}\,\overline{m}_4[\mathbf{B}_N]\\&=(\mathbf{I}+\mathbf{I}_{qk}\otimes\mathbf{K}_{qk,qk}\otimes\mathbf{I}_{qk}+\mathbf{I}_{qk}\otimes\mathbf{K}_{(qk)^2,qk})(\mathrm{vec}\{(\mathbf{C}\mathbf{C}')^{-1}\otimes(\mathbf{A}'\Sigma^{-1}\mathbf{A})^{-1}\})^{\otimes 2},\end{aligned}$$

and $\overline{m}_4[\widehat{\mathbf{B}}]$ *is obtained from Theorem 4.2.2 (iii).*

PROOF: The form of the approximation,

$$f_{\mathbf{B}_E}(\mathbf{B}_0) = f_{\mathbf{B}_N}(\mathbf{B}_0) + \frac{1}{2}E[(\mathrm{vec}'\mathbf{U})^{\otimes 2}]\mathrm{vec}\,\mathbf{f}^2_{\mathbf{B}_N}(\mathbf{B}_0),$$

follows from Theorem 3.3.1 and was also given in Example 3.3.1. As noted before,

$$\widehat{\mathbf{B}} = \mathbf{B}_N + \mathbf{U}, \tag{4.3.14}$$

which is identical to $\widehat{\mathbf{B}} - \mathbf{B} = \mathbf{B}_N - \mathbf{B} + \mathbf{U}$, and because of independence between $\mathbf{B}_N$ and $\mathbf{U}$ it follows from (4.3.10) that

$$E[(\mathrm{vec}\mathbf{U})^{\otimes 2}] = \mathrm{vec}(D[\widehat{\mathbf{B}}] - D[\mathbf{B}_N]),$$

since $\mathrm{vec}(E[\mathrm{vec}\mathbf{U}\mathrm{vec}'\mathbf{U}]) = E[(\mathrm{vec}\mathbf{U})^{\otimes 2}]$. By using Definition 2.2.1 and Corollary 2.2.7.1 (i) $D[\mathbf{B}_N]$ is obtained. In Theorem 4.2.1, $D[\widehat{\mathbf{B}}]$ was established, and $\mathbf{f}^2_{\mathbf{B}_N}(\mathbf{B}_0)$ was presented in Theorem 2.2.10 (iii). To obtain the error bound via Example 3.3.1 we note that according to (4.3.9), $\mathbf{B}_N - \mathbf{B}$ is normally distributed with the characteristic function (see Theorem 2.2.1 (i))

$$\begin{aligned}\varphi_{\mathbf{B}_N-\mathbf{B}}(\mathbf{T}) &= \exp\{-\tfrac{1}{2}\mathrm{tr}\{(\mathbf{A}'\mathbf{\Sigma}^{-1}\mathbf{A})^{-1}\mathbf{T}(\mathbf{C}\mathbf{C}')^{-1}\mathbf{T}'\}\} \\ &= \exp\{-\tfrac{1}{2}(\mathrm{vec}'\mathbf{T})(\mathbf{C}\mathbf{C}')^{-1}\otimes(\mathbf{A}'\mathbf{\Sigma}^{-1}\mathbf{A})^{-1}(\mathrm{vec}\mathbf{T})\}.\end{aligned}$$

Here it is important to observe that

$$\{(2\pi)^{\frac{1}{2}qk}|\mathbf{A}'\mathbf{\Sigma}^{-1}\mathbf{A}|^{k/2}|\mathbf{C}\mathbf{C}'|^{q/2}\}^{-1}\varphi_{\mathbf{B}_N-\mathbf{B}}(\mathbf{T})$$

is a normal density where the expectation and dispersion equal $\mathbf{0}$ and $(\mathbf{C}\mathbf{C}')^{-1}\otimes(\mathbf{A}'\mathbf{\Sigma}^{-1}\mathbf{A})^{-1}$, respectively. Thus, from Definition 2.1.4, where the central moments are defined,

$$\int_{\mathbb{R}^{qk}}(\mathrm{vec}'\mathbf{T})^{\otimes 4}\varphi_{\mathbf{B}_N-\mathbf{B}}(\mathbf{T})\,d\mathbf{T} = (2\pi)^{\frac{1}{2}qk}|\mathbf{A}'\mathbf{\Sigma}^{-1}\mathbf{A}|^{k/2}|\mathbf{C}\mathbf{C}'|^{q/2}\mathrm{vec}(\overline{m}_4(\mathbf{B}_N))$$

and $\mathrm{vec}(\overline{m}_4(\mathbf{B}_N))$ is obtained from Corollary 2.2.7.1 (iv). Finally, when obtaining $E[(\mathrm{vec}'\mathbf{U})^{\otimes 4}]$, the independence between $\mathbf{U}$ and $\mathbf{B}_N$ is utilized once again:

$$\begin{aligned}\mathrm{vec}\,\overline{m}_4[\widehat{\mathbf{B}}] &= E[\mathrm{vec}(\mathbf{U}+\mathbf{B}_N-\mathbf{B})^{\otimes 4}] \\ &= E[(\mathrm{vec}\mathbf{U})^{\otimes 4}] + (\mathbf{I} + \mathbf{I}_{qk}\otimes\mathbf{K}_{qk,qk}\otimes\mathbf{I}_{qk} + \mathbf{I}_{qk}\otimes\mathbf{K}_{(qk)^2,qk}) \\ &\quad\times(\mathrm{vec}\,D[\mathbf{B}_N]\otimes E[(\mathrm{vec}\mathbf{U})^{\otimes 2}] + E[(\mathrm{vec}\mathbf{U})^{\otimes 2}]\otimes\mathrm{vec}\,D[\mathbf{B}_N]) \\ &\quad+\mathrm{vec}\,\overline{m}_4[\mathbf{B}_N].\end{aligned} \tag{4.3.15}$$

∎

Corollary 4.3.1.1.

$$|f_{\widehat{\mathbf{B}}}(\mathbf{B}_0) - f_{\mathbf{B}_E}(\mathbf{B}_0)| = O(n^{-2}).$$

PROOF: Consider the right hand side of (4.3.13). Observe that

$$\int_{\mathbb{R}^{qk}} (\mathrm{vec}'\mathbf{T})^{\otimes 4} \varphi_{\mathbf{B}_N - \mathbf{B}}(\mathbf{T})\, d\mathbf{T}$$

equals a constant and thus we have to study

$$E[(\mathrm{vec}\mathbf{U})^{\otimes 4}].$$

Once again the canonical representation (4.2.14) is used, i.e.

$$\mathbf{A}'\mathbf{\Sigma}^{-1/2} = \mathbf{H}(\mathbf{I}_q : \mathbf{0})\mathbf{\Gamma}, \tag{4.3.16}$$

where $\mathbf{H}$ is a non-singular and $\mathbf{\Gamma}$ an orthogonal matrix. Furthermore, let

$$\mathbf{V} = \mathbf{\Gamma}\mathbf{\Sigma}^{-1/2}\mathbf{S}\mathbf{\Sigma}^{-1/2}\mathbf{\Gamma}' \sim \mathbf{W}_p(\mathbf{I}, n - r(\mathbf{C})).$$

In the subsequent the partition (4.2.16) is applied. Now

$$\mathbf{U}|\mathbf{S} \sim N_{q,k}(\mathbf{0}, (\mathbf{A}'\mathbf{S}^{-1}\mathbf{A})^{-1}\mathbf{A}'\mathbf{S}^{-1}\mathbf{\Sigma}\mathbf{S}^{-1}\mathbf{A}(\mathbf{A}'\mathbf{S}^{-1}\mathbf{A})^{-1} - (\mathbf{A}'\mathbf{\Sigma}^{-1}\mathbf{A})^{-1}, (\mathbf{C}\mathbf{C}')^{-1})$$

and

$$\begin{aligned}
&(\mathbf{A}'\mathbf{S}^{-1}\mathbf{A})^{-1}\mathbf{A}'\mathbf{S}^{-1}\mathbf{\Sigma}\mathbf{S}^{-1}\mathbf{A}(\mathbf{A}'\mathbf{S}^{-1}\mathbf{A})^{-1} - (\mathbf{A}'\mathbf{\Sigma}^{-1}\mathbf{A})^{-1}\\
&\quad = (\mathbf{H}')^{-1}(\mathbf{V}^{11})^{-1}(\mathbf{V}^{11} : \mathbf{V}^{12})(\mathbf{V}^{11} : \mathbf{V}^{12})'(\mathbf{V}^{11})^{-1}\mathbf{H}^{-1} - (\mathbf{A}'\mathbf{\Sigma}^{-1}\mathbf{A})^{-1}\\
&\quad = (\mathbf{H}')^{-1}\mathbf{H}^{-1} - (\mathbf{A}'\mathbf{\Sigma}^{-1}\mathbf{A})^{-1} + (\mathbf{H}')^{-1}(\mathbf{V}^{11})^{-1}\mathbf{V}^{12}\mathbf{V}^{21}(\mathbf{V}^{11})^{-1}\mathbf{H}^{-1}\\
&\quad = (\mathbf{H}')^{-1}\mathbf{V}_{12}\mathbf{V}_{22}^{-1}\mathbf{V}_{22}^{-1}\mathbf{V}_{21}\mathbf{H}^{-1}.
\end{aligned} \tag{4.3.17}$$

Using Corollary 2.4.12.1, we have

$$\mathbf{Z} = (\mathbf{H}')^{-1}\mathbf{V}_{12}\mathbf{V}_{22}^{-1/2} \sim N_{q,p-q}(\mathbf{0}, (\mathbf{A}'\mathbf{\Sigma}^{-1}\mathbf{A})^{-1}, \mathbf{I}), \tag{4.3.18}$$

which is independent of $\mathbf{V}_{22}$ and

$$(\mathbf{H}')^{-1}\mathbf{V}_{12}\mathbf{V}_{22}^{-1}\mathbf{V}_{22}^{-1}\mathbf{V}_{21}\mathbf{H}^{-1} = \mathbf{Z}\mathbf{V}_{22}^{-1}\mathbf{Z}'.$$

Thus,

$$\mathbf{U}|\mathbf{Z}, \mathbf{V}_{22} \sim N_{q,k}(\mathbf{0}, \mathbf{Z}\mathbf{V}_{22}^{-1}\mathbf{Z}', (\mathbf{C}\mathbf{C}')^{-1}).$$

Since

$$E[(\mathrm{vec}\mathbf{U})^{\otimes 4}] = E_{\mathbf{Z},\mathbf{V}_{22}}[E[(\mathrm{vec}\mathbf{U})^{\otimes 4}|\mathbf{Z}, \mathbf{V}_{22}]],$$

where $\mathbf{Z}, \mathbf{V}_{22}$ in $E_{\mathbf{Z},\mathbf{V}_{22}}$ indicate that expectation is taken over $\mathbf{Z}$ and $\mathbf{V}_{22}$, it follows from Corollary 2.2.7.4 that $E[(\mathrm{vec}\mathbf{U})^{\otimes 4}]$ is a constant matrix multiplied by

$$E[(\mathrm{vec}\{(\mathbf{C}\mathbf{C}')^{-1} \otimes \mathbf{Z}\mathbf{V}_{22}^{-1}\mathbf{Z}'\})^{\otimes 2}],$$

which in turn is a function of

$$E[(\text{vec}\mathbf{V}_{22}^{-1})^{\otimes 2}],$$

which by Theorem 2.4.14 (iv) is $O(n^{-2})$. Since $\mathbf{Z}$ is normally distributed and independent of n, the corollary is established. ∎

Since the order of magnitude of the upper bound of approximation error of $\mathbf{f}_{\widehat{\mathbf{B}}}(\mathbf{B}_0)$ is $O(n^{-2})$, the approximation is fairly accurate. For example, in many cases, such as in Theorem 3.1.4 (i) and (ii), the law of large numbers gives an error of order $O(n^{-1})$.

We are going to investigate under which assumptions $f_{\mathbf{B}_E}(\mathbf{B}_0)$, given in (4.3.12), is a density. In general, the sum of the first few terms in an Edgeworth-type approximation is not a density. First observe that

$$\begin{aligned}\int_{\mathbf{X}\in\mathbb{R}^{q\times k}} f_{\mathbf{B}_E}(\mathbf{X})d\mathbf{X} =&1 - \tfrac{1}{2}skq + \tfrac{1}{2}sE[\text{tr}\{\mathbf{A}'\mathbf{\Sigma}^{-1}\mathbf{A}(\mathbf{B}_N - \mathbf{B})\mathbf{C}\mathbf{C}'(\mathbf{B}_N - \mathbf{B})'\}]\\ =&1 - \tfrac{1}{2}skq + \tfrac{1}{2}skq = 1.\end{aligned}$$

It remains to check whether $f_{\mathbf{B}_E}(\mathbf{B}_0) \geq 0$ for all $\mathbf{B}_0$. Since

$$\text{tr}\{\mathbf{A}'\mathbf{\Sigma}^{-1}\mathbf{A}(\mathbf{B}_0 - \mathbf{B})\mathbf{C}\mathbf{C}'(\mathbf{B}_0 - \mathbf{B})'\} \geq 0,$$

it is required that

$$1 - \frac{1}{2}skq \geq 0,$$

which is equivalent to

$$n \geq (p-q)(1 + \tfrac{1}{2}kq) + k + 1. \tag{4.3.19}$$

Hence, if (4.3.19) is true, $f_{\mathbf{B}_E}(\mathbf{B}_0)$ is a density. Moreover, in this case the density represents a matrix elliptically distributed variable. This is concluded in the next theorem

Theorem 4.3.2. *Let the distribution of $\mathbf{B}_E$ be defined via its density in (4.3.12) and suppose that (4.3.19) holds. Then $\mathbf{B}_E$ is matrix elliptically distributed.* ∎

There are other properties of the approximation which are worth observing.

Theorem 4.3.3. *Suppose that (4.3.19) holds. Let $\widehat{\mathbf{B}}$ be given by (4.3.1) and the distribution of $\mathbf{B}_E$ be defined via its density in (4.3.12). Then*

(i) $$E[\mathbf{B}_E] = E[\widehat{\mathbf{B}}] = \mathbf{B};$$

(ii) $$D[\mathbf{B}_E] = D[\widehat{\mathbf{B}}] = \frac{n-k-1}{n-k-p+q-1}(\mathbf{C}\mathbf{C}')^{-1} \otimes (\mathbf{A}'\mathbf{\Sigma}^{-1}\mathbf{A})^{-1}.$$

PROOF: To prove (i) we note first that since $f_{\mathbf{B}_E}(\mathbf{X})$ is a symmetric function around $\mathbf{X} - \mathbf{B}$,

$$\int_{\mathbf{X}\in\mathbb{R}^{q\times k}} (\mathbf{X} - \mathbf{B})f_{\mathbf{B}_E}(\mathbf{X})\,d\mathbf{X} = \mathbf{0}$$

and thus (i) follows, because $\int_{\mathbf{X}\in\mathbb{R}^{q\times k}} f_{\mathbf{B}_E}(\mathbf{X})\,d\mathbf{X} = 1$.
For (ii) we have to calculate

$$(1-\frac{skq}{2})D[\mathbf{Y}] + \frac{s}{2}E[\mathrm{vec}\mathbf{Y}\mathrm{vec}'\mathbf{Y}\mathrm{tr}(\mathbf{A}'\mathbf{\Sigma}^{-1}\mathbf{A}\mathbf{Y}\mathbf{C}\mathbf{C}'\mathbf{Y}')], \tag{4.3.20}$$

where

$$\mathbf{Y} \sim N_{q,k}(\mathbf{0}, (\mathbf{A}'\mathbf{\Sigma}^{-1}\mathbf{A})^{-1}, (\mathbf{C}\mathbf{C}')^{-1}).$$

Now, let $\mathbf{\Psi} = (\mathbf{C}\mathbf{C}')^{-1}\otimes(\mathbf{A}'\mathbf{\Sigma}^{-1}\mathbf{A})^{-1}$. Then, by applying Proposition 1.3.14 (iii),

$$\begin{aligned}&\mathrm{vec}E[\mathrm{vec}\mathbf{Y}\mathrm{vec}'\mathbf{Y}\mathrm{tr}(\mathbf{A}'\mathbf{\Sigma}^{-1}\mathbf{A}\mathbf{Y}\mathbf{C}\mathbf{C}'\mathbf{Y}')] = (\mathbf{I}_{qk}\otimes\mathbf{I}_{qk}\otimes\mathrm{vec}'\mathbf{\Psi}^{-1})E[(\mathrm{vec}\mathbf{Y})^{\otimes 4}]\\ &= (\mathbf{I}_{qk}\otimes\mathbf{I}_{qk}\otimes\mathrm{vec}'\mathbf{\Psi}^{-1})(\mathbf{I}_{(qk)^4} + \mathbf{I}_{qk}\otimes\mathbf{K}_{qk,qk}\otimes\mathbf{I}_{qk} + \mathbf{I}_{qk}\otimes\mathbf{K}_{(qk)^2,qk})(\mathrm{vec}\mathbf{\Psi})^{\otimes 2},\end{aligned} \tag{4.3.21}$$

where Corollary 2.2.7.4 has been used in the last equality. Since, according to Proposition 1.3.14 (iv),

$$(\mathbf{I}_{qk}\otimes\mathbf{K}_{qk,qk}\otimes\mathbf{I}_{qk})(\mathrm{vec}\mathbf{\Psi})^{\otimes 2} = \mathrm{vec}(\mathbf{\Psi}\otimes\mathbf{\Psi})$$

and

$$(\mathbf{I}_{qk}\otimes\mathbf{K}_{(qk)^2,qk})(\mathrm{vec}\mathbf{\Psi})^{\otimes 2} = \mathrm{vec}(\mathbf{K}_{qk,qk}(\mathbf{\Psi}\otimes\mathbf{\Psi})),$$

the expression in (4.3.21) is equivalent to

$$\begin{aligned}&\mathrm{vec}E[\mathrm{vec}\mathbf{Y}\mathrm{vec}'\mathbf{Y}\mathrm{tr}(\mathbf{A}'\mathbf{\Sigma}^{-1}\mathbf{A}\mathbf{Y}\mathbf{C}\mathbf{C}'\mathbf{Y}')]\\ &\quad= (\mathbf{I}_{qk}\otimes\mathbf{I}_{qk}\otimes\mathrm{vec}'\mathbf{\Psi}^{-1})((\mathrm{vec}\mathbf{\Psi})^{\otimes 2} + \mathrm{vec}(\mathbf{\Psi}\otimes\mathbf{\Psi}) + \mathrm{vec}(\mathbf{K}_{pq,pq}\mathbf{\Psi}\otimes\mathbf{\Psi}))\\ &\quad= (qk+2)\mathrm{vec}D[\mathbf{Y}].\end{aligned}$$

Hence, (4.3.20) equals

$$(1-\frac{skq}{2})D[\mathbf{Y}] + \frac{s}{2}(qk+2)D[\mathbf{Y}] = (1+s)D[\mathbf{Y}],$$

which according to Theorem 4.2.1 (ii) is identical to $\mathbf{D}[\widehat{\mathbf{B}}]$, and establishes the theorem. ∎

From Theorem 4.3.3 one can conclude that the basic statistical properties are kept in the approximation. This is, of course, also an indication that the approximation is performing well. However, when $f_{\mathbf{B}_\mathbf{E}}(\mathbf{X})$ is not positive, one should be careful in applications.
Next we turn to a different approach and will use a normally distributed variable with the same mean and dispersion matrix as $\widehat{\mathbf{B}}$. Let $\mathbf{B}_{\underline{N}}$ be the approximating normally distributed matrix such that

$$\begin{aligned}E[\widehat{\mathbf{B}}] &= E[\mathbf{B}_{\underline{N}}] = \mathbf{B},\\ D[\widehat{\mathbf{B}}] &= D[\mathbf{B}_{\underline{N}}] = (1+s)(\mathbf{C}\mathbf{C}')^{-1}\otimes(\mathbf{A}'\mathbf{\Sigma}^{-1}\mathbf{A})^{-1},\end{aligned}$$

i.e. $\mathbf{B}_{\underline{N}} \sim N_{q,k}(\mathbf{B}, (1+s)(\mathbf{A}'\boldsymbol{\Sigma}^{-1}\mathbf{A})^{-1}, (\mathbf{CC}')^{-1})$, where s is given in Theorem 4.3.1. Since the third cumulant of $\widehat{\mathbf{B}}$ is zero, and for $\mathbf{B}_{\underline{N}}$ cumulants of order higher than 2 equal zero, a straightforward extension of Corollary 3.2.1.1 and Definition 2.2.2 yield

$$\begin{aligned} f_{\widehat{\mathbf{B}}}(\mathbf{B}_0) &\approx f_{\mathbf{B}_{\underline{N}}}(\mathbf{B}_0) \\ &+ \frac{1}{4!}\mathrm{vec}'c_4[\widehat{\mathbf{B}}]\mathrm{vec}\mathbf{H}_4(\mathrm{vec}\mathbf{B}_0, (1+s)(\mathbf{CC}')^{-1} \otimes (\mathbf{A}'\boldsymbol{\Sigma}^{-1}\mathbf{A})^{-1}) f_{\mathbf{B}_{\underline{N}}}(\mathbf{B}_0), \end{aligned} \tag{4.3.22}$$

where the Hermite polynomial $\mathbf{H}_4(\bullet, \bullet)$ is defined in §2.2.4. Let us approximate $f_{\widehat{\mathbf{B}}}(\bullet)$ by $f_{\mathbf{B}_{\underline{N}}}(\bullet)$ and let the second term on the right hand side of (4.3.22) indicate the order of the error term. By Theorem 2.2.12 it follows that

$$\mathbf{H}_4(\mathrm{vec}\mathbf{B}_0, (1+s)(\mathbf{CC}')^{-1} \otimes (\mathbf{A}'\boldsymbol{\Sigma}^{-1}\mathbf{A})^{-1})$$

is $O(1)$. Now $c_{2r}[\widehat{\mathbf{B}}]$ will be studied and we are going to use the characteristic function. From (4.3.10) it follows that

$$E[e^{i\,\mathrm{tr}(\mathbf{T}'\widehat{\mathbf{B}})}] = E[e^{i\,\mathrm{tr}(\mathbf{T}'\mathbf{B}_N)}]E[e^{i\,\mathrm{tr}(\mathbf{T}'\mathbf{U})}],$$

where $\mathbf{B}_N$ is defined in (4.3.9) and independent of $\mathbf{U}$ in (4.3.19). We are going to show that $c_{2r}[\widehat{\mathbf{B}}]$ is $O(n^{-r})$, $r = 2, 3, 4, \ldots$, from where it follows that $\overline{m}_{2r}[\widehat{\mathbf{B}}]$ is $O(n^{-r})$, $r = 2, 3, 4, \ldots$. Since $m_{2r}[\mathbf{B}_N]$ is $O(1)$, $r = 2, 3, 4, \ldots$, we have to consider

$$m_{2r}[\mathbf{U}] = i^{-2r}E[\left.\frac{d^{2r}e^{i\,\mathrm{tr}(\mathbf{T}'\mathbf{U})}}{d\mathbf{T}^{2r}}\right|_{\mathbf{T}=\mathbf{0}}].$$

Observe, as when proving Corollary 4.3.11, that

$$\mathbf{U}|\mathbf{S} \sim N_{q,k}(\mathbf{0}, (\mathbf{A}'\mathbf{S}^{-1}\mathbf{A})^{-1}\mathbf{A}'\mathbf{S}^{-1}\boldsymbol{\Sigma}\mathbf{S}^{-1}\mathbf{A}(\mathbf{A}'\mathbf{S}^{-1}\mathbf{A})^{-1} - (\mathbf{A}'\boldsymbol{\Sigma}^{-1}\mathbf{A})^{-1}, (\mathbf{CC}')^{-1})$$

and

$$\begin{aligned} &(\mathbf{A}'\mathbf{S}^{-1}\mathbf{A})^{-1}\mathbf{A}'\mathbf{S}^{-1}\boldsymbol{\Sigma}\mathbf{S}^{-1}\mathbf{A}(\mathbf{A}'\mathbf{S}^{-1}\mathbf{A})^{-1} - (\mathbf{A}'\boldsymbol{\Sigma}^{-1}\mathbf{A})^{-1} \\ &\quad = (\mathbf{H}')^{-1}\mathbf{V}_{12}\mathbf{V}_{22}^{-1}\mathbf{V}_{22}^{-1}\mathbf{V}_{21}\mathbf{H}^{-1}, \end{aligned}$$

where $\mathbf{V} \sim W_p(\mathbf{I}, n-k)$, with its partition defined in (4.2.16), and $\mathbf{H}$ is the same as in (4.3.16). By assumption $\mathbf{Z}$, given in (4.3.18), is independent of $\mathbf{V}_{22}$ and

$$E[\left.\frac{d^{2r}e^{i\,\mathrm{tr}(\mathbf{T}'\mathbf{U})}}{d\mathbf{T}^{2r}}\right|_{\mathbf{T}=\mathbf{0}}] = E[\left.\frac{d^{2r}e^{-1/2\mathrm{tr}(\mathbf{Z}\mathbf{V}_{22}^{-1}\mathbf{Z}'\mathbf{T}(\mathbf{CC}')^{-1}\mathbf{T}')}}{d\mathbf{T}^{2r}}\right|_{\mathbf{T}=\mathbf{0}}].$$

Now the exponential function in the above expression is expanded and it is enough to consider the $(r+1)-$th term, i.e.

$$E[\frac{d^{2r}(-1/2)^r\{\mathrm{tr}(\mathbf{Z}\mathbf{V}_{22}^{-1}\mathbf{Z}'\mathbf{T}(\mathbf{CC}')^{-1}\mathbf{T}')\}^r}{d\mathbf{T}^{2r}}]. \tag{4.3.23}$$

By expressing the trace function with the help of the vec-operator, we obtain that (4.3.23) equals

$$\frac{d^{2r}\mathbf{T}^{\otimes 2r}}{d\mathbf{T}^{2r}}\mathbf{I}_{(qk)^{2r-1}} \otimes E[\mathbf{Z}^{\otimes 2r}]E[(\mathrm{vec}\mathbf{V}_{22}^{-1})^{\otimes r}] \otimes (\mathrm{vec}(\mathbf{CC}')^{-1})^{\otimes r}.$$

Here $\frac{d^{2r}\mathbf{T}^{\otimes 2r}}{d\mathbf{T}^{2r}}$ is a constant matrix, $E[\mathbf{Z}^{\otimes 2r}]$ because of normality is $O(1)$ and one can conclude from (2.4.46), with the help of induction, that $E[(\mathrm{vec}\mathbf{V}_{22}^{-1})^{\otimes r}]$ is $O(n^{-r})$. Thus it has been shown that (4.3.23) is $O(n^{-r})$. In particular, we have shown that the second term on the right hand side of (4.3.22) is $O(n^{-2})$. However, it also means that the overall error of the expansion is determined by a geometrical series, i.e. we get the terms $O(n^2)$, $O(n^3)$, $O(n^4)$, ..., and

$$\sum_{r=2}^{\infty} \frac{1}{n^r} = \frac{\frac{1}{n^2}}{1-\frac{1}{n}}.$$

Thus we may conclude that the overall error is $O(n^{-2})$. One advantage of such approximation is that it guarantees an approximation with a density function where the mean and dispersion are the same as the mean and dispersion of $\widehat{\mathbf{B}}$.

Theorem 4.3.4. *Let $\widehat{\mathbf{B}}$ be given by (4.3.1) and*

$$\mathbf{B}_{\underline{N}} \sim N_{q,k}(\mathbf{B}, (1+s)(\mathbf{A}'\mathbf{\Sigma}^{-1}\mathbf{A})^{-1}, (\mathbf{CC}')^{-1}),$$

where

$$s = \frac{p-q}{n-k-p+q-1}.$$

Then

$$|f_{\widehat{\mathbf{B}}}(\mathbf{B}_0) - f_{\mathbf{B}_{\underline{N}}}(\mathbf{B}_0)| = O(n^{-2}).$$

∎

Thus, Theorem 4.3.1 and Theorem 4.3.4 give two alternatives of approximating $f_{\widehat{\mathbf{B}}}(\mathbf{B}_0)$. It is by no means obvious which one has to be used. If the condition (4.3.19) for $f_{\mathbf{B}_E}(\mathbf{B}_0)$ being a density is not fulfilled, the choice is even more problematic. It seems reasonable to use a matrix elliptical distribution because this can have heavier tails than the normal distribution. This is what is expected from the distribution of $\widehat{\mathbf{B}}$ because $(\mathbf{A}'\mathbf{S}^{-1}\mathbf{A})^{-1}\mathbf{A}'\mathbf{S}^{-1}$ adds some variation to the normal variation induced by $\mathbf{XC}'(\mathbf{CC}')^{-1}$. The choice of an approximation depends on the situation and the interested reader is suggested to perform explicit calculations for various choices of $\mathbf{A}$ and $\mathbf{C}$ matrices in the model.

4.3.3 *Approximation of the distribution of $\widehat{\mathbf{\Sigma}}$ in the Growth Curve model*

The dispersion matrix $\widehat{\mathbf{\Sigma}}$ was given in (4.3.2), although from now on no full rank conditions are assumed. Moreover, we are not going to work with $\widehat{\mathbf{\Sigma}}$ directly.

Instead, $\widehat{\boldsymbol{\Sigma}}$ will be centered around its mean and standardized to some extent. Thus, instead of $\widehat{\boldsymbol{\Sigma}}$, the density approximation of

$$\mathbf{R} = n\boldsymbol{\Gamma}\boldsymbol{\Sigma}^{-1/2}(\widehat{\boldsymbol{\Sigma}} - E[\widehat{\boldsymbol{\Sigma}}])\boldsymbol{\Sigma}^{-1/2}\boldsymbol{\Gamma}' \qquad (4.3.24)$$

will be considered, where $E[\widehat{\boldsymbol{\Sigma}}]$ is given in Theorem 4.2.3 (i), and $\boldsymbol{\Gamma}$ and the symmetric square root $\boldsymbol{\Sigma}^{-1/2}$ are the matrices used in the factorization of $\mathbf{A}$ in (4.3.16). We do not have to worry about the fact that $\mathbf{A}$ may not be of full rank because the expressions including $\mathbf{A}$ are unique projections. The matrix $\mathbf{R}$ can be decomposed in various ways. The reason for decomposing $\mathbf{R}$ is the same as in the case when $\widehat{\mathbf{B}}$ was discussed. We have some information about the moments of one part of the decomposition. The idea is to use this and to perform an adjustment which is based on the knowledge about the moments. Instead of (4.3.24), we will now consider the equivalent representation

$$\mathbf{R} = \mathbf{V} - (n - r(\mathbf{C}))\mathbf{I} + (\mathbf{V}_{22}^{-1}\mathbf{V}_{21} : \mathbf{I})'\mathbf{Q}(\mathbf{V}_{22}^{-1}\mathbf{V}_{21} : \mathbf{I}) - \boldsymbol{\Psi},$$

where $\mathbf{V}$ and its partition are given in (4.2.16),

$$\mathbf{Q} = \boldsymbol{\Gamma}\boldsymbol{\Sigma}^{-1/2}\mathbf{X}\mathbf{C}'(\mathbf{C}\mathbf{C}')^{-}\mathbf{C}\mathbf{X}'\boldsymbol{\Sigma}^{-1/2}\boldsymbol{\Gamma}'$$

and

$$\boldsymbol{\Psi} = E[(\mathbf{V}_{22}^{-1}\mathbf{V}_{21} : \mathbf{I})'\mathbf{Q}(\mathbf{V}_{22}^{-1}\mathbf{V}_{21} : \mathbf{I})].$$

Observe that $\mathbf{V}$ and $\mathbf{Q}$ are independently distributed and that

$$E[\mathbf{V}] = (n - r(\mathbf{C}))\mathbf{I}.$$

Since according to Corollary 2.4.12.1 $\mathbf{V}_{12}\mathbf{V}_{22}^{-1/2}$ is normally distributed and independent of $\mathbf{V}_{22}$, it follows from Theorem 2.2.9 (i) and Theorem 2.4.14 (iii) that

$$E[\mathbf{V}_{12}\mathbf{V}_{22}^{-1}\mathbf{V}_{22}^{-1}\mathbf{V}_{21}] = r(\mathbf{C})E[\mathrm{tr}(\mathbf{V}_{22}^{-1})] = a\mathbf{I},$$

where

$$a = \frac{r(\mathbf{C})(p - r(\mathbf{A}))}{n - r(\mathbf{C}) - p + r(\mathbf{A}) - 1}.$$

Furthermore, since $E[\mathbf{V}_{12}\mathbf{V}_{22}^{-1}\mathbf{Q}] = \mathbf{0}$,

$$\boldsymbol{\Psi} = \begin{pmatrix} a\mathbf{I}_{r(\mathbf{A})} & \mathbf{0} \\ \mathbf{0} & r(\mathbf{C})\mathbf{I}_{p-r(\mathbf{A})} \end{pmatrix}.$$

The matrix $\mathbf{R}$ is a function of a Wishart matrix and it is Wishart distributed in the special case $\mathbf{A} = \mathbf{I}$. Therefore the density of $\mathbf{R}$ will be approximated by a Wishart density. We are going to apply Theorem 3.2.3. The reason for using a centered Wishart distribution and not the ordinary Wishart distribution was motivated in §2.4.8. There are at least two natural and fairly simple ways of approximating $f_{\mathbf{R}}(\bullet)$. Both will be obtained as applications of Theorem 3.2.3.

Since $\widehat{\boldsymbol{\Sigma}} - \frac{1}{n}\mathbf{S} \xrightarrow{\mathcal{P}} \mathbf{0}$, we will first use the density of $\mathbf{S} \sim W_p(\boldsymbol{\Sigma}, n - r(\mathbf{C}))$. This means that we are ignoring the mean structure $\mathbf{ABC}$, as $\mathbf{S}$ reflects only the distribution of the difference between the observations and the "mean" $\mathbf{X}\mathbf{C}'(\mathbf{C}\mathbf{C}')^{-}$. Remember that we are considering symmetric matrices in the following and therefore we use the upper triangle of these matrices, which means that $\frac{1}{2}p(p+1)$ elements are used. The next theorem is immediately established by using the first terms of the expansion in Theorem 3.2.3.

Theorem 4.3.5. *Let $g_{\mathbf{W}}(\bullet)$ denote the density function of $V^2(\mathbf{W})$, where*

$$\mathbf{W} = \mathbf{\Gamma}\mathbf{\Sigma}^{-1/2}(\mathbf{S} - E[\mathbf{S}])\mathbf{\Sigma}^{-1/2}\mathbf{\Gamma}',$$

with $\mathbf{S} \sim W_p(\mathbf{\Sigma}, n - r(\mathbf{C}))$ and $\mathbf{\Gamma}$ is the same as in (4.3.24). Let $\mathbf{R}$ be given by (4.3.24). The density of $V^2(\mathbf{R})$ is approximated by

$$f_{\mathbf{R}}(\mathbf{R}_0) = g_{\mathbf{W}}(\mathbf{R}_0)$$
$$\times \left\{1 + \frac{1}{2}\text{vec}'\{D[V^2(\mathbf{R})] - (n - r(\mathbf{C}))\mathbf{G}_p(\mathbf{I} + \mathbf{K}_{p,p})\mathbf{G}_p'\}\text{vec}\mathbf{L}_2^\star(\mathbf{R}_0, \mathbf{I})\right\} + \ldots,$$

where

$$n^{-2}D[V^2(\mathbf{R})] = d_1(\mathbf{I} + \mathbf{K}_{p,p})(\mathbf{M} \otimes \mathbf{M}) + \frac{1}{n}(\mathbf{I} + \mathbf{K}_{p,p})(\mathbf{N} \otimes \mathbf{N})$$
$$+ d_2(\mathbf{I} + \mathbf{K}_{p,p})(\mathbf{N} \otimes \mathbf{M} + \mathbf{M} \otimes \mathbf{N}) + d_3\text{vec}\mathbf{M}\text{vec}'\mathbf{M},$$

and d_1, d_2 and d_3 are as in Theorem 4.2.3,

$$\mathbf{M} = \begin{pmatrix} \mathbf{I}_{r(\mathbf{A})} & \mathbf{0} \\ \mathbf{0} & \mathbf{0} \end{pmatrix}, \qquad \mathbf{N} = \begin{pmatrix} \mathbf{0} & \mathbf{0} \\ \mathbf{0} & \mathbf{I}_{p-r(\mathbf{A})} \end{pmatrix},$$

$\mathbf{G}_p$ is defined by (1.3.49) and (1.3.50),

$$\mathbf{L}_2^\star(\mathbf{W}, \mathbf{I}) = -\tfrac{1}{2}\mathbf{G}_p\mathbf{H}_p\Big\{t(\mathbf{W} + (n - r(\mathbf{C}))\mathbf{I})^{-1} \otimes (\mathbf{W} + (n - r(\mathbf{C}))\mathbf{I})^{-1}$$
$$- \tfrac{1}{2}\text{vec}(t(\mathbf{W} + (n - r(\mathbf{C}))\mathbf{I})^{-1} - \mathbf{I})\text{vec}'(t(\mathbf{W} + (n - r(\mathbf{C}))\mathbf{I})^{-1} - \mathbf{I})\Big\}\mathbf{H}_p\mathbf{G}_p',$$

$\mathbf{H}_p = \mathbf{I} + \mathbf{K}_{p,p} - (\mathbf{K}_{p,p})_d$, and $t = n - r(\mathbf{C}) - p - 1$. ∎

Observe that in the theorem we have not given any error bound or indicated any order of the error. By copying the approach when considering the error in Theorem 4.3.4 this could be achieved, but we leave it to the interested reader. Unfortunately again it requires tedious calculations which are not of principal interest. The result shows that the first term in the remainder is $O(n^{-1})$ and that the following terms are of diminishing order.

A minor modification of the approach above is to approximate $f_{\mathbf{R}}(\bullet)$ with the help of a Wishart distribution having the same mean as $n\mathbf{\Gamma}\mathbf{\Sigma}^{-1/2}\widehat{\mathbf{\Sigma}}\mathbf{\Sigma}^{-1/2}\mathbf{\Gamma}'$. From Theorem 4.2.3 (i) it follows that $W_p(\mathbf{\Theta}, n)$, where

$$\mathbf{\Theta} = \begin{pmatrix} (1-c)\mathbf{I}_{r(\mathbf{A})} & \mathbf{0} \\ \mathbf{0} & \mathbf{I}_{p-r(\mathbf{A})} \end{pmatrix}, \tag{4.3.25}$$

$$c = \frac{r(\mathbf{C})(n - r(\mathbf{C}) - 2(p - r(\mathbf{A})) - 1)}{n(n - r(\mathbf{C}) - p + r(\mathbf{A}) - 1)}$$

can be used. If n becomes large, $c \to 0$, and in this case the next theorem will be identical to the previous Theorem 4.3.5.

Theorem 4.3.6. *Let $g_{\mathbf{W}}(\bullet)$ denote the density function of $V^2(\mathbf{W})$, where $\mathbf{W} \sim W_p(\mathbf{\Theta}, n)$, and $\mathbf{\Theta}$ be given in (4.3.25). Let $\mathbf{R}$ be given by (4.3.24). The density of $V^2(\mathbf{R})$ can be approximated as*

$$f_{\mathbf{R}}(\mathbf{R}_0)$$
$$= g_{\mathbf{W}}(\mathbf{R}_0)\{1 + \frac{1}{2}\text{vec}'\{D[V^2(\mathbf{R})] - n\mathbf{G}_p(\mathbf{I} + \mathbf{K}_{p,p})\mathbf{G}_p'\}\text{vec}\mathbf{L}_2^\star(\mathbf{R}_0, \mathbf{\Theta})\} + \dots,$$

where $D[V^2(\mathbf{R})]$ is given in Theorem 4.3.5, $\mathbf{G}_p$ is defined by (1.3.49) and (1.3.50),

$$\mathbf{L}_2^\star(\mathbf{W}, \mathbf{\Theta})$$
$$= -\tfrac{1}{2}\mathbf{G}_p\mathbf{H}_p\Big\{(n-p-1)(\mathbf{W} + n\mathbf{\Theta})^{-1} \otimes (\mathbf{W} + n\mathbf{\Theta})^{-1}$$
$$- \tfrac{1}{2}\text{vec}((n-p-1)(\mathbf{W} + n\mathbf{\Theta})^{-1} - \mathbf{\Theta}^{-1})\text{vec}'((n-p-1)(\mathbf{W} + n\mathbf{\Theta})^{-1} - \mathbf{\Theta}^{-1})\Big\}$$
$$\times \mathbf{H}_p\mathbf{G}_p',$$

and $\mathbf{H}_p = \mathbf{I} + \mathbf{K}_{p,p} - (\mathbf{K}_{p,p})_d$. ∎

The third idea for approximation is based on the decomposition of $\mathbf{R}$, given in (4.3.24):

$$\mathbf{R} = \mathbf{Z} + \mathbf{U},$$

where

$$\mathbf{Z} = \mathbf{\Gamma}\mathbf{\Sigma}^{-1/2}\{\mathbf{A}(\mathbf{A}'\mathbf{S}^{-1}\mathbf{A})^-\mathbf{A}' - E[\mathbf{A}(\mathbf{A}'\mathbf{S}^{-1}\mathbf{A})^-\mathbf{A}']\}\mathbf{\Sigma}^{-1/2}\mathbf{\Gamma}' \qquad (4.3.26)$$

and

$$\mathbf{U} = \mathbf{\Gamma}\mathbf{\Sigma}^{-1/2}\Big\{(\mathbf{I} - \mathbf{A}(\mathbf{A}'\mathbf{S}^{-1}\mathbf{A})^-\mathbf{A}'\mathbf{S}^{-1})\mathbf{X}\mathbf{X}'(\mathbf{I} - \mathbf{S}^{-1}\mathbf{A}(\mathbf{A}'\mathbf{S}^{-1}\mathbf{A})^-\mathbf{A}')$$
$$- E[(\mathbf{I} - \mathbf{A}(\mathbf{A}'\mathbf{S}^{-1}\mathbf{A})^-\mathbf{A}'\mathbf{S}^{-1})\mathbf{X}\mathbf{X}'(\mathbf{I} - \mathbf{S}^{-1}\mathbf{A}(\mathbf{A}'\mathbf{S}^{-1}\mathbf{A})^-\mathbf{A}')]\Big\}\mathbf{\Sigma}^{-1/2}\mathbf{\Gamma}'. \qquad (4.3.27)$$

This time we will present an approximation of the distribution function instead of approximating the density function. We start by showing that $\mathbf{Z}$ and $\mathbf{U}$ are independently distributed. The details are given in the next lemma.

Lemma 4.3.1. *Let $\mathbf{S} = \mathbf{X}(\mathbf{I} - \mathbf{C}'(\mathbf{C}\mathbf{C}')^-\mathbf{C})\mathbf{X}'$ and $\mathbf{X}$ be as in (4.1.1). Then $\mathbf{A}(\mathbf{A}'\mathbf{S}^{-1}\mathbf{A})^-\mathbf{A}'$ and $(\mathbf{I} - \mathbf{A}(\mathbf{A}'\mathbf{S}^{-1}\mathbf{A})^-\mathbf{A}'\mathbf{S}^{-1})\mathbf{X}$ are independently distributed.*

PROOF: As previously, $\mathbf{A}$ is factorized according to

$$\mathbf{A}' = \mathbf{H}(\mathbf{I}_{r(\mathbf{A})} : \mathbf{0})\mathbf{\Gamma}\mathbf{\Sigma}^{1/2},$$

where $\mathbf{H}$ is non-singular and $\mathbf{\Gamma}$ is an orthogonal matrix. Furthermore, let

$$\mathbf{Y} = \mathbf{\Gamma}\mathbf{\Sigma}^{-1/2}(\mathbf{X} - \mathbf{A}\mathbf{B}\mathbf{C}) \sim N_{p,n}(\mathbf{0}, \mathbf{I}_p, \mathbf{I}_n)$$

and

$$\mathbf{V} = \boldsymbol{\Gamma}\boldsymbol{\Sigma}^{-1/2}\mathbf{S}\boldsymbol{\Sigma}^{-1/2}\boldsymbol{\Gamma}' \sim W_p(\mathbf{I}, n - r(\mathbf{C})).$$

Then

$$\mathbf{A}(\mathbf{A}'\mathbf{S}^{-1}\mathbf{A})^{-}\mathbf{A}' = \boldsymbol{\Sigma}^{1/2}\boldsymbol{\Gamma}_1'(\mathbf{V}^{11})^{-1}\boldsymbol{\Gamma}_1\boldsymbol{\Sigma}^{1/2}, \tag{4.3.28}$$

where $\boldsymbol{\Gamma} = (\boldsymbol{\Gamma}_1' : \boldsymbol{\Gamma}_2')'$, and $(\mathbf{V}^{11})^{-1}$ is the left upper block of size $r(\mathbf{A}) \times r(\mathbf{A})$ of $\mathbf{V}^{-1}$, i.e. we use the partition of $\mathbf{V}$ and $\mathbf{V}^{-1}$ given by (4.2.16) and (4.2.17), respectively. Furthermore, let $\mathbf{Y}' = (\mathbf{Y}_1' : \mathbf{Y}_2')$, $n \times r(\mathbf{A}) : n \times (p - r(\mathbf{A}))$. Thus,

$$\begin{aligned}
&(\mathbf{I}-\mathbf{A}(\mathbf{A}'\mathbf{S}^{-1}\mathbf{A})^{-}\mathbf{A}'\mathbf{S}^{-1})\mathbf{X} \\
&\quad = \boldsymbol{\Sigma}^{1/2}\boldsymbol{\Gamma}'\{\mathbf{I} - (\mathbf{I} : \mathbf{0})'(\mathbf{V}^{11})^{-1}(\mathbf{V}^{11} : \mathbf{V}^{12})\}\boldsymbol{\Gamma}\boldsymbol{\Sigma}^{-1/2}\mathbf{X} \\
&\quad = \boldsymbol{\Sigma}^{1/2}\boldsymbol{\Gamma}'\begin{pmatrix} \mathbf{0} & \mathbf{V}_{12}\mathbf{V}_{22}^{-1} \\ \mathbf{0} & \mathbf{I} \end{pmatrix}\mathbf{Y} = \boldsymbol{\Sigma}^{1/2}(\boldsymbol{\Gamma}_1'\mathbf{V}_{12}\mathbf{V}_{22}^{-1} + \boldsymbol{\Gamma}_2')\mathbf{Y}_2.
\end{aligned} \tag{4.3.29}$$

According to Theorem 2.4.12 (ii), $\mathbf{V}^{11}$ is independent of $\mathbf{V}_{12}\mathbf{V}_{22}^{-1}$, and from Theorem 2.2.4 (iv) it follows that $\mathbf{V}^{11}$ is also independent of $\mathbf{Y}_2$. Hence, (4.3.28) and (4.3.29) establish the lemma. ∎

By rewriting $\mathbf{U}$ and $\mathbf{Z}$ given by (4.3.27) and (4.3.26), respectively, we obtain

$$\mathbf{Z} = \begin{pmatrix} (\mathbf{V}^{11})^{-1} & \mathbf{0} \\ \mathbf{0} & \mathbf{0} \end{pmatrix} - \begin{pmatrix} E[(\mathbf{V}^{11})^{-1}] & \mathbf{0} \\ \mathbf{0} & \mathbf{0} \end{pmatrix}, \tag{4.3.30}$$

$$\mathbf{U} = \begin{pmatrix} \mathbf{V}_{12}\mathbf{V}_{22}^{-1}\mathbf{Y}_2\mathbf{Y}_2'\mathbf{V}_{22}^{-1}\mathbf{V}_{21} & \mathbf{V}_{12}\mathbf{V}_{22}^{-1}\mathbf{Y}_2\mathbf{Y}_2' \\ \mathbf{Y}_2\mathbf{Y}_2'\mathbf{V}_{22}^{-1}\mathbf{V}_{21} & \mathbf{Y}_2\mathbf{Y}_2' \end{pmatrix} - \begin{pmatrix} b\mathbf{I}_{r(\mathbf{A})} & \mathbf{0} \\ \mathbf{0} & n\mathbf{I}_{p-r(\mathbf{A})} \end{pmatrix}, \tag{4.3.31}$$

where

$$b = \frac{(p - r(\mathbf{A}))(n - p + r(\mathbf{A}) - 1)}{n - r(\mathbf{C}) - p + r(\mathbf{A}) - 1}. \tag{4.3.32}$$

When proceeding we will make use of Lemma 4.3.2 given below, which presents an extension of the expansion

$$F_{\mathbf{Y}}(\mathbf{X}) = G_{\mathbf{Z}}(\mathbf{X}) - G_{\mathbf{Z}}^1(\mathbf{X})\text{vec}E[\mathbf{U}] + \frac{1}{2}E[\text{vec}'\mathbf{U}G_{\mathbf{Z}}^2(\mathbf{X})\text{vec}\mathbf{U}] + \ldots, \tag{4.3.33}$$

where $\mathbf{Y} = \mathbf{Z} + \mathbf{U}$, $\mathbf{Z}$ and $\mathbf{U}$ are independent, $F_{\mathbf{Y}}(\bullet)$ and $G_{\mathbf{Z}}(\bullet)$ denote the distribution functions of $\mathbf{Y}$ and $\mathbf{Z}$, respectively, and $G_{\mathbf{Z}}^k(\bullet)$ stands for the $k-$th derivative of $G_{\mathbf{Z}}(\bullet)$. The proof of (4.3.33) follows by using the equality $F_{\mathbf{Y}}(\mathbf{X}) = E_{\mathbf{U}}[G_{\mathbf{Z}}(\mathbf{X} - \mathbf{U})]$ and then performing a Taylor expansion. The statements of Lemma 4.3.2 can be proven according to these lines.

Lemma 4.3.2. *Let $\mathbf{Y}$, $\mathbf{Z}_{11}$ and $\mathbf{U}$ be symmetric matrices such that*

$$\mathbf{Y} = \begin{pmatrix} \mathbf{Z}_{11} & \mathbf{0} \\ \mathbf{0} & \mathbf{0} \end{pmatrix} + \begin{pmatrix} \mathbf{U}_{11} & \mathbf{U}_{12} \\ \mathbf{U}_{21} & \mathbf{U}_{22} \end{pmatrix},$$

where $\mathbf{Z}_{11}$ and $\mathbf{U} = [\mathbf{U}_{ij}]$ are independent. Partition the argument matrix $\mathbf{X}$ in the same way as $\mathbf{U}$. Let $F_{\mathbf{Y}}(\mathbf{X})$, $G_{\mathbf{Z}_{11}}(\mathbf{X}_{11})$ denote the distribution functions of $\mathbf{Y}$

and $\mathbf{Z}_{11}$*, respectively, and let* $h_{\mathbf{U}}(V^2(\mathbf{X}))$ *be the density of* $V^2(\mathbf{U})$*. Furthermore, the inequality " $\leq$ " in* $\mathbf{U}_{12} \leq \mathbf{X}_{12}$ *means elementwice inequality, whereas " $\leq$ " in* $V^2(\mathbf{U}_{22}) \leq V^2(\mathbf{X}_{22})$ *means that* $\mathbf{X}_{22} - \mathbf{U}_{22}$ *should be p.d. Then*

$$F_{\mathbf{Y}}(\mathbf{X}) = \int\limits_{\substack{V^2(\mathbf{U}_{11}) \\ \mathbf{U}_{12}\leq \mathbf{X}_{12} \\ V^2(\mathbf{U}_{22})\leq V^2(\mathbf{X}_{22})}} G_{\mathbf{Z}_{11}}(\mathbf{X}_{11} - \mathbf{U}_{11}) h_{\mathbf{U}}(V^2(\mathbf{X})) dV^2(\mathbf{U})$$

$$\approx \int\limits_{\substack{V^2(\mathbf{U}_{11}) \\ \mathbf{U}_{12}\leq \mathbf{X}_{12} \\ V^2(\mathbf{U}_{22})\leq V^2(\mathbf{X}_{22})}} \Big\{ G_{\mathbf{Z}_{11}}(\mathbf{X}_{11}) - \mathbf{G}^1_{\mathbf{Z}_{11}}(\mathbf{X}_{11})\mathrm{vec}\mathbf{U}_{11}$$

$$+ \frac{1}{2}\mathrm{vec}'\mathbf{U}_{11}\mathbf{G}^2_{\mathbf{Z}_{11}}(\mathbf{X}_{11})\mathrm{vec}\mathbf{U}_{11}\Big\} h_{\mathbf{U}}(V^2(\mathbf{X})) dV^2(\mathbf{U}).$$

∎

Note that the lemma deals with the case

$$\mathbf{Y} = \mathbf{AZB} + \mathbf{U},$$

where $\mathbf{A}$ and $\mathbf{B}$ are known matrices of proper sizes and ranks.
Thus, when approximating $\mathbf{R} = \mathbf{Z} + \mathbf{U}$, Lemma 4.3.1 states that $\mathbf{Z}$ in (4.3.26) and $\mathbf{U}$ in (4.3.27) are independently distributed. Furthermore, from (4.3.30) and Theorem 2.4.12 (i) it follows that $\mathbf{Z}_{11}$ is centered Wishart distributed, where $\mathbf{Z}_{11} + E[\mathbf{Z}_{11}]$ is $W_{r(\mathbf{A})}(\mathbf{I}, n - r(\mathbf{C}) - p + r(\mathbf{A}))$ and the other elements in $\mathbf{Z}$ equal 0. Therefore, Lemma 4.3.2 implies that the distribution function $F_{\mathbf{R}}(\bullet)$ of $\mathbf{R}$ can be approximated by

$$F_{\mathbf{R}}(\mathbf{X}) \approx \int\limits_{\substack{V^2(\mathbf{U}_{11}) \\ \mathbf{U}_{12}\leq \mathbf{X}_{12} \\ V^2(\mathbf{U}_{22})\leq V^2(\mathbf{X}_{22})}} \Big\{ G_{\mathbf{Z}_{11}}(\mathbf{X}_{11}) - \mathbf{G}^1_{\mathbf{Z}_{11}}(\mathbf{X}_{11})\mathrm{vec}\mathbf{U}_{11}$$

$$+ \frac{1}{2}\mathrm{vec}'\mathbf{U}_{11}\mathbf{G}^2_{\mathbf{Z}_{11}}(\mathbf{X}_{11})\mathrm{vec}\mathbf{U}_{11}\Big\} h_{\mathbf{U}}(V^2(\mathbf{X})) dV^2(\mathbf{U}), \quad (4.3.34)$$

where $G_{\mathbf{Z}_{11}}$ is the distribution function of $\mathbf{Z}_{11}$, and $\mathbf{G}^k_{\mathbf{Z}_{11}}$, $k = 1, 2$, denotes its $k-$th derivative (for the interpretation of " $\geq$ " see Lemma 4.3.2)). The next step is to approximate the density $h_{\mathbf{U}}(V^2(\mathbf{X}))$ with the help of $\widetilde{\mathbf{U}}$ following a centered Wishart distribution, i.e.

$$\widetilde{\mathbf{U}} + n\mathbf{\Psi} \sim W_p(\mathbf{\Psi}, n).$$

Because of (4.3.31), we choose

$$\mathbf{\Psi} = \begin{pmatrix} \frac{1}{n}b\mathbf{I}_{r(\mathbf{A})} & \mathbf{0} \\ \mathbf{0} & \mathbf{I}_{p-r(\mathbf{A})} \end{pmatrix},$$

where b is given in (4.3.32). Let $k_{\widetilde{\mathbf{U}}}(\bullet)$ denote the density of $\widetilde{\mathbf{U}}$. Moreover, let $\mathbf{M}$ and $\mathbf{N}$ be as in Theorem 4.3.5. Theorem 3.2.1 yields

$$h_{V^2(\mathbf{U})}(\mathbf{U}_0) = k_{\widetilde{\mathbf{U}}}(\mathbf{U}_0)\{1 + \frac{1}{2}\mathrm{vec}'(D[V^2(\mathbf{U})] - D[V^2(\widetilde{\mathbf{U}})])\mathrm{vec}\mathbf{L}_2^\star(\mathbf{U}_0, \boldsymbol{\Psi})\} + \dots,$$

where

$$D[V^2(\widetilde{\mathbf{U}})]) = n\mathbf{G}_p(\mathbf{I} + \mathbf{K}_{p,p})(\boldsymbol{\Psi} \otimes \boldsymbol{\Psi})\mathbf{G}_p', \tag{4.3.35}$$

$\mathbf{G}_p$ is defined by (1.3.49) and (1.3.50). With the help of Theorem 4.2.3, and from the independence of $\mathbf{U}$ and $\mathbf{Z}$,

$$\begin{aligned} D[V^2(\mathbf{U})] &= n^2\mathbf{G}_p\{D[\boldsymbol{\Gamma}\boldsymbol{\Sigma}^{-1/2}\widehat{\boldsymbol{\Sigma}}\boldsymbol{\Sigma}^{-1/2}\boldsymbol{\Gamma}] - D[\mathbf{Z}]\}\mathbf{G}_p' \\ &= \mathbf{G}_p\Big\{(n^2d_1 - n + r(\mathbf{C}) + p - r(\mathbf{A}))(\mathbf{I} + \mathbf{K}_{p,p})(\mathbf{M} \otimes \mathbf{M}) + n(\mathbf{I} + \mathbf{K}_{p,p})(\mathbf{N} \otimes \mathbf{N}) \\ &\quad + n^2d_2(\mathbf{I} + \mathbf{K}_{p,p})(\mathbf{N} \otimes \mathbf{M} + \mathbf{M} \otimes \mathbf{N}) + n^2d_3\mathrm{vec}\mathbf{M}\mathrm{vec}'\mathbf{M}\Big\}\mathbf{G}_p', \end{aligned} \tag{4.3.36}$$

where d_1, d_2 and d_3 are defined in Theorem 4.2.3, and by Lemma 2.4.2 and Theorem 2.4.16 we have $(t = n - p - 1)$

$$\begin{aligned} \mathrm{vec}\mathbf{L}_2^\star(\mathbf{U}, \boldsymbol{\Psi}) &= -\frac{1}{2}(\mathbf{G}_p\mathbf{H}_p \otimes \mathbf{G}_p\mathbf{H}_p) \\ &\times \Big\{\{t(\mathbf{I}_p \otimes \mathbf{K}_{p,p} \otimes \mathbf{I}_p) - \frac{t^2}{2}\mathbf{I}\}\mathrm{vec}(\mathbf{U} + n\boldsymbol{\Psi})^{-1} \otimes \mathrm{vec}(\mathbf{U} + n\boldsymbol{\Psi})^{-1} \\ &\quad + \frac{1}{2}\mathrm{vec}(t(\mathbf{U} + n\boldsymbol{\Psi})^{-1} - \boldsymbol{\Psi}^{-1}) \otimes \mathrm{vec}\boldsymbol{\Psi}^{-1} \\ &\quad + \frac{1}{2}\mathrm{vec}\boldsymbol{\Psi}^{-1} \otimes \mathrm{vec}(t(\mathbf{U} + n\boldsymbol{\Psi})^{-1} - \boldsymbol{\Psi}^{-1}) + \frac{1}{2}\mathrm{vec}\boldsymbol{\Psi}^{-1} \otimes \mathrm{vec}\boldsymbol{\Psi}^{-1}\Big\}. \end{aligned} \tag{4.3.37}$$

Note that it is much easier to integrate over $k_{\widetilde{\mathbf{U}}}(\bullet)$ instead of $h_{\mathbf{U}}(\bullet)$ when performing a correction in (4.3.34). The result is summarized in the next theorem.

Theorem 4.3.7. *Let $\mathbf{G}_{\mathbf{Z}_{11}}$ be the distribution function of $(\mathbf{V}^{11})^{-1} - (n - r(\mathbf{C}) - p + r(\mathbf{A}))\mathbf{I}$, where $(\mathbf{V}^{11})^{-1} \sim W_{p-r(\mathbf{A})}(\mathbf{I}, n - r(\mathbf{C}) - p + r(\mathbf{A}))$. The distribution function $F_{\mathbf{R}}(\mathbf{X})$ of $\mathbf{R}$, given by (4.3.24), can be approximated as*

$$\begin{aligned} F_{\mathbf{R}}(\mathbf{X}) \approx \int\limits_{\substack{V^2(\widetilde{\mathbf{U}}_{11}) \\ \widetilde{\mathbf{U}}_{12} \le \mathbf{X}_{12} \\ V^2(\widetilde{\mathbf{U}}_{22}) \le V^2(\mathbf{X}_{22} - n\mathbf{I}_{p-r(\mathbf{A})})}} & \Big\{G_{\mathbf{Z}_{11}}(\mathbf{X}_{11}) - \mathbf{G}_{\mathbf{Z}_{11}}^1(\mathbf{X}_{11})\mathrm{vec}\widetilde{\mathbf{U}}_{11} \\ & + \frac{1}{2}\mathrm{vec}'\widetilde{\mathbf{U}}_{11}\mathbf{G}_{\mathbf{Z}_{11}}^2(\mathbf{X}_{11})\mathrm{vec}\widetilde{\mathbf{U}}_{11}\Big\} \\ \times \Big\{1 + \frac{1}{2}\mathrm{vec}'(D[V^2(\mathbf{U})] - D[V^2(\widetilde{\mathbf{U}})])&\mathrm{vec}\mathbf{L}_2^\star(\widetilde{\mathbf{U}}, \boldsymbol{\Psi})\Big\}k_{\widetilde{\mathbf{U}}}(V^2(\widetilde{\mathbf{U}}))dV^2(\widetilde{\mathbf{U}}), \end{aligned}$$

where $D[V^2(\mathbf{U})]$, $D[V^2(\widetilde{\mathbf{U}})]$ *and* $\mathbf{L}_2^\star(\widetilde{\mathbf{U}}, \mathbf{\Psi})$ *are given by (4.3.35) – (4.3.37).* ∎

Observe that by conditioning we can simplify the computations in Theorem 4.3.7. We may consider three independent variables and the corresponding densities, from which one is the density of

$$\widetilde{\mathbf{U}}_{1\cdot 2} = W_{r(\mathbf{A})}(b\mathbf{I}, n - p + r(\mathbf{A})),$$

where b is given in (4.3.32). Since $(\widetilde{\mathbf{U}} + n\mathbf{\Psi})$ is a linear function in $(\widetilde{\mathbf{U}}_{1\cdot 2})^{-1}$, we may integrate out $\widetilde{\mathbf{U}}_{1\cdot 2}$ from the approximation of $F_{\mathbf{R}}(\mathbf{X})$ given in Theorem 4.3.7. For details see Kollo & von Rosen (2000).

4.3.4 *Approximating the distribution of the mean parameter estimators in the* MLNM($\mathbf{ABC} + \mathbf{B}_2\mathbf{C}_2$)

In Theorem 4.1.9, the maximum likelihood estimators of $\mathbf{B}$ and $\mathbf{B}_2$ were given and we shall write them out once again. Under full rank conditions the estimators equal

$$\widehat{\mathbf{B}} =(\mathbf{A}'\mathbf{S}_1^{-1}\mathbf{A})^{-1}\mathbf{A}'\mathbf{S}_1^{-1}\mathbf{Y}\mathbf{H}'(\mathbf{H}\mathbf{H})^{-1}, \tag{4.3.38}$$
$$\widehat{\mathbf{B}}_2 =(\mathbf{X} - \mathbf{A}\widehat{\mathbf{B}}\mathbf{C})\mathbf{C}_2'(\mathbf{C}_2\mathbf{C}_2')^{-1}, \tag{4.3.39}$$

where

$$\mathbf{Y} =\mathbf{X}(\mathbf{I} - \mathbf{C}_2'(\mathbf{C}_2\mathbf{C}_2')^{-1}\mathbf{C}_2),$$
$$\mathbf{H} =\mathbf{C}(\mathbf{I} - \mathbf{C}_2'(\mathbf{C}_2\mathbf{C}_2')^{-1}\mathbf{C}_2),$$
$$\mathbf{S}_1 =\mathbf{Y}(\mathbf{I} - \mathbf{H}'(\mathbf{H}\mathbf{H}')^{-1}\mathbf{H})\mathbf{Y}'.$$

Since $\widehat{\mathbf{B}}$ in (4.3.38) has the same structure as $\widehat{\mathbf{B}}$ in (4.3.1), Theorem 4.3.1 can be used immediately when presenting the approximation of the density of $\widehat{\mathbf{B}}$ in (4.3.38). Let

$$\mathbf{B}_N =(\mathbf{A}'\mathbf{\Sigma}^{-1}\mathbf{A})^{-1}\mathbf{A}'\mathbf{\Sigma}^{-1}\mathbf{Y}\mathbf{H}'(\mathbf{H}\mathbf{H})^{-1}, \tag{4.3.40}$$
$$\mathbf{U}_{\mathbf{B}} =(\mathbf{A}'\mathbf{S}_1^{-1}\mathbf{A})^{-1}\mathbf{A}'\mathbf{S}_1^{-1}(\mathbf{I} - (\mathbf{A}'\mathbf{\Sigma}^{-1}\mathbf{A})^{-1}\mathbf{A}'\mathbf{\Sigma}^{-1})\mathbf{Y}\mathbf{H}'(\mathbf{H}\mathbf{H}')^{-1} \tag{4.3.41}$$

and

$$\mathbf{D}[\mathbf{U}_{\mathbf{B}}] = D[\widehat{\mathbf{B}}] - D[\mathbf{B}_N] = \frac{p-q}{n - r(\mathbf{C}' : \mathbf{C}_2') - p + q - 1}(\mathbf{H}\mathbf{H}')^{-1}.$$

Theorem 4.3.8. *Let* $\widehat{\mathbf{B}}$, *and* $\mathbf{B}_N$ *be given by (4.3.38) and (4.3.40), respectively. Then, an Edgeworth-type density expansion holds:*

$$f_{\widehat{\mathbf{B}}}(\mathbf{B}_0) = f_{\mathbf{B}_E}(\mathbf{B}_0) + \ldots,$$

where

$$f_{\mathbf{B}_E}(\mathbf{B}_0) = \left\{1 + \tfrac{1}{2}s_1\left(\mathrm{tr}\{\mathbf{A}'\mathbf{\Sigma}^{-1}\mathbf{A}(\mathbf{B}_0 - \mathbf{B})\mathbf{H}\mathbf{H}'(\mathbf{B}_0 - \mathbf{B})'\} - kq\right)\right\}f_{\mathbf{B}_N}(\mathbf{B}_0),$$

$$s_1 = \frac{p-q}{n - r(\mathbf{C}' : \mathbf{C}_1') - p + q - 1}.$$

∎

It is also possible to find an upper error bound of the approximation similar to Theorem 4.3.1, but we shall not present it for technical reasons. Now the density of $\widehat{\mathbf{B}}_2$ in (4.3.39) is approximated. Let

$$\mathbf{B}_{N2} = \mathbf{X}\mathbf{C}_2'(\mathbf{C}_2\mathbf{C}_2')^{-1} - \mathbf{A}\mathbf{B}_N\mathbf{C}\mathbf{C}_2'(\mathbf{C}_2\mathbf{C}_2')^{-1}, \tag{4.3.42}$$

where $\mathbf{B}_N$ is the same as in (4.3.40). Furthermore,

$$\widehat{\mathbf{B}}_2 = \mathbf{B}_{N2} + \mathbf{A}\mathbf{U}_{\mathbf{B}}\mathbf{C}\mathbf{C}_2'(\mathbf{C}_2\mathbf{C}_2')^{-1}$$

with the important property that $\mathbf{B}_{N2}$ and $\mathbf{U}_{\mathbf{B}}$ are independent. This is a consequence of the fact that $\mathbf{X}\mathbf{C}_2'$ and $\mathbf{Y}$ are independently distributed, as well as that $\mathbf{U}_{\mathbf{B}}$ is independent of $\mathbf{B}_N$. Before stating the next theorem it is noted that $\mathbf{B}_{N2}$ is normally distributed with mean $E[\mathbf{B}_{N2}] = \mathbf{B}_2$ and

$$\begin{aligned} D[\mathbf{B}_{N2}] =& D[\mathbf{X}\mathbf{C}_2'(\mathbf{C}_2\mathbf{C}_2')^{-1}] + D[\mathbf{A}\mathbf{B}_N\mathbf{C}\mathbf{C}_2'(\mathbf{C}_2\mathbf{C}_2')^{-1}] \\ =& (\mathbf{C}_2\mathbf{C}_2')^{-1} \otimes \mathbf{\Sigma} + (\mathbf{C}_2\mathbf{C}_2')^{-1}\mathbf{C}_2\mathbf{C}\mathbf{C}_2'(\mathbf{C}_2\mathbf{C}_2')^{-1} \otimes \mathbf{A}(\mathbf{A}'\mathbf{\Sigma}^{-1}\mathbf{A})^{-1}\mathbf{A}'. \end{aligned}$$

Theorem 4.3.9. *Let $\widehat{\mathbf{B}}_2$, and $\mathbf{B}_{N2}$ be given by (4.3.39) and (4.3.42), respectively. Then an Edgeworth-type density expansion holds:*

$$f_{\widehat{\mathbf{B}}_2}(\mathbf{B}_0) = f_{\mathbf{B}_E}(\mathbf{B}_0) + \ldots,$$

where

$$\begin{aligned} f_{\mathbf{B}_E}(\mathbf{B}_0) &= f_{\mathbf{B}_{N2}}(\mathbf{B}_0) + \tfrac{1}{2}E[(\mathrm{vec}'\mathbf{U})^{\otimes 2}]\mathrm{vec}\,\mathbf{f}^2_{\mathbf{B}_{N2}}(\mathbf{B}_0), \\ E[(\mathrm{vec}'\mathbf{U})^{\otimes 2}] &= \frac{p-q}{n - r(\mathbf{C}' : \mathbf{C}_1') - p + q - 1}\mathrm{vec}\mathbf{M}, \\ \mathrm{vec}\,\mathbf{f}^2_{\mathbf{B}_{N2}}(\mathbf{B}_0) &= \{\mathbf{M}^{-1}\mathrm{vec}(\mathbf{B}_0 - \mathbf{B}_E)\mathrm{vec}'(\mathbf{B}_0 - \mathbf{B}_E)\mathbf{M}^{-1} - \mathbf{M}^{-1}\}f_{\mathbf{B}_{N2}}(\mathbf{B}_0) \end{aligned}$$

and

$$\mathbf{M} = (\mathbf{C}_2\mathbf{C}_2')^{-1} \otimes \mathbf{\Sigma} + (\mathbf{C}_2\mathbf{C}_2')^{-1}\mathbf{C}_2\mathbf{C}\mathbf{C}_2'(\mathbf{C}_2\mathbf{C}_2')^{-1} \otimes \mathbf{A}(\mathbf{A}'\mathbf{\Sigma}^{-1}\mathbf{A})^{-1}\mathbf{A}'.$$

∎

4.3.5 *Approximating the distribution of the mean parameter estimators in the* MLNM$(\sum_{i=1}^3 \mathbf{A}_i\mathbf{B}_i\mathbf{C}_i)$

We are going to find density approximations of the parameter estimators $\widehat{\mathbf{B}}_1$, $\widehat{\mathbf{B}}_2$ and $\widehat{\mathbf{B}}_3$ given in (4.3.5) – (4.3.7). The technique of finding relevant approximations will be the same as in §4.3.2. Let us first consider $\widehat{\mathbf{B}}_3$ given in (4.3.5). By using the results and the notation of Lemma 4.1.3, it follows that $\widehat{\mathbf{B}}_3$ is identical to

$$\widehat{\mathbf{B}}_3 = (\mathbf{A}_3'\mathbf{G}_2(\mathbf{G}_2'\mathbf{W}_3\mathbf{G}_2)^{-1}\mathbf{G}_2'\mathbf{A}_3)^{-1}\mathbf{A}_3'\mathbf{G}_2(\mathbf{G}_2'\mathbf{W}_3\mathbf{G}_2)^{-1}\mathbf{G}_2'\mathbf{X}\mathbf{C}_3'(\mathbf{C}_3\mathbf{C}_3')^{-1}. \tag{4.3.43}$$

Thus, from (4.3.43) it follows that $\widehat{\mathbf{B}}_3$ has the same structure as $\widehat{\mathbf{B}}$ in (4.3.1). Therefore,

$$\widehat{\mathbf{B}}_3 - \mathbf{B}_3 = \mathbf{B}_{3N} - \mathbf{B}_3 + \mathbf{U}_3,$$

where

$$\begin{aligned}\mathbf{B}_{3N} - \mathbf{B}_3 = &(\mathbf{A}_3'\mathbf{G}_2(\mathbf{G}_2'\boldsymbol{\Sigma}\mathbf{G}_2)^{-1}\mathbf{G}_2'\mathbf{A}_3)^{-1}\mathbf{A}_3'\mathbf{G}_2(\mathbf{G}_2'\boldsymbol{\Sigma}\mathbf{G}_2)^{-1}\mathbf{G}_2'\\ &\times(\mathbf{X} - \mathbf{A}_3\mathbf{B}_3\mathbf{C}_3)\mathbf{C}_3'(\mathbf{C}_3\mathbf{C}_3')^{-1}\end{aligned} \tag{4.3.44}$$

and

$$\begin{aligned}\mathbf{U}_3 = &(\mathbf{A}_3'\mathbf{G}_2(\mathbf{G}_2'\mathbf{W}_3\mathbf{G}_2)^{-1}\mathbf{G}_2'\mathbf{A}_3)^{-1}\mathbf{A}_3'\mathbf{G}_2(\mathbf{G}_2'\mathbf{W}_3\mathbf{G}_2)^{-1}\\ &\times\{\mathbf{I} - \mathbf{G}_2'\mathbf{A}_3(\mathbf{A}_3'\mathbf{G}_2(\mathbf{G}_2'\boldsymbol{\Sigma}\mathbf{G}_2)^{-1}\mathbf{G}_2'\mathbf{A}_3)^{-1}\mathbf{A}_3'\mathbf{G}_2(\mathbf{G}_2'\boldsymbol{\Sigma}\mathbf{G}_2)^{-1}\}\mathbf{G}_2'\mathbf{X}\mathbf{C}_3'(\mathbf{C}_3\mathbf{C}_3')^{-1}.\end{aligned} \tag{4.3.45}$$

Because of Theorem 2.2.4, the matrix $\mathbf{X}\mathbf{C}_3'(\mathbf{C}_3\mathbf{C}_3')^{-1}$ is independent of $\mathbf{W}_3$ and $\mathbf{A}_3'\mathbf{G}_2(\mathbf{G}_2'\boldsymbol{\Sigma}\mathbf{G}_2)^{-1}\mathbf{G}_2'\mathbf{X}$ is independent of

$$\{\mathbf{I} - \mathbf{G}_2'\mathbf{A}_3(\mathbf{A}_3'\mathbf{G}_2(\mathbf{G}_2'\boldsymbol{\Sigma}\mathbf{G}_2)^{-1}\mathbf{G}_2'\mathbf{A}_3)^{-1}\mathbf{A}_3'\mathbf{G}_2(\mathbf{G}_2'\boldsymbol{\Sigma}\mathbf{G}_2)^{-1}\}\mathbf{G}_2'\mathbf{X}.$$

Thus, Theorem 4.3.1 establishes

Theorem 4.3.10. *Let $\widehat{\mathbf{B}}_3$, $\mathbf{B}_{3N}$ and $\mathbf{U}_3$ be defined by (4.3.5), (4.3.44) and (4.3.45), respectively. Then*

$$f_{\widehat{\mathbf{B}}_3}(\mathbf{B}_0) = f_{\mathbf{B}_{3E}}(\mathbf{B}_0) + \ldots,$$

where

$$\begin{aligned}f_{\mathbf{B}_{3E}}(\mathbf{B}_0) =& f_{\mathbf{B}_{3N}}(\mathbf{B}_0) + \tfrac{1}{2}E[(\mathrm{vec}'\mathbf{U}_3)^{\otimes 2}]\mathrm{vec}\,\mathbf{f}^2_{\mathbf{B}_{3N}}(\mathbf{B}_0)\\ =&\Big\{1 + \tfrac{1}{2}s\Big(\mathrm{tr}\{\mathbf{A}_3'\mathbf{G}_2(\mathbf{G}_2'\boldsymbol{\Sigma}\mathbf{G}_2)^{-1}\mathbf{G}_2'\mathbf{A}_3(\mathbf{B}_0 - \mathbf{B}_3)\mathbf{C}_3\mathbf{C}_3'(\mathbf{B}_0 - \mathbf{B}_3)'\}\\ &- k_3m_2\Big)\Big\}f_{\mathbf{B}_{3N}}(\mathbf{B}_0),\end{aligned}$$

where k_3 is as in (4.1.43),

$$s = \frac{p - m_2}{n - k_3 - p + m_2 - 1}$$

and m_2 is defined in (4.1.63). ∎

Now we turn to $\widehat{\mathbf{B}}_2$ given in (4.3.6). As before, we are going to split the estimator and consider

$$\widehat{\mathbf{B}}_2 - \mathbf{B}_2 = \mathbf{B}_{2N} - \mathbf{B}_2 + \mathbf{U}_2,$$

where

$$\begin{aligned}\mathbf{B}_{2N}-\mathbf{B}_2 =&(\mathbf{A}_2'\mathbf{G}_1(\mathbf{G}_1'\boldsymbol{\Sigma}\mathbf{G}_1)^{-1}\mathbf{G}_1'\mathbf{A}_2)^{-1}\mathbf{A}_2'\mathbf{G}_1(\mathbf{G}_1'\boldsymbol{\Sigma}\mathbf{G}_1)^{-1}\\ &\times \mathbf{G}_1'(\mathbf{X}-\mathbf{A}_2\mathbf{B}_2\mathbf{C}_2)\mathbf{C}_2'(\mathbf{C}_2\mathbf{C}_2')^{-1}\\ &-(\mathbf{A}_2'\mathbf{G}_1(\mathbf{G}_1'\boldsymbol{\Sigma}\mathbf{G}_1)^{-1}\mathbf{G}_1'\mathbf{A}_2)^{-1}\mathbf{A}_2'\mathbf{G}_1(\mathbf{G}_1'\boldsymbol{\Sigma}\mathbf{G}_1)^{-1}\mathbf{G}_1'\mathbf{A}_3\\ &\times(\mathbf{A}_3'\mathbf{G}_2(\mathbf{G}_2'\boldsymbol{\Sigma}\mathbf{G}_2)^{-1}\mathbf{G}_2'\mathbf{A}_3)^{-}\mathbf{A}_3'\mathbf{G}_2(\mathbf{G}_2'\boldsymbol{\Sigma}\mathbf{G}_2)^{-1}\\ &\times\mathbf{G}_2'\mathbf{X}\mathbf{C}_3'(\mathbf{C}_3\mathbf{C}_3')^{-}\mathbf{C}_3\mathbf{C}_2'(\mathbf{C}_2\mathbf{C}_2')^{-1}\end{aligned} \qquad (4.3.46)$$

and

$$\begin{aligned}\mathbf{U}_2 =& (\mathbf{A}_2'\mathbf{G}_1(\mathbf{G}_1'\mathbf{W}_2\mathbf{G}_1)^{-1}\mathbf{G}_1'\mathbf{A}_2)^{-1}\mathbf{A}_2'\mathbf{G}_1(\mathbf{G}_1'\mathbf{W}_2\mathbf{G}_1)^{-1}\\ &\times\mathbf{G}_1'\boldsymbol{\Sigma}\mathbf{G}_2(\mathbf{G}_2'\boldsymbol{\Sigma}\mathbf{G}_2)^{-1}\mathbf{G}_2'\mathbf{X}\mathbf{C}_2'(\mathbf{C}_2\mathbf{C}_2')^{-1}\\ &-(\mathbf{A}_2'\mathbf{G}_1(\mathbf{G}_1'\boldsymbol{\Sigma}\mathbf{G}_1)^{-1}\mathbf{G}_1'\mathbf{A}_2)^{-1}\mathbf{A}_2'\mathbf{G}_1(\mathbf{G}_1'\boldsymbol{\Sigma}\mathbf{G}_1)^{-1}\\ &\times\mathbf{G}_1'\mathbf{A}_3(\mathbf{A}_3'\mathbf{G}_2(\mathbf{G}_2'\mathbf{W}_3\mathbf{G}_2)^{-1}\mathbf{G}_2'\mathbf{A}_3)^{-}\mathbf{A}_3'\mathbf{G}_2(\mathbf{G}_2'\mathbf{W}_3\mathbf{G}_2)^{-1}\\ &\times\mathbf{G}_2'\boldsymbol{\Sigma}\mathbf{G}_3(\mathbf{G}_3'\boldsymbol{\Sigma}\mathbf{G}_3)^{-1}\mathbf{G}_3'\mathbf{X}\mathbf{C}_3'(\mathbf{C}_3\mathbf{C}_3')^{-}\mathbf{C}_3\mathbf{C}_2'(\mathbf{C}_2\mathbf{C}_2')^{-1}\\ &-(\mathbf{A}_2'\mathbf{G}_1(\mathbf{G}_1'\mathbf{W}_2\mathbf{G}_1)^{-1}\mathbf{G}_1'\mathbf{A}_2)^{-1}\mathbf{A}_2'\mathbf{G}_1(\mathbf{G}_1'\mathbf{W}_2\mathbf{G}_1)^{-1}\mathbf{G}_1'\boldsymbol{\Sigma}\mathbf{G}_2(\mathbf{G}_2'\boldsymbol{\Sigma}\mathbf{G}_2)^{-1}\mathbf{G}_2'\\ &\times\mathbf{A}_3(\mathbf{A}_3'\mathbf{G}_2(\mathbf{G}_2'\mathbf{W}_3\mathbf{G}_2)^{-1}\mathbf{G}_2'\mathbf{A}_3)^{-}\mathbf{A}_3'\mathbf{G}_2(\mathbf{G}_2'\mathbf{W}_3\mathbf{G}_2)^{-1}\\ &\times\mathbf{G}_2'\mathbf{X}\mathbf{C}_3'(\mathbf{C}_3\mathbf{C}_3')^{-}\mathbf{C}_3\mathbf{C}_2'(\mathbf{C}_2\mathbf{C}_2')^{-1}.\end{aligned} \qquad (4.3.47)$$

In the calculations it has been used that

$$\begin{aligned}\{\mathbf{I}-\mathbf{G}_1'\mathbf{A}_2(\mathbf{A}_2'\mathbf{G}_1(\mathbf{G}_1'\boldsymbol{\Sigma}\mathbf{G}_1)^{-1}\mathbf{G}_1'\mathbf{A}_2)^{-1}\mathbf{A}_2'\mathbf{G}_1(\mathbf{G}_1'\boldsymbol{\Sigma}\mathbf{G}_1)^{-1}\}\mathbf{G}_1'&\\ =\mathbf{G}_1'\boldsymbol{\Sigma}\mathbf{G}_2(\mathbf{G}_2'\boldsymbol{\Sigma}\mathbf{G}_2)^{-1}\mathbf{G}_2',&\\ \{\mathbf{I}-\mathbf{G}_2'\mathbf{A}_3(\mathbf{A}_3'\mathbf{G}_2(\mathbf{G}_2'\boldsymbol{\Sigma}\mathbf{G}_2)^{-1}\mathbf{G}_2'\mathbf{A}_3)^{-1}\mathbf{A}_3'\mathbf{G}_2(\mathbf{G}_2'\boldsymbol{\Sigma}\mathbf{G}_2)^{-1}\}\mathbf{G}_2'&\\ =\mathbf{G}_2'\boldsymbol{\Sigma}\mathbf{G}_3(\mathbf{G}_3'\boldsymbol{\Sigma}\mathbf{G}_3)^{-1}\mathbf{G}_3'.&\end{aligned}$$

We are going to show that $\mathbf{B}_{2N}-\mathbf{B}_2$ and $\mathbf{U}_2$ are uncorrelated. Unfortunately independence does not hold. First observe that $\mathbf{X}\mathbf{C}_3'(\mathbf{C}_3\mathbf{C}_3')^{-}\mathbf{C}_3$ is independent of $\mathbf{W}_2$ and $\mathbf{W}_3$, and $\mathbf{X}\mathbf{C}_2'(\mathbf{C}_2\mathbf{C}_2')^{-1}$ is independent of $\mathbf{W}_2$. Furthermore, Theorem 2.2.4 (i) states that $\mathbf{A}_2'\mathbf{G}_1(\mathbf{G}_1'\boldsymbol{\Sigma}\mathbf{G}_1)^{-1}\mathbf{G}_1'\mathbf{X}$ is independent of

$$\{\mathbf{I}-\mathbf{G}_1'\mathbf{A}_2(\mathbf{A}_2'\mathbf{G}_1(\mathbf{G}_1'\boldsymbol{\Sigma}\mathbf{G}_1)^{-1}\mathbf{G}_1'\mathbf{A}_2)^{-}\mathbf{A}_2'\mathbf{G}_1(\mathbf{G}_1'\boldsymbol{\Sigma}\mathbf{G}_1)^{-1}\}\mathbf{G}_1'\mathbf{X}$$

as well as $\mathbf{G}_2'\mathbf{X}$, while the last condition also implies that $\mathbf{A}_2'\mathbf{G}_1(\mathbf{G}_1'\boldsymbol{\Sigma}\mathbf{G}_1)^{-1}\mathbf{G}_1'\mathbf{X}$ is independent of $\mathbf{G}_2'\mathbf{W}_3\mathbf{G}_2$. Now, according to Problem 1 in §2.4.9, the covariance

$$\begin{aligned}C\Big[&(\mathbf{A}_2'\mathbf{G}_1(\mathbf{G}_1'\mathbf{W}_2\mathbf{G}_1)^{-1}\mathbf{G}_1'\mathbf{A}_2)^{-1}\mathbf{A}_2'\mathbf{G}_1(\mathbf{G}_1'\mathbf{W}_2\mathbf{G}_1)^{-1}\\ &\quad\times\mathbf{G}_1'\boldsymbol{\Sigma}\mathbf{G}_2(\mathbf{G}_2'\boldsymbol{\Sigma}\mathbf{G}_2)^{-1}\mathbf{G}_2'(\mathbf{X}-E[\mathbf{X}])\mathbf{C}_2',\mathbf{G}_2'(\mathbf{X}-E[\mathbf{X}])\mathbf{C}_3'\Big]\\ =&\,\mathbf{C}_2\mathbf{C}_3'\otimes E[(\mathbf{A}_2'\mathbf{G}_1(\mathbf{G}_1'\mathbf{W}_2\mathbf{G}_1)^{-1}\mathbf{G}_1'\mathbf{A}_2)^{-1}\mathbf{A}_2'\mathbf{G}_1(\mathbf{G}_1'\mathbf{W}_2\mathbf{G}_1)^{-1}]\\ &\qquad\times\mathbf{G}_1'\boldsymbol{\Sigma}\mathbf{G}_2(\mathbf{G}_2'\boldsymbol{\Sigma}\mathbf{G}_2)^{-1}\mathbf{G}_2'\boldsymbol{\Sigma}\mathbf{G}_2\\ =&\,\mathbf{C}_2\mathbf{C}_3'\otimes(\mathbf{A}_2'\mathbf{G}_1(\mathbf{G}_1'\boldsymbol{\Sigma}\mathbf{G}_1)^{-1}\mathbf{G}_1'\mathbf{A}_2)^{-1}\mathbf{A}_2'\mathbf{G}_1(\mathbf{G}_1'\boldsymbol{\Sigma}\mathbf{G}_1)^{-1}\mathbf{G}_1'\boldsymbol{\Sigma}\mathbf{G}_2\\ =&\,\mathbf{0},\end{aligned}$$

since $\mathbf{A}_2'\mathbf{G}_1(\mathbf{G}_1'\boldsymbol{\Sigma}\mathbf{G}_1)^{-1}\mathbf{G}_1'\boldsymbol{\Sigma}\mathbf{G}_2 = \mathbf{0}$. Furthermore, the linear functions $\mathbf{G}_3'\mathbf{X}$ and $\mathbf{A}_3'\mathbf{G}_2(\mathbf{G}_2'\boldsymbol{\Sigma}\mathbf{G}_2)^{-1}\mathbf{G}_2'\mathbf{X}$ are independent, since $\mathbf{G}_3'\boldsymbol{\Sigma}\mathbf{G}_2(\mathbf{G}_2'\boldsymbol{\Sigma}\mathbf{G}_2)^{-1}\mathbf{G}_2'\mathbf{A}_3 = \mathbf{0}$. Finally we observe that

$$(\mathbf{A}_2'\mathbf{G}_1(\mathbf{G}_1'\mathbf{W}_2\mathbf{G}_1)^{-1}\mathbf{G}_1'\mathbf{A}_2)^{-1}\mathbf{A}_2'\mathbf{G}_1(\mathbf{G}_1'\mathbf{W}_2\mathbf{G}_1)^{-1} \\ \times \mathbf{G}_1'\boldsymbol{\Sigma}\mathbf{G}_2(\mathbf{G}_2'\boldsymbol{\Sigma}\mathbf{G}_2)^{-1}\mathbf{G}_2'(\mathbf{X} - E[\mathbf{X}])$$

and

$$\mathbf{G}_1'\mathbf{A}_3(\mathbf{A}_3'\mathbf{G}_2(\mathbf{G}_2'\boldsymbol{\Sigma}\mathbf{G}_2)^{-1}\mathbf{G}_2'\mathbf{A}_3)^{-}\mathbf{A}_3'\mathbf{G}_2(\mathbf{G}_2'\boldsymbol{\Sigma}\mathbf{G}_2)^{-1}\mathbf{G}_2'(\mathbf{X} - E[\mathbf{X}])$$

are uncorrelated.
We also need $D[\mathbf{B}_{2N}]$. By Theorem 2.2.4 (i), $\mathbf{A}_2'\mathbf{G}_1(\mathbf{G}_1'\boldsymbol{\Sigma}\mathbf{G}_1)^{-1}\mathbf{G}_1'\mathbf{X}$ and $\mathbf{G}_2'\mathbf{X}$ are independent and we have

$$\begin{aligned} &D[\mathbf{B}_{2N}] \\ &= (\mathbf{C}_2\mathbf{C}_2')^{-1} \otimes (\mathbf{A}_2'\mathbf{G}_1(\mathbf{G}_1'\boldsymbol{\Sigma}\mathbf{G}_1)^{-1}\mathbf{G}_1'\mathbf{A}_2)^{-1} \\ &\quad + (\mathbf{C}_2\mathbf{C}_2')^{-1}\mathbf{C}_2\mathbf{C}_3'(\mathbf{C}_3\mathbf{C}_3')^{-}\mathbf{C}_3\mathbf{C}_2'(\mathbf{C}_2\mathbf{C}_2')^{-1} \otimes (\mathbf{A}_2'\mathbf{G}_1(\mathbf{G}_1'\boldsymbol{\Sigma}\mathbf{G}_1)^{-1}\mathbf{G}_1'\mathbf{A}_2)^{-1} \\ &\quad \times \mathbf{A}_2'\mathbf{G}_1(\mathbf{G}_1'\boldsymbol{\Sigma}\mathbf{G}_1)^{-1}\mathbf{G}_1'\mathbf{A}_3(\mathbf{A}_3'\mathbf{G}_2(\mathbf{G}_2'\boldsymbol{\Sigma}\mathbf{G}_2)^{-1}\mathbf{G}_2'\mathbf{A}_3)^{-}\mathbf{A}_3'\mathbf{G}_1(\mathbf{G}_1'\boldsymbol{\Sigma}\mathbf{G}_1)^{-1} \\ &\quad \times \mathbf{G}_1'\mathbf{A}_2(\mathbf{A}_2'\mathbf{G}_1(\mathbf{G}_1'\boldsymbol{\Sigma}\mathbf{G}_1)^{-1}\mathbf{G}_1'\mathbf{A}_2)^{-1}. \end{aligned} \qquad (4.3.48)$$

It is somewhat unfortunate that $\mathbf{B}_{2N}$ is not matrix normally distributed, which means that the inverse of $D[\mathbf{B}_{2N}]$ is difficult to express in a convenient way.

Theorem 4.3.11. *Let $\widehat{\mathbf{B}}_2$, $\mathbf{B}_{2N}$ and $\mathbf{U}_2$ be defined by (4.3.6), (4.3.46) and (4.3.47), respectively. Then*

$$f_{\widehat{\mathbf{B}}_2}(\mathbf{B}_0) = f_{\mathbf{B}_{2E}}(\mathbf{B}_0) + \ldots,$$

where

$$f_{\mathbf{B}_{2E}}(\mathbf{B}_0) = f_{\mathbf{B}_{2N}}(\mathbf{B}_0) + \tfrac{1}{2}E[(\mathrm{vec}'\mathbf{U}_2)^{\otimes 2}]\mathrm{vec}\,\mathbf{f}^2_{\mathbf{B}_{2N}}(\mathbf{B}_0),$$

and

$$E[(\mathrm{vec}'\mathbf{U}_2)^{\otimes 2}] = \mathrm{vec}(D[\widehat{\mathbf{B}}_2] - D[\mathbf{B}_{2N}]),$$

$D[\widehat{\mathbf{B}}_2]$ and $D[\mathbf{B}_{2N}]$ are given in Theorem 4.2.11 (ii) and (4.3.48), respectively. ∎

Now we turn to $\widehat{\mathbf{B}}_1$, which is more difficult to treat than $\widehat{\mathbf{B}}_2$ and $\widehat{\mathbf{B}}_3$, because $\widehat{\mathbf{B}}_1$ is a function of $\widehat{\mathbf{B}}_2$ and $\widehat{\mathbf{B}}_3$. First let us introduce some notation:

$$\begin{aligned} \mathbf{P}_{\mathbf{A}} &= \mathbf{A}(\mathbf{A}'\mathbf{A})^{-}\mathbf{A}', \\ \underline{\mathbf{P}}_{\mathbf{A}} &= (\mathbf{A}'\mathbf{A})^{-1}\mathbf{A}', \text{ if } (\mathbf{A}'\mathbf{A})^{-1} \text{ exists}, \\ \mathbf{P}_{\mathbf{A},\mathbf{W}} &= \mathbf{A}(\mathbf{A}'\mathbf{W}^{-1}\mathbf{A})^{-}\mathbf{A}'\mathbf{W}^{-1}, \\ \underline{\mathbf{P}}_{\mathbf{A},\mathbf{W}} &= (\mathbf{A}'\mathbf{W}^{-1}\mathbf{A})^{-1}\mathbf{A}'\mathbf{W}^{-1}, \text{ if } (\mathbf{A}'\mathbf{W}^{-1}\mathbf{A})^{-1} \text{ exists}, \\ \mathbf{P}_{\mathbf{A},\mathbf{B},\mathbf{W}} &= \mathbf{B}'\mathbf{A}(\mathbf{A}'\mathbf{B}(\mathbf{B}'\mathbf{W}\mathbf{B})^{-1}\mathbf{B}'\mathbf{A})^{-}\mathbf{A}'\mathbf{B}(\mathbf{B}'\mathbf{W}\mathbf{B})^{-1}, \\ \underline{\mathbf{P}}_{\mathbf{A},\mathbf{B},\mathbf{W}} &= \mathbf{A}(\mathbf{A}'\mathbf{B}(\mathbf{B}'\mathbf{W}\mathbf{B})^{-1}\mathbf{B}'\mathbf{A})^{-}\mathbf{A}'\mathbf{B}(\mathbf{B}'\mathbf{W}\mathbf{B})^{-1}, \end{aligned}$$

where it is supposed that $(\mathbf{B}'\mathbf{W}\mathbf{B})^{-1}$ exists in $\mathbf{P}_{\mathbf{A},\mathbf{B},\mathbf{W}}$ and $\underline{\mathbf{P}}_{\mathbf{A},\mathbf{B},\mathbf{W}}$. It follows that $\widehat{\mathbf{B}}_1$, given in (4.3.7), can be written in the following form:

$$\begin{aligned}\widehat{\mathbf{B}}_1 =&(\underline{\mathbf{P}}_{\mathbf{A}_1,\mathbf{W}_1}\mathbf{X} - \underline{\mathbf{P}}_{\mathbf{A}_1,\mathbf{W}_1}\mathbf{A}_2\widehat{\mathbf{B}}_2\mathbf{C}_2 - \underline{\mathbf{P}}_{\mathbf{A}_1,\mathbf{W}_1}\mathbf{A}_3\widehat{\mathbf{B}}_3\mathbf{C}_3)\mathbf{C}_1'(\mathbf{C}_1\mathbf{C}_1')^{-1}\\ =&\underline{\mathbf{P}}_{\mathbf{A}_1,\mathbf{W}_1}\mathbf{X}\mathbf{C}_1'(\mathbf{C}_1\mathbf{C}_1')^{-1} - \underline{\mathbf{P}}_{\mathbf{A}_1,\mathbf{W}_1}\underline{\mathbf{P}}_{\mathbf{A}_2,\mathbf{G}_1,\mathbf{W}_2}\mathbf{G}_1'\mathbf{X}\mathbf{P}_{\mathbf{C}_2'}\mathbf{C}_1'(\mathbf{C}_1\mathbf{C}_1')^{-1}\\ &- \underline{\mathbf{P}}_{\mathbf{A}_1,\mathbf{W}_1}(\mathbf{I} - \underline{\mathbf{P}}_{\mathbf{A}_2,\mathbf{G}_1,\mathbf{W}_2}\mathbf{G}_1')\underline{\mathbf{P}}_{\mathbf{A}_3,\mathbf{G}_2,\mathbf{W}_3}\mathbf{G}_2'\mathbf{X}\mathbf{P}_{\mathbf{C}_3'}\mathbf{C}_1'(\mathbf{C}_1\mathbf{C}_1')^{-1}.\end{aligned}$$

As before, we are going to split the estimator and consider the difference

$$\widehat{\mathbf{B}}_1 - \mathbf{B}_1 = \mathbf{B}_{1N} - \mathbf{B}_1 + \mathbf{U}_1, \tag{4.3.49}$$

where $\mathbf{B}_{1N}$ is a matrix obtained from $\widehat{\mathbf{B}}_1$, when $\mathbf{W}_i$, $i = 1,2,3$, is replaced by $\boldsymbol{\Sigma}$. Thus,

$$\begin{aligned}\mathbf{B}_{1N} - \mathbf{B}_1 =& \underline{\mathbf{P}}_{\mathbf{A}_1,\boldsymbol{\Sigma}}(\mathbf{X} - E[\mathbf{X}])\mathbf{C}_1'(\mathbf{C}_1\mathbf{C}_1')^{-1}\\ &- \underline{\mathbf{P}}_{\mathbf{A}_1,\boldsymbol{\Sigma}}\underline{\mathbf{P}}_{\mathbf{A}_2,\mathbf{G}_1,\boldsymbol{\Sigma}}\mathbf{G}_1'(\mathbf{X} - E[\mathbf{X}])\mathbf{P}_{\mathbf{C}_2'}\mathbf{C}_1'(\mathbf{C}_1\mathbf{C}_1')^{-1}\\ &- \underline{\mathbf{P}}_{\mathbf{A}_1,\boldsymbol{\Sigma}}(\mathbf{I} - \underline{\mathbf{P}}_{\mathbf{A}_2,\mathbf{G}_1,\boldsymbol{\Sigma}}\mathbf{G}_1')\underline{\mathbf{P}}_{\mathbf{A}_3,\mathbf{G}_2,\boldsymbol{\Sigma}}\mathbf{G}_2'(\mathbf{X} - E[\mathbf{X}])\mathbf{P}_{\mathbf{C}_3'}\mathbf{C}_1'(\mathbf{C}_1\mathbf{C}_1')^{-1}.\end{aligned} \tag{4.3.50}$$

Observe that by Theorem 2.2.4 (i), the three terms on the right hand side of (4.3.50) are mutually independent. Thus,

$$\begin{aligned}D[\mathbf{B}_{1N}] =& (\mathbf{C}_1\mathbf{C}_1')^{-1} \otimes (\mathbf{A}_1'\boldsymbol{\Sigma}^{-1}\mathbf{A}_1)^{-1}\\ &+ (\mathbf{C}_1\mathbf{C}_1')^{-1}\mathbf{C}_1\mathbf{P}_{\mathbf{C}_2'}\mathbf{C}_1'(\mathbf{C}_1\mathbf{C}_1')^{-1} \otimes \underline{\mathbf{P}}_{\mathbf{A}_1,\boldsymbol{\Sigma}}\\ &\quad\times \mathbf{A}_2(\mathbf{A}_2'\mathbf{G}_1(\mathbf{G}_1'\boldsymbol{\Sigma}\mathbf{G}_1)^{-1}\mathbf{G}_1'\mathbf{A}_2)^{-}\mathbf{A}_2'\underline{\mathbf{P}}_{\mathbf{A}_1,\boldsymbol{\Sigma}}'\\ &+ (\mathbf{C}_1\mathbf{C}_1')^{-1}\mathbf{C}_1\mathbf{P}_{\mathbf{C}_3'}\mathbf{C}_1'(\mathbf{C}_1\mathbf{C}_1')^{-1} \otimes \underline{\mathbf{P}}_{\mathbf{A}_1,\boldsymbol{\Sigma}}(\mathbf{I} - \underline{\mathbf{P}}_{\mathbf{A}_1,\mathbf{G}_1,\boldsymbol{\Sigma}}\mathbf{G}_1')\\ &\quad\times \mathbf{A}_3(\mathbf{A}_3'\mathbf{G}_2(\mathbf{G}_2'\boldsymbol{\Sigma}\mathbf{G}_2)^{-1}\mathbf{G}_2'\mathbf{A}_3)^{-}\mathbf{A}_3'(\mathbf{I} - \mathbf{G}_1\underline{\mathbf{P}}_{\mathbf{A}_1,\mathbf{G}_1,\boldsymbol{\Sigma}}'\underline{\mathbf{P}}_{\mathbf{A}_1,\boldsymbol{\Sigma}}'.\end{aligned} \tag{4.3.51}$$

After some calculations the term $\mathbf{U}_1$ in (4.3.49) equals

$$\begin{aligned}\mathbf{U}_1 =& \underline{\mathbf{P}}_{\mathbf{A}_1,\mathbf{W}_1}(\mathbf{I} - \mathbf{P}_{\mathbf{A}_1,\boldsymbol{\Sigma}})\mathbf{X}\mathbf{C}_1'(\mathbf{C}_1\mathbf{C}_1')^{-1}\\ &- \underline{\mathbf{P}}_{\mathbf{A}_1,\mathbf{W}_1}(\mathbf{I} - \mathbf{P}_{\mathbf{A}_1,\boldsymbol{\Sigma}})\underline{\mathbf{P}}_{\mathbf{A}_2,\mathbf{G}_1,\mathbf{W}_2}\mathbf{G}_1'\mathbf{X}\mathbf{P}_{\mathbf{C}_2'}\mathbf{C}_1'(\mathbf{C}_1\mathbf{C}_1')^{-1}\\ &- \underline{\mathbf{P}}_{\mathbf{A}_1,\boldsymbol{\Sigma}}\underline{\mathbf{P}}_{\mathbf{A}_2,\mathbf{G}_1,\mathbf{W}_2}(\mathbf{I} - \mathbf{P}_{\mathbf{A}_2,\mathbf{G}_1,\boldsymbol{\Sigma}})\mathbf{G}_1'\mathbf{X}\mathbf{P}_{\mathbf{C}_2'}\mathbf{C}_1'(\mathbf{C}_1\mathbf{C}_1')^{-1}\\ &- \underline{\mathbf{P}}_{\mathbf{A}_1,\mathbf{W}_1}(\mathbf{I} - \mathbf{P}_{\mathbf{A}_1,\boldsymbol{\Sigma}})(\mathbf{I} - \underline{\mathbf{P}}_{\mathbf{A}_2,\mathbf{G}_1,\mathbf{W}_2}\mathbf{G}_1')\underline{\mathbf{P}}_{\mathbf{A}_3,\mathbf{G}_2,\mathbf{W}_3}\mathbf{G}_2'\mathbf{X}\mathbf{P}_{\mathbf{C}_3'}\mathbf{C}_1'(\mathbf{C}_1\mathbf{C}_1')^{-1}\\ &+ \underline{\mathbf{P}}_{\mathbf{A}_1,\boldsymbol{\Sigma}}\underline{\mathbf{P}}_{\mathbf{A}_2,\mathbf{G}_1,\mathbf{W}_2}(\mathbf{I} - \mathbf{P}_{\mathbf{A}_2,\mathbf{G}_1,\boldsymbol{\Sigma}})\mathbf{G}_1'\underline{\mathbf{P}}_{\mathbf{A}_3,\mathbf{G}_2,\mathbf{W}_3}\mathbf{G}_2'\mathbf{X}\mathbf{P}_{\mathbf{C}_3'}\mathbf{C}_1'(\mathbf{C}_1\mathbf{C}_1')^{-1}\\ &- \underline{\mathbf{P}}_{\mathbf{A}_1,\boldsymbol{\Sigma}}(\mathbf{I} - \underline{\mathbf{P}}_{\mathbf{A}_2,\mathbf{G}_1,\boldsymbol{\Sigma}}\mathbf{G}_1')\underline{\mathbf{P}}_{\mathbf{A}_3,\mathbf{G}_2,\mathbf{W}_3}(\mathbf{I} - \mathbf{P}_{\mathbf{A}_3,\mathbf{G}_2,\boldsymbol{\Sigma}})\mathbf{G}_2'\mathbf{X}\mathbf{P}_{\mathbf{C}_3'}\mathbf{C}_1'(\mathbf{C}_1\mathbf{C}_1')^{-1}.\end{aligned} \tag{4.3.52}$$

It is of advantage that $\widehat{\mathbf{B}}_1$ and $\mathbf{B}_{1N}$ have the same mean and now this is verified. The calculations depend heavily on the definition of the matrix $\mathbf{G}_i$, $i = 1,2,3$, given in Lemma 4.1.3:

$$\begin{aligned}E[\mathbf{B}_{1N}] &= \underline{\mathbf{P}}_{\mathbf{A}_1,\boldsymbol{\Sigma}}E[\mathbf{X}]\mathbf{C}_1'(\mathbf{C}_1\mathbf{C}_1')^{-1} - \underline{\mathbf{P}}_{\mathbf{A}_1,\boldsymbol{\Sigma}}\underline{\mathbf{P}}_{\mathbf{A}_2,\mathbf{G}_1,\boldsymbol{\Sigma}}\mathbf{G}_1'E[\mathbf{X}]\mathbf{P}_{\mathbf{C}_2'}\mathbf{C}_1'(\mathbf{C}_1\mathbf{C}_1')^{-1}\\ &\quad - \underline{\mathbf{P}}_{\mathbf{A}_1,\boldsymbol{\Sigma}}(\mathbf{I}-\underline{\mathbf{P}}_{\mathbf{A}_2,\mathbf{G}_1,\boldsymbol{\Sigma}}\mathbf{G}_1')\underline{\mathbf{P}}_{\mathbf{A}_3,\mathbf{G}_2,\boldsymbol{\Sigma}}\mathbf{G}_2'E[\mathbf{X}]\mathbf{P}_{\mathbf{C}_3'}\mathbf{C}_1'(\mathbf{C}_1\mathbf{C}_1')^{-1}\\ &= \mathbf{B}_1 + \underline{\mathbf{P}}_{\mathbf{A}_1,\boldsymbol{\Sigma}}(\mathbf{A}_2\mathbf{B}_2\mathbf{C}_2+\mathbf{A}_3\mathbf{B}_3\mathbf{C}_3)\mathbf{C}_1'(\mathbf{C}_1\mathbf{C}_1')^{-1}\\ &\quad - \underline{\mathbf{P}}_{\mathbf{A}_1,\boldsymbol{\Sigma}}\underline{\mathbf{P}}_{\mathbf{A}_2,\mathbf{G}_1,\boldsymbol{\Sigma}}\mathbf{G}_1'(\mathbf{A}_2\mathbf{B}_2\mathbf{C}_2+\mathbf{A}_3\mathbf{B}_3\mathbf{C}_3)\mathbf{P}_{\mathbf{C}_2'}\mathbf{C}_1'(\mathbf{C}_1\mathbf{C}_1')^{-1}\\ &\quad - \underline{\mathbf{P}}_{\mathbf{A}_1,\boldsymbol{\Sigma}}(\mathbf{I}-\underline{\mathbf{P}}_{\mathbf{A}_2,\mathbf{G}_1,\boldsymbol{\Sigma}}\mathbf{G}_1')\underline{\mathbf{P}}_{\mathbf{A}_3,\mathbf{G}_2,\boldsymbol{\Sigma}}\mathbf{G}_2'\mathbf{A}_3\mathbf{B}_3\mathbf{C}_3\mathbf{C}_1'(\mathbf{C}_1\mathbf{C}_1')^{-1}\\ &= \mathbf{B}_1.\end{aligned}$$

This implies also that $E[\mathbf{U}_1] = \mathbf{0}$. Next the covariance between $\mathbf{B}_{1N}$ and $\mathbf{U}_1$ is investigated.
Since $\boldsymbol{\Sigma}\mathbf{G}_1(\mathbf{G}_1'\boldsymbol{\Sigma}\mathbf{G}_1)^-\mathbf{G}_1'\mathbf{X}$ and $\mathbf{A}_1'\boldsymbol{\Sigma}^{-1}\mathbf{X}$ are independent, and $\mathbf{I}-\mathbf{P}_{\mathbf{A}_1,\boldsymbol{\Sigma}} = \boldsymbol{\Sigma}\mathbf{G}_1(\mathbf{G}_1'\boldsymbol{\Sigma}\mathbf{G}_1)^-\mathbf{G}_1'$, the expression $\underline{\mathbf{P}}_{\mathbf{A}_1,\boldsymbol{\Sigma}}\mathbf{X}\mathbf{C}_1'(\mathbf{C}_1\mathbf{C}_1')^{-1}$ is independent of all expressions in $\mathbf{U}_1$. Moreover, $\mathbf{X}\mathbf{P}_{\mathbf{C}_2'}$ is independent of $\mathbf{W}_2$ and $\mathbf{W}_1$,

$$C[\underline{\mathbf{P}}_{\mathbf{A}_1,\mathbf{W}_1}(\mathbf{I}-\mathbf{P}_{\mathbf{A}_1,\boldsymbol{\Sigma}})\mathbf{X}\mathbf{C}_1', \mathbf{X}\mathbf{P}\mathbf{C}_2'] = C[\underline{\mathbf{P}}_{\mathbf{A}_1,\boldsymbol{\Sigma}}(\mathbf{I}-\mathbf{P}_{\mathbf{A}_1,\boldsymbol{\Sigma}})\mathbf{X}\mathbf{C}_1', \mathbf{X}\mathbf{P}\mathbf{C}_2'] = \mathbf{0}$$

and

$$\begin{aligned}&C\big[\underline{\mathbf{P}}_{\mathbf{A}_1,\mathbf{W}_1}(\mathbf{I}-\mathbf{P}_{\mathbf{A}_1,\boldsymbol{\Sigma}})\underline{\mathbf{P}}_{\mathbf{A}_2,\mathbf{G}_1,\mathbf{W}_2}\mathbf{G}_1'\mathbf{X}\mathbf{P}_{\mathbf{C}_2'}\mathbf{C}_1'(\mathbf{C}_1\mathbf{C}_1')^{-1},\\ &\qquad\qquad \underline{\mathbf{P}}_{\mathbf{A}_1,\boldsymbol{\Sigma}}\underline{\mathbf{P}}_{\mathbf{A}_2,\mathbf{G}_1,\boldsymbol{\Sigma}}\mathbf{G}_1'\mathbf{X}\mathbf{P}_{\mathbf{C}_2'}\mathbf{C}_1'(\mathbf{C}_1\mathbf{C}_1')^{-1}\big]\\ &\quad= (\mathbf{C}_1\mathbf{C}_1')^{-1}\mathbf{C}_1\mathbf{P}_{\mathbf{C}_2'}\mathbf{C}_1'(\mathbf{C}_1\mathbf{C}_1')^{-1}\\ &\qquad \otimes E[\underline{\mathbf{P}}_{\mathbf{A}_1,\mathbf{W}_1}(\mathbf{I}-\mathbf{P}_{\mathbf{A}_1,\boldsymbol{\Sigma}})\underline{\mathbf{P}}_{\mathbf{A}_2,\mathbf{G}_1,\mathbf{W}_2}\mathbf{G}_1'\boldsymbol{\Sigma}\mathbf{G}_1\underline{\mathbf{P}}_{\mathbf{A}_2,\mathbf{G}_1,\boldsymbol{\Sigma}}'\underline{\mathbf{P}}_{\mathbf{A}_1,\boldsymbol{\Sigma}}']\\ &\quad= (\mathbf{C}_1\mathbf{C}_1')^{-1}\mathbf{C}_1\mathbf{P}_{\mathbf{C}_2'}\mathbf{C}_1'(\mathbf{C}_1\mathbf{C}_1')^{-1}\\ &\qquad \otimes E[\underline{\mathbf{P}}_{\mathbf{A}_1,\mathbf{W}_1}(\mathbf{I}-\mathbf{P}_{\mathbf{A}_1,\boldsymbol{\Sigma}})\mathbf{A}_2(\mathbf{A}_2'\mathbf{G}_1(\mathbf{G}_1'\boldsymbol{\Sigma}\mathbf{G}_1)^{-1}\mathbf{G}_1'\mathbf{A}_2)\mathbf{A}_2'\underline{\mathbf{P}}_{\mathbf{A}_1,\boldsymbol{\Sigma}}']\\ &\quad= (\mathbf{C}_1\mathbf{C}_1')^{-1}\mathbf{C}_1\mathbf{P}_{\mathbf{C}_2'}\mathbf{C}_1'(\mathbf{C}_1\mathbf{C}_1')^{-1}\\ &\qquad \otimes \underline{\mathbf{P}}_{\mathbf{A}_1,\boldsymbol{\Sigma}}(\mathbf{I}-\mathbf{P}_{\mathbf{A}_1,\boldsymbol{\Sigma}})\mathbf{A}_2(\mathbf{A}_2'\mathbf{G}_1(\mathbf{G}_1'\boldsymbol{\Sigma}\mathbf{G}_1)^{-1}\mathbf{G}_1'\mathbf{A}_2)\mathbf{A}_2'\underline{\mathbf{P}}_{\mathbf{A}_1,\boldsymbol{\Sigma}}' = \mathbf{0}.\end{aligned}$$

We also have that

$$(\mathbf{I}-\mathbf{P}_{\mathbf{A}_2,\mathbf{G}_1,\boldsymbol{\Sigma}})\mathbf{G}_1' = \mathbf{G}_1'\boldsymbol{\Sigma}\mathbf{G}_2(\mathbf{G}_2'\boldsymbol{\Sigma}\mathbf{G}_2)^{-1}\mathbf{G}_2'$$

and since $\mathbf{P}_{\mathbf{A}_2,\mathbf{G}_1,\boldsymbol{\Sigma}}\mathbf{G}_1'\mathbf{X}$ is independent of $\mathbf{G}_2'\mathbf{X}$, the second term of $\mathbf{B}_{1N}$ is also uncorrelated with all the terms in $\mathbf{U}_1$.
Turning to the third term in (4.3.50), note that $\mathbf{X}\mathbf{P}_{\mathbf{C}_3'}$ is independent of $\mathbf{W}_1$, $\mathbf{W}_2$ and $\mathbf{W}_3$. Therefore, since $E[\underline{\mathbf{P}}_{\mathbf{A}_1,\mathbf{W}_1}] = \underline{\mathbf{P}}_{\mathbf{A}_1,\boldsymbol{\Sigma}}$ (see Problem 1 in §2.4.9), $\underline{\mathbf{P}}_{\mathbf{A}_1,\mathbf{W}_1}(\mathbf{I}-\mathbf{P}_{\mathbf{A}_1,\boldsymbol{\Sigma}})\mathbf{X}\mathbf{C}_1'(\mathbf{C}_1\mathbf{C}_1')^{-1}$ is uncorrelated with $\mathbf{X}\mathbf{P}_{\mathbf{C}_3'}$. Now $\mathbf{X}\mathbf{P}_{\mathbf{C}_3'}$ and $\mathbf{X}\mathbf{P}_{\mathbf{C}_2'}$ are independent of $\mathbf{W}_1$ and $\mathbf{W}_2$, and

$$E[\underline{\mathbf{P}}_{\mathbf{A}_1,\mathbf{W}_1}(\mathbf{I}-\mathbf{P}_{\mathbf{A}_1,\boldsymbol{\Sigma}})\underline{\mathbf{P}}_{\mathbf{A}_2,\mathbf{G}_1,\mathbf{W}_2}\mathbf{G}_1'] = \underline{\mathbf{P}}_{\mathbf{A}_1,\boldsymbol{\Sigma}}(\mathbf{I}-\mathbf{P}_{\mathbf{A}_1,\boldsymbol{\Sigma}})\underline{\mathbf{P}}_{\mathbf{A}_2,\mathbf{G}_1,\boldsymbol{\Sigma}}\mathbf{G}_1' = \mathbf{0}.$$

Thus, the third term in (4.3.50) is uncorrelated with the first two terms of $\mathbf{U}_1$. The same arguments yield that it is uncorrelated with the third term of $\mathbf{U}_1$. Now we have to consider the last three terms of $\mathbf{U}_1$. First it is used that $\mathbf{XP}_{\mathbf{C}_3'}$ is independent of $\mathbf{W}_1$, $\mathbf{W}_2$ and $\mathbf{W}_3$. Moreover,

$$\mathbf{G}_2'\underline{\mathbf{P}}_{\mathbf{A}_3,\mathbf{G}_2,\mathbf{W}_3}\mathbf{G}_2'\mathbf{\Sigma}\mathbf{G}_2\underline{\mathbf{P}}'_{\mathbf{A}_3,\mathbf{G}_2,\mathbf{\Sigma}} = \mathbf{G}_2'\mathbf{A}_3(\mathbf{A}_3'\mathbf{G}_2(\mathbf{G}_2'\mathbf{\Sigma}\mathbf{G}_2)^{-1}\mathbf{G}_2'\mathbf{A}_3)^{-}\mathbf{A}_3'.$$

Since

$$\begin{aligned}
&(\mathbf{I}-\underline{\mathbf{P}}_{\mathbf{A}_1,\mathbf{\Sigma}})(\mathbf{I}-\underline{\mathbf{P}}_{\mathbf{A}_2,\mathbf{G}_1,\mathbf{W}_2}\mathbf{G}_1')\underline{\mathbf{P}}_{\mathbf{A}_3,\mathbf{G}_2,\mathbf{W}_3}\mathbf{G}_2' \\
&\quad = \mathbf{\Sigma}\mathbf{G}_1(\mathbf{G}_1'\mathbf{\Sigma}\mathbf{G}_1)^{-1}\mathbf{G}_1'\mathbf{W}_2\mathbf{G}_2(\mathbf{G}_2'\mathbf{W}_2\mathbf{G}_2)^{-1}\mathbf{G}_2'\underline{\mathbf{P}}_{\mathbf{A}_3,\mathbf{G}_2,\mathbf{W}_3}\mathbf{G}_2', \\
&(\mathbf{I}-\mathbf{P}_{\mathbf{A}_2,\mathbf{G}_1,\mathbf{\Sigma}})\mathbf{G}_1'\underline{\mathbf{P}}_{\mathbf{A}_3,\mathbf{G}_2,\mathbf{W}_3}\mathbf{G}_2' \\
&\quad = \mathbf{G}_1\mathbf{\Sigma}\mathbf{G}_2(\mathbf{G}_2'\mathbf{\Sigma}\mathbf{G}_2)^{-1}\mathbf{G}_2'\underline{\mathbf{P}}_{\mathbf{A}_3,\mathbf{G}_2,\mathbf{W}_3}\mathbf{G}_2',
\end{aligned}$$

we have to consider the expectations $E[\underline{\mathbf{P}}_{\mathbf{A}_1,\mathbf{W}_1}(\mathbf{I}-\mathbf{P}_{\mathbf{A}_1,\mathbf{\Sigma}})(\mathbf{I}-\underline{\mathbf{P}}_{\mathbf{A}_2,\mathbf{G}_1,\mathbf{W}_2}\mathbf{G}_1')]$ and $E[\underline{\mathbf{P}}_{\mathbf{A}_1,\mathbf{\Sigma}}\underline{\mathbf{P}}_{\mathbf{A}_2,\mathbf{G}_1,\mathbf{W}_2}(\mathbf{I}-\mathbf{P}_{\mathbf{A}_2,\mathbf{G}_1,\mathbf{\Sigma}})]$. However,

$$\begin{aligned}
&E[\underline{\mathbf{P}}_{\mathbf{A}_1,\mathbf{W}_1}(\mathbf{I}-\mathbf{P}_{\mathbf{A}_1,\mathbf{\Sigma}})(\mathbf{I}-\underline{\mathbf{P}}_{\mathbf{A}_2,\mathbf{G}_1,\mathbf{W}_2}\mathbf{G}_1')] \\
&\qquad = \underline{\mathbf{P}}_{\mathbf{A}_1,\mathbf{\Sigma}}(\mathbf{I}-\mathbf{P}_{\mathbf{A}_1,\mathbf{\Sigma}})(\mathbf{I}-\underline{\mathbf{P}}_{\mathbf{A}_2,\mathbf{G}_1,\mathbf{\Sigma}}\mathbf{G}_1') = \mathbf{0}
\end{aligned}$$

and

$$E[\underline{\mathbf{P}}_{\mathbf{A}_1,\mathbf{\Sigma}}\underline{\mathbf{P}}_{\mathbf{A}_2,\mathbf{G}_1,\mathbf{W}_2}(\mathbf{I}-\mathbf{P}_{\mathbf{A}_2,\mathbf{G}_1,\mathbf{\Sigma}})] = \underline{\mathbf{P}}_{\mathbf{A}_1,\mathbf{\Sigma}}\underline{\mathbf{P}}_{\mathbf{A}_2,\mathbf{G}_1,\mathbf{\Sigma}}(\mathbf{I}-\mathbf{P}_{\mathbf{A}_2,\mathbf{G}_1,\mathbf{\Sigma}}) = \mathbf{0}.$$

Therefore, the third term in (4.3.50) is uncorrelated with the fourth and fifth term of $\mathbf{U}_1$. Finally it is observed that since $\mathbf{XP}_{\mathbf{C}_3'}$ is independent of $\mathbf{W}_3$ and

$$\begin{aligned}
&E[\underline{\mathbf{P}}_{\mathbf{A}_1,\mathbf{\Sigma}}(\mathbf{I}-\underline{\mathbf{P}}_{\mathbf{A}_2,\mathbf{G}_1,\mathbf{\Sigma}}\mathbf{G}_1')\underline{\mathbf{P}}_{\mathbf{A}_3,\mathbf{G}_2,\mathbf{W}_3}(\mathbf{I}-\mathbf{P}_{\mathbf{A}_3,\mathbf{G}_2,\mathbf{\Sigma}})\mathbf{G}_2'] \\
&\qquad = \underline{\mathbf{P}}_{\mathbf{A}_1,\mathbf{\Sigma}}(\mathbf{I}-\underline{\mathbf{P}}_{\mathbf{A}_2,\mathbf{G}_1,\mathbf{\Sigma}}\mathbf{G}_1')\underline{\mathbf{P}}_{\mathbf{A}_3,\mathbf{G}_2,\mathbf{\Sigma}}(\mathbf{I}-\mathbf{P}_{\mathbf{A}_3,\mathbf{G}_2,\mathbf{\Sigma}})\mathbf{G}_2' = \mathbf{0},
\end{aligned}$$

the third term is uncorrelated with the last term of $\mathbf{U}_1$. Hence, $\mathbf{B}_{1N}$ and $\mathbf{U}_1$ in (4.3.52) are uncorrelated and the next theorem has been established.

Theorem 4.3.12. *Let $\widehat{\mathbf{B}}_1$, $\mathbf{B}_{1N}$ and $\mathbf{U}_1$ be defined by (4.3.7), (4.3.50) and (4.3.52), respectively. Then*

$$f_{\widehat{\mathbf{B}}_1}(\mathbf{B}_0) = f_{\mathbf{B}_{1E}}(\mathbf{B}_0) + \ldots,$$

where

$$f_{\mathbf{B}_{1E}}(\mathbf{B}_0) = f_{\mathbf{B}_{1N}}(\mathbf{B}_0) + \tfrac{1}{2}E[(\mathrm{vec}'\mathbf{U}_1)^{\otimes 2}]\mathrm{vec}\,\mathbf{f}^2_{\mathbf{B}_{1N}}(\mathbf{B}_0),$$

and

$$E[(\mathrm{vec}'\mathbf{U}_1)^{\otimes 2}] = \mathrm{vec}(D[\widehat{\mathbf{B}}_1] - D[\mathbf{B}_{1N}]),$$

while $D[\widehat{\mathbf{B}}_1]$ and $D[\mathbf{B}_{1N}]$ are given in Theorem 4.2.11 (iii) and (4.3.51), respectively. ∎

4.3.6 *Problems*

1. Present the expression of the density $f_{\mathbf{B_E}}(\mathbf{B}_0)$ in (4.3.12) in the case when $k = 1$. To which class of elliptical distributions does it belong?
2. Find the dispersion matrix and the kurtosis characteristic of the elliptical distribution in Theorem 4.3.2.
3. Show that for $\widehat{\mathbf{B}}$, given by (4.3.1),

$$\begin{aligned}
&\overline{m}_4[\widehat{\mathbf{B}}]\\
&= (1+2c_1)(\mathbf{CC}')^{-1} \otimes (\mathbf{A}'\boldsymbol{\Sigma}\mathbf{A})^{-1} \otimes \mathrm{vec}'((\mathbf{CC}')^{-1} \otimes (\mathbf{A}'\boldsymbol{\Sigma}\mathbf{A})^{-1})\\
&\quad + (1+2c_1)(\mathbf{CC}')^{-1} \otimes (\mathbf{A}'\boldsymbol{\Sigma}\mathbf{A})^{-1} \otimes \mathrm{vec}'((\mathbf{CC}')^{-1} \otimes (\mathbf{A}'\boldsymbol{\Sigma}\mathbf{A})^{-1})(\mathbf{K}_{qk,qk} \otimes \mathbf{I})\\
&\quad + (1+2c_1)(\mathbf{CC}')^{-1} \otimes \mathrm{vec}'((\mathbf{CC}')^{-1} \otimes (\mathbf{A}'\boldsymbol{\Sigma}\mathbf{A})^{-1}) \otimes (\mathbf{A}'\boldsymbol{\Sigma}\mathbf{A})^{-1}(\mathbf{K}_{qk,qk} \otimes \mathbf{I})\\
&\quad + c_2(\mathbf{CC}')^{-1} \otimes (\mathbf{A}'\boldsymbol{\Sigma}\mathbf{A})^{-1} \otimes \mathrm{vec}'((\mathbf{CC}')^{-1} \otimes (\mathbf{A}'\boldsymbol{\Sigma}\mathbf{A})^{-1})\\
&\quad + c_3(\mathbf{CC}')^{-1} \otimes \mathrm{vec}'((\mathbf{CC}')^{-1} \otimes (\mathbf{A}'\boldsymbol{\Sigma}\mathbf{A})^{-1}) \otimes (\mathbf{A}'\boldsymbol{\Sigma}\mathbf{A})^{-1}(\mathbf{I} \otimes \mathbf{K}_{qk,qk})\\
&\quad + c_3(\mathbf{CC}')^{-1} \otimes \mathrm{vec}'((\mathbf{CC}')^{-1} \otimes (\mathbf{A}'\boldsymbol{\Sigma}\mathbf{A})^{-1}) \otimes (\mathbf{A}'\boldsymbol{\Sigma}\mathbf{A})^{-1},
\end{aligned}$$

   where the coefficients c_1, c_2 and c_3 are given in Theorem 4.2.2.
4. Derive the order of the error in the approximation given in Theorem 4.3.5.
5. Verify (4.3.33).
6. Prove Lemma 4.3.2.
7. Are the first remainder terms in the expansions in Theorems 4.3.5 and 4.3.6 different?
8. Find an upper error bound to the approximation in Theorem 4.3.8.
9. Find an approximation to the density of the maximum likelihood estimator $\widehat{\mathbf{B}}$ for the MLNM($\mathbf{ABC}$) with restrictions $\mathbf{GBH} = \mathbf{0}$.
10. Find an approximation to the density of the maximum likelihood estimator $\widehat{\boldsymbol{\Sigma}}$ for the model in Problem 9.

REFERENCES

AFRIAT, S.N. (1957). Orthogonal and oblique projectors and the characteristics of pairs of vector spaces. *Proc. Cambridge Philos. Soc.* **53** 800–816.

ALALOUF, I.S. (1978). An explicit treatment of the general linear model with singular covariance matrix. *Sankhyā, Ser. B* **40** 65–73.

ANDERSON, T.W. (1951). Estimating linear restrictions on regression coefficients for multivariate normal distributions. *Ann. Math. Statist.* **22** 327–351.

ANDERSON, T.W. (1963). Asymptotic theory for principal component analysis. *Ann. Math. Statist.* **34** 203–224.

ANDERSON, T.W. (2002). Canonical correlation analysis and reduced rank regression in autoregressive models. *Ann. Statist.* **30** 1134–1154.

ANDERSON, T.W. (2003). *An Introduction to Multivariate Statistical Analysis*, Third edition. Wiley, New York.

ANDERSSON, S.A. (1975). Invariant normal models. *Ann. Statist.* **3** 132–154.

ANDERSSON, S.A. & PERLMAN, M.D. (1993). Lattice models for conditional independence in a multivariate normal distribution. *Ann. Statist.* **22** 1318–1358.

ANDERSSON, S.A. & PERLMAN, M.D. (1998). Normal linear regression models with recursive graphical Markov structure. *J. Multivariate Anal.* **66** 133–187.

AZZALINI, A. & CAPITANIO, A. (1999). Statistical applications of the multivariate skew normal distribution. *J. R. Statist. Soc. Ser. B* **61** 579–602.

AZZALINI, A. & DALLA VALLE, A. (1996). The multivariate skew normal distribution. *Biometrika* **83** 715–726.

BAKSALARY, J. K. (1987). Algebraic characterizations and statistical implications of the commutativity of orthogonal projectors. In: *Proceedings of the Second International Tampere Conference in Statistics, 1-4 June 1987*, Eds. T. Pukkila & S. Puntanen. University of Tampere Press, Tampere, 113–142.

BAKSALARY, J.K. & KALA, R. (1978). Relationships between some representations of the best linear unbiased estimator in the general Gauss-Markoff model. *SIAM J. Appl. Math.* **35** 515–520.

BAKSALARY, J.K. & KALA, R. (1980). The matrix equation $AXB + CYD = E$. *Linear Algebra Appl.* **30** 141–147.

BARGMANN, R.E. (1964). Matrices and determinants. In: *CRC Handbook of Tables for Mathematics*, Ed. S.M. Selby. Chemical Rubber Co, Cleveland, 146–148.

BARNDORFF-NIELSEN, O.E. & COX, D.R. (1989). *Asymptotic Techniques for Use in Statistics.* Chapman and Hall, New York.

BARTLETT, M.S. (1933). On the theory of statistical regression. *Proc. Roy. Soc. Edinb.* **53** 260–283.

BASILEVSKY, A. (1983). *Applied Matrix Algebra in the Statistical Sciences.* North-Holland, New York.

BELLMANN, R. (1970). *Introduction to Matrix Analysis*, Second edition. McGraw-Hill Book Company, New York.

BEN-ISRAEL, A. & GREVILLE, T.N. (2003). *Generalized Inverses. Theory and Applications,* Second edition. Springer-Verlag, New York.

BENTLER, P.M. & LEE, S.Y. (1975). Some extensions of matrix calculus. *General Systems* **20** 145–150.

BERAN, L. (1985). *Orthomodular Lattices. Algebraic Approach.* D. Reidel Publishing Company, New York.

BHATIA, R. (1997). *Matrix Analysis.* Springer, New York

BILLINGSLEY, P. (1999). *Convergence of Probability Measures*, Second edition. Wiley, New York.

BILODEAU, M. & BRENNER, D. (1999). *Theory of Multivariate Statistics.* Springer-Verlag, New York.

BIRKHOFF, G. (1967). *Lattice Theory,* Third edition. American Mathematical Society, Providence, R.I.

BOIK, R.J. (2003). Principal component models for correlation matrices. *Biometrika* **90** 679–701.

BUTLER, R.V. (1998). Generalized inverse Gaussian distributions and their Wishart connections. *Scand. J. Statist.* **25** 69–75.

CAMPBELL, S.L. & MEYER, C.D. (1991). *Generalized Inverses of Linear Transformations*, Corrected reprint. Dover Publications Inc., New York.

CASALIS, M. & LETAC, G. (1996). The Lukacs-Olkin-Rubin characterization of Wishart distributions on symmetric cones. *Ann. Statist.* **24** 763–786.

CHANG, T.C. (1973). On an asymptotic distribution of the characteristic roots of $\mathbf{S}_1\mathbf{S}_2^{-1}$ when roots are not all distinct. *Ann. Inst. Statist. Math.* **25** 447–501.

CHIKUSE, Y. (1992a). Properties of Hermite and Laguerre polynomials in matrix argument and their applications. *Linear Algebra Appl.* **176** 237–260.

CHIKUSE, Y. (1992b). Generalized Hermite and Laguerre polynomials in multiple symmetric matrix arguments and their applications. *Linear Algebra Appl.* **176** 261–287.

CLINE, R.E. (1964). Representations for the generalized inverse of a partitioned matrix. *J. Soc. Indust. Appl. Math.* **12** 588–600.

CONSTANTINE, A.G. & MUIRHEAD, R.J. (1976). Asymptotic expansions for distributions of latent roots in multivariate analysis. *J. Multivariate Anal.* **6** 369–391.

CORNISH, E.A. & FISHER, R.A. (1937). Moments and cumulants in the specification of distributions. *International Statistical Review* **5** 307–322.

DAVIS, A.W. (1977). Asymptotic theory for principal component analysis: non-normal case. *Austral. J. Statist.* **19** 206–212.

DEEMER, W.L. & OLKIN, I. (1951). Jacobians of certain matrix transformations useful in multivariate analysis. *Biometrika* **38** 345–67.

DEMPSTER, A.P. (1969). *Elements of Continuous Multivariate Analysis.* Addison-Wesley, New York.

DWYER, P.S. & MACPHAIL, M.S. (1948). Symbolic matric derivatives. *Ann. Math. Statist.* **19** 517–534.

EATON, M. (1983). *Multivariate Statistics, a Vector Space Approach.*. Wiley, New York.

ESSEEN, C.G. (1945). Fourier analysis of distribution functions. *Acta Math.* **77** 1–125.

FANG, C. & KRISHNAIAH, P.R. (1982). Asymptotic distribution of functions of the eigenvalues of some random matrices for nonnormal populations. *J. Multivariate Anal.* **12** 39–63.

FANG, K.-T. & ANDERSON, T.W.(Eds.) (1990). *Statistical Inference in Elliptically Contoured and Related Distributions.* Allerton Press, New York.

FANG, K.-T., KOTZ, S. & NG, K.W. (1990). *Symmetric Multivariate and Related Distributions.* Chapman and Hall, London.

FANG, K.-T. & ZHANG, Y.-T. (1990). *Generalized Multivariate Analysis.* Science Press, Beijing, Springer-Verlag, Berlin.

FARAUT, J. & KORÁNYI, A. (1994). *Analysis of Symmetric Cones.* The Clarendon Press, Oxford University Press, New York.

FARRELL, R.H. (1985). *Multivariate Calculation, Use of Continuous Groups.* Springer-Verlag, New York.

FELLER, W. (1968). *An Introduction to Probability Theory and Its Applications,* Vol. 1, Third edition. Wiley, New York.

FELLER, W. (1971). *An Introduction to Probability Theory and Its Applications,* Vol. 2, Second edition. Wiley, New York.

FEUERVERGER, A. & FRASER, D.A.S. (1980). Categorical information and the singular linear model. *Canad. J. Statist.* **8** 41–45.

FINNEY, D.J. (1963). Some properties of a distribution specified by its cumulants. *Technometrics* **5** 63–69.

FISHER, R.A. (1915). Frequency distribution of the values of the correlation coefficient in samples from an infinitely large population. *Biometrika* **10** 507–521.

FISHER, R.A. (1939). The sampling distribution of some statistics obtained from non-linear equations. *Ann. Eugenics* **9** 238–249.

FLURY, B. (1997). *A First Course in Multivariate Statistics.* Springer-Verlag, New York.

FUJIKOSHI, Y. (1977). Asymptotic expansions of the distributions of the latent roots in MANOVA and the canonical correlations. *J. Multivariate Anal.* **7** 386–396.

FUJIKOSHI, Y. (1980). Asymptotic expansions for the distributions of the sample roots under nonnormality. *Biometrika* **67** 45–51.

FUJIKOSHI, Y. (1985). An error bound for an asymptotic expansion of the distribution function of an estimate in a multivariate linear model. *Ann. Statist.* **13** 827–831.

FUJIKOSHI, Y. (1987). Error bounds for asymptotic expansions of the distribution of the MLE in a gmanova model. *Ann. Inst. Statist. Math.* **39** 153–161.

GANTMACHER, F.R. (1959). *The Theory of Matrices,* Vols. I and II. Chelsea, New York.

GERBER, H.U. (1979). *An Introduction to Mathematical Risk Theory.* University of Pennsylvania, Huebner Foundation for Insurance Education, Homewood, Ill.

GIRI, N.C. (1977). *Multivariate Statistical Inference.* Academic Press, New York.

GIRSHICK, M.A. (1939). On the sampling theory of roots of determinantal equations. *Ann. Math. Statist.* **10** 203–224.

GLESER, L.J. & OLKIN, I. (1970). Linear models in multivariate analysis. In: *Essays in Probability and Statistics.* University of North Carolina, Chapel Hill, 267–292.

GOHBERG, I., LANCASTER, P. & RODMAN, L. (1986). *Invariant Subspaces of Matrices with Applications.* Wiley, New York.

GRAHAM, A. (1981). *Kronecker Products and Matrix Calculus: With Applications.* Ellis Horwood, Chichester.

GRAYBILL, F.A. (1983). *Matrices with Applications in Statistics*, Second edition. Wadsworth, Belmont.

GRÄTZER, G. (1998). *General Lattice Theory*, Second edition. Birkhauser Verlag, Basel.

GREENACRE, M.J. (1984). *Theory and Applications of Correspondence Analysis.* Academic Press, London.

GREUB, W.H. (1978). *Multilinear Algebra*, Second edition. Springer, New York.

GUPTA, A.K. & NAGAR, D.K. (2000). *Matrix Variate Distributions.* Chapman & Hall/CRC, Boca Raton.

GUPTA, A.K. & VARGA, T. (1993). *Elliptically Contoured Models in Statistics.* Kluwer, Dordrecht.

HALL, P. (1983). Chi-squared approximations to the distribution of a sum of independent random variables. *Ann. Prob.* **11** 1028–1036.

HARVILLE, D.A. (1997). *Matrix Algebra From a Statistician's Perspective.* Springer, New York.

HASSAIRI, A. & LAJMI, S. (2001). Riesz exponential families on symmetric cones. *J. Theoret. Probab.* **14** 927–948.

HASSAIRI, A., LAJMI, S. & ZINE, R. (2004). Beta-Riesz distributions on symmetric cones. To appear in *J. Statist. Plann. Inference.*

HAYAKAWA, T. (1973). An asymptotic expansion for the distribution of the determinant of a multivariate quadratic form in a normal sample. *Ann. Inst. Statist. Math.* **25** 395–406.

HELLAND, I.S. (1988). On the structure of partial least squares regression. *Comm. Statist. Simulation Comput.* **17** 581–607.

HELLAND, I.S. (1990). Partial least squares regression and statistical models. *Scand. J. Statist.* **17** 97–114.

HERZ, C.S. (1955). Bessel functions of matrix argument. *Ann. of Math.* **61** 474–523.

HOLMQUIST, B. (1985a). Moments and cumulants from generating functions of Hilbert space-valued random variables and an application to the Wishart distribution. *Research Report*, No. 3. Department of Mathematical Statistics, University of Lund and Lund Institute of Technology.

HOLMQUIST, B. (1985b). The direct product permuting matrices. *Linear and Multilinear Algebra* **17** 117–141. Corrigendum; *Linear and Multilinear Algebra* **22** 211.

HOLMQUIST, B. (1988). Moments and cumulants of the multivariate normal distribution. *Stochastic Anal. Appl.* **6** 273–278.

HOLMQUIST, B. (1996). The $d-$variate vector Hermite polynomial of order k. *Linear Algebra Appl.* **237/238** 155–190.

HORN, R.A. & JOHNSON, C.R. (1990). *Matrix Analysis.* Cambridge University Press, Cambridge.

HORN, R.A. & JOHNSON, C.R. (1994). *Topics in Matrix Analysis.* Cambridge University Press, Cambridge.

JACOBSON, N. (1953). *Lectures on Abstract Algebra.* van Nostrand, New York.

JAMES, A.T. (1960). The distribution of the latent roots of the covariance matrix. *Ann. Math. Statist.* **31** 151–158.

JENSEN, D.R. (1985). Multivariate distributions. In: *Encyclopedia of Statistical Sciences, Vol. 6, Eds. S. Kotz, N.L. Johnson, C.B. Read.* Wiley, New York, 43–55.

JOHNSON, D.E. (1998). *Applied Multivariate Methods for Analysts.* Duxbury Press, Pacific Grove.

JOHNSON, N.L., KOTZ, S. & BALAKRISHNAN, N. (1997). *Discrete Multivariate Distributions.* Wiley, New York.

JOLLIFFE I.T. (1986). *Principal Component Analysis.* Springer, New York.

KABE, D.G. (1975). Some results for the GMANOVA model. *Comm. Stat.* **4** 813–820.

KALMBACH, G. (1983). *Orthomodular Lattices.* Academic Press, London.

KANG, C. & KIM, B.-C. (1996). The $N-$th moment of matrix quadratic form. *Statist. Prob. Letters* **28** 291–297.

KANTOR, I.L. & SOLODOVNIKOV, A.S. (1989). *Hypercomplex Numbers. An Elementary Introduction to Algebras.* Springer-Verlag, New York-Berlin.

KARIYA, T. & SINHA, B.K. (1989). *Robustness of Statistical Tests.* Academic Press, Boston.

KATO, T. (1972). *Perturbation Theory for Linear Operators*, Second edition. Springer, Berlin.

KÄÄRIK, E. & TIIT, E.M. (1995). Higher moments of random vectors: Some problems of representation. In: *Multivariate Statistics and Matrices in Statistics. Proceedings of the Fifth Tartu Conference, Tartu–Pühajärve, Estonia, 23–28 May 1994*, Eds. E.-M. Tiit, T. Kollo & H. Niemi. TEV/VSP, Vilnius/Utrecht, 313–324.

KELKER, D. (1970). Distribution theory of spherical distributions and a location-scale parameter generalization. *Sankhyā, Ser. A* **32** 419-430.

KENDALL, M.G. & STUART, A. (1958). *The Advanced Theory of Statistics*, Vol. 1. Charles Griffin & Co Ltd, London.

KENWARD, M.G. (1986). The distribution of a generalized least squares estimator with covariance adjustment. *J. Multivariate Anal.* **20** 244–250.

KHATRI, C.G. (1968). Some results for the singular normal multivariate regression model. *Sankhyā, Ser. A* **30** 267–280.

KHATRI, C.G. & SRIVASTAVA, M.S. (1978). Asymptotic expansions for distributions of characteristic roots of covariance matrices. *South African Statist. J.* **12** 161–186.

KOLASSA, J.E. (1994). *Series Approximation Methods in Statistics.* Springer, New York.

KOLLO, T. (1977). Some notions of matrix calculus with applications in mathematical statistics. *Proceedings of the Computer Centre of Tartu State University* **40** 30–51 (in Russian).

KOLLO, T. (1984). Asymptotic distributions of eigenprojectors of covariance and correlation matrices for testing hypotheses. *Acta et Commentationes Universitatis Tartuensis* **685** 3–13.

KOLLO, T. (1990). Investigation of the convergence of functions of sample means and covariances to limit distributions. In: *Probability Theory and Mathematical Statistics. Proceedings of the Fifth Vilnius Conference, June 25 – July 1 1989.* Eds. B. Grigelionis, Yn. V. Prohorov, et al. Mokslas/VSP, Vilnius/Utrecht, Vol. 1, 638–646.

KOLLO, T. (1991). *Matrix Derivative in Multivariate Statistics.* Tartu University Press, Tartu (in Russian).

KOLLO, T. (1994). A note on patterned matrices with applications in multivariate analysis. *Acta et Commentationes Universitatis Tartuensis* **968** 17–27.

KOLLO, T. (2000). Chapter 15. Asymptotic inference based on eigenprojections of covariance and correlation matrices. In: *Innovations in Multivariate Statistical Analysis, a Festschrift for Heinz Neudecker.* Eds. R.J. Heijmans, D.S.G. Pollock, A. Satorra. Kluwer, Dordrecht, 207–222.

KOLLO, T. & NEUDECKER, H. (1993). Asymptotics of eigenvalues and unit-length eigenvectors of sample variance and correlation matrices. *J. Multivariate Anal.*, **47** 283–300. Corrigendum; *J. Multivariate Anal.*, (1994) **51** 210.

KOLLO, T. & VON ROSEN, D. (1995a). Approximating by the Wishart distribution. *Ann. Inst. Statist.Math.* **47** 767–783.

KOLLO, T. & VON ROSEN, D. (1995b). Minimal moments and cumulants of symmetric matrices: An application to the Wishart distribution. *J. Multivariate Anal.* **55** 149–164.

KOLLO, T. & VON ROSEN, D. (1995c). Multivariate minimal moments. In: *Symposia Gaussiana. Proceedings of 2nd Gauss Symposium, Munich, Germany, August 2.7, 1993, Conference B: Statistical Sciences* Eds. V. Mammitzsch, H. Schneeweiss. de Gruyter, Berlin, 15–23.

KOLLO, T. & VON ROSEN, D. (1998). A unified approach to the approximation of multivariate densities. *Scand. J. Statist.* **25** 93–109.

KOLLO, T. & RUUL, K. (2003). Approximations to the distribution of the sample correlation matrix. *J. Multivariate Anal.* **85** 318–334.

KOLMOGOROV, A.N. & FOMIN, S.V. (1970). *Introductory Real Analysis.* Englewood Cliffs, Prentice Hall, New York.

KONISHI, S. (1979). Asymptotic expansions for the distributions of statistics based on the sample correlation matrix in principal component analysis. *Hiroshima Math. J.* **9** 647–700.

KOOPMANS, T.C., RUBIN, H. & LEIPNIK, R.B. (1950). Measuring the equation systems in dynamic economics. In: *Statistical Inference in Dynamic Economic Models*, Ed. T.C. Koopmans. Wiley, New York.

KOTZ, S., BALAKRISHNAN, N. & JOHNSON, N.L. (2000). *Continuous Multivariate Distributions, Vol. I: Models and Applications*, Second edition. Wiley, New York.

KOTZ, S., KOZUBOWSKI, T.J. & PODGÓRSKI, K. (2001). *The Laplace Distribution and Generalizations. A Revisit with Applications to Communications, Economics, Engineering, and Finance.* Birkhäuser, Boston.

KOTZ, S., NADARAJAH, S. (2004). *Multivariate t Distributions and Their Applications.* Cambridge University Press, Cambridge.

KRISHNAIAH, P.R. & CHATTOPADHYAY, A.K. (1975). On some non-central distributions in multivariate analysis. *South African Statist. J.* **9** 37–46.

KRUSKAL, W.H. (1975). The geometry of generalized inverses. *J. Royal Statist. Soc., Ser. B* **37** 272–283.

KRZANOWSKI, W.J. (1990). *Principles of Multivariate Analysis. A User's Perspective.* Second edition. The Clarendon Press, Oxford University Press, New York.

KSHIRSAGAR, A.M. (1972). *Multivariate Analysis.* Marcel Dekker, New York.

KSHIRSAGAR, A.M. & SMITH, W.B. (1995). *Growth Curves.* Marcel Dekker, New York.

LAURITZEN, S.L. (1996). *Graphical Models.* Oxford University Press, New York.

MACRAE, E.C. (1974). Matrix derivatives with an application to an adaptive linear decision problem. *Ann. Statist.* **2** 337–346.

MAGNUS, J. (1983). L-structured matrices and linear matrix equations. *Linear and Multilinear Algebra* **14** 67–88.

MAGNUS, J.R. (1988). *Linear Structures.* London and Oxford University Press, New York.

MAGNUS, J.R. & NEUDECKER, H. (1979). The commutation matrix: some properties and applications. *Ann. Statist.* **7** 381–394.

MAGNUS, J.R. & NEUDECKER, H. (1999). *Matrix Differential Calculus with Applications in Statistics and Econometrics*, Second edition. Wiley, Chichester.

MARCUS, M. (1973). *Finite Dimensional Multilinear Algebra, Part 1.* Marcel Dekker, New York.

MARDIA, K.V., KENT, J.T. & BIBBY, J.M. (1979). *Multivariate Analysis.* Academic Press, New York.

MATHAI, A.M. (1997). *Jacobians of Matrix Transformations and Functions of Matrix Arguments.* World Scientific, Singapore.

MATHEW, T. & NORDSTRÖM, K. (1997). Wishart and Chi-square distributions associated with matrix quadratic forms. *J. Multivariate Anal.* **61** 129–143.

MAXWELL, J.C. (1860). Illustration of the dynamic theory of gases. Part I. On the motions and collisions of perfectly elastic bodies. *Phil. Mag.* **19** 19–32.

MCCULLAGH, P. (1987). *Tensor Methods in Statistics.* Chapman and Hall, London.

MCDONALD, R.P. & SWAMINATHAN, H. (1973). A simple matrix calculus with applications to multivariate analysis. *General Systems* **18** 37–54.

McLachlan, G.J. (1992). *Discriminant Analysis and Statistical Pattern Recognition.* Wiley, New York.

Mitra, S.K. (1973). Common solutions to a pair of linear matrix equations $A_1XB_1 = C_1$ and $A_2XB_2 = C_2$. *Proc. Cambridge Philos. Soc.* **74** 213–216.

Mitra, S.K. (1990). A pair of simultaneous linear matrix equations $A_1XB_1 = C_1$ and $A_2XB_2 = C_2$ and a matrix programming problem. *Linear Algebra Appl.* **131** 107–123.

Mitra, S.K. & Rao, C.R. (1968). Some results on estimation and tests of linear hypotheses under the Gauss-Markoff model. *Sankhyā, Ser. A* **30** 281–290.

Moore, D.S. (1977). Generalized inverses, Wald's method, and the construction of chi-squared tests of fit. *J. Amer. Statist. Assoc.* **72** 131–137.

Morrison, D.F. (1967). *Multivariate Statistical Methods.* McGraw-Hill, New York.

Muirhead, R.J. (1982). *Aspects of Multivariate Statistical Theory.* Wiley, New York.

Murnaghan, F.D. (1938). *The Theory of Group Representations.* John Hopkins Press, Baltimore.

Nel, D.G. (1980). On matrix differentation in statistics. *South African Statist. J.* **14** 137–193.

Nel, D.G. (1985). A matrix derivation of the asymptotic covariance matrix of sample correlation coefficients. *Linear Algebra Appl.* **67** 137–145.

Nelsen, R.B. (1999). *An Introduction to Copulas.* Springer, New York.

Neudecker, H. (1969). Some theorems on matrix differentiations with special reference to Kronecker matrix products. *J. Amer. Statist. Assoc.* **64** 953–963.

Neudecker, H. & Wesselmann, A. M. (1990). The asymptotic variance matrix of the sample correlation matrix. *Linear Algebra Appl.* **127** 589–599.

Nordström, K. (1985). On a decomposition of the singular Gauss-Markov model. In: *Linear Statistical Inference (Poznan, 1984).* Springer, Berlin, 231–245.

Nordström, K. & von Rosen, D. (1987). Algebra of subspaces with applications to problems in statistics. In: *Proceedings of the Second International Tampere Conference in Statistics, 1-4 June 1987*, Eds. T. Pukkila & S. Puntanen. University of Tampere Press, Tampere, 603–614.

Olkin, I. (2002). The 70th anniversary of the distribution of random matrices: A survey. *Linear Algebra Appl.* **354** 231–243.

Olkin, I. & Sampson, A.R. (1972). Jacobians of matrix transformations and induced functional equations. *Linear Algebra Appl.* **5** 257–276.

Ouellette, D.V. (1981). Schur complements and statistics. *Linear Algebra Appl.* **36** 187–295.

Parring, A.-M. (1980). General asymptotic distribution of noncomplite multiple correlation coefficients. *Acta et Commentationes Universitatis Tartuensis*, **541**, 18–26 (in Russian).

Parring, A.-M. (1992). About the concept of the matrix derivative. *Linear Algebra Appl.* **176** 223–235.

Pollock, D.S.G. (1979). *The Algebra of Econometrics.* Wiley, Chichester.

POTTHOFF, R.F. & ROY, S.N. (1964). A generalized multivariate analysis of variance model useful especially for growth curve problems. *Biometrika* **51** 313–326.

PRESS, S.J. (1972). *Applied Multivariate Analysis.* Holt, Rinehart and Winston Inc., New York.

PRINGLE, R.M. & RAYNER, A.A. (1971). *Generalized Inverse Matrices with Applications to Statistics.* Charles Griffin, London.

RAO, C.R. (1958). Some statistical methods for comparison of growth curves. *Biometrics* **14** 1–17.

RAO, C.R. (1973a). *Linear Statistical Inference and Its Applications*, Second edition. Wiley, New York.

RAO, C.R. (1973b). Representations of best linear unbiased estimators in the Gauss-Markoff model with a singular dispersion matrix. *J. Multivariate Anal.* **3** 276–292.

RAO, C.R. (1974). Projectors, generalized inverses and the BLUEs. *J. Roy. Statist. Soc., Series B* **36** 442–448.

RAO, C.R. & MITRA, S.K. (1971). *Generalized Inverses of Matrices and Its Applications.* Wiley, New York.

RAO, C.R. & RAO, M.B. (1998). *Matrix Algebra and Its Applications to Statistics and Econometrics.* World Scientific, Singapore.

REINSEL, G.C. & VELU, R.P. (1998). *Multivariate Reduced-rank Regression. Theory and Applications.* Springer-Verlag, New York.

REINSEL, G.C. & VELU, R.P. (2003). Reduced-rank growth curve models. *J. Statist. Plann. Inference* **114** 107–129.

RENCHER, A.C. (2002). *Methods of Multivariate Analysis*, Second edition. Wiley, New York.

ROGERS, G.S. (1980). *Matrix Derivatives.* Marcel Dekker, New York.

VON ROSEN D. (1988a). Moments for the inverted Wishart distribution. *Scand. J. Statist.* **15** 97–109.

VON ROSEN D. (1988b). Moments for matrix normal variables. *Statistics* **19** 575–583.

VON ROSEN, D. (1989). Maximum likelihood estimators in multivariate linear normal models. *J. Multivariate Anal.* **31** 187–200.

VON ROSEN, D. (1991). The growth curve model: A review. *Comm. Statist. Theory Methods* **20** 2791–2822.

VON ROSEN, D. (1993). Uniqueness conditions for a maximum likelihood estimators in a multivariate linear model. *J. Statist. Plann. Inference* **36** 269–275.

ROY, S.N. (1957). *Some Aspects of Multivariate Analysis.* Wiley/Indian Statistical Institute, New York/Calcutta.

RUDIN, W. (1976). *Principles of Mathematical Analysis*, Third edition. McGraw-Hill Kogakusha Ltd., Tokyo.

SCHOTT, J.R. (1997a). Asymptotics of eigenprojections of correlation matrices with some applications in principal components analysis. *Biometrika* **84** 327–337.

SCHOTT, J.R. (1997b). *Matrix Analysis for Statistics.* Wiley, New York.

SCHOTT, J.R. (1999). Partial common principal component subspaces. *Biometrika* **86** 899–908.

SEARLE, S.R. (1982). *Matrix Algebra Useful for Statistics.* Wiley, New York.

SEN, A. & SRIVASTAVA, M.S. (1990). *Regression Analysis: Theory, Methods and Applications.* Springer, Berlin.

SHINOZAKI, N. & SIBUYA, M. (1974). Product of projectors. *Scientific Centre Report*, IBM Japan, Tokyo.

SIOTANI, M., HAYAKAWA, T. & FUJIKOSHI, Y. (1985). *Modern Multivariate Statistical Analysis: A Graduate Course and Handbook.* American Science Press, Columbus, Ohio.

SKOVGAARD, I.M. (1986). On the multivariate Edgeworth expansions. *Internat. Statist. Rev.* **54** 169–186.

SKOVGAARD, I.M. (1987). Saddlepoint expansions for conditional distributions. *J. Appl. Prob.* **24** 875–887.

SPIVAK, M. (1965). *Calculus of Manifolds. A Modern Approach to Classical Theorems of Advanced Calculus.* W.A. Benjamin Inc., Menlo Park, California.

SRIVASTAVA, M.S. (2002). *Methods of Multivariate Statistics.* Wiley-Interscience, New York.

SRIVASTAVA, M.S & CARTER, E.M. (1983). *An Introduction to Applied Multivariate Statistics.* North-Holland, New York.

SRIVASTAVA, M.S. & KHATRI, C.G. (1979). *An Introduction to Multivariate Statistics.* North Holland, New York.

SRIVASTAVA, M.S. & VON ROSEN, D. (1999). Growth curve models. In: *Multivariate Analysis, Design of Experiments, and Survey Sampling*, Ed. S. Ghosh. Marcel Dekker, New York, 547–578.

SRIVASTAVA, M.S. & VON ROSEN, D. (2002). Regression models with unknown singular covariance matrix. *Linear Algebra Appl.* **354** 255–273.

STEIN, R.A. (1972). *Linear Model Estimation, Projection Operators, and Conditional Inverses.* Ph.D. Thesis, Iowa State University, Ames, Iowa.

STEYN, H.S. & ROUX, J.J.J. (1972). Approximations for the noncentral Wishart distribution. *South African Statist. J.* **6** 165–173.

SUGIURA, N. (1976). Asymptotic expansions of the distributions of the latent roots and the latent vector of the Wishart and multivariate F matrices. *J. Multivariate Anal.* **6** 500–525.

SYLVESTER, M. (1884). Sur la solution du cas plus general des lineaires en quantites linare, cest-a-dire en quaternions on en matrices du second ordre. *Comtes Rendus Acad. Sci. Paris* **99** 117–118.

TAKEMURA, A. (1983). Tensor analysis of ANOVA decomposition. *J. Amer. Statist. Assoc.* **78** 894–900.

TAKEUCHI, K., YANAI, H. & MUKHERJEE, B.N. (1982). *The Foundations of Multivariate Analysis, a Unified Approach by Means of Projection onto Linear Subspaces.* Halsted Press, Wiley, New Delhi.

TAN, W.Y. (1979). On the approximation of noncentral Wishart distribution by Wishart distribution. *Metron* **37** 50–58.

TAN, W.Y. (1980). On approximating multivariate distributions. In: *Multivariate Statistical Analysis, Proceedings of the Research Seminar at Dalhouse University.* North Holland, Amsterdam, 237–249.

TJUR, T. (1984). Analysis of variance models in orthogonal designs. *Internat. Statist. Rev.* **52** 33–81.

TOLVER JENSEN, S. (1988). Covariance hypothesis which are linear in both the covariance and the inverse covariance. *Ann. Statist.* **16** 302–322.

TRAAT, I. (1984). Moments of the sample covariance matrix. *Proceedings of the Computer Centre of Tartu State University* **51** 108–126 (in Russian).

TRAAT, I. (1986). Matrix calculus for multivariate distributions. *Acta et Commentationes Universitatis Tartuensis* **733** 64–84.

TRACY, D.S. & DWYER, P.S. (1969). Multivariate maxima and minima with matrix derivates. *J. Amer. Statist. Assoc.*, **64** 1576–1594.

TURNBULL, H.W. (1927). On differentiating a matrix. *Proc. Edinburgh Math. Soc., Series 2* **1** 111–128.

TURNBULL, H.W. (1930). A matrix form of Taylor's theorem. *Proc. Edinburgh Math. Soc., Series 2* **2** 33–54.

TURNBULL, H.W. (1931). Matrix differentiation of the characteristic function. *Proc. Edinburgh Math. Soc., Series 2* **2** 256–264.

TYLER, D.E. (1981). Asymptotic inference for eigenvectors. *Ann. Statist.* **9** 725–736.

TYLER, D.E. (1983). A class of asymptotic tests for principal component vectors. *Ann. Statist.* **11** 1243–1250.

VERBYLA, A.P. & VENABLES, W.N. (1988). An extension of the growth curve model. *Biometrika* **75** 129–138.

VISKOV, O.V. (1991). Some remarks on Hermite polynomials. *Theory Probab. Appl.* **36** 633–637.

WATERNAUX, C.M. (1976). Asymptotic distribution of the sample roots for a nonnormal population. *Biometrika* **63** 639–645.

WATERNAUX, C.M. (1984). Principal components in the nonnormal case: The test of equality of q roots. *J. Multivariate Anal.* **14** 325-335.

WATSON, G.S. (1964). A note on maximum likelihood. *Sankhyā, Ser. A* **26** 303–304.

WATSON, G.S. (1983). *Statistics on Spheres.* Wiley, New York.

WILKS, S.S. (1962). *Mathematical Statistics.* Wiley, New York.

WISHART, J. (1928). The generalized product moment distribution in samples from a normal multivariate population. *Biometrika* **20A** 32–52.

WONG, C.S. & CHENG, H. (2001). Estimation in a growth curve model with singular covariance. *J. Statist. Plann. Inference* **97** 323–342.

WONG, C.S., LIU, D. (1994). Moments for left elliptically contoured random matrices. *J. Multivariate Anal.* **49** 1–23.

WONG, C.S., LIU, D. (1995). Moments of generalized Wishart distribution. *J. Multivariate Anal.* **52** 280–294.

WOOLSON, R.F. & LEEPER, J.D. (1980). Growth curve analysis of complete and incomplete longitudinal data. *Comm. Statist. A, Theory Methods* **9** 1491–1513.

WROBLESKI, W.J. (1963). Extensions of the Dwyer-MacPhail matrix derivative calculus with applications to estimation problems involving errors-in-variables and errors-in-equations. *Technical Report.* Department of Mathematics, College of Literature, Science and Arts, University of Michigan.

WU, L. & PERLMAN, M.D. (2000). Testing lattice conditional independence models based on monotone missing data. *Statist. Prob. Lett.* **50**, 193–201.

ZHANG, F. (1999). *Matrix Theory. Basic Results and Techniques.* Springer-Verlag, New York.

ZYSKIND, G. & MARTIN, F.B. (1969). On best linear estimation and general Gauss-Markov theorem in linear models with arbitrary nonnegative covariance structure. *SIAM J. Appl. Math.* **17** 1190–1202.

SUBJECT INDEX

GPSR Compliance
The European Union's (EU) General Product Safety Regulation (GPSR) is a set of rules that requires consumer products to be safe and our obligations to ensure this.

If you have any concerns about our products, you can contact us on

ProductSafety@springernature.com

In case Publisher is established outside the EU, the EU authorized representative is:

Springer Nature Customer Service Center GmbH
Europaplatz 3
69115 Heidelberg, Germany

www.ingramcontent.com/pod-product-compliance
Ingram Content Group UK Ltd.
Pitfield, Milton Keynes, MK11 3LW, UK
UKHW020225290926
15470UKWH00001B/390
9781402034183